The Reunification of Science and Philosophy

Organised Project Research Notes

Volume 1

ISBN 978-1-8380618-1-4

A catalogue record for this book is available from the British Library.

Published by Merops Press
websites: www.scienceandphilosophy.co.uk
www.cosmicconnections.co.uk
www.michaelpitman.co.uk

Acknowledgements: Suzanne, my wife,
Françoise, my daughter,
Dave Gant, Rick Pulford.

TOTAL CONTENTS

Contents

In keeping with the nature of this book's framework contents are, except for the continuity of chapter numbers and illustrations, assigned binary (0, 1) digits.

Volume 1: Playtime

Book 1: A Player's Body

Part 1a: Biological Society/ Ecology

Volume 1: Playtime

Book 1: A Player's Body

Part 1b: Human Society

Illustrations

TOTAL CONTENTS

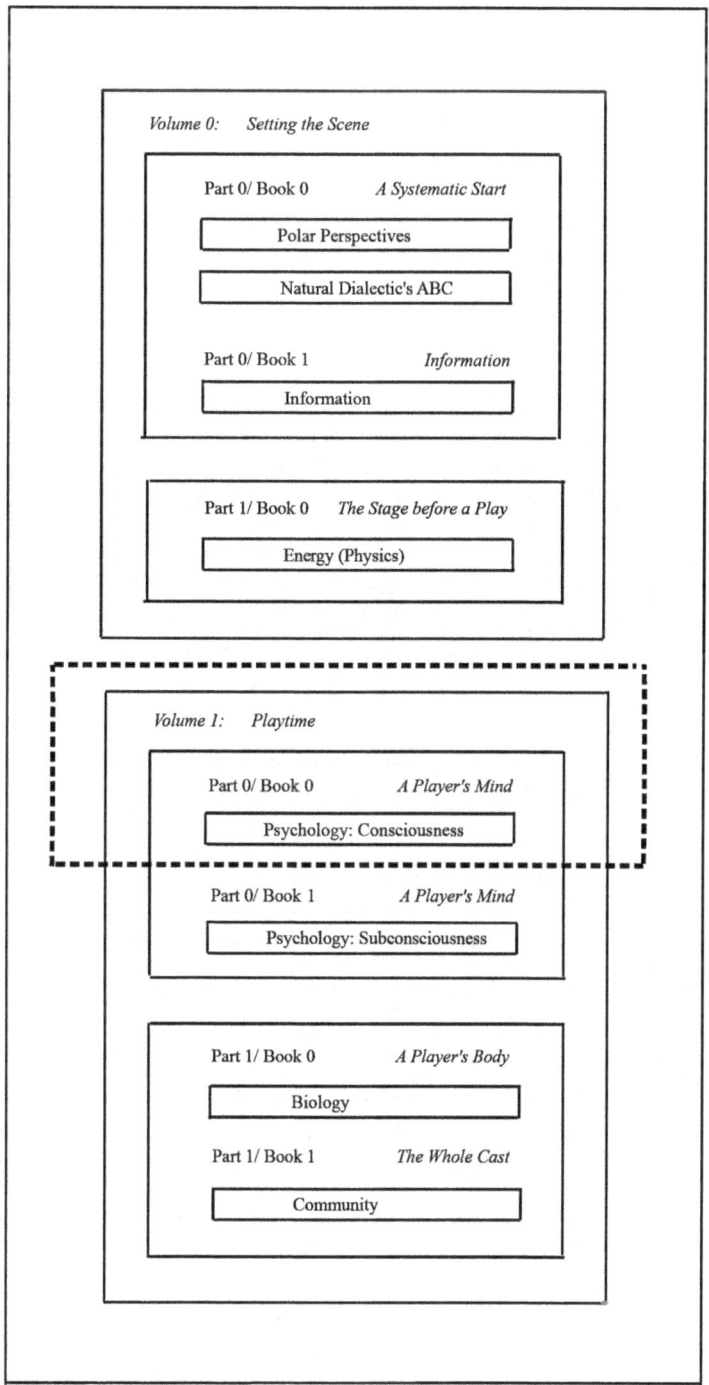

Volume 1: Playtime

Book 0: A Player's Mind

Part 0: Active Information (Psychological)

13. Consciousness

There are apt and ancient metaphors for life. Theatre is one, odyssey another. If you are acting in a vast theatre you also sail an ocean of context, gist, detail, currents and association. We explore the surface of this deep. Your life navigates the waves, tides and superficial storms of circumstance; it is driven before fateful winds but also rides out strong, internal cues that issue up as currents from your memories. For each human vessel there is a captain who, hopefully, knows where he is going and the purposes of his command. There are navigators, commentators and reverse-engineers who, curious persons, are trying to answer their own questions, understand and optimise the situation; call such explorers philosophers, scientists, mystics and, simply, you and me. There is also a lower crew (of various habits and duties) and, if you consider the whole biological fleet, vessels of various capacity and purpose according to the type of life you take. Each kind of body has its own specific attributes, not least its structure, its engines and, to guide the driving forces, navigational equipment in the form of different kinds of sense and sensibility. Such a navy, along with the atmosphere and ocean, is the subject of material science. Thus sails the biological community through life.

Who writes dramas? Who builds ships? Science wants to understand the reasons and the method that give rise to everything that navigates or rolls within the cosmic swell. Its broader sense of 'knowledge' must embrace all time, every space and place and interaction on the journey of the universe. It must, therefore, include the metaphysical as well as physical. Just take the body of a single merchantman. An inspection does not miss the role of captain and its crew as well as study of the nervous navigational kit, biological body of the vessel and the other chemo-physical aspects of the 'system'. Ships carry cargoes called desires and these, beyond the wild incident of wind and waves and weather, are their real burden. Do you remember Mozart's composition ((from Chapter 5: Information, Messages, Arrangement)? You cannot just describe a system without drama otherwise your partial truths will miss the point. What is the port of origin and where, upon a dream, the further shore? Is this shore in fact the goal of odyssey or is the business circular, out and return where home is London Town? What is the Climax of a sailor's plot? Survival? Sure. But is that all? Is anybody Harbour Home? What is the nature of that Port of Origin?

You can see, immediately, that a human cannot expect full and satisfactory answers to first order questions of purpose from scientism (a belief in the omni-competence of material science); and that most modern philosophy, having bowed at the altar of scientism, is now awash upon external ebbs and flows. Has it not lost an expedition's sense of destiny and so, with helmsman missing, is blown by prevailing trends of air or putters rudderless in ineffectual circles? Whatever happened to the captain? Surely schools and universities have not, forgetting where they came from and a navigator's overriding sense of destiny, lost touch with the larger, metaphysical plot?

It was declared on the first page of Chapter 0 that a scientific world-view not profoundly and completely coming to terms with the nature of conscious mind can have no serious pretension of completeness. Science and the Soul hopes, using Natural Dialectic, to generate a scientific (but not just materialistic) frame of reference by which to understand your metaphysic, mind. Base-line needs underlining. ***If the axiom that informative mind and energetic matter are two different kinds of element is true the logic of this volume is entirely unassailable***. Thus we include and not, as neuroscience might intend, diminish or exclude the primal factor of psychology, life's consciousness.

Do you remember the basic existential dipole of energy and information? Book 1 elaborated both constituents of cosmos; it covered what occurs before the curtains rise and a drama starts unfolding. It involved author, text and stage.

Do you also remember 'The Diamond Capstone' (*fig*. 7.1), Consciousness, its triune aspects of Infinity, Nothing and Unity, and its Purposive Potential? The emphasis in Chapters 7 through 12 was on explaining how these so-called principals are crystallised, especially in the world of energetic physics and material cosmos.

The emphasis now swings to information. No doubt information is embodied, in its passive form, in codes and coherent systems. It can be crystallised in bodies and machines. The interest of this immediate section (Book 2) is, however, in its active phase, in mind. *It is playtime*. We involve life's cast of actors and the First, Essential Principal of Consciousness with animating attributes - attention and enthusiasm.

What is the difference between what we call 'life' and consciousness? The essence of a mind is consciousness; and consciousness-in-motion is a way of describing existential relativity and changefulness of mind. **A man perhaps about to lose it, not in sleep but death, soon reshuffles his priorities. He grasps for sure that immaterial consciousness is life and life more precious than the universe - because without it *physicalia* forever disappear**. In this sense life is naturalism's grave. What, therefore, does a seeker find who holds its mystery to the light? If it is immaterial then surely nothing; and if material then why has science failed to weigh it up? Indeed, its immaterial quality is scientific atheism's tomb. Philosophy aside, however, with awareness and a person's way of thinking comes the public, heavily researched domain of psychology. Chapters 13-14 consider flexible, conscious mind, that is, active information; and Chapters 15-17 consider its inflexible, subconscious mode. Subconscious mind involves, like biological material, passive information. Chapters 19-24 review biological circumstance and Chapter 25 the nature of social environment. In short, this whole volume treats the mind and character of players as they weave embodied drama on an earthly stage.

Psyche and Psychology

Microcosm of the Macrocosm

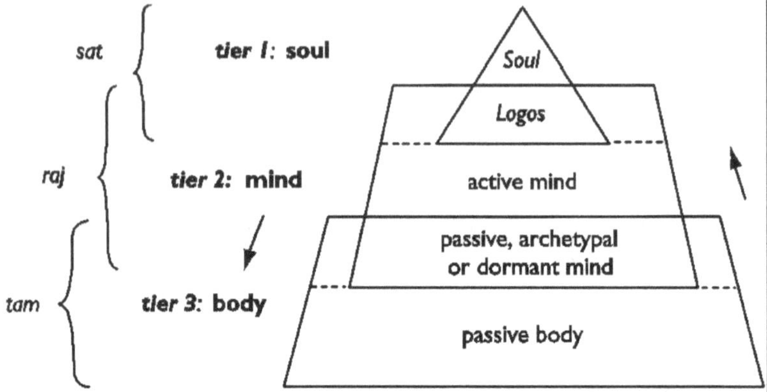

Do you reflect creation as a whole? Is your constitution in the image of a three-tiered universe? In this analogy the anti-parallels run in (for knowledge), out (for action); input, output - their two tracks lead to and from an information-centre at the head. In this respect they form an image representing you. This is how your nervous system is arranged; indeed, as we'll see, they represent the way that your whole body is disposed, that is, the structure of biology.

fig. 13.1 (see also figs. 0.4, 1.3, 2.3, 7.8, 13.4)

	tam/ raj	*Sat*
	tier 3/ tier 2	*Tier 1*
	body/ mind	*Psyche/ Soul*
	lesser selves	*Self*
	conscio-material range	*Conciousness*
↓	*tam*	*raj* ↑
	physic	*metaphysic*
	energetic	*informative*
	material	*immaterial*
	non-conscious	*states/ grades of mind*
	body/ tier 3	*mind/ tier 2*

Microcosm of the macrocosm - a simple, three-tiered cosmic ziggurat represents the overview. Physical awareness! Incorporated cogitation! What about your information-centre, mind; and at this point's human centre, where you think and have your conscious being, the third eye - the axis of your microcosmic personality? Whole libraries record how humans think about their thoughts, about subjective mind and its objective, correlated nervous systems. It is not the object of the next few chapters to summarise this corpus but to rephrase psychology in the simple but dynamic framework of Natural Dialectic. As in the case of non-conscious energy (Book 1) such translation from the phraseology of science can only, in a book of this scope yet manageable length, represent a broad-brush initiation. But why add 'soul' to 'science'? If no discrete, immaterial element exists then isn't science by itself enough; if 'soul' is somehow physical or simply non-existent then what, to scientific paradigm, is there to add?

Perhaps, however, there is more to 'soul' than meets objective eye. It is not a question of a mass of detailed scientific data but of principle. It is not a question of upsetting established nomenclature but incorporating it within what is possibly, even probably, a broader context. *Do electromagnetism, gravity and forces of the atom totally explain experience? Can their materialistic combination sum to your entirety?* **Be absolutely sure that materialism nothing more than guesses hopefully they do**. If, however, there is light beyond what meets the lens and retina of eye what perspective best systematically explains *all* facts of life, including subjectivity of consciousness itself? Perhaps turning to this angle would involve a revolution in materialistic lines of thought.

The word '*psychology*' means study of *psyche* (from the Greek word for 'soul'). What is currently called psychology were perhaps better termed '*noology*'. This is because the discipline studies mind, for which the Greek is *noos* (from the verb 'to notice' or 'to think'), and not *psyche*. Or does it? Perhaps not even 'noology' goes far enough. Contemporary research and, as a consequence, journals, educational material and the popular press, swing even further towards a hands-on, neurological commitment. ***Top-down*, psychology is metaphysical; but scientific materialism can, by definition, only allow physical composition**. Has psychological science therefore lost patience with its own psyche and externalised itself from mind to the behaviour of nerves; and thus, embroiled with objects made of atoms, died?

Soul to Sell?

Is your soul for sale? To whom will you sell out? If you sell to Mephistopheles then what is he? The devil, sure, but what about immoral pacts with yet another buyer - Mammon? Do you believe, excessively materialistically, that cell or cells 'evolved' a 'soul' whatever 'soul' might be? If you believe it's just imagination by material mind, a figment of the brain, then you've sold out to soullessness.

Try soul to cell instead. From a *top-down* angle soul is paradoxical. Psyche's a restriction of Infinity; it is Central Essence; and the Absolute of Pure Awareness woven, in creation, round with existential relativity. Weave mind and body round a soul and find a triplex entity; but, with constant motion of the lower couple of this tripartite combination, Central Self is lost from view.

Fig 13.1 illustrates this triplex which applies the same to cosmos as to you. Separate bodies each conjoin their fraction of non-conscious, material energies; mind's composed of grades of immateriality; immateriality's informative and mind's Transcendent Root is Axial Immateriality. Such Axial Soul is single; *Atman-Brahman*, One God and Unity of Consciousness are different words that mean the same; and since it's single, you must be essentially one with it. No difference! You know your complicated self but do you know your Axial Self?

This Essential, Central Self wears round it changing, existential selves. The Individual One expresses individualities. Potential Information generates all kinds of active, psychological and passive, physical behaviours; purpose, meaning, rules and regulations all derive from information. Soul has, therefore, special 'status' since it's at the heart of personal and universal orders. It is Most Natural and Real.

Oh no, it's not! Physic, metaphysic - philosophical duality becomes a punch-and-judy show. 'About turn!' Now the opposite, materialistic angle seems what's true! Where 'immaterial' translates to 'spooky' how unlikely some sort of divine infusion seems to be. According to this *modus cognoscendi* soul must have evolved from cells. Since cell evolved mind must have too. Brain evolved and brain is mind and so whatever you call soul is at root cellular; and is not cellular the same as chemical? A chorus from molecular biology agrees. That's therefore what brain scientists have to think. Embodied 'soul' might, as a vague description of a person's 'individuation process' or 'core uniqueness', still verbally exist but only squeezed beyond the actual world's periphery. **Materialism thus thoroughly degrades Creation's Central Factor into vagueness whose reality is now diminished and dissolved among mysterious properties of hormones, genes and neurons in grey jelly**. Soul's are not for test tubes; thus decline in use and understanding of the 'soul' word correlates with the ascendancy of scientism. Concept of soul, once immaterially dissolved away, leaves but the fact of brain. *Thus from soul to cell is the direction laureate thinking has evolved.* Has not progress almost managed to erase what used to be an insubstantial but a widespread superstition? An objective outlook on psychology has rather binned than sold its soul.

The Neurological Delusion

Thus psychology emerges as the study of a neurological phenomenon. Well-educated and evolved professors tilt their weights towards neuroscience as the guru. Does not consciousness 'emerge' from molecules grouped in some special way? Is mind not generated by the workings of a brain? Just sling sufficient atoms, in the form of nerves, together - they'll become no less than self-aware! The atomic property of knowledge, which some educated *cognoscenti* realise as truth, awakes in nerves. Consciousness and mind are therefore thought of as activity of brain. They can't exist apart from it. They amount to just an 'efflux' of neuronal circuitry, an intangible illusion that's dependent only on what you can sense. Not only is oscillation between two states of brain (waking and sleeping) dependent on some clever chemistry but, in this view, the ultimate reality of subjective life - that will o' the wisp called consciousness - is no more than an unforeseen biochemical capacity. By chance the chemical reactions making nerves, perhaps including subtler quantum interactions, began to by-produce a new invention. Call it mind or soul or 'virtual reality'.

What sort of 'almost-ness' is that? A virtual world is one that registers an electronic image on a screen. Does such an image feel a thing? Are you an image registered this way upon your brain? How? Who sees and feels and knows? **That mind is an illusion is the neurological delusion. To mistake its neural correlate (as, say, registered in brain scans) for a thought itself is error prime as taking electronic pulses in a wire for all there is to telephonic conversation. It is a prime, elementary fault, a first category philosophical mistake**. There is no evidence for it; no-one has ever seen or entered into someone else's thought. To identify consciousness as an illusion is itself, denying the reality of one's own experience, a pernicious - even dangerous - delusion. Yet, lemming-like, many intellectuals leap into this dialectically false conclusion. Some are scientists. The discipline of neuroscience dives routinely into such a hole. Who, however, cares for error when materialism's promissory faith - that atoms make a brain and thence make thought - caps all?

Do you remember (Chapter 5) Mozart and the scientist - a pair who measured music in a very different way? In the light of their division should lopsidedly materialistic ideology be left to arrogate Sole Answer or to trump the Dialectic's kind of explanation? No doubt the universe is made of energy. Vast amounts reside in every atom. 'Wall-to-wall' the cosmos is an 'ocean made of motion'; the universe is spinning wheels of energy; and everywhere you find translation by vibration. Might you therefore carry such translation and its quantum physical description through perception's doors and enter into mind? After all, sensation is an oscillatory vibration in and out of mind; and, in the brain, you can measure frequencies that thoughts express electrically. The question is, though, whether vibes and frequencies are actually thought or simply thought's reflections set within a skull? Is observing thinking's simile the same as measurement of subjectivity? *If energy is not the whole part and subjective information is 'the life and soul' then brain is not the thought but, rather, thought's reflection by materials - electric ions, chemicals and, equally, essentially lifeless cells.* You think to run and science measures body running; you only think and chemical effect is gauged. Quantum physics may be able to define the operation of mind's mirror, brain, but not life's feeling and experiential subjectivity. **To think so were delusion writ again.**

Visualise a white horse. How do nerves, and masses of them, 'represent' it? **Nervous particles and atoms aren't, like atoms anywhere, alive. Therefore, if life is made of them it shouldn't be alive!** How, therefore, does a nervous system 'come alive'? What's 'alive' if chemical alone? No matter when it comes to subjectivity, consciousness and understanding what it is to be alive - there are, wise men aver, three steps unto scientific psycho-heaven. After neural 'representation' you simply 'represent the representation', that is, become aware of it. Then 'represent the previous two representations' and become aware of awareness. Now you've got wise; you are fully conscious. Does this solution to the nature of your consciousness sound like a rational brainwave or breathtaking beggary - of the question, that is? To describe is not the same as to explain. Has this description 'scientifically' explained or just explained away our most prized possession - awareness? Is it not, actually, an effete form of waffle? How hard is 'soft' consciousness to understand and who is in a muddle under an illusion?

Natural Dialectic proposes that Descartes, who wrote 'I think, therefore I am', turned truth about. Essence precedes active existence. 'I am therefore I think'. But modern psychology, without the benefit of misconceived Cartesian duality or entirely preposterous dialectical trinity, writes that 'There is a brain, it thinks, therefore I am'. So mind, having been wrapped in nervous electricity, thereby falls under the influence of biology; from which materialistic science of bodies it is but a logical, conceptual hop-skip to a genetic basis of thought, the evolution of brains and evolutionary psychology. Is the circle now complete? Have visionary minds uncovered new, transparent truth of which previous generations were completely ignorant? Or just reversed the wheel of truth's rotation so that, new myths for old, cutting-edge psychology now ploughs such a furrow of confusion as has not been seen before?

You have to laugh. Do nerves and hormones really run the show? Do endorphins or serotonin crack jokes that make you cry? Or do they just make making jokes a laugh? Hormones aren't a joke; nor are ions. So doesn't mind come into humour so that metaphysic has both first and last laugh?

Therefore, does mind or brain continually screen perceptions and secrete your inmost thoughts? Are they the same or different? Where does soul, if it exists, come in? Practical as well as principled confusion reigns to an extent at which our forebears would have been amazed. Do, for example, embryos have a soul, potential soul or not? Is a zygote or stem cell alive; and is life sacrosanct or just a pennyworth of common chemicals? The British government, made up of fallible MPs, has allowed creation of chimaeras (hum-animal hybrids) to pacify religiously deluded souls when they object to reason-in-experimental-mode impacting on the soul-less molecules that make up wholly human embryos. Be scientific, please! Does an embryo, child or even legislating politician really have an immaterial soul? Are not spirits, psyches, ghosts and suchlike gross irrationalities? What, conversely, might the brain of a chimaeric youngster, one not killed by law while embryonic, feel? What, indeed, *is* feeling? In short, the neurological delusion claims that 'brain is mind' and that 'unless you call an electronic flicker soul, no such reality exists'. So close your eyes. Are you just nervous in the dark?

For neuro-non-sense in its darkness abstracts such as faith, hope, charity, creativity and love become, in reality, nothing more than excitations in an electronic jelly. What humanitarian diminishment; from a *top-down* perspective what lamentable reductionism, what unwholesome pseudoscience swells with arrogance sufficient unto branding science, reason, art, imagination and experience at root no more than various inflections of the brain! Neuro-superstitious atheology (there are academic experts in this discipline) suggests you're only nets of tissue shuttered in a calcium box. Such rumour weighs a gram of truth but, with gravity of error heavy as a black hole, crushes weightless spirit out of the equation. This kind of mind-set has expelled the soul. It is a full atomistic stop; it is the neurological delusion's 'meaning' and materialism's final logic brought to bear upon your pointless life. Is this the message neuroscientific lectures and the textbooks truly spread? If so, who is preaching darkness in the dark?

Line of Mind

The line of mind is how you look at it. Already it is clearly drawn, cutting in between *top-down* and *bottom-up* antitheses. Let's summarise the anti-parallels this far.

Bottom-up, what might be termed the megalomania of 'promissory materialism' denies, in spite of evidence, immateriality. The line becomes one-track, bone-headed when it comes to nerves and, therefore, brain-mind identity. Since matter - forces, atoms, particles - is witless then intelligence was witlessly evolved. You are 'a bunch of chemical reactions in a bag'. And some of them, called brain, make mind. Brain evolved and then somehow exuded consciousness. Nerves create a phantom called intelligence; soul-less mind is an illusion mindless atoms make. If, standing on your head, you feel the right way up then God did not make brain but brain excreted the illusion of a god. The latter, with religion, must therefore be found in brain - some gene or genes, spot, program, switch or nervous flicker. After all, aren't you no more than such a quantum flicker? Perhaps electric pulses, sick delirium, drugs or an epileptic lobe of mind are you need to throttle up and turn The Grand Delusion on! If this sounds silly and misguided perhaps it is. But do not underestimate the guidance scientism chafes to give; do not underrate materialism's willingness, using science and the media as tools, to reduce life to a vision of non-conscious bits. The current of this one-way swell that, given Darwin's impetus, generates the most just-so imaginations is evolutionary psychology. Of course, the scientific bull pronounces evolution is a 'fact'. Feel justified and free, therefore, to simply add a qualifier ('may', 'perhaps' etc.) to untestable, non-falsifiable hypothesis and you can spin an answer that will always put religion back where it belongs - inside the brain of brain-box; and your qualifier's the escape route if you're later trumped by someone else's plausibility. Such whimsical psychology is thus an effort to explain experience and behaviour in the terms of hominid survival; which, since our forebears must have felt like us, is partially fair. If brain evolved, as neuroscience would confirm, psychology and subjectivity boil down to nerves in skulls. An evolutionary line of psychological hypotheses must, therefore, claim the path to truth. What, however, if materiality is not the whole caboodle? Evolutionary psychology becomes a bag of tales. Since when was scientific value anecdotal? Therefore, God help its clinician's patient!

Top-down, brain is a medium, an intermediate, an instrument panel that allows your immaterial, subjective self to drive the vehicle of your body and, as sure as possible, survive on earth. In this view your mind is no illusion; you identify with it; it is closer and more real than arms and legs and all the other objects and appendages that tie you to the world.

If mind is mediated by the human brain then brain's a filter, valve or template that conditions Consciousness in a specific way. It is seen as a delimiter of possibilities, an ordinator of the way humanity is programmed to experience and understand the physical department of reality. The same logic would, of course, apply to ape or antelope or bee.

It is a fact that you can change your mind; you can control emotions too. An immaterial act of will power, purpose or intent is impressed upon grey matter. Various kinds of scan can register the neural correlates of thought and show how it affects the brain. Mind informs, by nerves and muscles, its material surroundings. And (Chapters 5, 6, 16 and 17) the reverse is true; through the agency of sense we absorb external information, through those organs we experience materiality. In Chapter 14 (see especially *fig.* 14.2) we examine five main 'gears' by which a brain reflects or else affects the correlate of nerves, a state of mind. This line of mind is clearly one that separates itself from brain.

Does Brain Originate or Mediate?

Do you remember (from Book 0 and the Glossary) a couple of different Primary Axioms - the PAM, the PAND and their corollaries? I say there are two fundamental components in existence - subjective consciousness and objective matter; immaterial and material. You say material alone.

One party, it is clear, believes brain *causes* subjectivity; thought (and therefore belief and all the purposive effects of will and faith) is part and parcel of nerve chemistry. And what is the experience of consciousness? The essentially robotic view of neuroscience holds that nerves *are* consciousness. We just don't yet understand, the faithful purr, how brain 'squeezes out' experience or how the juice that's 'you' must be exuded from its molecules. A revelation is, however, prophesied. Materialism's scientific certainty decrees that life will be reduced to chemistry and mind experimentally identified as simply due to complicated ionicity! You are a product of your physiology and so, at root, your genes alone. Life has, hasn't it, to be an electronic afterthought?

A laptop's screen displays, according to its program, images. But is the system physical? Who has decided what to show? It's physical but minds invented and now operate its passive show. Screens, radios and brains reflect mind's immaterial activity; its forethought governs your material creativity; brain is a brilliantly conceived connector, two-way link or coded medium that interfaces mind with matter. *Thus the other party's line suggests, conversely, that brain isn't an originator but a mediator.* It is a **transducer device** that reads environmental signals and translates them 'upward' into mental perception; and issues orders 'downward' into body chemistry (including muscular, hormonal and, most probably, developmental patterns of response. As we'll see in Chapters 15 and 16, the brain's complex interpretation and regulation employs, as well as molecular biochemistry, vibratory electromagnetic/ energetic signals. As an organ of 'cockpit control' its 'dashboard' is extremely well constructed to connect immaterial mind to a material body and, thereby, physical conditions. Thus we and other kinds of creature travel, in the vehicle of body, various sagas on the senseless stage of matter. Molecule-constructed brain is certainly a very specially developed interface, a superlative design for animate experience of the inanimate and purely energetic cosmic zone. Sense perceptions change the mediator's chemistry as messages are passed up into the 'psycho-space' of mind but, equally, by rational thought or instinctive response, mind hands down control of body's action. Electrochemistry is the body part of this mediation process. In this view mind and brain, although compounded, are quite separate entities - the former metaphysical and latter physical. And brain chemistry is identified as a design that expedites informative mediation between the pair. How can this be scientifically true? How can an immaterial element exist? Can materialism solve a problem hard as this; is it a 'naturalistic', scientific one at all? Or do we need, as Natural Dialectic clearly shows, a drastic and yet easy change in paradigm concerning how we view the universe?

Your life is intimately identified with what you know - your subjective consciousness. Like the eyes, ears, brain and other superbly fashioned instruments that contribute to its whole, is 'knowing' just material? Is your personal experience constructed only from four basic forces and a few kinds of particle? *If not and it is essentially immaterial then psychology and neuroscience miss their own main point, popular materialism is illogical and atheism irrational.* This section (Chapters 13-17) reconstructs accordingly. Of course, if science includes only what is known, materially knowable or testable by physics and chemistry then, by definition, what is immaterial is 'unscientific'. If subjective, informative mind falls into this category it is 'unscientific' but certainly exists. *Natural Dialectic now describes a viable, natural, logical format for the dynamic both of individual and universal mind. It correlates such construction, long familiar to humankind, with modern scientific findings. It maps, as one overhead projector transparency fits square over a second, the maximum overlap between the contours of 'scientific' and 'unscientific' paradigms. This dualistic premise is developed.* **Making no material difference by adding immateriality, the Dialectic simply reconstructs creation on the basis of a 'conscio-material' duality.** *It explores the revolutionary consequence for thought and, therefore, all our lives.*

Do you remember (from Chapter 0) the two pillars of faith? Narrow is the gap, fine the line between these two great citadels of human thought. There is, in material terms, nothing in it, no gap at all. From the seeds of these two psychological positions (whose difference is therefore immaterial) world faiths are grown. Yet the current of each strain of logic seems, while often crossing in between, to flow from poles apart in opposite directions. What crops ripen, what fruits of consequent behaviours mature in each one's field of mind? Which, material or immaterial, will bear the Nuclear Grain?

This chapter addresses both what's called the 'binding problem' of psychology viz. how the 'seamlessness' of waking experience is achieved; and the 'hard problem' of philosophy and science viz. how a physical object, the brain, can have subjective experience. It suggests that the holistic solution might, in fact, be softer than anything physical. It might be nothing whatsoever physical but something metaphysical. **In short, perhaps brain neither does nor ever did enjoy a seamless, subjective experience**. The implications of this seminal idea are so extensive that the entire book explores them.

Whether or not you think they are essentially the same thing, mind affects body and *vice versa*. There is psychosomatic or mind/ body interaction mediated by nerves and hormones and, grand in principle and purpose, there is the exercise of mind affecting body called medical science. In its quest for complete understanding biological science must eventually tease out every previously unknown or misunderstood molecule and place it in the light of reason. What about a cauliflower-sized jelly in which the mysteries of mind, memory and consciousness are immersed? Is consciousness immaterial, a material zero? Or can wobbly pinkishness invoke the spell of thought and thereby, from its bony hat, conjure nothing less than you? How can such grey matter be devoted to ideas; how can neurons want to kneel in awe or love? But there is fascination and, for some, enchantment with the notion there is nothing but matter, nothing immaterial and nothing to thought but brain. The materialistic aspect of psychology is rooted in neuroscience. In this view our sophisticated mental appreciation of the physical world is built from complex electrochemical patterns. *Such perspective is unproven; it is born of hope, belief and faith. Let not materialism's pot call holism's kettle black!*

Neurologically, therefore, it is believed that brain makes mind and the structures for its different functions can be studied two main ways. By inspecting damaged brains it is possible to correlate their loss of function with specific lesion. It is also possible to watch reflections of yourself in thought. *EEG*s (electro-encephalograms) and *CT* (X-ray computed tomography) have been complemented by exciting, new and powerful technologies. For example, *MRI, PET* and *fMRI* brain imaging techniques offer a composite picture of cerebral structure and function. While *MRI* (magnetic resonance imaging) indicates its structure *PET* (positron emission tomography) and *fMRI* ('f' for 'functional') can, by detecting blood flows, glucose/ oxygen concentrations and metabolic response levels, indicate the nervous correlations of experience in real time; computation translates the feedback into almost point-accurate three-dimensional 'video-maps'. These maps are not an abstract model but physical impressions that, as ripples reflect gusts of wind, reflect the winds of thought; they locate mind on the panel of its instrument, the brain. *In other words, we watch as thoughts impact on brain*! The study is exciting, intensive and its net is closing. Soon the cerebral atlas of mind will be complete. Every neuron's molecules will have been fully fished but will neurology have caught the cellular substrate of an idea, conceptual circuits or some nervous shadow shaping consciousness? Indeed, materialism cries, it will and at the same time will have squeezed soul absolutely out of neuroscience. It will have rendered mind, spirit, ghost or any other superstitious metaphysic non-existent. It will have proved that they don't objectively exist and, not being scientifically there, are an extraneous irrelevance, a subjective immateriality. Victory will be complete.

Did you forget already? Don't confuse a viewer with his television screen - a screen as lifeless as a brain alone. Don't confuse his goals in life with the scheduled traffic a computer program generates - passive, electronic motion that is lifeless as a nervous impulse. **Who confuses a flow of blood with mind itself? You might have mapped the last electron or ionic pattern in grey matter but missed consciousness.** Crushed in materialism's fist there's absolutely nothing left - just what you'd expect if information's immaterial and just what thought's metaphysic physically is. The serious neurological delusion is that, while it's there but immaterial, you'll find material mind in brain.

Francis Crick, the *DNA* man, hated 'vitalism', 'entelechy' or any notion of an ordering principle in nature. Thus he saw that for his atheism to prevail you had to 'kill' consciousness; you had to materialise your mind. He therefore proposed that subjectivity was nervous and sought 'neural correlates of consciousness'; such correlation would both stand for and would actually *be* your thoughts. The Dialectic perfectly agrees that thought, emotions and mind's process have objective correlates in nerves but not that, *à la* Crick, these are the same as biochemical constructions. His hopeful coupling of confusion and delusion marks the common scientific one - equating mind with brain thereby conflating subjectivity and objectivity. It is one-tiered philosophy and flat-earth rationale at work again.

Does paint paint paintings, petrol drive a car or electronic motion conjure creativity? How could consciousness 'emerge' (save by materialistic animism's philosophical necessity) from a non-conscious multitude of carbon, sodium, potassium and other atoms? Material particles and forces are devoid of purpose and you don't expect to find mind in machines, even those as wondrous as a brain. No doubt, Baird is found in every television set and yet, even if each atom in them all were charted, he (or his concepts) would be found in none; nor do televisions have a clue what they convey. What purpose has the actual image that's projected on a screen? What set of chemistry could bubble up experience, why should a mass of active but oblivious neurons feel a thing? What is anger or affection when considered atom-mystically? **If purposes are immaterial then - repeat it - you will never find mind in a brain.**

You use a screen of drum-skin or a sheet of silk. One man in front, called 'involuntary body-side', taps patterns then, with 'mind-side man' unseen behind, exclaims the 'thoughtful diaphragm' is able to respond in voluntary, unexpected ways. Thus one who thinks that, by accounting for each biochemical or recording every motion in each neuron, he has proved life all-material is to this end embarking on a waste of time. He is under the self-reinforcing, neurological delusion that consciousness is an illusion solely born of cerebral electrochemistry. He probably believes that somehow, like the most profound yet popular illusions of abiogenesis (Chapter 20) and Darwinian 'design' (Chapters 20-25), it is made of chemicals; subtract them and what real is left? Nothing made of 'natural reality', that is, of energy or matter.

Such a line of thought follows the logic of materialism but not, if consciousness is immaterial, actual fact. In which actuality, when nothing of vacuum or its load is left unregistered, the immaterial bird of subjectivity flies just as free as ever. Which of the two, materialistic monism or holistic dualism, is existentially correct? If metaphysical is not physical, the mind is simply not material. Why should it be? What's wrong with that idea? Why, except to suit a working practice and conceptual framework (including the angle on psychology), does metaphysical have to be physical? Why, due to its own intrinsic constraints of operation, should material science ever find an immaterial object or event? *How could it ever reach inside subjectivity? Or is your experience unscientific?*

Is not your nut worth cracking? Scans can take a powerful peek; up and very personal, these tools can take a trip inside your nervous head. Estimating such a prize is surely worth another shy. Imagine that an invisible pilot moves controls according to his whims, his purposes and, informed by sensors to the dials, the context of the body of his plane. You could work out where the sensors, automatic processors and effectors were attached; you could track, like blood flows or ionic twitches in a brain, their operations and their states; and you could, thereby, by such 'body language' (not of the whole plane this time but inside its cockpit) gauge to some extent the pilot's thoughts. Objectively. But not subjectively. **His controls, which you watch moving as if operated by a ghost, act as the medium by which he flies the body of his plane. They are its material, instrumental medium exactly like your body's brain.** A pilot turns and lands ten thousand times in different circumstance and, including calculation and emotion, with ten thousand variations of experience - yet the same instruments are used each time. The same 'lobes' of cockpit are engaged. Inside your head your nervous hotspots represent the linkage of your mind with body through the latter's own 'computerised' controls. Brain is how you delicately correspond with your own vehicle to fly it through terrestrial life. Same controls, different experiences. Where exactly are the pilots if their states of brain for a manoeuvre are the same but their experience differs? Where are *you* when your speech centre lights up saying very different things? When you remember or experience different stresses brain lights up the same. Control compartments can, of course, be mapped (see *fig.* 24.4); weren't they designed to be composed like that? But scans, precise or vague as they may be, reflect yet cannot read experience; nor can non-conscious, material nerves enjoy it. Immaterial mind is made of firmly different stuff.

Tracking such a line of thought the founding fathers of science did not expect a direct hit on God - either to reveal or to explode him. Happy with the notion that 'nothing physical' does not exclude the presence of mind or whatever they conceived of as spirit, they no more expected to find a Creator under a microscope or looking back through the lens of a telescope than we might expect to find Baird in his invention, the TV set. By contrast modern reasoning decrees that what cannot be found by scientific instruments does not exist; or else exists as an illusion born of evolutionary necessity. Such a 'reasonable' idea permeates the intellectual atmosphere. Professors toast it, pedagogues make echo and, with secular enthusiasm, media correspondents hokey-cokey down the line. Or lose their jobs. Thus, as every propagandist knows, by dint of repetition fiction is converted into uncritically accepted fact. If you cannot materially prove a fact is wrong - because you speak of immateriality - what hope is there? What, if you speak of nothing, can you say? In the High Court case of 'materialism v. holism' the prosecution simply claims defence lacks any adequate defence. Is your logic not of scientific style? Materialism, having disbarred all the cards it does not hold, would seem to win hands down. This lop-sided viewpoint dominates the case today.

Look at things another way. Freedom of thought and dogma, especially the forceful imposition of religious or political dogma, never mix. Modern science therefore takes visceral (and, dialectically, correct) exception to any notion that might seem 'leading backwards' to infliction of a medieval 'thought-cage'. Yet its own 'forward' counterthrust will happily place faith in untestable imaginations, in 'reasonable' fantasies that include multiverses, cyclic universes, colliding p-branes, strings, abiogenesis, biological transformism and so on and on. What does 'reasonable' mean? It means 'feasible within a wholly materialistic and therefore mathematically amenable scenario'. This position is sacrosanct. It is both promoted and defended by secular auxiliaries who, if provoked, rapidly descend from highbrow academic philosophy to lowbrow polemic. The issue of truth, physic and metaphysic, becomes a battleground. What can a seeker after truth do but, casting aside dogma and fundamentalism of all persuasions (including atheistic), press along alone? Are you a pioneer?

It has even been suggested (see Chapter 8: Conjuring a Multiverse) that what appears our particular universe might be a Grand Cosmic Simulation. It would, therefore, have to be the work of a Super-intelligence or super-intelligent race of computer specialists. Did matter come from Super-minds or did they themselves evolve to shape eternity? *Notwithstanding the origin of such 'angels', is it more outlandish simply to suggest (which is the thesis of Natural Dialectic) that subjective consciousness is metaphysical? And that, by implication, so is intelligent design - all cases of?* Does such a simple suggestion open a greater chink in the armour of scientific realism than other 'reasonable hypotheses'? Its apostasy certainly attracts a wasp-like attack from the neurological division. Indeed neuroscience, molecular biology, anthropology and evolutionary psychology combine forces to limit the damage, recharge the spell and keep unbroken faith with *mantra* - 'it is all material, mind *is* brain and brain like all the rest evolved'. A molecular biologist might rehearse abiogenesis (Chapter 20) or an evolutionary psychologist re-educate you in the ever-changing tales of hunter-gatherer 'ape-men' - because 'it *must* in principle be true, don't even question that'. Who dare 'irrationally' say otherwise? Who dare contradict the repetition popularity recites, question the Darwinian anti-holy book or resist spellbinding propaganda of a dominant mythology?

To discredit or, still worse, disprove the extrapolations made by biological evolutionists has knock-on effects. Perhaps the most severe occur in the correlated mind-set and imaginary (that is, wholly theoretical) 'expertise' that inhabits evolutionary psychology, sociology and (see Chapter 25) politics. How might you feel if competition rendered wares upon your stall set in the academic market of philosophy devalued? What if the pedigree of such a doctorate became *passé*? Never mind that revolution, in its swing from *bottom-up* to a reverse, *top-down* perspective, might still include you in a three-tiered universe; adherents of materialistic faith respond as priests might when the tables in their 'scientific temple' are completely overthrown. In what way might that be?

How can you authenticate an *OBE* (out-of-the-body experience, Chapter 18)? A single such authentication would, slighter than a pinhead, puncture big thought bubbles; it would cripple a delusion's dominance. It would imply that subjectivity exists apart from brain and reduce scientific materialism to, at best, a working hypothesis. Nor, if you or I had never had one, would it preclude the possibility of such an immaterial phenomenon. Even the entirely normal attributes of mind (purpose, will-power, manipulation of symbols, memory, desire, imagination, feeling etc.) are quite different from those of matter. Mind is natural enough but its natural laws quite different from those of non-conscious materials. Subjectivity is you. While the material zero of consciousness may seem like nothing it is, being everything without which we are nothing, our most powerful and precious factor in the universe. Consciousness-in-motion is mind and mind is life, life that in the case of human physicality employs the most powerful psychosomatic instrument in the universe, a brain, to mediate upon this plane.

Phenomenal! Can you see the wood for trees? Rainforests foliate luxuriously but is the foliage an accident except for how its seeds first fell? Nerves are mostly made of water but with architecture that's atomically precise. Fifty thousand of them every second teem to make an embryonic brain. Their root-like journeys sprout great distances to favoured destinations forging links that work. There will, at maturity, be one hundred billion each connected to between 100 and 250,000 others summing to perhaps a hundred trillion synaptic links; your brain's nerves would, unraveled, wrap four times round the world! Now add plasticity; each link involves capacity to modulate intensity of its transmission, change its shape and redefine its connectivity. Such dynamism does not tolerate mismatch, mistake or accident. The system fires, overall and 'sensibly', at billions of times a second. Phenomenal accuracy develops into healthy brains as different as their owners - very little! Ask a surgeon.

Electricians understand what problems faulty wiring makes. Neurons are wired according to genetic plan and can, in what is perhaps the most complex and deliberate arrangement of atoms anywhere, make a million relevant connections in a second. Don't think mutations in such plan will make a better brain; bugs make brains defective. Don't think those connectors are so simple either. Every synapse, as such relays are called, trades in nothing less than the cyclic cooperation of many precise chemicals. Each cycle's purposive complexity can modulate the passage of a nervous impulse in just thousandths of a second! The computational power of such a multi-layered grid is vast; no doubt also that coded transmissions within a system of electrical impulses form a physical basis for the subjective elements of thought, perception and personal memory; they link mind with its body and they do so very subtly, very well. Indeed, perhaps the whole array reflects (see Chapters 15-17) the informative sub-routines of archetypal memory. *None of this objective phantasmagoria is, however, any more subjective than a telegraph whose code is tapped along a copper wire in morse.* It is neither source nor sink but as transmitter *represents* mind's messages. In this view, therefore, human mind is 'snared' within a cage of body but is not that cage; subjectivity, like a sailor in ship or pilot in a plane, is closely locked for a life's journey-time and, using a control called brain, learns to navigate as if the corporeal vehicle were simply an extension of its psychological behaviours.

In short, brain is an information panel wired up to instruments that make the vehicle go. It exchanges information from the outside to the inside and is often set to automatic pilot; it also records life's learning process and relays decisions from the pilot to the outside. You can see it in the way a vehicle behaves. **In a phrase, brain mediates and not originates**.

Therefore the single, simple, holistic premise of the following discussion, one consistently followed from start to finish of the book, is that subjectivity is not made of matter. It is immaterial. If the subjective experience of consciousness (which includes purpose, will, knowledge etc.) is an immaterial entity, then it is not the product of matter; metaphysical information (Chapter 5) is made, despite the neurological delusion, neither by nor of electrochemistry. Sensation certainly affects the brain but so, from 'above', do purpose, thought and will. **Materialism's view, which sees the molecular activity of nerves as the generator of subjective 'illusion', is topsy-turvy. It inverts effect (nervous activity) into the place of cause (the experience of information, conscious mind)**.

Isn't the presence of an invisible event commonly inferred from its effects? Examples include planets that circle distant stars and electrons orbiting atomic nuclei. *Why, therefore, should the presence of invisible, metaphysical mind not be inferred from its informative effects*? And why, if mind is a separate element from matter, should some aspect of immaterial, universal mind not shape materials.

A Unified Theory of Psychology and Science therefore allows, according to Natural Dialectic, a metaphysical dimension in the form of active information (mind) and its source. Its primary focus will thus be to elaborate the nature of subjectivity (*aka* the nature of material nothing) and the electrochemical mechanisms of its interaction with matter. The explanation rolls from pole to pole. Chapters 13 to 17 roll out divisions of psychology from 'top' (its head, conscious or positive pole) to 'tail' (its lower, sub-conscious or negative pole). Discussion naturally includes the sub-psychological integration of mind with special, 'biological' forms of matter, bodies called organisms. *From soul to sole, from top to bottom does microcosmic man reflect the gradient of Mount Universe and, therefore, the pattern of creation?* How, in addition to the ways discussed in Chapters 5 and 6, could archetypal human bio-logic accord with the triplex pattern (see *fig.* 0.5 or 13.3) of cosmic construction? Exactly who or what is one of life's dramatic players, you? First and foremost, though, what is the basis of it all, that is, of subjectivity?

Consciousness

This is what it's all about. Without it you are nothing. The star of every play is mind; the kingpin of psychology is consciousness. What is the 'thing'?

Since by 'life' is generally meant awareness, sensitivity and cognizance you'd think it was simplicity itself. Consciousness is your subjective heart. It unifies disparate parts and is the only single, seamless way - a waking, feeling state in which experience and thought take place - that you can know the world. What have you, what are you without it? Absent, nothing, less than even ghostly. It is, therefore, the universe to you and at the same time central to your own existence. Call consciousness important!

On the other hand, the last two centuries of psychology and fifty previously have each given different answers when you ask the nature of the beast. Is it simple or in fact a very complicated thing? Is it a separate entity or 'just' emergence from a complication called a brain? If so you'd have to ask, as Chapter 15 does, the nature of *un*consciousness as well. Myriad experiments involve non-conscious neurochemistry of nerves; thousands more devices prod and probe the way that consciousness is modulated as it operates a human body (or other bodies that are more a vet's affair). Brain is an intellectual's mirror house whose light is bounced around to various effects. Ingenious experiments test this and that response. Theories slap as thick and fast as theses on an academic table but, when all is said and done, what *is* the 'electric' that drives all the mechanisms of the mind (or brain)? What is the light, the cause, the consciousness that underlies a quagmire of confabulation and, equally, the clear and seamless 'knowing' we call life?

We've been here before. Check, for example, Chapters 3 (Truth, Appearance and Reality) and 6 (Mind Machines). Is your life molecular illusion or is what you think is you, the subject, a reality? You can see brain as a 'computer' with perpetual input/ output and internal flows of electricity. But computers are for users; who is the user of a brain? And is grey matter a 'machine for generating consciousness'; or, influencing but not generating mind, is it one because it mediates between a body and the immaterial subjectivity you know as Self or Inner You or Consciousness? If life can be derived from physical equations you evolved; experience emerges from non-consciousness and the former answer, mind-from-matter, ticks your box. *Conversely and perhaps counter-intuitively, the holistic vector points to 'matter-out-of-mind'; mind hierarchically precedes material phenomena and its transcendence seeds our mindless world. In this view consciousness, if immaterial, is not an 'object' that's available for worldly, scientific test.* It is irreducibly subjective. There is, indeed, no *direct* scientific evidence that it exists - although you anecdotally confirm that you're aware! You're the owner! Your experience can only ever be, objectively, inferred. Thus consciousness, an uncreated and essential element, strips all materialistic theses out of skull and dumps them in a bin. How can you work out the way that consciousness and mind evolved when you are clueless what they really are? By whose authority, therefore, can anyone assert for sure that consciousness is brain while nescient of its nature? Therefore, wheel in a bigger bin or, better still, for evolutionary psychology a paper pulp.

In fact, if you don't know how you know then what's the basis of psychology? You need to since can't, unconscious, know a thing; consciousness substantiates intelligibility. Intelligence and information weave upon its immaterial screen. Are this screen and hence all world-views, including scientific, accidentally derived from particles and their uncertain motion? Reduction of cognition to the action of non-conscious fleck and force is itself unthinking; it elicits such asymmetric and thereby unsound psychology as afflicts much modern thought. Is, on the other hand, mind's core forever out of scientific reach? **Maybe, for the sake of argument, Descartes and Galileo excluded experience from experiment; such exclusion has been powerfully successful but it's wearing thin**. Subjective activity is an inescapable, central component of reality; the triumph of their ideology over subjectivity must be reversed and common sense redeemed. And yet the re-inclusion of immaterial consciousness threatens to unravel the entire materialistic paradigm. From *top-down*'s point of view it will.

The collapse of intellectual empires can't touch natural reality. It won't affect your liveliness a jot. You are not robotic and you need re-find your core; but, as Alan Turing, saw it, are you strictly computational? Is mind to brain as software to computer hardware? There is similarity and difference in operation but your computation needs prior mind to generate prior programs and a power source from the outside. Or is mind's power from within? Is computer-ware enough? Do you remember (Chapter 6 again) that we took a painting of yourself and upgraded to the point we'd made a futuristic robot even friends mistook for you? Has 'as if conscious' been transformed to 'is'? Is this robot as alive as you?

The existential dualism of Natural Dialectic (*figs*. 1.1, 1.2, 2.6 etc.) points up mind/ body, informative awareness/ oblivion or consciousness/ matter. If this is right it means mind machines and robots must be actively, intelligently conceived; their reason is their maker's not their own; the information vested in them is all of the passive and unthinking kind. 'As if' does not turn into 'is'; the material simulation, howsoever complex, still is not subjectively alive; and the convoluted ramblings and sometimes fantastical confabulations of professional philosophers and AI scientists are, if they take materialism's line, based on false premise and thus incorrect.

Energy (like matter) is non-conscious. In this respect it's *pure non-consciousness*. Its variability involves (as also does that mix of form and information, mind) such properties as concentration, state or reactivity - but what about antipode, a formless opposite formed from *pure consciousness*? Materialism doesn't like this turn of phrase at all. Yet, high or low, the grades of form in consciousness involve such mental qualities as mood, intensity of focus or intelligence. Could, like that transcendent burst of motive energy we call the 'bang', a highest purity of consciousness engender all the lower grades of mind and, when non-consciousness is reached, solely energetic patterns of behaviour? From radiance to utter black, from purity to gross impurity could Essential Potential realise a universe? You don't believe it. Yet holistic faiths endorse it! The call's to purify the mind. Distilled Purity's a common aspiration. So is Communion of Friends. The orient calls this stateless state *sat chit ananda* (truth, consciousness and bliss). *There are, in other and holistic words, no complexities of matter evolved or ever turned to consciousness.* Instead there is, *top-down*, original consciousness constrained by degrees in bodies made of various 'energetic subtlety'. There are 'mental sheaths' of increasingly materialistic expression. They range, as the 'super-positional' steps of *fig.* 14.1's ziggurat explain, from 'fluid' levels (as described in Chapters 13-14) to base, fixed and archetypal frames of mind (Chapters 15-17). Each such frame is in turn fixed to its physical expression - a biological sheath. This is the outermost, coarsest body; and from the range of life on earth perhaps humans are endowed with optimum potential, that is, informative freedom. Humans are, by nature and if given half a chance, knowledge addicts always interested in a fix of learning. A brain (2% of body's weight) uses 20% of its energy; this indicates an emphasis on information business, the hallmark trait of man.

In short, energy and information are distinct elements that compose the existential dipole (Chapter 1). The idea of non-conscious matter transforming itself into consciousness is incorrect; but (*fig.* 0.14) the couple interact in various ways and different extents. On the other, materialistic hand you would wish to entertain the notion of sophisticated robots slicing up the 'as if' barrier until its wafer disappears and computations merge with our humanity! No matter the original intelligence required to make robotic form; if it made you then, you believe, natural selection can manage any trick. This form of non-intelligence is credited with brains that now can strive to simulate or even overtake themselves. A further evolution of material consciousness! How perspectives generate antipodal interpretations of the facts and lead the evidence to opposite extremes of which, if truth is single, one must be illusion.

Do you want to materialise and make grey matter mind or not? The two sides seriously head-butt. They beat each other's brains up but, when it comes to consciousness, whose is the real headache? Do neurons cause the 'emanation'? *Are* they awareness? Or is it a separate entity impinging and impinged upon by them? In short, is awareness an objective or subjective thing? Of course, if it's subjective and apart-from-matter you will have to make connection. If the brain is like a console of control then how exactly does an incorporeal pilot fly his craft or psychonaut by thought sail in his ship of body? What is the nature of the borderland, psychosomatics and the way that metaphysic meets with physic? This stinger we shall (Chapters 15-17) have to grasp.

Physical and Metaphysical

You say there is only matter. Since the time of Bacon, Galileo and other founders of the quantitative, materialistic methodology of science, psychology has gradually crystallised mind and soul into the confines of matter. From soil to 'soul', did life's awareness not evolve? Flesh made soul: mind is brain. You are a body whose brain (and therefore mind) is meat that may, if so inflicted, suffer from delusion - spiritual experience - or belief in super-natural power. The clinically objective observation is that brain 'somehow' generates the mind; 'higher levels' of consciousness emerge as the brain evolves. States of mind are states of brain; they are simply products of electrically coded biochemistry and can be measured using biochemical assays, electro-encephalograms, various kinds of brain scan and so on. According to this 'mind-is-meat' brigade the organ generates ideas. Such presumption might logically revere a venerable 'relic' such as Einstein's brain for character and calculating capability locked in its grey stuff. Sure enough, in 1955 his brain was pickled, sliced up and preserved by Thomas Stoltz Harvey. By 1996 it had ended up in Princeton Hospital since when and where it's been inspected to explain its genius. Does lack of a wrinkle or excess growth in parietal parts confer an excellence of thought? Did a glut of glial cells once support the mathematician's nervous troops as they stormed cosmic riddles? You might contemplate the furrows and gaze on the marinated essence of his mastermind, his genie bottled in formaldehyde, his saturated genius jammed in jars. Reliquary in glass! Not for worship but to take apart! A perspective with only place for matter is the studious, professional bent of physics, chemistry, biology and, subsumed under their auspice, 'soft' psychology. Such objective viewpoint was assigned (in Chapter 0) to the 'anti-church' of scientism, that is, the pillar of materialistic faith.

Is active, metaphysical information (Chapters 5 and 6) therefore simply a subtle, dependent kind of electrochemical illusion whose consciousness results from rates at which nervous networks in their circuits oscillate? Feeling, thought, memory and other features of a mind must be factors scattered through the composition of that master storyteller, brain. Self could be seen as a computer chip made, instead of silicon and metal, out of nerves upon a board of different tissues. Simple, isn't it? There is no such thing as 'inner world'. Free will is the function of a mechanism called *anterior cingulate sulcus*; morality is a function of genetic make-up - we are behavioural puppets walked upon the leash of nervous and hormonal chemistries; and life 'pops' irresponsibly from nerves. *If thought is reduced to brain chemistry and a meaningless firing of neurons what is the point of life? Is such modern psychology (with its impoverished application, clinical psychiatry) not blinkered to the point of powerfully missing it?*

I say material <u>and</u> immaterial exist. The latter element is information, knowledge, consciousness. **And, complementing physical, mind is a metaphysical body.** A forms-of-thought container.

Consciousness might not appear considerable to material science. Being in all senses immaterial it might even, from the outside, seem like nothing. **From the inside, though, it is what life's all about**. Shall I therefore beware or greet the implications of objective vision when it turns upon my mind? Certainly the holy grail of biopsychology with its associated neurological disciplines is a unified theory of mind and brain. If, however, mind *is* brain and consciousness made up of molecules, clearly material and immaterial cannot dissociate - because the latter can't exist. In this case there are no such 'paranormal' things as telepathy, an out-of-the-body-experience (*OBE*) or, since you identify yourself with skin and bones, 'life after death'. 'Super-natural' is nonsense, 'miraculous' is delusory and life dies with its life-giver, a nervous system. It depends, of course, what you mean by 'super-natural' or 'metaphysical'. If you equate 'natural' with 'physical' then anything immaterial must be extracted from the equation. *Psyche* becomes, as noted above, a figment of the imagination (whatever imagination is); it is, with its twin ghost *noos*, exorcised from the machine. That pair are abstractions. Psychology's main components have been annihilated. They are nothings, empty and impotent absences. The subject has been reduced by its own adoptive materialism to a study of organic information processing. Of course, if the evolution of neurological systems is really the same thing as the evolution of consciousness, then evolutionary psychology must, as teacher, mentor and healer, preside as the official 'church'. Who dares question its authority (see Chapters 13-25)?

It depends what you mean by 'miracle'. What is 'miraculous' is, literally, awe-inspiring. This is because its operation is outside the ordinarily perceived operation of cause and effect. For example, two hundred years ago computers would have appeared miraculous to an observer. Thus 'super-natural' gives way to 'natural' as understanding grows. A problem arises only if, according to materialistic creed, you equate the word 'natural' with 'physical'; or, to turn the issue round, if you deny that 'super-natural' is a natural state, that mind is 'informative' or 'metaphysical' and the essence of a natural universe could be a natural non-physicality - in fact, Most Natural a Source of every other entity.

No doubt that scepticism is a healthy mental trait; and the filtrate of experimental science, clean of such impurities as superstition and the dross of charlatans, is truth. Material truth. Is, however, study of the immaterial by material means the way to find out more? Is state of mind predictable or the ability to think affected by a hostile or devoted crowd? Are analytic test, insistence on repeatability and 'double blinds' a proper net to lay across the field of thoughts? Is that how you will understand emotions and the paranormal capabilities of mind or, seeing that your method cannot even verify the contents of a dream, will you dismiss the element of subjectivity? The double bind for scientific method is that you cannot consistently experiment nor yet accept an anecdotal kind of evidence. *In fact, perhaps subjectivity ring-fences scientific work; perhaps it marks a step too far, a barrier that material research, by its unsympathetic principle, can never breach.*

On the other hand, is scepticism a devoted, faithful, trusting frame of mind? If scientific doubt is rational then an emotion like devotion constitutes irrational empathy and faith plain gullibility. In this sense love and lab are not allied. Warm and positive is not the same as negative and cold; is reason cold or are there ways that you can warm it up? The immaterial testing-ground is mind; mind is the lab and exercise of focus is the gym in which practice of the metaphysical occurs. Another kind of truth is there distilled. Does not focused light elicit flame? Or enthusiasm's focus fire you up? The yogi, Patanjali, for example, considered mind (not brain) a natural instrument of paranormal, psientific capability. *The Sanskrit word 'siddhi' means a psychological power, a faculty of mental concentrate alone. A single 'miracle' from his standard list (telepathy, levitation, healing etc.) would demonstrate that subjectivity exists apart from brain and would, as already noted, reduce scientific materialism to, at best, a working hypothesis.* Of course a 'saturated materialist' denies and an agnostic doubts such possibility but neither can dispute that yoga as a practice works in treating common ailments (depression, headache, back pain and so on) but, more importantly, in optimising mind-with-body health.

It works in practice. Why, therefore, should its principles and most important thrust of practice, 'vaporising' mind in reaching towards the Cosmic Essence, be awry?

The 'US National Institutes of Health' is an agency of the US Department of Health and Human Services. It includes a National Centre for Complementary and Alternative Medicine. Studies funded by this centre (e.g. in Seattle) have suggested the superiority of physical yoga over standard western drug-based treatments or conventional physiotherapy and exercises. A basic premise of yoga is the precedence of mind over matter - the priority of metaphysical over its physical derivative. If physical yoga beats or even equals western treatments why deny its metaphysical foundations - *prana*, *chakras*, orderly creation and the goal of enlightenment? These translate, in dialectical terms, into sub-conscious energies/ potential matter, the caduceus/ *H. archetypalis* (Chapters 15-17) and a goal of Super-consciousness beyond the confines of the intellect that is, translated to an English word, Enlightenment. Evolution of a soul and evolution of a body are entirely different, anti-parallel concerns. Evolution of a soul from its confinements is, from Darwinian evolution's body-loaded point of view, sheer nonsense. It is something clever chaps don't practice but, socio-biologically, explain away.

Even in the staid, etymological sense of 'awe-inspiring' there is much miraculous in brain, mind and consciousness. Indeed, objective science has difficulty grasping, in its own measurable and repetitively testable terms, the immeasurable, immaterial quality of consciousness. Consciousness is subjective. Subjective experience tells its story in the first person; objective science, even psycho-science, tells it in the third. *Science separates itself from its objects.* The 'ideal' scientist is, theoretically, an unemotional, 'on-the-side-lines' reporter rather than a player, an observer rather than an active participant. Such detachment is fine when it comes to the study of universal non-consciousness, that is, the reflex world of physical science. It also plays a useful part in, for example, the diagnosis of physical cause for psychological upset, in the psychosomatic evaluation of how mind and body relate and in a clinical analysis of emotion, mind corresponds with brain. However, while analysis surrounds a circumstance with words and numbers, empathy in silence and without such calculation also, absolutely naturally, enters in. Nor is it clinical '*hauteur*' or 'distance' but an empathy that leads relationships into communion.

Simple psychologists, as we all are, strive to accentuate the positive and eliminate the negative. One seeks freedom from the pain of restrictions (such as hunger, illness or unsatisfied desires) and isolators (such as selfishness, fear and loneliness). *Central to subjective psychology is the unifying dynamic of sympathy, friendship and love; central to mystic subjectivity is kindling these to the extent of ecstasy, a fire beyond confinement of the body and its world. If clinicians professionally avoid this path, then the saint's powerful psychology is simply unprofessional. Does such an epithet change anything?*

While subjectivity may seem like nothing and as easy to dismiss it is in fact our central being. It is holistic essence without which, for its bearer, nothing else exists. From nothing it 'balloons' and seamlessly joins everything into a single whole. Such an immaterial unifier at the heart of life has to be well accounted for. If *bottom-up* accounts will not suffice can you rule out a *top-down* order of psychology? The whole of Natural Dialectic tries to come to grips with such a possibility.

Open Up the Brainbox!

Perhaps you're still unsure. Why doubt that consciousness evolved with brain and cerebral development? The scientific answer's elementary. At base life, science and life science all depend on elementary particles. Subjectivity at root is earth and what is brain but complicated earth? Therefore unhinge the cranium! Open up the brain's box! What is there to see?

Build, say, a modern day jet engine. Could a billion mindless years 'conceive' such ingenuity? Statistically, you cry, the numbers never yield impossibility. If, unlike jets, *DNA* can replicate then mindless 'logic' might build brains. An aggregate of happy accidents could easily evolve a mind! Job done! Of course, separate components of a system are part of its entirety. Nerves and other coded chemicals compound to *mediate* in processes of interaction with the world but, still, awareness is as difficult to explicate in terms of subatomic particles as is its evolution from them. Check *fig.* 2.6. In a physical universe where the dominant (*tam*) tendency is down towards the blind dead-ends of entropy how could an instrument of order just be ordered up? *Thought is an anti-chaotic creator of purposeful patterns. It is logically surprising that, for no reason, non-conscious and illogical chaos should have constructed order to a very highly systematic climax in the most complex working system of the cosmos, an information processor whose whole, sole, negentropic business is order - a central nervous system and associated brain.* Did matter, getting far more than it didn't bargain for, perchance 'evolve' a brain? Materialistically speaking, it must have. *In reality did it?*

And what is the relation of order-making mind to brain? Are they the same thing or is brain the *medium* through which active information (mind) is physically located and thereby interacts with universe? If not neuroscience then psychology, years after quantum physics, is becoming used to the admission of a metaphysic, mind; subjectivity is entering the equations. *None of such founding fathers of neurophysiology as Wilder Penfield, Roger Sperry and Sir John Eccles believed that mind was brain.*

If, however, mind is metaphysical, by what steps did atoms evolve a metaphysical program whose intent is making sense, whose purpose is creation of coherence and whose medium, the brain, is an immensely complex, accurate technology? If, on the other hand, it is entirely physical then how come anything became, from a non-conscious, absent state of mind, minded? How did a group of atoms come alive? If consciousness is not an element your problem of explaining it becomes much worse.

How do you *explain* at all? Using symbols such as words? Language expresses meaningful order. All language is a code that can reflect its speaker's rationale. This reflection is connected by way of very fine control through nervous areas to the muscles of tongue, lips and vocal cords for speech. Each language matches sound-to-concept in a different way you have to learn; but, as with computer logic or the language of the genes, the principles for speech are coded ready for expression's day.

A stored reflection of the speaker's rationale is called a memory. In this sense genetic code is, as surely as a book, a record and memorial - but what *DNA*, protein or other biochemical builds mental shapes or itself becomes a memory? How do you explain memory, perception or, even harder, communication, learning, understanding and belief in terms of electrons, protons and patterns of force? Of course, you can explain computer hardware or a brain in those terms but what about the all-important element - information? Their construction comes from information and all code (or language) is no more than software that reflects a metaphysical domain - ideas and purposes of mind. Do not, therefore, elevate the software over its creator; never elevate the genes above their simple status as a database.

Open up the brain's box. Can you find a thought or single memory? We think of thought but remember, for a thinker, memory is critical to hold the world together. Nor has brain feeling so that, if a portion of its skull has been removed and electrodes then implanted, the regions that correlate with different senses, functions and parts of the body can be mapped. Curiously, mapper Wilder Penfield found that in a small percentage of cases stimulation at certain points could generate intense and precise flashbacks of memory. These flashbacks were no more expected than a hoard of gold coins buried but detected in a field. Similarly, years later in 2007, during deep brain stimulation of fibres in the fornix/ hypothalamic area associated with the limbic system Andres Lozano 'excavated' a walk in the park. The Canadian researcher reported that an electrode 'hit' had, in an apparently random way, activated a long-lost memory of his patient in a park with friends - and the more Lozano 'turned up the volume of memory circuits' (by increasing current through the contacts) the more vivid and precise became recall.

Both Penfield and Lozano took on trust that these events were not imaginations but actual memories. Either way, such neurological excavation in unexpected places raises interesting questions. Why this particular memory in that location? What about its 'continuity', that is to say, what happens either side of its remembrance? Are a billion memories scattered round the brain? Why, therefore, do not probes 'buzz' them everywhere they're placed? And if they are not scattered are they all stored in a few localities? How big is each; why do crude probes at any point not activate them by the score? Very little is known about the 'circuitry of memory' but could it really be that memory as well as thought is made of nervous linkages? Could electrical circuitry somehow 'hold' a memory, thought or process of perception so that, in effect, your subjective experience is reduced to nets of special physicality called synaptic neurons? In other words, your experience would amount to complex flows of ions; it is electricity that makes you 'know'; and isn't 'cognizance' essentially what you are - electricity-in-motion? Never is such current conscious but materialism works on the hypothesis it sometimes is!

Brain Very Briefly

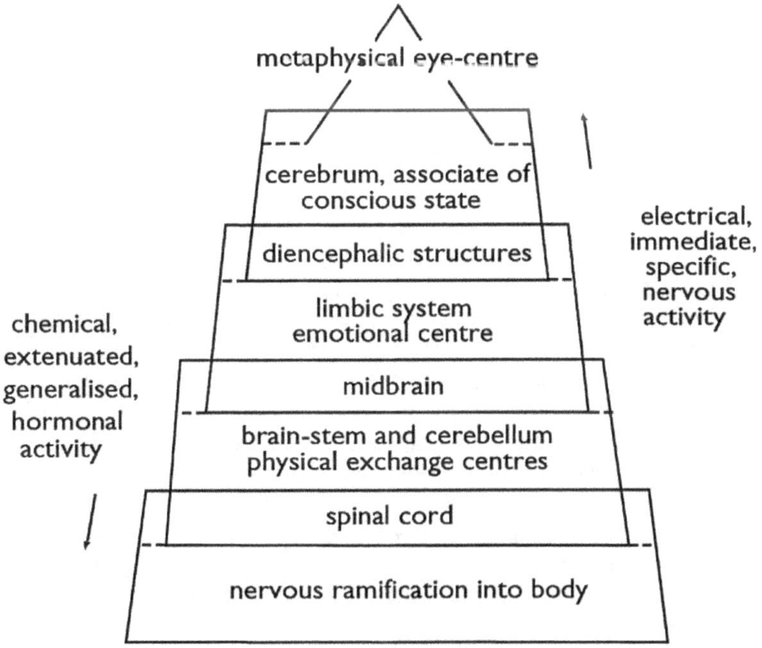

Can it be demonstrated that a nervous system reflects *top-down* order; and that various types of brain implicate degrees of flowering of capacity? In principle you always start (*sat*) top-centre with your personal cosmological axis, pivot, or eye-centre. In this subjective, metaphysical domain you would, as the psychology of conscious mind describes, actively experience the process of information. From here you would expect to drop back, down and out through more objective, reflex centres of control towards physical detail.

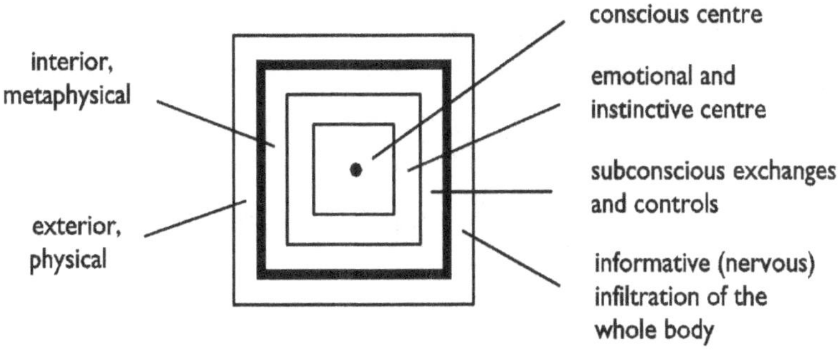

nested command and control

You would to fall through instinctive level and its sub-conscious, psychosomatic interface (or *PSI*) to a physical arena, that is, to the energetic, bilateral domain of (*tam*) body. **The ziggurat suggests that brain shows more than simple, physical, bilateral polarity. Its construction also, more fundamentally, reflects descent along an**

information axis (*fig.* 0.7 iii) called the conscio-material gradient from positive metaphysical to negative, physically-embodied pole. Such descent marks the order of psychology.

From subtle, psychological interior to gross, physical exterior there exists a clear motor materialisation of conscious cause as it affects bodily behaviours and activities; and, in the opposite, sensory direction, a drawing together of multiplex environment into a single, coherent, unified whole - conscious experience. The tiers of a ziggurat are nested. **The human information centre is mind but this nest describes the dialectical construction of its interconnection with the physical world, the 'dashboard' through which it pilots the vehicle of body. Mind drives through its physical environment using a panel of control called brain.**

As well as nested boxes a brain can be viewed (which better reflects reality) as nested or concentric spheres, one 'stage' cradled within the next. This and the ziggurat model allow the structural arrangement of brain to be viewed both *top-down* and, as reflects an emphasis on physical reality, from a *bottom-up*, inverted perspective.

A mammal's brain flowers in development from 'simple' egg (see Chapter 22) through neural tube to complex, triplex structure (hind-, mid- and forebrain parts); cerebral cortex is composed by cells migrating with the precision of a military drill from 'inside out' (the organic, Dialectic way) to form the structure in six layers. Such *top-down* program is the absolute converse of *bottom-up*'s imagination that a host of chances could evolve a single cell to cerebrum (and all the rest of brain and body integrated with it). Do not confuse these two developmental processes, the former very real the latter just a figment of the mind, in the slightest.

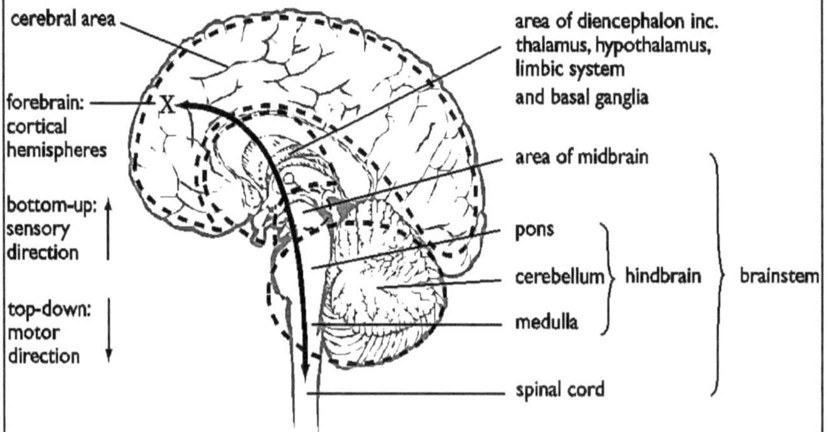

This illustration shows the main divisions of the brain that run from X in *top-down* order following the order of psychology.

The *top-down* 'motor vector' follows, inside out, the order of creation. Such an act informs the exterior world (including, mostly, one's own body). It translates into the way we execute decisions and rearrange materials; as such, it represents 'information out'. It also represents, dialectically, the way that information stored in such material potential as an archetypal memory, the *DNA* code or an egg is expressed in biological development. And, finally, it represents the original order of a brain's design (see 'Planning a Brain').

In a symbolic Star of David (*fig.* 15.6) two worlds interlock. *Top-down* and *bottom-up* are meshed; mind (↑) is linked with (↓) body.

The *bottom-up* 'sensory vector' follows, outside in, the path of perception. Such an act informs the interior world, mind. It represents the path of 'information in' from the exterior side, that is, the path of a return to *HQ* - headquarters of the brain and mind. It marks an ascent from non-conscious through sub-conscious to conscious condition. Such evolution of awakening reflects the psychological capacity of different organisms, ranged in propensity along the conscio-material gradient, for sensation and for cogitation. In man alone the 'full-blown' potential can be realised. Such evolution would involve return to Natural *HQ*, first principle of information, Pure Consciousness. This 'substance' is, for Natural Dialectic, the original immaterial factor; it is, at the Apex or the Heart of cosmos, Being; it is Essence.

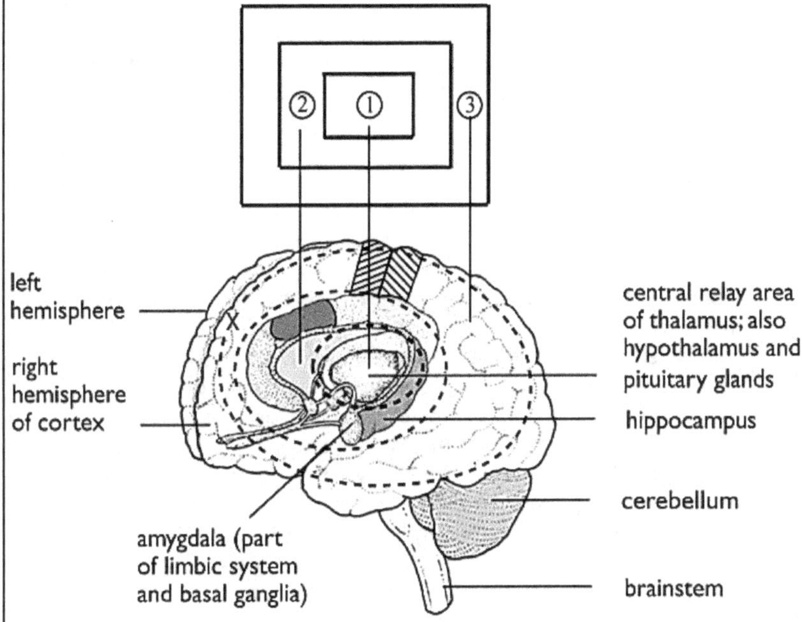

① diencephalon/ thalamus

② limbic system

③ cerebrum inc. prefrontal cortex

This next diagram is slightly left-inclined to show, by rings, the nested structure of the brain. The order of the superscript is (1 to 3) taken from a *bottom-up* perspective. Embryonic brain develops (↑) to engage sub-conscious, autonomic and then conscious states; it blossoms as the flower of **cerebrum**. Cerebrum marks a 'maturity' of consciousness. X marks the spot. Front-centre of its cortex at the full unfolding of the flower is the place you think and therefore are. In other words, you inhabit a metaphysical point of reference, an axis or a link through brain and body to the material world.

Let's trace *top-down* order from your conscious place, your thinking space called metaphysically '**third eye**' or 'eye-centre' and, as its physical co-location, the **prefrontal cortex**. Travelling back and down from here you pass the motor, sensory and other media centres of enfolding cerebrum; here nerves from sense receptors or to muscles interface electric impulse with the faculties of conscious mind - perception, thought and purposing. Below this you fall upon a second and interior 'cradle'. On each side of this cradle's outer edge find a **hippocampus** next to an **amygdala**. These organs are, it seems, involved with data storage and retrieval; they log experience in the manner of a record/ playback head; they catch or release a moment that, in fixity, is called a memory. As such they localise a connection, they address a psychosomatic interface between conscious mind, sub-conscious memory and the non-conscious, nervous complex of brain. The whole cradle is called the **limbic 'system'** (if it really is a self-contained system) or, sometimes and prejudicially, 'the reptilian brain'. Along with a further structure nested inside it and called the **basal ganglia** this area involves basic instinct and thus reflex motivation by the so-called four Fs - fighting, fleeing, feeding and f**king. Nested inside the basal ganglia you find a further 'Russian doll', a central relay station that translates electrical messages to and from the right place in the cortex. It is called the **thalamus** and, within and below its compass, is revealed a triplex nest of master command-and-control glands. These electro-physiological lynch pins 'mate' nervous with chemical, hormonal *IT* (information technology) and are called **hypothalamus**, **pineal** and **pituitary glands**. Below this 'arch' connection is established with the psychosomatic control units in charge of reflex, physiological functions; the 'primordial issue' of a **brain stem** links conscious and sub-conscious to non-conscious dimensions; in company with the master glands it ties metaphysic to physic, it bonds 'heaven' to the body's earth. This stem comprises a **midbrain** and a **hindbrain**. The latter is composed of the **pons** and **medulla oblongata**. These factors are linked internally by a 'slot' called the **reticular formation** which is probably a device involved with sleep and waking. Pons-with-medulla (site of first line, automated regulation of thermal, cardiovascular and respiratory systems) is also connected with a rear-disposed unit called the **cerebellum**. This sub-cortical workstation, often called the 'auto-pilot', receives copies of all traffic to and from the body. Its dedicated function is fast processing and refinement of the sensory/ motor responses that govern physical orientation and motion. From the medulla neural structure falls along a spinal trunk-route; from this trunk nerves ramify, as 'branches from an upturned tree with roots in heaven', into the body's earth.

The whole nervous system is material. Its physiology is as intrinsically reflex, inanimate and chemically 'dead' as a vehicle's dashboard or its wiring, dials and linkage systems.

Intelligence and not oblivion creates the vehicles of intelligence - all sorts of technology and art. Did Natural Intelligence create the natural vehicles of intelligence - cells and organisms? Don't think of evolution but development. Think of a system of development that in its constructions increasingly allows the liberation of intelligence from its obliteration in material coverings. From this perspective you start with wholly automated, archetypal systems such as cells and cell metabolism. In the case of more sophistication you might add some nervous ganglia and flexibility accorded from the context of a personal memory. From physical exterior you ascend towards the interior processes of brain and then, above the physic, metaphysic - mind. As seed develops into plant there evolve the spinal cord, brain-stem and higher, more flexible agents of connection and relationship. Full flowering would unfold into a symmetrical couple of hemispheres, the winged purveyors of consciousness, thoughtful 'mediators in between earth, man and the gods'. Except that these hemispheres are, *per se*, as chemically 'dead' as any other part of the body; and except that a seed unfolds into a flower according to a strict, pre-programmed regime. If you are thinking 'chance appearance', then forget it. Think, with brain, intelligence! A well-planned, well-rehearsed appearance! And if the constructions that permit the exchange of information between mind and matter are, in the human case, rationally and minutely programmed, it would suggest that there is also a pattern in the degree of development typically allotted to each one of a range of biological life forms. In this case a form would represent a particular point of arrest, a limit of unfolding or of evolution on the conscio-material gradient. It would, therefore, equally suggest that, from dormant through lesser into fuller consciousness, humans are extremely fortunate to find themselves gifted with the fullest biological flowering of informative potential. How they use or abuse the gift varies (can't you see it all around you?) greatly!

fig. 13.2

Certain circuits in live mode can somehow dream up neuroscientific life. Some thinkers even dare to call their key presumption fact.

Or is there more to life than nerves and, bundled by the billion, a mass of master-connectivity called brain? Does no concept underwrite development and ordered structure of organic calculation? Is there no metaphysic to the mind? For Natural Dialectic mind including memory is, both in origin and operation, metaphysical. *Brain is a medium.* Of course, embodied mind must have connection with specific input (called sensation) and control the neuromuscular output. How might its impressions be appropriately stored? You might expect to find a neural filing system correlating their committal. Such systems always need addressing codes. Couldn't a neural 'algorithm', set up in accordance with input received, act as an efficient 'point of recall' when a certain memory was prompted in response to present circumstance? What else is learning? Memorisation/ committal is followed by appropriate remembrance/ recall. Specific nervous combinations are assigned to index and, like pin-codes unlock a given memory. Nor are cerebral systems naturally as random as a surgeon's probes. External stimuli must trigger relevant stored information; and such 'remembrance' be included in the calculation leading to a physical response.

Language is a case in point. Is memory of a language physical? Are words found as nerves, nouns declined, verbs conjugated or a sentence parsed by synapses? Touching the speech area can produce a temporary loss of that facility (*aphasia*). A patient may understand something but fail to find words for it; but he has not forgotten his vocabulary. People may return from coma or amnesia and find that they can speak. These instances suggest that language, speech and its mechanisms are, although mediated by nerve cells, not identical with them. And is experience itself, including the context of memory and recognition that envelops it, simply the electrical stimulation of a specific, local net of nerves?

Imagine, simplistically, that brain is like a tape recorder but the tapes are metaphysical. Hit the button, hear the song. A record player mediates between recorded information and the body of the world. This does not mean that memory itself is physical or made of nerves. It may lie buried but, as Lozano's 'increased-current-gives-you-higher-resolution' seemed to shout, perfectly preserved. You might, in other words, damage a play-back system or lose access to your songs but these don't therefore disappear. Committal and/or recall are impaired. They might be lost but the 'reel' of memory's film is perfectly preserved. Behind the scenes our archives stay intact; personal memories are linked to our requirements by addresses in the brain. And after death? No brain, no mind? No nervous link with memories but could (see Chapter 18) emotional attachments carry through? Could the 'reel' itself be metaphysical; could our psychology endure?

By now the thrust of argument is clear. Nervous flux *reflects* our subjectivity. It mediates between the 'outside world' of our own bodies with the circumstance that they inhabit and the inner world of memory and conscious mind. The flux connects our metaphysic with its earthbound episode; it interacts (*fig.* 15.8 and Chapters 15-17) with our subconscious structure and, through this, the ship that skims our mortal surface, consciousness.

In other words double translation, whereby a stimulus is translated by a sensor into a sequence of electrical impulses that are, in turn, translated into a pattern of activity in the brain, is the function of nerves. *Nervous objects, however intricately coded, no more explain the subjective aspect of sensation than pulses down a telephone explain the conversation.* How does the third and final translation from cerebral patterns into what we experience occur? *Physics and chemistry lead to the threshold of sensation but it is not reducible to them alone.*

Similarly, responses that are recognised by a patient as involuntary rather than voluntary can be caused by electrical stimulation in some regions of the cerebral cortex. The person might say "I swallowed" or "my leg moved" but add that the electrode caused it, not his own deliberation. At no point does stimulation cause belief, decision or understanding. Indeed, if electrodes can move a patient's body without being able to make him *want* to move it, the indication must be that human will, intellect and learning have no bodily organs. *The implication is that the brain is an interface or connector; it is a conduit or medium whereby information is exchanged between two worlds.* It is a control panel through whose 'neural morse' its owner communicates with the physical world but it is *not* the organ of thought or consciousness.

You see, hear, remember, can't remember, think and so on. You may persuade yourself all kinds of 'psychological' events are in fact physical or, conversely, that awareness is 'knotted', in the form of mind, with body. Either way animation involves an exchange of information. If you watched a plane but could not see the pilot what might you expect? That flight behaviours involved not only sensory read-outs (orientation, altimeter, fuel etc.) and moving instruments but also the invisible pilot's attention and decisions? In the same way it is possible to interpret scans which identify various sorts of 'hands-free' psychological manipulation at specific locations in the brain as an accurate map of the operations of a sophisticated control panel and its 'nervous' connections with the body. Every time a turn is made or flight-level adjusted you see the same instruments move or light up. This indicates nothing about the specific instance either in terms of context or reasons. You can turn hundreds of entirely different corners with the same panel routine.

Remember (Chapters 5 and 7) the white horse you visualised? How do nerves dream such imagination up? Let alone a 'video' sequence of it galloping, a variable sequence into whose detail you can zoom at will? It is conceived of remembrance. What is a memory? If, in an act of remembrance, limbic hippocampus, the amygdala, prefrontal cortex and other cerebral lights shine, do they signify any particular memory? *They certainly indicate neural activity, a biological appurtenance of psychological condition and, perhaps, resonant aspects of an associative address system. If they are damaged then, as a damaged tape-head faultily records a song, memorisation is impaired. In other words, they may be necessary correlates of storage and retrieval but does this add up to the memory itself?* Is your history just electric patterns written in grey matter like information on a hard-drive or CD? What about an image of your mother's face? Is that just sparkling hippocampus and the atoms of some nerves - or is what's subjective something more? **To reiterate, the question of what memories consist of still remains.**

The brain is about mediation. It makes connections, by way of body, with outside circumstance. Is memory of a perfume or reminding by a song not immediately associated with physical as well as emotional response? That is, with nervous and hormonal patterns mediated by the brain? Would you not, therefore, expect that memory be tied to physical address? To some tag that, at least locally, reconstituted the brain-state of the original experience so that in remembrance the old physical response is 'refreshed'? In this case if you press the physical button (unnaturally using an electrical probe inserted through the skull rather than the natural sensation of an outside influence) the mental side is switched on; and *vice versa*. Such connection need not be interpreted as other than a link. *Evidence from electrodes or brain scans no more makes thought physical than a study of telephone wires or computer operations brings their chemical machinery to life.* Physical systems such as brain can store, connect and transmit informative patterns but not, without metaphysical aspect, experience them. **To confuse electromagnetic pulsations with the subjective message they carry is, from a *top-down* perspective, the standard error of materialistic perspective**. *It is to mistake message-bearer for message, to confuse priorities and take effect for cause.*

Because this error crops up continuously, like a bad but unchecked habit, in scientific interpretation of the working of our minds it is worth rephrasing. To confuse a configuration of magnetic particles on a tape with the purpose, origin or import of a song is as mistaken as to confuse neural patterns with subjective realities. *Standard neuroscientific explanations always start, from a dialectical perspective, upside down.* They run from the wrong place *bottom-up* and, due to the primary philosophical error of dismissing immaterial factors, take objective representation for all there is. They take the carrier (be it code, silicon chip or neural circuitry) as if it were the whole and not the lesser, passive half of mind's reality. There follows an inevitable attempt to squeeze consciousness into electronic passages and mind into synapses and other nervous parts. Is a radio its program? Whose purposes does a computer serve? **Such material confusion of physical and metaphysical, to the point of denying the latter exists, is a cardinal error**. *Moreover, whether or not you are an atheist (and many neuroscientists think in an atheistic, evolutionary way) the treatment of mind and memory as brain is pure hypothesis. Such speculation is, from a dialectical perspective, misleading. In spite of textbooks and best-sellers written wearing such a style of coat, it is best stripped off.*

Doubting Thomas

No doubt, tell Thomas, that your body with its brain is a miraculous vehicle. Its control panel, which includes nervous and hormonal systems, trades huge amounts of information each second. No doubt, too, that the controls of a vehicle can go awry and adversely affect the driver who, until its journey's end, is locked inside. A mind snared in a chassis with fault-prone controls has a desperate time. Therefore, because accidents and faults happen, someone might even open your cranial bonnet and, having identified specific psychological events tripping specific neurological and hormonal events (and *vice versa*), decide to tinker. This is the basis of a 'brain mechanic's' drug, hormone, electrical and even surgical appliances.

No doubting also, Thomas, thought or the metaphysical aspects of your mind seem able both to direct the chemical details of brain and provide an enveloping unity of conscious experience. Intention, for example, impacts the brain and thereby continually changes its electric patterns and, using the attachment of a system of muscles, the physical structures of the world. From outside, on the other hand, mind continually translates the sensible world into a dazzling multi-tech, hi-fi movie whose qualities multi-billion-dollar Hollywood would love to emulate. It is able to access memory, exercise more or less emotional control, contemplate and issue highly accurate, coordinated motor responses. Waggle the little finger of your right hand slightly twice. It was certainly, as any damage to them would make clear, nerves and muscles that caused your finger-dance. What of the conscious part? If consciousness *is* immaterial why should it not be able to act in ways matter cannot? To exhibit, for example, free will? Purpose? Understanding? You would, however, no more expect to find such capacities in brain than you would a programmer in a computer. Are you still absolutely sure that mind is brain; and. therefore, it's matter over mind?

Mind over matter - is that true or just a step too far for Hesitating Thomas to advance? This cannot, understandably, be materialism's logical position. But mind *suggests* and the concentrated power of suggestion is immense. For example, every doctor knows it cures but, because the western ethos fosters confidence and a dependency of patients on its type of medicine (with drugs and surgical procedures to the fore), it is not surprising that a patient's own power, resting in his mind, may be insufficiently aroused and exercised. A potentially strong and voluntary cause of health is played down as involuntary effect - 'placebo'. This point of view turns, as materialism does, 'mind over matter' on its head. Drug companies might suffer loss of profit but health schemes save millions by re-turning 'matter over mind' to the right way up. They might promote a natural remedy at least as 'pleasing' in potential and powerful in parallel effect as a molecular prescription in a course of pills.

Why not, in other words, promote the great good will already vested in medicine by drawing out another kind, the patient's power of positive suggestion? This efficacious 'trick' already claims astounding cures so why not maximise its capability by swallowing 'useless' pills, enhancing the amplifying power of ritual but most of all by systematically embracing positive relationships with confident healers and teaching the practice of positive auto-suggestion by the patient? There's nothing, Thomas, new to human history here. For example, three internal practices of consciousness - the focus of mind on a specific object (internal or external), the development of will and power of suggestion - are fundamental to the instrument of yoga. And in this hallowed practice perhaps suggestion is the greatest of the three.

Every moment, theory says, prompts in new ways. These prompts arise from both inside and outside mind. Humans at all ages learn from and actively respond to constant cues from others, the environment and their own mental world. This world includes creativity, that is, suggestions in the form of ideas, planning and imagination. Such immaterial creativity is as real as any other part of universe. The law of action and reaction *(karma* or causality) adds that all suggestions are realised sooner or later, for better or for worse and with intensity proportional to the concentration of their impulse. If indeed there is an element of information, mind, would not projection of the world involve suggestion by a great but natural power?

There is, therefore, no trick. Is there doubt that a subject's independent will-power, concentration and determination to be healthy, fit and well work, in combination, wonders? Trust is not the same as gullibility nor is faith 'suspension of belief'; their focus fosters superhuman strength. The question is from where? From molecules, electromagnetism or a circuit made of nervous cells? Or is mind operating over matter?

The objective, conceptual framework of western medicine, interminably reiterated in the culture of its schools, emphasises body over mind. Could this be the reason that it fails in its promotion of subjective faculty, that is, the overriding capability of mind? Does it not need to turn, take in and then develop both sides of a man's equation equally? Neuroscience even shows a physiological effect reflecting a suggestion. Dopamine and other chemicals associated with the expectation of reward and its enjoyment are released in the brain of a patient in the throes of faith healing. A dose of positive emotion is the subjective equivalent of electric current. 'Focused suggestion' by another healer or the sufferer himself causes material, physiological effect. Secondary effects ensue; they cascade through the body and, where sickness is illegality, help to restore its normal, lawful state. Faith and determination make a powerful doctor surging through the system in a quest to help it mend. Such 'miracles' help set the record straight and prove that Doubting Thomas lacks a faith in faith.

From even these obvious observations it might well be claimed that the riddles of awareness, will-power, thought and sensation are, answered in purely materialistic terms, not merely unsolved but devoid of a basis of their very beginning. **Their evolution, in the sense of subjective experience as well as the objective origin of neural structures (neurons, synapses, glial factors etc.), is a Darwinian black box.** *Because the evolution of consciousness and the evolution of mind are unexplained evolutionary psychology (the most logical form of materialistic psychology) has neither clear context nor explanation of its primary subject of research. In which case what validity has it?*

How can you call psychology material science if you can't touch what you're talking about? Yet if the life of mind is more than a material process what does the subject name as immaterial? The problem is that our experiences are, in subjective aspect, scientifically intangible. They are objectively inaccessible. We can *infer* something of mind's operation from outward neurological and other behaviours; and we know from our own experience that things like consciousness and memory are true. Science, self-manacled to philosophical materialism, is bound ascribe them physical status and, accordingly, couch its explanations in terms of atomic automata such as electrochemical interactions and neurons. **How, in this case, do the properties of cells and circuits result in consciousness? Quite possibly they do not nor ever did**. *Natural Dialectic, noting the profound explanatory shortcomings of one-tiered materialism, simply and reasonably proposes an addition to physical biochemicals - non-material, metaphysical mind.* What greater inherent improbability, wrote the founder of modern neurophysiology, Sir Charles Sherrington, than that our being should rest on two fundamental elements than on one alone? Does Sir Charles mean Natural Dialectic's elements of consciousness and matter?

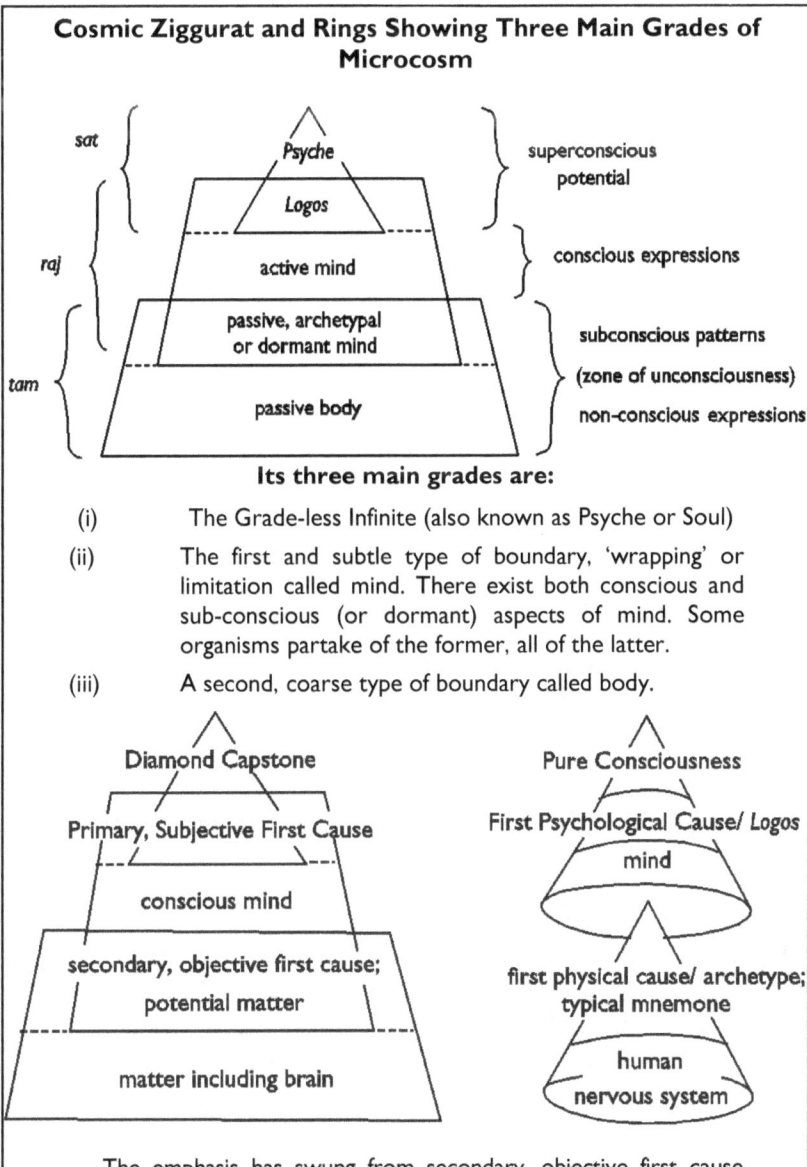

Cosmic Ziggurat and Rings Showing Three Main Grades of Microcosm

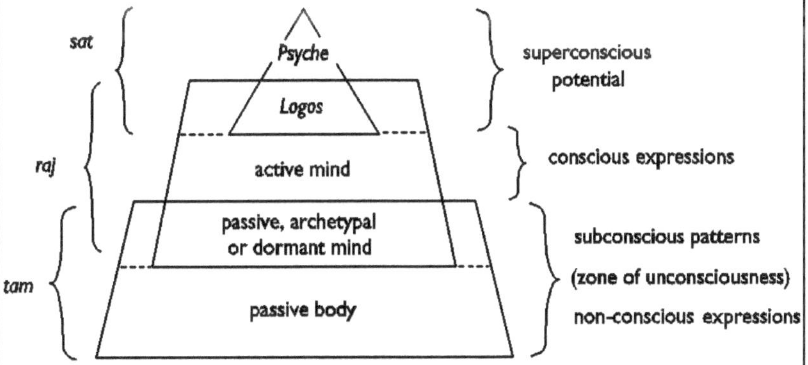

Its three main grades are:

(i) The Grade-less Infinite (also known as Psyche or Soul)

(ii) The first and subtle type of boundary, 'wrapping' or limitation called mind. There exist both conscious and sub-conscious (or dormant) aspects of mind. Some organisms partake of the former, all of the latter.

(iii) A second, coarse type of boundary called body.

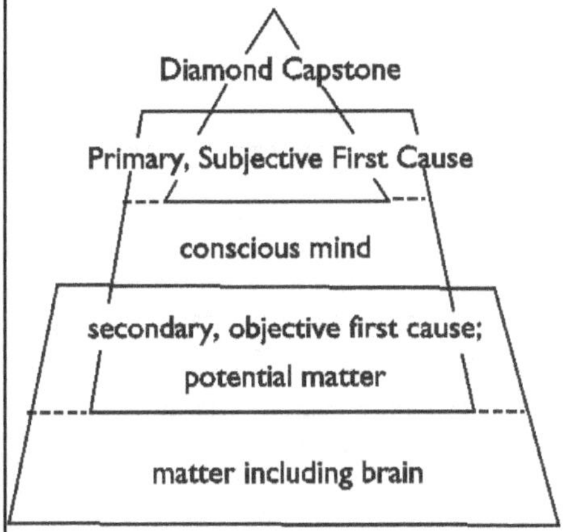

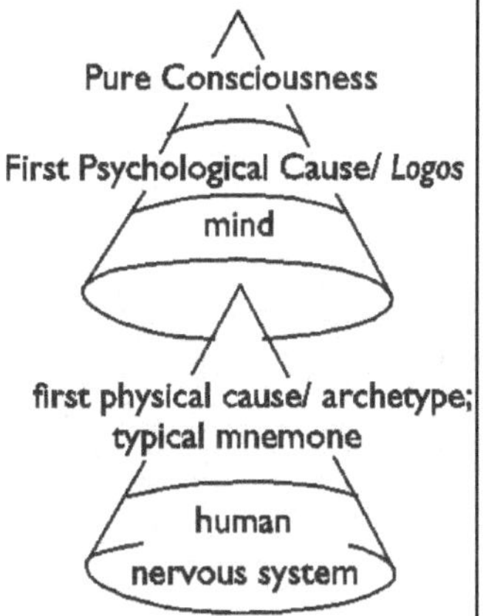

The emphasis has swung from secondary, objective first cause (Chapters 7-12) to Primary, Subjective First Cause of the Diamond Capstone (*fig.* 7.1). The characteristics of its Essence (right-hand column of Main Dialectic) include Infinity, Unity, Nothing and Pure Consciousness. The latter is called Psyche or Soul and, in its universal capacity, Universal Soul. Using this terminology Universal Soul (Pure Consciousness) is the diametrical antithesis of the material universe (pure matter). Between these poles run shades of mind which are, as we shall see in figure 14.2, reflected in the neural correlates of brain.

Although it is usual to think of a biological body as animate rather than inanimate in fact both terms apply. All objects or things are bodies. Do you remember the equation of Essence with No Thing or the (N)One (especially in Chapters 0-2, 7 and 10)? No Thing could also be thought of as No Body; or as Pure Consciousness that precedes and constitutes the central axis of a couple of bodies - one psychological and the other physical. Mental is nested within material. A couple of forms restrict the formless; this couple of bodies cloaks Infinity. As a reflection of the cosmos, your mind would operate within the objective crystallisation of your biological, strictly physical body. In this view you (yes, you personally) are bindings placed, most paradoxically, round Boundlessness!

If existence issues from The Essential Centre, it is in this sense the offspring of No Thing. As cosmos (made of motions and incorporated fixities) issues from No Body so we, minded and embodied microcosms, are the sons and daughters of The Pure Potential prior to any individual body. We are, in other words, the issue of No Body.

Concentric Sheaths of Mind.

Just as you need a diver's suit to explore the ocean so you need a human body to interact with the 'normal' physical world. Each universal level needs a natural body of its type to interact. Has your mind no limit? Mind is also a container, much more fluid than material body, correlated with the world of mental possibilities. Such 'levels of containment' can be modelled as 'sheaths', Sheath-localities in time and space are 'nested layers' that veil Central Essence.

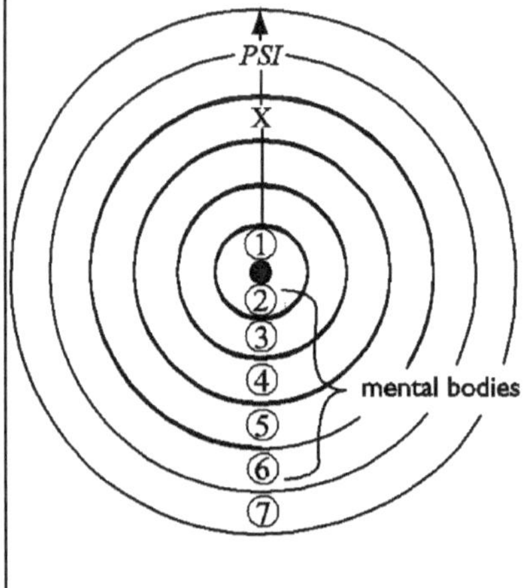

②-⑥ mental bodies from fluid to fixed

① Psyche, Soul; Self; Central 'Body'; Sanskrit: *atma* Latin: *anima*

② First Cause

③ causal mind

④ 'astral' mind

⑤ conscious, embodied self (3 levels)

⑥ fixed, subconscious mind; reflex habit, instinct; personal mnemone, typical mnemone (archetype)

⑦ physical body inc. nervous system; physical/ non-conscious; the special case of 'no mind'

The dialectical creation is a coherent unit incorporating three levels (see, for example, *figs.* 1.3, 2.11 and 13.3 (i)). It is three-tiered with sub-divisions.

Individuals 'live' their normal existence at different tiers, bands or levels of the conscio-material gradient. This gradient drops from (1), where Essential Consciousness or Psyche is, in its 'naked glory', unaware of created existence. It falls through 'hardening' degrees of separation, each with its level's relative lack of 'cosmic consciousness', until (5 and 6) the dark zones of sub-consciousness (potential matter) and non-consciousness (physical matter/ universal body).

These individual minds and bodies (microcosms) are graded according to the level of subjective activity they can achieve.

(7) Material body achieves none and is therefore graded lowest. Its oblivion is closed to all subjectivity.

(6) The archetypal level of potential matter; personal and typical aspects of sub-conscious mind (Chapters 15-17). 'Lower life', that is, unicellular, fungal, plant and some animal forms appear to function at this level.

(5) The lowest level of conscious mind, physically embodied mind. It is what, with differing conceptual capabilities, conscious animals experience as normal. Human grades are (see Chapter 14) three - higher, intermediate and lower.

Levels (5 and 6) of embodied mind and (7) mindlessness are unaware of the super-conscious conditions of detached or disembodied mind. These conditions transcend intellect. Such 'post-intellectualism' transcends all verbiage and is, as such, outside the scope of more than mention here. From a super-conscious viewpoint lower states of mind are 'dormant'. From their opposite point of view the higher is unknown and, even if suspected, unrealised. A disembodied state (Chapter 18) is involuntarily provoked by, possibly, severe trauma and, definitely, death; and in voluntary, controlled way by profound meditation. This lucid separation of consciousness from physical contact is like a higher gear, an overdrive of mind. It is often called 'an astral projection' or *OBE* (out-of-the-body experience). Super-conscious divisions 1- 4 are grouped together in *figs.* 13.5 and 14.2 as a single Transcendent Phase (1).

(4) Energetic, free as opposed to 'fixed', embodied mind. This 'sheath' is not restricted by physical incarceration. Its region of experience includes so-called heavens and hells, 'astral travel' and direct interaction with other minds, that is, thoughts. Such interaction is called telepathy.

(3 & 2) Mind-in-principle. This is 'causal' mind whence the order of existence derives. The inmost sanctum of this region is Potential Mind also called First Cause, *Logos* or *Aum*.

(1) Essence.

fig. 13.3 (see also figs. 0.5, 0.11, 12.1 and 17.6 - 10)

In short, while metaphysics is 'scientifically' debarred its subject matter may in fact be as true as your own mind and memory. Consequently rational steps must be taken to explain psychology and psychosomatic linkages within metaphysical terms of reference. <u>*Such terms are presumed and therefore the*</u>

next four chapters are written, Thomas, as if you had overcome the subjunctives that accompany uncertainty. They prefer indicative to a conditional mood.

Is not psychology already scientific? Since mind is brain, thought is nervous and experience a kind of electronic motion psychology's a science. The pair are unified by common physicality! Materialism's line, with theory and its practice well advanced, is without doubt the only reasonable one. Surely, as a naturalist, you'll claim such thrust as 'towards' implies is as redundant as it's ill-conceived.

Well, first, is mind amenable to mathematics? Can the scientific language made of numbers and equations fully satisfy a study of psychology? No doubt that in the case of observation and description of the various physical expressions of behaviour (hormonal, nervous, muscular and so on) it can help. Material operations in a given context are, when all impinging factors are accounted for, computable according to objective, natural law; and, if some factors are not known or understood, in imprecise or very complex cases operations are reducible to a statistical analysis. Such calculation can apply to brain and subconscious, automatic reflex of the mind, that is, to archetype or instinct; it applies as far as conscious thought (say, choice, purpose or desire) does not affect or override such reflex. Although the external, environmental context of an organism's body changes, reflex response is calculable; but if the internal genius of a human mind is changed by *choosing* a response, this 'override' is mathematically incalculable. The less conditioned, freer choice the less amenable it is to sums. Nor do numbers ever catch the feeling. Ever phrased equations for a dream?

Yet active mind may also act 'in character' by choice according to, say, code of conduct. In this case response can intuitively, perhaps even statistically, be calculated - unless the rules are 'scrambled' by minded body; but it becomes increasingly irrelevant as the subjective calculus of experience is entertained. Here psychology is not reducible to a numerical analysis. Numbers are irrelevant to feeling. They miss the central, knowing part of mind. They are especially impotent where positive, non-numeric feelings such as love and friendship rule supreme.

Secondly, if there is any metaphysic to the subject, then no - 'towards' were never ill-conceived! If a key element of mind has been reduced to a ghost of its real self and the central, immaterial dimension of psychology's reality ignored then how far past hormones and the neurological has science penetrated? Do genes and biochemistry explain the lot? I say there are matter *and* mind; mind is, as opposed naturalistic axiom, not a sort of matter. Just as a geocentric view of the universe needed its Copernican revolution, so a cerebro-centric view that mind is restricted to brain needs its constraint lifted. On this basis we can take a standard dialectical appraisal of the *psychological* circumstance.

The above diagram shows (as elaborated in *figs*. 17.6 - 10) a gradual embodiment of man according to the conscio-material gradient. Do you remember (Chapter 3) the equation of psyche with soul? Top Centre is (*Sat*) Essential Psyche; the cosmic balancing factor is its Pure Consciousness.

tam/ raj	*Sat*
existence	*Essence*
objective/ subjective effects	*Subjective First Cause*

	lesser levels of consciousness	*Pure Consciousness*
	body/ mind	*Psyche/ Soul*
	passive/ active expressions	*Potential*
↓	*tam*	*raj* ↑
	non-conscious/ objective	*subjective/ conscious*
	reflex	*voluntary*
	passive information	*active information*
	material body	*mind*

Top-down it has been supposed (Chapters 1-12) that creation is a polar exteriorisation or projection from this Infinite Potential. Potential comes before its own expression. As information precedes action or plan construction, so Psyche precedes existence. The first existential sheath is (*raj*) mind, an informative domain; and the second a body, a (*tam*) physically energetic domain.

You might check the stacks in Chapter 7's Order of the Infinite and Finite. From these you gain an inkling of the nature of our triune, cosmic home. *The same principles as applied to matter apply now to psychology and will apply to animated figures of biology* (in Chapter 19). The conscio-material gradient which vectors signify is called Mount Universe. What is the nature of its peak?

	tam/ raj	*Sat*
	lower	*Highest/ Peak*
	outcomes	*Potential*
	duality	*Unity*
	periphery	*Centre*
	more or less	*Completion*
	range	*Absolution*
	hierarchy	*Transcendence*
	spectrum	*Transparency*
↓	*tam*	*raj* ↑
	down/ lower/ outer	*up/ higher/ inner*
	passive	*active*
	negative	*positive*
	inertial	*dynamic*
	division/ diversification	*union/ unification*
	complexity	*simplicity*
	tension/ pain	*relief/ pleasure*
	confinement	*decompression*
	gravity	*levity*
	constrictor	*liberator*
	separation/ isolation	*communication/ relationship*
	particle	*wave*
	discontinuity	*continuity*
	materialisation	*dematerialisation*
	crystallisation	*evaporation*
	creation	*dissolution*

On the one hand (*tam*) inertial principle puts on the brakes. It 'descends' through executive and motor mechanisms until it grinds to a halt in solidity. Its materialising, motor tendency links Psyche with earth. In dialectical terms a kinetic phase proceeds towards exhausted, finished product. From potent start to impotent end, from infinite possibility into finite restrictions called forms and fields of influence the universe is rolled out from (*Sat*) Potential. It falls through (*raj*) kinetic mind to (*tam*) exhausted precipitate called matter. Such a scale reflects the three levels of involvement of Consciousness with forms of energy (*fig.* 0.14). The levels are - as previously explained - None (no activated energy), mixed and none (no consciousness) - Transcendence, mind and matter. On this basis you could call Consciousness-in-motion mind. (*Raj*) mind informs and is informed; it is the informative aspect of creation. Matter is, in this view, a special case of mind, that is, mind with a complete absence of consciousness. Non-mind. 'Crystalline' mind. Its totality is informed energy.

On the other hand (*raj*) kinetic principle ascends through the physical senses and contemplation towards (*Sat*) Psyche. Consciousness integrates. It unifies. Thought 'brings together' otherwise disparate entities. Education's motive is to heighten comprehension. And in this case, if the right-hand mystic path is of extreme ascent, it is no mystery. It is simply one of total focus and commitment towards togetherness - simply the path towards Ultimate Understanding and Complete Communion. This is a sure-fire way of 'getting naturally high'.

In the first diagram (13.3 (i)) man is modelled as a reflection of the cosmic pyramid. It is also useful (13.3 (ii)) to think of him as, rather than an aggregate of chemicals, an energy complex comprised of a series of concentric rings. Such rings are an image of discontinuity in continuity; they portray structured power in the simple form of waves spreading from a central commotion. Viewed in this way the construction of man and woman radiates from a Centre, Psyche, outwards through subtle, subjective mind to the gross, objective, outer sheath of their respective bodies. (S)he is seen more as a multi-layered, flowing event, less as a rigid, biological phenotype.

Secondly, more importantly, if your human axis is the cosmological point X, the eye-centre, it must lie outward and below the source of all expression - Central Axis. Pure Consciousness or Psyche is the Axial State of Grace and point X marks a fall from it. Equally, turned in the other direction, your own pivot marks the lintel, the starting-point of a *return*, a contemplative return to Origin and Original Grace. Creation floods outwards so that, without such a U-turn, such a re-think or repentance, there can be no inward re-ascent. Only an individual mind can decide which way it turns. In other words, you did not ask to be here but must, if you want it, ask for a return. Ascent from 'fallen state' is voluntary; a return to Rightful Mind will require a focusing of dispersed attention until, from paradise lost, the condition of Pure Awareness is regained.

From Potential issues motion, from Inaction action. This explains why a mystic sees both mental and physical action as 'interference'. They are kinds of 'noise' that degrade a full and clear Awareness of Self. His practice therefore seeks to minimise arousal and maximise quiescence. (S)he sits in a quiet place and blocks, as far as possible, external sensory input; then begins a 'battle royal' against the restless, outward tendency of mind. A lesser self, ego and its assertions, wanders continually unless calmed until a point of coherence is obtained, until no further interference disturbs the peace.

What has such extraordinary, unworldly 'extremism' to do with the routine of factory, office, street or home? How is it brought down to ordinary earth? There are a couple of final, key realisations to be drawn from the 'concentric rings' perspective. They are derived logically from a *top-down* perspective and are simple but also, set in a contemporary context, revolutionary. Firstly, a metaphysical archetype of primordial man must have preceded in time and must still - in the Dialectic's informative sense - hierarchically precede his biological expression. Secondly, if the 'organic' materialisation of successive sheaths stems from an Essential Centre then this radiation will incorporate an 'umbilical string', that is, an ability to connect back to the Cosmic Core.

Thus the state of man involves two forms of transcendence - one conscious, one unconscious. The former is the source of mind, the latter of material behaviours. Higher, psychological Transcendence goes beyond the mind; it precedes existence; and this higher form of archetypal order is (as Chapter 14 demonstrates) revealed in 'super-conscious' states of mind. Lower, physical transcendence involves (as Chapters 15-17 elaborate) archetypal memory. Such memory, whose imperceptible habitat is the sub-conscious zone of mind, precedes our energetic universe. Also called potential matter, its programs are reflected in material behaviour and shape; and, if creature animation is involved as well, instinct and morphogene. As such it is reflected in the architecture of the physical structure closest in operation to subjective mind, objective brain. For example, the tracks of super-conscious state impinge on brain as alpha waves, endorphins and so on. Neurology becomes, seen from this materialistic angle, 'neuro-theology'. You can study how a nerve or neurochemistry might resonate with an ecstatic state of mind or conjure images of light or paradise or gods. It's Mozart and the physicist again. Don't confuse analysis of sound with creativity or thoughts or feeling; don't muddle such reflection of experience with experience (perception, thought or feeling. In short, don't reify the bliss deifying matter as first cause. Electromagnetism and associated chemistry lie at the *bottom* of the cosmic class.

Essential Psychology

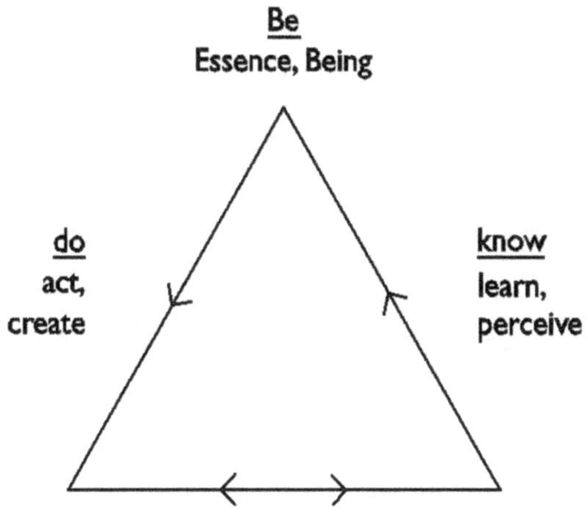

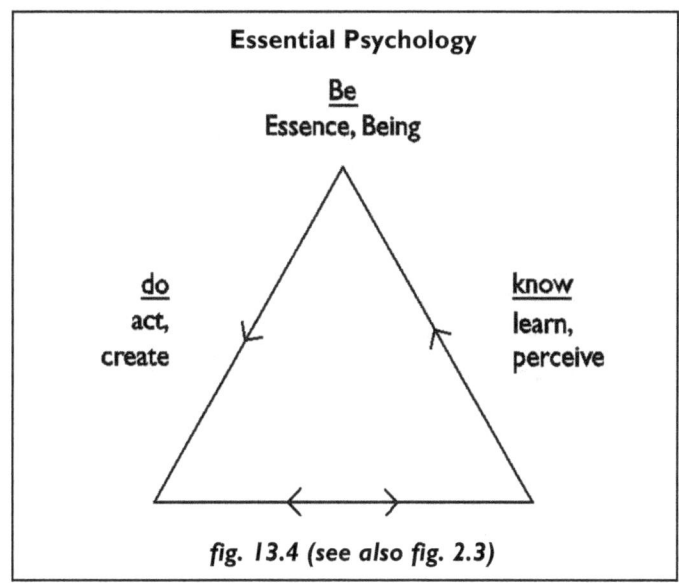

fig. 13.4 (see also fig. 2.3)

tam/ raj	*Sat*
existence	*Essence*
do/ know	*Be*
less conscious	*Super-conscious*
body/ mind	*Soul*
relative knowledge	*Knowledge*
separate disciplines	*Whole Science*
lesser truths	*Truth*

↓	*tam*	*raj* ↑
	external/ outward-bound	*internal/ inward-looking*
	do	*know*
	objective	*subjective*
	info. out	*info. in*
	motor/ action	*perception/ sensor*
	energy	*information*
	informed	*informer*
	involuntary	*voluntary*
	storage	*processing*
	memory	*thought*
	unaware	*aware*
	body	*mind*

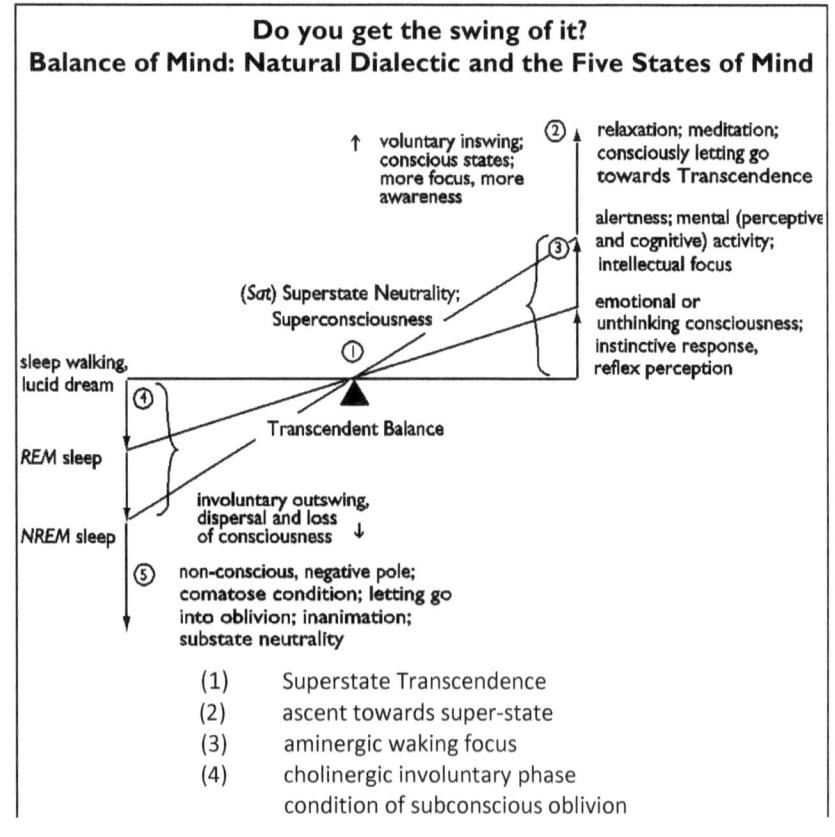

Do you get the swing of it?
Balance of Mind: Natural Dialectic and the Five States of Mind

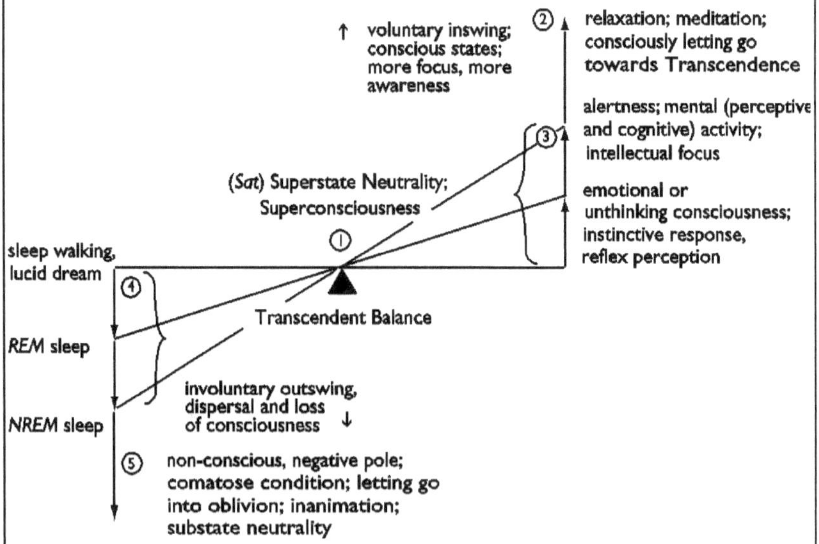

(1) Superstate Transcendence
(2) ascent towards super-state
(3) aminergic waking focus
(4) cholinergic involuntary phase
condition of subconscious oblivion

(5) shut-down, 'flatness', sub-state condition of material oblivion

This diagram shows, in terms of dialectical gradient, an oscillation between poles. Neuro-modulation occurs between waking (2, 3; voluntary) and sleeping (4, 5; involuntary or dormant) conditions. While all psychologies recognise waking, dream-sleep and deep sleep as altered states of awareness, the Dialectic also includes a Transcendent Superstate (1). Check *figs.* 7.6 to 7.8. The relationship between the four humanly normal states of consciousness is analogous to four states of matter with a fifth, archetypal potential. In the psychological case four levels of mind ascend from none (oblivion) towards the Primal Superstate. Such Prior Potential is Pure Consciousness or, for an experient, Enlightenment. Do you get the swing of it?

Although states of mind are well defined their grades 'slide', in the manner of spectral wavebands, continuously into one another. Mind also oscillates daily, elastically, almost metronomically between the reciprocal poles of awareness and unawareness. In its mid-range 'hallucinatory' dreams complement sensible, waking awareness of the world. While the super-state extreme of Pure Consciousness is uncommonly attained it is voluntarily approached in meditative states of internalised focus. Such 'deep thought' is diametrically antithesised by an involuntary 'drop' into deep or comatose sleep; attentive stimulus is opposed by the slide of psychological entropy towards 'scatter-brain', diffuse consciousness, and dormant exhaustion. The entirely internalised, subjective and central extreme of Pure Consciousness is set against its anti-pole of the entirely externalised, objective and peripheral sub-state of non-consciousness - matter.

In Pure Consciousness, which can be obtained voluntarily, mind has been transcended. Nor, on the other hand, is mind in physical energies; no 'life frequency' is possible in the absolute, involuntary 'flatness' of material being. Psychology ranges, therefore, between two extremes of void. One is fully alive and unconditionally free, the other psychologically impotent, inanimate and permanently locked away. Human psychology involves a brain and allied nervous systems but does all sensation need a cerebral interface? Many animals are capable of wakefulness but could some (e.g. coral, sponges and unicellular amoeba, paramecium or euglena) live permanently dormant - as apparently comatose as plants, fungi or bacteria?

***fig.* 13.5 (see also *figs.* 7.6 - 7.8 and 14.2)**

Consciousness comprehensively is, knows and does. The diagram (13.4) is therefore a simple foundation on which to build a psychology of the Great Integrator. It summarises mind's two basic directions of focus and the vectors involved in any cycle of information. An initial stack that can be derived from its pattern raises questions whose answers the next ten chapters try, within the same dialectical framework, to answer.

Oscillation. Ups and downs; ins and outs; waking and sleeping. It is also useful, in order to obtain a broad overview of the business, to view psychology in terms of polar oscillation.

	tam/ raj	*Sat*	
	range of consciousness	*Super-state*	
	normal states	*Conscious/Super-conscious*	
↓	*tam*	*raj*	↑
	inattention	*focus*	
	unaware / oblivious	*aware*	
	unconscious	*conscious*	
	material sub-state	*mind*	

The motion of *dynamic equilibrium* is vibration. Such oscillation at the heart of biological, psychological and also cosmic processes is also called *homeostatic*. There is no psychological oscillation in a dead body or in 'flat', non-conscious matter but for us the diagram illustrates in dialectical terms a metronomic swing. It cycles between our complementary, polar states of mind - waking and sleeping.

	tam/ raj	*Sat*	
	below	*Transcendent*	
	relatively aware	*Aware*	
	ignorant/ informed	*Knowledgeable*	
	partial waking	*Fully Awake*	
	degrees of sleep	*Never Asleep*	
	motion	*Poise*	
↓	*tam*	*raj*	↑
	entropic	*negentropic*	
	focus downward to earth	*focus upward to Source*	
	increasingly tiring	*inc.reinvigoratin*	
	dark	*bright*	
	flat/ slow	*vibratory/ fast*	
	less alert	*more alert*	
	less focused	*more focused*	
	incoherent	*coherent*	
	forgetful	*learning*	
	involuntary	*voluntary*	
	towards inertial equilibrium	*towards dynamic equbm*	
	more sleepy	*less sleepy*	
	sleeping	*waking*	
	sub-conscious	*conscious*	
	lower/ mid-brain	*higher/ frontal register*	
	unconscious/ non-conscious	*experiencing*	
	automatic	*flexibly*	
	response	*responsive*	
	inanimate	*animate*	
	unawake	*partial waking*	
	matter/ body	*mind*	

The first thing that emerges from these initial stacks is how well a representation of the world by reciprocal processes fits psychology. The second is, therefore, its potential for developing a thorough, rational framework within which to sift and sort psychological facts. A third prompt gives rise to the clear understanding that, while all psychologies recognise at least three states of mind (waking, dream and deep-sleep), there exists two more. The first of these, Transcendence or Enlightenment, is well-known and accorded pole position (both micro- and macrocosmically) by all except materialistic versions of the subject. The second, material oblivion, is understandably ignored. From a dialectical point of view, however, subtendent matter is located at the base of the conscio-material gradient; it forms the 'coccyx' of a universal spine; and the enervated, inanimate psychology of non-consciousness is called physical science.

Top-down, any attempt to reconcile science and psychology must logically start at the top (Super-consciousness) rather than the bottom (matter). It must start at the Axis, Centre, Source or Pivot rather than the effects of subsequent motion. For this reason Chapters 13 and 14 deal with the two waking states, super-state transcendent and ordinary. Note how 'Balance of Mind' (*fig.* 13.5) involves homeostasis, a highly teleological, centralising process which obtains to a preconceived norm, law or target. *In the case of conscious mind **voluntary**, flexibly-conceived and personal targets of desire are the subject of negative feedback until either satisfied or abandoned.* On the other hand Chapter 15 falls to cover, at the sub-conscious end of mental balance, the conditions of dormant mind, dreaming and deep-sleep. *Of special interest here* (Chapters 15 - 17 and 19) *is the operation of homeostasis whose norms are **involuntary**, inflexible and impersonal.* These pre-set targets exist in the psychosomatic domain of instinct, memory and archetype; they account for the psychosomatic connection between mind and its sub-state, non-conscious material body. Description then falls outside the actor's mind to (Chapters 19 - 25) his gross body or biological phenotype. Still further from the Centre, outside his own physical periphery, there lives a community of other actors. These comprise, set in the local scene, a network of interactive organisms playing on an ecological stage (Chapter 26). The whole flexible set includes, in principle and in practice, any player at any time in the universal drama.

The Ingredients of Psychology

We can collate what's gone before:

	tam/ raj	*Sat*
	experienced/ experiencing	*Transcendent Experience*
	range of lesser being	*Superstate*
	existential conditions	*Essential Self*
	objective/ subjective	*Subjectivity*
	body/ mind	*Psyche*
	personae	*Self*
	awareness of things	*Awareness*
	expressions/ changes	*Potential*
	prakriti	*Purusha*
↓	*tam*	*raj* ↑
	objective/ outer	*subjective/ inner*

non-conscious	*conscious*
what is known/ other	*knowing/ knower*
creation	*creator*
physical self	*mental self*
body/ object/ event	*mind-stream/ subject*
informed specific	*informant*
substate physicalia	*limited experiences*

As regards the whole creation Natural Dialectic posits mind and matter as two existential aspects of an underlying Essence. Two-in-one is also one-in-two. Essence is potential from which forms, higher psychological and lower physical, are willed into appearance. Cosmic 'ground-state' is Pure Consciousness. Consciousness is thus the root ingredient of psychology. It unifies; though immaterial it integrates; its interest 'gets the act together'. On this basis a *GUE* (Grand Unified Experience) might qualify as ultimate communion. A *GUE* is not a scientific formula. The equation of its balance is ignored without having started let alone completed the methodical research and practice needed to attain it. Is to dismiss without experiment the scientific way? And although such subjective 'integration' may be scientific nonsense, it is worth asking which of the complementary pair has 'got it all together' in the most profound way - mystic or scientist? Which part of me resounds with and which rejects an expression of psychology's top principle as 'Love the Infinite with all your heart and love 'not-you' (your neighbour) as your Self'? Which is the light of psychology - nerves or love? Or is each part of the same?

At the core of Consciousness is Awareness. There exist many qualities of awareness. They range from Super-state Purity through all different kinds of experience down to the special sub-state case of oblivion. A focus of awareness is called attention. Life is, if you like, a beam of attention. Such psychological light, focused through the glass of mind, kindles fire. This fire is called interest. Interest is a kind of love. What I am interested in I love. Interest awakens. Positive attention 'lifts'. It revitalises. The mind rises and raises. In a dualistic world, however, there also exist disintegrative, depressive, distractive forces. They flip interest inside out; they stress, scatter, repel and petrify. Negativity darkens and drains life away.

At this point a delicate disentanglement is required. A fundamental distinction is drawn between subjective experient (*purusha*) and the non-conscious object of experience (*prakriti*). They are (*figs.* 6.2, 7.5 and 13.6) subjective awareness and objective energy. Informant and informed; and exchange between them is the bread-and-butter of mentality. It occurs as a direct engagement between *chitta* (attention) and *manas* (a 'wrapping' of subtle, mental energy). In conscious mind *purusha* is active and called attention (*chitta*). Attention is my single conscious attribute, my life, with which I weave my world. It purposely designs patterns of thought and perceives images. As a potter uses clay to realise a shape so attention needs material to work with. Focus of attention can directly influence, develop and shape passive *prakriti.* The latter is like film that is developed by attention into the 'chiaroscuro' of mind. It is generally easy to distinguish between self and non-self, potter and pot, observer and what is observed. *It is impossible, using the relatively weak concentration*

Sanskrit Terminology

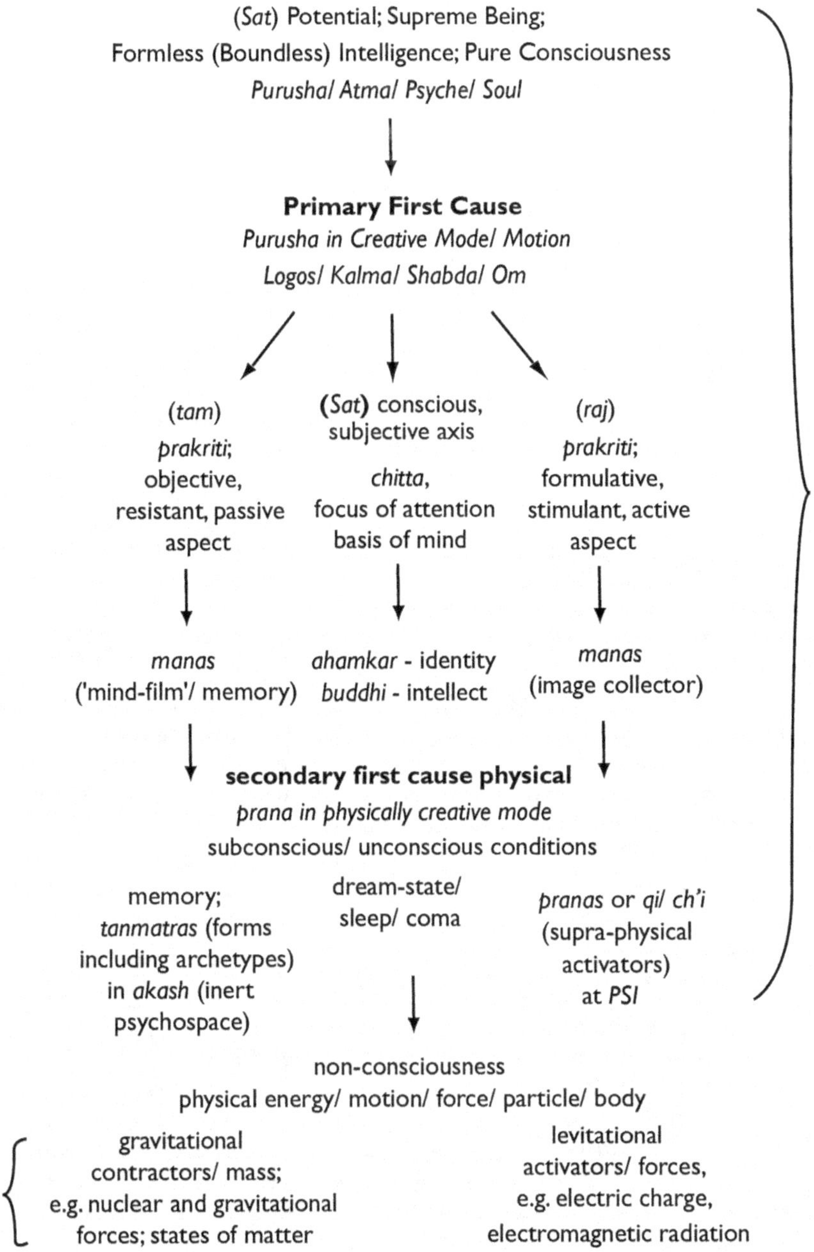

In this diagram the right-hand bracket includes metaphysical components and the left-hand physical.

For oriental analysis mind, a generic term, is a metaphysical body. The subject is broken into four main, cooperant parts (but often many more sub-categories). These four are *chitta, ahamkar, buddhi* and *manas*. It is not, however, possible to analyse these and the complexity of other pre-scientific oriental terms into definitions with which all schools of thought agree. Nor is this a book of Buddhist or Vedic explanation. Rather, natural Dialectic engages, as far as possible, general consensus about the meaning of Sanskrit terms and relates them both to its own binary system and to science. This generates, as this diagram begins to show, a broad sweep of simplified, systematically organised consensus. For information that will allow many further associations and connections to be made please consult the illustrations of this book and its Glossary.

fig. 13.6 (see also figs. 6.2, 7.6 and 14.1)

of intellect, to distinguish between chitta (conscious element) and manas (thought-form); that is, between Essential Self and the existential whirligig of mind and its physical attachment, body. So, lacking sufficient power of resolution, it is easy to blur the subtle distinction and, as actors, identify with the mental drama rather than, as seers, resolve the difference. Nor, for the most part, does occidental psychology even disentangle metaphysical thought from neural patterns. It identifies mind with body so that its drama, behaviour, is first and foremost physical. Thus it discards The Central Term, Self or Psyche, from its dissertations and equations.

Below the Source or, if you like, the Transcendent Origin of Consciousness are scaled the main, hierarchical subdivisions of psychology. *Purusha* in combination with *prakriti* creates information. Predominance of *purusha* (the positive or creative pole) is called active information or mind; the predominance of *prakriti* (the negative or created pole) is called passive information whose forms are sub-conscious mind (memory) and non-conscious matter. Thus, through an increasing involvement of *purusha* with *prakriti*, the levels of psychological and physical phenomena appear. *Viewed this way, existence is a gradual precipitate of information; it is a conscio-material spectrum whose lowest frequencies appear as 'heavy wrapping', that is locked, oblivious physical objects and events.*

If the hierarchical subdivisions of mind can be likened to a spectrum of light, each band with different properties, what are the *components* of that light?

Energised information is the same thing as interested attention. It is 'charged' with relative intensity of will and desire. The push-and-pull of this couple is the 'volitio-magnetic' driver of the mental spectrum; particular images and symbols are their vehicles. Mind-film is, therefore the *instrument* of awareness. It carries information in the way of software; it is a tool like paint, brushes and canvas to an artist; and an effect in the way that, for example, the operation of machines is an effect. Of what cause is it the effect? The cause is its inventor. Changes in the 'shape' of mind-film are either caused by inventive, imaginative attention or are data received on its invisible screen. Either way, in learning or creation, such information is active. Its information centre, mind, is aware.

In the sub-conscious state of dream (Chapter 14) attention is residual; it is practically overwhelmed and, in deep sleep or coma, completely so. This dormant, inactive condition is experienced as 'nothing'. The false *nirvana* of materialistic oblivion is in diametrical opposition to the experience of the No-thing. It is sub-state absence as opposed to Super-state Presence, endless sleep set up against Awareness. At the very lowest end of mind's scale active information is extinct.

Yet, as we'll see in Chapters 15 to 17, sub-consciousness is critical. It logically links the dual aspects of existence, mind and matter. Where a conscious moment passes only memory remains; memory can snag a thought and fix experience. Its zone, sub-consciousness, is a storage warehouse, the library of learning and the remainders of experience. Memories are fixed in mind; memorials, like an artist's portrait, film or vinyl record, are released in matter. Either way, only a remainder passively reminds. What of universal mind? If existence were reflected in a hemispheric brain then between its aspects, mind and matter, sits an immaterial kind of commissure, a communicator called psychological attractor or an archetype. The nature of connective archetypes will be explored.

At base of the cosmic pyramid coarse, physical expression of *prakriti* shows in various ways (such as light, heat, chemical bonds etc.) that can be described according to the rules of physics and chemistry. Levels of organisation nest hierarchically; no doubt, every element of matter is a system shaped by vibrant streams of energy; the energy informs and is itself, the dialectic gradient indicates, informed; the informant is its archetypal rule. Nothing at this level knows a thing. Physic is non-conscious, inanimate, without sign of life. When, as with a cinematic movie, the celluloid shapes are unchangeable, we call it passive information. In the 'real world' it is the rules of change that are unchangeable. **The world is composed of inflexible, passive information and we could, in this further sense, call physical science the psychology of non-consciousness**.

The shopping list for mind so far ticks consciousness (awareness) and focus of attention under the heading of *purusha*; also objects of attention, images and memories (aspects of *prakriti*). Both combine into an entity called self.

The basic unit of the psychology, its cell, is 'self', 'ego' or 'me'. And, since 'self' keeps changing, selves make up the 'body' of an individual's mental life. The issue of your self runs deep. What is, behind the mask of personality, True Self? Although there are a billion stations down the line you might call the esoteric start and terminus of mind 'Psychologie Centrale', the station of your Psyche. Origin and end of every journey, source of existential drama, this 'top' station is 'a fraction of Transcendence'. It is *Om*'s *bindu*, soul - a finite individual's assignment of the Infinite. As all things appear to be divisions of it, so it is the Most Coherent Unity connecting up existential multiplicity. Existence is, depending on this view, the way that Psyche differentiates itself. The true core of psychology is, therefore, the cosmic Axis. This Hub is neutral and precedes polarity. Its point of origin, Subjectivity, is beyond subjective images or material, objective shapes. Its Infinite Presence is like smooth water which, when moved, oscillates with cycles of existence. *The primordial motion of mind is a current that combines will, purpose and desire. These constitute a focus of attention. They localise and limit potentially infinite consciousness.*

The shopping list has now included 'self/ selves' and 'Self'. The Axial Ingredient is named; round this essential factor you can cook existence up.

'Self' and lesser 'selves'/ personalities show how the Dialectic logic works. Unless the Symmetry of Unity is broken there can be no other; there can be no second part and, therefore, no creation. Such a fundamental starting-point is reflected in the basic laws of any logic. First is the Law of Identity. A is A. A cloud is a cloud. I am that I am. The second is the Law of Differentiation or Distinction. A is not B. A cloud is not a tree. I am not that; it is not me. *From the first law flows the ability to identify and from the second the ability to classify.*

It is interesting to note that, in mystic terms, creation involved this order. From the polarisation of Pure Consciousness issues the first, subtlest dipole of finite existence. This is composed of subjective awareness *(purusha)* and non-conscious object of awareness *(prakriti)*. Duality composed of energy and information has appeared. *Essential Purusha* is characterised by the identifier 'I am that I am', *prakriti* by the separator 'I am not that'. *There is primal self and otherness. This is the very root of nature, logic and the cosmos.*

A Hindu monk in saffron robes will ask you who you are. '*Neti, neti*'. Not this, not that. When all the otherness is done what have you identified as you? Is what you think is you the bottom of the matter? Have you identified with body? Mind? These 'selves' change but is there any constancy? '*Neti, neti*'. Not in this world. If you are looking for The Self you will not find it in the world. It is hidden from sensation. When all the otherness is done how many can I know but one? This Axial Ingredient is what the monk is looking for; knowing it he calls *samadhi*. *Samadhi* is Communion.

Existence is a fundamental duplex; it is not chaos or confusion wherein all is mindless, energetic substance of which mind is just a sort. Energy and information are the polar twins. Model them as each strand of a double helix - separate, separable but entwined and complementary as mind is to body and a body to a mind. They may be unified in Essence but existence comes about because they are apart. *Such a logic of polarity blocks, at root, any notion of the biological evolution of awareness; it completely inverts a materialistic point of view.* How could 'other' precede 'self'? Original identity is only self-aware. From Infinite Awareness are devolved lesser, variously limited 'experiences' known as individual minds. From Self, by virtue of their personal thinking, are issued selves. Through principle to practice, from *noumena* into *phenomena*, from mind to matter information shapes and governs energy. This is standard, *top-down* procedure in any manufacturing or IT industry. It is how ideas are realised. Evolution of the mind is therefore not seen as a biological phenomenon at all. That never happened; consciousness was never absent from the scene. True evolution of awareness does not deal in existential gain or more complexity but loss. Mind and body are dropped off; such subtraction leaves you only soul. The process of 'subjective purification' is obviously, therefore, one of withdrawal from detail through principle to Original Pure Consciousness, Psyche or Self. The current of a real cure, the stream of true psychology is one that ends up, Omega in Alpha, with Identity in terms of Cosmic Soul.

Is it not ironic but consistent? The very entity materialism wants to squeeze out of existence is the very immaterial element that cosmos issues from!

The end is the beginning. Starting from its Central Station Self moves out. It weaves a net of drama ever growing in complexity. Psyche spins a web with energy; you could say *purusha* and *prakriti* dance. The Infinite is not divided but, it seems, it is; Soul is not divided but, it seems, it is. As Soul spins finite selves, so plot thickens and involvement complicates. Each 'separate' soul becomes encased in changeable *personae*, mind and body, and through these masks finds circumstance with other persons. So fascinating and attractive (or at least histrionic) are these *lower selves* that we identify with them. The funny thing is that such changeabilities don't feel they change; and also that, unianimously, these personalities vote they are real. And in our own, foreshortened sense it's true. I am no longer simply 'I' but 'I' the subject of a predicated drama. 'I' and 'not-me' strut the stage together forgetting only that our act is an appearance dressed in costumes, minds and bodies, that will change from scene to scene. Now is not for ever. What is the sense in which it is? Do you remember how a prince dressed in disguise was wandering in an alien land until, by one fortune or another, he retrieved his rightful state, a state of mind; and with this haloed coronation authority that ruled his lower, subject selves? Your Self is above your selves. What happens if the citizens forget their king, pretend that law has been suspended or believe the order of their government is myth? Even if you have not reached the rank of Central Government it is best to understand the hierarchical bureaucracy by which your own community of individual selves coheres - simply because imbalanced incoherence leads to chaos called ill-health. What expulsion from paradise is an underworld of pain! What fall from grace! Sense-based human consciousness is unaware, that is to say, ignorant of its Central Psyche. Levels that are super-conscious relative to the low point of a normal waking state ascend inwardly towards life's peak, a transcendent peak that's psychological as well as universal. Therefore normal man, even one as astute as a leading academic, is not the measure of mind. **What light is to physics Transcendence is to psychology; in other words, the only absolute measure of consciousness, and therefore psychology, is Transcendence**.

How, though, can you find 'True Self'? How can you purchase The Ingredient or take a scale and measure out Transcendence? In the battle to survive are High Ideals what we have time for? Will they repay the effort, what's the *quid pro quo*? Casualties of all kinds suffer; psychiatry and science offer triage. You need pragmatic hands to patch up symptoms or to operate upon the throes of abnormality. Although the quirks often involve involuntary bad habits and faulty patterns of thought, scientific emphasis is on genetic, biochemical and physiological malfunctions and their medical cure. No doubt, in part, such relief is right. Darkness cured is light; but what of Light?

There is, therefore, an ingredient of psychology none can afford to miss. Without a light how can escape from darkness know its way? Ideals are beacons. Where would partial-sightedness direct itself without ideals to let it see - ideals of honesty and trust and selfless service, ideals that balance in a dualistic world the weight of sickness, criminality and other dark sides of the human moon? What are your ideals? Who, as ideal person and top object of your estimation, love and admiration, embodies them? Can you see what religion is about? The best psychology is one that rallies round high, positive ideals. If you want a voluntary lift from social quagmire, if you want relief from pain then the psychology of mystics offers you The Answer. His is key psychiatry. At least that's what The Buddha, Christ and so on say. What do you call a person who denies you answers, ideals and drugless addiction to a path that leads from criminality? Is he an expert fool, a graduate of material muddle whose advice or government will cost you dear? Ideals are not, at root, a secular and changeable commodity. They are ropes pinned to the Apex so that you can climb; they are reflections of the nature of The Immaterial One.

Vehicles go wrong; you take them to the garage. Human bodies do the same; you take them to a human garage called a hospital or surgery. 'How practical a cure I have prescribed' said Dr. Buddha 'the prophylactic eight-fold path to lessen suffering.' 'How impractical,' retorts a busy doctor 'I have patients with specific, urgent ailments and know the treatments I can valuably use.' Matter over mind, body out of shape needs physical restorative; yet, fixed with treatments mind devised, mind over matter also plays both prophylactic and remedial part. What is, for example, lifestyle but a function of a person's mind? Healthy rhythms, vegetarian diet and deliberate positivity of thought - an inner surge that brightens body just as well as mind - add up to an elixir in themselves. It is not in every patient's case a doctor's option but thought can, well shaped, ignited and directed, be a doctor's helper. How wisely does one swap ideals for pills, inject stimulants or tranquillisers or rake over psychoanalytic history instead of driving towards a future filled with aspiration - especially when mind needs medicine as much as any other organ. Could you say the world is sinful? Fallen? Or, in humanistic terms, that individuals and society are wrestling with their negative propensities? In this tense and costly, daily struggle you need strength. Determination born of psychological ideals. *Faith, hope and love are the great motivators. Mark success of psychiatric treatment in proportion as it generates these three.*

There is one more ingredient of mind that links up with ideals. The other name for grasp of principle and the ability to handle information with efficiency is called intelligence.

How can it be argued man excels the other forms of animation here on earth? Considering their superb constructions it cannot be true. Man cannot see, hear, smell, taste, touch, move, reproduce or even survive 'in the wild' as robustly as a multitude of 'lower' organisms. Indeed, from the *bottom-up* perspective man is just a different kind of animal. Some evil specimens confirm that man is, due to immorality, inferior to any animal. If, however, a superior capacity exists it is not biological but intellectual. It is not physical but metaphysical, informative and (if the word is linked to purity of consciousness) spiritual in capacity. It lies in the ability to contemplate. Man manipulates with symbols, grasps hold of principles, craves information and, understanding somewhat, is always curious to know more. In this search man's mind, with brain to match, is formulated to engage with abstracts and employ conceptual logic in the forms of grammar and arithmetic. Thus, blessed with fingered hands on arms and balanced feet on legs, as well as senses and a tongue, you're also granted psychological superiority, above all other creatures of the earth, with which to work intentions out.

In other words, as well as having instinct and sensation Adam contemplates, anticipates with purpose and interrogates the world. Although other organisms have the capacity to feel emotion, communicate with sound, touch or gesture, problem-solve and (like chimps, crows etc.) invent tools, it may be agreed that only man is capable of contemplation and the sort of powerful, sustained intellectual activity that combines complex language with conceptual thought to produce libraries, complex technology, civilisation and, beyond even these, a thirst for more and more unto infinitely extended information. *It is this reasoned sharpness, systematic grasp of principle and resultant creativity we call intelligence.* **On the psychological basis of intelligence, therefore, it might be proposed to classify man in a separate phylum if not domain from all other life.**

Intelligence is *datum.* Why be proud about what you were given from the start? Moreover human sense-based consciousness has not realised its full potential. What is there above a mountain peak? From the perspective of Transcendent Consciousness all down below is sub-Conscious; conversely, from a lower standpoint on the slope where humans live what is above is super-conscious and, relatively, what is below sub-conscious. *While educated man is far from the top of Mount Universe he may at least, if his education pointed him the right way, aspire. Such persons, through books, institutions and the manner of their lives, pass on this aspiration. As long as one keeps looking up....*

Which psychological ingredient is physical? The anatomy of mind is metaphysical, conceptual and wireless. Where 'bio' is wired, 'psycho' is radio. The key simile is radiance and therefore, with respect to gradient, spectrum. The Transparency of Pure Consciousness is dispersed into 'colours' that vibrate, resonate and interact in different ways at different frequencies. A mind exchanges both information and energy with its own metaphysical and physical environments. Its *modus operandi* is association, resonance and, in the way of 'birds of a feather', the attraction of like to like. Do you remember, from Chapter 0, the idea of a hologram? A hologram is a 3-d photograph made with the help of lasers. Unlike a normal photographic image each part of it contains the image held by the whole. 'The whole is in each part and each part in the whole.' Perhaps brain's operation, over and above each nerve cell's *DNA*, uses holographic principles; at any rate mind needs brain, nervous system and all the integrated mechanisms of body so that you and I experience this great gift of life. These physical components need to pull together; nervous and hormonal systems and, as far as Natural Dialectic is concerned, archetypal memory interlock to see they do. If any part is pressured, damaged or malfunctioning then the result is 'interference', 'noise', signal impairment to the point its healthy message may not register; that is, it throws the system off its balance into relativities of pain. Material components are the objects of biology; and medicine is there to fix the breakdowns on that side of things.

Fixing cars is quite another thing from making them. And fabrication is a massive step removed from an invention. You take your brain for granted and it is. Whence? Could chance, without a figment of imagination in its empty headlessness, have raised it up? Or, as in everything that's been invented, was conceptual purpose the real order of the day? Purpose, plan and program. Brains are not cooked up but the ingredients of mind need bringing down to earth. How, replacing chance with your intelligence, would you proceed?

To Build a Brain

From a *bottom-up* perspective evolution doesn't think; if you credit billions of unseen years then you might imagine that this bank of time can knock up principles, organise coherent, working systems and perchance make anything - including brains - for which you need an explanation. Mindlessness gives rise to mind. Matter conjures up a very special kind of matter, a molecular complexity called brain that *is* the mind. In this view brain has evolved not by replacement but addition to its parts. To primitive nerve nets were added brain-stem, mid-brain and finally, beyond sub-cortical inflexibility, the flexibility that cortex, with its hemispheres, adds to the nature of a creature's consciousness; and this cortex has grown, in man, to size supporting all the intellectual activities with which, transcending every other animal, we lead our psychologically very active and creative lives. It's all down to accidentally constructed, multiply connective interaction of a powerful, computer-like machine, the human brain!

If you believe that story you bet on that guess. On the other hand, precisely what genetic pickle generated a capacity, totally superfluous for predators or gatherers as many creatures are, for learning, reason and detailed manipulation of such abstract principles and codes as language and mathematics? Nobody knows. Some, according to a Theory of No Intelligence, guess chance lurks under everything. Thus molecules developed love of art, appreciation of what's beautiful and, in the mind of such religion and philosophy, an engagement with eternal truths and love. Where mind is brain and actually nothing more than a complexity of molecules could atoms really have evolved a sense of purpose; could particles and forces logically construct plans that anticipate and obtain ends? One marvels at the power of an unconscious molecule joined up with friends! Mindlessness gives rise to mind; but (from Chapter 6: Mind Machines) you might remember that, although they're into planning 'brainy' robots, AI experts aren't themselves a mindless crew. Intelligence makes simulated mind. And dialectically the scheme's reversed as well so let's, instead of chance, deploy intelligence. Let's take an informative approach that makes sense, one a systems analyst might rapidly and skilfully employ - deliberate elaboration from first principles to detailed practice according to the goals and reasons laid out in that 'spec'.

Is matter really as extraordinary as mind? Is special matter mind and just brain 'experiencing itself'? And do development, structure and function of this, information's organ, reflect irrational an origin? Such is the fashionable guess. The *top-down* view, however, is that definitely they do not. Bio-logic that conforms to Natural Dialectic's scheme infuses brain as thoroughly as grammar language or scheme its machine.

A brain so simple we could understand would be so simple that we couldn't; but, *top-down,* we can enumerate the principles round which its great complexity might gather. Thus, before examining the different qualities of consciousness (Chapters 14 up to 17), let's deduce a system that will generate the faculties through which to feed its principals and thereby link mind with physicality. Let's sketch a definite and natural development and, eliminating chance and error, outline the tasks and their associated mechanisms leading to what lets you read this book - a human brain.

First, remember, you're not just going to construct and then connect a complex coded ball of nerves with chemicals. There's a body to attach but, even if you put this to one side for now, there's that old immaterial element. How will you work to conjugate a mind with brain? Yet again, do you recall Mozart's response to scientific measurement? Mozart didn't understand the mathematician's dossier describing instruments and sounds so well. And, though it rigorously describes brain's parts and, apparently, associated functions, neuroscience misses the composer's point as well. In the same way you could measure up a church and still completely miss the point. Indeed, we've all been sold a story by the school of natural history but, on the contrary, perhaps 'unnatural', that is, immaterial history also had a part to play. That part would be the primary contribution of all metaphysic, information. Intelligence and its designs. And is the greatest mind in cosmos really a professor's in his college rooms? If not then perhaps a human brain reflects its archetype; perhaps it was, in fact, planned as a template for humanity. If so then, contrary to a Theory of No Intelligence, brains stand to reason rather than to chance. They would line up with intelligence and, never born from lines of accidents, continue to eliminate them. Ask yourself which viewpoint makes better analytic sense. *From a top-down, holistic point of view it's clear that a cerebral 'panel of control', whose structure is gifted and whose basic functions are soon instinctively grasped by a freshly embodied infant, is indeed a very highly ordered medium whereby ounceless information is plugged into energy, mind into body or 'light heaven into massive earth'.* In a machine like, say, a motor car conceptualisations are localised on a control panel and connected up within the vehicle's body. For example 'stop', 'go', and 'change course' have brake pedal, variable accelerator and steering-wheel which correlate with brakes, engine and wheels in the body. The biological way that information locks with an energetic body is just the same. *The conceptual 'wish-list' is materialised in a way that operates like a cybernetic control system with, if necessary, conscious override.* **You might well infer that complex brain is planned but, more importantly, that any information linked to life forms and their cosmos did not shape the dust by accident.**

In this case, let's get metaphysical! Then afterwards, having established the principles involved in the flexible interaction of mind with matter, one might check (using *MRI/ fMRI* or similar scans) how this plan correlates with the design of a *deluxe* 'control panel' in the form of brain. Indeed, what Natural Dialectic logically predicts neuroscience often, from lesions, split brains, anaesthesia and other studies, has uncovered. This book has neither time nor space to relay copious research but you might start your confirmatory search (replete with notes, bibliography and references) in The Master and his Emissary by psychiatrist Iain McGilchrist.

	tam/ raj	*Sat*	
	forms of mind and matter	*I-ness*	
	foci of attention	*Central Awareness*	
	lower forms of consciousness	*Top Level*	
	elements of experience	*Experience*	
	matter/ mind	*'Ground-State'*	
↓	*tam*	*raj*	↑
	down and outward	*up and inward*	
	objective locality	*subjective locality*	
	objective focus	*subjective focus*	
	sense perception	*contemplation*	
	nervous system	*image collection*	
	chemical signals	*body feelings*	

If your machine must let subjective 'I-ness' operate in a dynamic, dualistic world what conceptual faculties would you mandate? What would be your shopping list for parts to make a brain? Would it need to feel, analyse, compare, perceive, interpret, purposely organise and act in both the physical and metaphysical domains? To question, search for answers and make sense? What constructions might allow exchange of information with objective circumstance - one's body and, as far as you can tell, surroundings? Brainstorm. Take focus for a start. Call it attention and, if there is will-power to create in correspondence with creation, allow construction purpose too. Such interest will need to differentiate, quantify (in terms of space, time and number), qualify (in terms of conceptual category), be able to discern priorities (such as cause before and separate from effect) so as to understand and thus anticipate behaviours in the future; and you'd need memory of past experience for recognition to occur - otherwise you'd be locked in the present never having learnt a thing. Information storage, we shall learn, is absolutely critical at every level, step and stage of life - so don't forget some manner of recording and referral whereby you can store the information you will later use to make decisions. How about inventing data storage/ retrieval and instinctual mechanisms? You'll need memory, short and long-term, giving you the context of a 'fixed' but metaphysical surrounding in which mind can orientate, a databank to which it constantly refers. Thus recognition can occur, comparison be drawn and, according to your learning, you can choose the next step down your life-line. You'd need to analyse (to understand the parts of things in order to control) but also synthesize and understand the whole. Is it worth it? You have to be able to evaluate, that is, assign a worth in terms of investment of attention. Are any of these conceptual functions physical? Yet a brain will also need to integrate aforesaid operators as a whole identified as 'self', to reconstruct an image of events in order to align with self's criteria of what is 'good' and then, again, decide how one should act to 'bring things into your symbolic reconstruction's line'. Not least, information must be gathered from the world of matter and, subsequent to conceptual manipulation, fed back by a motor system in response.

Symbolic operations known as thought, will power and purpose all need translation into nervous, muscular and metabolic process. If a single, main message springs from information technology and computers it is one of complex wonder-worlds built, like nature's, out of binary simplicity. 'Yes', 'no', 'either/or', possibilities and certainties - what instruments of nature better reflect the Boolean gates of digital revolution than opposites? You could use 'on-off' electrical polarity and so, computer-like, drive connectivity. And, as we've started out above, you could stack other sorts of dialectical polarity. A binary operator that distinguished both extremes (such as self/ non-self, general/ specific or isolation/ relation) and the covalent, compromising interactions ranged along a sliding scale between would be a useful tool. This is how the Dialectic 'ranges' and in cerebrum such modulator-in-between the hemispheres is there, your *corpus callosum.* Indeed, 'lateralized polarity' could well represent distinct modes of cogitation which would act, in the way of negative feedback, as cybernetic counter-balances and thus promote psychological homeostasis. Are not peace, poise and balance healthy? Is not comfort in the form of health and happiness what, in their own terms, creatures always strive to gain?

Phew! This is just an outline sketch, a basket of coherent tasks a living vehicle's dashboard and controls must satisfy before it has the slightest chance to flower, flourish and survive. It does not even start to generate efficient mechanisms (eyes, ears, muscles and so on) to facilitate the schemes of mind; nor does it consider how you might develop these and brain itself from clues left in a bag too small to see, an egg. We have glimpsed just how complex this informative project would be. Give credit where it's due. I'm sure your first recruit would not be Lady Luck.

Systems analysts may work *top-down.* Brain engineers, the *IT* specialists *par excellence*, might correlate the architecture of their mechanism with the conscio-material gradient (*figs.* 13.2 and 13.7 below); and let the whole, as seeds all do, develop from the inside out. From 'primitive' to 'modern' you don't need to think in terms of history. Instead of 'time and chance' employ 'intelligence' to best express, in physicality, intelligence. 'Primitive' would then mean 'furthest from reason' or 'closest to unconscious, automatic traffic'. Brain-stem, cerebellum and the mid-brain lead, as a plant develops from its seed, to fulfilment in the final flowering of nervous comprehension physically represented by two hemispheres dividing up a single, superintendent function of the brain. Furthermore, if human cortex *does* reflect full flowering of informative manipulation (what improvements does an evolutionist suggest?), one might regard the animal array as, in comparison with this respect, variously retarded down a scale of psychological development until we reach, say, coral, jellyfish or even cells called protozoans. There are many ways for life to know the world.

Having developed a theoretical mechanism that should deliver according to specification, an engineer seeks to devise an efficient, economical model from the materials to hand. Are not the biochemicals involved a very special mix - whose coherent properties and generation might be somewhat beyond your powers to invent? In the case of brain or biological life in general the problem of packaging and reproductive development has also to be solved. All information has to be capable of miniaturisation in a form that can re-develop on its own, time and again, material copies of an original. You might expect the *top-down* 'vector' of such brain-design (whose potential stems from information) to run counter to its *bottom-up* antagonist (chance evolution due to matter's non-existent, impotent imagination called by some 'design').

Records how to make a 'simple' biochemical or complex, integrating looms of nerves are harboured where? Where are files, packed much more intricately and accurately than parachute or even microchip, kept orderly in store? This potential, far more powerful than any egghead's dreams, is found in eggs. An ovum packs a picture of an adult made of *DNA*; and *DNA* from both forms is essentially the same. Zygotes are 'unwound' along a gradient that runs from physically simple to complex and, mind-wise, non-conscious (or at most sub-conscious) up to conscious operation. Once unwrapped and working in its preordained, informatively active mode, mind's 'hierarchically networked computer' reveals a clear metaphysical gradient. Adult brain structure reflects the logical path of its own original conception, whose consideration dropped from metaphysical to physical. This is also the vector, mind to matter, of its own creative, motor operations. Such shocking *top-down* logic does, of course, upturn the evolutionary direction and, heretically, the angle of a mind-set. Construction drops from top-front (the cosmological axis of the third eye) backwards and down. Its three sectors are cerebral cortex, mid-brain and lower brain stem. They involve (*sat*) conscious, (*raj*) central eye made an invisible connection with their Origin. As waking excels sleep so super-conscious acts above sub-conscious and (*tam*) physical or reflex regulation of basic life-support systems. Such an order reflects cosmic principle as we see below.

Russian dolls, Chinese boxes, nested eggs. Experience is localised at the front of your cortical hemispheres. Topmost, firstly, what is 'super-consciousness'? If the human mind and correlated brain are an expression of intelligent design, you might expect the former would incorporate a metaphysical ability, a life-line leading back, an immaterial umbilicus that, from a frontal locus called the third, excels conscious, body-bound normality.

For neuroscience such a state does not exist. Nor physically does it though brain's nerves might register, as wind ripples wheat, a ghostly shadow of its passing. Let's drop, therefore, do what most evidently is real - brain-channelled human waking state. When you're awake the prefrontal cortex 'buzzes'. Here are gathered sense impressions, thoughts and 'live' remembrance, that is, the experience of animation. To support this active information you need passive. Subconscious memories are filed in support. Their context guides (according to your species) the way you think and/or behave (see Chapters 15-17); and, in remembrance or anticipation, lends a life's plots a sense of continuity. The cortical lobes of your mid-brain cradle various sub-cortical relay systems. These 'transducers' are often termed, as noted, primitive' but might as well be termed 'a step-way from earth to heaven's realm of thought'. On duty at the top step a system called limbic (whose central business is the mediation of instinctive response, emotion and memory) mans the gate to consciousness. Here 'whirr' read-write 'recording heads' (emotional amygdala and defining hippocampus); these heads address, retrieve and integrate life's data-files with nervous correlates and thus the body. This system, cradled at the heart of the brain, constitutes the psychosomatic point of contact; here, about the threshold of the thalamus, occurs sub-cortical exchange of telegraphic information with brain's hemispheres.

Brain in Principle - Logical, *Top-Down* Construction; An Organic Representation of the Levels of Mind

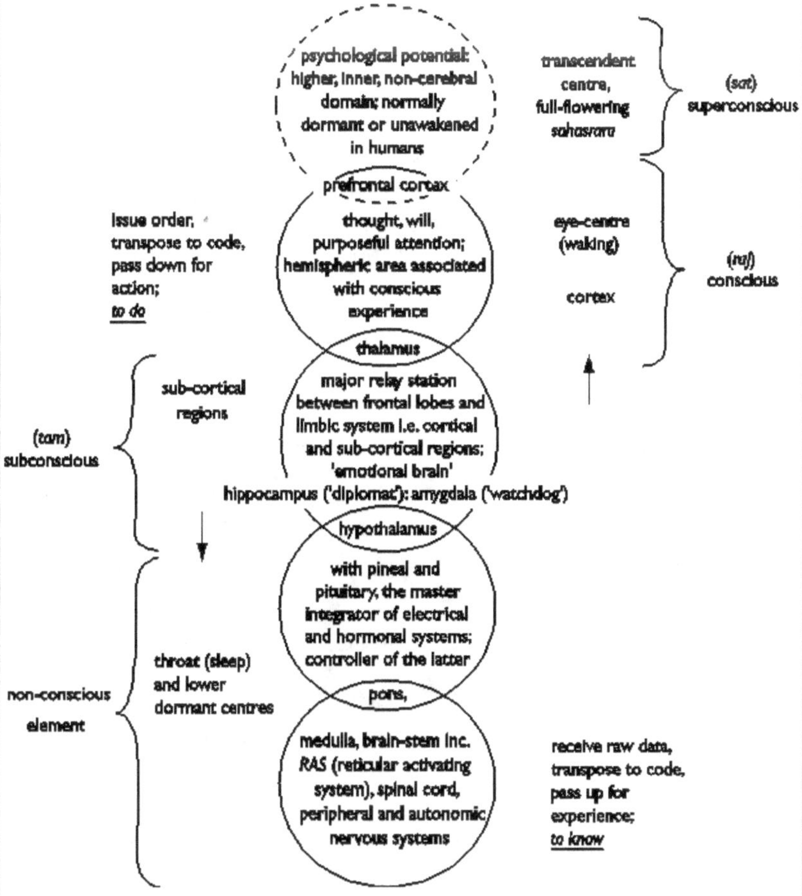

This construction starts top-front with areas mediating physical awareness and falls back down a conscio-material gradient to those representing sub-consciousness and non-conscious areas of homeostatic control. The structure and mechanisms of brain reflect the conceptual principles and information of mind; brain itself is composed entirely of non-conscious, chemical elements.

fig. 13.7 (see also 13.2 and 13.4)

Beneath the thalamus and in their turn are cradled three key master glands, the pineal, hypothalamus and pituitary. This triad links nervous and hormonal pathways that cooperate in the homeostatic maintenance of overall poise, 'tone' and physical health. They govern form's dynamic equilibrium. Behind the brain stem is a sub-organ that neither initiates motion nor is involved in conscious sense perception; yet the cerebellum, which has connections with the higher cerebral cortex, midbrain and the lower desks of brain stem, plays a crucial role in timing, positioning and overall sustenance of bodily equilibrium; it is, therefore, the office of agility. A reticular formation of nerves runs the entire length of the brain stem and connects (*raj*) conscious, (*tam*) sub-conscious and physical, reflex processes together. The *pons* houses an arousal system that influences the sleep/ waking oscillations of diurnal life, alertness and other basic mental supports. Hence we descend to the medulla oblongata situated in the inferior part of the cranial cavity. It is here that nerve tracts from the left side of the body 'decussate' to the right side of the brain and *vice versa*. This portion of the brain, which sustains vital bodily functions such as heart and respiratory rates, blood pressure, coughing and sneezing and so on, is continuous with a spinal cord that, in descending, ramifies to impact reflex interaction with the physical creation by way of your earth-suit, your body.

Natural Dialectic and Brain's Cortical Hemispheres

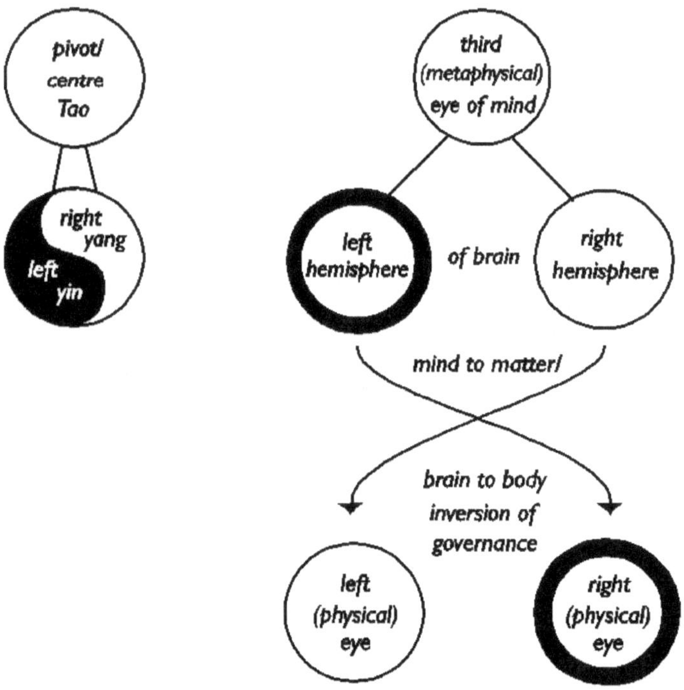

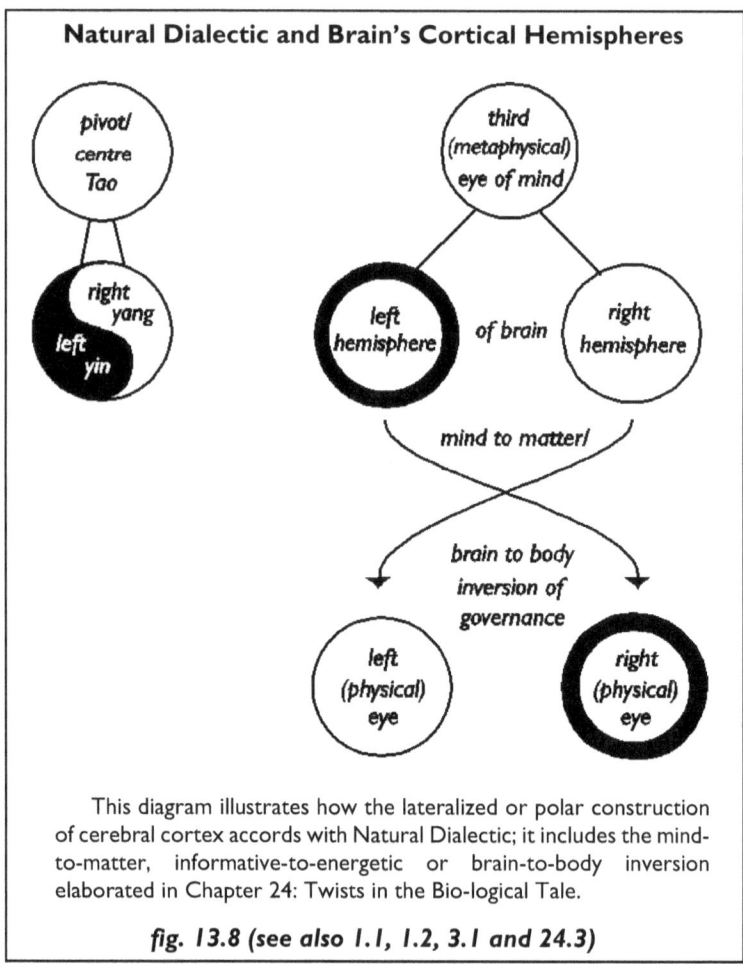

This diagram illustrates how the lateralized or polar construction of cerebral cortex accords with Natural Dialectic; it includes the mind-to-matter, informative-to-energetic or brain-to-body inversion elaborated in Chapter 24: Twists in the Bio-logical Tale.

fig. 13.8 (see also 1.1, 1.2, 3.1 and 24.3)

How does Natural Dialectic fit brain's scheme - or *vice versa*? To explain this we'd best turn to the top, to consciousness and hemispheric segments of a two-way world.

Natural Dialectic is built round a 'two-in-one' or 'one-in-two' form of polarity. Its stacked columns represent unity-in-duality (Main Dialectic) and the swings of such duality (represented by the anti-parallel vectors (↓↑) of Secondary Dialectic); and the whole represents paradoxically complementary, polar opposition that is ubiquitous in nature's operations. While consciousness is primal, prior and unitary the polar hemispheres represent two ways of attending to the world, two major modes of cognisance and thus two directions of experience. We'll now explore how these principles, as they impact in practice on construction of a brain, stack up. In short, does it look as though brain happened accidentally or, more likely, was constructed in accordance with the cosmic plan?

	tam/ raj	*Sat*	
	foci of attention	*Global Awareness*	
	lower forms of consciousness	*Top Level*	
	polar/ binary expression	*Unitary Consciousness*	
↓	*tam*	*raj*	↑
	left hemisphere	*right hemisphere*	
	separating tendencies	*unifying tendencies*	
	picking out	*generalising*	

Non-conscious matter hasn't any cognitive ability so let's start with psychology's ground-zero-physical, the state of consciousness and focus of it called attention. Immaterial consciousness is, as high-brow professors say, 'phenomenologically and ontologically unique' - which, in translation, means there's nothing like it. And yet everything we know depends on it. What is knowledge but, with thought, feeling, sense and memories combined, experience? Experience amounts to the attention given to 'not-me' factors, that is, whatever forms appear upon the 'ground-state' screen of 'I-ness' we call consciousness. **This combination of attention, knowledge, experience and consciousness amount to life as we subjectively conceive it.** Unitary consciousness, which joins all things in one, thus forms the primal datum; and 'round' this datum levels of constriction, sheaths or bodies can logically accrue that together we call mind and, top-central in the body, brain. This sheath, cortex, represents our conscious thought, especially towards its front. Its electronic action is divided into hemispheres themselves divided by a modulator-in-between, a thick-skinned body (*corpus callosum*). This separator, at the same time a communicator, consists of between 300 and 800 million fibres to which about 2% of cortical nerves are attached. Its channels can, like and/ nand logic gates, inhibit or transfer a signal or a group of signals through. It filters. It refines relations and control between the asymmetric, hemispheric modes of operation.

Conventional psychology identifies several kinds of attention. You may be alert, vigilant and remain in this 'detached' manner; and you may, as opposed to all-round vigilance, focus your attention on a specific item (whether a train of thought or an event things). **In fact, you may operate in a range between the general and specific, so that consciousness is divided between the two.** For example, you may be engrossed in music but still hear a door open or be so engrossed that you do not; or be so alert to the environment-in-general that particulars of background music scarcely register. **These modes of attention are, at root, complementary, polar ways we need employ to know the**

world. Both are essential and a brain would have, at a fundamental level, to reflect them both.

↓ *tam*	*raj* ↑
	specific	*general*
	division	*unification*
	isolation	*relation*
	separation/ singling out	*communication*
	local	*global*
	detail	*context*
	part	*whole*
	practice	*principle*
	analysis	*synthesis*

These kinds of attention are not only directed outwards to the physical environment and individual factors of which it is composed. Another polarity of attention is required (Chapter 0: Opposite Directions of Mind and Two Pillars of Faith), this time to flip fluently and practically synchronously between 'internal' thought (data processing) and 'external' sensory and nervous reflex (data reception and expression). *In this case the range would run from total involvement in external, reflex interaction with the physical world (including one's own body) through to total detachment from it; such detachment means total involvement in 'internal' contemplation.* We say 'lost in thought'. Subjective contemplation is either 'localised' by specific intellectual pursuits, artistic creativity or emotional 'day-dreaming'; or it is globalised in the form of attention paid to attention. Paradoxically, as mystics explain, such concentration involves one-pointed focus towards the source of mind, ground-state or consciousness itself. In dialectical terms this would mean an approach towards the Inmost, Original Unity of Nature. *In short, Natural Dialectic would add the modal divisions of 'attention turned outward to the senses and observation of the world's objects' and 'attention turned inward in contemplation of principles, purposes and other invisible connections' to its shopping basket for a brain.* This division is, essentially, one between physical and metaphysical involvement.

↓	*tam*	*raj* ↑
	sleepy	*vigilant*
	dim/ dull	*bright/ alert*
	unaware / ignorant	*aware*
	disinterested	*interested*
	bored	*enthusiastic*
	uncomprehending	*understanding*
	scattering of focus	*gathering*
	towards sub-consciousness	*towards lucidity*

There is, as *fig.* 0.3 illustrates, also intensity of attention to consider. As well as ability to vary width and direction of focus a range of intensities of beam, as with a dimmer switch, would need to be supplied. Intensity would relate to level of awareness, that is, ignorance as opposed to knowledge; and to level of ability to concentrate and, thereby, grasp and comprehend.

tam/ raj	*Sat*
two/ many	*One*

	counterbalances	*Balance*
	arms	*Pivot*
	asymmetry	*Symmetry*
↓	*tam*	*raj* ↑
	left hemisphere	*right hemisphere*
	down (to material world)	*up (into mind)*
	specific grasp	*global grasp*
	objective tendency	*subjective tendency*
	motor	*sensor*
	creation of form	*comprehension of form*
	analytical emphasis	*synthetic/holistic emphasis*
	sense craving	*information craving*
	thing-bias	*value/ meaning-bias*
	body-side	*mind side*

It should be clear by now that, as cosmos is the embodiment of a range of energies, so mind incorporates a range of information; its blacks and whites are, in operation, mostly ever-changing shades of grey. In conceptual reflection it is easy to express polarities by way of their extremities - black/ white, male/ female, isolation/ union and so forth - but in practice the construction of a brain must reflect the scale of nuance, tendency preference, bias and emphasis. To label 'areas of cognition' and identify 'zones of function' where in fact only networks actually exist is to make an artificial distinction; indeed, a useful mechanical buffer would occur if one part of a brain could compensate for another's injury/ deficiency. Yet it is also true that brain functions are associated with specific areas of brain and lateralised by 'tendency' (as opposed to absolutely) so that compensation can and does occur to the extent that in the case of loss of a whole hemisphere the other can (still with its own nuance) stand in as understudy. Thus, although easy stereotypes are drawn when, for example, male/ female or left/ right hemispheres of the brain are discussed, these are more properly construed in dialectical terms of range, sliding scale and dynamic oscillation. Asymmetrical opposites are complementary; they join to make whole human being or information's single most important organ, brain. The latter's complementary hemispheres *both* serve all human activity but each deals with events in a different way. They act as counterbalances; each attends to the world in its appointed way. They act antagonistically like two complementary muscles or, in that ever-close analogy, man-and-woman making up the human archetype. And they reflect action at metaphysical (mental) and physical (electronic and chemical) levels.

Brains did not evolve by way of semi-working craziness, hemiplegic discord or, except according to imaginative guesswork, mutant pointlessness. While nerves then brains developed what a bedlam would have been! Sparse sanity at every level as life's senselessness couldn't even try to get a grip! For even trilobite or early worm any imprecision in its chemistry would, drug-like, induce bizarre, unfit behaviour; mutant madness cannot build organic logic but, left to its own devices, mighty soon destroys. On the contrary, compartmental systems integrate towards an anticipated goal. Brains work bio-logically; they pack purpose and their shapes reflect it. Such shape impacts the way its owner - spider, bird, bee, you and me - sees the same environment. One might deduce, *top-down*, this is the way that we're each *meant* to see the world; and thus, in our own terms, are not half-hinged or quarter-cut but operate aright Where do you spot ineffective instinct or nerve nets due to imperfection still evolving? Just as bodies suit, in structure, the expression of their habits, so the information organ suits, to realise peculiar potential, what its creature needs to know. Brain serves, it is argued, as a medium. It connects mind with matter; it channels archetypal architecture and, in our hemispheric case, consciousness in ways appropriate to the asymmetrical, polar modes by which we pay attention to the world.

↓	*tam*	*raj*	↑
	left hemisphere	*right hemisphere*	
	detail	*general context*	
	analysis	*integration*	
	local interest	*global interest*	
	processing/ classification	*exploration*	
	assistant	*leader*	

No doubt the hemispheres are a composite, two-in-one, that act together in a way that Natural Dialectic easily includes. This does not mean each working may not have its own agendum or that a gradient (always implicit in the dialectic scheme) does not apply to operations of the brain as well. Balance we call homeostasis and gradient we call hierarchy are central aspects at the heart of cosmos and cosmic control. What, after all, are brain and mind but our own mini-cosmos? It may be argued, in the cyclic case of information exchange, that the right-hand hemisphere is hierarchically prior.

sensor	→	*checker*
effector	←	*balancer*
input	→	*processing*
output	←	*hemispheric reinterpretation*

From right to left to right again - the right-hand hemisphere engages what is new before it's passed across for left-hand re-cognition, re-presentation and, having been thus classified and placed within our individual world, shuttled back right for a response and thus, in the present circumstance, such decision as reflects a purpose. Right-left-right, from global to specific and the brief then passed back to the boss. *From sensor through left-hand processor and back to right-hand determination - this perception process involves a homeostatic loop. Equally, such psychological homeostasis or dynamic resolution in cerebral action amounts to a cycle of dialectical synthesis.* **Indeed, this cycle might be termed a basic 'unit of embodied mind' - in turn reflected as a 'unit of cerebral action'**. The nerves involved summate; they are not, but they compose the basis for, a physical experience. They serve, like electronic chips, to unify all matters on a single, conscious screen.

The hierarchy of process, also alluded to in Chapters 5 and 14, is at once temporal, spatial and informative; and its pattern, executed by the complementary agendas of the two hemispheres is easily represented within the template of dialectical stacks. From conceptual to physical, such orderly construction would, rather than haphazard appearance, indicate that mental function is not only the product of chemical preordination in the form of *DNA* (from whose program twin hemispheres were developed) but of conceptual archetype (whence patterns of human instinct, character and aspirations are derived).

While the directions of agenda antithetically oppose they also complement, modulate and balance out each other. Do men and women never fight? Yet how could life go on without their synergy? Operations in the two hemispheres, one whose centripetal tendency is towards self-consciousness and narrowness of focus and the other whose involvement is with global context and appreciation of fresh possibilities, act in dialectical synergy. The right hemisphere's vigilance scans our 'experiential' horizon while the left zooms in to fit its details to our schemes, our prejudices and ways we might exploit for personal gain; then the right contextualizes, fits left's theory into circumstance and launches plan of action - be it all in a split-second or a long-term and complex campaign. Such lateralisation of function, such ingrained polarity well reflects the way the cosmos works. Brain structure is embedded in both dialectical polarity and, as a whole, its trinity. Not only, we argue, does it symbolise the world in thoughts but is itself a symbol, a metaphor that serves as a reflector of the universe.

	tam/ raj	*Sat*	
	bilateral parts	*Brain*	
↓	*tam*	*raj*	↑
	left hemisphere	*right hemisphere*	
	isolation	*connectivity*	
	language/ speech/ analysis	*empathy/ context/ music*	
	attention to detail	*overall perspective*	
	shorter attention span	*longer attention span*	

Black and white is a crude, stereotypical kind of stroke. Mind is rarely driven to extremity. Balance, modulation (push-and pull), and relativity of emphasis is how the real world carries on. Not in warring opposition but with interference, inhibition, negativity of feedback; not in just promotion, resonance, coherence or short-lived, climactic positivity of feedback but in perpetual, changeful oscillation in between the two. In this respect let's draw up hemispheric stacks that nuance what experiments and scanners also show. Let's start physically with an asymmetry called 'Yakovlevian torque'. Mind's bent is (↑) towards immaterial information gain - in mankind to such extent that we might, according to our craving, re-classify ourselves uniquely as *Phylum cognoscens*. Such psychological torque reflects (↓↑) dialectical antagonism tugging between the two directions of every human centaur's mind - downward towards the sensible world and upward to the inner world of ideas, principle and plans. This tension is physically expressed in hemispheric asymmetry the informative direction of whose torque is, as opposed to (↓) energetic/ bodily, (↑) up-front-right. By 'Yakovlevian torque' the brain appears 'dialectically' twisted slightly about its central axis such that it is wider left rear and right front; it also extends forward to the right and rearward left. Muscles grow with exercise and, similarly, cognitive exercise may actually affect the size of correlated parts of cortex; it may therefore, also be that left-hand hemispheres - or regions within them - are sometimes larger due to cultural bias, that is, overemphasis on analytical, linguistic and serial reasoning skills in a 'scientific' society. It is certainly also true that brains reflect other dialectical asymmetries. For example, male and female structures differ (e.g. women have slightly less nerves cells but greater connectivity between them, a larger *corpus callosum* leading to faster inter-hemispheric transmission, and proportionately larger language-dedicated regions). In both sexes there is more white matter, apparently greater interconnectivity between nerve endings (dendrites), less fluid-filled ventricular space and more diffusely organised but better interconnection across regions in the right hemisphere; whereas functions in the left have less fast-transmitting white matter, are more focalized, well-connected within individual focus but less well interconnected with each other. The hemispheres differ in sensitivity to hormones and depend in degree on different neurotransmitters (e.g. noradrenalin and less easily fatigued noradrenergic neurons on the right side).

Cortical hemispheres are, as well as tendentious, complementary in mode of operation. A final stack summarizes how left and right modal tendencies affect our sensation, emotion and cogitation; and how they impact on our selfish and social leanings. The latter aspects are expressed in science, politics, art and religion so that hemispheric division reaches to the roots of our being. Brain structure needs, to be complete, both parts *in balance*. The cortex is an extremely subtle mediator between consciousness and physical creation. These sections of stack, overlapping in their spheres of influence, illustrate such polar and yet balance-serving governance.

	tam/ raj	*Sat*
	balancing	*Balance*
	left/ right	*Centre/ Whole*
	hemispheres	*Cortex*
↓	*tam*	*raj* ↑
	left hemisphere	*right hemisphere*
	down and out to earth	*up and in to heart*
	secondary/ follower	*primary/ leader*
mode:		
	narrow focus	*broad beam*
	detail	*detachment from detail*
	specifying	*generalising*
	practice-to-principle	*principle-to-practice*
	bottom-up	*top-down*
	particular	*global*
bits and all:		
	analysis	*integration*
	break up	*continuity*
	apart	*a part*
	extraction of part from whole	*inclusion of part in whole*
	spots difference in change	*spots sameness in change*
	picks out different groups	*encompasses different group*
	divisive	*integrative*
	separation	*relation*
	antagonistic	*holistic*
	body-as-thing(s)	*body-as-me*

one-by-one and one:

serial order	*parallel processing*
logic	*insight*
stepwise appreciation	*intuition*
'sticky'/ nit-picking	*fluid*
category recognition	*individual (face) recognition*
tendency to split/ classify	*tendency to gather up*

symbol and reality:

re-presentation	*presence*
image-making	*event (physical or visionary)*
explanation-in-self's-terms	*environmental inclusion*
processing	*receipt*
symbolic manipulation	*non-manipulative phase*
theoretical construction	*perception-as-is*
numerical calculation	*numerical relationships*
language/ local code	*universal code/ music*
verbal	*non-verbal/ visuo-spatial*
phonogram	*pictogram*
prose	*poetry/ song*
intellectual activity	*intuitive activity*
literal/ explicit	*implicit/ humour/ metaphor*
science	*art/ religion*

creativity:

creature of habit	*explorer*
rules	*boundary-pushing*
likes certainties	*happy with possibilities*
repetitive/ formal	*informal/ novel*
tendency to reinforcement	*tendency to balance*
positive feedback	*negative feedback*
'sticky'/ obsessive	*free*
mechanical response	*fresh approach*
technological interest	*interest in nature*
interest in mechanisms	*interest in people*
physical construction	*metaphysical creation*
rigidity	*creativity*

self and emotion:

ego	*super-ego*
egotistical perspective	*self-in-context*
me	*me and others*
grasping to self	*social self*
attachment to sensory self	*detachment from ego*
neg. emotions/ anger/ fear	*pos. emotions/ unifiers*
act of will	*act of submission*
calculation	*empathy*
observation of 'things'	*feeling*
earth-ward involvement	*heaven-ward transcendence*
cut off	*together*

disconnected/ lonely	*connected*
impersonal	*personal/ empathetic*
cold	*warm*

conflict and community:

thing-side	*life-side*
survivor	*co-survivor*
competition	*community*
denial of other position	*openness*
against	*with*
controlling/ forceful	*encouraging/ allowing*
inequality	*sense of justice*
self-centred law-breaking	*law-abiding*
exploitative/ utilitarian	*benevolent*
utility	*value*
economic/ survival value	*moral value*

morality:

formal/ fixed	*informal/ responsive*
controlling side	*informal side of morality*
punishment	*forgiveness*
religious canon	*mystic heart*
political law	*inspiration*
husk	*seed*
commandments	*love*

Why has such radically conserved bilateralism, from earliest organic times, been brought to bear upon the information organ, brain, itself? Is it all by accident, or does construction purposely reflect a sliding scale of balance in between the fundamental constants of our universe? Is it by accident that, for example, emphasis towards complex speech with grammar engages more to left side while musical appreciation is nuanced to the right? Clear signs of a rational division of information processing at all levels emerge from close inspection of the metaphysical necessities of life (as illustrated in the stacks) married to the hemispheric architecture of material brain. This conjugation represents paradoxically complementary, polar opposition that is ubiquitous in nature's operations. Indeed, common sense and scientific scans now both confirm the dialectical polarity of brain-function, mind and, which both reflects mind and of which mind is the paramount reflector, cosmos.

Such informative refinement in the balance of the operation we call life is not, according to a rational interpretation, down to jiggling molecules and incoherent chance. Evolution is supposed to replace by means of accidental and therefore irrational improvements and yet here, in brain's hemispheres is not replacement but creation of an integral addition; informative potential has been, through an array of animals, increased until - where do you go from you? Alternatively, the brain was shaped and meant to shape our world. We are the coherent way we're meant to be.

Yet some, amazingly, with symptoms of left-hand materialism in ascendancy (see Chapter 5: Intelligent Design), would currently, competitively ban these rational, dialectical conclusions from consideration in classroom, college or, by media, the public forum! Such exclusion, we shall see, unhinges human rationality!

In conclusion we might note that, should one want to build a brain, then its conceptual architecture might well be designed to best reflect the inner architecture of the universe itself. Polar ways of thinking would combine - with every nuance catered for - within the whole, most intimate but temporary conjugation of a human mind with matter. Look only to yourself. How ingeniously is concept earthed; how intricately fashioned is life's window on the world.

We return to discuss the rational symbolism evinced by architecture of the brain in Chapters 17 and 24. Now turn the page to qualities of consciousness.

14. Hierarchical Psychology

Bottom-up and *top-down* really head-butt when it comes to mind and brain! Materialism has maybe a harder nut but, on the other hand, when you butt at nothing how does that hurt immateriality? Your state of consciousness can alter (waking, sleeping, excited, peaceful, happy, sad and so on). Do you define the alteration from subjective or objective point of view? Does brain create the state of mind, mind the state of brain or can the traffic travel either way?

Of course, you have a human mind and, set with your body, its connective template, brain. Thus, ineluctably, you realise only human and humanity's potential and can but imagine what another kind of organism, with a different kind of template, feels. Is there, in this case, a healthy suite of states of mind that everyone can recognise as 'normal sanity'? Recognising that imbalances can happen how might you define stability and natural cycles of psychology? Certainly the brain, a chemical concoction, conjoins your state of mind with body by acting as a filter of sense input and a local panel of control. Brain *mediates*. Your headset brings ideas and images to earth as bodily behaviours; and passes earth into the mind as an experience made up of myriad sensations. Disturbances in consciousness, like water, vary as you pass them through a different shape of grid; you modulate the taste of food with additives and so you might expect that chemical additions and subtractions, natural or otherwise, affect the grid of brain; and thereby colour mood and state of mind. In other words, genetic difference, mood-altering drugs and hormone imbalances affect the template to produce a variable experience. Indeed, even an environmental change can shift a mood. It happens naturally all the time. Is there a state that mind is always looking for?

Take a special case of alteration when you feel restrictions in the network lift - hot interest, fluidity and even ecstasy. Is wisdom or the understanding of a state of ecstasy best induced by medicine or meditation? Can natural or unnatural addictions generate identical experience of bliss? At the other end of spectrum a second special case appears - unconsciousness. What happens here? Brain activity and what is called sub-conscious process no more cease than heartbeat but subjectivity's away; where is all awareness gone? Sleep, knocks and anaesthetic drugs induce the same effect; natural and abnormal each produce the same. If you disturb cerebral chemistry do you inhibit *generation* of 'electric' emanation (thought) or, conversely, disrupt the flow of information to and from subjective mind? Are you brain, is mind you or are they both the same?

When you are awake, like now, states of mind, mood and will affect your body just as body can affect them too. Therefore let's start on the basis that embodied state of mind is, as you might expect where two components interlock, a mutual business of informative exchange. Do physiological variables, however slight, affect an *ASC* (altered or altering state of consciousness in which perspective shifts)? If states of consciousness are mirrored in brain structure and its patterns of activity then couldn't we work back, reverse engineering from scan data to compute subjective states of mind? States and not personal, detailed content of experience would have to be the limit of our goal. Brain waves should be especially informative. We might suppose their function is coordinative; they would serve like gears in motor-cars and each modulate or predominate, in conjunction with a sleep/ waking oscillator found in the brain stem, over a particular background state of brain - or mind or consciousness since aren't all three the same? Each would have its own particular usefulness. Every instance in a psychosomatic exchange mechanism both affects and is affected by 'the other side'. You can voluntarily change your brain waves (e.g. deliberately act calmly in a psychologically stressful moment) but so, involuntarily and in compelling circumstance (e.g. shock or fatigue), can your reflex body chemistry. Mode, mood and quality of thought are affected from one side intentionally and from the other by the play of neural chemistry. Exhaustion, pain, drugs, illness, happiness, morale, motivation and health all play a part. You might call changes exerted 'from within' by mind on brain and body 'internal stimulation'; and those exerted on mind 'from without' by brain, body or any environmental event 'external stimulation'. In normal, waking consciousness modulation between these two forms of stimulus is practically continuous. The adjustment is, however, always homeostatic and, as far as possible, directed (↑) one-way - 'accentuate the positive, eliminate upsetting negative'. 'Positive' means, as desire cycles with fulfilment, satisfaction. Fulfilment of personal wishes or instinctive biological needs leads to the target-state - contentment.

So far, so good; both parties are agreed. Mind-body interaction is a two-way street. Polar divergence then occurs. *Which is cause and which effect*?

Bottom-up, materialism singles matter as the cause. There's no horse put before the cart. 'Emergent' mind depends on brain and brains derive from genes. It's all genetic; mind's the gift of special molecules. *DNA* makes nerves and thence, correctly wired, incalculable consciousness rises ghost-like from incalculable connections. Thus chemistry and physics drive psychology and thoughts are made, effectively, of atomic multiplicity! Who can argue with such guess? In this view grades of mind or scales of immaterial consciousness are disavowed; the notion of a Buddhist void or Christian light is, since it's merely brain-work, relegated to the shelf of nonsense. Hierarchical psychology is, in a rationalistic science classroom, out.

Top-down it's a two-way street as well - but this time there is something on the other side. Atoms don't *feel*, nerves do not *experience* nor brain *understand* a thing. The latter mediates between the body (close but still external agency) and mind (internal, immaterial agency). Embodied mind affects and is affected by brained body. Fearful, lustful or contented thoughts are each reflected by cerebral motions; and, the other way, sensations drive the grey machinery and thence affect the mind. Such two-way traffic makes allowance for genetic trait, that is, predisposition towards a chemically affected mood or bias in behaviour. Varying expression of alleles and their controlling factors (for example, more or less testosterone or dopamine) impacts each individual's tendencies. Such chemistry no way explains what set the trait, instinct or nervous framework up; nor what exactly *feels* experience; nor that, against the flood of instinct's physicality, rationality and curb of body by its mind cannot occur. It happens all the time. The concentration of a deeply meditating monk is simply an example of the metaphysical extreme. **Is, therefore, his state a product or the cause of secondary cerebral effect? Does purpose or molecular biology attain transcendent ecstasy?**

Top-down, Hierarchical Psychology

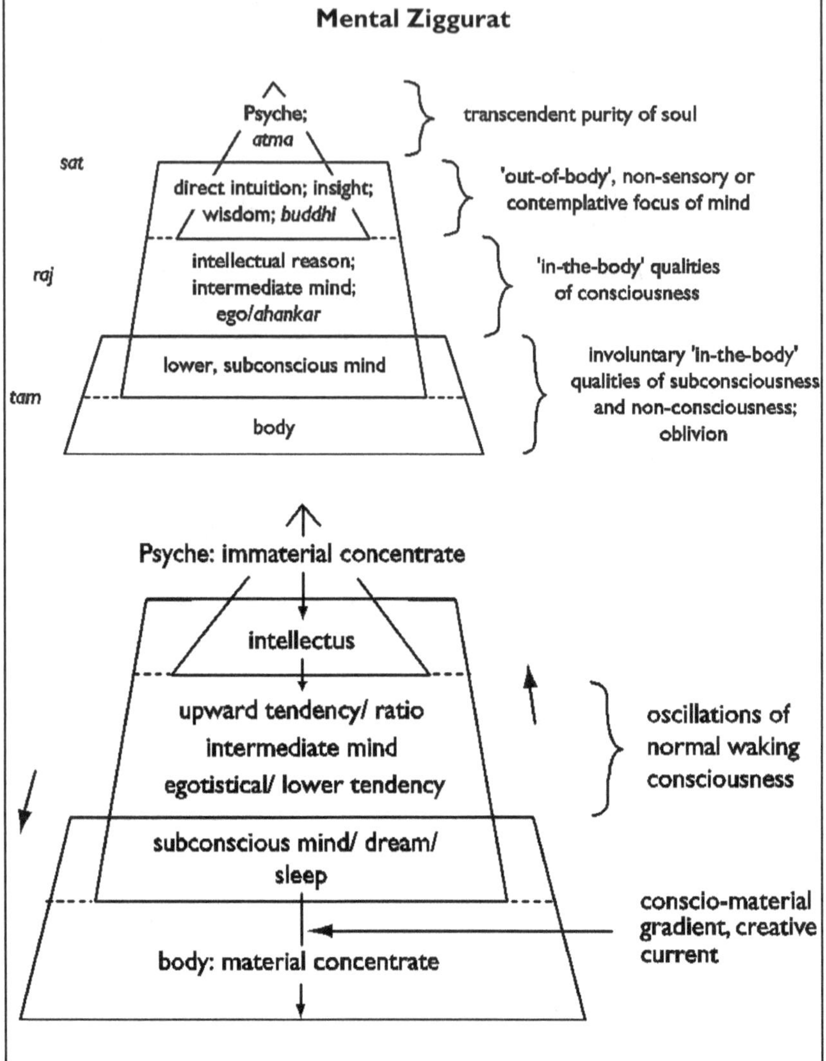

fig. 14.1 (see also fig. 0.14ii)

Steps; but the grades of hierarchy are not set in space. They are vectored in accord with energetic power and comprehensive scope of information and control. Frequencies my, for example, rise and fall; so, by degrees, may subjectivity. In a conscio-material gradient, therefore, you can identify, as well as states of energy and matter, states and moods of mind. *Top-down*, hierarchical psychology explains the levels and directions to which operation of the mind is geared; and, with oscillations in between these levels, nervous correlation in the human brain. In which gears does your own life predominantly run?

tam/ raj	*Sat*
info. out/ info. in	*Information Centre*
lower/ higher	*Third Eye*
↓ *tam*	*raj* ↑
lower	*higher*
material bias	*spiritual bias*
spiritual anaesthesia	*material anaesthesia*
nervous system	*contemplation*
detail	*principle*
diversification	*unification*

States of mind condition consciousness. They restrict it until, gradually, its light fades out. Over the next two chapters let's examine them. Check *fig.* 14.2. *Top-down*, as you'd expect, we start at grade 1 and continue with its phasing down into grade 2, the highest state of normal human mind. Such normalcy, your waking state, includes within grades 2 and 3 the operating subdivisions we call higher, intermediate and lower mind. Finally, we turn in Chapter 15 to involve sub-consciousness at levels 4 and 5.

Reformulation of *fig.* 13.5
(Five Main States of Human Mind)

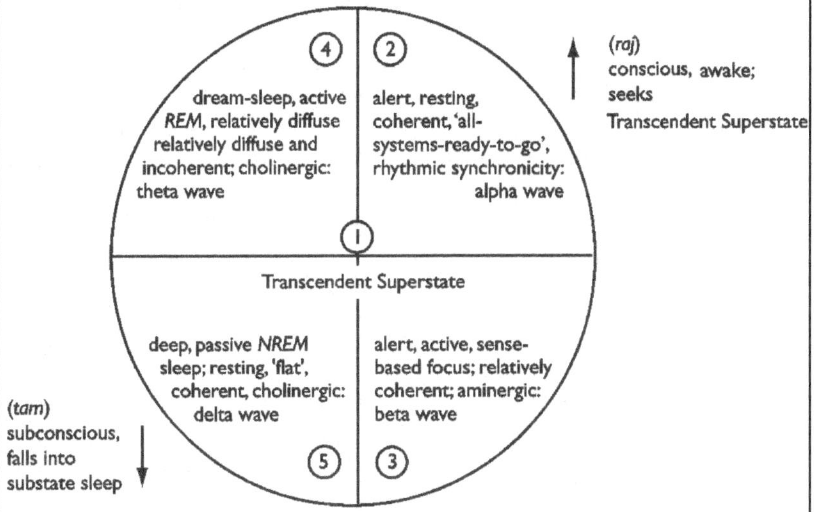

Diagram 13.5 is reformulated in terms of correlating predominant brain wave to a state (or gear) of consciousness. As cycles progress they constitute periods. There is, therefore, an additional correlation (although it includes sub-cycles) with our diurnal life cycle.

***fig.* 14.2**

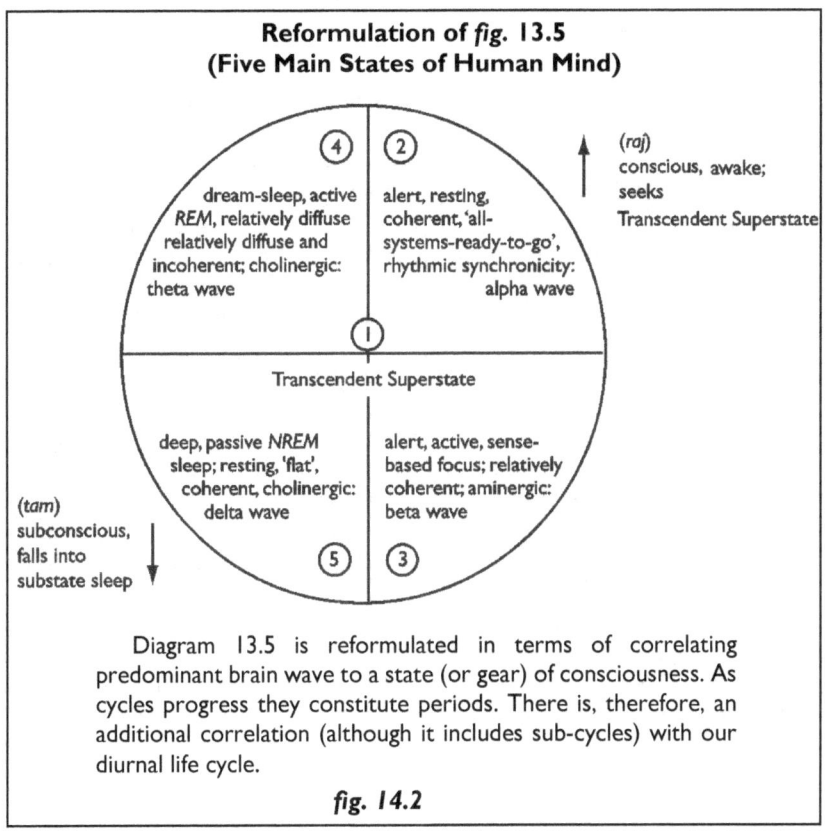

tam/ raj	*Sat*
relativity	*Absolution*
matter/ mind	*Pure Subjectivity*

below	*Transcendent*
motion	*Poise*
range/ spectrum of consciousness	*Super-conscious/ Supra-rational*
(2),(3),(4),(5)	*(1)*
↓ *tam*	*raj* ↑
downward tendency	*upward tendency*
passive/ exhausted	*active/ alert*
unconscious	*conscious*
sleep	*awake*
incoherent	*coherent*
sub-rational/ irrational	*rational*
involuntary	*voluntary*
inertial equilibrium	*dynamic equilibrium*
deep sleep/ coma/ delta (5)	*contemplative/ alpha (2)*
busy-in-sleep/ dreaming theta (4)	*busy awake beta (3)*
subjective absence	*various subjective motions*
complete objectivity of matter	*rel. subjectivity of mind*
sub-state material oblivion	*levels of mind*

What are gears for? Gears allow a vehicle to function best when different sorts of work need to be done. In this case the two fundamental states (or gears) of consciousness are the polar reciprocities of wakefulness (on-line action) and sleep (off-line inactivity). Within these main conditions there exist subdivisions most clearly illustrated in *fig.* 14.2 which is itself a simply reformulation of *fig.* 13.5. The major gears, with which most brain waves are correlated, comprise alert/ resting (alpha wave), alert/ rationally active (beta wave), asleep/ irrationally active (theta wave) and asleep/ resting (delta wave). These four waves interweave normality's diurnal cycle. Measuring only a few millionths of a volt they refresh the areas of brain they influence with frequencies that normally range from about 1.0 to 30 hertz, although 'high beta' or 'gamma' frequencies (up to and over 100 hertz) are said to reflect bursts of 'hyper-awareness' and the exercise of exceptional or, some say, paranormal powers. Delta waves are, conversely, slow and of high amplitude; they range from (abnormally in coma) dc. or 'flat' but normally between 0.6 and 2 hertz; and they occur only in the deepest states of rest and sleep. Theta waves are less coherent, more variable and range from 3 to 7 hertz; they predominate in so-called *REM* (rapid eye movement), dream or light but 'active' sleep. There exists at about 7.5 hertz (curiously the resonant frequency of earth and ionosphere) a so-called alpha-theta (α-θ) border, a semi-lucid dream-state. From 8 to 12 hertz vibrate alpha waves, reflective of a psychological condition of well-being, coherent alertness, calm and creativity. A 'groove' of alpha waves is the one we can best align with intense interest, meditation and the dialectical vector that extends towards an increase in right-hand characteristics. This is not news. From out of prehistory our ancestors' first historic signals are in books whose main topic is the way to increase a share of alpha waves or, more to the point, engage the bliss that these waves physically represent. The Vedas, Shastras and Puranas are concerned with prayer and meditation. The message is repeated in modern techniques (such as transcendental meditation and other forms of contemplation that target the same goal) and, less pointedly and therefore successfully, in religious ritual and ceremony. Silly us! Against alpha we find, from 13 to 30 hertz, the beta waves that predominate in an action-packed, stressful daily environment of work and business - our chosen, secular routine for life.

Top-down, the boss is at the top. The start's state (1) is as illustrated in *fig.* 14.2. This alpha-theta border is of special interest because, resonant with earth's natural frequency, it appears to neurologically reflect a window between waking and sleeping through which attention may slip either involuntarily into sleep or voluntarily from personal into moments of self-less universal consciousness. Such consciousness transcends the physicality of incarnation and our day-to-day experience.

First State of Consciousness - The Psychology of Transcendence

Bottom-up, we already know reductionism has squeezed mind out of mind and into a material deposit, brain. Nothing metaphysical remains. The first, foundational state of consciousness is, it is claimed, non-conscious matter. Consciousness is a mysterious but material property of brain; thought and free will are each illusions super-posed upon genetic make-up, neurological connections and electric flickering - except if quantum factors govern brain and then complete determinism might not steer your puppet life.

In this case life's *real* secret is, according to the zealous atheists who in 1953 discovered how it coiled, *DNA*. In their view mystic truth is actually a 'God Delusion' that's evolved from genes, their proteins and electric fields of charge. *Material truth must, therefore, involve what is otherwise identified as the prime category error of philosophy - a failure to distinguish energy from information, mechanism from its operator (or creator) and material from immaterial; and, on top of this, the treatment of systematic effect (say, written code) as if it were its own cause (that is, the author of itself).* If you once construe mind and brain inseparably the same then neurologically you cannot distil a thought or separate a feeling from each one's solution, physicality. Scan brain, scan mind. Thus naturally, from this perspective, neuroscience now asserts that, in reality, the seat of wisdom is the activation of certain sets of neurons as opposed to the ignorance shown by others. Electrical disturbances cause thought; they somehow *are* that thought. If, therefore, consciousness, divinity and its religions have evolved as an effect of brain then you assume their metaphysic has to be a secret that is, sub-scientifically, unreal.

Is not theology therefore an illness, folly or a bit of both? Is other-worldly fiction not concocted by a mess of hormones, intercranial tuberculoma or the buzz of tinnitus? Epileptic lights and voices, oxygen deficiency or drugs each seem to let you know, by strange and yet specific chemistry, divinity. You assert that ions here or there let logic flow; and by such charged logic you reduce fantastic deity to monoamines and the rest, you squeeze godhead to the volume of a skull. Such reduction unto magic mushrooms or disease is a measure of the literal ridicule in which atheism effectively holds saints, their teachings and all metaphysic that relates mankind with its Creator. Haven't scans identified the

Godhead's vibratory spot and thus - the ultimate reduction to absurdity - nailed deity inside some nerves and micrograms of polypeptide? The hope of mankind wouldn't fill the bottom of a vial!

Forget such infantile religiosity and turn instead to check exactly how nerves, hormones, transmitters and the evolution of a cerebrum could generate delusion on such pervasive and traditional social scale. Mystics seem to be the root cause of the problem. They personify, by their delusional excess, the error into which they lead mankind. Indeed, you'd never say it to a contemplative's face but he deceives or is deceived. In this case surgery should help your monkish patient. Why not try 'godectomy' to sort his problem out! Don't those epileptic abbots need lobotomy?!

After all, God's different sorts of houses - temples, churches, mosques - boil down to psychiatric hospitals, a Freudian refuge for the feeble-minded and a province where the blind support the blind. If it's not a pretty clinic for neurosis, then is not worshipful assembly just, according to Freud's atheistic fellow-traveller Marx, an opiate of mass addiction, an oppressor stupefying the oppressed? Yet emperors and kings, most powerful in the land, in less intellectual and scientifically unenlightened times held sway as regents and as servants of elusive deity. How absolutely out of kilter hierarchical society appears to meritocracy. And even if religion 'gets you high' then drugs can also raise you, this time without effort or a trace of moral grief!

A pilot at his panel flies the plane; are not brain, senses and our muscles simply instruments of flight of thought? Materialism thinks, however, that the motion of a steering wheel is just the same as a decision made to turn; representation is construed as if become its mental picture's fact; and then 'as if' construed as 'is'. As seismograph to a geologist so brain-scan to neurology - except that rocks are mindless objects and don't *feel* a thing. No doubt that neural correlates reflect each mood and motion of an immaterial mind; why should scientific instruments not capture neural 'fall-out' from a 'peak experience'? The psychology of mystical transcendence is, in these terms, nothing but biology. Chemistry will lead us out of error and deliver us unto the scientific truth.

And that's the rub. For a materialist some geni or another must rub mind from matter's lamp; and worst of mind is to account for mystical, religious or spiritual experience - such a widespread yet insane delusion. You are a body suffering spiritual delusion not a spirit suffering physical experience, aren't you? You might believe in body, brain and brain's mind but don't believe, do you, that the soul-piece, as attention could be called, is other than electrons, protons and, perhaps, some interior apparition made of light? Therefore the logic of monistic 'matter-only' mind-set is obligatory - you're driven to the answer that some god-spot, module or divine secretion must exist.

No doubt a pin pricks psycho-physiological effects. Fear generates adrenaline, pleasure dopamine and each emotion prints dynamic patterns on grey matter you can scan. Drugs, physical sensations or electric pulses can induce entrainment with a brainwave and/or change a mood. This is how, in passive mode, you are treated by the world. It may be that such products randomly or inadvertently impinge upon the neural correlates of mystical experience and generate a shadow of its bliss. Indeed, mankind was never slow to chase intoxication. The very word 'enthusiasm' means 'the god's in you'. If alpha waves reflect upbeat experience then why not force the superconscious pace. Consider psychodelic substances. Or do you remember, with his cap and wingèd sandals, Mercury the messenger who travelled fast as thought between the gods? Instead of magic mushrooms Michael Persinger of Laurentian University, Ontario constructed a hermetic cap to induce very small electrical signals in the temporal lobes. Sitting in a 'hell and heaven chamber' wearing such 'deific' helmet you should be transported to the heights of Helicon; four sets of solenoids should fly you to Mohammed's seventh heaven! The 'neuro-theological' idea was that specific pulses of electromagnetism might massage your lobes to ecstasy. The 'brain-box' might induce an alpha wave or propel you into alpha-theta regions of profundity. Was this pop science or the proper thing? Although the one-way scepticism of materialistic journalists suspended critical analysis actual research proved seriously less gullible. Replication using control groups in some Swedish double-blind experiments (by Pehr Granqvist and collaborators) found that the mercurial helmet did not seem to work. Perhaps, they surmised, suggestibility had done the trick. Indeed, as *fMRI* and *EEG* scans of meditators demonstrate, entranced experience is not associated with a temporal lobe alone but many regions of the brain. Why, therefore, any more and perhaps much less than *LSD* should the misconceived conversion of a skiing helmet generate hallucination of divinity? In fact, no study of the brain can prove or disprove subjectivity, transcendence or the quality of God.

Psilocybin, mescaline and *LSD*! If monoamines set the mood then perhaps what mimics them might help you feel divine. You might argue, cart-before-the-horse, that chemicals *create* divinity but, in reverse perspective, might conceive of brains as 'windows' that restrict an immaterial potential, consciousness, in specific ways that their design requires. This intended filter lets an animal behave appropriately and, in the human case, allows the case of full connection with Essential Super-state or Primary Divinity. Mental with physical effect; a lustful dream can change brain chemistry. Physical with psychological; external agency can do the same and thus cause sexual perception. You might even cut sensation out and chemically induce such sex-perception at the inmost of external agencies, the brain. A meditative monk drives neurological, although irrelevant, effect; but, if external agency (perhaps chemical induction by a drug) could generate due opening of perception's door, why shouldn't a programmed capacity for such divine experience be stumbled accidentally upon - this time without sage preparation, focus or the monk's intent and orderly control? How wide our window opens! You'd infer that the capacity for 'peak experience' was written into human functionality.

Potions aside, perhaps the closest, most intense of physical sensations naturally spring from the union of two made one materially, that is, (see Chapters 22 and 23) from complementary, polar sex. But dialectically you find (*fig*. 0.5) material body at creation's outside edge. Involvement with the physical is relatively superficial; it is peripheral regarding Central Truth. You could say that carnal lives skid over Life. Thus, although intense, a sexual episode is poles apart from spiritual ecstasy, the pole opposed to love divine, the base antithesis of immaterial transport out of physicality. Yet, even so, is the aura of emotion that surrounds a transient reproductive game unreal? Is lust's addiction madness? And, with its sex-free love, is not the alpha moment of transcendence also a subjective fact? Subjective states of mind most certainly exist. Healthy alterations up and own include the 'gears' of sleep, dreams, wakefulness and (arguably) hypnotic trance. Is it, therefore, in delusion evolutionary psychiatry (the latter word means 'healing of the soul') conflates transcendence, rapture, visionary episodes and apparition with psychosis, crazed hallucination and a raft of mental instabilities into a single, pathological division? Arts and culture are defined as the creation of some neural blips! 'Enlightenment' is filed as a 'disorder'! What depth of ignorance confuses heights of meditation with orgasm or an epileptic fit! *Who is, although he may have passed the board's examinations, fundamentally confused? Who's lost the plot'? How wrong can you get?*

tam/ raj	*Sat*
lesser being	*Supreme Being*
existence	*Essence*
lesser selves	*Non-self/ Self*
lower states of mind	*Transcendent Super-state*
relatively sane	*Enlightened*
imbalance	*Balance/ Equanimity*
turbulence	*Mens Sanissima*

↓	*tam*	*raj*	↑
	negative	*positive*	
	egocentric	*Self-centring*	
	bonding by ego	*liberation from ego*	
	mind-contracting	*mind-expanding*	
	darker/ dimmer	*lighter/ brighter*	
	ignorant	*knowing*	
	bad	*good*	
	depressing/ dragging down	*uplifting*	
	asymmetrical imbalance	*dynamic equilibrium*	
	impotence/ fallen state	*up-and-going-strong*	
	inwardly numb/ dead	*joyful/ all alive*	
	stress-induced	*calm-induced*	
	disintegrative	*integrative*	
	garrulous/ gibberous	*succinct/ accurate*	
	disorderly/ disruptive	*harmonic*	
	angry/ explosive	*calm/ collected*	
	incoherent	*coherent*	
	chaotic	*focused*	
	draining	*shining*	
	ugly	*beautiful*	
	proud/ hard	*yielding/ humble*	
	hateful	*loving*	
	hellish	*sacred*	
	demonic	*angelic*	
	unwanted/ repellent	*wanted/ longed for*	

incoherent	*purposive*
down to sub-state	*up towards Super-state*
paranoid	*ecstatic*
psychotic	*enlightening*
mens insana	*mens sana*

The Essence of creation is Stability; stability-in-motion means dynamic equilibrium, that is, constant restoration of a balance; and the homeostatic balance of material cosmos is inherent in those natural behaviours called laws of chemistry and physics. Life's central instrument of biological health is homeostasis (Chapter 19). *And psychological stability is equanimity; equanimity means least or no disturbance cramps the space of mind; our minds naturally and constantly desire, in ways they variously imagine best, peace, contentment and serenity. Liberation!*

Top-down, mind over matter, you put horse (mind) before corporeal cart. Immaterial information can, with effortlessly cool control, balance matter's turbulence. A vehicle, including *VAMT2* genes and nervous chemistry, is servant of life's journey's show. Psychological - suggestion, purpose, faith and hope - affects the physical. A body's subject to its mental fuels and information chemistry. Even *placebos* (where you think you're healing so you do) and *nocebos* (where you believe that something makes you ill so, like a voodoo curse, it does) each show mind substantially affecting head's pink jelly. Treatments quelling phobias or quashing obsessions by positivity of thought-power demonstrate it too. Every mother uses substitution and suggestion to divert her distressed child's attention to another line of thought. Mind over matter is, in fact, the basis of all education, human progress and the methods metaphysical of, if not neuroscience, spiritual psychiatrists. It is not brain that formulates the targets, generates the will-power and thus realises its ideas. It is your *mind* that, like driver separate from car, is separate from brain and metaphysical.

So does the diagnosis have to be that gurus need lobotomy? Are monks and nuns the pilgrims of deluded paths? Or are they on a track that's cosmically correct? *fMRI* scans show that prayer and trans-religious meditation change brain chemistry. The neural networks realign. Recalibration of, especially, parietal and prefrontal action reduces or eliminates their 'worldly' type of exercise. It works the same in every human brain. That's it! Materialists become excited. No blood flow to the parietals somehow generates 'divine' experience. You see how universally some nervous juices can evolve delusion; God is no more there than fizzing in your lemonade.

Oh dear! Standard error: zero probability! Who finds light while scrabbling on the floor? *Top-down* you can allow more than material reality. **Brain is a brilliant medium between the mental and material worlds. Thus its dynamic shape reflects the thoughts (or lack of thoughts or dreams) of mind as well as, from the body side, sensations**. Brain chemistry reflects but does not, being made of atoms, think or feel. It's the medium whereby active metaphysic meets the physic of your body and thereby the world. Thus Adam's heaven-conscient brain would have become the shape that certain chemicals promote and monks by exercise try to regain - but are they after parietal blood flow? Like sweat from running that's a negligible side-effect. Would not the link, the channel to divinity have been, like eyes or ears, designed for man's Communion? **It would not, therefore, be the case that chemicals or meditation stimulated an illusory effect but that, conversely, certain shapes of brain reflect detachment.** Release from physical locality, the body, opens contemplative gates of glory and what's always there is freed to enter in. How open is your contemplative gate?

What, in this case, is the inmost nature of the so-called 'spiritual' mind? What is its primal state of consciousness? *The purpose of all psycho-practice such as prayer or contemplation is to answer this.* Devotion to a teacher, worship of an icon or concentrative disciplines are all to develop focus and prolong expansion unto full communion with selfless, supra-mental consciousness. They are to climb the cosmic ladder and obtain Full Knowledge. Such enlightening experience ranges from novitiate glimpses to sustained and voluntary entry into Subjectivity or, in Dialectic terms, ascension to the Apex of Mount Universe. The way of Mental Mountaineering is one that's clean and clear beyond confusion of all man-made faiths. It might be science of the soul but definitely not the science of materiality. Pure of religion, is it not a natural capacity, a mechanism built into the human formulation, a connector wired back to the centre of this existential labyrinth? We are wired by design to understand Reality. The life-line grasped is Logical and called The Reason, Word and many other names. What kind of Natural Impulsion might, from the Top down, drive creation?

	tam/ raj	*Sat*	
	duality	*Unity*	
	lesser loves	*Pure Love*	
	self-contamination/ ego	*Self*	
	love's spectrum	*Transparency*	
	divided into parts	*All-Inclusive*	
	peripheral	*Central*	
	conditional	*Free*	
↓	*tam*	*raj*	↑
	inverted love/ hate/ repulsion	*interests/ attraction*	
	opaqueness/ darkness	*colours of emotion*	
	blindness/ incoherence	*purpose/ coherence*	
	cacophony/ chaos	*harmony/ coordination*	
	imbalance	*balance*	
	out-of-step/ haphazard	*synchronised*	
	interference	*affirmation*	
	discomfort/ stress	*relaxation*	
	isolation	*relationship*	
	confinement/ obstruction	*release*	
	ill-health	*health*	

In which case it follows that an 'alpha hit', whose brainwaves are a downstream and exterior reflection of the spring of Inside Information, is a fundamental source of hope, strength and health. Although it's indefinable in intellectual terms what kind of active metaphor might bear on better understanding of the super-conscious side of mankind's subjectivity?

The Master Analogy of Natural Dialectic is (Chapter 0: A Complementary Course) music. Song and music easily transcend loquacious intellect. It involves an integration of harmonic patterns, resonance and inflections that include different qualities and combinations of structure; its rhythms and cadences are, far from random or chaotic, orderly, purposeful and therefore meaningful; and its vibratory effect, outpacing language, can immediately touch, like contact by the eyes, our inmost nature - what is sometimes called the soul. Music moves. It does not just reflect but also seems to constitute the integral and integrated heart of mind from which less rhythmic, daily grind and its relationships detract. Love, music and its dance are 'playtimes' central to our subjectivity while the objective concentrations - detail, analysis and chores - seem more like work and in as much peripheral. The food of love is in another world. With how much other-worldly sense of music do you swing through all the irritating trivialities? Vibrations help give scientists a clue. Sounds and lights are what we learn by. If metaphysic sources physic could there be, above the quantum level and in mind, internal lights and subtle sounds whose natural excitements shape phenomena and cause material appearances? If so then no wonder mystics seek the inner light and unstruck music on whose 'thermals' they ascend to paradise! Perhaps, in fact, their lights and music are The First Cause of an orderly creation.

Science, by perspective, does not recognise this kind of language. Nonetheless all kinds of music stimulate response and it may be that some kinds (such as Bach or Mozart) can excite a specially heightened focus of awareness, increased sensitivity and the ability to learn. They foster an improved ability to appreciate more readily the intricacy, subtlety, beauty and kind of measured order found in integrated systems - such as language but also such as science and the other subsets that our education system strives to have us grasp. Could this be because the dynamics of such strength and power map to coded and therefore pre-programmed states of brain? After all, unless it's dead a brain can't lack activity; and so a certain state of mind induces such a state of brain as 'up-beat' alpha wave formations. If the construction of a brain is intrinsically harmonic, rational and resonant with order might not other types of music excite other sorts of nervous correlations and their states so generating consequent behaviours? In other words, you might conclude that the original template for neural construction and its various resonant modes of operation is harmonic; that from an Operatic First Cause springs the panoply of scores that, etched as memories and filed in universal mind, give rise to physics, chemistry and bodies of controlled and 'legal', lawful energies; and that such orderly and purposeful construction of a nervous system (including the teleology of its development) might reflect the nature of its most rational, orderly and musical beginning. Would not such an origin befit the paragon of information centres, the quintessential crystal of intelligence in all the universe, your ball of connectivity called brain? Is not psycho-somatic brain, that works upon an edge, the ultimate translator of mind's 'music' down to matter and of matter's up to mind? Are not nervous systems an expression of a higher plan that, in conjunction with genes, hormoes and the other kinds of bio-signal, pluck the human score called archetype (Chapters 15 - 17)?

The instruments of chance don't band together and compose a song. The only sort of evolution music undergoes is, like any system's, intellectual. A musical perspective therefore starkly contradicts materialism's version of our origin.

Nor does Darwin's theory predict, explain or otherwise illuminate the nature of your consciousness. Why therefore, you might ask, should such material delusion as 'a spiritual experience' and subsequent religious worship ranged about it ever have *evolved*? Perhaps untestable, inconsistent, incorrect or unsupported guesses are the best materialism, faced with immaterialism, makes. Convoluted theories abound. Could the primary purpose, for example, of the theocratic business of religion really be a cunning form of commerce or, by means of a bureaucracy of priests, a method of control and social order - one that benefits the lord and serfs alike? Exploitative, perverse and cruel – this is how religious despotism often ricks its face. If, moreover, an intelligent capacity for such perversion is dependent on your size of brain, is woman less perverted or deluded than her man? What use is cerebral capacity if over-brained 'ape-man' just nourished and developed superheated theo-fantasy? It is indeed a mystery when and why in ape-hood, hunter-gatherer or caveman-ship imaginations lacking value for survival came to rule survival's roost.

Maybe what appears, to scientism, mystery and superstition might be true. Why otherwise should the transcendent (or, perhaps, faked) delusions of the Buddha, Hafiz, Rumi, Sarmad, Saint Augustine or the whole congregation of scientifically ignorant mystics have led to such order, consistency and clarity of message, an informative message whose resonance so deeply inflicts huge numbers of equally deluded humans that they also mistake reality, take falsehood to heart as truth and seek, wrongly, to improve their lives in this strange way? Nanak analysed no followers' genes but was his Truth a figment of uneducated lies; or did he represent a more important, central sort of truth than bodies or their genes? Did Christ miss Crick's real secret? What was Lao Tzu's Alpha Moment or Ramakrishna's Great Idea? From ideas ripple powerful effects and from Idea ripples the most powerful of them all. Because he's touched The Bell a mystic's message resounds with the ring of Truth; perhaps it powerfully radiates because he's really entered, haloed, into Light.

Materialism stutters to explain Tzu's Moment. Its reflex is denial. It discredits, charges with deceit or plain explains it right away. Erase all immateriality! Identify it, for example, with some imagination like a so-called 'meme'. Although hypothetical and scientifically untestable an entity, a 'meme' has been identified as an immeasurable unit of consideration - an idea, attitude, belief, or instinct - any pattern of behaviour. A meme's a 'cultural analogue' of gene and no more cares for carriers. It's anything transmitted non-genetically from one mind to another. Such 'unit', being something taught or learnt, has been proposed as a materialistic measure of psychology. It might evolve like Chinese whispers or, when copied faithfully, remain the same. Its imprint's stamped upon informative a nucleus - your brain! It is an inflection of your nerves because, at root, that's where your mind emerges from. Art and science both evolve 'memes' but, you reckon, scientific ones reflect the truest excellence of neural rigor. They will be the fittest, survive best and, as representatives of lasting truth, mutate the least. *Indeed, the idea of a meme is perhaps a meme itself.* Since, however, the memetic paradigm is from an arsenal of materialism's missiles aimed at the destruction of 'irrational' gullibility then the most powerful and enduring memes in practice - the religious ones - are viral to your clarity of mind. These 'holy demons' are unscientific and although, unconscionably, they've won the immaterial competition for survival up till now the purely scientific meme will root them out. Its demonology will exorcise them. And, of course, the Alpha Moment round which all religions are arranged is most reviled or else explained away by evolutionary psychology.

The bottom line, in *top-down* view, is that non-conscious genes and conscious memes are simply used as ropes knot you up with atheistic explanation. When it comes to mystical transcendence you might rate inanimate but 'selfish' genes recessive. Where evolution's strength is rated by fecundity and 'fitness for survival' in a stereotypical caveman sense, you note that neither factor entertains the measure, goal or outcome of a mystic's practice. What use had altruism, lack of sex, celestial visions, prayer or surrender to an Immaterial Essence in the Pleistocene? Indeed, such qualities specifically negate the Darwinian ones; and this elimination shines from any uncorrupted faith (or viral meme)!

If, however, memes, the bits of brain that commandeer free will, are not gene-spreaders what are they? Is each one a neural circuit, the same or different connectivity of nerves in every infected brain? What is its synaptic correlation with electrical transmission? What is memetic origin except '*eureka*' moments when an idea springs to mind or, rather, brain? That would make the Great Idea the Great Meme - a Virus corresponding to the Supramental State of Rationality and Central Beacon round which men have always built their various social worlds. In what sense, by denouncing Psychological Transcendence as viral and rejecting our First State of Consciousness as just delusion or duplicity, is the memetic explanation neutral? Physics never clocked dimensionless a meme nor do scanners register their length or strength so in what sense are these imaginative wisps a scientific more than pseudo-scientific entity? Or are they just canned metaphysic of a ghost in atheism's grand machine?

Perhaps Tzu's Moment indicates a mental illness. Malady can, naturally, be coloured either by religious or an irreligious context. Such psycho-pathology does not suggest that sanity, coloured by religious or by irreligious beliefs, is other than society's majority condition. Nor does it invalidate the positive and therapeutic impact faith is shown to enjoy concerning health. However, isn't prayer like a sugar-pill and religion swallowing The Great Placebo?

It seems that drug-based medicine may have failed to understand the scope and power of suggestion and suggestibility. Scientific studies have been carried out into the benefits of meditation, that is, into the production of brain waves termed alpha and alpha-theta. These are correlated with approaches to first state and the first state of mind itself. Transcendence. It needs be emphasised that, although nervous response is being measured by *EEG, fMRI* or whatever other method, it is not nerves but the awareness of a subjective entity that is both decisively causing and experiencing the metaphysical condition of which these neural patterns are a physical correlate. *What do the studies claim to show?* Firstly, they demonstrate an increased stability of rhythm in brain waves. Waves from each cerebral hemisphere tend to fall in step, to synchronise. The effect is a reduced sense of anxiety (and amount of stress-related hormones), an increased ability to resist shock or disorder, sharp alertness and an increased sense of well-being (generating associated biochemicals such as endorphins). Secondly, such internal calm as is associated with, for example, concentrated focus of attention unsurprisingly shows a corresponding quietness in the lobe involved with external awareness and orientation, the parietal. Thirdly, alpha waves associated with the posterior cortex spread coherently and synchronously to the frontal region thus including the whole brain. The 'message' is one of order and integration rather than the converse. Fourthly, there appears an increased coherence of communication between the higher cortical and lower sub-cortical regions of the brain. The latter are enabled, in an orderly and parasympathetic fashion, to implement their government of major physical support systems such as *BMR* (basal metabolic rate), respiration, heart rate etc. Alpha waves are in fact associated with sporting as well as intellectual, creative and meditational peak performances. It sounds like pristine man regained! It may be predicted that scans taken at moments approaching enlightenment will emphasise the abovementioned characteristics but also, having blocked either internal or external input to the focus of attention, will show a highly defined point of activity. May metaphysic, at transcendent climax, be physically reflected by a strongly 'orgasmic' flowering of refreshment that engulfs the body too? It would be strange if the shape of such a scan turned out to resemble the petals of a lotus flower!

Neuro-physiological coherence and stability is a reflection of state of mind. The 'alpha' or 'Adamic' state translates into measurable personal, educational and social benefits. Whence came education in the west but through the monasteries? And in the orient the goal is, as was, union with *Sat Nama* or *The Tao*. The practice of *wu wei* is one of masterly inaction to achieve (*sat*) perfect equilibrium; the wisdom of the sage, transcending all the fluctuations of polarity, is found in poise. In peace above. Strange, therefore, in a scientific age that immaterial wisdom in our schools is lost. No curriculum specifically teaches how to best develop and maintain essential equilibrium and harmony for health and happiness in life. Yet scientists involved with *TM* (a specific form of Vedanta-based meditation) graphically illustrate research which, it is claimed, can be tested to show the sort of improvements that should make any parent, teacher or politician's mouth water. Mental potential is realised in the form of increased intelligence, creativity and problem-solving ability. Reduced stress leads to a reduction in psychosomatic illness, muscle tension and a corresponding improvement in all-round health. There also accrue greater emotional stability, positive motivation, enhanced moral reasoning and, as a consequence, reduction in anxiety, aggression, hostility, crime and recidivism. The latter are expensive failings for us all; the former personal benefits can, of course, be extended to local, national, international and world socio-ecological scenes. *Think of the savings in money alone! Isn't it about time, for economic reasons if no other, that the boys at the top woke up?*

Top-down, perspective views Darwinian confabulation of all kinds as starting from the wrong point (*PAM* and *PCM*) and thereby plain, lock-stock-and-barrel wrong. The structures underpinning socio-politics in even secular societies have as their basis information 'tipped and taught down' from the pivot of an 'Alpha Moment'. You find the worldly shelter of this Moment housed in the enduring stone of churches, mosques and temples but its pulse beats in the derivation of man's law and politics. This is because the better side of humankind is rational and in that Moment is, if wisdom is construed a rational frame of mind, found Rationality Itself. Yet, if matter's all there is, then unto what blind folly atoms have evolved! This book plumbs the heights but profundity of saintly benefaction surely marks dementia's peak. Of course, a rational atomic aggregation knows it's only made of molecules. And there's unfortunate materialism's wrong-way rub.

Man first, cities and civilisations afterwards. Mind first; religion, commerce, politics and books follow later. Indeed, our universities were seeded in a climate of monastic life geared to produce a contemplative's alpha-theta waves. With the abovementioned cornucopia of benefits on offer would anyone be surprised if, in future, our educational institutions made a daily period of properly monitored, religion-free meditation, in a quantity appropriate to age and maturity, a precondition of study in their classes? The uplifting goal of such psychology is clear. Psychologists (clinician, priest or other kind of tutor) would teach the immaterial way that Truth and Wisdom are achieved. *Indeed, according to this line of argument Supreme Psychology is taught by an enlightened mystic; and this saint's healing constitutes Supreme Psychiatry.*

Perhaps, therefore, the metaphysic of illumination is not all scientific darkness or irrationality; perhaps consciousness did not evolve; and perhaps 'mystic' and 'mistake' are not the same. **It is no exaggeration whatsoever to say that human civilisation is constructed from and around a supra-religious tryst with the eternal moment. Human faith, hope and ideals are derived from the materially meaningless experience of transcendence.**

Top Teleology or Communion is beyond the second state of normalcy. It is altered 'out' of self-bound human man and thus, although entirely natural in its right, termed supernatural. Such Natural (Super-) Consciousness is the quintessential and thus first, original state of subjectivity; indeed, it is The Essence that substantiates existence. Beneath this Unconditioned State of Union are ranged First Cause (see Chapter 5's Transcendent Information and Top Teleology) and what mystics grade as lower zones of super-conscious condition (see *figs*. 6.1 and 13.3 (iii)). Such zones are variously known as Paradise, Arcadia and, of the Oriental Heavens, *Brahmand*, *Swarg* and so on.

In its experience a snail is awake but cannot dream of yours. A cat is wide awake but cannot think like you. And you are wide awake but relatively dreaming; you are dormant set against the fullness of awakening. The state is only super-conscious while you are below it; is not waking from level of sub-consciousness quite natural? Thus stretches through creation hierarchy; a scale ranges through the rungs of mind that consciousness is clothed in; a conscio-material gradient extends from Pure Consciousness to mindless lack of it called matter. In conscious creatures there exist, as educators and the courts will equally aver, high and low expressions of the human mind; and as biologists observe, the wider world displays phyletic grades of knowledge and intelligence. You can rank the type and quality of information used. *If consciousness that's physical in its experience can be divided into species it is possible and dialectically logical that super-conscious and discarnate levels might be hierarchically graded too.*

According to the rank of creativity (*figs*. 5.3, 5.4 and 6.1) the higher, 'nested' regions reach like Cantor's tower towards Infinity. Thus it is logical to posit states of consciousness detached from earth. Incarnate consciousness, your present state, is an appendage of these realms. It is an extrusion that, by volition you were tempted to experience and, through the vehicle of parents, do. In any bout of creativity you find an architect, executive and overseer of the minions that make it work. Creation is a Grand Design, a Business Projection *sans pareil*. Thus interior cosmology denotes, by different names, the levels through which principles devolve to practice. And super-conscious are the regions through whose denizens the concentrated current of creation flows. Call, if you want, the 'wavelengths' undiluted pleasure and their riders Elohim, the Seraphim or just angelic hosts. Check Psyche and concentric sheaths of causal mind (numbers 1 to 4 of *fig*. 13.3 iii). The possibility is logical but scarcely matters since, like the snail with us, we cannot understand. The super-conscious levels whence you truly trace subjective genealogy are hidden from below.

When the Apex of creation moves then this emergence has a Name. The inmost 'phase' of causal mind is called First Cause, Archetypal Mind or *Logos* (see also Chapter 5: Potential Information and Top Teleology); from such *Logic* all the orderly arrangement of the world derives. Ultimate experience, communion with this Cause and thence with Essence, is known to all mystics as their heart's desire - enlightenment. **This gives a whole new meaning to the exhortation 'Realise your Potential'!**

How might you reach and check such superhuman consciousness? Can anybody struggle to this state 'debugged' of passions, ego and all vice, an elevation stepped above humanity's normality? Such transcendence, gaining height, is closer to the Cosmic Hub than intellect and thus, dialectically, more and not less rational, symmetrical and real. How unreal or irrational would you expect The Principle of principles to be? Who, to solve the sphinx's riddle, takes bearing in the right direction, who in the wrong? How irrational or rational is the mystic quest to reach Life's Source?

In this *top-down* respect the method of transforming intellect with all its hive of reasoned conflict into wisdom's honey is the way of meditation. And, if the logic that mind precedes non-conscious matter is correct, there is no chance that the human ability to contemplate is some evolutionary add-on. The reverse. The rest of mind, body and the world are sheathed additions to the core of life. Its instruments and method of a reconnection with Essential Axis were reeled out as the world was spun. A thread was laid within the labyrinth of cosmic possibilities, a trail that leads back to the summit whence you came. There exists, in other words, a natural belay that's attached to Central Origin.

Second State of Consciousness - The Psychology of Normalcy

If the first state of consciousness is absorbed with Top, Essential Principal, the second is involved with intellectual understanding, manipulations, feelings and the pragmatic business of negotiating life's sub-principles (such as healthy life-style, cultural regulations and apprenticeship); and, not least, dealing with the satisfaction of its needs and physical desires. The flow of normal mind is, despite its experient's unawareness of the fact, relatively unfocused, sluggish and clouded with such 'mud' as living on the bottom of a stream stirs up. From a *top-down* point of view average human psychology below the brilliance of super-conscious potential falls naturally into three sections. These are *conscious*, *sub-conscious* (psychosomatic) and associated *physical* (or non-conscious, biological) factors. We deal now with the first.

	raj/ tam	*Sat*
	duality	*Unity*
	polarity	*Neutrality*
	restrictions of awareness	*Awareness*
	lesser selves	*Self*
	player	*Supra-rational Seer*
	lesser truths/ shadow games	*Clear Truth/ Reason*
	restrictions of energy	*Potential*
	balancing	*Balance/ Axis*
	conditioned free will	*Free Will*
↓	*tam*	*raj* ↑
	negative/ no	*positive/ yes*
	passive	*active*
	effect/ outcome	*cause/ input*
	balanced	*balancing*
	passive info.	*active info.*
	less conscious	*more conscious*
	sub-rational/ instinctive	*calculating/ reasoning*
	more instinct	*more thought*
	necessity/ non-will	*conditioned free will*
	puppet/ body self	*thinker/ mental self*
	reflex reaction	*process of reason*
	involuntary	*voluntary*

There is logic to psychology and it is caught, like existence, in between infinities. Mind oscillates between Illumination and the vacuum of non-conscious oblivion. Like a tidal volume drawn by normal breathing in between extremities of full and empty lungs, normal human 'self' pulsates. This 'self' is understood as a *persona*. It is a mask, a map or web of tales; called 'me', it is the personality that is developed by each actor passing through an incarnation. Who, if millions of previous selves impinged upon your present one, would you become? This is, in fact, how information's stacked to make you what you are today. Aren't life's adventures stored in memory (see Chapters 15-17)? Where, like 'self', sits memory? Could you prise, isolate, excise or even surgically create one from the nervous circuit board of brain? In other words, is memory (apart from its reflection in your physiology) physical?

Immaterial memories record life seen uniquely through your eyes. They record the contours and locations of a mental landscape, of a context often faded into generalities but which retains the influence of history. Specific moments, even if forgotten, still influence your actions with their inclinations. You are, if you like, a ship of continuity upon a voyage of discovery; and the ocean that you sail is information.

Close your eyes to think. You find yourself inside your forehead in between your eyes but can you physically isolate and surgically excise it from the nervous matter of your frontal lobe? Is the single-handed sailor's trip (as distinct from its reflection in his nerves and muscles) physical? Grey matter and physique can be, not always easily, observed but 'self' is never there. So neuroscience makes it disappear! You disappear. You melt into thin scientific air because you're really an illusion brought down onto inter-nets of nerves. A brilliant and sophisticated trick, mind you! Although we've never seen a thought or memory, they're somehow banked by limbic systems and pop up from their locations on the web of brain. Reality is genes and brains. Apparent metaphysic of the self does not exist because naturalism, humanism, atheism or whatever else can't prove, in its own materialistic terms, that immateriality exists. How rational is a mind-set that denies your own foundation for that sort of reason? *No doubt, the isms worry endlessly how matter might make consciousness; they guess how molecular arrangements might deliver life but, categorically, refuse to countenance that subjectivity's a datum, something nothing makes or takes.* **Hypothesis 'explains' the inexplicable. Materialism conjures explanations that explain away. And, cleverly, such promissory escapology is what sustains its mind-set in the universities!**

Does it matter self's all matter since in spite of and, somehow, because of this you're here! What is must scientifically be! You're like a computer linked to sense so that, as long as you're embodied and the links all work, you will survive. Software and hardware co-exist. Genes, patterns of their proteins and the functional structures proteins somehow generate combine; and from that combination there emerges rational self! None other than the thinking man that's you! How could a brainless mind, like software lacking all material support, ever run? The answer is, as far as life incarnate is concerned, it can't.

How could life be otherwise? Body's a material hulk that holds the mirage of a puppet captain in command. Is there plainer, more persuasive evidence of that mind is inextricably annexed to nerves than sleep, mind-bending drugs or, when a person's mental map disintegrates, old age or brain disease? The intricate machinery, 'devised' and developed mindlessly, is in decay and thus informative capacity can't make connections, nervous links fall into disrepair and so the neural map, self's context, memory is lost. What more proof than an autopsy do you need to show the problem's simply nerves? When brain dies how could 'brainless self' survive? Forget your deprivation tank and contemplative ecstasy since you need brain for these. A brainless mind is an absurdity.

How could a holist possibly reply? For whom a theory of mind's relativity such as these chapters psychologically elaborate is based around the premises of hierarchy and a conscio-material gradient. In parallel with Einstein's physical absolute, light, a holist understands that metaphysic's absolute is immaterial illumination, consciousness. All is natural; super-natural transcendence is completely natural when it's yours. And, as opposed to this Illuminated *Self*, the relativity of consciousness resides in the diffractions of a prism, mind. Mind's perspectives differ in respect of level on the gradient. Their shadow varies as they're filtered through the clouds of thought and feeling known as humanity's experience. Thus normal *lesser selves* are ones whose frame of reference varies due to the diffracted knowledge (information) they obtain. According to its 'filter' or its 'wavelength' their perspectives change. Mind is, in this view, a many-coloured, psychedelic coat.

Rephrased, the base of human relativity is an incorporated *self* (as opposed to absolute, *The Self,* a Deathless Concentrate of Elemental Consciousness). Through senses it perceives non-self (other minds and objects) according to its particular frame of reference. Logically, the higher this point of reference the relatively more profound the answers, the better solutions and the greater fulfilment will be. There are, however, many physical things (including operations and constructions of its own body) of which *self* is unaware. Similarly, because it is distracted by a deluge of sensation and the emotive tug of desires, a human mind only rarely 'goes within' far enough to glimpse its own higher capacities - capacities that scientism must, materialistically, deny exist. For this reason the invisible regions of upper mind remain as unknown and (unless revealed) as unsuspected as physical secrets. Nor can the revelation ever be of scientific Galilean kind; you have to trust the mystic Galilean kind and, following such contemplative practice as prescribed, discover mind's truth not by bench-proof calculations but by supra-rational experience. Unless and until the normal, lower self evolves into alignment with transcendent consciousness the Central Psychic Self stays veiled by clouds of mundane thinking.

Of course, materialism doesn't like the tone of this! It absolutely disavows unscientific logic. 'Anti-scientific' holists, even if they practise medicine, chemistry or engineering for a living, are deluded since non-mystics of materialistic faith decree they are. Nor naturally, despite what every faith proclaims is right, is this the way that scientism sees the world at play. Instead, while you're in a body 'mind' is knotted to a bodily experience. This knot is a neural lock because you *are* your life-span's nerves. Self's link with world is through the brain and when this fails the linkage crumbles. Is not matter all there is? Animation is, when the appearance of its flickering phantom fades, revealed for what it truly is - material in-animation, dead as stone! Chaps who imagine there is more are, technically, wrong! Therefore, ignore religion, faith or transcendental meditation. Don't be fooled by such pretenders as your rabbi, guru, imam, priest or pope; don't be gulled by such misinformation as these ignorant, mendacious fellows put about!

Yet we're still here. Our 'selves' remain. What do you think? Do you identify yourself with nerves alone or not? *Top-down,* consciousness exists *per se* and is expressed, in alien non-conscious zones of which the earth is one, through mind, the senses and coordinating brain. By these media Conscious Self can come to know non-consciousness, that is, the non-self that we call the natural world. Where brain's a medium that is physical the mind is one that's metaphysical. Psychological. *And we can simply cut our band of psychological 'normality' three ways.* Your mood and state of mind is always oscillating in between three grades. Mind's vectors are simplicity itself to understand; and Natural Dialectic well describes the tensions tearing humankind between its biological and psychological ideals. Its polar terms have likened men, half-angel and half-beast, to centaurs. These 'two-faced', bi-directional creatures have to reconcile two basic quests in life - stress reduction and provision of secure survival with desire for learning who they are and why they're here on earth. The first looks outward to the detail of material circumstance; the other inward and perceives in mind those general, unifying principles by which the cosmos works. Such quests normally combine in practice to an oscillation in between ideals and imperfect compromise; they sum to an internal war between higher principled and lower selves. The higher reach engages positive and upward (↑) tendencies towards union, understanding and transcendence. The intermediate step is where we live our normal ups and downs. The lower reach engages negative and downward (↓) tendencies towards disruption, incomprehension and suffering.

In short, *fig.* 14.1 shows 'normalcy' as intermediate mind; and in *fig.* 14.2 it represents states 2 and 3. Now we cut 'normalcy' in three. On the basis of mind's spectral relativity three clear 'colours' of normal mind can be identified; these bands correspond to three familiar divisions or modes of operation. (*Sat*) higher mind or '*intellectus*', called in Sanskrit *buddhi*, is closely related to what we term the First State of Consciousness or Psychology of Transcendence. It tends towards the super-conscious or illumined state; development of its 'thoughtful focus of attention' is the staple of education. The lower pair involves *ahankar* (*ego*). They include the egotistical, rational/ emotional modes of mind. Below conscious normalcy extend sub-conscious states 4 and 5. These, the subject of Chapters 15 and 16, constitute reflex instinct (perhaps the Freudian '*id*'), the zone of memory and a psychosomatic medium between mind and body.

(*Sat*) Higher Normality

	tam/ raj	*Sat*
	imperfection	*Perfection*
	less than ideal	*Ideal*
	far from/ not quite	*Climax*
	striving	*Satisfaction*
	desire	*Contentment*
	levels of turbulence	*Serenity*
	lack	*Completion*
	apartness	*Communion*
	lesser loves	*Love*
↓	*tam*	*raj* ↑
	down/ lower	*up/ higher*
	downgrade	*elevate*
	division	*unification*
	discontinuity	*continuity*
	differentiation	*integration*
	separation	*relationship*
	break-down	*synthesis*
	incoherence	*coordination*
	resistence	*flow*
	isolation	*communication*
	confinement	*release*
	self	*other/ non-self*
	ego/ I want	*self-sacrifice*

This aspect of mind tends (*raj* ↑) *upwards towards unification*, universal principle and, thereby, relative release from diversity and limitation. It best approaches and therefore reflects the characteristics of Transcendence. These include, as well as *unification*, continuity, coherence, integration, relationship and communication. *Such positive, right-hand principles educe a well-balanced, focused and attractive personality. Such a side is idealistic and strives for wisdom, beauty and love.* In dialectical terms, an increase in the predominance of (*raj*) right-hand characteristics '*expands*' consciousness. Its direction is one of increasing comprehension of general or universal principles.

Medieval philosophers were careful to distinguish between serial reason (*ratio*) and *intellectus*, a word they used more in the sense of intuition, grasp of principle or illumination. Reason is something used to unpack the condensed information of intuitive realisation - rather like the way that Albert Einstein unpacked and elaborated the information from his 'mind experiments' into mathematically described theories.

Higher mind is often likened to clear, radiant, limitless space of sky above the clouds. Highness. Conscious light. It is associated with translucency, enthusiasm (which means 'god-in-you') and joy. Its wisdom is the fruit of such contemplative focus, interest or devotion as ripens not with age but with perspective. In ascent self-consciousness is rarefied and, as specific individual better integrates with universal Self, self withers into selflessness. With this expansion rises, as does cream on milk, the full range of positive characteristics that inevitably emphasise what is good and right. In other words, higher mind uncovers a natural, cosmic moral dimension whose principles centralise towards higher truth, understanding and the communion of love. In exercising judgment by these principles it shows as blessed, generous and therefore sanctified behaviour - the best in man.

Such a mind-set is called, as intimated, *wise*. It is, in a human frame, the closest reflection of the divine.

A healthy body is 'transparent' in that it acts as an extension of the mind. No disability intrudes to drag mind down into the lower mind-set of frustration, pain and negativity. Thus health facilitates a higher or a rising state of mind.

(*Raj*) Intermediate Normality

Of course, we sense and think simultaneously but the Dialectic distinguishes between aspects of what seems a single experience. Waking experience is composed, in constantly varying proportions, of a triplex balancing act - sensation, making sense and then deciding what to do. It is drawn out to the body-world by the senses, back into the mind by processing sense information and then out again into physical reaction. These are the Opposite Directions of Mind dealt with in Chapter 0 and illustrated in *fig.* 0.9.

↓	*tam*	*raj*	↑
	outward focus	*inward focus*	
	objective	*subjective*	
	body	*mind*	
	sensation	*thought*	
	centrifugal sensation	*centripetal meditation*	
	intellectual anaesthesia	*intellectual appreciation*	
	sensory aesthesia	*physical anaesthesia*	

A bombardment of the senses by stimuli is called hyperaesthesia. Under abnormal conditions (including the use of hallucinogenic drugs) nervous channels may even be linked or excited sufficiently to produce 'synaesthesia', a condition where the senses seem to coalesce and, for example, lights are 'heard' as sounds. It seems pre-coded mechanisms sometimes more than integrate. They overlap.

The reverse, where the world is cut off, is called anaesthesia. The sub-conscious state of mind, sleep, is a natural anaesthetic; and drugs may block the junction boxes (called synapses) in nerve circuits or interfere with other aspects of electrical or hormonal signal transmission. Clinical anaesthesia uses various chemicals (ether, 'laughing gas', ketamines and so on) to, depending on your point of view, block the formation or inhibit the reception of waking consciousness. They knock you out but chemically induced oblivion, unconsciousness, is not the only case. Mankind has always used drugs like opium, cocaine or even alcohol as a passive, involuntary way to neutralise the pains of the world and become 'comfortably numb'. The analgesic, anaesthetic influence of drugs is temporary, variable and risks negative physical, psychological and therefore social side-effects. The cost of pain-deleting crops is often high. However coded signals can also be blocked by sheer concentration in another direction. Development of just such powerful focus is a by-product of deep contemplation. While physical sensations block contemplative insights the reverse is also true - physical anaesthesia, psychological aesthesia.

In this respect it is remarked that thought itself, an interior concentration, is a kind of anaesthetic. In the normal case our thought is sufficiently weak or combined with sensation for us not to think of it as such. But daydreams and times when we have become engrossed in thought (sometimes even talking to ourselves in the process) are reminders that, in thought, we can and do lose track of the outside world. This is partial, voluntary anaesthesia. We could delve further. Do you remember the sense-deprivation tank (Chapter 0); and how the experience resembles a first step on the subjective exploration of mind? The contemplative exercise is not down at the gym muscularly pumping iron. Nevertheless it involves the development of an effortless but Olympian inward focus of attention. Such a natural concentration leads not towards sleep but heightened, relaxed inner wakefulness, that is, it leads past higher mind towards Transcendence. While sleep is a passive vacation of mind, this is active; intentional vacation leads, eventually, beyond cogitation to wholly anaesthetic 'out-of-the-body' experience and with its detachment an automatic understanding of the post-mortem condition, that is, an experience that, in principle though not identical in practice, everyone will have at death (Chapter 18). When contact with the physical world is no longer possible even if desired such detachment is undergone as simply as walking into a previously unseen room. The whole business is natural; it may occur at will and, when practised according to controls, with no negative side effects. Indeed, the reverse. The goal of a psychiatric mystic (or subjective scientist) is radiance, enlightenment and knowledge of Infinity. It is to reach the Central Peak of Mount Universe, the Cosmic Cause. It is very healthy and, where its effect *is* the Cause, very effective.

Anaesthesia is, in other words, even in its extreme forms, a natural and not a bad thing. Loss of sense does not just mean sleep. It is also the product of inward focus and produces enhanced awareness and comprehension. Material numbness seeds the roots of scholarship and learning; and, on the other hand, the world's economy is spiritually numb.

body emphasis	*mind emphasis*
adrenaline rush	*psychological 'high'*
sporting 'high'	*creative 'heat'*
sexual heat	*intellectual 'heat'/ learning*
hedonism	*beatitude*

Which just points up the problem. Which way should a centaur go? There's nothing wrong with body beautiful. It's natural, a gift and needs utilities and business to survive and thrive. We all employ both tendencies so which, when body down below and higher mind above both call, is best to emphasise? Or is there, as we mostly tend to intermediate thinking, nothing in it?

↓ *tam*	*raj* ↑
ignorance	*knowledge*
heaviness	*lightness*
taking	*giving*
pride	*humility*
oppression	*nurture*
rapine	*donation*
greed	*generosity*
competition	*cooperation*
egotism	*altruism*
hate	*love*
turbulence	*equanimity*
war	*peace/ serenity*
pain	*pleasure*

Self-centred, grasping *ego* is clever more than wise. Here's an atmosphere of ups and downs. Winds blow, bouncy interference bugs the broadcast and clouds fuzz the sky of higher mind. This weather's full of highs and lows. *Ego* (self-first) is the boss of wheeler-dealing at life's office but not at its heart.

Ego is self's intellectual executive and weaves with serial logic (*ratio* or today's 'reason') to achieve its ends. As a wilful, manipulative survivor it is a driving instrument of problem-solving and achieving goals. Like any middleman, however, *ego* has to balance both sides of equations and in so doing cuts both ways. Its operation may involve higher (↑) or lower (↓) moods, tendencies and intents; and it employs in various persons varying intelligence. Just as it can establish political, technological and commercial constructions for betterment it is easily capable of descent into passionate scheming, oppression, exploitation, destruction and war. Just as the glories of this intellectual level, science and the arts, advance the cause of man they also excavate conditions for collapse - overpopulation, ecological disaster, global warfare and the running fire of an excitement known as depravity. This illness, once called sin, is contagious; severe eruptions show as the carcinogens of violence and, while clever as a devil's advocate sin's reasons make excuse, crime.

Indeed, the invisible barrier between (sat) higher and lower (raj/ tam) mind appears as passion in the forms of greed, lust, pride, attachment and anger.

Such sins are simply swollen deformations of a need to operate within the bare necessities of life. They are extrusions of a calloused selfishness arising when corporeal considerations (body and its instincts) overwhelm and thus pervert the reasonable physical needs for circumstantial control, food, shelter, sex etc. Such needs are nevertheless, even in modest form, non-monastic drivers that busy mind and bind it down; they are motivators that, like weights, drag mind off its higher qualities. As a source of continual dissatisfaction that needs to be satisfied, as a drift towards imperfection (*dukkha*) always needing some correction the bare necessities and associated passions lash the beasts that bear them through their lives. They are the stake of a dilemma to which everyone is bound, the dilemma of embodiment. In fact, it is the goal of contemplation to vacate the mind of their motion. *The real battle is internal*. It is business in a soldier's mind. His business - handless, footless - is to extinguish ever-bubbling egotistical desires and thus 'behead' himself! A cowardly crusader or a mad *jihadi,* on the other hand, cannot face inside or, in proper imitation of Mohammed, Christ and the other mystics, take the sword of sharp attention to his own deluded mind; instead he charges out imagining that chosen 'infidels' are demons and, therefore, his enemy. How far from higher mind are muscles and grenades! How far such pilgrim is estranged from Bunyan's mode of progress! Fight the good fight - against the devil in one's self; defend against the rearing *ego*; help another to the centripetal path. This is true mission; this is, sat in meditation, *jihad* ranged in battle on the field of middle mind. In fact, all outward forms of 'holy war' are just excuses for barbaric bullies to indulge in misdirected brawls for power and control. They reverse and thereby, often in religion's name, pervert the real, peaceful meaning of the word.

Wise men do not walk with threatening weapons. They throw pain away. They cast the weight of curses out. 'Reduce to rise'. Upon the scale of levity you watch the weight of passions, dropping them like stones. Is not the goal of such athletic sacrifice, having worked out all the burden of desire, free run of contemplative ecstasy? Logic of the mystic practice (meditation) involves intentional release free from the gravity of *self.*

As well as finding cures and answers middle mind can worry, fuss and drag you down. You could say that it blows hot and cold; storm and spring wind both blow through this weather of the head. Its atmosphere is not a peaceful one. It can be overcast.

(*Tam*) Lower Normality

↓			
↓	*tam*	*raj*	↑
	lower/ negative/ dim	*higher/ positive/ bright*	
	limbic	*cortical*	
	ignorant	*knowledgeable*	
	reflex/ instinctive	*thoughtful*	
	self-centred	*other-centred*	
	superficial understanding	*profound understanding*	
	stunted/ vicious	*evolved/ virtuous*	
	idle	*diligent*	
	bad and worse	*good and better*	
	sinful/ criminal	*sage/ supportive*	

If there's an upside there's a downside too. Now comes a dark day. The other side of mind is a (*tam*) *separator*. The main static or downward principle of psychology is *confinement* with its correlates of structural discontinuity, differentiation and individuality. *Negative, left-hand principles educe disruptive, distracted and unattractive personality.* This subjective isolate is self-centred, restricted, aggressive, turbulent and as demanding as unpopular. The lowest cast of conscious mind shows (*tam*) inertia, depression, laziness and stultification. It is heavy, brooding, more inclined to force than reason. In as much as lacking comprehensive goal or strategy low-level mind responds in involuntary, reflex ways. The light of this mode is dim, its comprehension shallow, slow and crude - of low intelligence and sensitivity. In lack or loss of intellectual brightness it is overwhelmed by force of instinct and its passions. A human in such common state is 'animal' because an animal is much involved with 'static' interference of sensations and consumed with body's urges. It is inertia's and exhaustion's correlate - a deadweight drag as heavy and as tiring as 'high maintenance'. An increase in predominance of (*tam*) left-hand characteristics '*contracts*' consciousness. In other words, it isolates and structures an individual in an increasingly lonesome, localised self-centred way.

The life of lower mind is not simply shadowy. Its swollen, turgid deformations are what pull your highness down; its sky is the one in which those black, electric clouds of brutal passion ride. Worse than this its zone includes mind fallen into hurt - times of fear, anger, cruelty and despair - the *(tam)* dark night of souls.

It's just my luck. After all, what's the point of slaving on a treadmill lashed by various sorts of weariness and pain? Where's the meaning for a race of orphans bred by chance, flickering sparks that for no real reason now endure through long, cold hours of space and, soon enough, will be rubbed back into the faceless emptiness of earth? Pass the Prozac, dish me pills to cure depression that materialism of a heartless, atheistic world leaves a heart evolved from eyeless, hopeless earth to struggle for a short survival. And if you're a victim of your own perspective there's even worse a possibility as victim of another's. There lurk around the shaded fringes of an atmosphere of suffering the basest brutes of all - sub-animalian ghouls whose pleasure is in others' pain. Victims from a slave's perspective suffer from the careless evil of a slaver. Have you, even for an instant, lived within the dark cloud that composes either frame of mind? Have you been violated by a criminal? Or done as you would not be done by - voluntarily communed with devilry, most vile and wretched folly of a soul? Just look around to see that sin and prisons, even outside prison, thrive.

To summarise, three grades of normal, waking consciousness are correlated with a tension between the cortex and mid-brain. Is your focus primarily involved with the excitement of learning and intellectual endeavour? Is it living in the market place of calculation how to profit and, from some palanquin of influence, command attention and obedience? Or is it chained to a raw instinct for survival underneath the cortex in a limbic system that, as a guard dog, triggers life's alarms? Who has not engaged or even oscillated in between all three? Where, over time, does your particular balance lie? What is the norm that other people recognise as you?

Focal Loops

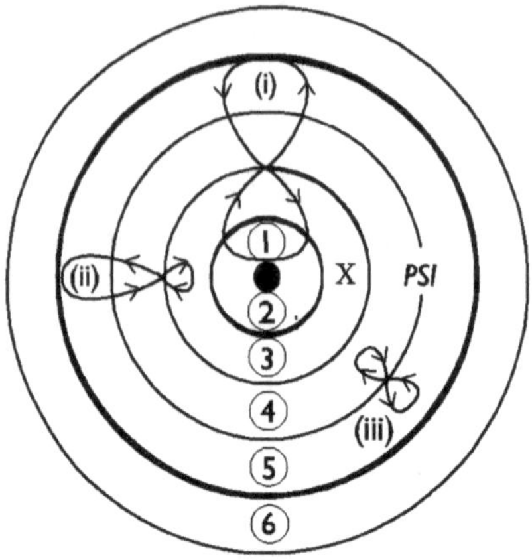

① Centre; Cosmological Axis

② superconscious state

③ conscious state/ personal cosmological axis/ third eye (X)

④ subconscious state/ *PSI* (psychosomatic interface)

⑤ physical body

⑥ external world

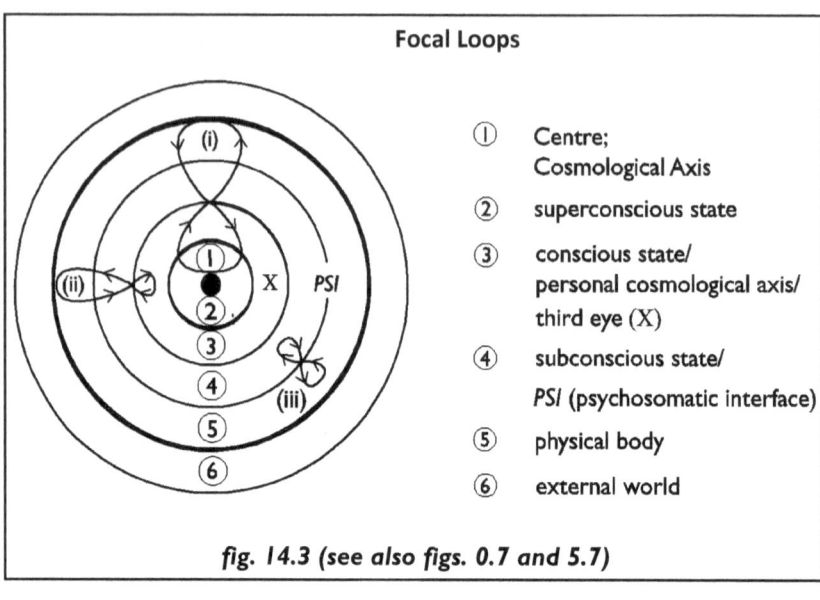

fig. 14.3 (see also figs. 0.7 and 5.7)

	tam/ raj	*Sat*
	polarity	*Essence*
	do/ know	*Be*
	info. out/ info. in	*Information Centre*
	swing	*Balance*
	process	*Precondition*
	oscillatory change	*Homeostatic Norm*
↓	*tam*	*raj* ↑
	to do	*to know*
	info. out / output	*info. in/ input*
	volition	*cognition*
	creation	*knowledge*
	behaviour/ response	*understanding*
	actor/ action	*preceptor/ perception*
	downward loop	*upward loop*
	down from centre	*up to centre*
	outward focus	*inward focus*
	exteriorisation/ mind > matter	*interiorisation/matter > mind*
	spiritual anaesthesia	*material anaesthesia*
	symbol to thing	*thing to symbol*

The tendency to equilibrate, whether reflex or by choice, is natural and universal. Do you remember (Chapter 5) the cyclical order of information and (Chapter 2) the unity and duality of information exchange, that is, the triune nature of a balancing act called homeostasis? Your nervous system, for example, is an informatic homeostat.

The homeostatic nature of informative and energetic systems is triune. It involves two vectors oscillating about a balancing factor, axis or norm. Such systems always involve causal input (by sensor), process (by controller) and, by effector, appropriate output. Kinetic bio-systems (such as thermoregulation), being passively informed by code and natural law, equilibrate reflexively. But unlike entropic matter that 'seeks out' least tension in exhausted states, animation's tendency is towards a vibrant norm of happiness, fulfilment and contented peace.

In a psychological system this amounts to a desire. Waking mind will try to actively inform its circumstance according to a choice from possibilities chosen in order to fulfil a psychological interest or biological need. In this case the goal/ norm is, like a missile constrained to target by negative feedback, a one-shot propulsion until satisfaction is obtained. Waking decisions can also engage subconscious and non-conscious physical response. A conscious organism can be influenced by past experience; learning and memory are involved. And in proportion as its mind is dormant an organism's reflex behavioural response is passively informed by 'instinct'. Finally, physical systems at all levels embody the principle of balance in the form of homeostatic trinity. It is, as Chapter 19 elaborates, their primary mode of operation. Negative feedback informs not only muscular and nervous but hormonal and metabolic systems; bodies are at root a bundled combination of such norms. That is one reason to identify the basis of biology as information.

A key question arises. Can the origin of all this information possibly be down to chance? If not, then what (see Chapters 17 and 19-24)? Not only is such original information required to establish coherent sets of norms in the first place but also to develop structures that monitor continual feedback and are able to maintain them. Such maintenance is called 'dynamic equilibrium' and its achievement 'satisfaction', 'well-being' or 'health'.

The nervous system, for example, is a mechanism constructed in principle of sensory input, top-centre cerebral processing and motor output. In the brain itself all parts are interconnected; they operate in a localised yet, at the same time, delocalised, unitary way. They cooperate as a (*sat*) single whole. Physical interlock with metaphysical information systems each whose purpose, whether conscious or subconscious, is the same - weighing up a situation. If, in this information-based balancing act, psychological precedes physical, how does the conscious trinity of knowledge work? *The business of con-sciousness ('knowing-together' or 'knowing-as-one') is the unification of disparate data.* Within this (*sat*) unity do you remember (Chap 0) opposite directions of focus? A *(raj)* inward vector and a (*tam*) outward? Psychological hierarchy is 'dynamised' by a two-way flow of information in cyclical, feedback loops. *In the first case attention is concentrated 'inwards' towards the centre*; contemplation is rewarded with insights according to its depth. *In the other direction attention runs 'out' towards our external, physical periphery.* This loop involves physical sensory and motor mechanisms; it is rewarded by survival and sensory pleasure. Both directions involve an exchange of information so you might call them Information Loops (*fig.* 14.3).

Each loop involves the coded transfer of information. In the bottom loop either a thought is translated via electrical and hormonal signals 'down' into muscular or other motor physiology or, from sensory perception 'up' to that symbolic imagery called thought, *vice versa.* In the top loop principles are grasped; details are classified into categories and the more general a principle the more powerful its exercise. Thus details are 'synthetically composed', that is, are 'raised' into abstract groups; data is subsumed within some personal formula or universal law. On the other hand, a metaphysical abstraction can be applied to circumstance; it can be brought, with words and numbers to describe particular events, specifically down to earth. Do you remember (Chapter 5) the gradient from active to passive information? Different codes support each stage in the transduction of an idea down this gradient into its physical outcome; they support the concretisation of symbol. Conversely, they support the conversion of physical fact into symbol, that is, the 'ascent' of physical objects or events into subjective perceptions. The medium of messaging is always, in one form or other, code.

The Basis of Exchange is Code

The basis of psychology is consciousness but the basis of communication's code. What do minds do except communicate? They exchange information always in the form of code. Mind is the world's inventor and manipulator; it pictures everything; it symbolises all there is to know. Its communication to and from a body and the universe around it is transmitted by electrical or chemical encryption. Code. *The prime business of life is information, communication, signals and code - all subservient to a purpose.* The purpose is to avoid pain, fulfil wishes, obtain pleasure and survive. To learn; anticipate; and not to die. No bunch of atoms ever wanted that! How could molecules evolve a code embodied in a complex system that depends on it? In other words, communication is conceptual business. **All media that inform, from computers and books through lasers to digital *DNA*, derive from mind. To think otherwise is Darwin's great illusion**. Who'd die for that?

Mind's relationship with the nervous system is its mode of interaction with the physical universe. In this case there is no doubt that certain biochemicals of precise three-dimensional and electrical configuration affect thought and *vice versa.* They are ciphers, media or symbols standing in for various kinds of it. They are relays. Information exchangers. They help to transmit the general nature of each class of thought (fright, fear, desire, contentment and so on) from side to side, from physical to metaphysical and back. Neurons relay encoded images of 'outside' to the point of psycho-space; and from 'inside' they decode thought into muscular action. They operate as channels to and from the central processor, brain. In this symbolic way thought rides on matter - indeed it is also carried on many non-biological media such as computers, tapes, books etc. *However, because a medium transmits a message it does not mean it generates or thinks or knows a thing.* Brain and nerves stand in the place of a radio or TV set that, as a conduit, mediates but is neither source nor sink for the program itself. A body is a non-conscious medium (albeit one of breathtakingly complex yet purposeful, integrated design) animated only by the presence of mind.

The Information Plug

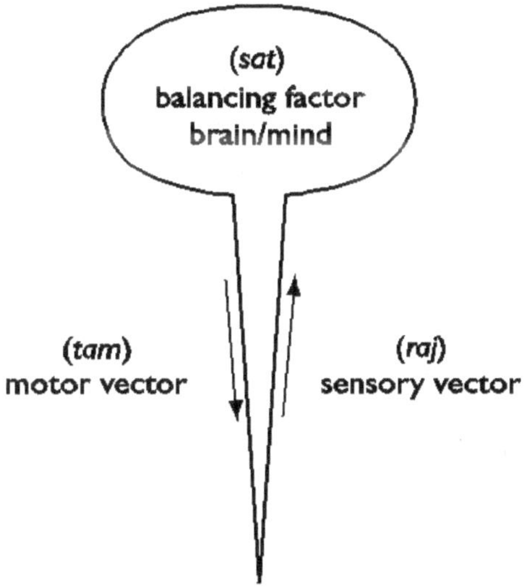

(i) **Illustration of earth-based, 'physical' mind plugged, through the media of nervous system and body, into the physical cosmos.** Voluntary 'heaven' and involuntary 'earth' thus interact. In this two-way exchange of information the non-conscious world impacts on our bodies and, in return, our minds through the agency of body respond and re-arrange the world.

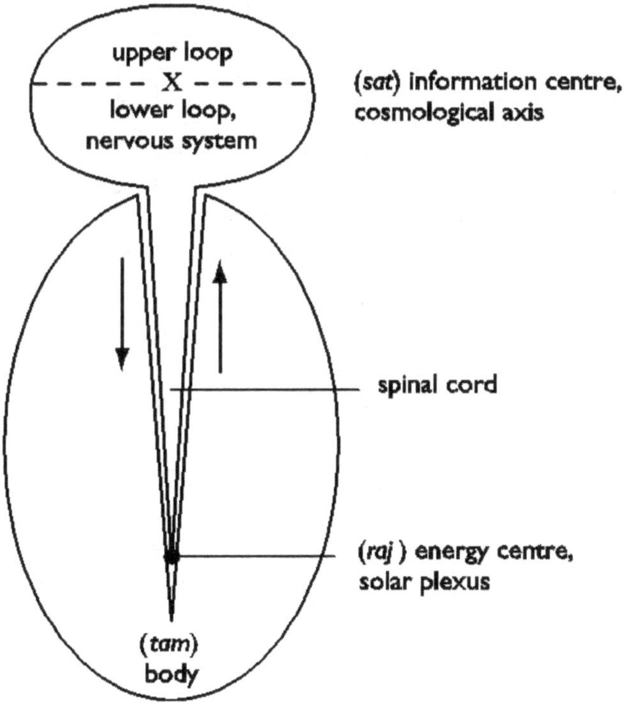

(ii) The plug itself involves fundamental Dialectic:

	tam/ raj	*Sat*
	informative motion	*Balancing Factor*
	motor/ sensor	*Regulator*
	nervous system	*Brain*
↓	*tam*	*raj* ↑
	output	*input*
	action	*knowledge*
	body	*mind*
	motor	*sensor*

fig. 14.4

You identify with body: you live in your mind. Mind is plugged into body, it is knotted to sensation and responding so that you can interact with the physical universe. It animates the body for its lifetime. Is mind the same or a different entity from body? Is the identification of self with a complex of chemicals only partly accurate? Of course, emotions affect physiology. A person's thoughts are affected by their mind-set (including subliminal instinct, memories etc.) and environmental setting; also by physiological and genetic factors, especially in the case of imbalances or errors that lead to discomfort or ill-health.

Chemicals affect quality of thought in the way that wires and modulating devices affect the homeostatic norms designed to deliver optimum efficiency, tolerable limits of oscillation and least interference in the operation of a two-way radio. There exist, for every aspect of human constitution, co-operative management when, for whatever reason, its in-built limits are exceeded. It is, therefore, possible to be born with or otherwise experience (e.g. by genetic fault, illness or drugs) either defective nervous 'wiring' or a congenital imbalance with respect to chemicals such as hormones or neurotransmitters that predispose to a particular quality of thought. The quality of 'in-the-body' thought will be affected in the same way as by anatomical stress, imbalance or pain. That is, pristine information that delivers health is interfered with.

Take sensation. Sense organs transmit a coded electrical signal. This is 'passed up' to a corresponding part of an integrator, brain. From here it is relayed to a decoding centre in 'psycho-space' which, in turn, builds up a simulation of the objective reality outside itself. For example, photons falling on the retina are 'seen' by the mind as dots of light; billions build into the seamless picture you are looking at. It seems to be your eyes that see but actually the picture is in your mind. The whole world is outside but you only know it in your head! In other words, no organ of sight is involved in the inner vision; and various non-physical associations can be made with other sensory information, memory, language, reason and so on to produce a rounded but personalised interpretation of what is going on outside. The inward vision of a dot of light is pre-programmed. Eyes are for seeing and not for synaesthetic hallucinations; therefore the signal from a retinal cell or the optic nerve, if artificially stimulated by an electrode rather than light, will still be interpreted by 'psycho-space' as light. Auditory nerves likewise stimulated will register sound and so on. In other words the whole system is pre-programmed; the body is like a multi-sense camera that reflects the 'real world' and decodes, according to syntax, what is not necessarily the real world itself. *Your universe is in the head; it is in psycho-space so that well-placed electrodes could play tricks on you or some laboratory animal.*

In brief, the nervous system in its sensory mode converts the universe into an experience. It is a process whereby the world is dematerialised into consciousness, of whose nature we know objectively little or nothing and yet, subjectively, potentially everything. And it is a process ordered by precise and accurate transfer of biochemical, electro-dynamic and symbolic code. Coded algorithms are the infrastructure shuttling information that forms an infrastructure critical for earthly life.

The reverse process, from symbolic to chemical and bulk, muscular body, is designed so that we can project our wishes into the world. We can, in ways that range from mundane to exceptional, vent our creativity. A motor system and bodily interfaces such as hands or feet are able, with a transparency that is proportionate to their optimum working efficiency (called health), to rearrange the world according to our taste. They allow a man to realise his imperfect visions and impose his particular order on a fraction of the whole stage, that is, materialise his personal mini-cosmos. Still, this whole business is, from top to bottom coded. *It is no exaggeration to say that life is replete with information wrapped in codes.* What is experience if not communication by exchange of information? Furthermore, each organism incarnates sheer information in the way it's coded, built and operates.

What, now, about the information loops (*figs*. 0.10, 5.7, 5.8 and 14.3) that radiate from your third eye?

<u>Upper, Metaphysical Loop</u>

↓			
	tam	*raj*	↑
	downward loop	*upward loop*	
	down from centre	*up to centre*	
	outward focus	*inward focus*	
	motor vector	*sensory vector*	
	exteriorisation	*interiorisation*	
	creative action	*understanding*	
	decision	*contemplation*	
	response	*sensation*	

Check *fig*. 5.8 (i). *The inward, 'upward-facing' loop of interiorisation is contemplative.* Information is exchanged within the metaphysical domain of mind. Although this loop (known in its partial blossoming as human intellect) involves every level of thought its *voluntary* focus involves no physical organ; it uses neither sense nor muscle but, as ships pass, you might spot nervous waves relating navigation of a thought to body's brain. Attention (*chitta*) manipulates *manas* (mind-stuff) using the organs of thought to create images or data symbols.

The *purpose* of engaging the upward loop is to improve the quality of experience. It rises through emotional imaginations to problem-solving manipulations. It involves intellectual capabilities such as language, reading, music, mathematics, creativity, entertainment and the whole intellectual pleasure zone. Since humans crave information it is continually used to learn and to understand things in terms of patterns and principles. By thinking you 'make sense' of an apparently diverse jumble, you unify what superficially appeared different, you comprehend. For example, a student studies a machine and '*returns*', in the sense of knowledge and emulation, to meet the source of its mechanical ideas, an engineer. Intellectual manipulation of symbols is refined by practice and the exercise of concentration. Greater internal consistency, harmony and understanding 'transport' the practitioner towards unification. *For a physical scientist this means an empirical approach and, after this, a body of material knowledge built on consideration of the data that his tests alone confirm. For a seeker after psychological truth the focus is inverted. It is solely switched inside.* Consideration seeks the principles of mind and, in deeper contemplation, ascends through higher mind towards Central Psyche. *In this way the upper, contemplative loop reflects a return towards the Highest Self. This is also referred to as the full realisation of human potential.* ***So the upward loop spans from earthly possibilities to Central Potential.***

↓	*tam*	*raj*	↑
	left-hand	*right-hand*	
	analysis	*synthesis*	
	division	*unification*	

Metaphysical organs of the upper loop include an ability to focus attention, collate material and develop will-power at the eye-centre; also cogitation (an ability to imagine, visualise, juggle and compare alternative, symbolic scenarios), an aptitude for the refined and orderly representation (either linguistic, mathematical or pictorial) of mental and physical events and, in an outward, executive direction, the capacity to make and oversee the implementation of decisions.

In keeping with the dialectical structure of complementary polarity, cogitation of the upper, metaphysical loop seems to involve, in its association with the lower sense/ motor loop, a curious, reflective asymmetry of brain mentioned earlier in the chapter. Although the left and right cerebral hemispheres in many respects duplicate each other in appearance, function and capability they act as mirrors in others viz. gender bias and local emphasis with respect to 'unifying/ synthetic' and 'separating/ analytical' tasking - the former on the right side, the latter on the left. In our culture, which places emphasis on linguistic, analytical and scientific skills, the left-hand, 'masculine' hemisphere cross-linked to the right-hand side of the body is 'dominant'. In fact the sides cooperate to solve problems. A 'masculine' side seems to convert speech into a conceptual apprehension; the 'feminine' half adds emotional tones and inflections. Does the pair compare nuances through their interconnecting commissures and, when their 'mock-ups' click, pronounce a problem dialectically solved? Or is this simply what the nervous waves appear to show? Conceptual operators are associated with different areas of the brain but, because thought, character and subjective experience are not measurable in terms of SI (International Scientific) units, the measurement of metaphysical 'output' can only be *qualitative*.

Lower, Physical Loop

On the other hand an outward, 'downward-facing' loop of exteriorisation is involved with the physical domain of body and environment. *This involuntary exchange of information involves, according to Natural Dialectic, not only metaphysical psychosomasis by way of a typical mnemone (Chapters 15 and 16) but also, lower down, the physical mediator brain with associated nervous, hormonal and muscular systems.* The latter are referred to here.

In terms of its voluntary nervous system we can write:

	tam/ raj	*Sat*
	effector/ sensor	*Balancing Factor*
	expression/ impression	*Modulator/ Brain*
	output/ input	*Functional Coordinator*
	substations/ lines	*HQ*
	somatic nervous systems	*CNS (central nervous system)*
↓	*tam*	*raj* ↑
	downward	*upward*
	outward	*inward*
	centrifugal	*centripetal*
	materialisation	*dematerialisation*
	reflex loop	*voluntary*
	sub-conscious	*conscious*
	output/ info. out	*input/ info. in*
	counter-effect/ response	*stimulus*
	effector	*sensor*
	motor/ muscle	*receptor*
	informs matter of consc.	*informs consc. of matter*
	symbol to thing/	*thing to symbol/*
	subjective 'deactivation'	*subjective 'activation'*
	action/ doing	*perception/ learning*

and of the involuntary autonomic:

	tam/ raj	*Sat*
	involuntary ramifications	*CNS*
	output/ input	*Homeostasis*
	motion control/ body balance	*Cerebellum*
	automatised functions	*Brain stem*
↓	*tam*	*raj* ↑
	stop	*go*
	quiescence	*arousal*
	suppression	*promotion*
	inhibition	*excitation*
	brake	*accelerator*
	parasympathetic	*sympathetic*

This section includes not only autonomic, reflex government of functions such as respiration, heart-beat rate and digestion but also the mass of proprioceptive information (concerning the movement of limbs and muscles) that floods in and out of the cerebellum but is rarely consciously recognised. In principle, as the next stack shows, hormonal systems are also based on the homeostatic, complementary activity of antagonists:

	tam/ raj	*Sat*
	exterior glands/ target organs	*Mid-brain*
	controlled	*Controller*
	informed	*Informant*
	hormonal cascade	*Pineal/ Hypothalamus/ Pituitary*
	swing	*Balance*
	vibration	*Norm*
	negative feedback	*Homeostasis*
	upset/ problem	*Solution*
↓	*tam*	*raj* ↑
	off	*on*
	inhibition	*excitation*
	end	*activity*

Information is exchanged between metaphysical and physical domains i.e. between mind and central nervous system/ body. Its natural 'language' includes sensors, codes and effectors which govern what sort of information is exchanged and how. The loop involves decision-making, usually without much contemplation and often taken in immediate response to external changes, events and behaviours. Its practically involuntary, reflex circulation is composed of instincts and habitual practices and uses, as well as higher and lower brain controls (e.g. in the medulla), organs of sense, appropriate autonomic and hormonally induced physiological and metabolic responses and, last but not least, muscularised motor organs to move and manipulate materials in a precise and orderly way. *Its five organs of sense* are sight, hearing, smell, taste and 'all-over' responses of skin to touch, pleasure/ pain and heat/ cold; *and its five of action* are mouth, arms and hands (five fingers), legs and feet (five toes), genitals and the organ of egestion.

In short, the 'left-hand', downward loop involves the transfer of psychological data to body; mind to muscle, it *materialises* thought or instinct. On the other 'right' hand, upward motion involves the transfer of material data to mind. In other words the sensory half of the loop *dematerialises*. It collects environmental data (including from the biological body), filters, codes and passes it up for dematerialisation into symbols within an individual field of consciousness called mind. Information from the exterior is physically assembled in three main stages. Sensors pick up the raw data, translate into electrical code and forward it for further refinement. In the medulla 'wires' from the left-hand side of the body switch to the right hemisphere and *vice versa*. The wires then pass to sensory centres where a second transduction from electrical code *via* typical mnemone into psycho-spatial 'meaning' occurs. Each centre interprets what is a single kind of signal in terms of its own faculty. We saw, for example, that the visual centre interprets signals as light; if, while the subject was blindfolded, you tweaked an optic nerve he would 'see' light. Associative areas (attention/ frontal lobe, orientation/ parietal, verbal-conceptual/ temporal and visual/ occipital) integrate these signals with memory and emotion to create an observation. In this way many different items are dematerialised, symbolised, collated and *unified* as a single whole. Look around. **The world you are 'seeing' is actually symbolised inside your head - so the system is very, very efficient.**

Dialectical System Analysis: Nervous/Hormonal Homeostasis

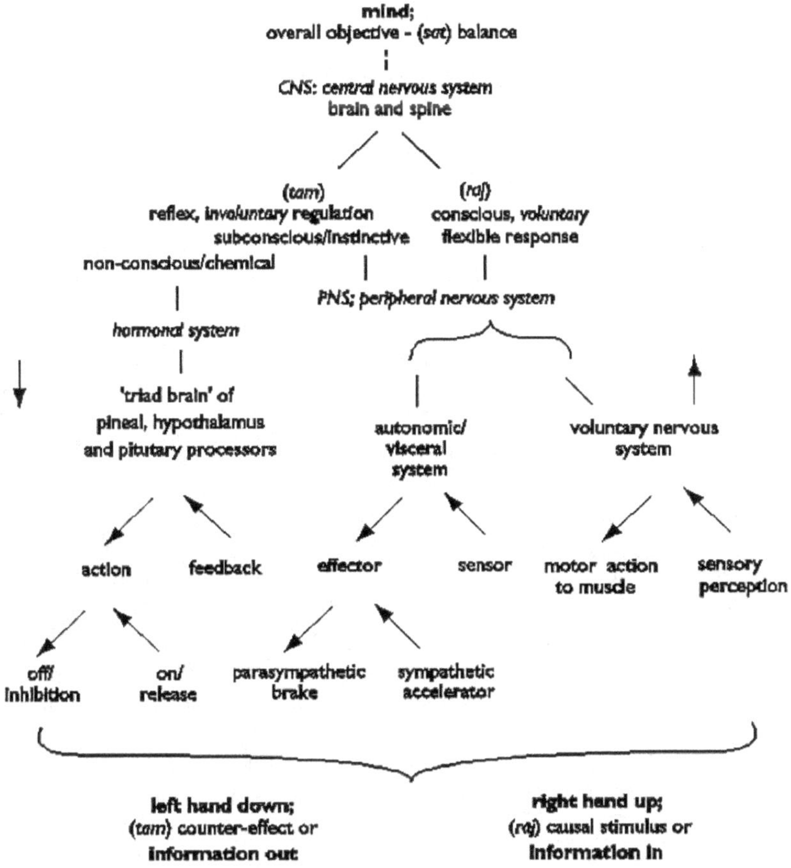

Just as a driver drives a car, so mind is the real driver of its vehicle, the body, using a control panel called the brain. **Just as a single dashboard light indicates the same function in myriad different circumstances, so the lights of brain indicate the same processes of perception, thought and memory used in myriad different circumstances.** Mind is an invisible master that, as its plots dictate, pulls the strings and animates the body of its puppet. Since mind is the cause and physical behaviour its effect, the dematerialisation of objects into symbols is classed as 'active' and placed to the right-hand, upward side of conscious regulation. So, in the upper loop, are learning, understanding and knowledge. Mental activation by 'information in' is complemented by an opposite, motor process - 'information out'. Since this involves the end of the decision-making process, the fixity of a thought in some material effect, it is paradoxically classed as 'passive'. The increasing deactivation or formalization of thought and instinct into physical terms is placed to the left-hand, downward side of regulation.

fig. 14.5

As with any equilibrium reaction, things work both ways. Perception is evaluated, a process of thought balances the issue and a decision is made. Now the loop's left-hand, downward motion influences matter; mind acts to make body act accordingly. The motor half of the loop *materialises* the patterns of mind. It devolves the order of mind 'down' to matter; by way of neuromuscular output it literally makes a difference. Both input and output involve physical, measurable data and, as such, the loop is called *quantitative*.

Of course, science is all about quantitative measurements but how do you judge *quality* of mind? This section states the textbook obvious but materialistic science struggles since it leaves the whole great immaterial side of mind a blank! How, without immaterial subjectivity, can character be estimated and as a result the quality of action called behaviour ruled for worth?

Quality of Mind

Homeostasis is regulation according to norms. Circumstances are continually monitored. Information is compared with a norm. Does it measure up? OK, do nothing. Does it deviate? OK, correct the deviation. The process is simple and fundamentally dependent on pre-set, programmed norms. A problem is a question to which we need to find an answer; in biological terms it involves deviation from a norm that is in consequence corrected; such correction, leaving the organism again 'feeling right', is an answer. Instincts involve the same kind of homeostasis; the problem is a need or desire and its satisfaction the answer. Accordance with physiological norms is a stability called health. Health 'feels right'. Feeling right is also certainly a quality of mind.

However, the business becomes more complicated and plastic in the case of conscious imaginings. We establish a desire. This is a problematic frustration until we can find a way to satisfy it. The desire, which may have no connection with biological reality, becomes a temporary, self-set 'norm', a target from which deviation must be corrected. In this way mind, an information centre, is a voluntary regulator, an instrument of psychological homeostasis whose 'norms', according to which we regulate ourselves, are our wishes. Instead of automatic programming that regains a pre-set balance, conscious mind develops schemes to satisfy individual wishes and, thereby, regain the balance or contentment that the wish itself disturbed. Wishes, which drive behaviours, are made within the overall framework of our thinking, context or character. Therefore in judging the quality of a mind we judge the criteria according to which its priorities are established and controlling decisions made. Quality of mind becomes a question of the 'psychological infrastructure' within which wishes (or aims) are developed. What is the best way to think? Which thoughts are best entertained, which desires best deliberately overcome or left to fade away? What criteria, standards or ideals should constitute an individual's set of psychological norms? What thereby regulates his business and best leaves him 'feeling right'?

Is individual 'feeling right' the same as universal 'good'? Does it secure morality? If quality is judged according to the criteria on which mind makes decisions, psychology immediately finds itself transported to a hallowed philosophical arena. Nor is this area a scientific one. What is 'good'? What is best? *What is true; of what nature is **top** information and the key to its disclosure*? **You ask, in other words, in what or whom exists the Highest**

Truth; and what, therefore, are the best criteria or is The Best Criterion on which to base our lives. The quality of psychology's central 'object' of research, mind, involves character and, in turn, a clear moral stance.

Physical science, on the other hand, dealing with automata, is not philosophical except in its first step. It shies from moral fisticuffs and its present-day materialism generally leads a retreat to the evolutionary corner. It tends to call upon life's central code which, according to its own criterion, is genetic and concerned with practical survival. From this stricture it pronounces on contentious issues that can involve morality and quality of mind. *Nevertheless, whilst it may be objective and scientific, psychology that omits a clear code of ethics at the same time misses the crucial point. It becomes, in real, subjective terms, relativistic, vacillating and vacuous. It leaves a moral vacuum and thus lacks, in this important sense, quality of mind.*

If an explorer travels a path and, because the initial bearings and psychological preparation were wrong, suffers a multitude of unnecessary adversities, wastefulness and error, what do you conclude? That proper orientation is critical? That it is smarter to logically anticipate and avoid a problem rather than learn by physical or psychological suffering? *Now suppose you are the explorer and the journey is your lifetime.* Abstract preconditions and high morale cannot be packed in knapsacks but, first things first, for an expedition to run smoothly must not its leader understand the game? How does a doctor, having checked a healthy body, tick the mind? What is its operational optimum? It needs be finely tuned and (which is education's real import) operating according to good purpose and right principles. What are these?

A decision concerning 'right' psychology involves, according to dialectical principles, three cosmic fundamentals and two main directions of focus.

	tam/ raj	*Sat*	
	exterior	*Centre*	
	action	*Poise/ Potential*	
	informed	*Informant*	
	controlled	*Controller*	
	negative feedback	*Regulatory Norm/ Rule*	
↓	*tam*	*raj*	↑
	from centre	*to centre*	
	away	*return*	

The cosmic fundamentals translate, as we have seen, into levels of consciousness - (*sat*) higher, (*raj*) intermediate and (*tam*) lower mind. Moreover in the upper, internal, metaphysical loop (*raj*) principled information will predominate and in the (*tam*) lower loop personalised and often reflex, unthinking and passive information. That is to say, the thinking loop evolves upward towards (*Sat*) Poise whereas the downward, nervous loop devolves into sensori-motor activities - what we call 'living in the real world'. From which loop, therefore, do you tend to control events? From a 'detached' seat of principle, from an 'other-worldly' chair of government; or by dramatic, passionate involvement in your own 'reality show'? Of course, you use both perspectives. You marry principle with daily practices. The question's one of emphasis. Does practice tend to rule your principles or principles rule practice? Which order mostly qualifies your thinking?

A lower, physical direction of response tends to put practice before principle, that is, exploit the moment, use the hour to best advantage and to hell with consequence. Don't ends justify the means? This is a mode with which scientific materialism is prone to identify. Evolutionary psychology will emphasise sensation, bodily response and animal behaviour in issues that involve quality of mind; and, correspondingly, downgrade morality based on metaphysical logic. If body is the ultimate, common-sensible reality, then metaphysical logic will be a delusive figment of its brain chemistry. Morality is pragmatism; what is an object or an action's use to me? If you read the world by such a book the message of the atoms is 'survive'! The real message of the genes is 'win'! Is nature not as cruel as natural selection? The goal is winning never mind the means - only winners hold aloft a prize. If altruism and morality are actually imaginations what have you to lose by casting, in the name of rationalistic realism, what is only an imagination out? Ashes to ashes, nothing to nothing, religion to the scrap heap! We sorted this at university! Isn't nature's real truth anarchy?

Yet the problem does not fade away. Do you remember half-beasts by the name of centaurs and (Chapters 3, 5 and this one) the centaur paradox? For such a half-man as a centaur tension tugs between himself and others - his dilemma is the individual's in society; and, just as acute upon the upper loop, it pulls inwardly between the self and Self - an individual and his 'State of Grace'. How about Utopia - both physical and psychological? Won't one kind automatically create the other? Don't you want to live in paradise or maximise at least a peaceful, pain-free state of human happiness? What stands in the way? What laws lead to liberation? What rules best restrain disorder and what kind of order leads, as anarchy cannot, to liberation? What is, you ask, the best philosophy to underpin all politics, law and religion? You may like or may resist self-regulation. You may like or may resist another's regulation of you. You may never even want to think of regulation until it disappears or turns on you - when anarchy or war has strangled moderation. Nor does it matter whether soul- or matter-centred canons modulate your thought. For an atheist or a theist his religion is the first and foremost point of jurisdiction. What, therefore, is religion at its best? What is the truth? How can I mould to it?

The word 'religion' means a 'binding-back'. Man cannot live upon the nerve's edge of uncertainty but thrives on straight stability; dynamic equilibrium and not some jarring unpredictability is what feels rhythmically right. Thus consciously or otherwise he constantly refers back to his central themes. These themes are criteria. From them his personality derives and thus the character of his behaviour. Your own quality of mind and therefore courses of behaviour are derived from your religion. It is unavoidable. A student of materialism's 'atom-speak' is reduced by his own argument to claim genetic predisposition steers 'free will'. A Darwinian philosopher explains an act of altruism as a cunning social but self-interested ploy. And the teachings of the prophets often undergo distortion at the hands of priestly mongers. In a mind-world where self-interest rules the waves religion is political and politics, behind its silky, diplomatic glove, a fist of power-games. Empire. War. Defence. Attack. Original intention in the hands of centaurs often gets reversed. You only have to trace the history of such 'faithful' as divorce themselves from their religion's purpose - inward quest for personal transformation and spiritual transcendence - and seek instead to purge those people they have cursed as infidel. In such a case is God or Godlessness the greater curse?

Yet regulation is the necessary weight to balance individuals with each other, that is, with society. Upon an ideal scale the imperfections of the world are weighed. Who can escape the boomerang of his decisions and, therefore, what is your specific choice? Which pillar of perspective, *top-down* or *bottom-up*, seems all-inclusive and therefore basically true? Which coherent canon of beliefs shines brightest in your eyes? Are its ideals incorporated in the fabric of existence or are they men's inventions, that is, not ideal enough? Which ones should the frame of government adopt and lead us where we're going? Which ones, in the name of God or Godlessness, would you adapt and thus tend to corrupt? What kind of purge eventually scrapes out corruption's rot? Although the cynic may deny it, these are questions asked of mental quality that every man and his society need reasonably answered.

	tam/ raj	*Sat*	
	lesser truth	*Truth*	
	colour	*Transparency*	
	shades	*Light*	
	apartness	*Communion*	
	difference	*Love*	
↓	*tam*	*raj*	↑
	down	*up*	
	deceitful	*honest*	
	miserable	*happy*	
	dirty	*clean*	
	slack	*diligent*	
	disloyal	*loyal*	
	hostile	*friendly*	
	selfish	*altruistic*	

The trend is clear. Metaphysical ideas and physical events, in either loop the quality of information can (*raj*) stimulate and elevate; or else (*tam*) depress. Descent translates to things gone 'wrong' and ascent to such swell of happiness, coherence and serenity as may eventually approach stability of radiance. The Dialectic just confirms our intuitions when it comes to highs and lows of mind.

Humans are, however, born to imperfection and dilemma. We wake up riding on a sea of oscillations; we grow up to experience relativities of mood and morals. What do mother, father, teacher or the policeman want? What do fellows judge acceptable? How runs the wind, how roll the waves, how to set the sails? Any captain will agree to tensions while he's sailing circumstance.

The upward drive towards knowledge, principle and Central Self extracts you from your selfish clamour. Such ascent of man is dealt with later.

Descent leads down into your lesser self, your ego-centric, body-centred and sensation-seeking cave. Attention drops to a surrounding moment and locality of vessel, body, and its launch-name - yours! '*Mens sana in corpore sano*'; 'a healthy mind in healthy body.' Instruments cannot be moral and if mind's instrument is body then the human body and its health are clinical, amoral matters. The quality of body is of medical concern. Objective body is an organ of subjective mind whose 'apparent absence' - when we're almost unaware of it - we diagnose as health. Illness or suffering cloud transparency; they cast shadows over any kind of earthly paradise. And though the simple, healthy needs of body are imperative, their satisfaction may involve another person and, therefore, morality and moral stress. What is, in principle, the way to carry on?

It is *use* of instrument that touches on morality; morality is mind's and not, as ascetic anchorites in desert caves sometimes confused, the body's. Why torture amorality? What, otherwise, could cast a shadow or snake shivers into healthy paradise? Could anything go wrong in mind? Could anything reduce its quality of state? *Difficulties rise when body-minded self becomes ambitious, passionate and, as an effect, sometimes frustrated.* Quality of mind, both of the 'selfish isolate' and those within its adverse sphere of influence, is compromised. Since quality of mind affects its organ, body, material health may also be affected. Thus, surely, it is right to call a free indulgence of the driving passions, especially if it degrades the quality of other life, psychologically and ecologically 'wrong'?

'Self' is, like magnetic influence, invisible but really there. Useful psychology identifies its immaterial influence. It cannot weakly vacillate, 'detachedly', around provision of morality. What aspect of your 'field', the positive or negative, predominates? Whence do you radiate your power and persuasiveness? Do you identify with body, with the lower loop and a continual search for sensual pleasure? *Or with mind, in which case quality of mind is the prime consideration*? Mind's upper loop involves subjective need for information, relationships and understanding. The '*thinking loop*' perceives connections, solves life's problems and seeks principles; it seeks patterns, order and the underlying truth. Higher man dissolves the selfish box. Apart from egotistical constructions he wakes into a bigger picture or some greater whole.

This awakening treats the world of 'not-self otherness' in one of two valuable, complementary ways. You can divide yourself from it; in this detached, objective case you analyse its many parts apart from yourself. Or you can subjectively identify with 'not-self' and realise that, eventually, everything is networked in relationships and, in this sense, 'all is one'. According to dialectical logic the latter perspective is 'higher'. The purest attitude occurs at the extreme (*raj*) upward right - the climax of unification in (*Sat*) Communion. This is holistic aspiration's goal; yet the highest quality of thought is, paradoxically, in thoughtlessness - the 'non-condition' of enlightenment.

Such proposition is, especially to an intellectual rationalist whose thinking gear is set in overdrive, pure nonsense for an answer! What quality exists in mindlessness except material oblivion? No quality of mind at all. Let us therefore try again. Objects exist but, subjectively, we have a living experience. *You and I relate the quality of an experience to the sense of well-being it bestows or robs.* Such mental health and happiness is related closely to a mind's 'transparency'; the more I'm free of pain and worry, the more engrossed in hedonistic or in intellectual pleasures, the more comfortable and therefore happy that I feel. Conversely, such transparency is badly blotted by the shades of self-awareness that anxiety, guilt and pain create. Mental health reflects a loss of lesser self, of ego and its fantasies. The Greek word 'ecstasy' means 'standing outside' petty ego. Loss of this self occurs when, on the one hand, I am merged in 'not-self' or, on the other, in The

Self. In plain language and on either hand this means interest, enjoyment, friendship and in merging with another. Love. *According to Natural Dialectic such altruistic enthusiasms are not an evolutionary quirk, a deceptive method in order, in social circumstances, to get your own way and propagate your genes. Nor are they a subset of psychosis but a natural exuberance of love. The Dialectic reasonably identifies love as the pinnacle of self-transparency and, therefore, psychological health.* The real questions thus become 'with whom shall I best fall in love', 'can I obtain a permanent, intense and deepening mood of love irrespective of material circumstance and if so, how' and 'what is the nature of such lovely Super-Self?' Do not hold your breath until the science of psychology has asked them in its lab. You would die before since these are not laboratory questions. How could material science prove or provide the answers of a mystic?

A blind, atomic maelstrom never made up teleology. Mindlessness did not create the loops of mind. Although the cause of visible behaviours, mind's interior, upper loop is of itself invisible and, in numerical terms, immeasurable. It *evaluates* an object, event, behaviour, plan or character in a *qualitative* (aesthetic or moral) rather than a quantitative, mathematical way. What is worth most? One bargains with the world, that is, with mammon; the other discriminates between 'good' and 'bad' characteristics and, according to one's own best criteria, makes value judgments. This sort of calculation is derived from quality. *Indeed, mystic practice is pursuit of a commodious utility, the highest quality of life.* It tracks its progress in the light of best intelligence; it measures up against the Ideal of Pure Consciousness. There is no mystery. Pushing such limits, an idealistic venture to which most people weakly, spasmodically and intellectually subscribe, the practitioner lays emphasis on the importance of developing the right-hand dialectical characteristics of unification, goodness and truth.

How far removed from the sweat and grind of workaday is such abstract psychology? A busy, pragmatic person, exhausted by the demands of a hard, hectic day, might query its relevance. He might also ask why saints ('enlightened beings') rather than rulers, chiefs-of-staff or businessmen have generated the major social and historical waves of influence. *Is it because they have emphasised the value and quality of worthwhile life?* From a *top-down* point of view existence itself derives from the Highest Quality Intelligence; and according to its Professors Logical Information is the first step in a creative cascade that leads to biological life like yours and mine. There is method in what scientism stigmatizes mad. The holist is not out of mind; indeed, a mystic might be quite intelligent!

<u>Quality of Information</u>

	tam/ raj	*Sat*	
	lesser truth	*Truth*	
	shades	*Light*	
	grades	*Importance*	
↓	*tam*	*raj*	↑
	negative	*positive*	
	dim	*bright*	
	ignorant	*knowledgeable*	

wrong	*right*
incorrect	*correct*
inefficient	*efficient*
useless	*useful*
trivial	*important*
error	*truth*

Information is often called by its users 'intelligence'. *In fact, intelligence is a function of attentive focus and ability to spot the principle behind a practice.* Quality of body (a good, healthy physique) is matched by sharpness, brightness and an intellectual sense of wit. All educators are aware of a scale of intelligence from lower (less contemplative, more sense-based and concrete) to a higher (more focused, accurate and subtle) ability to manipulate symbolic information. Intelligence is therefore a quality of mind inferred from the relative ability to absorb, process and create information. Life is intelligent, always goal-oriented (*teleological*) and orders things according to its designs. Purposeful design is the top psychological motivator behind each of the myriad information loops within which mind keeps cycling. All goal-oriented loops depend on a prior source of intelligence. Bright minds 'shine brilliantly'. *The greater an intelligence the faster, more accurately, intricately, thoroughly and efficiently its goals are achieved. In short, the more teleological is its nature.*

Mind actively informs; it creates information by its schemes and plans. For this it uses passive information, that is, messages it gathers, sorts and filters from the world outside. **If quality of mind is a function of intelligence and mind itself an information-centre, then what about the quality of information that it filters and then operates according to?** By what criteria can we judge its value? The importance or triviality of information is rated relative to various desires and sets of priorities. Such priorities and the wishes that sequence them change. Instinct, education, advice and cultural experience help shape such changes and, in life's relative muddle, provide answers and directions. What are religions and political creeds but attempts to encourage coherence, develop *esprit de corps* and emphasise idealistic priorities? Their purpose is to orient, create a certain quality of mind and thereby improve the way it handles information.

Everybody operates according to their own variable, personal set of priorities but do there also exist permanent, natural and universal priorities? Is there a hierarchy of information, a set of principles? If so, what is nature's 'top set'? Is there Prime Information? Is there anything of Absolute Importance?

Both subjective and objective scientists recognise, in pursuit of their objectives, different qualities of information. *They recognise that important, principle or fundamental information is the key to power and understanding.* It is most sought-after. Next, within a strategic framework of principles, comes the detailed, tactical information necessary to achieve specific goals. Lower comes trivial or insignificant data that has little relevance, meaning or power. Such categories at best promote and at worst do not impede the search for happiness. Indeed, any information that leads to subjective happiness is judged relatively apt. What, therefore, constitutes happiness? Whence, logically, might you draw upon the highest quality of information? It depends on your perspective. For example, the mystic answer and consequent, '*TOP*-set' priority is plain -

communion with the Source of Information; physical science seeks it in mathematical abstraction of a *TOE* (a theory of everything); and the world of daily business scours the market for a venture that will profit mightily.

You might seek but perfection but how often do you find? What about imperfect truth or error? The quality of information also shows a downside full of negativity. Information that misleads you includes accidental, ignorant and deliberate error. Illusion, delusion and deception twist the truth into a catalogue of suffering. Misinformation, treachery and blunder savage what is left. What is truth's least part but error?

Error lacks alignment with the facts. If facts are right and right is truth, then show me what The Fact is. Is it scientific? Matter, of itself, is never 'wrong'. Therefore objective, scientific error must be caused by accident, incomprehension or inaccuracy of measurement. For example, theories can be tested and found wrong. Errors in program or process are the bane of computational life. And bugs in technological machinery cause stress and catastrophic breakdown but their real import, what they really mean, is that a purpose has not been achieved. *Error is a function of the mind but, if so, how does this apply to health and sanity, that is, to information coded into bodies of biology*? False information in a system spoils; it blemishes perfection. For example gene mutation, damage to a nervous or hormonal system and ugly memories can each contribute to a physical or psychological malfunction. Other faults derive from physical attrition. The questions then become, 'what was, in the first place, right?', 'was there perfection when the system that is compromised was first produced?' and 'if I had primal information could I not apply its yardstick to correct such entropy as broke the system down?'

A blueprint would be helpful but, from Theories of No Intelligence, do not expect a plan. There was no discriminating mind so accidents were never errors; in evolution there is neither right nor wrong. According to a Theory of Intelligence, however, you had better check the patent. You better check the manual if you want to rectify a fault. Primal information lies in archetype and from archetypal mind extends the logic that informs all fact. This logic is what medicine tries, by teasing out the principles, to understand. The program is, in concept, metaphysical although what wears its product out is not.

Apart from its subconscious blueprint mind involves the conscious element of subjectivity. Is intent subjective? Is a purpose unimportant or is it what informs an action? Ask a court of law. Subjective error is computed on a scale against Top Subjectivity, Transcendence. It is therefore measured in the balance of good will and love. Unintended error does not carry penalty but reckless does; and more so a premeditated error (such as lies, cruelty or crime). There is no crime without a corresponding law. Is there, supporting the morass of legislation humans generate, a natural bedrock of morality? Are there commandments issued by the Logic of the Universe and, as such, errorless? Is the source of regulation where there's none - in Pure Subjectivity, Communion of Love? Such regulation would then have to harden in proportion as minds egotistically do. Sin would not be simply something you, materialistic individual or government, decide.

If matter can be explained as a distortion in the otherwise 'pure' fabric of space-time so mind can be viewed as distortion in the fabric of Consciousness. In this sense body and mind are both distortions of Pure Truth. They are crumpled, lesser ruths. The Buddhist word '*dukkha*' means imperfection. When set against the absolute criterion of the Infinite everything is relatively imperfect; when set against Freedom all experience is more or less limited and painful. Relative 'suffering' is a translation that, although often used, obscures the larger implication of the word, that is, the nearer you approach Enlightenment the more that *dukkha*'s shadow is dissolved; and when you hit the bell that is, according to the holist camp, information's acme then a cloudless radiance rings. Funny that, by absolute contrast, such judgment is inverted by materialism and so called 'the ultimate delusion'! Such perversity would try the patience of a saint!

If to ascend is lightening then to descend is darkening; to reject the lightening principles of the right-hand levity is, therefore, to sin. Such commonplace metaphysical error is qualitative, caused by ignorance, faulty perspective or selfishness of one kind or another. Such mistakes are effectively self-disqualifications; they range from trivial to serious and attract a *karmic* (i.e. appropriate and proportionate) penalty. To deliberately accentuate the negative is, however, known as evil. You won't meet evil's fellow on the path of unity. On the contrary, such a character aggressively promotes division, alienation, domination, rape or the destruction of another's inoffensive happiness. The moral regression of evil is an agent of personal, communal and national strife and, in spite of technological advances, appears as common as ever. *Turning downward and away from the light generates the worst quality information, constitutes a perversion of the goal of happiness and results in serious disqualification.*

What is information that is not communicated? Is it useless, latent or unused? Communication is intrinsic to the cosmos and a central, guiding feature of all life. Communication can be conscious or unconscious. Forces of physics cannot know the patterns they non-consciously convey; nor does their automatic message vary in effect. Conscious transactions and unconscious but purpose-bearing mechanisms constitute a different game. In this case quality of communication needs to rapidly and accurately reflect the quality of a mind's emotional/ intellectual information or a system's stepwise process. *Clear, precise communication needs a sufficient vehicle, a linguistic medium of information exchange. This means* (see also Chapters 3, 6 and 19) *code.* Mind uses code to communicate ideas and purposes in a meaningful and comprehensive way. Sub-conscious faculty (*fig.* 15.9) communicates with body using reflex signals coded in electrochemistry. Conscious faculties (*fig.* 15.8) use neurological and anatomical apparatus as well. Animals can sing, dance, screech and otherwise express their instincts with, in some cases, an ability to think. Only humans, though, have capacity to learn and use complexities of language and mathematics, that is, an ability to create images and use symbols in theoretically unlimited yet logically, meaningfully limited combinations.

In accordance with a seventeenth century world-view a twentieth century sect of science known as 'behaviourism' used (assuming mind was brain) to pursue a belief that animals behave robotically. No doubt cabbages converse in dim and reflex manner using gaseous communication (ethyl jasmonate) but there is no doubt such science radically underestimated animal intelligence. Its counter-intuitive rare sense fooled no farmer or pet-owners but its scientific folly led and leads to cruelty in cages. Belatedly psychology has caught up common sense. It recognises that animals express emotion, problem-solve and have, in short, minds of their own. It's obvious that in the wild all kinds of organism have to creatively solve immediate puzzles and plan survival strategies. They put their mind to it; they don't mindlessly evolve a physical solution. In some tasks their genius outclasses ours. Studies now confirm the intelligence, if not highlight the cleverness, of 'feathered apes' like crows and jays, squirrels, fish and rats along with bees, fruit flies, octopi and (did I forget?) apes, monkeys, whales, dogs and even, lower down the class, more woolly-thinking sheep and goats; they also show not apes but dolphins ranking intellectually second to our own abilities. Surely we, next up in line, did not evolve from dolphins?

Surely, like St. Francis of Assisi centuries ago, we should think of dolphins and a host of other creatures as *non-human persons*? Children and pet-owners also have no problem here. The previous brand of hard, 'scientific' objectivity that treats a dolphin, ape or atom just the same is, thankfully and none too soon, receding. Are, for example, fish as stupid as an evolutionary judgment rates? They have, in fishy ways, intelligence and memory to match with any chimp. While frogs don't talk some birds don't only parrot but can use the English language. Apes are able to relate a symbol to a particular object and string symbols together. Indeed, you won't be half as fast as Ayumu, a chimp, at number memory-and-sequence games. A gorilla (Koko with Penny Patterson) communicates using over 1000 signs and 2000 words. The vocal cords of apes do not permit them to speak or, by Koko's standard, they undoubtedly would. But African grey parrots (with Irene Pepperberg) and N'kisi (with Aimée Morgana) don't just parrot; they have mastered basic grammar including verb tenses, sentence construction and large vocabularies (with N'kisi 1500 words!) and can hold undeniably intelligent conversations in English. They make pertinent comments using sentences of up to fifteen words in a creative way. When N'Kisi first met Jane Goodall, the renowned chimp expert, after seeing her in a picture with apes, he reportedly (Times Supplement 27-1-04) asked "Got a chimp?" A question that, strange for evolution, puts great apes as well as the behaviourists to shame! No doubt that there exists a spectrum of intelligence. Reason, mimicry and creativity are not just human attributes; but only humans can write volumes on the objects of imagination or research and thus extend our creativity. Our own kind's flexibility of mind, its intellectual capacity, is in another league.

Pre-programmed chemical stimulus and response developed from bacterial beginnings until, for a *bottom-up* evolutionist, the ability to reason followed a fluent ability to manipulate symbols in the form of language. Since language seems to be the product of innate neural circuits and reason the product of language, then the whole 'illusion' of thought is based on randomly evolved brain structure and the motion of electric charge. *Yet inanimate matter never knows nor purposes. It evolves neither reason nor thought; nor generates 'virtual' symbols and informative code to carry them.*

Top-down the picture is reversed. Information is immaterial, metaphysical and conceptual. It is the product of intelligence and the purposive, meaningful order of its patterns needs, in the course of communication, to be preserved. This means being translated, moved and stored in code. Psychological, genetic and nervous codes are obvious examples. Information is communicated through forceful pattern and chemical messengers. The structure of brain itself must reflect the intention to define and generate a quality of experience we take for granted - 'human'. No doubt other kinds (tiger, spider, spider-hunter and so on) reflect their owners' type and mode of thinking. Some other codes inscribed in non-purposive matter, the province of 'hard science', obviously reflect the purpose of a mind though you can sprain your logic and deny it. Printed language, *DNA* codes and the sounds of speech are some examples. Others are less obviously purposeful, more arguable. Are subatomic particles the letters of inanimation's code? And the forces they connect with pattern makers in the mould of archetypal law? What purpose of itself has any stage without a play? Its only logic is that it supports the play. This perspective is, predictably, the reverse of *bottom-up* interpretation.

Involved communication. Conversation. Books. You have to ask if complex language, base of reason, is a predetermined mark of man. Over a hundred muscles in the tongue, lips, jaw, throat and chest exquisitely cooperate with various areas of the brain and instantaneous memory to let you say a word. Word-sounds are strictly made and ordered to impart a meaning; they generate responsive and creative social oils, our metaphysical semantic, speech. It is known that a foetus begins to register qualities of sound and patterns of language while still in the womb. Like walking talking is an innate ability that children learn without being taught; but how can genes frame the infrastructure of language or nerves encode a yearning to communicate? **Meaning represents another evolutionary black box**. Is it, in fact, unreasonable to judge that human mind, with its teleological communicative faculty, was intended from the start?

As for speech, so it is for literacy of number and of measurement, the wholly abstract language of mathematics.

The laws of nature don't evolve; nor does, as far as we can tell, metabolism (Chapter 20) or electrochemistry and codes of nerves. Minor changes don't affect, in principle, the job that's done. Of course, linguistic vehicles of human thought are varied but the rationale is just the same. Sounds with a self-consistent infrastructure differ over time and place but does plasticity explain their evolution from grunt-whistle; should we expect the evolution of still more sophisticated grammars, syntax and vocabulary? In fact there is neither 'primitive/ primal' language nor any evidence of one. Some stone-age tribes speak very complex languages and the earliest known languages involve richer inflexions than, say, the relatively 'primitive', devolved grammar of contemporary English. All known languages can be grouped in less than twenty 'families' of which there is no known common ancestor. The enormous complexity of language appears to spring intact. Written (historical as opposed to conjectural) evidence is only about 6000 years old. This is about 0.1% of the time man was, conjecturally, evolving; what happened during the other 99.9% of the time? There are certainly no signs of 'evolving' Palaeolithic or Pleistocene languages on clay or stone and the progenitor of Indo-European languages, Sanskrit, is one of the subtlest, most complex of all. If you can reason, express your reason orally and fabricate tools, why not also write? Why did writing not 'evolve' in tandem with the language capability? No clues exist but almost obviously it did; you should ask, 'why should it not?' Is not predisposition of a child to keenly learn a language and to count part of the human archetype, inscribed precisely at sub-conscious and non-conscious levels? Are instinct, nervous correlations and the vocal instruments of language not all predisposed and ready waiting, at about two years, to flower? If in principle and practice it were so then the combination would amount to a linguistic program. This program's communicative purpose would express the nature and the quality of mental content (ideas, feelings, purposes) with detailed, high fidelity. In such circumstance you might expect an infant to be born with an innate, incorporated 'blueprint' for acquiring linguistic and numerical skill. Such archetype would be expressed as aptitude for learning and for solving problems. It would involve capacity for organising symbols in support of purpose - reason. There would exist no need to resort, for reasons of philosophical materialism, to evolutionary tales of accidents, grunts and groans to grammar or (against the evidence but with the speculation) the development of complex from simple languages. Instead the whole business of language makes the reasonable sense it is supposed to - informing and informed communication from the first word. Go!

Quality of mind, information and communication are a function of your focus of attention and intelligence; and the latter is, in conscious organisms, a function of the relative ability to think, reason and contemplate. *In normal waking practice attention oscillates (often rapidly) between the upper, contemplative and lower, physical information loops but they are rarely equally attended.* Some vertebrates use both directions but, except for humans, the upper loop is stunted (see *fig.* 5.8 (ii); the rest of life on earth, locked into reflex patterns of stimulus and response, seems to lack or almost lack self-consciousness. It asks few questions and the ones it asks concern pragmatics and not principles. They are bound up in the details of the moment and what the instincts of the body are demanding. Even a majority of humans adopts, in deference to the overriding currents of sensation, this weakest, least disciplined and most diffuse grade of contemplation. Its shallow sort of reason supplies sufficient scheming for daily, physical survival. Rare intellectuals and contemplatives habitually practice a less superficial, more single-minded, inward absorption. Of these only scholars of a contemplative discipline sharpen up their focus with a purity of single-mindedness. Their experience is therefore uncommon and, as uncommon often is, misunderstood. Because it is subjective it can at best just be inferred. Do you believe the things that mystics always say? Are they not inevitably, being of the inmost subjectivity, anecdotal - repeatable but not in a laboratory way? To trust an invisible, immeasurable and paradoxical inference is not in this sense common-sensible. Nevertheless to reach the very depth of things a contemplative may have to storm the Top, the citadel of Truth of truths. To obtain the Answer needs no sophisticated technology. Its voluntary (and therefore not necessarily followed) path and associated mechanisms are part of a design laid, like capacity for language and the thread of reason, into archetypal man. *This is why, in his potential to complete the cosmic loop, upwardly mobile man is judged top of creation; he stands a creature once created in the image not of ape but of Creator. That is the inside information leading to the highest quality of mind.*

The Ascent of Man

What, after all, is quality of information or of mind but cerebral arrangement? *Bottom-up*, it's all determined by the evolution of your genes. Now ebbs and flows of monoamines vary moods; thalamocortical and limbic linkage weaves attentiveness and egotistic sense. Perhaps 'spiritual illumination' is a product of posterior shut-down in our cerebral reactor; perhaps transcendence jettisons the ego in proportion as the 'back-brain' doesn't fire! Thus a redirected nano-voltage ought to push you, front-ways, Home. Tot all the chemicals and still God isn't worth a lot. Truth's value is as meaningless as its creator - brain's electrochemistry.

	tam/ raj	*Sat*	
	existence	*Essence*	
	expression	*Potential*	
	duality	*Unity*	
	limited 'me-it' experiences	*Peak Experience*	
	periphery	*Centre*	
↓	*tam*	*raj*	↑
	down	*up*	
	descent	*ascent*	
	compress/ diminish	*grow/ liberate*	
	passive/ involuntary	*active/ voluntary*	
	created	*creative*	
	effect	*cause*	
	materialisation	*dematerialisation*	
	entropy	*negentropy*	
	devolution	*evolution*	
	body	*mind.*	

This stack is not, of course, how materialistic evolutionism sees ascent. A current generation always thinks that it is most advanced and, for an evolutionist, there stands upon the present peak of all progressive time the culmination called a human brain; here stands, at the apex of all history, the most evolved of minds - his very own. This is the way, from molecules to man, the latest scion of original bacterial slime shines forth. Top of the cosmic class! The intellectual head of creativity heads up creation! This is the notion of ascent for which, in our age, prizes are assigned.

Is such a notion born of professorial puff or simply mortar-board of common sense? What, if mutant apes with PhDs are cosmically incorrect, might even more evolved, enlightened brains in future see? If they agree materialistic man was right or partially right will they have nothing more to add? And if they add what, with respect to consciousness and life, might their addition be?

Top-down, on the other hand, such logic's upside-down. Though taught by clever dons in universities it stands upon its head! It should be clear by now, through 180°, what the ascent of man involves. It is not biology.

It is psychology. Towards the end of Chapter 3 mention was made of the achievement, following ascent on the right-hand path of knowledge to its Extremity, of Complete Information - Enlightenment. *Do you remember that, by negentropy of information, mind evolves? By self-purification it is sanctified and in such sanctity ascends towards the Principal of principles. This discipline was dubbed 'the science of the soul'.*

Does 'enlightenment' mean the experience of (*Sat*) Essence? If so it involves experience of the full potential from which cosmos is derived - Pure Consciousness. What is *that*? The mystic path means growing up (↑) and, with full potential of enlightenment achieved, the coming of an ageless age, eternal youth, pure life and ultimate maturity. The question is - what's your perspective? Can such Potential, if it's real, be realised. Would you recognise your Self?

Check back to *fig*. 4.1. *Top-down*, Natural Dialectic reflects a cyclical, two-way traffic of creation, information and focus of mind. The (*tam* ↓) outward or motor aspect of materialisation is construed as devolutionary. Cosmos is devolved from top-to-toe; its inversion runs from Inward Centre to outer circumference, from Potential to its final expression in matter. If bodies are material then they are devolved.

Conversely the (*raj* ↑) motion of dematerialisation is construed as evolutionary. A creation, having been discharged from potential to impotent condition, is recharged. Conditions quicken; lost potential is regained. Such stimulus depends, in the involuntary case of matter, on an injection of energy. As solid is (↓) devolved from gas, gas is (↑) evolved from solid. Ascent is towards simplicity, homogeneity. The injection promotes ascent towards flux, freedom and, in the form of subatomic particles or plasma, a return to 'seed condition'. You might deduce that this was very hot; but matter can't evolve to feeling nor energy spark subjectivity.

'Seed condition'. Recharge. Liberation. In this view Potential is creatively materialised from top-to-toe. It is devolved but, conversely, toe-to-top, evolved from impotence back to original form - First Cause. Lost potential is, in the voluntary case of mind, regained by an injection of attention; enthusiasm is a traction sucking information up. What fire best ignites improvement, what stimulates a person's mental progress fast? Could evolution of the mind embrace the (↑) right-hand path of union and, at the same time, liberty? Mind burns with enthusiasm, interest and love. What, therefore, is the nature of the origin of mind? What, by this logic, is the scientifically senseless character (*figs*. 2.11, 7.3, 7.4 etc.) of First Cause, *Logos*, at the Apex of Mount Universe?

No doubt that, from a *top-down* point of view, all creation - artistic, technological or simply routine in its business - runs the gradient explained in Chapter 6 as top-to-toe. No doubt, equally, that in the voluntary case of mind 'recharge' involves an injection of affection, a sharpening of focus of intent, and, in consequence, more understanding. Such evolution involves ascent from mundane practice to considered grasp of higher principles. It is the goal of higher and still higher education. But for man, a cosmic mountaineer, the climb presses on and upwards towards Olympian heights, towards a pinnacle of success, the summit of Mount Universe. Creators 'top-to-toe' disburse their information into making things; students looking toe-to-top gather information up. They recreate the maker's mind. This is the way that in creation creatures, students in the form of men, can understand Creator. What crampons, ropes and practice does the inward science of ascent involve?

Look at it another way. Bodies are non-conscious, purposeless and passive. They are (see Chapters 19 - 25) capable of variation-on-theme but incapable of origin by physical evolution. *On the other hand not biological but psychological evolution is possible.* **Such evolution might be called a lowering of entropy**. Of course, wholly sub-conscious, dormant organisms, such as plants, have neither more aim than sleep nor will than minerals to 'improve' their condition; and among conscious organisms the only one with the capacity to really break out from the physical and instinctive loops in which he is snared is man. Man is, in this respect, unique. His mind is a most active, purposeful information centre. It evolves plans to further its aims, projects to achieve its goals. If such a goal is to understand the nature of the world, its life forms and the cosmic place of man, it is called science (Lat. *scientia* - knowledge) or philosophy (Greek 'love of wisdom'). The excellence of science may be outward in its quest to understand all physical phenomena or, in the contemplative focus, an inward quest towards the metaphysical source of mind. Either way, a man seeks disentanglement from mundane detail and, instead, progressive involvement with general causes, principle and knowledge. *In this case mind evolves, phoenix-like, from material fetters.*

Intellectual evolution is the happy lot of academe but, if you seek the very origin of mind, there is a snag. It is not easy for 'thought junkies', addicted as they are to complex intellectual games of mind, to contemplate the precondition of ascent towards a supra-rational state. This precondition is a process that will climax in annihilation of what the *ego* holds most dear - itself. It seems suicidal and, therefore, both irrational and dangerous. Indifference to honours won't obtain the seal and scroll of a degree. Is not the only opposite of waking sleep and that of reason dim irrationality? How distinct from its sub-rationality is supra-rationality? And how on earth might any centaur, any 'human animal' by voluntary *loss* of reason delve the formulations of existence, paradoxically annihilate the mind and still retrieve the insight of Original Cause? Yet such is the psychological practice, anathema to busy worldlings, that prises man from his animal parts and in which psychiatric saint or mystic is absorbed. Absorption claims it yields up Self-awareness, final Knowledge and the Truth. What Socratic and yet intellectual nonsense that may seem. Yet if, as the contemplative claims, the acquisition of such knowledge is the fundamental purpose of our life on earth then who has missed his vital point? And, in missing, missed The Point?

In summary, it is not possible, in a dialectical framework, to understand psychology without grasping its origin. Psyche is mind's source and substrate but the place of human intellect is at a lower level. Its potential is bound up in process - a process of weighing up whatever kind of family it finds itself embodied and embedded in. Your mind is at this moment working as a balance at a pivot called your cosmological axis, third eye or eye-centre - where you think.

This balancing factor is complemented by a two-way dynamic - 'information in' and 'out'. Life incarnate is constructed round the information business and, as your own example shows (Chapters 17 and 19), this means its broad-brush, triplex principle informs the basis of biology. *Biological form is clustered round both mental and material nuclei. The former nucleus sometimes includes a conscious factor but in all organisms a dormant, archetypal one; and the latter DNA.*

A two-way traffic of communication flows. Immaterial information is exchanged. In the case of conscious mind (*fig.* 14.3) it flows around two loops stapled at the axis called 'third eye'. One heads 'upward', 'inward' towards the metaphysical; we call it contemplation. The other, a reaction to sensation, spreads 'down outwards' to the world. Below the third eye is the body-of-the-world's domain, a world of detail and of difference; above it is the inward universe, the place of unifying principle. The paradoxical (opposite yet complementary) nature of these loops highlights the distinction between experimental and experiential paths towards Truth. Each approach is valid; each seeks to elucidate a different aspect of our wholeness and thus weights its pan upon the scale of things. They are complementary. Which, however, is more fundamental? If both yield fundamental truths about life and the universe of things is one more important than the other? It depends on your perspective.

A teacher chalks up principles; his pupils carry out experiments. Framed in the *bottom-up* perspective of a scientific discipline psychology sheds manifold and fascinating lights on how life-forms process information and notably mind's passive, physical reflections (neural correlates of consciousness) in the brain of man. Having philosophically rejected the real nature of its subject, however, it misses the main, immaterial point.

Top-down, its theorem is simple. Your absolute and inmost being suffers, with embodiment, the relativities of ups and downs, that is, the joys and burdens of a physical domain that's prone to entropy and death. But what is its inmost 'spirit'? Is it real? Once the experience of humanity is past what, if anything, survives the grave? You test empirically within a natural laboratory (your body) using natural instruments (your senses and your mind). The method of repeated inspection is one technique or other of meditation; and the results of due diligence are, while often poorly resolved by inexperienced technicians, universal. The object of research is reconnection. It is the conversion of sensuality to spirituality. This, not physical or biological at all, constitutes the true, voluntary evolution of ascending man. When the experiment is life-on-earth (and perhaps some more) you have a science of the soul.

From sole towards soul - improve your quality of information. Rise (↑) towards order, evolve towards the highest kind of information, the most powerful knowledge in the universe! When the waves of fear, *ego* and the passions settle then the peaceful 'residue', the central concentrate of consciousness is closed upon. *Such union with Natural Essence reigns Supreme; and the ascent of man, in seeking such Supremacy, is voluntary. It is wanted.* How can you force freedom? It is chosen. A seeker purposely engages the return leg of creation's gradient and ascends the right-hand path to its Extremity. Humans are soft-wired for Communion with Truth. **Such final re-connection is intrinsic to our gifted circuitry; the work is so-called science of the soul, objective in its subjectivity, whose practice leads you to the origin and apex of the world.**

TOTAL CONTENTS

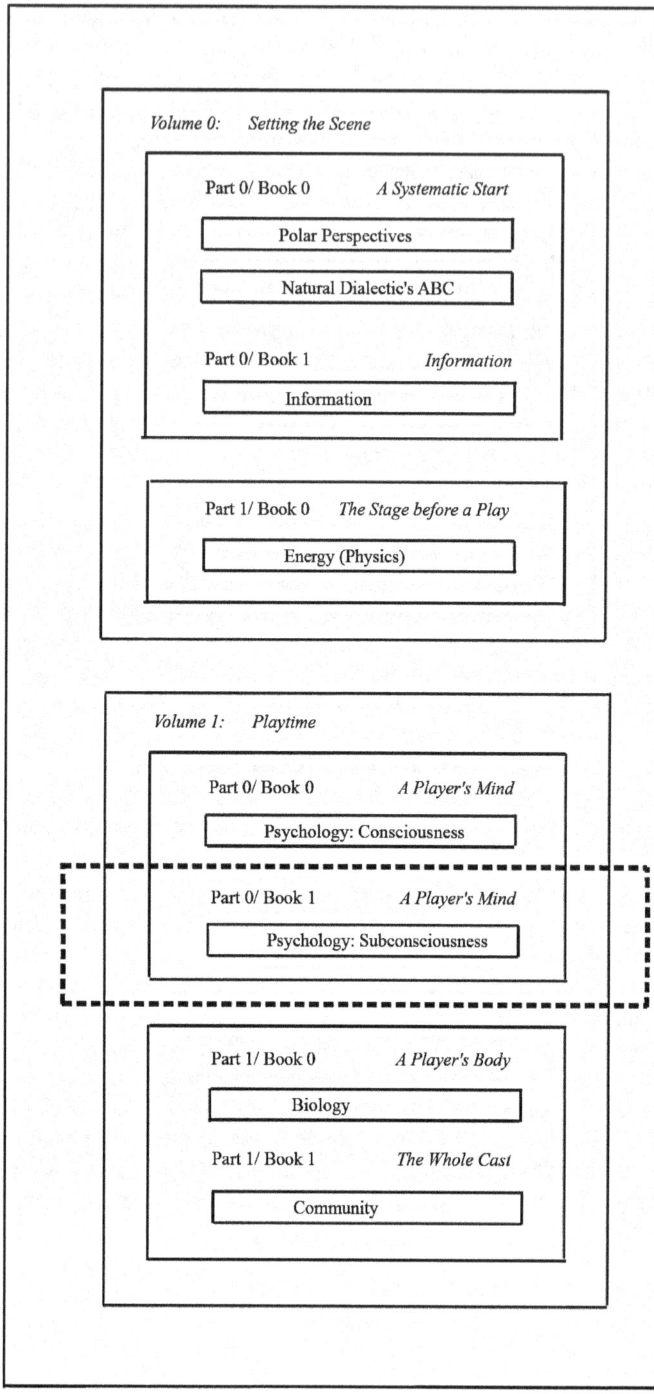

Volume 1: Playtime

Book 0: A Player's Mind

Part 1: Passive Information (Psychological)

15. Subconsciousness

Forget the top of the world. **For a naturalist, whose guessed idea of consciousness is of an 'outcrop' of brain chemistry, the hierarchical position of the next four chapters is as irrelevant as their thesis; thus you will not have met their content in the context of a modern scientific course.** They are superfluous because in its paradigm, called in the jargon 'materialistic monism', no dualism of separate awareness and matter exists. Information and energy, mind and matter are of wholly the same substance. And if this substance is material it is non-conscious. Consciousness is therefore a peculiar effect of certain formulations of non-consciousness. How strange are mind and subjectivity! How queer the 'scientific' formulations of electrons in a brain!

The Sub-conscious, Psychosomatic Sandwich

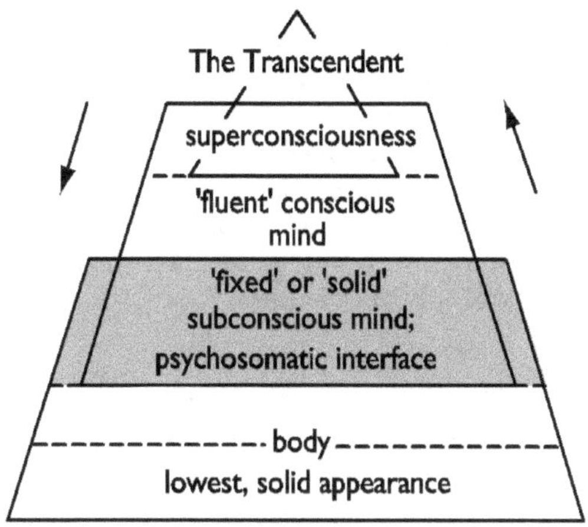

fig. 15.1 (see also 2.8, 2.9, 13.3 and 14.1)

Professionals are paid by results. They need sufficient tools and a profitable paradigm within which to ask questions and devise accurate, working solutions. Neither tools nor paradigm currently exist to conflate physical with the metaphysical part of a holistic package. It is therefore natural for science and technology to adopt as sufficient and attractive a philosophy that tries to explain the world exclusively in terms of what it can get its hands on - the physical. Having once squashed everything into the 'safety' of a framework composed of physical energies and neurologised (physicalised) consciousness, it is logical to dismiss the 'holistic dualism' of Natural Dialectic as an irrelevance. No matter that, in the last analysis, 'materialistic monism' might be less than half the right answer and, by the same token, inclusive holism wholly right. If this is fact then truth, in its entirety, is a scientific casualty.

From a *top-down*, hierarchical perspective the subject matter of the next three chapters, sub-consciousness, is sandwiched between the perceptible grounds of cognitive psychology above and biology below. *It deals with issues mentioned in Chapter 4 that, following the top-down acceptance of a metaphysical domain of mind, need to be addressed at the key level of its psychosomatic interaction with matter.* Just as mathematics can (sort of) describe music so it may be able to describe archetypes. But their ambit is above a physical 'glass ceiling'; and below where the so-called 'scientific flakiness' of the psychology of consciousness enters in. As gaseous matter, losing energy, drops through liquid into solid state, so mind, losing informative awareness, drops from active into the passive, disabled conditions of ignorance and sleep. Consciousness drops into sub-consciousness. You, as a mind, experience the effect of both conditions. They are linked to an atomic instrument through whose agency you can inform and be informed about the physical world - your body. What is the protocol of this psychosomatic linkage? What is the nature of the intermediary or 'medium' between metaphysics and physics we call 'solid', sub-conscious or unconscious mind?

Just as atoms found universally are specially arranged to compose your body, so a special composition of universals may be involved in the dynamic arrangement called your mind. Solid, gas and liquid are combined to make your body up; why should mind not have its phases too? Its awareness and energy are certainly, in your experience, capable of transformation; it is, as opposed to the relatively fixed operations of your body, a 'fluid' and imaginative information centre. Let's deal with the role of the psychosomatic level of mind in the way you *experience* your body and its environment, that is, the way you know the world. What about the subjective psychology of sub-consciousness?

<u>**A Black Box**</u>

	tam/ raj	*Sat*
	existence	*Essence*
	finite	*Infinite*
	below	*Transcendence*
	expression	*Potential*
	periphery	*Axis*
	swing	*Pivot*
	range	*Super-state*
	conscio-material gradient	*Pure/ Perfect Consciousness*
↓	*tam*	*raj* ↑
	downwards	*upwards*
	less informed	*more informative*
	diffuse/ unclear	*focused/ sharp image*
	dull/ tired	*bright/ alert*

sleep	*waking*
involuntary	*voluntary*
purposeless/ undirected	*directed/ purposeful*
sub-state non-consciousness	*mental spectrum*
subtendent matter	*levels of perception*
atomic body	*sensitivity*

Material bodies are hard, scientific ground; what, however, is mind's mystery about? Physical science is uncomfortable with mind's consciousness because, by definition, these are *not* robotic in the way of non-conscious force. Their flexibilities involve, even at the base, involuntary levels of instinct, memory and sleep, some degree of unpredictability, uncontrollability and unrepeatability. Such relative automatism decreases in proportion to the advent of wakefulness and voluntary thought. Of course, mind's natural enough but if it's insisted 'natural is material alone' then unnatural immateriality is placed conceptually out-of-bounds. Whatever science-as-we-know-it can't explain becomes 'abnormal', 'paranormal' or just plain impossible. **Thus, when only brain is treated, mind becomes an obvious mystery - a psychological black box**.

Is 'paranormal' just another word for 'weird' or 'metaphysical'? Mind is, by Natural Dialectic's definition, neither 'brain-meat' nor 'abnormal'. If mind *is*, despite equivocation, metaphysical then psychology becomes, except for its 'body-side' or neurological interface, a study in metaphysics. Still, despite voluminous documentation, widespread personal experience, Institutes for Paranormal Research and, more clinically-oriented, Psychological Research around the world, mainstream materialism (in the form of 'hard', scientific atheism) is disinclined, sometimes fervently disinclined, to accept the proposal that mind and matter are, although interlinked, fundamentally different entities. It has the nerve to deny, without evidence, that information and energy are essentially different; or that consciousness and non-conscious, physical phenomena are two basic components of our single creation. No dualism, no immaterial *and* material, inhabits such one-track, bone-headed certainty. **This attitude is especially disinclined to accept that the origins of cosmic 'deep structure' are, in fact, in mind**. *Nevertheless if, as well as matter, there is a familiar, metaphysical domain of mind, then saturated materialism is but only partially right.* There really would be a 'ghost' in the machine. *And in the interest of truth as opposed to orthodoxy it is therefore reasonable and right to consider the consequences of logically broadening its partial paradigm.*

Please forgive me here! An important question has to be re-asked. Numbness is not easy, anaesthesia is, if not opaque, unclear and sleep - what is sleep? Unconsciousness takes different forms and grades; by various causes, much researched, loss of feeling overtakes a local part of body or its generality including mind. Then what is lost? Where go 'I'? Why can't I remember absence and, perhaps, even things that passed before the period of forgetfulness?

Lack of sensation means a pattern of informative chemistry has been affected. Maybe sodium channels have been blocked, neurotransmitter or receptor action changed or even fat dissolved in tissues of the brain. Theories, each plausible but unconfirmed have not yet nailed the complex business to its mast. Plain sailing hasn't crossed the sea of mind. The question is: what senses, what can be desensitised, who is conscious that becomes unconscious? Is it nerves or chemical anatomy? All agree such matters play a part. Molecular biologists research the details that pertain to both conscious and unconscious modes of 'being'. Reversible unconsciousness is sleep; anaesthetists induce a coma hopefully not final; injuries and drugs block, variously, sensibility; death sweeps reversibility away. The question stays alive. What, held and released like breath, flips in and out of knowing; who oscillates, healthily or otherwise, in and out of consciousness?

From a *bottom-up* perspective consciousness is wholly due operations of nerves, molecules, electric charge or other mechanisms. It 'emerges' out of patterns so materially complex simple minds can't understand.

Top-down, what's soul; what *is* the simplex, unitary, immaterial element of consciousness? Computers, wireless or machines each offer an analogy; program, broadcast and force-field are models; forms of words and images of concept may as much mislead as reveal. Incorporated in a 'transmitter/ receiver set' it's possible that brainwaves modulate our major states of mind (electric pulses of some rhythmic kind might modulate every organism's 'states of being'). While entrapped in body's 'set' states can be varied or switched on and off; and when the mechanism of attachment, mind-with-body, fails irretrievably in death then only mental body - made of immaterial memories - is left. Without the 'diving suit' with which you contact matter's ocean thought and feeling in a real dream constitute experience. Subjective mind was anchored in objective vessel made, like a wireless, of material parts; once the vessel breaks down it's released, without restriction, into purely mental atmosphere. Your body is a part of the material world; your mind is part of metaphysical entirety.

Are you surprised? This view does not equate you with your body. Body's only the non-conscious 'cage' - a fraction of tri-partite self. You may be unconscious, in a form of ignorance, of your super-conscious part; your mind is subdivided into conscious and sub-conscious regions; and your body's irredeemably oblivious.

Thus from the top to tail of psychology we drop from waking into the zones of sleep. Life tips (*fig.* 13.5) down into the negative, involuntary zones of mind. These are active and passive sleep - the residual consciousness of dreams and unconsciousness of deep sleep. After discussing this pair we shall turn to inspect the place of memory, instinct and the mechanisms of exchange, called remembrance, between active and passive information. This is, you may remember from Chapter 4, one of the 'stingers'. It is the sub-conscious interface through which symbols of mind are traded with coded signals from body and its surrounding environment; and the place mind meets matter, the psychosomatic medium of life on earth.

The psychology of sub-consciousness involves (*tam)* lower, involuntary or dormant mind. If consciousness is translucent, like light, and at the same time full of colour then the nature of sub-consciousness is colourless, opaque or, at best, is 'coloured' by the flickering shadows of residual awareness - dreams. A grey area. Or is the experience of such mind, in dream or deep sleep, more black than grey? It is, in principle, ranged between conscious and non-conscious conditions and hides the critical psychosomatic border. Of course, you can use the same instruments to observe a sleeper as a waking person, that is, you can treat him as a sophisticated object of research; you can then try to correlate objective data with what he tells you, later in the case of a sleeper, about his subjective experience. *In subjective practice, it is difficult to investigate sub-consciousness because you can enter it neither voluntarily from above (you lose awareness) nor involuntarily from below (it is non-material).* **Both sub-consciousness and the proposed evolution of that state are, as with consciousness, black boxes**.

Not the only ones. How did instinct, language, dreams or sleep evolve? How did and do non-conscious atoms rise into unconsciousness and surface as your conscious state? **No doubt your brain is in your white box, skull; but mind's whole business is a physical black box - perhaps because it isn't physical at all.** In fact, science has no real clue how its white or black box came to be. Is not a scientist trained, in such a circumstance, to set aside prior beliefs and keep an open mind? Is this discipline adhered to when materialism's basis, neo-Darwinism's *PCM*, is questioned? A neurologist might lie-detect cerebral twitch.

Natural Dialectic notes two aspects of sub-consciousness - the *subjective* experience of sleep and its *objective* aspect of 'thought object', memory or 'frozen time'. What is memory but thinking or impression fixed? What is memory-in-action but the automatic vehicle of person, the subliminal working of unconscious mind? At unconscious level the conscio-material gradient approaches the psychosomatic border (*figs*. 0.7, 0.8, 0.11 and 15.1). Subjective is delivered to objective; thought is scaled into a physical event. Thus, to neurology's external view, the non-conscious, automated aspect takes up pole position. Chemistry seems dominant. Nevertheless, for individuals like you or me the fact of subjective experience takes precedence. Let's first deal with this subjective aspect of sub-consciousness. It has two levels. These appear in the (*tam*) downward, sub-conscious phase of diurnal oscillation as residual waking called dreams and an abyss that holds deep sleep and coma. On falling asleep you first plummet to the abyss (*figs*. 13.5 and 14.2). You then bounce back through phases of dreaming into wakefulness. However, in order to retain hierarchical order we'll call up the lower states of consciousness by level.

<u>Sleepy Head</u>

	tam/ raj	*Sat*
	range	*Super-state*
	conscio-material gradient	*Pure Consciousness*
	more or less sane	*Sane*
↓	*tam*	*raj* ↑
	passive	*active*
	sub-conscious	*conscious*
	sleep/ dream	*waking*
	sub-state	*intermediate levels*
	involuntary	*voluntary*
	uncontrolled	*directive/ informative*
	unperceptive	*perceptive*
	illogical	*logical*
	incoherent	*coherent*
	random recall	*specified recall*
	purposeless	*purposive*

insane	*sane*
meaningless	*meaningful*
mental blank/ physical business	*physical blank/ mental business*
cholinergic system	*aminergic system*

Biological homeostasis, whose purpose is to achieve dynamic balance, works in cycles. Action drains, inactivity 'recharges batteries'; tension tightens but, when you relax, you heal. So it is with a day's business. Forgetfulness allows the tensions to dissolve and, as necessary, the natural currents of health to reassert themselves. Sleep is fundamental in biology. Most people spend a third of their lives in a condition whose deprivation kills quicker than starvation. Even cultured neurons in a Petri dish, while never actually awake, seem to oscillate between active and passive, 'sleeping' modes. Is sleep just relaxation? Why, exactly, when you wake up do you feel refreshed?

The pulse of life is, like a metronome, oscillating to and fro. Reciprocity between sleeping and waking poles of consciousness occurs all the time. You can, for example, daydream while awake and wake (in dream) while asleep. It is a question of tendency and predominance. For the sake of simplicity we can divide the scale into waking awareness (states 1 to 3 of *fig.* 14.2) and sub-conscious sleep (states 4 and 5). The former includes both normal perception of the external world combined with a controlled, voluntary but internalised focus of attention called thought; and, which is not considered by materialistic psychology, an attainable state of Pure Awareness. As waking consciousness blacks out the sub-conscious influence of sleep swings into play. How?

Sleep is elastic. You fall deep asleep and, from this nadir, 'bounce' up into dreams. Thus, carefully controlled by chemical switches, each of four or five 'bounces' is in two parts. At last you 'surface', that is, wake.

Of course, science in the form of neurochemistry wants to explain the physical reasons and effects of these oscillations. It intends, as in the case of thought or sensation, to identify the emergence or loss of waking consciousness with neurochemical activity. How are the principles of waking or of sleep facilitated? How might an 'on-off' sleeping/ waking toggle work?

There certainly exist genetic, metabolic, hormonal and nervous switches that, without any conscious directions, regulate various kinds of physical homeostasis. Why should brain-function, which modulates embodied consciousness, not also respond to switches? Why should its diurnal oscillations not amount to cybernetic control or, as the homeostatic process of staying on target is sometimes known, negative feedback? The swing into sub-consciousness marks a global change in state. It marks a swing from voluntary to involuntary, from logical order to uncontrolled, illogical associations and from 'store-and-remember' to 'don't-store-and-forget' modes of operation. Mind is fallen from its day's personal business into an involuntary, passive 'blank' - although the blank includes a shadow-land of dreams. In sleep the influence of body can, without voluntary interference, balance exhaustion and restore unstressed health. If sleep, in its homeostatic capacity, marks a global change in mental state, you might predict this would be triggered by a global change in brain state. If so, what chemistry controls the cycle? What modulation regulates the 'tone' of brain and thereby affects the state of mind? Maybe brain needs sleep and mind does not; in order that exhausted brain and body are refreshed their restless stimulant, the mind, is cooperatively put to sleep.

From a *bottom-up* perspective mind is simply a 'functional state of brain'. It is not that consciousness, as a separate 'spiritual' entity from matter, is 'squeezed out' of this picture. It never existed. Therefore the difference between waking and sleeping is simply one of chemistry. Just as more or less of the hormone called testosterone affects the subjective as well as objective sides of sexuality, so chemical instructions program the global states of brain and, therefore, mind - sleeping and waking are 'simply' biochemical effects. In this view matter is prior. It is everything. And, as such, must be the first cause of all things. Mind is no more than the rare but secondary effect of an extraordinary atomic system - brain.

There is indeed no doubt that biochemicals mediate states or modes of mind. They structure 'fluid' consciousness but is a group of them capable of becoming aware? Can such a collective think and therefore be? In other words, is the world simply a sequence of neural and hormonal patterns representing images? Are these chemical 'images' in combination somehow the substance of consciousness itself?

If this is so it needs be asked how all these images and parts of brain (which are known to correspond to different functions of mind such as sensation, instinct, volition, memory and so on) can generate your seamless subjectivity, your unitary mental operation. Is there some encompassing and unifying neuro-modulation? **If so how might things cohere; who, in the centre of this composite, are you?** And how accordingly might an 'on-off' toggle switch for waking work?

	tam/ raj	*Sat*	
	degrees of awareness	*Awareness*	
	body/ mind	*Consciousness*	
↓	*tam*	*raj*	↑
	passive	*active*	
	body	*mind*	
	instinct	*intellect*	
	emotion	*rationality*	
	dream state	*waking state*	
	cholinergic system	*aminergic system*	
	in ascendancy	*in ascendancy*	
	psychotic mind-stream	*controlled mind-stream*	

You cannot switch electrical devices on by mind; if brain is physical you need a switch that's physical. Switched oscillation sways the brain and its embodied mind between contrasting states of waking/ alertness and sleeping/ oblivion. Slumber severs sensory and motor contact with the world. It swings a subject in between coherent thinking and psychotic dream - because internalised psychosis is our private state of madness every night.

In fact, two systems might reciprocate. The autonomic nervous system is responsible (at least in vertebrates) for countering different sorts of change and stress and thus maintaining the stable, balanced operation of the internal environment of a body. Its relaxant, parasympathetic nerves employ a neurotransmitter called acetylcholine while most of the sympathetic (excitatory) system uses noradrenaline. You have potential, if you can embed the switches, for a global cholinergic system and its antithesis, a global aminergic one. Here is a basis for polarity, complementary opposites and a dialectical approach.

Neuro-modulation derives from localised nuclei in the brain stem. In brief, a cholinergic system is active 24/7. Its projections upwards to the cortex and downwards to the body are sourced from the thalamus, mid-brain and brain stem. This system excites the limbic system and is therefore associated with emotion, instinct, memory and primordial lack of rational control. It also mediates parasympathetic impulses and a fundamental sort of relaxation is, of course, refreshing sleep.

What does sleep refresh? The brain stem also involves cells containing aminergic neuro-modulators - noradrenaline (associated with excitement, stress, adrenaline and the four 'Fs'), serotonin (associated with alertness, focus of attention, learning and rationality) and dopamine (associated with pleasure and reward). This aminergic system sways 'upwards' towards the light of conscious, waking state while an absence of serotonin (by 50% in deep sleep and 100% in REM or dream sleep) casts it 'downward' into slumberland. The brain is as electrically active in slumber as in waking but its chemical activation (and therefore the patterns of electrical activity) varies according to circadian rhythms. Could sleep allow refreshment of an 'off-line' aminergic system?

Let's rephrase. Sleep toggles involve complicated neurochemistry. The brain is 'switched on' in sleep but its sensory and motor systems are blocked at thalamo-cortical and spinal levels respectively. Thus its activity is internalised and no 'waking behaviour' normally occurs. 'Waking behaviour', on the other hand, includes electrical activity but adds order. Critically, the presence of serotonin (curiously, also found in plants and fungi) inhibits the cholinergic system; it promotes a sense of order, purpose, attention, logical sequencing and general cognitive control. If serotonin is in turn inhibited then the so-called primordial emotive system swings into the driving seat. In the cyclical absence of serotonin the suppression of the cholinergic system is released and the latter's excitability spills over. Mind is swept passively. Sleep may overwhelm its consciously directed thoughts and wash them into a psychotic stream, a hallucinatory course of dreams. Sleep would be, in this context, a clearing and a cleansing process - reason's busted flush, psychological evacuation. However loose sleep's rationale it's one we can't survive without.

How could a sleepy head evolve by accident? By mutating over aeons out of coma? If anything in the oscillatory process (*fig*. 13.5) might seem chaotic, such 'chaos' were an illusion. Disorder and haphazardness are dreams. In fact, the borders between wakefulness and sleep are, like each state itself, closely regulated through a network of neural pathways and biochemical relays. The whole system, which appears dialectically designed rather than cobbled up by accident, is one of balance, global reciprocity and circadian sway. What, therefore, controls the pacemakers? What triggers the metronomic swing between serotonin and its lack, that is, what switches on and off the neuro-modulation of two global chemical conditions of the brain that we experience as sleep and waking? Orexin (or hypocretin) is a protein connected with basic 'house-keeping' such as thermoregulation, appetite and sleep. Orexin neurons in the hypothalamus project this peptide throughout the central nervous system but it has a special effect on the aminergic system. It promotes production of serotonin in brain stem nuclei and thus, unblocking the voluntary system, switches 'waking' on. Conversely a lack of orexin induces narcolepsy. You could see orexin as a 'toggle' between the two fundamental conditions of embodied mind, wakefulness and sleep. No orexin, no wakefulness. Too little orexin and a tendency to narcolepsy. Therefore, is not the diurnal swing of orexin a toggle that controls our balance of wakefulness, part of a pacemaker system that can switch between two separate neurological circuits - a holistic one for waking and the other, just involuntary, for sleep? Some creatures are, compared to you, master togglers. Dolphins, fish (including sharks) and birds can keep half a brain awake and half asleep at the same time. They take shut-eye open-eyed. They toggle, left to right, as they doze awake by night or day! Swifts navigate with half a brain, not half a mind. Some migrating songbirds may not even toggle; they fly by night, refuel by day but their 24/7 exertions show no effects at all of sleep deprivation! It is not consciousness but body (including brain) that tires. It is not that an electrical switch generates but simply facilitates the flow of electricity; or that a molecular switch such as orexin produces but *facilitate*s consciousness. Does your TV switch create the program? The presence of orexin is a switch that 'turns us on', that overcomes the 'little death' of sleep and from its burial resurrects us.

Top-down there is no problem with machinery. A busy reticular system may promote alertness by inducing the production of orexin; it may work in antagonistic concert with a sleep promoter called adenosine. However, neither they (nor any other systematic chemical) produce but rather facilitate consciousness. Wheels neither invent nor produce a car - they are simply a function of its business. Similarly orexin is part of the vehicle in which we physically drive; it is part of the controls that vehicle needs; it acts as a gear-switch, tuning fork or connector. If chemicals are coded to induce a resonance with mind, then orexin may transpire to be *the* psychosomatic wake-up molecule. Conversely, the purpose of its cyclical inhibition is to promote cyclical refreshment for the physical body (including its nervous systems) and defuse psychological stresses before the next 'reincarnation' of awakening. A radio works using an on-off switch and so, with orexin and adenosine, might the brain. Ons and offs and ups and downs are one thing. What about their source from above? What about their archetypal origin? Did chance inform the information system and, in most of us most of the time, its automotive bug-free operation? What's the real source of wakefulness?

The Third State of Consciousness - The Psychology of Dreaming.

'Sleepy Head' took the objective side. What about the subjective one? Are you ready for a drop to the diffuse conclusion of psychological entropy, for a descent into the underworld? You know what it is to be mentally as well as physically exhausted. You've often fallen off into mind's flat, dark condition we might call inertial equilibrium. The 'little death' is not a real end and so let's simply fall asleep.

Morpheus was the son of Hypnos (sleep) and god of dreams. The halls of

Morpheus amount, objectively, to the upper fraction of sleep's elastic 'bounciness'. The top of each vibration is characterised by a lack of the biochemical serotonin, lengthening periods of *REM*, theta brain waves (4-7 Hertz) and an increase in bodily twitching, perspiration, genital erection etc. Subjectively these periods consist in part of residual, involuntary streams of internalised consciousness called dreams.

It was noted (Chapters 0 and 14) that there exist two directions of focus, internal and external. External focus involves, by way of sensory and motor nervous systems, the operation of physical body in the physical world; internal waking focus is either purposive (called thought) or undirected (called day-dream). In contemplation voluntary, internalised focus of attention is refined; the subject is awake and able to control psychic events. In dream, on the other, contrasting hand, attention is attenuated and diffuse; the subject is unable to control a stream of experience of which he is only more or less aware. In sleep exchange of information with the exterior world (including one's own body) is blocked but this time involuntarily. Wakefulness is overwhelmed and, even if residual consciousness flickers internally, it lacks focus or direction. A subjective effect of sleep's 'shut-down' is to relieve the tensions of purpose, wishes and frustrations.

At this point it's fair to ask. If nerves alone evolve, secrete or otherwise emit an image how, in consciousness, do they perceive themselves? The same applies to dreams. The answer, *bottom-up*, must be that 'knowledge' is an electronic figment and 'experience' shows how atoms come alive! Matter lives! As objective is transformed into subjective seer is seen, thinker thought and both the product of a complex and intrinsically oblivious nature - state of brain. Mind is a material illusion. Life's turn-on is molecular, grey jelly is our one and only source of life. In this case don't nerves so obviously perceive themselves? *Top-down*, this is nonsense. In each of waking consciousness or dream, the answer is, 'No, they don't in either case; and never did or will!'

For dreamers dreams are real; but the experience is uncontrolled either by external trammels or the ability to think. A wave washes equally over whatever is in its path; a torch shone randomly around picks out disconnected or illogically connected objects and events. The files are scattered, narrative is blurred. Although dreams (the uncontrolled, internal generation of psychological events) can occur at any stage in sleep, their illogical disconnections stream in vivid profusion at periods of rapid eye movement (*REM*) sleep. Movements in *REM* periods seem to want to correspond with a dream's drama but are inhibited by induced and natural paralysis at spinal level. Successive 'bounces' become less pronounced as the moment of reawakening approaches. You then break surface into a relatively logical, voluntary association with physical reality.

Of what is the subconscious world composed but memories? This world is not *per se* irrational but, carried passive on an incoherent dream-stream, the observer's slumberous vision of it is. He's 'awake'. He sometimes recalls the jumbled narrative. There are even memories of dreams.

No doubt, like Darwin Freud was drawn to dislodge reason and then focus on irrationality - the former to explain our bodies and the latter mind materialistically. Chaos; unconscious planlessness; mutant forms arising from a dark and uncontrollable abyss. Subconscious memory and motivation may incite irrational, emotional behaviour or, in dreams, irrational experience but, as we'll see, this does not mean the architecture of subconsciousness itself is wild, incoherent or irrational at all. Indeed, the functional principles and structure of such objects and events as make up mind are just as real as those composing matter. Like bio-logic psycho-logic may be part of orderly and thereby rational design.

A foetus dreams almost continually. What of? Does the objective 'nervous physiology of dreams' require a dream or is it simply part of sleep? Surges of electric roam and activate, where they exist, personal memories; perhaps they even rouse the slumbers of an immemorial, common program, archetype. Does this mean you can dream an archetypal dream or memories are physical? Does it imply that an arousal of the metaphysical by a borderline of matter (known as electric charge) is how psychosomatics work? At any rate the various motions generate involuntary, confused episodes trammeled neither by the real world nor any systematic, intellectual organisation. A free-range, plot-less psychodrama is released. Not quite a random one; an element of free association swirls and in its stream half-stirs a trail within the pool of memories. Still more directed, now that salient memories are coupled with desire, is the streaming of imaginations everybody knows as daydreams. A sleeping dream is 'mad', daydreams try to dodge some psychological or physical frustration but, unless full-blown hallucination overrides the senses, such withdrawal to an inner world is checked by interruptions from the outer one. Dreams are experienced, in hindsight, as irrational.

While they are a 'safe', restricted sort of madness that derive from the associations of a sleeping mind, the projection of waking imagination onto the outside world also involves degrees of sanity. If, for example, we project a currently acceptable social or religious construction onto what we see, we are called 'sane'; and if we integrate an accurate response to circumstance with this construction we are also judged 'sensible'. In each case sanity is defined by reference to what a community believes is 'good' or 'true'. A 'sane' thinker allies more tightly than an unreasonable or eccentric person with an acceptable world-view but is there, beyond contemporary beliefs, an Absolute Perspective, an Original, 'Ground-State' Truth? Is the Greatest Good, against which sanity is naturally measured, a Single, Metaphysical Reality? If so then isn't knowledge of this Good the Absolute in Sanity? Or is the ultimate in truth a scientific kind? In which case everything is relative and the so-called Peak Experience, Enlightenment, is simply absolute insanity?

It is a question first of substance then degree. How do you see the nature of Truth and Reality? For a materialist reality resides in forces, objects and their physical events. To identify with these (including your own body) is the height of scientific rationalism's sanity. For a holist, on the other hand, it is materialism dreams. It takes illusion for the truth when, actually, elimination of illusion happens at enlightenment.

If materials are the primary reality then thought (or subjectivity) is secondary and, where it does not align with things, illusion-prone. But, on the other hand, are dreams, ideals and emotions such as love or hate less real or engaging than the body that they occupy? Are they less meaningful than scientifically measurable quantities? Materialism's affirmative, 'yes they are', inverts holism's estimate that they are not. No, objects are in fact subsidiary. They are important but, as means to ends, ancillary. The ends they serve are ideals or imaginations of one sort or other. Such purposes are powerful motivators; they, not objects, are life's engines and, in this sense, are at least as real as anything outside - and far more meaningful.

So dreams are real. They are coherent or incoherent imaginations but a dreamer is 'under their spell' to the extent that he is helpless to control them. He is lost to the extent that he is led behind their apparition. What, therefore, tells a phantom from the truth? What really constitutes the Truth and therefore Knowledge? What constitutes ecstatic truth? Is spiritual enlightenment a dream you wake from in laboratories? Or is the world a dream you wake from in the immaterial substance of enlightenment? As dream to waking, so is waking unto super-state. How do psychiatrists distinguish life from its oblivious, material apparatus or define its sanity from madness? How does psychology rate psyche or an element of soul? And if it's chemical then what would represent its state of Wakefulness, its grade of Analar?

Maybe one could understand that wakefulness on earth, including dreaming, is a question of both oscillation and degree - one that ranges in between the poles of a coherent logic and an incoherent lack of it. If so, repelled by chaos of the nether pole, what sort of logic is it best to choose?

While a person is awake, a combination of reason and physical reality tends more or less to check, correct or override the elements of chance or incoherence. When he is asleep, neither factor reins them in; they run amok. Deranged dreams are normal but, if he lives them while awake, the dreamer is unhinged from social and, maybe, physical reality. He is declared insane. More than occasionally, in relatively restrained degree, the tendency to day-dreaming fantasy cuts everyone adrift. Indeed, the need is profitably expressed by books, computer games and the whole entertainment industry. The end is always a release and, at best, an ecstasy. The question's one of substance and degree. What constitutes ecstatic truth? What is the right direction and to what degree? It is direction and degree that constitute the range of fantasy and its effect - because the truth of dreams is realized by living them. The scale extends from 'Peak Experience' down through shadows of frustration, irrationality and, at the wild and dark nadir, sleeping dreams or else the waking anarchy of bedlam.

Is this a scale of heavens and hells that, sited with awareness of a body or without it, make up the experience of life? Is life-on-the-run a struggle to elude its captors, to escape from different sorts of pain and prison, to stay free? We wake from nightmares to the sanity of a familiar circumstance; mystic reveille calls beyond the confines of existence into the concentrated form of Freedom and its Ecstasy. Is this the utmost sanity or, in the polar swing of Natural Dialectic, an insanity? In dialectical terms moods, morality and 'social sanities' swing round The Axial Sanity. Around this Pivot of the Self, the centre of psychology called Psyche, there are clustered lesser selves, *personae*, flawed characters at more or less remove. Rationality is not a clutch of wishes or opinions vent as 'rationales'. The more aligned is reason with Reality, then automatically less flawed and saner character becomes. It is, then, as with all philosophy, down to a Truth. An Absolute. Is it oblivion or utter wakefulness with which a rational 'insider' most profitably associates?

Are these two exclusive truths or complementary ones? Could you have the best of both worlds? If truth is, at base, material then everything derives from a non-conscious oblivion. Why should objective science not explore the nature of oblivion, the way material operations play, the 'psychology of non-consciousness'? If, on the other hand, the basic truth is wakefulness then awaken, sleepers, out of partial into full-fledged Sanity! It is the aim of a subjective science, using meditation as its tool, to staunch the flood of thinking that arises either in connection with the world or, when you disconnect the senses, in the mind alone. The object of this sanctified and practical psychology is to awaken, from the state of either sleep or waking dream, into full perception of the Greatest Good. In brief, you have two species of research, two kinds of knowledge, one sweet and crunchy apple in each hand - but ones of very different quality.

Let's return to dreams while sleeping! Mind is able, by way of cerebral controls, to interpret the signals your senses pick up from the outside - including your body. It can, equally, translate your commands into refined anatomical responses and purposeful rearrangements of that world. The topmost part of brain, the cerebrum, facilitates thoughtful response; the 'lower' thalamus, cerebellum and brain stem control thoughtless, involuntary transactions with your body. In a dream-state *PGO* (paroxysmal) waves originate in the brain stem and are pulsed 'up' through the mid-brain and cortex. Here they activate parts of the brain associated with sensori-motor, emotional and memory retrieval centres. The consequent internally generated images constitute chaotic films in the cinema of psycho-space. The random association of dreams is reminiscent of modern rather than classical art. Perhaps this is why some critics think that modern 'scribbles' are degenerate; they break the orderly, coherent reflection of a classical reality. Others think that, by blurring logic and the rules, a dash of incoherence makes a fresh and welcome style.

One thing is sure. Only reason develops physics-with-a-purpose; only discipline creates fresh and logical working constructions; and only sane inventors and a plan of action surely bring designs to satisfactory conclusions. This is why wisdom, in accordance with perennial truths of physics and psychology, is at a premium. Cosmic logic is the golden rule. Natural science, physical and metaphysical, must learn and educate accordingly. For the eccentric and unwise, of course, this is by definition not the case.

What use, therefore, is an experience of randomness, of the apparent chaos found in dreams? Are they a valve for pent-up emotional steam, agents of prophecy or premonition, humdrum 'organisers' of the previous day's events or none of these? If they need remembering then why does waking-up so soon forget? While their content is personal dreams seem, in general, to trail behind an archetypal process of detachment in a stream of letting-go. If, on life's shoreline, night's dark tide washes daytime's sharp-lined 'castles' into ruin, aren't dreams simply part of a constructive deconstruction - wiping clean tomorrow's slate? An ancient 'loosening' process that would represent collapse of edgy yesterday and, in this case, refresh, cleanse and renew. You could think of them as leaves upon the tree of memory that nervous wind sets aimlessly a-fluttering; or 'tummy rumblings' of the mind. In most prosaic terms, are dreams not for the most part a subjective waste, psychological excreta from yesterday's digested, or perhaps undigested, fare? Are they not simply disinformative as opposed to energetic waste? Their character, lacking rational direction and coherence, suggests this is the case. Their dialectical position on the conscio-material gradient would equally suggest that involuntary, residual flickering of consciousness within the chamber of subconscious state reveal no Freudian profundities. Their jumble augurs nothing. You may enjoy their opiate kind of relief but they are simply pointless cinema; they are, like yawns and stretches, a forgettable, mostly forgotten part of gradually waking up.

To the extent that it is uncontrolled or overwhelming any association or interpretation of the world, physical or subjectively metaphysical, resembles 'psychotic' chaos. This is why a young child, faced with such circumstance, naturally and continually tries to establish order, make connections and re-orientate by asking questions. The inevitable, involuntary trammel for 'wild' imagination is body and its exterior environment; the voluntary, internal trammels are emotional feeling (for an overwhelming ride) and reason (for a calmer, plotted one). The 'better' your reasons the smoother the ride. This is why scientists have always been interested in reducing adverse impacts of physical and biological events; and it is what propels philosophers on their pursuit of 'highest goodness'. It is also why common sense clicks with the dream-free, realistic point of view; and why the ultimate criterion of any mental shadow's quality is the source-less, shadow-less illumination of Enlightenment.

Of course, states of mind are more complex in detail than in the broad-brush principles that underlie them. Asleep, awake; internalised, externalised; voluntary, involuntary and so on; conditions of mind involve the combination, in various proportions, of absolute opposites, the poles of Super-state Transcendence and sub-state coma. It is the changeful range of reciprocal interaction between these opposites that makes our mind, according to tendencies and characters listed in the dialectical stacks, such an exciting metaphysical experience. *This, at least, is the theory of psychological relativity.*

If 'natural madness' is dream, delirium or an inward vision projected on the outside world then we are all to some degree insane. The issue is, at root, control. Purpose, will-power, reason and physical necessities all focus, more or less, potential anarchy potential anarchy into peaceful, coherent, productive orders called sanity. No doubt, for reasons psychological or biochemical or both, that nightmares, sleep disorders, waking disorders and irrational habits or life-style to some extent afflict every imperfect waking shadow, every human ghost. The question is, as always, how to make and keep a heaven free of what could easily become a hell. This, as the mystic knows, applies with full force in the world of dreams. Is not dream, and you in it, as much a fact as any 'outside' object? In the realm of Morpheus, the area of disembodied happenings, reality belongs to mind alone. In this case, to be subjectively 'on top of things' assumes importance that extends past sleep into the realms of sleep's big brother Hades, death. Is death a zone as dark, bizarre and uncontrolled as dreams? If mind does not exist apart from brain then what is '*post mortem*' but a fantasy, a dream of wishful thinking that does not in fact exist? If, on the other hand (as Chapter 18 shows), awareness is a thing apart then what does he, who in a body dreamt, wake to? You wake from the encasement of a dream; into what kind of morning, if the bodily encasement of this life were but a passing dream, would a dreamer when his dreaming had past awaken? In other words, why do the prophets cry 'Dreamers, you have lost the plot, wake up to what beyond the dream you've got - a stake in Wakefulness'?

The Fourth State of Consciousness - The Psychology of Deep Sleep.

Or is your truth oblivion? What about the anti-pole of Wakefulness; what about - instead of rising, dreaming sleep - the falling or abyssal fraction of a boundless unawareness? What about a nether zone of utmost darkness, Erebus, that hovers over death? What, that is, about an inner outer space, pitch lack of mind without the flickering shadow of a dream still less bright forms of day – profound, deep sleep?

In deep, non-*REM* (or *NREM*) sleep the 'upper', voluntary structures of brain are cut from the loop. A sleeper's movements, including eye movement, are much restricted; sensation is dull or absent. Brainwaves, the overall coordinators of the central nervous system, slow to between 0.6 and 3 hertz. These are so-called delta waves. Maybe deltas drop to zero. Brain death. If, by head injury, stroke, tumour or poison, the sleep/ wake toggles fail or signals cannot reach the forebrain then the patient drops into oblivion. The curtain falls but drama does not start again. Coma is an open tomb, an unpinned shroud or wake-less sleep but in its stillness deeper grooves of mind, archetypal constructions such as long-term memory, habits, instincts and archetype, remain unburied, frozen but intact.

Normally, however, after the first abyssal plunge from waking-state into the 'slow delta' of deep unconsciousness there follow, at intervals of about ninety minutes, four or five more 'bounces'. Their quickening is sufficient to buoy the sleeper back to the surface. The *NREM* fraction of a 'bounce' is low. It drags on buoyancy and shows as almost in suspended animation. Along with the restricted eye movement and virtual dreamlessness of this state breathing-rate, pulse, blood pressure and body temperature also fade. Consciousness but not memory is lost; and at the third bounce or there-abounce a sleeper sloughs the delta wave. (S)he is sufficiently unwound to have risen from the darkest shadows and to start to spring towards dawn.

Sleep is a Darwinian black box. Organisms sleep in different ways and yet, without dropping off into their 'little deaths', they'd die. Survival is insistent on it. How, though, did the physiology of *REM* or *NREM* sleep evolve? Rest assured, genes, other chemicals and switches are involved but sleep's a metaphysical affair as well. It knits up *mind's* ravelled sleeve of care. That's its purpose. Hormones and genetics might be triggers but they're not the whole affair. What, beyond a good night's kip, about the even more profound de-animations that extremes of season sometimes bring? Complex, synchronous controls accompany the processes of aestivation and hibernation. Instincts appertain thereto. Each package is a single one and less spells not survival but, as natural selection wordlessly avers, death. How could the myriad parts of package therefore gradually and randomly accumulate?

Sleep over, now we shift from the subjective to objective aspect of sub-consciousness. Do you remember (Chapter 13) the 'delicate disentanglement' between *chitta* (subjective thinker) and *manas* (objective thought-matrix)? The way that nature freezes time is, as you know, memory.

Frozen Time

You sleep but your past does not disappear. You wake and your past has not disappeared. You think you have forgotten, you may even suffer amnesia but the memories remain. Memories are the way we freeze time. **A memory is frozen time. It symbolically encodes the past.** *A memory is a thought object and, as such, has no life of its own.* A disc encodes music once recorded 'live'; it's a memory that, when replayed, affects the present and, from its effect the future. Thus mental memory is an encoded image; it is a record and, on conscious recollection, becomes a presence of past action that may affect the future. What applies to music may apply to plan or any form of creativity. Could you call memory indefinitely suspended animation?

Bottom-up mind is thought to be some aspect of dynamic brain. Not that numbers sum to consciousness but 100 billion nerves, each with 1000 branched connectors called synapses (each synapse involving with several functional states), mean your hemispheric brain is awesomely constructed; and it's suggested that, somehow, dynamic changes of connections constitute a 'nervous trace' of which the databank called memory's composed. You'd confirm that physical construction was the case because your brain seems to embed a small recording head; a hippocampus is about the size of a USB memory stick. How does this 'memo-phone' relay impressions into ever-changing nets of nerves? How does it cut a record's grooves; and, when you remember, how does cortex draw correctly from your library and splice the information into consciousness? It's hard to see how memory could be composed by nerves and yet dementia's a medical condition caused by faulty constitution of the brain. Loss of function leads to loss of memory and even body function; surely this proves that mind is made of nerves. Nerves are material. Why, therefore, need immateriality exist? Such disease means immaterial God does not exist; a kind of madness seems to yield the very proof that atheism craves!

Is this true? *Top-down*, brain modulates. It exchanges nervous body-code with mind; and, as a string is plucked, mind plucks nervous current in the brain. Symbolic thought or instinct's pattern is translated into signals that evoke (or don't evoke) a chemical or muscular response. Brain is like the cockpit's panel as we fly our world; it is designed to 'screen' events; its prescribed function is communication with material circumstance. Perhaps, in this case, nervous quivers are not memories themselves but indices established when a memory is stored or reflecting its impact, through brain to wider physiology, when recall occurs. Library addresses aren't the same as books; but if the linkage system is corrupted this, in proportion to the size of fault, impairs storage or retrieval systems and maybe, in severe instance, both. Similarly, if a panel of controlling dials and switches is compromised or fails no wonder that the pilot's 'screen' goes haywire and the flight is forced off course. Not that one faulty plane portends the end of planes; the blueprint of its type and its personal black box persist. Personal and archetypal memories persist beyond a broken brain; while archetypes embody natural law the personal imprints will, according to the strength of impress, fade; but brain ties consciousness to body and, until the knot is cut, there's no way to escape the cockpit of a life-time's flight.

The fact is that nobody's ever found a nervous trace of memory. Over a period of 30 years Karl Lashley failed to find material trace in rats and chimps but found that they remembered what they'd learned even when he surgically removed large portions of their brains. Hydrocephalic persons left with only 5% of brain have taken university degrees. No doubt there's some recording head where sensory impressions are recorded (perhaps the hippocampus) and if this is damaged then filing into memory stops; but if memories themselves are not specifically located nor destroyed by brain dynamics how are they engraved? Are they split and spread, as in computer storage, into different locations? Of course, if memory isn't physical it well explains why it has not been found, nor ever will, in flickers of the nerves.

There are different kinds of memory. Habituation means that an impression drops below your conscious threshold; sensitisation means the opposite - more reactivity. Since even one-celled animals do these you don't need brain for memory. Learning certainly saves memories and recognition uses them. Indeed, storage and retrieval (recall, recollection) are thought's staples and they happen all the time. There is as yet no answer as to how remembrance physically works; though aspects that connect with body need a 'read/write head' to make connection with perhaps the whole memorial structure never will be scientifically found - because the main part of its system isn't physical.

Recollection is one thing, endurance another. Memories, whose species include short-term, personal and long-term, archetypal forms, endure. The latter, it may be argued, are innate, read-only files that are laid into the fabric of a being; they are like carrier waves of an archetypal broadcast. And the former are like individual modulations on the waves of such stations; they are read-write overlays, filters, records of varying endurance and intensity ranging from the ephemeral scratch-pad that supports moment-to-moment orientation through time to sharp, deep emotional impressions that colour a lifetime's responses. Either way, the past is present. It is, quantitatively, the airs and ocean to a boat; it is 'larger' than the superficial presence sailing in such eddies as it makes. Subconscious memory is the context, undercurrent or 'deep structure' that buoys up each moment, sways the vessel and generates a portion of the vagaries with which our conscious voyage has to deal. Yet, at the same time, it also acts as helmsman's helpmate; it is the decisive captain's first lieutenant on a ship as it forever falls across the world's edge, surfs the world's wave, skims the boundary of time and flies into a further presence that had just been called the future. Memories endure beneath the stream of voluntary operations and through the dark, involuntary night of sleep. They await the call to live again and, unless unprompted and forgotten, will. 'Fade' of a personal memory is inversely proportional to the depth of its original impression or emotional importance; and conversely, like any drill, the more rehearsed it is the more clearly etched and more quickly retrieved - until the point that it becomes a habit. Memories constitute a personal file of experience, a reference library, your database of learning without whose recollections you would forever be, like a slate from which impressions were immediately erased, waking from a lifelong sleep. No recognition. Blank. You would, if nothing were retained, feel always at your story's starting line.

A memory can be activated from above or below. From above, a thought can 'bring back' memories; from below, sensations trigger them. Either way, recollection may trigger instinct and its associated nervous, hormonal and behavioural effects. Both thought and sensation can also create memories. Either way, such records inform. They constitute a reference database for purpose and intelligent response. They are the latency without which an organism would be just a reflex object lacking guidance. It would also be, in proportion to its lack of personal record, short of any detailed, individual personality. Indeed, as a log of past mental constructions memory forms the context within which the present is lived and, as its result, the future is shaped. In Indian psychology memories are called *sanskaras*. *Karmic sanskaras* are impressions, good and bad; their sub-conscious grooves create tendencies and patterns in the current of your thought and thence behaviour. They affect your actions (*karma*) and from these fresh circumstances that perpetually (and fatefully) arise.

Remembrance is not, of itself, a state of mind. Rather its re-cognition is an integral part of all cognition. Whether archetypal or personal, the 'static' patterns of memory give conscious motion shape. Out of mind or in it, stored or reactivated, memories are manifold, essential and ubiquitous controls. They are reefs, banks and channels in the shoals of mind; or, if you prefer, they are mind's vehicle of state, the carriage of its presence or containment of its stream. The imprints of memory constitute no less than a metaphysical world, the internal universe.

How efficient is your video-recorder or, if you like, your hologram? You press a button to record; no fingers in a head but press again and play it back. Whatever the particular tape, you press the same keys on a control panel. Just as you can see many different videos, all looking outwardly the same, on the same screen, so a multitude of memories are stored and then replayed through your own customised, 'hard-wired' studio. Analogous storage and recall buttons are built into the brain. For example, you wouldn't want a bilateral medial temporal lobectomy because you would lose your hippocampus, parahippocampal gyrus, amygdala and adjacent portions of each temporal lobe. Although you would still have frontal lobes that seem to facilitate the retrieval of memories to consciousness, the effect of abovementioned surgery might be to cripple your capacity to file new long-term records - although those filed previously wouldn't be destroyed. Ever-present, archetypal instinct would remain intact but new personal memories not be logged. In computer-speak a 'write' failure would have occurred. As already noted, the standard neurological explanation of memorisation is that impressions are 'encoded' by modifications in the connections (called synapses) that occur between neurons; indeed, such changes might reflect fresh allocations of address.

Repetitious use of pathways will 'habituate'; disuse may cause a memory to fade or corruption interfere with its retrieval. This all fails, however, to explain how nerves can 'hold' subjectively experienced information; how the same 'brain-buttons' can, as if they were also tapes in a recording studio, not only record and play back but also hold myriad different facts, songs, records or whatever you wish to called stored information; how or where the 'read-write' recording head works; or how the excision of sections of brain tissue can leave memories intact. It would be interesting to see, if an accurate excision denied access to a memory, whether its reinstatement caused the memory to 'return'. Would the brain simply resurrect access to a metaphysical memory that had never disappeared? Or would you reconstruct the memory by reconstructing a nervous circuitry? In which case is there a circuit for every kind of memory? Or is everyone's circuit for the same memory (e.g. the letter 'A') the same? Would a poorly reconstructed memory (of, say, a parent or child's face) render it unrecognisable? Forever lost?

From a *top-down* point of view memories themselves are metaphysical but correlated nervous circuitry acts as a storage-and-retrieval system that, by association, allows the library of remembrance its efficient, selective operation. If either 'storage' or 'playback' buttons are inoperative, an embodied system fails. Either records are not made in the first place or become poorly accessible or irretrievable. But, once made, a metaphysical memory is, as a mind-object, physically indestructible. This does not mean it can't be reinforced by repeated similar experience or 'learning by heart'; or that, conversely, it can't, by retrieval and re-storage, be gradually falsified. Who can fully trust their memory - not least when, by retrieval's absence, impressions fade and will be forgotten anyway? Secondly, it does not mean the instrument of brain itself, as well as interfacing personally constructed memories, is not constructed as a medium for archetypal instinct and related, type-specific records. Indeed, you might definitely expect it was.

An analogy with biological archetype is apt. Individual bodies, in our case human, derive from an aboriginal, archetypal program associated with an egg; and from cleverly encapsulated information stored on a database made of a chemical called *DNA*. Do not genes code for enzymes? Are not enzymes agents of the application programs we call metabolic biochemistry? If chemicals are involved in memorisation then so, of course, are genes. The components of neurons and neurotransmitters need to be present and correct in the right quantities for 'ignition'. For example, a protein called *NMDA* (N-methyl-D-aspartate) is supposed to help orchestrate the flow of nervous morse from one neuron to the next. And if the materials or components of your video machine were faulty you would not expect its faultless operation. The information it conveyed would be disrupted, its message inaccurate. So mutant genes and disease may, like weak weariness or age, affect remembrance.

A psychological analogy is also apt. If mind is prior then brain would correlate with how a humanly embodied mind should work. In other words, the construction of sub-conscious mind is a 'chip', framework or archetype (see *figs.* 15.8 and 15.9) through which mental energies are directed This unsleeping archetypal memory is, because the body parts are built around and integrated with it, non-falsifiable; it is as if a principle round which expressions automatically coordinate; and (see Chapters 15-17) is continuously in subconscious motion with the body's correlate, its biochemistry. *It is suggested that 'signal translation', instinct, facility for language and the psychological aspect of orientation skills are, for example, prepared formats into which personal data is filed.* Indeed, it is further mooted that this general 'psycho-physical operating system' includes the facility to store personal data out of which the context of an individual 'self' is composed. 'Personal' is, as already noted, the nurtured extension built onto innate mental patterns, that is, onto a 'typical' archetype.

Of course, such layered construction cannot even have occurred to nerve-chocked materialism's brain - of which mind is but intangible, subsidiary excrescence. Therefore, let's rephrase. We are designed to remember, learn and behave in certain ways called human. Personal experience is laid into the framework or, if you like, the carrier-wave of human generality. Natural

Dialectic terms the subconscious archetype 'typical mnemone' and its personal overlay 'personal mnemone'. We'll meet the word 'mnemone' (see Glossary) quite a lot and it needs be made clear at the outset - the Dialectic's usage is totally distinct and unconnected with evolutionary psychologist D. Campbell's (a so-called 'cultural replicator' like R. Dawkin's memes).

Let's underline. A reasonable analogy for the couple of dialectical mnemones has already been derived from radio. As separate radio channels are each 'carried' on the fixed wavelength of their own carrier wave, so the psychological details of a type of organism ride on an 'archetypal carrier'. This wave broadcasts 'read-only' files laid into the fabric of a life form, say, a human being; and the 'general program' is overlaid with specific, personal 'read-write' records - memories of varying endurance and intensity. Individual, ephemeral variations (personalities like you and me) may come and go but the channel broadcasts on; it amounts to a continual reverberation or refrain. Time is not exactly 'frozen' by this fluid channel but the background data and the wavelengths are fixed; they remain, as with general form of body so with form of mind, the repetitious same. Nor need this channel, archetype or carrier wave be only human. Such a circumstance would stand in every organism's case; and (Chapters 7 - 12) in that of physics' basic entities as well.

No doubt that, looking darkly through the glass, brain scans have shown that unconscious perception stimulates parts of the brain that conscious perception might similarly activate; no doubt, also, that all unconscious, autonomic action passes underneath the bar of notice but is noted by the nerves of brain. Chemical connections tick in dormant organisms just the same. Information systems do not necessarily require a brain. How do such continually sub-conscious organisms operate in rational ways? Are plant tropisms (such as growing towards the light or water) purely chemical? Cannot amoebae hunt and farm? Problem-solving slime moulds forage; like an external form of memory their trails impart the knowledge 'do not visit here again'. **Since when did *chemicals* embody will-power to survive?** Yet such a will is basic to biology. *We shall return to brainless 'brains' because it is assumed that every organism has one; that is, an archetypal channel is what every cell of each body is built round.* This view argues with not one material finding; but it revolves the paradigm of pure materialism through one-eight-nought degrees.

From a *bottom-up* perspective, you remember, everything derives from matter, body and therefore, in the biological aspect, nuclear code. Therefore airbrush out the pilot; delete any non-existent ghost in the machine. Watch switches click and levers move without an operator and yet, it all seems, purposely. Automatic motions of control! You probe the vehicle's cockpit; you inspect a flying brain. But can you see what I am thinking? Can scientific pincers grip a thought or an electric probe translate what pulses in the wiring mean? To try and corner subjectivity objectively is 'looking darkly through the glass' but what else can you do? Ghosts are not research realities; factors metaphysical and immaterial navigators slip beneath the scans. Therefore all you have and ever will have as you take apart the cockpit and make tests to see how it must work are hippocampus, neural congregations, *NMDA* and so forth. A pilot-less machine; design without a systems engineer. This is a contorted story; this is the conceptual vision that materialism presses on the atoms of your brain. These atoms, you are told, can think. Their thinking's yours; and yours is actually theirs. Mind's strong-bonded to its molecules. No real difference exists. These, where brain *is* mind, are objects that by virtue of their orderly construction sum to the experience of life. No ghosts fly this machine!

***Top-down*, on the other hand, brain is not mind.** Its role is, like the captain's instruments, one of interface. Do you remember Penfield's probe? Of course, sensation makes its nervous mark. So does motor traffic moving, mind-to-muscle, out the other way. A human brain is preconditioned to 'compute' in human mode; and you need storage to compute. Why should nerves not link this storage to the body's action? You need to index memory so that on cue call-up happens; you need to label information and, as cells are indexed to produce the proteins that they need, it seems that brain is indexed with the stamp of neural pattern linking it to memory. And you need, in addition to the preconditioned pathways, to register experience and build a library of context that's your own. You need to register the input with specific and appropriate links to all the rest in order to survive. This database you live in and you call 'myself'.

In this view nerves are only half the game. They are the half that's physical but psychological capacity, events and data-items (such as, respectively, intelligence, feelings and memory) are metaphysical. But if they're neither *DNA* nor biochemical then what's, for ghost's sake, left? And is this sort of nothing unimportant? What, for example, *are* the memorial records, the videos, the library of data items which, by the million, a cerebral video-machine incessantly accesses? What commands which tapes, where is the library filed and what, or who, is the subjective screen? *Who* remembers? Who scans potential input or retrieves relevant output; by what system are the necessary filtered from unnecessary data entries; and by what mechanism does the search, association and selection of required information accurately occur? What, in other words, is clicking switches, pulling levers and responding to requirements of a flight through life? Fact: nobody knows.

Discrimination is critical. The brain/ mind blur needs resolution into two clear parts. Therefore, if you don't like half-cut answers, let's revise the way the issue is presented.

You will have received an invitation to believe that, over billions of years, chance has produced your body from terrestrial clays. You will also, in analogous manner, have been invited to believe that thought somehow 'exudes' from neural networks and that each record in your database of memories is, by definition, just a different pattern of connections stored in a few of your hundred billion brain cells. Each thought and each sensation, weak or strong, ripples changes through that computational web. Is not memory therefore, so the reasoning runs, simply an ability of *brain* to store the learned effects of its experience? In this view memory, like consciousness itself, is simply an electromagnetic by-product of special atomic configurations. You are the same robot that your brain and nervous system are. Just as the purpose, perceptions, feeling and meaning of your living experience are reducible to a network of electrical pulses percolating a few dollars' worth of chemicals, so is the presence of your past. It is as if the whole meaning and contents of a video had been vested in the plastic, metal or electricity of its tape and mechanical operation. What other invitation can a neuroscientist send? There is no way to materially prove that anything immaterial does not exist nor, however strong the inference, that it exists. *Therefore, unscientifically, such a philosopher will have decided that matter is the lone component of existence and that neither the immaterial consciousness of mind nor sub-consciousness of memory can, for all their informative importance, exist independently of brain.*

This is not, of course, the logical inference Natural Dialectic draws; the reverse, it is a deluded misinterpretation of the facts. Top-down, brain acts as a mind/ matter linkage system. It also integrates past with present both fluently and, according to the kind of information required, incessantly. If brain were a library's hardware (paper, shelves, storage and retrieval systems or books themselves), then software (the actual information conveyed by books) and the readers constitute its immaterial mind. Brain is part of a psychosomatic information centre that, as well as linking past to present, links mind to matter. Information is stored in code and is only meaningful to a reader or writer, that is, to a conscious or at least subliminal, sub-conscious involvement. Yet something, in this most dynamic of informative enterprises, must process and order the volumes of data. Books are classified, catalogued and shelved according to a predefined program. Similarly, something must smoothly link current sense perceptions, emotions and thoughts with associated memories in order to make comparisons and decisions. Such cybernetic processing is central to the evolution of immediate, balanced, 'sane' responses to exterior events. It occurs, according to Natural Dialectic, across a psychosomatic border, that is, it is *partially* a nervous exercise according to a system of classification reflected in the way different 'shelves' of the brain are hard-wired. The hippocampus, for example, is associated with the memorisation of details, various parts of the cortex with different sensory patterns (e.g. the secondary somato-sensory cortex for tactile patterns), the amygdala with emotional experience and the cerebellum with the storage of records involving learned sensori-motor skills such as balancing, catching etc. In this view the brain does *not* store memories. It tags them. It provides the sorting-addressing-storage-retrieval tag necessary to handle any particular datum according to a pre-arranged (archetypal) operating system. It acts for human or other biological organisation as does a physical library with its linkage system called a catalogue. This is *not* the same as information itself. Modern science tends to confuse 'library' (with its physical hardware designed to store coded materials catalogued and arranged according to software called 'classification') with 'information'. *Perhaps the distinction is subtle but it is also real and fundamental.* It does not mean that neuro-physiological research is misguided or its answers wrong, only that such study concerns the objective apparatus of *IT* and not the meaningful, subjective part. Do you remember Mozart's concerto (Chapter 5)? Bio-psychology, as the books call it, involves only a material part of the whole, greater truth.

Top-down, therefore, mind is a nucleus for the immaterial storage of informative creations; and what are books or other kinds of coded pictograph but materialised memories? They are storage vessels, passive records of such active information (in the form of ideas, purposes etc.) as first created them; and round these collections, for safe-keeping and availability, the body of a library is built. In other words, metaphysical core is reflected in physical containers and round these accumulate further 'storage and transmission shells' that make up library, data bank or other information fix. *Books and libraries are made by men but memory is the natural way of storing, without physical agency, information.*

Where is the 'place' of storage? Can you see how, as opposed to more energetic liquid and gaseous 'levels' of behaviour, energy is 'stored' in solidity; it is locked up in objects. Where, however, is the 'place' of *information* storage? It is in passive records that amount to deactivated but informative objects. In physical terms this means inscribed paper, vinyl, tape or (Chapter 6: Machines) any purposely organised material; no doubt, in this sense, you (that is, your body) are a book that carries print called *DNA*. In psychological (or metaphysical) terms storage means memories. *Memory's 'level' is the place of frozen thought, 'solidified mind' or sub-consciousness.* Its structures constitute a type of psychological glue that, time-wise, fastens past to present. Just as it lends time context so its psycho-space affects both conscious thought (above) and physical behaviours (below). Its blueprints, archetypes or instincts are a form of code so that, sited at the psychosomatic border (*figs*. 6.1 and 15.2 - 4), it is assigned to the passive 'rules and coding' level' of creation. Such plans are, as Chapter 6 describes, written in potential matter and the architecture of sub-conscious mind is (with *fig.* 15.8 and 15.9) described below.

Metaphysical files, sub-conscious objects and mental logs are the 'stuff' of memories. An item stored in your own mind is, by definition, inactive. Prompted by events or reactivated in recall, it is only resuscitated by an act of remembrance. In this act it can be reformulated and, as such, re-stored - thus our memories may undergo shifts over time.

A hologram is, you may remember, a 3-d photograph made with the help of lasers. Unlike a normal photographic image each part of it contains the image held by the whole. The whole is in each part and each part in the whole. This is, dialectically, the way of memory and, importantly, of generic, archetypal memories in universal mind. This mind's records, its relevant images, its holographic but passive information is accessed by individual cells and bodies. Mind's records serve as implicate order that's expressed explicitly in how things physical behave. Could it be that individual minds reflect the structure of an archetypal cosmic holograph; and their implicit information is exchanged through a translator, brain, with body and its physical surround? In short, the interface involves both physical and metaphysical components.

It needs be re-emphasised that an archetype is a memory and, as such, a thought object. And, if sub-conscious mind is memory, neither it nor archetype think or feel anything. *An archetype, as lifeless as a stone, has no life of its own. A memory is, whether universal or individual, a mental structure that of itself no more contains consciousness than a test tube or a church.* It is, simply, like the record of a song, a frozen piece of time - one that is played or, if you will, accessed incessantly by as many users as there are physical operations or cells of its type. Its prescribed vibrations transmit order. Its programs specify both the fundamental shapes of matter and, in a more complex series, forms of life. You may think of archetype as any organism's brainless brain. Such various, energetic records activate the cosmic 'dance' but if you ask when any specific disc was recorded, what the song means, who made the record player or what drives it you are moving away from, not towards, the roomful of sensational music - which musical box contains the sole physical evidence for anything. This is the position of science when it analyses orchestrated sound waves, studies molecular biology and takes all kinds of measurements but also asks questions about the cosmic box that may have metaphysical answers. It's Mozart and the scientist again!

It is not, as already intimated, only you who can remember anything. Birds do, bees do. Worms, sea-slugs and jellyfish do. The ability to focus on one object alone is called selective attention. Primates have it, so do dragonflies to track and intercept their prey. It can't take much of a brain. How, though, can a neuron or small bunch of them exhibit concentration and experience the hunt? A fly also functions in broadly the same way as a human. It senses, eats, sleeps, navigates, learns, memorises and remembers using a brain the size of a mustard seed, that is, a 'control panel' of only 250,000 neurons. Whether or not such a compact operator can think of, dream of or remember banana mush, one doesn't actually need any brain at all. If you thought you did then you're in for a big surprise - and a profound problem. Single-celled microbes aren't nerve cells yet organisms like amoeba do not move randomly. They optimise their hunting/ gathering chances; indeed, one species, *Dictyostelium discoideum*, can farm bacteria as a crop. Computer modelling has shown the angles at which *Daphnia* forage as optimal for success. And single-celled protozoa like *Paramecium* and *Stentor* can respond, learn and presumably, as part of the process, remember. What brain-partless sort of subjectivity is theirs? Who can learn without consolidating memory? Perhaps instinct and experience do not depend on nerves. What, therefore, substantiates behaviour; and what, if any, state of matter could produce it?

Many animals show conscious sensitivity and volition; they can respond according to memory, develop habitual response and, if they can recall, also forget. On the other hand, all organisms show involuntary (sub-conscious or unconscious) reflexes of one sort or another. Both fleeting and enduring sculptures of sub-conscious mind are called, collectively, memory. All organisms have archetypal, instinctive memory that includes their strategy for survival. No thought is involved. The psychology of many, such as fungi, plants (with their complex responses to many signals) and bacteria, may appear dormant. Nevertheless, while they never 'wake up', they may respond in less than entirely robotic ways to stimuli, communicate and perhaps even take initiatives. Sensitive, variable response sets life apart from the absolute automata of physics and chemistry. A great part of mind is bound up in sub-conscious memory and it is, therefore, instructive to try and understand its nature.

Bottom-up you are bound, of course, to reiterate that memory is the product of atoms, neurochemistry and therefore, ultimately, genes. You cannot possibly believe it in the case of unicellular, nerveless *Stentor* but, in the complex case of humans, you might be persuaded that a neuron is the basic unit of consciousness. Neurons join (or synapse) with one another to cause electrical patterns. It is true that, just as muscles grow when I decide to exercise, so sensitisation to a certain course of physical (i.e. muscular) or psychological behaviour may cause a change in the number and connective patterns of synapses. Conversely, habituation to a stimulus may involve a reduction in transmitter substance released across a synapse. Is such synaptic variability the mechanism that constitutes the physiological basis of memory? Is it the cause or, as muscle growth derives from the exercise of mind driving body, simply the effect of mental exercise? Again, the thin-lipped answer is: nobody knows.

It is possible that science today understands the metaphysic of mind in the same fraction as, three hundred years ago, it grasped chemistry? *It is certainly easier to explain away than explain memory in terms of nervous relay and storage.* How is information recorded (or memorised) in a nervous unit? Hum a song. A tape is nothing without playback mechanisms but, most importantly, a listener. If, as Penfield suggested, the brain is like a tape recorder would a particular cell remember a particular note or a set of nerves a refrain? Could, as his electrical probe might seem to have suggested, hundreds combine to compose the whole song, any 'commemorative pictogram' or, in transmission, a thought? Think of a well-loved face. How could will-power or imagination have created the image out of neurons? Or, if faculties of mind are simply sets of neurons, could they have 'boot-strapped' to first memorise and then subjectively remember it? Could you rate some neurons more 'intelligent' or calculate intelligence by quantity of neurons actively combined? In this case if you chopped a neuron or a neural circuit could you cut thought's process up, sever the components of belief or quarter memory? It might be sensible for a surgeon to see the brain in terms of grey matter; but it is not subjectively, psychologically or philosophically sensible to see the human information centre as physical alone.

It seems, as K. S. Lashley showed, you cannot quarter memory. He found it impossible to destroy physical access (by electrical stimulation) to memory if any cortex is left. In other words, complex habits and conditioned reflexes can survive great damage. This discovery does not sit well with Penfield's suggestion that memory might be wholly composed of nervous circuits. *In fact Lashley abandoned his original notion of a reflex theory of learning. So cortex looks like a channel or medium and memory more like a metaphysical hologram.* If one part of a hologram is snipped off and illuminated the whole of the original scene is reconstituted. It is also possible to superimpose images by making holograms in a solid. Each image is present throughout the solid yet can be extracted separately. In this way billions of data items have been stored in the volume of a cubic centimetre. Moreover a hologram acts as a kind of coding/decoding agent, a lens that resolves an apparently meaningless blur of light into a coherent image. Could 'holographic action' sub-consciously convert 'electronic morse', itself a translation of a range of sense perceptions, into the mental picture that you see and even seem to be part of? In this analogy each memory is linked so that, subject to the 'angle' of a prompt or the degree of similarity, effectively immediate association could be made. What light might lighten memory's hologram? Could it be coherence, focus or a resonance of mental energy?

The full suit of individual archetypes would constitute a 'super-hologram' called universal mind. Metaphysic does not need the space and time of physic. Imprints in mind-stuff (*manas*) no more need a correlate in matter than a broadcast on the media waves needs listeners; but antennae will locate and earth the programs to a well-tuned set. This is, in fact, the object of the exercise; what broadcaster creates a corporation and broadcasts his programs if he does not have an audience? Transmitters need receivers. In this model what are bodies (or their molecules) but mind's antennae? Aerials. *Brain is seen as a connector in between dimensions. It is a 'set template' onto which physical patterns are mapped during their exchange with metaphysical.*

Conscious mind projects, films or, to vary the communications model, acts as a two-way radio. Recording is continuous. Its archive is memory and memory is myriad clips of film or tape filed accessibly. Its operation of perception includes an archetypal 'cognitive' structure that decodes physical (nervous) signals into a form that can present them to consciousness; or, conversely, translates thought into nervous morse. Such a mechanism of linkage (*fig.* 15.8: signal translation) would involve an association that is able to exchange information between subtle matter (electromagnetic behaviours) and sub-conscious mind. In such construction mind is the informative film, tape or broadcast and brain is the physical instrument of two-way translation - a recording/ play-back head. A single 'co-mechanism', although its complex operation may be spread over different parts of the brain, simultaneously services both image-creation and the storage/ retrieval of myriad 'clips' or 'memorised moments'. Thus memories themselves are neither filed in nor perceived by brain. Neither, although they are channelled through and (in our case) humanised by it, are sense perceptions or motor directives.

It was proposed that, as biological bodies are constructions employing universal matter, so individual minds involve principles of universal mind; and that the wireless nature of mind, either individual or universal, can be likened to a spectrum of light. It ascends from complete obscurity (the 'black flatness' of non-consciousness) through infra-conscious to 'visible' and ultra-conscious wavebands. These modulations involve both state and quality of consciousness and, at any given level, its specific content (e.g. the events of a dream, objects of a perceptions or depths of a feeling).

Natural Dialectic locates memories in sub-conscious mind. It explains the image of memory (either in static or dynamic 'video' form) as a creation sculpted from a relatively coarse 'waveband' of *manas*, a record imprinted on 'mind-film', footage 'canned' on inert qualities of mental energy; and it identifies this sub-conscious, psychosomatic energy as potential or transcendent matter. You may have already made the dialectical equation of potential matter with (*figs*. 6.2 and 13.6) oriental *prana/ qi* and with sub-conscious 'holograms' of archetypal memory in universal mind. Are these (as Chapter 10: The Order of Nothing suggests) closely related to or just the same as that ethereal substance, *ZPE*?

Memory, the only form of metaphysical information storage, is the shape of infra- or *sub-consciousness.* Indeed, objectively it *is* sub-consciousness: sub-consciousness is made of memories. It is an organism's library of precondition and conditions that comprised its past, that is, its personal experience. The precondition is its archetype, the basis of its sort. We might call this sort of memory 'typical' while 'personal' experience includes both active (created and transmitted) and passive (received) information. Files are 'written' to the personal library with a sharpness of imprint proportional to their impact or importance to the owner. And they are filed according to their association with other data in the owner's bank. Whilst focused, deliberate access is called *remembering*, unfocused and involuntary awareness of memory is called *dreaming*. The latter includes physical response to dreams such as sleep-talking or -walking. Below these dreamy shallows there plunges imageless and inconceivable abyss - to levels of deep sleep and even sub-sleep, coma. The latter are still metaphysical as opposed to physical non-conscious conditions. At this level reside general, archetypal or 'read-only' impressions, that is, the body-linked archetype explored in the following section and, especially, Chapter 16.

Memory is not, by this account, a 'frozen' medium like a photograph. Nor, although it comprises a concept or basic expression of an idea, is an archetypal memory. Memories operate like a movie and store programs that, once triggered, may unfold like stories in a sequence to their completion. Any stored plan is such a memory. The aspect of biological archetype called a morphogene is a striking example of such dynamic memory. *We have all heard of instinct but morphogene, instinct's 'downward' correlate, is yet another important, revolutionary derivative from the proposal of a logical, conscio-material spectrum. Its nature will, as a part of archetypal memory, be elaborated over the next eight chapters of the book.* It can apply, as routine or subroutine, to various biological levels. Its main or 'master' routine expresses the typological form, such as dog, cat or human, to which various sub-types (say, genera or species) may belong. From the master are spun a suite of subroutines that define the different aspects of a type's overall body such as energy metabolism (including digestive and respiratory systems) or reproduction with associated organs. *In this sense you might, in computer terms, dub a typical mnemone an archetypal program.* Programs are, of course, although dynamic still a frozen form of mind; and they're replete with information. They specify, without error, the most efficient means to a well-defined end. **You might argue in this vein that biological structures are codification incarnate; and that the concept they express is an archetypal program**.

Let's further inspect the specific psychology of recorded time, sub-conscious memory, and after that its mechanism of storage and retrieval, remembrance.

Psychosomatic Linkage

Top-down sub-consciousness, sandwiched between conscious mind and physical systems, is practically inaccessible from either side. It is, in the first case, subliminal and in the second, above which it is ranged, sublime. *However, just because something is hard to access or measure scientifically does not mean it neither exists nor impacts our reality.* If (see Chap 0, The Light in Your Eye) there is more to the information business than electrochemistry, what is it? We noted (Chapter 13) the profound explanatory shortcomings of materialism with respect to conscious experience. Check, however, with Natural Dialectic's suggested explanation of archetypal memory as given in Chapter 8: Are You Certain? We've also noted the correlation between Natural Dialectic and the suggestion by neuro-physiologist Sir Charles Sherrington that there is no greater inherent improbability in the human consistence of ***two*** fundamental elements, both mind and matter, than matter alone.

Objective and Subjective Components of Mind

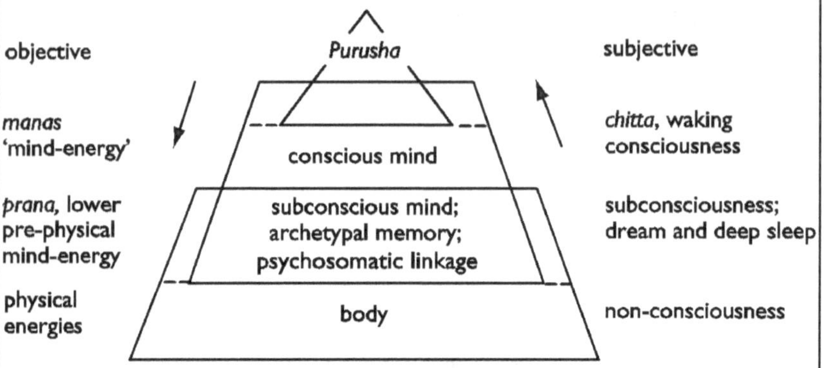

Check this ziggurat against another formulation of the two informative and energetic aspects of mind (*figs*. 4.2 and 6.2). Check also Chapter 8: Are You Certain? to review the way psychosomatic link might work. Archetypal memory (or archetype) is bio-logic's upper tier.

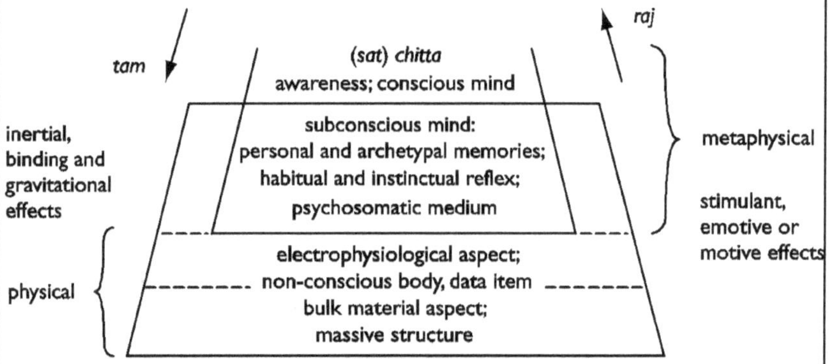

This ziggurat shows memory as the link between conscious mind and body. It includes both subjective and objective aspects of sub-consciousness. The former is sleep but with respect to the latter the diagram introduces a logical but revolutionary derivative from the conscio-material spectrum - the 'dormant' position of archetypal memory. It shows the hierarchical place of its synonym, typical mnemone. At the same time it identifies this channel of order with both the passive, 'coding or grammatical level' of (↑) sensation or (↓) creation (Chapter 6); and with 'top' or potential matter. Potential matter (Chapters 7 - 12) is the universal archetype whence issue the various basic patterns of physics and, with respect to complex, coherent 'bundles-of-laws' called organisms, biology.

Archetype is a generic term. For example, a single biological archetype (e.g. for a human) may nest hierarchical layers of integrated subroutines. These are archetypal sub-routines (or sub-archetypes) in their own right.

fig. 15.2

In the past a systematic explanation of sub-consciousness was elaborated from above (*top-down*) through the practice of psychological disciplines of concentration such as *pranayama yoga.* Its most experienced authorities claim complete voluntary control of normally involuntary functions such as heartbeat and metabolic rates along with Patanjali's so-called 'psientific' or 'psychotronic' powers of remote-viewing, healing, telepathy and so on. They describe consciousness and sub-consciousness as a spectrum of informative capacities. Everyone is familiar with the focal power of a *plexus* called the eye-centre because we are awake at it. It is the source of human consciousness. Light is spectral and its visible component shows about five colours; similarly five lower plexuses are described (see Chapter 17) as receivers/ transducers of five 'frequencies' of sub-consciousness. Matter is not spectral but shows about five states - vacuum, gas, liquid, solid and, as with light, purely energetic. Thus five 'frequencies' correlate with states of matter and, in corresponding spheres of predominance, five functional areas of the human body, five organs of sense and five of action.

From below (*bottom-up*) an objective assessment is completely different. How can you 'get a physical handle' on mind? Access ascends as far as electronic or electrochemically framed patterns of charge in motion. We can peer from outside through a glass darkly. Voltage, current and associated electrochemical effects carry us to a glass ceiling. They carry us to the threshold of sensation but we cannot enter the subjective house of another's mind. Indeed, to a materialistic rationale the exercise were futile. *Prana, qi,* sub-conscious and conscious energies do not exist. A 'conscio-material gradient' is the height of mumbo-jumbo. Flat-earth consciousness is all a matter of chemicals or, at least, subatomic particles and force! Yet what, at quantum root, are *they*?

A materialist can neither perceive nor conceive of a 'gap' which separates psychological from physical worlds. If, however, what he terms illusory exists, this 'gap' is crossable. Indeed, it is crossed at every waking moment. The crossing (*psychosomasis*) takes place with perfect ease and without any discernible lapse in time. Hand-in-glove. If mind and body synchromesh like gears together it is a prime necessity to narrow the 'gap' between our *top-down* and *bottom-up* versions of their interaction. In brief, we need to identify and clarify the nature of the point at which 'I' exchange information with the physical syntax of a nervous system - where mind meets matter. *What is the structure and function of the psychosomatic interface?* Is it possible to integrate a yogic system with the discoveries of biology, chemistry and physics, especially the quantum physics of subatomic particles, energy and forces? *How might the integration of psychic and physical approaches to the relationship of mind with matter affect our understanding of human construction, function, health and healing?*

A dialectical stack that describes mental body (mind) might read as follows:

tam/ raj	*Sat*
existence	*Essence*
body/ mind	*Soul*
conscio-material gradient	*Pure Consciousness*
prakriti	*Purusha*

↓	*tam*	*raj* ↑
	objective correlate	*subjective experience*
	what is experienced	*experient self*
	non-conscious context	*conscious centre*
	akash/ 'psycho-space'	*attention/ focal beam*
	manas/ 'mind-energy'	*chitta/ awareness*
	mind-film	*projector*
	image	*image-maker*
	thought	*thinker*
	memory	*remembrance*
	dormant pattern	*instinct activated/ felt*

and lower down, as elaborated in the next stack:

objective correlate	*subjective non- experience*
fixed mind/ memory	*subconsciousness*
dream-images	*sleep*
prana/ infra-conscious psychosomatic medium	*subliminal psychology/ habit/ instinct*
morphogene/ physical schema	*psychosomasis*
materiality/ physicalia	*oblivion/ lack of subjectivity*

How *do* mind and matter interact? To subject the psychosomatic interface to microscopic analysis we need try to understand its constituents. In order to anticipate semantic problems, however, we need first to open the black box. Its 'field' of sub-consciousness is unfamiliar territory. It has not been systematically mapped. Where exploration has occurred different systems of psychology (e.g. Freudian, Jungian, Buddhist or yogic) have used different terminology. As the words woman, female, lady, wife, mother, daughter etc. can be connotations of a single person so different words are used to describe aspects of the sub-conscious mind. In such description various words (in the same and different languages) are sometimes used to describe the same thing. Furthermore the same word is sometimes used to describe different aspects of the subject and, most confusingly of all, used inconsistently by different authors. In science fashions change but at least the words used (e.g. corpuscle/ particle, vapour/ gas, ether/ space or element/ state of matter) are English. However, there exists a fine heritage of oriental psychology and philosophy whose relevant vocabulary needs, on top of English words, to be defined and integrated within the terms of Natural Dialectic. For example, you may find *prana* or *qi* used in some contexts, incorrectly, as a synonym for *Logos*, *Tao*, 'breath of life' or animating spirit; in others as a vaguely defined 'universal energy' or 'vitalising air'; and in others again, especially Chinese, some almost animistic 'genius', emotional characteristic or 'spirit of place'. Concordance is required. And in this case the Dialectic, for reasons mentioned above and those elaborated over the next three chapters, restricts its definition to the Glossary's.

Check *figs.* 6.2, 7.5 and 13.6 and their contextual notes to revise the dialectical way the words in *fig.* 15.2 and its subsequent stack are used to help describe the duality of mind as it oscillates between its positive (conscious) and negative (non-conscious) poles.

Like atoms, every item in existence has its vibratory signal. In an analogy with wireless broadcasting mind's energy is not electromagnetic but involves a spectrum of *manas*. If you analogise human awareness-of-the-world with the waveband of visible light, you might liken its 'super-conscious wavelengths' to the 'transparencies' of UV or X-ray beams; and the sub-conscious 'black light' of unawareness-of-the-world with 'sub-luminary' infrared or lower wavebands. In other words, while the psychological activator is awareness and the physical activators are excited fields of force, the psychosomatic activator shows as what oriental literature names *prana* or *qi (tai chi)*. *Prana* is therefore identified as Natural Dialectic's sub-conscious energy; also, at this point, bear in mind the *PSI* (psychosomatic interface) and the further correlation of *prana* with 'electro-dynamic' *ZPE.*

Two people may describe the same physical object in ways that barely match. It is not surprising that a degree of precision is lost trying to describe a feeling, an emotion or even a dormant metaphysical entity like *prana*. It is this entity which, in order to optimise health, the breathing exercises of *pranayama yoga* seek to exploit. This is because *prana*, translated as 'breath' or 'vital air', is considered to drive the flywheel on whose motion living, in its aspect of biological function, depends. Respiration is also likened to a bellows at the fire of life. Indeed re*spir*ation derives from the same etymological root (and therefore idea) as *spiritus*, whose sense is a current of air. '*Animation*' derives from a Greek verb meaning 'to breathe' and is possibly related to a noun (*anemos*) that means, again, a 'current of air' or 'wind'. Life's an *animated* breeze! Beware, however, confusion if metaphor is taken literally. *Prana* is not air; nor is it spiritual in the sense of elevated consciousness or soul.

A *physical* proximity of oblivious, metaphysical *prana* is often identified as *electrical charge - the automatic informant of physical shapes*. In yogic literature *prana* is also generalised as 'universal energy' and associated with life, sunlight and the visible light spectrum as well as negatively charged ions. Electrons, currents of electricity, light and electromagnetism are aspects of charge. The 'spark' of polar charge is latent, locked or active in all things. In its association with the electronegative charge of oxygen *prana* seems to represent stimulus, vitality and the reactive side of biochemistry. *Wired lightly to the sun life is, of course, dependent on the sophisticated transformation of electromagnetic to biochemical energy (ATP).* Electrons (in the phosphorylation process), hydrogen and oxygen are active components in the recharging of biological batteries. Oxygen is not, however, just the 'breath of life' in a respiratory context. It is a polarity-creator not only in water, the special medium of life's metabolism, but also in phosphate and the functional groups of biochemistry. If you want sparkle, insert oxygen!

Does this mean that *prana* is simply a pre-scientific formulation of electricity or of biological energy? Is it a generalised description of what modern analysis has resolved into much more precisely identified components? Or is the concept more profound? In Chapters 6 and 7 a correlation was made between (*sat*) pre-physical, potential or transcendent matter and the 'grammatical' level of passive information. This level was identified with the laws of nature, the virtual ground in which the harmonic oscillations of phenomena are rooted. As four dimensions define the extent of objects and events in space-time, so a fifth and perhaps more define the character of its inner, metaphysical extent. This fifth dimension was further correlated with (universal) sub-conscious mind, memory and the same laws or patterns of nature. These correlations were developed in Chapters 5 - 12.

Could *prana*, therefore, be a single, archetypal instrument from which issue the patterns of nuclear, electromagnetic and other sorts of energy that constitute our world? In this case the pre-physical energy, the fifth-dimensional source from which those forces 'emerge' is identified as immaterial *prana*. Its sub-conscious, psychosomatic energy is conceived as a low-grade 'diffusion' of *Om,* that is, low voltage on the cosmic grid. As sleep to waking so *prana* to *Om*. While *Logical Om* is conscious spirit and the First Cause of all existence *prana*, at the base of mind, is the first cause of its physical part. It is the 'breath' of universal body and in this case 'breathes life' through archetypal memory. Memory is 'fixed' or 'solid' mind. *Om*, transduced to the frequency of memories, vibrates with the underlying harmonies that are broadcast into physical quanta and, by aggregation, classical forms of matter. Mass and energy, forces and events - all issue from the same orderly source. Therefore sub-conscious energy (*prana* or *qi*) sends the order, gives the rule, informs the universal field. It is the intangible 'extra' that controls and, by the coherence of its order, generates a cosmos greater than the sum of forceful parts.

The nub of the issue resolves to the way in which so-called *prana* interacts with matter. Where does it touch our physics? Check *fig*. 2.13. This suggests, with *fig*. 15.3, the *PSI* (psychosomatic interface) is 'entangled' with the zone of physics we call 'quantum'. **Through quantum physics science comes in touch with universal mind**. Potential matter is the archetypal phase of immaterial fields. These fields are (as Chapters 7-12 describe) universal; but specific, complex archetypes compose the substrate of biology. You naturally, therefore, press the vital question - how are these translated into forms of life? **Natural Dialectic logically (as developed in this and the next chapter) and decisively connects at the level of least mass and therefore subtlest matter - electrons, electromagnetic fields and therefore light**. Radio. Aerial linkage. What might constitute antennae for informative exchange?

Conscio-material Ups and Downs

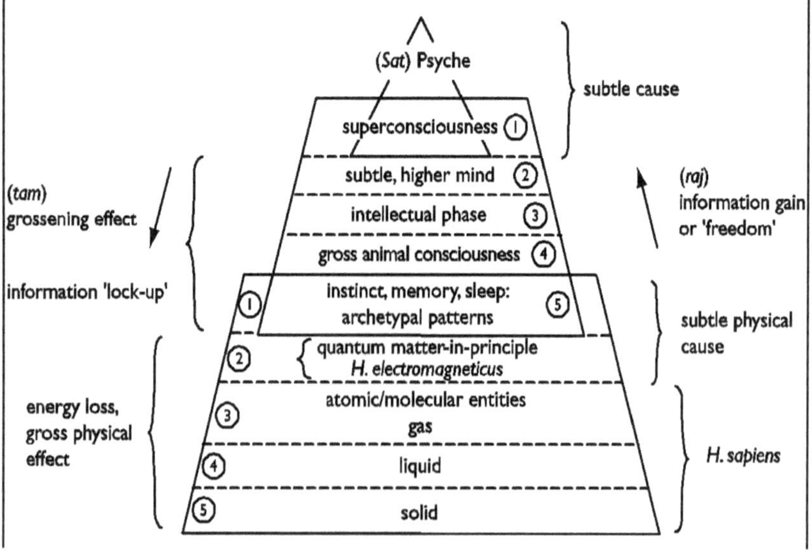

In the case of information, 'gain' refers not to the quantity but the quality of information involved. The more 'important', principled or fundamental is the information, the greater its power and, therefore, gain.

In the case of matter, 'gain' involves not the quality but quantity of energy supplied for stimulus, 'freeing up' or release. This includes both large-scale impetus and collision but also, 'higher' on the scale of matter, non-massive, 'subtle' stimuli such as light, heat and electricity. Although they are neither large-scale, visible nor violent, subtle quantum effects are dialectically rated 'above' gross external changes. In other words both gross and subtle materials interact but in the aboriginal order of creation subtle precedes gross.

In your case *H. sapiens* refers to the gross aspect of your body. Could there exist subtler, internal, informative aspects of this body called *H. electromagneticus*, *H. archetypalis* or even, when it comes to waking 'mind-set', *H. conscious?*

fig. 15.3

Its sub-conscious nature is why *prana* is sometimes, where animated bodies are concerned, vaguely and maybe confusingly called 'bio-energy', 'life-force' or 'etheric nourishment' and its sphere of operation identified as the 'etheric or spectral body'. The dialectical view is more precise. **Just as a continuous supply of electricity powers the 'nervous system' of modern civilisation, so at root electro-dynamic interactions, reflecting archetype, inform and power biology**. A wireless 'etheric body' coincides, hand-in-glove, 'within' a bioelectrical *Homo electromagneticus* (see *figs*. 15.3 - 5 and Chapter 16); and the 'secret' of archetypal *prana* is its ability to optimally align a body's electrical coherence and, thereby, its associated gross physical shape against a sub-conscious template. Such resonance charges it with what we perceive as health and energy. The *pranic* or sub-conscious body amounts to a subtle, organising infrastructure. It is an internal 'shell' or wireless anatomy. *It is the memory that constitutes an organism's archetype, more commonly known as its 'vis medicatrix naturae'. It is therefore perceived not only as an instrument of information exchange that operates through the appropriate nervous code but also as the dynamic of metabolism and morphogenesis.*

↓		↑
tam	*raj*	
physical	*metaphysical*	
light of sight	*light of perception*	
oblivion	*life*	
matter/ body -somatic	*psycho-somatic/ mind*	

The first principles of psychosomasis are thus already clear. *Mind precedes matter; quantum energy of the vacuum mediates all quantum and electromagnetic processes whose particle is the photon; and, where electro-dynamics describes the effect of moving electric charges and their interaction with electric and magnetic fields, biological electro-dynamics precedes, in the sense of underlies, all bio-molecular and bulk, classical considerations.* **Every biological process is electrical; and the flow of endogenous currents is the rimary and not secondary feature of physical life. Not biochemistry but**

quantum biochemistry heaves to the fore. Natural Dialectic lifts perspective from molecular to vibratory, field perspective. It is at wireless level patterns of subconscious mind meet and influence matter; archetypal information is relayed to chemistry by electricity and light.

tam/ raj	*sat*
physical/ objective correlates	*metaphysical governor*
non-conscious effects	*sub-conscious cause*
classical/ quantum matter	*potential matter*
material expression	*archetypal memory*
physical cause/ effect	*psychosomatic medium*
sensation/ action	*psychosomasis*
physical body	*morphogene*
H. sapiens/ H. electromagneticus	*H. archetypalis*
↓ *tam*	*raj* ↑
informed	*informant*
classical level	*quantum level*
structure	*function*
gravitational influence dominant	*radiant/ levitational influence uppermost*
base level	*PSI/ interface*
massive body	*'network of light'*
external body	*behavioural 'driver'*
anatomical shape	*e-m/ biochemical patterns*
H. sapiens	*H. electromagneticus*

It is useful to subdivide the wireless e-m spectrum (X-rays, microwaves etc.) and the bulk, bonded conditions of matter (gas, liquid and solid states). Similarly (*figs*. 15.3-5), running from 'high to low', wireless mind can be subdivided into a spectral continuum. Its lowest division, the 'solid' zone of memory, organises physical expressions. It 'tops' or sources the bodies of our starry universe. A binary dialectical stack describes biological bodies in terms of <u>*pranic*</u> archetype (*H. archetypalis*) and its control exerted through two lower 'sheaths' (electrodynamic-level *H. electromagneticus* and bulk-bodied aggregate called *H. sapiens*).

The Grades of Man

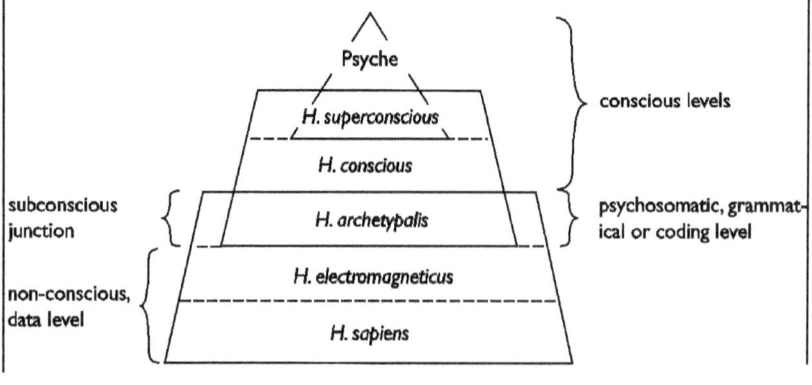

Man is seen as a series of sheaths (or bodies) nested one within the other. The subtlest is at the centre; potential information is finally realised, the construction is completed by the outermost, gross shell.

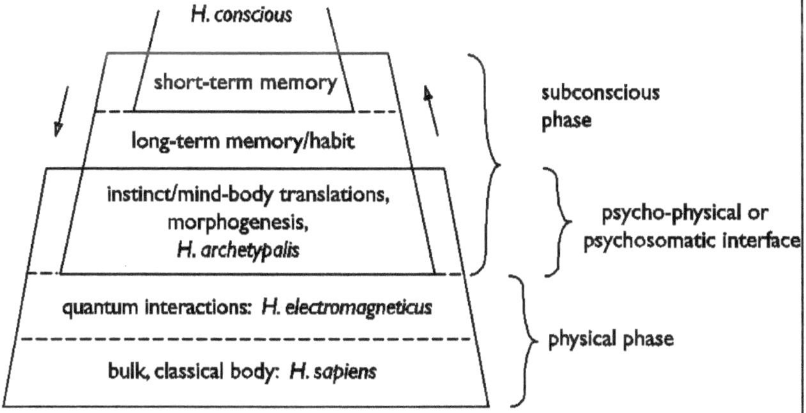

The sub-conscious phase is divided into a 'light' personal archive and a 'dark', indelible archive. The latter is like a *ROM* (read-only memory) compared to a cache or *RAM*, an active 'read-write' form of memory. It is to the indelible, instinctive aspects of mind that cellular, biological 'resonances' are linked. You could think of cells as transducers of mind (in the form of recorded programs or memory at the fixed archetypal level) into biochemical and thence bulk, orderly objects and events. Some organisms (including humans) are built to express an added, more flexible level - consciousness. In this case the exchanges of information between thought and body occur through the medium of sub-conscious mind (see *fig.* 15.7).

fig. 15.4

The three bodies represent, effectively, phase changes whose special, lowest case is a bulk, sensible aspect of the others. That is, the outward form of body whose hands you can see holding this book and which is called *Homo sapiens* is, in fact, an expression of inward layers of command and control.

Between the inner, informative and outer, energetic domains of existence there occurs an effortless transmission, an ever-open exchange of signals, a communication that keeps the whole show (which includes every part at each level) in orderly business. The source of physical order has been identified dialectically, as archetype. An archetype is one of a series of memories that collectively comprise the lowest zone of universal mind. Like personal memories, an archetype is a 'solid' or 'lasting' feature of mind. It both binds the moment of its creation with any subsequent, present moment and links mind with matter. *In other words, the past in the form of universal (not personal) memory has been identified as the sub-conscious channel from mind to matter*!

Wireless Man

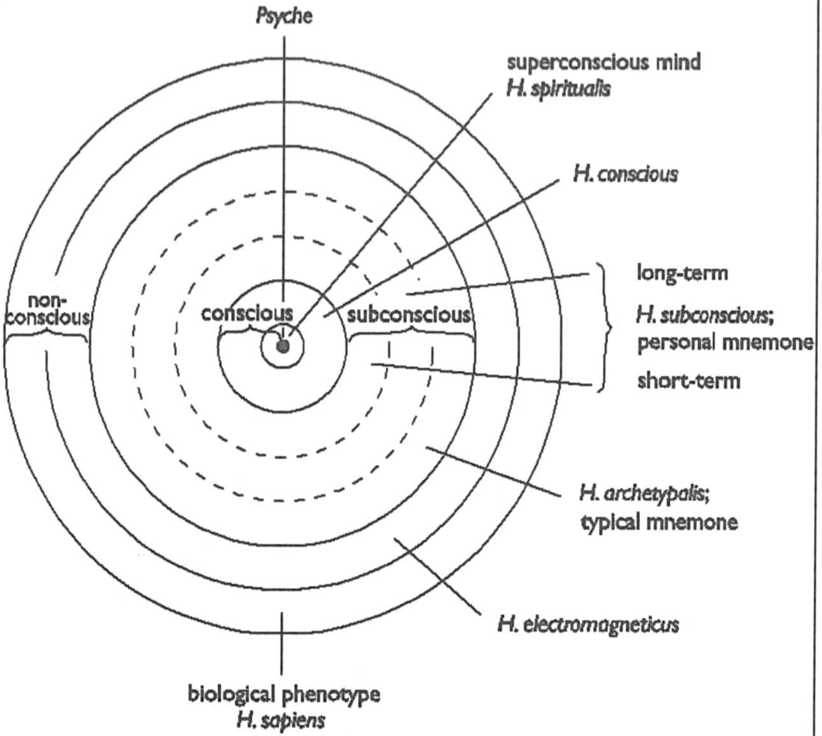

In this energetic, spectral representation of human grades, sheaths, shells or bodies only *H. sapiens* is wired, fixed or made of classical matter. The others are wireless or radiant.

Thus you might visualise yourself (and any other organism) as both wireless and wired anatomy, that is, as a composite of fixed and fluid sheaths. As discrete electron orbitals surround a nucleus these sheaths surround the Nuclear Psyche. Separated by 'an exclusion principle' their phased 'energies' materialise towards the periphery of creation. There the most dense, wired level - called your body - is obtained.

fig. 15.5 (see also 16.1)

Such an assertion raises a host of questions. What *is* memory? What are its components and how does it work? What, as well as its constituents, is the structure of the psychosomatic interface? There exist neurological aspects of memory; of what psychological material, however, is a memory or, in universal terms, an archetype made? What is the archetypal relationship with instinct and, possibly, biological development and morphogenesis (the structural development of an organism from a single cell into its adult shape)? Conscious instructions are translated into nervous morse to which muscles respond; incoming morse is translated into conscious perception and thought. What has the sub-conscious medium to do with this normal to-fro (sensory and motor) traffic between mind and body? How, in other words, is psychosomatic energy channelled to articulate with matter? How do metaphysical broadcasts combine sub-conscious with quantum interaction? What physical feature could act as an antenna through which mind's 'radio' information is exchanged with electromagnetically 'wired' matter? And by what protocol do its wireless broadcasts interact with local wired (i.e. receiver-transmitter) sets called materially bonded bodies? Is there any kind of answer in the air? *Such questions constitute the foundation of psychosomatics, that is, the new psychobiology.* The next few chapters will suggest a perspective and framework within which detailed query can be pursued.

	tam/ raj	*Sat*	
	mind	*Pure Consciousness*	
	life form	*Soul*	
↓	*tam*	*raj*	↑
	lower	*higher*	
	sub-conscious	*conscious*	
	a memory	*its creation/ recollection*	
	dormant	*awake*	
	instinct	*reason*	
	unlearned	*learned*	
	typical	*individual*	
	universal	*specific*	
	impersonal	*personal*	
	rchetypal	*experiential*	
	body-linked	*soul-linked*	

One day, after prolonged and concentrated research, science may fully understand the physical subtleties of bio-information technology. Yet saturated materialism excludes all metaphysical aspects of awareness and ascribes the whole construction to chance and chance's butcher, natural selection! Holism does not. A *top-down* perspective takes us back to the drawing board. Its emphasis on non-chance (i.e. design) starts from the subjective point of consciousness and refreshes a number of hoary questions. From this view it is obvious that at least the initial outline of an explanation needs be sketched from which to develop a logically consistent theory of 'the sandwich medium', to shed light into the black box and to develop a detailed picture of its inner, opaque operation - the working-area of sub-conscious mind. Indeed, it amounts to a key prediction of Natural Dialectic that science will one day discover the mind-matter border mediated by an orderly exchange between immaterial, archetypal patterns and material patterns of photons and charge (in the form of electrons and ions). The latter couple link, of course to molecular and large molecular aggregates.

Variation derives from theme, specific from general and practice from principle. In this case the 'soul-principle' is enmeshed with 'body-principle'. Soul (pure life or consciousness) is everywhere one and the same but individual bodies are numerous and different. Each 'drop of the Infinite' is wrapped, sheathed, embodied and thus separately personified. This embodiment is serial. Each case involves a series of interactive, meshed 'sheaths' each of which is successively 'tighter', 'lower' and grosser than the one above; and each such 'localiser' is constructed of the material of the realm (or level) at which it is to interact and work. Thus 'soul-linked principle' is identified as a conscious *ego* operating within the context of memories. And 'body-linked principle' is recorded in archetypal memory; it is from this generality that specific bodies are derived. You, therefore, amount to a life form composed of specific 'life-principle' enshrined in 'body-principle' specified by archetypal, genetic and environmental factors. This means, in practice, a particular embodiment of mind; and, translated into commonplace terms, the trinity/ tri-unity of a soul in a mind in a body.

Two people can, as previously noted, describe the same thing differently. Part of the business of science is to sharpen descriptions into a quantifiable, testable, universally acceptable format. Instead of using the ziggurat you could translate man into an energetic model, the concentric rings of a conscio-material continuum.

Today's physics teaches that human flesh can be viewed both as a solid *and* a lattice of force fields. And biology teaches that cells and tissues function as chemical batteries that all sustain a potential across membranes. This creates body-wide field patterns. Now, therefore, imagine the nuclear mass of your biological body, *Homo sapiens*, air-brushed out. Imagine a material framework stripped except for electrons, electronic interactions, current and mobile ions. From this perspective biochemicals are electrical patterns rather than objects. Such patterns are associated with electrical disturbance that can, with motion or the energetic 'jumping' of electrons, show as radiation. Of course, as every soldier with night-vision goggles knows, warm bodies radiate heat as an infra-red 'aura' of light. There exist subtler, less obvious weak radiations and, where the whole spectrum of electromagnetic radiation (not just the tiny band visible to our eyes) is generically known as 'light', we might call the pattern of your 'larger dance', that is, of the quantum level of your body's operation *Homo electromagneticus*. In this sense the relative abstraction called *H. electromagneticus* constitutes 'a body of electricity' or, with its biophotons, a 'network of light'. This is the lower half of the psychosomatic equation. The other, metaphysical half with which the 'network of light' interacts is a record of the natural idea according to which an incarnation's purposive pattern is predisposed. It is a framework, a broadcast, an archetypal memory that in the human case can be called *H. archetypalis*. Your subtlest physical pattern is thus understood as a vibrant reflection of programmed information.

It may be remarked that the level of *H. archetypalis* in *fig.* 15.5 corresponds to a factor called, in *figs.* 15.7, 15.8 and 15.9, the morphogene or morphogenetic template. The question first tentatively raised in Chapter 7 and now addressed in force is how this morphogene (or biological archetype) interacts with its physical counterpart. Is Mandelbrot's fractal geometry a form of mathematics that helps to describe it? Where and how does morphogenetic resonance react with chemistry, especially with information inscribed on a chemical, *DNA*, and thus called 'the physical book of life'? Is it a double helix alone that generates not only protein but, perfect to the position of each proton and electron, such functioning metabolism, organelles, cells, organs and structured systems as compose an integral form of life?

Synchromesh 1 - Awareness and Memory

Instincts for home-building, reproduction, offence and defence are critical for survival. They are ubiquitous. Always allied to physical sensors, processors and effectors they are expressed in forms as varied as the bodies they support. Are they physical? Is instinct subconscious and subconscious mind material? Is it a facet of the atoms, a dependency upon an organism's genes alone? *Myriad examples might suggest there's more to tricks than matter.*

Star of David: Psychosomatic Linkage by Vector

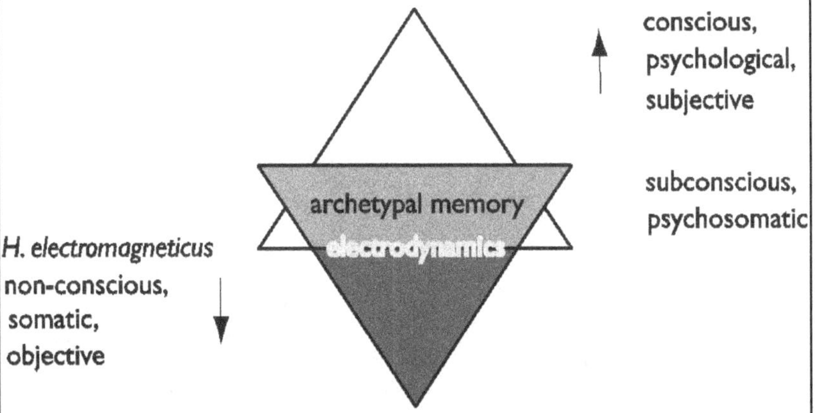

The Star of David, interpreted as a dialectical symbol, represents two interlocked arrows. It shows the (*tam*) downward and (*raj*) upward fundamentals in balance. If the top section is taken as the domain of conscious mind and the bottom as physical body, then the grey portion corresponds to subconscious linkage. *Fig.* 15.8 is an expanded Star of David. Memory (including the *typical mnemone*) corresponds to the light grey/hatched portion of 15.6.

The psychosomatic linkage between mind and body is of special interest. The subtlest, lightest form of matter involves electromagnetic fields, massless light and almost massless charge-bearers, electrons. **Thus Natural Dialectic logically, and therefore decisively, identifies the mind-matter interface with (bio-)electro-dynamics**. The nature of this connection - between *H. archetypalis* and *H. electromagneticus* - forms the substance of Chapters 16 and 17.

fig. 15.6 (see also 2.6)

You need all factors to cooperate at once. Less is fatal. How could gradual mutations leap and instantly rebuff mortality? How could blind graduality grind out an aggregation of complex, specific patterns of behaviour associated with coherent physiology (on all its levels from a molecule-in-cell to system) at once? Without which any species would, at its first generation, die? Even species number one. Nor can learned skills, even when a parent's present, be passed on by genes!

Psychosomatic Linkage by Domain

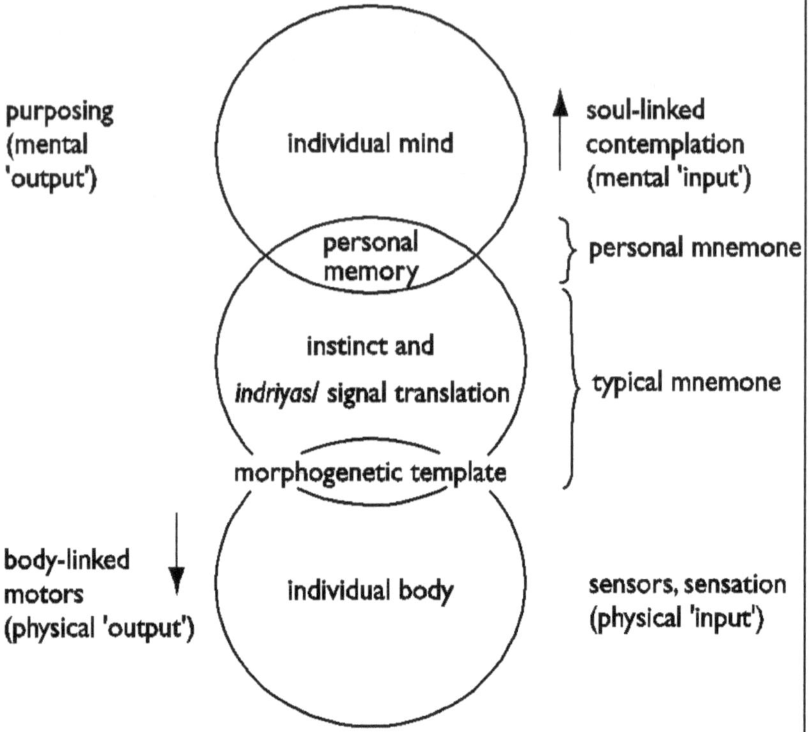

Three domains of human being: conscious, sub-conscious and non-conscious. This diagram indicates the order of linkage, through sub-consciousness, between mind and body. Various *personal memories* (such as language, habitual behaviour, familiar faces and objects) are each inactive until reactivated (brought to consciousness) to provide context for conscious mind; such memories can be short-term or long. On the other hand the three aspects of *typical (or archetypal) mind* are sub-consciously active; they are, like genes in a genome, permanently associated with every cell, all development and all multi-cellular bodies. 'Signal translation' is the subroutine that provides for an exchange of information between mind and body; they comprise the channel that translates information (for example, nervous information) into perception or, conversely, decision into the morse that culminates in specific muscular action. *Instinct* works upwards into conscious mind; or downwards through the faculty of signal translation and morphogenic subroutines to engage with cell-constructed body.

fig. 15.7

Top-down, information is metaphysical and body physical. *Linkage divides into two sections - an upper and a lower.* These two basic sorts of memory are called, for the sake of dissection, *personal and typical (or archetypal) mnemones.*

Synchromesh 1 involves the upper link between conscious awareness and memory. It is elaborated in the next section - The Personal Mnemone. This mnemone is specifically 'you and yours'. Its composition derives from individual mental or physical experience. **This higher, more inward cog gears memory, in terms of storage and retrieval, to conscious mind.** It includes short, medium and long-term records of mental activity. Its conjunction is therefore with your cosmological axis, point X, the 'power-point' at which you think.

For saturated materialism the only obvious form of information is genetic. **How, therefore, did mental, immaterial and not material instincts evolve? In fact, their complexities comprise another major black Darwinian box.** Why, on the other hand, try and squeeze it all into just genes? It doesn't fit. *Within Natural Dialectic instincts are easily explicable as immaterial, psychological programs.* So it remains to suggest the architecture of these programs (their 'chip circuitry') and the nature of their linkage to the body in the form of mnemones (see Chapter 16: Synchromesh 2; Psychosomasis and How Does the Connection Work?). If, in this context, archetypal program is materially developed by a co-operative, genetic one then it is incumbent on a *top-down* explanation to suggest the form oflinkage between such mnemone and associated body. *Materialism deals in chemistry; such quaint metaphysic as mind-matter linkage is axiomatically discarded as irrelevant.* But if there's linkage Natural Dialectic would suggest that this runs, Janus-like or anti-parallel, between subconscious and non-conscious media. The subtlest relevant forms of non-consciousness (i.e. matter) are, it has been suggested, light and electricity.

Suggested Architecture of the Subconscious

Two analogies are useful when thinking of the psychosomatic interface. Using the analogy of a mind machine, the computer, its '**conscious** *CPU*" (central processing unit) is not the machine itself. It is the inventor or user who employs the machine to serve his purposes. A computer as sold (or born) already involves immutable logic expressed in the form of operating system and various chip circuitries. **The algorithms of subconsciousness are likened to the programs of an operating system**. This informative software, varying with each type of organism, constitutes the typical mnemone; instinct is physically expressed by typical behaviour and morphogene by the typical hardware of a body-system. On the other hand, personal files (of the memory cache called personal mnemone) accrue through individual usage.

Using the analogy of radio a typical mnemone is the immutable 'carrier' frequency or broadcast wavelength (e.g. *H. archetypalis* or *Drosophila archetypalis*). As individual transmissions are carried on a general frequency so personal rides on the back of typical mnemone. It comprises a unique store of records for any particular organism with sensibility.

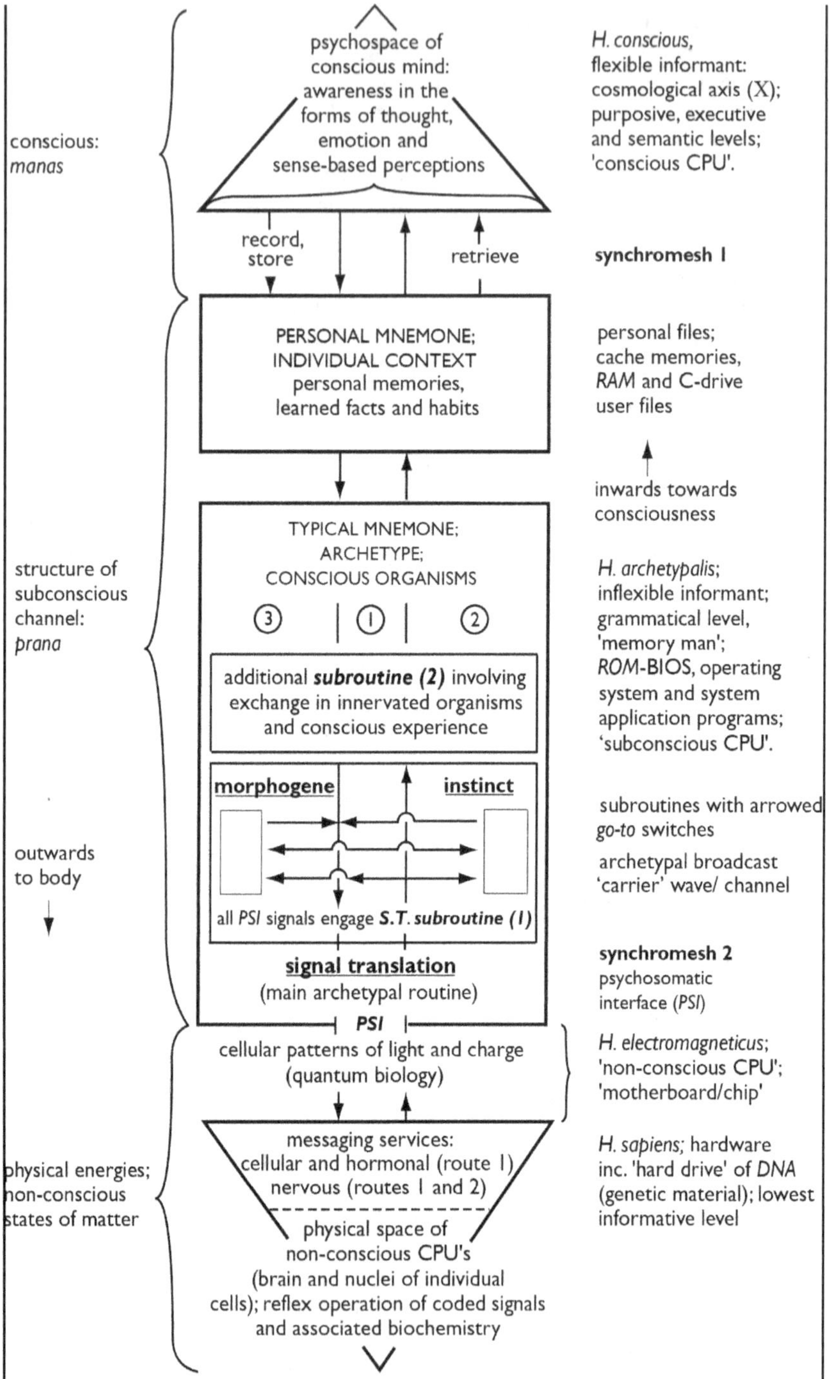

Mind is an information centre. Although it may be represented by material vehicles such as nervous pulsation, ions, hormones and so on, information itself is a metaphysical arrangement. And the natural mode of metaphysical storage is memory. The two main divisions of memory are called, dialectically, mnemones. The flow diagram above illustrates a possible 'synchromeshing' interaction of such mnemones with awareness, body and each other. It draws together the 'upper' personal and 'lower' typical characteristics of man. The upper 'personal mnemone' is an aspect of memory where images specific to an individual are stored. Such images provide a context continually and sub-consciously referenced by association. In fact, a personal mnemone might be viewed as records built up by an individual. Such files are 'read-write'. Data may be stored, retrieved (or remembered) and manipulated before re-storage. It is, in effect, a library of information against which all incoming impressions entering consciousness are 'weighed' and against which outgoing commands are 'steered'. Personal mnemone can therefore be viewed as a lens, filter or context built from a person's experience which 'colours' the way he or she thinks.

You can also view a second layer of sub-consciousness, the lower 'typical mnemone' or 'archetype' in this way - except this lens is, no more than a body, built from personal experience. If personal is 'weakly bonded' to typical mnemone, the latter is strongly bonded to every cell in every body that it represents. This 'typical mnemone', called in your case *H. archetypalis* (Chapters 16 and 17), is an archetypal memory. It can be visualised, metaphorically, as a dynamic blueprint, plan-stored-as-a-video or *ROM*-Bios operating software. It comprises, in the form of instinct, the natural, dormant sub-structure of mind; and the 'read-only' architectural component, called morphogene, responsible for the patterns of development, maintenance and repair of the physical body. In this capacity it interacts at electronic or electromagnetic level with chemical messenger services and, thereby, body. You might think of these services as 'buses' transferring information on a wired, material 'motherboard'.

Look at the 'circuitry' implied by the diagram. **The typical mnemone is an unconscious attractor, specific field in universal mind or archetype.** It is useful to think of it in terms of three major components, interlinked associates called (1) signal translation, (2) instinct and (3) morphogene (also called the morphogenetic subroutine or biological archetype).

The combination of personal and typical mnemones amounts to a complete psychological, psychosomatic and physical record of the organism it represents. Files are for storage but also communication. *H. archetypalis* is, sandwiched between psychological and physical domains, a transmitter. It resides, on a hierarchical scale of information (Chapters 5 and 6, *figs.* 5.3 and 5.4), packed as a 'grammatical stencil' between semantic and physical levels. It acts, in other words, as a signal translator, code formulator or transducer in the exchange between psychological and physical events. In this exchange instinct reaches up to consciousness and the morphogene down to non-consciousness. Both are referenced; physical response is various but they remain unchanged.

If, as in the case of animals, there is nervous sensitivity, then signal translation interfaces between conscious mind and body.

fig. 15.8

In dialectical terms the lower cog is archetype; this form of memory is, with respect to an organism's psychology, called its Typical Mnemone. Such a mnemone includes three major subroutines that will be explained in due course - instinct, signal translation and the morphogene. It gears mind to matter at the *PSI* (*figs.* 0.7 - 0.9 and 15.8) and is therefore a low cog in the transmission of information. Its smooth, psychosomatic conjunction, called Synchromesh 2, is dealt with in the next chapter.

Biological *ROM-BIOS* for Organisms Without Nervous Sensibility

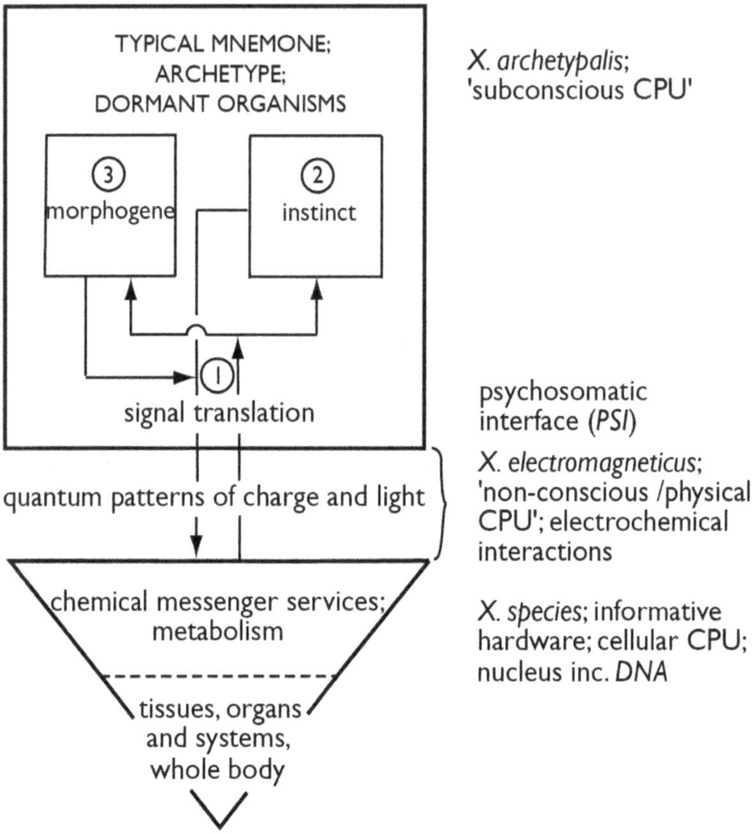

What about biological operating system for organisms without nervous sensibility? In the case of plants, fungi, bacteria and so on, the 'circuitry' is reduced as shown; only dormant, typical mnemone and cell (or multi-cellular body) is involved.

Either way, this mnemone is a psychosomatic communicator whose sub-conscious mode of operation is association. *The method by which its bio-patterns are continually referenced and employed, either psychologically from above or physically from below, is association by resonance.*

fig. 15.9.

Just as individual bodies are detailed within a general plan so specific memories are filed in archetypal format. Together general and specific mnemones furnish mental definition, architecture or, if you like, the context within which a person sustains both a sensible, conscious and orderly, sub-conscious interaction with their body and, thereby, the world of other people and things. The ins and outs of nature and nurture - what is innate and what is due to personal circumstance and, therefore, experience - are both represented. Typical mnemone is innate; its archetype is body-linked. Personal mnemone is, on the other hand, 'nurtured'. As such it is relatively 'superficial', that is, sits 'above' its deeper, changeless counterpart. Yet it serves as the context through which, as they 'rise to the surface', innate, instinctive responses are individually modified. It is the medium through which, according to context and circumstance, each person interprets instinct. This interpretation personalises its expression and, from feedback, allows such expression to be further modified. 'Instinct' can therefore be seen as a suite of 'general' urges; its preordination is locally expressed as various local, individual patterns of behaviour based on standard themes.

The Personal Mnemone

In an analogy with radio it's a carrier-wave or frequency; with a computer it's the chips and operating system; these are likened to an organism's archetypal memory or typical mnemone. A radio frequency is personalised by messages, by various broadcast schedules; coded transmissions need a local context to perform what minds that use them want; and so, with archetypal memory, a context must be added, an environment be integrated with its framework so that local life makes sense. The input of these contextual files constitutes an individual's history. They are the sum of what he has, to any point, experienced and learnt; they constitute the other, personal half of that subconscious sea of influence on which we sail our waking hours. After all, if nothing were remembered, how could anything be recognised or acted on? Subconscious whispers influence present wakefulness continually. Files of memory compose sub-consciousness and, in remembrance, let you move beyond step one. Without them learning could not be increased; in their absence nothing could be done.

Nurture as opposed to nature's element. Thus, superimposed upon the instincts of all human history, your personal history exerts its local colour and effects. Historic memories are ghosts that, in recall, are more or less revived; they are infiltrators mingling with experience; and they are like seeds that, sprouting in the present, influence the future. This chain, whose oriental name is *karma*, is composed of memories. You store them now; sooner or later but inevitably, each according to its own intensity, impressions germinate as destinies and flower as fates. This is the way life cycles on.

In other words, a *personal mnemone* is associated (through the process of Synchromesh 1) with conscious mind, sensitive interaction with the environment and an ability, by use of psychological filtration, storage and recall, to create variations on archetypal theme - a program called a person's individuality. Individual character is the product of experience. It involves learning, that is, appropriate storage and recall. The creator of his or her own drama uses this memory to help make decisions. The less aware an intelligence, the less discrimination and individuality it shows in its decision-making. Personal memorisation often but by no means always involves neuronal circuits - even unicellular organisms can apparently 'learn'; it involves forgetfulness as an 'unworked', 'cobwebbed' image gradually fades; and it involves upward linkage with consciousness. Could some degree of awareness, however residual, exist in all organisms? Or are many never more than dormant?

Each thing experienced, each dab of memory builds your picture up. Suppose, however, every human had a photographic memory, one whose 'dabs' or 'snaps' of every moment stayed with crystal clarity. Internet! Information overload! It would flood awareness with extraneous detail. Thus, just as focus of attention filters present circumstance, personal memory is a filtrate of the past. It incorporates a natural method of selection and of 'fade'. Mental images are impressed on mind-film with a level of contrast proportional to the emotional charge, focus of interest or dint of repetition that created them. In other words, circumstance is sieved and files are written with a 'weight', depth of imprint and sharpness of definition proportional to their impact, relevance or importance to the owner. Shock, fear, enthusiasm and love are typical concentrators of the mind. They gouge 'phosphorescent dabs' or, if you like, grooves that endure. Therefore as well as being colourful and cinematic such memories can be emotive too. We call them highly charged, acute or evocative memories. They are 'hot' thought-objects; each emits its own emotional 'radioactivity'. Pedestrian repetition wears its grooves less rapidly; habituation bores less notably but just as deep. It is as if important or repeated actions were, by 'force of impact', consolidated. 'Strength' or 'salience' of storage is therefore a function of attentive focus and repetition.

As already noted, past events are filed according to their *association* with other data in the owner's bank. Such classified association is facilitated by physically different 'reception desks' (areas of the brain) for information from different senses (e.g. eyes, ears) and homeostatic detectors (e.g. hypothalamus, medulla, pineal); and, conversely, by points of origin for the transmission of motor response to different parts of the body. Call it a library, a computer or an information centre. Whichever simile you use mind works, both in its informative and physical components, as parts cooperative within a single whole and towards a single goal - the satisfied survival of its user. The latter's focus of search, access and retrieval also works by association. **Association, whether psychological or bioelectrical, involves resonance**. On the physical, side a composite address of causal nervous stimulations would suffice. Such original pattern, on re-stimulation, would address the memory; and an incalculable search formula for a memory states that *ease of access is proportional to original 'depth'* (of imprint), *quality* (positive, negative or neutral emotional 'charge'), *intensity of seeker's interest and association factors* (according to the owner's perspective) *divided by fade* (the time between imprint and retrieval). Brain is an algorithmic point of linkage, a medium between two worlds.

If resonance is a search engine, does this help to explain famous anecdotes about inspirational ideas derived from dreams (as in the case of chemists Mendeleev and Kekule) or relaxation in the bath (physicist Archimedes)? *Eureka*! The Dialectic would suggest that, once the components of a problem or a suggestion have been clearly defined, committal to the sub-conscious 'deep' may, as in the case of the proverbial 'letter in a bottle', throw up an answer by return. Resonance of a search engine left to churn undisturbed through relevant files may frame its collection of 'resonant associations' in a way that answers you. On going to bed simply try telling yourself, firmly, that you will wake up at seven o'clock. Your internalised alarm is likely to work.

There are two main divisions of the first, personal mnemone. The first is an 'historic present'. This 'cache' connects immediate history with current affairs; its 'lightness' supplies comparison, orientation and continuity to the moment I am moving in; and its fresh and mostly evanescent images need little, if any, focus of retrieval. *This scratchpad of 'short-term' or 'working' memory is an ever-changing read-write 'dynamic' that allows you to function from moment to moment without losing track of time, location and the course of your plans.* It might be likened to the cache or working memory of a computer; it is some reverberating circuit telling me what just went leading into what's now going and may well go on.

A question occurs. Where is such mainly sense-based memory located as it impinges on the brain? In the medulla, thalamus or cortex? Or, like conscious thought, everywhere and nowhere just the same? To the extent that sense and motor information are both processed by the brain this question is materialliy relevant; to the extent it is derived from a belief that neurons somehow 'think' and 'memorise' it is immaterially irrelevant. A memory itself is metaphysical.

You might also question whether the focal content of every waking moment is not in fact recorded on 'mind-film' - with its clarity of image proportional, as noted above, to intensity of interest or emotional impact. Salient features of short-term focus are, in accordance with their perceived importance, automatically filed to a second division, the 'historic past' or longer-term '*RAM*' storage. Whatever pulls attention to its centre is, for that moment, relatively 'important'. The key positive factors that direct attention and therefore, equally, remembrance, are interest, association and purpose; and the key negatives are shock, pain and fear. Their log *is* personal memory. Of course, as well as one-off 'hits', sheer repetitious grind impresses deeply too. In either case no particular 'long-term memory switch' is needed to translate short-term into long-term '*RAM*' files. You do not need to quote frontal cortex (the physical site of mind's 'third eye'), hippocampal bulge or anywhere else as an agency of transfer - simple depth of emotional intensity (including the desire to learn) or repetition does the trick. It is noted that between its birth and death a living organism's 'computer' and, therefore, its 'software/ hardware complex' is never switched off. Therefore *RAM* becomes, effectively its database of personal or 'learned' files - the record of its life experience.

Memories can, like interests, also fade away. Lack of use disintegrates a phantom gradually; or maybe to forget is just to suffer 'fade' in the retrieval process, 'fade' which intense focus (and perhaps also hypnosis) can restore. Sometimes, in the interests of economy or peace of mind, it even seems that links are decommissioned - decommissioned, that is, selectively, subconsciously or, in some accidental cases, by brain damage. Amnesia appears to be a problem connected with storage or retrieval of memories rather than memory itself. People may forget either what happened before, after or during a trauma; or even anything that happened before or after their forgetfulness began. But it is cured when, for whatever reason, memories 'resurface'. They have not disappeared but appear, rather, to have been suppressed, submerged or veiled. The storage/retrieval mechanism has, in its physical component, been disrupted; contact has been lost. Of course archetypal memory, having no such access point, remains intact.

Remembrance can, as well as reinforce, distort. A memory remembered can, while in the conscious arena, be 'massaged' or manipulated before re-storage. Such rehearsal can occur times over so that, in the manner of 'Chinese whispers', distortion increasingly occurs. Tricks of the mind. Self-deceits. This is how we each, insidiously, rewrite our own life's histories. And old people can, sometimes perhaps distortedly, remember scenes from their youth clearer than things that happened an hour or two before. *Maybe nothing is lost, although the tapestry is dyed with different intensities of shape and colour.* Maybe mind-film is a good description for the capacity of mind to 'fix' whatever it has undergone, to record a lifetime's every thought and action; then, by remembrance replay and re-record the most affecting ones. Who, however, healthily inhabits history when there's the present and a future always eager to the fore?

From where's a memory restored? The question is 'Where is it memories endure?' Only certain areas of the brain (e.g. frontal lobe and amygdala) appear linked with the operation of personal mnemone, that is, with the storage and retrieval of personal memories; yet these seem available from everywhere but nowhere in particular. One answer is to beg them lost in the complexity of a cerebral cortex. *However, maybe in this case physical storage systems remain unidentified because they do not exist.* The nervous order is involved, as microphone or tape-head to a tape recorder, with the receipt and transmission of the sense-based, bodily aspects of a memory; it may act as an electromagnetic address log to the actual memory; but its is not the memory itself. It is not, therefore, surprising that stimulation of the appropriate brain cells will generate a correlated psychological response. You are meant to see with your eyes but stimulation of the 'correct' cerebral neurons will artificially create an involuntary vision of meaningless shapes and colour. What applies to the senses also applies to memory except that, in this case, the 'image' is already internalised. Artificial remembrance may therefore, as in the case of the Penfield probe, elicit an actual, meaningful memory. Yet, paradoxically, this does not mean that memory's image is located anywhere in space. Of what is metaphysical projection made?

No doubt that, if the brain and its associated nervous system is a control panel which facilitates the operation of sense-based consciousness, damage will affect its accurate operation. Hormonal imbalances, drugs or injury will affect its healthy working. No doubt 'the earthly experience' involves, for a biological lifetime, the incarceration of localised 'ego' in a wholly specific, local body. But, it is argued, this 'self' is actually metaphysical; although normally cuffed to a sense-based level of information it can transcend physical, even cerebral, activity. Normally the *excarnate* condition (complete detachment of mind from body) only occurs involuntarily, rarely and unpredictably from circumstances such as shock, weakness or chemical interference. This is in contrast to the case of yogic or mystic science, whose purpose is proper preparation for the moment of death and whose most advanced practitioners claim it as a voluntary, common and well-understood exercise. 'I die daily'. On occasions prior to biological death there is a return to the physical body; at that time there is not. Finally the material veil (including brain) is removed. Then, the experienced mystic avers, the unprepared are unceremoniously thrown into a post-mortem mindfulness undistracted by physical sensation. They are, as Chapter 18 describes, absorbed without respite in memories, desires, guilts and so on. Only if you have decided to equate 'natural' with 'physical' is metaphysical mind 'super-natural'. Otherwise it is, although immaterial, the most natural of things. Like memory. And, if the physical is stripped away, the closer, inward and metaphysical embodiment of mind remains. By no means is this a new idea; it is as old as Adam.

From death to birth, from old to young. What if there are no personal memories to fade or lose? Are the minds of newborn babies blank? Or impressed with archetypal human instinct? Aren't they imbued with *typical mnemone* in its triplex form of signal translations, instinct and morphogene? No doubt all organisms, made of cells, import their type of mnemone from the very start of life - but is the infant mind-field contoured with no personal fluctuations, tendencies or inclinations? Some would say the blankness of a new brain, like a clean slate, makes fresh mind. But though the brain is fresh and physical self's memories are not. Could, therefore, its mind be coloured with the tints of deeply stored experience, with an amalgam of much less than half-forgotten memories - an abstract of the pictures 'dabbed' in previous lives, a précis of its characters in history? Such materially unacceptable but sub-conscious burial of personality now modify the general instinct in a vague but individual way.

The argument can be rephrased. In the Greek underworld you drank of Lethe, river of forgetfulness, before metempsychosis, that is, rebirth into another form. Oblivion does not strike ascent towards The Central Truth; its strikes on the descent into materiality. What is a body but materiality? And what confusion if a prior life's specificities invaded that fresh incarnation, that separately embodied chance! In other words, although excarnation may wash the cloth of a previous incarnation's memories, stains remain. These stains, called *sanskara*s, represent the major 'hot-spots' of emotional experience; tendentious likes, dislikes and morality's exact account are carried over from a previous incarnation or, in series, incarnations. After all, a man who's conscious knows what he has thought and (even privately or undiscovered) done; it is in the mind-world that his record is held to account. If the imprint of an individual mind-set clusters round its central principle of soul, could the *karmic* or *sanskaric* inclinations from its previous incarnations have transmigrated to this one and, therefore, personally predisposed it from the very start?

In short, you may believe life is a one-off - come from nowhere, going nowhere, just your first and last. By this account an infant's mind is absolutely blank before specific memories accumulate; 'its all, therefore, in the genes'. Conversely, you may take a larger view - your present incarnation is one in a series. Thus the new-born infant is inhabited by an archetypal mnemone combined with a collection of *personal sanskara*s. From physical conception nature and nurture develop the brain's physical freshness; and pre-natal psychology (Chapter 18) contributes to the shape of mind. Neither mind nor body is conceived of as a prior blank; much information is incorporated in both of them from the start. For the metaphysical, mental part innate outlines nature and a soul's own past experience will predefine and channel response so that individual character, preferences and tendencies soon emerge. Just as the first sketches on a canvas shape the following detail, so childhood experiences (including substrate memories) colour the framework or 'texture' of a mind within which the complex brushstrokes of an adult are created. Either way, one life or many, blank or previously sketched, early lines will set the course for detail; early grooves will set the course for rills or canyons to develop. The intensity of early, informative impressions powerfully engraves the outlines and thereby affects the way the current of a mind will flow and the landscape it will create.

Perhaps everything is inscribed on mind-film. At least *'long-term memory'* is less ephemeral, deeper grooved than 'short-term'. It is also read-write but, as in a library of many files, requires precise access triggered by a sensation or concentrated focus of remembrance. Of course, as the abovementioned access-formula indicates, the more salient an imprint the greater its influence and the easier it is to remember.

Can a memory influence a consciousness unaware of it? Memory is a substrate within whose framework conscious operations are organised. Memories are, as noted, like underwater shapes that affect the way currents on the surface flow. They are like charged springs each of which, even if it does not break through, contributes to the overall 'bounce' of consciousness-in-motion. Just as an electrical effect can propagate across a molecule, so memories can trigger associations right across their storage field. In dreams this may occur in an incoherent and, therefore, logically incomprehensible way. Just as there exist different conditions of matter, so the dynamic of individual memories differs. Some are more, others less 'solidified'. Some rest more like 'objects', others smoulder charged with emotional potential. This is all another way of saying 'context'. Memory, made of thought objects, *is* a personal, internal landscape. It is also (Chapter 16) a subjective universe.

Long-term personal is not the same as archetypal memory; you no more interfere with that than with genetic make-up or your body's metabolic operation. Long-term memory is the cache of habits that have been acquired. It constitutes the personal basis of character. Its deep grooves are scooped, you might say, by the sudden torrent of intense emotion or worn by the repetitious, daily flow of mind-stream. Its shape can, however, also be remoulded. Determined self-help, character improvement is opposed by, involuntarily, the dissolute reverse. Buddhists have another metaphor. They liken the sub-surface cache of memories to a '*karmic*' bank or '*sanskaric*' storehouse of seeds. Each seed, triggered according to a 'resonant' web of circumstance, germinates and comes to fruit as fate.

Fate? Life is like a game of chess with you, its player, born into a state of play - location, parents, body, circumstances and so on. From this initial deal, by response or plan, you gain experience and, at the same time, seal a set of possibilities into a fresh condition every time you move or are compelled to move a certain way. Move after conditioned move, this is how we play the game until its natural check-mate, death.

What, in this complexity, shapes the quality of moves unless a will to win within the context of intelligence, desire and our experience? This powerful context is aligned with instinct, personal impressions, lessons and emotions - memories that constitute your mnemones, undercurrents that with motive bind you to a personal pattern of response. Motives and responses bind. Shaped to various extent by both external and internal cues, they create your course of thought and thereby action; they set up fresh circumstances all the time. Every moment is a fate, each response creates a destiny, that is, fresh future fate. Thus you can easily see past triggers future circumstance, memory holds information whose remembrance or subconscious stirring triggers personal response. From move to move we tread the wheel, we're whirled on an unending, giddy carousel. The undercurrents of our minds link past with present and the future; they link what's fixed and known with present possibilities and the unknown; personal mnemone links our moves with destiny.

How does a chess-master play? Can anybody neutralise false moves, obtain perfection in their judgment, 'burn their *karmas*' and against the ever-present rival, circumstances, always win? Might this ultimate endeavour not require that, with infinite relief, you find how to escape the existential wheel of time? If so, with promise of assured and happy destiny, what might be the substance of such highly rational and rewarding science of the soul? In Chapter 18 we shall check that out. For now it's time to study archetypal memory.

16. Natural Memory

The Typical Mnemone

A notice at the very start of Chapter 15 warned you. **If, on the one hand, you equate neural networks with the whole of sensation, memory, thought and consciousness then 'typical mnemone' is a brisk irrelevance.** If you equate molecular biology or genes with the whole of morphogenesis then **'memory man' is superfluous nonsense.** You might well, if *vis medicatrix* is something other than chemical, deny its clinical influence. What cannot be physically explored does not medically exist. You state that the wrong questions are being asked on the wrong premises and obviously, therefore, the answers will be senseless; and industry, in terms of pharmaceuticals, will certainly applaud you.

However, though materialism may pooh-pooh mnemonic archetype where does its absence leave us? Mind not molecule anticipates. We know, for example, cells use mechanisms for reliably encoding and detecting their locations. These interactive mechanisms (such as chemical 'morphogen' gradients, switches and cell-cell signalling) transmit information on behalf of interactive strategies. In no mechanism, though, does any part *sense* anything. Automatic biochemistry of cells preserves, repairs, reproduces and precisely orchestrates developmental patterns. How? The chemistry will one day be entirely defined. Once understood, however, does a complex mechanism thence invite a notional elimination of its engineer? Rather, it elicits praise. Why? Without attractor there's no reason for the maintenance of health or biological development. An attractor is a plan; and purpose plans how it will draw you to the future and, thus, expedite design. Such design, with all its sub-designs, might be, for organisms, their survival. Purpose, program and design - all need conception. How, otherwise, could shape-making strategies arise; and with them circuits of control, switches and 'instinctive' sense of time and place and goal? Cells 'sense' established gradients like safety valves 'sense' steam but what set the system up? Why should a coherent system of construction self-construct from incoherent molecules? How could encoded pathways, each step irrelevant without the rest, gradually appear? For what reason might a step, any step, emerge at random and then wait - since only with full muster registered can any step coordinate in profitable work? Don't stop at metabolic pathways leading to a target molecule; what reason is there for each stage of a development inexorably leading to anticipated adult, reproductive form? Mark you, moreover, manufactured shapes need moulds or cuts but forms of life build inside out. Their internal mould is program, inside information carried out on codes and transferred over into bodies beautiful in all their works. Is such informative programming squeezed out from mindless twists of serendipity? If chance is an irrational answer it's still the best materialism has and ever will have when it comes to source of information (as opposed to information minds collect about the mechanisms information builds). **Might, on the other hand, an immaterial element conspire with the material components of biology; might not subconscious control (that is, the exercise of a mnemonic archetype including cast of morphogene) inform all forms of life**?

Now, therefore, check back to *fig.* 15.8. In this section focus shifts from individual to universal aspects of the psychology of sub-consciousness, from personal to typical/ archetypal mnemone. Do you remember the analogy, by radio, of a variety of programs broadcast on the channel of a fixed and steady carrier wave? You might liken individual programs to a file of personal memories and the mnemone of a type (say, human type) to its fixed, archetypal frequency. This carrier mnemone is, like any memory, a mental structure. It is a 'psycho-type'. In this case the structure resides in universal mind. It is time to take a closer look at the psychosomatic connection of a 'carrier' archetype, composed of three cooperative components, with its body.

Must everything be physical, even if you wish it so? **Is it not the fundamental error of science (or at least many scientists) to deny any element or force dissociated from matter?** Perhaps the indulgence that it's 'all in genes' or 'simply chemical' is a professional deformation and, to boot, a philosophical delusion. And what about the quality of a dream or (whose nature is physically nothing) something as real and important as your imagination? Is memory, purpose or instinct material or immaterial? They have neural correlates but are, according to the *top-down*, dialectical perspective, psychological, metaphysical items. If you consider information, consciousness and mind to be material-only entities when in fact they're not - what of materialism then? Metaphysic that cannot physically be tested lies at the root of naturalistic (and therefore scientific) denial of an archetypal mind-matter linkage, independent morphogene or memory; but if, say, immaterial subjectivity *is* actual and if your interest is finding out the factual truth, then tunes will need to change. No nonsense! What follows tries to make sense of your own internal workings.

Suppose you are, instead of a physical body that may or may not imagine a 'spiritual' experience, a temporarily embodied element of immaterial consciousness having a physical experience. You are not, *bottom-up*, a human with cosmic delusions but, *top-down*, a soul cloaked in mind and body. No doubt, all bodies are exquisite kits for exploration in the ocean of non-consciousness. As deep-sea divers need their suits to dive in alien worlds so your mind needs a body to encounter physical events. Body interacts with bodies, earth can contact only earth. Direct, immaterial, telepathic interaction with oblivion is impossible; therefore, mind needs to communicate indirectly through the medium of a body. Self Incorporated is the business. Life Ltd. Muscles drive material rearrangements, sensation logs all changes in your outer world. You need a very special suit. You need to slip a body on to live this earthly life!

From a *top-down* point of view the 'general field' of cosmos is composed of many spheres of influence each with its kind of excitation and results. Indeed, field theory constitutes the summit of materialism. Fields organise matter. Modern physics (but not yet biology) explains its subject matter in terms of them. In fact, electromagnetic, nuclear and gravitational fields are entities more fundamental than material particles themselves. Furthermore, non-conscious, energetic nuclei, shapes called suns or stars, radiate great fields of influence. Why, therefore, should a nucleus of mind not also radiate its influence consciously? Why shouldn't even immaterial memories unconsciously exert their force? Body is your personal abode but is there any element, as body's natural elements are chemical, that's psychologically natural in you? Is there anything of universal mind; and, if so, what's its 'aerial linkage' with your 'psycho-chemistry' - a subject that we'll soon discuss.

The memories of universal mind are 'archetypes'. Such 'typical mnemones' are unthinking but, although the records of a universal rather than a personal plan, must be (like any memory) products of a conscious moment. In the case of universal mind memories are ordered logically beCause and not illogically perchance. As is the case with any file or record, recreation or replay of an original idea lies in its moment of recall; and excitation, that is, resonant association with these underlying, immaterial and archetypal fields keeps natural phenomena behaving orderly. In short, they induce material patterns of behaviour we call natural law.

At this point the work of Benjamin Libet may help to illustrate the way unconscious/ subconscious files of memory work. It was mentioned (Chapter 13) that we 'sail an ocean of context, gist, detail, currents and association'. On its surface consciousness is tossed although, of course, a captain tries to keep course on an even keel. This elemental ocean is composed of memories of shallower and deep effect. Memories, you recall, are records; they are files fixed (as a photograph) or dynamic (as the flux of program or a film) in the 'solid' phase of mind. Thus, as a solid object is composed of energies locked in a certain way, so memory locks information up. Its structure acts as 'psycho-storage' and, although the rules of immaterial dimension may not be the same as matter's, still mind-fields act as governors of shape, as inward organisers of events they outwardly express.

The abovementioned analogy for memory of two major kinds is carrier wave with input modulation; that is, for a given channel (e.g. human, oak or bee) a specific supplement (such as yours, mine or each individual bee's) accretes by memorised events. Thus a personal program is developed; a unique drama is composed on general theme. Check *figs.* 15.8 and 15.9. General, archetypal instinct (part of the broadcast called typical mnemone) is modulated by habits learned throughout a personal lifetime. The involuntary 'will' of such subconscious/ unconscious composite is known as 'force of habit'.

Do you get the natural drift? A body's archetype is individualised by genotype and the pressures of environmental circumstance; a mind's psycho-type is individualised by its associated body and then personal experience. The localised expression (e.g. gesture, manner and phrasing) of such composition is called personality. Of course, like any object or event, personal 'modulations' vary in endurance, strength and quality. So does the quality of carrier wave. A bacterial genome might be compared as 'basic', 'simple', 'coarse' or even 'primitive' against a 'subtler', 'more flexibly responsive', 'complex', 'higher' information-carrying capacity. Thus bacterial, quercine, bee or human psycho-types are 'on different wavelengths'; channel frequencies and the capacity for modular refinement vary. **You can see, in such an explanation, physically and metaphysically, coherent program and not chance writ large.** Since, however, neither teleology nor metaphysic is within materialism's frame, modern scientific *zeitgeist* won't be looking this holistic way.

Currents, eddies, waves; calm seas and storm. Against the force of ocean's habit and the winds of fickle, physical desires a conscious captain tries to steer the body of his ship. Or does he? Libet's experiment suggests, to some, that he does not exist. Libet asked subjects to make voluntary finger movements and monitored nervous activity by *EEG*. He found that a second before movement there is a blip called 'readiness potential'. This you might expect reflected a decision but, perhaps surprisingly, the conscious urge to move a digit *followed* the blip by about 0.2s. How might you interpret that? Nerves twitch, it seems, before your mind's made up. Stimuli pulse nerves before awareness surfaces. Does this suggest determinism rules, there's no free will and mind is brain? It might seem that matter can anticipate or go ahead of thought in time. In other words, nerves fabricate a thought. Consciousness 'emerges' out of neurons. Thus the 'captain' is a robot made of molecules whose consciousness is an electrical illusion. If so, materialism (and its subset of philosophies) seems psychologically justified. It almost seems it's proved. Check-mate, metaphysic; 'matter first' means physic's victory.

How might you otherwise interpret this 'strange' order of events? If brain's a medium and mind (including memory) is metaphysical how do subconscious conditions impact on the brain? *Top-down*, you firstly ask what any blip might actually represent. Of course, embodied thought relates, by way of brain, to physical accord. You'd expect *incoming* signals - in the order of sense to brain to conscious perception - to show a blip *before* perception. It would relate an innervated body to appropriate parts of centralised control. Would it also demonstrate subconscious relay? Would nervous twitches constitute your *actual* consciousness? If mind (including conscious thought and memory) is immaterial then definitely they do not. In fact, you might expect experience to be delayed by the time required to jump from sense receptor to the brain and thence, through subconscious structures of translation, to cognisance; conscious perception must, although it sees things as they were, fractionally lag behind the world.

In deep contemplation mind's detached from all sensation; and it's the same in sleep. In both conditions body needs to operate and requisite subconscious operations ply; corresponding nerves and hormones still communicate the body's various necessities. The hemispheres of brain cooperate yet each (as was explained in Chapter 13: To Build a Brain) 'weights' with intrinsic tendencies. Where right is global and synthetic left-side details, analyses, localises an event. General principle engages local practice. You might, therefore, expect *outgoing* signals to be translated through subconscious routines to where a blip of nervous excitation would occur in an appropriate department of a brain; and from this office be despatched to correlated parts of body such as Libet's digits. 'Revs' of expectation and of calculation might (as a checking captain 'feels from experience what's right') precede a conscious choice. Thus right-hand hemispheric fitment with the psycho-type would precede a left-hand correlation with the local elements of circumstance. Time and place are plugged into the program. *EEG* registration would occur. Post-feedback decision, conscious vote or veto, is reflected in the frontal lobe; it determines what the body's going to do. In this case conscious decision fractionally anticipates its world. Roger, action stations. The motor zone of brain will physically fire and, milliseconds later, digits move. All that's physical will register on scans but immaterial activity will not. It is not, therefore, a case of hemispheres (each a bunch of nerves) 'believing' or 'not believing' each or both originates a thought. They are simply stations of transduction in a process. A single operation, cognisance, partakes of several different steps; various modules, physical and metaphysical, reciprocate. **In this case why should psychosomatic oscillation, with probes that only catch somatic fraction, demonstrate that mind derives from matter?** It only proves that men interpret facts according to prior theory or belief.

Why, you may ask, should holy books be hardly interested to explain this cog of metaphysic in creation's wheels? Is psychosomatic linkage, mind to body, not a crucial one? Maybe it's because faith, first and foremost, turns 'face upwards' from the state of human mind and physical connections towards Ideal Psychology. It turns to super- not sub-consciousness. So why look down towards a place you do not want to go - oblivion? Why, unless your science is neurology, try to penetrate the dark space in between a conscious mind and its ephemeral, made-of-atoms body?

Perhaps we've already referenced the immemorial world. For a Jungian psychologist archetypes may constitute 'a store of collective unconsciousness', an immaterial element shared subconsciously by every human being; but while Jung's 'mythological memory' or a Hindu's '*akashic* records' may be thought of at the same psychosomatic level as a typical mnemone, they do not seem in principle to be the same. An archetype is not, any more than a technical drawing, an 'historical accumulation of parts'; it is a blueprint, a clear definition, part of (see *figs*. 5.3 and 5.4) a hierarchically purposed plan. Indeed, a Platonic 'blueprint' or idea is closer to a typical mnemone; perhaps the latter also includes archetypal bio-form, a metaphysical species of shape (Greek: *eidos* with plural *eidea*) that the founder of biology, Aristotle, conceived as the 'principle' or unchanging template behind each physically variable form of life that he and his botanist friend, Theophrastus, were used to study.

From below, on the other hand, a physicist explores the metaphysical abode of changeless rules by using certain mathematical descriptions. Why, for example, should an inverse relationship occur in metaphysical abode 'nature's language', calculus, whereby differentiation concerns (*raj*) velocities and rates of change and integration deals with (*tam*) areas and volumes? In a much broader sense you might rate the whole of physics an appreciation of the archetypal generalities. Pythagoras with vibratory scales, Plato with 'ideas' (*eidea*) and Euclid's geometry seem to combine with modern studies and help define the 'numinous' immanence that shapes physical phenomena. Physical systems include, of course, systems specifically encoded by symbolic *DNA* and called bodies biological.

It is worth emphasising that subconscious memory is uncreative. Its record is, of itself, passive, purposeless and fixed. And yet its reflex is, like that of any technology, originally the product of creative purpose; and such purpose is the motive behind the otherwise motiveless government of biological bodies. It is the origin of meaning and rationale in biology. Let's take a peep, therefore, at the distinct rationale behind the psycho-type of *H. sapiens* - you. It has already been identified (*figs.* 15.3 - 6) as a typical mnemone called *H. archetypalis*.

***H. archetypalis*, the Image of Man**

tam/ raj	*sat*
subsequent order	*archetype*

H. sapiens/ H. electromagneticus	*H. archetypalis*
physical bodies	*memory man*
non-conscious	*sub-conscious*
data level	*syntactical level*
expression	*potential*
classical/ quantum biologies	*transcendent biology*
effects	*cause*
oscillation	*norm*
deviation	*standard*
particular structures	*overall morphogene*
↓ *tam*	*raj* ↑
gross	*subtle*
H. sapiens	*H. electromagneticus*
visible body	*'network of light'*
classical biology	*molecular/ nano-biology*
periphery	*inward 'shaper'*
informed shape	*informant*
protein	*genetic code*
phenotype	*genotype*

Check *fig.* 16.1. There are (Chapter 7 and *fig.* 2.13) three main levels of matter - (*sat*) potential or transcendent, (*raj*) kinetic or quantum and (*tam*) bulk or classical. In biological terms the classical level studies organisms in terms of cells, organs, tissues, whole bodies and their ecology. Atomic and quantum nano-biologies are the province of molecular biology, biochemistry and the subatomic orders of life. Biology that deals with the reflex patterns of mind (e.g. instinct) might, being on the level of transcendent matter, be called transcendent. Such immemorial, archetypal memories are not necessarily single-frame photographic or even holographic 'stills'. The cinematic capacity of memory can include a sequence of events. Such a video can develop like the steps in a plan. It would then amount to a program, a recorded program which, like a song on tape, can be replayed, when called up, endlessly.

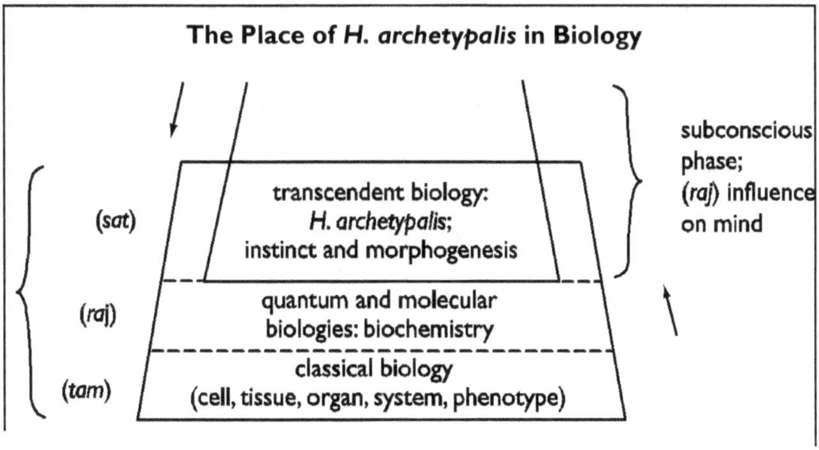

Physical prototypes (direct expressions of wireless archetype) include type of basic particle (photon, electron etc.), their behavioural routines, field and type of influence on other particles. Biological archetypes such as *H. archetypalis*, 'the image of man', are *not* the same as the current taxonomic definition of species. In the dialectical view such an archetype (see also Chapters 17, 19 and 21) broadcasts information to its physical specific. Such locals are, biologically, cells or multi-cellular bodies. Generic program acts as a homeostatic regulator, a reference-point or 'norm' that influences purposive developmental sequences, chemical algorithms (such as metabolic pathways or processes such as mitosis) and the morphogenetic arrangement of pre-coded building supplies in growth, maintenance and repair. An archetype is, in effect, a lawful program. Each biological archetype represents a statute (the organism as a whole) with its bundle of clauses (subroutines). Informative principles or laws derive from a level above. The level above matter is mind - in this case a sub-conscious store of logic called an archetypal memory. As your own behavioural patterns derive from a context of memories, so do those of nature. Indeed, are you not this very moment part and parcel of a very natural pattern, the real expression of a deep and natural memory??

In short, sub-conscious mind *is* the passive level called (Chapter 6) information's infrastructure. Its archetypal code informs, at the lowest level of existence, material energy. Every cell partakes, as an antenna, of archetypal as well as electromagnetic broadcast. The archetypal factor is a program that interacts with physical ciphers ranging from simple subatomic particles to complex, code-constructed molecules each performing its own intended task. The latter are built into *DNA*, protein, chemical signalling, cell structure and so on. As such, sub-consciousness helps regulate both biochemical and physiological activity. The morphogene of a typical mnemone acts as a regime within which correlated subroutines correspond to, guide and control the form and function of physical counterparts. It is a program whose output is the generation and maintenance of physical form; but why, any more than a computer program looks like pictures that it generates on screen, should archetype have bio-form?

If you think that 'archetype' is just a 'cop-out' in explaining how things are then I suppose you think conceptual plans for any working mechanism are a 'cop-out' too. However, your objection to purposive program in no way mitigates the impact of its natural possibility and, if an immaterial element of information exists, natural fact.

fig. 16.1. (see also fig. 15.5)

From a *bottom-up* perspective memories must somehow be encoded in the nerves; and instincts must be patterns whose development was somehow etched in neurochemistry by all-controlling genes. *Top-down*, on the other hand, a perpetually dormant but wholly coherent aspect of mind, archetypal memory, directly affects both the body and behaviour of every organism.

Check stingers 3 to 5 from Chapter 4: Stingers. *The psychosomatic mechanism whereby principle is realised, idea expressed or, if you like, conceptual design installed in physical activity is through the operating system of a typical mnemone, that is, through the agency of archetype*. From a *top-down* perspective different yet coherent archetypal memories inform not only components of the energetic stage but also patterns of its animation both in their body-shape and instinctual behaviour. Players, organisms, are incarnate information; cosmic actors (such as you, I, fungi, fish or pine-trees) comprise intelligence incorporated.

In dialectical terms the (↓) flow of physical creation is described by entropy; it runs from potential through stimulus and action to exhaustion. In this respect biological archetype is prior. It is hierarchically above its physical expression, its inexhaustible store of information is recorded metaphysically in archetypal memories and it constitutes the *potential* for organic structure. In other words, it is the (*sat*) causal, morphogenetic field whose influence in all organisms interacts with the (*raj*) electro-dynamic elements of their chemistry. As such this morphogene would amount to prior information, to preordination and the template for (*tam*) structural/ functional physical expression of patterns. It amounts, for every organism, to its 'norm', 'type' or 'law'. A 'geno-centric' picture of life elevates 'the gene' to the position of material jewel in its explanatory crown. No doubt genes are passive information and a chemical expression of life's database but what is a computer *sans* its conceptual software? An automated production line would well describe a cell but hardware, especially if it involves hierarchies of instrumental control, depends on software; and software depends on purpose for the order it transfers. A 'totipotent' morphogene, like *ROM*/ bios software, oversees life's automatic operations. So how, as we shall see, does *DNA* cooperate?

Check the legend for *fig.* 15.8. A typical mnemone is a modulator, a translator, a go-between. Its storage amounts to a context within whose frame of reference both conscious mind and physical body cohere, orientate and operate. While the personal mnemone contextualises in terms of an individual, the *ROM*-type inflexibility of typical mnemone contextualises in both general and specific ways according to the life form involved. This is another way of saying that classes contain individuals so that, for instance, spiders-in-general spin webs but spiders-in-particular spin particular geometries of web; this is obvious but the form of words indicates a mnemonic, non-evolutionary origin of class or particular type.

Another way of describing memory was, you may remember, in terms of holographic principles. In this case an individual life form's typical mnemone would amount to a snippet of the cosmic 'super-hologram'. It would amount to a lens to which you and I are, by virtue of our bodies and not least our brains, attached; or, conversely viewed, to a lens whose focal image, whose resolution is in your case human being. It would constitute a specific restriction on the way we are and how we see things; indeed, on the way that any organism is informed and how it interacts with habitat. What light shines through life's lenses? What light projects the image that for you is real?

Of the mnemone's three components one (signal translation) is a two-way code transducer, a psychosomatic channel translating between the orders of conscious mind and body; it communicates sense either way. A second, instinct, conflates 'upwards' with personal memories to influence, where such exists, conscious mind. There is no direct connection between the third, called the morphogene, and the personal mnemone or thereby conscious mind. On the contrary, universal or (in your case) human-in-general morphogene acts 'downwards' in resonant association with *H. electromagneticus* and thence your specific body. This said, mind is as comprehensive an 'organ' as the body; its metaphysical parts are not rigidly isolated and each, either directly or indirectly, modulates the others. For example, the sexual instinct is archetypal but, as far as individuals are concerned, its theme is modulated, understood and expressed in myriad ways in association with their personal mnemones and body chemistry. Variation-on-theme. And sexual thought or physical experience is transmitted by way of signal translation - whose passage is associated with instinct and morphogene. The latter is, in turn, associated with reflex metabolism and physiology e.g. hormonal triggers and physical organs. Parts of your body are interrelated, parts of your mind are interrelated and so are both coherently. Lets take a closer look, using a computer rather than a hologram as the analogy, at each typical subroutine.

A typical mnemone is composed of three main subroutines; or, if a protocol is a standard procedure for regulating the transmission of data between two end-points, three protocols. Together translative, instinctive and morphogenetic programs comprise an individual's archetype. If, as in the case of plants, fungi etc., there is no conscious component then the translator element is, as shown in fig. 15.9, reduced from nervous to chemical (e.g. hormonal) messaging alone. ***H. electromagneticus* has, in the human case, been identified as the physical side of translation and *H. archetypalis*, with its routines, the metaphysical correlate**.

If short-term memory is thought of, in computing terms, as a cache of working memory and long-term memory as *RAM* or a database, then archetype is an in-built, customised set of instructions, an inflexibly hard-coded chip known as Read-Only Memory (*ROM*). It is an overall operating system. The analogy from physics is a 'memory field'. **Fields organise and regulate**. They do not actively plan but passively direct events according to their innate characteristics. They dictate type of behaviour. *Such archetypal fields, called typical mnemones, are the staple of (sat) transcendent biology (figs. 2.15 and 16.1)*. Revise *fig.* 15.5. In this chapter the human archetype has been introduced as, for ease of distinction, *Homo archetypalis*. Its electromagnetic or 'quantum' aspect is called *Homo electromagneticus* and the well-known, gross expression *Homo sapiens*.

Two important aspects of '*ROM*' archetype are marked. **The *first* is that, just as the genetic 'book of life' is found as a nuclear genome in every cell, so every cell accompanies its typical mnemone. The *second* is that such generic '*ROM*' survives the death of any individual cell or body**. Genetic material is passed from generation to generation with its accompaniment, a cell; the physical medium of transmission is, of course, a single fertilized egg or product of asexual division. The metaphysical mnemone is not thus passed; you might liken it to a generic broadcast-channel of which any attuned cell is the receiving apparatus; destruction of a local receiving set does not affect the broadcast. Nor can archetype become extinct.

Biological archetypes, in their 'video' complexity, constitute the filter, template or typically restrictive, collective lens through which all organisms of a kind experience the world. This filter includes the generalised 'file' or 'program' from which individuals partake their shape. Its theme will, therefore, differ for a human, a horse or a moth. It is a layer not of habit (semi-fixed behaviour which can be consciously changed) but of deeper instinct. Immutable stereotype. Lying at the seam between physical and metaphysical, it is an unconscious, indelible psychosomatic memory. In the human case we dub this 'memory man' *Homo archetypalis* or *ROM* ('read-only-man') for short.

You might analogise *H. archetypalis* with a language. We can discern a subtle difference of dialect and usage in the speech of our friends but other languages sound like babble. Not, however, to their speakers - Frenchmen, herds of springbok, colonies of bat, bird or whatever. For each of them the same precision and subtle variety occurs within their linguistic theme. A local dialect becomes a species of the mother tongue. Translate this principle of grammatical code into biological terms and the same variation on theme, in practice as opposed to principle, occurs. This is because the universal aspect of an individual cell (its metaphysical archetype) interacts with matter in the form of genetic script (in the form of slightly varying *DNA* sequences), metabolism and environmental feedback to produce specific variations on its archetypal scheme. Physical construction is seen as, simply, the end product of an orderly suite of commands. It is a program, a preordained line of events.

Surely this can't mean that genes are not 'top dog' and proteins do not formulate the shape of organs, systems and, instincts included, entire bodies? It was, for example, claimed - misleadingly - that two different organisms, man and chimp (*Pan troglodytes*) have a 98.5% *DNA* sequence homology where proteins are concerned. In fact, protein-coding fragments (about one three-thousandth of the whole chimp genome) were chosen because of their similarity to ours and assembled in an order based on the notion of evolutionary relationship. Roy Britten, the original estimator, has lowered his equation to 94% but the similarity may well turn out as low as 90% if you count non-protein-coding 'junk' as well as genes. In fact, it may be 'junk' that critically counts. Such non-coding elements are now recognised as critical regulatory and structural factors in the operation of *DNA*, ones which may well help to explain the differences between the two types of body. The mode of sequencing, where protein-coding sections comprise less than 5% of the total *DNA*, also excluded mention that over 50 human genes are partially or wholly absent in a chimp; it excludes both 'indels' (large sections of complete dissimilarity in non-protein-coding sections of *DNA*) which exist in the cells of humans but not the monkeys they 'came from' and *vice versa*; excludes mention that in evolutionary terms women are more like chimps than men (the Y chromosome has 'diverged' markedly more than the X); and fails to mention that a chimp genome is about 10% larger than human. In fact over 99% of mouse genes (about 30000 like us) have human counterparts; on the simple basis of genetic sequence data the banana, a plant, nears 60% human; and 75% of human genes have counterparts in a nematode worm which is not three-quarters human just because of that.

Finally, a maximum 95% sameness equates to nearly 100 million specifically required *SNP*s (single nucleotide differences). Moreover only if a mutation involves a sex call can it be inherited. In the postulate of only a couple of million years such a chasm therefore rules out the gradual evolution of an ape to man.

Still, can it be that only genes and proteins make all difference? They underwrite materials but how do very varied forms of life emerge from standard biochemicals or 'morph' from similar genetic codes? Lacking notion of an archetypal morphogene we've almost no idea. Even if the genetic identity were almost 100% would this mean a chimp *is* a human? It might absolutely instead of marginally mean that 'it' should not be subjected to abuse or torture (as in some experiments) that you would not legally inflict on a human. The two organisms might be made of practically identical materials but if, as in the case of two houses made of identical materials from different plans, their archetypal template differed, they would be shaped accordingly. In other words your involuntary body is, by virtue of its typical mnemone, universal in type. Types (or archetypes) never blur; they do not transform or 'evolve' into each other but their plan is specified (or personalised) to generate an individual ape, human or whatever else by interaction with slight variations in typical *DNA*, local surroundings and experiences. Such variation-on-theme (*figs*. 5.1 and 21.1) is the norm, the order of the biological day.

↓ *down*	*central channel*	*up* ↑
morphogene	*signal translations*	*instinct*
body-link	*< two-way >*	*mind-link*
physical pattern	*conscious being*	*dormant being*

Re-check *fig.* 15.8. Its typical mnemone can be expressed in this tri-logical stack. A material archetype (e.g. light or electrical charge) may be restricted to a sparse, simple morphogene, a kind of 'one-liner' law. A coherent set of rules or laws can also constitute a program; the equations of science map such processes. The natural law of biological archetypes involves such bundles of principles/ norms/ programs; it involves coordinated processes of regulation. In waking animals, which show different degrees of consciousness, all three components of the triplex form of memory are active. Dormant organisms (bacteria, fungi and plants) have instinct too but they are, as any sleeper, unconscious of its expression. Each typical mnemone is, like the computer's *ROM*, a sub-conscious hive of industry. A broadcast from the catalogue of universal mind, it is type-specific rather than individual. A mass of TV programs, each broadcast on a specific frequency, do not interfere with each other; each single one can be clearly received on a suitably attuned channel. As a TV broadcast seems, at least locally, to be everywhere (on every screen) at once, so archetypal memory functions ubiquitously and simultaneously - which makes the particles and cells of the world receivers. They are like information reception stations or antennae that pick up the 'frequencies' on which are broadcast different memories of universal mind. For example we each receive, in the manner of myriad TV sets, the 'human program'. In this respect we are all part of nature's grand archive - no matter how many or how few sets there are. *H. sapiens* might become extinct but his metaphysical archetype (*H. archetypalis*) would remain intact. This mnemone, neither a program created by any individual nor body reproduced by any couple, constitutes the read-only archetype found in every organism's mind. **This subconscious, dormant form of mind extends to every organ and its cells**. And in this manner individuals are microcosmic variations on the macrocosmic principle.

To the notion of broadcast now add another powerful metaphor that evokes a feeling for the operation of universal mind. A hologram is a storage mechanism, a 2-dimensional pattern usually made on photographic film by 'freezing' interference patterns from two or more lasers. It can produce a 3-d image. Within this image various parts reflect the whole in many ways but, it is worth re-emphasis, any part or pixel can be used to project an image of the whole. It is, like the *DNA* genome in every cell's nucleus, both a part and apart. Each autonomous pixel is, in other words, also a 'sub-station' that both connects with and reflects the whole. Could you not liken each cell in a body to the pixel of a hologram? In which information for the whole picture resides? Each 'pixellated' cell incorporates both kinds of universal information (mnemone and its *DNA*) so that the phrase 'multi-cellular hologram' well encapsulates the informative nature of a body. This means every cell of every body would associate with its whole organism's archetypal memory, its typical mnemone. And, as has already been suggested, local variations on the physical side are due to 'acceptable' genetic variation (e.g. sexual, hybrid and viable mutant) and differing circumstance - nature and nurture.

A natural hologram is not, like a photograph, 2-d and static but emits, from a focal, metaphysical source, its coherent, space-time message. Indeed, you might conceive of all nature existing under the auspice of universal mind. That is, you might see the 'pixellated' subroutines of its memory as combined into a cosmic whole. A cosmic blueprint. In this sense appropriately evoked events are much less separate, much more related than they appear. And in this sense also the physical world is a memory of its own creation, the 'crystalline precipitate' of a psychological projection. Cosmos is a vast remembrance, a continual reactivation of its first conception, its original idea.

The holographic principle is of 'the whole in every part'. Such principle, wherein parts are automatically 'aware' of other parts and of overall function, is readily applicable to biological form and function. But, of course, if this is true the Theory of No Intelligence and with it that of evolution slip straight down the pan. They would be history.

With respect to holographic integrity it might be logical to understand that the human body represents an integrated, synergistic and co-operative whole; that different organs incorporate an imprint of the whole; and that the potential for this integrated whole is gathered in a single germ cell. In this view, one exploited by holistic forms of medicine, glands and organs represent subroutines; they are microcosms whose features are laid out in a manner that intimately reflects arrangement of the macrocosmic whole. One in all and all in one, health becomes a matter of dynamic balance between the parts. It springs from within as an intrinsic part of the way archetype, a *top-down* program, is electro-dynamically expressed in matter. This archetypal correspondence is known in its operation as 'deep homeostasis' or '*vis medicatrix naturae*'. Such orderly presumption, as opposed to the chaotic notion of origin supplied by Darwin's theory, underlies all medicine of the orient and, previously, the occident as well; and disruption of bioelectrical communication, that reflects a body's archetypal triggers, would emerge as a potent cause of ill-health.

In a holistic perspective differences are not so much created as disappear. Blocks are obstructions in a flowing plan, false notes jarring archetypal score. Indeed, Chapter 17 will elaborate how informative mechanisms (nervous and hormonal systems) reflect typical mnemone and link the body's operation into a functional whole. While acupuncture is a treatment based directly on expression of subconscious, archetypal whole, others use hands, feet, iris of the eye, tongue, lips and other parts to diagnose and work in practice back from imbalance to a whole system's dynamic state of health. Are not, for example, your feet your point of contact with this earth? In medicine's international market the stall of reflexology, a system that employs the feet, is worth a call. Each foot mirrors conditions in half the body's coherently reflective whole; and each represents five zones, channels for five different 'notes' or 'modulations' of sub-conscious energy (*prana* or *qi*). Such 'pedal longitudes' are crossed by horizontal 'latitudes'; these grade from head (the toes) to heel (the pelvic point of balance). The entire body conceptually fills this grid; the organs are thereby easily mapped as points within its lines and, therefore, the foot. Connection has been made. Hands-on (not drug-based) treatment can ensue.

Such medical logic, based on a Theory of Intelligence and employing fundamentals of its scheme, is clearly more than 'complementary' or 'alternative'. It reverses an objective, analytical angle that leads to clinicians specialising in their own disintegrated 'bit' of the body; and to drug-happy pharmacology that's focused to exclusion on the role of life's machinery's tiniest bits, the maelstrom of its molecules. It is obvious that, conceptually, the *top-down* dialectical and *bottom-up*, Darwinian perspectives are at odds. Therefore synthetic treatments, however ancient or widely employed, are sometimes deemed 'unscientific' and tend to be demoted by mainstream, western medicine.

Physical variations on such archetypal theme are wrought, perchance, by local influence and, not perchance, by nuclear code. This code's *DNA* bears biochemical determination. Slight differences in such 'intelligence' dispose specific individuality. The genome is a 'pixellated' version of the physical whole found in each separate, cellular part. Each cell is, in this sense, a microcosm of the macrocosmic phenotype, a whole body. The realisation that every cell contains a formulation of the whole, multi-level body of which it is part drops a heavy, holographic hint, largely ignored by materialistic philosophy unprepared for this perspective and its implications, as to the real source of bio-logic. It is, for example, still not fully clear how the expression of *DNA* coils in the form of appropriate type and amount of proteins is orchestrated. There must certainly exist a chemical and, therefore, electromagnetic level of control. *Natural Dialectic would suggest the molecule is a conductor, an aerial whose electrogenetic communications precede and govern molecular operations; and it would, of course, intimate a further transcendent or 'radio-biological' level of cohesion, one uncountenanced by current research - a psychosomatic resonance with the morphogene.* This morphogene is (*fig.* 15.8) part of the triplex, archetypal mnemone; which mnemone is itself a subroutine, the human subroutine, of universal mind.

In this view typical mnemone is the metaphysical blueprint for a body; it is body-linked at the level of a cell. Cell not soul has biological memory. Each and every one also contains genetic code to produce appropriate materials for itself and, where applicable, for chemical communication with the rest of the multi-cellular operation called a body; and different pixels of a hologram appear different, suppress or lock out parts of the whole picture but still contain its whole potential. *Similarly, in mind, a cell or body's typical mnemone is the morphogenetic 'intelligence' that moulds substance and thereby coordinates the production of functional shape - shape whose function relates specifically and accurately to its purpose.* **That is to say, all cells partake of memory. *DNA*'s material book of life is correlated with a book of immaterial information in the form of archetype. Each cell responds to its subconscious archetype.** For example, human cells each resonate with *H. archetypalis* and each cat cell with *Felis archetypalis*. So, within the whole picture, does each tissue or organ made of cells. Each pixel partakes of and corresponds to the whole. Just as a cell can express only its own relevant fraction of the genome, so it resonates with its relevant fraction of archetype. Individual subroutines or linked groups of them are accessed within the framework of a coherent whole. In this way the archetype is a complete context, a governing template whose different subroutines are accessed as parts of a coherent master program. An egg has, therefore, 'knowledge' of both itself and its potential adulthood; its 'daughter cells' will be of every kind incorporated in the adult form. Stem cells also have flexible identity that is by degree reduced until cell type becomes 'determined'. Determined, somatic cells do not transform but each has a holographic 'idea' of its own typical identity and its role in the multi-cellular whole. Body cells and unicellular organisms alike 'know' their behavioural patterns. **This knowledge is not conscious; it is unwitting as sub-conscious means and is.** *But all cells have dormant mind that interacts with their chemistry, chemistry that includes physical correlates of their 'dormant intelligence' in the form of information systems (cell signalling, hormones, nerves etc.) and a database of DNA-written coils (or solenoids?).* Nor is this passive form of mind less separate from atomic action than a dye from its cloth or a solute from its solvent. It is closer, close as broadcast is to the picture on its TV screen. Subtler than a nanometre, mind is within or behind material expression. It is, in all its aspects, an information centre. If for philosophical reasons science, especially neuroscience, tends to dislike the idea of an immaterial governing factor, that is one thing. Luckily for you, however, such distaste changes nothing. And most luckily so for the savant who wields his metaphysic to deny its own existence. Has he lost his mind or, in spite of all his efforts, not an ounce of it? Otherwise where, lacking thought or instinct, would he be?

Indeed, it is worth repeating that mind is not simply 'apart from', 'on the other side' or 'beyond' any force or body. These are simply conceptual devices, models of limited use. *In a paradoxical way mind is no more psychosomatically separate from a body than a broadcast from a radio; it is apart yet a part.* You do not think of a broadcast as entering or leaving the body of a receiver. Its wireless existence does not depend on such receiver but the latter materialises its otherwise insensible information. It links a program to the outside world. Moreover different thoughts can, like photons and broadcasts, occupy the same space - except that it is not physical but metaphysical, inner 'psycho-space'. Such space is immeasurable and unlimited in extent. Just as quantum properties interrelate with but differ from large-scale ones, so properties of the mental aspect of things differ again. The qualities of a psientific 'fifth dimension of psycho-space' strain the comprehension of an intellect designed to work in only four - as such quantum principles as superposition, non-locality and entanglement might start to demonstrate.

Dormant mind, which is called the sub-conscious archetype, operates in two directions. Upwards it 'emerges' as the organism's awareness of an instinct's mood. Instinct meshes with, where it is present, conscious awareness. *Downwards it acts as a morphogenetic template* that meshes with cellular biochemistry and, therefore, the expression of genetic code. A morphogene allows a few kinds of amino acid and relatively similar proteins to produce nervous, muscular, hepatic and all other often very distinct but functionally apt kinds of cell and, from the cellular substratum, organic, systemic and whole-body shapes. And different morphogenes with similar substrates 'catalyse' all bodies of the natural world.

Differences are 'gathered up'. The same sorts of cell can be found in different organs, the same organelles in different cells and, in the presence of water, the same biochemicals (carbohydrate, vitamin, protein, lipid and nucleic acid) everywhere. Parts relate to a whole that is certainly greater than their sum. Is a house different in weight or chemistry from the pile of its materials? In the same way a body is greater than the sum of its parts in respect of their order, coded arrangement and working design. Its passive intelligence is, prior to physical, metaphysical. The basis of biological business, a reference point for physical routines, is a natural holo-program run in every cell and called its morphogene.

A branch of scientism called evolutionism involves a 'religious' necessity in the belief that the theories of abiogenesis and macro-evolution are in fact fact. Its followers assert, in the name of science, the same dogmatic kind of exclusion to contrary argument that is expected from the adherents of any fundamentalist faith. Naturally evolutionism cannot accept that life forms are unable, like houses, cars or circuit boards, to self-construct by electronic bonds alone. Chemical evolution, genetics and developmental biology, all at the cutting edge of coding, especially pursue the 'self-construction' line. What else, given a materialistic philosophical framework within which to constrain your thought, can you say? Natural Dialectic agrees with self-construction but not at the level of chance and gene alone; its view includes an accord with the natural, sub-conscious blueprint of archetypal memory - a product of intelligence. This sinful view, expelled from scientific Eden, therefore states that genes alone are not the causal answer but rather, like a computer program, a secondary effect. An effect of what cause? Can intelligence be causal? Might there exist original intelligence of which all plans, fixed either in memory or matter, are the effect?

In sum, controlled access to a genetic database triggers the right construction of the right pile of materials at the right time and place. But a house is greater than the sum of its materials; neither house, chip nor motor-car 'self-constructs'. Building is intensely programmed and prepared. Forethought pre-arranges every single part. And the less there is human intervention the more completely robotic a production line must be; the more 'intelligent' its automated system has to be. Enter life forms. Mindless molecules cannot, without the precise framework of a preconceived order, 'self-organise' such complex mechanisms. The game needs rules. It needs conceptual mould, a construction line, an intelligent program that, while immaterial, is yet more than the sum of its own system's parts. Thus, *top-down* we say that metaphysical infrastructures called morphogenes precede and subsume the different types of life form. So that, as each type of organism is a function of its typical mnemone, it can be reasonably speculated that this mnemone is part of a larger suite, an ecological wholesomeness, a coherent conjunction of coordinates. As such you might think we are all part of a grand correspondence with the conscio-material gradient, an orderly expression of characteristics whose principles are 're-gathered up', eventually, into their Central Origin. Biology is resolved into a mind-field, a logical and coordinated set of plans. Not, therefore, that *H. archetypalis* is the image of man. The reverse. While this is certainly not the view of scientific materialism it may well be true (as this book henceforward explains).

Signal Translation

Is everything material? If everything, including perception and emotions, is electrochemical alone then there is no need for signal translation. If, however, an immaterial element of mind exists then so does the need for psychosomatic junction.

Life's key aspect is the exchange of information. A great deal of such exchange occurs in our bodies without us knowing about it. It is possible that, except in the case of some animals, no exchange of information is ever consummated in awareness. This is not to deny any biological unit, including a 'dormant' plant, fungus or bacterium, sub-consciousness. *Each type of organism with its appropriate natural protocols of information exchange senses the world, and responds to it, through the structure of its own chemistry and archetypal memory, that is, in its own peculiar, automatic way.*

It was noted in the first five paragraphs of Volume 0 that, in the human case, the outside world (including our own bodies) communicates with mind through a blitz of nervous 'morse'. Exactly how are such impulse-blips translated into either accurate muscular response or seamless awareness of it? From a *top-down*, dialectical point of view the sub-conscious mind is a channel, 'medium' or link between such awareness and non-consciousness (the physical world). Unity of consciousness as well as unification of governance demands a corresponding structural and functional unity. Only in this way can the coherent propagation of hierarchically organised signals occur. Therefore a typical mnemone (or archetypal memory) must include, as an important subset of its whole, the construction of an organism's *'transduction channel'*, its two-way translator of signals between mind and body, a psychosomatic translator. This is the type-specific mechanism through which environmental information is filtered to 'make sense' and, in return, an appropriate behavioural response is issued.

Thus the first subroutine of archetypal mind involves the translation of signal information between mind and body.

In the case of conscious mind such transaction is by way of a 'modulating agent', a metaphysical faculty, which shuttles coded information to and from conscious 'psycho-space'. Check *fig.* 15.8 and find this facet of archetype called 'signal translation'. Its psychosomatic channels translate information between nervous or chemical code and the symbols of perceptive mind. They facilitate meaning and action; their subconscious resonance-structures mediate between subtle matter (e.g. electromagnetic radiation) and conscious knowledge. The energy that activates these instruments is called *prana*; and they are composed of a specific 'tuning' (or *tanmatra*) of potential matter. Both these factors are often described (see the legend of *fig.* 6.2) as a rainbow of frequencies. This spectrum is, in the manner of visible light, 'split' into five bands is also, in relation to sound, likened to notes or harmonics. In Samkhya (Hindu) and Buddhist philosophy each 'band' or 'harmonic' of potential matter not only relates to five expressions of psychosomatic *prana* but also five senses and five states of matter. And the fivefold *tanmatra* modulates both anti-parallel sets of *indriya* - one upward, sensory and the other downward motor.

We need conceptual pegs; and, from Chapter 0, we know that images (called models/ metaphors) can only do their best. Thus in *fig.* 7.5: Universal Menus a preliminary attempt was made to iron out anomalies and correlate archetypal (and therefore immaterial) systems with modern science; and in Chapter 17: The Information Domain the sense connection is developed. It is as if you thought gross gas, liquid and solid states described the cosmic action. You think gross electronic impulses make mind; but just as subtler levels - atoms, molecules and particles - were delved so, unseen, deeper layers mediate between (↑) a body and its mind's perceptions; or (↓) between mind's orders and their physical effects.

In the *microcosmic*, animate case five *tanmatras* and their *pranas* are also associated with signal transduction of a non-sensory, non-motor kind. The 'transducers' are called *chakras*. From such focal points are distributed, by harmonic or resonant association, the programs that give rise to physical form. How, you may wonder, can a 'quality' (a 'causal ideal' or *tanmatra*) also involve harmonics and a morphogenic program? Think again of melody. Musical code. To repeat, the radio code is not chemical but vibratory. Each *chakra*, as *figs*. 17.9 and 17.11 relate, is associated with the expression of a material state. This is, in turn, related to the hierarchal construction of the human body (and, indeed, all others). Physical and metaphysical are interlocked; from the latter is evolved, in orchestrated concert with its *DNA*, a body.

In parallel *macrocosmic* case each focal point or so-called 'sun' for the dispersion of psychological energy (*prakriti*) at a particular 'level of frequency' on the conscio-material gradient is also called a *chakra*. Theset universal *chakras* are known as 'lords' or 'power-houses' of internal regions. Since the exhausted, physical zone is a special case, non-consciousness, it has no *chakra* of its own. Instead its power is derived from potential matter in the form (as described by the human lexicon) of a 'radiant sun' or 'thousand-petal' lotus; this star is the power-house of our universe; from this psychic concentrate pulsate harmonic streams whose order subtly underwrites the basic forms of our material cosmos. By now it's clear that external suns (or stars) radiate physical energy and that the function of internal suns (*chakras*) is to radiate metaphysical information. They do so in the form of archetype. The function of an archetype is signalling; it is an information-donor to the world of physics (which includes biology).

In the case of *H. sapiens* the physical correlate of sub-conscious archetype, nervous and hormonal co-systems, ramify into each part of the body. Each 'string' of signal translation correlates nervously with a specific organ of perception or an organ of action (made of muscle, bone etc.). In describing the connection Natural Dialectic diverges slightly from errors in the old systems. For example, it used to be thought that 'ether' (vacuum) carried sound but this is easily proved incorrect. Thus the association of space with sound, hearing and the organ of ear is incorrect. Natural Dialectic associates space with subconscious archetype or potential matter. Electromagnetic fluctuation is a vacuum we call light. Light and electro-dynamics are the medium of *H. electromagneticus*. Light, heat and electric current are also stimulants. Moreover the Samkhya relates air to pressure (rather than sound) and smell (molecules in air) to solid earth; whereas dialectical logic prefers that solids be related to pressure, touch, external temperature and pain; and in action, exhaustion of a solid, that is, egestion. There are thus passive or sensory facets of signal translation expressed physically, in hierarchical order, as sight (eyes), hearing (ears), smell (nose), taste (tongue) and skin (all-over touch, temperature and pressure-pain sensations). Also active or motor faculties expressed physically as speech (tongue), manipulation (hands), locomotion (legs), procreation and elimination (anus). As we'll see (*fig.* 17.10 and Chapter 19) the whole body is divided into three main sections. At the top comes the head or (*sat*) information centre; below but above the diaphragm is the passively informed zone dedicated to (*raj*) energy production; below the diaphragm is the zone of (*tam*) exhaustion (liquid, solid and procreative). Such orderly, non-haphazard metaphysical/physical construction, along with the whole sub-conscious structure of archetypal memory and including signal translators, is unknown to modern science and psychology.

Signal translation amounts, in conscious organisms, to an encoded connection between consciousness and the non-conscious world of body and its environment. In different creatures the quality and emphasis of their sense and motor faculties as well as their consciousness varies. For example, dogs live in a scent world you are absent from; and they are absent from the intellectual sphere. Not all organisms, however, have apparent organs of sensation or activity. Nor, in the case of sleep or knock out, is a form awake. Do bacteria, plants and fungi ever wake to sensitivity? What do the earless hear or eyeless see? In our terms do they have a life at all? Yet these dormant lower classes can detect and accurately respond to stimuli. They don't have faculties of signal translation since these connect non-conscious entities with consciousness. Micro-organisms, fungi and so on are not mindless but lack wakefulness and, thereby, choice. In this case (see *fig.* 15.9) electro-dynamic signals, borne on ions, currents and electromagnetism, interface with instinct and the morphogene alone. We can also by-pass consciousness in their defaulting kind way. An example of such signal translation occurs when a psychological phenomenon (e.g. excitement or fear) generates the production of adrenaline and associated effects; or when chemical factors or stimulating forms colour a train of thought and influence a mood. This is not to say that biochemistry is without its own micro-hierarchies of control and cycles of purely physical government. Indeed, dormant systems are practically reduced to cybernetic, coded chemistry. What control did sleepers ever have? It is, however, to affirm that each part, mind or body, may trigger events in the other. Reflex, chemical signal from body, reflex archetypal signal from subconscious mind and, whenever dormancy awakes, purposive signal from conscious mind each play a part in life's overall balancing act, homeostasis. But brainless brains and natural no-brainers each robotically react by instinct and entrained electrochemistry.

In fact, the whole ecological suite of life on earth shows different permutations, proportions and proclivities of psychological and physical qualities. Lower in the conscio-material scale various possibilities are reduced or excluded. In plants, for example, consciousness is excluded. Of archetypal faculties emphasis seems swung towards practically sole use of body-linked morphogene with its associated physical forms and biochemistry. Single cells also operate in this reduced condition of sub-consciousness. Conversely, the higher on the scale an organism appears the more it is 'released' to realise the potential of consciousness. At the high end of the scale all five two-way signal translators are expressed in balanced proportion along with a sixth faculty, awareness. These factors are coherently reflected in the 'full-flowering' hierarchy of metaphysical nodes found (see *fig.* 17.11) in the human morphogene. Such a system of construction and operation, linking ranges of biological feasibility with environmental preference, instinctive behaviours and psychological appreciation, excludes chance but exudes logic. Bodies are (Chapters 17 and 19) information incarnate.

Instinct

The other two subroutines of archetypal mind are, rather than a channel whose main function is translation, repositories of behavioural and morphogenetic information.

The former involves mental routine, *instinct*. Instinct is a term that covers, in broad terms, reflex actions that are neither taught nor acquired. Such actions are always purposive in scope and involve behavioural strategies of varying complexity. Purpose, strategy and tactic anticipate an outcome; they all involve, as well as the requisite biological tools, conceptual know-how and, therefore, mind. The issue is not the operation but the origin of programs of behaviour.

Some instincts, such as blinking in bright light, the epic journey of a tiny joey from its mother's birth canal to pouch or every baby sucking its mother's nipple, are immediate devices for healthy survival. Others involve complex sequences of activity woven into the fabric of an organism's behaviour. For one example out of many possible observe spiders closely. Not only are they abundantly endowed with multiple pairs of spinnerets but seven glands to spin different kinds of light, elastic silk that is, size for size, much stronger than steel. Tarantulas can even exude silk from spigots in their feet. What use a tool without the skill to use it? Are web-weavers and high-wire hunters not intelligent? If not in what does their 'intelligence' reside? Some spiders vary their superb designs each night; others emit appropriate pheromones to lure different kinds of prey. There are myriad brilliant variations on the 'spider theme'. From what source does a spider draw the skill to use its tools - appropriately?

What about the nests that most birds feel compelled to weave? And songs they sing while working? Even if they are isolated from the time their egg is laid, many kinds will develop to burst spontaneously, untutored and exactly into their own species' song. In another experiment similarly deprived warblers, blackcaps, were exposed at the time of their normal migration to a clear night sky. They became agitated and flew off in the right direction. How could these tiny creatures orient without guidance or practice by starlight? Have they in their birdbrains 'an innate image of the firmament'? If not, how is their response explained? If so, then how is such a mental map explained by genes, proteins or even so-called 'hard-wired' nervous circuits? How, if they exist, did such 'recognition circuits' first evolve and link into the program for development so that they were preserved? Is predisposition, ours for language theirs for song or orientation, programmed variously into archetype?

Not only birds but bees, ants, beavers and a host of other organisms construct complicated homes. With butterflies, fish, turtles, whales and so on they can navigate and harvest food in intricate, specific ways. Their programs are not blurred. They are distinct and accurate.

Instinct is innate and unlearned ingenuity; it is problem-solving often in sophisticated ways. Whence does the obvious and natural intelligence, bestowed by instinct and essential in one way or other for an organism's life, derive - especially in tiny brains? An ethologist, a specialist in animal behaviour, can point out tens of thousands of examples of procedures fixed in all sorts of creatures. But he cannot begin to explain (except in terms of 'scientific' myth or just-so stories scooped from materialism's only licensed information-centre, genes) the origin of their precise patterns, the 'unique biological adaptations' that accompany them or even the nature of a memory. How, for example, could blink or flinch from an approaching missile be by chemistry alone (especially if it misses)? As well as life on the web instinctual memories include call-signs, food-finding, hunting strategies, migratory navigation and complex, distinct breeding patterns such as courtship, home-building, nursing and rearing; and, when life bodes ill, what about a creature's instinct for effective natural medicines (zoopharmacognosy)? Instinct's indelible gift, translated into motivation, is survival. Are instincts only the preserve of animals? Or are these automatic choice-makers, that lend security in proportion to a lack of conscious thought, fundamental to every kind of organism and, therefore, basic to biology?

Bottom-up instinct is, by philosophical preference, only chemical. In the 'official, licensed view' it is recognised as a behavioural 'program' somehow 'hard-wired' into an organism's genetic constitution, a context that evolved by chance. Where else, if nerves are absent, might you identify memory or the storage of information? Of course, chemicals react with each other but how could this generate subjective perception or variable response? If it's all material then how do *DNA* and proteins make it work? How can chemicals promote complex and accurate behaviour? How could oblivious atoms, even a large collection of them, ever know the world and why, however intricate their congregation, should they 'want' to survive? No doubt some cognitive neuroscientists, having professionally and philosophically deleted the concept of metaphysical mind from their repertoire of response, persuade themselves that such motivation, along with will-power and 'free will' itself are simply illusions. Who is in reality deluded, hypnotised by his own materialism? Vague, unsatisfactory answers concerning instinct include all the usual animistic components of materialism.

Bottom-up genes make and modify all instinct. Genes alone can make a spider spin its web and even genes, at root, will let you think! Maybe, however, you don't buy that half-truth. Maybe thought and instinct aren't just functions of your *DNA* and a magnificent and as-yet unexplained development, the brain; maybe that, *top-down*, there's a prior place for immaterial mind. How, after all, can only protein build the 'nervous hard- and software' of behavioural complexity? How can instinct 'pop up' in a baby spider or a warbler ready-made, untutored and complete with no mistake? If such routines are not genetic you must ask (unless, as regularly in science, its philosophy prevents) whether they are 'made' of immaterial mind's sub-conscious part, a part of what the Dialectic has in mind as archetypal memory. Just as the broadcast of generic '*ROM*' survives the death so it precedes the birth of any individual aspect of biology. Cell, organ, organism are like aerials that resonate with mind-borne information; and, if this aspect of an archetype makes sense, why not the other couple of mnemonic channels? ***DNA* sequences are already known as a chemical data storage system that is incredibly responsive, specific and precise. Could not an immediate, radio system with its conceptual broadcast, archetype, be prior, proactive and equally precise?** Why shouldn't an essential link, mind-body psychosomasis, engage communication of a subtle, highly programmed kind?

The *top-down* point first made in Chapter 15 needs hammering home. **The evolution of instincts is a Darwinian black box because their exercise is not only physical.** Indeed, at root, they're immaterial. Metaphysical. You need a framework, a broader box that's outside but inclusive of the biochemistry. Within this box the instinctive aspect of mnemone works as an ancient, indelible and often complex inscription. **It is a psychological datum, immaterial software, an 'under-writing' of intelligence, a common theme whose various strands are woven in the warp of archetypal memory; it is, therefore, an aspect of a universal mind**. Instinctual information is exchanged 'upwardly' with conscious mind and 'downwardly' with body and its chemicals by way of signal translation and (Chapter 17) a combination of causal power points, the morphogenetic *chakras*.

Take any form of life. Just as the influence of environment or genes evokes a slightly variant set of phenotypes, so these factors modify expressions of an instinct; and if you add the factor of controlling, conscious mind then further variation on the archetypal theme is able to occur. 'Learnt' versions of behaviour accrue. General instinct, as it rises into consciousness (see *fig.* 15.8), is modified by patinas of personality; it is converted into habit or intrigue. Indeed, the more that conscious input plays a part the more that strategies to satisfy instinctive drives proliferate. For example, in the human case reflexive instincts surface 'upwardly' as passions for wealth, power, adoration, sex and sybaritic tendency. Each evokes the relevant behaviour whose emotions swell easily to forms of lust, pride, greed, attachment to the pleasing objects and sensations of the world and, in case of a desire's frustration, anger. In this aspect our sub-conscious motives, tarted up by a cosmetic of sophistication and civility, were grouped by Sigmund Freud into an '*id*'; maybe, in conjunction with the morphogene, they are what C. G. Jung meant by the 'collective unconscious' or 'mythological memory' of mankind. At any rate, instinct affects and is affected by sensory information. Its passions affect metabolism and *vice versa*; and it drives many forms of self-centred behaviour.

'Downwardly' and body-wards its dormancy interacts with correlated power points of the morphogene; or, if you like, it resonates with localised '*chakras*'. These in turn correlate and, Natural Dialectic logically suggests, electro-dynamically interact with a physical body's glandular hormonal and autonomic nervous agencies or, to coin a phrase, its 'informative biochemistry'. The sub-cerebral cradle of instinct and associated emotions is, in vertebrates, the limbic system. This system, which links cerebral, diencephalic and midbrain structures above the brain stem, is the physiological basis for pleasure, pain, instinctive drives and their accompanying emotions. Its sub-conscious remit is closely allied with the body's chief biochemical regulator, the hypothalamus. From this 'royal' gland, which is also closely linked to a sense-intermediary called the thalamus, hormonal directives cascade through the body's chemical 'bureaucracy.. They are tasked with the satisfaction of practically every major homeostatic exercise and its corresponding physiology (e.g. escape from the tensions of thirst, hunger, cold etc.). Thus sub-conscious commands are expressed through a physical hierarchy that involves the appropriate nerves, lower hormonal glands, organs and, eventually and most outwardly, muscular behaviours. *In both concept and practice the whole information system is (see Chapter 17 for detail) obviously, pervasively top-down.*

Take, for example, sexual stimulus and response. Either hormone levels or a variety of sensations can trigger varying intensities of response; so, on the other hand, can memories and creative imagination. The patterns and purpose of sexual behaviour are affected by genetic constitution, chemicals and other physiological agents; on the other hand neither instinct, thought nor experience of sex is physical. Does this sound strange? The fact is that all experience is subjective, metaphysical, in the mind. And the design of body has defined what mind will feel. No more than a TV program is *caused* by the controls of an individual set do neuron or hormones *cause* the programs of sexuality. They no more cause instinctive behaviour than, say, a penis or an ovary. If they did a single hormone, testosterone, would in different creatures somehow cause all sorts of different patterns of male behaviour; similarly oestrogen or even a complex of hormones do not by themselves create the various kinds of female apparatus and emotion. Similarly, how can a nucleic acid database for making protein create a mental pattern? How could genes or hormones alone cause a weaverbird to construct its nest, the urge for some birds and butterflies to accurately migrate or a spider to spin its own complex sort of web? Whence, therefore, derives the pattern of behaviour?

No pattern is half-formed, no instinct semi-skilled or bungling. No doubt, to the degree that conscious awareness is present, adaptive variations on innate behavioural potential (i.e. instinct) can be conceived and learned. Instinct can be steered, let loose or blocked up. Thought is voluntary and, by definition, makes non-reflex decisions. Thoughtfully or not, all animals register the topography of their local environment and note changes that occur in it. This applies to bees, ants, birds, frogs and monkeys. But to the degree that voluntary thought wanes involuntary instinct waxes. Order stays on top. As reason fails so instinct prevails; without a second thought pre-constructed program runs its automatic way. Fixed rigidity, operating in association with the morphogene, takes expertly over to ensure survival. Where conscious perception is virtually or wholly absent (as in plants, fungi etc.) only instinct working *downward* through the morphogene is left.

Such reflex process is observed in chemical responses known as tropisms, nasties and so forth. Why does that fact eliminate instinctual control, subconscious process or the presence of an archetypal government? Recall (Chapter 8) the 'crescograph' of Jagdish Chandra Bose and his view that plants have, although dormant, an emotional life. 'Green fingers' know that this is so. Do you not chemically respond to voice or music? The human voice has caused the rate of sweet pea growth to double because of change to gibberellic acid flows. Of course, a chemical is just one measure of emotion or intent - do you think, as science might, of excitement as a form of your adrenaline? Plants can 'taste' the earth for chemicals. If there's a gene that lets plant roots detect soil nutrients or a source of moisture and grow thence, what does that mean? A gene means protein. What does this protein activate? Of course it triggers biochemical response but 'knowing where' or 'growing towards'? How does 'chemo-taxis' work? How do gradients of chemicals impel unconscious sperm towards a goal or dry roots make for water? 'Lower' is less conscious on the scale of life; it means less flexible intelligence, more emphasis on instinct and the rigid mechanisms made by chemistry. But even in the 'lowest' sort of organism, say bacteria, is emphasis entirely mechanistic?

A sleeper moves his tickled foot but needs a larger frame of reference (including physical development and instinct) to complete the dormant job. Are autonomic or instinctive actions taken without reference to overall control? A 'brainless brain' in plants can also use a sense of touch to coil or feel the wind; or dormant 'sight' to track the sun or know intensity of light. In your own case embodied mind employs a subtle biochemistry to implement its orders and to sense the outside world; and you are unaware of it. Can reflex cart of biochemistry, when placed before the horse of mind, thereby deny the horse exists? Does adrenaline *cause* your excitement or is it an *effect*; is the experience itself just chemical and if so how do atoms garner instinct or chemicals raise subjectivity? Crucially, was the whole biological 'corpus' of reflex mechanisms wrought without mind; or was and is it shaped by mind to work with it?

Of course, any program's expedition involves physical kit. In case of instinct genetic, hormonal and other apparatus trips the 'job'; but is the job derived from them or do they serve its plan? If so, whither did this plan, a pattern of behaviour, evolve? There is nowhere evidence that an instinct evolves through stages of inferior skill. **Natural Dialectic therefore proposes that, because it is the cause of such an obvious effect as behaviour, instinct is our closest sensation of the immanent, numinous presence of archetypal memory**. In accordance with its systematic, hierarchical description of existence instinct is recorded as a subroutine in the sub-conscious, metaphysical archetype of any given type of organism. In man it constitutes a subset of the mnemone *H. archetypalis*. **In the case of gradual neo-Darwinian evolution, on the other hand, instinct and associated physiologies are yet another unexplained black box**.

Morphogene 1

So too is morphogenesis. While instinct is the functional subset of archetypal memory, morphogenesis is the structural. From its super-state, teleological top we have now dropped to the sub-state, passive tail-end of psychology. *This is realised as the third, base subroutine of a typical mnemone, an aspect of mind closely associated with electrochemical function and our earthly coil, the morphogene.* The human morphogene is in turn identified with the medical profession's caduceus (Chapter 17) and its sequence of plexi (*figs.* 17.5 - 10).

From a *top-down,* dialectical point of view an interpretation of morphogenetic activity as chemical alone is deficient. A hierarchical structure of control would suggest a move 'upward' from the tertiary molecular structure, which acts as an aerial for the controls of a secondary, electro-dynamic level. Such quantum biochemistry it names in humans, for convenience, *H. electromagneticus*. Such a wireless body with its endogenous flows of current, fields and associated photons, is in turn a medium - as are electromagnetic waves for radio - for the primary, archetypal broadcast. *H. archetypalis* is, in its context of biological operations, the morphogene.

In short, the subtlest physical entities, massless photons and almost massless charge, are identified as the physical agents on which, in the appropriate molecular location of an organism, the broadcast 'force' of mind can interact. A morphogenetic template thereby acts as a 'higher control', 'guiding principle' or 'informative field' with respect to the shape of an organism's physical body and, of which it is a composition, integrated sub-shapes. Caducean plexi each operate at different levels of sub-conscious energy; such 'frequencies' activate subroutines of the archetypal program. That is to say, plexi act as distributors, resonators and, thereby, section address tags. In terms of a library or database they act as access-points for particular 'sub-broadcasts' of a typical mnemone. Call them callers, activators or facilitators of the subroutines of morphogene. These subroutines are controllers of a body's levels. As will become clear such levels are associated both with characteristics of mind and the various states of matter. Each of the latter (e.g. gas, liquid, solid) obviously finds its own level; and the suite of biological bodies represents permutations and emphases on this basic, hierarchical theme. Biological life forms are, if you like, images that reflect psychological conditions. They represent a range of models staked off in order down the conscio-material gradient - which might make bacteria, at the material bottom of life's list, last and least in size but not in value!

Without direct access to consciousness, a morphogene is the most closely body-linked of sub-conscious formulators and, therefore, the one that should most closely concern a biologist. It is the preordained framework within which development and adult forms are sprung; and within which physical variations of any typical organism - bird, bee, flower or human - are, by genetic nature or environmental nurture, derived. The elaboration of its human shape is the subject of Chapter 17's 'The Logic of Embodiment'.

Meanwhile go and closely examine, of thousands of possible subjects, a Venus fly trap. Explain exactly how its predatory mechanism was 'developed' by haphazard steps. Next cup a bloom of orchid in your palm. The flower will be part of a whole plant. It will also, relatively speaking, be one of nature's simpler but evidently purposeful 'designs'. Yet how is it just the product of *DNA* base or amino acid sequence? How was its dynamic development actually registered in the seed? Are genes and proteins themselves capable of such obvious, large-scale teleology? Is chance? Or, as in the deliberate arrangement of processed materials to construct and perform computing operations, are chemicals only the tertiary, tail end of a whole, conceptual story?

A morphogene is the generic form of storage for a biological program. It is, if you like, a 'control satellite' off which specific *DNA* and other molecular signals are bounced. Such reference integrates the operational hardware of a unit called a cell or, cohesively, a multi-cellular body. In the human case it is a template through which, using the medium of *H. electromagneticus*, the construction and maintenance of *H. sapiens* as a conceptual whole is orchestrated. It will, in other words, include a developmental sequence with, as a final 'still', the adult form. The morphogene's adult subroutine of *H. archetypalis* remains an ideal Apollo or Aphrodite. It is not in archetype but our genetic nature and the circumstances of our nurture that we differ. Archetype's for all the same but around this negentropic, metaphysical matrix any individual's imperfect, damaged or aging physical embodiment, subject to entropy and death, declines. It is, by the same token, a psychological lattice around which a biological body will (unless disrupted by such external factors as 'catastrophic' injury, genetic mutation or nutritional deficiency) self-repair. As already noted, in this respect the template is well known to doctors and vets as a '*vis medicatrix*', natural healing power or regenerative capacity of biological bodies.

If it is supposed that the electrical configuration of a molecule (with its various electromagnetic forces) acts as an address tag, then its presence could act as a signal. If you pulse a specific 'barcode' to a reader you would expect an appropriate response. So, precisely with computers and their algorithms, one key hit gives routine A, another routine B and so on; and if the biochemistry beamed, from *DNA* and/or associated protein, configurative signals to its 'satellite' morphogene you'd expect this 'device' to bounce back appropriate, program-integrated instructions. If, therefore, you manipulate a key-signal you might expect to trigger either an error or, in its own right, an alternative routine. Stem cell research might illustrate the case in point. If you learned, using your technology, to manipulate the signals and could trigger routine A instead of B; and if routine A was at the natural program's start then you might expect to recreate stem cell markers and consequent cell reformation. From this stem cell you'd guide, manipulating further signals, chosen redevelopment.

If the morphogene is an aspect of archetype; and if archetypes are memories in universal mind; then its 'level' in creation serves as an information store. Passive information. Language is the ability to use symbols in unlimited combinations and archetypes amount to a 'grammatical and syntactical construction' according to whose order physical nature is derived. If the language of physics and chemistry is built of simple sentences the language of life is bundled into complex texts. Norms, patterns of behaviour and the laws of energetic operation are derived from archetypal memory. Indeed, all memories including your own personal ones constitute context in which your present physicality is shaped. The generic term for such reference files is sub-conscious mind and the subliminal energy of its level is called *prana* or, by Taoists, *qi*. Such current is metaphysical; it is subjective and therefore inaccessible to objective science - which typically responds by denying its existence and effects. Nevertheless it underlies such clinical systems as acupuncture, yoga and reflexology, whose principles are based on the patterns of subliminal energy. Oriental massage, for example, emphasises balancing and strengthening the flow of *qi* in the physical body, thereby promoting resonance with archetypal body. This means, in effect, the promotion of self-healing. Also, after many years of mainly Russian research, there has now emerged a routine understanding that electromagnetic appliances can, in their secondary, sub-archetypal role, stimulate 'stagnant' or 'blocked' bioelectric flows. They revive such archetypal dynamism as is, of course, synonymous with the abovementioned '*vis medicatrix*'. This 'bio-force' (as '*vis*' means here) speeds growth along morphogenetic lines - which reversion to health we call healing. It also reduces certain kinds of pain in a way that some, having no notion of archetype or its electromagnetic influence, have called 'miraculous'.

Archetypal program is a magnet, attractor or prior organiser. Thus the real psychosomatic question asks precisely how connection is made between informative mind and forces, atoms and molecules of the phenotypic composition that confronts you in the mirror - your body. A simple magnetic field exerts influence; it organises iron filings round the magnet. A mind-field tries to organise the world to fit its own thoughts and desires; and both conscious and sub-conscious patterns can exert direct influence over heart rate, breathing, body chemistry etc. In a similar way, as an electromagnetic broadcast is transduced into pictures by a TV set, Natural Dialectic is suggesting that a 'template of health' exerts influence over physical agents such as the genetic code, particular materials and higher level phenotype. Neither will act within the format of a coherent program without this animating influence. *DNA* or protein outside the context of a living cell is impotent, useless and inanimate. The bioelectric and morphogenetic constituents of its composition no longer resonate. The tertiary, molecular level has lost guidance and the organism's dead.

An organising template must exist independent of the content that it frames. A machine is made of parts but coherence, integrity and purpose are its frame. *The whole is greater than the sum of its parts.* Similarly, chemicals and their interactions constitute the divisible parts of an organism but they do not sum to its structural and functional whole - the coherence, integrity and purpose reflected in its shape. The role of a morphogene is the unification, coherence and thereby governance, in terms of resonance with electrical patterns, of a life unit, an organism. Biological *H. sapiens* is, as well as the phenotype of a genotype, the gross expression of electrical interactions that constitute *H. electromagneticus*. Both are physical, inanimate and essentially insensitive. Their reflex patterns are themselves a gross reflection of an independent organiser, the sub-conscious frame called *H. archetypalis*.

A molecule or cell's capacity for resonance has already been analogised to the aerial of a radio. Natural Dialectic logically connects mind with matter at the level of electric charge and light. For example, *H. electromagneticus* is the antenna/ electrical motion; it is the receiver/ transmitter of which *H. sapiens* the set's hardware. *H. sapiens* has an overall shape that consists of a hierarchy of subsets or subroutines called organs, cells and sub-cellular factors. *Each of these has its natural electromagnetic configuration that acts as an aerial to a subroutine of the 'broadcast' of ROM (read-only-man); down the energetic chain all cellular processes and physiological functions are subsequently determined by orderly electrical currents, the transport of charge and associated electromagnetism.* Physical damage or malformation at any level compromises information exchange with this overall homeostat in its role as the *vis medicatrix*; a specific part or even the whole form is 'stressed' in a way that tries to reassert the norm, that is, to heal. The morphogene, *H. archetypalis*, is immortal. Its incorporeal blueprint never changes, never fails. It is physical components that suffer wear and tear. Indeed the 'human frame', *H. sapiens*, may suffer accident or illness to the extent that it no longer responds to its own morphogene and, irreparably, ceases to work. It will certainly, according to the rules of entropy that govern this material dimension, grow old and die. *In short, the causes of disease, breakdown and mortality are physical not archetypal.*

Everyone knows that mind affects body and *vice versa*. Just as sub-conscious processes serve the master function, waking consciousness, so the psychology of sub-consciousness overlaps with the electro-physiology of its nervous, hormonal and other biochemical servers. *Another more surprising and controversial way of saying this is that the nervous system (and the rest of human physiology) is an expression of its sub-conscious patterns.* In dialectical words, the latter constitute a standard morphogene for various physiological norms; and, integrated with these, a standard for developmental morphogenesis. They provide, with the cooperation of *DNA* and biochemical metabolism, a type's healthy biological structure, function and instinctive response to its circumstances.

Morphogene and *ROM*? In this analogy the software's output is its hardware. Software oversees the automated lines that make and keep the bio-product, body, running. The business is entirely self-contained. And while bodies can malfunction the morphogenetic program exerts a continual 'homeostatic' pressure to conform to its norms, to keep fit. This psychosomatic link can be actively promoted by positive thought as well as physical disciplines. On the other hand, health can suffer from negative psychology, infection, structural defects, injury or, in life's 'psycho-neuro-immunological' department, the prescribed operations can be distorted by signal malfunctions (as in a defective radio set) caused by toxins and drugs. While doctors are fairly clear about what constitutes optimum performance by the physical system of information exchange (the nervous and hormonal communication systems) psychologists debate what constitutes the thinking behind optimum mental and moral health. Certainly, though, health and the most rapid return to its state of grace is not only a question of the physical frame but sub-conscious 'exertion' of *vis medicatrix*, positive psychological attunement with its ideal and conscious will-power.

In short, it would seem that the source of biological shape is above the scientific glass ceiling. It can be inferred but not, like its physical agents, touched. *Top-down*, it rests in archetypal memory.

Synchromesh 2 - Psychosomasis.

In a classic experiment the current from neuro-physiologist Wilder Penfield's probe in a patient's brain evoked a clear, repeatable subjective experience in the form of a memory. This experience involved the patient hearing a song as if it was being played at that instant. What you might almost call a 'track' or a program therefore extended far beyond a barely disturbed arrangement of atoms and electrons. How does electricity become experience? Mind and body interact but what is the *modus operandi* of this interaction? Before we take a closer look at the physical side of the equation let's ask again how the psychosomatic transmission of information actually works. How do mind and matter correlate?

Synchromesh 1 dealt with personal and typical mnemones and the way they mesh with each other and with conscious awareness 'above'.

Are you surprised that, when you think, imaging techniques can signal correlated brain activity? You're not just a head. Thought must connect with body to achieve material goals. Thus there's psychosomatic synchromesh. Synchromesh 2 deals with the way psychological meshes, through the medium of brain, with physical 'below'. It deals, in this case, with the junction of all three aspects of typical mnemone with the physical side of the psychosomatic interface - an electro-dynamic, resonant medium called variously your 'network of light' or *H. electromagneticus*. Psychosomasis is simply a word that identifies the process of translation, a transduction of information between (*figs*. 15.4, 15.5 and 15.8) *H. archetypalis* and *H. electromagneticus.*

Two hundred years ago John Dalton, the founder of modern atomic theory, could not see atoms. We still can't and use analogies to picture them. The analogies of a television or a computer with mind are useful but limited, possibly as limited as the analogies used to picture an electron or describe quantum behaviours. We have to use analogies and, until science can measure mind-film, it is the best we have to describe the medium of thoughts, feelings, ideas, memory, dreams and sleep - all of which exist, don't they? If you say they don't, then whose illusion are they?

An electron gun projects prefabricated pictures onto a television screen. What are perceptions and on what 'screen' are they 'developed'? The science of psychology is, subjectively, one of empathy and inference so how do you properly measure or understand its 'bits and pieces'? Consciousness is bad enough. Your own mind is as incapable of photographic description as an electron and immeasurable as a dream. Its physical science is dubious but the 'flaky' psience of sub-consciousness, *prana*, mind-film and so on seems far worse. Impatient materialism determines not to lose its nerve and to stick with only 'hard' parts such as the molecular biology of neuro-physiology; and to reduce the physical 'vagueness' of mind to electrochemical formulae. On which basis it could be a product of evolution. Is this not retreat to safe and solid ground? Absolutely not because, for example, you will never track a broadcast by fiddling with the set's components; nor, regarding memory, will you learn anything of past broadcasts by grinding up components into chemicals. Thus, engaging matter, you miss transmission and its meanings. You miss practically the whole point of the kit. Thus, ignorant of any unconscious morphological attractor, archetype, biology can't find the root of nature's purpose. Therefore, it attributes this, essentially, to genes. Genes, dialectically, are secondary.

At least we experience consciousness. Sub-consciousness is a murkier inexperience. Can three familiar analogies help tease out the possible nature of psychosomatic operations? The first (Chapters 0, 5, 6 and 13) is with radiant energy characterised by concentric rings or a conscio-material spectrum whose 'electrical' awareness interacts with 'magnetic' mental fields. In this way the world is seen as a suite of wavelengths, a layered series of 'fields of action' in each of which specific patterns of excitation (called behaviours) can occur. This layered series, which includes regions of mind, ranges from psychological luminosity to physical light's total loss - the absorption of blackness. Between these extremes the heavier, darker or more negative is a particular creation, the lower its 'frequency'; the greater the involvement of consciousness with its own creations, the more its 'purity' is wrapped up and lost in specific diversions. Just as physical descent involves loss of energy psychological descent involves loss of awareness, loss of information, ignorance and darkness - psychological entropy. Chapter 13 and *figs*. 13.3 and 15.5 describe the psychological hierarchy of 'wavebands' that appear as consciousness is lost. At a certain point the 'rainbow' of mind's spectrum is benighted. It is benighted either in the moral quality of its content, in its retreat from principle into trivial or perverse practices or in its condition of sub-consciousness - sleep. Although unaware of itself subconscious mind still computes in the dark. Of its activities dreaming is a trivial, subjective superfluity. *Its main, incessant and important operations are archetypal, reflex but crucial for engagement with a body*. Sub-consciousness is a condition of mind not matter; but light, electromagnetism and electricity have been identified as candidates for the physical side of two-faced psychosomatic exchange.

The analogy of spectrum is incomplete because (*raj*) energy needs to interact with (*tam*) substance in order to create shapes. *Therefore the second analogy, characterised by the ziggurat, is with states of matter*. It is exemplified by *figs*. 15.1 to 15.4.

The third analogy, in which the first couple are combined, is with resonance. Resonance, whose orderly aspect is characterised by the main dialectical analogy of music, is the tendency of a body or system to oscillate with larger amplitude when disturbed by the same frequencies as its own natural ones. It therefore involves the vibratory transfer of energy. Such transfer is an integral part of all vibratory systems wherein waves interfere with/ destroy or cohere/ amplify each other. Its vectors of (*tam*) division and (*raj*) unification well illustrate and are embedded in the dynamic of Natural Dialectic. Science is familiar, for example, with nuclear magnetic, electromagnetic, electron spin, acoustic and mechanical forms of resonance. Acoustic (musical) resonance involves the sympathetic vibration of stretched strings or air in pipes; thus it is easily understood that, as well as tuned circuits, quartz crystals and so on, musical instruments are resonators. Indeed, the motion of waveforms is closely associated with harmonics. Sound waves are governed by the law of harmonic relationship whereby notes of the right frequency combine into chords, and chords and notes in time into harmonies. Resonance occurs when one note or chord vibrates in harmony with components of notes or chords in a different octave. One body vibrates in sympathy with another. A similar sympathy can occur in the oscillations of electrical and mechanical phenomena. Energy is amplified and transferred by resonance and attunement. Common examples of electromagnetic resonance also include tuning a transmitter/ receiver and photo-electric initiation of the photosynthetic process, that is, the first step in life's chemistry.

Are atoms harmonic oscillators? Are quantum harmonics reflected in larger visible dimensions? Are crystals an example of this coordination? The answer's yes; and the effects of harmonic vibration on jets of gas and flame in air are dramatic. You can also make 'electromagnetic plants' whose flowery patterns arise in a dish of ferrofluid placed over a single iron-core AC coil; the fluid accords with the lines of the field. Figures produced by the resonance of sand particles with given frequencies of sound transmitted from tuning forks to the medium of a metal disc reveal patterns clearly related to energy (see Chapter 11, Chladni's dish). The ancient architecture of leaf, fern, wood grain and many other kinds of morph appears. This makes frozen music; it is Pythagorean sound and archetypal geometry combined.

Rhythmic beat, coherence and harmony. Their influence draws the entire world into order. It rings bells, chimes, dances together. It vibrates things into shape and its songs, each in their own way, feel right. The heart of morphogenesis is buried deep in resonance. **A key phrase in Natural Dialectic's suggested explanation for the wireless, psychosomatic transfer of information between subconscious structures of the mind and the physical plane is 'resonant association'.** Energetic frequencies can carry information; antennae are tuned for resonance with specific frequencies; and, since the emission/ absorption of electromagnetic radiation by atoms and molecules peaks at each's natural frequency, they act as antennae. Aerials for what? For the receipt, transmission and transduction of passive information between mind and matter.

In this natural view inner action precedes and governs outer structure. The visible expression of energy is the inverse of the actual vibratory pattern; visible mirrors invisible. Between the 'dead' ground of material shapes course channels of energy. **The invisible rules. Oscillation between polarities, cycles, vibratory rhythms, the interrelationship of waveforms and resonance are at the heart of Natural Dialectic.**

Do you remember (Chapters 6 and 10) that radiance and resonance were identified as agents of law enforcement? Of entrainment and control? Natural Dialectic identifies resonance as a key cosmic actuator. Coherent, orderly vibration is, in this view, not only a function of matter but mind. Indeed, thoughts and impressions in mind are sometimes referred to as vibrations, patterns of waves (Sanskrit. *chitta vritti*) in a lake of pure consciousness. And, lower down the scale, what are the brain's key activators of the four main states of consciousness (including lack of it) but different waves?

Just as sound and light are key manifestations of physical energy, so the first, topmost impulse that gave (and gives) rise to creation is said to be a living radiance. This radiance manifests in two basic forms - sound and light. These primal ripples in the 'lake' of Pure Consciousness resound like a gong and shine with the brilliance of suns. They are called *Om* or *Logos*. The 'tonoscope' built by Hans Jenny is a device that transforms sounds into two-dimensional shapes using crystal oscillators to precisely vibrate metal plates or membranes. Thus sounds uttered into a microphone yield visual representations on a screen. When vowels of ancient Hebrew and Sanskrit (but not other languages) were pronounced the vibrated particles took the shape of their written symbols; and when the Hindu sacred syllable for primordial creation, *Om*, is correctly intoned it produces a circle (representative of the Infinite Void from which all things issue). This becomes filled with concentric (nested) squares and triangles. Finally, as the last humming traces of 'm' fade, a *yantra* or meditational *mandala* takes shape. A *mandala* is an archetypal symbol, a formal, geometrical expression of sacred harmony, structured vibration and the orderly, kinetic basis from which the solid state, physical universe derives. In this case the *sri* (venerable) *yantra mandala* symbolizes the entire, sound-drummed cosmos. It is a perfectly symmetrical picture radiated from the Sound of Silence or the Nature of Nothing. *Indeed matter is simply, from a dialectical point of view, stresses, strains or tensions in nothing and in Nothing; but there is nothing random in emergence and creation from first acoustic principles is highly orderly.* **Here, even at the frayed, weak edges of such central influence, you might concur with those who have experienced the light and joy of such cosmology when they report 'creation is in order'.**

At this point recall a rule of synergy stating that in order to interact with any particular level (that is, 'frequency band') of creation a soul must be sheathed in a body of similar constituency. In fact a nested series of sheaths 'binds' the original light until, at base, a physical body is required to interact with a physical universe. In this case we probe the question yet again. By what method is information transmitted from psychosomatic morphogene to physical form? At this radiant point we can, conversely, ask by what aerials we receive morphogenetic information. In such a scheme what place have *DNA* coils and various other biochemical considerations? How is cellular 'knowledge' transmitted to its own biochemical anatomy and physiology? How does a single cell holographically 'sense' its identity as part of a whole, say, a human being? And how might the differential expression of genetic material work?

How Does the Connection Work?

It works by transduction. What? Archetypal broadcast is 'grounded' by an aerial composed of electronic dispositions in trans-membrane voltage fields, cytoskeleton or molecules like *DNA*. It is thence translated into bio-form; or, in the dynamic case of thought, is 'grounded' by aerial oscillations made of light, electrons and special dispensations of molecular constructions (e.g. nerves or brain). As physical so also psycho-physical transduction, given the precise equipment, exchanges information; it exchanges patterns of the mind and body worlds. Thus Natural Dialectic identifies the mechanism of connection as *resonance* of 'subtle', quantum forms of matter with potential matter; and potential matter is (*figs*. 15.8-9) synonymous with triplex archetype. Such psychosomasis is very common; it occurs continuously.

A psychic or psi-phenomenon is, out of kilter with this view, commonly defined as the case when information is exchanged with a physical or another psychological system without the use of any known form of physical energy. Such phenomena are notoriously subjective and difficult to force, control or repeat in the manner demanded by a physical experiment. If they cannot be so nailed and proven could they ever happen? You might invoke improbability's base end - impossibility. Is this because they are not physical but metaphysical in origin and their source is, therefore, not susceptible? At any rate, they are classified in two ways. *Firstly*, there occur phenomena that employ the agency of sub-conscious energy and are thus related to electromagnetic influence, that is, the electro-dynamics of how light and matter interact; the place of archetypal memory related, perhaps, to quantum physics was related in Chapter 8: 'Are You Certain?' and referred earlier in the previous chapter (Psychosomatic Linkage). *Secondly*, the direct transfer of a conscious formulation (a thought, a wish etc.) from one mind to influence another. *ESP* (extra-sensory perception) includes both species.

As an example of the latter case, have you ever felt attention, swung around and found the eyes on you? Sitting at the mouth of a cave high in the Himalayas I used to try this 'trick' on monkeys. As a troupe ranged through the jungle below browsing leaves I would, from my vantage point, focus on an individual with the thought that I was looking at it. Almost always the langur (such was the species of monkey) would immediately look up and stare in my direction - whether actually spotting me or not I do not know. No doubt that some 'mind-reading' member of a magic circle will pop up telling me he easily manipulates an audience like this! Or, on those sunny days, I was deluded and imagined it!

Have you ever thought of someone just before the phone rang, letter dropped or e-mail screen was opened? Or checked and found that, as they thought of you, you thought of them? Dogs seem to know when owners will return and, in times of crisis, many people have felt the 'projected' presence of a loved one speak or be with them - either as they die or afterwards. Of course, the sceptic frowns. Do not mistakes occur, emotions fog and hope triumph over fact? Cannot quacks abuse imaginations of a system no technology can reach? What if the sceptic, as materialist, positively wishes 'paranormal' thought projection was untrue? Negative projection acts like 'jamming' in the mind-set; 'no' wishes that a project fails; a sceptic by his attitude diminishes the possibility. Like a probe affecting the behaviour of a quantum particle that it was inserted to observe, 'negative electric' influences mind. Malevolence wreaks upset, interference and, it hopes, the failure of an aspiration it decries. A stage magician must, on the other hand, spirit up success infallibly; he will have to turn his 'psychic' trick mechanically; methods of deception have to be deployed most dextrously. Does this mean that truth and trick departments are the same? In fact, experience of telepathic interchange is common. The neuroscientific adage 'brain-is-meat' or 'thought-is-brain' denies thought transfer happens but since when were lack of explanation or a philosophical rejection reasonable reason for denial? Or, where any metaphysic is concerned, for always crying 'anecdotal foul'?

The dynamics of both species of projection, voluntary and involuntary, originate in 'psycho-space'. Their action is, therefore, metaphysical, psychological and not physical. Both have been the subject of serious secular use (e.g. military espionage and attack) and research by government agencies as well as 'unofficial bodies'! They include action such as psychokinesis (affecting distant objects by thought), television (seeing at a distance but as if present), teletransportation (moving objects with mass by mind), *ESP* and mind/ mind telepathy. According to the hierarchy Natural Dialectic draws, on the conscio-material gradient the field of information (mind) precedes all energetic (or material) fields. Nor are the properties of 'psycho-space' and physics' space the same. The universal field of information, universal mind with archetypal memory, does not suffer physical constraints in terms of distance or velocity. Nor is apartness of body any bar to telepathy; its salient operator is intensity of focus of projection.

If, in a mind-world, two minds wanted to communicate then, with nothing in between, they could only do so by telepathy. But how mind/ mind connection works bypasses how mind/ body interaction can occur. Despite their similarities, however, telepathy is not the subject of this piece. It is mind/ body interaction. *In fact, psychic exchange between mind and matter is, with the appropriate devices in place, common and continual.* **For the purposes of Natural Dialectic it is defined as psychosomasis. As already noted, on the mind's side of the interface its device is called a mnemone and it occurs internally whenever thought affects muscular activity or sense data is transferred (or signally translated) for symbolic appreciation in mind; not only this but also, by the morphogene, wherever archetypal template acts on biochemistry or governs systematic shape. And on the body's side it can occur when the appropriate molecular/ cellular bioelectrical devices are in place.**

light of sight *light of perception*

According, therefore, to its hierarchical logic Natural Dialectic suggests that psychosomatic linkage is obtained by means of electromagnetic transduction through a sub-conscious channel (↑) to conscious mind or, in the motor (↓) direction, *vice versa. In such localised, specific way mind can associate with and excite the relevant physical (nervous, hormonal and muscular) effects.* Both dynamic transduction and the 'static' storage of information in memory involve code; but such code is not, in the case of either personal or archetypal mnemone, one of alphabet or number but **resonant association**. This instrument of business operates, the Dialectic would suggest, at the least massive end, that is, the top of the physical gradient - photons and electrons. In other words, at physical level the association involves electromagnetic interactions, one of whose aspects is 'matter's holy ghost', light. Vibration carries information, information travels in the wave and light, bestriding real and virtual worlds, packs plenty in. What message does an animated broadcast send; what can you learn from an excited atom or the radiant stars? Electromagnetic spectra, 'chimes' or resonant associations are as specific as address tags, fingerprints or barcodes and, at the same time, made of virtually nothing. They are only patterns on the 'surface' of a vacuum state, keys in a quantum field from which physical energies are spun but which is, at the same time, linked to sub-conscious energies. **Vacuum state, which is by definition nothing physical, is identified as the psychosomatic border, the point of linkage between mind and matter**. In this case brain appears as much a vacuum-state computer as an electrically operated one. It is a complex crossroads for the inter-conversion of physical and metaphysical signals where molecules can act as antennae or, if you like, inter-converters. But you don't need to have a brain to engage, at least in its sub-conscious aspect of a body's mind, its archetype. Any single cell can vouch for that. And, as has been explained, the other face of archetype is a subtle, wireless substrate, an electro-dynamic medium whereby immaterial meets material, metaphysical controls the physical, psychology locks to biology and heaven really links with earth!

Thus, fallen into place at base, there emerge 'wired' molecular and aggregate, bulk forms. It is not that one sort of cell (for example, those of the cerebral cortex) is more capable of sensitivity than any other. In fact *no* biological component is sensitive. No doubt, neurons bear coded data (that is, passive information) to and from the threshold of mind but they know, in themselves, nothing more than any atom. Central *HQ*, your brain, is a physical instrument of signal exchange but the conjunction of an exchange, no matter the complexity of its nervous wires, is no more sensitive or knowledgeable than the wires leading to and from it. Is *DNA* not the enduring, physical repository of information cradled at the heart every cell. What about its metaphysical analogue? While conscious mind is centred where we think, sub-conscious mind extends its repetitious, 'holographic' or even 'fractal' archetype from every cell. *This top-down version of events precisely contradicts one saying that sub-consciousness dwells in brain alone or that mind is simply a side effect of neural connections; it states that the archetype of every cell is interacting everywhere with the electrical dynamic of an organism's textbook form*.

If all this is difficult from a non-hierarchical, materialistic perspective it is because, firstly, the nature of immaterial mind and memory are not well understood and, secondly, the standard view of chemical interaction is not one that involves specific, sometimes complex resonance. Neither memory (and, therefore, archetype) nor resonant association are easy subjects to tackle in a laboratory. A conclusion of this section is, nevertheless, that science will have to grasp its church anathema of metaphysical influence and accept, in a controlled, succinct and logical way, its presence. It is one reason why the whole book, in attempting to understand the operation of active (metaphysical) on passive (physical) information has identified its likely location and mechanism - for humans it is *H. archetypalis* (memory man). A will o' the wisp, *H. electromagneticus* has always 'ghosted' the physical scene but metaphysical *ROM*, the morphogene, is unlikely to be, in terms of scientism, an acceptable candidate in helping to explain both our substance and our origins. It is, however, crucial to a coherent explanation of the body of *H. sapiens*, biological organisms and the transcendent aspect of biology.

To summarise, such connection across the *PSI* interface (*figs*. 0.7-11) trades continually throughout your body. It is through this psychosomatic process that you live an earth-bound life. Physical semaphore (most obvious in nervous systems) involves the motion of electrical charge. *It is therefore proposed that a 'diaphragm' of information exchange resonates between the wireless energy of sub-conscious mind and the electromagnetic aspect of quantum physical component. Since electro-dynamics describes phenomena involving electrically charged particles interacting by means of an exchange of photons, we dub this resonant interface 'a network of light'; we call it the 'body' of H. electromagneticus*. Bodies and their organs are the large-scale expression of this electrical template. Intra- and extra-cellular flows of charge, by electronic current or by ion, also need micro-components analogous to those found in solid state circuits. These would include such bioelectrical devices as receivers/ transmitters (aerials), conductors, semiconductors, insulators/ resistors, capacitors, transducers and so on; and items such as cell membranes, membrane receptors, *DNA* and protein coils, specific molecular structures with unique oscillatory signatures, extra-cellular connective matrix (*ECM*), nano-structured aqueous water clusters, cytoskeletal tubules and so on may well, it is claimed, discharge their primary duties in those informative, electrical capacities. *The modus operandi of psychosomatic broadcasts is therefore, in a word, attunement. Resonance. It involves transduction between recorded information (memory) and physical energy.* **In short, mind is linked to matter by a wireless anatomy whose instrument is resonant association.**

At this level, therefore, there is no single 'life-force'. Instead *vis medicatrix* is a broadcast composed of various kinds of information. We have called these 'archetypal subroutines' that are filed together in a typical mnemone, not least in the morphogene. Morphogenetic information is reflected in chemical, physiological and anatomical systems; exchanges occur in the form of electromagnetic fields and the ionic motions of electrical charge. What species of mathematics best describes the resonance that governs morphogenesis and the geometry of outward forms of life is as yet unclear. **But the key concept that underpins psychosomasis is programmed order relayed through a resonant process of command and control**. *And a metaphysical point of reference, H. archetypalis, is physically realised by its electromagnetic correlate dubbed, dialectically, H. electromagneticus.*

H. electromagneticus.

If you could airbrush out its mass and thus reduce a body to fundamental factors (the presence (-) or absence (+) of charge in space), fundamental particle (electrons, free or bonded) and a fundamental force (electromagnetism and associated photons) you would have the subtle, electro-dynamical infrastructure. You would grasp the radio skeleton that Natural Dialectic has, according to its logical 'peg' on the conscio-material gradient, called *H. electromagneticus. H. electromagneticus* is an 'energy body'. It is the physical medium through which psychosomatic connection works.

Hints have been around awhile. Faraday showed that a changing magnetic field is accompanied by an electrical field. Maxwell showed the reverse and that interaction between the two generates electromagnetic waves (and, therefore, particles). All electrical fields therefore have a magnetic component. And the fact is that all living cells generate their own fields complemented by ionic motions and charged interactions. It might therefore, given this electro-dynamism, be reasonable to describe a cell as an electromagnetic unit.

Do you remember (Chapter 8) that Sir J. C. Bose, with a long-abiding interest in electro-physiology, published 'The Life Movement of Plants' in 1931? In 1935 Dr. H. S. Burr, Professor of Neuroanatomy at Yale University, and Dr. F. S. C. Northrop established that all 'living matter' from a slime-mould to an elephant, from a seed to a human being, is surrounded and controlled by electrical fields. Indeed, every single cell pumps ions to maintain a healthy electrical potential across its lipid membranes. Since the exercise burns 30% of its fuel (called *ATP*) you might infer importance. The membranes act as a leaky kind of dielectric - a battery that generates potential; and this source of voltage maintains the endogenous electric field that surrounds all healthy cells.

It is proposed that Burr's L- (for life) fields coincide with the description of cells as electromagnetic units and so with the idea of a mass-free lattice first identified in the last chapter as *H. electromagneticus.* Pinball machines, neon signs and so on may emit mobile patterns of light. If you represented each electrical fluctuation as a spark of light you would see the whole lattice of a body as a sort of shimmering array. The relative brightness of each sparkle would symbolise the energy involved in any shift. Of course, the whole network would flicker incessantly but certain areas pulse brighter and others (such as the central nervous system and especially the brain) glow in a consistently brilliant way. A photon is the only particle that, both real *and* virtual, bestrides the border of physical reality. *Natural Dialectic would logically identify radiant systems, whose weightless particles represent the purest form of physical energy, as prime candidate for the physical side of the psychosomatic exchange of information.* If this is right, we are illumined physically as well as mentally in our experience of the world!

If the hierarchical supposition of Natural Dialectic is correct then it is easy to understand how the wireless patterns of *H. electromagneticus* are ranged 'above' and regulate 'wired' or bonded molecular behaviour according to archetypal protocols. The *top-down* perspective of its logic would expect to find bioelectric activity facilitated by such circuit components as have begun to be revealed. This informative potential would extend to a body-wide field of electrical patterns unique to each biological type of organism. It would, on this basis, be unsurprising to find that specific fluctuating voltages, ion flows and thereby various hydration levels together 'electro-genetically' controlled such vital processes as mitosis, meiosis and the expression of genes. In this case disruption or distortion of archetypal/ electromagnetic field lines would tend to tension and illness; indeed, bioelectrical faults (with respect to membrane potentials, membrane permeability, charge distribution and thus water potentials, pH and so on) are known to manifest malfunction and pathologies including cancer. Whereas, on the other hand, the health-promoting role of *vis medicatrix* is known to coordinate re-alignment along archetypal lines. For example, the level of expression of repair-cell genes can be influenced and thereby modulated by electrical fields and in this case could not the abovementioned principle of electrical influence of genetic expression be a universal feature? Thus *H. electromagneticus* would, coming down to earth, involve chemicals. It would be transduced through specified and complex channels into body. At any rate the careful introduction of electrical fields around damaged tissue has recently been shown to expedite the healthy growth of skin, cornea and nerve cells. Care extends to the correct strength, positioning and orientation of such polar fields. In clinical trials (for example, at The School of Medical Sciences, Aberdeen University) the technique claims dramatic improvements in patients paralysed by spinal injuries.

Are you surprised? Each kind of molecule is an oscillator able to absorb, resonate with or emit its own unique frequency of radiation. Such radiation, although weak, nevertheless amounts to a signal, a fingerprint or, where resonance is involved, a switch; and the molecule involved amounts to a resonator or an aerial. Not only molecules but cells can be thus characterised. In the radio view a cell might be seen as scanning incoming frequencies while transmitting its own, local signal. In this case signal transduction is the conversion of an electromagnetic, electronic, ionic or chemically configured signal into a specific cellular response or *vice versa.* In the bio-informative network fluctuating fields act as messengers and the motion of charge (as, for example, in nervous impulses, electron transport chains and perhaps semi-conduction along spiral *DNA*) initiates and controls biological processes. Could, biochemist Albert Szent-Gyorgyi suggested, the sugar-phosphate backbone of *DNA* act as such a conductor? Or the molecule work as a resonator, that is, an aerial or aerials for incoming frequencies whose orderly nature might amount to a vibratory code? This code might, at resonant frequencies, act as the key to unlock certain sections of the molecule for expression. *A radio 'baton' would orchestrate the accurate retrieval of information from life's database. Electro-dynamical fields would, at root, command the expression of genes; and would themselves be archetypally entrained.*

In all cells currents, fields and ionic gradients form one half of the basis of biological signalling, that is, of information exchange. The other half, equally important, involves the pre-programmed, structural consequences of fixed or relatively immobilised electrical charge. Examples include the hydrophobic and hydrophilic interactions that result in accurate protein construction, *DNA*/ histone complexes, enzyme catalysis and, of course, the trans-membrane voltages that every living cell sustains. Furthermore, take the physical components of neurophysiology. These include neurons that, involve the same intra- and inter-cellular electrical signalling, trans-membrane voltages and current flows as other cells. Although a neuron can be crudely analogised with a conventional copper electrical cable, its signals rely for their function on mobile ions. Such charged particles not only create but interact strongly with both static and varying electrical and magnetic fields. Whether mobile or fixed, in this place or that, an aerial can receive or transmit at its natural frequency. Electrical fields, artificially induced by clinicians, act as a substitute for missing patterns; they constitute, a subtle kind of splint or 'wireless bridge'. And so, in mimicking what injury or disease destroyed, they resonate with morphogene and, reciprocally, by electro-dynamic control of biological function, encourage regeneration.

Each physiological process in the body has an electrical or electromagnetic component. Every event in every organ no less than every charged cell produces electrical changes and corresponding alterations of bio-magnetic fields in spaces round the body. Such ripples, a tiny fraction of the strength of earth's magnetic field, can be measured using extremely sensitive *SQUID* (super-conducting quantum interference device) magnetometers; and electrical fields of the brain, much weaker than the heart's, can be measured with an *EEG* (electro-encephalogram). The subtlest, organising biological interconnections are wireless. *H. electromagneticus* is, in effect, your wireless anatomy.

Does this anatomy, which organises morphogenetic fields and provides coherent, short- and long-range command and control, involve a component even subtler than the overt, physical photons and electrical charge that instruments can register? Some who favour archetypal fields and quantum virtuality (see Chapters 7 and 10) propose that a vacuum structure of, say, 'scalar electromagnetism' is in fact the primary expression of archetypal memory. Its virtual patterns hierarchically precede and therefore govern the expression of a body's actual energy. In short, just as it might be argued that two levels (atomic and large-scale, visible) comprise a crystal so two levels (quantum vacuum and physical electromagnetism) comprise the wireless anatomy dubbed *H. electromagneticus*.

What is today's eccentric salient is, if correct, tomorrow's textbook study. Electromagnetism involves photons. Is light not in fact the language of life's cells? Fritz Popp, founder of the International Institute of Biophysics at Neuss, Germany, proposes that low-level light emissions are a common property of all cells. Such weak luminescence ranges from thermal (infra-red) to ultra-violet. Not only do electrochemical forces across the membrane help control its permeability. Colleague Marc Bischof believes that weak, coherent e-m fields combine to regulate not only the cell's surface but its internal members. Thus a 'web of light' (correlated, perhaps with *ch'i* or *prana*) might harmonise cooperation of nucleus, mitochondria and other organelles with each other and with chemical substances throughout the cell. Moreover its cytoskeleton's coiled, semi-conducting filaments and tubules, whose other jobs include structural support and transport track-ways, compose a network for the conduction of charge and generation of electromagnetic fields. These 'wires' of electro-dynamic propagation may power up other structures such as protein alpha-helices and coiled/ solenoidal *DNA*. Indeed, Bischof postulates that *DNA* pulses as a 'light pump', that is, like an aerial both emitting and absorbing light. As brain-waves control the CNS such a web of light would, with its various frequencies, control cellular operations. Cannot radio carrier waves incorporate signals we tune into and call programs? Not only internal operations but also cytoskeletal components, where they attach to the surface membrane, may form electrical and possibly 'fibre optic' channels to the exterior matrix. This extra-cellular matrix (*ECM*) provides a medium for body-wide bioelectrical linkage.

For *bottom-up* biochemistry such radiation is simply a waste product of metabolism; it is not so much critically informative but incoherent, chaotic and meaningless. Popp, on the other, *top-down* hand, has interpreted these 'bio-photons' as electromagnetic signals that coherently transfer information from cell to cell throughout a body. Natural Dialectic philosophically frames his notion of hierarchical order and meaningful cooperation; it agrees with his idea of language or, seen as resonance, the harmony of health. Will the future swing towards an image of light-organised as well as light-driven biology? Which interpretation of a cell's emission of its own rainbow will eventually prove correct?

However that may be, anyone who has suffered muscle damage knows how complex is the interaction of muscular looms and how 'compensation' spreads across the body to disturb patterns far from the original injury. Every physiotherapist is aware that finely tuned balance between interconnected parts is upset. It is not difficult, therefore, to conceive of an organ (such as a brain or liver) as, rather than a particular object in a separate place, a function whose influence extends throughout the body. Indeed, each part can be regarded as a local domain, a particular subdivision of the whole matrix. Brain waves, for example, pulse throughout the central nervous system and, therefore, affect associated structures. The fields of all organs, oscillating in the *ELF* (extra-low frequency) waveband of the electromagnetic spectrum, spread information throughout the body; they change from moment to moment according to events happening both within the organs themselves and further afield. Biological electricity, ubiquitous in all cells, is a result of the movement of charged ions (sodium, potassium, calcium etc.) and electrons (as in an electron transport chain). It also involves, as noted, a voltage (potential difference or electrical polarity) across cell membranes that, in at least the nervous case, can oscillate. And injury potentials trigger tissue repair; in dialectical terms they are gradually able to override the injury or disturbance by 'imposing' the underlying archetypal pattern. There also exist flows of information that reflect subtleties as fine as the changing positions and chemical interactions of protons and electrons. Metabolic pathways are an example, hormonal and nervous systems another. The blood vascular system also carries a great deal of electrochemical information throughout the body. Holistic biology therefore takes the line that complementary parts, specific interconnections and intelligent arrangements amount to programs whose coded complexity far exceeds the creative capacity of non-consciousness. A factor more than mindless matter is, in this equation, reasonably inferred.

What about life's basic unit? Is a cell just a crude mix of chemicals that spill out when you burst it with a centrifuge? Wrong. Is its cytoplasm mostly normal water? Wrong. Do its metabolic programs proceed by reactions due to random collisions? Wrong. The old-school idea of a separate cell interacting in a spasmodic, chancy sort of way with its neighbours and, even less certainly, with far-distant reaches of a body is misleading to the point of error. Not only neurons and hormones but fibrous matrices link every cell, tissue and organ with every other. A global communication system reaches from nucleus and cytoplasm through trans-membrane connectors to connective tissue (extra-cellular matrix or *ECM*) that, permeating throughout an organism, creates a thorough, hard-wired information web - a body-wide web.

Let's start inside a cell. The components of this highly ordered structure are constrained by an external cell membrane, internal membranes and a dense cytoplasmic network of microtubules, filaments and, in some cells, fibrous elements that constitute, effectively, arrays. They might all be thought of as crystalline so that, rather than a haphazard sort of place, the cell is discerned as a highly organised affair. Phospholipid lamellae, myelin sheaths and muscles are examples of 'soft' arrays. So is cytoskeleton whose filaments are composed of a protein called tubulin. Many connectors such as actin, myosin, collagen and keratin are helical molecules that, along with informative *DNA*, act in living systems as piezoelectric semiconductors able to emit, absorb and convert light into vibratory effects. Indeed, Szent-Gyorgyi pointed out that all proteins are semiconductors, 'devices' through which conduction can be precisely controlled. Semiconductors are, therefore, signal or information controllers.

Polar (i.e. electro-active) water has a tendency to form highly fragile, local and evanescent arrays; but even pure, inorganic water is more chaotic than ordered in its flux. Not so cell water, which tends to 'aggregate' in an organised, hexagonal (or liquid snowflake) condition. It is claimed that, at 37°C, clusters of such 'structured' water pass more easily through membranes and have a dissolving power as high as extra-cellular water at boiling point. A 'liquidly crystalline' matrix also facilitates cellular communications, interacts in the form of a 'coat' with dissolved macromolecules and thereby mediates, though a 'laminated' film about ten layers thick, all their electrochemical/ electromagnetic reactions. It also affects protein's configuration and its electrical properties. Clearly, adequate hydration levels affect the operations of *DNA*, protein folding, mitosis and, as opposed to dehydration, the general viability and well-being of all cells. And where a 'crystalline array' supports electron flow, there charge moves and electromagnetism is generated. The medium also supports the diffusion of plus charge (in the form of protons). Not only are its clusters 'latticed' but many enzymes are attached to various scaffolds so that efficient, immobile enzyme catalysis in such an orderly circumstance is the norm. Cell chemistry is far from a haphazard affair. In bioelectrical fact, the coherent structure of a cell begins to look more like a special kind of liquid crystal and its operation one of solid-state circuitry.

In fact, as previously intimated, studies may lead soon to common knowledge that a large proportion of the DNA's workload is, with its helix acting as an antenna, engaged with bioelectric signalling. Water layers pack the chromosomes; electromagnetic fields can change the structure of the water and in turn the shape of *DNA*; such subtle prompts may be what, primarily, regulates expression of the genes. **In this case *DNA* would become a central station for a cell's network of electrochemical communications. The principal genetic operant would be radio!**

There's more. Cytoplasmic interconnections are wired to trans-membrane proteins (called integrins, connexins, anchor fibrils, desmosomes etc.) and thence directly out to collagen-rich connective tissue, also called extra-cellular matrix (*ECM*). *ECM* is, effectively, a gel-state reticulum that surrounds and links up cells, tissues and organs. A highly ordered continuum integrates, from inmost cell nuclei through cytoplasmic to extra-cellular networks, all parts of an organism. Indeed, connective tissue determines shape, is sensitive to deformities (e.g. by compression) and, as a consequence issues electrical signals that define every body movement. The *ECM* may, in this context, be viewed as a resonant organ. As with external skin, so internal *ECM* is a whole-body organ. Its three-dimensional matrix is composed of a complex assembly of proteins such as collagen, sugars, glycoproteins (including laminin), proteoglycans, hyaluronate, chondroitin, mineral ions and so on. Also, importantly, the constituents of this negatively charged 'zone' allow for organised water clusters; and its 'liquid crystal', semi-conducting matrix involves an electrical field whose fluctuations occur in response to proteoglycan concentrations and whose currents are engaged in cell signalling, that is, the passage of information. As mentioned earlier, the cell is directly anchored by trans-membrane proteins to, on the inside, internal cytoskeletal 'wiring' and, on the outside, to the *ECM*. It is noted that, as well as *DNA*'s aforementioned duplex coil, the *ECM*'s laminin triplex (see *fig.* 17.5) may also act as an electrical conductor. Emphatically, a pervasive, bioelectrical network of communication is established; its command and control directly connects innermost, nuclear *DNA* with exterior modulations throughout the body. It is thus easy to see that problems with 'holistic' *ECM* composition, hydration/ dehydration or injury can lead to electrical resistance, disruptions to electron, ion and nutrient flows, impaired ion permeability, incorrect pH and other factors that may cause illness or inhibit the healing process. Such electro-dynamics may also, as previously noted, be involved in carcinogenesis.

In the way described *H. electromagneticus*, the form of bio-electrical energies, is based in physiological fact on nervous, hormonal and protein-built connective systems that unite a body so that individual parts become local functionaries in a purposive community. Biology is, from top to toe, a study in fluid, homeostatic coherence. Indeed, the bioelectrical image of *H. electromagneticus* reflects archetype and forms no less than the basis of a fresh medical image of the body.

It is energy that, within a system, shapes external appearance according to both physical and metaphysical codification. Informed energy gives rise to purposeful, functional constructions. In dialectical terms *H. sapiens* is the (*tam*) external, solid or massive body and, with mass air-brushed, *H. electromagneticus* (Burr's 'L-field'?) is the (*raj*) internal, dynamic, morphogenetic one. The latter's electrical effects govern shape and process - government that accords with predefined parameters more intricate than a modern supercomputer's. But physical energy is purposeless and, therefore, carelessly incoherent unless 'rationally' contained. What, therefore, prefixes *H. electromagneticus*? With matter the ordering system involves mindless forces and, of these, charge is the definitive agent of chemistry. Electromagnetic charge is also an agent of psychosomatic exchange but, while matter is purposeless, mind is the only cosmic factor that *is* an information, command and control centre. In the biological case, therefore, the prefix of electromagnetism is mental. For Natural Dialectic such prior 'super-norm' is a metaphysical morphogene, *H. archetypalis*, a generic memory. Body is the glove of archetypal hand.

The physical system is not fixed. It is labile. Connections at all levels in the matrix can break and reform. Within such flux electrical fields maintain stability. Indeed, *ECM* and cytoplasmic arrays are field sensitive and can themselves, under tension, act like the aerials of a molecular broadcasting system. They generate coherent oscillations which occur in different forms at different frequencies. In fact each and every molecule, whether or not part of an array, oscillates. It 'rings' with its own 'chime' or 'ring-tone'. It emits a characteristic energy spectrum, its 'fingerprint', the unique signature that is used by spectrographers to identify it. *It was noted that each molecule, cell, tissue and organ has its ideal resonant frequency.* If molecular arrays act as aerial sensors/ transmitters for such radiations, could it be that each bio-molecule has its own distinct 'song' with which it interacts in resonance or otherwise with all others? Could the whole business, in all the complexity of a million-pieced puzzle, be integrated at the subatomic level of protons, electrons and light? And work like a living book whose letters (atoms) and words (molecules) make sense and cooperate in a way as meaningful as the words in a story? When two objects have similar frequencies they can interact without touching. Their vibrations become coupled or 'entrained'. *They resonate. As such they associate and affect each other.* All sorts of motions within molecules emit electromagnetic radiation. Orbital shifts, bond stretching, bending and rotations can as equally absorb as radiate. Could the effect of such resonance not pass across 'crystalline' arrays of relevant molecules? An antenna works best if its length corresponds to the wavelength of a signal transmitted or received. It is suggested that, at the nano-biological level of metabolism, biological interactions are as much a matter of vibratory resonance as physical contact; and that such signals are absorbed and conducted through the abovementioned inter- and intracellular matrices, affecting or not as the case may be every molecule in their path. It is further suggested that this information complex, whose basis is electromagnetic, entrains with the sub-conscious archetype. Molecular resonance 'chimes' with the patterns of 'solid' mind. Broadcasts from *H. electromagneticus* trigger not only chemical reactions that underlie metabolic programs but also, 'above', they trigger metaphysical associations in memory. Not unlike Wilder Penfield's probe. It works both ways. In both cases a subtle association is created through entrainment, resonance and coherent behaviour throughout the system. At the same time a subtle recognition of or disassociation from foreign 'chimes' automatically occurs. A 'chime' therefore acts like a password. Such ionic and molecular resonance might be how nerveless organisms, even single cells, interact with instinct and their morphogene.

Resonance is how memory (personal or typical) and mind-body exchanges work. The brain, for example, acts as a complex trigger. Its own brain waves, supplemented by myriad particular signals from specific molecules, local patterns of charge and changing electromagnetic fields, combine to trigger response from the morphogene and associated instinct or *vice versa*. Your typical mnemone, *H. archetypalis*, is activated in one, two or all three of its capacities. Such associations are also how, by way of relevant sensory locations, personal memories are filed or accessed. Impressions from outside or mental events (thought, feeling etc.) are filed according to type and that type's association with locations in the brain. Indeed, consciousness is linked with physical information by way of sub-conscious signal translation. As noted, for example, a beautiful woman or even the thought of her can trigger a cascade of physiological effects in a man; conversely, scents, pheromones, a touch or even artificial stimulants can bring sex to mind, induce personal memories, trigger imaginations and in this opposite way spin the cycle round.

All bio-information exchange involves phenomenal subtlety (miniaturisation, precision and operational velocity). *DNA*, for example, is a storage medium, a physical form of memory. It is an archetypal, read-only database for protein structure whose packing density is 10^{20} bits per cubic centimetre. Similarly, the brain is a complex organ whose physical operations (let alone their metaphysical extensions) are not yet fully understood. It has been estimated that, if each of 10^{11} human neurons transmits 10^{3} signals per second, optimally 10^{19} operations might occur per day. It has, however, also been estimated that information flow in a human eye might reach 10^{9} operations per second and overall bodily flow daily top 10^{25} bits. Such traffic exceeds the capacity of nerves. It would lead to jams that palpably do not occur. So how could things work? Just as 'crude' electrical pulses in a telephone wire convey a much subtler intricacy of message, may not waves of depolarisation that sweep along nerve processes activate more numerous, subtler and varied signals?

We met the intracellular part of a body's communication network, that is, a cytoplasmic matrix of microtubules. This cytoskeleton, made of hollow tubes that consist of tubulin, occurs in all eukaryotic cells where it functions as the plastic agent of internal scaffolding or structural support, locomotion, transport and cell division. This section logically suggests that its microtubules also act, even in the case of an amoeba, paramecium or sponge, as agents of information exchange. Indeed, since such networks exist in neurons they may be construed as comprising a first-order informative system with the secondary nervous sub-system; the fact is *all* cells use cytoskeleton.

A cytoskeletal tubulin dimer can exist in either of two conformations. If, as seems likely, the difference between conformations involves the shift of an electron from one position to another, it may be that these electrical polarities of the conformations act as on/ off switches. There are perhaps 10^7 tubulin dimers per human neuron and if their elementary (on/ off) operation is 10^6 faster than the creation of a nerve impulse, then an optimal 10^{27} signals per second is achieved. If this weren't enough then signal patterns might be multiplied by such breakages and reconnections as constantly occur between lengths of tubulin and its anchor points (called centrioles); and, in the case of innervated organisms, ctyoskeletal changes that redirect synaptic connections. The potential speed of transmission might enter the range of quantum computation. Quantum effects such as 'coherence' occur when large numbers of particles cooperate in a single, isolated quantum state. The action of lasers shows coherence. Coherent light (composed of photons) can pack a lot of information. Indeed, a key feature of high-tech communications is fibre optic cable for the lightning, hi-fi transmission of large amounts of 'electromagnetised' information. Such fibres often crack and break but the optical properties of a superior brand of fibre are found in the siliceous spicule lattice of a sponge, *Euplectella.* Could such minuscule glass architecture double up as a porozoan signaller? Your own brain is neither siliceous nor calcareous but it contains plenty of nervous tissue. Could its 'lattice', the presence of charge, membranes and cytoskeletal arrangements called *H. electromagneticus* mediate, as an agent of quantum coherence, the large-scale, coordinated exchange of information between mind and matter? Outbound signals from your typical mnemone, *H. archetypalis*, would involve the electromagnetic lattice and, consequently, trigger organised cascades of information through the body's nervous, hormonal and muscular systems. Inbound signals, collected from sensors, would ascend to distort the cerebral lattice (which is itself composed of tissue lattices and individual, cytoplasmically-networked cells). The signals would translate patterns to *H. archetypalis* and, where appropriate, images to conscious mind. In such a multi-layered yet coherent hierarchy different permutations of conformational differences in sheets of tubulin dimers could carry much more information than the linear morse of nervous messaging; they would offer much greater resolution of a message passed to or from 'psycho-space'. There is no doubt about the high-speed, high-fidelity definition in which our information system, including its nervous fraction, trades.

Look around. Your physical awareness involves soft machinery of phenomenal complexity. Yet is this all there is to know? At this point adherents of quantum computation, mathematical (perhaps chaotic) modelling of mind, cytoskeletal coherence and similar notions may commit the classic materialistic error of equating mind with matter and thus imagine psychological emergence from atomic stuff. **From a *top-down* perspective, however, it is not that all these signals *are* part-less mind or consciousness.** They simply carry mind as words or wired morse does; they transmit perceptions and intentions, Physical structure serves the prior purposes of metaphysical mind in both conscious and subconscious phase. Astounding electronic specificity is useless without higher, psychological conjunction; indeed, lacking such conjunction we call a body 'dead'. *So, although it is crucial to know how coded information is transferred between physical systems, we need also grasp how the psychosomatic transduction of electrical and perhaps also luminous patterns to and from the mind-field work.*

To summarise, Natural Dialectic suggests that psychosomatic interaction occurs between a typical mnemone (with its three parts of signal translation, instinct and the morphogene) and highly organised electrical and electromagnetic patterns. These patterns are framed, as the active site of an enzyme is framed by the conformation of its protein chain, by gross matter; but they are, equally, framed by archetypal program. In other words, the interaction of mind with matter is seen as a dynamic, coherent association of two anatomies, wired and wireless. And wireless orchestrates the wired. In your case fixed psychological harmonics called memories interact with the subtlest natural phenomena, photon and electron. In this transductive association molecules of your visible body, *H. sapiens*, act as oscillating resonators; the broadcast medium is *H. electromagneticus*; and the latter's archetypal associate, *H. archetypalis*, is the program's broadcaster. *Thus interlocked, Janus-like psychosomatic bodies can be seen as sub-conscious information, in the form of memory, acting in concert with complex electrical fields. Sub-conscious mind is the hand, electricity the glove.* **There is nothing in biology to suggest such concert does not exist, indeed there is much to support it.**

In this view, therefore, cerebral and all other cellular activity becomes a kind of aerial, a multitude of various antennae whose 'settings' exchange information with wireless, radiant mind. The physical agent at the root of this psychosomatic exchange is identified as light. A human brain is probably the most exquisitely programmed, complex, ecstasy-capable information agency in the cosmos. Its purpose is the concentration, centralisation and conscious awareness of that cosmos. We take it for granted but consciousness, wherein the physical world is 'turned outside in' and, *via* motor expression, the inside metaphysic of mind 'turned out', is an incredible experience. **It is likely that the human information system is the crowning glory of physical creation**. *It therefore seems unlikely that brain, whose very raison d'etre is the exchange of information with an active (conscious) source, should have evolved in a Darwinian sense from the reverse of information - chance and mindless 'selection'.* **Extrapolation downwards, from mind to matter, leads to the conclusion that integrated information networks, the precision of cell operations and *DNA* hard coding are no accident.**

Such a view may well press huff and puff from the lips of at least a few molecular biologists; and some neurophysiologists may in contempt dismiss. The problem is that denial of an immaterial and informative element leads, as demonstrated in Chapters 19 - 25 and especially clearly in Chapter 20, to absurd conclusions for biology. *If pure materialism is insufficient to explain both psychological phenomena and irreducible, physical complexities then a discussion of psychosomatic relations becomes as sensible as it is inevitable. This chapter, whose principals are H. archetypalis and H. electromagneticus, has begun to address the issue.*

In summary, *bottom-up* materialism theoretically reduces consciousness along with both conscious and subconscious forms of mind from metaphysical to physical status. The presence of metaphysical (psychological) parts is denied.

From a *top-down* perspective there is, however, no such denial. The metaphysical/ immaterial factor is (see Chapter 11) information. And from the hierarchical perspective of holistic Natural Dialectic the subconscious phase of mind is sandwiched, as computer software, between conscious programmer and physical, bodily hardware. For the holism of Natural Dialectic it comprises two integrated parts - metaphysical archetype (*H. archetypalis*) and, at the psychosomatic interface (*PSI*), physical *H. electromagneticus*. The subconscious amounts to an inner, psychological context composed of memories both personal (uniquely individual) and typical of a kind of organism. The latter, typical mnemone is, in turn, composed instinct, morphogene and, where consciousness is involved, signal translation. Two of these inhabit, like *DNA*, every cell in every body. The third, signal translation, occurs in conscious, nervously endowed animals.

In this case two-way flows of inward sensory and outward motor information are translated from nervous into conscious, experiential forms and *vice versa*. These anti-parallel, in-out vectors characterise the whole information system but in this case (dubbed synchromesh 1) only signal translation and instinct impinge, via the 'library' of personal memory, on conscious mind; and, in the outward, motor direction, a purpose instructs body by way of signal translation and morphogene. The system amounts to an inner, psychological context which, in conjunction with a biochemical 'chip' called *H. electromagneticus*, autopilots mind and body (including brain); its bio-cybernetic circuits keep the plane flying while its conscious pilot concentrates on his chosen, narrow screens of data. In other words, it constitutes an exquisite, involuntary operating system backing up a voluntary operator.

There is no doubt the whole business is highly, exactly programmed. Its metaphysical and physical components inhabit, like *DNA*, every cell of each body. That is to assert, mind with electromagnetic fields inhabits *every* organism even if many (i.e. plants and bacteria) are wholly dormant. *H. electromagneticus* involves electromagnetic fields with their electrons and photons. If you could strip cell structural material, tissue, organ and a whole body of all atomic and therefore molecular and bulk mass, the residue would constitute the quantum biology of signalling and connectivity.

As in physics, the field is primary. Its ubiquitous bio-electrical activity informs and controls secondary, passive structures. Of the latter the cell 'brain' (or membrane) with its two-way, in-out flows of message and response is one; the behaviour of intracellular biological machinery in the form of changing protein shapes, the specific motion and interaction of biochemicals, the operation of organelles and the dynamic data bank of chromatin (that is, genetic blueprint and epigenetic signalling) also act in response to either external or internal prompts. In short, the whole homeostatic process of living bodies is instructed by the field(s) we label *H. electromagneticus.*

As regards conscious mind-to-body communication the body's overall controller, brain, appears to concentrate psychosomatic transfer between *H. archetypalis* and *H. electromagneticus* at its centre, called the 'interbrain' or limbic system. This system exchanges information between the forebrain (or prefrontal cortex, the immediate 'wire' onto which a pilot's thought messages are, like morse, tapped out) and midbrain, hindbrain, brainstem, spine, peripheral nerves, glands and immune networks. It involves organs associated with emotion (the amygdala) and memory (the hippocampus); its diencephalon includes a crucial triad of glands - pineal, pituitary and hypothalamic - which link electrical (nervous) and chemical (hormonal) systems of information carriage.

Finally, chance and rudderless necessity could not originate the wired side of life. Bearing this in mind we are now in a position to turn from metaphysical mind and, at the physical side of the psychosomatic border, *X. electromagneticus* to, firstly, the energetic patterns of physics and then gross, codified bodies of biology.

17. The Logic of Embodiment

Faced with an opinionated wall of 'no!' you have to keep on asking. Is there an immaterial element, information? Not material vehicles that carry information passively - speech, inscriptions, drawings and so on - but immaterial conveyance. Not brain or nerves but metaphysic passive (memory) and active (consciousness, intelligence and thought). For materalism's atheistic secularity the idea is heretical. Philosophies, mind-sets and scientific paradigms depend on 'no!'

If 'yes!', however, then a link exists. Mind and material dimensions must be joined. A psychosomatic *PSI* needs be identified. Here, therefore, we go again.

Books 0 and 1 dealt with creation in terms of authorship, conceptualisation and preparation of the stage for a theatrical drama, a play. It covered (Chapters 7 - 12) the construction of an inanimate platform, its stage. *Top-down*, an act of creation is hierarchical. Its serial order runs from potential ('gathering' or focus of attention) through kinetic planning to the orderly rearrangement of non-conscious materials. In personal terms this means putting an idea into practice through the physical medium of body. In universal terms it means, if the starting-point is Pure Consciousness, a gathering of Causal Concentration. From this First Principle, Idea or *Logos* there are expressed, in a consistent way whose logic Natural Dialectic tries to best reflect, the principles of mind, universal memory and finally, according to the latter's archetypal files, the substrate of a physical cosmos. And microcosm reflects the macrocosm; universal order is reflected in the characterisation of individual, microcosmic players.

Book 2 deals with the animation of different, personalised actors as opposed to universal principles. A *top-down* look at psychology drops (Chapters 13 - 16) from conscious through to sub-conscious mind. The latter phase is identified in, for example, *figs*. 2.12 and 4.1 as 'potential matter', the place of archetype. Therefore, although Book 3 (Chapters 19 – 25) will detail a *top-down* appraisal of physical biology, in this chapter we approach the expression of a player's part, his body, through the structure of its metaphysical potential, its morphogene. Later (Chapter 25) we can consider the exteriorisation of a body from physical potential (a zygote) through kinesis (growth) to stasis (its mature adult form) - a process that follows the dialectical pattern of creation from (*sat*) through (*raj*) to (*tam*) and is called development. In this way we gradually link metaphysical principle with its exteriorisation to a biological conclusion and thus call this section the logic of embodiment. Its emphasis is, not least because we are human, on the human form.

Are you not an incarnation? Are you an 'assumption of the flesh'? The word 'incarnation' means 'made flesh' but its use is, in one-tiered country, politically incorrect. Because of its three-tiered implication this word for embodiment fell from grace. Now, because of Natural Dialectic's three-tiered implication, does it rise phoenix-like again? Conceptual resurrection?

The Logic of Embodiment

	tam/ raj	*Sat*
	expression	*Potential*
	outworking	*Plan/ Morphogene*
	classical/ quantum matter	*Potential Matter*
	gross/ subtle bodies	*Archetype*
	informative agents	*Informative Source*
↓	*tam*	*raj* ↑
	H. sapiens	*H. electromagneticus*
	classical biology	*quantum biology*
	gross constraint	*subtle motions*
	fixture/ container	*stimulus/ drive*
	structure/ anatomy	*physiology/ function*
	external appearance	*internal informants*
	informed product	*informant code*
	seeming sameness/ slow change	*rapid changes*
	bulk expression	*molecular embodiment*

Is there logic of embodiment or is there the embodiment of logic, of functional, dialectical logic? What is the purpose of a body?

Bottom-up, there is none. No logic, no design. The whole business of bodies is a matter of 'evolutionary' chance. The 'design' attributed to a biological complex is due to incremental accidents 'clinging on' to themselves. 'Clinging on' by natural selection may seem to involve a notion of 'improvement' or 'progress'; but if prescient, purpose-driven organic chemicals seem odd then biochemists from Friedrich Wohler onwards assure us they are not - because no such prescience exists.

Nevertheless life, which is simply a complexity of chemicals, seems to 'want more of itself'. The molecules 'want at very least to keep complex', that is, to survive. May we suppose that for some lack of reason biochemically-employed atoms (and especially a subset that can be falsely represented as replicating itself) 'want to live'? Can we, therefore, reduce the basic unit of life to a cell? Cell for soul? Soul sold for cell? In which case a bunch of biochemicals with the ability if not the instinct to survive intact is at the heart of a soul-less mythology. Tough love tells the way you think it is. And, of those key concatenations, there exist sub-cellular polymers whose 'molecular urge for self-preservation' (bizarre a phrase) dominates life's troublesome domain. Such an imaginative subset might translate into a myth of 'selfish genes' and a whole subsequent fresh way of seeing life. In this case bodies become 'purposelessly purposeful' packaging. Your own sense of importance is a deceit and your reality an illusion. You simply encapsulate the 'real purpose' and, unwittingly, protect genetic material. You serve as hive to nuclear queen bee, your *DNA*; you are paper round a precious gift; you are reduced to just a dish that cultures cunning, Machiavellian genes. You think it's all for you but such conscious selfishness is simply spun to serve the genes' non-conscious version. Sex, reproduction, family and family lineage are simply 'cover' for the humourless, genetic 'will to live'. In this case the 'real purpose' of life (whatever that means biochemically) reduces to mindless reproduction 'in order to' obtain a potentially endless survival, an *ersatz* immortality through successive purposeless bodies. Of course, genes are unaware of package-schemes called body-plans. The latter grew by accident from round a most successful replicator, a bacterium, into the panoply of life on earth. So we're only bits and bytes, odd bobs of digits in genetic crucibles of chemicals that have occasionally, strangely bubbled up as 'life'. No purpose whatsoever animates the whole, grand genocentric show! *This is twisted, dismal animism; it is a story not overly rational and by no means necessarily more than a small fraction of the whole truth.*

Is such convention in the slightest right? Though information's carried in them is an organism really, wholly, solely down to genes? Suppose instead that a hierarchical, *top-down* construction exists. Material order's crystallised around prior plan. Such archetype informs a physical projection; and within this reproductively enabled bio-form an individual combines to undergo adventure, a physical experience of drama on this plane. The combination play out an allotted part; and this objective part, a single character from the whole social and ecological cast, seems to its subjective player like a personal, leading role. We're all in leading if not starring roles! From this angle an incarnation is not construed as an unusual, essentially purposeless arrangement of subatomic particles. *On the contrary, from molecule through cell and organ to whole construction it must work as intended.* It is constructed according to archetype *in order* to be able to act the bidding of its author, that is, to act out a bundle of inclinations, desires and consequently, plots. You individualise the general potential of your type - our type. *A character represents a principle or combination of principles. If bodies are made at the behest of mental principle, this casts their actions in the light of purpose. It casts them in the drama of a real philosophy!*

Are you not, in this view, an incarnation of supremely functional logic? Is your body not a brilliant birth-right? And, if this were not enough, is that all? This individual gift reflects the generality of man; it extends beyond the atoms and the stars to shape of mind, archetype; and then, beyond its archetypal framework, into conscious mind and at its heart Pure, Natural Consciousness. Is this heart not aboriginal, first in the cosmic line, your heritage - one making you the object of a Great Idea? How is the plan devolved, how the fabric woven? By what path came you here and now?

Put this prosaically. Do you express ideas with symbols? Don't you use either fluid, energetic code called speech or language fixed in written words? Biological ideas are, similarly, expressed through symbols. Their physical expression is genetic code. This language is fixed on ink and parchment known as *DNA*. Its 'book of life' is stored on a nuclear database. **A *DNA* database is, in short, a symbolic expression of ideas.**

Biological life is, first and foremost, a synthesis of mind and matter. Information, you remember, is the potential for all subsequent rational action/creation. The *top-down* logic of embodiment is marked, starting from Potential, by the left-hand, downward sweep of materialisation. From a dialectical perspective life forms are the physical expression of an image, a prefabricated plan lodged in universal memory. Such metaphysical prefabrication exists, as previously described, in the form of a typical mnemone. No organism lacks the archetypal elements of sub-conscious mind. Mind comprehends; it invisibly substantiates every purposive construction. Without inventiveness and the informatic storage of inventions in a program or a plan no such construction could exist. No organism, equally, lacks correlate biochemical provision in the form of a 'typical cell'. This phrase describes a generality that may involve multiple subroutines - there are, for instance, over 200 different sorts of cell in a human body. Each comprises its own database, a genetic library according to whose inscribed plans pre-programmed metabolic responses and other physical components can, in a most orderly yet automatic way, be 'machined'. Is it the extreme ease yet complexity of execution that makes us feel no mind is behind the self-perpetuation of a cell? In dialectical terms metaphysical and physical constructors co-operate.

In some organisms a degree of consciousness, whose quality varies considerably from type to type and even from individual to individual, is correlated with a nervous information system. Perhaps the fullest flowering is recognisable in yourself, a human being. *Top-centre is the seat of consciousness, the cosmological axis. The central nervous system with spine is the root and trunk of your information centre; from this inverted tree ramify nerves into the surrounding body.* Short-term electrical are combined with longer-acting hormonal networks. Your biology is replete with all sorts of informative structures. All serve, at the centre, you - that you may, in living, experience and affect the world. This is a reversal of the materialistic (and therefore scientific) view. 'You' are not the product of random effects but inhabit a construction *granted* you for the dramatic expression of principles and anti-principles. The whole theatre is deliberate. Information dominates matter, the puppeteer the puppet. *Lives are meaningful information centres.* Your mind, for example, is conscious, your body electrically animated but unconscious and between them is sandwiched a sub-conscious template, a translator, an information exchange.

Information involves purpose. *The voluntary psychological purpose of life is what, as best you are able, you decide.* What you want is, for that moment, your truth. The importance of such truths, according to which you are constantly orienteering, varies. What, in the hierarchy of truths, is the most sensible decision you will ever make?

What you cannot or may not wish to decide is decided anyway - by instinct and physical processes. Instinct is life's fail-safe, its substitute for thought. What thinking lacks instinct provides. *The involuntary physical purpose of life (which relates to its biological embodiment) is to keep living; to stay alive, keep playing and, in so far as it is granted, enjoy the experience; the first premise of bio-logic is to satisfactorily survive.* If you react two chemicals the reaction runs to completion or, as we say, inertial equilibrium. It 'dies'. Biological forms are not like this. They 'want' to keep living and in order to survive they utilise energy in a precisely programmed way. *So, not quite as effectively as a fountain of perpetual youth, they evade death and for a while sustain a dance, a dynamic equilibrium of life.*

Dynamic equilibrium, rather like walking a tightrope, involves balance.

Life's balance is vibratory in the form of cycles round a norm. Such norms include temperature, the correct concentrations of chemicals, heartbeat rate and so on. Many periods, wheels within wheels, are integrated into each whole-body norm - a type of organism. In this sense each organism is analogous to a watch with meshing cogs whose healthy tick depends upon coordination. Each cog represents a homeostatic cycle; bundled up such cycles constitute its 'statute law'. Integral balance codified in scales of law - this is the great biological idea.

Central (*Sat*) Homeostasis:
Whole-body Balance based on Information.

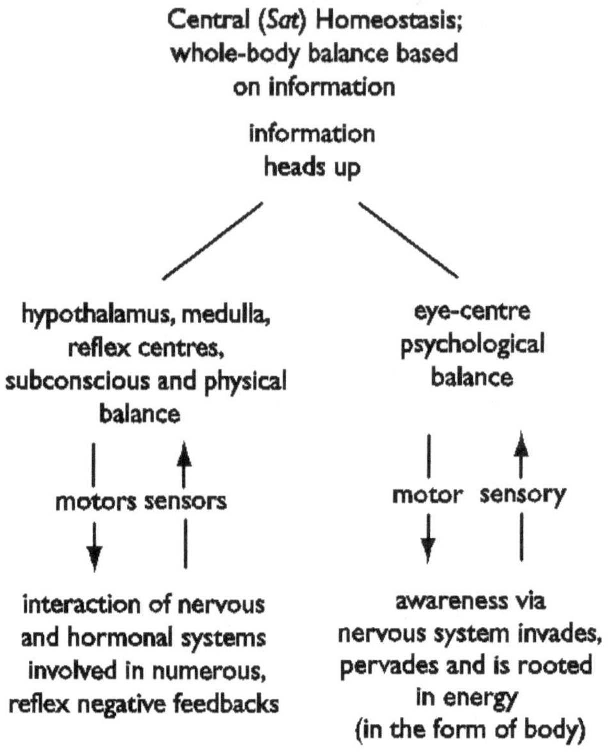

In the case of a machine information controls the energy of its action. Homeostatic systems, whose conscious pivot is the eye-centre, inform and control energy according to preconceived 'targets' or 'informative norms'. A voluntary, conscious target is fulfilment of a specific desire. An involuntary, instinctive or physical control (e.g. thermoregulation) makes cybernetic comparison against a preordained 'norm'.

fig. 17.1

tam/ raj	*Sat*
energy	*Information*
periphery	*Centre*

	imbalance/ motion	*Balance/ Equilibrium*
	implementation	*Plan/ Norm*
	effector/ sensor	*Regulator*
↓	*tam*	*raj* ↑
	down	*up*
	involuntary	*voluntary*
	physical	*psychological*
	info. out	*info. in*
	effector	*sensor*
	external	*internal*
	body	*mind*

The process by which life's vibrant forms survive is, in its aspect of control, called *homeostasis* (Chapter 19). Its game amounts, in a word, to balance; it amounts, in principle, to questions and answers. A problem (or question) arises and a response is formulated and implemented. In other words a stimulus (a loss or lack of equilibrium) causes a disturbance. The system loses track. There follows an attempt (reflex and/or voluntary) to re-balance, to regain the pre-set norm. It is irrelevant whether the stimulus was an internally generated desire or an external, body-related event. A sense of balance or the peace of satisfaction is regained. **Homeostasis** is the core activity around which all other psychological and biological processes are constructed. Its pivot, its basis is information. (*Sat*) information precedes (*raj*) energy. Information swings the way events turn out. But if bio-logic pivots on the centrality of balance and begins with vibrant homeostasis, biology textbooks rarely do. This is because their mind-set is dialectically off-key; it is 'out-of-tune' and called 'reductionist'. Thus most begin with a broad, archetypal (or typological) register which defines, in the form of classification, the scope of their topic. Thence they often switch to an atomistic, *bottom-up* perspective. Rather than primary function (homeostasis) primary structure (the cell or, even more atomistic, its molecular parts) is first emphasised.

But 'logic of embodiment' is a phrase astride two worlds. Mind and body each act in homeostatic capacities. Just as a plan is developed from seed inspiration, so a body develops from a zygote. *Could the plan for biological development have arisen as an inspiration, the seed of an idea? Of which the memorial is a typical mnemone, an archetype?* In this case each new body is the re-staging of an ancient, dramatic pattern of events. If you had to invent a self-reproducing, intelligent machine, what flowchart would you develop? What *top-down* set of modules would you link, in the most efficient way possible, to a master control routine? In this chapter we deal with logic or principle, centred in homeostasis; later (Chapters 19 - 24) we deal with its physical expression - working bodies. On the basis that information precedes the arrangement of energy and that principle precedes practice, what follows is an outline of the special, purposive case of natural logic that precedes its expression, a biological body. Does such logic accurately reflect the method of creation, making man a 'microcosm of the macrocosm' - a complete symbol of the universe?

Core Principles

The potential that precedes any bio-systematic activity or construction is its plan. Information. **The basis of homeostasis is information.** Its operation involves the triplex of sensor, processor/ regulator and effector mechanisms. Messages inform, organs are informed. *The remit is survival. Survival in the present involves energy metabolism* (with associated organs and organelles) to 'finance' sensation and action. Bodies resist (by the dynamic equilibrium of homeostasis) but eventually succumb to the universal process of entropy. They die so that *survival into an indefinite future involves reproduction* and its associated apparatus. The whole business is thoroughly informed, coded and works to plan. No chance.

Psychologically, your conscious or voluntary centre of homeostasis is that of thought, the eye-centre. Its metaphysical presence is incorporated, as befits a ruler, just above the organs of sense in the centre of the forehead. Its organ, the cerebral cortex, is pole-positioned as a port of information exchange. Top central, it is an expression both of the integrating nature of consciousness and homeostatic control. *The sub-conscious, involuntary centre of homeostasis is reflected, physically, behind and below the eye-centre in the form of central diencephalon, brain stem and nervous system.* Core agents, detailed later, include a 'transducer' called the pineal gland, a neo-cortical/ autonomic/ glandular interface called the hypothalamus and the latter's close associate, an executive secretor of 'nationwide' commands, an endocrine gland called the pituitary. Their 'throne' is located in the centre of the brain above a stem that, in its spinal descent, ramifies throughout an innervated body.

(Sat) information controls (raj) energy. The psychological (information) centre, with its cortical and sub-cortical centres, controls peripheral, physical energy in the form of the organs of a coherent, functioning body. Mind is plugged into body's socket (*figs.* 14.4 and 19.2 (iii)) by means of the *CNS*. It penetrates body like hand in glove, pervades from above and is fixed like a cerebral root branching down into earth by means of the nervous and associated hormonal systems of information exchange. This is how the two worlds, physical and metaphysical, are met.

Physically, the body's non-conscious centre of (raj) energy processing and distribution is the aptly named solar plexus. And its centre of (tam) support, foundation and fixed embodiment is, around a bony basis, spine's coccyx and the trunk's pelvis. **Such an order of descent reflects the order of creation.**

No doubt, the logic of biological embodiment begins with conception and develops into an adult form. *While its physical aspects are dealt with in the section on biology* (Chapters 19 - 25), *this section deals with the psychological aspect of sub-conscious archetype.* Because the example used is human it is therefore about the (*sat*) transcendent biology (*figs.* 17.4 - 9) of *H. archetypalis*. It is about both the structural as well as the functional (instinctive) aspect of this memory which, like *ROM* memory in a switched-on computer, is in a constant state of activation. In other words, it treats it like the memory of a song or a video that includes both developmental stages as well as a climax in the form of an adult frame. At the same time it recognises the possibility of interaction between individual, physical parts and the metaphysical, archetypal whole. The latter's shape of mind registers and responds to each of its 'holographic', physical compartments known as cells, organs and their linkage into systems.

While each physical body is specific the archetypal memory is shared in common. Think of it as a musical score that, accessed by a thousand musicians each with his own instrument and personal circumstance, is rendered fresh each resurrection. In this case you might understand the rendition that each concert brings to life as a slight variation or fresh genetic possibility of, in our case, the human song.

From a bottom-up, objective point of view the abovementioned paragraphs are nonsense. Mind, a material outcrop, is dependent on matter and not *vice versa.* Neither logic nor symbolism can appear in a structure thrown up piecemeal by chance. There is, therefore, clearly no logic in the construction of bio-logical bodies. Atoms have no mind. Where there is no mind there is no logic. Paley's watchmaker is more than blind. He is wiped out, non-existent. Accident, in the form of evolutionary mutation and natural selection, is the best you can suppose. **Although the evolution of protein synthesis, mitosis, sex and so on and on are Darwinian black boxes, there seems no alternative.** Therefore, a compendium of *interim* 'just-so' hypotheses has been devised. For example, developmental embodiment (Chapter 25) is supposed to follow a course of evolution that may have involved the mutation of genetic switches, neoteny, mathematical transformations (as per D'Arcy Thompson), mathematical 'attractors', arcane chaos theories etc. etc. *Uncertainty about the evolution of morphogenesis is, however, just a subset of the certainty that such 'progressive' process must have happened unintentionally.*

On the other hand, from a *top-down*, subjective point of view the logic of embodiment is clear. It reflects a gradient of creation as it falls, from top to bottom, through the basic dipoles of Natural Dialectic - information and energy. *The higher logic of mind (information) and the lower logic of body (energy) each extend influence from their respective centres in a mutually integrated, harmonious way. Both are symbolically unfurled to best effect in an adult human form (see this Chapter 'Caduceus: the Human Morphogene').* **The construction of human body, by reflecting the conscio-material gradient, also reflects the construction of cosmos. Not just the order but the balance, elegance and beauty of adult human form symbolically accord with it to such degree that it entirely eliminates chance as its originator. This is why it is said, accurately, that 'microcosm reflects macrocosm' and 'the universe is in man'**.

At the Axis dwells Central Consciousness, Awareness or Life. This Awareness is, in existence, snared in mind and lost in matter. Its Logical Spectrum has 'dispersed' from the Centre and is constrained (*fig.* 13.3 (ii) and 15.5) by the sheaths with which it is involved. In this view an organism's framework, as well as associated instincts, are viewed as a construction within which dwells awareness. Therefore we obtain the *yogic*, *sufic* or mystic view, objectionable to a materialist, of bodies as architectures (say, temples) of the Living Illumination. *It is possible to commit mental constructions to memory.* The memory of biological architecture constitutes an archetypal part of universal mind.

While both top-down and bottom-up perspectives see embodiment as a process that involves cell division or development from a zygote (fertilised egg), their bio-logic is quite different.

Bottom-up, the 'logic' is annulled by using the formula of inverted commas with the word 'design' to indicate that it does not mean what it means; instead it is supposed to mean (Chapters 5 and 23) a haphazard genetic mutation 'rationalised' by a culling process called natural selection. Development from seed to adult is programmed but the program was inexplicably derived from accidental incidents. No logic, only 'logic', generated its fine bio-logic.

Top-down the psychological influence of mind is prior; its order is according to its purpose. Although influence at the psychosomatic interface is sub-conscious, this does not eliminate the possibility of the conscious *origin* of sub-conscious patterns. We do it ourselves. We commit inspirations, sense impressions, emotional responses, thoughts and schemes to memory; and deliberate repetition of a desired scheme (such as a language) or pattern of behaviour sinks into memory and becomes a reflex habit. Indeed, this is exactly the view of the oriental 'law of *karma*' or occidental 'rules governing behaviour'. Repetitious thought (*japa*, *zikr* or *simran*) creates deep grooves (*sanskaras*) in the mind. The grooves of such mental behaviour (which is entirely natural) amount to good or bad habits; these in turn lead to physical behaviours and their due consequence. The trick behind deliberate repetition, perhaps using prayer beads, is to eradicate negative thought and, at the same time, establish an overriding memory, a salient context within which mental life is moulded.

Could involuntary repetitions of the archetypal suit of memories be what uphold the order of the world? Could 'reverberations' of continual remembrance quicken subtle (subatomic) patterns and thus, as 'meta-law', shape the way things have to be? This sustaining current has been named as a transformed, lower species of the stream of First Cause; the metaphysical reverberation is known as a 'stepped down' form of *Om*, sub-conscious energy called *prana.*

What, therefore, about the lines within which physical life is moulded? *Recall that any man-made object is a physical form of memory, a memorial.* It embodies the plan of its author. This fact applies with special force to complex, purposive machines. It is worth re-phrasing such a key psychosomatic concept. Take, for example, the cybernetic production-line of a chemical factory. *The basic biological robot, a cell, is just such a factory.* Its physical plans certainly involve the precise production, maintenance and operation of hardware such as nuclear code ('memory chip'), biochemicals, cytoplasmic architecture and, higher than cell, tissue, organ and whole-body structures. Perhaps even more startling than a mature form is, in the case of all multicellular organisms, development from gametes (egg and male missive whose delivery initiates creation) through zygote and intermediate configurations. *However, just as a robot is not made from the delivery of materials so correct production of materials is alone insufficient to trigger the biocybernetic emergence of specifically differentiated, purposive cells and body shapes.* According to *top-down* logic, morphogenesis involves a metaphysical dimension in the form of archetypal memory. Memories can be algorithmic; they can involve sequence and steps. And they can, like language or a suite of drawings, represent an orderly construction. *The morphogenetic mnemone is a kinetic, cinematic template. Such memories or archetypal 'shells' are 'engraved' in lower, sub-conscious mind.*

Factory operations may accede to conscious, sub-conscious or non-conscious influence. At the control panel of an industrial complex a supervisor may check, direct and override the robotic operations of both software and, through the software or by physical inspection, the hardware. Every cell has such a metaphysical aspect. Each cell in a multi-cellular organism has, as well as its *DNA* database, a *typical mnemone* which is, according to its location and developmental 'commitment', differentially accessed. A software engineer or a librarian establishes a sufficiently logical address system *before* any physical correlate (in the form of computer chips or a library) is built. It was suggested (Chapter 16) that specific memories are unlocked by molecular resonance; that such resonance amounts to an electromagnetic address system; and that its two-way, mind/ matter resonance constitutes an attunement. If music involves resonant harmony then so does interaction between typical mnemone (archetype) and body. The language of life is, one might therefore suggest, intrinsically 'musical'. Such harmonic attunement is, effectively, an attachment. It is the way so-called *H. archetypalis* and *H. electromagneticus* are bonded.

Of course, organisms sport metabolic similarities (such as respiration and *DNA* manipulation) so that they all, even if as far separated as yeast and human, display some *DNA* sequence commonalities. A human is, with respect to gene sequences, somewhat less than 50% banana. Organisms with over 90% similarity (such as primates, cats or mice) produce very similar protein 'building blocks' but, just as the same bricks can generate different buildings according to architectural designs, these blocks are built up into different yet fully coherent and competent body-plans. According to gene theory morphogenesis is by regulation and transcription factors alone. Dialectically, shape does not come from biochemicals. It is due to these factors combined with the rule of morphogene.

A chemist with a spectrometer is able to uniquely identify any molecule because each has its own 'fingerprint'. *In this respect bio-molecules, each with its own vibratory fingerprint, are seen as hardware. They are seen as agents of sub-conscious software.* The overall electrical 'fingerprint' of any cell-type is also unique. So, therefore, is an organ's or a body's. Together, in man, they constitute the subtle antenna called *H. electromagneticus.* Information is exchanged. Not only does *H. electromagneticus*, an integral part of biological constitution, receive the psychological archetype's signal but transmits its own condition back. Due to the holographic nature of memory's program, overall consistency, conformity, cooperation and control are sustained. Although it involves suites of subroutines grouped under a typical or main routine, an organism operates as a single, integral, sensitive whole. In other words, the morphogene of *H. archetypalis* is specified as the repository of human metaphysical code.

This sort of pre-programmed, vibratory interaction sounds no more accidental than a symphony - one in which the morphogene is score and individual body a particular recital. In this case it should be possible to describe the structure of the score. According to what order is the sub-conscious expressed? Using the device of *H. archetypalis* and *figs*. 17.5-17.10 we can further probe how the logic of embodiment works.

The Whole Works

All 'logic' resides, so the genocentric story goes, in the genes. These are construed as ultimately responsible for the characteristic structure and function of protein and, therefore, the quality and quantity of all biological morphs. They are the material vehicle of basic information; could not, must not information come about by mindless chance? No doubt being fat, weak, dim or, more subtly, having neurotransmitter or hormone imbalances, can affect behaviour. Therefore a materialist sometimes traces even moral or immoral tendencies to this or that gene on this or that chromosome. Highly sexed with much too much testosterone? With autistic or psychotic tendency to ignore another's circumstance? What about a criminal tendency to rape, extort, bully, thieve or murder? 'It's not my fault, it's my amoral genes'! *Except that morality* (Chapters 3, 13 and 25) *is a product of decision and is as distinct from chemicals as cars from the way they are driven.*

No life form and its associated bio-logic is simple. 'Simple' and 'primitive' are relative terms. For example timing, as in the production and maintenance of any complex mechanism or machine, is critical. The quality, quantity, carriage, location and time of delivery of materials employed must be calibrated, organised and, with necessary modulation, switched on and off. Biochemicals, whole cells and even whole constructions come and go. Genes (what else?) are the operators of such precise, biocybernetic flows of work. Genetic and perhaps also cytoplasmic programs control production lines. 'Program' is a word loaded with the notion of underlying purpose; it is vitalism, entelechy or deliberate direction by another name. Such a notion cannot be countenanced in any evolutionary sense so that a biologist, Jaques Monod, cleverly invented the word 'teleonomy' to describe a systematic process guided not by purpose (which would be 'teleological') but by natural constraints. Might one, now armed with this sophisticate's diversion of attention from the truth, more happily presume that 'survival', 'selection for survival' and the purposeless agglomeration of atoms are notions sufficient to explain the origin of coherent 'teleonomic programs'; and presume that chance, necessity (of natural law) or else chance and necessity combined comprise sufficient agent of 'design'? This, it must be emphasised *pace* molecular biology, is simply a belief, a necessary axiom of materialism. It is termed a scientific belief but the addition of a contrived adjective transforms neither necessity nor belief into proven fact. Could, as Monod hoped to show, mindless forces really code such complex programs as, in every single cell, biology surveys?

Because sensitivity and mind are assumed to be products of neurochemistry, genes are also invested with responsibility for patterns of behaviour. Instinct is presumed hard-wired. All process, transmission and storage of information, physical and psychological, can be traced back to gene sequences and an order of amino acids that arose (and can still arise) in a mutative, random way. The whole works is down to a selfish, microscopic anti-deity, the genome. *Such 'teleonomy', although targetless in any rational sense, is central to the whole, modern biological evaluation of the evidence. In this sense science lacks mind and thus higher reason.* Observing high-grade, coded order it therefore invests all explanation in the clearest physical carriage of order it can find, the genetic database made of *DNA*. It is as if, hearing a radio program, you thought you could realise its whole character from the wires of the set; or as if, finding a book, you presumed there were no immaterial factor and it wrote itself. As usual the problem is one of exclusively materialistic goggles; the goggled omni-answer is (Chapter 24) mutation and natural selection.

Top-down there is complete agreement that physical is physical. A body, including its intricate hierarchies of switching and control, is *per se* entirely physical - as physical as CDs, TVs or DVDs. Its entire, informed structure will eventually be understood in terms of molecular, cellular and higher level interactions. It is, however, the *arrangement* quite as much as the substance of components that is under scrutiny; and the logic of Natural Dialectic suggests that hard-wiring of the brain is not the *cause* of instinct, thought or purposing but rather an *effect* of construction coordinated by the morphogene. As the 'informato-centric' story goes, both instinct and nervous structure are part of a distinct conceptual blueprint for distinctly human perception of our earth and universe.

Analysis and criticism are admirable tools to break a whole into small, individually comprehensible parts. Break-down and build-up. Analysts and critics rarely work creatively. They are not originators. Their conceptual mode is the reverse of developing a purposive, working whole in the first place. Organisms show all the signs of being such synthetic wholes.

Philosophical Pin Pricks

Could pricks of pins be systematic? Could they add up to a cure for individuals but evoke a systems failure, a crisis and an inability to understand?

Acupuncture is widely used in China. It is used alongside western medicine in state-run hospitals as an analgesic and cure for many forms of illness. It regularly replaces the risk-laden form of western anaesthesia in surgical operations as serious as heart transplant. The patient remains conscious throughout, suffers no pain at the time and makes a rapid recovery. Discharge times are reduced to a few days. Over at least two thousand years millions of patients have benefited from its courses of treatment. Has the method any basis in western clinical science? Is it taught or practiced by most doctors here? Absolutely not - because it is 'scientific' neither in practice, perspective nor conceptual framework. It is only half materialistic. Therefore it's not on the syllabus.

Oriental *qi* (*ch'i*) is identified with Indian *prana; prana* (Chapters 6 - 12 and 15, 16) is in turn identified with potential matter, sub-conscious energy and the activation of patterns laid in archetypal memory. *Ch'i*/ *prana* is often called 'universal energy', 'life energy' or 'bio-force'. Acupuncture needles are used to clear blockages in its normal healthy, unobstructed flow throughout a patient's body. These blockages will have been 'pin-pointed' with the help of maps marking channels, meridians or paths which *ch'i* follows. The diagnosis accepts that such body-wide meridians do not correspond to nerves and have, in fact, no physical presence. They are metaphysical (this is the unacceptable rub) but at the same time clearly connected with the body's electro-physiology. Powerful *fMRI* scanners have charted the 'deactivation' effects of variously placed acupuncture needles on areas of the brain that include the 'pain matrix' and limbic system. These are associated with how a person 'modulates' or handles pain. In this respect definitive large-scale tests have, in a scientific manner, confirmed what China's billions and millions of others already knew - the efficacy of acupuncture in the relief and cure of illness. Nor is the practice restricted to humans. Veterinary doctors also apply its techniques to sick animals. The results have often seemed 'miraculous' after standard scientific treatments failed.

Tai ch'i is another well-known exercise which, as well as *pranayama yoga* (Chapters 7 and 12), seeks to promote the flow of pre-physical and thus 'unscientific' *prana* and, in accordance with holistic but not mechanistic perspective, treat body and mind as a single entity. In this endeavour its central principles accord with those of Natural Dialectic's (*sat*) homeostatic balance in respect of vibratory norms and (*raj*) dynamic flows; as blood through arteries or current along nerves, so psychosomatic *prana* along metaphysical meridians. No doubt that sceptical science has a problem if required to treat a metaphysical energy as 'real'. Some conventional drug-minded practitioners may even scorn the use of physical magnetic therapies, available on the NHS (or National Health Service), for ulcers, arthritis and other cures. However an increasing number now agree that China's got it right and acupuncture works. The occidental explanation is that needles stimulate a nerve or endocrinal response. East and west may both be right. The interaction between *H. archetypalis* and *H. electromagneticus* is close and the familiarity of science with electrophysiology, electromagnetism and the nervous system has also become close and extensive. Why, in this case, should it be difficult for science to admit that, for example, studies in Berlin involving over 700 patients have firmly indicated treatment by acupuncture relieves the chronically painful symptoms of arthritis and significantly complements the use of drugs (Times 30-10-06 p.16)? Nor should it prove impossible to translate the metaphysical, causal aspects of the study of acupuncture into measurable physical effects.

No doubt also, however, that if a system is seen to work then the conceptual framework whence its physical expression emerges deserves investigation. The system of acupuncture exerts its effect on body parts but the study of its immaterialistic part identifies actual psychological factors and is called, in dialectical terms, transcendent biology. Such biology involves not only archetypal instinct but also the medium of sub-conscious energy, potential matter or *ch'i/ prana*. The pattern of this energy is called, in terms of biological construction and operation, the morphogene. The structure of this part of archetype is called a caduceus (see figures 17.5 - 10). It may be that *H. archetypalis* and his physical correlate, *H. electromagneticus*, cooperate but physical pin pricks still appear to have shafted a whole philosophical system. Their effect is to provoke a reaction that unblocks the flow of clinical interest and releases a new perspective, a new layer of information that might accrue to mechanistic, materialistic paradigms of medicine. You can easily pipette your entire genome onto the sharp tip of a silver plated pin. Do you want to squeeze its whole conceptual system into yours or will you allow the element of a fifth dimension, one just as informative as *DNA* but even more elusive than the apex of a needle?

Morphogene 2

Top-down there's more to things than meets the eye - or any scientific instrument of register. Mind, even in its 'solid' form, is metaphysical. It is 'an information cloud'; it is an immaterial field. Memory is a particular, persistent pattern or construction in that field; and archetypal memory is more a broadcast than an object which, like a film or computational subroutine, communicates information. The focus here is on typical mnemone and, within it, morphogene.

In this case there are two reasons why *H. archetypalis* and not, say, *Equus archetypalis* (horses), *Scyphomedusa archetypalis* (jellyfish) or *Hymenomycetes archetypalis* (gill-bearing toadstools) is used as the exemplar. Firstly, it is part of ourselves. Secondly, the range of organisms is held to represent a complex scale of grades whose 'chemistry' is based on particular proportions of cosmic fundamentals and hence particular combinations and emphases of associated characteristics. At least in 'higher' animals a sixth state, conscious awareness, supplements its sub-conscious and non-conscious, physical parts. Each organism is more or less attuned to certain environments (e.g. water, air, darkness) and imbued with salient characteristics (e.g. aggression, cunning, acquisivity). However, in each construction except man these are more or less imbalanced. For example, in birds the 'elements' of air, energy and water are ascendant, in insects energy and earth and the botanical emphasis is one of association with water and earth. Five elemental states express the *tanmatric* qualities of potential matter. These are (from top free to bottom locked) spatial/ electro-dynamic, thermal, plasmic/ gaseous, liquid and (last and energetically least) solid. The most balanced expression of all six universal constituents, including psychological capacity, is found in man. This expression will soon be traced in a pattern called the caduceus.

Caduceus can be related to 'ziggurat' (*fig.* 17.9) but do you also remember the energetic model (*fig.* 13.3 (ii)) of organisms as concentric rings? And how (*figs.* 0.7 - 0.10) this notion reflects such a view of the cosmos as a whole? 'Organic' development means development from a concentration of information, a centralised potential. As well as an 'organic' this is a dynamic, vibratory view of things.

Bodies, which in all cases are at least chemically dynamic, are not seen as solid objects but flowing events. In this case an archetypal key is the association and amplification (or malignant interference and reduction) of vibrations - *resonance*. In psychological terms 'resonant association' means 'feeling right'. Things chime. They click. We resonate with preferred energies (e.g. sound waves of particular music) and behaviours (e.g. ways of thinking expressed by concept, language and behaviour). This is how we feel what, for us, is right. Click 'n' chime are friends. In the same way as a musician 'feels' the music or we conduct ourselves 'instinctively' through life's enchantment, so the archetype interacts with physical form. In the case of biological form the resonance is (*figs.* 15.3 and 15.4) between psychological *H. archetypalis* and the physical set's radio broadcast, *H. electromagneticus*.

What models, what metaphors can help explain the '*top-down*' operation of morphogenetic archetype? A tuning-fork? A three-dimensional diaphragm? A tuning fork is changeless and yet, set to vibrate against different materials (sand, powder, water), the shapes it creates vary. Less than perfect materials might, as Stradivarius would agree, stress or to some extent distort the tones and shapes in air an instrument emits. Communicant vibration dances into creativity.

Both elastic bands and habits resist change. You can deform elastic but, if stretched too far, it snaps. Habits are a healthy part of normal, stable patterns of behaviour but, if their elastic pattern is distorted into overblown passion or obsession, soon become deformities. What else is vice? Is malformation, when it comes to body, any kind of virtue? What good does either do?

Mind and body interact. As reflex nerve arcs occur, so material stimulus will trigger reflex chemistry. As, however, mind can also influence nerves and muscles it impacts on body chemistry. The dialectical suggestion to explain such interaction is a link between the 'virtual' archetype and electronic patterns; it is sub-conscious resonance with the position or the motion of electric charge and field. Mind meets electrochemistry. And of the chemicals involved proteins play a major role - informatively (as signallers), energetically (as metabolic enzymes) and structurally (skin, muscle etc.). They derive from *DNA* and differ with its order of command. In similar types (e.g. hereditary blood-lines, parents, children) it may differ very subtly; and between yeast, oak and buffalo a good deal. Therefore the physical materials which 'bounce against the diaphragm', 'stress the original elastic shape' or 'resonate against the tuning-fork' of a particular archetype will produce subtly different 'tones', 'sounds' or, in terms of interaction with material body, shapes.

Tensions geometrically Related in Two Dimensions
By D"Arcy Thompson

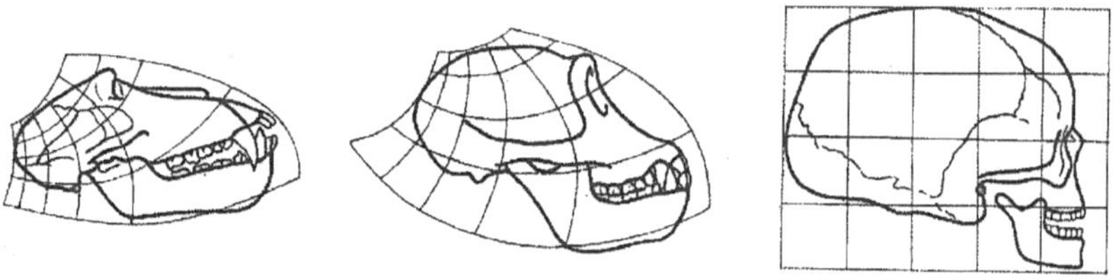

D'Arcy Thompson showed geometrically how regular transformation of shape might occur. Radical changes, such a fresh body-plan or organ, would exceed typical boundaries. However, he never implied such macro-evolutionary transformation could or did occur. Nor did he supply any mechanism for the translation of such 'mathematically elastic tensions' into forms of life. Nor, finally, did he explain the biological reason or origin of his 'transformable archetypes'.

The substance of each skull involves a similar amount of the same minerals, proteins and so on. Do 'rate genes' compute both the amount *and* different forms involved? How (see also Chapter 22) can only proteins extruded from a ribosome mould, for example, your opposing hands or act as templates to achieve the finely accurate, integrated and functional construction of a three-dimensional life-form? In informative, dialectical terms no doubt genetic programs play an automated part. What, however, is the origin of complex computation? What orchestrates the shapes of cells or organs or a body like your own? Don't such major changes as illustrated synchronise with many other anticipated checks and balances? In other words, it is suggested that the mould, an archetypal plan, is metaphysical.

fig. 17.2

We saw (Chapter 16: How Does the Communication Work?) the archetype as a harmonic whole, a signature whose tune for any given type of organism stays the same. Its score conjoins with each slightly different physical rendition. In this case physical differences (including genetic variation and its consequences) slightly distort, stress or 'blur' the notes; they are like various overtones, timbre or a particular artist's rendition of the song. Vibrant shape is shifted, habit varied. As different fingers 'bounce against' different instruments with varying effect so different materials 'bounce against' different plans with similar variation. Such 'bounce' is, at every level of an archetypal program, a resonance effect. Signal association. There is strong or so-called 'dominant' and degrees of weak, so-called 'recessive' reception. The morphogenic score itself is well tuned but circumstance, fretting and plucking a little differently, varies the interpretation of its 'song'. Various possibilities are realised. A morphogene remains the same but bodies differ, by genetic nature and their nurture, every time a tune is played, a version born and a fresh organism presses into life.

You are, physically, vastly more similar (about 99.9%) to every other human, especially your parents, than dissimilar. Sex delivers variation-on-a-theme. Sperm and egg involve two sets of chromosomes (whose gene expression will determine protein); and through special chemical arrangements archetypal 'fingerprint' is pressed. The former interacts with the latter, *H. archetypalis*. Sperm and egg coalesce to form a zygote. When they fuse it is as if, in coalescence, two similar versions of the same record were played together or two singers combined to mix a single species of their song. Or two photographs of faces with slightly different features were superimposed. Variation-on-a-theme. Is the product more like dad or mum? A new 'diploid' superimposition appears whose genetic expression, in conjunction with *H. electromagneticus* (or electrical form), represents the dominant traces from each parent. Such likenesses are reflections of physical not archetypal feedback. Generic morphogene is changeless; local contents ring specific changes. Thus the fine structure of a child's features 'resounds' with its parents, grandparents and, increasingly diluted in effect, further ancestors. Maternal, paternal or neither line may dominate. The outcome depends on the purely physical genetic recombination of meiosis and the differential strengths of resonance with archetypal 'criterion'. The archetypal memory can accordingly be seen, both with respect to sexual and asexual reproduction, as a probability-generator. It is, a 'tympanum', 'habit' or 'instrument of form' whose physical reflections vary while at the same time it acts as a strongly conservative 'attractor', or a template whose physical consequences vary with circumstance but always tend, if circumstance allows, to rebound towards the preordained norm.

In general the more powerful a material reflection, due to pristine resonance with archetype, the stronger its routine, the more dominant a particular influence and the greater the probability of its expression; and, conversely, the more 'off-key', then the weaker its resonant association and the greater chance that erroneous transfer will occur. Over generations some biological characteristics are locally amplified (become dominant in a population) and others fade out. It may even happen that a particular physical resonance 'breaks away' to form a recognisable but distinct version of the same basic song. It certainly happens that suites of the same musical genre (say, blues, operatic, Bach etc.) sport their own signature of style. They fall into distinct orders, classes, families; but though they may share notation patterns, cadences or other commonalities one style does not transform itself into the next - and *DNA* with mutant hiccups is a lot less fluid than a tune. Such intra-specific variation within genre is called, in biological terms, a new race or species. Various genres (genera) will make up different types; perhaps even distinct and separate families can radiate within a type. Where does a classifier draw his line? Does nature set impregnable a bar; no change beyond a definite extent; no unlimited plasticity? No, cries evolution, though its music lacks composer. Yes, retaliates the notion of design.

Now see the body as a rubber sheet or three-dimensional 'tent' which, pegged at different tensions by its physical genes, is subtly transformed. Each 'tent' pegged out, each musical recital or tumbling throw of the genetic dice is, except for identical twins, subtly changed. A persistent skew might generate separate 'races' or even what you classify as species (witness *figs*. 5.1 or 21.1) but this in no way implies that such changes are endlessly progressive. Such variation does not bear imaginary extrapolation or unbridled transformation into all the phyla that exist on earth.

Vibrant Morphogene

Information Loop between Physical Body and Morphogenetic Mnemone.

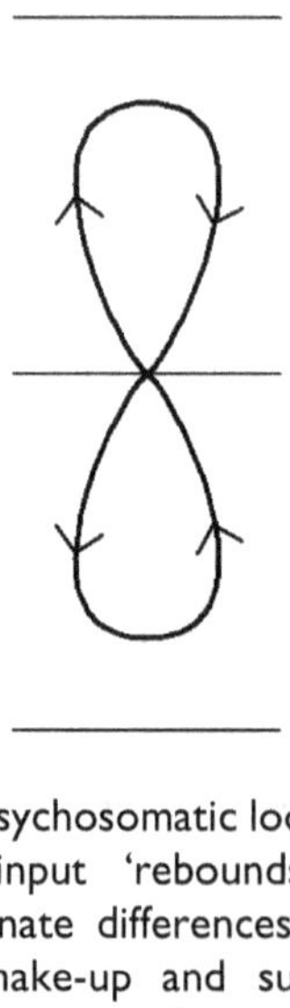

This psychosomatic loop engages rings 4 and 5 on *fig*. 14.3. Different physical input 'rebounds' from its particular morphogene with proportionate differences of form. In other words, slightly different genetic make-up and surrounding circumstance (commonly called 'nature and nurture') generate physical variation on archetypal theme.

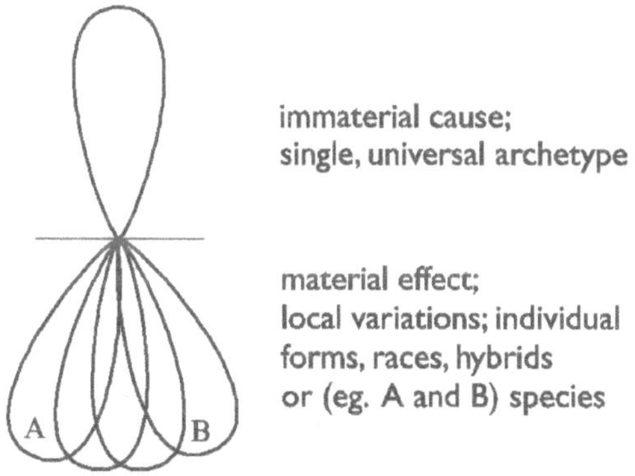

Sexual organisms are 'dimorphic viz. polarised into male and female forms. While these represent complementary opposites of a single type, the issue is not simply one of emphasis. A sexless primordial 'gonad' is caused, by a pre-coded signal, to develop into male testis or, lacking this signal, female ovary. And further definite, deliberate, predefined switches trigger either male or female external reproductive organs from a 'bipolar precursor' - an embodiment of polarised information intended to create one of two forms of genitals. In short, particular physical 'chimes' will elicit the appropriate morphogenetic subroutines. These will 'bounce back', in the continuous manner of radar, the signal framework that reflects either male or female 'morph'.

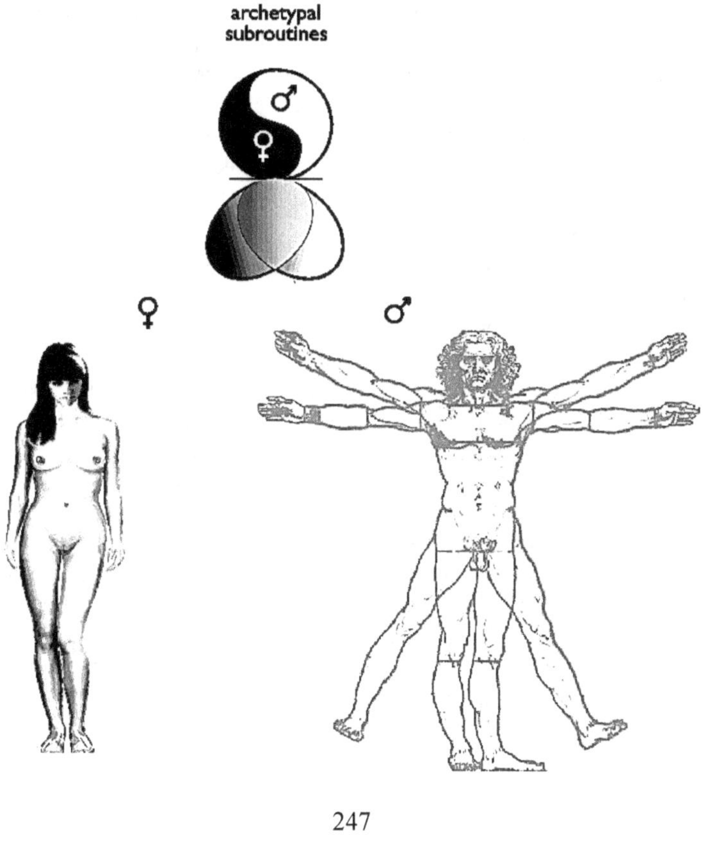

Can you tell the style of one guitarist from another? Different combinations of musician can play endless variations on the same song. Similarly different cells will, according to their physical constitution, 'resonate' differently with the archetypal score. Their rendition will show individual style. What if, as in fertilisation, two combine? In this case each parental body, in its quintessential form of egg or sperm, represents a slightly different recital of the archetypal score. Each plays a slightly different way so that theirs, of all possible duos, plays the single song in a recognisable but new way. Genetic recombination 'blurs' or, we might say, 'blends' the archetype's expression to a sexually slanted composite of parental characteristics. A new version of the old song is born, a resonance reflecting its musicians.

With respect to sex expressed (Chapter 24) male and female each evolve from their united, archetypal root with such fitness, beauty and emotional charge as to eliminate all chance; how gladly Lady Luck dissolves into the arms of Lord Deliberate's great love.

fig. 17.3 (see also 5.7 and 14.3)

D'Arcy Thompson, a zoologist, mathematician and classicist, showed in his theory of transformation that regular, mathematical relationships exist between certain forms; in other words, you can derive different morphs from a basic, geometric archetype (see *fig.* 17.2). Industrialists don't hesitate to manufacture different models of a popular machine. Why should morphogene-by-theme (say, fish, mammalian or sub-types) exceed the wit of engineering? But morphing themelessly? Mindful Thompson never sought to mathematically transform a single cell into an oak or buffalo but even if he had there's more to bodies than their overt form. Mindlessness and coded form are chalk and cheese.

Do pressures from the outside change an instinct or confirm it - even though each individual circumstance is slightly different? External 'pressures' (say, a different individual's genome) on a morphogene do not deform it. A drum skin is not changed but just reflects a change of beat, a different rhythm up against itself. And you can only vary tensions to the point where something snaps. Skin splits, cloth rips and cacophony's a killer. You cannot convert the self-consistent melody of one opera into another by random scribbles on the score. The reverse - you're stuck with minor variation on original score because close resonance with archetypal 'melody' is what will manifest as your coherent health, that is, a best performance. Wander far and you grow sick and die. Archetype is pure information. Information is potential and adherence to its lines shows up as optimised potential. Call such optimum a 'wild type' if you want. A limited cacophony, even if a highly-strung or sickness-prone departure, might sound unpleasant but remain endurable. But the further players stray from an original coherence of their tune the worse it gets. Out-of-tune, off-beat - too much noise provokes a lethal sickness. Many inconsistencies do not a fresh and whole consistence make. They kill the song.

The rational idea of a biological archetype used to be the norm. The materialistic notion of transformation from type to type of organism by small distortions (evolution) has so distorted this original mode of thought that it is worth rephrasing. General characteristics (e.g. facial features, organs, body proportions) occur throughout a biological type. A metaphysical, sub-conscious archetype (which includes, where appropriate, male and female sexual programs) is a kind of 'resonance structure' inflected in physical practice by genetic composition and circumstance. *Its image is no more shaped like a mature organism than an egg or electromagnetic frequencies of broadcast that will be 'grounded' by an aerial and thence transduced to a 'mature' and physically comprehensible end-product on a screen.* Different bodies work the changes round this median, seed-form or source of original 'perfection'. There is, in other words, detailed variation driven from the physical or genetic side and which appears as different individuals, races or even species (figs. 5.1 and 21.1 again). *Such variation is sometimes called micro-evolution but no archetype is morphed into another.* At any particular time and place, in any particular body or population, a certain stress or asymmetry may be accentuated. Genetic entropy of information (by mutation) may occur and less healthy bodies be produced; but natural selection will 'negentropically' sort the fittest, that is, those most resonant with archetype. Such resonance is called, by geneticists, 'dominance of the wild type'. There will exist a tendency to gravitate, an elastic pull towards standard versions of the archetype. The lines of least resistance are the ones that will tend to be expressed. Archetypal trace is reinforced. This is the same as saying that the memory is resilient. Its 'elasticity' (*vis medicatrix*) will tend to correct even such excessive stress or deformity of its field as might result from severe genetic or physical damage. *Such 'elasticity' is strongly conservative. It is anti-evolutionary and is what we find in practice.* Evidence of such 'elastic' correction has come from genetic experiments with, for example, fruit flies, cress weeds and so on (see also Chapters 21 and 23). The universal morphogene of, say, a fruit fly (*Melanogaster archetypalis*) is dynamically expressed. Are notes of music not dynamic too? Just as acoustics change according to container or chords are variously and subtly damped for different effect, so chemical environment of body rings the changes set against a constant archetype. Propensities, probabilities and (weaker) possibilities derive from interaction with its physical containers. Variation-on-theme crystallises due to the physical medium; slight variations are produced, as snowflakes in the air or slight distortions on the sands of Chladni's plates (Chapter 9: High Wires), by the archetype's reflection in a complex biochemistry. Air flowing over hindrance eddies; even subtly different snags will change the shape of water's flow. Same 'chime' but slightly variant; same 'type' but changes subtly rung. As the genetics of a population shows, such 'distortions' ebb and flow and cycle over time but, like swings and roundabouts, never can exceed 'acceptable integrity'. Either physical disturbance is overcome or an organism stretched beyond its range of coherence aborts. Monstrosities that survive miscarriage will be scythed by incapacity, disease or predators called collectively natural selection.

No doubt, therefore, genes play a crucial physical half. They 'express' specific proteins and are therefore a link that can switch biochemistry's automated production lines on and off. *Top-down*, they are part of the hardware employed to express software but are not the *source* of morphogenetic information. They are not shape's template but a genetic database that interacts with the cytoplasmic architecture of egg, consequent cell-types and, hierarchically, the larger constructive influence of *H. electromagneticus*. Bodies are organised according to a metaphysical component so that, as in the case of a machine, both the metaphysical component of planned architecture and the coded, physical applications that support it interact. In short, the integrity and accuracy of a genetic database affects the application programs that use it and the physical materials, specially processed to construct hardware, that derive from it. **However, the primary, all-powerful gene is a materialistic myth**.

Is a metaphor or model that cannot be accurately verified of any scientific use? How, for example, might one model claims for acupuncture or a morphogene? The latter can be described more precisely using a caduceus but first let's check the logic of development.

The Logic of Development

Could you infer the presence of a morphogene from acupuncture's working or, more powerfully, from the bio-logic of development?

As a packed parachute is a different shape from its expanded use, so the (*sat*) potential of an egg is different both from the (*raj*) unpacking process and its full expansion into the (*tam*) target product - finished, adult form. *In each case the whole construction and its operations are preconceived, preordained and, from the very first, work perfectly*. They had better!

Just as a metabolic starter is separated from its finished product by steps on an intentional, codified pathway, so a zygote is separated from adult form by a suite of developmental subroutines. The origin of each kind of routine will be discussed in Chapters 20 and 22 respectively. Let's rehearse the bio-logical construction of *H. archetypalis*, both in development and adult frame, in a slightly different way - from the perspective of its interaction with *H. electromagneticus.*

As mentioned in the previous chapter, in 1935 Drs. H. S. Burr and F. S. C. Northrop first established that all cells and bodies are permeated and controlled by electrical fields. Burr showed that salamanders possess such a matrix complete with positive and negative pole arranged along the longitudinal axis of the body. When he traced the development of the field back through the growth of an embryo he found that it existed even in the unfertilised egg. He marked with blue dye the pole of the egg where there was a noticeable drop in voltage and noted that the head of the salamander always grew opposite that point. Eggs, which also involve chemical gradients (or differentials), are electrically polar. Burr's embryo cells were arranging themselves according to the pattern of an electrical field that was present before the individual came into existence.

Development by Branching:
Systematic *Top-down* Construction of a Human from Germ Layers

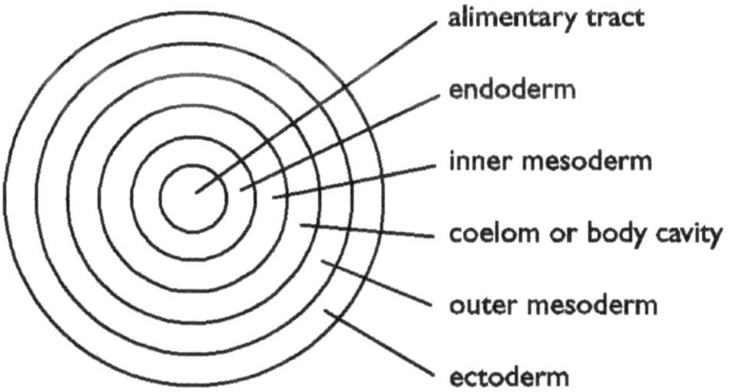

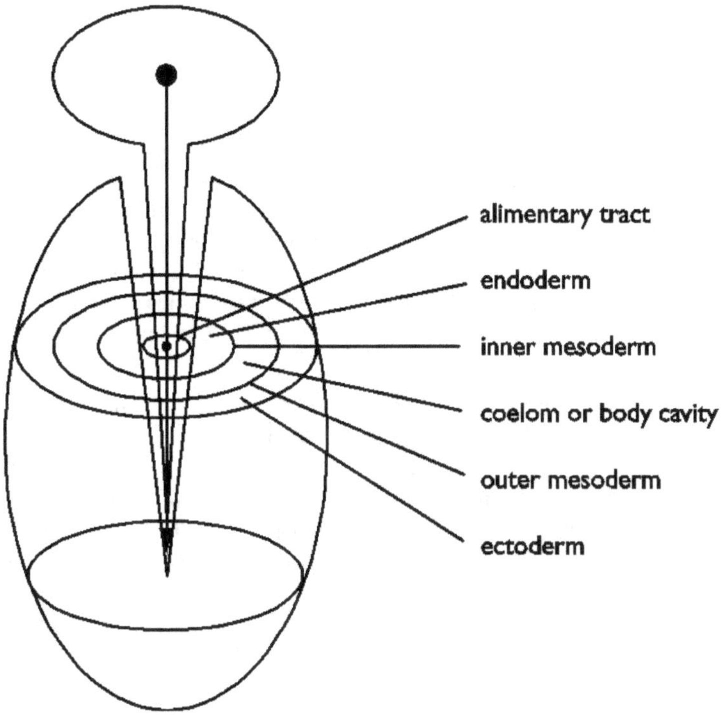

Top-central from the primary single-celled zygote binary division initiates developmental cascades. These flow through branching systems of archetypal logic whose goal is an adult, reproducing form. One of these branching systems is the way major structures are derived from three primary germ layers - the endoderm, mesoderm and ectoderm. The schematic concentric circles represent you. Your anatomical features (e.g. intestines from the endoderm, muscular and connective tissue from the mesoderm and nervous tissue from the ectoderm) have 'flowered' from them. Microscopic analysis of the process clearly illustrates high-level programming. A similarly high-level of illogicality, reason-based from the framework of materialism, denies an intelligent information source, preferring to believe that chance, evading all lethal bugs on the way, can 'program' even simple let alone highly sophisticated systems. Others, flowing with the *top-down* logic, assume that intelligence originates goal-oriented constructions and that, once created, chance is a simply their degrading detractor, a bug-creator.

fig. 17.4

To measure such voltages a platinum electrode is vibrated between two points outside the organism and marks the voltage (in millionths of a volt) between them. However, the work of Burr, Northrop and (from Chapter 8) Sir Jagdish Chandra Bose has not formed much part of the trends and fashions profitably centred in biochemistry, molecular biology and genetics that have shaped research over the last three quarters of a century. As a result the modern study of electro-physiology is practically restricted to the study of nerves and muscles. Even though some startling discoveries concerning the morphogenetic properties of electrical fields have been made, no general theory of electro-physiology has been developed. *There arises, moreover, when the complex shape of any mechanism is built, the question of what moulded it according to what. In a biological case the question is what moulded, moulds and controls the electromagnetic matrix itself - since its purposeful and systematic shapes cannot be the product of chemistry or physics left to their own device.* **The linkage of causal bio-electrical patterns with genetic, physiological and anatomical effects will be further confirmed and pursued in Chapter 25: Bio-logic and Logically Expressed**.

The first step in any construction is foundational. How is the approximately spherical symmetry of an egg translated into the bilateral, radial or other symmetry of a developing and developed body; and, concurrently, into the reflective and other asymmetries (such as the position of heart and liver in a human) that the same body may exhibit? From 46 cm. (Chinese dino-dragon *aka* dinosaur) through 39 cm. (extinct giant elephant bird, *Aepyornis maximus*) to 0.109 cm. (Vervain hummingbird), 0.014 cm. (human) and still smaller, eggs vary almost as much as the adults which derive from them. However the polarity of an embryo depends on the polarity of its first step, the egg. The animal pole is the point on the surface nearest the nucleus and the vegetal furthest from it. There exist asymmetries in the form of chemical gradients and cytoskeletal ('anatomical') arrangements that influence the positional axes of head-to-tail, back-to-front and left-to-right. In the case of the toad *Xenopus*, for example, yolky vegetal cells give rise to central gut formation while the animal cells will form the peripheral body.

Another kind of polarity is present in every cell. It is drawn between (*sat*) informative potential of the nucleus, (*raj*) cytoplasmic kinetics and the 'static' anatomy of confinement represented by (*tam*) membranes, tubulin microskeleton and compartmental organelles. It is suggested (see *fig.* 19.2) that each cell contains, in miniature, functional apparatus that mirrors the equipment found in developed form in multi-cellular organisms. These functions descend the conscio-material hierarchy from (*sat*) central potential (that is, immaterial information carried on the medium of *DNA*) in a gradient through to their manifestation with (*raj*) energetic, gaseous, liquid and (*tam*) solid cooperants. They involve the storage and transmission of information, intake of nutrients and energy metabolism, osmoregulatory and reproductive mechanisms, elimination of waste products and, finally, physical definition in the form of anatomical supports and limiting 'skins'. In this respect each cell is an antenna that oscillates with the same elementary principles as the multi-cellular body of which it may form a part. *From principle to practice; such principles and their physical associations are elaborated in the next three sections of this chapter.* Sub-consciousness, composed of memories, is of itself a reflex zone. While its physical state is grounded in the algorithmic operations of biochemical metabolism, a cell's psychological state is seen as sensitive but ever comatose oblivion...

Do you remember (Chapters 15 and 16) the analogy of a multi-cellular body with a pix(c)ellated hologram, wherein each part contains a representation of the whole. Nuclear *DNA* is such a representation and so, psychologically, is the psychosomatic morphogene. In humans the outer limit of a full body has, by the blastula stage of development, been defined. Horizontal endodermal/ ectodermal and vertical head/ tail polarities are established followed, at gastrulation, by differentiation into outer ectoderm, mesoderm and endoderm. These germ layers constitute three primary subroutines of a branching structure according to whose 'cascade' the organs of triploblasts such as you can systematically emerge. The main spatial parameters are now organised. Simultaneous triplex development builds from around a central tube called the alimentary tract. It gives rise in concentric order (*fig.* 17.4) to endodermal, mesodermal and ectodermal features. *Highly intricate biocybernetic routines coordinate the correct production of materials with the correct position and shapes of the subunits of an integrated whole that works while it is still being built.*

The bio-logic has not been exhaustively detailed. However, many research institutes are in hot pursuit and the principles are clear. *An artist, having sketched outlines and proportions, fills in the details; after the establishment of a framework composed of the main spatial and constructional polarities, a body is observed to develop, both in space and time, along top-down Natural dialectical lines.* That is, the emphasis runs from (*sat*) potential through (*raj*) kinetic to (*tam*) exhaustive, finalised components. It descends from 'before' (informative preconception or plan) through developmental stages (of determination) to 'afterwards' (the adult result). It drops from a spark of conception (fertilisation) through growth and maintenance to death. Or, seen in a structural way we shall shortly elaborate, it falls from informatic head through energetic trunk and its appendages of action, the limbs, down to base organs serving the physical 'hereafter' - anatomical recreation (reproduction) and the exhaustion of waste materials. From top to tail, from (*sat*) highest through higher (*raj*) energetic to lower (*tam*) regions of gross materialisation such direction follows the cosmic fundamentals through an act of creation. The bio-logical sweep brings intention to our world; it delivers bodies to earth. It realises the pattern of cosmic descent or, if you like, cascades down a slope of the conscio-material creation gradient. *Bottom-up*, genetic code and genes deliver everything; but whence the code? Are 'code' and 'goal-directed system' meanings you ascribe to chance? *Top-down*, bio-logic is the order of caduceus.

The main existential principles are information and energy. Survival of a (*tam*) biological body demands (*sat*) coordination, control and (*raj*) energy supplies. In humans a primitive streak, notochord, neural tube and development of the (*sat*) central nervous system are triggered. A beating heart and other organs of the cardiovascular nutrient supply system also take shape. Eyes and ears (the top senses) soon begin to appear. We are unfolding *top-down*. During a highly (*raj*) kinetic initial phase the brain develops rapidly. The instrument of metaphysical manipulation evolves in an order that reflects both structure and quality of consciousness - from sub-conscious brain stem and medulla through semi-conscious, reflexive cerebellum to, in its cerebral hemispheres, the flowering of consciousness. The instruments of physical manipulation, hands and feet, also sprint ahead. Nutrient-processing equipment (pancreas, intestine, haemo-regulatory liver) begin to form in the lower area of solar plexus. These deal with energy in its grosser form. They break down, dissolve, treat and circulate food. Urogenital osmo-regulatory organs (such as kidney and bladder) start to form. Pre-muscular masses appear in the head, trunk and limbs. An outer boundary, epidermal skin, encapsulates the whole evolution of form - *evolution from a predefined clutch of instructions and thus totally removed from the hypothetical, Darwinian concept of instructionless chance as the informant.*

Soon (after about 8 weeks in a human) sexual differentiation of a 'unisex' or sexually neutral genital ridge occurs and the major organ systems approach completion. It is now a question of (*tam*) con*solidation*. From the outlines of principle details such as fingers, toes, nostrils, final eye and ear shapes and so on are sculpted. Muscular and neuronal relationships are established and the foetus enters a growth phase. Skeletal tissue ossifies, chubbiness accrues and external features such as hair and nails are visible. A life form grows towards the moment of its chute when, as a result of consistent heavy pressure, it is laboriously expelled down a canal into the sea of an exterior, material world. From a *top-down* point of view there is a real sense in which this launch is not a birthday but a death-day. Out-swing from the centre. Incarnation. Mind 'canned' in a physical shell. Life fixed on the outermost edge of the universe. The reverse is, therefore, to experience in-swing - in the process of either voluntary mystic or involuntary natural death (Chapter 18).

Biological morphogenesis always involves a precise and intricate plan. Equally clearly, it evokes variation-on-theme. Its complexity will, like abiogenesis (Chapter 20) offer years of head-scratching research to those who want to work out how purpose-like and integrated specificity accrues by chance. *Top-down*, it never does. Information rules coherent integration of forward-thinking plan (is not an adult *target* of an egg?). How, in this case, does *H. archetypalis*, through the agency of *H. electromagneticus*, impress its top-to-tail sequence on the adult logic of *H. sapiens*?

Caduceus - the Logical Shape of Man and Symbol of the Medical Profession

The caduceus is a representation of basic human structure. It is an outline-in-principle, a matrix, a part of typical or archetypal mnemone also known as *vis medicatrix* or morphogene of *H. archetypalis*. Wings represent the conscious element, made physical as the two cortical lobes of the brain; the knob or bulb represents sub-conscious elements manifest as diencephalon, cerebellum and brainstem; staff and snakes represent three major channels of information exchange - central, inward (sensory) and outward (motor) - whose physical correlate is the nervous system. As glands distribute chemical information, so the caducean crossovers represent central nodes for the distribution of sub-conscious energy. These morphogenetic power points are variously called foci, plexi, nodes or *chakras*.

How do you see a wood for its trees? Notice the description is, unlike the wealth of detail found in a research paper or textbook of clinical medicine, rudimentary. That's how, at a higher level than the detail, principles are. They are infrastructure - the basic outline from which complicated practices are logically derived. They are the source of self-consistent order; immaterial order of the mind is, as caduceus, embedded in material form. We are dealing with invisible but comprehensive infrastructure of a body called a morphogene. Morphogenes lend different types of life-form their 'generic shapes'.

Caducean Structure Reflected in a Cohesive Protein.

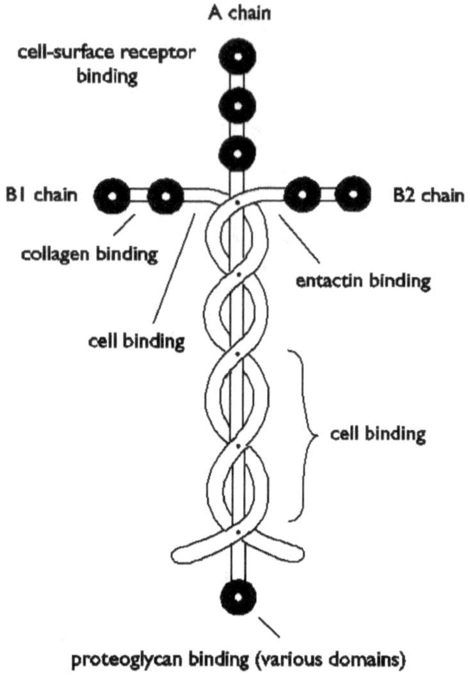

Curiously, the archetypal structure of our physical being, caduceus, is reflected in a cross-shaped cell-cell adhesion molecule called laminin. Found in basement membrane this glycoprotein forms a networked foundation for structural cohesion. It binds to cell membrane and, at the same time, extra-cellular matrix (*ECM*). Without it the tissues of most animal cells and organs would fall apart. No glue, no you - without it you'd disintegrate! Caducean coils and crosses are thus crucial!

fig. 17.5

The *caduceus* is an ancient symbol. It was the staff of Hermes (or Mercury) whose power was thought and who was, thereby, the winged messenger (information exchanger) of the gods. It is still the Hippocratic totem, even though its metaphysical import has been forgotten or erased, of scientific medicine man. From its central bulb spread two wings which represent the higher, flying capacity of conscious thought as expressed in the form of the cerebral hemispheres of the brain. In biological terms the bulb itself is said to represent sub-cortical, central forebrain. This structure, called the diencephalon, links conscious to sub-conscious and totally automatic, physical regulation of the body. The diencephalon contains both the 'seat of emotion' (limbic system) and a triplet of neuro-hormonal governors in the form of pineal, hypothalamic and pituitary glands; from their 'throne' at the spatial centre of the brain, these masters regulate all autonomic and hormonal information and, thereby, the body. They are psychosomatic nuclei. The bulb, representing all sub-conscious, psychosomatic aspects of mind/ brain, also includes the cerebellum, core processing units of the brain stem and reticular system, midbrain, pons and medulla. In this region reside the orchestration or government of balance, an on-off toggle for waking-state or sleep, the coordination of autonomic functions and a kind of switchboard, point of exchange or crossing-over for the routes of sensory and motor traffic between mind and matter. The staff (central, spinal cord) and the double helix that surrounds it represent the triplex distribution of sub-conscious energy. At each of five coils from the top this wireless energy is transformed in a kind of spectral, *tanmatric* correlation with the 'energetic frequency' of states of matter. Spectral *ch'i* or sub-conscious energy grades from higher to lower frequency; from freer, fluid to more sluggish or even fixed states; or from lightness towards the heavy darkness of solidity. At base it is coiled, like a spring or a crouching tiger ready to leap, at the tail end of the spine. Such hidden, generative power of earth was symbolised by the Chinese as dragons and known to the Hindus as *shakti*, *kundalini* or snake-power. The whole staff, head to toe, represents not a luminous but a psychological rainbow. Is the traditional caduceus just a vague, pre-scientific way of describing neuro-physiology? Why use such an antiquated framework to describe the human morphogene? Has it not been elucidated *top-down* mind to mind without material instruments; and if material instruments can't reach it is it really there? Modern science replaces caducean construction with an anatomically precise, subtle, hierarchical arrangement called the central nervous and correlated hormonal systems. In this neurological view consciousness is itself a physical product of the brain. 'What is going on in your mind?' now really asks 'What is happening in your brain?'

Dialectically, on the other hand, consciousness at the eye centre, located at the front of the cerebral cortex, is transformed through semi-conscious to sub-conscious and the lowest, physical or neurological levels. From the spine of this psycho-physical information system branch both the influence of psychosomatic meridians and the peripheral voluntary and involuntary (autonomic) neurons. Through their all-pervasive agency coded information is passed up from sensors to *HQ* and back down, through motors, to muscles. Sensation and locomotion; knowledge and practice; life in the material world. *Is matter all that matters? Does matter mind?* Or do the caduceus and biological information systems run in tandem resonance, in psychosomatic co-activity?

How can matter set up systems? How can it, 'any which way', ever *think*? Yet isn't it, materialists say, doing just this at this moment in your brain? Aren't you 'thinking matter'? Conversely, from a holist point of view there must exist (unless mind-power could act directly on whatever kind of matter) a psychosomatic linkage system of which nervous or other systems form the somatic side. The other side of this informative linkage, one 'sandwiched' between conscious mind and non-conscious physicality, is sub-conscious mind. A persistent mental shape or record is called a memory; memories, whether personal or archetypal, are the 'stuff' of sub-conscious mind. An archetype is, in these terms, a complex mental shape; if it is created personally we call it an idea, if universally it is part of the blueprint for physical creation, a file of archetypes called universal memory. In ours, the human case, it is a 'virtual' matrix whose 'archetypal program' dictates psychological behaviour (instinct) and governs its own reflection in the localised shapes (both in development and maturity) of an associated body. In other words, you represent the projection of a plan upon an earthly screen.

Is metaphysical, psychological structure impossible or not? Is it immaterial yet a most important fact? Or is it arrant fiction? Another name for a wireless 'fiction' that, whether rightly or wrongly, is entirely unacceptable to scientific materialism is (Chapters 15 and 16) *H. archetypalis*. *Figs. 17.5 - 10, with accompanying text, elaborate the logical shape of man or, in dialectical terms, his human morphogene.* They show, in other words, how a long-delineated system of sub-conscious power points represents the inner functional logic from which (wo)man's outer shape derives.

Human Extent: the Conscio-material Gradient

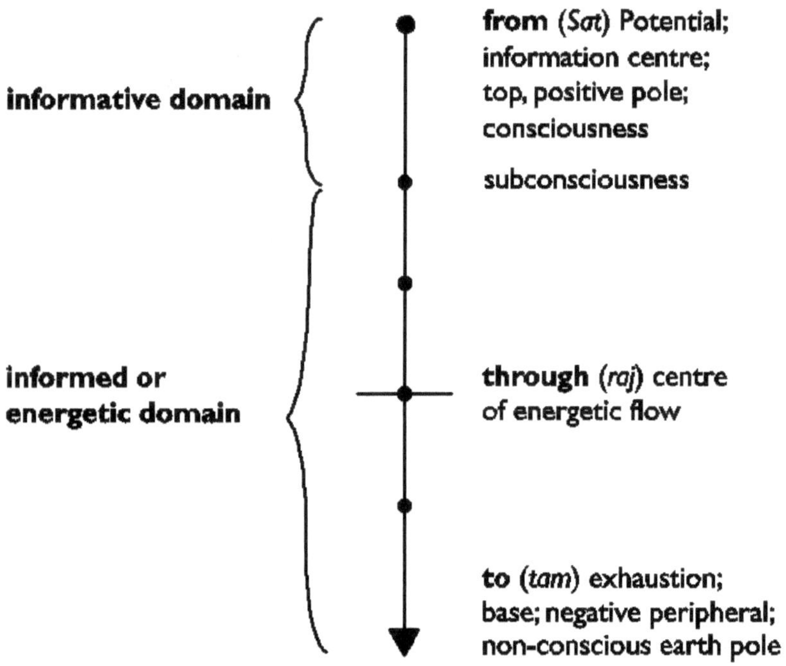

Human radius amounts to a microcosmic conscio-material gradient. A radius from the centre represents cosmic polarity - basic to Natural Dialectic - that drops between conscious and non-conscious levels. It falls from (*sat*) informative potential through (*raj*) active function to the (*tam*) static, structural phase of gross form, that is, bulk shape and containment. In this sense, with its roots in Essence, human form spans the whole of existence.

Note, in this respect, that the major function of the (*sat*) informative domain is command and control to maintain balance. Conscious command is flexible, sub-conscious archetype is inflexible.

The informed or energetic domain also involves two divisions. (*Raj*) energy's centre of balance is at the solar plexus. In a human the systems of the upper trunk involved with energy (*ATP*) production include, lungs, digestive system, heart and blood transport system, pancreas and liver. These break down nutrients to 'loose' chemicals that are irradiated to support each cell's metabolism, growth and maintenance.

Solidity swings low. (*Tam*) exhaust drops from organs of the lower trunk (colon, kidneys and the urogenital system). From reproductive organs falls the reproduction of a solid form.

In keeping with the non-conscious nature of energy lower controls are (except for an occasional awareness of the need for air, drink, food, excretion or egestion) reflex and inflexible.

fig. 17.6 (see also fig. 19.1)

Do you remember (*fig*. 2.1) concentric spheres of energy and atomic 'shape' were derived from a *top-down* axis?

Now consider (*fig*. 17.6) 'the human extent' as a wireless anatomy with conscious mind as its positive pole and pelvic foundation as the negative.

And now check *fig*. 17.7. Lacking any detailed systematic account in the west, we henceforward trace the archetype through oriental clues derived from practice. This practice has been described throughout historical millennia. Its yogic experiment is personal and psychological.

The primary *top-down* frame for man is, in its most basic orientation, represented by three major channels of the spinal axis. Together they link the major domains and plexi. The first, top or 'royal' connector runs *central* between these second-in-command plexi. Called *sushumna*, this central channel is the same '*sesame*' that a mystical Ali Baba called open to reveal, within the cavern of the cranium, riches of the soul. Ali's 'nerve of bliss' passes down the spine where its physical correlate, CSF (cerebrospinal fluid), circulates through the central canal and sub-arachnoid spaces. Pulsations in this fluid will, logically, have electromagnetic or other periodic controlling influences; they may well resonate in the duct that surrounds the pineal, a master gland that could be sensitive to or even create such modulations. It is certain that the fluid acts both as a protective buoyancy against shock for the delicate central nervous tissues and (in conjunction with the blood-brain barrier) as an agent for the homeostatic exchange of cellular nutrients and wastes. As a composite of three main meridians, one representing each fundamental quality, the (*sat*) *sushumna* is balanced or neutral.

At the base end of its line, coiled like a compressed spring, is the most confined, locked up energy of sub-conscious mind called *shakti*, *kundalini* or, simply, 'snake-power'. This is an allegorical way of expressing the 'least metaphysical' pole, a concentration of the lowest frequency of *pranic* archetypal memory, a correlate of solid earth. There are two ways to vacate the body, voluntarily or involuntarily - usually the latter. However, mystic practice seeks to attain this condition in a voluntary, coherent way. Death's mystery is revealed (Chapter 18) by an orderly exercise, its sting is extracted in a precise, controlled manner. Focus of attention is locked to the point where physical distraction disappears; the '*pranas*' are withdrawn from the base plexus upwards until all physical awareness is lost; life is literally wound up. Allegorically '*kundalini* is activated; the snake rises until it pierces the third eye'. Such practice is sometimes called, unsurprisingly, *shakti* (power) *yoga*.

Wireless Framework of *H. archetypalis* (Memory Man)

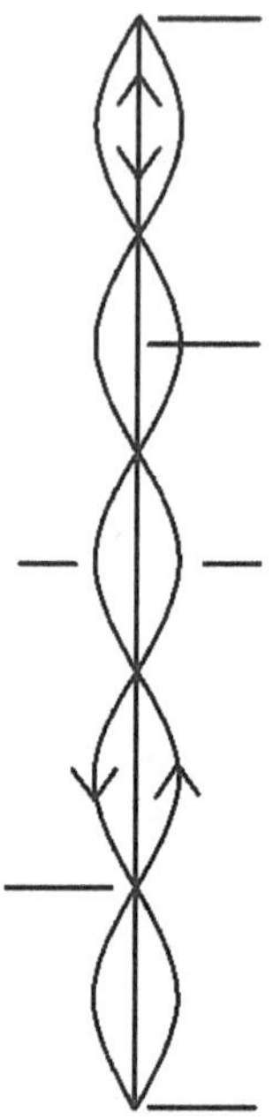

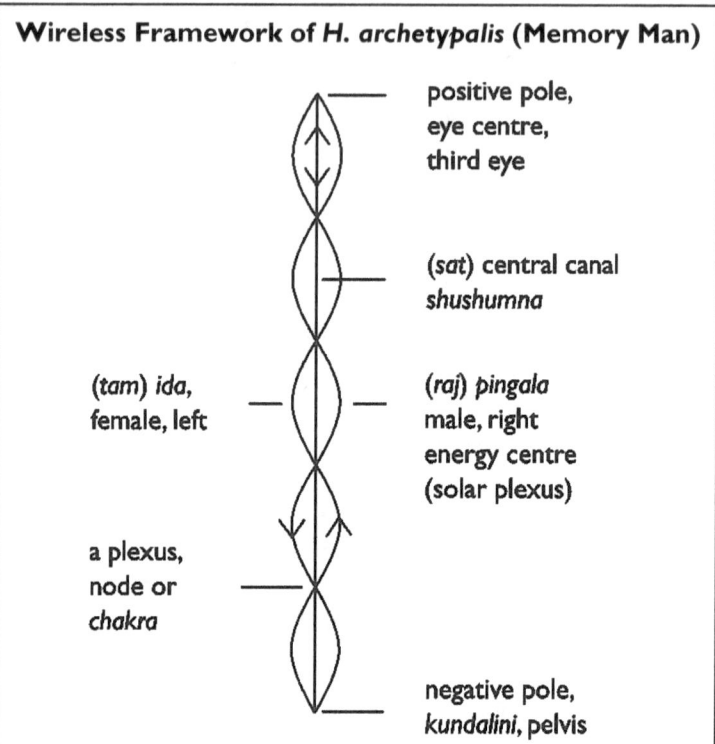

This framework, elaborated through *figs*. 17.8 - 10, forms the psychosomatic basis of oriental medicine including, as already noted, acupuncture.

It might appear, in animals at least, that the informative caduceus is physically reflected on chromosomes by developmental clusters (see Chapter 25). These code for proteins that control the highly coherent, hierarchical cascades of events that sum to algorithms of development. Such proteins are, effectively, switches. Distinct clusters of them enjoy both spatial and temporal 'co-linearity'; this means that there is a correlation between their physical arrangement on the chromosome, their activation pattern and the anterior-to-posterior expression of the body segment patterns they have triggered. In short, they represent top-level triggers that initiate the various sub-routines of body plan. Therefore archetypal man is, in informative terms, in the first and principle instance brought to earth in a coordinated way by a chemical 'bureaucracy'. Its gradients and sequences encapsulate, by thorough integration with all other sub-routines (that is, parts) of a body, conceptual code and anticipatory forethought. If the critical, conservative role of a cluster in a region of the body has been properly identified then what, upon its disturbance or excision, would work there? And what conversely, upon its inclusion but without pre-programmed connections to all necessary sub-routines, could ever make it work? Could serial chance mutations (Chapter 23) find the master sequence of a homeobox and its correspondent protein domain that in turn works computationally by exactly fitting *DNA* sequences used to switch a gene on and off - in hundreds of different super-coding permutations per organism? No use just waving arms and puffing that an archetypal cascade just evolved! Since when must any concept - not least the systematic, integrated concepts that combine to build life forms - have risen out of molecules without the bearing of a mind?

fig. 17.7

Wireless Framework: Information Channels and Distributors

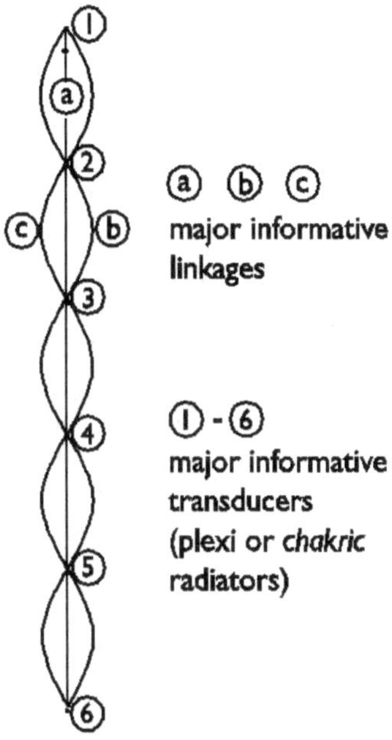

Between the two major poles (1 and 6) a spectrum of *pranic* values ramifies out through transducers (called plexi or *chakras*) and a network of channels (called meridians). The idea is that with descent the degree of materiality expressed increases.

A star is a focus of energy that radiates in space. A plexus is viewed as a sun that radiates subconscious information.

This illustration excludes a seventh, supramental focus of energy - the crown plexus - located in the next three diagrams. Thus transducers 1 to 6 here correspond to 'b to g' in *fig.* 17.9 and '2 to 7' in *fig.* 17.11.

Remember that this schema does not represent the physical nervous system. It represents the higher-level principle, infrastructure or basic blueprint from which physical detail is devolved. It represents, in other words, the archetypal memory of Adam (which is the middle-eastern word for 'man').

fig. 17.8 (see also fig. 17.11)

Lay your index and second fingers to the point on your forehead behind which you think. This point marks the first (or third) eye, eye of the mind or pivot of human awareness. It is your cosmological axis and the information centre from which (*fig.* 17.6) the radius of human extent begins. Voluntary penetration of this 'eye of a needle' is, in the simulation of death, where the upward, mystic path really starts. It is the product of meditative practice, that is, of laser-like focus of mind. At this point the logic and consequences of disembodiment begin to show. Victory over grave and gravity of earth begins.

Neutral *sushumna* is polarised by a double helix whose anti-parallel coils link the six major plexi. Because motion along a helix spirals in a snake-like way they are called 'snakes'. In accordance with sexual stereotype (Chapter 23) polar 'masculine' and 'feminine' counter-currents animate embodied mind. Down and up. One snaking helix of these currents descends from the eye-centre (the conscious pole called *ajna*) to the earth pole's base plexus called '*mul*' or '*muladhara*'. Its (*tam*) 'moon' current is associated with parasympathetic 'cooling down', relaxation and detached observation. Conversely the (*raj*) solar coil ascends 'from earth to heaven'. Its 'masculinity' is associated with arousal, excitement and involvement; its 'hot action' engages and compels because mind issues motor commands down the chain, identified with sympathetic and somatic motor systems, to muscles. In short, this central, informative triplex constitutes the core of a psychosomatic framework, the caduceus. The '*pranic* fuel' of this caduceus, sub-conscious energy, 'vibrates' non-life (in the form of a special biochemical disposition called body) into life.

Now (*fig.* 17.8) consider the wireless anatomy in a little more detail.

Having noted the archetypal structure that substantiates your human body let's elaborate (*figs* 17.9 - 10).

Did you, do you have any control over the way your body grew and works? Why, therefore, should you consciously know any more about your mind's construction and its archetypal operations? Just as a TV's 'life' derives from electromagnetic broadcasts, so the sub-conscious 'caducean archetype' broadcasts its program to molecular arrangements. These, having been established in a physical prototype, thereafter act as 'antennae' or resonators with the psychological matrix. Because this prototype was reproductively enabled, the system proceeds automatically through time.

Pyramidal Man.

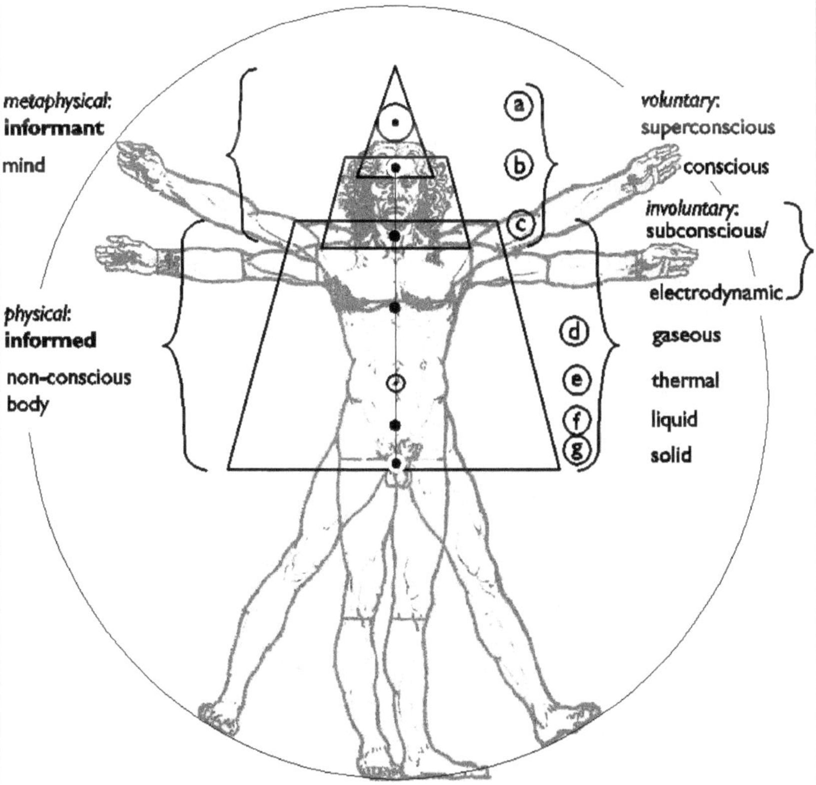

This ziggurat simply rephrases and extends information from *figs.* 17.6 to 17.8.

(b) what is above this cosmological axis is construed 'of heaven' (a) and below 'of earth'; subconscious archetype and the senses (which are junctions with the earth), all subtend this point; the hierarchical arrangement runs from most subtle to most gross.

(c) subconsciousness represents the psychosomatic interface, 'sensorium', zone of archetype, '*akashic ether*' and physical void; it is nothing-material, potential matter or the sub-quantal vacuum of space.

Do you remember (Chapters 6 and 10 and, especially, the legend of *fig.* 6.2) that five *pranas* were, through the five strings of archetypal instrument (*tanmatras*), associated with five plexi? These 'levels' of psychosomatic vibration might be likened to a rainbow of energies; or theoretical Higgs fields that each confers mass on its own kind of particle; or the expressions of various fields of the quantum vacuum each of which supports its own discrete kind of force or particle.

Perhaps the best way to view the scale of pyramidal man is as an harmonic scale or set of frequencies; seven notes that compose life's melody include (a) superconscious and higher states of mind (called *buddhi*), (b) egotistical or lower mind (*ahankar*), (c) subconscious, archetypal mind subdivided by the five *tanmatras* (see Glossary) and non-conscious body subdivided by five states of matter (space, energy, gas, liquid, solid). From the association of *prana* with the five *tanmatric chakras* also springs the character of biological form and function. Classical subdivisions of sub-conscious energy (*prana*) associated with the five biological plexi (c - g) are:

(c) *upana* (associated with sensitivity and overall physical control)

(d) and (e) *vyana* and *prana* (a secondary use of the word associated with the influx and circulation of 'vitality', that is, with oxygen and 'sun-born' or photosynthesised nutrients)

(e) *samana* (associated with assimilation, the 'furnace' of the respiratory process and energy metabolism)

(f) and (g) *apana* (associated with materialisation, reproduction of material forms, efflux of sexual products and the elimination of wastes).

Tanmatras are associated with the modes of signal translation in respect of sensation (↑) and motion (↓), action or creation; these modes are in turn expressed, as illustrated in *figs*. 17.10 and 17.11, in the biological arrangement of glands, organs of sensation and organs of activity. Thus, in nature's clearest and most elaborate reflection of hierarchy, human biology is displayed as a microcosm that reflects the scale of creation.

fig. 17.9 (see also figs. 6.2, 7.5 – 9 and 13.6)

A little revision is due. Sub-consciousness is not seen, dialectically, as a state of brain but a range of elemental mind; it is identified (Chapters 7 - 11, *figs*. 2.11, 7.2, 7.4, 10.1, 10.2 (iii) and 11.1) with physical first cause. A further, critical link is made (Chapter 5, *figs*. 5.2 and 5.3) between potential matter and the archetypal memory of natural law. As a gas may also, at a lower range, be expressed as a solid, so the subliminal energy of archetype, called *prana* (*figs*. 6.2, 7.5 - 9 and 13.6), is expressed as subtle (quantum) and thence gross, material forms.

In this view psychosomatic connection (Chapters 7, 10 15 and 16) is made through fields in space whose activations appear in the form of fundamental particles (or quanta) and, thereby, patterns of physical behaviour. These latent fields were once treated as a single, immaterial entity, so-called 'ether'. Of special interest here are nuclear and electro-dynamic behaviours, the latter involving light (electromagnetism), electronic charge and electricity. They compose the physical side of linkage between metaphysical and physical domains. Here, therefore, are found the levels of *H. archetypalis* and *H. electromagneticus*. Physical expression then drops from this (*sat*) high, informative grade through a (*raj*) energetic phase symbolised at its top by radiant, free light and lower down by bonded, grounded light called heat. Thermal energy is the internal energy of a body measured as temperature; and heat is the process of energy transfer from one such body to another. Loss or gain of energy may show as elemental change of state. The (*tam*) bulk, macroscopic states of matter are gas/ plasma, liquid and solid. The latter are, however, visible 'externals'; nowadays a great extension of scientific insight includes 'internal' components of bulk matter viz. forces, quantum subatomic considerations, atoms and molecules. These microscopic elements might be disposed, in pyramidal man, under 'energies' - radio or 'wireless' and, as in molecules and atoms, 'wired'.

Five *pranas* (*fig.* 17.9) are therefore related in accordance with dialectical logic to conditions of physical phenomena and, through five specific plexi, to the construction of biological form. The higher, subtler 'levitational aspect of *prana* is, in its subtlest physical manifest, expressed as electromagnetic radiation and electricity - hence its yogic association with sunlight, lightning and clean air, especially the negatively ionised oxygen found near waterfalls and rushing rills. Such a fresh, clean, healthy place was advocated for the sunlit practice of 'breathing exercises' called *pranayama*. It is equally logically believed (Chapter 16) that the interface between mind and matter is, on the material side, at the level of a 'subtle', field materiality, electromagnetism.

The top-to-tail axis of wireless anatomy reflects, like the cosmic ziggurat, a conscio-material gradient. This gradient is punctuated by plexi. Neither plexi nor their interconnecting meridians are objects like organs, vessels or nervous 'wiring'; nor, any more than the antennae, wiring and radiation of TV broadcasts look like the information they carry to a screen, are they based on physiological or anatomical correspondences. Just as atoms are oscillators and suns are concentrations of energy, so plexi are visualised as radiating oscillators, modulators or transformers. In this view each plexus corresponds with a different frequency of sub-conscious energy and each frequency is the 'tag' that calls its own set of subroutines from the morphogene. It therefore acts like a filter that allows certain archetypal memories (or programs) through; that is, each plexus is related to specific functions of the body's economy and, as such, its orders guide the metabolisms involved with particular sorts of maintenance and regeneration. These particulars are specified by their position in the conscio-material gradient; the gradient is itself reflected in the 'internal', wireless structure of metaphysical *H. archetypalis* and physical *H. electromagneticus* and, as a consequence, the external, 'wired' structure of *H. sapiens*.

The question is not whether you believe mind can interact with body. It can and does continually. Psychosomasis is fact. The question is whether you believe mind *is* body or not. If, as a 'saturated materialist', you subscribe to the Theory of No Intelligence and believe it is, then the structure of psychosomasis described here is as fictional and irrelevant as non-physical consciousness. It is, like all your thoughts and feelings, a neurological illusion of the atoms. You may not, on the other hand, feel that you are simply a complex electromagnetic pattern. Science cannot gainsay this immaterial feeling but, as an important *quid pro quo*, you have to explain the manner of interaction between the two cosmic fundamentals of informative consciousness and informed energy, that is, life and matter. A TV set depends, both in concept and in fact, on a desire to transmit messages, to communicate information. Just as the body of a TV set, broadcast and the creative desires that produced this physical couple are each useless without the other, so metaphysical and physical forms interlock. You have to describe the immaterial operation of this natural kind of interlocking, a ligand called psychosomasis. You have to explain how, using its antennae, the system works. Let us be clear, the psychosomatic plan we describe as *H. archetypalis* is the sub-conscious interfacing system for humans. Archetypes vary according to biological form but in our case the logical form of (wo)man is called, traditionally, the medical profession's caduceus.

In yogic and Buddhist literature each plexus (or *chakra*) of the caduceus is assigned the shape of a blossom with a different number of petals. Each 'lotus blossom' is associated with a different colour of light, quality of sound, geometrical shape and, even sounds that, when spoken, constitute letters of the 'god-given' or sacred Sanskrit alphabet. *Combined as a whole the plexic energies appear like a rainbow or a suite of musical notes, an octave.* It is said that the dimmer the 'glow' of a plexus, the more likely is physical malfunction within its sphere of influence; conversely, brilliance is supposed to indicate radiant health. Both physical condition from 'below' and the influence of 'mind-set' from 'above' affect plexi and, therefore, health. For example, a current of will-power in the form of an enthusiasm or a wish to self-heal 'brightens the *chakras*' or 'galvanises' the body. It generates a positively electrical effect. Do you remember (Chapter 13) the power of auto-suggestion? Health is, through the medium of psychosomatic *vis medicatrix naturae*, promoted by positive thought and, even more powerfully, by specifically focused attention.

Although chemicals naturally assume shapes that can reflect the construction of their atomic interior, in the case of complex, purposive morphology every industrialist knows that he needs a mould or matrix with which to fashion his products. A cell or egg, with its informative nucleus, informed metabolism and membranous boundaries, constitutes an archetypal and a physical matrix from which further potential unfolds. *Figs.* 17.9 - 10 indicate, for a human, the unfolded 'wholeness' within which parts are expressed and harmoniously constrained. This development runs from principle to practice; it devolves from simplicity to detail, from prior information to physical consolidation.

An inventor's blueprint becomes a manual to which one refers to make, maintain or repair a machine; it amounts to a record, a memory of invention. It is 'an ideal' which describes and controls the invention's 'working life' but is not the machine itself. The latter must, in turn, contain its own materially expressed control systems. Archetypal programs, stored in memory, crystallise the body parts. The channels and plexi of *H. archetypalis* are reflected in a cerebro-spinal control centre; the major command centres of biochemical homeostasis are glands, inflexible 'brains' each of which, as you pass down the body, corresponds with its own caducean plexus in a way that, in turn, reflects the cosmic logic of a conscio-material gradient. Last but not least the materials for all body parts, composed of cells, are coaxed like music from the 'piano' of *DNA*, each key a gene and its note a protein.

Do you remember the cosmological axis (Chapter 0) also called the eye centre, third eye or (*fig.* 17.11) *ajna chakra*? It is the axis or pivot whence you swing between two dimensions, the physical and metaphysical. It is at the same time the top, central plexus of the *primary, informative domain* where your own consciousness 'dwells'.

Information Man

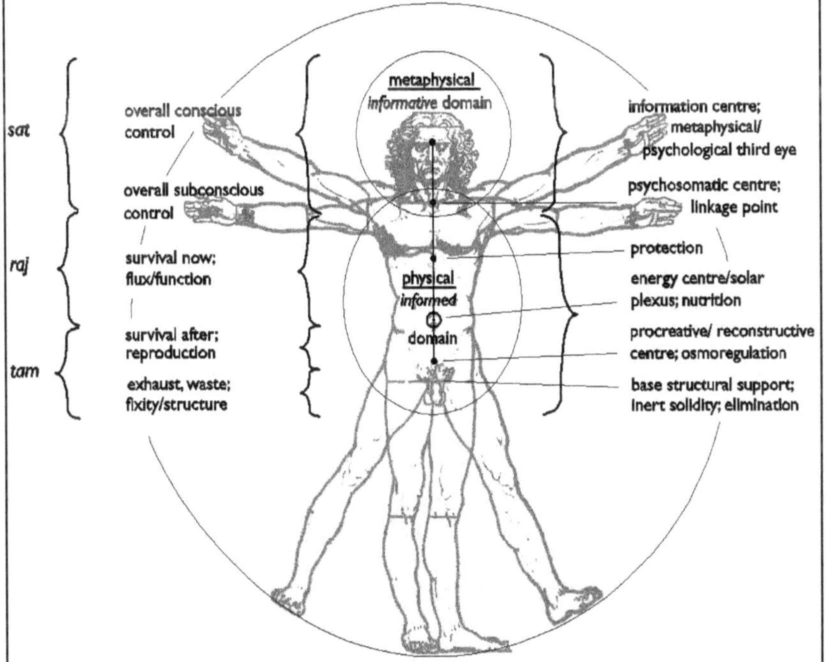

Pyramidal man can also be represented in the form of two domains, the *metaphysical informative* and *physical informed* or *energetic* domains. These domains involve active and passive information respectively (see *fig.* 5.2 and Chapter 6).

The higher a biological organism the more concentrated becomes the expression of form towards its (*sat*) information centre, the head and especially the eyes. This top/ front '*HQ*' is polarised against the base/ rear 'zone of expulsion'. Man walks upright, that is, with maximum uplift and thereby symbolic emphasis upon his information factor. His head is raised furthest from the ground.

fig. 17.10

It is an obvious no-brainer that sensitivity in organisms that have no brain is not confined to brain. An implication drawn from *H. archetypalis* is that, despite the current view of psychological and physical medicine, within the creature form mind is not confined to brain. Indeed, the informative domain links with a *secondary, energetic or informed domain.* Psychological connects with physical through the medium of sub-consciousness. The substance of this psychosomatic level is archetypal memory or, as the obverse 'tail-side' of its coin, potential matter; and its intermediate plexus is sited, suitably, at the neck and is called the 'throat centre' or (*fig.* 17.11) *vissudha.*

You might imagine (*figs.* 14.4 and 19.2 (iii)) that a cell or a body represents information plugged into energy. Energy gives motion but, without direction, motion were chaotic. Information gives direction; it instructs behaviour; it informs energy with the result, not of chaos but cosmos. Thus cytology, physiology and anatomy involve immaterial information in the form of an unconscious but materialistically unacceptable plan. Such informant plan is passive; it is delivered by an informative factor called a morphogene.

Of physical parts genes, which operate in both electromagnetic and chemical capacities, are at the top of the informative pyramid. In a cell the informed domain descends hierarchically from a central source outward; its cellular gradient radiates from (*sat*) informative nuclear management through (*raj*) fluid cytoplasmic dynamics to (*tam*) structural 'fixities' such as cytoskeleton and membranes. In the human body as a whole it descends from a (*sat*) metaphysical consciousness through (*raj*) physical energetics to the (*tam*) base levels of solidity, encasement, exhaustion and elimination.

Figure 17.10 elaborates how the informed domain links at its zenith with the base of the immaterial, informative domain; it also shows how the axis of this informed, material domain is at the point naturally represented by your original, umbilical point of nutrition, the navel of your solar plexus. This axis of energy is sometimes called (see also *fig.* 17.11) the *manipur chakra*. Meanwhile at the nadir of the informed domain rests an 'earth centre'. Its polarity antithesises *ajna* and is located at the coccygeal base (or '*mul*') of your spine.

There is, in this perspective, no argument with neurological mapping. It is simply a question of broadening the psychosomatic context in which the biological storage and transmission of information occurs.

The functional logic of man is now inspected. The caduceus is analysed in terms of the two domains of influence that involve the component basis of its logic - information and energy.

The Informative Domain

You, of either male or female polarity, express universal humanity. Your archetype, of all organisms', best expresses the conscio-material gradient in a full and balanced way. How?

Drop a pebble in a pond. Note that concentric rings of influence are radiating from an epicentre. You might, expanding a flat surface into three dimensions, think of nested spheres of sound from gong or light expanding from a candle flame. Take the source of knowledge or the point of power that's concentrated in the middle of such games. Coiling downwards from this centre spiral outwards as a growing cone (*figs.* 0.4, 0.10 and 0.11). Square the vortex and become a pyramid or ziggurat - Mount Universe. Check *figs.* 0.5, 0.7 ('layers of the cosmic onion') or 13.3 (ii). Take the centre, peak or epicentre as First Cause and, further out, point X as your cosmological axis, *ajna chakra* or in common parlance personal, subjective self. In other words, relate the grades of universe to universal man and measure up a conscio-material cosmos by yourself!

Universal Man

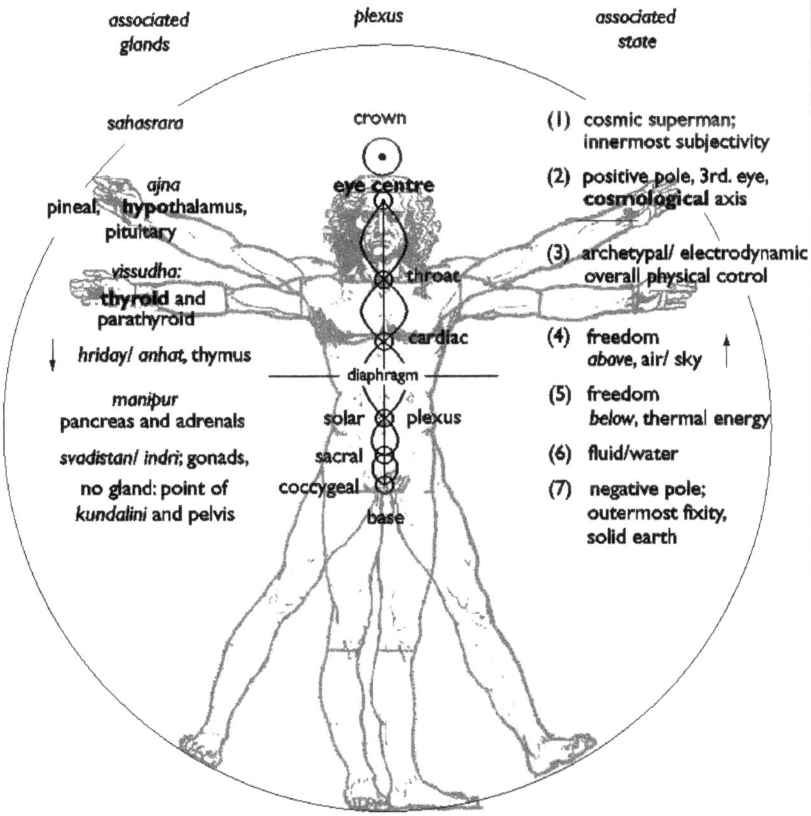

Vitruvian man is often used as an expression of bilateral or polar symmetry in the human body and, by extension, the natural universe; and of the golden rectangle or divine proportion. As well as such 'horizontal', mirror symmetry in the physical plane Natural Dialectic represents 'vertical' or hierarchical polarity (*fig.* 0.8iii) from apex to base along the conscio-material coordinate.

Major internal organs correlated with the associated states. Allow, as with the colours of a spectrum, certain overlap:

(1) none

(2) consciousness: cerebrum/ voluntary nervous system; note especially the correlation of the eye centre with your point X or cosmological axis (*figs.* 0.7 - 0.11).

(3) subconsciousness: subjectively, the dormant zone of sleep; psychosomatic communication/ systematic electro-dynamic expression of archetype or idea: mid and lower brain/ involuntary nervous system/ throat: also, energy input: by mouth and oesophagus

(4) energy related irradiation: lungs/ heart protective: liver/ spleen

(5) nutrient absorption; umbilicus; digestive system except entry (mouth) and exit (colon/ rectum)

(6) osmoregulatory kidneys and water reabsorption (colon); reproductive system

(7) waste output: urinary tract/ colon/ rectum

fig. 17.11

How abstract and yet informative are numbers, ciphers that exist nowhere but as expressions of the mind. Geometry is 'line-some'! Does arithmetic not count? What is it, you might ask, about some numbers? Like 1.618. Strange but so good-looking! Take (Chapters 8 and 23) the Fibonacci series and its yield, phi (or φ). This ratio, an apparently meaningless number, crops up from the roots to the leaves of biology. Root, stock and branch it evolves leaf arrangements, the uncurling of fern fronds, the distribution of branches on tree trunks, shells, insect segmentation and so on. It is especially evident, as Alan Turing noticed, in sunflower blossoms. Mathematical features embedded in growth patterns! Immaterial patterns beautify material flowers! How wonderfully wind and rain work atoms into architecture - not to mention simultaneous inscription of the immaterial codes that both substantiate and generate such ideal formulation! Greek architecture and medieval art, our cultural heritage, were based around the math of 'golden mean', of φ's 'divine proportion'. So how and why do you, as universal man or woman, embody it? *DNA* is drawn, in a ratio of width to length per spiral turn, according to φ. Why, like that molecule, do you incorporate proportions of a 'magic number' that, like Eve or Adam, leaves you looking so divine? Because, from nuclear helix spun into a visible consequence of size and shape, you do.

Measure up. Top to toe/ umbilicus to toe, shoulder to fingertips/ elbow to fingertips, finger joints and spinal vertebrae - all φ for your perfection! Why did such perfection, based around a top-tail axis, spiral from a zygote into adult form? Is your spine with its divisions a reflection of an archetypal ratio; is it too a set divided up in Fibonacci style? It is, indeed, so informed. This, with all the other indications of a systematic order, is how your personal geometry developed and now stands. It is, furthermore, a way coherent reason might inhabit things. How, as well as symmetry, could 'golden ratio' systematically, aesthetically but accidentally evolve proportions? Perhaps, on the other hand, archetypal logic can employ them to express an elegance and natural consistency of form. Beauty is a hallmark that φ's immaterial number well describes. Indeed, is this the sort of figure Einstein had in mind as a 'real constant'?

Is there revelation of a universe in universal man; can you now relate your microcosm to the macrocosm; is there a nodal metaphysic of the human and, if so, what are the power points in our archetypal kind? From 'inside out' or 'topside down' the highest authority in a hierarchy involves the most conscious information and wields the largest overall influence. Influence is, as in a civil service, progressively limited and localised down the ranks. It is as if the conscious seed of an idea had ramified, like an upturned tree of life, down into physical behaviours on this plane. From centres of a psychosomatic influence unfurl You (or at least your plexi and thence your physical shape) unfurl developmental patterns, molecular configurations, glands, organs and the systems science now can analytically describe. No chaos. Just a tight-knit, out-sourced order, a conscio-material gradient that's orderly from top to bottom, head to toe. Each higher level exercises overall coordination and control upon the lower ones. The influence of this exercise cascades through the electromagnetic and biochemical 'fingerprints' of *H. electromagneticus* out to informed molecular and made of molecules, gross structure of a body.

How is information for constructions signalled and relayed? It was proposed in Chapter 16 that archetype exerted influence electrodynamically. It is, for example, still not fully clear how the expression of *DNA* coils in the form of appropriate type and amount of proteins is orchestrated. There must certainly exist a chemical and, therefore, electromagnetic level of control. *Natural Dialectic would suggest the molecule is a conductor, an aerial whose electrogenetic communications precede and govern molecular operations; and it would, of course, intimate a further transcendent or 'radio-biological' level of cohesion, one uncountenanced by current research - a psychosomatic resonance with the morphogene.* And, with respect to earthbound code, biochemist Albert Szent-Gyorgyi suggested that the sugar-phosphate backbone of *DNA* might act as a conductor? Or the molecule work as a resonator, that is, an aerial or aerials for incoming frequencies whose orderly nature might amount to a vibratory code? This code might, at resonant frequencies, act as the key to unlock certain sections of the molecule for expression. *A radio 'baton' would orchestrate the accurate retrieval of information from life's database. Electro-dynamical fields would, at root, command the expression of genes; and would themselves be archetypally entrained. In fact, as previously intimated, studies may lead soon to common knowledge that a large proportion of the DNA's workload is, with its helix acting as an antenna, engaged with bioelectric signalling.*

It is thus suggested that the vibratory nature of an archetypal program carries, just as light can carry, information. It can signal using subtle and specific resonant associations with the spectrographic 'fingerprints' of molecules. Charge, light and shapes build up the complex forms of life according to photonic and genetic codes. The influential nuclei of these photonic codes are the archetypal plexi, *chakras*, whose holographic image radiates (like *DNA*) in every cell as well as (unlike *DNA*) the whole body. Thus the plexi (or *chakras*) of informative caduceus act as distributors that filter different energies. And these bands of pranic 'rainbow' associate, in turn, with different constructive subroutines. This (*figs.* 17.6 - 11), from *H. archetypalis* through *electromagneticus* to *sapiens*, is how your body really is evolved - not accidentally evolved but purposely devolved. Such caducean logic, whereby radiant, sub-conscious but still subjective contact is established with the objective, lifeless physical cosmos, is now elaborated.

Therefore at this point we turn to the uppermost 'roots' of control, the informative domain of mind and psychosomasis.

Is a good mood not a 'high' and a dark one 'low'? Remember, prepositions used with mind are not used spatially. *Sahasrara* is above you so let's start with an ascent. Forward, upward and inward focus rises towards the ***first*** *or highest* *plexus*. Some assign the plexi colours, harmonic notes, numbers of lotus petals and so on. In the case of colour an analogy with the visible, body-related waveband of electromagnetic light makes *sahasrara* ultra-violet (or clear); and it is called, as well as *sahasrara*, crown or glorious 'thousand-petalled lotus'. Constraints of individuality are gradually eased on such deliberate ascent which, given the normal earthbound bias and egotistical nature of human attentions, is unusual. Normally, therefore, the (*sat*) superior, super-conscious *sahasrara chakra* remains unknown except to rare, highly-evolved humans, saints - not necessarily canonised saints (who may have done excellent humanitarian works) but 'technical' adepts (who have achieved enlightenment).

No evolutionist suspects or anticipates such nonsense as a *sahasrara plexus* or spiritual evolution towards the Infinite. *Yet because it is wholly irrelevant, as are the other plexi, to materialistic psychology and biology, it is interesting to ask what, bottom-up, is expected as the next step in physical evolution for the human brain.* The latter has already, at the head of an erect, vertical stem, flowered into the consciousness we now take for granted. Has its evolution stalled, stopped or, with a processing capacity that has achieved the theoretical maximum, found perfection? Or, on the other hand, is not man as presently informed and formed an insufficient fellow in so many ways? A duncy ape compared to what his future holds? Surely, well before the sun burns down, there is time for betterment? What, therefore, might involuntarily evolve? Radar? Infrared sight could be useful in the dark; stun-gun tactic or a sensibility involving electricity could shock your enemies. Or, lower down, perhaps extra limbs, wings, knife-like nails, retracting wheels or submarine capacities might each lend super-human edge to your survival. Is not man a 'super-ape' that apes turn into when they have evolved? Super-human just turns into what is normal when you have evolved to it - exactly what the mystics say concerning mind not body. What could be more natural a state of consciousness than super-conscious Natural Enlightenment?

And, of course, you know it! Brain itself will certainly transform old evolution's ways from slow, involuntary to rapid, voluntary advance. A brave new world of science, one entirely unanticipated by aboriginal chemicals or slime, will be conceived from this grey springboard; brain will bounce us purposely into trans-human species of intelligence. Who would not grasp the opportunity that life this far has granted and so gear his nervous motions and non-mystical intentions to overdrive? And thus, according to committees and some ethical agreement (perhaps), let rip an exponential curve of evolution more extravagant, ingenious and exuberant *in vitro* than old-fashioned mother, nature, ever managed to conceive? Such drive should logically evolve into the flowering of what the Greek Tithonus long since craved - ultimate survival in the fittest state, corporeal immortality. He was warned! All that is mortal ages. Do you want to age forever? He ignored the signals. Arrogance rebuffed this wise advice and so, with nemesis of gift received, he only craved for its withdrawal! If consciousness is immaterial or rebirth happens what a wholly wasteful exercise dragged out, unending life on earth would be! What folly to prolong a youthless life when you could, naturally, receive a fresh one each time round! What dramatic error is a play without an end!

	tam/ raj	*Sat*	
	below	*Top*	
	ramifications	*Info. Centre*	
	effector/ sensor	*Regulator*	
↓	*tam*	*raj*	↑
	outward	*inward*	
	sub-conscious	*conscious*	
	towards physical	*towards psychological*	
	less conscious	*more conscious*	
	involuntary	*voluntary*	
	motor	*sensory*	
	effector	*sensor*	
	arousal	*quiescence*	
	sympathetic	*parasympathetic*	

After crown above the down-sweep comes upon your forehead. Here is located your cosmological axis, frame of reference, point of balance or third eye. Even if most people are unaware of superconscious *sahasrara* everyone is familiar with the ***second** plexus*. After all you live in it. *Its level is (sat) conscious awareness at HQ.* You think at *ajna chakra*; that is what the *chakra*'s all about. All above this focus is entirely subjective in its operation, all below objective in pertinence i.e. built for function in this world alone. *Ajna* is, as closely as can be localised, where 'above' meets with 'below' and each reflects the other. *This is the place of integrated personality, wakefulness, the intellect and conscious, overall control of information.* It ranks below top *sahasrara* but the second node is where you actively inform the physical domain. Is that not what you do continually? Thoughts are the work of your intelligence whose (*raj*) function is to optimise, decide, answer, govern, satisfy and from the lashings of desire and its frustrations seek the peace - in a word to find one's balance. *Ajna individualises universal principles and purposes*. As boss of the biological system it works in its own best interest to manipulate and coordinate symbolic data, to 'get an intellectual grip'. It is the throne of your awareness and the bud which concentration, like a light beam, brings to flower. It is, equally, the seat of egotistical self and selfishness, whose passionate deformity is arrogance. The colour of *ajna* is violet; and the two petals of its lotus bring resolution of duality to the single focus of mind's third and inner eye.

'I am, I understand, I act' (*fig.* 13.4). The physical correlate of *ajna's* 'inner eye' or 'sixth sense' is the cerebrum, especially the prefrontal cortex; this cortex is the register of psychological potentials - purpose, thought, remembrance, will - at point of primary engagement with the body. Note the concentration of detectors clustered in the boss's office, organs of perception localised in head - eyes, ears, nose and mouth. Note symbolism, logic of positions and the dialectical, hierarchical order of face. 80% of physical information is collected from the light (pure energy) in our eyes. These two luminaries shine each side and just below the metaphysical third eye. The visionary trio constitute an arrow-like triangle that points upwards. Slightly lower, set at each side towards the rear, the ears capture sound (or energy-in-air). They register the power of spoken energy, an intelligent medium into which physical nose, mouth, tongue, throat and lungs combined translate the metaphysical work of comprehension and construction. As well as breathing air for informative purposes nostrils are involved with energy (gaseous exchange for respiration) and sense of molecules-in-air called smell.

Lower down but still 'up top' supplies are loaded through a single, bilaterally symmetrical orifice. The mouth actively ingests potential energy in the form of materials called food; it moistens solids, imbibes fluids and involves a sense of molecules-in-water, taste. Finally, as a part of your overall protective, sensitive sheath, head's skin can evaluate different textures and degrees of pressure (molecules-as-solids, touch), thermal energy (temperature) and pain. You can know the world with your head alone - but you can't *do* much with it.

At the top mind impresses and is impressed. Its domain is one of information and desire. Conscious mind generates the current and negotiates fulfilment of desire, wish or whim. It is equally, therefore, an instrument of balance. Its voluntary courses of thought and behaviour vary according to what an individual thinks is 'good' or 'in best interest'. Mind generates imbalance by desire then acts to cancel out the tilt. It swoops and veers and then, in satisfaction, finds the straight again. It is, in this respect, a balancing factor, an agent of control you could refer to as a metaphysical homeostat.

Looks easy, doesn't it?

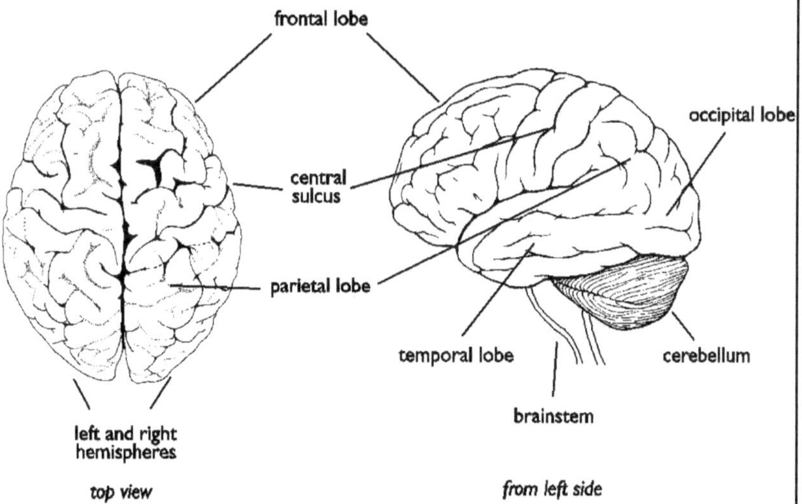

fig 17.12

It is, on the other hand, also a primary instrument of inflexible, biological homeostasis. Its organ is the brain. The greatest difference between humans and the animal world is encapsulated in this organ but if you want to know about its evolution you will have to guess. 75% water, containing 100 billion neurons with trillions of connections yet weighing only 1.5 kg., your brain exhibits perhaps the most exquisite physical complexity in the cosmos. *Developed purposely, awesomely, from a single cell, it is designed as the critical link between mind and the body's world.* It is the physical side of a psychosomatic bridge between subjective and objective, the basic modes with whose distinction this book started. Such a relay acts as a multiplex switch and gate or filter; it organises the two-way exchange of information between psycho-space and body. As such the brain integrates sense data from eyes, ears, nose, mouth and skin and orientates them within an all-important referential framework born of memories. As well as perceptions the interface called brain also facilitates the actions born of motivation and response. These include, in *top-down* order, speech, manipulation, locomotion, procreation and elimination each with its associated organs. *For all this, although it exchanges information with psycho-space, brain chemistry is not mind.* **Although brain and mind are aspects of a single 'self', biochemistry is not the same as mind; nor is the single, seamless experience called awareness just a complicated show of atoms.**

Does a piston or steam *cause* a steam engine to move? Of course; but without the *purpose* of both engineer and driver nothing would move. So what drives the hard machinery? Is it steam, piston or purpose? A steam engine or space rocket is kinetically active and yet passively informed. What is non-conscious, that is to say, physically energetic is informed. Thus physical and biophysical fixity and flux are just the same – static and dynamic form of information. *Nerves and muscles are no more atomically alive than a steam engine; they are, like DNA, a coded form of passive information.* Similarly, it is true that chemicals play a critical *part* in human psychology; but to assert that they alone *cause* the motions of cerebral soft machinery is not the whole truth, just a secondary, lesser half of the whole truth. In other words, neurological and hormonal agencies no more *cause* consciousness than a hand or foot. They are simply top-level informants. They are physical instruments used to help inform and express mind's metaphysical information; or, conversely, press the imperatives of body on the mind.

A body is an organ of information exchange between the two elemental coordinates of existence -subjective mind and objective matter. It represents a hierarchical system both for the 'step-down' of conscious to non-conscious energies and vice versa. In this view different biological organisms represent composites of these energies; each type expresses a particular permutation of emphasis and association. For example, in birds a gaseous association is paramount; for fish the liquid and for plants the earth. Such associations are sub-divisible according to metaphysical characteristics such as aggression, libido, stealth, sociability and so on; according to habitual way of life (as shown, for example, by avian sub-divisions such as ducks, waders, warblers, hawks, owls etc.); and thereby according to a work regime that makes best living from a certain habitat - a combination called its niche. A human, uniquely, involves all six main grades of association (conscious, sub-conscious and the physical states of energy, gas, liquid and solid) expressed in a balanced proportion through a complex of six power points. Each hierarchical grade exercises power over those below it.

Therefore what is inferior to conscious *ajna*? All the world, the whole physical universe is inferior to *ajna*. And which grade is immediately inferior? *After consciousness the system drops back and out from the eye-centre into the sub-conscious influence of the* ***third*** *plexus, visuddha, the throat centre.*

What I do not know I cannot disturb or be disturbed about. It would be impossible to have to think about every physical need so that lower operations are mostly automated. This automation is both sub-conscious (through which a conscious override is sometimes possible) and physical. Therefore in keeping with its role as psychosomatic link *vissudha* binds upward with the conscious phase of *ajna* and, downwards, with the body. Its link zone, the 'sandwich' of Chapter 15, is pre-physical. Its state is dormant, its matter potential and its governance reflex. As such its plexus is the 'residence' of typical mnemone or archetypal memory (*figs.* 15.8 and 15.9); and is a channel from which patterns of the lower four 'peripherals' unfurl. *Vissudha's* sphere of influence includes the office, in the brain stem, of a sleep/ waking toggle and it is the subjective link-level at which circadian consciousness oscillates with its lack, that is, with sleep. In the colour analogy of sub-conscious energy with visible light its colour is deep, dark blue. Its associated condition is *akashic* 'sky', 'etheric' space, void or vacuum. Vacuum is, as a reflection of Nothing (Chapter 10), the physical absence of matter 'resident' at the interface between sub-conscious mind and non-conscious body, that is, between metaphysical and physical factors. The brain itself is seen, primarily, as a computer whose secondary 'hardware' is composed of biochemicals, cells and sub-organs (such as cerebrum or cerebellum); its main business is between the implicate dimension of *H. archetypalis* and, framed within the obvious, extrovert form of *H. sapiens*, *H. electromagneticus*.

Thus the top half of *vissudha's* sphere of influence includes mid- and hindbrain of the central nervous information exchange system whose organs manipulate all nervous traffic to and from the body. These include the diencephalon, cerebellum, medulla etc. Reflex reactions are mediated through the cerebellum and the sub-conscious regulation of respiratory, blood vascular and other systems is by the central brain stem. *All functions of the nervous system converge on the mid-brain.* There are connections with cranial and spinal nerves, both voluntary and autonomic. Here a *corpus callosum* connects the polar hemispheres of the brain and branches of the optic, auditory, olfactory and taste tracts are found. On the inner core of the cortex at the thalamus of the brain are found master glands; in the very centre of the head a pineal, hypothalamic and pituitary command complex conjugates the 'grammatical' traffic of nervous and hormonal systems.

How wrong can you get? Nineteenth-century Darwinists counted nearly two hundred 'vestigial organs', organs that for one reason or another they supposed evolution had evolved then outgrown any use for. Now, because cellular bodies are lean, mean, efficient machines, all but less than a handful of these 'remnants' have been declared purposeful. Of this small number the pineal gland (or epiphysis), now the object of worldwide research, is emerging as the diametrical opposite of vestigial. *In a body packed to the hilt with integrated cybernetic systems it is a regulator of regulators.* So-called because of its superficial resemblance to a pinecone, the pineal is 'enthroned' in a position of symbolic significance, right at the centre of the brain. It has been identified, understandably but incorrectly, as 'the seat of the soul' or *ajna*, the metaphysical 'third eye' whose positional significance is front central. This single 'eye' is our localised receptor of internal light (that is, consciousness).

Meanwhile the pineal gland includes a physical kind of third eye, a photoreceptor that responds to variations in external, physical light. In certain lower vertebrates (species of lamprey, frog, lizard, fish) this parietal eye functions as a directly photosensitive homologue of the pineal. It. In mammals, the 'eye' is internalised; its pigmented cells show some superficial resemblance to retinal cone cells and indirectly receive light through nerve impulses from the retina. As well as light it is sensitive to the earth's geomagnetic field and seems to be a kind of psychosomatic transducer, a conduit or aerial that translates electro-magnetic influence into physical and psychological effect. While *ajna* operates as a mental command and control centre above and between the two physical eyes it might be supposed, by analogy, that the illumination-sensitive pineal, set above the hypothalamus and pituitary glands plays a key role in non-conscious regulation; using neuroendocrine secretions it governs the system of automated biochemical response called hormonal.

In fact, as a unitary body in conjunction with its adjutant, polar pair, the hypothalamus and pituitary glands, the pineal contributes to a triad that directly regulates the whole body's autonomic, hormonal and immune systems and coordinates, in the manner of a command and control centre, all biochemical 'intelligence'. Its photo-periodism employs the neurotransmitter serotonin and its hormonal analogue, melatonin to mediate circadian (sleep/ waking), seasonal (reproduction, hair growth, coloration) and life (growth, sexual maturation, aging) cycles. With respect to 'grounded light' (heat) it is perhaps involved with the hypothalamus in thermoregulation. Indeed, it hosts hormones from every major endocrine gland and must therefore be involved in the regulation of their rhythms.

The chemistry of melatonin is, while unlike any other hormone, very like serotonin. Such structural 'resonance' between neurological and hormonal systems also occurs with noradrenaline and adrenaline. Because of its different perspective modern biology has paid little or no attention to hierarchy, polar analogies and internal resonances (also called 'barcodes' or 'fingerprints') of biochemicals involved with the transmission, quite possibly psychosomatic transmission, of information. This is, however, a direction of research that Natural Dialectic indicates.

When a regulator malfunctions its system malfunctions. If such a regulator as the pineal loses its 'sense of rhythm', it is no surprise that both physical illnesses (due to sleep disorder, immune deficiency or carcinogenic potential) and psychological problems follow. Episodes of depression, convulsions and schizophrenic psychosis (Chapter 13 and note that some hallucinogens like *LSD* mimic the structure of serotonin/ melatonin) may be traced to the imperfect operation of 'king' pineal.

Separated from the pineal by a sub-cortical information exchange unit called the thalamus (see *fig.* 13.2) is the hypothalamic/ pituitary duo. All three glandular command centres are surrounded and overarched by 'emotional' limbic and 'rational' neocortical architecture; below them lie the brain stem, spine and corporeal ramifications. The hypothalamus, part of the thalamus and influenced by pineal secretions, commands many functions. Its inner section is connected with the quiescent, parasympathetic system while its outer edge is part of the sympathetic function of arousal. It is a link through which conscious cortical and sub-conscious sub-cortical systems interact; through which electrical and hormonal nervous systems interact; and through which fundamental regulation of the body's heat, food, water and hormonal sex level occurs. Such government is achieved using a clutch of neuro-hormones that, affecting pituitary secretions into the bloodstream, initiate hormonal cascades destined to activate lower glands (such as the thyroid/ parathyroid complex, adrenals, gonads etc.). *Each gland corresponds to a node on the caduceus and governs its associated functions.* In short, the hypothalamus exerts a hierarchical, body-wide form of informative potential (government) over the critical factors of heat, food and water (for energetic survival now) and sex (reproductive survival for the future).

While the posterior pituitary secretes neuro-hormones made by the hypothalamus, its anterior lobe is a cornucopia of command. Hypothalamic 'factors' either promote or inhibit the aforesaid body-wide cascade of tropic secretions. These trigger local glands affecting *BMR* (basal rate of metabolism or functional 'tickover'), stress regulation (cortisol) and sexual condition. Growth, milk production and neurotransmission are also promoted from the pituitary's 'desk'. If, in the sub-conscious section of the human republic, pineal is president then hypothalamus is cabinet and pituitary high-ranking civil service whose ministrations activate, using various missives, the mayoral hormones of localities and thus, eventually, a toiling populace of individual cells. Concept is, through this system, wonderfully expressed in the tight organisation of a healthy, cooperative nation - the body in its world. Conversely, chance construction by haphazard evolution is nowhere in any plan.

In short, sub-conscious activity localised in association with the central and lower regions of the brain (the limbic system, diencephalon and brain stem) organises involuntary, physical homeostasis and, where necessary, links it with conscious, voluntary requirements. All internal organs are thereby linked but conscious mind is unaware of their operation unless they signal stress or pain, that is, need for homeostatic redress.

At the base of the informative domain it is, symbolically, the neck that connects 'heaven with earth'. It links mind (head section) with matter (mid- and tail-sections). Hence informative *HQ* descends, by way of the cervical region with its vertebrae and plexi, into the informed domain. From its association with physical throat and neck it ramifies into the physical exterior of both the body and its universal surroundings. The traffic is two-way. Motor agency is balanced by sensory perception.

Just as, reproductively, male complements female so, informatively, mind complements and mirrors body. In this respect a couple of interesting inversions or 'neurological twists' (elaborated in Chapter 23) occur within *vissudha's* sphere of influence where sub-conscious mind plugs into body, where information is exchanged with energy, where the mirror-worlds mesh. Not only do the left- and right-hand hemispheres of the brain 'shine' at different, complementary sorts of task (e.g. left/ analytical, right/ integrative) but they reflect or mirror events on opposite sides of the body. In terms of dialectical polarity this 'inversion' or 'reflective asymmetry' between metaphysical and physical poles of existence is logical; but why should blind evolution have the left-hand sphere govern the right-hand side of the body and, with reverse symmetry, *vice versa*? Why should almost all motor and sensory pathways express a well-defined inversion or 'neurological twist' in their systematic logic? Why do they cross over (decussate) in the region of neck - brain stem and upper spinal cord - the psychosomatic region of *visuddha*? Why is there reflected, in the channel between heaven and earth called a neck, such a mind/ matter mirror?

The five lower plexi are (check *fig.* 17.9) associated with five 'qualities' or states of matter; physical expression drops from vacuum and radiant energy through the thermal energies of gas and liquid down to cold solid. Of these the cervical or throat plexus represents the psychosomatic union of metaphysical '*akash*' with physical vacuum, the 'emptiness' from which all particles emerged and have their being. Higher controls lower. Can the attributes and 'character' of this plexus be related to the four below? And, *top-down*, what can be learnt from this node's structural (biochemical) and functional (glandular) expressions?

Although educated you still have no moment-to-moment sense of most of the myriad biological complexities occurring at any of your levels. It is no different at this one - except that it is also as physically unverifiable as the content of a thought. Resonant exchange of sub-conscious energy with material equivalence is the province of quantum or 'nano-biological' *H. electromagneticus. In this respect it is important to conceive of molecules as vibrant energy. Each sort of molecule (as, for example, above-mentioned serotonin) is 'bar-coded' like spectral light; it emits its unique vibratory pattern, kinetic 'fingerprint', 'chime' or 'ring-tone'.* Changes in energy vibration are how we sense (or feel) changes in our own surrounding energy field, the environment of our bulk body and the world beyond it. *It is no different at quantum level. The crux of information exchange occurs* (Chapter 16) *by resonance between archetype and physical expression. Associative 'chime' is the dynamic address system. Such psychosomatic resonance constitutes the linkage between mental, emotional and physical energies.* Electro-physiology, neurology and endocrinology are seen as studies of a rapid interaction between archetypal as well as physical influences. In this view psychosomatic interactions are tightly controlled by the system of 'ring-tones'. You might expect to find groups of oscillatory tones (and therefore chemical configurations) common to the 'harmonic' or 'waveband' expressed by a given plexus. Social molecules will ring each other up! The order of the instant is overall linkage, coordination and cooperation whose objective is the maintenance of homeostatic balances specified by archetype in the role of *vis medicatrix* or 'health-plan' for its physical correlate. The subtlety and sophistication of precise, precoded signals extends far beyond the remit of mere *DNA*; it is intrinsic and integral to every three-dimensional molecule for whose fabrication *DNA* is simply a physical channel; it is, if you like, the systematic fruit each branch of *DNA* was made to specify. This is not, of course, a perspective adopted by reductionist molecular biology.

Biochemical Dialectic:
The Scales of Hormonal Homeostasis.

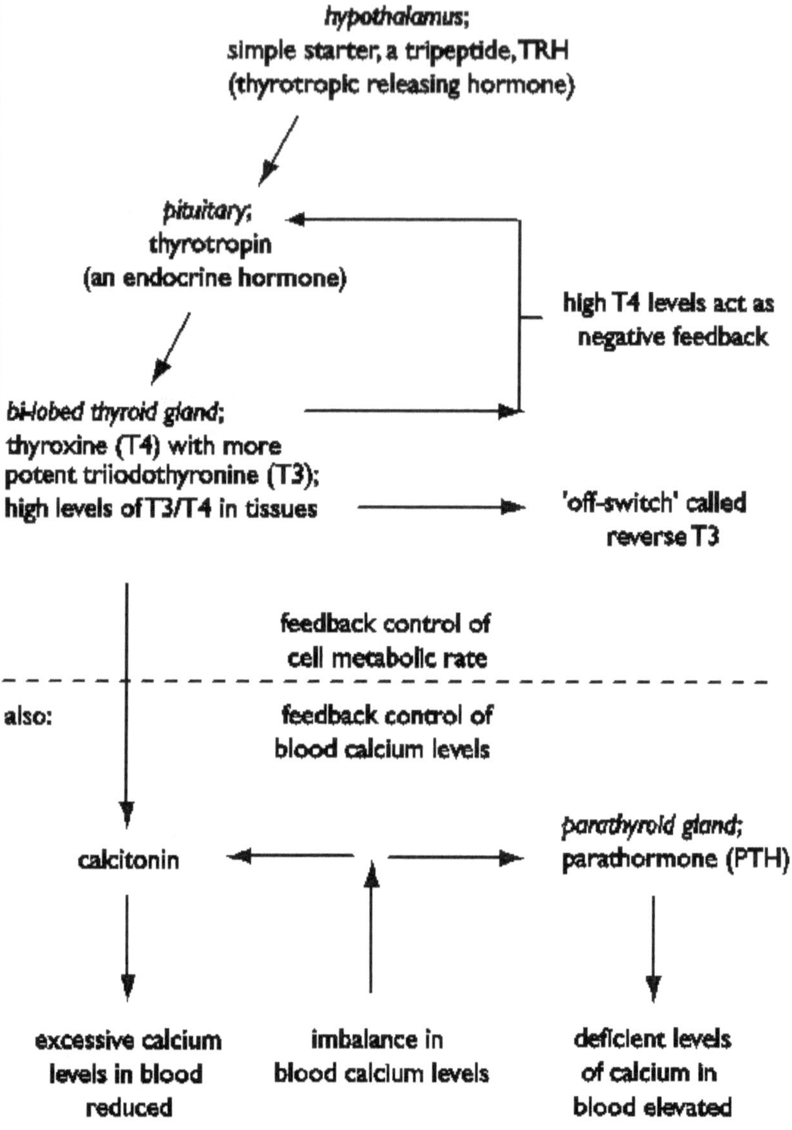

This is an example of a purposeful, duplex cybernetic mechanism. Such mechanisms are, wheels within integrated wheels, repeated over and over throughout all biological forms. The conceptual aspect of biochemical systems resembles that of, say, electronic circuitry. In fact, such 'informative or conceptual morphology' constitutes the very basis of these forms. In this case a systematic cascade involves a typical negative (homeostatic) feedback and involves thyroxine which exercises general control over the rate of cell metabolism. It is yoked to a dialectical polarity, that is, an oscillatory interaction between complementary opposites - calcitonin and parathormone. Such regulation is the last link in a critical chain of command. This is because calcium is itself a practically universal 'end-of-line' informant - a 'foot-soldier' whose concentration gradients control (often in conjunction with phosphate, another recipient of parathyroid government) many aspects of cell metabolism.

fig. 17.13

Rising from the heart emotions sing; and rising air vibrates within the throat to give you speech and song. This is the *chakra* of expression and communication - but the lower half of *vissudha's* sphere of influence involves more than you know. Metaphysic of the throat plexus involves the passively informative level of 'grammar' (Chapter 6 and *fig.* 6.1) that is called (Chapters 7 - 12) psychosomatic potential, transcendent or top matter. As its metaphysical source such matter controls the automatic patterns of which the tangible, physical world is entirely composed. It represents (see *figs.* 15.8 and 17.11 - 12) their archetypal level. We might therefore expect thyroid *visuddha*, representing both sides of the psychosomatic coin, to be upwardly associated with instinctive impulse and nervous response; and downwardly to exert, at the level of general manager, nervous, hormonal and thus overall sub-conscious control of what is below, that is, the energetic domain of informed body. *In fact the throat chakra is, at the base of its informative domain, associated with another, lower agency of overall control - the thyroid and parathyroid glands.* A complex negative feedback cycle governs the relationship between the two glands; and they are themselves regulated by a higher negative feedback. cycle that involves messages issued hierarchically from the hypothalamic and pituitary command complex. Thyroid and parathyroid together govern both the rate and quality of cell metabolism. The tick-over of this 'engine speed' (whose base-line or 'norm' is called basal metabolic rate) affects the rate of every biochemical operation in the body. If it is either sluggish or overactive various illnesses tend to occur. The throat node is also ascribed the emotional tendency of sensitivity whose deformity becomes either over- or under-reaction - excitability or apathy.

Biological nature is not rigid. She is flexible, pliable, responsive. All targets (i.e. end-purposes) are obtained by a dynamic, cybernetic means of control called negative feedback. Such control, keeping a missile of intent on course, is at the heart of biology - which subject is effectively a study of information incarnate. In such a psychosomatic world *visuddha* exerts (through the agency of diencephalic master glands and the thyroid complex) overall biochemical homeostasis. If this is the case we might also expect a negative feedback mechanism between polar antagonists. We might not, on the other hand, expect blind evolution to generate such complex, elegant consistency as we find exposed in, say, a flow diagram mapping the counter-operations of thyroid and parathyroid (*fig.* 17.13). A proposed gradual evolution of such integrated business involves profound problems (Chapters 19 - 24). These pervasive problems include the rationale, structure and function of metabolic pathways, glands and organs that are in dialectical terms derived from the influence of lower morphogenetic plexi (*fig.* 17.11), that is, archetypal plan. To illustrate the specific, purposeful and complex nature of their chemical derivatives let's briefly detail thyroid 'management'.

The thyroid is itself controlled from the 'regal' complex of hypothalamus (using TRH) and pituitary (using thyrotropin) at the central throne of chemical messaging. As a rule of thumb hormones secreted above the solar plexus are 'hydrophilic' protein and those below 'hydrophobic' steroids. However, in the thyroid a protein, thyroglobulin, binds two hormones that (with adrenaline) are the only ones neither true peptide nor steroid. These distinctive 'master-keys' are called T4 (thyroxine) and more potent T3 (triiodothyronine) to which, when necessary, the more plentiful T4 can simply be converted. Together they are involved in the control of general metabolic activity and the maintenance of the abovementioned healthy 'tick-over'. While other hormones affect nuclear *DNA* indirectly through cascades of intermediaries across cell cytoplasm, T3 is the only hormone with privileged access to a cell's inmost chamber, an indication of its importance as it communicates directly, control on control, informant on informant with the precious database of genetic code. As a result genes are transcribed with various, immediate effect over local biochemistry. A large proportion of T3 is made from T4 at target cells by snipping off an iodine atom and, thereby, subtly changing the 'ring-tone' of the molecule. Indeed, configurations of iodine atoms (which need specific enzymes to ring their changes) seem responsible for various 'ring-tones' that characterise the switch-codes of this system. Switched on, more T3 is made to stimulate - except in the *CNS*, spleen, lungs and testes - metabolism. While T3 and T4 are distinguished by the subtle redistribution of a single iodine atom, so is their antagonist, a simple switch called 'reverse T3'. This is manufactured in various tissues by conversion of either T4 or T3. The effect of reverse T3's 'ring-tone' is to switch the system off.

Calcium is an important agent of metabolic regulation involved in cellular gradients, blood clotting, milk production, bone construction, muscular contraction and, tellingly, signal transmission both in nerves and cellular messaging. It is heavily involved, in other words, in the traffic of information, switching and chemical government. It is in the role of metabolic management that the physiological connection between thyroid and parathyroid is most evident. One might yet again infer from both the logical, dialectical position of these critical managers and their actual, integrated duties, that they are part of an accurate, efficient *top-down* program rather than the unexpected outcome of wholly unintelligent and haphazard 'keyboard strokes' by Thomas Huxley's imaginary monkeys (see also Chapter 20: section xxiii).

An electrolyte is a compound that, in solution, conducts an electric current. While most organic chemicals are non-electrolytes, physiologically important electrolytes (or stimulators) mediate cell metabolism, contribute to body structure (e.g. bone), facilitate osmoregulation and help maintain the pH (acid-base) balance or neutrality of life's fundamental medium, pure but polarised water. Of electrolytes calcium and phosphate are two important general administrators of body chemistry. Both are involved in informative, energetic and structural roles. Their informative part includes intracellular messaging by concentration gradients and in conjunction with specific chemical messengers.

The critical junction from which life's energy is derived involves phosphorylation; and what links units of *DNA* but phosphate? Calcium is also, as noted above, a vital component in tasks ranging from bone construction and cellular messaging to microtubule assembly. The homeostatic control of blood calcium levels involves secretions from the thyroid gland whose calcitonin lowers blood calcium level; and from its antagonist, the parathyroid, whose parathormone along with an associated variant of vitamin D increases it. Dynamic balance. *Top-down*, overall control. *Parathyroid, with thyroid, helps to administer general, all-body electrical, metabolic and structural order; the pair promotes development, growth, protein synthesis and development of the calcium-rich skeletal system.* The finely tuned patterns of hormone construction need to be catalogued according to bio-logic.

The whole business of biological order and control is, even in terms of molecular biology, not fully understood. For philosophical reasons a materialist needs hotly dispute any *top-down* perspective, especially one that includes the metaphysical or, if you like, psychological element of psychosomasis. An adherent, howsoever elevated his academic and/or scientific qualification, may decide he does not like the ideas of holism or Natural Dialectic. **That is his opinion. It is an opinion based, as previously discussed, on a belief that knowledge is material and there exists no independent, immaterial element of information**. As one turns to fact, however, it is clear that the deeper and more subtly science probes the more obviously appears and the harder it becomes to deny a balanced, highly controlled and interactively dependent web of biological structure and function. As it digs, the more not less complex and subtle the wondrous organisation and, as in an automated factory, self-organisation it reveals. You might holistically predict that close inspection of the *PSI* will happen at the quantum phase (*figs*. 15.3 - 5 and 16.1) of *H. electromagneticus*; and that investigation 'rising' to this phase may well discover how resonance structures, in conjunction with sub-atomic motions of the brain, dynamically reflect the features of logical, archetypal design and clearly indicate metaphysical, mental purpose behind organic structure. A theory of accidental, haphazard evolution retards this understanding yet, as the latter grows, will be destroyed by it.

The Informed, Energetic Domain

The hierarchical order of human state descends from the level of informative mind to informed energy. If you *know* with your head you *do* with the body below. *Below head and neck subtends the province of energy, of torso, limbs and, of refined agency, their hands and feet*. In the descent of any hierarchy power and influence is gradually restricted. At the neck's throat centre *visuddha's* link-grade is vacuum or the potential matter of sub-conscious, archetypal memory. Below this point the order falls through (*raj*) inner, kinetic to (*tam*) outward, structural aspects of energy-in-matter; it emerges from quantum to classical levels of physics; or lengthens from pico-seconds of biochemical reactivity to the 'large-time' of classical physiology and anatomy; invisible networks of communication substantiate the 'solid', visible and textbook patterns of microbiology, botany and zoology. (*Sat*) inner plan (*raj*) orchestrates the generation and survival of a (*tam*) bulk body, a material life form.

In hierarchical grade force (such as the expression of electromagnetism as light) precedes molecular gas. You might therefore expect to locate the energetic, 'luminous' plexus above its gaseous, cardiac inferior; but you don't because the body's pivotal mid-point, its umbilical centre called the solar plexus, works from the middle of its domain. It is the axis on whose energy absorption (at the small intestine) the lower grades depend; warmth, nutrients and 'breath of life' are circulated from the heart around the body's 'universe'. Nevertheless (*figs.* 17.9 - 10) *top-down* convention treats the torso's four lower plexi by spinal succession rather than material grade. In broad terms the (*raj*) top couple tend to support flux and function (metabolism and dynamic physiology) while the (*tam*) bottom pair decline towards 'earthiness' - reproductive output (body-making), the exhaustion of waste and anatomical foundations.

Whereas the informative domain shows a couple of coordinated, inverted trees of information (the nervous and hormonal communication networks with their roots in brain) *the informed domain shows a couple of inverted trees of energy.*

In the 'higher' gases are exchanged with the environment through the nostrils, using a two-way trunk-route composed of the trachea (situated in front of the oesophagus), bronchial branches and 'leafy' alveoli of the lungs. 'Down here' gases are exchanged to and from a circulatory river of blood.

In the second, 'lower tree' (*sat*) potential energy in the form of solar-derived food, having been ingested top-front by the mouth, drops through the darkness of a second, one-way trunk-route set behind the first, the alimentary canal. It is dissolved and digested until, from a tube situated around the solar plexus at the centre of the body's (*raj*) energetic domain called the small intestine, its nutritious components are absorbed into the bloodstream. Here they are united with the aerial substrates of respiration and radiated throughout the whole biological sphere. As in a tree itself, life-giving saps are circulated from the air and earth.

Can you see, where plans are not by chance, the symbolic plan? Lighter in front, heavier behind. The pulmonary system is 'bouncy', light and levitational. Its highly branched system is 'leafed' with alveoli and sprung by intercostal muscles and a diaphragm. Its two-way business is completed in the body's 'sky'. On the other hand the apparatus of the alimentary tract, which deals with liquids and solids, runs behind the 'airy' (or aerated) respiratory 'tree'. It is branchless, leafless, less 'sprung' and more 'weighty' or gravitational in its operation. Stomach under spongy lungs and radiator heart; it dumps its wastage down to ground.

Lighter above, heavier below. The diaphragm, situated beneath the lungs, divides the 'sky' from 'earth'. *In the sky is fixed a* ***fourth*** *plexus, the anhat or hriday chakra.* Its colour is azure or a blue-green cyan and it is the oscillator involved with energetic molecular matter (gas), especially oxygen. Also called thoracic, cardiac or heart centre this power point supports life's levity, motility and lightness; and it supplies a body's 'fire', the respiratory dynamo, with fuel. Thoracic organs (lungs/ bellows, heart/ pump, spleen and female breasts) help build energy and resist any loss of strength. The secretion of milk from the breast is, of course, a life-giver; and from both breast and open heart flow love. The other organs, along with this centre's associated hormonal node, the thymus gland, represent respiratory, blood vascular and lymphatic (immune) systems. With red oxygenated circulate white defensive cells. Water is the fluid basis of life and this plexus dominates the dynamic, fluent, aerial section of its 'hydrocycle' - circulation of the blood.

Metaphysic of the breath of life is a stream of consciousness called spirit; its physic is the pulse of lungs, an ebb and flow of oxygen, the vital inhalation that you can't stay minutes in this universe without.

The metaphysic of a heart is love but even the material instrument is truly warm and like the sun. Beating out life's rhythm it irradiates the micro-cosmos of your body with star-kissed nutrients, a light-ignited combination of the elements for your assimilation and well-being. More prosaically, it is a pump that transfers 'energy-in-motion', heat, while also circulating gaseous, liquid and liquid crystalline states of matter through its medium. The emotion of *anhat* is love but psychological imbalance associated with this centre pulls as grasping, clasping attachment or jealousy.

Do you remember the hierarchical suite of organs of knowledge in the boss's office, your head? The top, sixth or internal, metaphysical organ of mind, called the third eye, is positioned behind the forehead above and between the physical eyes. Mind and its senses are related not only to conditions of matter but to systems of the body. The informative nervous system is associated with voluntary consciousness and involuntary (autonomic) sub-consciousness. Of sensory organs the eyes are associated with electromagnetic radiance, illumination and a thin waveband of vacuum-penetrating light. Below and behind the eyes ears are also linked with vibratory motion, this time energy-in-air called sound and, as sound's active organ, throat. In front and below the eyes nose breathes gas for energy and smells molecules-in-air. Systems correlated with the nose and found within the ambit of the *anhat* plexus are pulmonary (of lungs) and cardiac (of heart). Lungs fill the blood with 'special air', exciting oxygen, and heart, a pumping station, continually and energetically irradiates both air and nutrients (or molecules-in-water called the blood) to service respiraion. These nutrients drop from mouth whose moisture and whose sense of taste are at digestion's door; and are absorbed at the solar plexus by the small intestines into the bloodstream; delivery body-wide allows respiration, chemical reaction and the generation of thermal energy. Light is 'grounded' photosynthetically; you might call all electromagnetism locked in bonds (not least in *ATP*) a kind of 'grounded' form of light that, when released, evolves some heat. Light is flown into life by energy metabolism. Hence, from macrocosmic to a microcosmic sun, behold the fifth, umbilical and solar plexus, *manipur*.

As well as with digestion water links with the regenerative, reproductive system in the sixth zone down below. Finally there extends over the whole body an exteriorised, generalised field of perception, a 'global' sense of pressure, touch or 'feel'. While touch is an exquisitely localised sensation it is complemented by a general 'feel for oneself', a sense of solid, bodily existence, a localised position in relation to the rest of space, time and any other thing. Are you not a unified experience made of a billion trillion parts?

Below the neck, we'll travel down the *chakras* of the trunk in this informed domain. First, is there anything that you'd cut out of the construction? What don't you want to know? Is it a sense of negativity, ill health and what is wrong - the experience of pain? Who wants hurt, malfunction or an organ's operation that is only half or less complete? This is where I love an absence and so, it seems, does brain. Unlike the other welcome senses pain has no particular place in brain, that is, it is not represented by a local space in cortex. It is exiled to the wilderness. It is nowhere mediated like the other inputs but what, you ask, is its psychology? Pain does not think but forces thought onto its knees; it does not please but forces pleasure to submit. Who wants hurt yet, at the right (or wrong) time needs it? This sharp paradox of pain is a violation yet a necessary warning signal something's wrong and threatening your survival. Although its diabolical compulsion can, in varying extent, be blanked by mental focus elsewhere how without it would you know that anything was untoward? Though natural, opiate analgesics and perhaps 'blocking circuits' underneath the thalamus and in the spine can offer you relief you need a system of alarm. Without a guard-dog you could suffer injury. Pain is the bleakest siren, warning of attack, a negative exchange that always cries out 'Trouble!' It provokes your riposte 'I must scotch or mend its cause unless I want to suffer more!'

Within *anhat's* sphere of influence the internal, involuntary and energetic motions of bellows and pump are complemented by external, voluntary dexterity in the form of arms and especially touch-sensitive, five-fingered hands. Just as brain is the organ of intellectual (i.e. metaphysical) manipulation, so hands are the primary organs of physical manipulation. As brain is 'thinky-feely' so hands are 'touchy-feely'. And as brain is the physical organ of emotional feeling so skin (especially the densely-innervated skin of the hands) is the organ of physical interface, of general sensitivity to anything that's close. In short, mind and hands are, respectively, metaphysical and physical manipulators of desire, the willing and willed servants of intent.

The gland linked with this 'airy' plexus is the thymus. It is situated just above the centre of the rib cage and its hormone, thymosin, promotes the body's internal security system. While skin is an external boundary the immune system, patrols and defends its owner's personal commonwealth from alien violation. The health that it defends is elastic, bouncy and resilient. *If health is a result of interaction between physical and archetypal (sub-conscious) patterns, it did not evolve by accident.* The tendency of every system is to bounce back to its archetypal norm, its *vis medicatrix.* **This metaphysical form is the very basis of homeostasis, physical life's overriding principle. Health is the norm**. Wounds heal, infections are fought and cell debris cleaned up. Indeed, the central nervous and autonomic, endocrine and immune systems are integrated in such a holistic way that experts sometimes use the phrase 'psycho-neuro-immunological system' to describe their cohesion. Mind, as every doctor knows, affects matter. *The channels of 'psycho-somatic effect' are precisely what this chapter is describing - unless you neuroscientifically believe that mind is brain and psychological means, essentially, chemical.* This means you believe metaphysical *is* physical, will-power is a function of atomic configuration and, in dialectical terms, your world is upside down. This is why we call it *bottom-up.*

While the skull, rib-cage and pelvis shield internal organs from gross external blows, any would-be invader of the body's walls has to run the gauntlet of skin, mucous membranes, cilia, chemical complement, bug-eater cells, antibodies and 'hand-to-hand' cell-mediated battle. Information is an essential element of warfare. Command and control centres deal in nothing else. It is as if this particular security service was a mobile, global, non-conscious brain whose basis, in both natural and adaptive immunities, was information in the form of code, signal, memory, recognition, specificity and plan. What does not have this body's protein (and, therefore, its *DNA*) should not be here; what is not 'self', the system logically concludes, must be expelled, expunged, expurgated. The manual of physical immune procedure involves 'language' (stored on its carrier, *DNA*), discrimination, editors and all manner of identification tags. 'Intelligence' is at the heart of all control. Home defence deploys to guard you and its information's postal traffic circulates on psychological meridians and lines of blood and lymph.

Resonant power to protect needs selective power to destroy. A non-specific system of defence which tends to work on enemies in the bloodstream, tissue fluids and lymph is complemented by a specific system which includes long-term, protein-based association by 'memory cells' and a specialised 'lottery' mechanism involving rearrangement of gene permutations in order to generate diverse immunity. The lymphatic system collects tissue fluid, flushes it through filters called nodes and tips it back into the bloodstream near the heart. Plasma is continuously screened at such stations as the thymus, spleen and liver. Some white blood cells called B-lymphocytes mark the enemy for destruction with tags called antibodies. Others called T-cells are converted by the thymus into primers, killers or, antagonistically, agents that decide whether a battle is won and enough is enough. At the same time the body also invests in a massive staff turnover to maintain optimum strength. For example, ten million new bug-eaters are manufactured a peacetime minute and thousands of antibodies per second per cell in the case of attack. Even humble bacteria have antibiotics, toxins and restriction enzymes along with communications, recognition capabilities and a *DNA* text to underwrite the way they cut up foes. The human immune system is one of great complexity but its principle, like that of any army, is simple. Expel the antagonist from abroad. Kill threatening 'non-self'. Peace is healthy, health is undisturbed by pain. What more homeostatically healthy than to repel attack?

Below anhat is the plexus, manipur. This compartment, the solar plexus, marks the pivot around which the informed domain swings. The plexus is about both body's outer strength and inner, mental drives. Fire is its central sign and colour sunny yellow. Regarding mind, associated feelings are of physical excitement and fiery, passionate love; heat also powers the will; the character drives motion and loves to freely travel but imbalance or obstruction shows as friction of frustration and sees red with anger. Regarding body, *manipur's* 'vibration' is kinetic matter's - one of thermal energy; but its flames burn on behalf of vital purpose in the heat of action. Not only mid-point at the navel. Mitochondrion and chloroplast are, sited in between the nuclear centre and the outer membrane of a cell, stations that provide the energy for cytoplasmic business, for sun-driven chemistry.

Now you may have understood why some yogis sometimes contemplate their belly buttons. They're concentrating on the axial focus representing light and warmth and the original sources of their energy for life. They're lasering a beam of physical strength and calling up a sphere of psychic radiance. Don't laugh! Even biologically the power starts here. In the navel region nutrients (once called 'vital elements') are sieved and absorbed from the second, lower energetic tree into the blood for irradiation. This omphalic midpoint reminds you of the link through which your energy was first, primordially derived from a beloved goddess, mother. Similarly life on earth is wired, umbilically, through rays of light to a starry generator, to our source of nutrient, our heavenly womb. *This poetry is fact; it might be myth but it is science too.* Life's 'electrical circuit' (Chapter 20) depends on radiant, wireless, solar energy to recharge its batteries. 'As above, so below'; and 'from above to below'. If you had known the inner light you might notice its reflection on the outside. It pours from sky-light into earth-life. Softly and silently the sun, through light, is the candle to life's physical vitality. Photosynthesis is the primary bio-process, an initiator and sustainer. Its mechanism is the method of light's capture; it splits up sunbeams, freezes sunlight into pure white crystals, glucose that stores stellar power for life. From light to life - organisms use for fuel the transformed product of an atom bomb, the sun. And through the medium of a food chain nutrients are circulated in life's river, blood, to every cell and serve (*raj*) dynamic, metabolic and thence muscular designs; respiration everywhere serves standard units (*ATP*) on which cell chemistry can draw for 'fire'. In other words original solar power, having been radiated from the centre of the solar system, is now re-radiated throughout our microcosmic bodies. Central heating. Thermal radiation. We can feel and move! We can still survive! Such energetic output is complemented by the construction, maintenance and repair of (*tam*) 'static', solid, anatomical constraints. Finally exhausted waste is discharged down and out of the lower orifices - urethra (front, liquid) and colon with anus (bottom-rear, solid). In this way the informed pattern of energy moves from high potential (raw materials input at top-front) through mid-way kinetics (centralised processing, overall circulation and respiration) to low exhaustion (output of waste products) at the base.

Internal organs associated with firepower are those of ventilation, digestion and circulation. Digestive organs particularly associated with the solar plexus include the pancreas, stomach, small intestine and liver. *If information's central organ of homeostasis is the brain then the energetic domain's is the liver.* It is endowed with a rich supply of blood whose contents and multiple activities it regulates. It promotes carbohydrate, lipid and protein energy metabolism. In the case of lipids it governs the level of cholesterol from which bile salts and steroid hormones are both constructed. The former are used in the small intestine to help digest energy-rich lipids; the latter are associated with control of water-related functions (osmoregulation and reproduction) in the domain of the plexus below. By disposing of hormones whose message has been delivered and job done the liver promotes smoothly vibrating feedback cycles. It also produces albumin, an abundant plasma protein that regulates the water potential of the blood. Without it blood could 'run dry' at the same time as tissues swelled uncontrollably - with lethal effect. It could also, on occasion, run cold. The liver is a reservoir of blood and general body heat. On the other hand specific units of heat, coinage that slots into all life's metabolism, are the potential fire that respiration generates. Organelles of firepower, chloroplasts and mitochondria, contain suites of enzymes whose structures all support (as shown in *fig*. 20.3) the transformation of light to life. So marvellous is this metamorphosis that it is worth a second mention. Heat is released by breaking bonds in quantity that amounts, in this particular case, to standard scintillations of sun's ice. Strike a light. Biologists know life's match, its fitting coinage or power pack geared to every metabolic need as *ATP*. *The body's sun, its fuel distributor, is central but its influence is radiated carefully to every part of every cell. Here, just as exactly and appropriately in time and place, each single 'burst of flame' is struck.*

External organs associated with the power of fire's light are, of perception, eyes which catch movement, shape and colour from reflection; and, of action, legs and feet which grant locomotion. Light, sound, sensation and, at best, harmonic motion swing us through the sparkling dance of life.

Manipur's glands are pancreatic, intestinal and adrenal. Of these the pancreas contains 'polar' glands whose antagonistic effects (glucagon and insulin) control the proportions of active, available and passive, stored energy in the form of blood glucose and glycogen - a critical role in the operation of energy metabolism. Both pancreatic and intestinal glands also issue agents of digestion. Meanwhile lower, to the rear, an adrenal gland rides astride and above each kidney, an organ predominantly involved in salt and water regulation. It has strong connections with the two regions below. Both adrenal cortex and renal medulla are involved in energetic response to stress and, in the form of 'fright, flight or fight', immediate reaction. The cortex, whose generalised, dynamic influence as a 'survivor' spreads throughout the body, produces 'fatty' hormones derived from cholesterol and the medulla, closely linked with the sympathetic nervous system, adrenaline and its close structural and functional companion, the sympathetic neurotransmitter noradrenaline. In addition to medullary tissue there exist chains and concentrations of chromaffine tissue distributed all over the body but especially along sympathetic nerves (where it also tracks the spinal column and thereby links with nervous and psychosomatic plexi) and in the dermis of the skin. This tissue also produces adrenaline. The hormone is a general excitant that tenses muscles, causes cells to increase sugar uptake and promotes respiration. Given that the 'four F' connection (see also Chapter 13) also involves sex it is understandable that the adrenal cortex is a source (noticeably in post-menopausal women) of testosterone. Finally, two small hypothalamic proteins (*ACTH* and *ADH*) act on the cortex. *ACTH* is involved with the abovementioned stress responses; it influences the cortical synthesis of steroids that promote carbohydrate 'sugar-for-energy' metabolism. The kidney is a homeostatic organ whose functions span association with both solar and sacral (water) plexus. In this context *ADH* is involved with the main business of the next region down - water works; so, in terms of salt and water balances, is the cortical steroid aldosterone.

Space and gas are associated with sky-born levity. (*Raj*) energy drives. But the lowest couple of *chakras* are involved in the (*tam*) earth-bound gravity of materialisation. As such their sections occur in the lower part of the body and their main function is the reproduction of bodies, osmoregulation and the exhaustion of wastes. First down from gas is 'solid-flux' in the form of liquid.

Polar water with its extraordinary biological properties is your major physical constituent (about 60% but rising to 90% in the case of some organisms). It is the solvent medium on whose polar make-up the ebb and flow of biological life depends and is, as such, involved in the dynamic, fluid aspect of your internal being - the process of metabolic operations. All dissolved metabolites are wrapped in its coat, many reactions take place through its structured medium and it may (see Chapter 16) facilitate gene expression and in general act as a key component of bioelectric signalling. Water's functional aspect gives way to structural components, non-polar solids that act as fixed containers of the dynamism. Neither solid nor liquid, one might understand the body as a 'liquid crystal'. Together the two features constitute bulk form; they constitute the medium of subtle metabolism and its gross manifestation, a biological body. According to this logic the higher of the two base plexi (the sacral) is involved with liquid and the lower (the coccygeal) with solid dispositions.

The functions of the **sixth** plexus (also called *svadisthan* or *indri chakra*), are thus connected with the fluid/ water principal.

The higher aspect of the region involves osmoregulation, the regulation of water potentials, acidity and the excretion of liquids. Its main organs of homeostasis are the kidneys. Its organs of action are sweat glands (an overall 'excretion' of heat), the urogenital tract and colon; its organ of perception, which recognises solutes (dissolved solids), is the tongue and its sense, taste.

↓	*tam*	*raj* ↑
	yin	*yang*
	out	*in*
	inhibitor	*stimulator*
	absorption	*release*

The *lower aspect* of the sacral plexus is generative; it involves the polar attraction of sexes whose erotic communion results in fresh bodies.

	tam/ raj	*Sat*
	duality	*Unity*
	division	*Union*
	parts	*Whole*
	polarity	*Neutrality*
	subsequent order	*Archetype*
↓	*tam*	*raj* ↑
	yin	*yang*
	female	*male*
	absorptive/ receptive	*emissive*
	attractive	*outgoing*
	centripetal	*centrifugal*
	oestrogen/ progesterone	*testosterone*

The spectral colour of this 'sex *chakra*' is occasionally given as white (marking the pure, transparency of water) but more often and rainbow-logically, orange. Its emotional tendency is the fire of lust and its deformation sex mania. A water/ body relationship, reflecting that between the lowest elemental plexi, is clearly demonstrated in the combined urogenital tracts. Liquid is 'higher' than solid so, symbolically, liquid exhaustion in the name of urine issues 'from above'. That is, the bladder is 'above' the testicles and 'in front of' the female reproductive tract; and while a male's urethra is also his sperm duct the female urinary orifice is situated just above her vaginal birth canal. The region's organs of reproductive action are the primary male or female sexual apparatus including glandular gonads and various vessels connecting to its port, the penis or vagina. The female gonads, ovaries, are positioned inside the body and, on each side, apart; and, in a kind of polar inversion of location, the outgoing male's testicles are situated together and, in the scrotum, outside. The sex organs (female vagina inside and male penis outside) are, equally, organs of perception and of communicative intercourse. In their use for sexual conjugation sperm issues from a penis at the front of the base area. Life, which has a future, is passed forward. Sperm is 'bounced' into the womb and, on recoil, a child is bounced back out. From heaven to earth! Solid potential in the form of embryonic body is projected towards a future lifetime of experience. An infant is expelled and falls, symbolically, upside down from its mother's front-base orifice head first into a hard world. Down to soil! The curtains rise. Dramatic shock initiates another mundane yet exciting episode.

At base level solid waste, which is of the finished past, takes the dialectical role of subtendent neutrality. It is exhausted down from the rear.

According to the principles of Natural Dialectic polarity arises from neutrality. In this case two very similar but antagonistic hypothalamic/ pituitary hormones affect the sacral region. One, *ADH*, is concerned with the asexual retention of fluid and causes an increased re-absorption of water. The other, oxytocin, mediates reproductive contractions of one kind or another. It causes contractions in the female oviduct (to help the passage of the egg towards the womb), in orgasm (to issue sperm or draw it to the egg), in myometrium as labour pains and in the expulsion of milk while breast-feeding; orgasm and lactation involve the warmth of pair bonding and so it is sometimes called the 'cuddle and contentment hormone'. *ADH* and oxytocin illustrate a polarity that pervades the homeostatic nature of both nervous and hormonal systems. Molecules involved in allied but polar kinds of activity (in this case retention of water and expulsion of liquids and eventually a solid form through the urogenital channels) are derived from the same source, hypothalamus, through similar chemical precursors ending up with a similar shape to perform complementary work. Check male testosterone and female oestrogen for yourself. *A dialectical expectation might be that balanced pairs of hormones with active, polar effects will be derived by differently 'tweaking' a neutral (i.e. passive, non-functional) precursor. This would be a sign of efficient, systematic and intelligent design.*

Remember that informative and constructive bio-molecules are simply the frames for electrical positions, vibratory interactions and switching. Perhaps such frameworks are most evident in the case of enzymes, much of whose material is used to three-dimensionally frame a precisely active site, but they reflect the whole intricate, integrative network of *H. electromagneticus* and, therefore, *H. archetypalis*. Each of these dynamic patterns affects the other. The subtle search is, therefore, for vibratory signatures and electrical switching that induce interactions. In this respect the abovementioned sex hormones are interesting. As opposed to water-soluble, unisex proteins (GnRH, FSH and LH) from the brain 'above', sex hormones produced below the diaphragm, especially under 'watery' sacral influence, are often steroids based on the anti-water (or lipid) precursor, cholesterol. This is the case with hormones of the adrenal cortex, ovaries and testes. A testis produces male and an ovary female sex hormones. Because of their similarity, transmutations are possible. Indeed, after their cholesterol starter 'female' progesterone is a stage in the production of a number of steroids. And there exists in both sexes an enzyme, aromatase, whose specific role is to interconvert oestrogen with another very similar molecule, testosterone. Sex is not absolute but depends on a web of hormones, enzymes and other fine controls. It is a matter, in either sex, of emphasis and predominance. Such imbalance keeps things swinging.

Sexual chemistry is a good example of polar, dialectical interaction. It engages in complex, homeostatic relationships of stimulation and inhibition, stop-go, peaks and troughs. For example, the female menstrual cycle shows a polarity between its two halves. The first, oestrogen-predominant half is 'centripetal'; it draws in and gathers for the spring of life; it is the attractive, 'sexy' preparation whose 'heat' rises to climactic ovulation and penetrative input by the male. For both sexes this input involves, with its focus of physical heat, a rush of go-getting testosterone. The second, progesterone-predominant half of the month is 'centrifugal'; it induces an out-growing, maternal phase whose productive 'push' will, if sustained due to fertilisation, climax in the natal expulsion of a child.

Such examples appear to support an axiom that balanced pairs of hormones with polar opposite effects are built from a common precursor. How well mindless evolution didn't think things out.

The base (mul, coccygeal or earth) plexus is the **seventh** *and lowest. The 'darkness' of its under-world is located diametrically opposite the 'translucency' of the top, transcendent chakra's over-world. It most emphatically represents the negative, insensitive, non-conscious pole of existence - the physical domain.* As such it provides support, anchor or stability for the temple of earth body. Primal earth power's instinct is for physical security, endurance and survival. Its character is structural solidity, mass and heaviness, its colour dull red (or the 'black light' of infrared) and its emotion dull, dense and inert with an unpredictable, animal tendency towards brute force, rage and violence. This is the (*tam*) bony, bottom end, as far as materialisation can go or gravity can pull. Hard earth represents the last stage, an outer shell of exhaustion. Its foundational structure is the pelvis and its base function egestion. *Mul's organ of action* is therefore, in conjunction with the water-absorbing colon, only a couple of sphincter valves composing an instrument of gross exhaustion, the anus. Associated perception is vestigial. Informative function has 'rigidified' to the functionless point where neither coccygeal spinal nerves nor endocrine gland exist. As you might liken duplex brain on spine to an arrow pointing (↑) up so pelvis and supported spine are like an arrow pointing down (↓).

From positive to negative pole six 'elemental' plexi are active in humans and the top one, 'seventh heaven' of the *sahasrara*, is dormant. Do you remember, earlier this chapter, reference to a range of organisms held to represent a complex scale of grades? In this panoply each broad type of organism expresses different permutations of active as opposed to dormant plexi. For example, birds show the potentials (characteristic instincts, morphology, ecological preference etc.) predominantly associated with air, warmth and water, insects warmth and earth, plants earth/ water and bacteria perhaps simply earth. Within these broad spheres of influence exist many subtler degrees of action or restrictive dormancy. Each expresses different shades of colour, different combinations of 'elemental principles' that in turn correlate with permutations of the cosmic fundamentals *sat*, *raj* and *tam*. In this view there arise different criteria (cosmic fundamentals, plexi and elemental correspondence as well as homology) by which to classify a hierarchy of body plans. There exists a connection by cosmic coordination and design rather than haphazard evolutionary heritage that, it is alleged, accidentally lurched from a bacterium to a man. Different actors, different parts comprise the holistic drama staged with bodies here on earth.

After all, can you perceive no pattern in all this? Is there neither reason nor logic outside the wind, rain and hills? No creation bar mutation? Biologists look for reasons. Even though chance is believed to have created them, biologists (even Darwinian biologists) look for reason in every molecule, cell, organ or organism. Reason by chance! Intricate, coherent systems derived from incoherence; ones whose clear cooperation is hierarchically *top-down* whistled up by a tuneless jumble of atoms!! Yet if a structure or function (for example, plugs in a foetal human's nose) is unexplained, you presume ignorance rather than irrational chance. Judgment is rationally reserved.

Swing life round to the other side. Instead of reason seek its absence. Metabolic pathways and organs express, in different organisms, common physiological functions. These functions are, effectively, 'reasons' subordinate to the overall 'reason' that drives a biological body. This 'reason' is homeostatic survival in response to changes in different aspects of its own and surrounding physical conditions; and 'sub-reasons' include the establishment of individuality (a boundary between 'me' and 'not-me'), sensitivity, biochemical command and control, digestion, respiration, osmoregulation, reproduction etc. *Which of these functions is, where found, superfluous?* Which of its associated organs, metabolic factors or coded molecules, lacking a reason for existence, is therefore redundant? If any, explain which. *On the other hand, which can survive without the others?* **Calculate the minimum number of functions necessary to support life, any life at all.**

Of course, you find that in some form or other every single one inhabits, by necessity, even the simplest cells of complex life. Functions all need agents and, in the biological case, the main agent is protein; and structure of a protein is precisely and without redundancy encrypted in the genes. Molecular biologist Craig Venter has calculated (using *M. genitalium* as exemplar) that the absolute minimum number of genes necessary for the simplest microbe to function is over 350. How's that, as they say in cricket, for the evolution of a first cell? It's definitely (Chapter 20) out. Why, furthermore, should genes (sheer chemicals) create coherent suites of functions anyway? Cells are forms of passive information; materialistic incoherence bluffs by wholly missing active information out! Take any cell or body made of cells you like. Is there not an absence of redundancy, a plenitude of rationale? The question's not rhetorical and so, again, list the least number of *functions* needed to cooperate within a system and express their structured purpose simultaneously. How might each have 'happened' to combine at the same original moment into a physical system? In other words, explain how chance (which is by definition devoid of reasonable explanation) could throw a goal-based system together. How did such irreducible complexity jump into working order? How did such a mandatory minimum of purposeful, coherent physiological functions with associated biochemistry and structural definition (in the form of organ, cell or organelle), co-evolve 'at-a-go'? Reason is not present in a molecule or an accumulation of them. Is it really absent in a working body?

Reason and chance are like chalk and cheese. *Reason is meaningful, chance meaningless. Why should a highly rational system irrationally 'self-organise'? Chapters 19-25 will demonstrate the fallacy of a belief that any chance whatsoever could invade the initial, entirely purposeful construction of a cell, a human or any other type of biological body. Your own human microcosm underlines the utter feebleness of evolutionary explanation. Is it, with its mind and brain, a good expression of chance - no more ultimately explicable than by a shrug or, rather, a long series of them? Or is it inconceivable that such a logical, integrated, self-consistent embodiment, constructed with purposive complexity in accordance with the grades of the conscio-material gradient of creation itself, occurred by chance?* **If reason wins, chance and time's grand theory crumble back to their home ground - they bite their progenitor, the dust.**

18. The Logic of Disembodiment

One Thing Is Certain...

Will you never die? Do you suppress the thought or, as society around you seems to, live in death-denial and suffer from delusion? The end is nigh. The end is, naturally, upon us - for you in maybe fifty years, for me in thirty or, perhaps, a single hour. This whole world will disappear. It will be no more. Yet commentators turn the sense upon its head, evade the issue and thus think it's claimed the starry universe's time is up, that cosmic death is nigh! **For cosmos, in the near future, it's unlikely but for each life on earth one thing is absolutely certain - death.** Therefore cut all cant and superstition. Approach its moment in a steady, lucid, balanced way. Death is, in Natural Dialectic's terms, a very lively subject.

Life Cycles

Of course, if you believe that mind is made of nerves then death disintegrates. You, just atoms, will be ecologically rearranged. Dust to dust; oblivious immortality; death's void. That's it.

If consciousness, though modulated by, is not dependent on the brain another possibility appears. Check *fig.* 2.14. Oscillation, up-downs, ins and outs; see-saws, pendulums and springs vibrating till, exhausted, they stand still. An act of creation is hierarchical and yet at the same time existence is, from the regular wheel of stars to vibratory, biological periods, profoundly cyclical. Is there (Chapter 12) a cosmic life cycle? What about the human coil? Awake, asleep, the tide of consciousness rolls in and out. You spring awake but, gradually, vitality subsides to sleep; the vibrant tide has ebbed. In this view sleep re-coils the spring, refuels and 'knits the ravelled sleeve of care'. It is a phase of balance; from down rise up, from back go forward. Compare 2.14 with 18.2. What about a whole life cycle? Living and dying are each one a part of whole rotation. The previous chapter dealt with *out-swing* from the Centre in a process of materialisation. It framed the logic of embodiment. Life flows until high tide and, from maturity, it ebbs. A body, at the end of its life cycle, dies. Its physical components are recycled ecologically. Is there also metaphysical recycling? This chapter concentrates upon embodiment's reverse, *in-swing*, the logic of an organism's disembodiment.

	tam/ raj	*Sat*
	existence	*Essence*
	oscillation	*Poise*
	body/ mind	*Supermind*
↓	*tam*	*raj* ↑
	passive	*active*
	objective	*subjective*
	towards matter	*towards Supermind*
	sub-conscious	*conscious*

involuntary	*voluntary*
separator	*connector*
sleep	*awakening*
death	*birth*
out-swing	*in-swing*
materialisation	*dematerialisation*
physical birth	*spiritual initiation*
egress	*ingress*
exodus	*return*
end	*start*
ebb	*flow*
mass	*energy*
body	*mind*

You need to understand life's pendulum. *Out-swing* from the Centre is (*fig.* 15.3) a process of materialisation. It is the manifestation of potential gravitating through kinetic into exhausted phases. Pure Alertness dims through conscious into dormant mind. Sleep, in this direction, represents a nadir. It is a sinking from the Centre, out-swing towards mind's margins, a journey to sub-consciousness. Subjective knowledge of sub-consciousness - in sleep and dreaming - is as near as life approaches the non-conscious oblivion of matter.

Sleep is elastic. It is like jumping on a trampoline reversed. The first fall is steepest, the first recoil strongest; then oscillations lessen till you gently bounce awake. Waking marks an *in-swing* and propels us back towards a zenith of alertness. At least, in waking, darkness is turned into light, consciousness increased and the elasticity of life confirmed. Life surges, interest flows swiftly as in-swing raises the ability to focus and to concentrate. When someone does not understand we say 'Wake up!' When they do they have 'woken up'. In this sense the ascent towards Central Consciousness is a gradual birth or re-awakening and, at the ebb tide, you are dropping off. Descending you are nodding, increasingly enveloped by a fascination with gross objects or fallen spellbound under Hypnos, that is, sleep.

Sleep and Enlightenment

Incarnate Cycle.

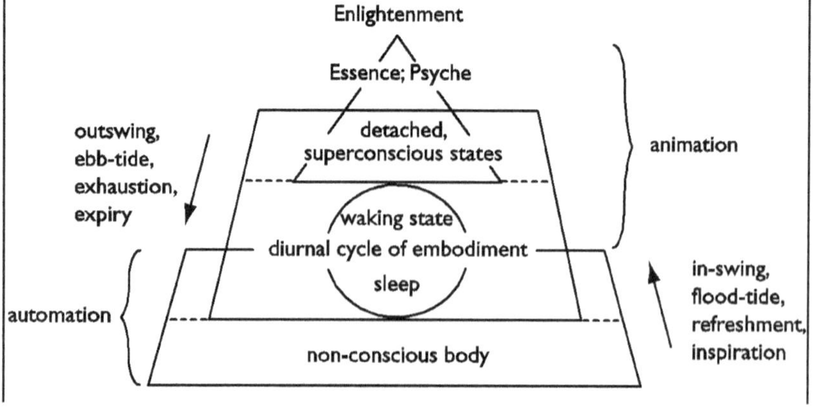

The incarnate cycle oscillates between waking and sleeping. Its nadir of awareness is dreamless sleep (unconsciousness). The zenith is a variable dependent on the evolution of a person's mind. But spiritual evolution's Zenith is Enlightenment; and an individual's best approach to elevation in this 'non-condition' represents his or her personal, relative zenith. You might understand that, in this swing of things, set against enlightenment our present consciousness is a relative sleep, a fault-prone dream: and that, while birth into a heavily constrained material body represents a relative death, the ascent towards Zenith approaches immaterial rebirth. Zenith is the Peak of Wakefulness - Full Life.

We'll check out the Excarnate Cycle a little later on (*fig.* 18.3: The Logic of Disembodiment).

fig. 18.1

In other words, the natural state of consciousness is wakefulness and of matter the reverse. *Pure Consciousness, permanent wakefulness; pure matter, permanent oblivion.* Non-conscious objects are inanimate. In this sense they are 'negative absolutes'. Combinations of each ingredient, called life forms, oscillate within the middle ground between extremes. They are neither super-conscious nor non-conscious. When, however, are you most like a material body? When dormant. Unconscious. Asleep. Hypnos was the twin of Death but Death never lived. *Even biological bodies are only alive when animated by mind.* Their constitution is, as a biochemist or molecular biologist correctly perceives, made of chemicals, wholly material and, when the influence of mind ceases to 'resonate', a dead body's atoms are recycled. Hypnos sleeps but the body of his brother, Thanatos, never rose above a grave of earth.

Grades of consciousness are manifest in different forms of life. Sub-consciousness marks the most diffuse and weakest flow, the lowest watermark of consciousness. Bacteria, plants and fungi are, as far as we know, dormant. Plants are, compared with beetles, permanently comatose. A beetle set against a fish or mammal walks in sleep. Sense-based awareness sleeps beside awakening contemplation. The higher an awakening, the freer of its physical encumbrance, the more it is (*fig.* 0.10) removed from world-periphery towards the Centre. This is why physical birth may be regarded as a downward 'out-swing'. It involves the drag of material embodiment. It is why contemplatives, 'in-swinging' towards liberation, have always regarded body's death-day as more potent and important than its birth-day. It is why a deity of death, Siva, is paradoxically most lively in the Hindu pantheon and why Christ could say to uncomprehending rabbis "You are unknowing, you are sleep-walking, you are as dead". Conditioned free will, choices that are bound by context and 'sleep-walking through the heavens although wide awake on earth' are normal human states of mind - even in the case of clever intellectuals. You think, as you read this book, that you are wide awake. You are *relatively* so. Equally, relative to Enlightenment or such grades of cognizance as clasp their praying hands in its direction, you and I lack comprehensive grasp and in such nescience are as dreaming.

The way you use the words birth, life and death depends, therefore, on the context within which you understand them. Is it *bottom-up* or *top-down*?

No doubt, the medium of body with its brain constrains and patterns consciousness. A *bottom-up*, objective view goes further and decrees life *is* a neural network and consciousness a special kind of electronic computation. It is, at least in mammals, just a state of brain. In this view an organism doesn't so much learn to use its instincts, body and its brain; instead, consciousness itself develops as it grows. Similarly death is just a catastrophic bug, an irreversible, systemic failure. The program's circuits are disrupted and consciousness ebbs, like a fading screen, into its final unawareness. There is no elastic buoyancy to catapult you back to life; there is no vibration only flatness, the end of an illusion, oblivion. All parties agree that death, like sleep, is a change of state. For a materialist it is a corporeal one; and bodies all disintegrate. It is time for ritual scrap, decay whence physical components (there is nothing else) are ecologically recycled. Ashes to ashes, dust to dust but life thereafter? No. No. No soul unto Soul.

	tam/ raj	*Sat*	
	existence	*Essence*	
	spectrum of consciousness	*Pure Consciousness*	
	embodied life	*Life*	
↓	*tam*	*raj*	↑
	non-conscious	*sensitive*	
	non-life	*life form*	
	death	*temporary lifetime*	

Of course, energy is always lost and motion apt to fall away. Does decadence called entropy not rule the whole material universe? For each embodied life its body's failure sweeps it to the world's end's bin. For you, at least, this end is nigh!

What, though, about negentropy? Can immateriality wear out? Is life composed of just material elements or is, as *top-down* holism holds, its better half an animated element apart? Embodied life combines the two; physical is married to the metaphysical. A dualism made of mind and body is presumed. A mesh. The conscio-material mesh involves a slope of interaction. On such a sliding scale (*figs*. 0.7iii or 0.14) you view increasing materialisation as the ebb of consciousness from its High Tide. Or, conversely, see it as a gradual predomination of non-conscious energy, information's ordinate dropped (down its vertical y-axis) to nought - a fading, flattening of subjectivity.

Out-swing involves increasing lack of awareness, sleep or death. *Thus entry into physical worlds involves an ebb tide. The outflow of creation is an ebb tide or expiration from Life.* In this process The Infinite is split into separate, finite parts; and an individual mind is locked, by physical conception and birth, into the tight locality of an incarnation; from the body focus of attention is predominantly outward and, at its *nadir*, can fall sub-consciously asleep. Sleep is a falling out of mind, a temporary material experience, an inkling of the non-subjective state of things. Matter is beyond this 'little death'; at the outer region of creation there is cast a desert of perpetual oblivion. This cosmos is an oceanic vale of death sparkling rarely with a star of life. The desert, a black hole in the light of animation, surrounds and, as your body, is you.

Can animation and oblivion separate? Is the marriage 'til death does them part? Can the elements of physic and of metaphysic be sundered by an act of death? This would mean you had it both ways but there's only one you'd know. Which 'you', mind or body, might survive catastrophe? Incinerated or interred and entered, incorporeal, into life beyond this world, could dying be one world's end and at once another's start?

D-Day

Are you ready for your own D-Day? Is it not the world's way bodies turn to dust? So are we, you and I, prepared for our respective D-Days? All-change at the end of our 'normality'? Why did the hermit contemplate a skull? What about the other end of birth-day, death-day, and its hour of passing?

What is non-conscious matter but oblivion's tomb? Thus, locked in a biological construction, soul becomes immersed awhile in lifeless, physical phenomena. It becomes an actor on the world's stage but if immaterial mind and energy of body are two different qualities then they can part. *What meshes can detach. When the pot cracks life may, as from tomb, fly. When the psychosomatic connection is broken you part with your pretty (or not so pretty) arrangement of chemicals; you part with oblivion and the incarnate experience.* The Bible, Koran, Kabbalah, Zend Avesta, Mahabharata, Ramayan, Granth Sahib, Buddhist Sutras and all other such human inspirations have distinguished, in sophisticated or simple language, between the sacred and profane, between soul and flesh, life and death. *Indeed, the whole point of mystic practice (and therefore tradition) is to experience and thereby understand the ways of dying and of death.* Natural Dialectic would suggest they are agreed on the Essence of Consciousness and its existential dualism with matter. If scientific materialism, having deliberately excluded all but profanity from its considerations, wishes to assign the central tenet and, therewith, all else of mankind's collective 'ignorance' to the dustbin, it should say it outright. It will do so without either physical or metaphysical evidence; and without having had the grace to consider the philosophical context, to balance the psychological, moral and consequent physical implications or even to complete the experimentation involved - holistic practice of the mystic path. How objective, how experimentally scientific were that? Or even irresponsible!

In-swing reverses the ebb tide and conditions of out-swing. Its incoming 'rise' represents conditions of either partial or full excarnation. These are *waking up* (a resurrection from sleep), thought (rising beyond instinctive response) and death. The strong inward focus of contemplation is itself conceived of as a partial detachment from body; and death a full detachment, a sloughing of the coil, what you might idealise as a phoenix flying from the ash. The physical materials of an outer sheath are shed and, incoherent, are recycled ecologically; but, equally, the inner sheath persists. Mind is recycled in a different but coherent sort of way. *If the objective, physical side goes but the subjective, metaphysical side remains then where, death, art thy sting?*

Psychological recycling? This, if we consider mind and its attention metaphysical, re-begs the hackneyed question of psychosomatic interface and the nature of the conjunction (at out-swing or birth) or disjunction (at in-swing or death) of body and mind. *How does subjective recycling work?*

Life Swings

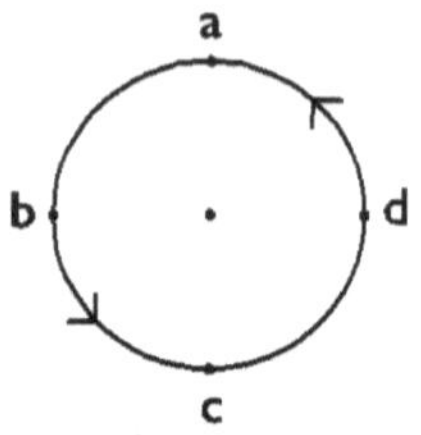

a wheel rolled through time traces
a period, wave or vibration

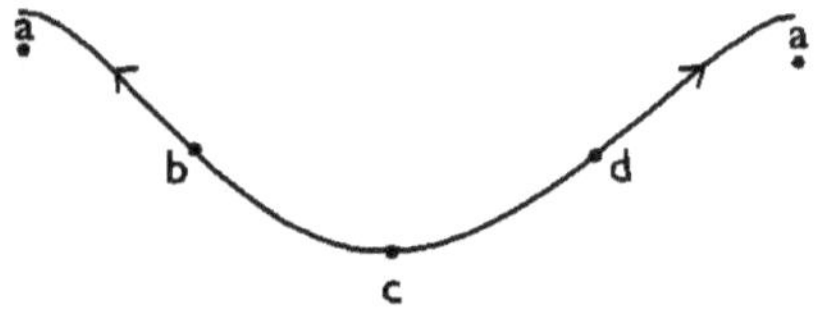

Check the following 'Information Cycle' against figure 2.14.

Information Cycle:

Wheel of Life

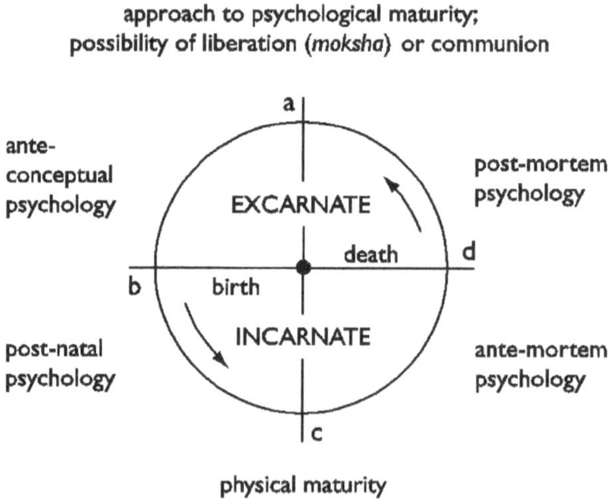

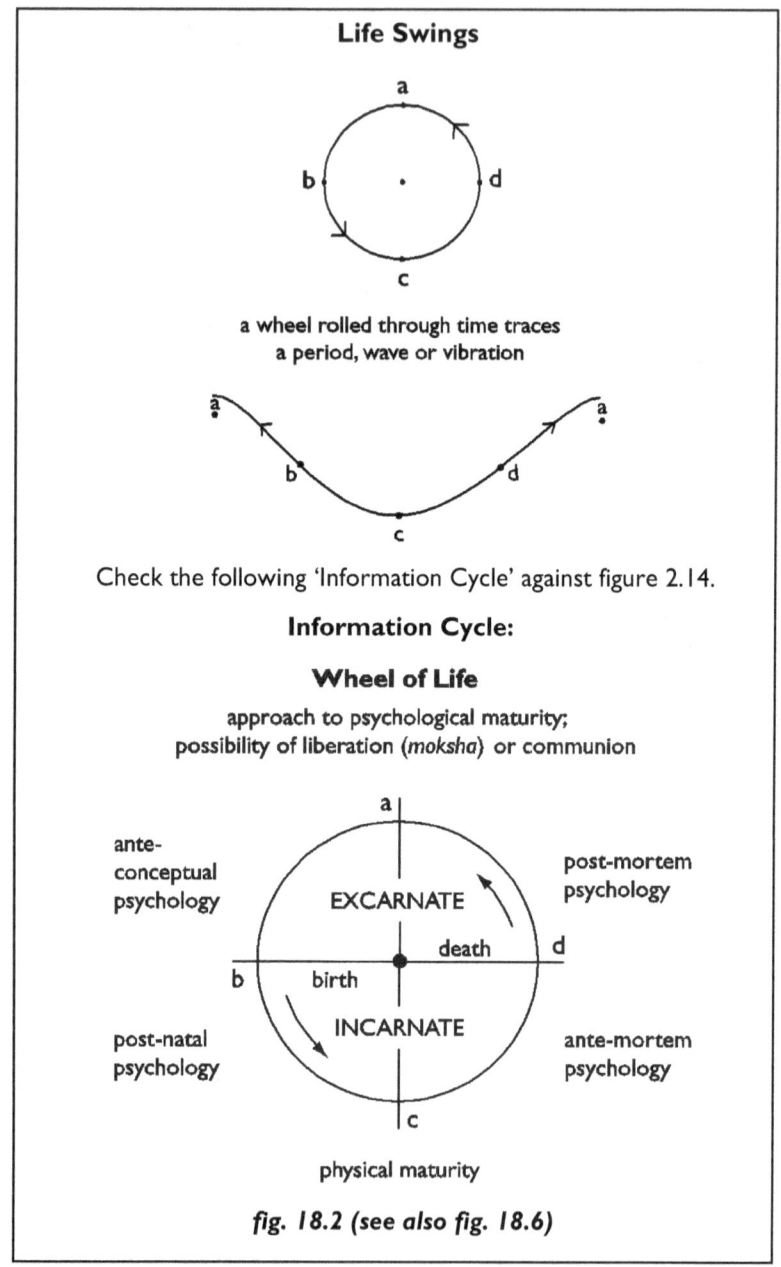

fig. 18.2 (see also fig. 18.6)

It is necessary, in order to understand the human life cycle, to consider both incarnate and excarnate phases of its pendulous swing.

Do you remember (Chapter 5) the Cyclical Order of Information? It dealt with nervous loops, contemplative loops and the major cosmic loop that 'lassoes' Enlightenment. If there exists a major cosmic loop of subjectivity then there exist within it lesser wheels; there exist all kinds of cogs and cogitations of the mind. And if mind is, in all its individual works, an information centre there exists informative recycling. Each separate, conscious unit of the 'substance', mind is always changing. Yours, mine and so on cycle through locations, times and their experiences (see also *fig.* 18.6) until, if ever purified and swung to zenith, they communicate with Source; they commune as one with Singularity. Recycling nature's immaterial component is a process called reincarnation.

The four phases of this cycle are, from a subjective point of view, transitions called birth (incarnation), (incarnate) biological life, dying (excarnation) and (excarnate) metaphysical life. To describe any cycle (whether female menstrual, cell cycle etc.) you have to decide where it begins. *Here the logical, top-down starting-point ('a' in fig. 18.2) is in-swing at its zenith, the moment of equilibrium before out-swing commences.* This moment occurs, however, in an excarnate condition.

As western psychology operates within a framework of materialism it denies the existence of such a state. Since mind itself is supposed, *bottom-up*, to evolve from neural patterns it is natural to assume that contemplative experience, inner vision, 'expanded consciousness' or any excarnate form of psychology is the result of chemical disturbance. Such back-to-front theory has, as noted in Chapter 13, bundled together *all* conditions of awareness (except 'normal' waking consciousness and dream) as states of malfunction. In other words, only 'ordinary' incarnate waking consciousness, dream and unconscious sleep were healthy 'electrochemical illusions'. A partial understanding of the nature of mind has, therefore, unfortunately pathologised both higher and excarnate states of mind. It has tried to reduce their cause to neural biochemistry, diagnose them as a form of ignorance or illness and so suppress or re-educate belief in them. Moral, spiritual and excarnate aspects of the human condition have been downgraded or dismissed. Indeed how, from an objective, clinical perspective could the unfamiliar study of post-mortem as well as ante-conceptual psychology be born? Anecdotal evidence is unacceptable so that, *bottom-up*, the starting-point is either fertilisation (the moment of conception) or birth. And the simple end is death. For a materialist death is a headache that doesn't even warrant aspirin; yet it marks a scientific black hole.

Top-down, excarnate psychology will be dealt with in the next three sections. Nor is it the thrust of this chapter to rehash the conventional psychology of infancy, child development and adult mentation in either healthy or diseased forms. It is noted that the human life cycle resembles a spring at first uncoiling very rapidly; the thrust of this first acceleration loses momentum until its full extension is reached, whence it recoils and eventually loses all vibrance. After fertilisation a zygote 'uncoils' its potential at so great a rate that after only a few weeks the full human form is discernible; after birth growth continues at a slower pace until sexual maturity after which a slow but sometimes, towards the end, increasingly rapid 'loss of bounce' sets in. Similarly, infancy is a period of phenomenal psychological development followed by a voracious youthful appetite for information and the knowledge of experience; the joy of such an appetite may, in a more measured, directed and less excitable form, remain life-long.

The gloss Natural Dialectic adds is that during the out-swing of youth, when life seems eternal and the body immortal, the zest for information and achievement is focused on technical skills, an acquisition of factual knowledge and the business of worldly affairs. Ambitions, sexual maturity, marriage, family and consequent financial responsibilities rule the mind. A common materialistic response to the consequent wane of age is an attempt to arrest it, recreate youth and cram in as much menopausal 'pleasure' as possible before you lose your puff, the whistle blows, the game is up and there is nothing left. If, on the other hand, mind were metaphysical, it could outlast the breakdown of brain, physical connectors (such as eyes and ears) and body as a whole. Body is not, in this view, seen as a total identity but a temporary address. *Dialectical vibration hints at in-swing and a possibility that the approaching mortality of skull and grave are, like broken shells or cast-off casing, an inconsequential fraction of the whole story.* If so, it would be rational to adopt a traditional, *top-down* agendum. From this perspective an older, wiser person turns increasingly from affairs of the world, actively prepares for death, treats it as a release from the physical 'roller-coaster' and, as for a voyage into a new land, works to ensure that the journey successfully achieves its goal - a healthy, incorporeal afterlife. What does this consist of?

<u>The Logic of Disembodiment</u>

It is clear that, contrary to the previous chapter's logic of incarnation, the top-down logic of disembodiment is dominated by the upward, right-hand path of Natural Dialectic.

	tam/ raj	*Sat*
	finite/ restricted	*Infinite*
	division	*Communion*
	existential episodes	*Life*
	body/ mind	*Soul*
	lesser attractions	*Attractor*
	surrounding rings	*Source*
	oscillation/ revolution	*Peace/ Stability*
↓	*tam*	*raj* ↑
	negative	*positive*
	passive/ involuntary	*voluntary/ active*
	out-swing/ descent	*in-swing/ ascent*
	more objective	*more subjective*
	exodus	*return*
	physical	*metaphysical*
	diversification	*unification*
	creation	*dissolution*
	incarnation/ embodiment	*disembodiment/ excarnation*
	isolated	*relating*
	apart	*part*
	bulk/ bonded	*radiant/ wireless*
	body	*mind*

The Logic of Disembodiment

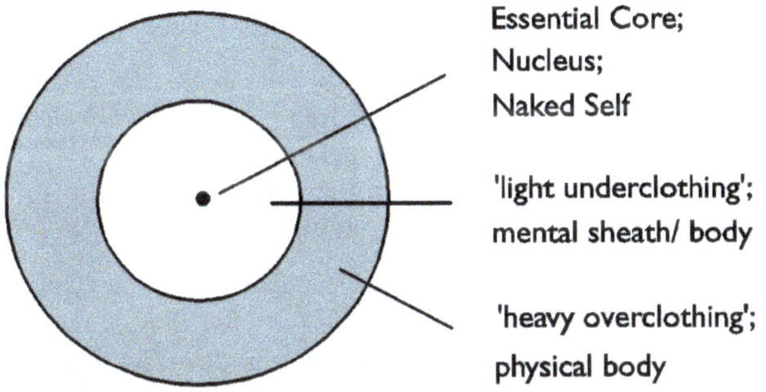

Death and Enlightenment: Excarnate Cycle.

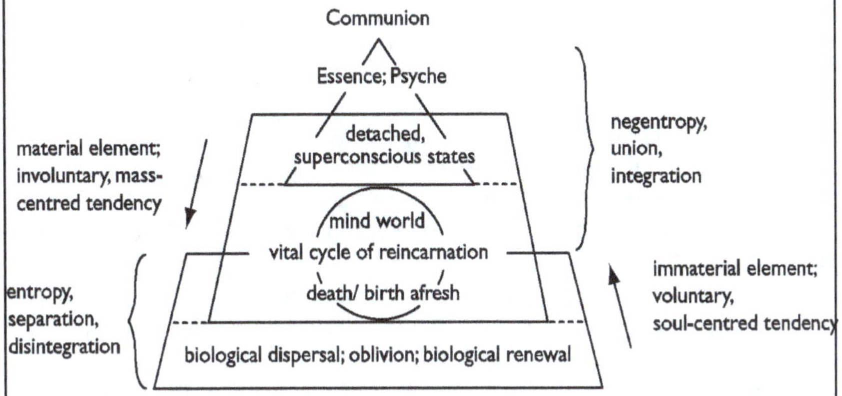

Death cracks the pot. Dust to dust the outer, lower case disintegrates. *Top-down*, however, death of physic is but birth into the mind-world. The 'great sleep' may be physically unaware but psychologically experience persists. Animation's factor tends, according to its purity, to rise. You *are* your mind. If mind is clear Communion is guaranteed. If not, what rises falls. Gravity propels the bundle of accounts (mind weighed in the balance of past actions and desires) towards incarnation. Sooner or later the excarnate cycle leads to rebirth in another body, one commensurate with the incoming type of mind. New clothes for old - death of body is the final cut but mind will live to fight its next embodied day.

What is a 'karmic cycle'? In 'Buddhist-speak' the Axis, Axle or Essential Pivot of the cosmic wheel (see also *figs.* 18.1 and 18.6) is Communion, *Nirvana* or The Ultimate Experience of Unity; and, equally and personally, the centre of our microcosmic wheel of life is known as Nuclear Psyche or *Nirvana* too. Round this the world, revolves; round Supremity of Being circulates existence.

fig. 18.3

Think of The Infinite as a Clear Broadcast everywhere. Think of existence, made of finite objects and events, as 'within it'. Think of psychological existence, mind, as 'underclothing' - context in which The Naked Self, Pure Consciousness, resides. And think of body as non-conscious 'overclothing', an outer third sheath inside which the other two are nested; body is your suit of physical existence.

Now think of the latter marvel as communicant. See it as an aerial that resonates with mental broadcasts of both personal and archetypal kind. Memory and awareness play their parts; and the 'receiver' and 'transmitter', brain, modulates experience of, for you, a human kind.

Each and every organism ticks towards its bomb. What happens when the crock of body breaks? Are corpses holy? Of course they're not. They are the same as when 'alive' - a chemical complexity that death disintegrates. Think of a carcass as a broken aerial, deficient antenna or a membrane of inoperant metabolism that earth pulverises, damp dissolves or fire swallows up. No last trump or bony resurrection there.

The set is 'dead' but has the broadcast died? The disembodied context called your mind still clothes your Naked Self. It still limits and defines Pure Consciousness. And now the drivers are not physic's energetic forces but subjective motivators - metaphysic's will-power, purpose and desire. You will seek to work your real, unveiled emotions, lusts and aspirations out. Mind motors through 'internal' landscapes of devotions, longings and intent. What is the quality of your intent? Do you (↑) yearn for Life or, unable to detach from your contextual web of love-and-hate relationships, want to (↓) obtain such body as continues them?

For each action there's completion, for each cycle end and for life death. Do you, therefore, yearn for work-out, action and incorporation in a worldly form of death - an atomic body that will fall apart? If so, the wheels of desire will run you there; your line of thought will track you down a birth and coat you biologically in cells of life. So spins the wheel of life and death, that seems so real but is not Real, round and around.

Thus, after disembodiment, the logic charges you with either upward disentanglement from mind or re-embodiment in physical impediment. You may wish to throw encumbrance off or, in continuation of your drama, take a new suit on. Is mind not always changing and the scenery of bodies too? *Status quo* is not an option in the theatre of the existential field.

Towards a Unified Theory of Life and Death

You say there is only matter and therefore after-life and reincarnation are impossible. Science has progressed beyond such superstition.

That is nonsense. How can material science 'progress beyond' the immaterial? Scientism simply ridicules its mention; but if immateriality is real what does ridiculing or denying it avail? Denial is a guess but not scientific one.

I say there are two fundamental components - matter and consciousness and therefore at least disembodied consciousness and probably also, according to dialectical logic, cycles of reincarnation. Such cycles of mortality, both physical and metaphysical in series, would be governed in metaphysical part by psychological imperatives; and, when constrained in the 'overclothing' of biological form, by psychological *and* physical imperatives. Both states would involve experience rather than oblivion, motivation as their motion and the forces of mental drama. They would involve, as their subjective element, intentions, associations, memories and states of mind; also, in the incarnate element alone, involuntary interactions and the kind of changes found to make up physical events.

Science, a study of what is mathematically or physically explicable, works from the best of motives and a high principle - a determination to find the straight, unadulterated truth. Its renaissance mirrors the healthy development of a strong-willed, sceptical individual. An adolescent rebels. He wants to slough old-fashioned parental creeds and, empirically, determine for himself what is right and wrong. He wants to learn by experience or, in this case, physical experimentation. He wants to test and push the limits. Over four hundred years of analysis, theory and testing have led to brilliant technology and a synthetic world-view that embraces the whole of physical time and space. *In its relative maturity science has now reached a couple of interesting limits - mind and death.*

	tam/ raj	*Sat*
	existence	*Essence*
	peripheral shells	*Central Nucleus*
	lesser consciousness	*Consciousness*
	lifetime	*Life*
↓	*tam*	*raj* ↑
	passive	*active*
	down	*up*
	descent	*ascent*
	body	*mind*
	less conscious	*more conscious*
	less keen	*keener*
	asleep	*aware*
	absence	*presence*
	objective	*subjective*
	involuntary	*voluntary*
	death	*life*

Death is a closed box; it is a critical unknown, the last horizon of ignorance. What is the nature of its inevitable 'black hole'? There are two views:

(1) the *objective* which looks, from the outside, at another's

(2) the *subjective* which, crossing the event horizon, experiences death for itself

Bottom-up only the former is possible. Objective science can't include subjective views of conscious life or death. There is no logic of disembodiment. Psychology post-mortem never starts because death is the end; death is the death of what never lived - a body of atomic aggregate. After all, life *is* its body; this body is made of non-life, that is, of atoms; if non-life *is* life and non-life is also death then life is a species of death. What is it but a fragile, ephemeral and illusory suspension over voids of non-conscious non-life? If this suspension is somehow 'subjective' then death's relapse would drain it into endless 'sleep'. Objectively, a corpse is disposed of by cremation, burial or natural means alone. Ashes to ashes, death to death - if incarnate life is just an evanescent flickering associated with cell chemistry then what on earth could your excarnate being be - except a flickering imagination that you have while still alive? The notion is absurd; 'life-after-death' is an anathema; and this chapter is (like any metaphysic) blank.

Faith is that mind is somehow a reflection *caused* by a molecular arrangement. **Death of body is annihilation - that's the end of it!** Some might tag such atheistic opt-out as 'the soft and easy option'. No imminent post-mortem judgment on behaviour, no code of inexorable morality, no sense of answerable responsibility to an authority that's higher than your local rule of law - if it can catch you. After all, you argue, everyone's eventual reward is the identical impunity of everlasting 'sleep'. No wonder materialistic Karl Marx castigated a non-materialistic church for offering the pretence of compensation after this embodied life has ceased. You don't prepare for nothing. Therefore instead of such absurdity he thought, according to materialism's rationale, that 'heaven' (in the form of communist utopia) should be prepared for here on earth. Prevalent philosophy 'confirms' the view that death is not to do with life. Self's illusion vanishes and after death you are disintegrated into nothing but molecular recycling. *If materialism (such as yours?) won't hold that there's an immaterial element then certainly the holist's preparation for a brainless future seems, well, brainless. Isn't it absurd?*

For a materialist, whose body makes his mind up, the sequence of 'there's-nothing-in-it' argument is natural and logical. Brainless equals mindless and it is therefore senseless to believe that mind persists in 'somewhere-land' beyond a cranial catastrophe. Mind construed as a product of brain is a mysterious but, to its owner, obvious 'emergent property' derived from patterns of electrical charge circulating in chemicals. Consciousness is an evolved product of biochemistry, emotions spring from hormone concentrations, behaviour from genes, 'self' is a construct of language and subjectivity an illusion. Modern psychology and psychiatry have almost squeezed the life from it. Have they not disproved the wisp of consciousness as a separate entity and, conversely, proved that life is really matter? Some hope that in the future theory will grasp the spirit in its fist, crush it - if you can crush the immaterial - and, certainly, find nothing physical. In which case, if life is an electronic film then death must be the end of it. A dead body means the absence of its mind; it means absence of life. End of materialism's story. Fact. This *must*, from an exclusively 'type-1' point of view, be the case. An end of illusion is not, in this view, the revelation of Truth. Truth is oblivion. It resides in the oblivion of non-conscious matter. A subjective void. Inanimate nature is an empty truth because it has no metaphysic, subjectivity or life. Not life but lack of it, death, is its Truth. Chemicals are rearranged but the grand delusion of 'a self' is dead. Then long live death! This is a world where life is certainly, at least in part, entombed in biochemical phenomena. The psychological part, which may include the hallucinating twitches of a dying brain, should in the end analysis consist of no more than its own disintegration - an obliteration and oblivion.

Is this not a half-truth masquerading as a whole one? **Can nothing inaccessible to scientific instrumentation, research and the logic of mathematics exist? If you believe in a subjective element that's immaterial consciousness then death becomes a door. Without physical sensation mind's environment will change. Naturalism may deny this kind of fatal transformation but, if it's real, to enter such a portal unprepared were folly. Do services that deal with life's emergencies not drill? Is to prepare for sudden, overwhelming change absurd?**

Top-down, to assert that mind can never leave or dwell apart from body is no more than simple posturing by materialistic creed; faith is that it's somehow caused by a molecular arrangement. True, there occurs disintegration, dispersal or disposal of a body. Mind is not, however, construed as the by- product of grey matter. The view has always been that mind is a separate component of the universe from matter. So that to seek a physical explanation of mind is as demented as physically and chemically analysing a piano in order to discover its hidden melodies. *In this view the brain is itself a product of an archetypal score. It is no more a producer of mind than a TV set of its programs.* Instead the nervous system is construed as a receiver/ transmitter complex that allows mind to interact with body and thereby the physical world. From this perspective the breakage of a TV set is not the end-all of a program. That is an illusion. While a physical broadcast remains 'on air', even radiates through our bodies and can be picked up by any intact set, metaphysical mind loses contact with its broken attachment and, therefore, the world. Thousands of mystics throughout recorded history have likewise unanimously stated that death is an illusion. A stage in their practice allows voluntary death, at will. That is to say, a mystic may expertly vacate the body in the same manner as happens clumsily at the involuntary circumstance of death. They are familiar with the process and claim to negotiate the mind-world as skilfully as we negotiate the natural events of our day. If to experience is to know, then they know the nature of death. It is, therefore, no surprise to note their textbook familiarity with type-2 descriptions reported by perhaps millions of people worldwide to whom *OBE*s (out-of-the-body experiences) and *NDE*s (near-death experiences) are unexpected, somewhat incoherent surprises.

This is the rub. Science depends on physical proof and it is impossible to physically prove to anyone that, for example, you have had the dream you say you have - even though it was entirely clear to you. Indeed, they will only believe you dreamt because they do too. Why should you lie? So they *infer* you dream the same. Does this make your dream, so real to you, unreal because a test tube can't contain its content? Similarly your thoughts, feelings and the whole gamut of your mental dynamic are - what - fantasies? These, if you think about it, *are* you. You are mind first, body trailing. Are you being told that body matters, mind doesn't? And, turning your subjective truth upon its head, that it's an illusion?

Science is an excellent seeker after repetitive, predictable, measurable facts. But even if you enumerate the electrical activity of a telephone line or wireless set, the signals *per se* have no purpose or meaning. They are simply code-conveying agents of thought. By measuring brain waves you can't know the metaphysical, subjective component of a person's consciousness; indeed, does consciousness even need neural activity to support it? *Until you obtain an instrument that can register the subjective component of thought as an object, whether embodied or disembodied, you will have neither objective evidence against post-mortem survival nor other than circumstantial or trust-based, anecdotal evidence in favour.* As well as lack of register, how experimentally predictable or capable of numerical definition is something as common yet intangible as a thought let alone an *OBE* or *NDE*?

It is not as if either experience is rare or the province of mental disorder. An *OBE* is nothing like a dream; the experient is wide awake and, where there was none before, no symptom of a mental illness follows after. No change in sanity. It has been calculated (Times 'World at the Millenium' supplement, Week 5, p.26) that perhaps thirteen million persons living in the USA at this time alone have had 'sneak previews' of what happens when the 'damper' of their body is removed. That equates to more than 5% of the world population; some estimate still more. When all sensation of the 'clamper' is removed, when the 'drag' of body is released then activity of mind is greatly accelerated; there is a sense of understanding clearly, time radically changes and the intensity of a 'heaven' or 'hell' can, according to the mind-set ('wavelength' or 'attunement') of the individual, be overwhelming. Intensely, overwhelmingly real. *It is not unreal to the insider. It is very real to the participant.* At death the black box, mind, is thrown out. A lifetime's flight recorder is ejected. *Even a conviction, born of some desperate circumstance, that death is imminent and inevitable can provoke such an ejection. No physical injury is required.*

While mind-brain identity is crucial for the sustenance of atheistic materialism, the separate identity of immaterial, informative mind and material brain is basic to all variations of holism. No doubt life-transforming revelation can't be scientifically verified; it's anecdotal as a dream or any personal experience. But if a single *OBE*, *NDE* or other 'paranormal' event were ever confirmed beyond doubt, it would demonstrate brain is not mind, mind not matter. *Therefore materialism would fall and with it the part of any grand synthetic world-view that involves materialistic interpretations of origins, mind, human purposes and capabilities, death and the nature of life and consciousness. A radical re-appraisal would become necessary.* **No less than a transforming shift of paradigm would have to overtake scientific materialism.**

If, in other words, the mystic interpretation is vindicated, death becomes a change of environment, disembodied life a fact and the rest of its message about an intelligence-based creation might well, by inference, be true. If those who claim this particular revelation of Truth are right, materialism is turned upside-down. Pure Consciousness, Life and lack of death are the Ultimate, Absolute Truth. No *fact* that science has discovered is changed; simply, having excellently mapped the exterior, it will have to realise that there is also an interior, metaphysical dimension to life. It will have to take full account of complementary, subjective science. **Any grand synthetic world-view will henceforward re-interpret the facts to include both the further dimension and metaphysical science in its new paradigm. This is what this book has, from the start, been doing.**

This is a much harder rub. *PAM* believers therefore always clamp down hard on any chance 'post-naturalism' might have of seeming true. It's not, of course. Naturalists explain away; and they dismiss, by dint of *PAM*, any 'super-natural seeming' as illusion. In this case what kind of illusion is an *OBE* or *NDE*? Is there life after life or not? In what way could this crucial issue be decided? Whose authority might sway agnostics? It's all in the mind, nobody disagrees; but is there, in between the two opposing camps, a happy medium?

No mediums, please! No doubt, the feeling of 'location' in one's body is a grounding mechanism of mind/ body mesh, that is, your incarnation. What, however, does that mean if both are made of matter and mind is itself some species of material illusion? If the neurological basis of 'connection' is weakened or destroyed you might expect that separation could occur - except that such 'umbilical' connection of the mind with its host mother, body, is - you claim - just another trick of mind. Virtual reality! Have you ever rocketed to space and back while riding in a simulator? For *OBE*s the nervous working is unclear but experiments have shown that, using special goggles, cameras and screens, you can simulate the weird experience! Science correspondents therefore clamour that this more than just explains away, it proves all 'super-natural experience' is a void illusion - except that in a 'real-life' *OBE* there are no instruments of simulation to apply! Is simulation really actuality?

If you have anticipated and thoroughly, correctly prepared for a life-threatening situation you can, like the expert stuntman in a movie scene, cope with it in a calm and measured way. If not things could rapidly unravel. But their experience is not scientific so forget what silly saints and mystics say. Might not a trip to the local cardiac arrest unit better serve research and, in the face of death, its truth? Involuntary *NDE*s and *OBE*s are often associated with trauma, severe shock, terminal weakness, drug overdose, anaesthesia or illness. They can happen anywhere but at least if the patient is in a hospital *NDE*s (which can also involve an *OBE*) are closely monitored by clinical staff with no professional interest in scoring philosophical points. The ambience is about as objective, methodical and scientific as you could find. While there is no test for an *OBE*, clinical death is recorded when every measurable output has stopped. Not only heartbeat but also brainwaves register a flat line; both are gone. There is no neuronal activity and so, from a materialistic point of view, no consciousness. Metabolism may continue for a while but death is pronounced. Life has ceased.

The issue is simple and clear cut. Either there is life after death or not. Either you are going to live after you lose contact with your body and, therefore, the physical world; or there is nothing more.

For a materialist there is, we've noted, nothing more. Death is, therefore, unimportant and, if it ends accustomed pleasures, an unwelcome guest. Neuroscience calculates that about eight seconds after breathing and heartbeat stop no part of the brain continues to function. No function, no consciousness, no experience. If mind is an illusion supported by the brain then *OBE*s and *NDE*s must be hallucinations created by brain biochemistry as it malfunctions or shuts down. They are blips that pip mortality. In this case either *NDE*s occur in the first few seconds of 'death' or in those immediately after resuscitation before the patient 'comes round'; or else, if they occur when brain is 'flat' and motionless nerves cannot physically generate a thought, mind and brain are not the same. The last is, according to materio-philosophical presumption, impossible. It is therefore vital to correlate the exact moment or period when an *NDE/ OBE* occurred with brain-wave patterns. This is as easy as locating the specific instant of a scene by watching brainwaves as they register its dream. Neither brainwave nor machine can enter into thought or dream.

What occurs? Perhaps the commonest clinically related form of *OBE* is when an anaesthetised patient 'pops' out of their body, 'floats up' and watches the operation on their outstretched body from above. Severe road accidents can also 'pop' a ghost-body in which an unaccustomed mind may take time to realise that, although metaphysical *indriyas* (signal translators) can sense without the physical agents of sense, others are unaware of its 'body'. Indeed, the excarnate person can pass right through these other persons and through solid objects like walls. Is such a body what we call a ghost? Is a ghost therefore as natural as the flesh, being just a lowly mental sheath? Is it the form of an unseen connector? Is it a conjoining medium inside the 'clothing' of biology but worn outside the 'psycho-space' of conscious mind?

A case of *NDE/ OBE* was encountered in the USA by Dr. Robert Spetzler, a surgeon, and corroborated by Dr. Michael Sabon. Pam Reynolds was undergoing surgery for a large and therefore dangerous aneurism located deep in her brain at the top of its stem. After the operation was over she reported having watched it from above the surgeon's shoulder. She accurately described events that had occurred, things that were said and technical instruments that were used while her body was flat out on the operating table. There was no way she could have known the things she related. Her description of one surgical instrument was such that the corroborating doctor did not recognise it and had to send away for further information. The response confirmed its presence in the theatre. **Such an event appears to categorically affirm that mind can separate from body; and thus subjective consciousness is, as dialectically proposed, an entirely separate element from matter. At a stroke it validates the super-positional hierarchy of Mount Universe.**

The subject of an *OBE* can also leave the 'real' world and become absorbed in such visions as have been reported universally. As in the case of any explorer and his new-found land, the visionary's interpretation is coloured by his education, culture and descriptive powers. Nevertheless a few distinct types of experience are commonly mapped. The catalogue includes a tunnel at the end of which is 'living, loving and compassionate light', a border, barrier or point of no return, the presence of loved ones, spiritual guides and holy persons, paradisiacal scenery, exchanges exclusively by means of telepathic or 'direct' communication (what body is now interposed?), a 'summary review' of the life just lived, a judgment of the conscience etc.

Does, it was asked, an aerial, electron gun or screen make TV programs? According to the laws of physics mind's impossible. Therefore materialism, looking downside up, believes that electrical networks, hormones and other factors *cause* our screen of thought; that non-conscious molecules and electrical currents can produce a persistent illusion called consciousness; and biochemical imbalances or drugs do not so much interfere with reception as themselves *create* hallucinations. In which faith the extraordinary episodes of *OBE* and *NDE* are construed as the result of systemic malfunction, a collapse of the nervous system and the immanence of death. As such they are as impossibly irrelevant as any subjective experience. Perhaps their fantastic nature is due to lack of oxygen, changes in neurotransmission or a rush of 'anti-shock' proteins; or, again, agents of anaesthesia, euphoria, incoherence or suggestion. They must sum only to the splinters of a breakdown, the residual fall-out of a life just prior to death. In no way does a vivid dream add up to life beyond it.

Such is materialism's scientific explanation but it may not capture the whole truth. *You could see the process in another way.* If its aerial is broken, out of tune or any of its components are malfunctioning, what sort of program could a radio transmit? The human brain-set is constructed to operate in a dynamic way and perceive the world as it normally, healthily does. However, drugs interfere; so does any other sufficient biochemical imbalance, illness or injury. And any breakdown of the brain affects its tuning and, therefore, it picks a 'crackling program' full of interference up. On the other hand, if death were as programmed as biological life certainly is, you might expect an orderly and intended sequence of 'gear-changes'. Its so-called hallucinations might be part of a suite of psychological events that are as natural and necessary a part of creation as biological operations. In an *NDE* such gears may engage only partially or not at all. After all, the circumstance is neither here nor there. It is fragmented and incomplete. And if it runs to its completion there is no way back to tell - except in the voluntary case of mystic practice. *Implicit in almost every funeral rite, however, is the assumption that death is not an end but a transition. A change of state. Therefore is not the greatest lesson in life to learn how to die properly, to use death's window of opportunity successfully*? To learn, in other words, how to avoid rebirth into the shocks and pains of physical embodiment; and to avoid expulsion, through birth or death, into hells. To learn instead how to attain enlightenment, to commune with the Infinite Heart of Nature, The Father? Books of the Dead accompany Books of Life. Guides and travelogues on the art of dying and excarnate adventures have always been popular. Well-known eschatologies include the European 'Ars Moriendi' and Egyptian, Mayan and Tibetan Books of the Dead. Have you read Dante's 'Inferno' or Milton's 'Paradise Lost'? And the Bible, Koran, Zend Avesta, Guru Granth Sahib, Bhagwad Gita and Buddhist texts show, among many more, how to live in order to die well. This is the theme. Getting a grip on Reality. Success at D-time! Successful victory over death!

The abovementioned is, while science remains exclusively materialistic, scientific nonsense. However it makes some sense to a lucid dreamer and perfect sense to a mind (or person) who has experienced even an involuntary, relatively clumsy *OBE* or *NDE*. The latter are real, radical and often life-changing events. This interpretation of the business is confirmed and extended by mystics whose voluntary, practised excarnation is performed as a gymnast performs a routine. *Indeed, they say that only once you have died does your perspective on life shift profoundly*. It is turned upside down or, rather, right side up. Why should these revolutionaries be believed?

The problem reasserts itself. If all brain functions have shut down and yet a person (whether patient, practitioner or anyone else) can still receive information then mind must act independently of brain. *The implication is that consciousness is immaterial; and if the dialectical premise is right and the scientific wrong then perhaps the consequent dialectical logic, as expounded in this book, is right as well.* No wonder that neurologists splutter and, with his materialism disappearing down the pan, an atheist vituperates.

In the last analysis the whole issue of thought is one of inference, faith, trust in witnesses and circumstantial evidence. I can understand what I have experienced. We infer in others what is 'normal' for ourselves. It is more difficult if I lack experience. Who understands a drunken, drug-induced or, indeed, a normal condition of mind unless they have felt it through? Intellectual, objective theories cannot understand in the way that a shared experience is a shared and certain bond. How many people have experienced *OBE*s or *NDE*s? How can they convey their understanding of that foreign field? Nor has science any instrument except its brain or, rather, mind - its own illusion. *It cannot share but why in consequence should it deny?*

An interested student would need to sift anecdotal accounts of subjective *NDE*s and *OBE*s and objective 'ghosts'. Many such anecdotes involve strong supporting circumstantial evidence. In the end, however, a magician could simulate the circumstance. He could deceive you into thinking his illusion was reality. He could exorcise the ghost of death - although nothing could prove what really happens except a carefully monitored physical resurrection or the actual metaphysical experience. Science cannot, by naturalistic definition, prove or disprove what is immaterial. That position calls for care but not denial. Until a 'sneak preview' or death itself is realised the issue must remain a matter of faith. In this case let us turn, bearing in mind the cultural restrictions imposed on metaphor, to see the business of death briefly described in 'type-2' terms by the Tibetan Book of the Dead.

<u>Post-mortem Psychology</u>

	tam/ raj	*Sat*	
	expression	*Potential*	
	relativity	*Absolution*	
	relative freedom	*Freedom*	
	oscillation	*Axis*	
	shells	*Nucleus*	
	lesser consciousness	*Consciousness*	
	lifetime	*Life*	
↓	*tam*	*raj*	↑
	descent/ down	*up/ ascent*	
	externalise	*internalise*	
	away from waking/ towards sleep	*away from ignorance/ towards awakening*	
	inc. physical aesthesia	*dec.physical aesthesia*	
	darker/ heavier	*lighter*	
	more objectively involved	*more subjectively involved*	
	enmeshed in body-world	*detached from body-world*	
	confinement	*release/ liberation*	
	gravity/ inertia	*levity/ kinesis*	

materialisation	*dematerialisation*
embodiment	*disembodiment*
birth of biological body	*death of biological body*

Physical health is awareness of the world without obtrusion in the form of pain, decay or disability. In this sense body-less, your mind's attention includes elements of cogitation, sensation and emotion. It oscillates (*figs*. 0.1-3) between the outside world and inward contemplation. Of these states the closest to embodiment is sensation and nearest to the mind-world are cogitation and emotion. Thus, even while embodied, which tendency would you say best approximates a state of disembodiment or, no longer equipped to sense physicality, death?

Death is, by 'type-2' subjective token, a release. The question of how a 'soul' enters or leaves a body translates (with the proviso against literalism that applies to all analogy) into how a TV set is switched on or off. When the correct aerial is constructed and the set biologically energised, it will receive the relevant program. *In a physical sense nothing 'enters', 'leaves' or 'goes' anywhere.* Immaterial mind, having quit its connection with the material universe, a connection called its body, is no longer in it. Nothing has flipped except some biological switch. Nothing has changed except a disconnection and a shattering. Membrane voltage disappears, body's currents cease to flow and the atoms of a corpse are re-arranged. The body of the TV set is broken; no sense remains with which to know the earthly plane; once 'unplugged' from its physical medium dematerialised mind is left in its own metaphysical place - the so-called information field. Its program is still broadcast; and, as filings round a magnet, its coherence round the soul remains. In this sense an absolute finality of death is, as all the manuals of 'truth metaphysical' agree, an illusion born of ignorance. It is a real and critical enough event but naturalism's trick is to persuade an audience that type-1's view is the full rather than just the lesser half of the story. That death is the end of life. Oblivion.

If, however, mind *can* separate from a body to which it has been joined then such junction must be 'temporary'. There must exist incarnate and excarnate conditions; depending on your current condition, you might call the other 'inter-phase' or 'inter-life'. From your current incarnate viewpoint 'inter-life' would therefore mean type-2's experience, 'a disembodied stage'.

What subjective incidents, post-mortem, might occur? They include salient episodes in the journey of a soul through 'inter-life' prior to final release (called Communion, *Nirvana* etc.) or reincarnation. Cyclical reincarnation (also called metempsychosis or the transmigration of souls) has, in this view, replaced the eternal states of 'heaven' and of 'hell'. These are states of mind. Both embodied and disembodied forms are capable of happiness and suffering but the logic of the Dialectic precludes an existential permanence of state. Is not existence and, therefore, the part called mind in flux? Is its basic nature not a changing state, a continual flood of transformations? Thus you might exclude the dogma of damnation that is unrelieved; no doubt that hell is a real psychological condition but everlasting fire and sulphur is a metaphor solidified. Medieval images of a perpetual roasting on the devil's spit are just as 'cardboard' as cartoons of fluttering, harping angels in a host that never fall from grace by straying off cloud nine.

As individual lives may vary greatly but as regards humanity stay generally the same, so common episodes of 'inter-life' are variously replayed in individual 'inter-lives'. And just as mode of perception and level of education differ, so a key distinction is drawn between the experiences of more and less spiritually advanced souls. Taking this into account you might sketch the nature of an 'inter-life' as well as its corresponding 'body-life'. Could we take a well-known text? What does the Tibetan Bardo Thodol say?

Bear in mind that this book was written down well over one thousand years ago by a monk, Padmasambhava, for an unscientific culture that used imagery and had customs different from ours.

A *bardo* is a suspended condition, a gap between critical boundaries. It means a lifetime between discontinuities called, in scientific terms, phase transitions. There are four of these in a life cycle (see *figs.* 18.2 and 18.5). If, in accordance with the turn of post-mortem psychology, we begin at point of death there follows a *bardo* in the voluntary field, luminous state or mind-world. After this, if the cycle has not been broken, sooner or later a phase of descent towards the *karmic bardo* of birth in the physical universe will start to pull. This *ante-natal bardo* is metaphysical. *It has nothing to do with the involuntary, automatic development of the specific biological body that will act as a physical receiver/transmitter associated with the mind in question.* You can identify the end of this *bardo* as the point when disembodied mind begins to resonate with and personalise its new 'home'. Bodies are like antennae tuned to specific, appropriate wavebands. The question of frequency, wavelength or, as we say, quality of mind is therefore the critical factor which determines the kind of body, parent and parental circumstance to which, by resonance, it attaches. Mind then drops out of consciousness, sleeps and forgets prior to the phase change into its physical adventure. Rather than the point of physical conception or psychosomatic linkage, you can identify this phase transition as the moment of exit into our universe - birth. There follows the psychodrama of the third and fourth *bardo*s we call post-natal and ante-mortem stages of embodiment; that is, development to physical maturity then, from that climactic prime, decline.

Death, like birth, varies from individual to individual. It can happen violently, suddenly, painfully; or peacefully in sleep or consciousness. The Bardo Thodol describes a normal, natural death taken in its own time and due to weakness or old age.

There are two phases of this first *bardo*, dying - external and internal. The process is described in general terms and, as in the case of biological birth and other life-processes, numerous variations on theme can occur. Death follows a theme that varies from person to person.

D-Day has arrived: and within the day its moments. Externally, as the system begins to shut down hearing fails. Sight loses focus, detail blurs and even outline dims. Likewise the other senses recede. The body weakens, the head cannot be held up and it becomes hard even to open or close the eyes. The cheeks have now sunk and pallor closes in. Visions are associated with each stage of withdrawal. As strength ebbs away there is a sense of heaviness, falling or feeling crushed. This is called the vision of earth dissolving in water.

Hence withdrawal ascends through the next, 'watery' stage. Fluids are lost by way of runny nose, dribbling or incontinence. Clamminess sweats and thereby dehydrates. The eyes feel dry; mouth and nose become sticky and clogged. The body may tremble or twitch and there is a feeling of thirst. The vision is of a mirage, being swept away in a river or drowning in an ocean.

As water secedes to fire bodily sensations taper off. Breath cools and the person can no longer take food or drink. At this point the residue of sight and sound becomes confused. Feelings begin to dissolve and the mind becomes hazy, irritable and nervous. It swings between clarity and confusion. The names of even family members cannot be recalled. After mild delirium drowsiness invades. The vision is of haze with wisps of smoke.

As fire subtends to air the heat of the body drains from its extremities towards the heart. It becomes harder to breathe. Intake is rasping, panting, short and laboured; expiry is prolonged. The body slackens and becomes immobile. The eyes roll upward, outside becomes a blur and the intellect becomes bewildered. This is the last contact with the outside. There is only looking back; a lifetime is slipping away like a dream. A vision of red sparks dancing like fireflies above an open fire may be glimpsed. The mind is 'going inside'.

Mind is being systematically withdrawn up the psychosomatic *chakras*. As the air plexus dissolves into the plane of consciousness a familiar world finally, irrevocably disappears. All feeling of body, not so much in the form of anaesthetic numbness as forgetfulness, has passed. The 'breath' of sub-conscious energy now rests in the heart, throat or eye-centre. Hallucinations and *NDE* visions begin. If the life has been negative these may be terrifying, if positive warm and welcoming. There is slight warmth at the heart as pumping stops, blood gathers and after several final expirations breathing suddenly elapses. This is the moment of physical expiry. The person is certified clinically dead.

At this point we can dovetail the process into the way that universal man (*figs.* 17.6 -10 and especially 17.11) is understood. We can click the retreat of death, at psychosomatic level, tightly into how our own psychology, archetypal format (Chapters 16 and 17) and biology (Chapters 19 and 22) are explained. Theirs was an outward, materialising order, this the inward swing. Do you remember that universal man is seen as a projection of the 'chakric' set? How the sub-conscious power-points 'spiralled' down along the spinal axis whence their influence spread into the materialising emergence of physical form? Conversely, if the caducean mnemone unfurled as a dynamic template around which associated nerves, glands and organs develop, at death you might expect a reversal. A winding-up. Previously extended as a dynamic template into the physical domain, it would be shut back. *Bottom-upwards*, there is drawn an orderly recoil. Sensation is withdrawn from the body; and node by node sub-conscious energies (*prana* or *qi*) is re-furled to the top *ajna chakra*, the point of consciousness called cosmological axis, third eye or 'located soul'. This, you remember, is your interface between subjective and objective worlds. Now, from the objective side, a tide is pulling back. It represents ebb from the shore of this world, an in-swing towards the inner ocean, the natural metaphysical. The single collection of nodes, now repacked like a folded telescope, detaches.

Cellular body loses contact with its integrating archetype and the process of disintegration, unless frozen solid, embalmed or otherwise preserved against decay, begins. All sense of body and thus material envelopment has been lost. Some have said the breath of life leaves through the nostrils. This is the physical reflection of an invisible, inward event. A TV set switched off is dead. The current and the programs both 'vacate' it. From the 'screen' of the third eye the bundle of mental energies and their carrier, called soul, 'vacate' the body. They collapse from the physical arena leaving, as far as life is concerned, emptiness. Ultimate detachment. The void of death is, from 'view-1', a cadaver. If the 'silver cord' is broken mind's 'umbilical' with body is unplugged. Active information (conscious mind) has departed from informed energy (the body and its world). Connection has been severed. Irretrievable death has definitely occurred - at least from the perspective of 'view-1'. What about 'view-2'?

Before embarking on subjective angles what about objective apprehension of an *OBE*? What kind of phantom is a patient, hovering in *OBE* above an operating table, able to sub-consciously project? Could a disembodied personality affect material circumstance or even, as an incidence of 'ghost', be sensed? Countless anecdotes attest to phantoms as a fact. Countless witnesses seem genuine but anecdote is, rightly, not the sort of test acceptable to science; and anyway a single ghost, if scientifically proven to exist, would disprove materialism's *PAM* as it is so far understood. No less than 'rational' hypothesis will do. That means you'd better formulate materialistic explanations, supply ghost-busters with an electronic kit and send them out to make some instrumental measurements.

What is the dialectical interpretation of 'a spooky visitation'?

Sub-conscious mind involves psychosomatic border. The mind/ matter tangent (Chapters 15 - 17) logically exchanges information through the agency of electromagnetism so that a disturbance on the material plane might register in forms of radiation. In other words, you can't say just because you found unusual radiation there is not a spectre - or that there definitely is. Sub-conscious mind is, equally, the zone of memory. It is the zone of 'frozen image' or of action cached on 'video'. As such it forms a 'shell' of context that envelops active, conscious mind. From such a view two possibilities emerge. What does the phrase 'jump out of your body' mean? If terrifying, petrifying and, therefore, rigidifying shock occurred at death then a 'mnemonic shell' might linger after physical remains had been destroyed. The shock might 'jam the works' and, as flashbacks and repeated nightmares do, 'root you to the spot'. Some minds can't escape the influence of very damaging experience. If attachment, anger or obsession dominated such experience then its gouged imprint, the 'body of its memory', might also 'haunt' its place.

This personal shell, this body of sub-conscious memory might still be occupied or not. If, for example, its occupant did not believe death had occurred, could not detach or felt in other ways compelled to stalk the spot then his or her projection from the base of mind might stay 'inhabited'. A common circumstance of incorporeal 'remainder' is that a person physically dying or just dead may feel the need to comfort loved ones or a reluctance, whether such a feeling is due to shock, untimely death, unfinished work or strong emotional attachment, to leave a scene of 'incompleteness'. In other words, a disembodied soul remains attached by strong emotion to some specific circumstance on earth; and such a ghost, a complex built of memory and feeling, can like that pair eventually fade away.

Neither case of a post-mortem ghost is normal. From physical to metaphysical, complete vacation from this world to mind-worlds is the norm. Crossing over is a metamorphosis that sheds the husk of worn out dreams. Snakes and spiders sometimes 'spook' susceptible observers but an actual 'spook', although the presence might involve its element of pain, is no more harmful than the casing of a moult. Nor, any more than subjectivity is scientifically measurable, are apparitions. With what instruments could you apply a test of repetitious regularity to such phenomena? Indeed, if it were so then, as the Dialectic keeps explaining, science would have changed its boundaries and its definition of what counts as 'physicality'.

The journey is now, having sloughed connection with the earth, purely psychological. In dialectical terms, it is a subjective soul-mind complex left travelling through the space of 'inner worlds'. This is because, as we'll shortly see, *Nirvana*, *Dharmakaya* or the Apex of Essence has not been attained, Liberation from existence has not been achieved. Soul has not shed its second, inner 'body' - mind. From here, therefore, different minds follow different patterns and trajectories.

The less aware, the more ignorantly reprobate or instinctively sub-conscious a life has been, the 'heavier' it is and the shorter its curve back to earth, the quicker its gravitation to rebirth. This is even the case with many humans. Their lack of development is such that a comatose condition may fall. Oblivion can also accompany a condition of denial ('I am not dead' or 'there is no life after death'). This short-circuits many of the bardo experiences and, of course, any chance of liberation. Such minds may not reawaken until their psychological tendencies have begun to sweep towards reincarnation. They exercise no control over the process.

A 'departed soul' may, on the other hand, be drawn through a tunnel, across a border or however final exit is explained. Many reports involve mentors, guides, friends and other associates. In their company appraisal and learning occur but the Tibetan Bardo account 'goes it alone'. Dissolution of internal consciousness follows withdrawal from the outside, sensible world. It is as if the coarse, ordinary mind is progressively stripped. Sheaths of anger, desire and ignorance are peeled away. The clouds of five passions - anger, lust, greed, attachment and egotistical pride - are dispersed. Emotional rigidity of mind turns soft, pliable and sweet. To an appropriately tuned mind there may arise the perception of sound and light and with it the key knowledge that harmonic vibration (the main, musical metaphor of Natural Dialectic) underpins all levels of creation. The process amounts, for those keenn and capable, to a crescendo of increasingly subtle understanding. *This dawning culminates in the critical vision of Dharmakaya or The Clear Light of Primary Reality.* As the shadows of negative tendencies dissolve there naturally appear the qualities of clarity, bliss and knowledge. It is not knowledge of specific facts or manipulations. These have dwindled into triviality and vanished beside the rightness, attractiveness and loveliness of this condition. Its radiant luminosity is the light of *NDE*, the thousand-petalled lotus (*sahasrara*) of the Hindus, heaven or lowest of the mystic's inner suns. It is variously called the Point of Entry, First Step on the Final Ascent, Door to the Infinite or, simply, 'The Way In'. The Bardo says:

"*Let the flame of the clear light burn up all ideas and recollections of materialism.*"

The Radiance represents a great opportunity. Surrender, entry or immersion severs the link with physical mind and matter. This breaks the cycle and means you will not reincarnate. The problem is, the texts warn, that the light may only be glimpsed, not recognised or, as quickly as a snap of fingers, lost in other distractions. Indeed, this is the easy and normal reaction. It happens because the dissolution of conceptual mind, which forms an invisible, agitating barrier, is incomplete. Disturbances (called *karmas*) 'pop up'. *Karma* means action; actions are recorded in memory and, unless the mind's account is balanced, they act like hidden springs whose discharge continually propels it into fresh destinies. These hidden memories are also likened to waves that disturb the mirror-like surface of a lake. Or seeds. As you sow you'll reap - the harvest's destiny. If thoughts, exciting or depressing, ruffle this critical moment they act, in proportion to intensity, as a distraction. Any keen focus is lost. Attention slips. They throw the mind more or less off balance.

The heart of meditative, mystic practice is, on the other hand, to prepare for this crux. It is to develop the stability of a state of minimum distraction so that self may merge with the light. Meditation is preparation. It is learning to die while not losing aim. It is to project precisely. *Dying to live.* As usual, mystics flip the body's viewpoint upside down; we live to die but, more importantly, we die to live. *Their goal is to experience death while living and thus know, and to that extent conquer, 'the last enemy'.* If a practitioner can 'pass through the eye of this needle' (s)he has learnt to exert voluntary control over the process. The agenda is directed. *Its goal is non-reincarnation or, put positively, Liberation.* It is release from the cycles of birth and death, an increase of intensity from a cloudy, unclear state of the mind towards full *Dharmakaya*, immaculate sky or Buddha-nature. This in turn is identified with the (*Sat*) Central Illumination of Chapter 12, Communion with the Father, Enlightenment, Nature of the Infinite or whatever terminology you use. This is why there is an exclusive emphasis on Natural Dialectic's (*raj*) positive, right-hand column, the column that rises from relativity towards such Transcendent Absolution. Although (*tam*) negativity is a critical part of polar existence its materialising drag needs to be overcome, its pull of gravity thoroughly escaped. In this case death is welcomed in due course as a window of opportunity, carefully anticipated, to escape the outward tow and to 'surf' a rising, inward flow towards Completion. In mystic terms you do not even have to biologically die to achieve this. It is, to reiterate, the point (or, rather, Point) of mystic practice.

In other words, the experience of *samadhi* or Enlightenment is one from which an adept can, while still with body, return to physical sensation. The *Dharmakaya* will appear during the course of voluntary death, mystic transport or the 'flight of the soul'. In that case return to the biological sheath is, until the connection is finally broken, equally voluntary. At the naturally appointed time, when that sheath is finally sloughed, *Dharmakaya* is re-obtained, this time without return. *In the case of the unpractised majority this is not the case.* The invisible barrier of egotistical mind is not sufficiently thinned. The strength of un-sprung *karmas*, potential of un-germinated seed or unfulfilled desire remains. They are the seeds of desire. If they involve physical phenomena their propulsion is down, their pull is like gravity, they will drag you down. In this way the escape route is sealed. At the critical moment, therefore, unready eyes cannot countenance the purity, distractions upset the moment of truth, the chance is lost. In its unsteadiness the 'soul' may slip, at least this time round, from its chance of Salvation.

Anathema

Before post-mortem psychology we encountered materialistic anathema - post-mortem psychology is denied. Any kind of life after death is denied. Before ante-natal psychology we encounter another kind of anathema. *No doubt reincarnation is not compulsory but orthodox Christianity denies its cycle altogether.* The final, official source of this denial proceeds from Anathemas pronounced by the Fifth Ecumenical Council convened in Constantinople for political motives by the Emperor Justinian on June 2nd. 553 ad. This synod was a dubious publicity exercise designed to drum up solidarity within the Eastern Church and to condemn anti-monophysite schisms that threatened a united religious and political front. Monophysites believed that the body of Christ was wholly divine and never for a moment combined human and divine attributes. Whatever the obscure theological import of this dogma, it was held by the Empress Theodora as she challenged western doctrines and especially those of Origen. On this rigged Council only 6 out of 165 bishops were from the western church. Through its Anathemas the teachings of Origen, which included the pre-existence and transmigration of souls, were condemned. The idea of reincarnation was expurgated.

Is no doctrine-ridden politician ever in denial of the facts? Does a politically inspired attempt to delete a doctrine from the canon of some orthodoxy make it inconceivable or render it immediately untrue? Whence, if mind and spirit aren't the same as brain, do ante-natal minds appear from? And whither, post-mortem, return? Of course, the one-stop shop preached by some stern churchmen is understandable - in the face of the opposite kind of eternity to make sure of heaven. One life on earth, one chance alone: no idle, spiritually debased Indian or Orphic circulation round and around a pagan universe. No life after different life, simply an eternal fixture - paradise or the abyss. Should a fear of hell not whip any individual into righteous living, respect and, don't forget, tithes and legacies for churches? Should fear of damnation not prick spiritual exercise sufficient to 'Home In' at the first and only shot? Nothing concentrates the mind like pain and fear of pain. No second chance is a doctrinal dose of salts, a fiery stick to discipline the wayward soul.

Nevertheless although it was anathematised, other early church fathers believed in the theory of reincarnation. The Gnostics and neo-Platonists including Plotinus, Porphyry, Iamblichus and Macrobius as well as Origen and the Alexandrine School, Saint Jerome, Saint Clement, Saint Gregory, Justin Martyr and others expressed sympathy with the idea. What did Christ mean when he announced that he and John the Baptist were Elisha and Elijah come again? The idea of 'metempsychosis', as the transmigration of souls was called, rumbled on in various sects until its last known 'death by anathema' in 1439 by the Council of Lyons; but for one and a half thousand years orthodox officialdom has emphasised Salvation and, due to a council meeting, excised and thereby forgotten the existential wheel of life. Dogma does not recognise its buoyant view. It's not a way directly to the Top and therefore a Primary Principle to which both Christian teacher and pupil might aspire. *Yet unless the heavenly Father is continually sending down new souls and then mercilessly committing large numbers of them to eternal damnation, reincarnation seems a better explanation.* Individual eternity in a stylised heaven or hell is a static, cardboard kind of imagination. One is as boring as endless ritual, the other hopelessly cruel. Both are rigid, both make literal metaphor and neither is realistic.

Reincarnation is, by comparison, a dynamic possibility whose metaphysical episodes are as natural as its physical. It harmoniously incorporates the fundamental principles of cause, effect, motivation, free will, the law of *karma* (action and reaction), fate, destiny, justice, equilibrium, oscillatory motion, reciprocity, hierarchy, recycling, (*tam*) gravity and (*raj*) levity, life, death and relativity set against a Central Absolute. In this view heaven and hell are textures of mind; they vary in degree but are intrinsic qualities of the mental spectrum. Mind's engine, on the other hand, is motivation; it is focus of attention and will-power concentrated by desire. *An individual's texture of mind fluctuates according to the quality of his strong or habitual desires, consequent actions and their results. His or her mental projector creates, in its flux, destiny which is realised, at the appropriate moment, as fate. Therefore kinetic fate, which is a result, can include the result of a desire to 'try again' as a human - rebirth.* The theory of reincarnation is reasonable. It does not preclude Salvation. The reverse. *Salvation is included as the special, highest, absolutely anti-gravitational, immaterial case - the positive goal of non-reincarnation.*

Let's rephrase this circumstance. Below the Essence of Salvation passes a continual flux. What are the rules of existential flux? They are, at root, equation and equilibration; they are vectored motion around norms or, if you like, principles. These principles are expressed as a set of physical conditions so finely tuned that (as we saw in Chapters 7 - 12) the slightest 'twiddle of any individual dial' would have wrecked the chance of cosmos let alone life's appearance. Could you therefore say the 'rules' of physics locked up energy's behaviour? Not only are there various kinds of prison (forces, atoms and so on) but process also is incarcerated in an ineluctable chain of cause and effect. *The prison is invisible.* Patterns of behaviour are conserved and thus called laws; and conservation of energy throughout their dynamic interactions is, as Newton's laws of motion indicate, one of rigid equation. 'Action and reaction are equal and opposite'.

What, however, about metaphysical conditions? What about psychology? Suppose that the intrinsic, fundamental balance of physics was a subset of a broader one, a law of action (*karma*) that applied to mind as well as matter, to information as inexorably as energy. 'As you sow you will reap'. Action circles in a chain reaction. You reap this moment fate that previous 'moves' have led you to; and by reaction at the same time sow the seeds of destiny. You commit to 'earth of memory' the learning that will, as your 'context', lead you into future incidents of fate. Do not decisions of a captain in the context of his past experience (his memories) influence a journey? Is not navigation of life's sea, with you the captain of your vessel, just the same? So that reincarnation simply represents another journey from the harbour where you on your latest D-day docked. It is part of life's equation, balancing the books; it is an agent of order, a factor in the precise adjustment of psychological actions and reactions. It is, where rigid balance of equation dictates the outcome of all physical transformations, the balancing factor that delivers 'conservation of psychological motivation' or, put simply, justice.

Incarceration in a body is an obvious restriction. *Mind's prison is invisible. It is constructed of a barrier of egotism and, as invisibly, of a context of karmic process (action with its consequence) and impressions deeply grooved in memory.* Rings and chains are binding; metaphysical recycling is a chain that, as with destiny, is forged by action - action in the form of thought, its execution and the consequences these events produce. You are in jail with locks and bars of iron or gold; detention in your body may, according to its pleasure/ hardship levels, be a high- or low-class shackle but you are caught up just the same. Reactions to your fate are binding you to actions that create your destiny; and thereby, round again, inexorably seal a fate called 'payback'. In this view a cell of circumstance rewards an inmate's previous deeds according to a legal system called morality. Or, if you prefer, it is the sprouting of a karmic seed that you have sown.

Fatalism? Is this perspective slack with fatalistic inactivity, torpid acceptance and inertia? Nothing could be further from the truth! *Karma* theory is not fatal but reanimating. Its equilibration neutralises 'bad' with 'good'; it directs you on the way to positively treat your present circumstance and thus improve your destinies/ fates; its proper use ameliorates imprisonment in mind and body. Such use can do so in today's life-cycle or in future lives. At climax it can lead, as all unshackling mystics teach and *figs*. 18.2 and 18.6 describe, to freedom from the existential wheel.

In the case of a physical event reaction is immediate but in the case of a psychological one various lapse of time may occur before orderly justice is administered. This happens because the context of memories (past attachments, accumulated tendencies and what amounts to the record of our personal history) draws us to specific conditions. The heart of these conditions is not so much a function of their physical surround as their subjective quality, their expression of characteristics such as loyalty, kindness, criminality and so on. What circumstance is not, in this respect, a fateful challenge? It waters a particular seed, called a *sanskara.* It activates a memory that, having germinated, sprouts a characteristic response. All memories migrate through time. These 'undetonated mines' can ambush any moment; or, as latent tendencies to talent, thrive with the water of encouragement. Circumstances trigger a *sanskara's* charge. And these tendencies of mind can transmigrate. Memories are the grist to metaphysical recycling's mill; they drive reincarnation's wheel. And thus both triggered action and inactive latency drive life's dramas forward.

The instinct for balance is adapted and defined by each man's own 'case law', a subjective record of his choices and experiences only wholly known to himself. In this view morality may vary during life; it may be distorted by one's egotism but in the end intrinsic balance weighs you up. 'You get what you deserve'. Is that anathema or fallacy? Is it wrong to say 'I am where I am by my own actions'? Could the 'context' of my mind, its character, have landed me where I deserve to start on this embodied cycle? Could this round's accumulation of impressions define the point of incarnation to the next? This is how mystics calculate. Life's equations show as endless balanced transformations. Wheels of desire recycle bodies through their plots. There is no cessation or escape unless the primary desire becomes escape or, from a *top-down* point of view, return to Completely Natural Freedom of the Boundless Origin. This axial point of balance, this eternal pivot is the *Dharmakaya.* Essential *Dharmakaya* is innate. It is pure conscience, the light of truth against which every mental action is compared. Indeed the light shines through such action or is overcast by swirls of viciousness, by storms of passion and dark thoughts. Behind the clouds, however, you may know the sun is always out. Indeed, there can be no end of things until all memories like mists evaporate, shadows of the habits of all lifetimes lift and winds, the driving forces of desires, have fallen quiet. Clear sky. The storm of mind has blown out. 'No more desire,' remarked the Buddha, 'sets you free'. There is liberation if the world is burnt like chaff in mystic fire; only from such ashes does a phoenix rise. Or, you could say, without such liberation life makes sail in serial, individual lives like passages with different ships. It journeys, port to port, around the ocean of existence. Its embodied episodes make endless passage through such waves of mind and matter as compose *samsara*'s sea.

You can no more see a memory than the force of gravity. Could you ignore the latter? Would you deny a memory its hidden influence? Reincarnation 'crosses' lines of birth and death as easily as space is crossed by gravity, as birds migrate or after flying on a plane journey you drop as simply into another culture as if stepping through a door from one room to the next. Psychologically it explains proclivities and traits of character emerging from an early age, well-defined talents and the apparently unjust distributions of wealth, health, pleasure and pain. It certainly explains some children's detailed and otherwise inexplicably accurate accounts of a previous life. As you sow you'll reap. This *is* karmic theory, the rule of cause and its effect. Such 'normalising' is as natural as the equilibrium of health. It is not a man-made form of justice though the latter, as a doctor treats for health, treats for social health. How can such a possibility as cosmic balance challenge or impugn instead of complement the basic Christian message? *Is a law of motion and of motivation with its consequences physical and metaphysical unacceptable? Is its dynamic linkage of the two worlds called reincarnation really an anathema?*

Immortalities

Immortality is packed in three main brands - the unconscious (sub-conscious and non-conscious), conscious and super-conscious. Of these the first two are apparent or existential immortalities; the third is quintessentially real.

Brands of Immortality

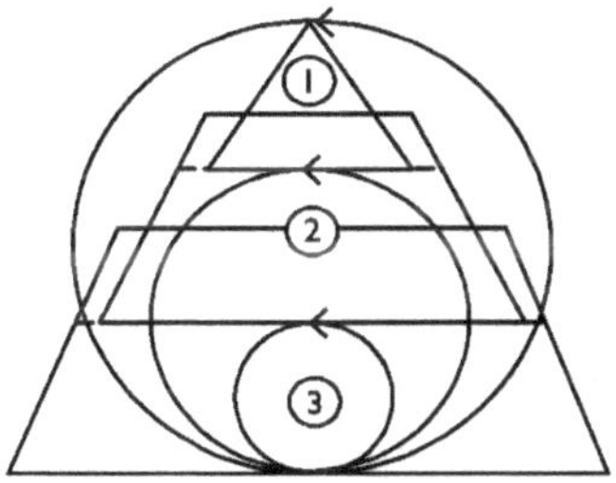

Immortality seen in terms of cycles of cosmos (1), mind (2) and material bodies (3).

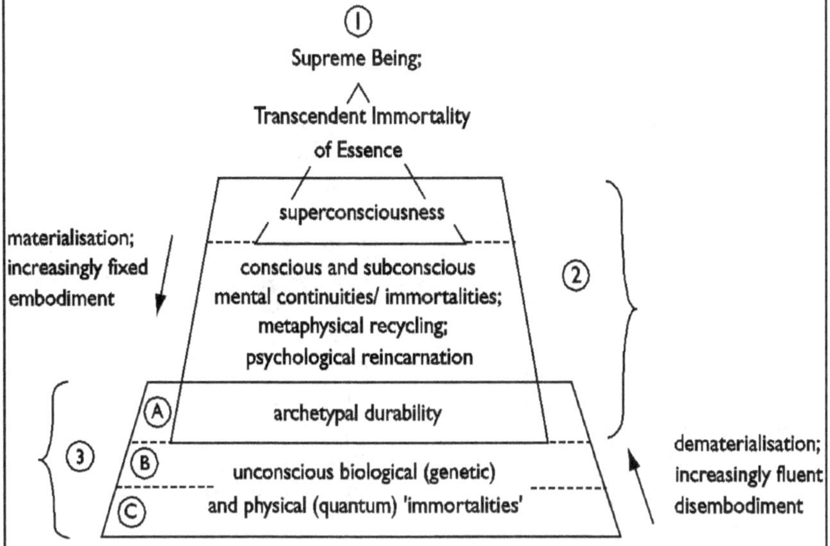

Three brands of immortality visualised in terms of the cosmic ziggurat. The lowest, solely energetic brand is itself subdivided (as *fig.* 0.11) into three parts; these are archetypal/ potential (A), quantum (B) and bulk or classical changes (C). Bulk, chemical bodies (C) 'transmigrate' endlessly through forms based on (B) and (A).

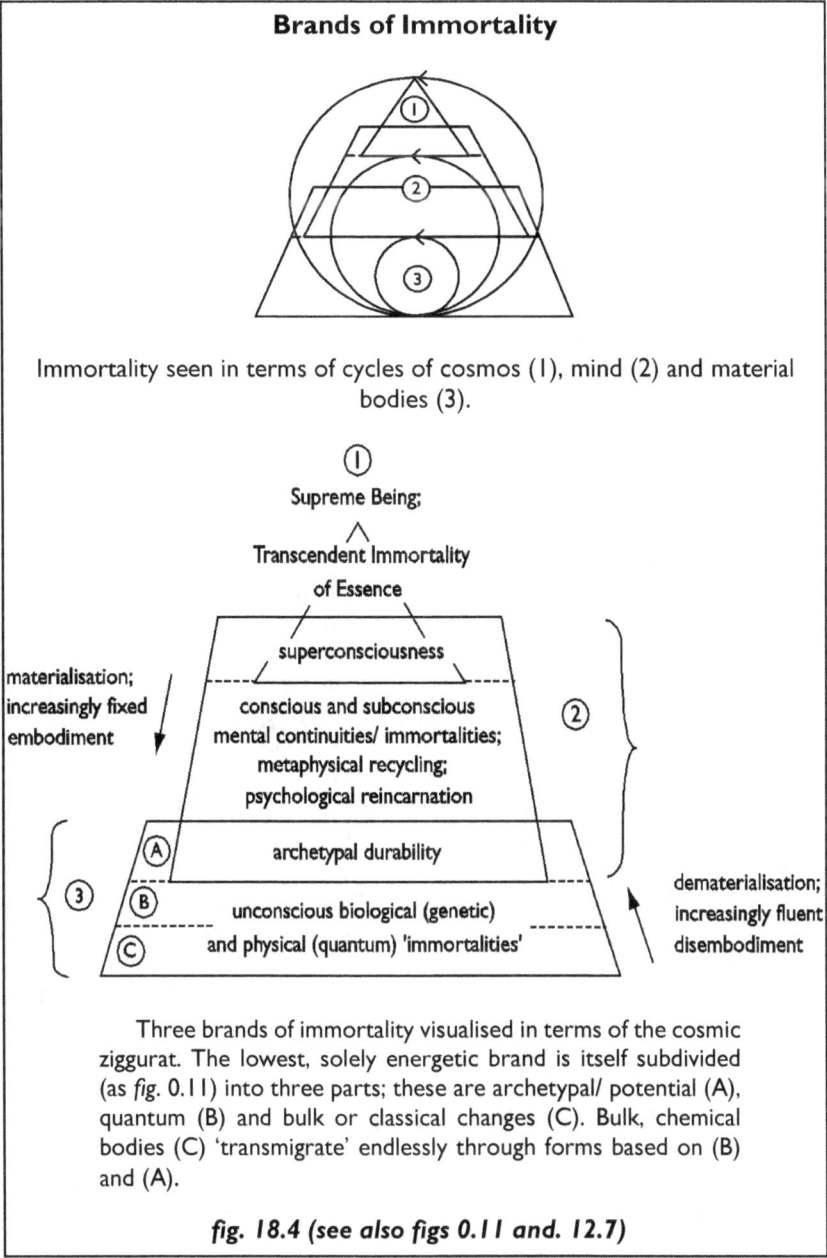

fig. 18.4 (see also figs 0.11 and. 12.7)

Let's start at the lowest brand - grade 3C - and work on up. A dialectical stack for the *non-conscious*, *physical* dimension of bodies physical (including biological) might be written as in *fig.* 0.11:

tam/ raj	*sat*
grades B/C	*grade A*
expression	*potential*
expression	*information*

	cycles/ changes	*axis*
	concentric shells	*centre/ nucleus*
	lifetime	*immortality*
↓	*tam*	*raj* ↑
	grade C	*grade B*
	passive/ inert	*active/ forceful*
	fixed/ bulk aggregate	*fluent/ inner constituents*
	discontinuous	*continuous*
	outcome/ conclusion	*program*
	informed	*informant*
	end-shape/ adult body	*construction/ development*
	run-down/ entropy	*build-up/ stimulus*
	senescence	*reproduction*
	towards nadir/ periphery	*towards zenith/ centre*
	aging/ oldness	*pre-maturity/ youth*
	death	*vitality*

Bulk forms (grade 3C) come and go; they do not last. Such transformations include the senescence of animate bodies (Chapter 25). Animate 'immortality', in its physical aspect, resides in the 'holographic', 'pixellated' cellular information bank of each individual life. This biological information is specified on chemical 'paper and ink' called *DNA* in the language of genetic code. It is at the heart of an intricate mechanism whose *purpose* is not only to support individual life but postpone its death and the extinction of its life's type as a whole. What might be called 'the computerised operation of digital bio-information' is a program for survival. It serves to 'outwit' the destruction of progressive, straight-line aging by recycling the physical libraries of information (genomes) required to reproduce 'photocopies' of the various forms of life on earth. Each birth is thus a kind of resurrection. Lifetimes are repeated. Each individual might presume (*tam*) *physical vicarious 'immortality'* through his or her descendents. You personally die but your ancestral genes live on.

Let's be clear. This is the 'least real' yet 'most apparent' kind of immortality. It is a reproductive continuity, although the chemicals continually change, of structure; it is a theoretically endless series of similar but different bodies. Biological continuity complements individual discontinuity. Biological continuity complements individual discontinuity. Specific substitutes compose a general chain. Life's line is a thread. Death is cheated by a connective strand as thin as chromosomal strings in a single cell that's called a zygote. From a zygote bodies of a genome are regenerated and the consequence of lack of offspring or extinction is postponed. For you, but not your genes, it is survival *in absentia*. In short, mortality ends immortality unless, by reproduction, genes have made escape. Reproduction by your 'selfish genes' might grant perpetual survival. In this sequential sense 'digital bio-information' is, subject to a congenial environment or lack of its destruction, deathless; with fit copy and without extinction there could theoretically exist an animate eternity, an 'immortality of genes'. Such creed is, indeed, the socio-biological essence of material religion.

In-swing and Out-swing: Biological

Genetic 'Immortality' through Offspring.

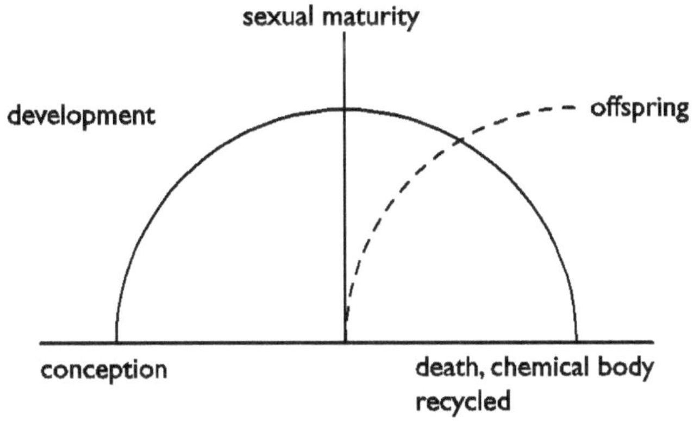

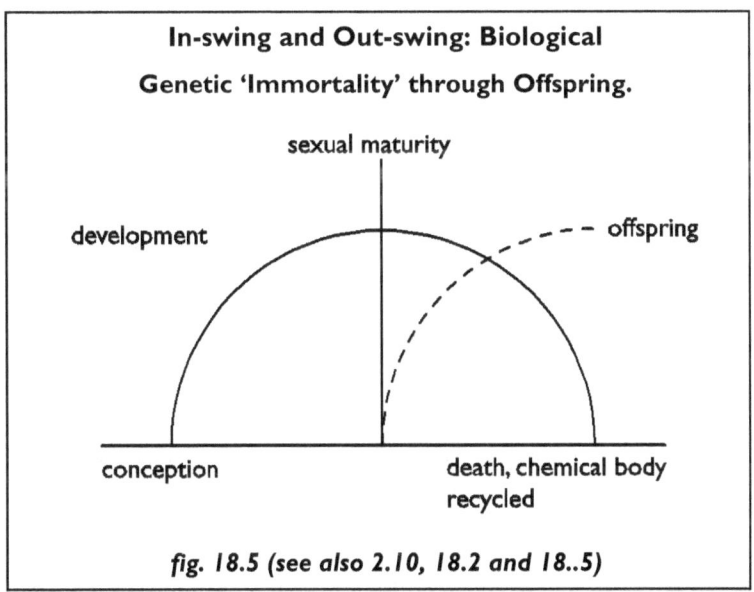

fig. 18.5 (see also 2.10, 18.2 and 18..5)

But isn't there an even closer, more continuous sort of 'immortality'? What about asexual inheritance by divisions of a single cell? Bacterial immortality is, theoretically, possible. Sea anemones and hydra might bud everlastingly. Bug or bud, aren't daughter cells just *doppelgangers* of yourself? Isn't this a schizophrenic heritage, a continuity of selves in which you actually become the whole extent of your successors? Divided parts of you survive. And, uniquely, the jellyfish *Turritopsis nutricula* is an inspiration for stem cell research. In this case an individual is able to revert from sexual maturity to immaturity and back again. The creature can 'trans-differentiate' back to a baby polyp then, from such regression, re-mature. Of course, predation and disease bite into cyclical eternity. No single longevity has been observed but, save extinction, you might still concede its 'theoretical immortality'.

In imitation of bacteria! Without their body human cells aren't keen to stay alive but might not cancer cells, if cultured properly, embrace a quasi-immortality? Aggressive Hela cells are cloned from Henrietta Lack, who died of vicious cancer many years ago. They can, subject to provision of a suitable environment, divide a limitless amount of times and, as international colonists, they populate the dishes of experiments on human tissues. Without a doubt such deathless cells evade the 'Hayflick limit' (the times a normal cell type divides before it stops); they don't wear out but are still totally dependent for survival on a '*deus ex machina*' consisting of histologists who care for them with special cultures in laboratories. Extract your '*deus*' and you pull the plug on 'Hela' or on any other kind of biological 'infinity'. Such 'potential immortality' is no more than conditional cipher for an unconditional deathlessness; it lasts till some disaster wipes a population, species or a higher grouping off the perch of everlasting life; you might call such exhaustion of potential 'immortality's extinction'.

A dormant organism 'lives asleep' and scarcely has awareness if at all. Subconscious 'immortality' is like an everlasting slumber, no more personally engaging than the life of stones; but at least its genes, made up of molecules and therefore as alive as any crystal, 'live' on. Aren't we really talking here about 'oblivious immortality' that could but won't last anywhere as long as its own atoms and their particles? This would be a lifeless sort of immortality, at best apparent and a temporary kind you might expect of particles if not the bodies they compose.

Can't anybody make it real and lasting? Bodies differ but the family genes remain the same. Why should only they turn death-defying tricks; why can't a single form accomplish medical eternity? Why not immortalise your present body using gadgetry? There is, of course, the risk of fatal accident or illness but why not otherwise freeze time? Put death on hold? Refrigerate the body in a tomb of liquid nitrogen until a nurse injects your wake-up call? You could, if you were really slick, engage with futuristic spare-part surgery and by gradually replacing pieces stop your youth from growing old. Rejuvenation! Scheduled resurrection! Why should simple nature flout my wishes? I'll grow even younger if I want! Why, therefore, should not I invest in ultimate refreshment - organ transplant from a personal bank of slaves, self-sacrificing stem cells? Perhaps a freshly culled or stem-cell cultured brain inserted just before each brave new life would slip the noose and let me vigorously strangle death. Or would it? Whose mind would be that brain's? In fact, if frozen process is a form of death then all appearance of solidity is death as well. Iced time. Material bodies are themselves a vale of death. If reincarnation is a natural phenomenon then serial, surgical replacement of one's body parts or procrastination in deep freeze were each absurd. What use is hanging on in different, captured bodies, timetabled death and surrogate rebirths? Artificially recycling nature's vital gift is an *ersatz* and a scientific sleight of life's reality.

Who, if you hope eternal life might be the grant of doctors, kids whom? Never mind if you forgot the life before and boredom born of repetition did not wind you down. What about the practicality? What about tremendous strain that such potential immortality and its attendant overpopulation might impose on food chains and the earth's resources? What about the social consequence for families and, survival after survival, never-ending sets and chains of relatives? Is not 'clinical reincarnation' the operation of insanity if only *top-down* truth were known?

So is this all? Is generality in genes alone? Are you just a figment of these chemicals? It was noted that indeed, from a *bottom-up*, materialistic view, you are. Genes, from which brain and therefore consciousness are held to be derived, are the primary blueprint; they are critical chemistry, the actual architecture around which some peripheral phantom, 'you', emerges. People used to believe that they, as people and personalities, were the climactic point of the creation's business. In the brave new world this anthropocentric view is as outdated as a geocentric one of cosmos. Man is a genetic variant. In a 'geno-centric' view of life it is genes, not you, that really count. They are the central chemistry, material essence (is there any other type?), the information bank and therefore nucleic logicality of life. The insubstantial, transient subjectivity you know as 'self' is, in comparison to a hard-coded, objective genome you can put in test tubes, well, frankly negligible. Immaterial 'you' does not conform to scientific standards of what makes for durability!

If animate eternity resides in 'immortality of genes' then an inanimate eternity seems, in a world of change, to reside in the endurance of protons, electrons and the forces of nature. These energetic forms (*fig.* 18.4: grade 3B: quantum) perhaps persist throughout a lifetime of the universe and are thus candidates, although they never lived, for the title of 'physical immortals'.

Is a bulk, creature body made of such immortals really any more alive? Mortality involves a body but is this corpus by itself alive? Is it not in fact a species of inanimation, that is, death? Physical that lacks its metaphysical is only half a story; and the Dialectic of Adam and Eve's immortality also indicates that, if anything seems 'immortal', it is their generalities. It is their typical part, their humanity recorded in the form of potential matter (grade 3A: archetype). In this conceptual (as opposed to genetic) respect the immortality of biological animation is metaphysical. It resides in archetypal memory.

From a *top-down* perspective immaterial information hierarchically precedes material energy; mind precedes matter. By this token a materialist is, of course, right to identify information as the essence of things and, in particular, humanity; but he is wrong to have chosen, according to his self-imposed injunction, passive information. Having denied an active originality he is thus forced to claim that exquisite biological information did not proceed, as the Dialectic current indicates, from mind but instead from a randomly self-assembled lack of purpose in the form of atomic genes. *This, denial of the real origin of information, is the root cause of the irrationality of 'rational, scientific' explanation. It is not born of the facts themselves but endemic in a professional orthodoxy of the mind-set treating them; it is born of habit. In short, interpretation of the meaning of the word 'rational' depends on your perspective; and each perspective has its complementary use.*

From a dialectical point of view a plan is documented in the form of blueprint, instruction manual or, in nature not in ink, memory. In this case you are descended from humanity's own type of immortality, the potential matter of its archetype. You are, in other words, a member of the universal human type. Your individual body is an expression of higher generality called, dialectically, *H. archetypalis*; its genetic database is part of a material reflection of this body-linked archetypal memory. Such memory is the immortal morphogene, an informative type of which you, either Adam or Eve, are a polar half. Is character for costume or costume for a character? You, your characteristically human body and its vital database are each an intentional part of the whole show. No doubt, however, that both archetype and genes are passive information. One is filed in universal mind as immemorial memory and the other in the nucleus of cells. Biological reproduction, born of the combination, is mechanical. 'Mortality Incorporated' (or is it 'Life Inc.'?) is a factory, a business whose turnover is repetitive and whose line is simply the fixed instrument of its founder's intention - vehicles of life not life itself. So much for body and objective, time-bound parts of immortality; so much for the *biological* dimension.

But the habit of identifying self with body is, by a lifetime's habit, well entrenched. The tough illusion doesn't want to budge. Thus it needs be asked again, "Are you just a body? If spare-part surgery had replaced your every part what would have formed undying, underlying continuity? Would it be your own mind or the other brain? Would you transcend your bodies?"

Not what but who, therefore, are you? There is 'quasi-immortality' of bodies by their reproduction. There is longevity of line but dubbing such mortality immortal is, we have seen, simply science in delusion. Leap from physic to psychology. Even were you able to watch the non-conscious machinery of genes grinding out your family's bodies, you are neither a genome nor a genetic process. Biology observes the process of corporeal reproduction but where do you live? In subjective mind - of which, if you think about it, objective body is an appendage. What of this subjective part? Might continuity of mind, if not of body, grant you immortality? We turn from biological to *psychological* dimension.

For grade 2 - this dimension's stack - we write:

	tam/ raj	*Sat*
	existence	*Essence*
	below	*Transcendence*
	finite	*Infinite*
	lifetime/ span	*Life*
	forms/ grades of consc.	*Consciousness*
	lasting	*Everlasting*
	lesser reality	*Reality*
	cyclical immortalities	*Immortality*
↓	*tam*	*raj* ↑
	lower	*higher*
	passive	*active*
	physical appearance	*metaphysical apearance*
	object/ body	*subject/ mind*
	informed	*informant*
	flat line/ lifelessness	*cycle*
	external	*internal*
	divided/ discontinuous	*continuous/ undivided*
	physical ' immortality'	*metaphysical ' immortality'*
	reproduction	*reincarnation*

Instead of reproduction, therefore, try recycling. Link discontinuity of a physique with continuity that's metaphysical. Ride upon vibration that weaves in and out of bodies and their tombs. Periodic cycling of the mind might constitute a (*raj*) conscious kind of immortality - reincarnation. It involves both sorts of scenery, physical and otherwise, upon the existential stage. Each embodied act of drama follows up a disembodied one and *vice versa*. Invisible, psychological plots and visible, physical costumes change as different scenes are acted out. Performance links up with performance, life visits 'outside' then, as death shuts doors behind itself, drops 'in' again. Such oscillating theatre casts a body psychological, called mind, but dresses it in different physical attire, that is, with costumes that are bodies biological. It revolves, one might reiterate, around uninterrupted, incorporeal psychology that visits physical disjunctions; it turns on mind's dynamic rather than the pre-fixed, chemically constructed carriages of life. *(Raj) reincarnation is, as opposed to (tam) physical reproduction, metaphysical recycling*.

What goes round comes round. Wheel, vibration or cycle - each imply returns. What goes up comes down. What is sown is reaped. Both physical and metaphysical scales of existence are instruments of balance - in one case mathematical equation, in the other justice. If cyclical reincarnation is a fact, it makes nonsense of attempts to prolong the life of any physical incarnation or seek eternal youth at any cost. Fear of death becomes as irrational as fear at the end of a holiday. To seek 'fixed immortality' (such as freezing youth, arresting age or endless survival in a body frozen between revivals) is deluded. It is thinking straight not cyclical. Perhaps scientism, if occasionally a little contemptuous of the poets and humanities, has read neither Greek mythology (starring Hybris and Nemesis in 'Pride before a Fall') nor Tennyson's verse (on the physical immortality of Tithonus).

If, on the one hand, reincarnation promotes a sense of detachment it equally attaches, with the respect to the quality of the next 'term of life', strong moral strings. After all, what animate but fondness, dislikes, honesty and guilt? And all to do with appetite. Is it the influence of passion for sensation and the motivation to possess, by fair means or foul, an object of desire? In metaphysical fields desire rebounds. A crop is reaped as sown. Will-power and wanting spin the mind-world round. They pull the strings. They recycle lives. In which case future 'trips to earth' will realise desires, conceive desires but also settle the accounts desires incur - for better and for worse.

The metaphysical pattern into which the process of life, death and rebirth is locked is simple. Interests vary and mind flits from show to show. It is not that an excarnate mind differs, intrinsically, from when it is embodied. You do not change immediately when walking through a door; or with your clothes on or off. The quality, texture or character of a mind does not suddenly change because one moment it is pre-mortem and the next post-mortem. Reincarnation links mind with matter, metaphysical with physical and life with so-called death. It links episodes, threads lifetimes and, most importantly, includes subjective motivation - purpose, will, desire, emotion and so on. These fires forge character so that, most marvellous, within the author's preordained constraints and using the aforesaid props, you are left to write your own script, project your own thoughts and produce your own inimitable play.

Patterns of body are fixed; they are involuntary, you cannot change them and they are called 'passive information'. The patterns of conscious mind, called 'active information', are flexible; with effort you can voluntarily change them or they can be, by circumstances, changed involuntarily. They can change for better or for worse. Upgrade or downgrade. You can take an escalator down (↓) into left-hand division and such divisiveness as springs with aggression, selfish passion and its forms of isolation - the character of materialising out-swing. Out-swing cycles, like a bob on pendulum, forever round and round. But there is another option open. Ascension towards the pivot is an upgrade; you swing less and less until, upon the axis, there is no swing left. And, as you well know, one can ascend (↑) on wings of right-hand union called friendship, interest and love - the dematerialising character of in-swing. In-swing/ upgrade is, therefore, of central interest. It leads us straightway to a second, higher level kind of death.

High-level Death

Evolution is a word that, with a sense of progress and development, you can use in many ways. It can mean simply 'change', 'process' or 'aging process'. Or it can imply 'improvement' and 'unfolding' towards some higher, better thing. You might relate this sense to 'program' or to what programs consist of, 'code' - both words loaded with implicit goal. Although the Darwinian theory of biological evolution implies 'target-less improvement', the sense of Natural Dialectic certainly includes a natural goal. Can you, therefore, mentally evolve? What kind of education stimulates such lightening? What kind of message personally evolved by you is broadcast just by 'being you' on earth's great stage? That is to say, what is the quality of your own theatre, 'soap' or psychodrama with the family of life?

Is its predomination to the right or left-hand column of the Dialectic? Towards what sun does your life, like a flower, unfold? Is there, indeed, an evolution of the soul?

Alright, you don't believe in 'soul'. Materialism, if not science, knows the mystic quest is a deluded one. What, however, are you looking for? Undiluted happiness? Is the source of bliss endorphins

In-swing and Out-swing: Psychological

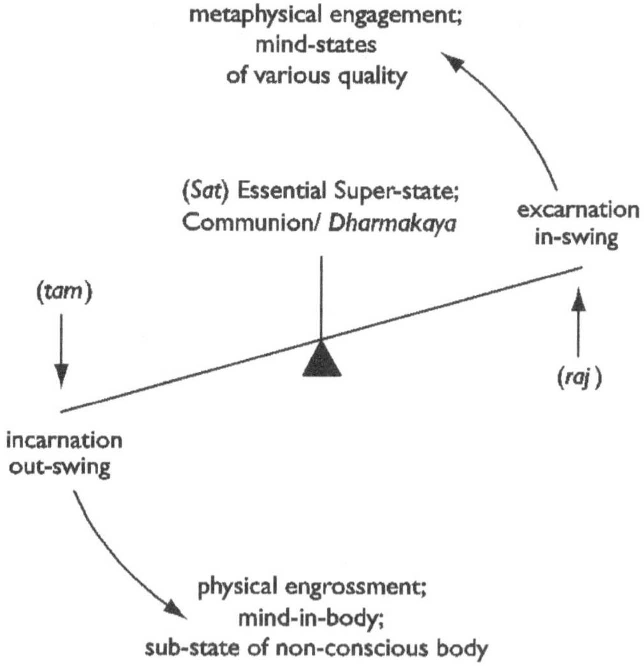

This scale shows the full cycle including both cosmic equilibria between which mind swings. It shows Essential Super-state, the *Dharmakaya* or Singularity of Enlightenment of Self; and it shows its anti-pole, the sub-state oblivion of matter. This sub-state includes your material body. Bodies are ecologically recycled; minds, if not centred beyond existence in Essence, are also recycled according to their constitution.

Full Wheel: Psychological and Biological

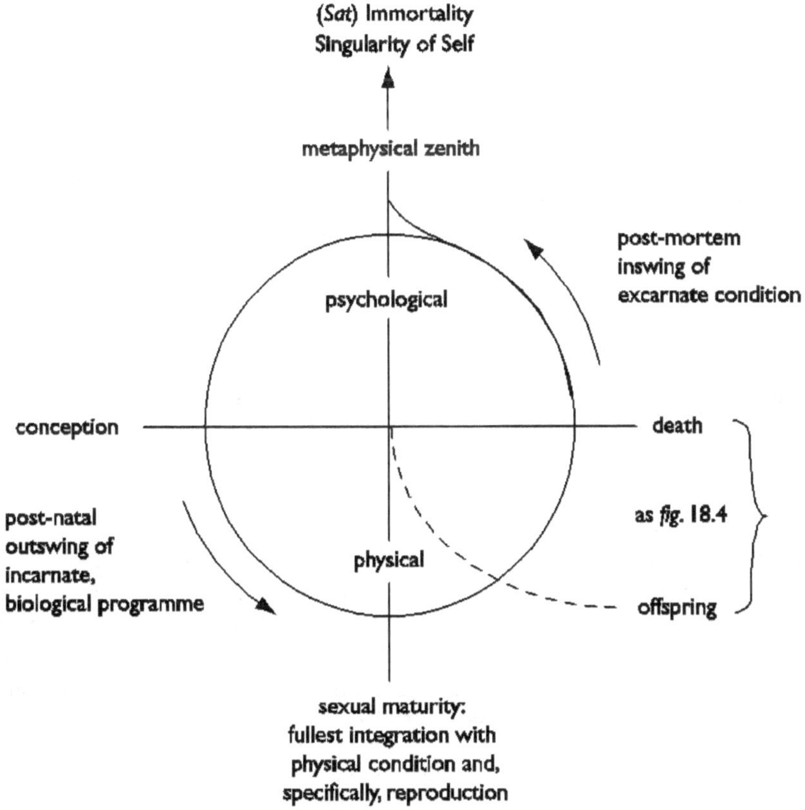

Within existence mind may cycle in and out of incarnate, that is, biological embodiment. The circle illustrates such cyclical continuity or *(fig. 18.4)* brand 2's 'psychological immortality'. (*Sat*) Immortality presides at the peak, centre or pivot (whichever model you use) of existence. In obtaining Subjectivity, therefore, one obtains the cosmic axis and is 'out of the cycle'.

fig. 18.6 (see also figs. 2.14, 18.2 and 18.5)

dopamine or drugs that lead to ecstasy? These are temporary resolutions, patches that don't cure the underlying cause of existential aches and pains. Might you therefore turn and find 'illumination' born of sexual pleasure, bliss that also only lasts an hour or two? Humans find a million sorts of interest but how long do any last? Can happiness that always craves another fix be other than dependency? Can it be other than a false appearance of the independent, real kind and, if so, what is this?

tam/ raj	*Sat*
existence	*Essence*
below	*Transcendence*
grades of error	*Illumination*

↓	*tam*	*raj*	↑
	descent	*ascent*	
	degradation	*improvement*	
	outward perspective	*inward perspective*	
	physical attraction	*metaphysical attraction*	
	body-centred	*soul-centred*	
	devolution	*evolution*	

Desire means discontent and, perhaps as well, its satisfaction. Does there exist immediate, permanent satisfaction or is it best to jettison desire itself? Would jettison obtain beatitude and is such bliss composed of peacefulness or boredom or contentedness? What is it like to fly in skies above the clouds? Is, leaving gravity of earth behind, any kind of joy sufficiently profound to pass beyond the gates of death?

Transcending clouds of mind, the mystic claims, is the way into the third, real (*Sat*) brand of Immortality. Brand 1, Grade 1 - this is the pellucid zone, a clear, lit space of concentrated consciousness, a life beyond all existential weather - what the Buddha called illumination is not religious but a natural experience each religion would, in its own way, enshrine.

Body's death is death of outer shell. **Death of *ego*, that is, death of ever-bubbling-up desires is the second, higher level liberation that crusading mystics aim to grasp**. And yet aren't passions and desires what colour life? Who'd exchange them for a colourless transparency; who'd slough this inner skin or take the second, higher kind of death as leading to a Preferable Reality? Mind and body represent existence. Without them what on earth in heaven's name is left? Are there, after death, eternal atoms only or unearthly Life as well? In such Life Absolute could you find transience or change or death? And thus, if they are not deluded and an immaterial element is central to incarnate life, the mystics seek Transcendent Immortality. Death of the mind by merging in First Cause, Natural Purity by bathing in The Spring's Stream is the way that anyone becomes Alive. Look round this extraordinary universe. To reach its Living Source were awesome as a goal and far past words when once achieved. **If, therefore, a mystic is materialism's clown still there is method in his madness. Might the universal fool not, breaking out of cycles of existence, come to consciously rest in Final yet Essential, Eternal Truth?**

Whose laugh were last? Death, wherefore art thy sting? Death stalks the first two kinds of immortality but not the third. In the centre, at the axis of an axle, turning stops. You cannot orbit at the axis of a wheel. Therefore turn towards the centre not the edge, move beyond the cycles to their Axis, coil round the cone and spiral to the Apex of Mount Universe. Beyond matter, mind and, as such, all trace of influential negativity find 'madness' that is actually the clearest, coolest sanity - (*Sat*) Axial Immortality.

The brief is short and simple but, in practice, long and hard. To evolve is to unfurl from darkness, unintelligence and (another way of phrasing ignorance) relative oblivion. Such flowering cannot, therefore, mean the totally oblivious operation of non-conscious matter. It applies to purposive construction (Chapter 6: Machines and Mind Machines) and the realisation of ideas. It may apply to physical skills, technological inventions or educational development but the mystic has, as you've just seen, another goal in mind - beyond mind's net. *For him evolution of the soul means throwing off existence in the form of mind and body; it means high-level death that reaches past existence into Central Essence.* Without existence Axial Immortality is Essential, Timeless and, therefore, Deathless. There is neither repetition nor recycling in the Absolution of Pure Life. **Therefore, beyond rebirth the final enemy is overcome, death is annihilated, only Nothing left. In seeking out this First Class Brand of Immortality the mystic's mind is clear and straight**. The path is single-minded, single-pointed and unwavering. To evolve thus means (Chapter 14: The Ascent of Man) to rise upon the right-hand path of levity towards Light without a shadow of illusion or of death. Death in Pure Life? After all, what's infinite lacks boundary; and, as grade 2's stacks above relate, Pure Consciousness (without attachment of a mind or body and transcending any form or relativity or personality) is Infinite. A synonym of immaterial consciousness is life; thus Life Immortal is the Absolute Reality. The nature of Essential Reality (which reincarnation just reflects a failure to achieve) is obvious from a glance at Primary Dialectic. The goal is, therefore, (*Sat*) True Immortality and, resting fresh at Heart, Unity with the eternal youthfulness of Origin. **High death is thus, in metamorphosis, the Highest Form of Life**.

Ante-natal Psychology

Mind blinked first and shrank from inmost climax with the *Dharmakaya,* from the zenith of its cycle. The moment was lost. The tide has turned. *Out-swing* has begun again. There follow episodes, either short or extenuated according to the driving strength of earth-linked motivation, of settlement. This 'settlement' is, in effect, the process of 'falling out' towards another birth. Psychological precipitation. In other words, what goes up must, unless it escapes gravity, come down. Having failed to escape the gravity of existence, having failed to obtain enlightenment, where will it be carried? A mind gravitates according to the type of its memories or, as they are called in Sanskrit, *karmas* and *sanskaras*. A *karma* is a physical or mental deed and a *sanskara* is a 'groove of habit born of repetitious thought', a reflex behavioural pattern of variable depth and intensity. This couple are the compound of character. They are an amalgam of inner context that predicates, in turn, the outer. Outer context shows as fate and destiny. Fate is what you are now undergoing. It is what you have brought on yourself. Destiny is what, according to response to fate, you will undergo. A mind's re-entry into the physical zone proceeds in proportion to its moral imperfections and material bias. The descent is buffeted by winds in mental air, currents of both fear and desire. At some stage, like anything that falls from the sky, it will have to land. By choice or otherwise metaphysical mind will assume another container. You will dress, according to the character of your debts and desires, for another physical adventure. According to this process you will 'drop through the dark tunnel' into physicality.

Why should I believe this? Apart from dialectical logic much research has been undertaken and evidence accrued from which the 'fact' of reincarnation might be inferred. One of the strongest involves childhood recall of facts it would otherwise be impossible to know; behaviour in terms of pronounced but inexplicable traits and responses; and advanced abilities (e.g. linguistic, musical, philosophical or mathematical) manifest an early age without prior exposure to a discipline in this incarnation. Because mind is metaphysical such evidence is always anecdotal; it cannot be measured against the previous (unknown) incarnation. As such claims of reincarnation are justifiably vulnerable to demand for physical proof or sceptical accusations (sometimes quite possibly well-founded) of suggestion, imagination, bias or fraud. They do not, definitely, pass scientific muster. Even if we all go through it nobody has observed the subjective process objectively, from the outside. How could you? Can you see your own experience from outside it, even now? You are only ever in it. Therefore the excarnate factors can, through an accumulation of carefully and dispassionately collected data, only be inferred. So that, in dialectical terms, the strongest compulsion to believe drives from the logic of hierarchy, existential cycle, balance, cosmic equation and equilibration set within the context of an individual - an individual personality composed of physical and metaphysical dimensions. Trust in the teaching of a saint cements such faith. You might, on the other hand, well ask a sceptic what evidence *would* convince him. Would the answer 'none' amount to more than inconsiderate and therefore negligible denial? What else do you expect from whom the whole of metaphysic is anathema - a materialist. After all, how can you tell a body as precipitate of mental state? Aren't mortal coils rolled by a shuffle of the genes?

Ante-natal out-swing begins before biological fertilization, conception or associated, typical psychological development has linked with the equation. It involves a sequence of experiences with 'birds of a feather' on the same 'wavelengths' in various regions of the mind-world. You might call these regions 'heavens', 'hells' or 'in-betweens'. They are 'other stages', other 'theatres of experience'. As on earth, each to his interests and abilities; does your mind travel everywhere? Many mental places therefore go unnoticed by an individual 'surfer' through not cyberspace but real psycho-space. The Bardo Thodol describes this part of the adventure in terms of encounters with various deities, other mind-forms and the lure of different lights. As a perfume or particular musical signature can evoke memories or imaginations so each spectral 'vibration' rings with its own significance, associations and consequences. Whatever we embodied intellectuals may think of inner worlds, disembodied vision, *NDE*s or *OBE*s, they are very real to the participant. Who does not participate in his own thoughts and feelings?

Do you remember a death simulator - the sensory-deprivation tank mentioned in Chapter 0? While our embodied cogitation is swamped with sense data, are ideas and thinking simply fantasy? Or are they not, at the heart of our subjective reality, in fact the reverse? Why, therefore, consider the excarnate self fantastic or negligible? Just as your waking dream is this instant constrained by sense perceptions and the rules of physics and society, so a waking dream bereft of these is a mind unconstrained, unburdened, liberated. It operates untrammelled by material drag. *Its activity is faster, its current more powerful and its 'sense' more vivid.* Just as you identify with vivid dreams, so disembodiment identifies with the associations and projections it conveys. Is this 'liberated' state as low as madness or high as enlightenment?

For non-practitioners it is more like the madness of a vivid dream. The *Dharmakaya* has been missed. No body's there to damp or ground mind's flow; sufficient self-control has not been learnt and so it is swept at the mercy of psychological weathers. A highly volatile mind is now at the mercy of its own accumulated tendencies, habits and memories - *sanskaras, karmas* and such trains of thought as memories remembered detonate. It experiences their effect in 'heavens' and 'hells'. Emotions of fear or attraction arise. At the same time as the negative influence of these passions grows stronger, the mind's old tendency to reassert control begins. It calls up habits, recruits old certainties and marshals forces in accustomed ways. Clouds are forming. Habitual modes of perception are starting to take hold. Partiality increases, familiar proclivities begin to re-emerge and, in descent, the excarnate episode is drawing to a close.

At this point the Bardo reiterates its key principle - the overwhelming power of thought. Focus; quell chaos; inner principle affects outer practice; mind governs physical action; metaphysical precedes physical. Decisions at critical moments may (and in the case of rebirth will) have long-term effects. For this reason it advises the necessity of a lifetime's practice in order to neutralise karmas, stabilise the mind, control it and familarise it with the ways of Liberation.

You can best exert positive control over a situation when you are detached or know the drill. For this reason the Bardo seriously exhorts detachment from the illusion of what it calls *samsara*. *Samsara* means the ocean of existence, a kaleidoscope of ever-changing forms. It includes, as well as physical, mental conditions. The excarnate experiences of the *bardo* cycles are a psychological part of appearances. Whether able to navigate or not, an individual is swirled on the river of his *karma*, that is, the currents of cause and effect activated by the 'underwater' context of his memories, the waves of fear and whirlpools of desire. The next job of the Bardo is, therefore, to indicate how to retain control, how to shoot the rapids and end up, voluntarily, where you want to be. The best option was *Dharmakaya*, the light of communion. Failing that you might remain as long as possible in the best psychological condition, the best visionary frame of mind. We all know heaven from hell. Each exists wherever minds do. You want, obviously, heaven over hell. At worst you may be caught in an undertow that sweeps uncontrollably towards a future that will include, if the gravity of material desires cannot be overcome, reincarnation. *You then want to know how to get sufficient metaphysical grip to choose an appropriate birth.* It is easy to see that, in the swings between higher excarnate and lower incarnate phases of life, the high-level decisions are taken from inside the outside, from above the dead world's pangs or graves.

The fact of the matter is that, in cosmic terms, we make and break ourselves. **We, as humans, are the authors of our destiny in this life and, in the disembodied interphase, the so-called 'next'**. In which case, the greater our knowledge and detachment, the greater the authority or voluntary control we can exercise over our future. As every politician knows, timing is also critical. This, says the Bardo, is the moment that big decisions are taken. This is the board level. Once a pattern of ideas is fixed in material implementation it will be too late. You'll have to live with it. There's no vacating that particular body until next, in its due course, death comes around.

For this reason the power of positive thought is re-emphasised. This will have been practised (drilled is not the right word) to the point where prayer or immediate call on a spiritual teacher is, in any negative or un-spiritual circumstance, habitual. This is not news. All faiths use prayer beads. The origin of profanity was in this powerful practice. When something goes wrong a Christian, with a sincerity and trust that is its energy, swears by "Christ!" A Muslim cries "Bismillah!" or "Al-hamdu lillah!" All faiths have their emergency number and, with it, the connotations that will see them out of harm. The Bardo asserts that such a response can destroy karmic illusions in the same way as, calling out, you may wake from a nightmare. Waking, in this case, is the presence of the teacher, the teacher's 'aura' or the *Dharmakaya.*

By this time, however, the resurgence of ordinary mind is increasingly strong. The sky is overcast with ignorance and desire. The various passions, fears or guilt with which a person is involved can lead to trains of gross thought that, in turn, lead to craving for an incarnation. In the terms of the Bardo Thodol images and visions are linked to passions which, in turn, are linked to different realms of existence. Each is associated with a different, dull light. They include demons, dissatisfied ghosts, animals, humans, demi-gods and gods. This sounds primitive until you realise that cruelty, anxiety, sensual ignorance, humanity and higher characteristics can each be personified. A rapport with any of these 'wave-lengths' acts in the magnetic way of an attraction, an allure. Once you have entered that line, immersion is easy, extraction difficult. The advice of the Bardo is, predictably, to use your presence of mind, 'think positive', obtain voluntary control and not become involved. The catch is that, without advanced practice, this is about as easy as taking control of a dream - but much more important. You are sliding towards a long-term future and had better, if at all possible, get it right.

At this point the Bardo issues another serious warning. *If you have to take a body, take a human one.* If you missed *Dharmakaya* this time, your best chance to hit the cosmic jackpot next time is as a human. This is because conditions for spiritual progress are only favourable in this form. While at this stage a traveller may still be aware of the 'old' or previous life but waiting for the right connection to start off afresh. A judgment, similar to one that sometimes occurs in an *NDE,* is made. It is immediate, telepathic and almost a self-judgment. The 'soul' knows, set against the light of Truth, what its previous actions were worth. Were they meritorious or not? It knows what it deserves and, consequently, into what fateful conditions it will inexorably descend. The Bardo, in accordance with Buddhism, Christianity and all other faiths, issues strong warnings against behaviour for which, irrespective of penalty exacted or not exacted by earthly systems of justice, full debt will be discharged in the course of the subsequent excarnate phase and/or the reincarnation that flows from it. There are, it is noted, plenty of hells on earth.

If there is craving for a body the Bardo counsels caution. Resist rushing like a Gadarene swine into the first available container. Take care regarding choice of realm and parents. Even now you can, by calling on the teacher and steering the mind away from the vision of your prospective parents' copulation, close the entrance to rebirth. If, by dint of one attraction or another, this proves impossible you should think of them in terms of holy figures. If even this does not occur to you, enter only a human realm.

Obviously none of this is the psychology of an unborn child. It precedes connection with a biological form. It is the psychology of excarnate mind as it is swept towards the next embodiment into which, along with its associated archetype, it will slot with precision. *To repeat, it is not energy but desire, not motion but motivation that drives the metaphysical cycle. This cycle includes reincarnation. It is a specific amalgam or 'composite wave-packet' of desires that locks onto a specific body. It is this that defines the pattern of landing, the resonance or 'click' of appropriate settlement - a particular reincarnation.*

The opening move happens well before we are born. Choice or lack of choice concerning the nature and circumstance of the new body is certainly critical for the whole, consequent incarnate session. Like the first move in chess or the opening scene in any drama, it sets the tone for all that follows. First and foremost our sex, parents and thereby family, treatment, country, culture and education have already been chosen, in effect, by the outgoing mind. The choice of channel, if not detail of the individual program, accords with the quality of its own vibrant character, its resonance. The weight and kind of its own momentum has voluntarily, by decision, but in most cases involuntarily, by sub-conscious default, driven it to the right kind of chassis and surrounding circumstance. A mind that resists the *karmic* cataract is rare. Of all organisms only man is sufficiently subtle, intellectual and thoughtful to create art, literature, science and technology. Of men only a few are sufficiently above the crowd to remain collected, focused and unmoved by the currents of their own metaphysical interior.

Even at this late stage the Bardo has something to add. Now internal dissolution has been reversed, 'the ball has fallen' and mind is 'caught' in its fresh, material body. It now normally drops into an unconscious condition of sleep and forgetfulness prior to birth. If not, the Bardo exhorts it pray, continue spiritual practice and bless the womb as a sacred environment, a palace of the gods.

In the meantime external creation rapidly unfolds. An archetypal memory (Chapters 16 and 17) is simply a low-level, intricate oscillator that, 'strong-bonded' to its specific biological correlate, mediates between mind and matter. This correlate is every cell that reflects its particular pattern, its type of organism. The same dormant, psychosomatic level of mind activates, with its sub-conscious energy, the processes that constitute an amoeba, a coral or a daffodil. All organisms. In this case *H. archetypalis*, particularly in conjunction with an egg, is broadcasting on a frequency of interest. Don't certain songs immediately appeal and others leave you cold? It is now possible to see how a particular 'resonance of mind' might be drawn to attune with an archetypal program. It is possible to see how an individual's 'experiential context' (sensation, thought and memories) might, voluntarily or otherwise, 'interfere constructively' with a given 'vibe'. Or, as relayed in Chapters 15 to 17, personal mnemone might resonantly 'weak-bond' with typical and, thereby, the latter's associated body form.

Like to like, resonance to resonance, a synchromesh of personal with typical mnemone is formed. As two magnets engage until a stronger force separates them, so a particular mind is linked to its own general, archetypal memory; and is at the same time and linked to (or located with) a specific zygote, its parental bodies and unique development. Such biological development is programmed;

it is automatic and no more a function of incoming mind than your own metabolism is of yours this instant. Thus animation docks with automation. An individual nexus of experience locks onto a fresh vessel for another natural trip on earth.

It 'clicks' with *H. archetypalis* and thereby *H. electromagneticus* and the psychological barrage of sensation from cellular existence and a body-world. Individual mind is, like electricity confined to the circuits of a computer, now confined in a close relationship with the general psychophysical restrictions of archetype and the elements of body. Unlike *Amoeba archetypalis* or the others, yours then engaged the potentials of a human 'set' (*figs.* 17.8- 10). Such linkage involves an antenna capable of operation on sub-conscious, conscious and, although normally unused, supernormal 'frequencies' - such as the overdrive engaged by mystics and called *sahasrara*. This antenna is, you may recall from Chapter 17, called the caduceus. Just as expansive ascent of mind was simply an unfolding, so the reverse descent into a body involves restriction to a local, sense-based horizon. This sensible confinement incorporates an increasing involvement with the exterior physical shell of the world. Therefore a physical foetus develops and grows according to its caducean framework (*figs*. 17.5 - 10), the developmental program of *H. archetypalis,* genetic constitution and the biological mechanisms of construction.

Haven't you forgotten many things? Yet those events have played a part in moulding you. The whole 'inter-life' normally involves a progressive amnesia concerning specific memories from the previous, receding life. These are distilled. They are fused from a personal into an impersonal composite, a 'wave-packet' of desires, the 'outline thrust' of a particular character. Such distillation, such filtration squeezed from detail to the juice of generality is necessary. Otherwise details of memories unconnected, irrelevant and alien to a new life form's circumstance would interfere, confuse and distort its potential for choice. Without such a wash in the waters of Lethe how could a 'blank' yet 'characterised' baby develop clean into a fresh experience of life?

If reincarnation is a fact you'll never prove it 'scientifically' by instrument or by experiment. Subjective metaphysic is not open to objective, physical investigation except with respect to its lower linkage with electromagnetism of a body's nerves and brain. This link is broken when you die; there is no link between a past physical body and its follow-up. Dead meat decays; a corpse is not reanimated nor will a last trump ever more than scatter dust. Reincarnation is not resurrection of old ash or bones. Its possibility is no more open to laboratory tests than to experience. What, anyway, is personal experience (a ride in mind) but intangible, invisible and anecdotal? What else but anecdotal, therefore, could you ever call a claim to life preceding birth (which would happen in a mind-world) or claims for life after after-life? What proof life on life? Yet an anecdote involving either usual or unusual experience may, though 'unscientific', still reflect the truth. The only tools to deal with a process such as reincarnation are inference derived from careful case analysis or personal experience. In the absence of the latter reputable psychologists have still persevered. Thousands of astonishing, corroborated cases of past-life memories have been recorded from all parts of the world - irrespective of local systems of belief. Many are drawn from a source least contaminated with worldly wisdom, hidden agenda or guile and one that shows, emotionally, the most transparency - children under five. *If only one case squares as real the sceptic is in deep trouble; a materialist abhors the implications, belittles the research and plays, according to his faith, the 'we-will-find-a-rational-explanation' card.* On whose terms 'rational'? Better go and sift the wheat from chaff yourself.

Consciousness might not appear much to objective science. Being immaterial it might even seem like nothing. **From the inside, however, it is the basis of everything**. The first, creative imprints of Inner Nature are now the universal memory channels through which, *via* sub-conscious mind, the passive, reflex world of matter is informed and sustained. It is vibrated. Are not the Buddhist notion of perpetual flux and the modern theory of kinetic matter just the same? In the latter constant motion and vibration of bulk matter's make-up - molecules and atoms - explain such basic features of our world as temperature, phase change and diffusion. In the former explanation is extended to include a medium of information - mind that also oscillates, radiates and can't keep still at all. Each thing, whether psychological or physical, packs a note, a chord, a song; it broadcasts a signature of oscillations ranging from a simple to orchestral kind of archetype. In other words, vibration spans the archetypal pattern of an atom, cell and complex organism; it also covers every level, state and machination of dynamic mind. Is it force or music holds the world together? Aren't they, where the going's good, the same?

Now, by the end of the ante-natal Bardo, genetic and morphogenetic combination have created a foetus; and there is linked with its typical, generic mnemone the personal mnemone of an incarnating individual. This hierarchical process of materialisation has produced the heart-beat of a child. At the end of the psychology of reincarnation we have therefore pulsed down to the subject of the next five chapters, bodily biology. It is time to be born.

TOTAL CONTENTS

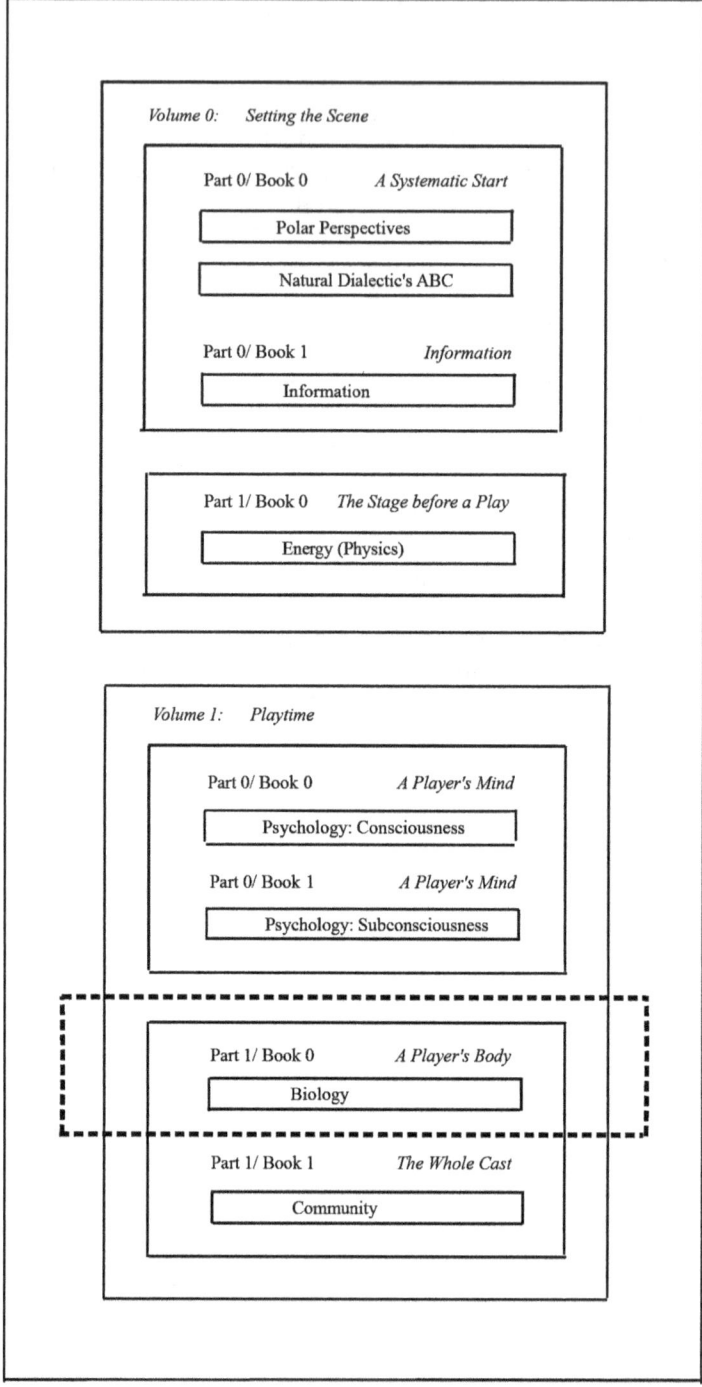

Volume 0: Setting the Scene

- Part 0/ Book 0 — *A Systematic Start*
  - Polar Perspectives
  - Natural Dialectic's ABC
- Part 0/ Book 1 — *Information*
  - Information
- Part 1/ Book 0 — *The Stage before a Play*
  - Energy (Physics)

Volume 1: Playtime

- Part 0/ Book 0 — *A Player's Mind*
  - Psychology: Consciousness
- Part 0/ Book 1 — *A Player's Mind*
  - Psychology: Subconsciousness
- Part 1/ Book 0 — *A Player's Body*
  - Biology
- Part 1/ Book 1 — *The Whole Cast*
  - Community

Volume 1: Playtime

Book 1: A Player's Body

Part 0: Passive Information (Biological)

19. Principles of a Unified Theory of Biology

Biology is the study of sensitive embodiment (as opposed to psychology which is the study of embodied sensitivity). From an initial study of archetypal biology (Chapters 16 and 17) let's turn to classical biology. Just as the study of physics and chemistry has delved subtly inward to identify unseen fundamental particles and four forces from which all gross, visible aggregates are expressed, so biological focus has now, as textbooks well illustrate, penetrated inward from the study of external phenotypes, taxonomic botany and zoology, physiology and anatomy to its subtle phase of molecular biology, dynamic metabolism and primal information written up on the four letters of *DNA*.

	tam/ raj	*Sat*	
	lesser consciousness	*Consciousness*	
	expression	*Potential*	
	order	*Preordination*	
	informed/ informant	*Information*	
	agents of purpose	*Purpose*	
	periphery	*Centre*	
	parts	*Whole*	
↓	*tam*	*raj*	↑
	passive	*active*	
	outer	*inner*	
	non-conscious	*conscious*	
	physical	*metaphysical*	
	objective	*subjective*	
	informed	*informant*	
	ordered	*ordering*	
	inanimate	*animate*	
	matter/ body	*mind*	

What surprise if biology's principal is life? Life and, which is survival, more life? What, however, *is* life? Of what exactly does a life or form of life consist? How differs animate from inanimate?

parts	*system*
analysis	*synthesis*
reductionist	*holist*

Bottom-up, the distinction is blurred. 'A whole's no greater than its parts', reductionism cries, 'and these are lifeless molecules'. If everything's material then can't you feel materialism straining at its leash to formulate that life is a molecular assemblage - one continually regenerating, replicating and evolving? That is all. That's essentially you and me! Atomic bits, strange bobs, composed spontaneously. Life, like mind, is construed to be a perhaps unlikely but 'emergent' property that 'happens' when some most unlikely chemical configurations accidentally occur, re-occur and as they do so self-complexify. Molecules self-organise; much more powerfully, they self-systematise. How else, if cosmos is a mindless entity, might one infer that patterns of embodiment are other than the product of contingency (that's circumstantial chance) constrained by natural law? Vitality is not defined subjectively but according to invisible molecular patterns and visible bulk patterns (such as organs) whose characteristics somehow include sensitivity, growth and reproduction. But if you defined a robot or an automated production line in terms of physics and chemistry alone would you have the whole truth? When it comes to their analytically correct definition of biological equipment, molecular biologists are especially prone to think so. Theirs is at the basis of a scientific, that is, wholly naturalistic version of events. And is your life not an event?

<u>The Principles of a Unified Theory of Biology</u>

	tam/ raj	*Sat*
	below	*Top*
	consequent	*Before*
	lifetime(s)	*Source*
	outer peripheries	*Inmost Core*
	shells	*Nucleus*
	informed product	*Message/ Purpose*
	sphere of influence	*Governor*
↓	*tam*	*raj* ↑
	informed	*informant*
	passive info.	*active info.*
	physical	*psychological*
	metabolism	*code*
	business	*instruction*
	phenotype	*genotype*
	oblivious/ purposeless	*purposive*
	division	*unification*
	separation	*relationship*
	isolation / part	*system/ communication*
	tendency to randomise	*thematic comprehension*
	matter	*mind*

Top-down, you have another view. Remember (*fig.* 0.5) a cosmos of three levels. Its central cosmic potential is (*Sat*) Essence of Pure Consciousness. This Essence is gradually confined; it is, in combination with energy, reduced along a conscio-material gradient through the (*raj*) psychological levels of mind and (*tam*) physical levels of body. Mind is a centre of specific information; body is a special nexus of energy/ matter.

Do forms of life reflect the cosmos in three-tier construction? Are they microcosms of the macrocosm? Mind may be active (or conscious) - which subjectivity is present in the higher animals; or inertialised (subconscious) in the inflexible forms of memory and instinct. Natural Dialectic identifies this lower, passive psychological level as the pre-physical level of archetype. Archetypal memories are part of universal mind and constitute the link of biological form with its initial conception. *A typical mnemone (Chapter 16) operates in every cell of every organism, that is, every form of what we call 'alive'.*

Many kinds of organism are almost ignorant of consciousness; they hardly rise to waking state and many others (dormant plants, fungi and bacteria) never wake at all. Can a bacterium, however, 'hear'? We know they 'smell' and 'taste' and 'see' and 'feel' but such response is chemically automated and, no more than a machine, needs mind - except at its inception. The question's not what a bacterium, any more than any sleeper, 'thinks' but what originally thought its system out. Nor do computer-like *DNA* operations and metabolism mean that it has no instinct or it lacks subconscious archetype.

At the third stage - the lowest, physical level of energy and atoms alone - all organisms are embodied. Inanimate matter has no life but bodies are, like any other kind of vehicle, composed of it. Matter has no knowledge. It has no sensitivity. Its only passive kind of information is embedded in behaviour called by physicists the laws of nature. Dialectically, therefore, your body is informed according to those laws but also, somehow, by a coded plan. Have chemicals, however, any sense or sensitivity? They have no mind and, of themselves, are dead as stone.

	tam/ raj	*sat*	
	3/ 2	*1*	
	H. sapiens/ H. electromagneticus	*H. archetypalis*	
	subsequent order	*archetype/ blueprint*	
	fixity/ action	*potential*	
	informed effects	*first cause/ prior information*	
	contingencies	*rule/ law*	
	material outcome	*top matter*	
	physical	*metaphysical*	
	non-conscious	*sub-conscious*	
↓	*tam*	*raj*	↑
	3	*2*	
	H. sapiens	*H. electromagneticus*	
	governed	*governing*	
	gross consequence	*subtle medium*	
	outward effect	*inner cause*	
	mass	*force-fields*	
	bulk structure	*wireless influence*	
	form	*function*	
	fixity/ anatomy	*physiology/ fluidity*	
	containers	*processes*	
	product	*reactions*	
	phenotype	*code and metabolism*	

Not perceived so not believed. This might be the epitaph yet, at the same time, clarion call of scientific progress. How could empirical experiment notice what is immaterial? Although the distinction between passive, bio-psychological mind and biological (i.e. physical) material is critical it is not mentioned by materialism because it's unperceived. It is therefore, because of its importance, worth re-stating. Although life has a psychological aspect, biology is the study of *bodies*, not minds or souls. *However, overlap between (metaphysical or subjective) psychology and (physical or objective) biology occurs across a psychosomatic interface.* It is between sub-conscious grades of mind (which involve memory) and subtle (e.g. electromagnetic) forms of non-conscious matter. In other words biological life straddles the border between transcendent (*sat* or archetypal) and active (*raj* or quantum) levels of matter (see *fig.* 5.2 and Chapters 7 - 12). It involves both biotic (psychological) and abiotic (physical) aspects. The former includes both conscious mind and (*figs.* 17.6 - 10) sub-conscious memory including instinct; the latter includes light, heat and gaseous, liquid and solid phases of matter. Biological interest therefore kicks in at the point of what is still, dialectically, metaphysical, that is, the repository of archetypal memory. It involves (*fig.* 0.11) three subdivisions of matter; and *fig.* 16.1 relates these to the triplex covering we call a biological body (the potential matter of *H. archetypalis*; the orderly reactivity of *H. electromagneticus*; and bulk *H. sapiens*). The object of this chapter is, however, body - biology and not psychology. Its focus is therefore on levels 2 and 3 of the previous stack - the physical pair.

But information originates; whether archetypal or genetic it is the cause that specifies material effects; it amounts to potential that directs any physical course of action or construction. From immaterial principles are devolved, as in the case of all technology, material mechanisms. **Thus top, potential matter, archetypal memory or typical mnemone is the basis of conceptual biology**. Because of this a brief recap is necessary in order to place stages 2 and 3 in context below 1.

Archetype constitutes the bio-psychological governance of all organisms. Indeed, in the majority its passive, sub-conscious or dormant mode of operation predominates. It involves their 'norms', it is their 'natural code' or 'law of nature' and, in each case, its complexity stands, as a machine to a crystal of salt, against the simpler programs that govern non-purposive constructions. A law is an inflexible program, principle or pivot around which the operation of a system sways. It amounts to preordination, to information that precedes action and is, therefore, biology's (*sat*) informative potential.

Mind, an 'attractor', tends to draw together and arrange things in the image of its purposes. In this case a typical mnemone exerts such tendency in the form of an overall scheme according to which parts are assembled as an integrated whole. With respect to any given type of organism one might call such influence its 'law of nature'. Its operation is like natural law of physics - involuntary, reflex and fixed. Archetypal memory stores information; it carries reason like a book, passively.

Is there an active principle that informs the passive, bodily expressions of life? Abiotic atoms do not crave to live or, which is to continue living, to survive. The complex molecular biology of a cell no more 'wants to survive' than building materials 'want to be a house'. Life, on the other hand, is distinct from non-life. No doubt as simple an object as a brick or a teacup has a purpose; it has no mind in its passively informed, purposely arranged structure but (Chapters 5, 6 and 16) each is a memorial of its creator's mind. Could a principle of purpose inform biology? Could the purpose be 'more life', a 'continuity of life' and 'life in a proliferation of experience and form'? In this case, while some forms are minded to *want* to survive such conscious wish is an 'extra' in the sense that all forms *are wanted* (i.e. designed) to survive. Survival is 'wished upon them' and the entire pattern of their animation is informed by the ramifications of such purpose. From sub-conscious instinct upwards life shows mind, from sub-conscious instinct downwards it shows bodies equipped with a hierarchical system of homeostatic norms. These norms correlated with their maintenance in the form of appropriate biological mechanisms.

In short, survival is the motive that informs life forms. All bodies are constructed around this conceptual core, this metaphysical axis, this purpose of obtaining more life. In this respect where mind is informative life is fundamentally purposive. Its purpose is survival and then more comfortable survival. Metaphysical drives physical and the purpose is translated into information incarnate - biological form. While mind's natural, subjective mode of information storage and transmission is memory, the natural objective method is code carried, symbolically, on a material substrate. From a *top-down* perspective the basic unit and most nearly non-conscious form of life, the cell, carries both metaphysical (archetypal) and scientifically accessible (genetic) information.

A Dialectical Plan of the Way Life Works

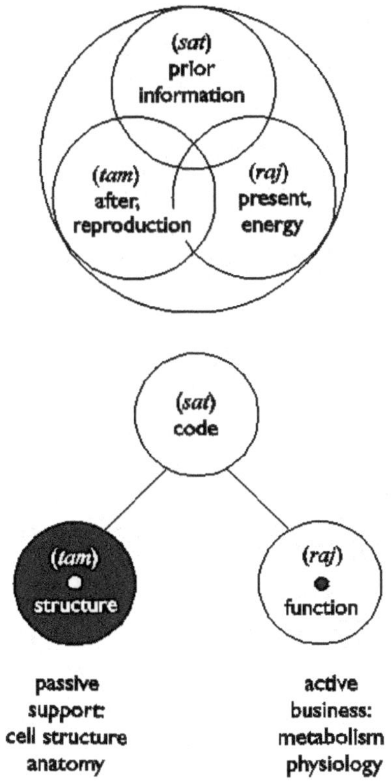

Information precedes. It is prior and anticipates. Information is the potential and *sine qua non* for the action that consequently issues orderly.

Information comes, as already explained, in two forms - active/ conscious and passive/ unconscious.

The **conscious case** involves sensation, intention and knowledge; it employs, in order to experience the body-world, a voluntary nervous system.

The **unconscious case** involves both psychological and biological components. These are (*sat*) archetype (instinct and morphogene); (*raj*) active but reflex homeostasis by, in some organisms, involuntary nervous systems and, in all organisms, pre-arranged, cybernetic metabolism principally controlled by genetic information coded up on *DNA*; and (*tam*) the physical organs of action and response. Life's process is one of dynamic equilibrium. Equilibration. Its goal is balance in accord with preset norms. Metabolism, being totally information-dependent, works with reference to precise, incoming messages and equally precise genetic response indexed by particular messengers and changing conditions inside its cell. Life is, in this way, an incarnate flux of order due to information.

Energy provides for survival now. It involves, metabolically, photosynthesis and respiration. It promotes cell biochemistry, physiological processes and, on the large-scale, (nervous) sensation and (muscular) motion.

Structure represents the outermost, fixed realisation of shape. Its base domain is energy's container, a fixed expression of internal, orderly flux. In other words, a 'phenotype' (see Glossary) is a peripheral aspect whose body both reflects and fixes the shape of inward information and energy. It includes the anatomical aspects of enduring physical support. If, for example, the human case is taken then its base caducean positions (*fig*. 17.10) indicate anatomical foundation that supports higher-level structures and processes, exhaustion of waste products and last but not least reproduction of the entire physical structure. Such reproduction averts extinction. It is the way that a typical life-form cycles forward through time.

fig. 19.1

The first biological hierarchy is informative. What is information but a central and essential source; hence streams guidance and to have reached back to source is to have gained original knowledge, core principle, the heart of any particular issue. In Latin 'nucleus' means 'nut' and, by extension, implies fruit and tree. Whence would you expect to find the kernal of cellular knowledge, the heart of biological issue, but at its nuclear core? And, indeed, every single sexual gamete or asexual cell encapsulates, at its nuclear core, a physical book of life. Its genetic information helps govern the time, place, specificity and, through cybernetic mechanisms, the quantity of bio-materials needed by itself or by any other part of its larger, archetypal 'self', the body to which it belongs. **You cannot contemplate biological life without, first, information. In this respect a whole precedes and, by archetypal virtue, is distinct from the sum of its**

separate and varying parts. In other words, a single and unchanging metaphysical blueprint is expressed, in the 'contingent' circumstance of individual genetic make-up, as myriad various bodies. Aren't you, like every other human, biologically unique?

	tam/ raj	*sat*
	divisions	*union*
	parts	*wholeness*
	ups and downs	*balance/ poise*
↓	*tam*	*raj* ↑
	inertial equilibrium	*dynamic equilibrium*
	fallen flat	*oscillating*
	irregular/ out of tune	*rhythmic/ regular*
	loss of balance	*equilibration*
	disunity/ illness	*homeostasis*
	differentiation	*integration*
	isolation	*communication*
	falling apart	*coherent parts*

The archetypal principle governing biological organisms is (*sat*) balance. Around this axis vibrate important (*raj*) dynamic principles. These centre round the notions of union, unification and wholeness. They include stability, continuity, coherence, integration, relationship and communication; and they show as internal constancy, dynamic equilibrium obtained by homeostatic regulation (or stable feedback cycles), finely controlled metabolism, a finely tuned interdependence of parts and the recuperative, self-corrective nature of all healthy bodies. This *vis medicatrix naturae* is a product of order, organisation and integration. Multiple, integrated feedback cycles are directed towards a single end - stability. *As such they are conceptual, they are anti-chaotic and each is represented by an irreducibly complex set of agents including sensor, regulator and effector.* Tailor-made electrical, hormonal and organic systems interweave. All bend to the order of preordained norms. Norms are laws or rules or regulations. The codes of their command are bio-universal. They apply from a single cell to whale or redwood tree; they apply to your own organic systems made up of perhaps 100 trillion cells; and they rely critically on 'potential action', that is, 'prior information'. *How, when these parameters are needed from the start of all and any life, could nature creep towards order out of accidents?* **The panoply of life's fully informed operations is as entirely consistent with deliberate design as it is entirely inconsistent with neo-Darwinian evolution's core co-creators, non-orderly (that is, chaotic) chance and non-creative 'necessity' (another name for inanimate order).**

In biology keeping the balance, that is, the maintenance of dynamic equilibrium is known as homeostasis.

All other biological functions are expressed as peripheral or subsidiary adjuncts to this central objective which in every case embodies the core informative (*sat*) principle of integral union or survival as an intact whole.

Homeostasis (Chapter 5 and Glossary) *is balance-in-action. If Essence is (Sat) Balance, then such balance is reflected in the (raj/ tam) vectored motions of existence. Balance-in-action is expressed as 'laws' and 'norms'. Such balance-in-action (also called dynamic equilibrium or regularity) is a key facet of existence.* In this case biological homeostasis is a reflection of a core cosmic process. It is useful to think of a biological organism as a spatial form of symmetry involving bilateral, radial or other form of anatomical regularity; and as a temporal form of musical harmony. This harmony involves the dynamic integration of a number of physiological vibrations each deriving from a separate instrument of homeostasis. To work healthily biological processes and productions need 'not too much and not too little' of any item in question; they all involve coordinated balance, the middling way, the centrality of preordained homeostasis. This informative process holds in place, according to provisions in the form of norms, the (*tam*) disintegrative tendency of disparate, undirected forms of matter. Vibrant health, that is, properly dancing homeostasis is the natural condition of a body. There exists a natural elasticity that attempts to ease aberrant, failing homeostasis back towards its archetypal norm. It is therefore easy to see how (*raj*) dynamic equilibrium or homeostatic ordering according to prearranged principles is central to the process of biology.

It is worth rephrasing this key principle. The order of creation involves a 'homeostatic triplex' of (*sat*) potential, (*raj*) action and (*tam*) fixity; and its overall process is one of continual equilibration. Biological order reflects cosmic; its microcosms reflect the macrocosm. The framework runs from principle to practice, concept to expression, purpose to the mechanism of its achievement. *Information precedes energy and its 'sticky' part, mattter.* It acts from a level above on one below. Hierarchically. The first two phases of creation (Chapters 5 - 12) are author (conscious mind and the committal of conception into memory) followed by stage (informed but inanimate matter). The third is animation or, if you like, characterisation (a combination of mind *with* matter). In this respect biology can also be defined as the study of (*raj*) active on (*tam*) passive information or of a highly informed, purposively programmed, 'machine-like' case of material complexity. Fixity of structure made of moving parts is, in this view, the final expression of prior potential in the form of information. How might hierarchical biology reflect the conceptual basis of its own created order?

Biology in Brief: Information Plugged into Structured Energy

(i) The Cell.

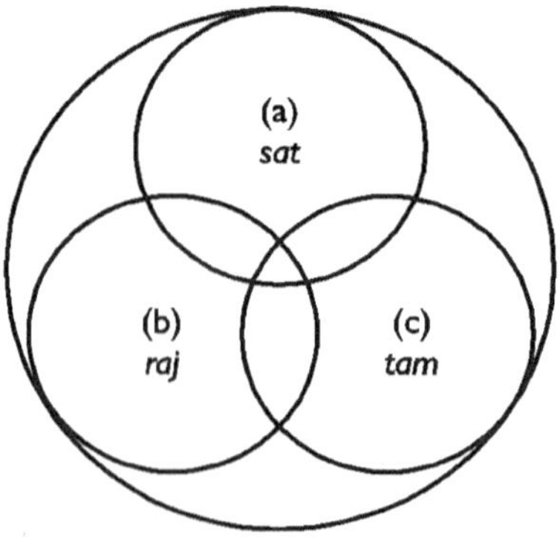

Smaller than the eye can see cells embody marvellous technology. And cell construction reflects the dialectical way life works. The 'colours' of existence are seen as a conscio-material spectrum, that is, the hierarchical dispersion of consciousness-in-motion (active information or mind) and variously stored images (passive information set in memory or in codified matter). Biological 'colours' are seen as a latter, lower waveband of that overall spectrum. Just as (Chapters 1 - 12) energetic existence is shaped according to the three cosmic fundamentals from its Potential, Essence, so the graduated characteristics of biological plan are based on the transparency of their potential matter, archetype. Cellular and multi-cellular composites are physical expressions of psychosomatic, archetypal records; expression can be seen in terms of subtle, electromagnetic interactions or their gross effect viz. atomic, molecular and bulk biological structures.

(a) (*Sat*) nuclear information is exchanged with active metabolism and passive cytoplasmic and membranous structures. A nucleus is a highly organised and sub-compartentalised organelle. Its signalling agents are specific biochemicals.

(b) Cytoplasm with energy-related organelles: the cell's (*raj*) business complex.

(b) (*Tam*) structural support: cytoskeleton and membranes involved in the stabilisation of intricate metabolism, packaging, intra- and extra-cellular import/ export duties, shape-making morphogenesis) and boundary constructions.

(ii) Link Scheme (with *figs.* 17.6 and 19.1).

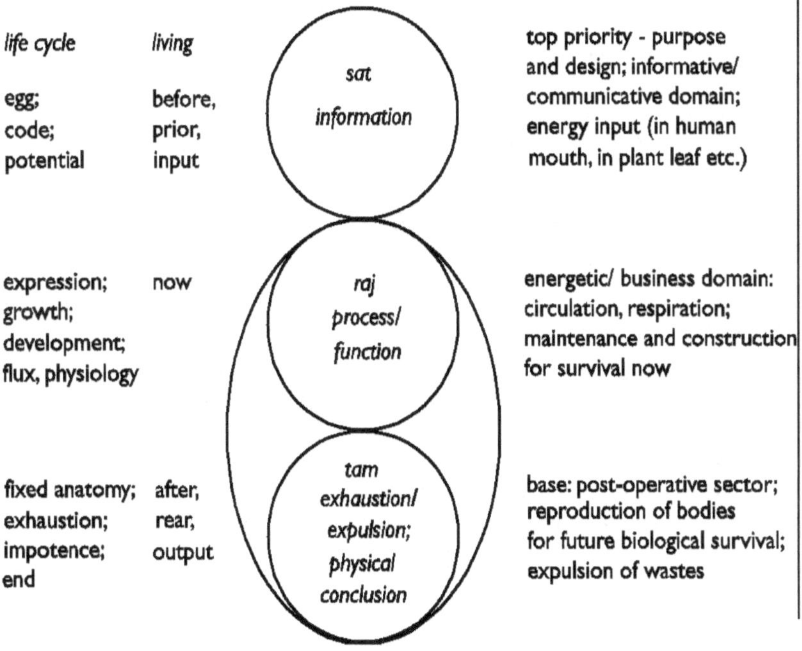

Note, as in *fig.* 19.1, how biological expression involves cosmic fundamentals with respect to both life cycle and day-to-day living in space and time. In such dialectical sequence (*sat*) information precedes orderly outcome, (*raj*) energy supports current dynamic equilibrium and its sub-cycles are 'housed' in (*tam*) structures that, while exhausting, temporarily endure. Continuation of such cycles, assured by whole-body reproduction, takes care of survival into the future.

Thus, in the manner of all natural action, information (or rule) leads to process (or behaviour) and finally information out as result, effect or end-product. Just as a chemical reaction could happen (has potential), reactants are fired to react and the process runs to fixed conclusion, so with life. Central potential (information stored in an arrangement of *DNA*) is employed in metabolic business; the final result of either metabolic or phenotypic development is a target form. This target is a structural container, a fixed or stable form. Approaching the end of its functional purpose/ life it decays into waste. The end of process is, therefore, on the one hand maturity (attainment of target) and on the other, worn-out (entropic) exhaustion from this target. The end comes with fixity, expulsion or death.

(iii) Relate these representations to *figs.* 13.4 and 17.9 - 10.

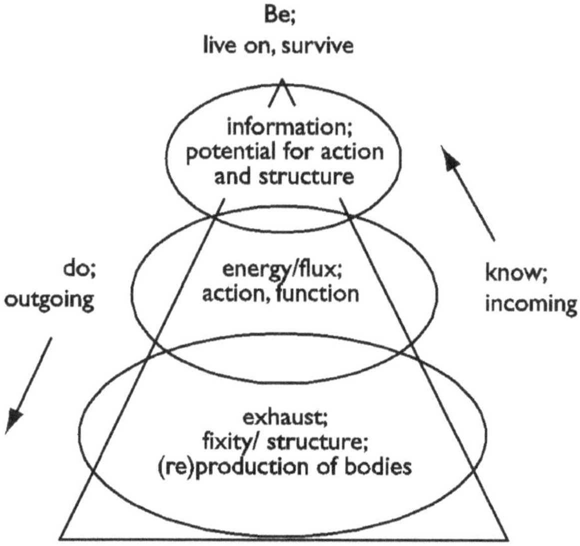

All cells and bodies symbolically represent Natural Dialectic. Yours expresses it particularly well (see also *figs.* 17.9–11). Top and centre, head and spine incorporate a concentration of information/ control systems. Its central portion (including trachea, lungs, oesophagus, stomach, pancreas, small intestine and liver) is involved with the uptake of materials for respiration; a 'warm' heart irradiates both gaseous and liquid elements. Energy means business. Arms are 'cranes' for hands that feel and manipulate; legs and feet are appendages that combine to give us balanced locomotion. Last and least but still a crucial part, the lower portion is involved with water regulation (kidney), the elimination of gaseous, liquid and solid exhaust (kidney, bladder and colon) and the reproduction (by way of sexual apparatus) of fresh, fixed forms. Solid bodies are elaborated here. The sweep of cosmic fundamentals (*sat* to *tam*) clearly falls from skull to pelvis, brain to genitals.

The Information Plug: Mammalian and Cellular Schemes (see also *fig.* 14.4).

(a)

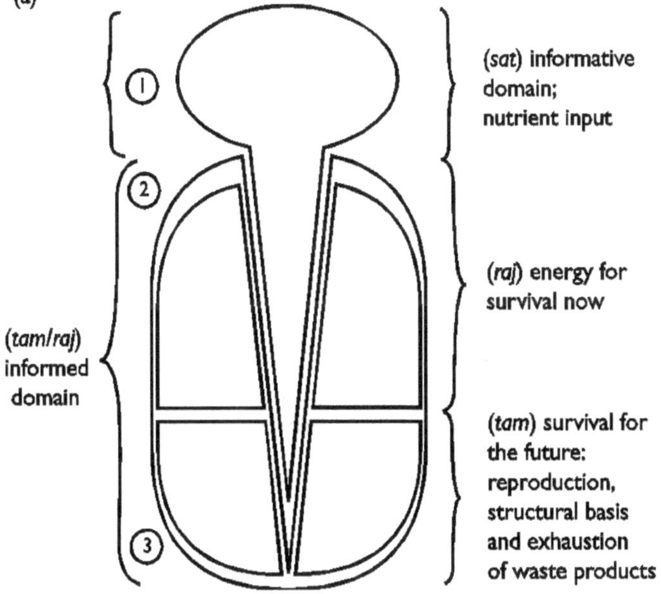

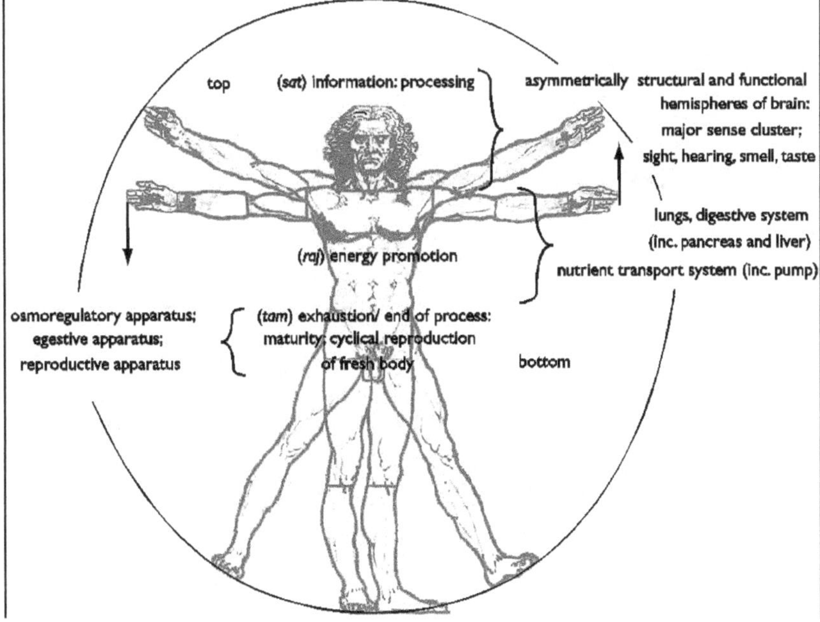

(b)

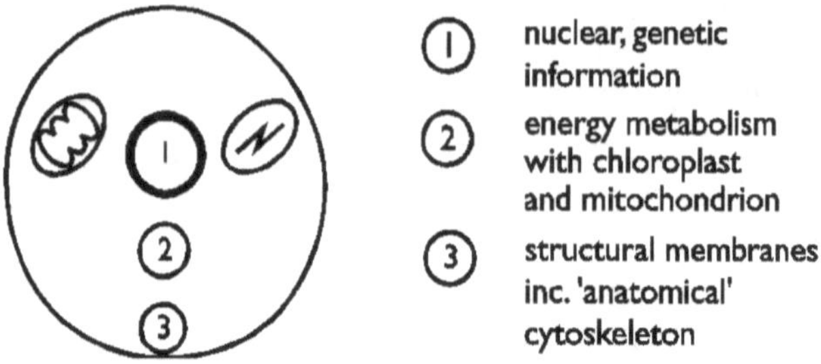

Information is plugged into energy/ matter in its active (physiological) and passive (structural) roles. The organic construction of bodies is organised around oscillators that reflect, in their different combinations, different shades of cosmic emphasis. In the human case (*figs.* 17.8 - 10) a body incorporates a balanced set of all such oscillators. As a result it is said to reflect the epitome of natural design - not least in its top element, an information modulator called the brain. The 'set' reflects a drop on the 'cosmic' gradient from potential through kinetic to exhausted phase, from inward (*sat*) informative code through the outward expression of (*raj*) energetic functions to, as in all good engineering design, (*tam*) structure that most efficiently accommodates their patterns of behaviour and their ends.

(c)

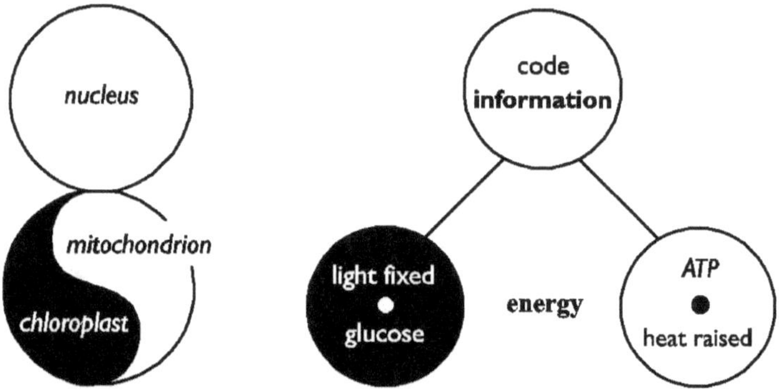

Compare this diagram with *figs.* 1.1 and 21.1. Information is the potential for orderly action; action depends on energy. Interestingly, the correlation of this basic existential dipole is reflected in the substructure of a eukaryotic cell. A primary division of cell function is distinctly represented by organelles of information (nucleus) and polar energy (chloroplast for (↓) fixing light and mitochondrion for generating the agent - *ATP* - of its orderly thermal (↑) release).

Indeed, if bodies are designed *top-down* a cell is *central outward*. Nuclear information leads the cytoplasm's business zone; its body is dynamically stabilised by membranes, walls and cytoskeleton.

fig. 19.2

From a *top-down* perspective the overall source of bio-cybernetic government is psychological. (*Sat*) immaterial, sub-conscious archetype is a recorded program whose routines exchange information with the electromagnetic and, thereby, atomic and molecular aspects of physiology and anatomy. These include, centrally, operational reference to a systems database, a codified instruction manual for the construction of physical materials, that is itself made of a chemical (*DNA*). Psychosomatic exchange is, whether sensitive or motor, continual. Its physical 'illuminant' is (*raj*) energy in the form of electromagnetism. As well as mediating informative processes electromagnetism in the form of light also substantiates biological energy through the mechanism of photosynthesis; then respired glucose mints a universal coinage, *ATP*.

With energy we reach stage 2. Expression follows its potential; after information comes energy. How might (passive) information be plugged into structured energy? How might it shape the matrix of a body?

Non-conscious matter has no tendency to create orderly, purposive patterns of behaviour. The reverse, in the case of non-living structures, whether simple or complex, it tends to disorder. Uncoded, non-purposive things are wholly prone to the 'influence of entropy'. In this 'dead' zone, therefore, the term 'equilibrium' mostly applies to 'inertial equilibrium', the end of reaction, flatness or exhaustion. Diffusion towards this static state is really breakdown, scattering and *(tam)* death.

Biology, on the other hand, is the study of life. Its processes involve potential and vibratory as well as inertial forms of equilibrium. Not only is there the informed, dynamic equilibrium of homeostasis but the energetic capacity to grow and to develop. Its 'waiting equilibrium' is represented, in the case of sexual reproduction, by an unfertilised egg whose closely controlled energies will, once ignited by the sperm, flare into living form and function. And, in the case of asexual reproduction, the latent reproductive capability of any particular clone; examples of such programs include binary fission and mitotic budding, sporulation, vegetative propagation and so on.

Once latency is sparked then nuclear information guides the action. Although always and wholly dependent on the potential of coded (passive) information, life's equilibrium is changed into (*raj*) dynamic gear. Its motion, that is, its disequilibrium would soon fall flat but programmed, energetic channels keep it juggled in the air. No doubt biological bodies are made of inanimate molecules but organisms 'bounce' and 'vibrate'. They 'cycle' according to the close government of pre-set, homeostatic norms. As long as the physiological and anatomical bundle of life support systems designed to promote this 'normal' oscillatory process stays intact, an organism will live healthily. Purposive, life-sustaining systems allow survival in varying physical conditions. They are complex. Such a degree of intentional order must be minutely programmed - even more so if it is to include both adaptive flexibility and the long-term survival strategy of self-reproduction. Of course, the agents of such program must themselves, like robots in an automated factory, be physical. Which elevates not detracts from the immaterial substance of intelligent forethought.

Forethought is very well but bodies need (*raj*) energy for survival now. **Following the informative, biology's second hierarchy is energetic.** On cue, there radiates reflection of divinity - prismatic rainbow or a wall of light that boundlessly supports the house of life on earth. Hence, first in line stands photosynthesis. Nutrients are 'frozen' out; a rain of nectar and a snow of sugar crystals is precipitated from the leaves of plants. Pure energy is joined by energetic matter (gas) in the form of oxygen; it fuels, like a bellows, the 'fire' of respiration. This regenerates the 'battery' called *ATP* - a standard, bonded amount of sunlight universally acceptable, like a coin in an international slot-meter, to the diverse metabolic processes of life. From this slip of 'grounded light' is struck a flame to drive metabolism and the chemistry of everybody's body. Energy metabolism is a key operation whose two aspects, anabolic photosynthesis and catabolic respiration, underwrite the 'buzz' of energy we know as life.

This 'buzz' involves knowledge, action and construction. That is, energy underwrites information (in the form of nervous patterns of sensation and thought), action (muscular response, growth or other changeful motion) and supports all life's anatomies. The biological world engages a whole range of energetic response using flexible conscious, relatively inflexible instinctive and inflexible biochemical programs. In humans, for example, the power game recruits all three kinds of program. Hunger and thirst are slaked by consciously finding and taking food and water; oxygen is breathed; thus three major systems (ventilation, digestion and blood circulation) supply respiratory power stations called mitochondria.

	tam/ raj	*sat*
	3/ 2	*1*
	peripheral	*central*
	outworking/ expression	*information/ program*
	shells	*nucleus*
	exhausted/ kinetic	*potential*
↓	*tam*	*raj* ↑
	outside	*inside*
	exhausted	*kinetic*
	info. out/ result	*program/ info. activated*
	compartment	*communication*
	discontinuity	*continuity*
	fixed/ static container	*fluid process*
	boundary/ buffer	*motion/ change*
	'lifeless' aspect	*reactive/ 'lively' aspect*
	structure	*function*
	shape/ morph	*operation*

To summarise thus far, an *information hierarchy* may involve conscious mind, certainly subconscious instinct and an archetypal morphogene; it may involve nerves but certainly involves the chemical messaging of hormones and a *DNA* databank. The *energy hierarchy* may involve muscles or other modes of manipulation, balance and locomotion; it certainly involves energy metabolism (photosynthesis and respiration), catalytic metabolism and bulk supportive organs.

↓ *tam*	*raj* ↑
massive	*energetic*
'frozen'	*dynamic*
fixed form	*process*
distinct/ separate	*interactive*
discrete	*cooperant*
unit	*system*
cell/ organelle	*metabolism*
membrane	*cytoplasm*
anatomy	*physiology*

After energy comes matter, the lowest precipitate of the entire conscio-material gradient. Non-energetic building material (whose state is solidity) ranges in phase from might be termed a liquid crystal matrix to hard shell and bone. It includes gels, membranes, organelles and various qualities of tissue, skin and hard materials. These constitute the (*tam*) board on which life's programmed circuits run; they provide form for function, a container for action, anatomy for physiology. **Thus a third (*tam*) *structural hierarchy* completes the purposeful range of dialectical logic.** Anatomy complements physiology, form facilitates function. (*Raj*) cellular functions, not least metabolism, take place in a fluid, flexible solvent - water; however, as well as functional fluidity an organism needs a robust (*tam*) structural framework at every one of its nested levels to contain, support and control the import/export economy of its activities. And cells employ rigidities such as walls, membranes and cytoskeletons, to contain, compartmentalise and shape their industry. For example, the nuclear 'mind' of a cell is, engraved on 'tablets' of *DNA*; this physical memory exchanges information with cytoplasmic industry, the mid-part where energy-related organelles (chloroplast and mitochondrion) are found and biochemical business is carried out; and the whole is held in place by membranes, walls and cytoskeleton. Multi-cellular organisms employ internal skeletons, connective tissues and external shells, skins, barks and analogous components to shape, support and protect the function of soft, internal and external organs. In the human body (*fig.* 17.10), organs of informative command and control are concentrated at physical headquarters - head and spine. In mid-torso are arranged your gross 'organs of energy' (lungs, heart, pancreas, liver and digestive tract); and, at base, those of exhaust and physical reconstruction, that is reproduction.

The question is whether this spectral order of biological operations fits the facts. Do mind and light combine with the states of matter to precipitate the swaddling clothes, glad rags and shrouds of life - bodies biological? Is this conscio-material description true?

At your head instruments of information also follow a *top-down* dialectical order. Eyes for light and ears for sound are positioned above those involved with the energetic hierarchy - nostrils for gaseous intake and mouth for the ingestion of liquids and solids. Sensitivity to touch, temperature and pain envelop the head but are spread to cover the whole lower structure. Instruments of informative manipulation (mind/ brain for thought and internal nervous/ hormonal processing) and energetic manipulation (arms and hands for external things) are arranged above contact with the ground. From heaven to earth underneath, you stand *on* locomotive legs and feet. Plant life, lower down on the oblivious horizon of subjective scale, is rooted in a lifeless matrix, soil. Roots suck the cold and watery breast of mother loam but, from this anchorage, stems reach towards the gaseous sky, light and the sun. Their aerial constructions are involved with leafy, light-related energy metabolism for energetic survival now or flowery reproductive patterns and behaviours so that other generations shoot into the future.

Indeed, life is not a here-today-and-gone-tomorrow, single generation effort. *It looks to the future, which atoms and inanimate matter never could.* More life *is wanted.* Organisms are *passively informed* and supplied both with subjective archetypal memories, objective genetic code and reproductive mechanisms by which original construction can be accurately reconstructed. *Reconstruction provides for maintenance and repair but its full provision is in the reproduction of a whole new body - continuity of the species, survival into the future.* Not only is the base domain of our structure (*fig.* 19.2 (iii)) involved with exhaustion, elimination and (as with the pelvis) containment and support; it also involves potential, in the form of reproductive apparatus, to fabricate a line of life's containers. Is not the base emission of fresh life, dropped in order that it rise, a supreme and lovely paradox? Earth to earth and earth to sky, vehicles are developed from a seed. Solid to solid, from the base or back-end of a body bodies are prepared and launched into terrestrial odysseys. Massive yet very subtle structures, the substance of biology, are robotically created in a potentially endless line.

In brief, the main (*tam*) static principle of both physical and biological matter is *confinement* with its correlates of structural discontinuity, separation and individual difference. Bulk matter, especially solid form, is the lowest, most exhausted aspect of creation. The biological restraint of flux occurs in functionally appropriate, 'customised' containers made of fixed matter. We call them membranes, tissues and organs that interact internally as systems with each other; such covers, skins or rinds protect, sense and enable physical response.

As well as intake there is output. A cell membrane is a highly gated system through which secretions and excretions flow. The (*tam*) hierarchy includes exhaustion of waste - gaseous excretion from nostrils, liquid from urethra and egestion from the anus are the way for you. Such expulsion of waste materials is, it was noted, from the same low-level vicinity as the gravid construction and laboured, gravitational expulsion of a lovely paradox - the 'anti-waste' of fresh and solid forms.

In short, the whole of every organism is, in terms of information *and* of energy, hierarchically arranged. As life evolves from inside outward all levels interact. The motion is in complete contrast to the Darwinian prospect of piecemeal and haphazard construction. *Can such reflection of cosmic order as described be, in its obvious biological purposes, an effect of chance?*

The Basis of Biology is Information

The biologist Theodosius Dobhzhansky is famous for coining a popular mantra: 'nothing makes sense in biology except in the light of evolution.' *This is utter nonsense.* **The actual, iconoclastic fact is: 'nothing makes sense in**

biology except in the light of information.' You may drag in evolution on information's tail but it is simply a fashionable word which (see Glossary) is used in biology to mean variation or development. It is not, therefore, an origin of species but, fundamentally, the origin of information that is at issue. **Not evolution but information is the basis of biology**.

Thus the basis of biology's an immaterial entity. The subject's root is metaphysical. Yet modern philosophy, trapped in a mind-set that exorcises 'universal mind', prevaricates, intellectually confabulates, long-windedly equivocates and agonises how sense *must* evolve from nonsense. Biological code, as clearly code as language in a book or music on a disc, is 'as if' intelligent; it is information 'by analogy'; it shows 'apparent' purpose which, like a bad dream, the clear light of physical analysis strives to dispel. But fact stubbornly endures: life, in both psychological and biological terms, is information-rich. The overriding questions are: when does an appearance, like the appearance of your humanity, become so overwhelming that you judge it actuality? Whence does what's inferred to be intelligent arise? Can information come by chance or not?

More and more molecular biology shouts metaphysic since, more and more, it illuminates the signals, switches, codes and carriers of complex operating systems. Of course, a book's as unintelligent as ink and paper; so are specific gigs of biomolecules. Thus, *bottom-up*, material information is as mindless as its agents, that is, the chemicals that science studies. It's passive; it's been garnered gradually, according to the *PCM*, from chance mutations acted on by variables of ecological or developmental constraint. The product is a trait, an adaptation that's 'selected' to improve its owner's lot. Natural histories are promoted to support this animated, evolutionary view; plausible narration presses to explain this kind of origin. A *post hoc* story that articulates a chain of possible historical events is, however, *not* a scientific exercise in the same sense as operational physics, chemistry or applied biology. Instead, it is scientism's method of informing you whence information, which was lacking, must have mindlessly appeared; best-guess speculation generates, *ex nihilo* since there was only lack before, such inference as substantiates the basis for an exclusively materialistic comprehension of biology.

Is such a method strong or weak? *Top-down*, there is no information without mind to understand; nor can information be created without mind's intelligence. Of course, such intelligence uses form to express itself. Such form includes rules of format (such as language and its rules), carriage (using sound, light or chemicals) and storage (memory, tape and so on). It includes communication (feedback) whether with animate or inanimate agent. Organisms all involve all these factors in abundance. Mostly their intelligence is reflected passively, as in machines; but active, creative, conscious response occurs as well. From a top-down point of view, however, matter has no chance of ever generating mind's core *sine qua non*, information. Active precedes passive state. Chapter 6 described in detail how this is. Mind's creativity rules matter's lack of it; mind's message gives rise to material patterns which, according to its purpose, other minds can understand. This frame of reference treats biological organisms as passive or 'rigidified' expressions of information. **The basis of biology is therefore (Chapters 5, 6 and 13 - 17) metaphysical**. Metaphysical leads, physical follows. The correct angle from which to view biology, whereby its parts fall into place, is therefore immaterial information's. Code, syntax and semantics - translation to and from a purpose - are the operators of information. The origin of life (and still its source) is abstract code; whence appears, by agency of wind and heat and rain, a set of abstract symbols? Yet current study, committed to scientific materialism, cannot see genetic code or language of the genes this way. It is therefore forced to bet on matter, energy and chance alone. It relegates the cause of code to randomness and thereby incorrectly raises evolution theory and its random generator to a pedestal of total rectitude.

At the causal heart of biology the modern discipline of genomics studies information systems whose purpose is homeostasis-for-survival and whose communications and data storage are, like any language, in code. Biology is replete with such systems. Their purpose is everywhere the same. It is, in a changing exterior world, to maintain a comfortable, healthy *status quo*; to maintain *dynamic equilibrium*. Dynamic equilibrium, we saw, is disequilibrium of the highly controlled kind that fluctuates around the stability of a pre-set norm. It is the product of mechanism designed to support its specific activities. In humans, for example, these include protein synthesis, hormonal and nervous controls, blood vascular controls, immune, hepatic and digestive systems, thermoregulation, osmoregulation etc. etc. etc. In the human case integrated feedback functions in a hierarchical system with *HQ* set up top in an information nucleus called the brain. All organisms own such central, informant nuclei or are controlled by balancing mechanisms (individual forms of homeostasis), each always integrated with the others to operate as a whole. It does not matter whether you speak of a multi-cellular body or a single cell. It does not matter whether you refer to maintenance, growth or reproduction. *Everywhere you find order, reason and coordination.*

Teleology and *teleonomy* are big words for saying something is directed. But is the mindless, teleonomic 'guidance' of natural law sufficient to imbue life-forms with their ubiquitous 'semblance' of reason and its messages? Signals, as on a railway system, control the steps by which biological reason or purpose achieves its destinations. An application program is a signalling system. Every line communicates its commands. *Biology is replete with signalling and code.* In higher animals these regulators form the basis of mental operation but also occur, subconsciously and physically, throughout a cell or organism as a whole. A book is seen as information first, paper and ink second; a computer as information first, plastic and metal second; **a cell is seen as information first, chemicals second.**

For example, the surface membrane of a cell could be regarded, as well as a container, as a communications screen full of addresses, labels, locks and keys, special entrances and exits, a control panel of specific 'bells, buzzers and winking lights' that communicate between exterior events and the chemical inhabitants of the interior. The screen is a two-way initiator whose business is programs of response. On the inside a cell's cytoplasm is a chemical factory at least as complex as any industrial combine. Its interactions are very effectively grouped and controlled. Control is, as ever, by position and signalling in code - by specific chemical pathways, concentrations and interactions. A code has meaning. It tells you what it means, what it wants and what it intends to do. The nucleus, of course, holds a database that is accurately and incessantly interrogated by an operating system - a system likely, as Bill Gates of Microsoft might surmise, to be highly complex, efficient and hierarchical. These are words that spell intelligence. They spell mind. All irreducibly complex mechanisms, all hard or soft machineries spell mind. Maybe, to some, the mind of Sir Nigel Gresleyis more teleologically evident in a streamlined steam engine than 'mind' in metabolism.

	tam/ raj	*Sat*
	energetic expression	*Potential*
	info. loss/ info. gain	*Information*
	consequent issue	*Archetypal Source*
↓	*tam*	*raj* ↑
	local detail	*general principle*
	'downstream'	*'upstream'*
	external/ exogenous factors	*internal/endogenous factors*
	lower/ controlled routines	*higher/controlling routines*
	physical energy	*metaphysical driver*
	physical forces	*purpose/ will/ desire*
	disruption/ info. loser	*integration/ info. holder*
	chaos-producing vector	*order-producing vector*
	thermodynamic vector	*negentropic vector*
	expression	*conceptual element*
	what it represents	*code*
	material aspect	*mental aspect*

The problem is, as we've remarked, the origin of the information, latent or expressed, in life's codes. It has always been obvious on the large-scale but now also on a molecular scale that biological bodies are all a product of intricate order. Proteins are produced and cell metabolism driven under the strict influence of coded instructions. Programs of instructions are the antipode of chance. Non-chaotic, codified information-rich chemistry exudes precision and excludes randomness from every formulation. Even lotteries as they occur in, say, meiosis or the immune system are deliberate and set within an algorithmic scheme. Yet information has no source in matter. Nor is there any evidence to think that things were ever different. Indeed, a systematic act of creativity might be defined as work evolved against the chaos-generating gradients of matter. What about the immaterial energy financing such conceptual work? It is will-power and desire. Actively-directed information. Intelligence. You pump an integrated system up with sheer intelligence.

But injection of information into a naturalistic system is, by *PAM*'s Clause 1, disallowed. Where cosmos is concerned you might infer (as in Chapters 5 to 12) an initial, transcendent projection of energy 'coded' or 'regulated' by minimal, archetypal information of the kind that we call natural law. You might equally, where life forms are involved, infer a transcendent projection of specifically complex and high-grade information that codes and thereby regulates the cosmically minuscule program of matter called an organism. Indeed, a prediction based on the inference of informed design is that research will, in fact, increasingly demonstrate the impossibility of blind, natural forces ever producing such intricate, complex and always specified constructions as even single cells. But, even if such inference and prediction well explain what random chemistry cannot, there is denial. For example, the scientific origin-of-life community is wedded to prolific speculation in which, despite the practical impossibility of its claims (see Chapters 20 and 21), only material solutions are allowed.

Such professional denial is serious enough to be rephrased. The intrinsic nature of gravity, which synthesises planetary orbits and orders the mindless cosmos, is unknown. It is known by its omnipresent potential and, in the presence of mass, its actual effects. Similarly, the omnipresence of coded information in biological material is an effect whose observation leads to inference and questions concerning the intrinsic nature of an informant, a coder. It is the purpose of artistic and not least literary criticism to gain a better understanding of an artist's mind. Does this not apply, because of a philosophical bar, to the 'book of life'? Why, highly illogically, is one required instead to place the origin of highly economical, effective information at the door of its anti-pole, chance? This is the unnecessary irrationality pragmatically ignored but aroused in principle when modern science takes a stance on where it came from - origins. *Because, faced with the obvious inference of orderly creation that is reinforced daily by its discoveries, the immediate response is a 'great denial' followed by reflex recoil to a theory of evolution*; *it is to a thin-lipped assertion that time, chance, coded laws of nature called 'necessity' and natural selection - in short that PCM - are sufficient to combine and generate the information that built up and supports your life.* You might accurately call this centrepiece of scientific explanation a 'Theory of No Intelligence'. It is a paean to unreason in the form of mindlessness. Is this, an essentially atheistic presumption, the intolerant huff of intellectual scholarship, examinations and, following up, the puff of popular journalism? Yet, if it is true that intelligence is measured by the economy and effectiveness of its creativity and true that coders create programs, must not such truth be biologically wrong? Must not the 'irrational' rationality of purposeful design be replaced with the 'rational' irrationality of accidental 'design'? The weasel couple, 'as if', claim that ubiquitous design appears 'as if but not really designed'. So that a 'Theory of Intelligence' is irrational! To be eliminated if introduced to scholarly curricula, to be cut off if seen to head that way! Uproar, resignations and capitulation with apology by heads of church! What this book conveys must be a kind of pseudo-science that no school can tolerate! Topsy-turvy whirls such missionary materialism through our halls of academe! **Indeed, denial is so divisive that 'origins' in science class means only mindless evolution; banned informative design had, therefore, better find expression, marshalling all the scientific facts, in those mindful centres of semantics, art and IT classes!**

The philosophical imperative of materialism is obvious because you cannot engage in scientific research except by manipulating materials and assuming, for the purposes of hypothesis, materialism. This is fine when dealing with the operations of natural constructs and machinery (a body is, we'll see, not a 'machine' but a real machine); but it is not necessarily sufficient when dealing with their origins - or with the subjectivity of meaning and intent. Indeed, it seems materialism has decapitated bio-cybernetics; it fails to admit a rational origin of omnipresent concept, code and purposefully working systems. It dismisses with the cavalier but weasel words 'as if'! **To treat this stark but important philosophical division according to biological origins Book 3 is arranged according to prebiotic chemistry (Chapters 20 and 21) followed by consideration of the three main (r)evolutions in evolutionary theory. *Firstly*, in the 19th century Darwin himself proposed evolution by descent and natural selection (Chapter 22).** This was highly unsatisfactory since the process lacked a mechanism. Therefore the theory evolved to stage 2. **In the 20th century mutation, a mechanism which might provide sufficient material for biological transformation, was coupled with natural selection to generate neo-Darwinism (Chapter 23).** This combination was, although protested ardently in the absence of anything better, still highly unsatisfactory if not trivial for reasons that we'll see. Some of these problems have now, after nearly a century, been relieved by a *third* revolution. **Stage 3 (also Chapter 23) involves so-called 'natural genetic engineering', that is, the mobilisation of coded modules sufficient, it is claimed, to support 21st century informatic as opposed to reductionism's failed 20th century mechanistic version of evolution. An important subset of Stage 3 (Chapters 24 and 25) involves reproduction, sex and especially, the genetics and evolution of biological development - which latter has been dubbed evo-devo. Could there even be, in years to come, a *fourth* stage? There'll be plenty here for arts, IT *and* science classes!**

It is not, however, the remit of this chapter to explain what higher purpose might inlay the structure of life forms, forms whose bodies are composed of lifeless molecules and inhabit what were otherwise a lifeless scene. It is simply to reiterate that, from a biological perspective, their overriding, prosaic purpose is the survival (or stable continuity) of their own highly-informed, individual patterns - in both an objective and, more importantly, a prior subjective sense. Such forms depend on information whose influence extends from subjective to objective, from psychological (in all cases sub-conscious but in some conscious as well) to physical and from mind to lower, visible aspects of body. *Their biological survival is based, in principle, on three core components - a (sat) precondition of precise information systems, (raj) energy metabolism for current survival and, as well as (tam) waste disposal, recycling in the form reproduction that grants long-term survival into the future.* In this respect biological construction also reflects (*figs.* 19.1 and 19.2) the hierarchical, dialectical order from (*sat*) starting potential (a single cell or zygote) through (*raj*) kinesis (growth and development to adult form) and (*tam*) decline towards an end-point.

Every kind of living organism must have, from its inception, all three core components whose operation is expressed through various sorts of coherent, interlocking mechanisms contained within its perimeter. Accordingly each factory-like cell, even excluding its metaphysical basis in archetype, comprises (*sat*) nuclear information, (*raj*) busy cytoplasm and (*tam*) structural components such as cytoskeleton, membranes, walls etc. Since even in a bacterium this triplex of mechanisms is complex, and involves self-regulation (homeostasis) how in principle could mindless molecules, against the universal thermal trend towards low-energy, degraded systems, synthesise such an upgraded, ordered, purposive structure? *In fact the central principle of integral survival is a juggling or balancing act supported by the balancing mechanism, homeostasis. Nowhere are biological purpose, information and control more evident than in homeostatic cycles.*

The (*Sat*) Central Instrument of Biology is Homeostasis

	tam/ raj	*Sat*
	lesser balance	*Balance*
	disequilibrium	*Equilibrium*
	less-than-ideal	*Ideal*
	oscillation	*Norm/ Law*
	comparison	*Criterion*
	practice	*Principle*
	information out / in	*Decision*
	controls	*Government*
	effector/ sensor	*Homeostasis*
↓	*tam*	*raj* ↑
	down	*up*
	info. out / effector	*sensor/ info. in*
	response	*stimulus*
	end-point	*action*

or in flat as opposed to cyclical terms:

absence of info./ comms	*info. exchange*
disintegrator	*integrator*
no response	*feedback*
silence	*signal*
flatness	*cycle*
inertial equilibrium	*dynamic equilibrium*
death	*life*

A painter wants a blank canvas and a musician silence. In order to completely control what he writes an author needs a blank page. The basis of biological structure and function is informed, efficacious cellular biochemistry. Because its neutral (aqueous) medium and low temperature are too mild for most reactions they present a blank page to such command and control whose agents are enzymes, specific proteins coded for and requisitioned using highly-regulated nuclear *DNA*. As catalysts enzymes lower activation energy. In concert with precise kindling by agents such as *ATP* an otherwise 'dormant' and perhaps highly improbable reaction can occur. Not just one but targeted sequences of them. Metabolic cogs, cascades of integrated cycles. At what a rate they work! For example, millions of *DNA* nucleotides can be copied in five minutes by a simple bacterium as just part of that short time-frame's business! Nor is material quantity the issue; it is quality of information. Quality of sequence, accurate arrangement of components as in all technology, that is, information set to purpose is what actually counts. Real value does not lie in number; biology's gold standard is the order of its chemistry. Consequent biochemical writing is, without any accidental surprises, both entirely grammatical and meaningful. Targets are as unerringly achieved as an editor's pristine text. Each cell has, theoretically, complete control of its internal chemistry and a multi-cellular aggregate of cells with appropriate interlocking signals can, in a pre-programmed cybernetic way, tightly regulate the industrial performance of the organism as a whole.

Physical systems tend to seek out states of lowest energy. They naturally diffuse into dead-ends like puddles or, like fag-ends, fade until there's nothing left to burn. Zones of (*tam*) death. The reaper haunts time's rule; this grim stalker scythes and scythes its crop. Of course, you can upset his swing by throwing in some fire or other reactivity; but such disturbance as raw energy kicks up invariably subsides into a further state of dullness. Lights go out. A reaction runs to its completion and is gone.

This is not the case with biological material. Such material appears 'machine-like'. In the case of a machine raw energy is needed but is introduced to the system through specific, precise, prearranged channels in order to achieve a specific result. In the case of a motor-car the purpose is transport. It would not work to douse the vehicle with fuel. You would simply achieve a purposeless, non-specific disturbance that, in running to completion, destroyed the machine it was meant to serve. It is the same with biological machinery whose fuel, derived from sunlight sliced up into standard units, is *ATP*. Nobody argues that sophisticated energy metabolism works like a crude, undirected blaze.

Just as the car uses fuel to drive a cyclically controlled engine of transport so bodies engage systems of homeostatic control in order to survive. In other words they resist stoppage, breakdown or the death of their integrity. In this case the purpose is to keep going, to stay alive. Whether amoeba, alga or human, survival is in order to fulfil various individual purposes.

Survival is *not* running to completion. There is no 'dead' and 'finished' equilibrium of inertia, mass and heaviness. Life is light, vibrant and purposeful in concept and in practice. It springs with the dynamic equilibrium of a fountain. It 'dances' in a choreographed balance between opposing tendencies, between interior control and exterior contingency, order and chaos, mind and senselessness, healthy harmony and the discordant rasps of war, exhaustion, disease and death. All its refrains involve set cadences, the norms round which its various cycles play.

Balance, scales, pivot, action swinging orderly around an axis made of information - this is homeostasis, the throbbing heart and central instrument of biological stability..It rules internal motions just as, simultaneously, it buffers changing circumstance. Is not the integration of conceptual cogs expressed as routes and organ-stations what makes bodies tick? Information, purpose and their various cycles seem, up against time's flood of entropy, to run survival's show.

Unlike abiotic matter, organisms appear sensitive, responsive and give clear indication of targeted behaviour. This behaviour applies not only to the whole organism but all its parts down to molecular, atomic and even subatomic patterns. In this harmonic way life delays the end of its reaction. It deftly dances past the inertial equilibrium of abiotic matter. Its quick, specific steps elude, for a while, the destruction of its patterns of behaviour. Where life the integrator draws together, death, disintegrator, splits apart. Biological life avoids, for a while, the tax of death. The instrument of this evasion, which allows awareness time on earth, is homeostasis.

Homeostasis is the cybernetic, computer-like maintenance of a special form of disturbance called dynamic equilibrium. Its process cycles around a stabilising centre, a pre-set quantity called a norm. It keeps peace, stability and balance. It is a missile's guidance system and it underwrites a target - poise that we call health. Such vital control involves a switching system called *negative feedback.* Your home's central heating system is (see also Chapter 5) a good, simple example of such feedback. *It needs, along with the associated plumbing, water supply etc., a minimum of three integrated mechanisms to work.* The first is a *sensor* which tells the temperature. The second is a *regulator* which decides if it is equal to, above or below a pre-set 'norm' and the third an *effector* which accordingly either does nothing, switches off or on the boiler, pumps etc. Such a system allows the temperature to fluctuate so that it is, in effect, oscillatory. Body homeostasis is also fluctuating or periodic in its various operations. Examples are your own thermoregulatory control set at 37°, allosteric metabolic feedback, pH control and so on.

It is worth emphasising that homeostatic regulation involving (*raj*) input, (*tam*) output and a (*sat*) balancing or regulatory processor has no use by itself. Its irreducible complexity works not in a vacuum but involves associated components or, in metabolic terms, pathways. Sub-systems have purposes and are, in complex cooperation, interlinked in a hierarchical fashion within an overall system, scheme, plan or, in this case, biological organism. All codified factors, whether at molecular, cellular or bodily level, are therefore necessarily fixed together in the right relationships at the same time in order to work. How (Chapter 20) might such a system originate?

The maintenance of life's dynamic (dis-)equilibrium involves concerted control over a large number of more or less individually controlled homeostatic subsystems. Each living body contains information, energy and recycling (reproductive) loops. It consists of interactive networks of homeostatic function whose mechanisms control aspects of its internal environment such as temperature, water potential, pH, pressure etc. and, in so doing, buffer against the effects of external environmental agencies. Such up-and-down, round-and-round routines govern the rhythms of life on earth. Are they flexible? What is their margin, tolerance or play?

The degree of tolerance varies. In the case of your temperature the controls are tight. The same sorts of norm and oscillations round them can differ in organisms of different kinds, the same kind or, as the female menstrual cycle shows, within an individual's lifetime. Yet, for all their flexibility, norms are norms. They are as critically pre-designed as an axle that can wobble just so much. After that you have to take it to the wagon-hospital. A norm could be called a theme or keynote around which minor extemporisations or errors still play on. Each homeostatic mechanism 'resonates' with a pre-arranged and more or less specific keynote. *Translate each keynote, harmonic or 'fingerprint' into a specific, correlated biochemical or set of associated biochemicals and you have the biological picture.* You would logically expect to find that the electrical 'fingerprints' of each molecule associated with a specific homeostatic task are mutually resonant; and that the keynote itself is integrated, like the harmonics of a piece of music, with all the other keynotes for that organism.

Do you remember (Chapter 17) the analogy of an organism with a watch's intermeshing cogs, a bundle constituting interlocking government? Each homeostatic cycle harmonises with the others, each vibration 'resounds' within the tune of all the rest. What do you call harmonious vibrations? Music. An instrument in tune is healthy; each kind of body has its own theme tune.

Balance is the key. Each individual, homeostatic cog cooperates in balance with the rest. Where health is balance discord calls the doctor, disharmony does not feel right and 'musical' distortions signal illness. Cacophony is death (though programmed death is not cacophony). The network of a body is pegged out on the basis of a pre-set 'score' - its 'power-points' (*figs*. 17.8-10), norms, codes and mechanisms. Its command and control, that is, data-processing centres in brain, nerve networks and each cell inform and thereby sustain its natural cadence.

In fact, so central and fundamental is homeostasis to life on earth that a biological body, whether uni- or multi-cellular, may be seen first and foremost as an invisible pattern of informative control. The pattern of this dynamic biology involves an irreducible minimum of working components. It needs pre-set but slightly tolerant (or resilient) homeostatic norms, integrated archetypal subroutines, electrochemical codes, sensors, regulators and effectors. Around this hierarchy of information, dropping from archetype through electromagnetic and molecular levels to that of macroscopic physiological processes such as renal osmoregulation or heart rate control, biological molecules 'fall into place'. Corporeal shape is automatically arranged within an incorporeal infrastructure. This infrastructure is informational; the basic necessity for the operation of homeostasis, and therefore biological life on earth, is information.

Each cell is, thereby, replete with balancing systems and, therefore, such ionic and molecular signals as its computerized procedure needs and generates. It is these intricate signals, their nucleic and other processors and the relevant specific output that today engages the full attention of the best minds in laboratories worldwide. Life is a giant, dynamic puzzle made of tiny pieces! Such coded hierarchies of micro-homeostatic complexity are not easy to unravel! Could chance accidentally have stumbled towards a very high degree of its own lack? Or, just as purposeless as time, have created such goal-oriented order? And do so all at once before cacophony could kill? If not how on earth (as the next two chapters show) could the basic subroutines of life have started out?

In summary, (*sat*) homeostasis revolves around cause and effect. A change in the environment causes a regulator to adjust and (from its point of view) maintain stability. The process is based on information and the purpose is, by keeping balance, a vehicle of life's survival. It forms the basis of biology so that its constraints and ordered guidance are implicit in each atom of organic form. *Program is expressed in signal, code, electrical and chemical gradients and in every sequential pathway.* At root, processes are switched on (1) or off (0); subroutines are 'gone to' or 'not'. The program of binary homeostasis is what controls the whole body and, wheels within wheels, its component parts. Cybernetic choices correspond to information and this 'substance' is passed 'up' to a control centre by a sensory mechanism and 'down' from a decision by effectors. *Such a loop is clearly, fundamentally, both hierarchical and cyclical: and it is born not of primal, senseless matter but, as a straightforward matter of inference, intelligence.*

Biology is Hierarchical and Cyclical

Bottom-up, genes rule. Genes are factors that store, in the manner of a database, coded information that can be translated, subject to cyclical feedback by various messengers, using a 'staff' of *RNA* and protein executors that are themselves the product of that code. According to the task in hand they 'employ' a staff of proteins as executors. The latter are the basis of both your 'bulk' (skin, bones, muscles, hair etc.) and your metabolism. Because genes sequence amino acids and thus help control the production of protein they are often assumed to underlie the shape (and thereby function) of the parts in a cooperating whole. *This is because, materialistically speaking, shape and not concept defines function.* It is thus that, as if words in a book were judged to render its author obsolete, the 'genetic program' is thought to generate biological function. In truth, the arrangement of materials may reflect intelligence (as in the case of books or machines); or it may not (as in the case of stones or stars - unless you recognise natural law to be originally inspired). Such arrangement is called *passive* information and, since matter is his ultimate, it is from such information alone that a materialist is obliged to derive his view of the ultimate origins of life. Genes and, therefore, protein are as central to evolution as evolution is to a materialistic (is this quite the same as scientific?) world-view. Just as a molecular biologist might assume that his bio-molecules are *the* basic component of life, so it might be assumed that genetic units underlie *all* aspects of biological construction. Do you remember (from Chapter 6) the objective analysis of a Mozart concerto, steam engine or TV set? Could you reduce these objects to their physical constituents alone? Does not arrangement also count so that, to constitute the whole truth, mind inhabits them as well?

Information is, on the other hand, invisible; its immaterial meaning and purpose intangibly but recognisably inhabit material shapes and behaviours. *Life, in all its aspects, is so obviously purposeful in character that genetic chemicals are ascribed all kinds of animistic properties.* They 'compete', 'organise', 'program', 'adapt', 'select', 'create form', 'engage in evolutionary arms races' and even, lyrically, 'aspire to immortality'. In inverted commas, of course, because like every other kind of chemical genes are non-conscious and have to leave any panegyric to their PR men. What kind of molecule can 'feel compelled to replicate itself'? What kind of atom or molecule has 'strategies for survival' on its lack of mind? The answer is, unless we engage in self-deception, none. Not even, any more than letters of an alphabet desire to make meaningful sentences, do base-pairs 'authorise' *DNA* or genes 'conceive' of proteinaceous organisation as complex as cells or a body like yours.

In reality, animistic language always implies, unequivocally, mind. Purpose in front: intelligence behind. Nevertheless you proclaim chance, an agent not only naked but by definition without regular, reasonable form, crowned the empress of life. Such contrary jubilee! Because bases like assembler and genes like a higher level of language definitely amount to coded instructions: there even seem, within genetic operations, to exist micro-hierarchies of *ROM*-like command and control. No doubt, like the integrated components of any machine, chemicals 'talk to each other' and 'tell each other what to do'. Complex, targeted programs are, in all other cases, the product of intelligent design but, within the exclusively materialistic paradigm of contemporary biology, must not bodies with their orchestrated chemistry have been created by 'experimental' chance? By a ratchet-effect composed of chance, environmental constraints and a mysterious 'understanding' of what is 'fit and proper' molecules must have contortedly hoisted themselves up to the province of 'life' (whatever that is) and even, most exciting for a crowd of molecules, emerged into 'consciousness' (whatever that is). Determined to 'survive' (whatever, without a mind, that is), must they not have straddled the gulf between the entirely separate 'languages' of *DNA* and protein? And then planlessly yoked anatomy with physiology to underpin the phenomenon of life and its behaviours? Of course they must. Molecular 'intelligence by chance', chemical animation - are you not the proof of it?

From a *top-down* point of view there is an element of truth in these confused claims. They are partially correct but overstated and, most importantly, wrongly identify the *source* of information, code, messaging and coherent operation that marks out all life. Inner precedes, substantiates and supports outer. Invisible creates visible. *Information precedes and orders energy/ matter.*

Biology is built from integrated forms of homeostasis that control the expression of plans; and networks of control are, by their nature, hierarchical. Genetic and phenotypic complexities involve chains of cause that order preordained effects. Code anticipates. Signals communicate. Information operates with feedbacks which ensure dynamic systems 'stay in line'. Therefore, to find the causes of 'downstream' effects you have to travel to their 'upstream' cause. This cause is not natural selection. Nor is it mutant lesion of a gene or chromosome. No doubt, encyclopaedic genomes bear on phenotype but so do the constraints of shape and so-called laws of form. 'Upstream' from even genes or tensions of morphology is the conceptual information that they represent. Whence derives their supervision; what's the source of biological machinery?

As in the case of a machine, in the biological hierarchy information precedes energy, mind precedes matter. So if your own conscious mind can't make you up, what can? A psychosomatic intermediate, a *ROM*-archetype is the memorable answer. Hierarchy is (Chapter 17) archetypal; such imprimatur is clearly expressed in its physical product. And coded information (in the form of *DNA* and chemical signals) is the recipe for subtle, molecular patterns called proteins from which are derived cells and cell interactions. On this basis, in turn, are constructed the large-scale organs and operations of biology. In other words, visible (*tam*) characteristics called phenotype depend on (*raj*) molecular action that, in turn, depends on (*sat*) genotypic and other pre-programmed information. Biological structure and function emerge from the exact chemistries of maintenance and development. And such complex metabolism is itself entirely information-dependent, information-structured and information-regulated. *Is chance ever a source of precise, meaningful information? Where is the chance in your computer? What part does chance play, except mistaken, in life's systematic arrangements?*

If it *were* chance you might claim life was the same as death. Indeed, some biologists go so far as to deny the possibility of defining life. They believe no essential difference exists between living and non-living specimens and nor, in purely atomic terms, does it. After all, what is a cell but a particular yet fortuitous arrangement of a farthing's worth of chemicals? Or a large body one of a few pence? Chemicals have no sense of survival or anything else. Nevertheless an informative agent is invoked, for better or for worse, called natural selection. It makes atoms know what is good for them. 'Life', whatever that is, is good for them. Increased complexity might seem to up your chances of survival; doesn't a sophisticated system seem to make more sense? It may even acquire a mind of its own. In this evolutionary view, however, everything is simply the result of purposeless adaptation (or new permutations) of the same atoms. Nothing is intrinsically better or worse, above or below anything else. There exists no preordained hierarchy in evolution's democracy of atoms. It is as if, remarking only its physics and chemistry, you 'cleverly' denied the intelligence behind a book or film and treated them like paper, acetate or rock or ocean wave. Since biology is, by definition, the science of life the obliteration of distinction between living things and other forms of matter ends its separate status. This sort of biological philosophy, which even Nobel prize-winning biologists have promoted, indulges in a form of self-annihilation. It is a tale as hopeless as destructive; but, of course, scientific atheism has decided it is true.

In truth a biological whole is not, as a reductionist view might try and persuade, the sum of individual parts. What is greater than the sum is their pattern, their juxtaposition and *order in order* to execute a purpose. Does chance work in regular cycles? Does it 'know' what it wants or vibrate according to pre-arranged norms? How, having the start (and therefore the end) of a homeostatic period in her lipid bag, did Caprice know all that has to fit between? How did she 'get it'? There exist, in the implicit hierarchy of mind 'down' to matter, explicit biological hierarchies of information, function and structure. These hierarchies always involve, through the exercise of (*sat*) balance, homeostasis or feedback, a cyclical flow of information between levels. Inside information is expressed and response collected for processing. Messages are passed, through chains of command, to and from information collected for processing. Messages are passed, through chains of command, to and from information

centres. *Whether in an explicit, conscious or involuntary, cybernetic way, these centres always execute orders in compliance with purposes.*

The origin of hierarchical order, cyclical flows of information and integrated function is always purpose. The origin of purpose is mind. We might, therefore, reasonably conclude that the basis of (*sat*) informational, (*raj*) functional and (*tam*) structural biological hierarchies is mind. What holds a scheme in mind but memory? A memory is a passive store and so, therefore, are sub-conscious archetypal memories like *H. archetypalis.* These archetypes are not products of the imagination of philosophers or, where conceptual design precedes their prototypes, of inventors. They govern biology. They inform it from above. They are life's records not fixed on paper, disc or other chemical but naturally archived in memory. At this point it may be instructive to elaborate on three or four metaphors that emphasise the conceptual, informational, non-random nature of biological forms.

Mechanism and Machine

The metaphor of biological structures as mechanisms or machinery is, we begin to understand, apt but insufficiently followed through. A machine (Chapter 6) involves a source of energy and operates according to the laws of physics and chemistry but, crucially, cannot be reduced to them. This is because its arrangement (at atomic, molecular or any other level of objectivity) is specially composed to execute a purpose. The structure and function of a machine are specifically informed, that is, designed accordingly. You can, for example, study a TV or a biological body to its last subatomic detail and find nothing but matter. In both cases, however, each part's order is directly related to the composition's purpose. Without that order, whatever its source, purpose could not be achieved. *The biotic components in a machine are its plan and the planned manufacture of specific parts and shaped materials derived from the abiotic component, raw materials*. It is the same with soft, biological machinery. *Implicit in both TV and biological organism, therefore, is the conceptual foundation of purpose, principle and plan.* Both machines and organisms exploit physical principles using ingenious mechanisms towards specific ends. They are replete with metaphysical purpose and information expressed as function, process, active signals, passive (stored) code and shape. The three-dimensional geometries that create the architecture of any mechanical structure are special because they allow its purpose (in the process of function) to be expressed. Where homeostasis is a special case of equilibrium, architecture is a special case of functional structure, of deliberate shape.

An engineer will confirm that care is taken to select or create the optimum materials to realise his architecture in physical fact. Indeed, the shape and quality nature's biotechnology is precise to the level of molecule, atom and position of electrical charge. Its architecture is encoded on the ink and paper we call *DNA* and thence automatically but accurately translated into operational reality! An organism is much more sophisticated than a television or any other machine constructed by man assembling parts according to a blueprint. Organisms develop from *within.* They mature according to a prearranged schedule of events. This is no accident. Their self-reproduction, development, self-repair and overall self-cohesion, whereby the parts coordinate with each other for the good of the organism-as-a-whole, are unlike any machine even brilliant humans have so far been able to build. This self-organisation is based in its entirety on information in the forms of which we have spoken - pre-programmed communications, codes and constructs. Life's biochemistry and physiology are neither more nor less metaphysical than the software and hardware of a computer into whose construction so much intelligent information (or consequential order) has been invested. If in this case life-forms not only precede but are beyond human wit to conceive and build from scratch, we might reasonably presume their metaphysical origin superhuman. Not in the sense of superman but certainly in terms of a super-intelligent plan.

Seasoned engineers gasp like children at the bio-show. They copy natural designs. Therefore, by what criterion are, for example, ribosomes, *ATP synthase* dynamos and dynein cargo transporters *not* literal machines? Are mitochondrial power stations, fuel factories called chloroplasts and the mechanisms of protein production not actual and excellent machinery? As well as oft-quoted thousands of examples of Darwinian black boxes (such as gecko's feet, woodpecker's head, stronger-than-steel spider silk, sex, eyes, feathers - the list goes on and on). *Bio-machines, including an example evolution's missionary preachers love to fulminate against, the bacterial flagellum, are not analogues.* **These specified complexities do not just give appearance of design but are really purposive designs.** They do not seem 'as if' machines but without doubt *are* natural contrivances whose excellence the engineers vie, clumsily, to imitate - all this with consequences academic atheists abhor.

Finally, the essence of a machine is its minimal functionality. Precise parts do not include, except as trivial and non-invasive decoration, any imprecise or non-functional parts. Similarly, a biological cell has precisely what it needs and also doesn't have precisely what it doesn't need. Could primal imprecision, incompleteness, extraneity and resulting malfunction have stumbled unselected-against for billenia until 'rightness' clicked and 'health' was eventually asserted? Such suggestion would, on today's overwhelming if not complete evidence against it, constitute a very poor, just-so guess.

Check Chapters 5 and 6. Mind machines we call computers give the game away. Nothing in the IT industry occurs except by springing first in mind. Abiotic atoms thence embody life's biotic plans. Mind expresses its intentions in the pattern of material constructions. Life lives clothed by death. If you think you are a most unusually configured form of atoms (which are not alive) you should ask yourself whether you think you are really dead or alive. If you are not dead, the same argument can be applied to every other biological organism. Body is in mind as much as mind in body. *Bodies are envelopes; they are the atomic outskirts of mind.*

DNA Computing

A memory symbolically encodes the moment of its making in the past. Memories sustain their histories in suspended animation. Thus, like a musical or written record, *DNA* might be construed as a memorial inscription - a dynamic one that encodes, with power of recall, an original intention.

A computer is (see Chapter 6) a mind-machine entirely dependent on the memory of its programs, chips and databanks. Its construction and encoded programs are teleology objectified. Computer operations are based on conceptual Turing machines. These consist, like homeostatic systems, of input, processing controlled according to a set of rules and output. Scientists (such as Leonard Adleman) have realised that proteins responsible for *DNA* manipulation act, with the *DNA* itself, as conceptual machines - input *DNA*, process, output regularly transformed *DNA*. Together these biochemicals operate as Turing machines; bio-information is computerised. From this realisation the scientific development of artificial *DNA* computing has developed.

No ifs or buts. Biological computing and its issue, biological machinery, are not 'as if'. They're not illusions or appearances but actually real. Natural is as conceptual as what is artificially engineered. In this case the excuse of analogy, beloved of evolutionists, is swept away. Call them alphabetical or digital (as machine code) but *DNA* molecules store digital information; and the nucleic acid/ protein system *is* a calculating machine. Information, abstract and metaphysical, is carried on biochemical 'speech'. This 'speech' is subject to the linguistic architecture of alphabet, syntax and grammar; its expression is tightly regulated. Sentences, phrases and whole routines - metabolic modules - are combined to most efficiently exact desired effects, that is, requisite end-products for their system to survive. No doubt, as chemicals, genes are hardware but just like hardware called a c-drive or a book they carry immaterial meaning, reason, purpose and program. Information is carried on the concrete arrangement of materials. **Genetic code is as designed as sentences you think and speak**. It represents, like any language, logic, reasoned meaning, program, purpose and intelligence; and, of course, if you can have one sort of natural real machine then you can have the others - all the kinds with which the world of living beauty is replete. If, on the other hand, you are a scientific atheist you can't. You will deny all sense in this. Intuitive appeal to order is suppressed if you deny design intelligence. Instead, appeal to so-called counter-intuition seeks, perhaps somewhat thoughtlessly, to bar the 'spooky' notion of an immaterial archetype. It's all 'design' not actual design. Why, however, should I buy this line?

Thus, *bottom-up*, the problem's to convince yourself (and others) that, granted a vastly complex starting-point, the computation of cell chemistry 'self-organised'. How could its chemical translation mechanisms have evolved? Did life's informative technology arise by gene mutation set against an organism's ecological constraints? What systems flow-chart could you build to demonstrate this possibility?

Crystals, clouds or rivers all lack specified intricacy, that is, coded or mechanical complexity. Mechanical constructions, on the other hand, *all* derive from concept; hard or soft dynamic systems that have purpose are informed, at least initially, by mind. Take, therefore, a simulated mind. Take an electronic information system. *Top-down,* above all else, are found the purpose and principles according to which the calculating machine is constructed. Around these is pragmatically geared its 'instinctive, programmed thought'. This is called software and is composed of four main parts. At the heart of all computers, first and foremost, is a predefined code, a language. Something *means* something. This code operates in the context of digital storage (a memory or database); of software (application programs) and machinery (hardware) for functional processing of stored or input data; and higher level regulation (an operating system) by means of which an external agent can interact with the computer and the intended outcome of programs be expressed.

The primary element of computer message, software, is binary or machine code. On-off switches regulate the flow of information. From such basic code complex (or higher) languages are built. The logic of a purpose, that is, program can then be expressed through lines of regulatory code. Expression has to be, in any automated system, both efficient and precise. Switches change the track of process in accordance with input from circumstance, that is, from data referenced by the program from 'outside'. If x is the case do this; else do nothing. If x is the case go here; else go there. *Top-down* programmers know that, from a main routine, switches branch to subroutines and, when a subroutine is done the process cycles back to start again. **These routines are modules**. This conceptual character of algorithm, this repetitious use of switches and of blocks of modular code is just what coded bio-systems show. It is how nature's life-forms, full of reason, always work. **One would, therefore, predict research will more and more reveal signs of bio-logic to the point that, in 'live' computing, such complex permutations, integrated combinations and hierarchical sets of regulation will squeeze then nullify the notion that celled systems ever came about by chance.**

It is not *ATP* but electricity that drives computers; and the primary element of physical hardware in a central processing unit is silicon. Of the same chemical family and with similar properties, the primary element of biological computing (including *DNA* and brains) is carbon. But in the analogy of life forms with computers more than an atomic couple is at stake. Information is up-front. No more than coded software can create itself can programs carved in *DNA*. No more than codes on discs can manufacture hardware or accessories can *DNA* make bodies. ***DNA* does not determine everything**. In fact, it's neither information's start nor end. Code starts in mind and ends expressed as target forms and functions. *DNA*'s a holding medium in-between, a passive carrier but not creator of life's messages, a manual for making and maintaining some specific body. The metaphysical organiser, an organism's 'typical mnemone' or archetypal plan, was described (in Chapter 15) in terms of informative software logic and routines. It is a bio-logical thesis of Natural Dialectic that, '*upstream*' from genetic chemistry, the source of animate incorporation works as the program called an archetype. Its medium in subconscious mind is reflected, chemically, in programs that are hierarchically arranged.

This begs a question. How could metaphysical connect with physical? For the moment put aside the issue of the origin of *DNA* or archetypal program borne by it. Look at its operation with a metaphor. A genome is, as well as c-drive, a cell's 'brain'. It is 'mtor' (in expression), 'sensor' (to molecular triggers) and its 'muscle' might be termed its *RNA*. What, however, by analogy with your conceptual reckoning, is its internal morphogenic link with archetypal program? It was suggested (Chapter 16) that 'upstream' connection with its mnemone was electrodynamic.

'*Downstream*' from the nuclear *DNA* devolve, through cascades of control, the protein products and their phenotype; from internal regulation are developed all the bodies that inhabit earth. It is only as a consequence that, as they run their course, phenotypes suffer the secondary, external and uncontrollable impact of ecological variables. **In other words, endogeny precedes exogeny.** Information causes, the shapes of bodies are effects. The inmost cause, archetypal architecture, hierarchically precedes inscription in the genes and, subsequently, its impression as a living individual of a certain body-type. This view has neither time nor space for chance but, like exquisite systems of technology, is squarely based on information, that is, prior working of a mind. This is absolutely not what Darwinism's mind-set wants to hear.

Sharpen a pencil and then dot the page. Smaller than this dot a cell contains a nucleus and, marvel of miniaturisation, a huge database, complex programs of genetic access, transcription and translation and then, according internal or external input, metabolic programs of response. Bill Gates may surmise (who better placed?) that *DNA* is like a computer program, one very much more sophisticated than human intelligence has yet produced. Is not the cell like a computer? Its dynamic software, using *DNA*, *RNA* and protein operators, transforms itself into its target, that is, realises meaning as a form of life! Symbol is translated into actuality! Hardware, on the other hand, is the infrastructure, framework or container of software without which the latter is useless. It includes metallic and plastic devices whose biological equivalent is a cell, tissue, organ or the phenotype of an organism. It also includes transistors, chips, disc drives, wiring, power supply and the realisation of code as binary electrical signals, multiple alphabetical signs or, in the case of a biological system, four genetic symbols in chemical form (G, C, A, T), triplex words, genetic sentences and so on. If restricted to a description of cellular components such hardware includes all physical parts including membranes, organelles, cytoplasm and nuclear material - all chemicals including energetic *ATP*. But *DNA*, which informs cell chemistry, is more than simply *like* a computer program; and the cell not just analogous to a computer. **Each, in the very real aspect of its purposive complexity, is *identical* to the other**. Again, therefore, we ask what the origin of purposive complexity, intentional improbability or specific, functional engineering could be. Is it chance? Is it the laws of nature left mindlessly to run; or do they constitute another program that cannot create a living thing? Alternatively, perhaps systems rich in teleological (and not just Shannon) information are well known to arise in agents of intelligence alone - in immaterial mind but nowhere else.

Progress through this book reveals how fundamental information's signals, code, messages and meaning are. We shall increasingly see that biological applications are, like their technological counterparts, 'written' *top-down*. Maybe they are engineered from a point of full comprehension. They certainly appear developed as branches or subroutines devolved from a main or core switching routine. **In this all biology reflects the *top-down*, philosophical structure of Natural Dialectic.**

Materialism's counter-song seems rasping from a withered reed. 'Half-baked' or unassembled hardware is a useless shell; a 'hard drive' on a dump is just as useless as nucleic acid on a pebble, in a puddle or in pools; nor is software anything without its holder, hardware. A disintegrated database is inaccessible, an isolated operating system nothing more. And, of course, nothing works without an intricately engineered and ever-ready power supply. All exist to serve the purposes that application routines incorporate. *In biology the applications programs all, directly or indirectly, serve homeostasis.* This is because informed homeostasis is the principle that, programmed, serves dynamic equilibrium; this in turn underwrites the integral survival of a life system. Where, in computational intention called a program, are the loose ends, clashes, cross-reactions, interference or redundancy? It will not work until the bugs that hamper its intention are eliminated. Clockwork, bug-free integration and encoded, systematic balance are not the kinds of property unaided chance evolves - although each healthy cell and every single body from the so-called simple to extreme complexity involves them 'top-to-toe'. Yet, simply on its whim, materialism excludes from any bio-logic a subjective origin, an element of mind, that same immaterial, informative intelligence that every *IT* specialist is pressing to his limit! This promissory faith proclaims, against the evidence (Chapters 20 and 21), that we might find some day how code incarnate first 'invented and inscribed itself'.

And then some more! Of course cells and computers can both suffer bugs in code. These run from mild, where process limps on, through a range to lethal, where the program dies. Is this to say, that either party 'built itself' by bugs? By dint of scientific faith, and it is faith (see Chapters 22 to 25), gene mutation must have been the way that life evolved. The 'fact' of evolution's set in mental stone. Yet, neither in their primal state nor 'evolved' into complex form are life's codes and their operations 'simple'. An enormous program of research is under way to understand expression of the 'simple' gene. Banks of switches, super-codes and editorial refinements are part of what combines to hierarchically control expression of the information that, in a most systematic way, mints every form of life. Who can derive a system's software by examination of its hardware? Science would laboriously decrypt. It coaxes, probes struggles to decode life's programs from a chemistry of transformation, sequences and flows - and finds incisive specificity. Does such a pool of keen intelligence engage original intelligence much sharper, surer and more comprehensive than its own? Or else, the more exceptional the integrations that it finds, the more ascribe the whole script to an awesome 'intellectual' or 'conceptual' fluke of evolutionary 'design'?!

Coherent forms of homeostasis, each with its norm, branch like sub-routines from main or master routine set at a program's core (one such program, being human, is your own); and these routines collaborate towards the same core goal - survival. Apart from any conjectural archetype, the main physical information storage centre that they access is called a genome. This genome acts like a library, encyclopaedia or database for any particular form of life. A whole encyclopaedia is found in the nucleus of almost every kind of cell. It is divided, in your case, into forty-six volumes called chromosomes - twenty-three from mother and twenty-three from father. The books themselves are divided into sections, chapters or digital files called genes. Each gene codes for a particular protein; proteins underwrite both physical structure and metabolic function so their correct presence, amount and ability to do their jobs accurately are crucial factors. It could be argued, furthermore, that complementary information is also coded into the structural and electrical factors of cell chemistry; into epigenetic 'super-coding' (Chapter 23); into 'promoter' switches that control expression of a gene; into editors, proof-readers and corrective agents; into splicing factors that can generate many 'nuances' of a single protein; and into all 'keynote' agents of message such as hormones, secondary messengers and so on. *This amounts to a dazzling array of informative power, to a steep hierarchical complexity of programming and control.*

The informative specificity of protein structure and messaging systems is dependent on the *DNA* library, operating systems (such as access and translate its instructions) and enzymes that, effectively, each represents a line of code in an application program. Such a program is a metabolic pathway that produces, from a specific starting-point, a specific product. Where is any chance or 'flexibility' in this? A prescription for each protein (and, therefore, each enzyme's 'metabolic line of code') is logged on the *DNA* database. It consists of three-letter words (or bytes) called codons. Sixty-four codons (life's entire vocabulary) are in turn made of permutations of just four letters (or bits). These letters are, in chemical terms, the abovementioned symbolic bases guanine, cytosine, adenine and thymine. Each word 'means' either a particular amino acid or a 'blank', a full stop. Such an alphabetical and verbal system far exceeds, in its efficiency, English or any other human language - yet it spells out the reality of all kinds of biological body.

Much *DNA* does not code for protein. In yourself perhaps only 2% is strictly genetic, that is, codes for protein. The other 98% consists of sequences called non-protein-coding *DNA* and is not thus expressed. As long as n-p-c *DNA* usage (including sections directly appertaining to p-c genes and called introns) has not been understood it has tended to dismissal by the name of 'junk'. When referring to organs whose use they did not understand Victorians tended to dismissal by the name of 'vestigial'; the vast majority of these are not 'vestigial' now they're understood. Similarly, for Elizabethans, 'junk' is more likely a description of man's current comprehension than the mechanism itself. Indeed, although Darwinism promotes the notion of 'construction by accident' and the parallel perspective of 'artless information', such misinterpretation of non-protein-coding *DNA* may prove to be one of molecular biology's few resounding clangers. We'll take a closer look at it in Chapter 23.The fact is, though, we can say straightaway, that a gene is not a gene if you think (as was the textbook tale) that it just codes for protein. **Many lengths of *DNA* code for regulatory *RNA* alone; and in all organisms specifically programmed control panels (introns) are attached to the protein-coding parts (exon).** The two parts together compose a genetic unit. This unit is a complex rich in programmed flexibility. Introns include regulatory elements (to variably promote or inhibit gene expression) and addressing/ coordinating factors of the genetic operating system. We'll see in more detail that still further 'epigenetic' factors add information hierarchically to this microscopic information factory. They include histone proteins and nucleotide methylation. Such epigeny allows genes to respond (without losing base sequence) to the environment. Thus they are not only causal; their expression can be affected (or, if you like, nurtured) by outside circumstances.

Such a complex operating system is hardly haphazard. How much junk, in terms of what is comprehensible according to their code (say, English), does a magazine or manual contain? *No junk signifies intelligence.* No junk - but not all genes-without-a-break-between - is what the Dialectic would predict you found in any freshly minted system. Of course, the world's in bondage to decay. Entropy of information follows hard at heel. Years take their toll. You might predict, therefore, that you would find all genomes loaded with mutations whose effect could range from lethal through to nearly neutral (Chapter 23). *You would predict that accidents and errors might afflict the sense of an instruction manual but, for heaven's sake, you'd never think it all went in reverse and accidents could stand for logic or misprints one by one accumulate a book replete with purpose and with sense.* The dialectical prediction squares extremely well with all we find about the language and the highly integrated, self-consistent logic of the genes.

Bugs disrupt smooth running of machine and coded text alike. What intelligent solution might a human best employ to rub them out? Repair shops and detective editors? Hey presto! Proteins maintain the integrity of *DNA* through their ability to execute running repairs related to sequence, that is, the order of letters that compose our books of life. An operating system of at least 120 factors, including enzymes, cooperates to reduce error in copying the human genome to about one in three billion parts. Extensive correction mechanisms exist to cover all kinds of mutation. These include cuts, pastes and 'proof-reading' that, for example, detect wrongly paired bases and repair them correctly. Electronic scans identify a broken strand. Together such factors may reduce mutations per nucleotide in bacteria to between 0.1 and 10% per billion transcriptions and ten times less for every other kind of organism. Why should evolution care if what it writes is 'wrong'? Indeed, bugs are supposedly the mechanistic spring of its 'progress' unto more intricate complexity of text, greater compass of abilities and, in addition, quality of genetic literature! *On the other hand, debugging systems certainly suggest the text that DNA is carrying is 'right' and must not be degraded into nonsense. It suggests conceptual care incorporated into protein and the genes.*

Nor is bug detection the only editorial role. What about spell-checkers or search engines? As noted, not all *DNA* makes genes. Just as not all of a book is made of words nor all of a car is its engine, it is likely that the other parts serve functions like wrapping or addressing. Certainly address fields are attached to data stored in every database and are cut off when the data is presented to the user. In a similar way protein editors called spliceosomes cut out sections of the m-*RNA* copy of a gene (which may or may not be used for other purposes and are called introns) and splice the remaining sections (exons) to form the template for a protein. And it has recently been learnt that with a high level of sophistication called 'alternate splicing' some genes can be appropriately modified by cutting and pasting different introns from their m-*RNA*. Such variation supplies versions (up to hundreds in humans and thousands in fruit flies) of the same protein for functions in different parts of the same body or perhaps even refinements of operation in the same cell. In other words a single gene can act as a kind of archetype and, by editorial cutting using complex splicing code, employ different permutations of exon to produce modifications of itself. Remember, in addition, spliceosomes are, like ribosomes, complex, encoded mechanisms synthesized, modified and partially assembled in the nucleus before being shipped out for further assembly in the cytoplasm!

Adjustable spanners; jigsaw customisation; super-coding. Epigenesis. Such flexible response is common and could underwrite most forms of adaptation even to the point of speciation. It might also explain why, on the one hand, organisms with similar genomes may appear quite different; and, on the other, why there exists no clear correlation between the complexity of a species and the number of its chromosomes or genes. For example, a salmon has more chromosomes and rice more genes (~38000) than you: a roundworm of 959 cells (*C. elegans*) has 50% more genes than a fruit fly; and sea anemones with about 25000 genes are similar in respect of genetic complexity and activity. But there is a huge disparity between gene sequences and actual forms that their proteins combine to make.

Molecular biology has by now catalogued careful housekeeping (e.g. ubiquitin proteins with the proteasome complex) and quality control procedures (for example, checkpoints occur at number of critical points during the fundamental process of protein synthesis). Variable ('volume') control of specific protein quantities, proof-reading, inspections (using short sugar sequences), help with folding (by chaperones) and editorial ('sense') control supplied by epigenetic switches will be mentioned later. As well as epigenetic super-coding other factors buffer the system, promote inbuilt potential for adaptations and (as, for example, using heat shock proteins) facilitate response to external shocks. Why, we asked earlier, should evolution have the least interest in quality and quantity control, efficient use of resources or conservation of a coherent symphony of molecular music? Of course, 'interest' is not the word but why should a system be of evolutionary use at all? How could bugs one by one compose it? The more is discovered the louder echoes emptiness of any claim that all this specified complexity arose by chance.

Organisms from bacteria to man employ sophisticated methods of repair. As organs so even molecules self-heal. Information maintenance systems (using, for example, cooperative enzymes like Mut Y and endonuclease III) can scan *DNA* 'wiring' by testing electronically, like an electrician, for faulty circuits. As well as this and other methods of repair (about 30 so far discovered) mistakes are actually pre-empted. There exist many prophylactic processes (including the aforementioned micro-*RNA*s) that regulate the crucial steps of replication and, in protein synthesis, the transcription of *DNA* to *RNA* and, thereafter, its translation into protein. There even exist 'post-translational' enzymes which, after a protein has been exuded from a ribosome, can modify a normal amino acid into a novel one (there are 20 normal kinds but over 100 found in protein chains).

The deeper we zoom in the more detail, and the more precise and coordinated the detail, we observe. If *DNA* represents an information system you would *expect* to find evidence of the conceptual apparatus of programming. **A prime definition of information is 'that which purposely holds things together, unifies them or is the principle from which their organisation derives' - a definition closely related to the principle of unity and the character of consciousness. From such principle you might expect to find the concept of inter-relationship expressed in interlocking, mutually informative systems - mechanisms.** The symbolic representation of interrelationship is called address and addresses involve the whole signal apparatus of postal systems. The Dialectic would suggest that, once the logistics of a system that can raise suites of enzymes to service metabolic programs and create the biochemicals that service life's demands, genetic expression will cleave still closer to the computer analogy than it does today, that is, in terms of density of information, further still from chance.

Would you expect a bacterium, with only one chromosome, to need an inter-chromosomal linkage by address? No. If 'junk' represents such a system, it has less than a eukaryote - but you'd still expect addressing that linked different genes for metabolic purposes. In eukaryotic systems, however, either 'junk' internal to a gene (called intron) or external may act in structural and functional capacities as well. It may help 'pack' chromosomes into their super-helices ready for mitosis. And sections may assume informative capacities such as link-points, controls that indicate the sequence of steps in a process, addresses and agents of parallel processing or, as it is also called, multi-tasking. Such features amount to part of an operating system. Then 'junk' would represent a system that, behind the scenes, enwraps each genetic file. And a geneticist, by 'keying in' his various probes, is trying to work it out. If, in addition, an aspect of introns is genetic modification then information's flexibility is increased by orders of magnitude. It was noted how much extra code highly evolved word-processing packages consume and therefore have to store - and how bugs disrupt proceedings. No wonder, if mutations occur to introns, so carefully excised by spliceosomes during m-*RNA* transcription, they can cause havoc as catastrophic as exon mutation. As an afterthought involving forethought, it is worth noting that all information systems are purposive, reasoned and thoroughly pre-programmed - so what would the first cell that evolved a spliceosome (the eukaryotic ancestor of all other protoctistans, fungi, plants and animals) have to splice? Or, without correlate system or spliceosome, was the first of accidental and non-integrated introns just the start of *DNA*'s proliferating junk? If, on the other hand, introns also contain information that permits different but appropriate expression of the same protein in different cells then they assume the logical status of programmed self-adjustment. The key phylogenetic imagination that sustains the neo-Darwinian theory of evolution is that complex, systematic micro-hierarchies can occur by chance. In fact, a cell *is* computational chemistry and the logic board of its genetic operation locked, like any complex algorithmic system's, into cooperative micro-hierarchies. Biologists routinely use such IT language of description with respect to biology's book of life but many, due to 'lazy' philosophical preconceptions, fail to acknowledge its implications. *No information technologist, however, misses the clear hallmark of design in such sophisticated, high-level programming, programming that involves the very hub, the operating system, of biochemical computation.*

Of course, with age some genetic variation due to 'dents', 'scratches', 'blotches' and other mutant imperfections is bound to be selected out, neutrally forborne or occasionally selected in. But, as we'll see, so-called beneficial mutations are observed (as with sickle cell disease or bacterial antibiotic resistance) to involve single changes and not the thousands of incremental benefits that, before being selected out, are needed to innovate at any level from molecular to larger systems. So what is the burden of the abovementioned observations? They demonstrate a cell (and higher organs of a body) act computer-wise. *Effectively coded application programs* called metabolic pathways run to fabricate each chemical that key switches trigger as required; *operating systems* comprising *DNA* manipulation, epigenetic 'super-coding' (Chapter 23), protein synthesis and suites of messengers (hormones, kinases, *AMP* etc.) support the running of these applications all subsumed beneath the program that conducts a body's 'song' on earth. And, of course, the software 'calls' stored information to support its actions; it calls precise instructions from a *nuclear database*, a system written on a medium called *DNA*. The burden of a computational observation is thus specific, purposive, ubiquitous precision. The previous perception of a 'junkish' origin by chance is, by intelligence, by far outclassed.

In all the informative precision of this micro-world, operative in every cell of the eyes with which you read this book, one needs ask what use is protein left to its own devices and without an overarching plan. A house uses bricks, wood, metal frames etc. Different shapes of house may or may not use the same or practically the same off-the-shelf, prefabricated components. Every builder's merchant publishes a catalogue listing different shapes and sizes of the same basic components. As builders use different well-tuned mosaics of the same basic ideas - wall, window, door, tile, pipe and so on, should one not expect to find evidence of genetic subroutines 'rephrased' for use in different bio-architectures. Biologists have long listed homologies - similar metabolic pathways and organic structures which perform the same or different jobs. For example, just as a petrol-engine is adapted for different uses in vehicles on land, sea and air so life's universal motor, energy metabolism, employs homologous variations on theme. In the same way mouse and other mammalian protein is 90% or more human. Indeed, it is as arguable that chimps evolved from man as the reverse. Although 30% of their genes are identical to ours they have 48 chromosomes (to our 46) and there are millions of single base differences, many non-coding but others involving genes, between them and us. Sponges and fruit flies are over 70%, banana mush 50% and even cabbage over 40% genetically the same as you. Nor, as noted, are chromosomal counts or numbers of genes much to judge by. *DNA* symbolises, like an architectural drawing, the building products that concept needs to realise itself in physical practice. What proteins, and therefore genes, cannot do any more than building materials is to spontaneously assemble the house for which they were delivered. How can a protein tell a left hand from a right one; or that it is part of a banana and not of a mouse? Indeed, you could theoretically make two different houses from *exactly* the same materials - same database and supplier but a different architectural archetype, same genetic database, same physical components but a different archetypal program.

Application programs signal their instructions. An operating system is a framework or facilitator that allows these communications to be carried out. A database (such as *DNA*) is important but secondary. It is the purpose of the application programs that is primary. Direction comes first. To set *DNA* at the informational centre is to put database before system, means before the end, the cart before the horse. Life is its own teleology. We don't ask what stones or clouds or water do. They do their 'own thing'. We *do* ask what a machine is up to. What's it for? How does it work? How does it work to serve its purpose? Everywhere biology spells *aim*. This aim is the irreducibly complex homeostatic survival of body. And body is a 'programmed container', a computerised appendage by means of which mind in conscious and/or subconscious form can temporarily engage the physical plane.

A computer is a mind machine. *On this basis it is established that the cybernetic operations of a cell, an object as thoroughly material as a computer, superbly meet the criteria that pass it as a mind machine.* **A cell is also a mind machine. Its instruction manual is a program written up as 'genome'; and its systems hold semantic meaning as modules or entities of bio-form like, for example, you.** In such a system you might expect the conservation of main or master routines; the possibility of simple, switching variations on the use of subroutines; and even the possibility of a 'universal genome' or a few such phyletic universals. Developmental biologists, especially, home in on such ideas. Causation with respect to life forms has shifted decisively 'upstream' away from neo-Darwinian appeals to exogenous, peripheral factors such as the impact of environment. It has, in many cases, uncoupled the 'selection' of mutations by 'downstream' ecological constraints and pressed up to the codes and master codes that supervise our form. This without an inkling how, unintelligently, these primal, masterful complexities first gradually arose - if ever it was actually so. If *DNA* embodies a computer system's bio-software a design theorist would predict the possibility of scanning, downloading or uploading it in order to create a novel organism, biochemicals or even, as regards the cybernetics of behaviour, some robotic artefact - say, a bionic bee, vicarious vegetable or, by some fraction, superhuman human. It is, further, Natural Dialectic's prediction that the source of bio-complexity, the metaphysic of an archetype, will inexorably force itself to man's conceptual view. ***It will be understood impossible that such biotechnological software 'self-organised' to make and operate its own hardware without intentional design.***

Government and Industry

There are a couple more interesting metaphors for biological order and economy. If more evidence were needed that fire, wind and water do not organise principled, targeted systems, government and industry supply it.

Top-down, the head of a political system is its source of directive power. The word 'governor' means 'helmsman'. King, president, pharaoh or whatever you call the prime minister steers an executive council. Active information resides in this cabinet. Intelligence agencies provide information, the privy cabinet or parliament debates it and hands down decision to the administrative bodies. Spheres of influence and implementation include organs (called ministries) whose coordination is intended to maintain a homeostatic, balanced state. They keep the peace, orchestrate the health and well-being of the body politic. Below the political head such Ministries include Agriculture, Food and Fisheries (for food), Health and Social Welfare (for health), the Criminal Justice System with its police to keep internal order and a couple of agencies to optimise external relations with the environment (Foreign Ministry to promote and exploit the homeland's interests and Armed Forces to exact immunity from foreign attack). You can easily relate such function to your own body. After all, you are the king or queen of it.

The industrial analogy is also instructive. What is a machine (see also Chapter 6) but a purposeful arrangement of parts, a functional complexity that serves a goal? What is the difference, in this respect, between machine and cellular machine? What, furthermore, is a factory but a suite of mechanisms and machines that serve an overriding purpose - to make the goods that let it profitably survive? *If its network of interlocking assembly lines serves the profitable purpose of survival is not a cell a chemical factory*? One whose machines, called proteins, manufacture microscopic parts down metabolic lines? Both cell and body work like factories. All three are subject to the same forces of decay and, in order to minimise their effect and maintain production, must control both internal and external circumstance. **In fact a cell is, marvellously miniaturised and of labyrinthine complexity, a computerised chemical factory.**

How do its structure and function serve their reason, their end-product - animation? A cell is built of flexible and moving parts. A membrane-barrier separates its internal activities from the different conditions outside and, thereby, defines its realm of jurisdiction. The 'wall' has gates that control its relationship with outside forces. This relationship includes orientation, identification and the exchange of specific information and materials between itself and its surroundings. Chosen raw materials are transported in, waste and finished products shipped out. The cytoplasm is its shop floor. In this controlled, compartmentalised and efficient place specific parts (called biochemicals) are lathed then delivered at the right time in the right quantities to the right places. Various departments (energy, post and packaging, cleaning etc.) combine to optimise the business of survival. Schemes of work often require the manufacture of items for which plans are kept at headquarters. In this case the boss's office is a nucleus and the plans filed in chromosomes. No master plan ever leaves the office to circulate, become crumpled and soon enough lost or destroyed on the shop-floor. Instead specific photocopies are issued (in the form of m-*RNA*) to guide assembly. This fantastically automated factory even has the capacity, under the appropriate conditions, to double its capacity. It can make a copy of itself and start again.

All cells and all organisms comprise both government and industry. They employ the same kind of informational, functional and structural hierarchies. *That is, they are conceptual in design and operation.* Factories and computers indicate an elephant, by name Not-on-your-Nellie-when-it-comes-to-Chance, is stomping round a theory's cage; nor will Darwinism ever recognise the beast exists.

Archetype

Is archetype a dangerous idea? For materialism it is fatal.

Bottom-up thus trenchantly denies it while increasingly the latest science points 'upstream' towards a chanceless source. Conservation of the main routines, mosaic recombination of their subroutines and pleiotropic (one-to-many) influence from 'upstream' or 'high-level' modules regulating detail that is forged below - these are discoveries in biology that involve genetic permutation and sdd to a *top-down*, computational terminology; increasingly they indicate a labour that will bear the dialectical perspective viz. logic that's derived from software called an archetype. Book 2 (Chapters 15-17) in some detail checked the nature of subconscious mind, potential matter or, in other words, archetypal memories in universal mind. It named an organism's subset 'typical mnemone'.

Archetype (see also Glossary) **may be defined as a conceptual template or, in dynamic terms, a program.** Software always issues from the purposes of prior intelligence. Purpose is not made of particles with forces; it is metaphysical. And archetype means architecture. It means plan. And plan implies intention. We've dealt with it in almost every previous chapter. Do you remember Archetype positioned at the top edge of existence and identified (Chapter 5: Top Teleology) as a pure stream of consciousness named *Logos*; and, lower down at the mind/ matter interface (*PSI* of *figs.* 0.8, 0.12 etc.), potential matter, the archetypal agency of quantum 'matter-in-principle' and its outcome, classical 'matter-in-practice' that we call sensible phenomena? Such 'transcendent matter' is (*figs.* 2.6, 2.15, 4.1, 5.2, 5.3 etc.) the top edge of material existence, that is, the physical universe. It is *physicalia*'s first cause and has been identified with archetypal memory in 'non-local' universal mind. The subjective/ objective 'instincts' of the world are also known as typical mnemones (Chapters 16 and 17). Such mnemonic archetypes are programmed patterns everywhere expressed in energy/ matter as simple, fundamental particles (including charge) and forces; and locally, in dynamic complexities informed through the agency of chemically inscribed code, as types of biological body.

Check *fig.* 13.3. The inflexible zone of unconsciousness includes both typical mnemone (or informative, 'attractant' archetype) and the non-conscious, physical automation of matter. It now remains, therefore, to link the ideas of Chapters 16 and 17 and 19; and thereby join, closer than breathing, metaphysical with physical biology.

If forms of life are wholly material then matter can only have 'organised' itself against the 'stream' of disorganising tendencies (called entropy) to produce animate sensitivity. If subjective experience, consciousness, reason, mind, instinct and memory are by-products of chemical changes and quantum effects, then any notion of mental design *preceding* the construction and operation of life forms is nonsensical. If the author of 'design' is chance with death as its only check then what chance is there that principles, which are condensed information, source the expression of biological form? Chance means lack of principle, inexplicable behaviour. This is why some scientists, finding the 'lucky' view of evolution just a tad too lucky, seek as-yet-unperceived physical laws of information in the universe. Their questions are good but starting-point poor. This is because a materialistic *(bottom-up)* point of view forbids the place of active, negentropic information (mind), archetype or purpose behind biology. *It inhibits a fully conceptual appreciation of the subject.*

From a dialectical point of view, however, the logic is reversed. Material bodies do not 'organise themselves' purposively; and immaterial subjective experience, consciousness, reason, mind, instinct and memory certainly conjoin and affect biological form. *In this case, according to the conscio-material order of creation* (Chapters 5 and 6), *negentropic mind precedes entropic matter. The question of its impact on chemical bodies then becomes a practical one of interface.* The nature of such psychosomatic interface - of which archetype is the immaterial, psychological side - has already been discussed.

If I find a necklace washed up on the beach does the object itself, in the absence of its maker, immediately imply or prove that (s)he existed? *In the same way, the closer we look, the more evident and exceeding becomes the immediate implication or proof of biological design.* How could chance construct mutually inter-locking hierarchies of form and function? Each new reason, every new inter-linkage, every new '*because*' that we discover reduces the chance that chance was ever reason's reason. Either design is merely apparent and purpose purposeless or they are not. Is there *any* chance that the inception of biological form was non-conceived - or is that inconceivable? Do you think that natural selection could perform the wondrous task? Then just read Chapters 22 to 25. In the last analysis, however, neither perspective can prove its negative. *Forensic biologists have only seen the reproduction, not the origin, of life forms.* Therefore, in the court of origins, they can no more than *infer* from a logical framework.

How, though, can frameworks that are 'logical', 'goal-oriented', 'reasonable' or 'purposeful' be implanted in biology? You may remember, from Chapter 1: Causality, *energetic* and *informative causation.* The former pushes you from past to future. **The latter is goal-oriented causation; and goals are in the future pulling you their way. They pull you *from* the future; they are metaphysical attractors, guides that govern your behaviour as they lead you through the world**. Of archetypal structure biology is most concerned with instinct and the morphogene. Morphogene means a specific field in universal mind, a metaphysical but also morphological attractor. Attractors pull things forward the way they have to go. Push and pull cooperate. **Energetic and informative causation, running anti-parallel, are the way the world proceeds; and, in the context of biology, primary archetypes pull, secondary genes and other physical components push.** We discussed the way, by electro-dynamic connection with biological matter and perhaps especially the coiled 'antennae' made of *DNA*, that creation's vibrant, radiant archetypes can operate. Thus, nested hierarchically within the structure of external form, they ply their mental but unconscious trade; hence biology derives its obvious reasons.

The basis of a conceptual approach to biology is, therefore, such archetype. It should be crystal clear by now that an unconscious archetype is, like instinct or memory, real but not physical. Indeed it *is* a memory, a natural record; it is a sub-conscious mental, metaphysical construct, an idea or rationale physically realised in a variety of similar forms. *Biological archetype is identified as* (Chapters 15 - 17) *typical mnemone or archetypal memory. This memory constitutes a lens or filter through which each individual type of experience, that is, organism is projected; and, therefore, through which the parcel of its parts has coherent being.*

No wonder Thomas Huxley expressly hated the idea of archetype; it more than leashes, it exterminates the idea he pugnaciously 'bull-dogged' - Darwinian evolution. Evolution as a macro-concept is dissolved.

Archetype is 'nature' to genetic 'nurture'; variation in genetic and environmental circumstance will lead to various expressions of its metaphysical potential. Such endless, local variation-on-a-theme is called individuality. For the most part such plasticity arises systematically by sexual mixing and (see Chapter 23) epigenetic switching, that is, pre-coded adaptive potential. If thereafter, one step down creation's scale, genetic code is construed as 'nature' then circumstantial 'nurture' also generates variety-on-theme. But the environment can also be destructive. Is, for example, damage to gene sequence called mutation your creator? This is what neo-Darwinism of the *PCM* avers.

By contrast, a Theory of Intelligence construes such mutant factor at the far periphery of creativity as no more constructive than a crash or crashes making motor-cars. Genetic code, recombination and environmental pressures each promote variety; expressions of archetype may vary in location, time and specific detail but not (*figs.* 5.1ii or 22.1ii) in purpose, type or theme. *An archetype is a thematic pattern or norm.* It is, indeed, the (*sat*) informational potential, the norm (consisting of a hierarchy of sub-norms) that specifies a given biological type.

To repeat and clarify, oscillation that shows as genetic variation may occur around this homeostatic 'norm' but the theme itself will never change. In other words, each type of organism is a natural law unto itself. This archetypal law is, in its chemo-physical expression, elastic but stable. While realising variation-on-theme it cannot 'morph' into a different body-plan; such variation is therefore ill-conceived as micro-evolution. There is always a tendency to harmonise as far as possible, within the context of other components or of the whole body of an organism, with the (metaphysical) archetypal pattern. There is, equally, a tendency for extreme divergence to 'rebound' or revert towards the inner stability of its 'native' or unstrained normality. We recognise the realisation of optimal or strong alignment with archetype as a dominant 'wild type' and elastic resilience towards such innate strength as a health-giving force known to doctors as the *vis medicatrix*. The substance of such a master control and communication system for cells and the body was discussed in Chapters 16 and 17.

For the information theorist an archetype is an idea realised as a main, core routine (or type of organism). Around this core are linked subroutines - some universal, others more specific - which are appropriately tailored to integrate with each other under the control of any particular main routine. These main routines and subroutines serve as modules; and they are recognised as homologies.

This central *top-down* concept is worth rephrasing. **Various tailored subroutines are called from a Main Routine** (see also Chapter 22: Mosaic Subroutines). Suites of such modules combine as permutations around which different bodies are, under the coordination of such an Archetypal Master Routine, expressed.

An architect's plan governs construction completely. If such construction should incorporate industrial cybernetic and mechanical complexities (say, an automated factory) he would not have exercised his rigour any less. Plans rule. They are the 'egg' reflected in material lines of code; they are conceptual 'seed' from which dynamic shapes of life unfurl. Omnipotent, totipotent - archetypes precede and wholly govern physical biology. An engineer actualises an archetype using sub-modules dovetailed into the main project of his design. Just as, for example, there exist myriad versions of the archetypal steam or petrol engines, used in machines of different power built for different purposes, so appropriate or bespoke versions of archetypal subroutine occur in specific organisms. There may also occur mutually exclusive modular designs. These represent taxonomic gaps or unbridgeable structural differences between which intermediate structures are unfeasible. How, as a mechanical example, could a series of viable engines exist between an oscillating and a rotary engine; or, as an optical example, between the distinctly different optical principles that underwrite at least seven different types of eye? Do we label such a phenomenon clever co-evolution (of different expressions of some bio-quality); analogous evolution (say, different wings for flies, birds or bats); or convergent evolution (which is the independent acquisition of similar characteristics by unrelated organisms, that is, similarity *without* descent - for example, a torpedo shape in fish and whales)? Could randomness keep innovating in this 'intentional' kind of way? Or are these words simply neo-Darwinian ploys, shells, vapid vagueness that would lead you off the scent of something more intelligent?

In the *top-down*, hierarchical view of Natural Dialectic, tiered sets of modules compose, at top level, the range of life-forms and, lower down, their various coded and coherent parts. We'll see the clearest example of this later in the way that biological development is controlled. In such a scheme a module's distribution across the living world is, in different degree, mosaic. If, for example, it is critical (like respiration) it may appear, appropriately attuned, in every organism. In other cases it may occur in scattered and sometimes unexpectedly different locations (as for example, haemoglobin in humans and various kinds of plant). While evolutionists term such apparently unconnected, mosaic appearance 'convergence', from the perspective of archetypal computation it is called 'modular programming'. In Chapter 22 we'll check the matter further.

It was previously noted that molecular biology's great strides are heading straight towards the notion of an archetype. A similarity of genetic instructions is indeed found to extend across a great variety of forms. For example, as well as the same homeotic modules (see Glossary) in different organisms coding for the construction of completely different kinds of eye, a normal gene from a mouse can replace its mutant counterpart in a fly; now an eye, a fly's eye not a mouse's, is produced. In other words, such genes act as 'go-to' switches in an archetypal program of development. The discovery of profound genetic conservation in critical root sequences responsible for the timely production of proteins starkly contradicts the hypothesis of gradual, randomly generated adaptive 'solutions' to ecological challenges. Down at the roots, at the sharp end's cutting edge molecular biology would seem to start to shout that modular arrangement rules and patterns are designed. *Top-down* it is predicted that the volume of such call will grow.

In this case, if gradually tweaking modules can't innovate, could macro-mutations conjure macro-evolution up? Could large-scale genetic 'crashes' shake biology's foundation by conceiving hopeful monsters or by springing missing links? Such disruptions are observed as catastrophic. Thus whence could macro-evolution ever rise to life? In the modern evo-devo view radically different shapes of body might surface from small and homeotic variations, that is, from changes in the high-level sets of genes that regulate organic shape. Profound conservation must have, so the theory runs, have varied slightly and occasionally. 'Downstream' these random switches must have rippled and reverberated into larger phenotypic differentiation. Did such morphogenesis turn macro-evolution's trick? We'll check in Chapter 25.

If mice and chimps were even 95% genetically human (which is doubtful) then how can such large-scale difference in creature shape occur? Genes alone can't be the template that gives shape to both development and adult form. In concentrating on the only information source that it can measure and manipulate, material emphasis on genes might thereby miss informative potential that operates from a forbidden plane. **The template is, Natural Dialectic clearly claims, beyond (although within) the chemical domain. It is in mind as archetype.**

Variations-on-a-theme are wrung, by breeders and mutation, on the shapes of dog, horse and many other organisms. Yet the direction of natural selection, though supporting trivial differentiation, can't *create* a metabolic pathway, tissue, organ or a body-plan. It operates the wrong way - (↓) down towards *loss* of alleles and genetic information. Variants, as Cruft's shows, are mainly due to loss of information and not evolution's necessary gain; and death always freezes up a limit to plasticity. **In fact, by artificially and fiercely driving such selection to its breaking point you'd soon empirically discover whether nature could snap barriers of type and *innovate*. There is no evidence, not even nascent evidence, it can**. In contrast, '*deliberate*, not random, changes rung upon the archetypal subroutines' is how the Dialectic would interpret facts. It would informatively explain convergence, co-evolution or mosaic evolution as the engineering application of a theme to local detail, of a principle routine to different practices. *It would be found no accident that a mosaic conservation of both form and function pervaded somehow, somewhere, every form of life. It would be found no accident that a mosaic conservation of both form and function pervaded somehow, somewhere, every form of life. These routines and subroutines - reflected by material genes, molecular conformities and thence the larger structures that entirely constitute biology - would reside in immaterial mind.*

Primary (but not 'primitive') processes involve fundamental archetypes. For example, both origin and survival depend first and foremost on economically and efficiently organised (*sat*) *information*, that is, on a prior and pre-set encapsulation of directives in a well-oiled system of storage and transmission. Present survival is also based on (*raj*) *energy metabolism* and future survival on (*tam*) bodily *reproduction*. It might be expected, therefore, that the metabolic computer, a cell, would show universal 'sub-archetypes' with respect to informative operating systems (such as genetic code and cell signalling systems), energy routines (such as respiratory metabolism) and reproductive patterns (such as asexual mitosis). It does. Principles such as these would be elaborated according to organism.

Specialist archetypes for sight, fight or flight involve eyes, wings, weaponry etc. All are, of course, subsumed under the top, informative principle whose archetypal, and thence physical, expression includes apparatus suitable to engage, in different organisms, various degrees of sensitivity, mind or awareness.

Could you call your mind 'an information field' whose shapes or particles of influence are thoughts and memories? An archetypal form is metaphysical, a memory in 'the field' of universal mind. Discrete archetypes (Chapter 9) constitute the natural law, modulation or behaviours of photon, electron and proton. They control the quantum operations of forces and particles. But each type of organism is a discrete part of this same 'rainbow range', this conscio-material spectrum of existential forms. Its physical expression will vary according to local conditions but only within the potential of its own unvarying theme. Chapter 17 was a study of the human 'set', an arrangement of plexi around whose characteristic themes (*figs.* 17.8 - 10) your own outward form is built. A nexus of specialised or specific versions of subroutines compose the range of internal processes which, interwoven, unite as the specific body of a type of organism. In other words the perception of archetype is not restricted to the *creatura* as a whole. Hierarchical levels of a computer program are called subroutines; nested levels of biological program, wherein interact electrical (e.g. *H. electromagneticus*), molecular, cellular, organic, systematic and whole body (phenotypic) subroutines, may be called sub-modules or sub-archetypes. Biology generally calls fundamental similarities at any of these levels '*homologies*'. Forms of life can be seen as incorporating permutations of such homologous architectures or archetypal subroutines. *In this view it is archetypal theme rather than mutation that underlies the exquisite engineering with which biology is replete.*

The biological range of organisms expresses mental character (by instinct and, where consciousness exists, in quality of thought); and each kind plays a definite role in the ecology of life on earth. There are types with critical recycling roles (saprophytes, detritivores and scavengers) that live along with photosynthesizers, herbivores and predators in solid (soil), liquid and gaseous (climatic) media. *Each type's rich variety changes round within the circle of its own typological idea* (*figs.* 5.1 and 22.1). Family trees were seeded and will end within their 'type'. It is the type or super-species, not the species, which is fixed: and the type (somewhere round the genus/ family level of classification) is itself not a single material specimen but an idea whose expression changes within its genetic circle of influence. Looked at this way, to explain the origin of species is a trivial and sidetracking exercise. It misses the main point. *It is the origin of types (or archetypes) by macro-evolutionary relationship or by design - both of which are unseen, inferred methods - which is the real, vexed question.* And those who deny metaphysic won't, of course, allow suggestion of an archetype!

If organisms are viewed as archetypes that incorporate different permutations of subroutines, tailored as necessary for optimal coherence, then there is no need to invoke phylogenetic/ ancestral relationship to explain the diversity of biological forms we see today. **Instead of an illogical, accident-based explanation with death as its blind executor we have a logical, consistent overall expression of purpose and of plan**. Immaterial information and intelligence exists in the universe. They are real. Just as energy can be concentrated in a particular body or be broadcast free, so information can be concentrated in a particular experience, projection of purpose or fixed in memories. If there is universal matter as well as human body is it impossible there is, as well as brain-attached mind, universal mind? And, by an extension of the same logic, you might expect to find life at any other suitable location in the universe. Yet (Chapter 26) physical constraints mean that it will probably consist of archetypal expressions like those expressed on planet earth - variations on the theme of carbonated water making yet another kind of fizz!

So we have arrived in the arena. This is the nub of a gladiatorial clash of Titans, the collision of two pillars of faith. *Materialistic logic, whose basis in biology is the theory of evolution, would exorcise the anathema of archetype from its frame of reference. Top-down dialectical logic, on the other hand, asserts that biology better expresses anti-randomness than its reverse.* It identifies an informative, functional and structural hierarchy of archetypes combined in all cellular and multi-cellular life. An obvious example of archetypal, structural hierarchy is the biological system of classification. In the attempt to track this natural template we construct our own. We place this man-made trace over specific organisms to try and determine their place in the overall, natural scheme of things. Being human it is not infallibly accurate and taxonomists argue various pros and cons. Large-scale disturbances occasionally occur such as the reassignment of blue-green algae from plant to bacterial kingdom. Generally, though, we work from four or five basic archetypes (say, prokaryote, protoctistan, fungi, plant and animal) down through refinements until we reach the detail of an individual organism. While *species* is a physical expression of archetypal variation it is, as was hinted at above, unclear which level of classification represents the lowest level of whole-body archetype - a Master Routine incapable of transformation beyond that point. It is certainly well below the level of kingdom and includes physical possibilities such as hybrids, polyploids, sexual and other sorts of variation. As previously noted, it probably rings a catchment area in the region of the man-made categories of *genus* or *family*.

The strength of the archetypal against the weakness of the evolutionary idea will be scrutinised over the course of the next five chapters. Right now we turn to inspect the three levels of archetypal hierarchy - informational, functional and structural.

(*Sat*) Informative Hierarchy

	tam/ raj	*Sat*	
	matter/ energy	*Archetypal Information*	
↓	*tam*	*raj*	↑
	passive product	*active/ reactive*	
	inertial stability	*dynamic stability*	
	run-down	*cycle*	
	completion	*continuity*	
	death	*life*	

It is not materialism's fashion and yet so critical is the previously encountered dialectical principle of biological hierarchy that we here reaffirm and formalise it.

Just as a synthetic, ecological approach contrasts with a combative, warring attitude towards life's environmental struggle to survive, so the primary dialectical emphasis on conceptual integrity (that is, on the unified, whole-body plans of informative homeostasis) contrasts with an analytical, reductive study of biological components.

The central feature of homeostasis, an information loop, exemplifies the fundamentally hierarchical, cyclical and regulated nature of biology. Passive (received) information is processed and a reasoned, active and/or reflex, passive response issued. *Turing machines. Computation. Signals, code and communication are life's central business.*

The minimal, irreducible requirement for any homeostatic function is three integrated, co-operative mechanisms. This triplex must be simultaneously present and operative from the start. Missing any part the system cannot operate. They are, as stated, sensor, regulator and motor. Let such irreducible complexity of associated mechanisms, situated at the informative heart of biology, serve as a trailer for the 'film' of Chapter 20 (Modern Alchemy).

	tam/ raj	*Sat*	
	motor/ sensor	*Control*	
	output/ input	*Centre/ Pivot*	
↓	*down*	*up*	↑
	output/ info. to exterior	*input/ info. from exterior*	
	motor	*sensor*	

Biology's central, essential (*sat*) factor is regulation. Homeostasis exemplifies the tendency of government. Its mechanisms serve the principles of control, balance, poise or harmony. Its reverse is, of course, imbalance, disharmony or disease. Cacophony is jarring noise; homeostasis orchestrates the right chords at the right place in the right time. In this view the natural, normal state is health and, where consciousness exists, contentment. Each biological organism is seen as the passive materialisation of a melody - a logical, many-phrased melody that is its causal, archetypal score.

A (*sat*) regulator is an organisational centre. It is occupied with the receipt, processing or transmission

of information to buffer the fluctuations around a pre-arranged norm or stimulate and coordinate any preconceived (or pre-programmed) sequence of events. What is the difference between a program, purpose and design? S*at* qualities are awareness, information and responsiveness (whether conscious, sub-conscious or through physical mechanisms designed for the purpose). Conscious, psychological desires (and their homeostatic neutralisation through satisfaction) are flexible; and the central elements of biology are homeostatic norms and the coded information accessed in order to obtain their involuntary, inflexible purposes. Mutually interactive information systems include, as well as the overall archetypal and electrical patterns, nervous, hormonal, nucleic, cytoplasmic and membranous sensitivities. Animate, as opposed to inanimate, material is seen as information-rich, specifically guided, code generated. Its teleology is passive but not apparent; it is real.

energy	*information*
output	*input*
movement	*sensation*

Top-down, the source of order is mind. *Your own most obvious homeostatic function, central to your existence and situated top centre at governmental headquarters, is your mind.* Its choices, your choices, will carry you in a direction you think may optimise your sense of well-being. The physical correlate of mind is its (*sat*) regulatory brain and from this central nervous system there extend (↑ *raj*) sensory and (↓ *tam*) effector/ motor systems. These operate at electrochemical level to construct and control, in an integrated, unifying way, the metabolic and larger-scale, physiological activity of other organs. The immediacy and pinpoint accuracy of nervous response also coordinates with the longer-term, more generalised influence of hormonal cascades. Both nervous and hormonal systems are knotted together in the centre of the head by a triplex comprising pineal, hypothalamus and pituitary glands. Both systems act like governments; they gather information and hand down commands. For example, a few hormones released into the bloodstream from central executive glands are amplified by target organs to organise the desired biochemical and correlated physiological order of effects. And out in the field such 'lower ranks' as, for example, progesterone report for duty. In the course of office they send signals back to high command. Military, commercial, political or administrative - every chain of human responsibility ever established to achieve its goals acts like a body or a cell. Is not the function of the latter, where no human lapse or frailty of judgment enters in, superior? A two-way flow of information keeps the missile right on track; strings at the heart of life are plucked in tune; dynamic equilibrium of faultless homeostasis keeps a healthy form alive. The circle, called a cycle, is complete; and it keeps completing what its purpose is. Where is chance in this? How, by chance, did its raw, triplex starting-point conglomerate?

How, more precisely, might an irreducibly complex series of interconnected cycles and components (which occur in single cell metabolism as well as whole physiological systems) arise piecemeal by chance? Despite a philosophical hope it might be possible how could such purposeful order arise from purposeless, unguided chemistry - chemistry that prior to its own coded and intact integrity would have no working value? Do chickens precede eggs? *Information, reason's twin, is evolution's major problem; surely plan-less lack of reason did not make a cell or, composed of cells, your body? Indeed, it could never and we'll see, in Chapters 20 and 21, why.*

Life is vibratory. Its rhythms are, from top to toe, coordinated. Order is tabulated in the form of chemical code (e.g. micro-hierarchical *DNA*); the program is dispatched by signal (e.g. electrical, hormonal, ionic interaction or by concentration gradient) to receptors that 'know' automatically what to do. Subroutines reference higher routines, higher routines instruct lower as information is continually cycled through the correct channels. Large-scale, whole-body cycles range from birth through circadian, monthly and annual to death. They gear with small-scale, rapid and specific sub-cellular feedback cycles. In processes that start with protein synthesis and metabolism these reference their own database and construct, regulate the flow, appropriately transport or destroy the requisite biochemicals. All such exchange operates according to a pre-arranged and 'understood' language. Information is smoothly materialised. Feedback closes the loop of omnipresent biological purpose and control. Does linguistic incompetence damage a conversation? **Lack of fluency is unhealthy if not lethal to the performance and, similarly, there is absolutely no place, nor ever was, for random chemical events to create multiple, informed networks of coherent activity - even in a single cell.**

(*Raj*) Functional and (*Tam*) Structural Hierarchies

	tam/ raj	*Sat*	
	concentric zones	*Centre*	
	explicit expression	*Implicit Order*	
	informed/ informant	*Information*	
	structure/ function	*Code*	
↓	*tam*	*raj*	↑
	outer/ gross	*subtle/ inner*	
	passive component	*active component*	
	informed materials	*genetic informant*	

matrix	*metabolism*
vessel	*reaction*
container	*process*
limit	*flux*
instrument	*role*
structure/ anatomy	*physiology/ function*

A *bottom-up* perspective might conceive that structure, once accidentally obtained, will generate its own function. In this case the function would, like a chemical reaction, be a product of happenstance and inherently purposeless. Of course, at all biological levels natural selection regulates variation upon functional theme - but is incapable of the innovation and integration of *fresh* theme (such as nervous or transport systems). Before cartloads of variation can accrue thematic horse must first be harnessed to the task. However, if the function is purposeful the perspective is incorrect and exponentially incorrect if, as in a machine, separate functions are required to coordinate towards a single, overall reason - such as the (homeostatic) continuity of life.

Do you remember (Principles of a Unified Theory of Biology) the triplex hierarchy of information, energy and structure? Biology is not chemical structure out of which emerge function and control. The reverse is true. **All the hierarchies run *top-down*.** No doubt function and structure, under the direction of information, are two sides of the same physio-anatomical coin but structure serves function and function serves purpose. An engineer runs from principle to practice; only the empirical student, who studies an invention, runs from practical observation to principle. Science is, like the student, empirical. This is no problem unless, because of its starting-point, *top-down* possibility is denied and the holistic order of creation thus reversed. In this case a product has to evolve itself because, for a philosophical reason alone, intelligent engineering is put 'out of bounds'. But it is not, in truth, as if haphazardly generated structure like a cliff face, waves at sea or any pool of chemicals will ever give rise to function. Chapters 20 and 21 make absolutely clear why.

Right from start (Chapter 20) you will need to make breath-taking guesses how 'black boxes' we shall mention ever started! Nor do the odds against materialistic/ design-less bets decrease when placed on tissues, organs and their systems. Ask an engineer or programmer. **The horns of hierarchy always lock with any element of randomness**.

Organic structure is a 'box' in which some function can occur. All engineers design their products so that structure best permits their preconceptions to appear. Not *vice versa.* No creator blindfolds himself and thinks of nothing (not even the reason for his project) while he scribbles his 'designs'. You cannot produce working systems like this. Biological systems *must* work. And they do - from the word go. Failure to work means death. From life's presence rather than non-presence, from its survival rather than its death we may reasonably infer they always, evidently did.

Biological design is therefore not only a matter of gross (or phenotypic) form; nor even coded homeostatic principles that regulate the structure and function of a life form. *It is at root an inner question, one of matching, in all the applications that coordinate to support survival, code with molecular and gross structure of the mechanisms with their specific functions*. This match must occur in a way that, whether or not it is deemed optimal by human critics, translates the concept of homeostatic survival into working practice. We have already examined the *minimum* requirements (isolated code storage, reproductive facilities, fuel supplies and isolating, environmental buffer systems) for even a single cell! We shall (in Chapter 21) elaborate on this evolutionary road-block.

Functional hierarchies involve the operation of systems (nervous, muscular, skeletal etc.) geared to each other at different levels but under the overall control of a master-processing unit. *The basis of functional hierarchy is pre-arranged sequence.* The basis of sequential coordination is, as every producer knows, communication in the form of information based on code. Timing, signal, switch, correctly composed instrument and position are each critical. Life's choreography, even in an archaeo-bacterium, engages a minimum but large number of complex, directed steps. Matter, because it is going nowhere, engages none.

What metabolic steps or mechanisms do you think that you could, for any given organism, harmlessly extract? How many are superfluous? Or is the equation set just right? (*Raj*) energetic function executes a 'dance' prescribed by (*sat*) regulation to achieve targets. The targets show at four physical levels from molecular to phenotypic. They include completion of the homeostatic cycles and developmental pathways necessary to survive both now and, reproductively, in the future. Perhaps development, which embraces the trio of informative, energetic and reproductive hierarchies, is the most powerfully obvious staged program of all. The creative archetype of a construction industry runs from potential (preordained information carried in, say, an egg) through a kinetic, shaping to a static, shaped phase. From within outwards, it follows the creation gradient. It originates; then it follows the cosmic pattern whereby inside (seed or potential) information is exteriorised to realise the physical reflection of its original idea. Are you not based on an idea?

Psychological process is imprinted on materials. **Purpose is the real nucleus or centre of any construction.** Around its core are woven concept, design and, finally, the construction itself. In biological construction the parts are not made separately and 'bolted' together. They develop in synchrony and differentiate as parts of a whole. They are not built from outside but 'organically' from within. Plan is realised. Mind is at the heart of things. A biologist does not, however, examine the mental precursors of his machinery. The first sign of plan or coded information that his searching 'radar' pings is a molecular database and nuclear control. Programmer actively informs, system is arranged (i.e. passively informed) accordingly. Such genetic information is, even when in computer-like action, passive. It is stored at microscopic level in a vessel called a cell or, for starters in the case multi-cellular vehicles, a specialised top-level information pack called an egg. At a signal the cell and its code embark on precise algorithmic but dynamic exchanges of data whose pathways lead, eventually, to the full realisation of the original idea's potential. In this case the idea is a self-reproducing, self-maintaining machine. A zygote (or, asexually, a clone) initiates a program of changes until its final realisation as an adult form which, while able to engage in cyclical procreation, is subject to the material, entropic process of aging and decay. *In this view the architecture of an adult, by which any multi-cellular species is normally judged, is actually only the last, special (final or non-developing) case in the series of a dynamic, hierarchical, developmental archetype.*

You might argue that life's evolutionary prototype, a 'simple' proto-bacterium, never needed a developmental program. Yet what, after all, is such a program at a molecular level? **Is not a metabolic pathway but a biochemical developmental archetype?** All organisms are wholly dependent, for their visible form, on molecular developmental programs - metabolic pathways - all of whose precise, preordained, intermediate chemicals are, except in the context of a targeted series of steps, functionally useless. And each intermediate is itself created by a pre-coded functionary, a catalyst called an enzyme that equally, except in the context of a targeted series of steps, is useless. In other words, these otherwise useless proteins have to be coded by a co-expressed series of otherwise useless genes which somehow, because they 'know' their uselessness might one day be useful, evade natural selection wholesale. Can ten, fifteen, fifty useless genes wait like sculptured ice in sun until the last piece in the jigsaw, the last member of the set that triggers what was 'wanted', 'happens'? From happy zigs to final zag, by luck the set is frozen into perpetuity. But didn't you forget control genes, useless until the day of reckoning, gradually 'building up' to orchestrate the other genes' expression? How little is redundant on a metabolic track! Yet so much 'inefficient uselessness' there must have been to interfere before somehow it all coordinated and maturity, end-product, was like a prize achieved! Coordinated with, of course, a hundred other prizes too! **Coded metabolism is the very foundation of biochemical and therefore biological process yet its evolution is scarcely ever mentioned. Is not only the evolution of pre-programmed metabolism but also of multi-cellular development (Chapter 25) each an order too tall to place at the unreliable, unresponsive door of chance? Unless some prior assumption blocked the thought you might think so too.**

What about paths not of production but carriage? Materials in a factory do not wander; they neither err nor 'walk' in error. Similarly, the informed accuracy with which biochemicals are transported in correct quantities at the right times from site to cellular site along micro-tubular routes is another profound challenge to any theory whose basic rationale is chance. We'll take a walk with the answer in Chapter 21.

What about construction sites themselves? (*Tam*) structural hierarchies provide the 'passive' containers within or through which metabolic or large-scale functions such as breathing, sensing or eating (with their associated sub-systems) can occur. Examples of subtle (molecular) containers of metabolism include cytoplasmic gel, cytoskeleton, membranes, lamellae and organelles such as mitochondria. Examples of aggregates of biochemicals, highly ordered both in specification and relative positioning, include different types of cell, tissue, organ, system and, eventually, organism. Take the example of your own breathing. Such gross respiration (ventilation) is supported by rib-cage, diaphragm, lungs, circulatory system and, thereby, linkage with eating, digestion, the alimentary tract, pancreas, liver etc. These structures together, through (*raj*) energy metabolism, help realise the concept of animation and its survival as life. They alone are, of course, insufficient and have to coordinate with (*sat*) informative and (*tam*) reproductive hierarchies.

Order-in-complexity of supportive infrastructure occurs at every concerted level in both active (physiological) and passive (anatomical) modes. However, just as code is based on simple, alphabetical and grammatical principles from which complexity can be realised so structure is based on chemical simplicity from which great diversity develops. Simple chemicals are built up to serve intricate, preordained patterns. At root, both functional (metabolic) and structural biochemistry are based on water, carbon dioxide, a few mineral salts and, most importantly, the binary code of electrical polarity. Polarity is (Chapter 1) a fundamental existential principle; its binary character is as basic to Natural Dialectic as to nature itself. In this case polar 'fizz' is provided by mineral ions but especially by hydrogen (whose ion is positively charged) and electro-maniac oxygen (negatively charged). Polarity activates. Take, for example, the (reacted) union of polar hydrogen and oxygen. The (*sat*) balance of an acidic hydrogen ion (H+ or proton) and an alkaline hydroxyl ion (proton and oxygen, OH^-) yields the simple substrate of life - transparent water (HOH). Water is itself a polar molecule. The properties that make it the key medium for a (*raj*) functional flow of biological life stem from this 'magnetic stickiness'. On the other hand, the stable basis of (*tam*) complex biological structure is provided by non-polar carbon and its concatenating property - although a few organisms, such as lettuces and jellyfish, are forms of practically liquid architecture. Hydrogen, carbon, nitrogen, phosphorus and oxygen – ingredients as light and simple as can be - except that, as we'll see, no form of life is simple. Not only must a lot of polysaccharides, lipids, amino acids, long and specific polypeptides and very long and specific nucleic acid helices have fallen into the correct orders in the same miniscule areas of Darwin's 'once upon a warm and salty pond' but protein synthesizing complexes, ribosomal machinery, energy production stations and informative software (the operating system, application codes and *DNA* database) must, in communicating concert, 'know' exactly what to do and how to do it. Warm pond, mineral clay, volcano, oceanic black smoker - swapping around the hypothetical site of hypothetical origin is, as we'll clearly see, no help. The origin of any metabolic *status quo* is hard enough to contemplate. Add the expression of a reproductive archetype; add the origin of informative operating systems that seem to have some 'purpose' but, at the same time, fillet the 'pseudo' out of science, strip out all unnatural ideas of 'purposeful design'. Now the mind really boggles. Is this not the height of nonsense, the height of intellectual contortion dressed up as modern, naturalistic rationality? You could, however, toggle the boggle. You could switch and, turning, look intelligence square in the face!

Biological systems operate on a top-down principle. They are simple in principle, complex in practice. Top-level is psychological. Psychological, mentally informative grades are expressed outwards and downwards through various permutations of archetypal and then physical subroutines. Biologically omnipresent are the concepts of homeostatic control, coded information systems, energy metabolism and reproduction. The structures of different types of organism comprise variations on these central modules. This sort of variation is no more random than, say, an architect's variation on the theme of a house or, within it, a room, window or water supply. They comprise mechanical and associated chemical adaptations for operation in the three states of matter (gas, liquid and solid) under various conditions of lighting and heating. In other words, the systems are economically combined into an integrated, hierarchical ecological structure called life on earth.

On the one hand it is palpably *certain* that mind can create complex, purposive machines. The higher the intelligence the more complex and effective its device might be. Having conceived of his construction an engineer employs integrated functional and structural hierarchies to build a bridge, a ship or a jet-plane. They are, equally, the staple around which an information technologist elaborates his applications. Whether soft- or hardware they are powerful evidence of talent for planning and deliberate design. *Technology does not pretend it is accidental, purposely accidental or accidentally on purpose. It always evolves ideas. It prizes intelligence, inventiveness and ingenuity and employs them with maximum strength to solve the problems its purposes generate. From mind its products are devolved. Such product is a materialised idea. Its 'crystalline mind' is definitely on purpose.*

On the other hand, given enough hypothetical time working against the thermodynamic odds, could *anything* at all innovative just happen? Could time build a camera? A camera, you say, does not have *DNA*. But *DNA*, with all the system that surrounds its operation for a purpose in a vehicle called a cell, comprises a machine more powerful by far than camera or micro-chip. The truth is that the feeble conceptual thread on which the whole evolutionary faith hangs is as immeasurably slender. The odds are, beyond reason, against. Are you logical? Will you call the long-term, serial bluff? How much will you bet?

Perhaps it's time to supplement analysis. Did praise and wonder ever spoil an observation? They more enhance. The intricate designs, informed technologies, of organisms warrant such response. In fact, any other wise were stingy in extreme.

In summary, a *top-down* perspective is found logical, reasonable and data-compliant. Its order, of which biological creation is a subset, reflects what we find everywhere. It implies, however, the antithesis of a neo-Darwinian evolutionary model of origins.

20. Chemical Evolution

The notion of a fine-tuned universe is, no matter the interpretation that you place upon its origin, now part of mainstream physics. Darwinism takes this fine-tuned stage for granted yet, in the play of its biology, rejects the notion of fine-tuned, deliberate design. It deprecates such source of order as eccentrically out of order. We, on the other hand, shall systematically concentrate and see.

Chemical Evolution?

The issue is not one of religion or opinion but science and logic.

Do you think of Henry Ford every time you use a car? Does a doctor worry where the broken bone he is mending first came from? Does the meaning of life enter into our every move or is the question of origin directly relevant to the daily operations and the applications of biology? Biologists might, in the reflex way of received wisdom, agree that nothing in their subject makes sense except through the lens of either evolution or informed design but at the same time mostly work without particular reference to either of these background 'frames'. They are superfluous and yet, because each exerts its specific, powerful influence on our self-image, they are indispensable. As adults we are like a child trying to orientate itself. Where am I? What is it? Where did it come from? Where did *I* come from? The question recurs, personalises and to this extent galvanises the scientific study of life, biology, and its origin on earth or, 'exo-biologically', on some other planet.

One explanation for our beginning is that we were deliberately created. The other, which acts as base for the neo-Darwinian theory of evolution, is that we were not.

The former is, materialistically and therefore naturalistically, preposterous. Any birth is special but such a mind-set drastically interferes with the notion of an 'original transcendent projection' of biological phenomena. Allowing mind exists in intellectual or instinctive artefacts (Chapters 5 and 6) it nevertheless execrates any notion that specific information might pre-ordinate a naturally complex bio-form. The idea that such metaphysical 'intelligence' (as spies call information) could be, at root, derived from potential matter's archetypal patterns or from memories in universal mind is thus scientific anathema. It is entirely unacceptable.

Thus the latter explanation, which the next two chapters run ragged, starts with a process called chemical evolution. This phrase implies that lifeless chemicals 'evolved' to the point whence they could 'self-construct' the primary unit of life, a reproductive cell; it means the generation, perhaps gradually over a long period of time, of life from non-living components by physical means alone. This process is also called, in the jargon, abiogenesis, biopoesis or prebiosis. It is integrally part of, strictly not the same as, Darwin's consequent evolution. After all, it casts no role for natural selection, variation or mutation; Darwin merely hinted, hopefully, that once upon a warm pond....

Darwin and Huxley thought a cell was just an amorphous blob, a homogeneous kind of jelly they called protoplasm. Protoplasm, said Huxley completely incorrectly, was the physical basis or matter of life. He even took a gelatinous ooze of calcium sulphate to be a primeval biological form. These fantasies, although erroneous, are understandable; they were also very important because they allowed the notion of chemical evolution to grow without recourse to reality. Nowadays electron microscopy and very sophisticated biochemical techniques provide the modern sage with far greater scope for admiration of even 'simple' life's complexity; and, proportionately, far less and decreasing excuse for clinging to the notion of uninformed, unguided abiogenesis. However, the result of nearly two centuries of speculative prolixity has been to entrain a mind-set that has evolved a life of its own. The next two chapters are devoted to clearing the unchecked wilderness of theories that have choked the reality of what must have been original cell computation and construction.

Chance, necessity or both *must*, if you exclude informative design, have created a first cell. There's no alternative, of course, but are the pair enough? If not then, as best explanation, the projection of a single cell implies a mind that could as well project the rest of life's rich panoply. *E cellula omnis cellula.* Only from a parent cell does daughter come. **No exception has ever been found to this rule so that it is called The Law of Biogenesis**. In 1860, a year after the publication of the Origin of Species, creationist Louis Pasteur showed that broth in sterile flasks did not spoil. This experiment, which underwrites the use of sterile equipment in medicine, therefore showed that life does not spontaneously arise. It confirmed the Law of Biogenesis. *Nor in modern times does any process of abiogenesis, whether quick or slow, occur.* Therefore perhaps, you have to argue, things aren't what they used to be.

As opposed to purposive life, atoms are completely unaware. Matter is not alive. The materials of a body and, by extension, body itself is not alive - unless by life you don't mean what we, living beings, think we are. We seem to be rare sparkles that inhabit a vast vale of death. When homeostasis breaks down and a biological body dies, dust returns to inanimate dust. Is life just a complex form of death? Is the key issue sterility (no prior life) or others such as atmospheric content, saline water and so on? In spite of Pasteur's unchallenged experimental proof to the contrary, could bodies have spontaneously emerged from simple dusts?

Development from egg to adult is most often the result of sexual reproduction. Of course, evolution's biogenetic 'egg' is not a sexual gamete but an asexually reproductive proto-bacterium. The principles necessary for the practical creation of a bacterium or any other form of life were discussed in the previous chapter. *The prime necessity, the very cornerstone of a materialistic theory of origins is that such an organism must have arisen; and since, a priori, mind did not make it chance must have.* Thus Hope of Abiogenesis specifically contradicts the Law of Biogenesis. If such abiogenesis occurred the chemicals and conditions involved would have been relatively simple and reproducible; nor could the process have taken long, otherwise the requisite components of the 'pre-proto-organism' would suffer chemical interference or otherwise degrade before it was made. Yet, despite Pasteur's clinical conclusion and a complete absence of observed evidence, the philosophical imperative of naturalism and thereby scientific assumption (that life is 'self-organised' matter) is relentlessly, dogmatically and, in rational terms, thoughtlessly by dint of repetition reinforced. It has become, rightly or wrongly, part of a habit, a materialistic mind-set. But could undirected chemicals alone ever actually, spontaneously achieve the principled, purposive patterns of life? *The very basis of that hopeful theory, abiogenesis, is the unverified claim that they did*.

Darwin: Half Right, Wholly Wrong?

Perspectives on the Three Central Tenets of neo-Darwinism - a Tabulation

✓ true

✗ false

		Bottom-up	Top-down
①	Abiogenesis	✓	✗
②	Variation (microevolution) by mutation and natural selection	✓	✓
③	Transformism (macroevolution) by mutation, natural selection or any other means	✓	✗

Aren't half-truths the hardest ones to disentangle; and the ones with greatest tendency to lead astray? You tell half a story and leave the gullible or motivated to fill in the rest. **Indeed, as was already mentioned in Chapter 5, neo-Darwinian evolution (as opposed to biological variation) may be viewed in the light of a logical fallacy, a trick called equivocation.** *This semantic trick is to conflate two entirely different matters - firstly, variation on existing features (tenet 2) and, secondly, tenet 3's addition of complex, highly informed fresh features (such as coherent organs, systems and so on) in the first place. You might even try to slip in tenet 1!*

fig. 20.1

On top of Hope the Theory of evolution's process equally, specifically contradicts the Law of Biogenesis. Who, however, disagrees that Darwin's observations were in fact correct? Astute and well relayed? No problem with the science here. Who, therefore, dares to disagree that chance, at least apparent chance, plays an important role in guiding any body's life? Or that variation of life forms is not an obvious, incessant part of nature? Differentiation (by meiosis or mutation), origin of cline, race, sub-species, even species (though this depends to an extent on definition) happens all the time. No problem here either.

It is not Darwin's facts but his hints, suggestions and extrapolations one might take to task.

Non-Darwinian seed of a Darwinian tree of life must have evolved from barren pools or clays; from mud or, maybe hot or cool, saline solution. You might guess this happened or refrain. What inference do the facts (most of which Victorian science did not know) support? If they support a positive then Lady Luck has won the toss. If they support a negative then 'life-from-non-life' flips to 'modern alchemy'. An axe is laid upon phylogeny; the root is sliced; Darwinian seedlings never sprang to sprout a tree.

No seed, no tree. The way one kind of organism branches into others is still a mystery. Transformation, macro-evolution (in the way of mice to elephants or whales) is, if the first cell did not make itself, no option. If a tiny cell can't build itself how can large-scale body-building just 'self-organise'? Darwin, as the tabulation clearly hints, was in micro (small-scale) right, in macro (large-scale) wrong; in detail true, in general suggestion and extrapolation not at all so. And if you disagree let's, in the following few chapters, see.

Modern Alchemy

Medieval alchemy included dreams of producing gold from base metals; not gold but nowadays from base materials you seek to find the elixir of life. An even more fanciful successor involves the transmutation of inanimate to animate, the dream of generating life itself from non-life. Such modern alchemy, either the spontaneously immediate or drawn-out generation of life from sterile non-life, is chemical evolution. **Scientific orthodoxy invests heavily in such abiogenesis because, as cannot be emphasised too strongly, from a *bottom-up* materialistic (or atheistic) perspective, life *must* have come from non-life**. Its whole intensely intelligent laboratory enterprise is combined with plausible interpretation by philosophy inside the science. All is prefaced with the cast-iron concept: 'there is no metaphysic; there is no archetypal plan from which bioelectrical/ biochemical forms are precipitated; therefore the origin of coded life is a random process developed and constrained only by mindless physical elements; there's no acceptable alternative'. Thus science correspondents chime in concert; such unwitting prebiosis-unto-genesis *must* have occurred and set grand evolution in its train. It is required *fact* in a paradigm that lacks the dimension of active information. A Big Accident *must first have* happened (matter flew from nothing) then a Little One (forms of life evolved from matter) *must have* followed it. The goal is no longer necessarily to discover the *most* plausible explanation for the start of life but only to identify and then expand upon a naturalistic one. Many such have come and gone until now the golden grail seems further out of grasp than ever.

Earth, fire, water, air - could life thence spontaneously appear? Must have, must have. 'Must have' pops up everywhere. It is an 'auxiliary imperative', a coercive mode of verb that in this context means, lest any thought of prior intelligence arise, you're wrong; and, even though I can't see how, mindless evolution's right and will be proven so. An argument *ad baculum* warns, as shaken sticks are wont, 'shut up'. Yet is it true, as atheistic Jacques Monod expounded, that only 'chance' and the constraints imposed by natural law ('necessity') must have created life? Stirred by currents in some cauldron's'pre-biotic brew', which nature somehow locked in 'cycles of increasing chemical complexity', was animation's weak solution caused by random fluctuations? Could 'vital spontaneity' swirl unto consecration of a life upon such watery altar?

Although abiogenetic alchemy is the foundation stone on which the materialistic explanation of life (and thus of atheism) rests, it needs be emphasised that its process is purely speculative, very highly unlikely and, in practical fact, impossible. It is the first of a long series of black boxes ranged before us by naturalistic theories of life's origin on earth. It may very well have nothing in it and, if so, it represents a coffin for the atheistic faith. It can be verified neither by observation (no-one was there) nor experimentation (even if intelligent, highly directed contrivances produced 'life' from raw chemicals, this would not prove but only lend credence to the possibility it once happened in that way). In fact, the impossible geochemical complexity of such a mission might equally lend credence to the necessity of mental guidance - by investigating scientists at least. Attempts to simulate the subtleties of chance are, as we'll see, as far from coded cells as snowmen flying through fictitious air, ones bearing chemists as they soar. One can only repeat, with promissory faith, the mantra that chemical evolution *must have* happened. Natural Dialectic, on the other hand, notes codes and symbols are mind's strength; it sees life as incarnate information and would not, therefore, expect a single cell to live by chance. Atoms assembling by themselves into encoded bodies might, in principle, seem impossible - but in practice surely chemical evolution, atheism's parable of hope and life, isn't just a Just-So Story, a walk on theory's water or a liturgy sad in the greatness of its delusion? **What are the facts? Could you, knowing all that's so far known and yet despite the theories science of the day professes, rest entirely confident no cell (and if no cell no multi-cellular construction) was ever organised by chance and natural selection? We shall see.**

Biological organisms are not like natural aggregates with one piece 'slapped on' as aimlessly as the next. They both are and are not like machines built up, as a toy, from 'Meccano' parts separately made and fitted together. They are like this because mechanical constructions are purposive realisations of a plan; they are expressions of their creator's logical train of thought. They are not like this because every part, having developed organically from an original parent cell, is never separate and always coherently familiar with every other. In fact life forms are *all* highly integrated, complex entities whose being is instructed by their code; the *meaning* of a code is in the conceptual information it conveys, that is, in the reasonable, teleological arrangement of matter according to a preconceived idea. **The basis of biology is not, from this point of view, one of 'Meccano' biochemicals but comprehensive information; and the current scientific view that, given a set of biochemicals, enough random jumbling will inevitably cough up codified excellence is, simply, fundamentally misconceived. It is, ignoring immateriality, an error of first order.**

Biological material is never, even at its most 'primitive', other than complex. Very complex. Has anyone dreamt up *from scratch* a parallel chemical system that might generate biological forms? And, although it would probably resemble a single proto-bacterial cell and therefore be much smaller than a full stop's dot, built and tested it? Perhaps there isn't one but at least we never had to think things up. We start with the conceptual answer. We start by copying what we know already works. Even in this case, granted the complete plan in front of him, no scientist even approaches (without 'cheating' by using large prefabricated chunks from previously living organisms) the construction of that most successful, 'simple' organism, a bacterium. Nor can he fabricate *ex nihilo* the whole game's starting-point - atoms from which its bio-molecules will be synthesised.

Perceiving how great is the problem of explaining the creation of life in material terms alone, contemporary science has adopted a standard, rational and pragmatic approach you can apply to any large problem viz. to break it down into a series of small ones. Solve each of these, sum the solutions and you may have solved the whole thing. The whole is no greater than the sum of its parts. You can apply this method either to the analysis or construction of any complex project - so why not to the improvident genesis of life or, as Darwin, to the evolution of all life on earth? In either mindless, biological case you might suppose that complexity derives gradually in small, random steps from an initial simplicity. Is not biology essentially biochemistry and, so the logic runs, atoms bonded into simple organic precursors? These precursors 'organised themselves' into a Meccano-like construction of biochemical parts from whose watery jumble life must have arisen. Pasteur's experiment using sterile broths had seemed to disprove the medieval idea that life could spontaneously generate itself from matter but for a century and a half, despite his scientific proof to the contrary, this line of thought has stimulated a plethora of scientific, abiogenetic speculations. Whether hopeless or not, these imaginations are vital. They keep science 'science', that is, exclusively materialistic and thus manageable. They provide a research arena and, by the same stroke, eliminate pseudo-science and lock an enemy called 'the super-naturalism' out. The essay is not, therefore, blasphemed with the label 'modern alchemy', 'sophisticated animism' or 'scientific mythology'. Rather, 'chemical evolution' is resoundingly encouraged to revive vitality from sterility, grasp life from the jaws of non-life, resuscitate 'spontaneous generation' and thereby seed the lifeless to evolve an animated family tree.

After all, what's wrong with a good, medieval notion? What's the headache? Earth-born alchemists failed in their task of transmuting base elements into gold but, where chemistry ousts alchemy, stellar furnaces can forge new elements from old. Were and are not simple hydrogen and helium so transmuted into the 92 elements composing stable matter in the modern universe? And if stellar alchemy can complicate hydrogen why should biological alchemy not elaborate cells? If the hallmark of life is simply complexity, why should some hidden but 'inevitable' drive towards complexity not have to finish up with life? Never mind the sharp distinction between purposeless and purposeful complexity - a difference involving immaterial information. Never mind the trifle of an undiscovered law of physics, a principle of naturalistic innovation which, circumventing general rigidity of rules, might support 'an improbable but inevitable complication' of matter. Can you imagine purposeful atoms? But an evolutionary scenario has been so diligently constructed by myriad explanations, so repetitively rehearsed in concert that it *must* have happened, mustn't it? Even though the import of 'inevitability' seems to approach 'intelligence' and, from a *top-down* point of view, such a required *mode* of thought resembles an attempt to rescue impossibility into the preserve of doubt. 'There is no alternative: we are here so we *must* have evolved'. Against inconceivable odds, far beyond any normally considered conclusively negative, this doubtful improbability is expanded to a confident certainty that, given enough time, anything could happen and life certainly did. It did. So strong is this consensus that a consensual reflex burns any other book, freezes out any other suggestion and, however intelligent it may be, kicks out any other possibility. Time, chance and a cosmic laboratory compose evolution's habitat; what need for more? *Therefore any alternative is blanked out. Materialism is in denial. All interpretation of evidence is ranged to support (which both factual odds and the top-down approach might judge fiction) the fact of abiogenesis.*

This chapter will, adopting the standard, pragmatic approach, break the huge problem of such abiogenetic notion down into small requirements. It is argued, however, that any attempt to link these 'steps' makes Monod's famous offering of 'life by chance and necessity' cumulatively less rather than more probable. This is because his couple - each alone or both together - simply aren't enough to generate specific, detailed information that even the 'simplest' forms of life require. Could 'necessity' of natural law, whose patterns never change, develop not just such irregularities as so-called Shannon-information values highly (see Chapter 5: Information, Messages, Arrangement) but information of the functionally specific kind on which biology depends? Or, if you turn to Lady Luck, is there sufficient probability in all the universe of time and space to generate the molecules that, in some tiny volume at the magic moment of its birth, were integrated into complex systems of the alleged primal cell? It may transpire that Monod's couple sum not to solution but, in principle, to permanent impossibility. Tracking the scenarios of chemically evolved 'ascent' this chapter's argument gathers up the steps into their logical result. At this climactic point what might emerge as clear? If the idea of accidental life is as illogical as it is impossible, is a metaphysical resolution to the purely physical impasse inevitable? Does life need an intelligent and yet forbidden miracle to quicken it?

What needs be itemized on life's most elementary bill? **Minimally, membrane assembly, membrane customisation (presence of requisite channels, ability to grow and divide etc.), encapsulation of large polymers such as nucleic acids and proteins, replication machinery, protein synthetic machinery, alliance of code to protein synthesis (construction of genetic code, its operators and an essential gene set to create necessary proteins), evolution of requisite metabolic pathways, creation of energy capture and usage systems, elimination of interfering reactions and the clutter of useless, redundant or superfluous materials, and the correct localisation, that is, spatial organisation of components**. These minima could not occur in drawn-out steps; natural forces would destroy a stage before the next could link. A cell is a few microns (millionths of a metre) large; materials for construction would need to be about that localised. Time and space both militate against. Is such an order tall enough to start with?

In the next two chapters (Chapter 20 from a *bottom-up* perspective and 21 from a *top-down*) we'll rehearse over thirty major, cumulative reasons why, despite protestations to the contrary, the speculation of terrestrial abiogenesis, whether in extreme or benign conditions, is a species of wishful thinking and, indeed, an unbelievable form of modern alchemy. Within this framework we'll note plenty of black boxes, contradictions and chicken-before-the-egg scenarios. Is it science or philosophy you want? *The abiogenetic canon is, we reiterate, zealously guarded because the alternative (which includes an informative dimension the current scientific version lacks) causes the official doctrine - naturalistic materialism - to collapse. The standard, prebiotic thread is fuzzy: it is woven from indefinite time with, by definition, unpredictable and therefore incalculable chance. Anything might happen. Can you deny it?* **Blurred, invisible, indefinitely flexible yet infinitesimally slender is the thread on which a whole mode of popular thought chances to hang.**

Isn't it frustrating? For reasons of philosophy and, as some see it, science minds are shut. Intelligence, though perhaps 'best explanation', is expelled. Thus many models and hypotheses of life's beginnings wrestle with the consequence. **Each fails to explain the origin of such information as life's codes, business and bodies obviously and systematically employ.** Each, more or less subtly, begs the question by supplying at the start the factor they are seeking to explain. **Controlling order, not lack of control, is the answer that informs survival.** Does not the controlling intelligence of theorists, chemists or programmers plant initial conditions intended to reduce nature's mindlessness? 'Targets' inform directions, 'correct' chemical solutions are assumed and 'correct' conditions guide the way that theory's molecules could perhaps have interacted. One way or another, ample information is injected at the start to reap an output that sustains an evolutionary possibility. Yet all attempts have failed by far (and there is reason to suppose they always will) to generate a single coded and metabolising cell from scratch.

Arcane profundities, the origin of life and things, leave us perplexed. For things, this universe was not enough; its transcendent projection is forbidden by the *PAM* and so we speculate upon a multiverse. For life, blithe dreamers blowing bubbles in the air or, rather, in an early bath may reckon all it takes is 'vesicles' to self-assemble so a cell itself will hubble-bubble up. No doubt, however, that the accumulation of hurdles that will be probed in this section has led more pragmatic scientists to suggest, despairingly, that earth is not enough; but its intelligent projection is forbidden so we speculate abroad. What might, therefore, be your celestial ancestry? Perhaps relevant organic products grew in interstellar dust or 'simple proto-cells' (such as bacteria) first appeared elsewhere and then, propelled through space by photon pressure, flew in from the wide open doors of a great refrigerator, outer space. Welcome to the barren earth! We'll double check this flight of fantasy a little later on.

Quartz, iron ore and mineral dyes exist but no Ferrari self-assembles from them. Water and organic molecules exist in space but life, even a bacterium, is much more complex than a motor car; and coded for its maintenance and reproduction. Both cell and car are very well informed. OK, cars don't incorporate nucleic acid and, you claim, *DNA*'s self-replicating. That's myth. Without a precise battery of operators it does not. It's as useless sitting on a stone or floating in a pool as any other chemical (we'll deal with this grave problem - the origin of replicating and protein synthesizing machinery - later in the section Perplexity). Can you, anyway, pretend that any self-replicating cell is less complex than a computer or an automated factory? Code implies a meaning and a meaning purpose. High spec, high-fi. When do coded/ CAD-programmed structures self-assemble from the lifeless bubbling-jostling of a brew? When imagination orders so. Even a 'simple' bacterium possesses high-spec cybernetic, industrial principles (see Chapter 19) whose mechanisms, whether or not frozen or dehydrated in suspended animation, need to be in place from the start. An overall homeostatic system of survival that includes information transmission/ receipt (called signalling) and storage (code) must govern even bacterial parts. Bacteria demand an irreducible minimum of principles and practices. In fact, such bacteria as could survive the hazards of a voyage in space would need complex *DNA* repair systems and complex biochemical capacity - no easy going here. Extra-terrestrial origin simply removes the burden of their production to another place instead of another (internal) dimension. *Whether or not we evolved from aliens the cumulative reasons that are registered below apply, in practice, anywhere in the cosmos.* Removing them from mother earth in distant time or space no way diminishes their force. So let us keep our feet on her firm ground and stick to facts.

Finally, let's summarise the target model. Chemical and subsequent biological evolution are each a scenario that scientific explanations have, today, to fit - but do they fully fit the facts or fail but issue promissory notice for the future? Such notice serves as basis of a faith, the naturalistic faith, a non-religion. When it comes to organisms you need voluminous and specified complexity. Informative content would have to leap from brute material 'creativity' by reflex, mindless natural law to most sophisticated, intricate, anticipatory plans. It would spring from zero to a very high-value level of information.

Do you remember (from Chapter 8) Lady Luck, Lord Deliberate and betwixt this couple Fixer *aka* natural law? In physics modes of origin and operation danced between the agencies of Chance, Necessity and Archetype. Now, in biology, one executes a sidestep so that chance becomes mutation, necessity the natural selection of a fittest type and teleological archetype, as an idea, is scrubbed. Is, however, fluid randomness that's fixed at first by chemistry and then survival quite enough? Can The Fixer (*aka* 'inevitable necessity' of natural law) combine with Luck to impregnate the sterile earth and bear a cell? You hope so since this couple is your lot. Without their progeny your theory is not born. Chapters 22 and 23 check whether they're enough.

But there's an oxymoron, a tight paradox. Necessity, whose mode is repetition, can't anticipate or innovate. You don't expect predictability to generate the wholly unexpected - functional innovation. With multiplex components it can generate irregular, improbable complexities (say, coastlines, rock formations or the weather) but it cannot *specify*. It cannot target or foresee the need for parts that, outside an integrated whole, are useless and would decompose alone. Light-headed Chance is even less help here. She's hopeless. Could, therefore, matter really seem to animate itself or, if you like, 'self-organise' a specimen of high technology with its encoded, functional specificities, a cell? Could, in the oxymoron, lifeless chemicals engage to 'specify' a living cell? Or not?

If Luck and Fixer fail to generate first-class degrees of specified complexity (as life's coded systems everywhere demand and show) why can't we recognise the kind of choice that language represents? Choice between alternatives is what the Latin for intelligence ('inter': between and 'legere': to choose) denotes. Luck and The Fixer have no choice and therefore Lord Deliberate personifies a better explanation for the origin of text that writes the forms of life on earth.

Cars and cells are made and work within the framework of physical law. Does this mean such laws created either? Does it mean, moreover, once you know in detail how and why a mechanism works you understand its origin in terms of concept and inventor? Perhaps, on the other hand, you feel compelled on principle to categorically deny that any immaterial architecture or inventor's plan informs biology. If so, your nihilistic form of 'bio-logic' - the foundation stone of a materialist's presumption - fires a trajectory of thought that, with rebuttal, runs as follows. Let, in this course, facts and not hopeful fiction speak. Let's shoot and see if the Hypothesis of No Intelligence survives.

Not a Great Start

The origin of the earth has been dated at perhaps 4.57 billion years. Fossils are found at ages calculated at 3.5 billion years. What happened in between? What is the time-frame for a single life's ignition?

A 'Hadean' period up until 3.8 billion years ago was inhospitable. At first our youthful sun was highly unstable and for half the frame its radiation (particularly life-damaging X-rays) was fifty time higher than now. Hot. And impact craters precisely measured on Mars, Mercury and the moon show that violent collisions by planetesimals and asteroids battered the solar system - including earth - most intensely in the period between 4 and 3.8 billion years ago. If life had begun to 'coalesce' such pummeling would have finished it off. Moreover because the earth had not cooled sufficiently for solid crusts and oceans to form no rocks or marine deposits older than about 3.85 billion years have ever been found.

Stromatolite fossils and microfossils have been found in Australian, South African and Zimbabwean deposits dated (if correctly) between 3.3 and 3.5 billion years old; and in Greenland carbonaceous geochemical evidence from the oldest known rocks (3.85 billion years) may also suggest the presence of life. If such evidence is confirmed the window for the spontaneous generation is narrowed to a slit of about half a billion years. Could all the simultaneously required items listed in the previous section occur in such a frame? The probabilities, we'll see, are vanishingly small. Non-appearance not appearance is what the sums suggest. Not a great start. However, we'll plough on.

Atmospherics

From hydrogen to human, from molecules to man - energy 'injected' into the so-called nucleo-synthetic construction of elements from hydrogen can be construed as the primary resource on which all subsequent chemical reactions have to draw. This *sine qua non* is taken for granted.

Down to earth several sorts of site, each with its own drawbacks, are proposed for the pre-biotic synthesis of relevant organic compounds. They include the atmosphere, volcanoes, deep-sea hydrothermal vents and, from outer space, comets or dust.

Since Pasteur had shown that spontaneous generation does not occur under current oxygenated atmospheric conditions Alexander Oparin, hammer and sickle atheist and father of chemical evolution was of necessity led to hypothesise a strongly anoxic, reducing atmosphere rich in, say, ammonia, molecular hydrogen and methane. Without this assumption the evolutionary scenario fails because even simple organic compounds, the smallest biological building blocks, fail to form in weakly reducing air (say, oxides of carbon, water and nitrogen) and would crumble as soon as they formed if oxygen were present. Oparin's suggestion consisted of some mixture of ammonia, methane, hydrogen, nitrogen, oxides of carbon, water vapour and perhaps formaldehyde and cyanide.

This politically driven speculation was soon propagated in England by Marxist fellow-traveller 'JBS' (Jack Haldane). The Oparin-Haldane textbook model includes the notion of a dilute, marine 'soup' in which floated the precursor molecules of amino acids, nucleic acids and so on.

Neither sky nor soup was likely to have been the case. For example, the role of ammonia in abiogenesis is central. How, bearing in mind the temperatures, pressures and catalysts involved in the Haber process, could enough be synthesised from simple hydrogen and nitrogen? A mixture of reducing gases, even if one had existed, would have been rapidly destroyed. Ultra-violet radiation would have broken up methane and ammonia; and free hydrogen, upon whose presence the latter's stability (or resistance to decomposition) depended, would have escaped to the upper atmosphere or space at such a rate that it could have persisted for only a fraction of the time required by chemical evolution. With respect to methane, most carbon in rocks is in the form of carbonates derived from carbon dioxide; and there is little geological evidence for an antique atmosphere rich this gas. Therefore, was sufficient ammonia, methane or hydrogen cyanide actually ever present for a whiff of fore-life in their foul and early air?

Current speculation prefers a non-reducing, neutral atmosphere composed of nitrogen, water vapour, carbon dioxide and possibly carbon monoxide. However, only UV-radiation could break the strong bond of the latter apart; and when this happened it would most likely react rapidly with the products of light-split water to give carbon dioxide. The same photolysis would generate considerable amounts of oxygen. While hostile to chemical evolution this same oxygen, in the form of ozone, actually protects life from those unhealthy UV-rays. Its pre-biotic quantity is disputed and ranges from minuscule to amounts inimical to the non-biological construction of amino acids and other molecular 'bio-bits'.

Oxidation is reduction's opposite. If, for example, appreciable quantities of oxygen were present they would be incompatible with the coexistence of ammonia. Geological evidence can be interpreted, according to what you are seeking to find, both for and against such composition - although in the presence of an appreciable quantity of 'fresh air' such basic bio-compounds as amino acids and sugars are unstable. A little oxygen degrades them into carbon dioxide, water and other substances; or else shuts down the pathways to their production altogether. Natural generation of a cell could never happen in the air we breathe - hence the exile of oxygen from Oparin's enthusiastically atheistic, red-starred atmosphere. In support there are adduced supporting arguments (including 'red-beds' - thick deposits stained in bands by oxidised or rusty iron) to show that oxygen came later. In fact was the gas first evolved from organisms that could photosynthesise? Or, with photolysis of water by sunny ultra-violet radiation, was it always here? The evidence now 'waters down' reducing skies but has forensic rain washed theory right away?

In brief, this contradiction called the oxygen-ultraviolet paradox is acute. Its problem is severe. If present, oxygen inhibits the formation of pre-biotic chemical reactions; if absent, then no ozone layer could have formed to shield fledgling products from destructive, short-wave radiation. Indeed, in 1903 a Nobel Prize was awarded to Niels Finsen for the discovery that such radiation is lethal for bacteria; and, we now know, almost 100% of viruses and other kinds of microbe too. Would you treat pre-biotic life with floods of ultra-violet death? Or would this constitute a real false start? Thus either with or without oxygen chemical evolution could not have happened. *Such is the lose-lose dilemma for a 'life-is-only-matter' point of view.*

Another popular scenario invoked involves ancient volcanoes. They belched, their lightning flashed and some kind of life-juice was evolved. Lo! This is how life must have happened! There is, however, little to indicate that such exhalations ever differed from today's - nitrogen, oxides of nitrogen and sulphur, carbon dioxide, water vapour and a little oxygen that would be swiftly lost in conversions to oxides. These oxidising volatiles would have been hostile to the abiotic synthesis of biochemicals. Combinations of gases may, under artificially directed, monitored conditions (especially with respect to timely withdrawal of products from a reaction site), yield several amino acids but the consensus nowadays does not support Oparin's politically scientific sky. Could heavenly help from the beyond have flown in? Could passing interstellar clouds have dropped off urea; or sufficient cyanide and formaldehyde - substances normally toxic to life? If not, how on earth might sufficient concentrations have been supplied to act as life's protein/ nucleic acid precursors? Well check both ET (extra-terrestrial) and underwater options later.

In short, a reducing atmosphere is a *sine qua non* but most likely never existed. Nor, if it had, would its presence have lent inevitability to life's unlikely start. It may simply appear as a device taken in isolation to lift the odds minimally above nil. Let us, however, generously allow its possibility and at least the reasonable production of organic precursors such as water, methane, ammonia and so on.

Unnatural Interference

The classic laboratory attempt to simulate chemical evolution (Harold Urey and his student, Stanley Miller, in the early 1950's) assumed an 'incorrect' atmosphere based on the 'Oparin-Haldane model' and containing a mixture of ammonia, methane, hydrogen and water vapour. This they circulated through glassware. Power sources assumed to have driven proto-syntheses exist the same today - cosmic rays, radioactivity, lightning and volcanoes. Instead a continuous spark discharge was used to simulate discontinuous bolts of lightning when, in reality, U-V light would have been the commoner source of energy. Such energy is biologically unhelpful - in its presence yields of amino acids are low and *DNA* is disrupted - but supporters argue that hydrogen sulphide in an early sky might have acted as an absorbent buffer. However, thermodynamic considerations, highly anti-evolutionary, predict the breakdown of any complex but ephemeral molecules. Natural conditions, especially harsh ones, are destructive of complexity. They *randomise*. Therefore the chemists' carefully, intelligently-designed equipment employed a highly artificial device, a 'cold-trap', to isolate any products from the deleterious influence of the abovementioned UV-radiation, any electric discharge zone or raw heat. With this cold-trap a muddy-coloured product was found to contain some organic compounds including various kinds of uncoupled amino acids. Spirits soared! *But without the cold-trap chemicals would be destroyed as quickly as they were made; with it no energy could enter to drive the hypothetical evolution of molecules forward.* **Such intelligent experiment, such unnatural interference is, we shall see, a hallmark of the laboratory search for pathways that might naturally produce organic chemicals.** It defeats the purpose but, since the process *must* have happened, what else can you do?

Thus another lose-lose dilemma confronts a chemical evolutionist. **Not only is there the problem of molecular degradation by the same sources of energy that created them but there is one of separating useful from non-useful chemicals bundled randomly together; and also of unintelligently interfering products or competing side-reactions in a mixture. Thus, in the Urey-Miller slurry both non-biological and biological chemicals occur**. These need be separated lest reactions between the two types degrade the whole mess. You must also switch on only 'lightning' or short-wave U-V light because long-wave destroys your target molecules - and destruction of its building bricks would leave construction of life's palace absolutely <u>zero</u> possibility. No early atmosphere could mimic such intelligent control. Indeed, careful and strategic manipulation of chemical and physical ingredients amounts, effectively, to the injection of mind into life's re-creation. Such artificial interference, however much it is designed to 'copy' non-design (in the guise of 'chance and necessity') involves prior specification. Indeed, what synthetic rig does not, through the devising of its experimenter, introduce a 'hand of God' into the equation? In the end analysis the method has generated some amino acids, nucleic acid bases and carboxylic acids. Most of these are bio-alien if not bio-toxic. How *were* the interfering non-biological 'aliens' naturally sieved out? Why did life, informed by code, 'choose' just amino acids and the bases that it did? In fact such artificially optimised but still poverty-stricken production is, even if you refrain from including the issues of 'optical isomers', correct bonding or creation of genetic code, so catastrophic for chemical evolution that the experiment can as justifiably be claimed to have 'disproved' as 'proved' abiogenesis.

Variants of the original experiment using other chemicals (such as sulphur dioxide, cyanide, formaldehyde, iron, carbonates and others) have since produced more amino acids, bases and sugars. But later in life Miller wisely, humbly and gracefully admitted that his textbook experiment (along with a couple of other versions) had no biological significance. As a principle, however, evolution cannot die. Thus fresh steam has been injected into the argument.

<u>Evaporated Soup</u>

Organic molecules, some relevant to life and others not, have certainly been created in laboratory conditions that may simulate those imagined for a primitive earth. Suppose that, like Meccano parts, all those life needed had been built. The simplest viable organism is an arrangement of hundreds of different kinds and thousands of each linked in specific shapes and orders. Programmed, coded, masterful arrangements - and purposeful arrangement is the very business of mind! Nor could mindless ocean swell wash all of them - a massively improbable congregation - at the same time into a volume greatly smaller than a thimble. Yet, for survival of a theory's sake, suppose the necessary mutual associations had, at the climactic moment of a rapid and supremely lucky chain of chances, been washed together in a local or a global puddle that became the very crucible of life - including yours. Once, in 1871, Darwin let slip a dream that a warm little pond (say, an acidic, clay-filled volcanic spring) might constitute our primal womb. Nearly 140 years later research at hydrothermal vents in California and the Kamchatka peninsula poured cold water on his reverie. Added lipids refused to form membranes; added nucleotides, amino acids and phosphates were adsorbed onto the clays strongly enough to block any further reactions. That was that pool done.

No doubt you need some heat for the non-metabolic production of amino acids and so on. Therefore why not consider sub-atmospheric volcanoes belching in the sea? Newly discovered, lightless hydrothermal vents spew a superheated brew rich in metal ions, methane, hydrogen sulphide and other inorganic chemicals. Couldn't thermophiles (bacteria that inhabit the airless exteriors of these 'black smokers' at super-heated temperatures of up to 400° C) offer clues about the chemical evolution of life on a hot, early earth? Could hot, alkaline crustal water spewing into acidic brine have conjured amino or fatty acids; or biochemistry have started in swirling, boiling and supposedly oxygen-free water - perhaps perched on or even down inside its crust? It's a pity, in that case, about amino acids whose half-life at 350° C is measured in minutes, small peptides in hours and sugars (at 250° C) in seconds. Watery furnaces that produce at the same time destroy. In addition there would be, at most, very little ammonia and, even so, *RNA* hydrolysis would occur in minutes; and dispersal into colder waters simply thins the soup while for long-term functional stability *outside* a cell you need to keep nucleic acids in a freezer. Not only does informative *DNA* have a pre-formative problem. Both above and below sea level other combinations of difficulty need instant combinations of solution. For example, a physical lack of thermal or pH regulation coupled with lack of predefined, genetically coded homeostatic mechanisms would not improve a pre-cell's chances! Moreover, water is not a condensing agent! How could any polymer then form? Check, yet again, the list of bio-items that you'd need to coalesce in either superheated or cool water!

The fact that some bacteria are adapted to live by hydrothermal vents in no way means life started out like them. Thus life-cradling vents are, for some theorists, profoundly frustrating. The pressure builds and so let's surface from the deep and splash in shallows. Is soup on the menu? 'Primordial soup' is, in view of Pasteur's sterile broth, an incongruous misnomer. For theory's sake, however, one may speculate - for which speculation, according to the tools developed by geochemists to measure the quantities of requisite organic chemicals on early earth, there is no evidence - that a liquid treasure rich in life's precursors must have swelled somewhere. If volcanic pitches or warm pools can't take the heat then why not pour cold water into theory? Earth's oceans contain perhaps a billion cubic kilometres of water; thus for a substance to achieve even a minute fraction of molar concentration it would require billions of tons - after the great majority had been destroyed by ultra-violet radiation in the atmosphere before being washed down by rain. Even there it would not last long but, *ad hoc*, you assume it congregates in cool, fresh pools where water might evaporate and the concentration of its solute grow. You'd need strong evaporation since in this exceptional puddle the likely concentration of even the simplest amino acid (glycine) would be the equivalent of about 0.2 mg. in a swimming pool. Thin broth indeed! On the other hand it would also, unless new batches were continually washed in or synthesised, have to resist too much evaporation over many thousands if not millions of years. Although length of time is no issue with respect to the actual formation of chemicals (the more time you allow for formation, the more you simultaneously allow for destruction), it is an issue for consequent life. Would seawater (rich in sodium) do? Life is relatively sodium-poor, potassium-rich - thus your briny wouldn't have to be the sea. Moreover, if the shallow ever dried up or were inundated such that nutrients were too far scattered, if the pool's life-correlated salt/ mineral concentrations varied only slightly or if the pH changed then any chance of nascent life would be killed off. An uncannily delicate balance! Unwavering eons! A deliciously correct and stable soup indeed! Yet there remains no trace of its nitrogenous or carbonaceous deposits in Precambrian rocks. No evidence for such a massive, long-term feature - only faith in promise and slim plausibility.

Indeed water and its solutes are, within a cell, highly controlled so that uncontrolled water, with its destabilising motions and hydrolytic tendencies, seems an unlikely launching pad for life. Even more so, because of the need for a liquid 'reaction chamber', does the dry lack of it. Or are primeval soups and powders cooked and spiced up by imaginations that they need to feed?

Where next to cast a line for life? As in hot sea so a cool pool - water's problematic. You'd have to find sear spots or suppose condensing agents (such as cyanate or cyanamide) conveniently to hand to help life's all-important polymers to start to form and not fall straight apart again. Why not, therefore, fish upon a solid surface - one either submarine or at the water's tidal or evaporated edge? Mineral surfaces on special clays might help to serve as templates for nucleic acid and protein construction. Hypotheses proliferate. Could iron sulphide (precipitated in the form of iron pyrites or 'fool's gold') at hydrothermal vents have driven forward the synthesis of pre-biotic molecules (although they don't appear to do so now); could they also act as templates on which short chain peptide or nucleic acid might crudely be developed? Then you'd only need to conflate libraries of soft inscriptions in a single, lucky ark of life's first covenant - a lipid bag. In fact, as we'll see (Bags of Life), there's trouble there ahead as well. The only known source of phospholipids is cell metabolism and so what sort of aboriginal membrane floated up from where?

Vexatious problems need firm handling on all fronts. Suppose, therefore, that you provide simple organic starters flown in from space by comet or some other kind of taxi. Then again, if '*vita ex machina*' from the heavens is not your kind of style, raise curtains on another set. Provide molecular ovens and refrigerators working side by watery nano-side for billenia to buffer your non-biological biochemicals against external thermal 'irregularities' that would otherwise certainly destroy them. Such steadfastness is unexpected, such natural fidelity a boost. Even then... It has been calculated that if a library were compiled of all possible randomly-sequenced proteins each a modest 100 units long its mass would fill the known universe many times over - a bit much for a thin soup which, anyway, needs the relevant half a thousand clustered up together in no more than forty microns (millionths of a metre). Along with all the other so-called homo-polymers (polymers made of the same constituents) like sugars and nucleic acids!

Hold on! Talk of polymers has jumped too many links ahead. **Hot or cold, served in a bowl that's localised as pools or global as the oceans, primordial soup is in the mind. It's evaporating as we speak. All that's crystallised is myth.**

Chained Up Unchained

No biological polymer can be made naturally outside a living cell. *It is not, therefore, easy to account for spontaneous production - especially without enzymes and genetic protocol.* How, for example, does the 'random chemistry' of a pre-biotic scenario couple a base (say, adenine) to ribose and phosphate in a way that, of several possibilities, is inexorably the one found universally in nucleotides? In reality, this 'beta-1' connection, critical for the correct operation of *DNA* and *ATP,* is a product of coded metabolism. Its discussion outside this orderly frame of reference is essentially irrelevant to its production in a cell - the only place that it is ever found. The question of the origin of life always drives back to the question of the origin of orderly mechanisms such as metabolism.

Metabolism works by proteins; proteins operate the *DNA*; both are made of joined up units. Whether you obtain such polymers 'at random' in a lab or, as in biological practice, entirely non-randomly as synthesized by order of a code, you have a problem. *DNA* nucleotides in water no more combine to make up polymers than dissolved amino acids spontaneously join into polypeptides. There is no such combination. This is because polymerisation is the product of a freely reversible 'condensation' reaction that expels water during the bonding process. By le Chatelier's principle of chemical equilibrium such polymerisation is, therefore, highly unlikely to buck the strong counter-gradient that water represents. The matrix of water, such a bio-friendly solvent, inimically resists the formation of life's basic polymers; and if by any fluke a short biopolymer (of protein, *DNA* or carbohydrate) did 'self-construct' then the reverse reaction, hydrolysis, would straightway break its linkages and thus destroy it. Indeed, the purer the water added to a 'free' biopolymer (as one in an early, less salty sea, a pre-biotic puddle or freshwater lake) the more it forces hydrolytic disintegration. Catch 22. Hydrolysis not condensation dominates. In a laboratory 'peptide reagent' and in a cell enzymes are able to overcome this disability and build up chains, each to its biologically critical mass - but whence naturally 'bubbled up' a steady stream of building blocks? Time breaks; long ages seal all polymeric fate. Thus even if such monomers arose the chance of bio-scale polymerisation were non-existent. From what appropriate source, therefore, might energy be drawn to make the right, 'uphill' connections? How first arose the coded enzymes needed to polymerise themselves and thence all other types of bio-chain? Enzyme before enzyme, by what *chicken-before-egg* spectacular did cell metabolism first hatch into history? Was it, since primordial linkage is arrested by wet water, a ripple of dry water did the early, most unusual trick? The short of it is that by chemical rights no biopolymers, let alone specifically coded or translated ones, should exist; the long is, therefore, that you shouldn't either.

You are here by articles of great design. You are here because of code, program and coherent, integrated systems. The construction of life's large, complex and very specific chemicals is strictly metabolic, needs enzymes and, like the products themselves, is never found in nature outside a living cell. Even if it were a 'proto-cell', it would have to contain the specific nucleic acid formulation and associated 'application programs' necessary to generate the proteins that catalyse polymerisation. By chance? Pull the other leg, the one with jingle bells on!

Let's spell it out. You would need, say, 1000 chemical bonds to form, against the thermodynamic grain and without metabolic assistance (no known enzyme precedes its own code and protein synthesis), a small polynucleotide of 1001 units. This chain might, almost impossibly improbably, happen to code for a single, useful but 'ordinary' protein (although there would be no way to prophesy its use and it would anyway be useless by itself). And, of course, at this very early, fragile stage it would lack transcription and translation mechanisms that could make it any use at all.

On the other hand, just one case of the chemistry that insistently *wants* to happen, hydrolysis, would disrupt your emergent molecule and annihilate any slight informative potential that it may have accrued. In other words, a few short, meaningless polymers might arise whose chance of immediate destruction is much greater than their continued existence - let alone 'growth and development'. Unnatural bondage, spontaneous disruption - they would appear only to disappear. As we saw, for polypeptides the situation is impossible. Even if one formed it could be of any sequence. It is even 'more impossible' for sugars and nucleic acids, whose sugar rings are quickly hydrolysed in water. It is a chicken-before-the-egg circumstance. Without prior, pre-coded enzymes to make them no biopolymers could exist in water. They should fall apart. 'Life' would fall apart well before it started. Unchained life is chained from even starting. Yet protein and *DNA* are themselves biopolymers.

Wouldn't such biopolymers be contaminated by 'alien' chemicals whose effect would be to further degrade their already minimal capacity to correctly link together either inside or outside a cell? To counter this problem Sidney Fox 'specially' heated pure (analar) amino acids in a dry environment. He obtained simple loops of amino acids optimistically called 'proteinoids' but no protein. His product was racemate (see Reflections) and of little or no catalytic value. Indeed, if the temperature was sustained the loops duly fell apart. It is regrettable that, although long dismissed by professionals including Harold Urey himself, biology textbooks still robotically regurgitate accounts of Miller and Fox's early experiments as if they 'proved' or even supported chemical evolution. This is, exactly, the power of reiteration and myth-in-the-making.

'Which comes first?' is the refrain. Is it chemicals or the code they carry? Is it code or outcome it is worded for? For example, a non-specific polypeptide (protein) that was not the direct product of working *DNA* code would be useless - its un-reproducible type could last no more than its own brief lifetime. *In life peptides, which must be reproducible, are the product of code and complex associated machinery. And all codes are thought simulators. Coded information is a form of thought.* **The whole abiogenetic business is, in short, another chicken-and-egg paradox of the sort first referred to in Chapter 6: Machines.** Call this poshly 'causal circularity' but it's a vicious circle of the kind in which, for example, the enzymes that make the amino acid histidine themselves contain it. Again, inorganic sulphur is rendered organically useful by a bacterium using an enzyme which, with causal circularity, itself includes a metabolised, sulphur-containing molecule, cysteine. This is the kind of cycle in which books can write themselves and chemicals 'self-assemble'. It is as if every part of a prototype machine, a biological one which is the only known manufacturer of its own parts, had to precede both prototype and, therefore, its own production; and then, without knowing it, luckily fall into the right, very subtle combinations required to appear as the machine that henceforward secured each its own careless survival. Code comes afterwards. Its 'after-thought' precisely integrates with what's already there and then takes charge! After all, code's added bonus is essential so that reproduction can kick in! Henceforward, that is, all chemicals become dependent on immediate, precise and correlated code for each one 'happening' just right! What a systems switch chance radically developed and then handed over in straightaway - unless you've got your logic upside down. And when it comes to source of information naturalism definitely has. Such apparition is a brainstorm! Transformative code, from nucleic acid into working protein, just by chance! If logic is itself the symbolic embodiment of anti-chance, then to invoke chance as the source of genetic logic is a fundamentally illogical distortion. The fact is that without prior arrangement (or preordination) of such machinery, there is wildly imaginative but no actual chemical way that chemically distinct bases, sugars, phosphates, amino acids, lipids etc. float together, link up then go forth and multiply. *This is because every biochemical in every organism is the product of metabolism i.e. code i.e. program.* **In other words, a set of chains which can't be made without another set of chains which can't be naturally, chemically made at all is all you need to liberate a life-form from primeval sludge. In fact, the unguided production of biochemicals outside a cell, in which metabolism is all strictly guided by instructive code, is fundamentally irrelevant to the question of abiogenesis.**

Let us reiterate this lethal injection of truth into the 'necessity' of chemical evolution. Even if you presume a billion appropriate molecules they are useless unless within a cell membrane. Neither here nor anywhere else is there such a thing as a 'self-replicating molecule' to maintain your stock levels. You indeed need *DNA* to make *DNA* and make *DNA* work - but only via complex metabolic systems. For example, codified *DNA* makes protein to edit and repair the same *DNA* as in the real world tightly orders biochemistry. Over a hundred proteins are engaged not in code creation but anti-evolutionary code maintenance alone. You would also, at the same clueless, primordial time, competent protein synthetic, replicating and metabolic mechanisms in place to reliably sustain the accurate reproduction of relevant chemicals. **The real necessity is codified, code-generated and thereby thoroughly informed chemistry - controlled chemistry that operates, computer-like, at lightning speed with great efficiency. You might hope it but what brine could wash such flotsam up?**

The search for 'natural' origins in salt solution is conceptually all at sea. It should abandon ship since what is concept in a watery waste of chance? What, however, is a lost cause when your theory never had a real one? Evolutionary chemistry will never, therefore, yield to logic. It will not ditch assays because the mind-set that drives forward naturalistic faith and scientific funding won't permit. What is banned, except in experimental concept and the execution of intelligent, first-class design, is any sense of purpose in biology.

Purposeful plan is, we've seen, just what chemists have to use to synthesise a product in the lab. Experimental programs must be carefully conceived. Perhaps pure chemicals are obtained from a supplier. Program amounts, as lawyers would confirm, to forethought and forethought to intention. *All application programs are foresighted plans.* Indeed, the logical flaw in all pre-biotic experiments is highly organised control. How did pre-biotic purity of the required reagents occur? Could the early earth have been as carefully directed? Does lab production mean the product ever naturally occurred?

The genetic code, with its storage, translation and editing mechanisms, helps to express plans. It is extremely economical, efficient and specific. Suppose that a few nascent polymers in the form of two, three or even ten-unit peptide sequence occurred by chance. With them would occur a vastly greater population of irrelevant and useless sequences. How would non-life (as it would still be) discern the worthwhile molecules and code up their metabolism while simultaneously throwing out a far, far greater load of junk? If those competent occurred inside some proto-lipid bag how would pre-life, demonstrating sharp discrimination, eliminate such swarms of undesirable cross-contaminants and mobs of interference and irrelevance as swilled with them? And if such bio-toxic hordes lapped with 'the chosen ones' outside the sac of life, how would their diffusion be denied; and only those required be filtered and proceed to enter into life? Then would they survive, without prior reproductive capability, more than a single generation?

Let's, however, very generously as if by magic grant this impossibility. Your stock of parts, including every different spec. of chain, is now collected like prefabricated items in a fluid motor spares shop and, in some finely controlled way, stored there. Will such stock spring to life? Or have you forgotten that whole vehicles are, by reason of the informed order of their operational arrangements, far greater than the sum of their parts - even if some of these (such as phospholipids or ribosomal bits) are built to satisfy the role of self-interlocking structures? There is no chemical reason why chemicals should adopt conformations, roles and tasks that are, in reality, the specific endowment of code, translation, replication and other clearly metabolic mechanisms. **Of course, a physical form of life is dependent on behaviours described by the laws of physics and chemistry but it cannot be derived from them**. Chemicals never 'come alive' but pre-coded 'super-chemistry' (called biochemistry) certainly *serves* life. No chemicals, even the ones of which human artefacts are made, have a purpose of their own - they simply serve their user. *And the operational purpose of life is, against whatever odds, survival.* No chemical *wants* to survive, does it? In which case what is chemical homeostasis for? *What, in atomistic terms, does 'struggle for survival' mean?*

And what survival value has a code without the concepts that it represents? *Such point applies to the genetic as to any other code.* Which, if rational engineering is considered, comes first? Is it concept or construction, chemicals or code? If C, T and A fell together to make 'cat', what meaning has that to the letters themselves? So which comes first, the physical or metaphysical? *The production of a few 'Meccano' chemicals does not even as much as indicate abiogenesis.* You can liken them to loose, lead type, a pile of letters of the alphabet. As such they carry no predestination. No more can a cell allow such materials to 'self-assemble' than an editor the letters of his text. Nobody imagines the former arranges his columns by throwing handfuls of type into the air. Not even if, to increase the possibility of random formation of sequences that he understands - words and combinations of them - he weights, say, vowels. What press splutters out a few uncoordinated letters of which some may even look like words but most will not? No. Both editor and cell operate in fundamentally anti-chance mode. Both need to keep the presses rolling with intelligible messages, with signally self-consistent publications. Nucleotides are letters but genes are, like words, thought simulators. And a whole 'book of life', called a genome, is with its operating systems orders of magnitude beyond a few random concatenations that might look like words. It is part of a large-scale conceptual simulator that may well include an archetype. So what's the score for early earth? Life's melody is tied up in specific phrases and, without such chains can't start to be unchained. No start. Full stop. Bitter silence.

Not So Sweet

Carbohydrates represent a half-way chemical house between much reduced carbon (lipids) and fully oxidised carbon dioxide. They are stable in water yet, once activated, reactive. Their polymers constitute an excellent storage mechanism and, in the form of cellulose, chitin and mucilages, structural materials. Many short 'oligosaccharides' are also information carriers but of course glucose and its metabolic offshoots are also life's prime source of energy. Glucose is a product of photosynthesis whose process itself depends on the prior presence of sugars like ribulose bisphosphate - the commonest carbohydrate in the world - and its associated enzyme called rubisco. Single sugar molecules are classified by the carbon atoms they contain. **No five, six or seven-carbon members have so far been detected outside the biological sphere of influence; nor, since you need enzymes for hydrolysis, could un-sweet polymers of starch have sweetened theory**. However, very low levels of a three-carbon specimen were found in the Murchison meteorite (possibly due to contamination since atmospheric entry temperatures would have been destructive).

Cells and multi-cellular bodies are internally coherent, mutually informed and the subject of overall governance dispensed from a nucleus, electrical matrix and, in certainly some but arguably all cases, mind. Much intercellular information is exchanged using glycolipids and glycoproteins on the cell surface membrane; glycoproteins also work within the cytosol. Their subunits are drawn from the 'glycome', a body's pool of code-derived sugars. A single glycome may involve hundreds of them. Polysaccharides, like all other sugars built by enzymes coded for by *DNA*, are complex molecules that can contain many thousands of subunits linked in a variety of patterns. Even monomers are constructed and used with machine-tool precision. For example, 'homochiral' is a word that means 'optically exclusive' (see Reflections); it means all molecules are of the same 'handedness' (called chirality). In double-stranded *DNA*'s case only right-hand deoxyribose is used; and, differing by just one strategically-placed oxygen atom, single-stranded *RNA* uses only right-hand ribose.

Yet how these and other life-essential sugars actually arose and dovetailed into a working cell's coded metabolism remains a mystery. Starting with formaldehyde the 'formose reaction', using a base and divalent metal as catalysts, produces different sorts of sugar chemicals. However, a large proportion of these products include sugars not used in bio-systems; or, like ribose, whose production would have been inhibited or whose antecedent chemicals consumed by unwanted side reactions (for example, by nitrogenous chemicals needed for producing amino acids or bases). Moreover sugars decompose at other than a neutral pH range; and sometimes even in that range. At temperatures above only about 80°C the half-life of ribose is reduced to a few minutes. If you want to reconstruct nucleic acids, the bases that you'd want to join with them readily react with formaldehyde (and acetaldehyde); these reactions degrade and render them biologically useless. You might manufacture some adenine or guanine using hydrogen cyanide and ammonia; or make cytosine with urea and other chemicals but many other chemicals could interfere with these reactions. Also, while purines bases may survive a short while in inactive, sub-zero conditions, pyrimidines require hotter and thus mutually incompatible construction and storage sites. Several further problems occur when trying to simulate early-earth production of the units needed for *RNA*. One needs, to sort through all the incompatibilities, intervention, intervention, intervention by an unseen guiding hand - a chemist's. Indeed, realistic routes for the spontaneous production, simultaneous condensation and preservation of three-chemical nucleotides on an uninhabited earth have not been identified. It looks as if vital sugars, only found historically encoded for in cells, could never have evolved. *No sugars, no ATP, no nucleic acids and no book of code.*

Research may find the glycome (an organism's or all organisms' complement of sugars - many of which perform critical tasks in cell biochemistry) as complex in its ramifications as the genome. Parts of the kit are far too numerous, specific and functionally coherent to have formed by chance. They do not now and only special pleading claims they ever did. In its uniquely biological context each essential sugar is, in a timely and accurate way, synthesised by synthesisers themselves synthesised according to an application program that derives information, by way of an operating system, from a pre-recorded genetic database that has, crucially incorporated in its structures, sugars it will make. Such complex 'chicken-before-egg' informative and metabolic protocols often involve the use of *ATP* (including this molecule's own ribose sugar) and would have needed thousands of cumulatively 'correct' steps to approach operational capacity. If it's simpler perhaps an evolutionist could the algorithm on his slate. Could glucose have actually co-evolved at exactly the same time and place with respiratory metabolism and thereby triggered all life's energetic possibilities?

Bags of Life

Seas aren't good enough. Chemicals disperse not concentrate together there. Nor is a puddle sufficient. You have to drop beneath a drop of water. Chance had to group all necessary chemicals, without confusing or unnecessary ones, within a sac whose edge is but a few millionths of a millimetre wide and called a cell membrane.

Every cell has, as its boundary, such a membrane that separates 'inside' from the world 'outside' and thus defines the basic unit of an incarnation. It is made of molecules called phospholipids; and the fatty acids involved are constructed of chains between ten and twenty carbon atoms long - more would create molecules too insoluble to mobilise in water and less ones too soluble, slippery and disruptive. Its head group includes a negatively charged phosphate part. Charged and uncharged ends lend the molecule an important, electrically ambivalent (or amphiphilic) tendency. One end dissolves in water, the other stays away. This means that in water double layers - called bi-layers - automatically form; the head group faces outwards and the water-insoluble tails cluster, like the filling of a sandwich, together in the middle. One important result of all this is that membranes are relatively impermeable and another that they can behave as electrical insulators across which (in a living cell) a voltage is always raised.

Proteins of various kinds embedded in the membrane act as pumps, gates and, with attached sugars, messengers and guardian intelligence. Actually the inner and outer layers of a biological bi-layer differ in composition and structure; they are effectively asymmetrical. This difference is reflected in each side's functions (transport, detection of environmental changes, adhesion to cytoskeleton etc.). Although the membrane is fluid there is, however, no free diffusion of proteins. In fact, fluidity varies from region to region and proteins and different classes of lipid (such as cholesterol and, with various head groups, phospholipids with various fatty acid chains and so on) are confined in discrete, functional domains. For example, mobile lipid rafts engage proteins involved in signal transduction and seem to play a role in vesicle secretion. Membranes are, therefore, not only complex but functionally organised. This even goes for 'simple' bacterial sheaths.

In eukaryotes they are also used as internal borders to separate different 'compartments' of a cell (such as nucleus, mitochondrion etc.), to form platforms on which metabolic machinery such as ribosomes, chlorophyll and respiratory protein complexes can be stabilised in an orderly, close-knit way and to create water-free reaction chambers in which synthetic chemistry can occur. In short, every membrane is a complex, dynamic mosaic of carefully arranged molecular pieces.

Nowadays (and for the last three-and-a-half billion years) lipids are the product of chain reactions mediated by enzymes specifically coded and called for. Phospholipids are the major component of a cell membrane, the embrace of which is crucial to split 'inside' from 'outside' and thus buffer homeostatic biological metabolism from its non-homeostatic, non-biological and unpredictable, uncontrollable exterior environment. This applies the same to a pre-biotic vesicle as to a 'modern' cell. Such specialised lipids are made of glycerol (or ethanolamine) and fatty acids, esterified by an enzyme and finally phosphorylated (meaning that a phosphate molecule is added). The question is how, if at all, they might have first self-assembled and then kept assembling (or at least dividing) without the benefit of prior *DNA* coding, its operating system and specific protein products; then how they 'handed over' to a specifically coded metabolic program for their own 'survival'. A suite of enzymes would imply a host of irreducibly complex molecular mechanisms that could only be developed a lot further down the line than we have reached.

It is remotely possible that some fatty acids (for which a complex pre-biotic cycle of construction has been hypothesized) could have been produced in the presence of iron or nickel catalyst - but where on early earth, since it involves complete dehydration, could reaction with glycerine and phosphate ever have occurred - especially in relation to other dissolved requisites? Not only this but fatty acids and phosphates each bind with calcium and magnesium ions to form water-insoluble complexes (commonly known as scum); in other words, they'd precipitate out of any solution they'd need to form cells in. This removes them from reaction zone. Therefore, was no calcium in the soft-sea starter? Just as soap will not lather so fatty acids would be precipitated out in hard, calcium-containing water. We will deal with the similar problem of phosphate availability soon (see *DNA*). Another problem, especially in the sea, is salt. Were early earth's great oceans much saltier than those today? Model prebiotic membranes fall apart in brine; they're only stable in pure water. Pure water's not a 'soup'. What aqueous locations would be free of salt? So what was the mode of codeless, biochemical construction? How did membranes ever gain stability?

Easy! You just wave a wand. You assume that - once they have appeared together in sufficient quantity - they self-assemble. Is this true? Different phospholipids show diverse behaviours that vary with conditions. They may self-assemble as you'd wish into unstable 'liposomes' but mostly do so in a non-bi-layered phase or make bi-layered stacks or multi-layered spherical vesicles. Only under very specific conditions do pure phospholipids transform into single, stable bi-layers (like cell membranes). Of these temperature is critical. Critical temperatures vary according to the layer's phospholipid composition; and life-forms regulate that composition to buffer external changes in temperature in order that their membranes are maintained intact as single, biologically acceptable bi-layers. Deviation's deadly. Membrane tracks environment. Dynamic homeostatic regulation hugs the norm, locks to the critical condition. Without such regulation fluctuations in temperature, pH, salinity and mineral presence would each have destroyed not built a membrane. How, therefore, could the stable boundary of a proto-cell spontaneously occur? And if it ever did, phospholipids are notoriously impermeable to amino acids, nucleotides and so on. How, therefore, could it then encapsulate the chemicals of life? How, suspended in this special state of equilibrium, were the first cell's genetic and metabolic agents then encapsulated?

How did mindless nature select the correct compounds to facilitate transport across its membrane and how, when nowadays unsaturated fatty acids play their part, did stable growth and fission of primordial vesicles occur? Directed, skilful chemistry in labs can hint at answers which the undirected circumstances and coincidences on the early earth could very doubtfully supply. The devotee invokes a promissory faith in remote possibilities. *But even if you produced the first phospholipids by accident you would still have to generate (by accident of course) the coding sequences and mechanisms to make sure they are permanently fixed in earth's biological process. <u>If not then no cell lives and materialism drops dead.</u>*

On the subject of lipids, it is also interesting to note that steroids are synthesised (from previously metabolised cholesterol) in a wide range of organisms from bacteria through algae, plants, invertebrates and all kinds of vertebrates. Nevertheless, despite such wide taxonomic distribution, their molecular structures and associated modifications show (as with other bio-molecules) no significant patterns of evolution. These regulators have never much changed. Nor is there evidence that cholestane or cholesterol precursors could form in a pre-biotic 'soup' and, subsequently, no chance that uncoded steroid-protein complexes could have opportunely drifted within a thousand miles let alone a thousandth of a millimetre of each other. If their non-biological origin is out of the question, by what steps did their biosynthetic pathways develop in order, it seems, to turn out a useful end product, a metabolic regulator?

In short, the ability of a cell to live depends, among numerous other critical factors, on the ability of its 'skin' to withstand disruption. In fact, the molecules of a membrane cohere almost as delicately as those of a soap bubble. It will rupture and disperse if any of a number of conditions (temperature, humidity, salinity, acidity, trans-membrane redox potential, various ion concentrations etc.) are not, even for a few seconds, monitored and met. A membrane-protected cell can only cope with a very narrow range of external fluctuation in each of these conditions. It must neither burst nor dry up and its internal regime must (through the agency of ion pumps, molecular gates, electrical potential, regulated transfer of materials etc.) avoid dispersal and actively maintain a high-grade, dynamic equilibrium - called homeostasis. So what non-steroid regulators did a life-originator have? Or must you wish impossibly that, like a packaged saline drip, seawater or the puddle were immune to natural flux; and that no tide of change for absolutely ages changed life's delicately balanced flow?

Evolutionary contrivances are all designed, of course, to confound the notion of design. Yet, even at this 'simple', primordial level, any discrete biochemical would have to involve design. Who denies the complex chains of reactivity that science is unravelling in cells? In fact, each cell and all other levels of any living system are organised in a much more complex, hierarchical way than 'simple' components alone explain. Life is replete with instructions, information, integrated metabolism and purposive construction - all of which are completely absent from any interstellar cloud of organic chemicals, carbonaceous meteorites or the tarry gunk of pre-biotic experiments. *Nor do such abiotic samples hint at the way genetic protocols were built. <u>In other words, the experiments tell us nothing about how a route for making even simple monomers or other basic biomolecules first developed in a genetically governed cell (and</u>*

there exists no other kind). **In yet further but important words, they miss both the informative dimension and the metabolic point. So, it may seem, does all supposed simulation of abiosynthesis.**

In fact, natural, non-biological lipids are unknown. Nor have viable paths leading to them been identified.

And biological lipids fall into evolution-busting, chicken-before-the-egg syndromes. Neither cell wall nor membrane's lipid bilayer can be made without proteins and nucleic acids; but no codified metabolism, such as protein or lipid synthesis, can occur except inside such phospholoipids as it needs produce. **There can, in other words, be no molecular stability without a membrane to buffer and maintain it. So which came first - membrane or its makers? Of course, all are needed all at once - with a minimum of another 300 or so proteins and other chemicals. At the top of hierarchy, finally, all such chemicals depend on informative codes. Did code or chemical come first? Or, fully integrated, both together?!**

Reflections

Mirror-image Chemistry (Chirality)

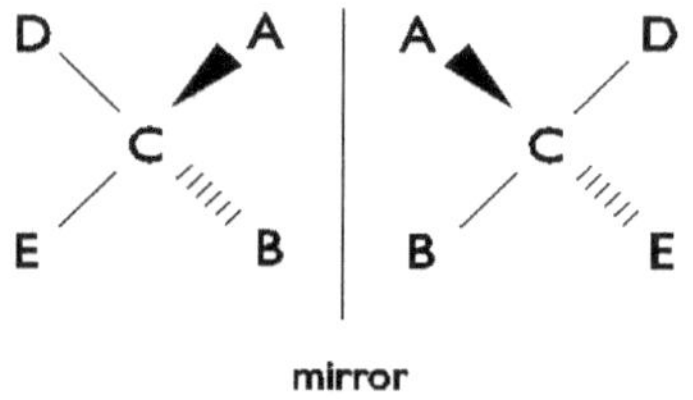

fig. 20.2

For the sake of argument, although neither experimentation nor an inspection of earth or outer space confirms it, allow the full range of biologically active amino acids. Delete all non-biological chemicals that would simply confuse and clog up a necessarily complex and efficient form of nanotechnology - a cell. Isn't that enough?

No, sir, not at all! The left and right hand of a pair of gloves are mirror images. Such 'handedness', dubbed from the Greek 'chirality', is exhibited by chemicals called optical isomers. All natural synthesis of amino acid and sugar isomers shows a 50/50 (called 'racemic') mixture of each hand. This even-handed result, with equal and un-segregated amounts of right (D-) and left (L-) forms of chiral molecule, resembles in its head/ tail probability the tossing of coins. It prevailed, for example, in the Miller-Urey experiment. And it must also, spelling ruin for the theory of chemical evolution, have prevailed in the natural, unguided chemistry of pre-biotic circumstance.

Ruin? Why? L and D amino acid are otherwise identical; so are the complementary forms of *DNA*. But life has, 100% obliterating 50/50 odds, 'selected' to control by codification the segregation of its major biochemical components into all L-form (left-hand) amino acids and D-form (right-hand) sugars. Because the ability to do their jobs depends on 3-d shape and because 'handedness' is an integral part of such shape, life's *DNA* and protein depend on rigid, metabolised segregation. Why these particular 'homochiral' left/ right complements have been chosen is as unclear as why we live in a 'material' as opposed to an 'anti-material' universe; but there is no evidence that 'isomeric purity' was ever biologically absent or unimportant.. The fact is that the strict structural requirements of proteins and nucleic acids means that the mechanism for generating homochirality must *precede* them; they are, critically, made by it; without correctly homochiral amino acids and sugars any naturalistic assembly of proteins is blocked.

The universal occurrence of 'analar opticality' must indicate some universal significance - some 'deep' feature of construction and its operational necessity. Uncontrolled reactions create racemic mixtures but the basic atoms and, therefore, molecules of each metabolised polypeptide and chromosome are, in their hundreds and millions of billions, all strictly controlled. For example, a strand of *DNA* evokes its smooth helical twist by using exclusively D-form sugars; nor could complementary strands bind into their famous, operational double helix without such use. If an L-shaped nucleotide were introduced it would disturb both local shape and destroy functionality. A 50% random spread, indeed anything less than a 100% correct constitution, would wreck coding operations; it would be as if, in a multi-cylindered engine, one piston were square. Nor could even partially D-form proteins (including enzymes) do their jobs. A consequent fact is that fully L-form protein editors scan replicated copies of D-form *DNA* and excise any error. Did such editors evolve before *DNA* (although in reality they are sequenced from it)? If not, how did any precursor or cellular, working *DNA* without 'anti-racemic' immunity hold out for billenia against the natural tendency to fall apart? What might, without a code or cell in sight, create homochirality?

One grasps at straws. Could it be an effect that links magnetism and chirality called 'magnetic anisotropy'? For various cogent reasons it is most unlikely. What about selective adsorption on mineral surfaces - which can occur with either handedness and anyway only fractionally select one hand over the other? So intractable is the problem that no less than Stanley Miller proposed that the first self-replicating molecules were achiral peptide nucleic acids (*PNA*s). In this case the bases are joined, not by sugar and phosphate but acetic acid and an amino acid not known to occur anywhere in nature called 2-aminoethyl glycine. There is no naturally supposed production path for *PNA*s. It is also doubted (and there is no evidence in any case) whether *PNA*s could engage in normal *DNA* activities or support blind, evolutionary transformation from a *PNA* to an *RNA* or *DNA* world (only the latter exists today). Such feeble guesswork is typical of the hope, or perhaps desperation, that must grasp to obtain a naturalistic answer. How hard die-hard naturalism, materialism and scientific evolutionism drive our minds to drive out other philosophical considerations!

To conclude, it is a fact that when a living form dies today its L-amino acids and D-nucleotides spontaneously degenerate into racemates. Any non-living, abiotic mixture of biochemicals is racemised. The laws of nature issue racemates but from code-informed control proceeds strict, biological segregation. If life's enzymes create chiral molecules, how did segregation happen before metabolism? If it first happened abiotically, how was sufficient coding, combined with application program, immediately engaged to make sure it happened thereafter on time, every time? Because without such handy handedness you couldn't, as already emphasised, make working *DNA* or, thence, proteins with their own specific three-dimensional shapes. *Metabolism couldn't happen*. **Metabolism neither occurs nor, speaking bio-logically, ever occurred without prior arrangement, without a purposive computational assemblage at whose heart is code**. Yet chiral rectitude depends on enzymes in the first place. **Thus another intractable, chicken-and-egg problem plagues wholly chemical, physical and even astronomical explanations of either single-handed, homochiral isomer production or *DNA* editorial processes.** The ragged response is a standard, arm-waving apologetic to the effect that 'life', in the form of such precise pre-metabolites, *must* have had a long-time lucky break to the nth, nth, nth degree. One's free to buy that hopeful bluff but, equally upon reflection, free to reverse it. Thus, turned round and down, no sale!

Join Up, Fold Up.

A 'primitive' cell contains many thousand *proteins*, an 'advanced' one millions. Each complex contains hundreds of types of component coded for in the language of genes. These genes often involve, including the contextual operating system for protein synthesis, replication and so on, many thousands more parts. Each of these parts down to a single nucleotide has, like the letters on this page, contextual meaning outside which it ceases to operate effectively, that is, it ceases to help bear a message. In the case of any genome such cessation means its own destruction. How, therefore, could an integrated genome ever build up piece by accidental piece?

As the number and complexity of its parts increases so, exponentially, does the number of precise interactions and the unlikelihood of a system's accidental origin. In fact a single gene is just a microscopic fraction of the possible arrangements chance would have to sample. Where n = 20 (for twenty biologically coded amino acids) a given polynucleotide (l for length) could show (n to power l) combinations. Thus, for an average protein of 300 residues (say, 900 genetic bases) the sample size would be 20^{300} - a far, far larger number than all the particles in the known universe or seconds since time began. What trick did evolution use to sample much, much faster than computers 'think' of sequences? And hit the jackpot time on happy time until the first cell formed?

This is just the one-dimensional starting-block. As well as joining chains you need fold the sequence up into a three-dimensional operator. Proteins need, like any tool, correct shape to work. Function is dependent on their folded structure; not only ion channels, bonding sites, active sites and infrastructure that frames working compositions needs to be precisely sequenced; then at its birth, with the help of ribosome and protein midwives (chaperones), every protein is precisely folded up.

A fifty-protein trick would be small beer compared with what is actually required by even the most minimal of cells. Even then, no metabolism works without all peptide bonds showing so-called alpha-linkage between the constituent units of amino acid. Such consistent linkage is the product of enzyme catalysis. Enzymes themselves show alpha-linkage. On the other hand NMR (nuclear magnetic resonance) studies have shown that non-biological, thermal synthesis produces mostly unnatural, unworkable, non-biological linkages. The issue's one of causal circularity. Did polypeptides not occur until *after* a genetic coding system that governed the consistent production of alpha-linkage in all protein? If they occurred before, how did 'pre-life' cope with the chaos of non-biological racemates and randomly-mixed, predominantly non-biological forms of linkage? The operation of *DNA* is inextricably associated with the 'right set' of proteins. How did the proto-operation of *DNA*, which itself codes for these proteins, manage to 'identify' this 'right set' from all those that evolutionists imagine swilling in the jumble of primordial whatever. Much, much harder, how did it then straightway manage to create the 'right and corresponding' genes from such previously uncoded debris of the early earth as it 'found useful'? Special pleading (without a shred of evidence) that protean 'life' might have worked without these apparently insurmountable constraints is as unscientific as it is unrealistic - and unacceptable. **These are serious 'chicken-before-the-egg' problems.**

A protein works according to that biologically active 3-d shape derived from multiple weak, electrostatic, hydrophobic and configurative interactions. These occur in the watery medium of its 'birth' and operation. A precise shape has a precise function and any deformation may dramatically decrease its operational efficiency. **In other words, incorrect folding means functional destruction. No function, no survival or, at very least, illness that a lack of proper protein generates.** There are myriad ways in which each specific chain could theoretically fold into its 'minimum-energy structure' but in a cell, whether folding for the first time as it leaves a ribosome or refolding after denaturation, the 'correct' order of fold leading to the same target shape always occurs within seconds. *It always must have because, as noted, the single correctly folded shape of any protein is critical to its operation.* Is this precision simply a result, as conventionally assumed, of amino acid sequence? In which case why do *IBM* reckon Blue Gene, a supercomputer completed in 2005 and 500 times faster than the present generation, takes about a year's round-the-clock processing to model the folding potential of a 300-residue polypeptide - which a cell can fold to target shape immediately, along with many others? How did such biological efficiency occur in the random, coincidental conglomerate of perhaps a minimum of two hundred and fifty proteins required (with *DNA* and its operating system intact) the instant a minimal first cell occurred? If sequence alone were the key why do, for example, haemoglobins found across a wide range of organisms and sequenced very differently all fold up much the same? How proteins fold to take their precisely required shape is therefore thought of as a 'grand problem' whose solution will be well within Nobel Prize range.

Yet folding's critical because a protein's function is dependent on its shape. How many changes are required, from all the 'sample' possibilities, to turn one function to another? If one amino acid changes you'd be very, very lucky to derive a novel function; no example has been found but maybe one could with rarity, occur. Two? Two steps would mean the 'broken gene' produced a less efficient intermediate but was not, perchance, deactivated or deleted. The nascent innovation might then need to explore, according to Doug Axe, up to 10^{77} further folding possibilities (including the few sufficiently recouping metabolic costs to avoid deactivation) until the pristine, unforeseen but efficient function was achieved. This number is, in innovative terms, far to low. Since most proteins operate as cooperative functional groups, complex proteins or proteins with co-factors and are integrated cornerstones of whole cell strategy, a two-step invention is, as well as statistically remote, functionally much remoter. Many gradual 'correct' steps would be required while 'incorrect' chances *en route* to the unanticipated metabolic 'target' were never deactivated or deleted. Derivation of correct re-folding of a single enzyme to perform a fresh and usefully integrated task - while leaving its predecessor's necessary performance biologically intact - were difficult enough; but cells are replete with protein machines (such as ribosomes, proteasomes, *ATP* synthases and so on) whose protein components all need exactly the right stereoscopic folding patterns in order to bind and work together. The larger a protein or protein complex the larger grows the sampling problem and smaller the innovative possibility. Constraints on the sequence of even simple proteins is severe; swaps from one kind of fold or folded structure to another would require unstable intermediates; for large proteins or complexes the improbability is orders of magnitude greater; and, on top of all this, strict editorial procedures are applied to ensure that, on translation to construction, the correct amino acids are delivered and the correct 'massage' to generate correctly folded three-dimensional configurations are 'grown'. Remember, a gene is the symbol for a protein albeit that genetic language builds its own material reality. The information required to accomplish this astoundingprocess is stupendous and still in the early stages of decipherment. No doubt, an 'elephant' called information (close ally of not-on-your-nellie intelligence) is stalking the spaces of molecular biology. The goads employed to tame this conceptual beast are weak; the arguments (such as 'consensus areas') forged to contain non-naturalistic interpretation of emerging facts strike with the force of grass.

'Massage' by chaperones requires a second glance. Proteins don't just fold into the right shape by themselves. On the other hand, inefficient, slow or incorrect folding presents various potentially catastrophic problems to a cell. The automatic answer is assistance helped by quality control. There exists a large family of molecular midwives and, included in it, chaperones of chaperones to render such morphological assistance as any protein needs. Dozens of these 'maternal' factors buffer any adverse influence while a nascent protein develops to such precise spatial maturity as will alone allow its operational association and interaction with other biochemicals. They also, perhaps helped by non-coding snips of *RNA* and certainly by a set of sugar signals that exact a quality control, guide the folding process. Such associations preclude malformation and malfunction; they promote coherence and stability in the workplace known as a cell. The sugars alone would require suites of enzymes to metabolize them; genes and non-genic *DNA* are both included in the recipe. Such orderly configuration indicates the involvement of much more information than is carried in simple, linear sequences of amino acids; it thereby argues influence beyond the nuclear database. Indeed, any protein needs translation machinery in the form of a ribosome created out of many specific proteins and exact *RNA* in order to be built in the first place. *Even a protocell must have sported such crucial machines.* Thereby we stumble across yet more chickens-before-their-eggs. Which came first? Was it ribosomes - that start every protein off? Or proteins, chaperones, *RNA* signallers, sugar quality controllers or requisite, integrated sets of genes? By what stages co-evolved these systematic factors?

It is further noted that quality control, applied at several stages in protein synthesis, here critically discriminates between an incorrect fold and a partial fold that will eventually flourish. The system uses sugar (oligosaccharide) signals. At least as subtle as *DNA*, such exact configurations act, using both exact morphology and charge, as 'programs' marking and controlling complex, three-dimensional shapes as they pass down the line. The exquisite precision is awesome; the logic light years ahead of chance. Here, down in the foundry of biological creation, essential micro-forms - without which none other would exist - are forged. Newly folded protein is inspected and, if still somewhat incorrect, marked up for further scrutiny by chaperones. The misshapen are, however, recognised and chalked for scrap. That is to say, if the *expected* state of folding does not match what the sugar signal indicates it should then appropriate action is taken. How could a cell before such controls existed ever, clogged up by mistakes, survive? The feeble answer 'must have' needs explanation with a flowchart and a plan. Chance was never logic's strong suite nor was reason's serendipity.

If the genes and proteins that compose a fly are 80% the same as yours, can it be that genes alone are responsible for the spatial organisation of biological tissues, organs, systems and whole, self-consistent bodies? Is not *DNA* simply a database accessed as necessary by an application program called 'cell functionality'? What has it to do with shape? Just as steam needs pistons for an engine's work might biochemistry not need polarities and shapes of cytoplasm for its functionality - functionality innate and initialised by single cells or eggs? In this case you might ask which comes first - a structure or its function? Does some shape develop its own reason - in concert with a hundred other previous amorphologies? *Does purpose precede implementation, book index or program its database?* If such considerations apply to larger organs and whole bodies they apply with equal force to the fundamental agents of biology, protein. Chicken-and-egg problems plague the issue of life's origin and resolution always involves, as in the case of any engineering, mind behind. You need, for example, a seventy-peptide string (coded for by over 200 nucleotides) for any folding to be able to occur; yet the first irreducibly complex cell would have needed many such proteins all correct and present from the start. One must, to maintain an evolutionary mind-set and keep the naturalistic home-front up, allow that stark impossibility.

In short, as part of the requirements for abiogenesis you'd need a wash of optically L-form (left-hand) amino acids, nucleic acid that could grab and join them up, the genes, machines and all the rigmarole of protein synthesis and folding. What good any integrant without its code in *DNA*? Yet whosoever, vaulting reason, next vaunts evolution's prowess in creating forms of life must by that token claim un-coded chains of both identical and different types of polypeptide would have had to accidentally appear together in the primal puddle; and then, in order to create some proto-life, encode themselves; and, finally, each single one fold to the accompaniment of chaperones (or, much more likely, twist without) correctly into its single biologically active shape - all at the same time. What an order - or tall order - but you're not there yet by far. Proteins rarely work but rather in spatial and/ or temporal complexes. A spatial complex means each protein has to stereotypically adhere correctly to compose the unit (as, for example, in the quaternary structure of a complex enzyme); or to work sequentially with others as, say, in a metabolic pathway. **Just one protein isn't nearly good enough.** *And if you begrudge 'design analysis' and claim the minimal, first cells were 'too simple' for this to apply then, again, please explain precisely and devise in detail how you'll show me what you naturally mean - or else, because the element of information speaks so clear, it might be fair to bottle up hot air, to stopper gas of an opinionated atmosphere.*

Number Games

Evolution touts the mists of time and chance (or unpredictability) to pull its unseen tricks, its naturalistic miracles. The process *must have* happened. That's your starting point. If you can't be precise, well, never mind - you opt to work statistically with probabilities. Remoteness hardly matters then. Doesn't *someone* always win the lottery? Evolution's won so *must have* won it countless times - a winner all the way! How can you argue with the vagueness of a 'proof' like that? OK, maybe by responding like for like. A probability of less than 1 in 10^{50} is considered statistically zero, that is, impossible. **Many calculations have been made that illustrate, by this criterion, the impossibility of obtaining by accident various sequences of *DNA* and protein singly let alone in multiplex cooperation**. Nor does statistical imagination of remote a probability make up for physical impossibility.

All sorts of 'odds against' are bandied about. There are four letters in the genetic code; they are disposed in three-letter words; the total possible number of words is therefore 4^3 or 64. By the same token if a full and sufficient stock of 20 amino acids were sitting waiting there would theoretically be, as Francis Crick once noted, only a chance in 20^{100} of randomly achieving a modest but specific polypeptide of one hundred residues. Such odds not only shout, they scream cellular impossibility by chance. Add some reality in the form of possible constraints and the probability of, for example, a small protein of 110 amino acid residues called cytochrome c occurring by chance has been calculated according to Hubert Yockey and William Dembski at about 1 in 10^{75} (there are about 10^{78} nuclear particles in the universe, not to mention massively less stars or grains of sand). By including requisite peptide bonding and left-handed optical isomerism Douglas Axe upped this to about 1 in 10^{164}. Put another way, given a vast and deep primordial soup containing 10^{44} amino acids it could take, at a chance per second, 10^{23} years to obtain a 95% chance of building that single protein. The universe itself has lasted only 1.4 x 10^{10} years! Noting that a minimal cell needs at least 250 types of protein this figure approaches Fred Hoyle's guestimate of 1 in 10^{40000}. You can quibble with impossibility but, given that intelligence achieves objectives 1 in 1, it is vastly more improbable that chance (in combination with physical law) engendered life. In other words, its counter to design is negligibly weak. Chance has no chance whatsoever. Zero means impossibility. All time and space are not enough to bolster evolution. How reasonably, in the interests of your philosophy and scientific methodology, can you still claw at evanescent straws?

Life's tolerance is very fine. High calibre. Hi-fi. Life's tack and tools, especially enzymes, are precise - specifically intolerant of imprecision. The forces that would, in theory, have created such a molecule are blind and would as soon destroy it. And you need thousands of them coded, wrapped together with at least a thousand other kinds that work together to begin to make a cell! **Is not the belief that life arose like this an exercise in absolute futility?** Don't aggrandise it in the name of science. Was it not noted (Chapter 8) that scientism sometimes flees from science? Here again the case is one of sheer, secular denial.

Time is no use. The reverse. Time's arrow is destructively (↓) entropic. Things wear out. Complicated things stop working. Biological survival is simply and, in the end, inevitably a losing struggle up against a passing gradient. On the slope of hours even 'simple' cells need make and constantly remake many copies of at least a hundred different protein molecules. Each one does this precisely many times a second.

What about the other biochemicals? In a thin primordial soup all such materials would have to have congregated in a space far less than a cubic millimetre. If all the particles in the universe (10^{80}) had interacted in Planck time (the maximum possible rate of 10^{45} times per second!) since the assumed start of time (10^{45} seconds) they could only have achieved only 10^{150} combinations. For example, could a haemoglobin molecule 574 acids long have been churned out? It has been calculated that the odds against any such chain 'happening' in the correct sequence, with the correct components (even granted a continuous, sufficient supply of monomers), with correct left-handed optical isomerism (see Reflections), peptide bonding and folding (Join Up, Fold Up) is about 10^{950}! But the chance is much, much less than that. There exist over 200 million such molecules per red blood cell, millions of red blood cells per mm^3 and thousands of trillions in your five-litre blood supply - a lot of optically 'immaculate conception' staining red. These exist because, in reality, the haemoglobin is the product of a template; this template is the code that's been inscribed in *DNA*. *What goes for proteins goes, hand in hand, for DNA as well.* The accidental production of one protein alone, without its coding and necessary synthetic paraphernalia that involves more than a hundred others, would be useless. While nobody suggests that haemoglobin was a 'proto-protein', every single 'chemically evolved' molecule without machinery for its own coherent reproduction would have been useless. Within a few hours the environment would have degraded what it chanced to make. What chance is there of accidental acme of cooperation - a hundred thousand million atoms coded and dynamically coordinated to exquisitely compose even the smallest kind of natural nanotechnology - a 'simple' bacterial cell? Is chance therefore a notional shot too long? A speculation, right off limits, turned to fantasy? The order of exactly specified complexity is basic. It is the very basis of biology. Statistics shouts aloud its quality of information could not arise by chance.

Mean-spirited parsimony! Why play Shylock with such wealth of speculation? Why, insists materialism, doubt such uncertainty? Give chance a chance. Extend the hand of possibility to evolution. After all, it might have, could have, must have and in fact did happen. You, fighting for your life, assure me so. Cut, therefore, from a numbers game to the experimental chase.

Take fatty and amino acids, glycerol, base, ribose and phosphate; however unlikely, you are granted any simplex chemical in any quantity you want; take stock that's analar and exclude racemates; stopper the tubes, seal every dish whose 'atmosphere' and solution are of any permutation that you wish; on an industrial scale incubate and agitate at various pressures, temperatures and any other kind of energy you want. Persist as many times and as long as anyone could wish until a codified, self-reproducing cell is washed up on the scientific shore. Such arrival is the thread on which materialistic promise is strung so why has this experiment not been systematically carried out? Given the complexity of even a bacterium, could it succeed? By Chance, it better had!

Then let imagination run its riot, lay your bets on anything except a certain but forbidden winner, teleology. Point out on this page an 's'; then 'p', 'o', 't', 'o' and 'n'. This simple exercise takes, due to the intelligence involved in execution, a few seconds. How long would it take a leaf in wind? Whereas a normal unicellular organism may involve millions of component parts, dare to consider a *very* simple organic aggregate of only two hundred. With how many did it start? Did a single chemical seed liquid-crystal life? Or, because you cannot deem a single bio-molecule an organism, how many 'self-assemblies' might accumulate into a vital mass? No matter that all organisms are a lot more complex than our proto-biotype. Grant possibility. Each addition has to 'super-integrate' in a positively functional or at least a neutrally, non-interfering way. Suppose that now, instead of just identifying letters on this page you had to write things up yourself. Could a leaf-in-wind compete with you and write, or even copy, sense? Compare the exercise with moves you have to make to build a stack of cards. As complexity increases so the odds of those additions that detract from rather than improve such teetering integrity increase. Each step becomes, statistically speaking, much less likely than its precedent. Each false move is dangerous, perhaps lethal to the stack's stability.

So is the hand that stacks just, all and only chance - chance constrained of course by rules of chemistry? No cause, no reason. Certainly not natural selection since chemicals aren't 'fit' and don't 'survive to reproduce'. **Materialistic faith is vested in a lack of reason**. Life's ingredients are very complex chemicals built by code and used by everyone in our globe's great family. The chance against them popping up spontaneously outside a cell has been calculated by physicists at 10^{229} against (New Scientist 24-5-97 p. 39). Do you remember Fred Hoyle's calculation? Paul Davies, an apparently ardent evolutionist, has mustered an even more generous number, about $10^{2000000}$, against abiogenesis. Its chances are less than winning a national lottery every day for thousands of years. *So never mind the odds. The issue is really one of 'don't want, won't have'. In this case you are forced to argue the toss - because the other, conventionally unwelcome side of 'no chance' is, of course, biogenesis.* Statistical long shots are far more mystical an explanation of life's origins than straight, rational, biogenetic design. The mind-numbing numbers cry out for functional logic to swiftly intervene, for informative intelligence to collapse improbability to certainty and execute the work.

Don't want, won't accept. It's not according to my own philosophy. From within the reasonable implications of 'intelligent design' might there not stride implacably unreasonable fundamentalism, all the suppression, pain and lies that missions in the wrong hands cause? Non-heaven forbid! *Don't want, won't accept*. The awful immanence of God must not exist. How could you bear an inane ghost like that forever lurking in your mind? Never mind the numbers game - there's certainly no God when all is chance and natural selection. Or, compounding the irrationality, if you insist on Him can't He illogically choose Lady Luck to make His mind up? Therefore, the unspoken thinking runs, even if an inference of 'intelligent design' were reasonable it's unreasonable; and if it's right it's wrong, if it's true it's false. Design's a dirty word. When confronted by his own calculations with the effective impossibility of life occurring 'naturally' President of the Royal Astronomical Society and atheist-with-intellectual-integrity, Fred Hoyle, said, "...it seems better to suppose that the origin of life was a deliberate intellectual act." Indeed Lord Deliberate, the 'mind of God', looks Davies in the eye but the latter seems, in signed-up contract with professional *PAM*, bound to systematically avert his gaze. Of course, denial in closed minds is where the thinking stops and where, because you can't be sure of what is sure, the betting carries on - raising astronomical odds against the birth of life in shallows and, therefore, against the panoply of those descendents risen out of Grandpa's marsh.

Pristine Instruction

Atoms of themselves don't have a purpose. No more do molecules or what is made of them. None can ever give instructions. But in matter of fact all biological proteins are the definite, unerring product of instructive code. Information is written on nucleic acid, *DNA*, which constitutes an organism's genome. Although chemically unrelated, each of these two classes of bio-molecule is uniquely fit for its purposes. And in terms of informative program each is closely related, through the medium of a translator, t-*RNA*, with the other. Protein succeeds *DNA* and, as products of metabolism, other chemicals succeed protein. **Information, in accordance with dialectical theory, is hierarchically prior. And just as thought and language precede alphabetic script, so purpose and genetic code precede the *DNA* in which their meaning is inscribed, stored and transmitted.** You get nowhere in the elaborate expression of an idea without its first principle, informative instruction. You get nowhere in life's micro-technology without essential structures which, materially viewed, are various sorts of code - application programs, operating systems, a database of chromatin and hardware. The signals are precise. How could a flux of micro-puddles throw life's messages together?

Consider for a moment the *meaning* of life's pristine instruction - messages called proteins wherewithal to make an organism up. A proteome is an individual's total complement of biological polypeptides called proteins. Proteins are made from about twenty sorts of amino acid 'chosen' for their size, polarity (or non-polarity), electrostatic and other properties useful in the construction of biological parts. Chosen? Chance alone could not 'choose out' and then ally with code a set of chemicals with properties combined to outperform life's chosen ones. And cells, of which all bodies are composed, are wonders of miniaturisation. Proteins, both as structural materials and functional enzymes, are at the heart of brilliant, optimal, natural nanotechnology. Precisely made and bound together, they compose all manner of molecular machines. Indeed, on this basis none of the chemicals of life are seen as haphazard but as optimal components deliberately machined by metabolic pathways for tasks such as information storage and transmission (nucleic acids), energy storage and release (especially carbohydrates, lipids and *ATP*) and material biosynthesis (especially protein). Engaging the limits of sub-atomic capacity they execute diverse structural and functional roles. Has chance optimised again or can you use your intelligence to push its luck? What greater intelligence is there, from an evolutionary point of view, than human? Yet no scientist has devised better materials to underwrite biological construction. Indeed, what *else* might you dream up that could substitute for them at work?

	tam/ raj	*Sat*	
	expression	*DNA Databank*	
	informative oscillation	*Balancing Factor*	
	off/ on switching .	*Regulatory Circuit*	
	negative feedback	*Homeostasis*	
↓	*tam*	*raj*	↑
	output	*input*	
	switch off	*switch on*	
	informed	*informant*	
	controlled	*controller*	
	products	*protein*	

To return to proteins in particular - once 'squeezed' from ribosomal manufacturing pads they fold up automatically and rapidly according to an in-built plan. How, after this, does a freshly minted protein know where to go to work? Nobel Prize-winner Gunter Blobel discovered that each sort has a unique address tag. This tag, whose shape is coded for, will fit only the spot at which it is destined to play its part. Such a computer-like, informative and at the same time vital attribute of addressing looks like high-level, systematic information. Is it the product, yet again, of chance?

How does a cell regulate the amount of any particular protein that it needs to make? Explanations for the origins of homeostatic regulation, a central, cyclical and teleological feature of biology, tend to provoke 'chicken-before-the-egg' puzzles. And they tend to involve different levels of hierarchical control. Functional enzymes, for example, are able to regulate their own activity by a form of homeostasis called 'allostery'. This means that protein they produce is itself able to inhibit further catalysis. Such metabolic feedback is conducive to the orderly operation of a cell. Its purpose may be assumed implicit in a code that subtly, simultaneously predefines the shape, catalytic function and catalytic regulation of its own molecular machines.

There are, as already noted and will be further noted, many other types and levels of bio-regulation in higher organisms. A single cell can chemically 'sense' its circumstance and metabolically adjust accordingly. If not, how could it survive a day? Just as a rheostat controls electrical current so a cell can adjust the quality and quantity of protein expressed. Even as a bacterium exercises such control so must have life's proto-bacterium, your putative Grandfather. How was the requisite supply of information first injected? What first tested and established regulation? What set up such 'intelligent', computer-like expression of his genes? Surely it was not that empty-headed Mother, Lady Luck, again - who started 99% of all wrong and gradually worked up to tolerable interference and redundancy? Regulatory sequences of *DNA* tell a cell when to switch production of a given protein on or off; but they only work in precise conjunction with other, specialised proteins or *RNA* snippets that are themselves the product of other sections of *DNA*. In what order did chance make the interlocking parts? If not all together how did natural selection piecemeal work the Tommy Cooper trick?

You kid yourself that systematic, coded information can be naturally built by chance and/or natural law. If this is not the case then every protein is, in the sense that it is only found as a direct consequence of cellular information, 'unnatural'. And although it may service both production and regulation of production at the same time, no protein is an island. Its work is cooperative with thousands more. It is an integral part of a larger, coherent whole. As any watchmaker will tell you, purposive wholes do not work if parts are misshapen, tolerance exceeded or the inventory is imprecise. *The sum of a machine is not the sum of its parts; the equation has to include information, arrangement and the exercise of purpose. In this case these are expressed with astounding technological forethought and precision.*

Of course, in the topsy-turvy, sterile world of life-from-lifelessness a micro-puddle somewhere did the breathless trick. Plans just happened; except an accidental plan is not a plan at all. Instead of a centralised plan, overall coherence and hierarchical cascades of information scattered bits and pieces blindly came un-jumbled. 'Self-assemble' is a more dramatic phrase. Nowadays a cell 'self-assembles' according to a strict regime, a coded, automatic schedule. It starts with itself. Abiogenesis is a start *without* itself. Codeless atoms 'boot-strapped' into coded systems. Non-biological biochemicals, if they ever lay around for long enough, are supposed to have witlessly 'self-organised' into finely-tuned parts that 'knew' (by virtue of their accidental code and operating system) how to integrate with one another to achieve a necessary end. The typically *bottom-up*, trial-and-error sort of estimation is that if pieces of kit are somehow provided, things will eventually and inevitably 'build themselves'. See the problem? It is so wilfully ignored that it needs hammering home - even if those pieces by some complex mystery gradually came to repeat/ replicate themselves! *You need an instruction booklet or, at least, intelligence and a sense of purpose. Biochemicals have none.* Coded 'booklets' are the way life works. Codes are the antithesis of randomisation or of chance. '*Getting things together', anti-randomisation or order is the intrinsic nature of mind. Not matter.* Therefore a high degree of codified order must be based on mind not chance. You can't see minds. ***Infer*, therefore, a mind behind**.

How do electronic circuits or computer programs gradually 'self-aggregate'? Cellular chemicals metabolise, not at random but according to electronically precise, integrated plans for the development of preconceived and functionally integrated products. These plans are called metabolic pathways. Is it for purely physical reasons that steam hits piston in a cylinder and thereby drives a wheel? Or electrons pass round integrated circuits? Is it for purely chemical reasons that nucleic acid and protein interact and thereby drive life forward? *Such interaction minimally involves division of labour, catalysts, transcription and translation agencies, ribosomal jigs and an accumulation of other multi-molecular mechanisms all co-operant under a communications network itself subsumed within a homeostatic system geared to well-defined end-results.* **As with any machine, if all factors are not present, correct and constructed into a precise and self-consistent whole, nothing will work. Nothing works unless everything does.** Materialism imagines proto-mutants, sickly precursor cells 'hobbling around' and only 'ahfl-alvie' until some 'happening' unctuously blesses them with 'you are now a fraction more fit than before'! *Top-down* holism, on the other hand, sees sickness as deterioration of perfection, a destruction of information leading from and not towards pristine health. *No doubt the molecular biology of the cell will eventually be understood fully in terms of the laws and processes of physics and chemistry. So are steam engines. Such understanding does not, of itself, explain their origin - which was an idea.* A person who had hypothesised a thousand different ways in which an engine might improbably have 'self-invented' could reasonably be forgiven for eventually including anti-chance, idea, thought or an inventor. *If that hypothesis were unscientific, had not science lost its thinking cap*?

DNA

So, therefore, to consideration of code's bio-carrier, a chemic elegance called *DNA*. Friedrich Miescher, who discovered it in 1869, did not realise its vital potential. Now we know this chemical embodies information; bodies are its book expressed.

Information precedes its expression. *DNA* makes *RNA* makes protein. Both conceptually and in synthetic process *DNA* precedes protein so, logically, it should originate *before* the protein that it makes and all the other chemicals that, down the line, protein enzymes grouped in metabolic programs make. *In fact it is the least simple, most complex chemical of all.* By the standard of complexity it should have evolved last but stands, in a form of life's construction, first and foremost. *DNA* is central and, because of its intimate association with protein, critical to the operation of a cell. It is unknown outside living cells; even if it were thus known, of what use is a data storage mechanism whose instructions are garbled and which lies on the ground detached from any computer? On the other hand, the tiniest organism trades all the time in complex, meaningful (i.e. coded) arrangements of its own deep structure, that is, information,. A cell masterminds thousands of rapid, simultaneous reactions all coordinated towards the same goal, its own homeostatic survival. Does a chip at the heart of a microprocessor happen by chance? Is chance a causal mastermind or not? Having derived a construction by conceptually working from principle to practice in an orderly way you can then build it at-a-go. **Chance creates chicken-before-egg syndromes. Logical mind annihilates them**.

We have, if materialists, to ask how nucleic acids (in nature never found outside a cell) came to float, language intact, into composition of line 1, scene 1 of life's amazing play. *How did DNA 'choose' such uncannily apt bases, exclude all other possibilities, find enzymes to polymerise and operate itself, work out a code, co-opt one system that linked it to ribosomal protein synthesis and, within the first cell's lifetime, another that allowed it to replicate? How, furthermore, did it adopt the intelligent-looking code sequences applicable to a second alphabet of twenty 'chosen' amino acids such that this second protein-language translates effectively from the first DNA-language to create a cooperative suite of structural and functional proteins that, in further turn, form metabolic pathways to create every other necessary (but no unnecessary) biochemical?* Only with the breathtaking completion of such accidental felicity is *DNA* any more use than an incoherent floppy disc without its drive. What can an atheist irrationally claim but, as did Carl Sagan, that pointless information takes a long, long time to make its point? If ever. No more than this page and ink is *DNA* alive but in imagination let it grow upstream, against the grain, against the tide of entropy. Let nucleotides accrue in bonded order till a code (for what?) appears. Sagan's appeal to timely chance is driven to conceive just-so technicalities and numb denial of a purpose anywhere in life. Therefore let us be pragmatic. *DNA* is a data storage facility; *RNA* is an agent of data transmission, translation and, in its ribosomal form (r-*RNA*), of catalytic function; and t-*RNA* modules constitute exact translation's vital key. All kinds of nucleic acid are polymers composed of nucleotides.

A *nucleotide* is a base, ribose or deoxyribose sugar and phosphate group. These are joined, although options are available, in a particular, always-the-same way. Linkage into chains uses exclusively what is called 3-5 linkage between sugar and phosphate molecules. In pre-biotic simulation experiments predominant 2-5 is jumbled up in a 'racemic' mix with more stable 3-5 linkages. Therefore short non-biological (pre-biotic) strands suffer the usual, predictable lack of sorting and selection. Without natural guidance (i.e. information) they have neither consistent 3-5 linkage nor base sequencing. Any short, unguided 'starter' polymer, unlikely in the first place, would as likely degrade as further construct long strands. Yet even 'simple' bacterial *DNA* is millions of nucleotides long. Polynucleotide strands are required to wrap around each other in a very regular twine. This means a 100% non-racemic fit! Chemist Dr. David Watts writes: 'Difficulties in the way of pre-biotic synthesis, however, are far, far greater than those which relate to the origin of polypeptides. Our own laboratory experience in the synthesis of phosphorous polyelectrolytes, which are comparatively simple analogues of the *DNA* main chain, leads to a vivid awareness of the need for rigorous control of monomer purity and of reaction conditions. Only with a well-designed apparatus and intelligent experimental planning can one achieve successful productions of these high-weight polyelectrolytes'.

The alphabet of genetic code is in the form of nucleotides whose bases are guanine (G), cytosine (C), adenine (A), thymine (T) and, instead of thymine for *RNA*, uracil (U). Four of these have been synthesised in a laboratory using unnatural concentrations of pure (analar) components such as urea, hydrogen cyanide or cyanoacetylene. It is unlikely these chemicals ever existed naturally, at least in sufficient quantity and, if they had, competing/ interfering reactions and decomposition would have rendered pathways manipulated by chemists irrelevant. No doubt the optical properties of bases minimize damage by U-V radiation - useful when incorporated into biological *DNA* because of buffering against mutation. But if a pre-nuclear molecule got through then ambush by formaldehyde or a fluctuation in pH or heat would quickly finish off its game. Sagan-time is not on offer. It is fantasy.

Another point - anti-parallel base pairing (G with C and A with T) lends stability to the very large and complex molecule. It is also fundamental to the fine structure and efficient function of *DNA*'s doubly helical purveyance of life's code. Since bases have to fit this way was such apt configuration struck upon by Lady Luck? Is *DNA* therefore a perfectly symmetric molecule? One that is stable yet, in order to transfer information, can precisely and rapidly unzip and re-zip? The pairs seem to have been 'chosen' from other possibilities for their complementary fit. The weak bonds between each exclusive pair are, because of the correct angle between each hydrogen atom and its acceptor, optimal for this job. Such fitness is the result both of 'correct' three-dimensional configurations, the right sizes to create a smoothly curvaceous double helix and a junction with sugar that produces the right weak bond connectivity and strength. Yet if this 'perfect elegance' was absent at the start how could a cell begin to work, that is, transfer data and metabolise?

Indeed which, replicator or metabolism, issued first upon the vital stage? Perhaps small molecules 'self-organized' in networks, cycles or whatever process ('auto-catalytically' or on a mineral surface) might first give rise to metabolic pathways. Then these pathways would, of course, have to 'build' an information system's grand complexity, complexity of code and replication that henceforward was going to make them. Does this contorted 'logic' sound ridiculous. It's what a naturalistic mind-set drives you to. In case such magic could not work then try 'homo-polymers' (like carbohydrate, protein or *DNA* wherein subunits are all of one class); but it won't work in early pools because, although a cell can specify a homo-polymer, outside a cell the codeless, random forces of the world would generate, at very, very best, haphazard backbone composition that just wouldn't work - and even if it did would 'live' just once and, on decomposing without congruent code, would never reappear. We'll take another look (Perplexity and *R- or DNA*?) at the metabolism-replicator quandary whose deep dilemma cleaves the heart of evolution theory before it has begun.

To base is joined the sort of sugar molecule called ribose. Ribose provides uniquely angled space to accommodate large, complementary bases. Deoxygenated ribose lends stability (which *DNA* requires) and ribose some degree of instability (ideal for a short lifetime and ease of degradation when *RNA*'s messenger work is done). Check 'Not so Sweet'. Tests with other sugars have been made. Some cause the strands of DNA to interact too strongly, others too weakly. Others don't allow a double helix to be formed at all.

While sugar production is quite easily manipulated in a laboratory, it is excluded when a volume of water containing nitrogenous proteins or ammonia is involved. So, as a consequence, are ribose-bearing *RNA* and *DNA*. Anyway, laboratory synthesis bears no relationship to the codified way such bio-molecules are made in cells and is, in this sense, unnatural and irrelevant. So let's focus for a moment on the natural, non-accidental, codified construction of the sort of nucleotide that has adenine as its base. It is called *AMP*. Its first form in the metabolic chain is a sugar called ribose-5-phosphate (itself the product of previous biochemistry). And its biosynthesis, under strict regulation that governs both method and amount of production, involves thirteen steps, twelve enzymes and some *ATP* (which, in a second chicken-before-the-egg situation, is itself constructed from *AMP*). Intermediates rarely have any use except, as in twists made with a Rubik cube, as stages towards a target molecule. Michael Behe rightly notes on p. 151 in his book 'Darwin's Black Box' (reading which perhaps the reasonable Darwin himself would have conceded defeat):

"The problem for Darwinian evolution is this: if only the end product of a complicated biosynthetic pathway is used in the cell, how did the pathway evolve in steps?"

Indeed natural selection, during the eons they would have to wait until every new recruit was 'correct and present', would not preserve each of the many intermediates, accessories and associated code through which useful chemicals are synthesised. The reverse. A part that doesn't have a system can't yet work or know that, if an as-yet-undefined new system were to invent itself, it might then gain a useful role. Lacking any target or agenda natural selection would therefore eliminate each novelty as a non-integral irrelevance before it had a chance. Decomposition would occur. Even if the impossibility occurred (and time alone does not make possible impossibilities), the emergence of any new, improperly regulated, unbalanced pathway would more likely be lethal than helpful. Yet such evolution of thousands of intermediate 'accuracies', all converging on the target operation of a cell and each as critical as components that are necessary before an engine works, is supposed to have happened at random over millions of years. *To repeat, the longer a period of time the less likely is the survival of a complex molecule.* This problem is ubiquitous in biochemistry. *What applies to energy metabolism applies to every other aspect of cellular (and therefore multi-cellular) biochemistry. Such biochemistry basically informs all 'larger' aspects of biology.* **Yet the central question of the gradual development of coded, polynucleotide *DNA* and metabolic pathways is answered with deafening silence.**

The third component of a nucleotide is its phosphate group. This group is important in membranes, polyphosphate energetic molecules (such as *ATP* and *GTP*) as well as nucleic acid. *In other words, its integrated presence is crucial in all informative, energetic and boundary operations of a cell.* It is perfectly suited to construct a stable backbone for nucleic acid chains. Phosphates join nucleotides 'elastically' as needed by a degree of conformational flexibility; because of negative charge they interact with protein regulators (such as histones) and at the same time protect the polymer against hydrolytic attack i.e. they slow down the disintegration rate of *DNA*. Surely phosphate is easily obtained? Curiously, it is not. Check Bags of Life. In the presence of calcium and magnesium ions, found in abundance in rivers and the sea, phosphate ions are, like fatty acids, precipitated out leaving only very small amounts in solution. Indeed, no known process is able to harvest sufficient quantities of phosphorous for biological use. Today polyphosphate minerals are rare. Plausible routes to obtain them from apatite or hydrogen phosphate require complete dryness so that natural formation from rock is impossible; and artificial synthesis of phosphate compounds (e.g. *ATP*), ribose sugar and cytosine has, despite skilful and unnatural manipulation, been plagued with gunk, low yields, instabilities etc. Nonetheless many nucleotides are supposed to have joined together by a condensation (or dehydration) reaction between sugar and phosphate. Such a condensation reaction, which is also the way polypeptides, polysaccharides and lipids are formed, occurs with the expulsion of water. Thus, as in the cases we've already met, it is always enzyme-mediated. Before the necessary pre-coded enzymes 'happened' it were far more likely that in a watery environment the reverse reaction would tear their would-be polymers apart. How on primitive earth, therefore, could the miracle of *DNA* or *RNA* occur? **It is yet another chicken-before-the-egg scenario.**

Mark nucleotide components as optimal. The phosphate is well suited, in conjunction with ribose, to form a stable backbone that, retaining negative charge, protects the molecule from cleavage by water and at the same time critically interacts with histone proteins and chromatin. Chromatin is involved with differential gene expression. Its 'nucleosomes' ('reels' or 'balls' of protein around which *DNA* is spooled) embody the basic organizing factor of *DNA*. Precise, repetitive positioning of nucleosome supports along the information chain create, with linker *DNA*, a bead-like structure that allows, as functionally required, for regulated coiling (for efficient storage) and uncoiling (for code transcription). Such 'balls' bind preferentially to specific nucleotide sequences that, placed at exact locations, effectively create a histone-binding super-code that helps control *DNA*'s 'opening and closing' operations.

Research into alternative forms of DNA using non-standard bases, various sugars and polymer linkages suggests they are less, mostly much less, suitable for the computer-like information management task that DNA itself so exquisitely performs. **It suggests that biological nucleic acids are, in their trade, the smoothest possible operators.** Another accidental master-stroke! A paean of praise is, then, due to chance that 'selected' and 'grouped' what are perhaps the only two pairs of optimally complementary bases in the universe, along with other specific components and the necessary linkage mechanisms, in quantity and at the same time in the same micro-location! Hallowed, O Fortuna, be thy scientific name - evolution.

Supreme Elegance

Do you remember (Chapter 6) that paper and ink constitute passive, loaded information? Ranked at the lowest, physical level of an information hierarchy, they simply carry message and meaning. Do you also remember (Chapter 10) particular 'shapes of ink' - not letters of the alphabet but sub-atomic particles in whose 'language' the physical sense of an automated world is written? *DNA is a form of paper and ink that transmits a biological message.* The paper, often called a 'backbone', is composed of a chain of phosphate and deoxyribose sugar molecules; and the 'shapes of ink' are, in this case, the abovementioned four bases (G, C, A, T). These carry, like letters of an alphabet, the sense of the biological world. They are *not* the message but its cast of type; they are just book-like purveyors of life's rationale. The actual code, cracked by Nirenberg, Khorana, Holley and others, involves 4 letters (2 complementary pairs) and 64 3-letter words (called codons) with 21 'meanings' (20 amino acids and a 'stop'). This is the entire alphabet and dictionary for the whole diversity of life on earth and probably, if it were found to exist, beyond. The sequential order of these words commands the way protein is built.

It is, in pertinent anticipation of the next section, at this point worth asking how translation's *code* evolved. 64 codons are translated by t-*RNA* molecules each bearing nuclear code at one end correlated with a specific amino acid at the other. Completely pre-agreed, accurate translation is essential for apt protein synthesis to occur. Therefore how, in spite of causing immediate, catastrophic damage to protein sequence, shape, function and thence metabolism, did the *t-RNA* components responsible gradually evolve and thus compose a code? How, particularly when considering mutant start and stop codes, might one conceive of *any* 'alternative language' gradually, randomly and non-lethally transmogrifying into fresh, non-canonical meanings? **In fact, since no answer whatsoever is on offer, this 'code-generating riddle' constitutes another Darwinian black box.**

Could there indeed have been built a greater conceptual, structural and functional efficiency in terms of information storage and transmission than the genetic code and its 'alphabetic' bearer, *DNA*? Is the quality of code optimal or can you find a more economical logic-bearer? Is there a more powerful agent of potential anywhere? Life seems healthy using only 20 coded amino acids; less would create problems but more seem superfluous. Leave the number of amino acids and, therefore, 'meanings' of genetic words the same. In this case 64 words standing for 22 'meanings' (including start and stop) means that some of the 22 are represented by more than one word. Such apparent over-representation is dubbed 'redundancy'. Can such 'redundancy' be improved upon?

With respect to letters of the alphabet try only 2 (1 complementary base pair) to write life's book. You could code all its 'meanings' using 5-letter words; you would need 2^5 (32) of them. The redundancy of such a code would be limited. Is it possible, however, that 'redundancy' represents not improvement but refinement? That it permits an extra layer of information, a further subtlety implicit in the code? Biological *DNA* may incorporate it to buffer the effects of printing error and mutation but possibly, more powerfully, to generate subtleties of language that relate to gene regulation. As 'female' and 'woman' are words that relate slightly differently to the same entity, so may TTT and TTC be nuances on the amino acid 'lysine', for example, a subtle command to leave or delete this lysine in different versions of the same protein. To achieve such flexibility in a 2-base system would require cumbrous 6-letter words; thus materials and energy for synthetic processes would have to be at least doubled while the process itself became more complex, error-prone and, at best, only half as fast.

You might query not only the efficiency but also the logic of a system that assigns six 'nuances' to some 'meanings' and only one or two to others. It is reasonable to assume that apparent anomalies and irrationalities will, as in the buffering 'backup' capacity of so- called 'redundancy' itself, be found to conform with biological nature's extreme teleological efficiency and turn out to be an integral part of economical plan. In fact, the commonest kind of change that could affect the code's accurate translation into protein is called a substitution mutation; 64 codons represent 20 amino acids and so there are repeats; and examination (Molecular Biology and Evolution 17 ps. 511-518) reveals that these repeats are systematically constructed to minimize the impact of error. Change may well result in the same or a similarly functional amino acid making the replacement. Computer studies have suggested that, of all possible genetic codes, your own is close to optimal in its capacity to minimize error. This suggestion might be interpreted as strongly as evidence of design as of coincidence.

No base pairs that fit as elegantly into the system as nature's couple have, despite plenty of research, been found. It is nevertheless possible to speculate that an extra, microscopic pair or two could be summoned from the vastness of space. Is it, on balance, more or less efficient to use 3, 4 or 5 base pairs than 2? 6 letters (3 pairs) could accommodate 21 'meanings' within 36 possible 2-letter words but with low mistake-buffering or flexibility from 'redundancy'. On the other hand 6^3 (216) 3-letter words would need 216 different t-*RNA* translator molecules. Not only does this mean extra coding but, if any were missing, the un-translated words would amount to blanks in a protein text. Where blanks mean full stops the structure and thereby functional intention of the protein would be split up and destroyed. This sort of mutation would inevitably and always be catastrophic. The more base pairs you add the more the outcome deteriorates. Speculation indicates that the system in use is as efficient and elegant as conceptually possible. There is nothing to indicate things were ever otherwise.

Such code is often judged as optimal. It is supremely elegant. And it is often argued, *bottom-up*, that the universal usage of nucleic acids, especially *DNA*, as a bio-language is a 'sure indicator' that all forms of life are related to a single primary program; bodies only informed by one chemical protocol have evolved from an ancestral cell. *Top-down*, this inference from the facts is readily disputed. Once you have developed a viable language, an efficient transmitter of code such as English, do you not write all your books in it? Why chop and change? Why complicate your alphabetic type supply? A single genetic language for earth's bio-program marks both an intelligent decision and a logical imperative. **And if code is universal then, since changes would be lethal to its sense, it must have been there from the very start.**

Code correct and present from the start is not a problem for a Theory of Intelligence. For a Theory of No Intelligence, such as contemporary intelligentsia prefer, it is. How could such a code *evolve* by steps when any change to its codon assignments would lead to changes in amino acids and, thereby, to the shape and functionality of protein that the change induced: or would cause stop and start commands to occur in the wrong places? Such quantity of defective protein would immediately be lethal - which is why Francis Crick argued that the code could not undergo evolution. It is why, in all its elements, the universal code sheds any rival and alone persists. **The code must have sprung whole in the ancestral cell!**

If this were so, the question arises how such excellence had already evolved by the time of this primitive proto-cell (called *LUCA*), a hypothetical common ancestor whose standard all subsequent life has inherited. The conceptual (not accidental) target of genetic information is the product of its translation; its *meaning* is life's basic, specific functional components called proteins and, lately described, *RNA* operators. Such clear purpose is linked to extreme elegance, economy and efficiency in the line of duty. It is worth noting that the discovery of extra-terrestrial life with a genetic code like ours would clinch a 'theory of intelligent design'. It would rise to the podium and, not by accident, grasp the same prize-winner's cup with us.

Supreme Density of Data Storage

If the code is conceptually optimal is it practically so? Is its purveyor, *DNA*, the fittest possible mechanism for the storage and transmission of complex information? It may predictably, according to the design principles of Natural Dialectic, transpire that terrestrial genetic code is the most efficient for protein synthesis and that its optimal coding chemical is *DNA*. Perhaps a more perfect material for the three-dimensional storage and transmission of biological data is impossible. Certainly nanotechnologists, who engineer molecular, atomic and subatomic systems as compact and efficient as they know how, know they are streets behind natural bio-nanotechnology; yet already they have developed nucleic acid hard drives whose storage capacity is 2.2 petabytes (million gigabytes) oer gram! Thus, where *DNA* will write the future of computing, a quantitative description of its information density beggars belief. Calculated at 10^{21} bits per cm^3, biological *DNA* embodies by far the highest known. A high-tech microchip might store 10^9 bits per cm^3 - which means *DNA* carriage is about a million, million times more efficient. All the information stacked in the libraries of the world is estimated at about 10^{18} bits. If it was placed on superchips they would, piled up, stretch from earth to past the moon. Registered in *DNA* it would cover, like the *DNA* that informs every different human in the world, about five pinheads! The *DNA* of every kind of organism that has ever lived on earth would need a fraction of this space! Yet the capacity of *DNA* for compaction is such that your super-coiled chromosomes, an encyclopaedic genome that contains all biochemical information needed for your construction, are routinely packed to occupy a tiny fraction of a cell's volume. Most cells are much smaller than this full stop. **There is no entity in all of physic's known universe that deals in high-grade information as effectively as *DNA* and its co-ordinated systems - not forgetting its controlling super-codes** (see Chapter 23). Yet, in order to support and satisfy an intellectual committal, atheism must assign its origins to chance! Pure chance, since how can natural selection carve the shape of molecules or codified metabolism? *Is this not the height of self-delusion in the cause of creed*? Can it be shown, even theoretically, how chaotic 'pre-code' might have built replicating 'pre-organisms' in a sequence of steps of which we have neither draft plans nor absolutely any evidence? It needs to be because such a highly miniaturised, nanotechnologically 'perfect' combination of grammatical code, correctly spelt-out (read 'sequenced') chemical alphabet and an operating system must have preceded the first organism. With such a precise flow-chart missing (as it is at present) any assertion of abiogenesis is rendered vague and baseless.

The *DNA* storage principle, with base pairs weakly linked as steps on a helical ladder, is dynamic. Precisely identified segments of information can be accurately transferred to cytoplasm, membrane or even other cells by complex read-edit-print mechanisms. A human body cell contains about 10^{10} bits of genetically coded information. This, translated into the written word, amounts to about 1000 books of 500 pages each. Even the humble *E. coli* bacterium, weighing 10^{-13} gram, contains nearly 10^7 bits of information all able to be rapidly manipulated. Its cell division takes twenty minutes and its information is completely and accurately copied, at a rate of about 30000 base pairs per second, in just a fraction of that time. *There is no indication that it was not always so.*

This is a mechanical account of how the genetic presses roll. It takes no account of the extra meaningful, semantic dimension of such information as is printed. **If you look to molecules you are looking at the wrong level of information for a full answer.**

Supreme Operation

DNA performs two critical and, in the sense that when it's doing one it's not doing the other, mutually exclusive operations. *They are protein synthesis and replication.* The function of protein synthesis is maintenance and preparation; that of replication duplication of information prior to cell division. Both are complex involving many co-operators and stages to completion. Every cell must perform them correctly.

Protein synthesis involves transcription and translation into protein using molecular industrial units, some, such as ribosomes, composed of many complex subunits. At the transcription end of operations (we'll deal with ribosomes at the translation end shortly) various co-components - large enzymes and complexes such as helicase, polymerases, ligase and so on - constitute a minimal collection.

This molecular printing press includes many specific, interactive kinds of nucleic acid and protein. It involves many functions, mechanisms and their switches in a truly remarkable manner. How, he asked, could any such system or sub-system work at all until complete; or enzymes be made without pre-existing ones to make them? Other *RNA* and protein molecules that service the *DNA* database are also, using systems at work within the factory, synthesised according to the cell's direction. The whole business, without which proteins cannot be made so life can start to roll, is intricately coded for. *Codes and codification represent the opposite of chance; so does any creation devolved from prior coded instructions.* Randomness in any code sequence destroys the code; and in any process of construction destroys its meaning, that is, its intended end-product. No *chemical* tendency to enforce non-random yet meaningless 'clumps' of nucleotides is known; if it exists it is certainly overridden or absent in the meaningful, software-driven non-randomness of biological *DNA*. *It is irrational to suggest that randomness could have spontaneously generated code sequences as super-specific as those of genetic code at the same moment as its integrated, self-consistent surrounding hard- and software operations.* This, however, is a basic claim of evolutionary science!

The heart of any purposive, decisive operation is program, command and control - an information centre. Let's take a closer look at the *DNA* database. Let's start with a 'simple' bacterium. Carl Sagan (New Encycl. Britannica 15th. ed. Vol. 22 p. 987) estimates the information - not only genetic - in a single cell at 10^{12} bits or 100 million pages of an Encyclopaedia Britannica. Thousands of genes reside in its circular databank. How does the system know what genes to switch on, to switch off, to forbid or generally orchestrate in the manner of a micro-hierarchy? How does the system respond in a balanced way (using homeostasis) to the demands placed on it? And how did the two *DNA*-to-*RNA*-to-protein conceptual categories called transcription and translation 'pop up'? With their associated machinery such as ribosomal 'tape recorders' and an efficient coding system incorporating exactly the right sixty-four translators to turn one chemical language into another, unrelated one? And so transform words into the very objects they represent! Could they evolve gradually, like a headless Microsoft, by chance? Or not (see Chapter 23)?

It is a question of *minimal functionality*. It transpires that specific forms of nucleic acid constitute an optimal information storage medium while three-dimensional protein is the optimal vehicle of function (metabolism) and structure. During protein synthesis transmission of information between them is mediated, minimally, by a multi-step procedure. Many factors, including primers and multiplex polymerase writer, proof-reader and repair faculties, engage in building either *RNA* copies of genes and regulators; small, exact sections of chromosomes are involved. In replication, at the time of cell division, the whole nuclear genome is copied at rates which approach 1000 nucleotides (letters of the genetic alphabet) per second. Firstly, in both 'simple' prokaryotes and eukaryotes, a large machine consisting of fifteen or more different proteins and called a primosome 'primes' the *DNA* at the point where copying/ replication will begin. Only then can a replisome, including large polymerase subunits each in turn made of multiple subunits, 'mount' both strands and begin work. Transcription is executed using a bidirectional at the site of replication called a fork. No organism lacks a protein motor, helicase, that cleaves the waves of double helix as a ship cleaves the surface of the sea; gyrase follows up to ease the strain. On the 'leading' strand a continuous script is produced and on the other 'lagging' strand discontinuous fragments that are then glued together (using another protein, this time ligase). After this yet another protein removes the primers and proof-reading services ensure correct copy. In eukaryotes several more steps (such as cutting/ splicing and topping/ tailing) must be correctly carried out. Flawed molecules are detected and destroyed.

At this point note that *RNA*, a key player in the synthetic process, is eminently fit for its work. M-*RNA* is a relatively unstable molecule but, since its work of carriage will soon be over, such instability as aids rapid degradation is required. R-*RNA* and t-*RNA* are, on the other hand, critical cogs in the longer term machinery; and their construction is more stable. Of course *DNA*, that needs remain intact through not only a single lifetime but by inheritance all the generations of a species, has to be ultra-stable. Each type of nucleic acid well fulfils its role.

Messenger *RNA* passes from the nuclear site to the heart of coding operations between the two main players. Although the pores it passes through are large enough to cope with macromolecules such as ribosome subunits and m-*RNA* they are not gaping holes. As usual with increased resolution, molecular biology comes to focus upon miniature technology of a kind that would have astounded Darwin. In this case the pores are mechanised gates, about 2000 per nucleus and each built of over 450 proteins of 30 kinds that interlock together in a precisely rational way. The reason is strict control over the entry and exit of heavy traffic; a gate facilitates wizard traffic-control at a rate, where necessary, of over 1000 translocations per second! Import and export of cargo marked with postal code is mediated by special proteins. Such sequentially, morphologically and functionally is highly informed. Its accurate business is critical for the sustenance of all eukaryotes but has nothing whatsoever to do with chance at origin/ conception or any other time.

Having passed a gate the m-*RNA* reaches a conveniently positioned translating machine called a ribosome. Ribosomes, central to protein production, have been the subject of intense study. They comprise a massive complex of larger and smaller subunits. These together involve three or four large *RNA* strands that act as an *RNA* reading tunnel and as scaffolds organizing between 50 and 80 proteins; that is about 350,000 atoms in precise order. *No ribosome, no life.* In bacteria thousands float free in the cytoplasm. In the case of non-bacterial organisms, each of whose cells may contain up to half a million, many are fixed on structures specially made for the job; they are strung like beads on nets (called endoplasmic reticulum) surrounding the portholes of a nucleus and, as an m-*RNA* molecule drifts through, it correctly latches onto one. Amino acids are joined at the rate of about four every second. Sometimes a hundred or more ribosomes are situated in a group called a polysome. The *RNA* tape can be pulled through the whole line of them all at once so that as many proteins as there are ribosomes can be synthesised efficiently, simultaneously. Then, as efficiently as they were assembled, ribosomes can disassemble as required. How came this dynamic translator, essential at life's very start, from a primeval scum? Along, of course, with all the other apparatus need for an operating system. Supreme coincidence, a miracle, gave rise to these with all the other apparatus needed instantaneously integrated for 'self-replicating' molecules to pull their vital trick.

Messengers attached to a ribosome are now translated into protein. Major operators here are three-dimensionally precise t-*RNA* molecules whose job is to match 'words' (called codons) with their specific 'meanings' (particular, prescribed amino acids). Just as *DNA* bases are almost chemically neutral with respect to each other so no chemical affinity favours any particular correspondence between nucleic acid and protein. However, specific *RNA*-codon and amino acid receptor sites, isolated well apart on their t-*RNA* carrier, match the physical expression of word with its meaning. The correlation is mediated by unique aminoacyl-t-*RNA* synthetase tools (or enzymes) needed to attach the correct amino acids to their corresponding t-*RNA* translator molecules. **Such complete lack of affinity precludes biochemical predestination so how do you explain not operation but origination of genetic code?** In other words, where and how did such un-predestined, unpredictable yet 'intelligent' code first evolve? Corporeal life hinges on the dialogue between them. **No theory has been proposed for the evolution of the critical, immediate and highly serendipitous linkage of protein with nucleic acid through the medium of such a suite of requisite, complexly-formed and yet precise t-*RNA* translators and synthetase tools.** Without this series intact nothing could start. So it goes on. *At least 100 different proteins, each coded for and synthesised by the very machine of which they are components, are used to convert DNA instruction into protein product-hardware*.

During translation proof-reading and editing, that is, quality control procedures, are applied by aminoacyl-t-*RNA* synthetase enzymes to check the right amino acid has been joined. They can discriminate between similar amino acids (such as isoleucine and valine) and delete mistakes. Loaded t-*RNA* is escorted to the ribosome translation site. The escort, a protein, can double-check the loading is correct. Obviously, this is critical because wrongly loaded amino acids would cause mutations. Mutation rate is thus reduced, in a system that has to balance rapid production with accuracy, to a minimum viable for life. And after translation inspection and control procedures ensure that proteins are properly stabilised (using disulphide bridges), folded (using chaperones) and assembled into complex machinery (such as ribosomes themselves). Goodness me! **Quality controls imply process towards a target**. They imply and end-goal, purpose, foresight and anticipation. What foresight nature therefore has; or, if it hasn't, how could such a high-fidelity, efficient system build itself by non-anticipatory natural selection? Immediately - because a cell needs all its operations up-and-running even in a skeletal and proto-form. It is not, therefore, too much to enquire how the translation mechanism could evolve by gradual, haphazard steps. What systems flow-chart can you build to start to actually get to grips with 'fact' and simulate blind building of its essential capability? Provide this flowchart, please! It might demonstrate how well protein synthesis has been designed.

In a final, extra twist of the synthetic tale even shut-down of a given gene's production is carefully controlled. The gene itself has, by some form of feedback, to be switched off; and, as already noted, m-*RNA* molecules have limited stability so that, by breaking down, their levels can be regulated. Furthermore, m-*RNA* messengers corresponding to the proteins of a particular metabolic pathway have very similar rates of decay; those specifying proteins for key activities have a slow rate; and those specifying transient duties break down rapidly. Unregulated protein levels would shortly disrupt a cell's operations leading, by natural selection against chaotic lack of control, to death. It's therefore fair enough to speculate that 'regulation all the way' is critical, complex and by refined design. Indeed, of the whole system the late Professor Malcolm Dixon, one of the founding fathers of enzymology (the study of enzymes) at Cambridge University, wrote, "*The ribosomal system under which we include DNA and the necessary co-factors, provides a mechanism ... for its own reproduction but not for its initial formation.*"

Full genome replication, employing replisomes, is no less a complex chip. A department that is half of life's printing factory it lacks none of the sophisticated components used by the department of protein synthesis. If only ultra-stupid chance could focus for a moment it would to enter into nightmares of confusion. How can mindlessness know order? How can the dimmest wit create what all the brightest sparks alive have only just now partially revealed? Why, from a comatose, chemical point of view, should such a multi-component operation ever have to happen? Reproduction's all about anticipation. It's about survival of the species. Why should nucleic acid or its protein stringers care a hang for that?!

Prior to cell division the whole genome is copied. The cell's databank is doubled. In this case of cell division and survival for the future a three-stage process of initiation (or priming), replication and termination, involving many tools, works in a way not dissimilar from copying to *RNA*. Action at the replication fork is also bi-directional. *DNA* is transcribed to *DNA* in a semi-conservative way, that is, the newly replicated strand entwines with its template to form a double helix. In *bacteria* replication is practically continuous and starts at a specific place called an 'origin'. Thousands of genes reside in the circular databank of a so-called 'simple' prokaryote and their duplication is about as complex as what occurs in other organisms (including you and me) with linear chromosomes. You can no more afford to lose information from life's book than any other. Therefore the system uses proof-readers, editors and repairers as well as separators, linkers, scribes, translators, controllers and other personnel (in the form of protein) to ensure its accurate and conservative, information-retaining reproduction. In favourable conditions extremely rapid printers mean division can occur every thirty minutes or so. Two replication forks travel at identical speeds in opposite directions from a predefined site of origin around the 7000 or so bacterial genes until they meet. Thousands of nucleotides are copied per second. Next one of the rings is peeled off and correctly attached to a new site on the cell membrane. This leads to a form of cell division called fission. In *eukaryotes* genome replication is timed in cycles and starts, for the sake of saving time, at many points along each chromosome - a process called bubbling. The coordinated regime involves many proteins including complex polymerase machines built of numerous protein subunits bound to each other with the correct geometrical configurations at correct strength and angles. Not least of operations very hard to ascribe to chance is, as described above, the bidirectional motion of polymerase transcription - on one strand producing continuous script and on the other discontinuous fragments that are then glued together (using a protein called ligase). Then of course, eukaryotes must engage mitosis or, in sexual mode, meiosis (Chapter 24).

No thread of DNA could function without myriad coordinates to operate upon it. Indeed, such molecular machines perform the function of an operating system in a computer. Geared to perform co-efficiently they copy, splice, join, mend, regulate expression and, at micro-level, are the basic reason forms of life are able to exist. **Yet, chicken-and-egg again, the information for this crucial protein machinery is carried on the very *DNA* that it manipulates. So which came first, print factory or coding for it**?

Both protein synthesis and replication are full-blown biochemical systems involving a plenitude of precise, encoded, interlinking parts. They are far too elaborate to have arisen more than once. **So there you have it - once by gradual, aimless evolution or once by targeted, conceptual design. Which, philosophy aside, looks the more rational inference drawn from the scientific facts?**

Supreme Flexibility

How came all this supremacy by chance? Do computers, though obeying laws of electronics and electromagnetism, 'self-organise' because of them? Researchers long ago despaired of chance alone to start life up; they re-pinned hopes upon the notion of Monod's 'necessity' - natural law somehow inevitably leads to life. 'Chemical predestination', using bonds, must forge the fabric of a code and then accumulate the immaterial, non-Shannon information that eventually makes mice and men.

Both *DNA* and protein are constructed so they best fulfil their working roles. Do you want to start with protein? Between various amino acids there exist affinities - but too few and weak to account for the extensive permutations and irregularity of sequence that, as a consequence of coding, proteins incorporate.

In principle amino acids can join up in any permutation. Just as an author needs flexibility of choice of words within his grammar, so the genetic code expresses necessary flexibility.

Why not, therefore enrol nucleic acid as progenitor; and try to extract some biochemical predestination out of *DNA*? In other words, could *DNA*, life's information-storage molecule, organise itself according to the properties and chemical affinities of its components? In fact, there's nothing preferential in the way its nucleotides are ester-bonded to compose the 'backbone' of a strand; nor in the way that different bases bind to this backbone. *Free as letters of an alphabet, nucleotides can join in any order*. It must be so or else, locked into blocks, you'd limit possibilities. No doubt base-pair weak-bonding, strand across to strand of double helix, occurs for the best of reasons - copying capacity; but this does not affect the sequence in which bases are arranged along a strand; and such succession, like that of letters in a book, determines what the genes spell out. *There's no preferential affinity along this lengthwise axis between any pair of bases. No chemical predestination is inferred; no repetitious 'clumping' or distortion from free order can occur.* Thus, as with any intellectual creativity, the choice of words is absolutely free. Such freedom has two outcomes. On the one hand, it vastly increases the theoretical probability of biologically useless sequences strung, like 'words' made of jumbled letters without any meaning, randomly together. On the other hand, complex, improbable arrangements may, like organised words and sentences, carry information that is accurately, expertly specified. You'd expect no less. High-grade, functional maxima can't be restricted. Affinities, by forcing repetitious orders, would reduce complexity, dramatically impair creative freedom and, in the jargon, seriously reduce (not, as needed for development, increase) the information content of a text, that is, an organism. They would cripple innovation so that, if you're thinking they might help self-organising evolution, they would actually rule it out.

The incapacity of nature to blindly generate any code, let alone the code of life, is critical to any explanation of biological origins. It bears rephrasing. Neutral chemistry, like plastic letters of the alphabet tossed on a kindergarten sheet, can only wholly randomly 'compose'; and everyone agrees that chance alone can't specify the book that's any body's order of biology. The unpredictability you find in *DNA*, as in the world of art, is of meaningful and specified intent - to develop plots and discharge schemes of work. Non-affinities in *DNA* bestow the freedom of unfettered choice that you see in the multi-coloured range of life. If, conversely, great or small affinities between life's letters (nucleotides) determined sequence then those letters would lack freedom (which they need) to create whatever order served their purpose. The informative capacity of *DNA* would be destroyed; its reasons could not be expressed. In short and fact, base freedom 'transcends' any determination by physics or chemistry. No chemical predestination but a requirement of neutrality and freedom to dispose life's texts in any way is satisfied. Thus 'self-organising' biochemistry, like alphabetic ciphers magically shaken into words, sentences and scripts, could not (except in dreams) occur alone. You'd need lots of information, an injection of specific order by intelligence to pump arrangement of the letters 'uphill', against the grain of entropy, into living books.

Thus there's no affinity but great affinity. Supremely flexible potential expressed by an excellent, fixed grammar; language capable of different meanings. There's no affinity between nucleotides; or between amino acids; or either for the other. But there's informative affinity between the carriers of signal. The correct amino acid binds with its t-*RNA* translator. A message-bearer binds exclusively with its receptor. And there are regulatory affinities galore as *RNA* is matched with *DNA*, as proteins on a regulatory region lock or release expression of a gene and as the genes which need to be expressed are recognised. Binding affinities of electrochemical precision are programmed. They are closely written as a product of life's script.

Scripts need authors. Nor can any machine, let alone a 'mind machine' that deals in information, exist without its egg (an inventor) and its chicken (its own parts clicked into a gloriously clucking whole). The conceptualising egg is metaphysical; the conceived, informed chicken is physical. Chickens-before-eggs are circular arguments and machines crack this ring. Neither egg nor chicken is, physically, prior; together the inventor and his invention form integral, inseparable parts of the same, purposive process. Neither does invention have to stay stuck at '*Eureka*!' or mind keep re-inventing the wheel. Plan, program, blueprint written out in code - mind simply stores the whole concept in memory with perhaps, as a physical reference, some sort of written code. We have already met one circular argument impenetrable to egg-heads who deny eggs-ahead. *Do you remember that any attempt to form a model or theory of the evolution of the genetic code is futile because that code is without function unless and until it is translated i.e. unless it leads to the synthesis of requisite proteins*? **This *impasse* is includes another. Even coded *DNA* is insufficient. *DNA* itself is just a string but its information's loaded as a multi-level library of code on code.**

As with a book you need to be able to access its meaningful information on demand. For this it needs hardware (the cell environment) and software (in the form of operating and applications code). It needs an operating system and any operating system needs, crucially, a complementary index. An index is, of course, an information-rich system of identification, signals, address, linkage/ relationship and specific retrieval - none of them the specialities of chance. Each form of code interlocks and supports the other. We turn full cycle back to amino acids and protein because, in intimate conjunction, these work to regulate the precise release of information from its store. **By itself this database is rigidly inert. It cannot operate without *RNA* or protein regulators which are themselves a product of the code.** *The simultaneous, spontaneous appearance of coding hierarchies (including control micro-hierarchies even within the DNA mechanism itself), which function together as one whole system, certainly reduces the probability of abiogenesis! Without full cooperation the system is meaningless, the information rendered as void as scraps of different books thrown together or parts jumbled in a knacker's yard.*

A genome is a dictionary of protein. It is, if you like, an encyclopaedia or a book of life whose protein-specifying sections are called genes. This genetic language expressly and directly translates to a second, proteinaceous one. Genetic words (called codons) translate into letters of the protein alphabet, amino acids. Sequences of these could make up nonsense 'words' (e.g. bkoo) but in a cell they never do. The nuclear dictionary dictates what's meaningful (say, book). So the *meaning* of amino acid sequence is a word-like protein whose rationale is obviously how it works. As words when integrated properly compose a sentence, so proteins compose functional complexes that have been informed, at root, by a coded scheme that 'rides' on *DNA*. Thus a general dictionary can, expertly computerised for access, build an individual, working model of its message; it can build, from life's nucleic acid book, the story of a body like your own. What is more, once granted body, as lead actor you inevitably write a further masterpiece, the drama of your three score years and ten. Thus objective language yields an instrument by means of which is harvested subjective meaning - the experience of earthly life.

Language is a triumph of the mind; comprehensive system is the gift of intellect and program the antithesis of chance. How, then, are the two semantic vehicles linked in their informative affinity? Besides protein synthesis (*fig.* 24.7) how is the gene to make a wanted protein found, operated and controlled? Apart from the idea behind it (which 'library' and 'databank' well express), its exquisitely apt construction and conservative stability, *DNA* has another couple of tricks up its coiled sleeve. In fact the first one, a major groove, is *in* the sleeve. This groove is of a shape and dimension that allows a very common conformation found in proteins, the alpha helix, to engage in a way that gives intimate access from outside into the veiled, coiled private sanctum of genetic script. Such access facilitates the second, economical trick. This is to double up its use of nucleotides to work not only for protein specification but also gene regulation through the agency of transcription factors that carry such *DNA*-binding motifs as 'zinc fingers'. A protein may carry multiple 'fingers' allowing sophisticated access to code. In the search process each nucleotide presents its own electrical pattern to the groove so that the penetrating alpha helix is able, like reading braille, to 'feel' and recognise about four adjacent bases. These, according to type and sequence, have their own signature 'chimes'. Since four bases can be arranged in 256 ways, could there be 256 'dial-a-gene' tones? In fact there are far more genes than this in, say, a human genome but, if you were to employ a sequence about sixteen bases long you might compile a directory of unique 'numbers' with which to call any particular gene. Stable yet reversible binding of a protein to a *DNA* sequence needs about sixteen weak bonds, the number that occur when contact is made with a sequence about sixteen bases long. Does such contact occur? Yes. Proteins are micro-machines. Two or more alpha helices, correctly positioned, are designed for a combined operation, on opposite sides of the *DNA*, to identify target sequences. They are probably helped in this exercise by small, informative distortions and other conformational 'imperfections' that snake along the length of the double helix. Less than sixteen weak bonds tend to inhibit identification, more lock the detective apparatus irreversibly onto its quarry. The reversible binding strength is in fact about right; probably allosteric feedback from the product of the search (another protein or other biochemical) causes the detective dimer to detach and so relock the previously active gene. It is always the same - the more subtle the message and more highly sophisticated the code, the less the chance of chance.

The organised, differential expression of genes by protein regulators built themselves from genes is key work. The identification kit issued to index, identify target sequences, unlock, express and then relock the right genes to make the right proteins in the right quantity at the right time is phenomenal. Your own genomic library/ databank of perhaps six billion bases (bytes) is accessed with fully informed precision in a split second. This second, millions of times over throughout the body, amazingly small machinery is rapidly and unerringly converting one-dimensional concept into three-dimensional reality. Together *DNA* and protein are awesomely, algorithmically, conceptually manipulating atoms and subatomic particles to build and sustain all life on earth. The couple with their multiplex relationship embodies biological reason. What more rational their origin than reason's reason, great intelligence; what less so than chance?

Code Rules Supreme

Nature has invented, from the start, a genetic code and bearers, the nucleic acids, that are supremely well constructed and, at the same time, reign supreme. **Biology depends entirely for its life upon the information carried in its nuclear books.**

Who could disagree, however, with the fact that molecules are mindless, non-conscious fragments of inanimate matter? A molecule has no 'sense of survival'. What goes for molecules must, of course, go for cells made of them. Just as it goes for the tapes on which a song is recorded it goes for coded *DNA*. What goes for a cell must go for multi-cellular collaborations. Look at bacteria, fungi and flowers. They are passively informed. No senseless biological body 'knows' survival, actively 'wants' to survive or makes 'improvements to the quality' of its 'life' - except mindful ones. And sub-consciousness came lately so it doesn't count; even less so later, clever consciousness. Evolution by the power of will could not have existed prior to minds, is dubbed Lamarckian and, very likely correctly, expunged from the canon. The radical solution has to be that 'chance happened to create its own order'. It must be that, while constrained by nature's forces, bio-systems self-assemble in an accidental way. They are, essentially, as meaningless as non-intelligence. They are based on the unreason of oblivion; and any breakdown of their working equals death.

The main thrust of atheistic argument against Big G whittles down, at the sharp end, to random chemical constructions never naturally seen and randomly mutated innovations (not just variations on an already innovated theme), also never naturally seen. What happens 'randomly' is, however, by definition and forever unrepeatable. It is not scientifically able to be proved; nor, experimentally, can it be falsified. So the thrust is blunt and brittle. Believe it if you need. It comes down to one tale of life's origins against the other. Which is more reasonable? Surely the irrational, illogical version is more plausible? That's exactly what you would expect that topsy-turvy thinking thought. Who in modern academe dares disagree?! Although it goes against the grain you have to be clever enough to abandon the idea of reason. You have to learn to trash meaning - any fundamental meaning or purpose for humanity, life on earth or, at the cutting edge, yourself. Take, for example, a jumble of letters from a child's toy alphabet that fell, by chance, into the word 'C-H-A-N-C-E'. This, if you spoke English, would mean something. It might even happen that a few more letters aggregated into a small sentence which their medium, unlike biological water, did not churn apart. What meaning have they except to the beholder? Why should aggregation *per se* represent meaning or purpose? Meaning needs translation, one of the most purposive (or teleological) processes known. Translation needs a preconceived code. So an evolutionist cannot speak of 'code' in a rational sense. He means 'code' in, with respect to its origin, an irrational sense. Randomness generates code. An alphabetical letter is a symbol, a nucleotide a 'symbol' of genetic 'code'. And pre-biotic roulette might, with enough spins, generate a starter 'meaning', a seed 'reason' for survival.

Evolutionists believe you can obtain non-random results from random variation by selection. You can. For example, by accepting or rejecting randomly-generated letters of the alphabet I can 'evolve' a message; but this, because the message itself is preconceived, is not Darwinian. It involves prior information, concept and the passage through useless stages to a working conclusion. Random evolution lacks all three. Its pointless nonsense isn't going anywhere.

To illustrate this point take soviet Oparin. Proved wrong by sky he engaged a second pre-biotic oxymoron. Animate inanimation in a pool! But such self-contradiction boxed him in. How might random fluctuations and some chance reactions in a volume of salt water generate informed sub-systems that allow cell replication to occur? Discoveries of very complex, algorithmic biochemistry based on digital genetic coding have stranded chance as too remote by far. Yet natural selection presupposes such complexity. If Oparin invoked the power of natural selection by assuming 'primitive metabolism' already operating in a 'membranous coacervate', he begged the question he was trying to answer - how and where informed self-replication just occurred. How, in any case, can mindless molecules 'compete', that is, somehow select a 'useful' code, combination or configuration in advance of any purpose for its integrated use? They'd be 'selected out', destroyed by forces of the elements well before the gist of any plan-less plan emerged. And, anyway, such 'plan-less plan' is only a conviction in your head.

As a second illustration take, although his hypothesis is now redundant, Manfred Eigen's 'hyper-cycles'. This idea still serves to illustrate the train of logic underpinning abiogenesis as, at any intellectual cost, the only - but quite possibly incorrect - solution acceptable to modern science. His hypothesis *assumes* a primeval 'soup' rich with huge quantities of protein, *DNA* and *RNA* parts. Thus assumption again provides the answer; Eigen begs the question. He then postulates hyper-cycles by which 'successive leaps of self-organisation' select, by some sort of Maxwell's demon, the 'fittest molecular assemblies'. This is, as already noted, biologically meaningless because the only generally accepted organising principle in biology - natural selection - cannot operate at a level below the cell. Eigen's pre-cellular 'fitness' is chemical. Neither chemical nor biochemical has any sense of 'survival' so that, fitness-wise, there is no distinction between one '*quasi-species*' (as Eigen hopefully prejudiced his molecular assemblies) and another - except that the *simpler* the molecule the less likely it is to be degraded. Moreover his hyper-cycles include the notion of 'auto-catalysis' although there is (see Chapter 8 and this chapter: Perplexity) no such thing as an auto-catalytic molecule - although, since Eigen, the blithe notion has been extended to include un-coded, non-metabolic citric acid cycles and a hypothetical model of primitive 'autocatalytic networks'. **Worse still for**

Eigen's theory is its assumption that reduced entropy is the necessarily same thing as information. It is not. Throwing letters of the alphabet in air or stirring nucleotides in water until, perhaps helped by natural constraints, they fall into 'vital' sequences, does not create informed, purposive systems. *It is at the heart of scientific error to assume it might.* In this sense Eigen's speculation (with every other purely materialistic one) is fundamentally flawed; the universal attribute of chemical reaction is not negentropy but, which increases overall disorder, entropy. Nor is it a question, for life, whether proteins or nucleic acids came first. *The fact is that translation mechanisms using both together produce meaningful biochemistry, organs and whole bodies. Meaning is manifest in codes translated in order that structures and processes develop and operate as a unified, self-consistent whole - an organism.* **Contemporary biology, studying material forms that may involve mental processes, is riven from top to bottom with purpose yet is, because of its materialistic philosophy, content to reduce life down to chemistry!**

I hear what you say. 'Purpose' and 'the will to live' are visions of the atoms. Mind is an illusion made of chemicals. And information is a trick of sense played by the non-sense of oblivion. The only ruler, whipping chance into a better shape, is the mindless, merciless dark lord named natural selection. Is such surmise correct? Can only an unscientific ignoramus disagree? Must 'untutored' reason bow before this brazen species of irrationality?

The sharp lens of molecular biology increasingly contrasts with the cataractic blur through which Victorians understood cells. The detail in life's house is being resolved into far more comprehensive and yet precise a teleology than the early Darwinian vision of a globule of slime. State-of-the-art biotechnological research has recently begun to mimic a cell's own nanotechnological genius but, far from the construction of an organism *de novo* from first principles or radical redesign of any biological body-plan or system, it only tinkers on the fringe. To *construct* a cell from inorganic chemicals is still like asking a child to build London. In the future, at a point when the modern cutting-edge looks as blunt as medieval, such a simulation might be achieved. *We could then calculate the massive injection of conceptual thought required not to invent but simply to copy!*

Some biologists yearn to copy, inject intelligence into otherwise unintelligent molecules and 'create' artificial life. How do you inject intelligence into matter? The syringe is you, the needle focus of attention and the solution injected is intelligence - intelligence that informs, arranges and manipulatively targets a material solution to your problem or desire. The logic of a book is cast in ink, the logic of biology is cast in *DNA* and, unlike any ordinary chemical it represents a language that is able to unfold and realise its narrative! It's incredible! You couldn't make it up.

Genomes are 'books of life'; and no doubt that ingenious genomes are a bio-builder's stumbling block. How do you construct what you presume pure chance first bubbled up? How might you arguably brew succinct, intense, efficient logic from an artless, primal stew? Because genomes capsulate the chemically written language regulating all the cell-based bodies we call forms of life. Such information stores are incomparably compressed in size, highly efficacious and exact. Their message is relayed dynamically. The product, if it's 'healthy', is spot-on. The whole is elegance incarnate. *Do coded systems start by chance?*

The origins-of-life conundrum - how to generate even the letters, let alone polymers, of *DNA*'s alphabet - has been discussed: nor is the context of any life restricted to its chromosomes. A cell is a self-organising, self-consistent, self-contained whole. It needs all it contains and it contains at least a million specialised, reproducible parts that combine to harmoniously produce an operation more flexible and sophisticated than any modern chemical factory, fighter plane or ocean-going liner. You can put ten cells in the space taken by an invisible speck of dust. Each single one is only possible within the constraints of a large set of 'mutually correct' cosmic factors as described in Chapters 8 and 26. These extend from 'life-friendly' natural law, its attendant material physics and chemistry, the state of the universe, solar system and planet earth down to the prefabrication of a genetic code which, in turn, can manufacture umpteen, critical biochemicals in a controlled and balanced way. Indeed, research laboratories everywhere are laboriously unravelling the details of code, communication and signalling that facilitate smooth function according to pre-set homeostatic equilibria. This was never the language of chance. Precisely the reverse. *The finer cytology's technological resolution becomes the higher the order of teleological accuracy revealed; and, in proportion, the more the idea of idea-less abiogenetic life is dissolved.*

By now we clearly see that the *DNA* enigma is not so much (although still greatly) with the production of complex chemicals as with the origin of code and information both conceptual features, both crucial and neither of which evolutionism rationally addresses. *It takes materialism's pirouetting flight of the imagination to o'erleap the element of information, one embodied by all codes.* **Codes are no less than embodied information; information is an immaterial element, a metaphysical component whose active centre we call mind.** Mind is anti-chance, the place of purpose; purpose is conveyed by messages in code; thus representing purpose code is always teleological. This fact includes, emphatically, the genetic code - embodied information that in turn embodies you. Yet still you harp how natural selection is enough!

Why not, therefore, cut the cackle and cut to the chase? Why not, as was urged before, administer a large-scale genesis by constructing an evolutionary roulette wheel made of millions, even billions of dishes or test tubes. Such a massive pre-biotic research project could, not even granting a start-point of *DNA*, systematically ring all the changes. You could test repetitiously using multiple, variegated and very large batches of 'initial chemicals' raised against a string of speculative micro-environments - different combinations of atmospheric gases, saline solutions, simple minerals, heating, electricity, incubation times and so on. Of course, you wouldn't be allowed to interfere with any process or isolate the products; you couldn't preserve nascent structures that you knew would later be required against the elements that made them or entropy's destructive time and tide. Would accidents have made life up? What, furthermore, would you expect to 'hatch' from your *in vitro*, pre-biotic 'eggs'? Time degrades at least as fast as builds up complex chemicals. Since they won't languish on an early shelf time is not the issue. You should be able to create required permutations in a year or two at most. What then, if rapid randomising found them, would 'life capsules' or 'the chips of genesis' have proved? Might any of a trillion prospects rise to expectations? We're not expecting Frankenstein. If chance could synthesize a single cell then that would be enough. Who's betting?

Perplexity

Paradoxical perplexity in biochemistry! *No chemical 'wants to survive', 'complexify' or 'get a life' let alone look to the future and reproduce itself.* But, as careless as computers grind away, one kind might appear to. Organic biochemicals seem to work, amazingly, in a machine-like concert that is highly organised. A single, unresponsive note or meaningless refrain is not repeated endlessly. Nor is their music like a jarring, mindless brouhaha but as if from a well-conducted orchestra with rhythm from coherent score. Chemicals in cells work now. As there were no previous composers to evolve the shapes of their informed, conceptual harmony, must they not have built themselves? Life 'self-organised', did it not? And since 'purposive' procedures pervade biology, then bio-molecules must lead in front-line productivity. The burning question, according to any 'theory of no intelligence', is how? Forget those speculation-spoilsports whose easy answer is a 'theory of intelligence'. How did the blind lead blind to sight? How did insensibility endow, how deadness bless the world with life?

After all, one pleads, life *only* needs a 'self-replicating molecule', a single seed about whose primer evolution crystallised. This, as the previous sections start to show, is more like a major assumption than a minor plea. Nevertheless *DNA* is, one asserts, such a molecule and therefore as soon as a few nucleotides condensed together (in water or on clay but without the help of enzymes) the rest was possible. No doubt the 'dramatic' cell is a very conceptual, structured form of information but, it is argued, aren't you proof positive that, given a pile of self-replicating 'alphabetical' chemicals, they will eventually, by trial and error, 'self-organise' into forms of information as grammatically meaningful and much more complex than newspapers or encyclopaedic books. Life is an author-free zone, an editor-less broadsheet but is it, nevertheless, worth the paper that it's written on? What, unless a scientific panegyric to the power of oblivion, does it all mean? What's the point?

So atoms churn until a lucky strike. Or, rather, strike on strike on lucky strike. What is there to 'naturally select' until the time they've struck up life? Why, therefore, should they ever reach the point of 'breaking into dance'? How could self-replicating chemicals acquire 'taste for survival'; why should unrelated chemicals associate with them and, around their core, evolve ever more sophisticated mechanisms to support that 'goal'? More than progress regress is the natural way. Ordered start-ups gradually decay. Entropy is a word associated with the Second Law of Thermodynamics that describes the tendency of a system to lose information, to become disordered and degrade. Therefore what agent drives, against this influence of disintegration everywhere, a scenario of integration prejudicially entitled chemical evolution? What drives atoms unto animation? In fact, neither raw information nor raw energy support the generation of irreducibly complex systems. Raw information in the oblivious, non-conscious form of natural law is devoid of purpose: it needs conscious directive to establish the informed complexity of art or technology, be it human or natural. Similarly raw energy, sunlight, no more supports life than petrol by itself fuels a car. It needs a car! And with the car a complex fuel injection system combined with an engine and associated parts (axles, wheels, chassis and so on) to establish, for a while against time's grain, a working vehicle. Or, in the case of the biochemical factory called a plant, photosynthetic and respiratory energy metabolisms with associated larger-scale structures (see Chapter 21: Energy Metabolism Perchance?).

So are you at root, as an evolutionary protestant protests, the offspring of blind chance, a child of mindless atoms and life from dark lifelessness? Or, on the other hand, is the speculation of abiogenesis plain wrong? *If it is wrong then why, from a speculative error, should a theory of the extrapolated transformation of all life forms be right?*

You claim that *DNA* 'self-replicates'. **The problem is, however, that *DNA* cannot by itself replicate itself.** *It is an evolutionary myth, commonly propagated, that nucleic acids are self-replicating entities.* Man has laboured long and hard but failed to design a single 'self-replicator' so why should atoms blindly chuck one up? ***The biological fact is that there exists no SRM (self-replicating molecule).*** *Nucleic acid cannot copy in a vacuum. Genetic mechanisms cannot operate without 'satellite' operating systems. Just as a database needs surrounding soft and hardware systems, so life*'s *metabolism needs satellite systems. Primary genetic material needs an operating system of specific proteins. Such systems are complicated, are themselves coded for by DNA and only found in living (i.e. post-pre-biotic) organisations called cells.*

So, again we have to ask which one came first - *DNA* or all the proteins that it needs to make a protein? Or, indeed, the proteins needed to metabolise their own genes, that is, to synthesize their own essential templates made of *DNA*?

Such paradoxical perplexity in biochemistry is critical. How did pre-biotic, 'animistic' atoms read the future? Why should pre-genes - no less chemical and lifeless as the genes-in-membranes that inhabit you today - animatedly seek to preserve and later complicate themselves, evolving all sorts of elaborate biological forms as vehicles for their own protection and propagation? How did or do they know they have both to survive now and for the future reproduce? Whence arises, in a sachet made of molecules, the inanimate 'will' or oblivious 'selfishness' for either program to develop ordered forms? You might well conclude such animistic story-telling is, in scientific terms, babble.

The issue is a nub of confusion. From neither philosophical perspective do inanimate atoms wilfully 'program' themselves or anything else. From a *bottom-up* view any apparent biological 'program' first arises by chance alone (how can natural selection works on molecules?). We presume no 'interference' by intelligence and, from this possibly incorrect but nevertheless materialistic presumption, try to imagine a scientifically plausible mode of construction. But, from simple to complex, the only contender is some unordered kind of evolution. It *must have* happened. We just have to 'make it work'.

Top-down, on the other hand, presumption includes a single further factor.

As well as matter there is information. Information input is called creativity. The process is not, therefore, one of simple start by chance to complex outcome. It involves devolution from simple first principles to detailed ramifications - the engineering process. No doubt chance occurs within the constraints of physical law - which includes neither principles of information-generation nor purposely-complex design. Therefore chance cannot be responsible for mechanisms that serve the anticipatory purposes of survival and reproduction. It can no more develop life forms than you or I, who build a bridge or even an accident-prone space probe, build either accidentally. In engineering and bio-engineering cases both hardware and software are *preconceived.* The *fact* is that matter has no more innate inclination to produce genetic databases (with their associated application and operating systems) than it has to produce a poppy or a penguin, a piano, computer or component protein. And it is a *fact* that the minimum free-standing biological entity that can reproduce is a complete cell. *Allow the cell, like any and every other machine, a metaphysical component.*

Whether shaped by mind or chemical necessity matter is indifferent to the forms it manifests. It is therefore also entirely unclear why it should, of itself, indulge in that most semantic institution - code. The genetic code shows 'grammatical' conventions well recognised by the textbooks. At root is the *notion* that three bases will *represent*, through a *translator* molecule because there is no necessary connection between the two '*languages*' of nucleic acid and protein, an amino acid. *Furthermore, that punctuated groups of 'words' will represent a conceptual end, a discrete section, an instructive sentence (called a gene) that plays an integral part in the logic of the whole book (called a genome).* Simulation. Symbolism - the heart of intelligent, informative behaviour. A chemically arbitrary storage system could not be called a code. *Codes carry information.* They correlate otherwise unrelated objects. The noise 'you' *means* something; it *stands for* something it is not. Because it is a meaningful noise that involves a notion we call it a word. You can take a few (or even millions of) letters of an alphabet and churn them. You will obtain, briefly, a few words of various languages but never, *pace* Thomas Huxley, a coherent manuscript.

How, therefore, arose life's most efficient, universal code? Its perfection must have been attained by *LUCA* (Last Universal Common Ancestor). If the code's not perfect then what improvement might there be? Even if imperfect (but it has not evolved since then) by what steps was its excellence achieved? How could such robustness happen accidentally? Some explanation is required but none scientifically forthcoming. Even stories that are plausible are not by plausibility alone made true.

One might have argued (Chapters 7-12) that 'the world stage' is itself a product of archetypal order; cosmos is a passively informed, mindless but dynamic, finely tuned automaton. In this case, a snowflake or a salt crystal falls within the predefined but meaningless dynamic of physics and chemistry. So, in the general sense of their components, do all technological and organic forms. It is the *arrangement* of these components that is special. They are special creations due to their purposive shapes and arrangements, mental origin and, prior to their specific physical expression, plan. The basic component of any informed system is its plan. A part of the biological expression of plan is genetic code. Players on the world stage are all made of coded parts. *At this point, as was glimpsed earlier, a real chicken-before-the-egg conundrum seals the way. A perplexing and impenetrable circle excludes abiogenesis from a rational account of origins.* Any attempt to form a model or theory of the evolution of the genetic code is futile because that code is without function unless and until it is translated i.e. unless it leads to the synthesis of requisite proteins. *But the machinery (or operating system) by which the cell transfers instructions or 'self-replicates' consists of about a hundred precise protein components which are themselves the product of the code.* **This alone renders a materialistic theory of abiogenesis, although central to atheistic faith, illogical and untenable.**

So which interpretation, *bottom-up* or *top-down*, best fits facts? Who lives by faith alone? *Top-down* certainly does not fit *bottom-up* philosophy and so materialism's hunt for resolution of perplexity ploughs on regardless. Which caused the other? So intractable is the conundrum of which came first - nucleic acid coder or protein operator - that compromise has been identified mid-way. Nowadays it's *RNA* that stands between the two. Perhaps *RNA* gave birth to each and then slipped coyly to its current intermediate role of go-between? Was life's ignition sparked in a lifeless, pre-biotic world of *RNA*? Let's see.

***R* not *DNA*?**

Catch-22s. There's intractability because **without its self-replicating molecule any atheistic doctrine of origins collapses**. To live you need to replicate; to replicate you need protein, *DNA* and *RNA* to work in concert; and specific proteins are the systematic product, using *RNA* and ribosomes, of accurately coded *DNA*. But, independently reacted, these would not match up. Translator-less, this is the fact. Such severe and serious obstruction blocks out any theory that, to try and weave a circumvention, throws up an accidental, serial appearance of chemicals whose interdependency is critical; and whose cooperative algorithms just won't work unless they're hemmed together in a cell.

Nil desperandum. Some new, hitherto unknown kind of proto-life can be invented fairly easily. *If you could persuade yourself that RNA alone could take on DNA and protein metabolic function (thus making a self-replicator that could catalyze its own synthesis) you might just convince yourself the deadlock's breakable.* Though single-stranded but with rare and very modest catalytic property might not this delicate and relatively unstable nucleic acid have primed all lives on earth? Perhaps it could store some information (*DNA*-like) and (as if a protein) catalyse self-replication too. To this end some research recruits exotic forms of virus to the flag. They incorporate reverse transcription (*RNA* to *DNA*) and modest double helices in genomes that, as yet, employ no *DNA*. However, a few *RNA* molecules (whose designated, dedicated business is replication) do perform specific, important, mainstream tasks likes mediating amino acid linkage during protein synthesis in a ribosome. Such catalytic *RNA* is known as a ribozyme. For example, *RNA* polymerase ribozyme can, after careful, non-codified, non-metabolic construction, can add about 20 nucleotides to a primer template before shortly falling apart. On this slim, coaxed sort of evidence it has been suggested that suites of natural ribozymes, made of nucleic acid but acting like protein, might have kicked life from the margins into play. Performing hundreds of necessary metabolic rites, could they have created cells and thereby life? When something has to happen anything is deemed a possibility. But, where other avenues are blocked, is this any more a corner turned by hope than cul-de-sac of hopelessness? Is it, when another simple but forbidden answer makes more sense, compulsive guessing to support materialism's logic or a counsel of determination that it's only molecules - without the organising principle called mind - that maketh man? Why not, however, since all else has failed, propose a pre-biotic 'world of *RNA*'? This is, presently, the favourite scenario.

No problem with hypothesis. Neither plausibility nor dreaming is against the law. What about reality? In fact, as already noted, it is unlikely that lab simulations of the production of ribose or nucleobases (such as adenine or cytosine) could have occurred on early lab-less earth. But, *if* they did, then they'd have to join up into homochiral nucleotides. In fact, because of their shapes it is very difficult to join a sugar to a base to make either the *DNA* or *RNA* monomer. If a first-class chemist is involved they still soon fall apart. Chancy life, on the other hand, supplies specialist, coded enzymes for the purpose (and must have from the very start). Even *if* chance churned out homochiral nucleotides at the same micro-time and micro-place sufficient unto chains of *RNA*, it is even less likely that they could have self-assembled into suites of catalytic, self-reproducing polymers; or suites of coding sequences (without storage mechanism) to make working proteins; polymerisation requires its own set of specialists. Therefore, it has been suggested, perhaps they might have grown from something simpler, say, a peptide nucleic acid (*PNA*)? These contain no sugars or phosphate, can form base pairs and helical structures but are stable to the drag-point of not releasing any daughter molecule. Thus if *PNA* were the alchemical case a whole new series of questions arise. They include: what created *PNA* (as *RNA*) in pure and sufficient quantity? How would swaps from *PNA* to *RNA* to *DNA*, for which there is zero evidence, invisibly occur?

Nor, any more than in the case of *DNA*, does 'chemical predestination' help with building coded order up. As with letters making words from ink, so sequences of *DNA* or *RNA* do not arise from any internal chemical property; no preferential affinities help *RNA* to generate the sequence useful for a single working protein, let alone its many necessary friends. So how do you, engaging in detailed experimental design, enjoin the nucleotides of *RNA*, correctly linked, to join up as informant code? Yet you insist that evolution, somehow, is a fact!

While pre-biotic waters could not generate a continuous supply of activated nucleotides you might pipette them into test tubes. Don't forget to add a specific protein called polymerase, some primer and, if ambitious, some reverse transcriptase for a snip of single-stranded *DNA*. However, even carefully manipulated polymers of artificially 'self-synthesised' *RNA* are short, fragile, tend normally to hydrolyze in water or curl into useless circles then soon fall apart again. So, like a house of cards, you have to build hypotheses to obviate these weaknesses - sure in the mind-lock that your evolutionary hypothesis is true. Nor, from millions of such random sequences, have any demonstrated powerful catalytic, self-synthetic capability. Engineers of ribozymes have orchestrated their experimental choreographies to generate nucleotides. They have selected the necessary sugars and other components, purified at every step (thus averting natural cross-contamination of the products) and obtained the desired result. *Intelligence in swathes came first.* They have then transformed the bio-friendly outcome of their algorithms into specific sequences needed for even limited self-replication. *Intelligence arrived there first as well.* Of course, couples conjugating have been caught in test tubes. Yet even then the ribozymal replication is not the normal kind. Such couples have been engineered. One of a prefabricated pair simply acts upon two halves of the other. No more than 10% melds. The halves are simply joined end-to-end to make a whole; which whole can similarly bind another pair of halves and theoretically (if you supply the energy, nucleotides and so on) so on. Above all, the prior 'information' problem of giving them a viable sequence has been overcome by careful minds. Yet those directing ribozyme experiments intend paradoxically to show that, in a medium of *RNA*-world, chance must have animated chemicals!

A bit of snip-and-splice is, however, far from triggering or running life on earth. There's far more to any cell than this. It is a hopeful and yet desperate apology for generating biological centrality - code <u>in order to</u> metabolise. The fact is *RNA* like *DNA* needs proteins to construct it in an independent cell. And the 'tentative' double helices that inhabit various reproductively dependent viruses are much shorter, wider, less regular and less accessible to protein manipulation than *DNA*'s. This would make it very inefficient if it worked at all. But say it did. *What about the necessary transformation into current, universal DNA?* Having created *RNA*'s uracil chance would have to systematically replace it with thymine, replace ribose with a deoxygenated form and expunge all 'non-canonical' nucleotides, such as inosine, from any regular length of double helix. Then how did genes evolve with ribosomes for protein synthesis? Could, due to faulty but evolving text, early burly cells have survived an abundance of redundant, interfering proteins? None can now so it's unlikely then. And, when you've done with *DNA*'s transcription, what about the other non-redundant part, translation - purposive, conceptual, meaningful translation? Without the lynch-pin of translation, *RNA* to protein, life's basic, coded exercise just meaninglessly self-aborts. You need enzymes, metabolic pathways fully up-and-running straightaway. Then, as well as a transition phase of *RNA* to protein code-storing *DNA* must supersede unstable *RNA*. Where, meantime, has ancestral *RNA* gone? The molecules have disappeared but, apart from blowing in imagination's wind, there is no jot of evidence that they were ever there.

No doubt, a few simple viruses use an *RNA* 'coding tape' but they employ the enzyme called reverse transcriptase. This unusual transcriptase, which was lifted from them and has become a basic tool in biotechnological engineering, hijacks host *DNA* and its suite of protein operators. However, no self-sufficient organism uses an *RNA*-based coding system. Further more such powerful catalysis as would be required to mimic life's replication is a huge extrapolation from what has actually been observed. *In myth RNA and DNA can copy themselves; the fact bears repetition - neither can.* **In fact, the myth conveniently forgets a complex system of protein operators that alone make any replication possible.** The theory of life by *RNA* still begs its own question - by what steps might such associated software have unpredictably self-synthesised?

An '*RNA* world' has proved so hard to sustain that yet another cliff-hanger, ribonucleoprotein (*RNP*), has been drafted to the front. But why should duplex *RNP*s (such as you find in ribosomes) grow and complicate itself against the tendency of natural forces to disperse; whence did the coded peptide part arise; and what would constitute their function outside the context of a cell (in which, like *RNA*, they exclusively exist today)? The same intractabilities rear up and block the tale. If things did start with *RNA* then the transformative genetic switch to *DNA*, which is universal in all but a few microscopic, viral, parasitic pests, is totally unknown and unexplained. For example, by what stages could it have occurred? Why is all sign of the transformation lost? Abiogenetic speculation twists and hunts and turns; maybe, however, that its fox is false. Storage scroll and notepaper: *DNA* and *RNA*. The fact is that each kind of information carrier is admirably suited to its own purposes. How exactly, therefore, could a chemical fanatasia (of *RNA*-worlds) turn into the real world of information that specifically, dynamically works in every cell?

For a more detailed demolition of the *RNA* tale read, say, Signature in the Cell by Stephen Meyer. Perhaps, in a world without chickens-before-eggs, you wouldn't have to try and tickle information from a watery grave. You wouldn't have to lift life out of the inertia of inanimation. Perhaps no 'boot-strapping' switch of program, *RNA* to *DNA*, ever had to happen. Or, except in imagination, ever did.

Raw Energy Destroys

You'd think, when talking to some naturalists, that energy was everything. Doesn't immaterial information count?

Evolution's based on build-up; unusual constructions increase in complexity. Its problem is run-down. Run-down, the way an energetic cosmos works, falls absolutely opposite to build-up's way. Time's arrow pierces Darwinism's heart with poison. Its dart is tipped, every moment of each day, with toxic entropy. Of course, raw energy can be injected into systems from outside but is that all an organism needs? Sunny weather's needed but is not, any more than air or water, sufficient unto animation. Crude radiation has to be refined before it can resuscitate, reinvigorate and flow through all the veins of earthly life. More than this, can energy alone initiate genetic code, metabolism and the systematic architectures that express vitality?

As noted (Chapter 8) initialisation of the physical universe appears to violate the First Universal Law of Thermodynamics. This states that energy/ matter cannot be created but is conserved. But without this post-original law physics and chemistry would be in disarray. You could not write an equals sign in any mathematical equation. The abstract foundations on which factual science has been built would crumble. In the flux you couldn't know a thing for sure. No outcome of any interaction would be predictable. In addition to physical violation an unintelligent biological start (abiogenesis and the theory of evolution) would appear to violate Axioms of Information Theory (e.g. information is neither matter nor energy) and the Second Universal Law of Thermodynamics. The latter states that within a closed system (like a universe whose energy is conserved) things occur under the materialising influence of (*tam*) entropy; and that in an open system they always tend to disarray. They tend, one-way, to diffuse into states of lowest energy and fall, as far as information is concerned, into disorder. For example, a drop into inertial equilibrium, can be disturbed by the introduction of raw energy (e.g. heat) but, as soon as possible, the upheaval subsides into a further predictable but purposeless state of reactive completion. Or some stimulus may cause a local aggregation to buck the trend but even then new 'order' or 'complexity' remains unspecified. Innovation, code or mechanism can't arise this way. Instead potential dissipates. The battery goes flat. Death's duty is a tax imposed on all material things.

A case of 'disorder' can, however, be 'disturbed' by an injection of intelligence. You could, for instance, meaningfully tidy up your room. Such energetic input would accord with the laws of thermodynamics and entropy; but at the same time its direction of intent would run against them. You could make a scientific discovery but, in that realm of physicality, is thought or immaterial reason barred? You might create a work of art but is intelligence taboo? Of course not, but intelligent circumstance as a reason for biological order is proscribed on the principle of 'no intelligence' in natural causes.

Forget intelligence. If the universe were eternal yet always functioning according to thermodynamic laws it would already be 'dead'. Therefore the Second Law indicates that it must have been created. But the First Law says it couldn't have created itself - unless the laws of physics didn't exist before the world and, therefore, there's no scientific answer to the spontaneity of its beginning. You might not want a metaphysic prior to physic but, despite endless hypothesising to the contrary, an archetypal frame of universal mind makes sense. In this case you must recall intelligence.

Zoom from the macro-system of the universe down to our terrestrial mini-system. Earth's biosphere is not a system that is closed. Raw energy from a molten core or external sunlight can disturb low-energy conditions on its surface. Zoom in to a single puddle or even a micro-environment within the puddle and the same circumstance applies. You can inject energy into it from outside, it is therefore an open environment and in this case might not a local instance of 'negentropy', that is, 'anti-entropy' occur? Could not complexity sufficient unto life 'self-organise'? To repeat: could sunlight work with chemicals and brew up life?

Time to inject some facts. Biological life is a highly unstable, low entropy process whose dynamic equilibrium needs a number of homeoststic engines to keep it balanced. Imbalance leads to malady which, unless rebalanced, falls to death. *Inanimate systems, however, tend to drive to high entropy and lower states of energy or both, while the highly improbable, codified order of life-forms needs the reverse.* **Then how could raw energy, without the device of pump or other engine, ever have begun? The premise of chemical evolution is, from a physical point of view, in principle an absolute non-starter**.

And so we live off light but not off light alone. The fact is life depends on the complex, codified *mechanism* of photosynthesis. Only in this way can solar energy be harnessed to the construction of life's chemicals from carbohydrate. Raw, radiant energy is a necessary but not, however, sufficient condition for such basic biochemistry; and if you're talking energy from chemicals instead of stars the same applies. Energy's relationship with life might be likened to the electrical power supply for a computer. A necessary component, even a driver, neither initiates nor creates a system itself. Photosynthesis (or chemosynthesis) drives forward life's operations but it needs a coded origin. It needs its own creator.

Think. Could sunlight ever turn a desert into cities or transform a salty pond into a brain? Time, where entropy's concerned, can't even build a complicated molecule. It never specifies progressive, coded hierarchies of complexity. Darwin's time-dependent theory runs up against its own necessity. Dead rabbits don't revive as bullets fly back to a gun; cars don't turn into trees and from that crash reverse into undamaged state; nor people grow young through their birth into whatever came before! Science works *with* energetic time; but evolution pumps *against* its grain. Raw energy destroys but mind alone builds specified complexity. Mind naturally provides informative negentropy. Information, basis of biology, crafts systems countering entropy with such negentropy. 'Active transport' that can carry concept to fruition or can code and work out purpose by inventing a device requires immaterial inspiration just as much crude materials. Only the intelligence of conscious mind can expedite refinement into operational subtlety. Yet operational subtlety is rife throughout biology's extent!

In short, in order to obtain constructive as opposed to a destructive biological effect raw energy must be specifically refined. Life forms need, indeed are themselves, open systems but they need specifically purified and not raw energy. They use sugar metabolically refined from water, gas and sunlight. The intricacies of such refinery (which must have been in place from earth-life's crack of dawn) will be detailed in the next chapter's section 'Energy Metabolism Perchance?' They well exceed any capacity of molecules 'without disturbance by intelligence' to purposelessly chance upon; but, without such orderly catalysis, life could not sustain its special, vital pistons of dynamic equilibrium, its clutch of orderly disequilibria defined as homeostasis. Nor does oblivious matter ever make these balancing mechanisms so as to maintain the other special, complex, interactive patterns that we know as bodies bio-logical. Matter's not alive. No non-biological molecular structure (is any, though perhaps organic, 'biological' before biology?) is known to persistently increase in complexity to the point of exhibiting, against the omnipresent influence of the Second Law, machine-like stability, coordination and, thereby, purpose. **Purpose, like machines, is the province of mind.**

Materialism won't give up. Never say die since matter isn't dead but never lived - yet isn't 'living matter' what, exclusively, one claims a life-form is? A mathematical model of almost any process can be made to work on paper as long as certain assumptions are made. Couldn't, therefore, 'dissipative structures' such as turbulent vortices perhaps gradually whirl up such self-organised complexity as cells? Ilya Prigogine writes: "Unfortunately, this (self-organisation) principle cannot explain the formation of biological structures. The probability that at ordinary temperatures a macroscopic number of molecules is assembled to give rise to the highly ordered structures and to the coordinated functions characterising living organisms is vanishingly small" (Physics Today Vol. 25, No. 11, 1972 ps. 23-38). Repeated attempts to explain the origins of programmed biological form in terms of the auto-catalysis, self-organisation of matter or (see Chapter 8: Metaphysical Evasion) Principles of Evolutionary Innovation are essential to uphold both a 'naturalistic' vision that is in harmony with the philosophy of science and the neo-Darwinian theory of evolution. *They are also, however, at root sophisticated, systematic attempts to ascribe the informative qualities of mind to matter. This is so much huff, puff and bluff. It is as wrong as it is misconceived.* The creations of both chance and mind are constrained by physical law. **But life is never 'self-organised' by external circumstance. It is controlled internally by a specific code.**

Topsy-turvy Logic

Raw energy is not the problem - except that it won't get you any bio-where. **Cut to the chase, the origin of information is what you need explain. The real basis of any evolutionary hypothesis is the mindless production of reduced entropy sequences bearing information**. This sounds difficult but boils down to the same problem as producing writing on a beach (which is grammatical and meaningful) as opposed to the patterns of the wind and waves (which are not). So basic is this problem of reversed logic (deriving information from mindless, atomic interactions) that, nothing daunted, speculation proliferates because, one is instructed, there is no scientific alternative. *No scientific alternative to topsy-turvy logic?* It is, naturally, the core philosophical wish of a physical scientist to explain nature as far as possible without the help of any super-natural agency but if by the word natural you mean physical then, unless you redefine it, explanatory mind is itself metaphysical. And therefore super-natural! And likewise (Chapter 5) immaterial information itself! Therefore, because biological information would derive unscientifically from beyond the remit of natural science, you argue (Chapters 13-17) that human mind is an aspect of material brain and universal mind is non-existent. Yet why do even expert humans err? Is it because of lack of information or an incorrect perspective? In this instance is the scientific *zeitgeist* blind to a core factor, one that's immaterial? Such blindness need not blind you too.

Non-conscious subatomic particles, atoms and molecules are certainly automata. They are, paradoxically, governed by clockwork necessity and, to our restricted vision, unpredictable chance but they cannot generate purpose. *Biological bodies, which use them and are governed by the same clockwork rules, exclude chance and entail purposes. Inanimate matter does not, at each of many levels, engage intricately coded patterns of behaviour.* Non-life never performs integrated routines that seem to *mean* that a particular, individual collection of atoms wants to best *survive*. What, in such terms, *is* 'survival'? What, in your case, *is* the awareness you are so aware of? To deny purpose is to deny awareness, life and thought. *It is thought that annihilates randomness and, speedily, focuses to generate specific, self-consistent and even self-operative technology. Machines are purposive but made of non-purposive automata. So is biotechnology. Your body part is a superb machine.*

Scientific materialism is, however, the philosophy (and naturalistic atheism its religion) that negates intrinsic reason or meaning in the origin of biological machinery. It is, accordingly, 'irrational' to propose any model for the origin of life based on other than solely physical or chemical processes. *Unfortunately, as we are beginning to see, this irrationality reaches to the philosophical heart of contemporary science.* Over the last couple of hundred years this Theory of No Intelligence has infiltrated and now permeates biological thought. Its determination persists despite a prime opposite directive in research, namely, to understand the *reasons* evident in biological form and function; and despite a prime opposite discovery viz. the ubiquity at all levels, however minuscule, of exquisitely high-grade design. How, therefore, does this jarringly contradictory condition of mind, this topsy-turvy inversion, survive?

To sustain the illusion three necessities arise:

(a) an emphatic *a priori* denial of Babbage in the computer

(b) an appeal to various theoretical diversions (e.g. chaos theory, complex mathematics or 'self-referential dynamics') to counter the thermodynamic considerations. *In fact, as noted in Chapter 8: Up Horseshoe Creek, no known natural/ physical law or event constructs detailed, self-reproducing, homeostatic mechanisms that support awareness or express purpose.* A surgical application of Ockham's razor allows, however, that such 'miracles' are the sole and common product of the only known anti-randomising, metaphysical agent in the universe - mind.

(c) an extrapolation that follows close and inextricably linked to (a) which argues that, lacking any source of information, an abiogenetically-primed organism must have evolved almost imperceptibly over eons from 'simple' to 'complex' through a series of 'beneficial' genetic accidents.

non-specific	*specific*
random	*self-consistent/ orderly*
oblivious	*aware*
targetless	*meaningful*
physical law	*purpose*
matter	*mind*

On the other hand, top-down Natural Dialectic proposes that active precedes passive information. Mind precedes matter. What happens in the human sphere of operations, namely, the outward expression of internal orders of thought, mirrors the cosmic process of creation. The higher or better an understanding the more effective and worthwhile is its product. The thrust of every educational system supports this dialectical line.

Do you normally put your faith in chance? Any information technologist knows that his intelligence is a much better bet than chance to set his system up. **Thought, a natural phenomenon, satisfies the criterion for an origin of teleological coding and design which the very basis of biology demands**.

Noise, Monkeys and Catalysis

Theatrical Jean Cocteau once facetiously exclaimed, "The greatest literary work of art is basically nothing but a scrambled alphabet."

Do you remember Shannon's nonsensical definition of information (Chapters 6 and 8)? At a statistical level Shannon and Cocteau are right and might equally have claimed that 'biological mechanisms are nothing but the evolved product of statistical improbability'. Do you think that you came from a scrambled egg? From the point of view of all superior levels of information (grammar, syntax, semantics, purpose and will) they are wrong but succinctly and unintentionally spotlight the basic flaw in all abiogenetic research and, unsurprisingly from a *top-down* perspective, lack of its development. It is the scrambling of reason, the presumption of chance and denial of prior purpose, intelligence and information.

From noise, it's said, evolved a perfect melody. A trancelike apparition merged (or was it emerged?) to the magic of reality! Forms of life - undirected, uninformed, uncoded - just arose. **To repeat, the fundamental error in a 'scientific' explanation of the origin and evolution of life is to assume that it arose due to the accidental production of genetic material; and then, however poor the quality of an initial script, mindlessly transformed its noise by entropy of information, that is, by mutation into many, many works of art.** Do ink and paper write a story? This is back to front. It runs against the stream of sense. 'Information entropy' degrades meaning. It bugs reason and distorts the purpose behind a message.

I know. You cry that life makes sense and death kills nonsense. We can all agree that natural selection weeds mistakes. You say mistakes that make survival sense survive and, after *LUCA*'s birth, gradually accumulate, innovate, produce and reproduce a code-informed complexity of life. Evolution works, against the natural grain, uphill.

I say each book was written with, in order to appeal and prosper with a changing audience, an inbuilt flexibility of script. In this case, nonsense makes no sense; mistaken reprints are deleted from each line of stock. Whose interpretation, given scientific facts, makes better sense? Which will, eventually, survive?

Could Humpty Dumpty fix himself or cows jump o'er the moon? **There is fostered a powerful myth that, given time enough, anything can happen if you but wish it so; and promissory faith that chance can manage even what's impossible. Not just mathematically, statistically improbable but physically impossible**. Impossibility is possible. Yet is an immaterial element of information so impossible? Check the section in Chapter 5 called The Lowest Physical Level. In dialectical terms constitution of this 'fixed' or 'automatic' domain is the final outcome of information; the physical world is passively informed. Nor does coded information arise at the statistical level of physics and chemistry. Chemicals and large-scale objects may carry but cannot, by any known law of the universe, generate such a packed and meaningful density of information as is, for example, found in *DNA*. The whole universe could never bake a tea-cup let alone a Wedgewood set - even if you asked it! Nor is time any use at all. Never in ten billion years have ink and paper 'self-composed' into the simplest pamphlet. Atoms *never* construct networks, mechanisms and sequential operations geared to the realisation of 'target molecules' or reasonable ends - because they do not think. So that Thomas Huxley's fantasy of monkeys at typewriters eventually scripting a work of art was as facetious as deceptive. How would they retain the 'right' keys clacked and not obliterate the 'work' with wrong ones? In fact, not artless monkeys but the mindlessness of water, wind and chemicals was what he really had in mind. Students and lecturers at Plymouth University recently put Arts Council money where Huxley's mouth was. A month-long experiment to test his hypothesis found that six Sulawesi crested macaques occasionally and incoherently pressed a few keys of their typewriter, amassed a literary output of no words at all, partially destroyed the machine and used it as a lavatory. They monkeyed with a metaphor and confirmed the *top-down* hypothesis. Such exercise in randomness is, on the other hand, symptomatic of a *bottom-up* hallucination. You can diagnose it from materialism's 'multiversal' class. In such trans-dimensional species of statistical illusion anything is possible and, if you believe that, nothing Real. This is the vision Darwinists insist is true; it is, sadly when the grandeur of life's plans is contemplated, naturalism's naturally selected state of mind.

To assert, as does neo-Darwinian theory, that 'information entropy' (Chapter 23: Entropy of Information) can work 'uphill' to create and then mindlessly reshuffle code-bearing material into 'improved code' reverses the *top-down,* dialectical direction of creation. The latter supposes that purposive plans, simple or complex, start in mind and, if requisite language and materials are brought to hand, are stored as code. The direction of creation therefore starts with (*sat*) pristine information. It is information that gives energy direction. Information is potential action. Such action is released through a creative phase and 'dumped' in physical expression. This expression may, as in the case of a book, involve code and code-bearing materials. Only after pristine production do individuals from a print-run suffer the vicissitudes of 'information entropy' such as, for a genetic book, mutation. How could corruption have created literature? It only gets books trashed or pulped or thrown out. Such mutant damage just creates imperfect genomes. This is the state of those we analyse today. Negative selection kills and not creates. If you think otherwise then Chapters 22 - 25 show how and why, taking up the arguments, an evolutionary resolution of the materialist's information paradox is simply a conceptual fiction. Are natural selection and mutation capable of body building on a specific, complex and extended scale? Or, despite the fact of Darwinism's central mechanisms, is a belief that they can develop purposive systems just a fiction, an illusion not born out in fact?

No chemical can multiply itself. And nucleic acids need a host of ancillary, coded molecules (enzymes and building blocks) to copy anything. It's critical but no-one knows if iterative chemistry like this aimlessly occurs at all. Let alone, as in any cell, on a large scale. The chances may well be absolutely none. But go on anyway! Estimate, from many billions of unknown planets, the chance that even one might churn an *SRM* (self-replicating molecule) whose absence so far shatters atheistic claims. Proceed with arbitrary assertions on which a faith's world-view depends. Imagine, since materialistic creed *must* have it, that the veiled-in-mystery deed is perchance done. Count up base pairs, genes and chromosomes of some first cause of life - some primordial cell or, indeed, any other kind of cell! Such statistical count would be as meaningless as the quantity of letters or words between the covers of a book. The reality of a book, not least a genetic book of life, lies in the quality of its expression and its *purpose.* It does not, as was previously noted, lie in paper and ink (or *DNA*) but in the *meaningful* information efficiently conveyed by an accumulation of prescribed shapes called letters of the alphabet. **The *meaning* of genetic language is protein. And the meaning of protein is, beyond itself, its job - the construction and**

maintenance of organisms, that is, of life on earth. In any rational text one word has to be taken in the context of the rest. Interactive programs, fashioned rigorously to serve well-defined purposes, abound. What about, for example, metabolic sentences that thread, under the syntactical direction of squads of enzymes, from start to target chemical? A Shakespearean play is less complex than such metabolic plots and sub-plots. They involve trillions of atoms in precise, architectural order and myriad, 'computerised' interactions. Anywhere outside a body, where time breaks at least as well as it makes, such reactions are vulnerable to a most un-lifelike, death-like condition - inertial equilibrium. They soon land up spent. On the other hand, the 'impossibility' of life's dynamic equilibrium (or ought it be called, as with a motor engine, 'controlled disequilibrium'?) is made possible by catalytically conditioned chains of exquisitely organised reaction. Unlike one fraction of monkey text that bears no relationship to any other or, indeed, anything at all (except perhaps by the interpretation of strange, human 'monkey businessmen') metabolic chains are involved in mutually homeostatic interaction. Their dynamic instability is contained within a planned, purposive framework. No nonsense, at least in health, exists. This flips their probability from 0 to 1. It makes all the difference. Inside information - preordained intelligence - makes all the difference.

This 'inside information' pivots about the genetic production of proteins. In bodies many of these are enzymes, molecules that promote cell chemistry. *Thus, all this leads towards a consideration of the catalytic quality of chemistry that DNA informs. From suites of such catalysis alone whole organisms will be able to exist. Let's therefore examine how such metabolic pathways may or must first arisen.*

21. Origins

We've checked the *bottom-up* approach to origins-of-life. It involves the unnatural build-up of precisely configured congregations of atoms into larger molecules, groups of such molecules into one or more extremely localised concentrations and then, most hopefully of all, bagged-up integration with a congruent and productive code. The practical impossibility of this has been indicated.

While *bottom-up* accumulated complex, cooperative entities without the notion of a plan, *top-down* is essentially conceptual. It deals with a cell or composite organism and, especially, the information that creates it as a whole. Could genomics, which combines the disciplines of genetics, biochemistry, molecular biology and IT, lend insight to the way we view first life? No doubt a cell is usually very small (between 5 and 40 billionths of a metre across). However, it's not size but intrinsic information that leads to its functionality which really counts. Is there a minimum databank and thus minimum subsequent functionality below whose genomic compass life cannot exist? We already examined, in Chapter 20, the minimal functionality and, with it, irreducible complexity of genetic operations; and hinted at such minima concerning developmental pathways, lines of bio-code called enzymes that work in metabolic steps to fashion any biochemical a cell commands be made. Since they anticipate the future in the form of target product and every step must be in place to reach predestination's goal metabolism's programs must be labelled teleological. They have, like any mechanism, purpose or at least, if you want to side-step actuality, 'appear to' have it. Let's therefore step back from a whole bacterium and check the possibility that vital parts, its metabolic pathways, might have 'come about' by chance.

Catalytic Philosophy

Let's examine the catalytic quality of the chemistry that *DNA* informs. Doesn't interaction happen unpredictably? Life's watery bodies are, intrinsically, impassive. Their temperature is low and pH so neutral as invites no unexpected chemical events. No unpredictable 'popping' and 'fizzing'. On blank pages you can write exactly what you want without any other marks appearing that could scramble your meaning. Life's chemistry, called metabolism, works like this as well. *Each cell is a chemically blank page on which the 'writer', regulator or homeostatic program can control exactly what goes on.* Using a computer analogy the same exercise applies. The cell is the hardware and its software (its internal information) controls, in a balanced, homeostatic way, the order of its chemistry. *In order that things run smoothly bugs are eliminated. 'Noise' is life's arch enemy.*

Against their background blank, however, cells write busily. They are 'control freaks' that regiment reactive text through the discipline of specific catalysts. A catalyst lowers activation energy and thereby increases the rate of a reaction without itself undergoing any permanent change. It may also, critically for life, permit a reaction that would not occur without it at lower activation energies; or a specific series of reactions that, due to their complexity, could not take place at all. 'Blank-page', water-based life depends entirely on the precise delivery of sufficient quantities of the right catalysts. These printers of an organism's script are called enzymes. They are accompanied, where necessary, by energy in the form of tiny, specially prepared 'standard units of packed sunlight' called *ATP*. Such script is 'come alive'. Life's minuscule, catalytic bio-machines are called enzymes. Enzymes are, to the three-dimensional position in space of an electronic charge, that is, to fractions of billionths of a metre, extremely accurately constructed.

What is minimal functionality? Enzymes (with their production system) and *ATP* (with its own metabolism) are absolute minima, along with code, membranes and the right raw materials, for any biological maintenance let alone creation to occur. Even in a 'simple' cell. All life's processes, including self-replication, depend entirely on them. Most biochemicals are sufficiently large and their conversions sufficiently complex to require suites of enzymes to turn initial input into an end-product. Rephrased, some starter chemical 'A' is fed through, say, ten steps but often many more during transformation to a target molecule (say, 'Q'). Each step involves a new substrate and another enzyme. Intermediate enzymes/ steps in the series are useless except for the critical part each plays in the chain. *Therefore some unspecified, currently unknown chemical selection would have had to have built up the correct sequence in order for it to work.* If any enzyme were absent its chain could not proceed. 'Q' would not be reached. In fact, no enzyme (or its code) would have *any* use before Q was reached; a useless candidate is deselected . How, thus, could life's pathways ever reach completion? **A metabolic pathway is, in other words, a molecular *developmental* pathway, one that develops towards a pre-arranged product. It is, with enzymes being like the lines in a highly structured computer program, anticipatory**. Each step is just means to an end. 'Means to an end' means purpose. In a word you call it teleology. With this you switch into a *top-down* mode.

Expressed the other way round, most proteins are enzymes and most enzymes useless except as specific intermediaries in a metabolic pathway; therefore most proteins must by definition constitute steps in such pathways and most genes must code for otherwise useless proteins. How come such great redundancy, prior to full complement and operation of a pathway, was retained while it was being built up step by step towards its metabolic target? Why does there remain no sign of any actual redundancy today?

The issue complicates. When different pathways cross by virtue of some common intermediate (and metabolic pathways do criss-cross) they use a common enzyme. When tracks diverge, however, what signals to the intermediate chemical which track it should pursue? More computational 'intelligence' is needed super-coded in the program-as-a-whole. But hierarchical layers of code, replete with integral complexity, never point towards chance; they firmly point the other way.

Was any metabolic pattern, just like developmental algorithms leading to a computer program's target product, ever woven accidentally? In short, by itself a lone enzyme is supposed to construct a physiologically useless intermediate chemical. Thus both enzyme progenitor and the gene that produced it are, by themselves, equally useless. **Use arises only at the moment a full and correct suite of 'colleagues' to complete the pathway fall into place.** Why did each 'struggling' and yet useless step survive the elements; why did any gene survive until the magic moment came? How could the cell survive without end-product until then; and, as theory says it must have, what advantage could a single (even if new) product-chemical bestow? Enzymes are all specified; a pathway amounts irreducible and specified complexity. How could non-conscious *DNA* or enzymes have attained a target they could never have conceived? Not once but many times with integrated pathways even in the simplest cell. How illogically materialists are swayed.

The fact is that every single metabolic pathway is, not only of itself but linked into its wider context, irreducibly complex. **Such pathways form the very foundation of life's continuity. It might therefore be concluded that if biology knows nothing of their evolution then it knows nothing of evolution as a whole.**

Because of its implications for the notion of abiogenesis and consequent resistance to its assertion, this point is well worth labouring. The whole point of an enzyme is its coded specificity, lack of redundancy and coordination within a rigorous bio-system. How many useless enzymes did or ever could any such system tolerate? How many impotent 'catalysts' are known? None. Life is efficient and, of its many interlinked pathways, few if any enzymatic and therefore biomolecular redundancies are even suspected. *It was noted that the first cell would have required at very least complete energy and reproductive metabolisms, with associated genetic hardware and software, in place before it had any 'survival' value at all.* **In short, multiple pathways including all intermediate enzymes, for which no 'selection' process could exist, must have arisen at the same time within the same ten millionths of a cubic metre of impossible puddle! Enzymes are themselves the product of code and so, with appropriate controls, must have been their genetic precursors. Is the enormity of the problem for evolutionary theory dawning?**

Information input into bio-systems is the cryptic factor that the blind-fold of a mind-set will not let its own professors see. Who generates strict protocols by chance? Genetic, metabolic and other signal protocols are manifold in every cell. How illogical, therefore, appears the rationalist's claim that chance made life!

To suggest that chance generated order of such sequentially-targeted, diverse-yet-integrated and hierarchical magnitude is a kind of madness. It is simply babbling self-delusion. To suggest, conversely, that mind-based reason is its cause is (as all technology confirms) wholly reasonable. Look at things another way. Consider a biochemist's complex, multistage synthetic flow sheet. At each stage of every reaction binary decisions ('yes' or 'no': 'and' or 'nand') are computed. By blocking or widening pathways the chemist coaxes the reactants through a maze to his target. He manipulates various conditions to best exploit the properties of intermediate chemicals. In life, such injections of information appear in the form of enzymes. In this sense enzymes are like 'crystallised' intelligence. There is meaning behind their structure. The code for that structural and therefore functional semantic resides in *DNA*.

As previously noted each metabolic pathway is, effectively, an application program. Suites of such 'bio-computation' are always perfectly (that is, very specifically) integrated into a larger 'bio-system' called, at least, a cell. In chemical terms they constitute an amazing biosynthetic flow-sheet that depends entirely on the correct space-time presence of selected, code-generated protein. Just as a production line where one component is absent stops, so a pathway cannot work until every enzyme is present and, by code, correct. It cannot start. It has neither survival value nor means of being selected naturally unless both it and every other enzyme needed to secure the *target* chemical have been made. **So intimate is the association between enzymes and biological life that the problem of the origin of life on earth can be seen as that of the origin of pre-coded enzymes.** *How could either cellular DNA or protein 'emerge' without the other - especially without their functional, linguistic interlinkage? How, in quantity, did they spontaneously arise together?*

Metabolism involves algorithmic chemistry. *Pathways,, it needs be emphasised, depend on suites of intermediate enzymes precisely tailored to the step-by-step transformation of a raw material into a target molecule. They are entirely purposive and, moreover, each intermediate enzyme is useless except within the context of its pathway. It is all or nothing.* Metabolism involves algorithmic chemistry. **Each route is an irreducibly complex system. Understandably, therefore, the silence that surrounds the evolution of metabolic pathways or, better defined, metabolic programs is deafening.** I once asked the abovementioned enzymologist, Professor Dixon, whether under the circumstances outlined in this section he believed that a cell could have abiogenetically evolved. He said no.

Energy Metabolism Perchance?

Because it must have been there from the crack of life's dawn, fully fledged, intact, let's examine energy metabolism. How well, therefore, does the case stand for this crucial, primal pathway being preconceived? What might a silk in court propose?

Cell metabolism is accurate, rapid, incessant, labyrinthine, responsive, purposive, homeostatically controlled and dependent on specific protein catalysts called enzymes. The presence of these is in turn dependent on relevant genetic hardware and software, on a sufficient supply of the correct raw materials and on energy. This applies as well to bacteria as any other kind of organism; even so-called 'simple'. 'primal' systems marvellously complex yet precise. *Energy metabolism, a superb example of metabolic parsimony and elegance of design, pumps at the heart of life's engine. You can improve on chance; thus how, if you were asked to invent such a conceptually beautiful system from scratch, how might you devise its operation?*

The Light to Life Energy Conversion Chart:

The parts of this chart illustrate how, like two inseparable sides of a coin, energy metabolism is a good example of dialectical polarity incorporated into symmetrical, complementary design.

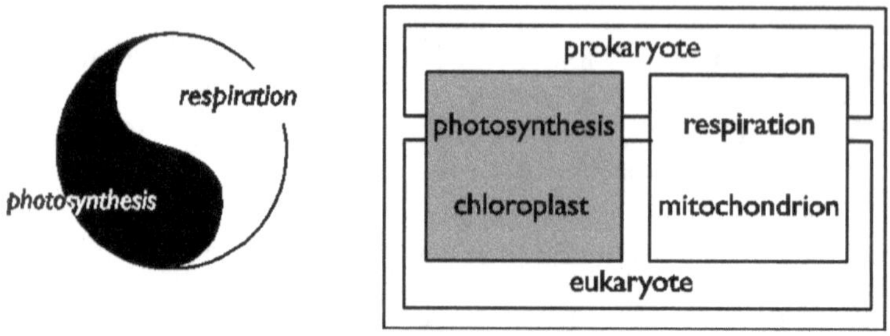

(i) Complementary energy units incorporated into life's phenomenally sophisticated engine express its dialectical polarity. Take the basic equation for photosynthesis (light + water + carbon dioxide > oxygen + sugar) and run it in reverse. You then have the basic chemical description of respiration - except that an input of light has been replaced by an output of 'units of frozen light' called *ATP*.

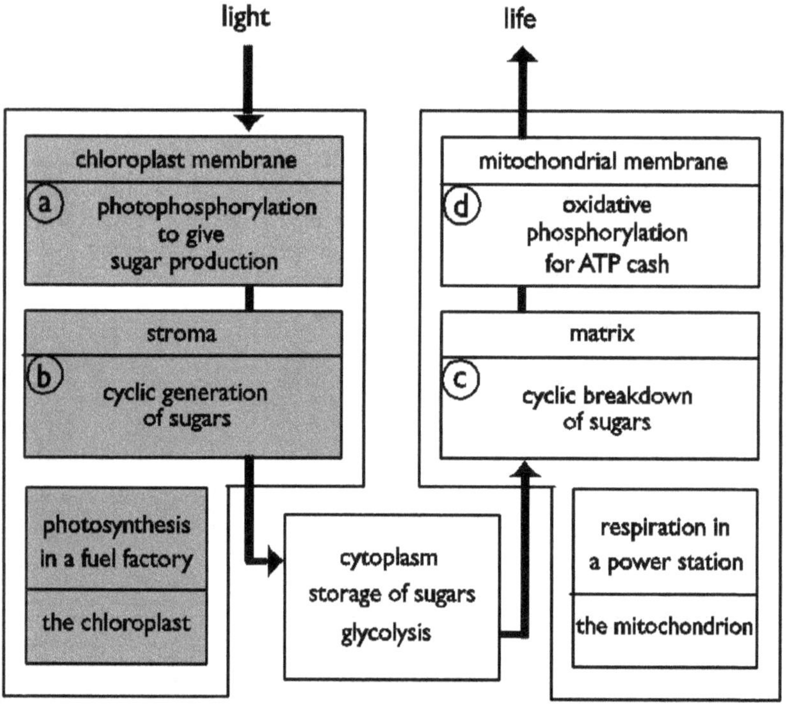

(ii) *Chloroplast* and *mitochondrion* are, like a dynamo and an electric motor, mirror-imaged opposites which interlock to drive life's engine. They are as complementary as right and left-hand gloves; and they amount to a fuel factory and a power station.

The conceptual point of a chloroplast, which is vested with its own *DNA* for the synthesis of crucial proteins, is to reduce carbon dioxide to glucose, that is, make life's fuel. Photosynthesis is a two-phase process. Products of the 'light phase' (a) are used to drive the sugar-productive 'dark phase' (b). The former occurs across proteins embedded in inner membranes called *lamellae*. Water is split into an oxygen atom (which is evolved as a waste product), two electrons and two protons. Chlorophyll with its porphyrin ring is the initial reaction centre in energy metabolism. Light falling on this molecule excites electrons for use on an *ETC* (electron transport chain) which photo-phosphorylates *ADP* making *ATP*. The two protons are pumped inside the lamellar membrane where they are prepared, with *ATP*, to promote the dark phase. Carbon, the kingpin of life's chemistry, is ushered through its 'dark' portal and sugars thereby biosynthesised. The process occurs in the colourless cytoplasm of a chloroplast called *stroma*. *ATP* and protons drive the Calvin cycle whose offspring are sugars, fatty acids, glycerol etc.

The reason for a mitochondrion, which is also vested with some genetic autonomy, is to oxidise glucose and other food materials i.e. to burn life's fuels in an efficient way. Sugars are 'stripped' by a Krebs cycle (c) situated in the mitochodrial correlate of stroma, called *matrix*. Their carbon dioxide 'wrapper' is thrown away and the precious fuel, hydrogen, conveyed to proteins embedded in the folds of an inner membrane called *cristae*. Oxidative phosphorylation (d) on an *ETC* generates *ATP* using a complex molecular 'dynamo' called *ATP synthase*. Two 'spent' electrons, two protons and an oxygen atom then reconstitute the molecule with which it all began - water. Oxygen is, in your case, carried to the site of electron pick-up by haemoglobin with its porphyrin ring. The waste product of the process is carbon dioxide which the haemoglobin helps carry away. In this way, where chlorophyll initiated, haemoglobin finalises energy metabolism. Where punctuated reproduction involves life cycles energy metabolism is a critical, continually spinning wheel.

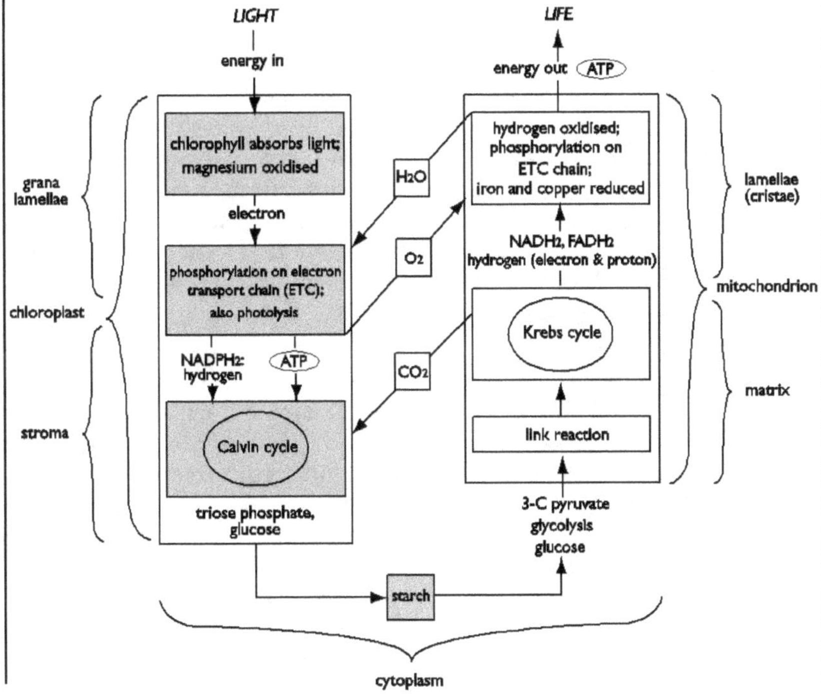

(iii) Aerobic energy metabolism: this is a more technical representation of mirror imagery shown in archetype (ii). **The presence of such an elegant, symmetrical and efficient solution to the problem of life's engine suggests that chance has been thoroughly outwitted. Indeed, not known for its ingenuity, it likely played no part at all.**

In fact, light to life can, in the electro-dynamic sense of Chapters 16 and 17, also be construed as light to light; and, in a few specific cases, light to light occurs when, from photosynthetic input, energetic output in the form of bioluminescence is derived.

fig. 21.1

↓	*tam/ yin*	*raj/ yang*	↑
	anabolism/ small-to-large	*catabolism/large-to-small*	
	light capture	*heat release*	
	bond made	*bond broken*	
	storage	*retrieval*	
	materialisation	*dissolution*	
	photosynthesis	*respiration*	

Do you remember, from Chapter 1, the existential dipole of information and energy? A match delivers a standard amount of heat. Biological energy comes in the form of phosphorylated nucleotides, mostly *ATP* (adenosine triphosphate). *ATP*, life's universal currency of energy exchange, can be seen as a frozen spark of sunlight, a latent flame or a standardised 'match' of heat supply. The general fabric of metabolism is geared to this unit which, with enzymes, makes things go. To humans, life is most obviously sensation and movement. Both of these depend on *ATP*. *ATP* itself depends, for fabrication, on complex and highly integrated programs called photosynthesis (except for a few bacteria with their own chemosynthetic complexities) and respiration. That's the rub. To make *DNA* you need protein and to make protein *DNA*. Both are needed to make *ATP*; you even need that molecule to to spark the universal respiratotry process called glycolysis. It's required to prime production of itself! **So which, with all this causal circularity, came first - energy metabolism, *DNA* or protein?** Which egg preceded chicken? *All* life, that is, *all* cells, including any hypothetical prototype, exact the precondition of fully-fledged, bug-free energy metabolism.

If animals constitute a range, with man at the 'high end', of awareness and ability to process information then unconscious plants are, at the 'low and dormant end', rooted in the energetic domain. **In this domain photosynthesis and respiration constitute complementary, interactive free energy devices, that is, metabolic machines** They compose two sides of the same coin, intimately linked (*fig.* 21.1) in a mirror image of each other. *Energy metabolism thereby illustrates, at the very heart of nature, a perfect example of natural, dialectical reciprocity and the balance of equation in action.* Plants, unlike animals or fungi, perform both tasks. They ground solar energy, they fix it in each glucose crystal and make the other biochemicals on which material forms of life depend. They also burn sweet fuel and charge themselves with energy.

They respire. Both sides of energy metabolism are played out in them (but only one in you and me).

Before investigating the exquisite way in which plants transform light to support life let us hear mean charges of their imperfection laid. For a start, does not chlorophyll poorly absorb green and orange lights? Various pigments, such as carotenoids, relieve the 'deficit' but doesn't the report read 'could do better'? On the other hand, you might query the effect of 100% efficient absorption of visible light. It might not, like 100% oxygen in the atmosphere, be for the best. Might 'souped up' plants not weave a tanglewood arresting other forms of life? In fact the biosphere is not unbalanced by a choking overgrowth of giant plants and there is certainly an aesthetic compensation in that plant-life is not coloured all-absorptive black.

It is also heard murmured that a few bacteria (called chemoautotrophs and methylotrophs) do not require the complex materials that derive from photosynthesis. Because of the 'simplicity' of their nutritional requirements such organisms used to be labelled 'primitive'. However closer inspection has revealed biochemical machinery at least as complex as found those organisms that need to ingest carbohydrate. So-called archaebacteria complex a sodium gradient with phosphorylation no less sophisticated than ours. Moreover, in evolutionary terms, aerobic chemoautotrophs and methylotrophs must have evolved as a consequence of the development of oxygenic photosynthesis. Even the simplest conceivable cell has complex phosphorylation and carbon fixation systems entirely dependent on suites of enzymes accurately produced by genetic hard and software. It also embodies precise information in order to expedite the precise patterns of its own reproduction.

In fact, photosynthesis, one of the most complex natural processes known to science, is no less than the gateway to life on earth. Not only does it 'freeze' light's energy to sugar crystals but is also the sole and single portal through which inorganic carbon is ushered to the biological temple, the sacred container of life, a body. Such fixation from fluent air to body chemistry is absolutely basic to biosynthesis. What therefore preceded bacterial or other type of photosynthesis? How did such crucial capture of light happen before this mechanism was in place?

tam/ raj	*sat*
complex	*simple*
practices	*principle*
effects	*cause*
conversions/ transductions	*light*

↓ *tam*	*raj* ↑
light-to-sugar	*thermal unit (ATP)*
'earthing'	*'de-earthing'*
light absorption	*heat emission*
storage/ fixity	*action/ flux*
fuel	*blast*
fuel factory	*power station*
chloroplast	*mitochondrion*

While a cell can wait a few days or even weeks for the molecular biology of its reproductive apparatus to evolve, it needs energy metabolism immediately! That is, it needs mirror-imaged photosynthetic (or chemosynthetic) and respiratory systems coincidentally, immediately.

This is too much for anyone to swallow. Thus many stories assume that organisms, somehow sufficiently kitted with genetic machinery and correlated respiratory metabolism, used organic materials that were conveniently lying around. Then, as a few bacteria do, they started to photosynthesise using hydrogen sulphide. Why not water? Because a process called photolysis, which is still not fully understood, releases a metabolic poison which, you may remember from section (i), could have wiped out nascent, anaerobic life: this poison, oxygen, is tolerable only because of complex biochemical mechanisms that *must have* evolved to suppress its harmful influences. Any kind of photosynthesis is an elegant, targeted, irreducibly complex mechanism for transmuting the pure energy of photons through electronic and protonic stages into the 'frozen' storage of a chemical bond. It reciprocates and is meaningless without its complement, respiration. *Such intertwined biochemical artistry is hard to explain as the product of chance.*

Do you remember an absurd equation and consequent considerations (Chapters 7, 9 and 10) of Triune Infinity, One and Nothing; and its linkage with a transparent, physical tri-unity/ trinity (Chapters 10, 11 and 12) of vacuum, time and matter's 'holy ghost', light? The second trinity links with a third still lower form of transparency that supports life on earth. The diagram shows how a triad of pure energy (the 'holy ghost' of light), gas (in the form of air) and liquid (in the form of water) are fixed together as the biological 'tri-unity' of a solid plant. This fixation supports the vast majority of life on earth because a plant's photosynthesis manufactures the precursor molecules from which pathways branch to make sugars, amino acids, *DNA* nucleotides, lipids and other biochemicals. We depend on plants and they on their transparent trinity.

First, take chlorophyll. Because this molecule only intercepts visible light and is insensitive to the UV part of the spectrum where the sun's rays are more intense, it might be considered an inefficient molecule. Alternatively, just as *DNA* may represent *the* optimal chemical template for the storage of coded information, so chlorophyll may represent *the* optimal design for the bio-collection of raw energy; as may its respiratory antagonist, haem, for the carriage of oxygen. Both haem and chlorophyll incorporate a key mechanism in life's energetic industry - the porphyrin ring. *Is it possible, in principle as well as practice, to imagine how the multiple steps required to produce such rings in full working order, mounted on appropriate protein frames, could have arisen from an incalculably freakish series of accidents?* There exists a base level below which no mechanism (which word implies, except in its loosest sense, purpose) works. Why should inanimate nature serially 'select' a multiplex pattern of atoms in complex, specifically shaped molecules that have, by its own lack of definition, no purpose? Even if a porphyrin ring had a purpose, how could the whole, precise environment of its construction occur if each stage, each biochemical or proto-biochemical did not work? Did not work, that is, because of its own unviable shape, its lack of relevant neighbours or the build-up of a minimal, coded metabolic system was incomplete? Due to the normal processes of chemical breakdown time offers no assistance at all. *The longer its period of 'natural development' the worse the problem of systematic, integrated complexity becomes.* How quickly do sand castles erect themselves against the tides? All dynamic complexities 'evolving by chance' against the implacable flood of time crumble. The same circumstance applies to each and every biopolymer irrespective of whether or not it can be produced singly and non-metabolically.

Note (*fig.* 21.1) the supremely balanced, reciprocal cycling of oxygen, carbon dioxide and water whose bio-philic efficiency is further elaborated in Chapter 26; the complementary processes that occur in life's fuel factories (chloroplasts) and power stations (mitochondria); and the way that photosynthesis 'grounds' and, in a very systematic way, recycles light originally derived from the solar fusion of hydrogen atoms. Throughout terrestrial energy metabolism hydrogen - the lightest element - can also be considered the key resource.

Light kicks off bio-life. **Photosynthesis starts with 'light capture' by a chlorophyll molecule** whose functional component consists of four aromatic rings arranged around a central metal ion (magnesium). From the lightest particle in the universe, a photon, energy is transferred in a trillionth of a second to the second lightest, an electron. This charge then runs the gauntlet of a phosphorylation system (called an electron transport chain) embedded in a membrane and makes some *ATP*. **Simultaneously (using a catalytic core composed of three manganese, one calcium and four oxygen atoms) a process - found properly located in all plants, algae and 'primitive' blue-green bacteria - splits water into hydrogen and oxygen.** Such precise, complex and critical photolysis replenishes the magnesium of 'used' chlorophyll with electrons, delivers the hydrogen round which glucose will be built to a carrier ready for the next stage and still returns some water to the system! The oxygen released is what you breath. **Both 'light-capture' and 'water-splitting' capacities are irreducibly complex, coded necessities for embodied life.**

Next proton and electron (in the form of hydrogen) with *ATP* are carried to the chloroplast's 'cytoplasmic' site called *stroma*. Here they are used in the so-called Benson-Calvin metabolic cycle to make glucose and, from this start, all the biochemicals a plant needs to grow and survive. At this point a further unhappy accusation of inefficiency is levelled against photosynthesis.

The cycle is kicked off by a huge, sixteen-part complex of proteins working coherently as an enzyme called RuBisCo. RuBisCo is most important in terms of biological impact because it catalyzes the primary chemical reaction by which inorganic carbon permanently enters the biosphere. It is carbon's usher at the portal of life's temple. Here its function is to yoke carbon dioxide to a sugar (ribulose).

How could it be that evolution has still failed to perfect this, a most abundant, life-critical protein complex? What's the carp? It certainly reacts 'slowly' with CO_2 and may seem to confuse it with another similarly shaped molecule, oxygen. Combined with O_2 the mal-evolved RuBisCo trips an apparently wasteful process called photorespiration. In fact, unfazed RuBisCo does discriminate and it is the transition state of this positive discrimination that slows things down. In other words, the trade-off for discrimination is a slower rate of catalysis. Moreover, refinement occurs in the form of different binding strengths adopted for CO_2 in different thermal and gaseous contexts. In hot, oxygen-heavy environments it binds tighter to its preferred molecule and *vice versa*. Clever stuff!

And in extreme cases, where CO_2 is at a low level or unavailable, the safety valve called photorespiration kicks in. It is as if, for every three pedal thrusts on your cycle, the gears slipped for two - a tiresome waste of energy; but this energetically expensive slippage releases the 'pressure' of oxygen and *ATP* build-up while (here lies the charge) inefficiently creating a molecule called *PGA* from which plant materials may be created. Since the waste product is CO_2 there is still an opportunity to retrieve a little glucose in the end. In fact, it may be that photorespiration promotes the assimilation of nitrates from soil. The pathway is certainly a major source of H_2O_2 and thus makes a key contribution to 'cellular redox homeostasis' - a process that controls and thereby balances the production potentially harmful free radicals. In turn, such activity influences multiple signalling pathways, in particular, those that govern hormonal responses controlling growth, environmental and defense responses and apoptosis (programmed cell death). It may be, due to such coupling, some level of photorespiration is found to occur even in relatively unaggressive conditions. Some plants engage mechanisms to reduce uptake of molecular oxygen by RuBisCO in the first place. Such 'patches' or subroutines appear, with no particular ancestral consistency, across the class of flowering plants. One complex 'add-on', found in a mosaic of diatoms and about 30 unrelated sorts of plant, is a 'convergent evolution' called C-4 photosynthesis. Most are grasses, almost none are trees and 11 genera include both C-4 and the 99% prevalent C-3 methods. If photorespiration is a fitness problem why do C-4 plants not dominate the earth? In arid places plants such as succulents and cacti use another method to keep 'carbon-fit'; it is called Crassulacean acid metabolism (*CAM*).

RuBisCo is half coded for by genes in the nucleus, half by others in the chloroplast; then dedicated proteins fit the parts together so it works. Has anyone devised an algorithm to precisely detail this enzyme's first appearance then its step-wise evolution? A photosynthesizing plant is superior to a human in its biochemical self-sufficiency. Indeed, human life is as irrelevant to plants as theirs is essential to us. No plant, no human. An excellent supremacy that leaves dependents in the shade! If I gave you mineral water in the daylight could you make yourself? Amazingly, a plant can. Moreover even the most brilliant engineer has not been able to simulate or excel its almost optimal energy conversion process. *To be able to translate such strategic information into the artificial conversion of solar energy to fuel would solve any terrestrial food problem at a stroke!* **And, if you could commercially mimic photolysis, the world's fuel problem too!** Very cleanly! Simply split water, use oxygen and hydrogen to 'oil' the wheels of transport and industry and dispose of their exhaust. The exhaust, same as the starting point, would not pollute. How could pure water dirty earth?

How could photosynthesis evolve? Please, don't cry 'simple'! Even the 'simplest', bacterial form of the process is amazingly complex. It involves a multi-enzyme complex that includes pigments, reaction centres, light harvesting complexes and the generation of electrochemical gradients. What is the point past which the mechanism could be pushed to lose its vital functionality? At what point does irreducible complexity, composed of much code specifying many precise parts, kick in? How could wind, rain and currents in some water gradually determine such a light-fantastic, life-supporting plan?

All this photosynthetic business is only *half* the set-up that is critical for any moment full of life!

Food is a packet of 'heavenly light' combined with earthly nutrients. Whether directly or indirectly (as non-vegetarians), we ingest plant-manufactured material from which to build or maintain our own bodies. To obtain energy (*ATP*) an opposite breakdown process called *respiration* mirrors photosynthesis. Various substrates (sugars, lipids and protein) are translated into forms from which *ATP* can be obtained. All organisms use an anaerobic routine based on glucose that needs *ATP* (its product) to prime it. It is called glycolysis. *Thus ATP is used but none produced (from a previously metabolised three-part ADP molecule) until the tenth step.* **This is, for any original organism, a chicken-before-the-egg situation.** *How, furthermore, could both such a complex, targeted metabolic pathways come into being together but gradually by accident? How could its individually useless intermediates, made up by enzy*mes like glyceraldehyde phosphate dehydrogenase (which is itself composed of four identical, exac*tly sequenced chains of 330 amino acid residues) be coded by chance and wait for each of the rest to 'happen'?* How, indeed, could any highly informed metabolic pathway pop up, interlinked with all the others, just by chance? And by what precise steps could regulatory mechanisms, which can include the presence of hormones like adrenaline, thyroxine, insulin, glucagon, somatotrophin and many others, evolve? What happened prior to any regulation? How did genes for construction of the critical 'hydrogen shuttle', *NAD*, without which 'simple', 'primitive' anaerobic respiration would not work, coincidentally aggregate at the same micro-time in the same micro-space of the abiogenetic proto-cell? Indeed, how were the ribose sugars that help make *DNA* nucleotides created before glucose oxidation.

Anaerobic glycolysis is the 'poor' relation. Its 'chip' reflects a flowchart of many integrated factors but it fades in complexity beside the 'cash machine' - aerobic respiration in mitochondria. After further unpacking by precise nanotechnology, 'primal' hydrogen is passed by practically the same carrier to practically the same membrane-embedded electron transport chain (or *ETC*) system as was used in photosynthesis. This chain is yet another protein machine - one that precisely handles electronic charge to a key metabolic end. Made of many parts its 'entry enzyme' is a molecule, called *NADH* dehydrogenase. This large and complex piston occurs in all kingdoms of life where it can also pump protons to generate essential membrane potentials. In the *ETC* hydrogen's proton and electron are split apart, steered through various components and used to power a molecular 'dynamo'. The dynamo is incorporated into the form of an amazing 31-part complex called *ATP* synthase. It employs a wheel-like component that rotates. Wheels are unusual in biology. In this case the wheel acts like a mill powered in most cases by a flow of protons. Controlled acidity is harnessed to drive a huge, refined and marvellous construction; in other words, the rotary motor converts a current of electrical charge through mechanical energy to generate (by means of a cam displacing the turbine's rotation) a universally 'agreed' biological unit of energy. You could fit millions of the whole molecule, turbine and all, within the volume of a pinhead! Trillions in your body create its weight of *ATP* each day and, if they failed to work, you'd be dead before you dropped! Not one of them could be useful unless (as a result of a highly accurate software system) it was made exactly right for its job in its particular environment. Rotary motors driven in a single direction by some current are the holy grail of all micro-devices. Yet the all the efforts of intelligent scientists have so far built up to results crude and weak in comparison to the operation of *ATP*-synthase. So how did the extraordinarily complicated, co-integrated but critical bio-mechanism, found in archaea, bacteria and eukaryotes, first 'pop up'? Was it by the laws of chemistry and physics? Each aspect of its physiologically-integrated structure and function needs be as sharply explained in terms of evolution (which its coded profile does not fit) as bio-engineered design (which it does). Only then we'll make a toast, we'll sing a scientific hymn - one in awe and wonder at this most unlikely fluke!

High-speed, high-fidelity performance - metabolism always works like this. As usual, therefore, the synthase construction only works when all its components are present and correctly assembled. An irreducibly complex set-up then mass-produces biological power. Idling at 6000 revolutions per minute the mini-turbine grinds out energy at a rate of three units per revolution! Each unit is called *ATP*. In such *ATP*, as opposed to glucose, chemical storage of energy resides in a bond between two phosphate molecules. This phosphate-phosphate bond, which constitutes a 'heat-potential', is the metabolic destination of the sun. It is the precise location of chopped sunlight, a thermal unit that is 'cut-and-dried' till called upon. When called to service it is biologically broken. Heat is released and, like an exactly placed and microscopic Bunsen burner, drives reactions. The word phosphorus' means (from the Greek) 'light-bearer' - the bearer, in this case, of starlight shed onto the subject of terrestrial life. Unlike 'photosynthetic' *ATP* 'respiratory' *ATP* is not tied to a single use. Every part of each body is geared, like a slot machine to a specific coin, to use this standard currency. All biosynthetic pathways need it. It is available for any regulated metabolic employment. How, as life's lamplighter, was it first 'chosen'? How did it plug in and supply, at the first instant and the correct 'grid voltage', power to its associated appliances? How were its many different metabolic purposes crucially, immediately coordinated in the first cell's larger scheme?

After *ATP* production things settle down. Electron and proton are reunited as hydrogen. At this juncture oxygen previously discarded in photosynthesis has been brought to the reaction site - in the case of many animals by a aromatic analogue of chlorophyll that, instead of magnesium, has an iron atom at its heart and is called haemoglobin. Hydrogen is now coupled with this oxygen and thus restored to its original form, water. The energy cycle is complete. You breathe to make water. You breathe in what plants breathe out. You breathe out the carbon dioxide and they breathe it in - to make your food; you excrete water, they absorb it. *How neat, ordered and efficient is the mirror imaged photosynthetic-respiratory cycle and, with it, natural plant-animal dialectic.* It is easy to retrospectively conjure inventions from imagination but their actual implementation, in working order, is another matter. Can those who claim the system *must have* 'self-concocted' piecemeal over, at a minimum, thousands or millions of years offer some serious thermodynamic and molecular constructions as to how?

An evolutionist argues that solar energy is sufficient to generate local increases in information represented by complex chemicals and, even and eventually, life forms. *This is untrue.* No doubt an open system (such as earth with sufficient incoming sunlight) is necessary but its raw, uncontrolled energy is destructive. You might as well throw some petrol at a car and expect it to work! Or, more accurately, ignite a long-term fire and expect it to assemble and then power the assembled vehicle! *In fact, as well as raw fuel you need, as a prerequisite and not an afterthought, energy conversion mechanisms and, hierarchically, a control system directing, maintaining and reproducing these conversion systems - that is, you need encoded photosynthetic and respiratory complexes as described above incorporated in a cellular context before you can even start.*

In short, to build the biologically complex (a cell) from the physically simple (a hypothetical mix of the right organic chemicals) not only energy conversion mechanisms and a control system directing, maintaining and reproducing the energy conversion systems would be needed. You also, critically, need the information, inscribed in code with its operating system, in place to coordinate the whole cell's balanced constancy of being.

Such concepts deposit you in a totally different, non-evolutionary ball-park - the real biological world. Electrical and thermal engineers are in awe at the highly miniaturised, optimally efficient operation of a self-reproducible system quite unlike their own technologies. The indirect and ultra-fine application of sunlight through the medium of *ATP* means that diverse kinds of work can be carried out through equally diverse but integrated, highly controlled and ubiquitous metabolic mechanisms. **To cap it all, the purposive instruments of life's universal engine are pre-coded. Both hardware and software are together prescribed.**

Bottom-up, the evolution of subroutines that support the coordinated elements of energy metabolism are supposed (despite the fact that a primal cell would have needed both intact and integrated from the very start) to have risen as oblivion strolled through time. In eukaryotic organisms its presence is encapsulated within organelles called chloroplasts and aerobic mitochondria. How could such complexities, the platform whence multi-cellular evolution must have been projected, first occur? One theory, of endosymbiosis, was the original 'brain-child' of Lynn Margulis, first and philosophical spouse of widely respected and missionary scientific atheist, Carl Sagan. Since the abovementioned organelles have, along with the nucleus, a double membrane she and associates vigorously argued that various bacteria have entered host cells, integrated symbiotically and, over time, lost their own autonomy. In other words, the circumstance is explained by the fortuitous incorporation of one kind of bacterium in another. For example, similarities of biochemistry between some photosynthetic bacteria and chloroplasts have understandably (since they do the same work) been identified.

The theory (see also Chapter 24: The Origin of Asex) underwent severe criticism but is still touted because, after all, until better idea turns up what alternative is there to explain a critical juncture in evolutionary history? 'Scientists have pinpointed the single tiny event that created all plants.' In the brash manner that media outlets typically promote a popular version of evolutionary origins The Sunday Times (9-11-08 p. 11) trumpeted the decades-old theory that a 'bacterium turned its attacker into the ancestor of every tree on the planet'. The Times simply cyclostyles classroom currency.

What problems are there? Some but not all mt-*DNA* (mitochondrial *DNA*) is, in bacterial mode, circular but there is lack of convincing evidence that one prokaryote can ingest another *and* keep it alive in order to allow symbiosis. How, furthermore, would mt-*DNA* from one bacterium migrate into the other and become integrated with it - especially as regards the couple's necessarily synchronous replication? How might Krebs cycle and *ETC* components have found their correct locations? Indeed, present-day evidence shows that mt-*DNA* functions are not self-supporting but controlled by nuclear *DNA*. Of course, an out-station for the on-site delivery of basic materials (such as a cement mixer on a building site) might be construed as a measure of efficiency. If so, though, how did the supporting eukaryotic nucleus, linear *DNA*, the correct controlling proteins and mitosis all occur together in a single prototype; or, since this is another black-box, 'dog' of evolutionary gap hypothesis, how did things gradually occur - remembering plausibility indicates no more than possibility. Furthermore mt-*DNA* code contains no introns and, anyway, differs from the universal protein dictionary. As a 'foreign' component paternal mt-*DNA* in an egg-cell is actively decomposed so that inheritance is matrilinear. How did this required mode of resistance come about? Since mitochondrial quantities are governed by cell needs and rails of cytoskeleton laid down to carry them about you could argue forethought as regards dynamic plan.

Another insertion exercise would have occurred (some argue several times) to generate plant chloroplasts; indeed, plants would have to have 'discovered' both kinds of energetic 'mate'. All eukaryotes would also need to have developed double membranes and their nuclear pores as well as to have developed organelle control genes. No *LECA* (last common ancestor of eukaryotes) nowadays exists but what, you guess, might such a nascent absence have chemically consisted of. On the other hand, from a *top-down* point of view, the theory accretes incredibility; it papers thinly cracks in history. Such gap is easily replaced by missing, immaterial but forbidden fruit - intelligence. The theory is thence replaced by one of pre-designated subroutines for which - as in the case of plans for a petrol engine - you might expect related coding adjusted to accord with each specific model. However, while it would be illogical to conceive of a petrol engine that did *not* incorporate the unseen mind of its creator it would, for a materialistic rationale be illogical to conceive that, for natural phenomena, it did! This attitude predicates evolution and debars intelligence as the necessary source of information.

In fact, not only the operational fact but the hierarchical symbolism of energy metabolism's extremely precise, purposive manipulation of subatomic particles is awesome. Pure energy (light) is transduced through an electrical system; mass, in the form of a proton, is co-opted so that having stepped down through three grades of subatomic particle life-giving photons are stored as a molecular hydrate of carbon, a soluble white crystal called glucose (or its polymer, starch). Iced light. The primary respiratory substrate is then, as required, restored to potential energy in a chemical bond. And when this standardised sliver of the sun, this *ATP*, is struck like a match at point of ignition it provides specific heat to drive metabolism, motion and sensation. Sunlit life, sunlit in all its parts, at your service, sir! If puddles cannot factorise life's chemicals, plants can. Plants are the world's Jeeves. They are biochemical plants that, from paths that branch off photosynthesis, fabricate carbohydrates, lipids, amino acids and *DNA* - a service on which the rest of life is thoroughly dependent. You need greens but they do fine without you.

Energy metabolism is fundamental and immediately necessary for every part of every organism. Is it possible to infer from this combination of facts (*fig.* 21.1) that such finely tuned interaction and mirror image operations comprise a single underlying concept - life's engine? Whose two modules (photosynthesis that grabs energy and respiration that systematically, usefully releases it again) combine with adjunct organelle, organ and whole-body systems to form a single, balanced power grid? Indeed, details of the clearly polar, complementary cycle of energy metabolism have been highlighted in order to demonstrate its elegance, symmetry and informed directions. It is not that photosynthetic and chemosynthetic bacteria do not exist, rather that the prototype cell would, from its inception, have needed energy metabolism complete - along with the correct genetic code to produce entire, minutely specific suites of necessary enzymes, reproductive system, cell membranes etc. etc. All would have to be in fully cooperative order. A tall order! Fine chance! Superfine chance! On the other hand, interactive, complementary design that is based on and packed with information typifies the perspective Natural Dialectic brings to biology. A to-fro dia-logical perspective! Such tight-knit integration of the Dialectic's main components, information and energy, evaporates every last drop of chance away from the power of life. *Is this why there has been devised no precise account of the binary system's co-origination nor detailed flowchart of how the program could have developed step by gradual, accidental, chemical step*? **In other words, the critical evolution of energy metabolism is, with that of general metabolism, the universal system of information storage and transmission and the technique of protein construction, another black box.**

Natural Nanotechnology

Technology involves machinery. A machine (Chapter 6) is a body or assemblage of bodies used to transmit and modify force and motion, especially a construction in which several parts unite to produce intended results. **Machines work according to natural law but the latter cannot explain them.** This is because machines are informed by purpose and, in obtaining the fulfilment of that specification, are irreducibly complex in a way that purposeless forces and particles can never be.

What does 'irreducibly complex' mean? Engineers develop, build and maintain machines. Professionals know perfectly well that a multipart machine must *start* its working life as a well-designed, functionally coherent unit; both then and thereafter, except perhaps for superficial faults, unless everything works nothing works. **'Irreducibly complex' therefore means, with respect**

to a system or a mechanism, one that must have several parts each made of the right material, the right positional relationships between parts (which may include quantum elements such as electrical charge) and, in the case of a biological mechanism, the right 'grammatical' instructions encoded in a manner that allows it to access data, perform tasks and repair and reproduce itself. Such criteria must all be met. ***The parts must be simultaneously present and built to perform to the standard of minimal function.***

Minimal function means a mechanism must fulfil its promise to the extent that an efficient, acceptable level of performance is achieved. Such a performance includes not only the correct shape and interconnection of parts but the correct materials of which they are made. In the interests of efficiency technologists strive to combine minimum complexity of working parts with maximum effect at lowest energetic price. This applies with equal force in nanotechnological engineering, that is, the development of functional machines at molecular scale. Such engineers are learning how to develop tiny devices such as wires, switches, valves, shuttles, rotary motors and, as they seek to mimic bio-materials, protein catalysts and *DNA* computers. In comparison with bio-machinery, however, their efforts are technically skilful but as yet unsophisticated. Natural nanotechnology is proving, with awesome excellence, to have got there first. Has ultra-sophisticated chance therefore preceded man in creating what are to all intents and purposes machines and not 'machines'? Could mutation honed by natural selection have produced, even in a trillion years, devices that life's cells must in their integrated circuits well combine to even start their metabolic motors up? You claim that, naturalistically, it must have been the case but how, from a *top-down* point of view, without prior concept could these technologies occur? Because they *are* technologies and not, as philosophers would have you believe, machines by definition. The word is *not* an explanatory analogy. Molecular bio-machines act as cooperative cogs inside the cellular watch. They mirror the diversity of designs produced by human engineers and increasingly, by breaking nature's patents, humans copy excellent bio-designs. The problem for atheism is that all machines (including bio-machines) need prior conceptual input. They require intelligent design. How can any machine and its blueprint 'happen' without such psychological component? You respond with a tired answer that the presence of chance-blown, 'self-replicating' *DNA* and, therefore, mindless code can change the whole equation. In Chapter 20 the cumulative, insurmountable difficulties that face this assertion were examined. **There's no scientific but only philosophical reason anyone should have to buy a story that's devoid of any pre-coded mechanism's vital, immaterial dimension, information.** Let's check what evidence can be adduced to post-materialistic interpretation of the facts.

We already peeped, in previous sections, at replication factories (called polymerases), translation machines (ribosomes), the RuBisCo complex and a motor called *ATP* synthase. Rotary motors are also used to inject *DNA* into host cells, transfer it from host to viral capsid (body) and, in bacterial conjugation, from cell to cell. Indeed, a class of *ATP synthase* molecules is often used to drive chemical reactions, pump elements across membranes, transport against concentration gradients and so on. Yet another classic example widely touted as evidence of psychologically engineered construction and, therefore, just as widely reviled is the flagellum. Many kinds of prokaryotes have this tail-like projection to propel them through fluids. Check the internet or descriptions in books. Take a close look at the motor that includes such components as stator, rotor (including a universal joint), bushes, drive-shaft, propeller and a housing of six or so electric motor. Together, as parts in a mechanism, these work to cause their flagellum to achieve, either backwards or forwards, a motion comparable, size for size, of you jetting through the water at 200 kph. Such practically weightless machines certainly 'revolutionize' the 'most primitive' of organisms. The minuscule technology of a rotary motor that can stop, start, change gear (including into reverse) and generate large propulsive thrust can also generate contention. Does pot not call kettle black? The definition of an irreducibly complex design ('in which all parts are indispensable to supporting a system's basic function') supports the notion of intelligent design. *This is a very dangerous idea.* Watches need watchmakers but atheism claims that such an explanation lacks, in the case of biological mechanisms, 'explanatory power'. More powerful, it is urged, is explanation that involves speculating (mostly with a waving gesture) how a vast improbability could accidentally, by mutation, create fresh and innovative 'designs'. Is it not fabulous how information-lacking chance with death (otherwise called natural selection) can engineer such wonders of the micro-world?

There is a fable here that, incidentally, involves 'black death'. Infections caught off germs are bad enough but you must never tolerate contagion that infects an evolutionary thrust of speculation. So it came about that after Michael Behe claimed the 'rotor motor' was evidence of 'irreducible complexity' an antidote was soon administered by Doctor Kenneth Miller. Because his rebuttal superficially appeared to counter Behe's teleology it both bamboozled a judge (Jones of Pennsylvania) and naturally excited a blatantly atheistic, neo-Darwinian biologist who enthused (Richard Dawkins in 'The God Delusion' p. 131) that Miller's argument 'is for my money the most persuasive nemesis of intelligent design'. Black death indeed! What was Miller's cut and Behe's counter-thrust?

The cut was that if one part of a system were also part of another (e.g. a battery could be 'co-opted' into a torch and an alarm clock) then neither system would be - despite its distinct structural design and entirely different functional purpose - irreducibly complex. A 'rotor motor' contains, with its propulsion device, a mechanism to transport some of the proteins that make up the device to their positions on the outside of the cell. This sub-system contains proteins similar to those also found in the secretion systems of plant and animal pathogens such as *Spirilla* and the black death's pest (*Yersinia pestis*). Therefore although a 'rotor motor' without these proteins is as inoperable as a torch without a battery Miller argues that it is not irreducibly complex! Indeed, he argues that the secretion sub-system might have been 'co-opted' and, jumbled with some other spare parts, accidentally have become a 'rotor motor'. Since plants and animals infected by the pathogens are supposed to have evolved after bacteria this seems even less likely.

Miller's cut completely misses because, counters Behe, the whole thrust of 'irreducible complexity' is that separate components are ordered to achieve a function beyond any individual one of them's capacity. For example, torches and alarm clocks have to be designed. And 'commandeering' a compatible battery from a torch might make an alarm clock work; but what about the efficacious origin of all the other integrated parts? Whether a co-opted battery fits or not, a power supply is an irreducible part not only of the operation but also the systematic conceptual make-up of these two complex although entirely different species of purposeful designs. Such 'irreducibility' applies to organ, organism, device or machine. It even applies, in this example, to sub-components of the battery itself. Add to this the fact that no detailed, step-by-step account of the Darwinian 'construction' of any biochemical or cellular system has ever been supplied. Thus, although intelligence is the only known factor to conceive (and even radically adapt or co-opt parts for) specifically informed, purposive designs, Miller denies it even while failing to supply counter-evidence. Therefore his battery of argument is flat, the nonsense of his pile laid low.

Indeed, although conceptually compatible (a rotary power source driven by proton flow) the actual components used in different secretion systems subtly vary. Many found in *E. coli's* flagellar motor are not, for example, found in *Yersinia's* hypodermic drill. Two-thirds of the proteins (some 40 parts) found in 'rotor motor' are unique to it. If, moreover, you knock out any single one of them the system fails to work; any integrated system missing pieces fails to work. Even with all correctly codified and present you need consider the origin of a complex assembly pattern. All said you might, pace Miller's incomplete and somewhat desperate account, infer that various sub-routines on the theme of 'secretion' add up to an economical, multi-functional and mosaic use of excellent engineering design.

On which interpretation shall we paste a 'black death' sign? A key aspect of Miller's argument is typically evolutionary. Emphatic arm-waving to the effect that one system *must have* evolved from another is standard practice. You had, unsupported by any detailed pathway, better believe it. What story can a good imagination not concoct? You might speculate that even the impossible were possible! Just such a magic wand has wavered over metabolic pathways, mitosis, meiosis, sex and a thousand other cases of innovative engineering. The fact is that Darwinians don't have a clue how 'irreducible complexities' evolved but only that, by scientism's incantation, they evolved somehow. *Innovation and not tinkering with innovation is, as Darwin himself recognised, the real problem.* Unless, of course, you stack the deck by craftily generating a computer simulation that incorporates, as its teleological end product, the very solution that evolution's chance and natural selection were supposed to have obtained, but craftlessly, from scratch! Hey presto - innovation! It is equally sad that his lack of balance failed to report any counter-thrust to Miller's fabulous remedy. If under atheism's glare you dare then simply 'google up' 'intelligent design', William Dembski's counter-thrust and other explanations. And, for example, you might care to study closely the biochemical operations of a single cilium on a single-celled alga called *Chlamydomonas*. It involves a complex routine called *IFT* (intraflagellar transport) that uses and exchanges many proteins in a highly controlled, coherent way. In The Edge of Evolution Michael Behe writes (p. 95) that "On the origin of the cilium/ *IFT* by random mutation, Darwinian theory has little that is serious to say. It is reasonable to conclude, then, that such theory is a poor framework for understanding the cilium." And, by implication, almost every other feature from integrated, hierarchical structures to genetic interactions as complex as those of computer logic circuits that we find everywhere in biology! One is left to ask whose logic has been infected by the consideration of unicellular spectaculars. Was it illogical chance or logical design first generated mechanisms (programmed to reproduce themselves) which have other functions as well as locomotion and dwell in organisms that range from a microbe through a mouse to man. *Who is it, you might ask, believes in fables*?

Smart operators don't take chances. The antidote for chance is anti-chance. The key to life is anti-chance. A cell is more responsive and keenly organised than any factory planned or built by humans in the non-organic way of a construction kit.

Who marches through that cell? Porters carry up and down its alleyways. In no factory do items 'wander' but are precisely moved from place to specific place. They are kept on the right tracks. Neither, in the militaristic order of a cell, do chemicals just 'diffuse' everywhere. Theirs is a strict world of coherent arrays, functional compartments, chaperoned transport and passport controls. Signals inform all activity as soberly and safely as on a railway system. Porters carry their cargoes of specific bio-molecules along predefined tracks. Both workhorse-porters and tracks are multiplex, being made of pre-coded proteins. The latter are called micro-tubules. They are arranged in a star-like manner spreading from a central cluster to the cell periphery. As well as acting as rails for the precise delivery of industrial components and serving a dynamic, structural function as cytoskeleton these tubules also (in the role of mitotic spindles) facilitate cell division. There are, on the other hand, two sorts of carter. The larger, called dynein, also plays a role in mitosis as well as generating the whip-like action found in cilia and non-bacterial flagella. It is composed of three domains. Stalk-like 'legs' attach to the walkways. The second part, a complex ring, mediates between stalks and the 'truck' or 'carrier' domain. Its motor is the ring whose 'stroke' causes an angle-change that makes the stalks 'walk'. The stroke varies according to load so that the motor is effectively geared. Dynein and its smaller collaborator each usually move in single, opposite directions, the former towards the cell's centre and the latter away. Kinesin is a simpler micro-machine and, with a single speed, not geared. This linear motor is either active or not. Its two legs are molecular characters made of upright, intertwined strands of twisted protein (like a body but called a stem). To each strand is attached, as to a leg, a 'foot'. The 'feet' are attached on their other end to the micro-tubular walkway. They carry, like a native lady, molecular luggage on the upper, head-end of their stem. Each bipedal porter uses a couple of shots of the standard dose of fuel (*ATP*) for each two-step cycle. In a non-equivalent way one leg is swung over in a 180 degree arc then the other slid horizontally past it. It is thought that release of carefully twisted 'chemical strain' in the stem causes the legs to spring, one after the other, in steps of 8 nm (over 100,000 per millimetre). Smart carriage!

The key factor in evolutionary theory is chance (acted on, in life, by death's selection). Yet, to repeat, the key activity describing life is anti-

randomisation - in other words, purposive command and control. How theory conflicts in essence with the facts! This conflict is not 'as if' or apparent. It is real. Doth Lady Luck protest too much? Should theory still override? As any viable factory needs the correct items delivered on time at the right places, so cells presently contain, as from the start they must have, detailed yet pre-coded transport, printing, construction and many other kinds of integrated system. A description of any aspect of the industry, such as tracks and transport, only starts to answer the questions that purposeful operation poses. How, at any given time, does a cell know which micro-tubular routes to construct and which to dissolve? How does a porter (of which there are many kinds) know what cargo to load when, what routes to map and when to discharge? What is the signalling system? Do such controls, which must have existed in life's proto-cellular blueprint or else it could not have survived, involve molecular resonance; or an electromagnetic exchange of information with an archetypal program? In the blur of questions it is at least clear that, as scientific focus improves, the blur of accident resolves increasingly into the fact of precision, order and purposive operation. Life has no positive role for error; chance degrades not creates biological forms. The implication is that, while one can dream that such irreducibly complex systems might have evolved, the hard facts bespeak a far stronger argument - that highly rational, logical and intentional operations are, and always were, informed. *Information walks the walk with resolve. Chance, staggering nowhere, has no chance.*

Your body contains over 50 trillion cells each of which contains, in addition to transporters, thousands of other minuscule 'functional units with interactive parts' called by some researchers 'molecular machines'. The word 'machine' implies design. More designs are being discovered all the time. Is their elucidation, multiplied throughout all the various levels of life, why Einstein, Planck, Eddington and other physicists could not ascribe complex biological order and harmony to coincidence? *Werner Heisenberg thought that if we transcend biology to include psychology the concepts of physics, chemistry and evolution would be insufficient to describe the facts*. This is if you add mind last, let alone put it first. **Some principles of biology are reducible to physics and chemistry, others are not**. Neils Bohr said, "A description of the internal functions of an organism and its reaction to external stimuli often requires the word purposeful, which is foreign to physics and chemistry." Niko Tinbergen echoed him, "Whereas the physicist or chemist is not intent on studying the purpose of the phenomena he studies, the biologist has to consider it." Even a doyen of 'hard-core', Nobel prize-winning molecular biologists, Jacques Monod, rightly but perhaps, as an atheist, surprisingly says, "One of the fundamental characteristics common to all living beings without exception is that of being objects endowed with *a purpose or project. Rather than reject this idea (as some biologists have tried to do) it is indispensible to recognise that it is essential to the very definition of living beings* (my italics)." Not only are molecular and larger physical structures incomprehensible without consideration of function but neither is any organism's behaviour intelligible without reference to intention. *Indeed the best biological explanations begin with a reasonable purpose whose teleological presumption leads to prediction and research.*

Minimal Functionality

Naturalistic scenarios place great emphasis upon simplicity. Primordial life *must* be simple. The peer-pressured assumption is that a last common universal ancestor (*LUCA*), with codes and basic metabolic functions now intact, gave rise to every other form of life. Thus keen interest is paid to the genomes of prokaryotes considered to be the oldest, ancestral representatives of our family. The question therefore is: how simple must 'simple' be?

No doubt bacterial morphology is simple but the chemistry is not. If, therefore, you argue proto-cells weren't complicated then just how simple did they have to be? If, for example, you propose that primal replication was somehow elementary could you produce a plan to explain what you mean? Remember to include the real primal necessity, code, and other metabolic cascades into your diagram. You will be doing life-from-chemicals a service since, as mankind has recently begun to understand, genetic code, chromosomes, gene control, protein synthesis and the programmed relationship between *DNA*, particular amino acids and, through the steps of protein synthesis, protein and metabolism are all crucial outcomes of an irreducibly complex, informative archetype. So too, further down your line, are bacterial *DNA* replication forks, mitosis and (about which books have been written) meiosis and the sexual archetype. Then, of course, there clucks the chicken these genetic eggs embody - or is it *vice versa*? Flowcharts should include the whole vexed issues of development and morphogenesis. They'll show how single-celled 'simplicity' leads, without weak or broken links, to many interconnected complexities. Accurate step-by-step accounts of the separate, accidental origin of any component let alone whole, integrated systems are absent. Volumes of silence speak volumes.

But there's no lack of chatter. Theoretical biologists are rarely short of seminal ideas. What boils down to the most plausible of possibilities? **Please don't, however, first and foremost in this quest forget that any vehicle of carriage isn't made of complicated chemicals alone. Life isn't simply complicated chemistry.** Vehicles that carry boxes, bags or life all derive from information and have been evolved from such materials in factories by man. Accidents or the arrow of time will eventually reduce your bearer to breakdowns and a heap of rust but neither can create it in the first place. **Similarly, life needs materials *and* information.** Its forms are vehicles that expire but, on a higher level of systematic complexity, can also resist decay, self-sustain and reproduce. They step above machines by using mind's anti-chance machinery - code, computerisation and a tripartite mechanism called homeostasis. Is mindless *IT* (information technology) come to life luck's child? Because according to the standard naturalistic answer all you need is a self-replicating molecule. This molecule is free to spin code-combinations generating 'lucky break' on 'lucky break' to innovate complexity. You only have to beg from Luck a chit of special acid that can trigger libraries of publication by the firm of Life & Co. After production of a proto-cell there automatically ensues its reproduction. Successors of Original Aggregate (*LUCA*) spin life's careless web. Thank your fates they spun the animated tapestry you look at in the mirror. Human code, like any other kind, is written up on *DNA*; this encyclopaedic book is called a genome; thus look to genome where your information is concealed. Genomics opens up the book. You might be complicated but we now can see just how 'simple' any prototype would have to be.

Could a virus just appear? Why not, for an ambitious start, consider as Fred Sanger first did, viral *DNA* - 11 genes in 5386 bases that constitute the information making phage virus *Phi X 174*? However a virus, although relatively highly ordered, cannot have been the 'simplest' prototype because it cannot reproduce itself. So you are not of viral descent.

How many bits in next best? An endosymbiont is an organism that lives within the body of another. Perhaps the smallest bacterial genome known belongs to a bacterium which inhabits a particular kind of mealybug. It is called *Tremblya princeps*. Having only about 150 genes, it lacks many considered essential for survival. However it succeeds, as a 'nested endosymbiont', by having an even smaller bacterium, *Moranella endobia*, inside it. Russian dolls? But although smaller in size *Moranella*'s genome is three times as large. Neither party has the complete genes for any metabolic pathway but together they make do. In this respect they represent a composite living being! And this composite has to live inside the mealy-bug. There's a problem for beginners!

Another tiny genome belongs to another obligate endosymbiont (perhaps a parasite), a proteo-bacterium called *Carsonella ruddii. Carsonella* sports perhaps 182 genes from about 160,000 base pairs. These are cast with highly sophisticated efficiency in the form of overlaps. Imagine writing a letter that incorporates its message by means of overlapping on the words! Another such bacterium, *Buchnera*, has genes solely for products essential to its life - 396 at smallest. And hot-vent *Nanoarchaeum equitans* with its 552 or so genes (from nearly half a million precisely organised, 95% coding base pair letters) includes *DNA* repair facilities. Parasitic *Mycoplasma genitalium* has ~520 genes that number nearly 583,000 base pairs (at one per second without cease they would take about a week to count) all in order. How was such combination informed accidentally? And these are merely dependent forms of life. In nature they cannot exist outside a host and did not, therefore, breed you.

Free-livers, though, roam independently. The smallest known are an Archaean called *Thermoplasma acidiphilum* (with its extraordinary membrane and about 1500 genes) and a bacterium called *Aquifex aeolicus* (a thermophile with 1512 genes from about one and a half million letters, bytes or base pairs). And what is to be made of the first sequenced Archaean, *Methanococcus jannaschii* from hydrothermal vents, 56% of whose 1738 genes resemble no known *DNA* sequence, let alone only 17% of the whole that resembles the two other main bacterial groupings? From these examples it can at least be deduced that the minimum genetic constitution for life exists somewhere between parasitic 180 and free-living 1500 genes. In any first instance the latter quantity must incorporate autotrophic processes, the capacity to produce varieties of lipid, carbohydrate, protein and nucleic acids, respiratory metabolism, genetic code, protein synthetic machinery and the capacity to reproduce. This represents a tall order for wind, rain and fiery force of nature to fulfil.

In order to estimate the minimum gene set necessary for independent as opposed to parasitic life genes of *H. influenza* and *M. genitalium* were compared. The results indicated a very bare, unlikely minimum of about 250 genes; such an organism could not digest complex compounds and would need a large supply of organic nutrients. Other tests lead to the consensus that such hypothetical bareness may range between at least 350 and 450 different protein and *RNA* coding genes (Venter's minimum would comprise nearly 400 protein-coding and about 40 RNA-coding genes required to cooperate to build a basic independent mode of life. This means that, from a *top-down* point of view, an absolutely minimal form of life would need perhaps 150,000 correctly sequenced nucleotides built up to form its book. This book would have, in micro-space and micro-time before degradation of its parts began, to be encapsulated and enter into synergy. By synergy is meant all functions mentioned in the paragraph above, properly organized and orchestrated (that is, regulated) in the proto-cell. The odds against producing a single, average protein (say cytochrome c of about 100 amino acids and 300 base pairs) is, as noted in Chapter 20: Number Games, at least 1 in 10^{75}. To muster a further 250 inter-functional proteins/ genes simultaneously this lack of probability explodes to incomprehensible numbers such as make the amount of atoms in the universe or time since it began look ridiculously small. To generate that testament to cellular complexity, an *E. coli* bacterium whose genome size (with 4288 genes) falls with the same order of magnitude as 'simple' microbes or the *LUCA*, you'd have to write trillions of zeroes (powers of ten, orders of magnitude) against the right code taking shape! This is fact. Even the most 'primitive' cell is, marvellously miniaturised and of labyrinthine complexity, a computerised chemical factory. **It is only science fiction to imagine less. Yet science fiction seems to rule the naturalistic, atheistic day!**

Biosynthesis

Smaller and smaller's normally the ingenious province of high-tech - except in nature where you choose to call it 'primitive'. The fact is, however, no prokaryote or any cell is primitive at all. All are fiendishly convoluted. For example, a minimal genome contains at least 350 genes and a viable analogy for *LUCA* (Last Universal Common Ancestor) about five times as many. Both parasite and free-liver contain, as well as their irreducibly complicated chunks of information inscribed on so-called self-replicatory *RNA* or *DNA*, the correct quality, quantity and locality of ions, proteins, carbohydrates, lipids and so on. Abiogenesis has absolutely no chance whatsoever in real life. What, however, if we add a dash of mind?

It is one thing to take a machine, say, a computer, and make a few changes or try and improve on its design; it is quite another to invent, from scratch, such a vehicle of purpose with all its conceptual novelties, pertinent materials, codes and interlocking parts. A synthetic biologist, granted natural biochemistry and mechanisms, falls into the former category. He does not *invent* a form of life but intelligently tinkers with the data. For years biotechnologists have created minimally 'unnatural' forms of life by the manipulation of genes. Such a *top-down* method of manipulating genomic information explicitly accommodates the teleological concepts of programming, engineering and technology. It artificially tailors, that is, reconfigures present forms of life in, as yet, a limited way. Indeed, the limits of such reconstructions remain to be seen. Certainly, though, reconfiguration amounts to the basis of a fresh, man-made or, if you like, unnatural biology. *This is, of course, not evolution by chance but, as in the case of any specifically constructed product, by mind. Mind and chance are chalk and cheese.*

Where to start more radically reshaping and constructing life? A human 'book of life' contains about 3 billion *DNA* base pairs that compose about 21000 protein-coding genes along with an important range of 'non-coding' regulatory factors. You definitely won't start playing there. Why not, for a less ambitious start, consider as Fred Sanger first did, viral *DNA* - 11 genes in 5386 bases that constitute the information making phage virus *Phi X 174*? Neither natural patent nor copyright exists and such bug-*DNA* has already been buglessly copied using artificially synthesised material purchased from a supplier. Viral replication works when a virus injects its genome into a host cell. Genetics expert Craig Venter injected stitched-up *Phi X 174 DNA* into its natural, component-rich environment, an *E. coli* bacterium. It was like taking spare parts for a bicycle from a junkyard and fixing them up. The fix worked. The operation of any system, biological or technological, must of course be wholly explicable in terms of its components. Could you not, therefore, pirate other more complex biological records? Or even, having grasped the language, assemble new ones?

Non-viral life is routinely cut to three domains - eukaryotic (including you and me) and two prokaryotic (*Archaea* and *Bacteria*). Although they look superficially similar the biochemistry of the prokaryotic microbes differs so radically that, in some ways, the *Archaea* as little resemble *Bacteria* as they do eukaryotes. Both sorts of microbe show great biological diversity and complexity. Nobody, however, claims that groups of degenerate bacteria such as *Ricksettias, Chlamydias* and *Mycoplasmas*, with various properties that once seemed to place them between viruses and cellular organisms, are indeed intermediate forms. This is because, under close inspection, these parasites are observed to have a typically non-viral, prokaryotic structure; and while cells that compose plants, fungi and microbes often have a wall those of *Mycoplasmas* don't. And they're small. The smallest known genome of any reproductively capable organism (*M. genitalium*, a parasite that inhabits primate genital and respiratory tracts) is probably close to the lower limit necessary to specify the properties of cellular life. Its genome comprises genes composed of more than half a million letters, all correctly sequenced and joined into a ring. Although a thousand times larger than Venter's cobbled phage virus it is, however, still unable to code for amino acid synthesis. This indicates either that it feeds off other organisms or that its genetic bank has been downgraded. Information loss is not life's gain; bankruptcy not profit is the result of lost or scrambled code. Nevertheless Craig Venter, having defined a minimal genome, identified its possible possessor as a *Mycoplasma.*

Despite their 'primitive' complexity Venter transplanted one type's genome into another's closely related body and so created, at the peak of a mountain of accumulated information and using great skill, tenacity and ingenuity, a novel life form. Lo! For sure a great achievement! *M. laboratorium*, a beauty of its kind and nicknamed Synthia! Synthia is 'pirate' Venter's baby, his booty captured out of life's high seas, the inventor's 'synthesis of life'! Of course, conceptual material - the basic molecules including *DNA*, membrane, metabolites, cytoplasm and so on - were given. They were taken. A simple bacterial genome was copied and machine-made segments added; a yeast cell stitched the bits together and the 'new' genome was transplanted into a second, equally simple bacterium where it took complete control. Although a technical triumph Synthia is hardly synthesised from scratch. Indeed, only a tiny fraction was computer generated. Nor is Venter's mind at all the same as Lady Luck's. He honestly admits his cell is *not* an artificial form of life. He did not create a life. Indeed, his biotechnological skills effectively confirm (though not admit) that the simplest possible kind of cell is far too complex for The Incoherent Lady to have flip-flapped up. Modification of fourteen genes and the addition of 'watermark' nucleotides add up to altered life from prior life. Intelligent design created a novelty but what created the intelligent designer?

Is the quality of scientific knowledge greater than oblivion by more than just some elongated quantity of time? Can you reduce intelligence, that's quick, to nothing more than time compression, compression whereby chance is simply squeezed out of the mix? Of course you can't. Even so, current research into abiogenesis is like a class of children playing with alphabetic letters or a sentence teacher first supplied. Using highly directed chemistry you may cobble together a few such preconceived letters, say bases or amino acids. You might even 'plagiarise' the teacher's sentences or re-jig new ones; but the construction of a whole new language is as far beyond your conceptual capacity as writing books or programs full of systematic meaning. No doubt, in its molecular analysis biology is prone to the materialistic view of 'life-as-body-chemistry'; and no doubt that trailblazing doctors advocate 'directed evolution'. If you can specify nucleic code, transplant it into host cells and then 'boot' the program of synthetic genome into life you will have generated a synthetic species. You will amazingly and accurately have manipulated data viz. code, specific complex chemicals, vital functions and integrated bio-systems but does this mean that you 'created life' from scratch? *You can synthesise an artificial gene, chromosome or life form like the copy of a microbe of minimal genomic size to support biological operations but this is not abiogenesis*. Billions of *intelligent* man-hours would have been invested by mankind in acquiring sufficient knowledge to create even this simple Synthia and thereby 'recast the originals' in a laboratory. **Indeed, the whole thrust of deliberation is to achieve what chance does not**. Nor, in physical terms, are genes the whole game - you would have to copy no less than a whole cell. *To copy's not to innovate or to create. Nor, even then, would such deliberate facsimile of life-form demonstrate that one might ever come by chance.*

Craig Venter is no understudy standing in for evolution's Lady Luck; nor does he embody time compression that would simply, quickly mimic her erratic ways. Perhaps quite soon man will understand each chemical particular in the reproduction, development and operation of organisms. Knowing precisely the multiplex, subatomic interactions orchestrated in each cell and larger body, he will understand biology in its entirety and then, maybe, build organisms out of chemicals in bottles. The creation of artificial life will become 'turn-key' and widespread. Man, the plagiarist, will proceed to construct forms much more intricate than unicellular Synthia up from scratch. Would such a brave, new world show, concentrating billions of man-hours of human education, research, planning and skilful manipulation combined with the ubiquitous use of highly-engineered machinery, that actually life's old-world caboodle was expressed by accident? **In fact, it is a wholly reasonable interpretation that such time-consuming, manufactured expressions of genetic code (with all their protocols, strategies, monitors and so on) demonstrate the necessity of direct involvement by intelligent agents; and that without large injections of such intelligence into their experimental rigs no such transformation as required by *LUCA* can occur at molecular level.** Can't and never could. **Finding out how something's built and works does not mean that it came about by chance; nor does re-building, even purposefully ringing variations on its copied theme, mean it ever 'self-assembled'.** The methodological assumption of physical science means you need to voice an explanation that casts the mindlessness of matter as creator. To invoke mindless 'inventiveness' is an axiomatic, essential ingredient of such a theoretical assumption. You *need* evolutionary theory to explain, on those terms, life on earth. However, in the last analysis this philosophical necessity may turn out false. If the explanation's chance then modern research seems to disagree. The innovative origin of coordinated information - on which all kinds of life depend - goes unexplained. Specified complexity of the genetic code, metabolism, integrated functions of the cell, development and every other process in biology are each pressed to the flag of Lady Luck. Instead, the obvious should 'out'. *Information and intelligence, not chance, are as rational an explanation as they are rational constructors of purposive, well-oiled systems.* And biological incorporations are, without doubt, the smoothest operators.

In Extremis

No bacterial constitution is a fraction as primitive as modern, humanly informed nanotechnology. *No contemporary organism is 'primitive'.* There exist only very complex or even more complex life forms. Take, for example, chemoautotrophs and methylotrophs (which lack photosynthetic complexity and derive their energy from inorganic or single-carbon organic substrates). They were thought 'primitive' until their biochemical machinery was discovered to be as complex as those of bacteria that need organic nutrients. Minimal such synthesis includes (after the specific assimilation of carbon, nitrogen, phosphorus, sulphur, hydrogen and oxygen) nucleotides, *DNA*, amino acids, protein, porphyrins, lipids, sugars and so on; which chemicals, when metabolised correctly, must be disposed in the right, cooperative positions having been compiled into the requisite structures! Does that sound haphazard? Could it ever happen bit by following bit?

Microbes, especially the slime moulds, can also aggregate and coordinate their behaviours; their genetic material undergoes recombination and mutation; and they show, according to the environment for which they are suited, diverse nutritive strategies - but classification remains difficult and even among the purple (or proteo-) bacteria there thrive many groups with distinct characteristics and no obvious linkages between. Of course, theoretically there exist phylogenetic relationships. In practice, for example in the case of the *E. coli/ Salmonella/ Shigella* group, variation (or micro-evolution) is routinely observed to produce different strains of a microbial type but, despite nearly 150 years of bacteriology involving continuous and often extreme forms of laboratory persuasion, no new type of prokaryote has emerged nor has one type been observed to assume any other identity. Indeed, so complex is *E. coli* (perhaps the most researched organism of all time) that after a century of work papers still flood in about newly discovered features. In spite of severe inflictions by generations of geneticists these bacteria have shown no sign of evolving into anything else. An appeal is therefore made, not to the fossil record where microbes resemble their modern counterparts, but to metabolic evolution. Proto-organisms *must have* preceded both contemporary prokaryotic domains. Their inexplicable mediation is needed. *It is inexplicable because, as noted above, all cells are dynamic, irreducible wholes.* Minimum components critical for survival in conditions mild or extreme include information processing and reproductive systems, homeostatic pH, water-regulating and other trans-membrane porting systems (a prokaryotic membrane is much more selective a barrier than the eukaryotic - the bacterium *Pseudomonas aeruginosa*, for example, has almost 300 porting systems), locomotion (including the fabulous flagellar propeller, a circular prodigy, a wheel that spins at speeds of up to 1700 revs. per second on a freely rotating axle set in a bearing), spore formation, metabolic programs (including energy metabolism and genetic control) etc. etc. *It's been clearly shown that the 'black box' evolution of metabolic pathways is logically inexplicable.* How can you leap, except in imagination, from nothing to all this? What is the simplest irreducible whole? *Who believes in miracles*?

For natural start-ups naturalism seems to have no choice. Neither earth nor life forms are eternal but, if you can't feasibly imagine how life here must have started up, frustration might well drive you to extreme suggestions - say, extremophiles; or Martian proteans; or spores from anywhere in space (and idea called panspermia). Let's not fly off into void just yet. Were proto-organisms on the hostile early earth extremophiles? Extremophiles are prokaryotes that thrive in very hot, cold, dry, acidic, alkaline or salty conditions. Earlier we met *Thermoplasma acidiphilum* and heat-lovers crowding sulphurous hell-holes called 'black smokers' deep beneath the sea. To pour cold water on hot, however, the hotter you go the worse the problem. A bench Bunsen burner demonstrates how heat stirs reactivity. Given the instability of *RNA* in such a crucible an *RNA*-world isn't where life's buzz began. The likelihood of *any* large, specific molecules surviving the destructive effects of high temperature is minimal. As this chapter shows the problem is vastly greater than the random production/ destruction of a few accidentally 'correct' macromolecules such as, in this case, thermophilic protein. For whole batches of constructions, each apparently interacting with the others as if they were parts of a whole, the chances are predictably zero. Try the life-capsule tests at 100°C to check the hypothesis.

And at cold temperatures the hydrophobic effect that contributes to molecular morphology ceases to operate - a phenomenon called cold denaturation that can adversely affect ribozyme activity and thus the idea of an icy *RNA*-world. Underground microbial ecologies might seem a good idea except they're rare. Where can these isolated colonies have found, if prototypical, organic chemicals sufficient unto aggregation as a metabolic form? It is doubtful if such organisms represent a cosmic loop-hole through which perhaps life slipped. Try briny brinkmanship in some dead, ancient sea; load molecules into extremely alkaline or acid pools; but the likelihood of complex, stably interactive organic configurations enduring long in an extremity are, unless as a bio-form endowed with the appropriate informative and energetic constitution, nil. How, therefore, does one reason that such codified organisations as extremophiles gradually 'self-aggregated' into the chemistry called 'life'?

If terrestrial extremes won't do perhaps we'd better slide off into freezing, airless space. Might a dust-cloud, if it harbours any carbon chemicals, throw life-on-earth a starting-line? Such clouds have been found but they're as far from life as rock or water by itself. What, therefore, about some extra-terrestrial incubation - life like plankton floating slowly through a dark and empty sea? Heavenly spores were thus, according to physical chemist and Nobel laureate Svante Arrhenius, the seeds of life. This *ET* idea, panspermia, has now creaked past its hundredth birthday. It has also been considered, in undirected and directed forms, by scientists as notable as Lord Kelvin, Francis Crick, Leslie Orgel, Fred Hoyle and Carl Sagan. Are you in fact an alien's ancestor (the directed form)? Or were you born from a wandering spore, a seed that rode a chariot of comets until one day it fell, dropped from the starry heavens onto earth (the undirected form wherein life's chemicals formed or life evolved naturally somewhere else in space)? Researchers point to the remarkable survival capacity of certain sorts of bacteria. They can endure conditions of vacuum, extreme heat, cold, lethal radiation and impact. Do you recall the excitement generated by carbonaceous chondrites - meteoric fragments found at, for example, Murchison, Australia and Lake Tagish in Canada? And, found in Antarctica after nine years on earth, ALH 84001 was supposedly from Mars. Through radiation, freezing cold and the fire of atmospheric entry some organic chemicals and lifelike forms survived on them - unless contamination here on earth occurred. Contamination can happen in less than a week (for example, the ALH 84001 amino acids were all, as against the mixture one might ordinarily expect, left-handed). Yields from the heavens are small and, as calculations by Sagan and Christopher Chyba indicate (Nature 355 (1992) ps. 125-132), insufficient to have supported even the thinnest of geologically-absent primordial consommés. Half a gram per year flown into all earth's oceans is not much; and such release of biologically relevant organics would not accumulate but soon be broken down. Moreover configurations seized upon as microfossils were too small for life and the rock itself too old (dated at 4.5 billion years in a solar system of about the same age). Aren't we therefore chasing shadows, chasing Martian ghosts? Did a plagued Nile not turn to blood and blood fall from the skies with rain on Kerala? Had it fallen out of space? When this 'rain-blood' from India was analysed its freight looked like red algae; if, however, dust, meteors and comets bus life here and there around the universe could carmine rain suggest how you, generations well removed, first came to be?

'Dusting' earth with life in some 'panspermic' way evades not answers how it all began. It simply removes the problem to an unqualified, unquantifiable and unknown arena somewhere else in space. *This is unacceptable.* Indeed, from a *top-down* point of view the wrong questions are being asked. It matter neither here nor there where you might choose to speculate. *There is in fact no evidence, only case-driven inference, that a 'proto-prokaryote' ever evolved into an prokaryote or that such microbes ever evolved* *into a eukaryote.* A 'theory of no intelligence' certainly demands that such transmutations *must have* happened. On which wall, therefore, is bio-logic chiselled? Is it, as serially interpreted according to various assumptions, chalked in rock fossils? Or does its script reflect, according to 'a theory of intelligence', an archetypal logic, coded information wholesomely conceived before the play began - in which case forms of life might be projected anywhere conditions could sustain them. *In extremis* might mean far removed from 'egocentric' earth but, as we'll see in Chapter 26, cosmic chemistry and specified complexity demand these 'alien' bodies will conform to universal scheme, a bio-pattern we would recognise. You would expect to find conceptual wholes and not just jumbled bits; a metaphysical character not accidental chemistry; ecological cooperation instead of competition. Microbes alone might possibly be found; microbes bio-systematically prepare life's cycling ground. Arguably the fittest of all survivors, they can be construed as essential agents of geo-chemical change. They can be seen not as precursors but participants; not precedents but, like plants and fungi, brothers in life's arms. They are workers, servants that sustain the 'higher' elements of ecosphere. They prepare and they sustain. From their lowly position at the base of an ecological pyramid microorganisms play a vital role in the recycling of matter. They help maintain a homeostatic stability (such as between the nitrogen content of air and the supply of combined nitrogen) on which life forms are dependent. Other critical cybernetic controls in air, sea and on land balance temperature, salinity, hydrogen ion concentration (pH) etc. Bacteria drive these cycles. They are unwittingly creators of the biosphere's dynamic equilibrium. Some live independent of eukaryotes but they can equally and easily be seen as the base platform of life's ziggurat, a chore labour force, an integral level of life's whole terrestrial economy. Is any life by all accounts a simple or a complex miracle?

Galilean Correction

Step many steps far forward. Suppose, for a moment and fantastically, we have arrived. Some magic alchemy has yielded *LUCA*. Whatever lonely *LUCA* was, prokaryotes have always excellently thrived. What, therefore, about survival of the fittest? Does more complex mean more fit? *No type of organism is fitter in terms of survival than a bacterium.* The rest are 'more evolved'. In other words, they are more specialised, perhaps multi-cellular and thereby fixed in and dependent on particular environments. More 'souped up', complex but fragile. Why, therefore, get tricky in the first place? Why should any organisms other than the fittest - bacteria - have appeared and survived? Wouldn't they have been selected against? Why should, how could bacteria ever have left their extremely tough, successful rut?

Out of the rut then what? Bacteria have circular *DNA*. Non-bacterial nuclei do not. They have chromosomal lines. Was it only once bacterial *DNA* split accidentally from loop to line and (along, coincidentally, with regulatory chromatin, mitosis and membranes) created the initial eukaryotic nucleus? Why is the cut so clean? Why aren't vestigial loops found in eukaryotic nuclei or loops split into chromosomal lines in bacteria or certain organelles? In fact, organisms except bacteria all have and appear to have always had multiple linear chromosomes. Cell division requires that the right information be delivered to each daughter cell. To this end we noted (Supreme Operation) the excellent replication mechanism all prokaryotes employ. How (Chapter 24: The Origin of Asex) should and could this effective mechanism have been extended to embrace the complex, eukaryotic ritual called *mitosis*? Mitosis packs, posts and delivers correctly enumerated linear material. An account of mitosis should append a step by step explanation of how its critical precision arose haphazardly, part by part, in concert with other differences between prokaryotic (bacterial) and eukaryotic (all other) cells; or how it must have fully orchestrated itself in the same comprehensive instant as the first eukaryote. Just as purposive, critical and somewhat more complex than the mitotic postman is a 'dance' called meiosis (see Chapter 24: The Origin of Sex). *Meiosis* packages the coded information for gametes or sex calls. *If evolution is scientific and if both processes evolved we might reasonably ask for an account. Not a vague just-so story but a cool, calculated and precise step-by-step molecular account.* The meiotic account would append a step by step explanation of how the shuffling hierarchy, associated sexed gametes, male and female sexual apparatus, instincts, feedback mechanisms and developmental programs implicit in meiosis all occurred before any of the incipient but so far useless stages had been selected against and eliminated.

If no rational account of possible accidental molecular steps is available what evidence is there? Biochemistry and molecular biology are at the base of all physiology and anatomy. **If Darwinism fails here, in accounting for any single, 'simple' molecular sub-component in the much more extensive system of, say, sexual reproduction, then it fails everywhere else.** Because 'everywhere else' is a component in one system or another and each system is in turn subsumed in fully integrated, healthy working order under the organism as a whole. Indeed, bringing a cybernetic approach to bear on the study of reproduction you might reasonably conclude that any intricate, sensitive and purposive hierarchy of feedback and control is not the product of random molecular agitation.

We have seen that massive and theoretically insurmountable problems face the notion of chemical evolution. These worsen not only as resolution improves but as the scope of structures and associated functions is revealed to be more and more complex and interdependent. **In these circumstances the suggestion of intelligent design appears like Galileo's correction of the Ptolemaic dogma**. As Galileo, an intelligent man, had the effrontery to overturn an earth-centred view of the universe, so Natural Dialectic overthrows a flat-earth, materialistic view of origins. What if mechanisms *are* purposeful? What if purpose *is* a metaphysical aspect of information? This is not, except in the eyes of a certain perspective, pernicious. It is heresy to the authority of that order alone. Secular law, unless painted red, forbids inquisitions but allows the 'liberal' burning of a guilty book. Far from being 'guilty', the judgment of this book could set you free. It's time to turn to *top-down* treatment of the modern alchemy.

Tick Tock

Count-down. It's time for summary.

What makes clockwork? Who makes clocks? William Paley argued in his

Natural Theology that the workings of biological organisms resembled those of the hi-tech, miniaturised intricacy of his day, a time-keeping watch. A watch, no less than a futuristic quantum computer, is a fabrication conceived with purpose. It has a maker, cannot be assembled in a haphazard way from imprecise parts and these parts, each with its own critical role, cooperate in a highly integrated way. Each part of any complex mechanism acts, along with the laws of nature, as a mutual constraint. Tensions, parameters and parts of such an instrument are orchestrated into a single, working whole. Although a metallic watch is less tolerant of tinkering and less resilient to injury than the dynamic of an organism, both systems work at their best when well-adjusted, highly-tuned and fit. *If you then change the plan of the whole or any part of the whole you will, unless you introduce a suite of compensatory changes across the system at the same time, degrade or destroy it.* A body trumps a watch because its well-balanced, smooth working, called health, is resilient; it can bounce back to its norm after the distortion of accident or illness. And its reproduction can generate restricted variation-on-theme. Nevertheless a watch is a fine analogy for what makes the physical basis of biology, a cell, tick.

Wait a tock! Some persons, namely most biological scientists and all atheists, do not like this analogy. Watches, like any other technological advance, have to be invented before they can undergo directed evolution in discontinuous steps (called 'new models') under the guidance of intelligent designers. These designers, having coped with the original development, can master the compensatory suite of adjustments each new model demands. The problem is they use their brains, brains are a product of evolution and evolution is the standard substitute for natural intelligence. Where intelligence conceives does not evolution occur by accidents (genetic mutations) that are honed by natural selection? So the argument of biological bodies from 'intelligent design' is both physically wrong (because there is no un-naturalistic, miraculous production of new models, only a naturalistic, 'miraculous' production of the first one) and metaphysically wrong (because there is no such thing as 'metaphysical', especially not in the form of a 'more-intelligent-than-human' mind). Atheists and accommodation theorists (see Chapter 23) therefore denigrate Paley's view using the argument that it is 'dated'. Biologically the word 'evolve' is used to paper over an oxymoron; its 'design-less design' smoothes not the crack but chasm between contradictory 'design' (by chance) and actual design (by mind). In this case one might update by arguing, for example, that a woozy watchmaker called Caprice evolved a prototype cell and, thereafter, everything up to your intelligently working brain. So numerous would be the unpredictable coincidences involved in this scenario that, if the show were re-run, you would as certainly as the first cell not be here to contemplate the universe, your navel or anything else.

One might argue, on the other hand, that such an analogy is curiously inept. Does reason age? Does mathematics change by the season or logic by the aeon? At the heart of physics is assumption that the laws of nature never change. How, therefore, can the natural principles employed by an inventor like, say, Marconi become 'out-of-date'? Neither the facts nor the logic of natural history are changed by contrary belief or re-scripting. In this case whence emerged the logic, the intrinsic, coded logic every organism ticks by? How, computer-like in process, could the regulating programs of life's chemistry just aimlessly accumulate? Could millions of years and solar systems throw up even such a glass as you sip from? You need swathes of time for chance to work what mind coordinates in moments. Therefore, of course, to one proposing accidental systems generation and improvement aeons of tick-tocking are '*sine qua non*'. Unseen billions of years are your prime datum. But is eternal churning-turning of a tick-tock any use at all? It counts for nothing if you count intelligence. Indeed, in proportion as concept and creativity arise length of time contracts into irrelevance. Information (purpose-with-designs) comes first. As always, Natural Dialectic inverts the conceptual hierarchy of materialism's explanatory scheme. Information is regarded as the foundation on which physical, and especially biological, reality is constructed. Information-centre (mind) → norms/ archetypes of nature → energetic patterns → material bodies; informed constructions and their operations follow down creation's line.

Chance neither targets nor schedules. There is nothing due. Unlike the guiding craftsman of a watch it is as craft-less as an aggregate of atoms, as unguided as the weather. The consort of 'necessity' - an immaterial anti-deity that pops up unexpectedly - lacks any shred of information. Intangible chance is lack of information's cipher. It is, by definition, absolutely incapable of evolving systems of predictable behaviour. Nor can natural selection (Chapter 22) actively create a thing; it just lets mutants live or die. Yet systematic patterns, programs, exactly constitute the coded logic that informs all life. They centrally, repetitively, ubiquitously express embodiment. In this case life's tick-tock is transformed by nuance to the sound of Paley's footsteps growing louder, clicking on the pavement through Christ's College lodge, moving closer as we learn. Is his skilled watchmaker with him as his figure re-appears intact?

Michael Denton, in his book 'Evolution: A Theory in Crisis' (p. 269), summarises the whole problem admirably. He writes "Eminent engineers and mathematicians, such as von Neumann, who have considered theoretically the general abstract design of self-replicating automata have shown that any automaton sufficiently complex to reproduce itself would necessarily possess certain component systems which are strictly analogous to those found in a cell. One component would be an automatic factory capable of collecting raw materials and processing them into an output specified by a written instruction. This is the analogue of the ribosome. Another component would be an automaton that takes the written instruction and copies it, a duplicator. This is the analogue of the *DNA* replicating system. Another component would be a written instruction containing the specification for the complete system, which is the analogue of the *DNA* (storage information system).

The fact that artificial automata and living organisms both have to conform to the same general design to meet the criteria for self-replication tends to reinforce the feeling that perhaps no system simpler than the basic cell can exist which could undergo genuine autonomous self-duplication.

The difficulty that is met in envisaging how the cell system could have originated gradually is essentially the same as that which is met in attempting to provide gradual evolutionary explanations of all the other complex adaptations in nature (my italics).....The problem of the origin of life is not unique - it only represents the most dramatic example of the universal principle that complex systems cannot be approached gradually through functional intermediates because of the necessity of perfect co-adaptation of their components as a pre-condition of function.....*The origin of life problem lends further support to the notion that the divisions of nature arise out of the necessities of life rooted in the logic of the design of complex systems* (also my italics)."

A Definite Flight from Science

The count-down reaches bottom-line. Let's be clear. Science has it, naturalistically, right. Evolution *must have* happened. Physic without metaphysic, specified complexity without prior information - both demand it. How else could things be as they are? Data is all cast, by presumption, into evolution's information-less mind-set.

Yet we've been using language drawn from information, code, semantics and intelligence. Its element of metaphysic profoundly and precisely contradicts chance and a dodgy theory of evolution. Do you remember (from Chapter 8) 'Big Accident', a multiverse and the irony of flight by scientism from its bio-friendly science? In science you want reasons and therefore try, on principle, to avoid appeal to flukes. *Yet when it comes to origins the very reasons are identified as flukes*! Great Accident or Small, the rule-of-thumb is wondrously reversed. And, since flukes are by definition unpredictable, the theory can't be falsified. Such slewed policy 'explains' a lifeless origin of life and all the evolution-by-mutation of life's sorts. Is, you ask, this shaky effort an entirely testing, scientific one?

By now you may understand that it is easier to wave a verbal wand and assert that chemical evolution *must have* happened than convince an open-minded critic that it really did. **The fact is that life only starts at the level of a functional cell**. And, as Chapters 19 and 21 highlight, the simplest such cell incorporates informative, energetic and reproductive hierarchies, sophisticated and accurate genetic language, various homeostatic norms with their triplex enabling mechanisms, hundreds of different macromolecules and many thousands of parts. **A 'simple' cell is not at all simple.** *Hundreds of thousands of scientists working over many years have still not completely understood its operations.* **Encapsulated in most cases in a sac much smaller than a full stop on this page, it is an irreducibly complex mechanism**. It is one that continuously and unwaveringly satisfies a huge requirements sheet by engaging highly coherent, rapid, coded and thereby accurately guided chemical programs. Have you, perchance, grasped the implication? It is worth reiterating that to live for a moment let alone survive it needs, as a bare minimum, a membrane, a nucleus storing accurate code to produce hundreds of proteins (including those needed for decoding, protein synthesis and the algorithmic creation of thousands of specific macromolecules), homeostatic sensors, processors and effectors attendant on the prior establishment of working norms and responsive to internal and external environmental cues, synthetic and respiratory aspects of energy metabolism and a capacity to replicate *DNA* and efficiently reproduce - all ticking over with each other and at the same time coordinating with instinct. Without such collection there is nothing to live and, therefore, nothing to mutate or naturally select. *Must* it not have sprung all at once by luck in a warm puddle of some description??!!

Whose logic therefore shouts that 'birth from earth' of a primal protoplasm (and, therefore, grand macroevolution from it) never happened? 'Not mine', retorts a Theory of No Intelligence, 'that sort of non-materialistic logic is irrational; whatever can be could be given ample time; I profess that Darwin's 'evolution', whatever facts might seem to say, is the logic of our lucky life on earth'. But, upon examination of metabolism and a host of other evidence, a Theory of Intelligence is bound to disagree. Two frames of logic here head-on collide. Each construes the other as illogical. **How far do you fly from facts to think that thought is made of atoms, faith and purpose spring from molecules or that non-consciousness bred highly systematic forms of life?**

This is life's issue's crux. How reasonable is questioning belief in scientism's root idea for life, its explanation of your being here - abiogenesis? How rational is it, on the other hand, to ignore a logic that might seem to undermine materialism's faith? Such irrational, literal ignorance is, concerning where the origins of systematic forms of life apply, ironically the sceptical and 'rational' atheist's! And if scientific rationalism is *by wilful choice* entirely 'naturalistic' in the sense of 'only-physical-exists' then be aware of fundamental bias. Don't be led astray by such a vote for mindlessness but include (*fig.* 0.7iii) a 'hidden variable', the vertical coordinate of information. Natural Dialectic is, you know by now, a framework to present a form of Naturalism that includes both active information and the passive, naturalistic kind that science gathers. No dissidence disturbs this pair until, when locating point of origin, irresolutely double dots appear; a blur; explanation bifurcates into those polar Theories of Intelligence and No Intelligence. Which, Naturalistic or a naturalistic view, best makes the case for origin according to a logical interpretation of the definitely known facts?

Let us be clear. Including a coordinated couple - energy and information - is not a vote against science but against its philosophical trailer, scientism. In a definite flight from science, scientism flouts the Law of Biogenesis by parading the Hopeful Hypothesis of Abiogenesis! In addition to flouting this First Law of Biological Conservation it proposes (as the Theory of Evolution) that entropy of information (mutation) and of energy is capable of 'designing' organisms. Quite apart from the informative implications of such stance it flouts the Second Law of Thermodynamics. In effect, scientism proposes the replacement of laws to which there is no known exception by hypotheses of which there is no known example! **The philosophy of materialism itself rests upon such cheap and cheeky impudence, such rabble-rousing protestation in the face of fact**.

Of course, the First and Second Laws apply with full force to the operation of biological structures but can a few ingenious but weak hypotheses of circumvention explain original construction? Scientism demonstrates in hope; materialists affirm their faith. This is invested primarily in a theory derived from the accurate observations and, in limited respect, explanations of Charles Darwin. *This theory is based on 'fitness of adaptation' and Darwin may indeed have partly explained the origin of some minor adaptations*. These can be 'non-purposive' (accidental) or 'purposive' (pre-programmed as specific buffers lending flexibility to life's response to challenges). **The appearance of an adaptation is, however, a relatively trivial event. Much more profound is the origin of organ, system, body-plan and organism as a living whole - all**

prerequisite to minor adaptation and none of which (except by an imaginative appeal to extrapolation) the theory of evolution starts to explain. Darwin's disciples still don't even explain the integrated, coded origin of tissues and, as this chapter demonstrates, types of cells. In short, Darwinism explains, to some extent, dynamic but minor variation. The assumption that biological variation is 'micro-evolution' amounts to a semantic nudge. However, to extrapolate from this prejudicial semantic, that is, to extrapolate from variation to 'macro-evolution' and a 'tree of life' is only arguably warranted. Indeed, it may well constitute, as we shall see, a third, unfounded flight from actuality.

Whose logic is, therefore, in the face of science strained? Pasteur's work, that disproved the spontaneous generation of life (*aka* 'The Little Accident of Abiogenesis'), is ignored. Spontaneous generation is dead but we are alive; therefore spontaneous generation *must have* happened! Long live abiogenesis! Indeed abiogenesis, ignoring conceptual principles, piles improbable practice (in the form of nearly thirty abovementioned starter problems - maybe you could think of more) on improbable accident. It stands or falls on the unpredictable antics of wobbly, don't-rely-on-me chance to the point that it becomes, in the balance, a hypothesis either for the credulous or obsessive who *must* believe because the alternative is, for them, unbelievable. *Who is kidding whom? Do people kid themselves? A dreamer converts abiogenetic impossibility into a fundamental fact. For a pragmatic realist, on the other hand, it becomes a false hope, a fiction.*

The problem is a psychological rather than scientific imperative. While its 'fact' is a story of happy imaginations, not just goggles but gas masks are testily scrambled at the appearance of any irredeemable alien who airs the infection of 'design without inverted commas'. The situation is wilful. To inhale the gas of metaphysic is, far worse than nitrous oxide, to rapidly dilute materialism's scientific sense of fun! Propagators of chemical evolution will therefore neither admit nor consider but only dismiss any alternative. Materialistic solutions, often bolstered by expensive research programs, are spun in any plausible way that might deliver passive order without reference to its active predecessor. *It is, with this spin, an attempt to reverse the entire order of creation.* This is because it cannot, looking 'earth-up', probe beyond self-referential physic and so reach to mind. From a 'worm's eye' materialistic point of view matter *is* the source of mind. Matter is primary, mind secondary. Having reversed the *top-down* view of things, hard-core materialism predictably resists any re-reversal. For example, the clear priority of Natural Dialectic is that active precedes passive information; mind precedes matter. Mind is the source of order. The logic behind the (unproven) theory of abiogenesis twists the Dialectic upside down. It is topsy-turvy because it states that what is ordered (passive information as found in physical bodies) *ordered itself* into increasingly specific, purposeful arrangements. If passive can create active information then, of course, no active initiator is needed. Passive's tail can wag the active dog!

Appeals are made to non-logical time (in the fog of which anything can happen), illogical chance and mindlessly 'logical' necessity. This produces a theory of history as unapproachable (because chance is by definition unpredictable) as it is unobserved, un-testable and incalculable. Nor is it falsifiable. How can you ever demonstrate a test-proof, guess-rich theory wrong? As such it is no more scientific, except in its rejection of metaphysical intelligence, than metaphysical inference. **Indeed, it *is* a metaphysical inference - one that prefers non-intelligence to intelligence!** Evolution happens in the absence of any immaterial element. It is a tautological refrain, one as circular as a rosary. Physic is and, it repeats, metaphysic isn't. In the predefined absence of informative mind, there is no intelligence. Indeed, the crucial moment of virgin birth (of a proto-bacterium) on sterile earth underlies the whole orthodox cosmology of scientific materialism, humanism and atheism. It is therefore definitely, from this point of view, the preferred and reputable answer to the way life started up - but is it right? *A student has to grasp the logic of nature and decide whether (s)he, a product of that logic, is governed by reason or unreason. So much faith is vested in abiogenesis that an axe applied to its root may, unhappily, provoke a redoubled determination to discredit dialectical rationale and, in the process, prove reason irrational.*

The facts are clear. **Chance and necessity engage, like natural selection, absolutely no creativity, anticipation, intent or intelligence - the latter being the only known generator, *by its very nature*, of specific, teleological information. These properties are, exclusively, the province of mind.** Thus not only chemical but computer simulations of how chemicals might, unaided, have accrued functional information that would raise them from dead waters into life are doomed to failure. Indeed, they disprove rather than prove Monod's bold thesis since at each attempt the provision of information, reducing randomness and edging towards a desired result, is hosted more or less surreptitiously by the theorist. Chemicals or programs are aided by choices, guided by decisions and helped along in one way or another towards a preconceived end. Each psychological intervention effectively adds anti-randomness (or information) to the brew. Such a negentropic 'spell' is established in the mind of experimenter or programmer even before initial or downstream conditions apply. Unmindful nature never worked this way. Thus a theory, like a course of action, that is ill conceived or wrong may generate a waste of public, let alone private, time and money. Would you still promote it?

The physical world is completely mindless and science really hasn't got a clue how life could mindlessly begin. *Instead you have, on pain of scientism's excommunication, to believe a wretchedly thin materialistic fable dressed with technical sounding jargon - chemical evolution.* Yes, it is true! **The myth of modern alchemy states that an unknown reacting mixture reacted in unknown reacting conditions to give unknown products by unknown mechanisms - against an accumulation of actual scientific evidence.** If this involves a flight from science what will you make of scientism's claims? Could it be, for reasons this chapter has outlined and Natural Dialectic confirms, that a study of 'chemical evolution' is based on fundamentally false premises? Could a believer's promise that such transformative alchemy - from leaden clay into life's gold - will obtain a wholly materialistic answer to the question of how life began be an empty one? Worse than that, could the catechism be a bluff? *Is it simply a philosophically driven, 'scientific' litany under whose illusion the naïve, gullible or acquiescent labour? Or under whose spell the more knowledgeable stand transfixed?* **Chemical evolution is a fantasy that the**

philosophy of naturalism and its atheology in their totality depend upon. Thus reason stands no chance! No god but future, perhaps undreamed of, science will close every gap. Once locked into such totalitarian necessity the faith of promissory materialism by continual re-telling keeps the lifeless creed alive!

Uncertainty, like imbalance, creates motion. It certainly drives science and is not at all the same as ignorance. Yet scientism lectures from a stance of certainty. Its podium rests upon foundations of the evolution of some chemicals. Is not the basic fact of life their special disposition? At present we can only guess but no doubt in the future we will understand how, smaller than the eye can see, serendipity created cellular technology. Or not. We have seen the problems and that, in the flight from science, a Theory of No Intelligence is not so rational. The very concept wobbles on some shaky ground. *The evidence is weak to the point of absence and sometimes, as detectives know, absence of evidence is actually evidence of absence.* Then, normally, they let the suspect go. **If abiogenesis is wrong the suspect wasn't there. The evidence for evolution theory never even leaves its starting block**. Is materialism actually, for this and other reasons, a weak faith built on shifting sands - the sands of chemical reactions and, by interpretation, evolutionary trends?

Is infinitesimal a margin - chance that's zero or statistically a wafer from impossibility - the abstract holdfast, great white hope and overriding faith of unbelief? The chemistry and maths (see, for example, Chapter 20: Number Games) confirm that it is so. If, however, scientific atheism wobbles on a very shaky altar of coincidence, by definition purpose and intelligent intent do not. The greater an intelligence, the more powerful in principle is its information and the less uncertain its designs. Knowledge and precision outwit accident. Chance is left no chance. **Life did not go live by accident**. The abovementioned list of observations serially reveals the presence of mind, as it stands behind every mechanism and each purposeful action, behind the systematic programming of biological patterns. The matter of irreducible complexity of mechanism, on which we focus more sharply in the next chapter, is one ubiquitous in the study of life forms. It is the final nail in chance's coffin. There is no reason for no reason. In the case of design chance were so remote and superfine a wisp that it is polished off. There is good reason to discard, as wishful thinking, a theory that irrationally proposes the completely accidental occurrence of purposive, pre-programmed soft machinery. **Indeed, if any mechanism be found truly irreducible in character then not only chemical but biological evolution as an originator of life's forms fails absolutely on theoretical as well as on the abovementioned actual grounds**.

This chapter has spot-lighted how death (in the form of matter) cannot, rising from its everlasting grave, be transmuted into life. It has disinterred a couple more black boxes - chemical evolution and abiogenesis. Both are dead in the water of Darwin's fancy, a pool of chemicals. Their nonsense buries any Theory of No Intelligence. *The notion of A Little Accident is definitely another flight, both physical and chemical, by scientism from its science; it is, at the same time, an epitome of error.*

22. The Origins of Species and of Type

Hold on! A materialist is (as noted in Chapter 7) wont to lapse into animistic phraseology such as 'genetic program' or biologist Haeckel's deification of matter as 'eternal and alive'. Such imaginary matter, while easily capable of evolution, is unknown to chemistry, to physics and indeed molecular biology. On the other hand, if the hypothetical chemical evolution of a proto-cell *is* (as illustrated in the previous chapter) simply a sophisticated form of animism or an alchemical myth, it becomes impossible to steam away to any evolutionary extravaganza.

Evolution is agendum-driven but its major props, The Big and Little Accidents, are both in quarantine. Can you start a cosmos out of nothing? Maybe physical constraints were absent but (Chapters 7 - 12) surely it is reasonable to argue that transcendent, metaphysical events and entities both were and are prior? Now, worse than quarantine, the Little One has, as last chapter's visit showed, suffered a fatality. But you can't start without a start. *The problem is profound because the whole edifice of materialism demands abiogenetic seed from which its tree of life can then branch up or, if you like, can commonly descend.* How, therefore, to seedlessly respond? Do you smile and simply gloss the issue? Then press forward regardless?

OK. It may be less rational to propose an abiogenetic than conceptual origin of biological information, exercise and machinery. Abiogenesis may be impossible in principle (Chapter 19) and practice (Chapters 20 and 21). Do you recall (from Chapter 20: Supreme Operation and Chapter 21: Cata;ytic Philosophy) that enzymologist Malcolm Dixon had, on the logic of his study's ground, denied a lifeless start to life? I probed further asking, "If there is no such 'seed' then how sprang life's tree and branches popular in books throughout biology?" As a Cambridge academic involved with senior molecular biologists he simply gave an enigmatic smile. Although his 'shrug' avoided controversy I sensed implicit confirmation that without a seed there cannot be a macro-evolutionary tree of life; therefore branching diagrams relating to phylogeny exist as evolutionary devices you will find in minds and books alone.

Harumph! This is not what scientific rationalism wants to hear. Without a point of origin whence does a theory of origins proceed? And if such diagrams are conjured by imagination from thin air then science if not scientific atheism will eventually submit to factual evidence. *There is much metaphysical at stake. If there was no abiogenesis then the bottom-up philosophical perspective is swept away. Scientific atheism catastrophically fails.* This is why the game is kept alive. Evolution *must* have happened so, without its starting-point, let us proceed to query the hypothetical branching that is supposed to have produced actual, leafy life. Ghostly seed, roots, trunk and branches! Let us notwithstanding, only for the sake of argument and to tease its problems from their woolly knot, assume Darwinian macro-evolution of all organisms from a common (but in truth a non-existent) ancestor occurred. This ancestor emerged from chemicals alone; are you, therefore, not a most evolved, esteemed reaction?

The Light Through a Lens

'I remember at an early period of my own life showing to a man of high reputation as a teacher some matters which I happened to have observed. And I was very much struck and grieved to find that, while all the facts lay equally clear before him, only those which squared with his previous theories seemed to affect his organs of vision.'

Lord

Joseph Lister (1827-1912).

Goggles, goggles, goggles! And not just for the chemistry experiment! From the outset (Chapter 0) we saw clearly that the lens through which we look affects our vision. It is not that, equipped with human senses, we see everything the same. The way that we interpret depends on education and, at chosen mind-set's root, whether we are looking *top-down* or *bottom-up.*

Science, in the excellent and entirely correct exorcism of ghosts, hobgoblins or any superstitious spiritualism, reduced its parameters to physical alone. Nevertheless, while there are no hard edges and you can't get a ruler to subjective experience it may, as we have seen, be wrong to throw the metaphysical baby out with superstitious bathwater. Only the empty tub remains. It represents a dry and lifeless physicality, a one-dimensional, flat-earth view of life whose best brief is scientific materialism. At the heart of this brief lies the notion that life is indeed entirely physical, evolved '*bottom-up*' from chemical reactions into the most complex and orderly arrangement of matter in the universe (a human brain). Brain is the sole cause and repository of mind. Matter therefore precedes and causes the metaphysical illusion of a mind (that may in turn conceive of evolution).

This earth-to-life application of an evolutionary lens will expect a seedless family tree to spring abiogenetically and branch upward from the barren earth. *It is called a phylogenetic tree.*

Because their remit is scientific most biological teaching and textbooks start with the application of such an evolutionary lens. They automatically adopt an evolutionary framework within which taxonomic relations start from an autotrophic ancestor. They have adopted an almost uniform 'cadre' of examples to explain their scientific case. Such 'icons of evolution' include a single bush or tree of life, common ancestry as evidenced by homology and fossils, embryological drawings, illustrations (such as bacteria, ammonites or Galapagos finches) of variation by natural selection, fruit fly genetics, peppered moths and so on. These are regurgitated edition after generation. If volume of print or dint of repetition were any criterion of truth then it is certain once upon an ancient time that energy-metabolising, reproductive chemicals 'abrupted' (as disputed in Chapter 20) out of sterile pools or barren earth. *The problem is, as Jonathan Wells and numerous other sceptics show (see, for example, Adam and Evolution), such icons sum to a litany of misinterpretation, unwarranted extrapolation, half-truth or, occasionally, downright deception. What a platform from which to preach interpretation of the evidence - especially if you 'scientifically' stamp on criticism!*

Nevertheless from this unfortunate educational starting point a student (who may later become a reputable teacher) is locked and locks all data into an exclusive, Darwinian frame of reference. Despite anomalies the facts are fed into a template of interpretation based on gradual evolution of a tree reproductively-linked life.

In this scheme nucleic acid information storage and transfer systems, ribosomal *RNA* and proteins that support core function (e.g. informative, energetic, reproductive) and structure (e.g. codified biosynthesis of the constituents of cytoplasm, membrane, extra-cellular matrix etc.) are interpreted to indicate fundamental ancestry. Thenceforward evidence is adduced from molecular and phenotypic sequence data. After an extremely hypothetical, primordial stagger from lack of biological principle to principles and principal (a first cell, axed in Chapter 20) it is presumed the cornucopia of life thence spiralled forth. Despite persistent, systematic anomalies between molecular, developmental and structural evidence evolutionary series (such as cephalopod ammonites or horses) are arranged and re-arranged according to schemes that might suggest how branching and complication occurred. How did the phylogenetic tree of life develop from its roots to current leaves? How, after a particular fully-working molecular, cellular, organic or whole-body plan 'appeared', did it morph into increasingly sophisticated successors? Chance is unprincipled. Because mindless unpredictability is difficult to track it's open season. *While the principle (evolution) is assumed unassailable the practical details are up for negotiation.* At least the single, simple guiding principle is clear - an unmotivated stagger from simple to sophisticated principles and practices, an accidental elevation from 'coarse' to 'complex' designs. Is that not, even if impossible, plausible; and if plausible, then why impossible?

Examples of progression are drawn from developmental patterns such as protostomy and deuterostomy, homologous organs, shell shapes (for diatoms), fossils, spore size, body symmetry, the evolution of germ layers, the coelom etc. etc. Divergence, convergence, parallel evolution from similar ancestors, functional loss and re-evolution, mosaic evolution and co-evolution are part of an armoury of ingenious and all-encompassing explanation. So are homologies (same apparatus developed in the same way but different function) and the analogous evolution (different apparatus, same function) of complex organs of manipulation (e.g. pentadactyl limb), locomotion (e.g. flight of bird and bee) and so on. Given axiomatic naturalism *something* like that must have happened!

Do you remember, from the last chapter, an example of convergent C-4 photosynthesis? Convergence is the circumstance wherein, in thousands of astonishing cases, the same or very similar genomic, molecular or morphological character is thrown up in different, sometimes completely unrelated animals, fungi and plants. In other words, blind evolution is supposed to have repeatedly recreated the same effective designs for operation in air, in water and on land. Examination of a catalogue of startling convergences and mimicries renders this hand-waving explanation (in which the small steps, genetic and morphological, are never closely defined) weak. *Indeed, 'convergence' is simply a semantic device that re-interprets 'archetype'.* It is for Natural Dialectic a 'just-so', 'evolution-of-the-gaps' story repeated every time there is no hereditary reason why two functional shapes, mechanisms or genotypic sequences should be similar in concept or appearance. It is the Darwinian reason why, for example, ruminant stomachs and relevant lysozymes occur in cows and relatively unrelated primate langurs but not in species 'in between'. Thus a Dialectician, simply and with strength, replaces the inexplicable notion of convergence with archetype. He replaces such 'mosaic' occurrence with the notion of variation upon subroutines or modular themes incorporated, according to a Theory of Intelligence, by design into the integrated program of an organism as a whole. Such archetypal record, the source of code, shape and biological functionality is thereby read, not as somehow blindly 'convergent' but conceptually deliberate.

Gaps not only invade between 'convergences' but, unless you start with many separate trees, occur between the branches of descent upon a metaphorical tree of life itself. In deference to a lack of hard fact phylogenies used to be joined at the base of domains, kingdoms, phyla, classes and orders by tentative sets of dotted lines. Perhaps from forgetfulness of the philosophical basis of the evolutionary framework, perhaps from an increasingly hard and emphatic assumption of rectitude, these lines are now often drawn continuous as if they represented (give or take a proviso or two) fact. An icon is an image, a symbol of faith and aspiration. Such representation amounts, in fact, to iconography. The image of a tree of life is certainly an icon. The real questions involve the nature of its seed, roots, trunk and branches; are these supporting linkages, quite apart from foliage composed of living organisms, real or thin as air?

As You Like It: Scientific Animism

Natural history documentaries are often brilliant in photography and, when they stick to current facts, narration. Is it true, though, with respect to history, that today's narrators are 'just-so' raconteurs? Aren't they story-tellers, as you like it, of naturalism's default tale? The answer, in the balance, if you stick with facts as they're script-written, has to be an arguable 'yes'.

Why? Conventional wisdom leads interpretation's way. Interesting, speculative and plausible, the scripts are based exclusively within a frame of mind set fossil-hard by its philosophy. Materialism: wherein the world's essentially a mindless place and therefore, critically for life, a Darwinistic zone. There's no alternative and so monopoly grants licence unto endless just-so stories and (which is fun) changeful hypotheses. You're allowed to think within this framework not without. After all, what's life but bodies? These *must have* evolved from single, simple cells; and how else to explain the history of life's progressive way but by guesswork that's informed. Our *zeitgeist* rules; such tactic brooks no argument.

So is it 'As You Like It'? How valid is it to presume that, just because a biochemical, a metabolic process, organelle, cell, tissue, organ, system or a body-plan was needed to some end, that animistic evolution almost 'willed it' to appear? You tell us cells 'make choices', 'engineering' evolution 'solves the problem' and nature always seems to 'find a way'. But this just-so shorthand is no more than animistic nonsense. Matter's mindless and such errant phraseology no more explains than it explains away the tacit *reasons* every bio-form displays. Reason is the gift of mind. How irrational, therefore, at the sharp end of informative development, to deny the source of information and assign the cause of life to energetic chance. Uneducated folk, whose rural circumstance might seem to harbour elements alive, animate their place with spirits, sprites and deities. Educated fellows plump, on the other hand, for chance. Since such cause is reason's absence 'reason without reason' is the frequent cry. Now a Great Sprite, evolution, chooses, solves, discovers and creates the mindless rationale of life. This is scientific animism's spirit - eyeless progress in the image of a human mind.

Could mutation bear creation's load; or billions of accidents, sorted out by death into survivors, have ever generated spritely life? You may wish so but, if not, how valid is it to compare a modern type with ancient specimens and thereby construct an arguable tree of evolutionary ascent? Could, for example, blue-green bacterial cells have bloomed to vent sufficient oxygen for aerobic metabolism; so that somehow (several hypotheses exist) dozens of metazoan types (e.g. fungi, plants and animals) separately evolved to take life's stage? Of course, many Vendian (pre-Cambrian) fellows are, like sponge and coral, evolutionary dead-ends. Still, from lineage that's unidentified, must have issued *Fractofusus*; complex animals with repetitious, fern-like branching segments that resemble sea pens first sprang.

Next *Dickinsonia* presented life with such bilateral symmetry as you and I inherit to this day. It seemed to browse bacterial beds of slime. Mobility! Actually mobility is of another kind! Controversy over this fossil has led to its meander across nearly every kingdom and many phyla of life. In 1992 seven authorities among paleontologists were asked by evolutionary biologist Rudolf Raff to specify the affinities of *Dickinsonia* and they gave seven different answers. The published alternative attributions include an independent eukaryotic kingdom Vendozoa or Vendobionta (Seilacher 1989, 1992) of quilted marine protozoans, giant protists like xenophyophore foraminiferans (Seilacher et al. 2003), terrestrial lichens or mushrooms (Retallack 1994, 2007, 2012), benthic placozoans (Rozhnov 2009, 2010, Sperling & Vinther 2010), comb jellies/ctenophores (Zhang & Reitner 2010), cnidarians (pelagic jellyfish or sessile corals or sea anemones) (Sprigg 1947, 1949, Harrington & Moore 1956, Erwin 2008, Valentine 1992, Buss & Seilacher 1994), platyhelminth/turbellarian flat worms (Termier & Termier 1968, Fedonkin 1981), annelid worms related to the recent genus *Spinther*(Glaessner & Wade 1966, Wade 1968, 1972, Glaessner 1979, Cloud & Glaessner 1982, Runnegar 1982, Gehling 1991, Jenkins 1992), an extinct higher animal phylum Proarticulata (Fedonkin 1985, 1990, Ivantsov 2007), or an extinct animal class Dipleurozoa (within Proarticulata) possibly related to chordates and/or nemerteans (Dzik & Ivantsov 1999, Dzik 2003, Fedonkin 2003), the sister group (together with the Ediacaran frond-like taxa) of Eumetazoa (Hoyal Cuthill & Han 2018), or a position among basal bilaterian animals (Gold et al. 2015), or an unresolved attribution to Metazoa *incertae sedis*(Hoekzema et al. 2017). All of these disparate attributions have been disputed by other eminent scientists! And as for body symmetry so for internal organs. Life's not easy!

Slug-like, grazing *Kimberella* scratched the surface as it moved and predatory *Sprigina*, needing energy to glide about, acquired a head-tail axis, mouth, teeth, digestive system and sensory organs to line up attack. Need it, get it! Evolution (Lady Luck and not the Good Lord) will provide! Repertory expands according to whatever functional 'improvement' needs. Perhaps *Funisia*, clustered reed-like, first exchanged genes and initiated that meiotic expedition into sexuality? Whence evolution could 'explode'? As you like it - every serial reconstruction is essentially the just-so same. Not only media raconteurs but academics also build imaginary bridges over gaps and skip in dotted lines from isle to functional isle. Kipling never made such doctored, learned links. Surely, though, there's no great flaw that would mutate such reasonable 'fact' to fiction or could lengthen stories into tall?

In fact there's a couple. Law not flaw, the first is fundamental to biology. It is the First Law of Biogenesis (see Chapter 20). Such law is to be confused neither with Haeckel's Theory of Recapitulation (his self-styled 'biogenetic law) nor infringement by the Hope of Abiogenesis (Chapters 20 and 21). No exception, except by mind's imagination, has been found to this First Law - although not for want of promissory trying.

Second, evolutionary tales are seen to break the Second Law, the Law of Heredity. Organisms, without any known exception, reproduce their own types. Variation (by mutation, sex, environmental pressures and adaptive super-coding) happens all the time but, by this law, with limited plasticity. This is what all observation shows. Unlimited plasticity has never ever been observed and yet the raconteur, who claims authority from science and in whose mind it is conceived, protests that such plasticity has made us, every one.

As well as blithely breaking these two laws our novelist omits an explanation of a series of 'black boxes'. These serial offenders, numerous and irreducible complexities, Chapters 19 to 25 identify.

If this were not enough an immaterial but massive barrier looms. By sequential 'logic' anything that evolution needs it gets. More energy required? Add organelles, fresh coding, head, mouth, stomach and the rest disposed coherently in an appropriate body plan. Fossils, hypothetically arranged, can prove what you believe. Yet the fact is evolution's world is mindless. It is without anticipation such as only mind supplies. Its irrational maelstrom can't, step by entirely groggy step, progress to build up integrated systems - not least since natural selection would cull starters that still, like any uncompleted mechanism, do not work.

Worse still for unconscious unintelligence, blind process does not generate 'advantages' informed by code. The basis of biology is information and this information is materialised by code. Signals stored or carried various ways (including *DNA*) inform specific, purposive complexity. From faulty metabolic algorithms through to incoherent parts of systems, natural selection kills what does not work. It kills evolvands off. So whence, in a way that's never been observed, did the initial information for each *innovation* (not just adaptation) come? *Raconteurs cry 'chance' but is chance, in the form of mutant code, sufficiently creative unto functions, targets and the interactive modes of complex life?* **There exists no natural Law of Innovation such as just-so plausibility requires. Nor jot of evidence, besides interpretation, that the progress claimed occurs.**

Do you protest? Is this interpretation of the evidence not what you like at all? You think that confident, enthusiastic raconteurs have to be right but Chapter 20 started and the next four will in greater depth explain just why the bards put Darwinism's cart before the horse; and why interpretations of life's history better add an element of immateriality to material forms and physiology. In short, photography is as you like it but narration, when it serialises fossils spanning multiplex conceptual gaps, misleads. Indeed, an engineer will tell you that design is a deliberate form of information; functional arrangement is mind's speciality; and neither an invention nor its explanation can be reasonably made without due reference to prior intelligence - all semblance of which just-so raconteurs abstemiously omit.

What You Will

You don't always get what you want but you may get what you don't. What animates? You want to say that from a microbic 'seed' there must have sprung the whole abundance of eukaryotic forms and yet by now you start to see that the iconic tree of life sounds simple but is not. A hundred and fifty years ago biologists had as broad and intelligent but less detailed an appreciation of the subject. Most Victorians saw biological organisms as we see cells today - expressions of consummate engineering and, it could be claimed, artistic skill. They did not shy (as Richard Owen's cathedral-like construction of the capital's Natural History Museum shows) from the philosophical consequence of teleology. In this view purposive information and mechanisms are each a product of design. *The Darwinian framework is one that has changed nothing in fact, simply the lens through which the facts are observed.* It is a lens that excludes and has therefore not developed a modular, typological and cladistic approach to biology and classification (see below). **This is in spite of the fact that everything able to be interpreted according to *bottom-up*, phylogenetic, evolutionary principles can equally and systematically be approached in a *top-down*, modular, typological way.**

No doubt from the lowest, statistical perspective a cell, like a TV, is simply chemical elements. No sensible engineer dwells on this single aspect. *Bottom-up* (and an engineer looks *top-down*) an informational hierarchy emerges. Biological order involves the transition from molecular activity to a super-molecular order of the cell. As we ascend an increasingly systematic unity is revealed. Specific molecular configurations give rise to metabolic pathways, such pathways support and relate to functions, which functions operate within specific structures to achieve definite ends. A cell was, to those Victorians, a 'black box'. Neither Darwin nor anyone else understood its order of complexity. In fact each unicellular organism is as different from mud as each of your own cells. No cellular mechanism, even the most 'primitive', could have been a 'homogeneous globule of slime' like Thomas Huxley's mythical *Bathybius haeckelii*. All that is 'primitive' is the understanding of that word's user. Far from ever being crude or random, the biological aspect of life is at all levels from molecular upwards exquisitely technological and almost inflexibly constrained. It is not as if such constraint applied only to large-scale systems such as renal osmoregulation, blood sugar level or heart rate. **Nor does life get fuzzier as you zoom to smaller parts**. Order is clearly defined, codified and executed within the molecular interior of a cell. Here, down to atomic and subatomic levels well beyond the scope of Darwin's vision, science is revealing precise, complex and inerrant coordination. Such management of minutiae is critical. *Up, down, all around there never was nor is a simple starting bio-point.*

Do you remember, for example, energy metabolism or the 'kinesin walkers' (Chapter 20)? Molecular biology 'buzzes'. It is a rapidly expanding, fascinating line of investigation into the chemistry of life. With ever sharper resolution it becomes harder to ignore the profound, systematic accuracy of information that underlies each integrated level of even sub-cellular components. **Of course, minor accidental variations occur; but in step with the clarity of revelation it becomes progressively more irrational to believe in the original production by chance of such exquisitely miniaturised, coordinated operations and their soft, organic machinery.**

It must, equally, be evident that there's no fossil record whatsoever for molecular evolution. You can just compare the molecules and metabolic pathways that we find today. Even fossilised soft organs, as opposed to petrified shell, scale or bone, are rare. Among the oldest are Precambrian sea pens (like, perhaps, *Charnia*) found in Newfoundland and the UK; also fossilised jellyfish from Ediacaran beds in South Africa and South Australia. The latter differ slightly but are at least as well formed as those found floating today in nearby Spencer Gulf: all the earliest multi-cellular organisms - not least those highly diverse and complex Cambrian eye-openers, trilobites - must have been in full working order. Actually, you *could* sequence an ancient genome; another couple of arthropods, crustaceans that are superficially similar to look at, the horseshoe crab and 'old three eyes', a freshwater tadpole shrimp called *Triops*, are 'living fossils'. They former is thought to have persisted for nearly half a billion years and the latter to have failed to evolve for well over a quarter of a billion; both organisms are as complex as trilobites (with different kinds of tadpole shrimp coded to employ a variety of unusual as well as normal male/ female sexual strategies including hermaphrodites and parthenogenesis) so that you could predict complex genetic operations then as now. *In this case, although there is absolutely no fossil record for histological or organic evolution either, you might also examine living fossils to find out how things likely were.* Otherwise you might have, for example, muddled stages of development for different species. Small-fry babies aren't a different species from their parents but, it seems, some junior *Triceratops*, *Tyrannosaurus* and duck-billed dinosaurs have been split into separate fossil species from their seniors. It has been suggested that as many as a third of such fossil reptiles might have been erroneously assigned; and many fossils similar to modern types have been assigned new levels of classification simply due, it seems, to separation in time. It would, on the other hand, be a brave palaeontologist who found a fossilised Saint Bernard, dachshund and greyhound and assigned even their hard parts to the same species. You could, with some imagination, concoct an arguable line from tapirs through Falabellas and racehorses to a Suffolk Punch. Plausibility is not the same as proof it happened. All except the tapir could live on the same farm. Nor does the tapir inhabit a different time but a different place; so bring him from the zoo to look at history in the field. Nor, even if his time was different, do a series of extinctions show that tapir, coney, deer or any other sort of 'likeness' was, by graduated evolution, ever turned into a horse. Equine evolution used to be a showcase but today the origin of *Equidae* and notion of 'descent' through 'intermediate' types of horse is challenged. Bets are laid and the dispute on course.

Speculation leans on inference but the coverage is not always binocular.

Fossil and modern types are carefully compared by monocle with evolutionary lens. Thus phylogeny implies an actual 'blood relationship', a common descent through ancestors of increasingly different types that lead back to a proto-cell. The leap to presume such familiar, familial linkage is, although a crucial thread in Darwinism's twine, perhaps more psychological than based on evidence - or are there really 'missing links'? There are claims of such fossil ancestry but they are few and far between; and, as in the case of reptile/ bird or ape/ man, in hot dispute. Nowadays a 'savvy' palaeontologist prefers to try and identify intermediate, transitional and perhaps even ancestral *features* of a fossilised organism. The inferential identification of so-called 'primitive' and 'advanced' features also invades the taxonomy of living specimens, especially plants.

It is a fact that working systems with interdependent parts do not appear, all the time working, while being formed by accident! Nor, before they are working and chance holds sway, is there any chance that natural selection can step in to 'order' things. What reason could non-reason have? You can drum up hypothetical sequences drawn from apparatus found in living organisms (the eye, as we shall see, is a popular resort); you can imagine the same teleological necessity, survival, lies behind selection; and that, therefore, some kind of pointless and oblivious competition to survive notches up improvements. Natural selection, progress, evolution but what about the evidence? *You may have noticed that, while both fossil and contemporary specimens are evidence for working body-plans, organic systems and other biological purposes, they offer no evidence for the myriad ancestral non-starters at every level of biology from molecular to whole-body that couldn't, because they weren't working yet, have led to them.* **In other words, nascent physiologies, incipient anatomies and provisional chemistry all tending towards a future, integrated working system don't exist today. Thus, if 'the present is the key to the past' they never did. Even if they had then such ambitious 'would-be innovations' would have simply fouled up what already works. Death not evolution would have cut a swath through all such bio-wannabes or, if anticipation's disallowed, such morphic accidents.**

Scientism assumes a universe comprehensible in its entirety by human reason. Does this mean that all processes must, as its *fiat* runs, be ultimately explicable in terms of only physical processes? How can discussion with a brick wall whose cement is called tautology occur?

Natural selection weeds out weaklings. It tunes the healthy population like a tracker fund. It tracks top business as its enterprise in tackling challenges succeeds; but, instead of variation tracking round a theme, what about successful innovations? Where is the evidence for gradual, all-round 'systems revolution', for radical restructuring by accident? Unless new plans and systems popped up fully formed then minor accidental steps must lead to them. Must it not be that mountaineer Caprice reduced an infinite negative (impossibility) to finite peaks of vast improbability? And yet no fossil series shows that, part by part without a goal, complex new systems (not just minor variations) self-assembled. Nor today is systems engineering seen to edge by stages towards a pinnacle of order without mind. Of course, you spot the lady skipping, step by selective step, among the shallow slopes of Mount Improbable. Is she, in the guise of variation dressed as micro-evolution, not the proof of theory? Is evolution (Chapter 23) not in full swing here? Or, on the other hand, isn't High Improbability just that? Isn't new design a 'winning peak' and aren't the heights of real achievement never left to chance? Revolutionary innovations, bannered up as macro-evolution, don't swan hand in hand with Lady Luck. It is a definite but unseen hand that, engineering systems, takes its orders and its information from the mind.

What a pity history's ambassadors, the fossils, do not offer any clues about the about the origin of all-pervasive information, embedded for example in controlling norms found everywhere like cogs in a machine; how did the multifarious forms of homeostasis, threads that weave the fabric of biology, come to evolve? It is as if, due to a complete loss of R & D documentation about ideas, preliminary sketches, detailed plans and trials of prototypes, working products have seemed to spring *ex nihilo* and fully formed. *Ex nihilo* (from nothing) physical, that is. No doubt that fully-preserved fossils have been found in millions by the phylum, class and order. Many resemble closely present-day descendents but what of all the intermediates, the stepping-stones that Luck skipped fairy-like across, between the separate banks of life? What about a million ghosts that should have filled the gaps in godless theory but seem to have evaporated from the rocks? If you suppose that, since no archetypal document exists, the evolutionary files amount to step-wise aboriginals then what if these intermediate steps do not exist? You could (if given half a chance!) link organisms that live in the world today into a just-so story that evolved as good a series as the fossils do. You wouldn't need a single common ancestor to link your range of products up.

If you can't build bridges over evolutionary gaps (as Darwin couldn't either) then, failing failed suites of prototypes, an urgent search for next-best evidence assumes priority. A single intermediate form might be enough, a gesture that implies a greater truth. The quest for such critical entities has therefore become a holy grail characterised by all the elements of mission - zealous conviction (they must exist, they are there, we only have to find them), honest toil, in-fighting, spin, even fraud and suppression of any rival doctrine. Galleries of portraits have imaginatively clothed a few bones and detective novels have been written about the monkey business involved in human palaeontology alone. Shady spectres such as 'embryonic Piltdown' (see Haeckel Chapter 25) and Piltdown 'ape-man' still haunt 'missing linkage'. In a recent case (according to The Times 1-12-04 p.34) one scientist alleged the other 'had secretly glued a wisdom tooth onto the jawbone' of a 'hominid' - or perhaps it was a female gorilla.

The whole issue of missing links and 'primitive features' is, especially where humans are concerned, vexed and overheated. So much (not least fame and funding) is at stake. And pride. No atheist is part of God's creation; he is a self-made man who worships his creator - or creatrix if Lady Luck evolved him. There is no room in all the star-struck heavens for Intelligence much larger than his own. How could you, therefore, tout a Theory of Superhuman but still Natural Intelligence?

For a son of Adam, on the other hand, the race of men did not first hop along with monkeys. It started with the archetypal fall of man into form physical. Man descended not ascended; he was devolved, did not evolve. Is metaphysic real? If you put information first you tout that theory of non-chance, The Theory of Intelligence - one that is officially taboo.

Which plaintiff is correct? Which sleuth's lens has skewed reality? Could you ever solve the riddle if you had omitted motive and design from the suspects listed in your file? Do you remember (Chapter 17) the exquisite disposition of the human type? What if there isn't need for 'missing links' you do not find so that failure to discover evidence merely reflects its non-existence? Each side re-interprets what the other has inferred. Whose appeal will be dismissed?

Upright Logic

The problem is (Chapters 5 and 6) *that anti-chance or meaningful order derives from active and not passive information.* It implies the reverse of an essentially meaningless *bottom-up* earth-to-life succession. Such upright logic stands like you with head on top and information to the fore. Natural Dialectic's bio-logic would suggest that only mind can generate high-grade, purposive information; and that, consequently, a functional whole is greater than the sum of its parts. This is because the whole includes the organisation of those parts. Organisation is the informative component. Information, even passive information such as material code and shape, is initially the product of active information, mind or purpose. Let's recap the theme that this book's polar, Dialectic system has developed until now. The last nineteen chapters have been spent outlining the self-consistent, systematic and therefore logical patterns that inform and organise physical, biological and psychological shape and behaviour. Let's remember especially the contents, as it relates to information, of Chapters 5, 6 and 19 to 23. Let's revise before we turn the upright logic upside down.

The Principles of a Unified Theory of Biology put information first. **It was affirmed that the very basis of biology is information.** The basic informative actions are development of form and maintenance of dynamic equilibrium - whose central instrument is homeostasis. The basic informative actions are development of form and maintenance of dynamic equilibrium whose central instrument is homeostasis. Both depend on a flow of appropriately coded signals - input, processing and output. And an organism's systems, molecular or otherwise, are built from coded information that delivers mechanisms capable of executing their conceptual ends - energy production, waste elimination, reproduction, information processing or homeostasis itself. Information is always purposive. Meaning is derived from its intent.

	tam/ raj	*Sat*	
	motion	*Balance*	
	up/ down	*Level*	
	oscillation	*Axis*	
	passive/ active	*Poise*	
	expressions	*Potential*	
	variations	*Criterion/ Norm*	
↓	*tam*	*raj*	↑
	passive	*active*	
	informed	*informant*	

matter	*mind*
entropy	*negentropy*
used	*user*
code/ communication	*idea/ concept*
outcome	*purpose*
product	*concept*

Do you remember the nature of immaterial information? Check *fig*. 2.6 and other stacks in Chapters 5, 6 and 19. The active (mental) aspect of intelligence chooses and arranges appropriate materials (which are passively informed). While matter is subject to the (*tam*) downward thrust of entropy mind is (*raj*) an order-making, negentropic entity. In biological terms and therefore involving three levels of matter the Dialectic reads:

tam/ raj	*Sat*
peripheral	*Central*
consequent order	*Archetype*
its translation	*Principle/ Order*
its conveyance	*Purpose*
inertial/ energetic	*Potential*

tam	*raj* ↑
informed	*informant*
effect	*language*
product	*code*
protein	*gene*
phenotype	*genotype*
inertial	*energetic*
bulk shape	*biochemistry*
task	*agent*
recipient/ container	*driver/ donor*
cellular feedback	*DNA*
its constructions	*metabolism*
structural protein	*RNA/ functional protein*
form	*function*
anatomy	*physiology*

Genes themselves, it needs be emphasised, are as entirely material and no more intelligent than nervous impulses, a larynx, sound waves or a TV set. All these convey code. They transmit information but in *passive* form. They create, like newsprint or a paint-box, nothing. *Genes are simply agents of intelligence.* What these primary instruments inform are secondary agents. A body's manual instructs and controls the fabrication of both tools (called enzymes), primary materials (called structural protein) and, through metabolic pathways, other secondary materials (such as chlorophyll, *ATP* or urea) that a given form of life will need to use. This is why protein (the word means first, primary or essential constituent of bodies) derived from informative genes occurs on both functional (organising) and structural (organised) sides of the stack. Both genome and its body-context are passively informed. They are passive information.

Passive information (Chapter 6) is not active. Upright logic states that active governs the behaviour of a passive state. In terms of conscio-material hierarchy physical effect depends on prior informant mind. The Dialectic's name for conscious mind is 'active information'; passive information is both mental in the form of what is stored in memory and also, through the laws of nature, all the shapes and textures of material phenomena.

Mind's potential (in the forms of idea, purpose or design) is the source whence issues action and its orderly results. You might construe 'results' as code, as meaning or creations that express their maker's mind. Non-purposive, inanimate complexity is coded, at root, by potential matter's archetype; the language of its patterns we identify as 'laws of nature'. Highly specific, purposive complexity as found in bio-logic's animation, that is, systems that support terrestrial life, is of much higher order. Anticipation, ingenuity and accurate integrity of each component with all others beams a message loud and clear to those who'll see and hear - information is incarnate in life units from a cell up to the size of whale or tree. All show, like machines, intelligence of mind behind. They carry passive information in the form of codes, of highly ordered language written out and worked specifically by chemicals. Designs are pressed on senseless elements in the way that men use ink or wood or speechful air. **The upright logic states that only mind makes symbols that make sense; and only sense makes tools that work.**

Therefore the major error of dysfunctional, 'down-left' reason is to confuse passive with active information. *It is, topsy-turvy with its head in sand, to reverse the logic and in this illogical procedure elevate creative capability of chance-within-an-automated-system over and above the active possibilities of mind. Do you rate blind turbulence ahead of clear foresight?* The error is forced. It is the directly enforced consequence of philosophical error, scientific atheism. It is to presume that passive generates active information, the expression of information precedes its potential or, in common parlance, matter creates mind. In biological terms it is to say that accidents (allied with passive information in the form of laws of nature) can create codes and these in turn become 'books' of genetic information. Non-intelligence can, by this inverted rationale, create a genome, contextual cell and, given time enough, conscious sensitivity with, in the final you might almost say climactic stages, lo, intelligence itself! At this point, if the intelligence is human and is high enough, the logic indicates that active information will take over and mankind start to evolve a brave new world! Surely this hopeful arrogance, judged by history alone, cannot be misplaced?

How, in short, can ideas be expressed without them? How can the irreducible complexity of cells occur without concept? The holistic answer is elaborated in a Theory of Intelligence. It states it can't. Codes and books, it claims, are derived as wholes. Each part is only useful in the context of the whole; indeed, its place in context adds up to the meaning of the whole. This context is conceptually informed by active thought or mind. The materialistic answer states, on the other hand, it can. Elaborated in a Theory of No Intelligence conceptual input is denied. The theory wheeler-deals in chance instead and, at least in imagination, lets the power of accident 'create' the natural ideas that we call organisms studied in biology.

Some persons loathe behavioural constraint. Others, rightly loathing various kinds of poison that perversions of theocracy inject, reflexively reject religion's other, human face. They refuse its salve, its logical intelligence and immaterial heart of love. Therefore they find the great conceptual flaw in evolution's answer most attractive. This allure, buffed with scientism's attitude of mind, shines irresistibly. Such authority is nurtured with tenacity; much time and effort are combined convincing both adherents and their student neophytes its literal nonsense makes the only kind of sense. In fact, it's fact! This is secular religion taught under other names in all our universities. Does pervasive, prolix repetition make it right?

<u>The Embodiment of Inverted Logic</u>

	tam/ raj	*Sat*
	imperfection	*Perfection*
	imbalance	*Balance*
	lesser truth	*Truth*
↓	*tam*	*raj* ↑
	towards illusion	*towards truth*
	loss of correct info	*gain in correct info*
	according to incorrect info.	*according to correct*
	failure	*success*
	incoherence	*logic*
	disorder	*order*
	accident	*design/ purpose*
	breakdown	*function*
	mistake	*correction*
	degradation	*improvement*
	mutation/ misprint	*accurate facsimile*
	disease/ falling ill	*health*

It sounds so obvious. Must we not have missed a trick? Yet *bottom-up*'s inverted logic starts at bottom left and runs to right-side towards its truth. Top becomes the bottom; top-right is your conclusion not your starting-point. Where disorder is your bottom line then informative potential becomes the last thing on your absent mind.

Is life just its body's form? Is its information content physical alone? The only information that inhabits matter is called 'natural law'. Within this frame of law the mindless incidents of cosmos flow. 'Chance and necessity' cohabit and construct the universe. Composite events of physics often turn out unexpectedly but each component of an 'accident' acts in accordance with 'necessity' of law. *'Chance-and-necessity' of physics is biology's 'mutation-and-inexorable-government-by-natural-selection'.* This is what we mean by *PAM* and *PCM.*

Materialism's inverted, dysfunctional logic holds that the pattern of physical operation (called 'natural law' or 'necessity') originated by chance; that, lacking any ingredient of purposive organisation, 'chance and necessity' are alone sufficient to produce both animate (biological) and inanimate systems; and that no such system is more than its atomic, molecular or other level of parts. *Blind factors can, in this view, generate purposive, high-grade information.* Such a line of thought inevitably produces and then rests upon the laurels of a Theory of No Intelligence. It therefore enjoins the *PCM* - organisms all evolve by means of random mutation and natural selection. But, you may demur, did mindless atoms really 'happen', without any logic of a plan, into such communicating minds as yours and mine? And if they did, is logic not itself illusion born of an illusion? Another twist in this strange tale is one that, as opposed to logic of embodiment, would deliver logical effects illogically, that is, at random. *That is to say, dysfunctional logic is twisted, topsy-turvy logic. It stands truth on its head.*

It is, however, no less than the key but illogical tenet of the theory of evolution that 'chance and necessity' can produce systematic, high-quality, purposive information. Take, for example, your heart. Is it purposive? No? Then all sense of reason has been lost. Yes? Then its order is necessarily, reasonably observed to dovetail with a larger, purposive systematic whole. It both informs and is informed by a self-consistent, bio-logical body. Such a principle applies not just to you but even the so-called 'simplest' organism - which is, in metabolic terms, very complex. Of course, you counter, Darwin plausibly invoked his scheme. *Indeed, but plausibility is not the same as truth.* Weak inference can be rejected in the face of strong. In this case assertion that purposive appearance and behaviour can evolve without a plan in mind may sum to just a vapid play on words. *Is it not smoke and mirrors to assert design without a trace of intent or intelligence? Are the key axiom and corollary of scientific naturalism, between them mustering no IQ at all, true or false?*

How can action start without a purpose? How on earth, if there is no intelligence in cosmos, could anything begin? Of course, you retort, chance rules. No higher information than the reflexes of nature weaves atoms into bodies far more complex than a jet plane, far more integrated than a space probe. They are conceptually but at the same time accidentally encoded in a language also accumulated out of accidents. Therefore why not promote, as the essential theme in volumes like The Origin of Species, 'purposeless purpose'? To minimise confusion wrap the oxymoron in a catchy phrase. For 'accidentally devised' write down 'evolved'. For 'blind filtration of intent' or 'mindless agent of life's quality control' read 'natural selection'. These two words combine to mean that, in a given circumstance, what form of life survives survives and what does not does not. That's true enough. Why dissent? It makes perfect sense. Why, for example, buy Charles Darwin's and not Edward Blyth's interpretation of the catchy phrase that (see next section) Blyth invented?

What does inverted biologic claim? Firstly let's rehearse the case of 'micro-evolution' - accidental variation on a theme. Yes, minor accidents occur but there is no necessity that any series of such accidental steps combine to innovate a different vehicle. Imagination can project long series of small steps into a process, namely 'macro-evolution', whereby major structural transformations can supposedly occur. Couldn't 'approval' of 'improvements' by natural selection sift and steer development along? Despite intense desire otherwise the origin of novelty remains a mystery. Black boxes blind, walls of irreducible complexity obstruct, dead-ends frustrate the thesis. In fact (*fig.* 22.1) circulation rather than concerted breakthrough is what genetics, breeding and the fossils seem to indicate. We've seen and will see in more detail that inference concerning 'Origin of Novelty' is drawn more reasonably the other way. No. Macro-evolution is not proven fact. It might be science fiction.

Now, secondly, let's inspect the strength and weakness of the proposed *creators* of biological evolution more closely.

Asexual and sexual kinds of reproduction create, without mutation, nothing original. Ever. The former simply and as faithfully as any printer copies a genetic pack: the latter shuffles (or recombines) the same cards simply dealing different hands. You can, as both common knowledge and genetics show, obtain reproductive variation-on-theme but, as the very word suggests, reproduction is not an originator but a transmitter.

Gregor Johann Mendel (1822-84), a creationist monk and the 'father' of modern genetics, established the conservative nature of 'factors', genetic units that are now called genes. These are passed intact from generation to generation. Occasionally, however, for one of a variety of accidental reasons, the base sequence of the *DNA* code can be miscopied or otherwise disturbed. This means that a change (normally construed as an error) occurs either in genetic operations themselves or the structure and consequent function of a protein. This error, called a mutation, can create novel *DNA*. As such it is the sole 'creator' of original material. *Such mutation as affects germ cells is heritable and, as such, the sole engine proposed for evolutionary innovation*. How, though, might slight changes made to proteins gradually innovate organs, coherent systems and their body-plans?

Natural selection, on the other hand, originates absolutely nothing at all. It is, simply, a fateful finger hovering above the genomic delete button. It is a term to describe differential copying or the way differences in the ability to reproduce are inherited or not. It weeds at the level of a whole genome and not any particular misprint it may contain. You might be lucky or unlucky. You might keep running with a load that runs to millions of minor problems or have one that, lethal, cuts you down. *Natural selection is no more than a name given to the lucky survival or unfortunate death of an organism or group of organisms in a particular environment*. A 'wild type' is, weighed in the balance of probability, presumed better adapted to thrive unless a catastrophe or major ecological shift occurs; but a 'mutant', on trial at the court of survival, is more likely to be selected 'against'. It is organisms not mutations that are chosen for the chop or not. But can any particular mutation cause an adaptation that 'works' and, furthermore, grants a competitive edge over its carrier's peers? Or is it, having failed to persuade the judge that it is 'beneficial', thus condemned? Beware of animation. You cannot imagine but must try to bear in mind the magnitude of senselessness, the total lack of judgment (except by mindless infertility or death) that in a Theory of No Intelligence makes nonsense of design.

Although the primary agent of evolution is mutation acted on by a secondary, natural selection, the twin 'creators' were historically presented backwards by the proponents of evolution. In 1859 Darwin proposed, without any mechanism for the creation of new material, The Origin of Species by Natural Selection; then in 1901 the Dutch botanist, Hugo de Vries, concluded from experiments that a new species of evening primrose had arisen as a result of a sudden change in its germ plasm that he called a mutation. A later repetition of his experiments (Nordenskiold E. Hist. of Biol. 1928 p. 588) showed that the change was not in this case due to mutation and de Vries' inference had been incorrect.

Nevertheless mutation is a fact and remains the first principle of evolution. Thus a historical twist in the tale or, rather, in the logical order of hypothesis, is preserved in the following inspection.

First Stage: The Judge (Natural Selection)

Natural selection was the first agency proposed for evolution and thus, as *the first stage in development of an inverted logic*, we begin with it.

Let chance contrive a first cell but, thereafter, given the very high improbability that the existence of any functional part displays, some biologists bar discombobulating chance from a 'creative' role! Even if chance made life it cannot systematically improve it! You can have it both ways! Cue, therefore, nemesis; intelligent design has met its match; unintelligent design, called natural selection, has led to its supposed extinction. Some naturalists raise natural selection to, it seems, the status of some kind of logic, agent of creativity or natural law. What delusion - unless death is such a regulation! Alfred Wallace (1823-1913), co-founder of the theory of evolution and whose paper 'On the Tendency of Varieties to Depart Indefinitely from the Original Type' preceded (1858) and precipitated publication of Darwin's 'Origin of Species' (1859), noted that natural selection not so much selects special variations as exterminates the most unfavourable. Such extermination is, in fact, a stabiliser; 'fit' traits blossom, 'unfit' wither as an organism's own ecology (its niche) dictates.

Ecological constraints (maybe changing, maybe not) cause natural selection to cull an unfit phenotype but a second course is also harnessed onto change's mill. Beneath the 'radar' of non-random cull random genetic drift acts on the genotype. Chance-based drift involves a directionless, selectionless swilling of alleles round a population's gene pool. Sometimes a given allele will predominate, at other times its sister; change might be visible or not; natural selection might bear down or not. Sometimes, however, such to-fro, dynamic equilibrium of the gene pool can be broken up. Rough lot may favour one allele to the exclusion of the other; or, when a small group splinters off, it may *lose* a proportion of the whole pool's genes. Such reductive 'founder effect' can, if one allele becomes 'fixed' at the expense of its fellow, cause phenotypic variation. The divergent spread of sparrow populations across America from their place of introduction, New York, is an example. Thus, by drift and/or selection, a different race or species may arise. However, no information has been gained. Allele permutations simply aren't enough. Indeed, such variation mostly involves not gain but (↓) loss of viable genetic material. Slow deer, slow fly - *reduction* in variety. Wild-type fitness wins the day. **Variation is not evolution (even if you call it micro-evolution).** There is no evidence, as we shall see, that it leads to macro-evolution, that is, evolution proper. This needs creativity. The only creativity an evolutionist can offer, we shall see, is something Darwin didn't know the details of. Mutation. Can mutation, we shall ask in detail, innovate enough? Or does it just dis-integrate?

In short, some plasticity occurs in populations. Change is rung, even speciation may occur but it's now generally recognised that there are limits, at low level, to the kind of innovation that gradual, exogenously dictated pressures can confer; nor is there direct connection coupling single genes (that might randomly mutate) to traits. Breeders minded to select a trait can only drive to limits; over this extremity an individual dies, a population will become extinct. Variation certainly occurs but nowadays the emphasis has shifted. It has sidelined Darwin's 'breeder', natural selection, as too weak a 'force'. Its method - cull - is uncreative and a lack of innovation cannot be the cause of evolution. Who, on the other hand, has ever witnessed nature drive a major transformation through? The evo- devo bet is, therefore, that natural selection conserves. It preserves the archetypal regulation of developmental patterns but, perhaps, occasional shuffling of its powerful routines produces missing links. Yet who, except in their imagination, has observed 'upstream' constraints (in Chapter 22 we called them 'bio-logic') flouted past extremity? No 'monstrous' missing links have been experimentally created. In fact, Natural Dialectic argues, so-called macro-evolution's an extrapolation way beyond the *PCM*'s capacity. It cannot cut transformist mustard; it is no more than vapid truism to explain that organisms are adapted well to where they live.

Yet the stakes are high. Darwin put his money on 'selection'. He replaced a 'supernatural designer' with a mindless kind of breeder - 'natural selection' - that is central to the atheistic faith. This faith's choir waxes lyrical on understanding how it can evolve 'prodigious heights of complexity and elegance'. A panegyric from secular fundamentalist Richard Dawkins even describes this key element of his philosophy as 'the champion crane of all time' that 'raises consciousness' (The God Delusion, pages 73, 79, 114 etc. etc.). What is it that this form of raising consciousness can show? If evolutionary theory is right (of course it is) and mind evolved from chemicals (of course it did) it shows the way that consciousness (whatever that is) indeed was raised by 'anti-God' from clay. You may thereby divine that obvious design in forms of life is an illusion. Purpose is an apparition airbrushed by the *PCM*. No doubt, this *illusion* of design is powerful to the point biologists assume 'natural reason' with its complex, rich designs as a working tool. The alchemy of Darwinism (as opposed to, for example, creativity or meditation) is that as a 'consciousness-raiser' it sort-of-reasonably allows design to seem 'apparent'. It actually converts the currency of mindfulness to fool's gold in the form of none. Is it, therefore, your natural reason or, actually, unreason cleanly cuts you to materialism's chase, a Theory of No Intelligence?

Instead of materialism's apotheosis let us treat 'selection' as it really is. Let us, at the start of an inspection of its 'mechanism', post three *caveats*.

The first is that natural selection is solely a process of elimination. It involves the disappearance of 'unfit' organisms. Deletion is not creation; such deletion is, as long as any individual life endures, suspended but it never makes anything new. To claim it creates the *emergence* of new form or function is, illogically, to use 'disappearance' to explain 'appearance'. However Darwin, with his belief in the Lamarckian theory of pangenesis (whereby the acquired features of parents are transmitted and blended in their seminal fluids), thought such acquisition bound up with selection's 'disappearances' was an explanation good enough. Prior to the advent of epigenetics (Chapter 23: Super-codes and Adaptation) the *PCM* believed random genetic mutations were the only mechanism of 'appearance'; now, though, you can add *epimutations* to the list!

Secondly, the effect of natural selection and random drift is, in the case of speciation, to have reduced the genetic potential of an original gene pool. Alleles are deleted or its complement split into separate populations. Information is not gained but lost. *For evolution, which needs not information (↓) loss but (↑) gain, this is entirely the wrong direction.*

Thirdly, it is false to conceive of natural selection as a kind of 'ratchet' that holds on to 'an extrapolation of improvements' at any level from nucleotide through protein to whole body shape. This is because without a prior plan to work towards there is no way for intelligence, let alone total lack of intelligence, to 'discriminate' a 'good' apart from 'bad' move as regards some novelty. *Indeed, it will breed out unwanted, nascent or non-functional characteristics.* **Only once an organ or a functional system (such as a beak or eye and associated factors) is complete can natural selection act on trivial, accidental variation to that system**. In short, natural selection, genetic drift and, we'll argue later, mutation are insufficient agencies to established informed, metabolic, cellular, organic or systematic 'norms' in the first place. These all need conceptual establishment.

Let's continue. From a *bottom-up* perspective mutation supplies the raw material for evolution but the real quality controller, sculptor or 'designer' of life, is biological selection. This judge, with IQ zero, has another name - Natural Selection. *Natural Selection is 'non-chance' in the armoury of evolution's argument.* Take a chance, the theory says, and selection gives you the result of it. It damns mistakes and condemns misfits. 'Sieved or survived; died or did not'. Such grand filter knocks out all that does not work. Whilst it may be categorically stated that the process of survival or of death is real, is such a 'God-substitute' really capable of the magnificent role ascribed to it by Darwinism - the creator of all forms of life? Can a eugenic executioner of sickness, ugliness and disarray have created health and beauty? Could destruction following malfunction have generated *in the first place* life's full range of body plans? You say so. Natural intelligence is not allowed and so you raise selection, while denying weakness any second chance, onto a pedestal beside the empty throne of Lady Luck.

I disagree. I would promote 'natural intelligence' as both judge and creator. This is the way that selection and decision work in mind - conceiving, planning, implementing preconception. 'Unnatural', 'artificial' or 'intelligent' selection in this manner is the way that breeders, from whom Darwin garnered inspiration, work. 'Natural selection', on the other, passive hand, no more than follows change. It is just another way of saying 'no intelligence'. Remember Chapter 20? How can such complete oblivion even mastermind the building of a single cell? *How can senselessness, without the least sense of anticipation or a goal, lock onto random changes that are 'positive' with respect to exactly what it has no idea of, future functional purpose? How can it 'design' and thereby construct such specified and purposive complexity as is evidenced ubiquitously in animate nature?* **Natural selection explains survival not arrival of the fittest.** It weeds the weaker but cannot create the fitter; it is as creative as a kitchen sieve. Both genetics and the fossil record witness that it creates nothing from nothing, that is, no pathway or organ from its start. Without concept or creativity it can only work on pre-existing, previously informed structures and their functions. It is, therefore, demoted to another name for non-survival, that is, death. Could death create all forms of life? Could non-intelligence inform the intricacies of every fine means to an end or is the idea (and with it macro-evolution) simply nonsense? In other words, is natural selection not in fact the form of an inverted, academic 'God delusion'?

A few more twists or, rather, untwists in the tale will help to set this particular record straight. The basic system of classification that biologists use to pigeonhole each organism in the panoply of life was established by the Swedish naturalist Carolus Linnaeus (1707-78). He grouped animals according to their parts (structure) and gave a distinctive name (binomial) to each species of organism. For example, he called the house-cat *Felis domesticus* and the lion *Felis leo*. Such classification begged, however, a key question. Just how are *F. Leo* and *F. domesticus* supposed to be related to each other and, by extension, other cats and non-cats? Linnaeus did not even bother to ask. This is because he was a *special creationist*, a species now extinct. Quite apart from the fact that his Bible authorised *kinds* but not species, the authority of Linnaean orthodoxy stated that what he called 'species' were immutable. Of course, Linnaeus may have had a broader definition than our own in mind; or, some say, he recanted later on in life. *Nevertheless his apparent conflation of 'species' with 'special creation' is today regarded by everyone as a mistake.* It established a 'straw-man' that became the target of Darwinian objection, a spark that lit the Origin of Species and, born of Linnaean error, the rise of neo-Darwinian counter-faith. Surely all the huff and puff has not been just because Linnaeus misconstrued or was misunderstood?

Edward Blyth (1810-73) was another creationist also involved in a twist in the tale. A keen naturalist and 'Father of Indian Ornithology', Blyth published essays that first appeared in The Magazine of Natural History in 1835, 1836 and 1837. *These, which were read by Charles Darwin who afterwards corresponded with him, introduced the ideas of a struggle for existence, variation, natural selection and sexual selection.* These four are central Darwinian tenets and we might ask why, since in large part Darwin's work was based on Blyth's ideas, the former hardly acknowledged the latter's 'intellectual copyright'. Blyth made no more of his notion of natural selection than the facts warrant. He, like Darwin afterwards, drew attention to its passive, conservative function, using it not explain how new types arise from pre-existing ones but rather why a healthy population is conserved. In short, the process conferred not change but stability by maintaining wild-type pedigree or archetype. For him such natural pedigree was a creation kept distinct, fit and 'in form' against the hone of natural cull. For Blyth, unlike Darwin, natural selection never made a thing. It was no *ersatz* creator. Instead, it acted as an inhibitor slowing any deterioration from original perfection. By acting on a less than fully healthy individual it merely kept a type's functional mechanisms tuned to peak performance. It put a brake on what we call today the entropy of information and, as such, it acted as a selector, corrector or, if you like, a robotic judge of error. Natural selection is not, in fact, 'an incredibly powerful mechanism, almost magical in effect' but a circular, tautological way of saying a population of organisms remains healthy. Nor is it a grand 'law of nature' but a trivial observation that an organism born with serious defect dies and/or does not reproduce. 'Weaker die, stronger live'. A truism states the obvious. 'Those that survive survive'.

'Those that survive survive'. Darwin agreed but turned Blyth's implication on its head. For him natural selection was the way original imperfection was led to 'perfection of adaptation'. It was conceptually transmuted from remedial cull to 'an incredibly powerful mechanism'. This is Darwin's magical, nay, conjuring effect. Those that survive might yet evolve.

For Blyth a perfect original script would, under natural conditions tend to decay, lose information and become nonsense. For Darwin original nonsense would, selected for by death, be turned into perfectly informed life. Which looks the more realistic a proposal?

Blyth proposed original creation that, like you and me in our life cycle, deteriorated. Darwin did not propose *any* origin of systematic information, just 'improvement' by chance on systems already present. For example you and I are, for no reason except 'a wish' of atoms to increase their 'chances of survival', a biological improvement derived directly from the very successful survivorship and reproductive capacity of a bacterium. **The Darwinian trick has always been to make non-conscious chance seem able to achieve what is normally recognised as the expression of conscious intelligence. Its handkerchief, behind which any jiggery can poke, is a fog of immense time in the course of which, the audience is invited to believe, nothing is impossible**. With quantum physical or Darwinian biological improbability what is impossible? Isn't an un-testable, un-provable yet 'scientific' faith in a Theory of No Intelligence better than so-called 'unscientific' faith in one of Intelligence? Such evolutionary prejudice might seem to eliminate the hypothesis of Blyth and the '*Natural Theology*' of another creationist authority with whom Darwin was well acquainted but barely if ever acknowledged, William Paley. It is possible that he failed to mention these or other of his teachers (such as geologist Adam Sedgwick, an avenue dedicated to whose name ironically lies opposite Darwin College in Cambridge) because he wished to eliminate all taint of a creationist framework from his new, topsy-turvy projection of their ideas.

Did Darwin specialise in twisting creationists' tales? In 1798 another creationist, the Cambridge graduate and reverend Thomas Malthus, argued against current 'systems of equality' as proposed by French republicans and, later, communistic socialism. His '*Essay on the Principle of Population*' demonstrated that competition for limited resources was a law of nature. Its anti-revolutionary politic may have stuck in the craw of Darwin's grandfather, Erasmus, but it struck a deep chord with the contemporary, Victorian spirit of colonisation as well as the country's own struggle against poverty and disease. Conservative Malthus observed correctly that in each generation more individuals are produced than survive. Whether they could possibly have survived is incalculable and irrelevant because they won't and don't. Predation is part of the ecological process. Malthus' implication was neither that birth sometimes involves miscarriage nor, in the whole ecological framework, eggs, seeds, fruit and youngsters comprise a large food-supply: it was that all living things are subject to an intense struggle for survival and therefore under severe and constant selection pressure. *Such pressure would favour healthy individuals and act as a stabilising force that prevented change in animals and plants.*

Darwin agreed with the first part of Malthus' argument. He correctly cited four factors that check a species' presumed tendency to increase without limit.

Like warring competition as opposed to ecological collaboration, they were negative - predation, starvation, climactic severity and disease are aimlessly upsetting or pointlessly destructive. In addition we now know that population growth is as much regulated by internal causes as external catastrophe. Initiatives, either behavioural or in the form of some intrinsic physiological feedback mechanism, are taken by the organisms themselves. Does not, for example, stress affect desire for sex or inhibit ovulation? Animals and plants may not, in adverse circumstances, produce as many eggs or seeds as possible. Brood sizes vary. Species no more strive for limitless increase in numbers than for size, they simply increase until some barrier obtains. In other words, adaptation of a positive kind is not necessarily a matter of chance. It is to some extent pre-programmed.

However by arguing that competition was the driving force behind the survival of superior organisms and, thereby, improvement of a race Darwin contradicted Malthus' conclusion. He thus turned him, like Blyth, on his creationist head. Later, in Principles of Biology (1864), evolutionist Herbert Spencer coined the famous tautology that only the fittest would survive. He describes plants and animals locked in a struggle for existence in which the weakest and least perfectly organised must always succumb. Survival of the fittest! Yet if fitness is measured in terms of survival sufficient to reproduce the greatest number of offspring and this is the criterion for success, why are all women not by now agelessly beautiful nymphomaniacs and men handsome satyrs? Why haven't rabbits or herrings inherited the earth? Why did our 'primitive ancestor', the bacterium, arguably the 'fittest' of all organisms, ever evolve from its extremely successful, undemanding rut?

chance	*intelligence*
antagonism	*cooperation*
predation	*symbiosis*
struggle/ competition	*niche/ harmony*
war	*peace*
death	*life*

Malthus' image of man at war with nature and nature at war with itself is a negative one. No doubt, as violent crime and warfare show, mankind can behave in a way red in tooth and claw, but does nature? Of course, existence shows both sides of the coin. *Modern ecology would, however, profoundly disagree with the Victorian bias as rendering the whole truth and nothing but the truth.* Nevertheless an epigrammatic 'struggle of man against nature' splashed the banner heroically lofted by both communistic and capitalistic politics. The frame of mind it engendered wrought and still wreaks massive human destruction of the environment. Take, for example, a Darwinian/ Marxist maxim along the lines of Spencer's fitness test. In complete contrast to the traditional Taoist/ Buddhist concept of man living respectfully in harmony with nature Mao Zedong unwisely exclaimed that modern, hard-headed red science 'must conquer nature'. The result has been widespread degradation and destruction of the environment with China's wildlife in many cases crushed to within a glimmer of life or death. Not only communist but capitalist officials are just waking up to the extensive damage wrought by 'progressive', scientific and yet aggressive, disrespectful human husbandry.

Is greedy over-exploitation really progress? Is sacred nature really 'red in tooth and claw'? Observation does not square with it. Although ruthless competition between species can be artificially induced in a laboratory, in the wild it is difficult to pinpoint any example of sustained mutual damage inflicted between or within species. There is competition for space and resources which generates ritualised signals and aggression with, occasionally, fatal results. The same applies to competition for females. Pressure but never, except by humans, genocide. Food specialisation is another simple method to avoid competition. Other species divide their habitat according to time - when the dayshift retires the nightshift takes over. Migration, sequential flowering and specific habits or equipment are other strategies adopted for a peaceful coexistence. Struggle is not even, as Darwin asserted, most severe between individuals of the same species because such mechanisms as dispersal, territorial principle, innate and learned behavioural differences, pecking orders and so on kick in to reduce tension. Bankers and flower-sellers both inhabit London but their 'trade' differs. Habitat and 'way of life' together constitute *niche. The elimination of competition by division of a habitat into niche is universal and has become an ecological principle of prediction and discovery. Peaceful coexistence, harmony not struggle, is the overriding rule.* The cooperative principle is stronger, even in harsh environments, than the competitive. **A key ecological as well as physiological principle is, as with Natural Dialectic, homeostatic balance**.

imbalance	*balance*
disease	*ease*
error	*rectitude*
accident	*control*
chaos	*homeostasis*

There is not even competition, unremitting struggle or war between predator and prey. There is alertness but never is the predator angry with its prey. It does not hate it. Wanton killing is abnormal, overkill (except by humans) not practised. For prey there is provision for escape, lack of imagination and reflex response to survive; for the greater part, suffering is swift and short. The inevitable moment of struggle is minimised as is, it seems, the horror or the pain. Natural selection's conservative nature sees the weak, sick or young more likely 'go to the wall' but accidents catch fit and unfit just the same. In fact, where non-Darwinian ecological principles hold visible sway, it is in the area of viral, bacterial and parasitic infection that Darwinian kill-or-be-killed principles are best expressed. Such infections cause prolonged and painful suffering from which, since they are invisible, there is no obvious reflex counteraction. A pathogen is simpler and fitter than its host. It should always win. It should always have defeated nascent immune systems before they could advance sufficiently to work.

The critical origin of immune defences is another Darwinian black box which, although interesting, there is no time here to explore.

death	*life*
isolation	*relationship*
strife	*collaboration*
sickness	*health*

We return to the ecological tack. Not in a single cubic metre of biosphere is a community of organisms restricted to a single population, a single total winner. The reverse. Diversity breathes easy. *And all organisms are dependent in many different ways on others for completion of their life cycles.* Indeed, as we shall see, there seems to exist a deliberate pyramid or hierarchy of types, each 'serving' those above. This pyramid (see also Chapter 26) seems to escalate from menial towards complex tasks that serve, at the top, intelligence. Intelligent life would, by this criterion, play lead roles that a cast of minor players and one-liners supported. Is 'theatrical' collaboration not the oldest, strongest strategy in nature's repertoire?

It is hard to over-emphasise the anti-Darwinian principles of peaceful coexistence and cooperation. *More than war symbiosis is the name of the game.* Such relationships abound, many affecting entire ecosystems. Organisms do not so much fight as collude or work round the environmental problems they encounter. Less pain, more gain is the rule rather than the reverse. *Such ecological and physiological coordination militates against the notion of 'creation by natural selection as a result of competition'.*

Nor is it at all clear that descendents are (or ever were) better adapted than their predecessors. Many a first fossil of its kind has been as well formed or looked the same as its later, even modern, counterparts. And many a maximally efficient design feature discovered by technologists has been either copied from or later found to exist in a biological adaptation. Indeed, faith in such engineering artistry is a principle adopted whenever a biologist inspects the structure and function of any part. In every case economy of design and efficiency of soft, biological machinery minimises effort, maximises effect. Surely the highly intelligent industry exploiting 'bio-mechanical design' has not been trumped by chance and death alone? **It is only because some biologists, believing it is exclusively scientific to do so, resolutely cling to the one-dimensional philosophy of materialism that we run into contradictory explanations and a wealth of 'just-so' stories concerning how such self-consistent code and its equally self-consistent, homeostatic, purposive chemical issues accidentally happened.**

The key phrase is 'self-consistent code'. Natural selection is, one is assured, a deterministic process that acts as sieve, as hone, as scissors that will snip the more untidy threads from life's great tapestry. It does Blyth's bidding but Darwin wanted more. How might one transform it into an instrument of transformation, the all-pervasive power of evolution and the central mover in a Cult of Chance?

When is a code not a code? When it's a random non-code, namely 'noise'. Therefore let there be noise. Generate swathes of meaningless rubbish from which quantity some scraps might just seem to make, somehow, some 'sense'. They might, although the wind and rain would not agree, resemble fleeting words of someone's language. How might you catch them as they came and went? If only you could isolate this sense from senseless storm. How might you, from the maelstrom, fix those fragments that would for no reason juggle up a real sense of comprehension called The Book of Life? You would have your cult's Great Fixer, a Designer Substitute. Its unholy word shall be a word of death not life; it shall be the word of life from non-life not from life. And it shall take the name of Natural Selection. Let local circumstance determine automatically what stays or goes. Let environmental justice execute its sentence with respect to changes. What is 'good' survives, what's 'bad' is handed down a penalty of death. *QED.*

Do you get the drift? Translate the picture into scientific terms. Darwin did not understand the fundamental source of randomness that generates the 'noise' whose worthiness selection judges but biologists have now identified this source of 'error' as genetic mutation. Natural selection never worked, for sure, on any pre-biotic chemical. It cannot work until the machinery of reproduction is in place. Therefore (as shown in Chapter 20) what great nonsense, unselected for, established this most complex, purposive of chemical contrivances, a cell? Did protein precede its *DNA*? Proteins do not arise without their coding and their cells. How did *DNA* arise in a primordial waterhole? Or did *DNA* and proteins make their entry arm in arm? None of these options is, biologists all know, feasible.

Simply *DNA* is not enough by far. Over and above the chemicals you have to generate the order for a reproductive system in a cell. *Before natural selection even starts to work you have to have the first instruction manual and with it a first specified 'machine'*. This is a logical impossibility that chance, you claim, has trumped. Therefore allow impossibility - a reproducing cell. If you start with randomness then was not *DNA* a 'noisy code' before it graduated into silent sense? How could a 'noisy code' spring, accidentally, into the meaningful coherence we call life? How could the first severely bugful bugs 'survive'? They could produce no single useful protein let alone coordinated cohorts of the stuff. For what reason should they float around and turn into a cell? Is not their ilk exactly what heat entropy if not selection operates against? Nonetheless, as Father Christmas to an evolutionary wish, grant them the presence of survival.

You have now reached the point to theoretically imagine that innumerable 'assays' on *DNA* by trial and error find you different ways to co-exist with your surroundings. Determination therefore builds on randomness. What works is scrambled into something 'better' - although the way blind chemicals assess best interest is not of interest to them. How interested is a numb, dumb stone in things? Are dissolved crystals any more perceptive than a stone? The substance of this mindless 'system', dubbed Darwinian evolution, is repeated accidents and 'lucky breaks'. They make a life form that, while highly systematic, lacks all reason. At this rate what is life? If mind and information are not elements in their own right then you have problems. How can consciousness arise from bubbling biochemistry? Have you a Theory of Intelligence or not?

Presuming mindlessness eradicates all chance of natural intelligence. It does not, however, begin to explain the origin of complex information, self-consistent code, structural archetypes (like seaweed, fungus or a vertebrate) and purposive design. Is it a 'fact' that only the best accidents are selected to design a car but most will break it down? No? But isn't evolution, permeating through a haze of years, promoting just this kind of 'fact'?

For Natural Dialectic order runs the other way. Active information, mind, makes order; passive information is its product, its material imprint. This includes all kinds of code, instruction manual or machine. Such passive information is, like everything material, subject to the force of wear and tear called entropy. It is subject to the forces of disorder. On this basis you would think genetic code, like a pristine copy of a book, was first minted perfect and then fell into decline. How many generations, you would ask, before the text becomes so garbled that a species, like the human species, falls into genetic meltdown and, by dint of entropy of information, extinction?

No doubt that natural selection, mutation and variation (sometimes speciously labelled 'micro-evolution') happen constantly. These are due to sexual recombination or to accidents. But they are trivialities compared with major issues like 'abiogenesis' or the maxi-morphs required to sketch new body-plans and their associated systems. These need the syringe of concentrated mind; they need injections of an immaterial juice - information from intelligence. An act of creation falls, *top-down*, from idea through to its materialisation. Randomness is its antithesis but anti-chance its thesis. Are accidents - the effects of faulty design, poor workmanship or wear and tear - what you seek out when things have got to work? Intelligence determines to eliminate such 'information entropy' and this determination springs, unlike natural selection, from *prior informative selection*. The basis of determination here is anti-randomness. This, we all know, is the way successful systems work. And what a system shows as bodies biological!

Water runs round obstacles or 'dies' into stagnation. Its creeping paths are, within constraint, entirely unconstrained. Hydraulic systems, on the other hand, channel water purposely and, unlike Darwinian pervasion round mutation's obstacles, obtain their end. The basis of biology is, equally, a channelled conduit called code whose information passes life precisely on its course. **The origin of code is always mind.** A guiding principle involves the way things swing along in line; homeostasis orchestrates the music of our chemistry. Unlike the random creep of damp its preordained co-systems fluctuate around their target norms. Life is vibrant, life is tolerant to obstacles and 'noise' but how much noise can information bear until its message fails to reach a friendly, correspondent 'ear'? **The origin of information, not constraint of babble, is the real issue that confronts a person asking after his full lineage.** Is the origin of high-quality information accidental? Is its 'corrector' only various degrees of 'opposition' from environment? If chance can't format properly then why should anti-chance not set the scene?

Anti-chance means mind and mind is, in materialistic context, not politically correct. Oblivion is unmindful, matter is non-conscious and so are you surprised that death is cited as the source of life's abundant cornucopia? The device for the delivery of the living world is, as opposed to the information-rich coherence of biomechanical design argued in the last three chapters from a base established in the previous fourteen chapters, promulgated as bug and death! This is in spite of the opposite working hypothesis that every biologist uses in seeking a *reason*, that is, a purposive design for every bio-molecule, cell, organ or system that (s)he studies. So that, rather than healthy biology, a case of mutant bio-illogy is diagnosed. Accidents are not creators but candidates for casualty.

You might, on the other hand, start with an archetype already programmed with adaptive potency. Such conceptual adaptation would occur at archetypal level and, as cells are determined in their operation, would be determined in original character of type. 'Type' would assume a physical immutability but (see *figs.* 5.1 and 22.1) there would be, within its bounds, a scope for endless flexibility; within constraint of standard form you would expect to find an orderly expressed plasticity. An organism's body could invoke prefabricated variation that, as well as sex-based individuality, would buffer different environmental pressures. Such buffering would involve the use of codes and mechanisms that, appropriately triggered, could evolve in practice principle inlaid on purpose. Adaptation that anticipates life's problems confers informed advantage so that chance cannot compete. Information generates potential for certain a path of action; this case of biological provision thereby prints its name below.

There you have it. How do you divide up life? What's a type? What's a species? Darwin started out with species. Let's do the same. Let's inspect the origin and state of species. Then let's see, as opposed to 'fluent' species, how fixed but dynamic type or archetype can be explained.

The Origin of Species

What is the smallest category into which similar individuals can be grouped? Biology needs the concept of species, or something like it, to identify what it is talking about. Carl Linnaeus, the founder of modern systematics, thought a 'species' was a group whose members could interbreed. This approximates to the modern taxonomic definition but diverges from the evolutionary because, whether or not he changed his mind later in life, Linnaeus is renowned for a belief that species are each created fixed and immutable. If he later raised the stakes to 'immutable genera' would this align more nearly with what Natural Dialectic calls a 'type'? It would draw the sting from any argument (such as Charles Darwin's) about a non-divine origin of species - a fact which anyway no-one disputes today (see *fig.* 20.1) and perhaps never did. *Nobody has a problem with an origin of species - although taxonomists still argue over what the label actually means.* After tallying homologies (as fundamental similarities of structure are called) or descending the archetypal levels of biological classification from, say, animal kingdom down to the individual detail of a man such as Darwin himself, where do you draw the line? Between ape and man? Caucasian, African, Patagonian or Chinese? What is the *lowest* level of classification to which an individual organism can be divided? Below which you cannot split the group and above which higher groupings are conceptual?

A presently preferred but occasionally blurred definition of species is a population of organisms reproductively isolated (by various factors) from a similar group. It is not without drawback. What, for example, about members of two different species (say, grizzly and arctic bears) which interbreed and produce fertile offspring? In the Galapagos Islands marine and land iguanas, of different genera, have been observed to hybridise. And why is a wolf (*Lupus lupus*) not placed in the same genus as *Canis familiaris*, the family dog, with which it may profitably mate? Look at the array of shapes, colours and sizes contained within the latter species or, say, that of the horse. How, furthermore, do you interpret a fossil species, especially one where the male and female have markedly different body forms? Darwin, as opposed to Linnaeus, preferred to 'plasticise' his definition of species to what he saw as its cause - gradual change. He wrote "*I look at the term species as one arbitrarily given, for the sake of convenience, to a set of individuals closely resembling each other, and that it does not essentially differ from the term variety*". A species, in blurred evolutionary frame, is therefore a 'variety'.

(i) Unlimited Plasticity

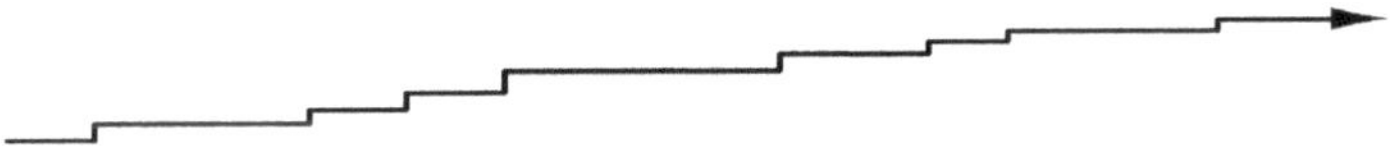

Darwinist view of potentially indefinite 'micro-evolutionary progress' by small, gradual steps (as 5.1i).

(ii) Limited Plasticity

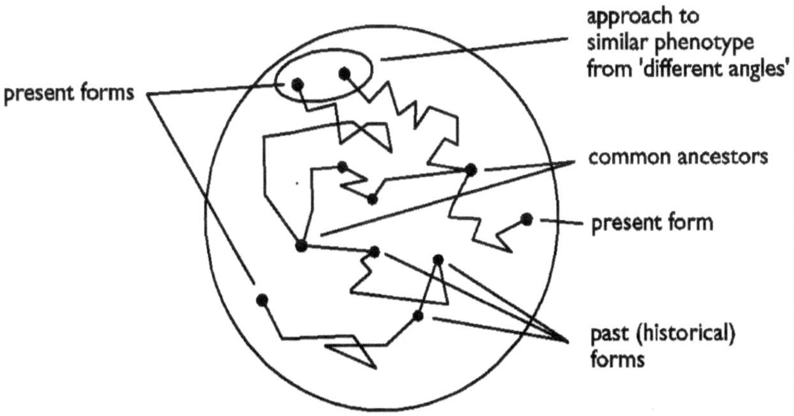

Do you remember, from Chapter 19, the (*sat*) central instrument of biological balance, homeostasis? The *top-down* view sees variation as oscillation round a norm, deviation from the mean or limited plasticity. **It proposes 'type' (see The Origin of Type) as the homeostatic norm around which limited differentiation may occur.**

Consequently this diagram enlarges on the notion of random mutation and genetic drift within a typical boundary (*fig.* 5.1 (ii)). It shows schematically how a number of species and sub-species can (as, for example, in the case of wolf, dingo, fox and all kinds of *Canis familiaris* or 'family dog') exist simultaneously within the same constraint; how constrained common ancestry occurs; how a given species may be 'approached from another angle', that is, derive from a different lineage; and how it may, as such, be approximated at other times and places.

In short, what is commonly called 'evolution' is dialectically described as 'principle-expressed-in-practice' or 'variation-on-theme' (see also Chapter 23: Evolution in Action?).

fig. 22.1 (see also fig. 5.1)

Darwin regarded varieties as incipient new species. Such limited plasticity is no big deal. For example, gulls breed in a continuum round the world. Herring gulls come to look more like the lesser black-backed variety until where the ends of the ring (or cline) meet. Here in Europe what you now call two species do not interbreed. Call it microevolution if you want. No way does this amount to macroevolution. There is no evidence, as we shall see, that speciation leads to macro-evolution, that is, evolution proper.

However, he not only regarded varieties as incipient new species but also proceeded to extrapolate on the hypothetical principle of unlimited plasticity. Because natural selection works on these varieties it would seem that the origin of species is what, in its flexible gradualism, in the end denies them any durability. Species are simply moments in flux. Snapshots of change. The classic model of speciation (see below) derives from his observation of thirteen or so finches on the remote Galapagos Islands. It shows that, by various mechanisms, one group of descendants from a single population may diversify and perhaps cease to interbreed. As with gull speciation, a kind of variation often termed *micro-evolution*, occurs. The question always is how far such variance extends. Some insist on a narrow, 'intra-specific' definition of the word. No-one, however, denies (*fig.* 22.1) various amounts and rates of change, that is, of flexibility to a type and therefore the unpredictable coming-and-going of new species; therefore others include 'intra- *and* inter-specific' in their broader definition. Is this, however, simply elasticity within an archetypal theme or can it be *extrapolated* past archetypal boundaries to the points where, say, a mouse can snap out of 'mouseness' into something else? On the principle of unlimited plasticity does speciation accumulate so that this stretch and snap of boundaries keeps happening until the mouse has become a whale? Is the integrity of archetypal 'rings' paramount or can 'broad' *micro-evolution*, against all experimental evidence, proceed unchecked into *macro-evolution*? Can it with unlimited plasticity 'break' non-existent archetypal boundaries and transform a bacterium into a human? This is the dogma. It is why Darwin's disciples recoil at the anathema of 'archetype'.

Is not the Victorian Sage's big mistake 'small variation written large'? *Because, conversely, on a genetically demonstrable principle of limited plasticity variation might be thought of as different permutations of a single genetic potential expressed at different times and places.* Genetic potential (Chapter 23: Super-codes and Adaptation) adds up to the amount of variation that a kind or type of organism can produce from genetic material already present and acting as a 'variation bank'. **In this way the definition of speciation might be seen as differentiation circling round a norm, wild-type or archetype.** Take, for example, the dog family. Many kinds of dog are found but the range of variation is finite: stretch the elastic too far and it snaps. *In other words, the intact survival of type is ring-fenced by impenetrable death. Ring-breakers are extremes selected out.* It might be argued that, even if no natural breeding occurred between two 'variants', artificial pollination or insemination might produce viable offspring - hybrids that could themselves reproduce. It happens. Our major cereal crops are viable hybrids. After a certain limit, however, variations and crossbreeds frustrate further attempts at 'improvement' by becoming sterile or reverting to the wild type. *In practice, when genetics* *decisively demonstrates the limits past which mutation does not survive it also demonstrates conclusively, through myriad experiments (e.g. with E. coli bacteria, fruit flies and mice), that the macro-evolutionary principle of unlimited plasticity (or unbridled extrapolation) is incorrect.* Ask a dog-breeder if he has bred other than a dog. Ask him why.

Darwin saw the problem - even if you give it bags of time. "The geological record is extremely imperfect and this fact will to a large extent explain why we do not find interminable varieties, connecting together all the extinct and existing forms of life by the finest graduated steps. He who rejects these views on the nature of the geological record will rightly reject my whole theory." In fact the fossil record (to which we shall return) is not nor ever has been in accord with gradualism. Where Darwin staked his claim on continuity, *discontinuity* (the primary prediction of an archetypal model) is what we actually find. Nobody is more aware than a palaeontologist that Darwin's 'interminable varieties' are missing from life's petrified family album.

Galapagos and All That

Meanwhile, importantly, Darwin's Malthusian causes for evolution also create nothing. They are neither informative nor constructive but anarchic and destructive. *If you add to their mayhem the notion of extinction, which is sometimes linked with evolution but is in fact its opposite, then you have a doctrine of decease. If you then add mutation and natural selection you have one of sickness, deformity and death. What a way to get a life!*

On the other hand, nature does not work so much by blind, brute struggle as by information and purposive cooperation. Its principles are, more than competition, economy and efficiency of design. It works, in the way of artist or engineer, 'smarter not harder'. **It seems, in respect of basic principles as opposed to trivial side effects, that Darwin really got it wrong**.

The Victorian era coincided with the end of a phase known to British historians as 'the agricultural revolution'. Darwin saw breeders impose *artificial selection* on sugar beet, sheep, pigeons etc. *This is an intelligent, planned process with a goal in mind. Its mechanism for selection is obviously the mind of the selector.* Even so, its remit never runs beyond accelerating variation.

No doubt unintelligent natural selection is an elegant and simple concept. It is, however, the complete antithesis of purposeful, teleological plant and animal breeding. Although it seems to complement artificial selection in the latter's evident, purposeful ability to generate special (or intra-specific) variation and even, dependent on your definition of 'species', hybrids and different species within the embracing ring of a type, natural selection has no goal in mind. Thus, an entirely different process, it simply cuts out variations that don't work. *No one has any argument with its succinct description of the way accident and disease - whether genetic, physiological or structural - terminate weaker or unlucky members of a species.* Is unguided variation sufficiently powerful, however, to allow the process of maxi-morphism, that is, macro-evolutionary transformation involving conceptual novelties and their associated organs (such as wings for flight) to break an 'illusion barrier' and bang into reality? 'Yes', cry materialists. 'No', cry holists, 'this is materialistic science blinding us'. Which side is baying to the moon?

If animals acquire some slight new differences (e.g. size, shape or coloration) and then mix with other, unaffected members of their species such interbreeding soon eliminates the difference. Darwin admitted that artificial selection may, if relaxed, lead quickly to the resumption of a typological norm or, in geneticists' jargon, 'wild type'. Natural selection is unconstrained; novelties are not weeded out according to a breeder's list. You might expect that 'relaxation' ruled, norms were rapidly resumed and variation but not evolution was the order of a natural day.

Whistle up the family dog again. Do you need proof selection doesn't work? Try Cruft's dog show. Many different-looking hounds such as boxer, whippet, sausage dog and poodle are 'cat-walked' but these inbred or restrictively, artificially selected specialities actually show loss of 'bred out' alleles, reduced frequency of others and by concentration round the lower number left, an overall diminution of canine information. 'Wild type' is diluted to 'recessive' oddities. Indeed, eugenic manipulation by the 'intelligent selection' of professional breeders has led to major suffering due to morphological deformities and other defects in many of the dogs concerned. Thus selective breeding, like founder effects and genetic bottlenecks, *reduces* rather than increases genomic diversity. *Reduced potential is, however, the exact opposite of what is required by evolution.* Rather than a dialectical reduction in practice from archetypal possibility evolution needs a hypothetical increase in information by inheritance from a common ancestor. Many of Cruft's winners aren't wild-type survivors! In the wild dogs don't evolve; and the competition painfully protests for anti-evolution.

Dogs will be dogs; and as with Cruft's so with the Chelsea Flower Show. You can 'evolve' a great variety of roses, dahlias or any other kind of plant - but such variation does not logically or typologically stretch into transformation of a dahlia to rose and so forth. Breeders know intense selection or inbreeding generates not strength but weakness in the novel stock; and that there's a limit of plasticity past which you cannot press. **For all that a rose remains a rose**.

Mutations, some malignant and others neutral or (perhaps) benign, leave things as they were or add to loss of information. Mutants are generally seen as blemishes upon ideal or 'wild type' potential. For some the effects of a genetic bug can be contained by the natural 'elasticity' of the *vis medicatrix* but in others cases it cannot. In this case the tension of 'speciality' becomes too much. Excessive inbreeding or artificial eugenics often leads, as every attendant vet knows, to deformity, incapacity, high sickness rates and, snap, death as the result. But death is not the sponsor of fresh biological ideas or their expression in the form of body-plans, homeostatic systems, organic mechanisms and associated biochemistry.

No doubt, cross breeding or even hybridisation within the limit of a type may sometimes *increase* (that is to say, *reinstate*) its general vigour. But if Cruft's inbred specialities were all left to their doggy devices the number of 'strange' variations would soon reduce. The winners soon start to re-assume a wolf-like character. From a *top-down* perspective this indicates that the whole, typological potential, in the form of the canine gene pool, has been split up and each boxer, poodle or sausage dog pedigree represents a *loss* of information, a restriction of potential in the form of *less* efficient overall performance than the wild type. That is, a split, detail or specification skewed from the norm occurs with artificial selection or inbreeding in a small, localised and separate population. *This is not evolution but its reverse. Therefore, from a bottom-up perspective, it is imperative that natural selection does not work like artificial selection.* After all, it has to generate not only radiations of variation-on-theme but also create, as parallel artificial selection never does, new themes themselves.

Although Darwin reported working models of natural selection, he proposed no mechanism by which its raw material (mutation of form) might be generated. His famous voyage to the Galapagos Islands revealed a variety of reptiles, insects and birds. Different forms inhabited different islands and, in the presence of this variation, he was struck to wonder whether the 'species' really was, as Linnaeus had at first proposed, immutable. As a result he extrapolated a concept of evolution by natural selection, a draught simple, materialistic and, to judge by its effect on some people's minds, potent. No 'bio-force' (except perhaps that of evolution itself) was needed. As an example of this he then described, citing Galapagos finches, barnacles etc., limited cases of variation that we now call micro-evolution. 'A Species is Born'. Instances of micro-evolution/ variation are regularly headlined. The *Anolis* lizards of the Caribbean, Hawaiian fruit flies, aforementioned cichlid fishes, snails, periwinkles, moths, apples, gulls and many other instances have excited academic interest in 'the roots of biodiversity'. Morphs, hybrids, clines, polyploids, races, sub-species and speciation itself is unexceptionally invoked. *Nobody disputes the facts of recombination, change and variation. They are the norm.*

Darwin himself was aware that artificial selection seemed to have limits. You could cause variation of characteristics in a species to a certain degree at which the limit seemed reached. Nevertheless he further and finally rolled out a general theory of evolution which, breaking those genetic and physiological limits, made the claim that his 'special' theory of micro-evolution could be generalised, that is, extrapolated boundlessly. *This, unlimited variation with direction towards increased complexity and awareness, is his 'general' theory of macro-evolution.* It is not micro-evolution, which *ID* claims is simply variation within archetypal boundaries, but macro-evolution that is alleged, by chance, to have conjured phylogeny from the first living cell to mankind and all other forms of life we see sprung around us today. In this view the family relationship between all organisms is one of 'blood' rather than conceptual design. In other words, your Grandparent was not an Adam or an Eve but an ape, a shrew, a reptile, fish and so on back to the first accidental, abiogenetic cell - the freak that is and was, according to Chapter 20, logically impossible.

By casual extrapolation many micro-evolutions are supposed to make a macro-evolution. *It is the primary article of Darwinian faith that, as noted in Chapter 21, macro-evolution is simply micro-evolution writ large*. **Viewed in the informative terms of *top-down* variation-on-theme this Darwinian creed is simply a great mistake**.

The argument for evolution therefore starts by proving a small step which, multiplied, becomes the large. *Although Darwin's attention was first caught by variations (species) of Galapagos mockingbirds his iconic evidence was drawn from the radiation of about thirteen species of finches descended from, possibly, a single mainland ancestry.* All the birds have similar dull grey-brown plumage, short tails, courtship behaviour, nest type and number and colour of eggs. They diverge in size, shape of beak and feeding habits. There are two main subgroups called ground finches (*Geospiza*) and tree finches (*Camarhynchus*). In fact it is difficult to dogmatise on their status as different species, subspecies or even races as opinion differs according to the definition of species employed. The commonest definition is one that says 'if two forms interbreed and produce fertile offspring they are the same species'.

The studies of ornithologist David Lack noted such similarity between birds of some species that a hybrid might be indistinguishable. However, he did not observe interbreeding such as was documented by Peter and Rosemary Grant, J. Weiner *et alii* between six of the thirteen species that were studied in the 1980's. Perhaps such hybridisation will be found to occur between them all. *If organisms interbreed in this way they are, by definition, still the same species. Neither speciation nor micro-evolution has occurred.* Genetic mutations, in some cases one in a hundred or so nucleotides per gene, have not changed the latter's functionality. Neither has new material been introduced nor explanation offered for an *origin* as opposed to adaptation of, say, beak or coloration. Indeed, droughts on the island of Daphne Major in 1977, 2003 and 2004 generated competition between large and small-billed varieties of finch. In the case of seed shortage what would be the pecking order? Would big beaks crack harder, tougher seeds or small ones better snitch the undersized remainders? Birds varied and the pundits cried 'strong evolutionary change'; but when 'a single drought can change a population' (P. Grant; Sci. American: Oct. 1991 ps. 82-7) is this really 'evolution in real time'? More likely, you might judge, simple gene dynamics wherein variation unequivocally demonstrated pre-coded adaptive response, that is, orderly epigeny (see Chapter 23) and not random mutation. It would thus be preordained 'variation-on-theme' rather than 'evolution-in-action' that the finch adaptations illustrated.

The question remains. *Is the variation in beaks and feeding habits the result of natural selection acting on random mutations or the expression of pre-coded adaptive potential for the beak subroutine?* In fact Great Tits can change their beak shapes twice a year; several Cameroon finches have optional beak shapes; and the Grants have shown that Galapagos finch beaks oscillate in size and shape according to rainfall patterns and hence available types of seed. If food source varies beak shape alters rapidly and accordingly. When conditions (say, drought) are over it can just as easily and in a single generation revert. Immediate, relevant modification; cyclic variation; adaptive but limited plasticity. *Just as cars are built with shock absorbers to buffer bumps so, it would seem, the adaptive potential of super-code can, in a useful but strictly limited way, buffer organisms from the shock of changing external conditions.* Laysan Isle finches, artificially introduced to isolated islands, adapted their beak shapes over just a few years; such changes likewise suggest not random mutation but the expression of in-built genetic potential, preordained possibility that is triggered by habitat and reinforced by ecological isolation (the natural equivalent of artificial selection). In spite of non-evolution at work 'these birds have become such a universal symbol of Darwin's process ...that their beaks now represent evolution in the way Newton's apple represents gravity'! Yet in this case the basic premise is unsafe and, *top-down*, the massive extrapolations prised from a partially correct observation are false. What a heavy downfall that should bring!

There seems, however, no doubt that variation and probably, despite the case of Darwin's finches, speciation does occur. The specialised beaks of Hawaiian honeycreepers seem to have diversified in a similar way from perhaps a single, original pair. And an especially lively example involves fruit flies separated into different populations by the islands' lava flows. There is, though, major doubt whether such variation (or micro-evolution) can be extrapolated into a grand, general scheme of things. For a typologist, who sees speciation as a *loss* of information in the artificially or naturally selected type, such elevation is wholly unwarranted. A study of genetics strongly reinforces this view. Large amounts of variation (as in the family dog or horses) can occur at the same time and place without any mutation. Galapagos cormorants have, by mutation or adaptive super-code, lost flight-in-air and gained 'flight' underwater better than their cousins elsewhere. And there exist on windswept islands flightless flies whose flight in gales were fatal. Indeed, despite whatever geneticists have inflicted on them, *E. coli* bacteria, flies and so on have never shown unlimited plasticity. The reverse. Do you remember Koestler's eyeless fruit flies (Chapter 21)? And four-winged versions, where the rear pair of aero-paddles aren't connected up with muscles and don't work, are just as bad. Who mates mutants? Kept in a lab they might survive awhile but in the wild such monstrosities soon face the chop. Natural selection axes them. Every stock or horticultural breeder knows that, if he lets slip his chosen, fettered lineage, then pedigree (let alone monstrosity) will revert as soon as possible to wild type.

Nor have fossils revealed a steady and enormous stream of intermediate fossil body-architecture, body-systems or organs. The reverse. Palaeontology's 'open secret' is the sudden initial appearance and subsequent persistence of archetypal body-forms - a *top-down* fossil record. Despite intense pressure from believers to find, interpret and popularise a few fossils in terms of evolution's missing links, it can as equally be argued that none has ever been found.

In short, for Natural Dialectic Darwin's observations on his voyage with the Beagle, especially the formative ones in the Galapagos Islands, demonstrate not evolution but adaptive potential. The Theory of No Intelligence was, in its materialistic (some say 'scientific') interpretation of 'random nature', from a *top-down* perspective ill-conceived from the start - a start that has now led several generations up a naturalistic garden path.

What about the moth myth? It is notable that nearly a hundred years elapsed before Darwin's theoretical arguments received experimental support. In the 1930's and 40's Professor Heslop Harrison, who passionately believed in Darwinian theory, made a series of fraudulent claims. For example, he claimed that feeding moths with chemicals found in industrial smoke could cause their offspring to evolve darker wings. In fact two inter-fertile morphs of the peppered moth (*Biston betularia*), dark and light, pre-existed any such conditions.

It remained for Bernard Kettlewell to perform, in the 1950's, an iconic series of experiments that still adorn school textbooks and public examinations as prime evidence for natural selection and, therefore, evolution-in-action. *If this is the best proof then there is none.* There was neither introduction of new genetic material (no new genes) nor evolution of a new 'melanic' species of moth from a light one. There simply occurred a shift in population from a predominance of light (*typica*) to dark (*carbonaria*) morphs in the industrial regions of England, a shift called industrial melanism that is now in a post-industrial process of swings-and-roundabouts reversal. *Kettlewell's experiment failed to explain the origin of the two pre-existent varieties of Biston betularia or a third, less common, speckled intermediate (insularia), let alone the moth species or moths as a whole.* It actually serves to illustrate that natural selection operates to *reduce* the gene pool, that is, the genetic potential of a species. Either one colour or the other is not created but wiped out. Wiped out means made (locally in this case) extinct. *It cannot be too clearly understood that a species' extinction has nothing whatsoever to do with its origin.* Indeed, Kettlewell's case seems to confirm Blyth's conservative rather than Darwin's (r)evolutionary, non-conservative conception. Blyth had said that natural selection acts as a mechanism to weed out mutant, weak or malformed stock and maintain a healthy species with, a typologist could add, morphs appropriately adapted to their particular niche.

More seriously, it may be argued that Oxford medical scientist Kettlewell gave to natural selection what Piltdown gave to 'ape-man' - a fraudulent confirmation of the theory of evolution well-timed to meet demand. How was Kettlewell a twister? His experiments have been criticised but neither replicated nor confirmed (Nature vols. 300, Nov. 11th 1982 ps. 109-110 and 396, Nov. 5th 1998 ps. 35-36) and there is evidence that feeding habits and migration rather than visual predation by birds was the true cause of the spread of the melanic form. This would invalidate the argument from natural selection by competition. Moreover his field notes have disappeared, he chose for his studies an area of woodland predominant in dark oak as opposed to neighbouring light birch and, key to the set-up, it is known that during the day these moths rest dispersed under higher leaves and almost never exposed on rocks or tree trunks. Yet tree trunks were the spot on which morphs, without camouflage if they had an inappropriate colour, were supposed to have been naturally selected for by dining birds. Kettlewell and other experimenters apparently made sure of this by either pinning or gluing specimens there (J. Wells, The Scientist, May 1999); or by releasing them in the daytime next to trunks onto which, in preternaturally high densities, they fluttered. Furthermore, original associates have since admitted a charge that the original photographs of moths on tree-trunks as depicted in our every textbook are *fakes*. They are of glued stock. Finally, whether unnaturally glued or not, no speciation has been shown. However, after nearly sixty years of heavy textbook promotion as a stellar evolutionary 'clincher', an experiment by Korean workers (Journal of Evolutionary Biology July 7th 2012) may help to straighten the record. In a parallel, 'real world' experiment it was found that, after landing (Kettlewell's 'frozen' position), moths very soon adopted inconspicuous, camouflaged positions.

In sum, there were dark and light moths before, during and after the industrial revolution. No origin of species (or even morphs) has occurred because the morphs interbreed. Kettlewell's experiment showed neither macro-evolution nor even variation (micro-evolution) in action. *Rather, it confirmed shifts in population; it confirmed dynamic equilibrium, according to changing conditions, swinging between races of peppered moth; and it confirms Blyth's creationist view of natural selection as a conservative mechanism operating to keep types of organism from, in the face of environmental change, becoming extinct.* Therefore were Harrison's claims and Kettlewell's results the product of poor science or motivated skew, that is, trickery? Why did the contemporary scientific community so vaunt, elevate and hype them as 'beautiful demonstrations' of evolution-in-action? Why does it still?

If it were needed then a weed's a further candle to the educational 'icon' of the moth. *Crepis sancta* produces *two* types of seed - one tufted, wind-borne and the other not. It was discovered that in paved towns the tufted kind prevailed, in earthy rural circumstance non-tufted thrived just where it fell. *This is natural selection (but not evolution) in action.*

Top-down, information is the immaterial basis of all technological and biological systems. Being immaterial it cannot originate from a material source.

Bottom-up, bio-illogically, simply substitute natural selection for intelligence, chance for thought and 'design' for design. That is to say, neo-Darwinian fundamentalists have elevated natural selection to the status of sole arbiter, nature's 'engineering' principle, an active, quasi-divine 'creator of the powerful illusion of design'. In its superhuman role of 'anti-thought' it non-consciously 'guides the accidental process of evolution' towards an ever more complex grandeur of biological 'design'. Yet, we shall see still more clearly, this elevated status is mythical. It is neither scientific - nature cannot think in terms of 'value' or 'progress' so that some biologists (notably Stephen Gould) rightly within an evolutionary paradigm make no hierarchical distinction between horse, fungus, fly or human; nor is its pretension possible - rude nature does not engineer, by death or any other means, machines. Science blinds that asserts otherwise.

Darwin himself, with many other biologists, was less extreme. Although he still emphasised the central importance of natural selection it was, in his view, 'the main but not the only means of modification'. *Primus inter pares.* Although he excluded sports (monstrously clear physiological defects due to mutation) from his account, modern Darwinists elevate them to the primary cause of change. While accepting their master's view of natural selection they realise it is, in fact, a passive agent operating in one-off, unrepeatable and therefore unique sets of circumstances. It simply sorts, as Blyth said, organisms best suited to a given habitat at a given time. It acts as nature's hone. Other means supposed to supplement this hone include sexual selection, 'random genetic drift' and, of course, unpredictable contingencies such as volcanic action, strikes from space, freeze, disease or even, simply, predators. Evidence from now and then (in the form of archaeology and fossils) is scrutinised. There is no argument about the facts but, such is the influence of philosophical perspective, the calculation diametrically differs according to our pillar of faith. Is it *bottom-up* or *top-down*? From a *top-down,* information-driven point of view neither natural selection nor mutation nor any other material agency could have *originated* life forms or pushed a random, macro-evolutionary stagger into today's natural world. The claim that Darwinism whistles in an uncreative and irrational breeze may represent a significant and unwelcome culture shock. What disproves it?

On the other hand, both fundamentalist and pluralist species of Darwinian hold natural selection as a central, inevitable and inalienable plank in the materialistic explanation of life on earth. Without an evolutionary interpretation of the 'power' of natural selection and, as Harrison and Kettlewell both understood, scientific proof to back its claim the case for such philosophy crumbles.

In summary natural selection, first proposed by the abovementioned creationist Blyth, is a notion plagiarised by Darwin, hijacked by evolutionists. In fact natural selection has neither creative power nor meaning except as a survival sieve. **It creates nothing and is simply survival by another name**. If the fittest survive, Blyth noted, they keep a species healthy but this by itself is far from explaining how a type (from which stem varieties that are, under certain circumstances, labelled as separate species) first originated. Nor does the maxim apply *between* species otherwise diversity would never flourish. The dominant, fittest group would forever wipe out its fledgling competitors. *Instead ecological science shows myriad positive, non-competitive, non-selective relationships.* It shows diverse ecosystems wherein the disturbance of one species more likely *adversely* affects others within its sphere of influence. This sort of natural selection would make weeds, pests or plagues of the stronger. A good example of this is the devastating impact of man-centred technology on natural communities and habitats across the globe. If destructive, competitive humans add up to a biological plague, when and how do you think their hone of natural selection might grind? Would such nemesis create a resurgence and redistribution of existent types or permit the evolution of new ones? What kind of humans might survivors be converted to? Your answer hangs upon a point of view.

While variation - or micro-evolution - is accepted by all parties as fact it places the theory of macro-evolution in a similar category to abiogenesis - unproven in fact, unacceptable by evidence and therefore, by many, unaccepted in faith. If that's the case then, in the mill of reason, where does it leave Darwin's central inspiration, the idea of a tree of life?

Are Bushes, Trees and Forests Just the Same?

An unacceptably non-Darwinian, information-rich life-to-earth construction ramifies *top-down* through mind from its source in the Infinite. Replete with purpose and its reasons such development is seen as an inverted, archetypal tree of life. Such a logical but immaterialistic view clashes head-on halfway down the stairs with its reverse, a phylogenetic tree of transformation climbing up from Puddles.

Plainly, here's a case of global warming if not warring in philosophy! The temperature is sweltering, no wonder tempers snap! Vituperous accusations start to fly. Is not 'creationist' a term used as abuse? 'Creationists peddle lies,' the media cry (e.g. The Daily Telegraph 28-2-08 p.16). These non-scientists (read 'non-materialists') exaggerate the size of gaps between the phyla, dismiss each month's new and short-lived 'missing link' and, topping even this, dot the bushy branches of phylogeny's great tree thus making, almost in the manner of a Cheshire Cat, tentative hypothesis of fact. 'It is you', the other, reviled camp retorts, 'make fact of an hypothesis. Whose arrogance joined Darwin's branching thoughts and modest, dotted lines? Subtract them and transform 'the tree of life' into a 'living forest'. This *top-down* image is much closer to the truth we find. From bird to beetle, frog to flowering plant and all the rest the missing links are in your mind. No doubt, permanence of gaps implies great heresy; they accord with archetypal program and are systematic; thus they refute conceptual macro- (but not all-party micro-) evolution as an organ of transition and diversity.

It has for decades been recognised but tardily (24-1-09 p.34) New Scientist set the public seal. The theory of evolution (on pain of academic death don't challenge that) is evolving. Indeed, a revolution has occurred. Darwin's inspirational doodle isn't right. The central idea of a tree of life has been uprooted, sawn and cast as dust by an 'onslaught of negative evidence'. At least, if it's not marked down for the chop, some heavy grafting has to radically re-forest it.

One problem is that genomics, bio-informatics and other modern disciplines have revealed a revolutionary non-linearity. Let's start with genetic code itself. Previously (in Chapter 20) it was noted that evolution of an optimal code is, for practical purposes, impossible. Yet deviant, non-universal codes exist. Over twenty eccentricities have been discovered (none in plants) and there may be more. Although the majority found to date involve small, discrete amounts of mitochondrial *DNA,* nuclear varieties are spread in a 'mosaic' way, that is, without apparent connection in bacterial, fungal and ciliate species. This plurality of non-standard codes occurs when alterations to t-*RNA* molecules affect their attachment sites and thus the assignment of start, stop and low-frequency codons. A codon translates, of course, into an amino acid but sometimes, in a diminution of the code, possible codons are not employed at all; but because all the main principles of genetic operation are adhered to it seems likely that non-universals are simply slight deviants from the universal standard. It is also possible, however, they have a reason that we don't yet understand; and if life forms were seeded from a range of archetypes rather than a mythical single tree then perhaps bridgeless evolution never did occur.

A second problem is that often genome sequences do not, as it was at first supposed they would, concur with suppositions drawn up in evolutionary trees. Visible characteristics appear similar but involve 'genetic discordance', anomaly zones' and 'anomalous gene trees'. In other words, parts of genetic make-up do not match a phenotype's supposed phylogeny. It seems, *prima facie*, that evolution has been compromised. Thus new models, such as coalescence model, try and patch the puncture up.

There's a third, mosaic problem too. It stems from modularity. For example, petrol engines, made in great variety, all need a piston and a cylinder; and by analogy some core processes in biology, say respiration, will require core genetic homology. What, though, about the anomalous presence of a rotary motor in the piston-engine scheme? Mosaic subroutines of *DNA* are, we shall see, found scattered round in organisms so disparate you would not suspect an evolutionary connection. A duck-billed platypus shows clearly what I mean.

The fresh permutations amount to numerous anomalies. They spin you contradictory tales. How did they get to where they, out of evolutionary kilter, are? How could such subroutines be scattered randomly (you can't say by design) in programs otherwise dissimilar? How can you account for the non-linear nature of these modules?

If 'vertical' inheritance won't do perhaps 'horizontal' must. If vertical transfer of genetic linearity from one generation to the next cannot explain mosaic distribution of chunks of similar *DNA* then 'horizontal gene transfer' (called HGT) must be proposed to patch the punctures in a theory of descent. Darwin's tree of life has branches that are disconnected and, to connect its myth back to reality, bacterial and viral vectors are invoked. No doubt, material is laterally swapped in several ways between such types but, where sex and mutations aren't enough, is this trade rich enough to fund the course of macroevolution? The idea's all the rage to fill up evolutionary gaps. Natural genetic engineering (Chapter 23) spins another web through which to guess.

Perhaps, therefore, microorganisms carry vital answers to the way new body plans and all their functions have evolved and work. About 90% of life is unicellular and, so the thinking runs, by swapping genes a complex 'web' of forms might grow; and then, occasionally, a multi-celled minority might sprout like bodies from the great mycelium. And this minority might further spread the net as they go forth and hybridise; branches fuse and thus confuse what once seemed reasonably clear. Except that hybrids, while they definitely exhibit variation from their parent stock, are never seen to innovate. No more than sexual intercourse do they produce fresh systems, organs, functions, metabolic or developmental pathways. Hybrids cannot cut the innovative mustard evolution desperately requires. Surely, though, germs drive evolution in untidy ways. They must seed mosaic distribution of homologous subroutines; they must be the cause of similar chunks of *DNA* in, say, mice, bats, lizards and frogs but not in fish, birds or humans. Where 'vertically' inherited mutations cannot do the trick bacteria have now been drafted in to horizontally drive the evolutionary process forward. Theory drives contrivance. Some theorists strive for conservation but modernists have sawn imagination's tree in separate pieces using fact. And if life's history cannot be represented as this tree then Darwin's crucial metaphor has simply failed.

Prokaryotic transfer of genetic information and eukaryotic hybrids are both facts. Each certainly involves a random element that, in any given instant, may or may not generate specific variation. But, as sections from Chapter 23: 'Entropy of Information' and 'Evolution in Action?' repeat in detail, such variation may be insufficient to explain the origin of novel features, body plans and macro-evolution. Geneticists have not observed such transformations in *HIV*, *E. coli*, fruit flies or any other organism in their three major categories of life on earth. Does, moreover, every aspect of an organism's structural felicity reside only in the biology of genes? Perhaps 'the onslaught of negative evidence' might be construed, *top-down*, as positive. The heretical ideas of archetype and an engineering/ computational theory of design stand with their living forest as intact as ever. Not one tree of life but many. Clearly, forests, bushes, trees and, perhaps, fungal networks of mycelia are not at all the same!

Forensic altercation! Litigation! *These are serious conflicts of interpretation that demand a systematic, point-by-point discussion of the fossil record link by, accepted or denied, each missing link.* Should, as was suggested earlier, the court of truth call the witnesses from both camps so the facts are marshalled from each point of view? Would open-minded science ever open up enough to tolerate an element of cosmic information for consideration by the bench? Neo-Darwinism and therefore materialism with all its progeny *depends entirely* on the *PCM*, that is, abiogenesis followed by macro-evolution as a result of mutation and natural selection. So would palaeontology seriously allow such flagrant, threatening breach of ethic as the notion of an immaterial archetype? We saw (Chapters 20 and 21) that abiogenesis is impossible and shall see (Chapters 19 - 25) that genetics supports limited plasticity, that is, restricted variation but not unrestricted macro-evolution; but without the former evolution cannot start and without the latter it would fizzle nowhere. **Therefore macro-evolution is critical, a central canon of the faith in fact.** Surely evolution's central process is a fact? Transformation from one type of organism to another just must have happened even if we have not found transition series, common ancestors or missing links - yet. **Thus materialist faith is wholly vested in the future, in the proofs that have to come but seem to be receding all the time.**

Homology: Common Descent or Common Design?

Archetype (Chapters 15 - 17) is a word that includes both conceptual rationale and its specific, physical expressions. The latter include homologous structures. *The science of homology, the physical expression of archetype, is called comparative anatomy.* Homologous structures are those of different organisms or parts of organisms which correspond in relative position and connections to the same fundamental plan. This plan must include the similar genetic code and developmental pathway as well as finished, adult form. Arthur Koestler described the presence of homologies as "the preservation of certain, basic, archetypal designs through all changes".

The theory of evolution is based on two main arguments. The first, hereditary variation or diversity by common descent, was briefly introduced in the course of last couple of sections and will be elaborated in Chapter 22. The other is homology. A basic example of homology might be body-plan, say, mammalian or chordate body-plans. Another commonly used illustration is the pentadactyl (five-digit) limb adapted as your hand for manipulation, as a whale's flipper, a bat's wing, a mole's trowel etc. **The concept of homology has now been extended from the morphology of phenotype (see Glossary) to the molecular biology of base sequences in *DNA* and amino acid sequences in protein.** Sequences that code for a protein like respiratory cytochrome are compared for different organisms. The number of base differences in, for example, yeast, cauliflower, cat and oak *DNA* is measured. This measure is not often interpreted as a matter of thematic design (as, for example, in petrol engines adapted to different vehicles) but, more commonly and 'scientifically', as the phylogenetic separation in an evolutionary relationship that, in ordinary language, means how long ago it suggests they branched from common ancestors.

Indeed, debate rages concerning the origin of molecular and large-scale homologies. In the rapidly expanding field of genomic research many biologists claim the evidence for evolution derived from a comparison of gene sequences is, especially when they appear to contain inherited but senseless mutations, 'utterly compelling'. In other words, molecular/ genetic homologies are often touted as confirmation of the phylogenetic interpretation of descent. Do the facts of gene mutation warrant such a claim for macro-evolution? The abovementioned high-school comparison of a single protein (say, cytochrome c) from primates through birds, reptiles, fish, yeast and bacteria yields insufficient data. A recent study (by Michael Syvanen) compared 2000 genes common to humans, nematodes, fruit flies, sea urchins, frogs and sea squirts; but the construction of an evolutionary tree from their homologies failed. Different genes related different ways. For example, about 50% of sea squirt genes are urchin-like and others chordate like a fish or frog or you. Such 'phyletic incongruities' are found in every group of organism.

Molecular homologies might be expected to parallel homologies of form (although they do not always so conform and in some cases characters controlled by the same, pleiotropic genes are not homologous); likewise the development of similar organisms would be expected to be homologous (but is not always so). But beneath such technicalities the question's simply your interpretation of the facts. *To what do you assume life's similarities are due*? You might, after all, expect the actual architecture of a building to reflect its plan; and individual components (bricks, tiles, sculptures) to reflect the context of both plan and the whole building that they represent. An architect or archaeologist will tell you each one's basis for interpretation is the same. Do you assume Darwinian gradualism rules or, conversely, that an architect has licence to commute his theme. Typology interprets biological homology as evidence of an efficient use of concept in design *and* accommodates exceptions to the planner's rule. **In short, phylogenetic interpretation from genetic or molecular homology is no more 'proof' of evolution than is such interpretation from a phenotypic counterpart. And, equally, these correlated types might be the typological product of intelligent design.**

Moreover the phylogenetic view takes no account of the *origin* of homologous structures or their part in evolutionary transformism whereby, say, ratty is transmuted to a whale. Denton (*op. cit.* p.155) summarises the point:

"There is nothing more deceptive than an obvious fact. The same deep, homologous resemblance that serves to link all the members of one class together into a natural group also serves to distinguish that class unambiguously from all other classes. Similarly, the same hierarchic pattern that may be explained in terms of a theory of common descent also, by its very nature, implies the existence of deep divisions in the order of nature. The same facts of comparative anatomy which proclaim unity also proclaim division; while resemblance suggests evolution, division, especially where it appears profound, is counter-evidence against the whole notion of transmutation."

Darwin defines homology as a "relationship between parts which results from their development from corresponding embryonic parts". It is, therefore, insufficient simply compare end-phase, adult homologous structures. *These need to correlate with clear genetic and embryological homologies before any phylogenetic relationship can be considered. In other words, to be considered homologous a similar organ needs to develop in the same way from the same genes.* Both circumstances spell trouble for evolution's argument.

In fact homologous structures have been found specified by non-homologous genetic code. In other words, the code has found a different way of saying the same thing. Most biological structures involve the circumstance whereby a combination of genes affect a single structure (*polygeny*) or whereby a single gene affects a number of structures (*pleiotropy*). How, for example, can a gene which in a chicken networks for both feather and skull formation be homologous for all vertebrates? *In other words, homology of phenotype does not mean homology of genotype.* This, if you are obliged by theory to evolve in gradual, consequent steps as result of genetic mutation, is not easy to account for. Alone it leaves the notion of a single tree of life in tatters. It is still more difficult if sensible, intermediate steps can be neither observed nor hypothesised; and most difficult if, because homologous organs are arrived at by different routes, homologous developmental correlation is weak or missing.

For example, the segments of fruit fly and wasp bodies are homologous but different genes account for their development; although they 'converge' to the same organ modes of gut development in lampreys, frogs and sharks are not homologous; and the eyes of a mammal and squid are superficially similar in appearance, remarkably similar in function and efficiency but built up in different ways from different elements. No Darwinian biologist dreams of 'homologising' these two kinds of eyes because they belong to two quite different kinds of animals. He would just say that these eyes have 'converged', that is, become similar. Fish and whales are supposed to have 'converged', coincidentally, towards a similar streamlined shape that gives them equal mastery in water. Can homologous organs, then, be matched only in animals that we believe on other grounds to be closely related? If so, the Darwinist stands within danger of the noose of circular argument. Homologies cannot be based on relationships and, at the same time, be considered independent evidence for them. *You cannot assume homologies stem from common ancestry and then use them, tautologically, as 'proof' for just that.*

A bone in the floor of the skull that lies between the eyes of amphisbaenian lizards makes this troublesome point. In other lizards this bone, the orbitosphenoid, is formed in the normal way from a cartilaginous precursor. It was presumed that this also occurred in amphisbaenians and that the bone's unusual thickness was an adaptation to their burrowing habit. Now it has been found that, in their case, it develops in the embryo in quite a different way - from soft tissue instead of cartilage. Because of this developmental difference it fails the test for homology, although it completely resembles 'normal' orbitosphenoid bones in other lizards. Could this type of phenomenon be more widespread than previously believed? An anatomist, Dr. R. Presley of University College, Cardiff, has written "...this apparently obscure finding seems to me in the light of my present knowledge of the subject to have shaken the philosophical and logical framework of comparative biology to a very serious extent and lots of people ought to be worried by it. I bet they aren't."

Symmetry is everywhere in nature. Its geometry 'falls into place' and lacks the tension of irregularity. It is, in shape and process, the kind of balance nature fundamentally conserves; and balance is vital for efficient function. Bodies show such capability for rhythm and, which is closely allied with beauty and aesthetic satisfaction, structural symmetry. For example, the muscular symmetry composing your own ability for balance and balanced agility is remarkable.

Now look at your hand. Ask what informed its purposive configuration. Is it evidence rude accidents evolved you from a proto-mammal, reptiles, fishes and the zones of life you're told 'preceded' them? Or, with the rest of body, was it informed by immaterial reason; was it obviously formed, exquisite as an arabesque, by what you never see with eyes - intelligence? Look at it a second time. What about the other hand? And feet? Evolution might suggest that these derived from, respectively, the pectoral and pelvic fins of a fish. Not that fins and limbs are (as you can see by purchasing a coelacanth at market) strictly identical structures; and fish themselves are not unbalanced! How did random sets of mutations so simulate each other as to produce the different fins; or the symmetry of fins and banks of muscle capable of efficient 3-d locomotion; or the deeply correspondent, serial logic of your mirror imaged hands and feet? Did hindlimbs and forelimbs evolve from a common source? Did your hand evolve from your foot? Why four feet, not three or two or one? What shapes the different limbs, made and muscled by the same genetically programmed proteins; and how was the profoundly coherent, hierarchical developmental program that governs their well-formed appearance in the right place at the right time first written on nucleic vellum? What long series of misshapen fossils do we find before 'life's authorship' hit on this 'winning formula'? How many accidents were needed to 'perfect' their staggering imbalance? Indeed, you ask how symmetry of complex systems that are purposive (and thereby meaningful) arise haphazardly; why should conceptual, mathematically ordered information add, by bits and starts, to working elegance and beauty? **In fact the order of balance and symmetry, a core feature of all life and of the Dialectic, is another Darwinian black box**. *And while the theory provides a possible mechanism of variation or minor adaptation of homologous patterns it provides no answer to their basic origin.* These patterns, such as *DNA* or the pentadactyl limb, provide optimal solutions to various problems encountered in the construction and maintenance of physical bodies.

In summary, while an adult body represents one sort of archetype homologies represent another. They are fixed, underlying, intra- or sub-archetypal patterns that represent functional solutions to various 'questions' raised by the environment. They are analogous to subroutines in a computer program or pre-assembled units that can be plugged into a complex electronic circuit. For example, appropriate variations in silicon chip homology are, as adaptations or attunements of the ideal, used in different computers. *A top-down perspective sees organisms as mosaics composed of functional units*. In each different type of organism its homologous units have been adapted both to perform particular tasks in air, water or on land and to work in concert, as an integrated and harmonious whole, with other units. Essentially each archetypal component of every body is integrated into the service of three main interlocking themes - information processing (including cybernetic homeostasis), the generation of energy (for sensation and movement) and reproduction (including development, growth and self-repair). *A specific permutation of these 'homologous solutions' to life on earth appears within the context of a whole-body archetype such as bacterium, fungus, plant or animal.* **Nothing of ancestry need be deduced from their possession.**

In other words, minor adaptations caused either by accidental mutation or some epigenetic switching system (Chapter 23) may superficially affect the appearance, shape, colour or operation of homologous organs. This does not mean that such mechanisms and associated systems 'emerged' by gradually self-assembling; or that intelligence cannot vary a good conceptual design to fit different habitats and life-styles. Small changes not a first cause make - especially if the cause is conceptual, informative and purposive in its expression. Such epithets click with the actuality of biological form and function. In such a case descent were not from common ancestors but, hierarchically, from mind. Denton has brought evidence to bear indicating true homology, which includes developmental homology, is rarer than analogy (where like function is implemented by a different factor). He writes (*op. cit.* p.154): "Invariably, as biological knowledge has grown, common genealogy as an explanation for similarity has tended to grow ever more tenuous. Clearly, such a trend carried to the extreme would hold calamitous consequences for evolution, as homologous resemblance is the very *raison d'etre* of evolution theory. Without the phenomenon of homology - the modification of similar structures to different ends - there would be little need for a theory of descent with modification."

<u>The Origin of Type</u>

If Richard Owen established the Natural History Museum in London his counterpart in the USA, Louis Agassiz, likewise a creationist, established the biological section of Harvard University's NHM. Indeed, pre-20th century biologists such as this couple, Cuvier and Linnaeus perceived *discontinuity* rather than continuity in biological form. *They perceived a principle of limited rather than unlimited plasticity.* They saw variation within an invariant theme, distinct boundaries within which intra- but never inter-typological changes might occur. Their crisp view of nature opposed what seemed a blurred Darwinian focus allowing endless, gradual, cumulative and unregulated changes of order. It seemed to them that empirical study and observation allowed a rational, non-religious case to be made for discontinuity while, on the other hand, Darwinian continuity seemed by the same observance the more irrational but still zealously promoted perspective. *<u>As a result they held to the immutability not of species but of type or kind.</u>*

Check *fig.* 22.1: Limited Plasticity. Since the notion of archetype (or super-species) effectively kills off the theory of evolution it meets implacable resistance. An accusation is first levelled that requires sharp definition of the limits of a given generality (called 'type' or 'kind' and not the less inclusive 'species'). The former, being outside the Darwinian mind-set, has not been researched; but lack of precise information is in turn regarded - ironically in view of myriad black holes in materialism's view - as sufficient grounds for summary dismissal!

For pre- and post-Darwinian biologists an archetype (Chapters 15 - 17 and 19) *is a general, abstract representative of a type.* It includes, as well as a purposive, irreducibly complex, minimally functional mechanism itself, the idea, purpose and principles behind it. That is, it includes metaphysical architecture. Blueprint is as good a word as ever. And doubly helical, symbolic *DNA* reflects an important aspect of *informative potential*. This code is charged with the storage and programmed (i.e. logical) transmission of instructions that comply with the commands for a cell's survival. Such potential is either restricted to meet specific needs (for example, a muscle cell's use of protein differs from a liver cell's) or, as in the case of recessive alleles, remains unexpressed. *Top-down*, at the psychosomatic border, physical and archetypal sorts of information are also exchanged. Archetypal commands are issued in terms of resonance (Chapter 16); a typical mnemone transmits particular vibratory patterns that relate to subsequent physical, visible shapes. *Archetypal routines are, in the way a motor or wheels are adapted for use with different vehicles, adapted and harmonised for specific coordination in different sorts of organism. In other words, individual, species and type each represent a restriction of archetypal possibilities, a constraint on potential, a particular outworking of the generality*. It is as if an archetype (e.g. the vertebrate construction) were a theme-tune, typological adaptations were rendered for different biological instruments (i.e. sorts of organism), genes were the strings/ keys and consequent 'musical' chemistry were exchanged with both internal (bodily) and external environments. Or, if you like, the genome is a 'boundary condition' varying the local, detailed outcome of theme. Each component, physical and metaphysical, limits and affects the others. Changed ecological pressure or physical abnormalities can provoke the expression of previously latent adaptations or else unlock inappropriate responses (e.g. hen's teeth, horse's toes or human tails).

In terms of Natural Dialectic the (*sat*) informative level of biology is (*fig.* 15.4 (ii)) the archetype, blueprint, plan or call it what you will. The (*raj*) level is expressed in a body's subtle or electromagnetic constitution and its (*tam*) outward structures are visible with either microscope or the naked eye. The three combine into an expression of archetype you can see - for example, yourself.

A typological perception that allows for innumerable yet limited variation is supported by three facts.

Firstly, no-one has observed, either now or from the past, either abiogenesis or the myriad concatenations of intermediates that must chain order to order of organisms. **E cellula omnis cellula; this is The Law of Biogenesis.** *And parents are observed, without exception, to reproduce their own kind; variation is limited not limitless;* **this is The Law of Heredity. <u>No evidence contradicts these most rigorously tested of all biological principles</u>**. And both contradict evolutionary theory.

Secondly, the fossil record (next section) - not least its Burgess Shales.

Thirdly, generations of extreme conditions inflicted by genetic engineers on *E. coli* bacteria, fruit flies and so on have produced genetic mutations and all sorts of deformities; new strains, races and, under the humanly imposed definition, species have also been observed *but no new types*. ***E. coli* remain *E. coli.* Fruit flies remain fruit flies.** The *Drosophila* type has been known for over 50 million years (since the Tertiary) and has undergone virtually no modification in that time; in this case both natural and artificial selections entirely and without exception contradict the Darwinian hypothesis. The list goes on. **Nor is diversification the same as evolution. As the section of Chapter 23 called 'Evolution in Action?' elaborates, the question is not one of limited plasticity but the origin of a type in the first place.** Diversification has been shown (e.g. between species of Hawaiian fruit flies, European grey rabbits and, over 150 years, races of house sparrow in North America) to occur naturally and rapidly. And take-it-easy natural is complemented by concentrated, purposeful, artificial selection. Refer, for example, to the wide contemporary variety in breeds of dog - alsatian, greyhound, St. Bernard, collie, dachshund, chihuahua and many others whose genomes are 99.75% or more identical; indeed, a plot of poodle and boxer genomes found they differed by about 0.1%. If their bones were fossilised how many species, genera and perhaps even families might a palaeontologist make of *Canis familiaris*? Analysis of mt-*DNA* (mitochondrial *DNA*) shows that British women (and their colonial descendents in Australia, Canada, New Zealand and the USA) are descended from about five original 'Eves'. In a similar manner the whole variety of dogs has most likely descended from an original wolf-like stock (Science 298 22-11-02 ps. 1610-3). As this stock spread worldwide, invading vacant niches, isolated populations with characteristics best suited for different habitats would have thrived. This is natural selection and adaptation at work but is it evolution? Evolution needs fresh information so that not only fresh minor but also major functional differences can arise. *On the other hand canine differences, dramatically enhanced by man's selective breeding, represent lost, redistributed or concentrated genetic material.* Poodles, for example, are one sort of degenerate result of corruption, deletion and selection from the original wolfish stock. You have obtained poodles from wolves but, because variety has been 'squeezed' out of the former's gene pool and it has become the 'specialised end of a line', you cannot breed wolves from poodles. Indeed, every vet knows that specialised forms of domestic dog show numerous congenital defects. *No new information has been added.* The process of modification is 'downhill' where evolution needs 'uphill'; what is it, after all, that dogs are now evolving into? Their descent into special sorts of breed is, according to dialectical principle, devolutionary, that is, non-evolutionary. It is the opposite of evolutionary. It is variation-on-theme, that is, different details devolved from typological principle. **But, for all known breeding, natural and artificial, now and in the past, dogs stay dogs.**

What applies to dogs applies to cats, rabbits, flies and (although genetic proof is not yet in the bag) probably to all other types of organism. In this case it will have applied to an original human stock. Therefore, although disorderly mutation may eventually cause the downfall of any biological expression of archetype, remixing in the form of inter-racial marriage would tend to reconstitute original potential and thus, where it occurred, strengthen a population. In fact constant shifts in biological form occur as a result of genetic shuffling, corruption and, according to a theory of archetype, recombinations that trigger different aspects of adaptive, archetypal potential. **In this view the first law of genetics would be conservation of archetypal information.** *There is no new injection of information only a kaleidoscope of original, resilient 'books' of code suffering, by now, a little wear and tear. There is no unlimited plasticity and no new types. Type remains true to type. As we find.*

If, however, typological axioms such as immutability do not apply at the plastic level of species where does irreversible determination set in? What exactly *is* a type, super-species or, as it is sometimes deemed, a kind? *The limit has not been determined since official mind-set has endorsed unlimited plasticity. The question's therefore not been asked nor research for an answer been engaged.*

Vertebrate Cladogram

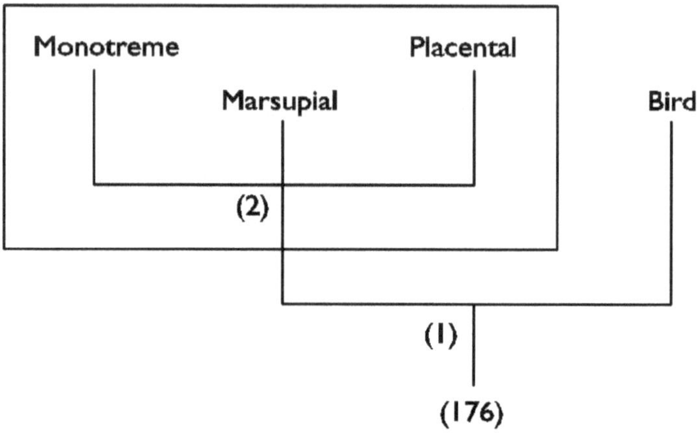

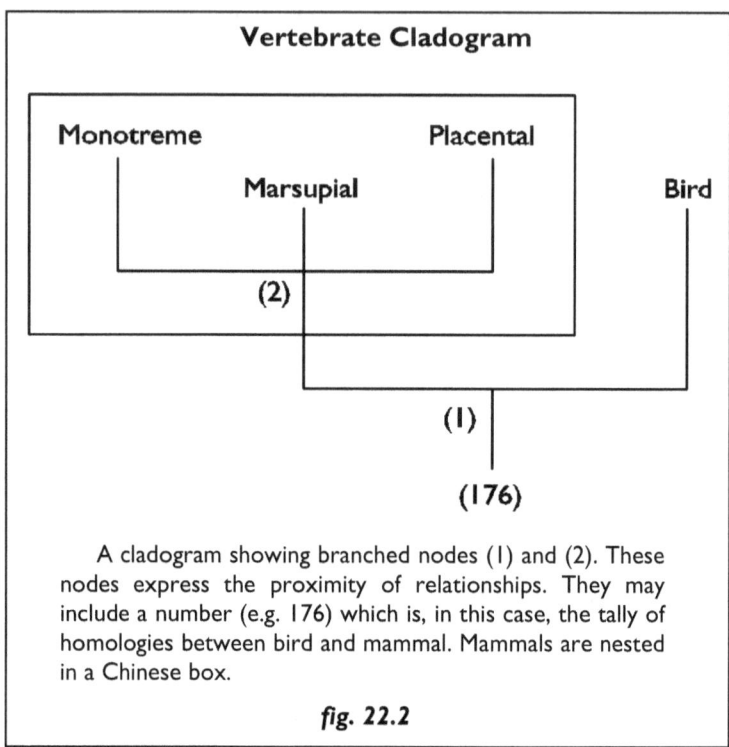

A cladogram showing branched nodes (1) and (2). These nodes express the proximity of relationships. They may include a number (e.g. 176) which is, in this case, the tally of homologies between bird and mammal. Mammals are nested in a Chinese box.

fig. 22.2

One possible definition might, however, be that a type is a group or class of organisms that possess a number of unique defining characteristics which occur in fundamentally invariant form in all the species of that class but which are not found even in rudimentary form in any species outside that class. Examples include birds with beaks, feathers etc. or mammals with hair, lactation, four-chambered heart etc. Types are, by this definition, exclusive and not approached gradually through a sequence of transitional forms. Indeed, close examination of the handful of conceivably transitional organisms such as the velvet worm *Peripatus* (annelid/ arthropod), lungfish (fish/ amphibian), *Archaeopteryx* (reptile/ bird) or duck-billed platypus (what indeed?!) simply underlines the principle of discontinuity. They constitute mosaic patterns drawn from otherwise distinct types. They embody unusual permutations of diagnostic features which appear distinct, not approached through gradual transitions and, as with homologies, fully characteristic of one type or another. In similar respect the proposed evolutionary series of plants (from algae through mosses and ferns to land-plants such as flowers) and animals (from fishes amphibians and reptiles to, on the one hand airy birds and on the other ground-based mammals) are as indicative of distinct adaptation to watery, semi-watery and dry environments asany cumulative transition.

Chinese box Classification

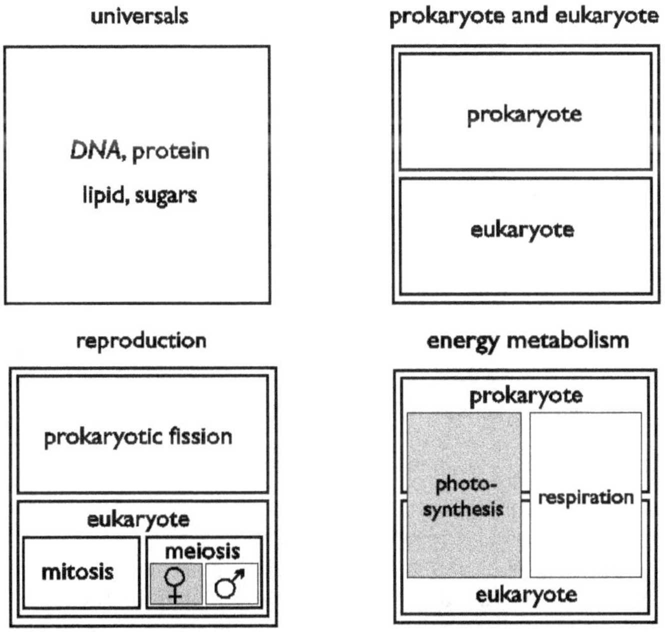

In this arrangement relationships could be by descent or, like a range of motor vehicles, by design. Chinese box classification anticipates a new perspective in biology. The idea of 'complementary opposites', most sharply expressed in Taoist literature, is one developed throughout this book.

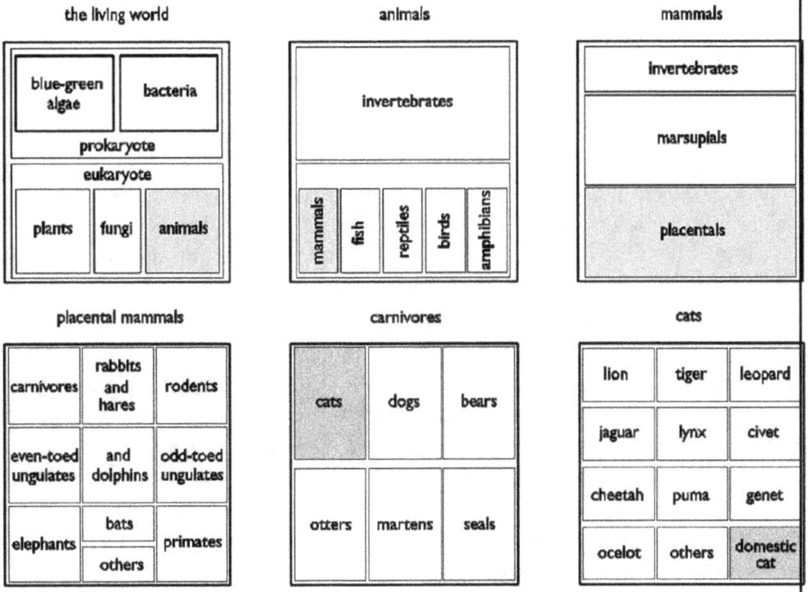

This example pursues the classification of *Felix domesticus*, the family cat.

fig. 22.3

Another possible definition is the possession of a unique set of characteristics whose distinction (in the form of functional homologies) was unapproachable by any gradual process. These characteristics are irreducibly complex mechanisms that include, as well as organic systems, specialised metabolic or developmental algorithms.

A third definition sets the integrity of type (as the basic unit of biological construction) at about the conventional classification level of family. In this respect you would expect to find distinct, archetypically adapted but physically unrelated types of fish, bird, flower, tree or mammal. You might, for example, expect to find hawks, herons, crows, waders, warblers, finches and so on each scattered world-wide in its niche. What test might be performed or criterion applied to sharpen our understanding? In nature plant compatibility and animal psychology, behaviour and coloration play a large part in determining whether or not male meets female gamete. However artificial pollination or insemination might demonstrate relationship within a type if fertilisation occurred and the first or more divisions of a zygote occurred. In other words, the instigation of embryonic development would be a criterion.

DNA is a database that governs the production of proteins for either functional (enzymic) or structural use. However while the shape of a house is determined neither by the builder's tools nor his materials, since 1953, when Crick and Watson in Cambridge famously discovered the structure of *DNA*, the focus on understanding *DNA* templates has meant that other aspects of morphogenesis have been relatively sidelined. Cells always result from division of a prior cell. *Multi-cellular bodies also start as a single cell (an egg, seed or spore) so that this critical structure must incorporate the trigger controls, indices and application programs used, in conjunction with the nuclear database, to initiate an intricate program of development.* Cellular information resides, of course, in *DNA*, chemical gradients in the cytoplasm, other biochemical messengers (such as hormones) and the external membrane with respect to trans-membrane voltages, glycolipid, protein and various other signalling fixtures - but also, it is argued, by an integration of this electrical configuration (in your case *H. electromagneticus*) with an overall metaphysical morphogene, a wireless archetype (called in your case *H. archetypalis*). This 'mass' of contextual, non-nuclear information engages with nuclear in a prearranged, interlocking dialogue that triggers both the (nuclear) directions and (non-nuclear) production, carriage and organisation of the right quantities of the right biochemicals at the right times in accordance with architectural blueprint.

The quality of a medium influences the passage of energy through it. Different conditions may occur in a single medium (e.g. air) or a quite different medium (e.g. water). Either way, the vibratory energy (e.g. sound) might be termed active and the medium passive; sound, light or heat would be 'dominant' and water, air or earth 'receptive' or 'controlled'. In a similar way archetype (prior order) and medium (the shaped energy that comprises a physical body) 'exchange' information. Archetype is dominant, body the 'controlled' receiver. Even a cell, each and every cell, is 'locked within' the influence of its archetypal memory. Nevertheless the condition of any medium will affect the way it conducts, transmits or 'expresses' an energy. For example, sound differs in hot, cold or turbulent air; and differs considerably in water or through rock. In a similar way the conditions of biological media (such as *DNA*, cytoplasm, organs etc.) affect and are affected differently. The analogy of wireless exchange (Chapters 15-17) with body as a 'set-with-aerial' is apt. Patterns of sub-conscious energy, the broadcast, interact with electrochemical features of the physical form. For example, small genetic variations separate every individual; they evoke a different 'interpretation' or 'expression' of the archetype, occasionally even one involving malfunction (genetic disease). Larger differences, both of archetypal 'tune' and medium (the shape and materials of a body) separate what we call higher taxonomic divisions such as type, phylum or kingdom. Much as the behaviour of sound differs through different elements, so organisms differ in their 'nodal receipts', psychosomatic plug-in points (see, for example, *figs*. 17.8-10) or templates for archetypal 'sound'. Atoms absorb and emit radiation of certain 'permitted' wavelengths. Compounds each have their unique, spectrographic fingerprint. It is how we identify them. In a similar way each type from the range of biological organisms has its own line spectrum, its 'permitted' permutation of archetypal wavebands, its combined norms and frequencies of operation - a dynamic program. **There is profound order behind the superficial, apparently chaotic differences in this creation**. It will, in this case, be argued that conceptual transformation of a program may be simple but material reworking, type to new type, model to new model is beyond the capability of naturalistic evolution. It cannot, even at a protein level, make the functionally innovative grades.

Relationships between species (and, as a consequence, higher categories) are represented by evolutionists as a hypothetical phylogenetic tree that has branched 'up' from a proto-bacterium. However to draw up a family tree we need to know, if we are not to indulge in just-so stories, precisely who the relatives were/ are. Neither modern species nor the fossil record tells us enough and it is thus unscientific (indeed, unsafe) to guess. *For a typologist a more useful construction derives from cladistics.* Cladistics is a method of classification using diagrams called cladograms. Organisms are collected into groups on the basis of shared (homologous) features. Homologies are tallied and numerical rather than speculative, historical links drawn up between organisms. Cladism is thus a powerful, neutral, objectively detached tool of analysis. For this reason the technique enjoys growing popularity among many of the world's taxonomists.

As well as cladistic trees typologists are free to employ a Chinese box arrangement of categories nested within categories. Species cluster within types and types show similarities that are contained in larger boxes (or higher taxonomic categories). Such a hierarchical form of classification emphasises the discontinuity or distinction between adult archetypes. It can also distinguish between functional homologies such as energy metabolism and reproduction and, at the same time, allow for the dialectical expression of complementary opposites such as photosynthesis and respiration, asexual and sexual, male and female etc. It depends entirely on complements and similarities that can be observed, measured and assessed and needs nothing of unobservable, ancestral phantoms that may or may not have existed long ago.

Denton (op. cit. p. 117) summarises: "All in all, the empirical pattern of existing nature conforms remarkably well to the typological model. *The basic typological axioms - that classes are absolutely distinct, that classes possess unique diagnostic characters and that these characters are present in fundamentally invariant form in all members of the class - apply almost universally throughout the entire realm of life...*(my italics). To refute typology and securely validate evolutionary claims would necessitate hundreds or even thousands of different species, all unambiguously intermediate in terms of their overall biology and in the physiology and anatomy of all their organ systems." Such intermediates may exist, as Darwin admitted to Asa Gray, only in the imagination.

Types of Fossil

We come to nature's book of years, life's archive stacked within the international library of earth, the fossil record.

Science works by observation. Facts are gathered and theories proposed according to these facts. Is this the case with the theory of evolution? Is not true science in reverse when facts, even contradictory anomalies, are squeezed by hook or by crook to fit a theory? What if such distortion happens wholesale?

The fossil record is amenable to models that argue either evolution or design. Indeed, the superficial appearance of evolution has, we'll see, to some extent been imposed on it by dating rocks according to their fossils and their preconceived relationships. How confirmatory - or tautological - is that?

Thus it needs be stated, in the interests of fairness, clarity and truth, that a major research project needs be established. **Palaeontology, whose interpretative mind-set is exclusively evolutionistic, needs root and branch reassessment. Thorough unpicking and radical comparative debate is long overdue. Such international program would involve a thorough comparative analysis of each and every fossil.** It would accept their obvious, objective and previously measured physical dimensions but would challenge, from the points of view of Theory of No Intelligence *and* of Intelligence, their relations in life's forest (or perhaps on its tree) throughout its presence here on earth.

Two questions to be asked are, simply, whether any bio-form might better be construed as integrally engineered from prior archetype or whether, without prior information, from an unplanned aggregate of add-ons; and, secondly, does the range of various homologies more exhibit seamless continuity, mosaic distribution or discontinuous clustering? With what mind-set might you best interpret the historical construction of life's many-splendoured tapestry; in what framework might you best cast hardware remnants of earth's bio-computational machines?

Top-down and bottom-up, 'design' *and* 'evolutionary' palaeontology would each apply their lens. Such comparative discipline has, dog-in-the-manger, been fiercely resisted by materialism, naturalism and the atheistic faith. Thus, as in the case of theocratic or of socialistic politics, sole public purview and publicity has rested with a single party's view. **To reiterate, the time for redress and for academic balance is long overdue.**

Tautology is circular argument. It is often employed to prop the presumption of evolution. For example, we are here and so we must have evolved; evolution happens therefore creator-less abiogenesis must have too - and abiogenesis proves evolution; homology is a product of common ancestry - therefore it proves common ancestry; we are here, evolution happens by chance and therefore we evolved by chance. Do lawyers not interpret facts? A documentary but common form of narrative is to decide an order of 'problems' which evolution must have 'solved' and then, from contemporary examples, arraign mute witness that it did so. You need legs? Try mudfish. Lungs? Lungfish. And so on. Problem? Problem solved this way or that by evolution, by the only answer - 'just so'. Whole courses, textbooks and TV series are built around this 'just so', 'need it/ get it' kind of 'logic'. Such analogy of teleological intelligence is, we're assured, 'shorthand' for the irrefutable assumption chance informed the living world.

Does fossil débris actually show such gradual transformations as a mindless problem-solver might 'contrive'? Yes, countless books on college shelves attest to them. They suggest, by drawing 'fossil series' up, a practical ubiquity of missing links and transitional features in the fossil record and in living lineage. This argument extends the 'problem-solving' paradigm from now into an undetermined past. Does not the fossil record seem, in this view, to support the case? *Having ordered a series of organisms of similar bony body-plan or perhaps even some 'soft' features, does not such 'progressive arrangement' prove they must have evolved from one another?* This *must*, indeed, from a materialistic point of view be true.

You could similarly range a suite of technological constructions (say, motor cars) to demonstrate their evolution; and, of course, they did evolve - by mind. How fair is that? Cars don't, like life forms, reproduce. Alright, take the orthodox but mythical evolution of the horse as described in an iconic 'tree' from a quasi-hyrax to modern horse. It has been transformed into a fuzzier bush by research but the textbooks lag. *In fact, any few fossils can be placed in a line and a story told about their evolution; then placed in another line for another story.* The *inference*, however, does not change; any series that an evolutionist concocts strikes out non-evolutionary objection. Other books with other goggles can, nevertheless, explain the evidence their way. How can such forensic confrontation happen? Do you, high-handedly, dismiss one party out of hand or do you commission reappraisal that compares, from both mind-sets, interpretation of the details of the fossil record case by case?

Modern geology is based on two major premises - the great age of earth and that James Hutton's principle of uniformitarianism (summarised in the aphorism 'the present is the key to the past') holds good. This idea, promoted by Charles Lyell in the 1830's, was never proved from the rocks but rather imposed as a bias in geology. It negates 'catastrophism', that is, it erases the notion of global tectonic, igneous or, by flood, sedimentary catastrophe in earth's prehistory. Of course, local catastrophes happen. The eruption of Mount St. Helens (USA) in the 1980's generated metres of layered sediments in hours. Wood in mineral-rich solutions is known to petrify rapidly. Many fossils, arguably most, indicate live burial. Closed clams, fish devouring other fish, polystrate tree fossils (trees buried upright though layers of sediment) and many other examples bear, as they are bared, mute witness to sudden disaster. Some episodes are catastrophic. Does this mean, however, that all are? Or that slow events (say. coral atolls layering or mountain chains upthrusting) do not occur? Modern geology, a critical pillar of support for the theory of evolution, espouses 'gradualism'. Changes, at least global changes, do not happen rapidly. Things, overall, look slow.

Geology's stratigraphic column is built of rock units, for example coal and chalk measures, and subdivided into beds and bedding planes. By using certain fossils as indicators a 'bio-stratigraphic column' can be divided into time zones. This column is defined by zone or index fossils. Such fossil types have been assigned an age based on the presumed age of strata in which they were found.

Anomalies occur. For example, some 'living fossils' (e.g. coelacanths, a small mollusc called *Neopilina galatea*, ratfish elephant sharks or the tuatara lizard) still flourish and many others (e.g. jellyfish, springtails, the nautilus or bats) bear uncanny resemblance to their modern counterparts. Take a flea, a spider, horseshoe crab, shrimp (e.g. *Aciculopoda mapesi*, 400 million years) or seed-shrimp (*Ostracoda* from palaeozoic times). After time-scales during which all land plants have appeared and an unidentified worm evolved into you they still look very like their modern counterparts (and perhaps, if their *DNA* was accessible, be found practically identical). Of course, no rule says a line that's branched from has to die; or that, despite mutations, it can't stay the same. But evolution is a theory of kinesis and not stasis. It moots continual change and if, from evidence and observation, the opposite is found then can a theory wholly contradict itself? If what fits and doesn't fit can be included just the same how then what is such a flexible and all-inclusive theory worth? How can you falsify what's always right?

Nevertheless, the presence of index fossils is used not only to date strata but any other fossil found in them. There is no one area where the whole stratigraphic series of fossils is found but an evolutionist, having decided that things evolved in certain ways (e.g. from simple to complex or sea to land) arranges finds from different areas on the assumption his decision is correct. Natural selection notwithstanding (Chapter 23) herein lies perhaps the most powerful tautology of all. *The assumption of evolution is the basis upon which index fossils are used to date rocks; and these same fossils are supposed to provide a main evidence for evolution. The fossil record, itself based on evolution, is interpreted to teach evolution.* A closed circle is a ring that binds. By this sort of reckoning, called begging the question, the main evidence for evolution could be the assumption of evolution!

Fossils show that variation has changed things in the past as it does now. They demonstrate neither how the variation happened nor that 'later' descended from 'earlier' fossil form; but because their evidence has been almost exclusively interpreted within the dogma of a Theory of No Intelligence a counterweight, a thorough reappraisal is needed. Palaeontology, the study of ancient forms of life, does not inevitably mean the study of evolution. ***So what does a top-down, archetypal view predict earth's graveyard will reveal?***

(i) *Since life's major subroutines will, at every level, concur with archetypal plan you might expect organisms to appear 'abruptly' and fully formed.*

(ii) *The archetypes will persist.*

(iii) *Systematic gaps (discontinuities) in the fossil record* will underline the obvious yet intangible order of biological form as vested in a metaphysical archetype. Intermediates will not exist. This does not mean that one type of organism may not resemble another or contain common features such as eyes, heart or legs. *It means one archetype (and hence its physical expression of 'type') never transforms into another.*

This is in stark contradiction to the *bottom-up* evolutionary idea that one type of organism must have morphed into another. According to this theory a metaphysical archetype is a figment of the imagination and there exist intermediate physical forms. Sooner or later, alive or long fossilised, they will be found. It may appear that some already have been. Intense energy is invested in the search and interpretation of finds. This is because 'missing links', either in the form of whole bodies or an evolving series of features, *must* be found. Every glint in sunlight seems at first like gold. Such confirmation of theory is critical.

(iv) *Continual variation on theme will occur due to intra-typical genetic exchange (including sexual reproduction) and mutation.* Changes (called micro-evolution) will include degenerations, circumstantial improvements, changes in proportion and complexity but no new types.

(v) *Extinction may occur but because it represents loss and not origin of information is irrelevant.* You may argue that the loss of one type gives another a chance but what has this to do with macro-evolution?

The fossil record displays, in fact, just these five predictions. You can fit fossil as well as contemporary taxa into Chinese boxes. Variations circulate within the perimeters of fixed archetypes. This is clearer now that, as the late palaeontologist Stephen J. Gould put it (Natural History 86; May 1977), 'the trade secret of palaeontology is out'.

Generalised Geological Records for Plants and Animals

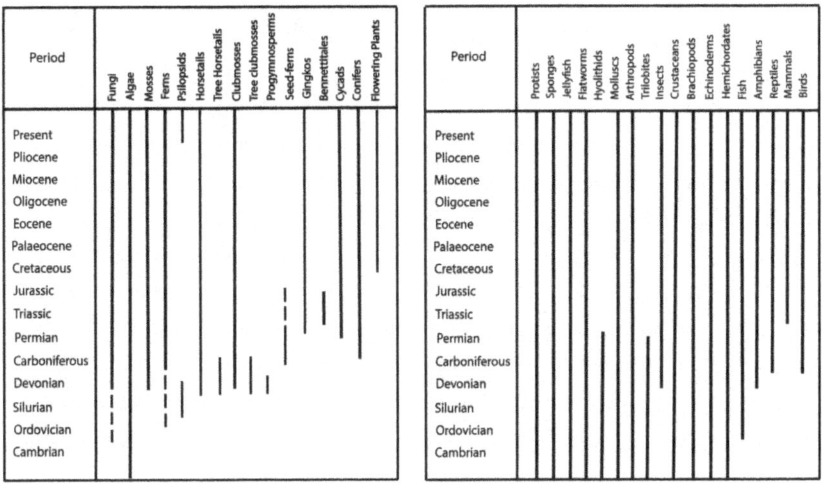

Solid lines represent the duration of existence of each group. Broken lines indicate doubt as to the earliest appearance of some groups. Common ancestors are uncertain or unknown.

fig. **22.4**

The fossil record reveals that distinct biological phyla spring abruptly, fully formed in typical 'radiations' of life forms. Austin Clark (The New Evolution: Zoogenesis) and Ernst Mayr (What Evolution Is) were also ardent evolutionists who both agreed. They exhibit the same basic characteristics as their contemporary counterparts. A mega-year (Ma) is a million years. According to the uniformitarian geological timetable Cambrian rocks show, about 560 Ma ago, a relatively clipped five-million-year 'explosion' of life that includes all domains, kingdoms and sub-kingdoms. It runs to well over a hundred kinds of diversity reaching down biological classification tables as far as phylum and class levels. Such a 'fossil big-bang' puzzled Darwin. How is it explained? Clark's gloss was 'concurrent evolution' from fantastic, simultaneous diversification by the first cells. Gradual build-up is the only other guess a Darwinist can give. Was there a gentle Chinese 'slope'? Fossil-rich Cambrian rocks at 'big-bang' Chengjiang in Yunnan show underlying strata that, as Precambrian, should house the answer. They do not. Nor does Ediacara's record of jellyfish, sea pens and other life as seen today. There is no key build-up to unlock the puzzle Darwin set. The question then becomes, 'Is Darwin's question a misguided one?'

The Long and Short of Geology

Era	Period	Span	Uniformitarian Chronolog	Ecological Zonation	European Revised Timetabl
uniformitarian only		Ma/ million yrs	possible period of origin	*non-uniformitarian*	*non-uniformitarian*
RECENT	*Quaternary*			post-catastrophic history;	received history begins
	Holocene	present		tapering of volcanic activity	
	Pleistocene	0.1		to localised, spasmodic	after some centuries the
				effect; glaciation; erosion	cataclysm's turbulence fades
					but more large-scale tectonic
CENOZOIC	*Tertiary*			organisms best able to flee	activity rapidly splits Pangaea;
	Pliocene	1.7	human	and resist catastrophic	repopulation and dispersal (at
	Miocene	5.4		inundation generally fossilised	different rates) of air-breathing
	Oligocene	23		in final phases of flood; initial	animals after exit from a refuge;
	Eocene	34	horse, whale	post-diluvian deceleration of	large-scale reforestation,
	Palaeocene	56	primate, bat	global upheaval	some reinundation.
					aquatic amphibians and reptiles
MESOZOIC	**Cretaceous**	71	flowering plants, ungulates	organisms less able to resist	from huge mats of vegetation
	Jurassic	151	birds	adverse weather or flee (e.g.	repopulate drying land; some
	Triassic	204	mammals, dinosaurs	reptiles and fish) buried in	fish buried in landslides.
				sediment and fossilised.	flood recedes; dry land appears,
					mats of pre-cataclysmic
PALAEOZOIC	**Permian**	254	cycads, gingkos		forests laid giving coal seams
	Carboniferous	304	reptiles, conifers, ferns	lowland vegetation (forests),	Pangaea formed
	Devonian	375	amphibia, insects, mosses	most swimmers, plankton and	marine invertebrates and heavier
	Silurian	419	liverworts, horsetails,	bottom-dwelling organisms	fish swept up in 'tides' of
	Ordovician	446	fungi, fish	overwhelmed; sea and shelf	continental inundations as
	Cambrian	498	Cambrian explosion; all phyla	deposits by surges	Cambrian sediments formed.
			of animals inc. vertebrates,		waters inundate levelled
			fungi		land surfaces; erosion yields
					the 'great unconformity'
PROTEROZOIC	**Precambrian**	635	stromatolites, bacteria,		sudden cataclysmic destruction;
			soft-bodied Ediacaran biota		volcanic/ tectonic convulsions
			jellyfish, bottom-dwellers		subterranean water chambers
			such as sea-pens, worms	earliest phases of geo-cataclysm	emptied; mountain systems
			algae, sponges	metamorphic and tectonic	stripped to crystalline basement;
				upheaval; rapid geo-transformations	land-dwellers wiped out.
				various protists (e.g. algae)	stable (as present-day) geological
ARCHAEOZOIC		2800	bacteria	origin of earth's crust	and ecological conditions

The Long and Short of Geology, a chart whose sharp abbreviation of a hot topic could be criticised, summarises *two* possible reasons for the fossil record. Either geologic processes acted in the past as they do now so that fossil-bearing sediments have accumulated due to *gradual* sedimentation, interrupted by brief and *local* catastrophic eruptions of activity, over hundreds of millions of years. This is the currently accepted academic norm into whose uniformitarian, evolutionary framework all fossil finds are by reflex placed for interpretation. Interpretation or hypothesis that deviates outside this professional norm is scorned o Deviation from this norm is dismissed. However, in the case that Lyell, Darwin and current students of life turn out wrong, a second reason for life's family history, the fossil record, is ascribed to the effects of a *sudden, global* catastrophe whose convulsions died down, despite a tapering series of local, lesser eruptions, over a few hundred or perhaps thousands of years. The former case supports evolutionary biology; the latter does not. Which, in careful, thorough comparison made against the facts, stacks up best?

fig. 22.5

No-one doubts the clear, consistent order of the first appearances of major fossil groups and their successions. However, interpretation of the facts differs (as it often does in science) according to the application of materialistic or holistic mind-set. The former allows unlimited plasticity (*figs*. 5.1 and 22.1), the latter does not. This latter is called The European Model and is a development of The Ecological Zonation Model. It proposes a *sudden*, huge and worldwide geological upheaval, caused by hydrological and tectonic forces, whose effects, with limited recrudescences, gradually diminished over years or decades. The facts are employed to test this scenario. In many cases its explanations are, although anathema to the evolutionary perspective of slow, local erosions and upheavals over hundreds of millions of years, at least as plausible. For example, a definitive book (Darwin's Doubt by Stephen Meyer) contests standard explanations for what was a mystery to Darwin, the Cambrian explosion wherein most of life's body-plans (phyla) seem to have emerged in a very short geological period, one impossibly short for evolution to have been their author. Although meeting strong resistance to his analysis from parties interested to refute him, these have so far been as unsuccessful as the European Model is potentially successful. *It is time that all fossils were revisited, interpreted and taught in the comparative light of both interpretations*. However, for some the uniformitarian interpretation of the fossil record constitutes, however imperfectly, a proof of evolutionary theory. Thus the 'politics' of their party, with over 150 years' academic and economic investment in its own narrative, does not want anyone to even consider any other view! A one-party state of affairs endures.

Note that *devolution* can occur. Loss of alleles by inbreeding and/ or speciation (of allopatric or sympatric variety) certainly occurs; so does atrophy of organs (e.g. functional loss of eyes or wings). Organs or systems may become 'vestigial' but, having lost importance, do not 're-evolve'. Such specialities diminish an original potential; they subtract and, as we saw and will see again (Chapter 23), this limits plasticity. Constructive information is not added. Therefore, the presence of wasted, useless vestigial structures exactly counters evolution's grain.

Novel complexities do not evolve but, like the 'Cambrian explosion', suddenly appear. Life-burst! Over a short period of Cambrian history, a plethora of life-forms, none simple and all complex, suddenly emerged. In 1909 the trowel of Charles Doolittle Walcott revealed one of that period's treasure troves. Similar formations to the Burgess Shale have since been found in Greenland and China. Together with 'simpler', soft-bodied Ediacaran fossils (many of which are very similar to modern organisms) almost every current phylum and as many more now extinct erupted and threw over Darwin's tree. A veritable forest of body-plans took root; trunks of phyla seem to have devolved until, today, we find in less of them (about 35) much more detail in the form of 'leafy' species. As the Dialectic would concur, simplicity devolves into complexity; variation branches from an archetypal theme; nor is extinction anything but opposite of innovation.

For example, there are currently over a million species of arthropod within four phyla; but the shales show more than twenty extra! An acorn worm (*Spartobranchus tenuis*) is very similar (in shape and thus quite possibly genetically) to those that inhabit today's watery depths. And our own chordate shape is present and correct, saluting in a fish-form called *Pikaia* (which seems very similar to extant branchiostomes called lancelets). More advanced than *Pikaia* is *Metaspriggina* found at Marble Canyon near the Burgess Shales. It has a backbone, well-developed eyes, nerves, nostrils, pharyngeal bars and strong fish-like muscles. It is, indeed, effectively an eel-like fish - in the Lower to Middle Cambrian! A similar couple of 'putative chordates' (*Haikoichthys* and *Myllokunmingia*) have been discovered on the other side of the world at another Precambrian site, Chengjiang in China. The almost mammalian features of *Metaspringgina* bring to mind Haldane's 'Precambrian rabbit' - the discovery of which would have definitively persuaded him that the theory of evolution was false.

Another fifteen or more phyla, long since extinct, include such amazing sea animals as *Hallucigenia*, *Wiwaxia* and *Opabinia*. Maybe there survive today less themes and, from each, more variation; less principle in the form of general body plans, more specialities luxuriously devolved from fewer founding shapes. In practice, though, shouldn't evolution keep creating more and not less novelties? While several explanations try to wriggle out of this unfortunate Darwinian reversal perhaps, due to constraint within an incorrect framework, they are all like the full-blown theory itself - only partially right or wrong. **General evolution (as opposed to special variation) is turned upside down. Developmental devolution of a product from its general concept better fits an engineering mode; the evolution of an innovative seed idea past blueprint into specific prototypes might be an inference that better fits the facts**. A single, self-transforming tree becomes, more sensibly transformed, a forest seeded by conceptual archetypes. Palaeontology yields evidence life is designed.

Constructive information suddenly, as with inventions, suddenly appears. To construct fresh body-plans, new hierarchically organised functional systems and their body parts would have required a set of hugely involved instruction manuals. Where did all the information come from? Genome, which specifies protein, cannot by itself create the higher-level, irreducibly complex novelties that life suddenly displayed. No known mechanism except intelligence can specify for purposive complexity derived from integrated working parts such as you find in a machine. Biological forms are very sophisticated 'soft' machines. Nor would intelligence need about ten million years though that is all chance in the Cambrian got!

In other words, 'explosions' mark a top-down fossil record. They uproot Darwinian family trees; they throw over genealogies. In dialectical terms principle precedes the variations on its theme. The main principles were struck. Once minted archetypal forms have given rise to detailed variation down the ages. These original principles have been conserved from then to now but all the while within their boundaries 'lower level' differentiation (call it micro-evolution if you want) rings endless minor changes. Not only the Cambrian but also a series of other 'explosions' have in fact marked the appearance of flowering plants, mammals and so forth. **In combination they exactly contradict the Darwinian theory of gradualism.** Darwin was fully aware of Murchison's discovery of the Cambrian explosion but it left him speechless. He had no answer for its challenge to his theory but to recognise its mystery.

Not only are there 'explosions' of major new arrays of form but once established they remain the same. Stromatolites, bivalves such as *Lingula*, the complex and beautiful geometrical structures of diatoms and *foraminifera*, bacteria, jellyfish, cephalopods, bats, bees, sharks - the list goes on and on. No transformation. You find fossils very similar to living forms or so-called 'living fossils', that is, forms that live now and are identical to fossils. Nor, it needs reminder, does extinction explain evolution any more than scrapping a car its invention.

A top-down record contradicts the course of 'macro-evolution'. Discount highly specialised deep-sea worms called *Poganophora*, recently discovered inhabiting hot-water volcanic vents in the depth of the ocean and for whom fossil evidence is absent and whose past is therefore unknown. Otherwise no animal phylum has arisen since the Cambrian. A large proportion of sub-phyla started then. All classes are found except hermaphroditic bryozoans, insects and, perhaps, vertebrates. Plant phyla (except Precambrian algae) spring from the mid to late Palaeozoic (horsetails, clubmosses, whisk ferns, seed ferns, conifers, cycads and mosses) followed by Jurassic angiosperm (*Archaefructus*) and Cretaceous ferns. This is strange reckoning. Food chains depend on primary producers (photosynthesizers) so it would seem that there was hardly any food for at least a lean 250 million years. There is, furthermore, a complete absence of transitional series between any two phyla. Orders, even families and most genera and species appear suddenly with no evidence of transitional forms. Indeed, the gaps between known orders, classes and phyla are distinct, systematic and large; and, as far as can be judged from fossilised indications of soft tissue, so are those between distinct biological systems (informative, locomotive, respiratory and so on) and the organs, tissues and even cell types that compose them. Moreover, as well as over fifty 'living fossils' (where living specimens are identical with their fossilised forebears), many of the abovementioned taxa show examples of types that have persisted, from Precambrian (e.g. algae, bacteria, jellyfish, sea pens, crinoids) through to Pliocene and Pleistocene with minor variations, that is, essentially the same as they first appeared.

We noted (Chapter 20) and will note (Chapter 25: Which End Is Bio-logic's Head?) the impossible leap from seedless earth to life. Now note especially the phantom leaps, supported not by evidence but speculation, from bacteria to eukaryotic protozoans (such as amoeba) and from protozoans to multi-cellular metazoans - many pre-Cambrian and Cambrian metazoans are of high complexity. These are serious, systematic gaps. They accord with dialectical typology. All Cambrian arthropods, except extinct but marvellously constructed trilobites, fall easily into modern classes. Insects are supposed to have evolved from a sort of worm, *Peripatus*, who can be found exploring leaf-litter in tropical forests. The earliest insect fossil (380 Ma.) is a springtail that, as already noted, almost exactly resembles modern ones. Add to this the abominable mystery of invertebrate to vertebrate evolution. Invertebrates have, in the main, soft inner parts with exoskeletons made of, for example, sugar-based chitin. Vertebrates have soft outer parts and calcareous endoskeletons. Who was it turned inside out and by what steps was the remarkable trick performed? There is no fishy invertebrate. Suspect transformees include acorn worms, sea squirts, echinoderms but, top of the list, a boneless 'fish' called a lancelet (or *Amphioxus*). A hollow nerve chord (notochord) is found in the embryo of all vertebrates but, in lancelets, persists throughout life. Lancelets are supposed, like *Pikaia*, to precede vertebrates; curiously, genome sequencing has revealed that sea squirts (which are sessile but with infant tadpole form) seem nearer to ourselves. However, this volume does not have time to deal in detail with the 'systematic gap' problem (in evolutionary terms) or solution (in archetypal terms). It is not, as Darwin yet again realised, in favour of evolution. *With vast, fresh banks of fossil data collected since his time, it could be argued that the cut-off not the linkage lines are clearer than ever.*

As opposed to *Poganophora* with no fossil counterpart, bizarre fossils such as *Hallucigenia*, graptolites or dinosaurs with no extant representative have been found. Of course, extinctions have occurred but these are the *opposite* of evolution. Conservationists entirely understand that extinction represents not gain but loss of biological information. Such anti-evolution illustrates decrement rather than gain in the whole body of nature. It records the destruction not the origin of life. Destruction, as any vandal knows, is much easier than creation.

And destruction is, by natural selection, evolution's 'method'. Yet frog, bat, springtail and a host of 'living fossils' differ little from the earliest fossils of them found; morphologically conservative priapulid worms haven't really changed in half a billion years. If such 'living fossils' lack transitional forms extinct types also lack Darwin's necessities. *The evolutionary tree is grown from inference but not from evidence.*

Over 100,000 fossil species are known and some, such as diatoms, micraster or ammonites show variation-on-theme (or micro-evolution). Such variations as occur are often trivial, reverse themselves and produce no new types. Indeed, while the complexity of ammonites fluctuated that of graptolites reduced towards simplicity. No partial forms leading to or from any of these types are found, nor even between genera within the type. An ammonite remained an ammonite and a graptolite a graptolite throughout their spans. *Both the distinct formation of major body-plans and the relatively trivial variations on their themes dramatically contradict Darwinian gradualism.* Instead of billions of transitional fossils that should have traced the paths between systematic gaps in phylogenetic classification we find only dotted lines. These lines, drawn by academic rulers, are tracks that acknowledge but at the same time severely challenge the theory of evolution.

Not only organisms as a whole but also organs, cells, organelles and even biochemicals are distinct in structure and function. The rocks show very little such 'soft', physiological material and arguably none at all which (any more than the contemporary array of distinct organisms) supports macro-evolution. Of course, everyone admits gaps in the fossil record; it is what they imply for your theory that grabs disputants. While a lack of such 'links' or transitional forms between groups of living creatures used to be attributed to lack of sufficient fossil material this excuse no longer holds. Indeed a Senior Palaeontologist at the British Museum of Natural History, Colin Patterson, wrote in correspondence with engineer Luther Sunderland that 'there is not one such (transitional) fossil for which one could make a watertight argument'. Is to claim there are no definite missing links to misinterpret what he wrote? Stephen Gould, a committed evolutionist, communist and atheist, added this postscript to our discussion:

"Palaeontologists have paid an exorbitant price for Darwin's argument. We fancy ourselves as the only true students of life's history, yet to preserve our favoured account of evolution by natural selection we view our data as so bad that we almost never see the process we profess to study."

This is serious. **Gradual transition from species to species is the engine of Darwinian argument yet if you seek a strict, detailed and continuous record from the rocks you may as well espouse a promissory faith in godless gaps**. Variation, yes. Transition, no. Distinction is the message ages old. That's just for hard parts we've got millions of; there's hardly any record of soft parts at all. Black box; black holes. Billions of links are missing. No God but gaps a-plenty; a theory of more holes than Gruyère cheese!

'If you can't see the creeps', runs a quip, 'was it jerks that did it?' Of course, evolution *must have* happened - it's just that you can neither see it nor find transition's frozen stills. Therefore Gould himself went on to develop a jerky device called '*punctuated equilibrium*'. In this account bursts of evolutionary change punctuate long periods of none, that is, of the '*stasis*' that we actually perceive. These bursts are, in geological terms, so fast that you might not expect to find any intermediate fossil links which, as you already knew, you have not. The macro-evolution of new body plans happened in the blink of a geological eye. You can pick no bones with that kind of argument. A loophole created in the law of evidence has, on a very serious charge, led to a postponement of the death penalty. The public did not seem to realise this fundamental import of this kind of resuscitation's trick. Gould's plausible if jerky jinks were well received and, with commensurate relief, the theory has revived.

In short, the fossil record shows that distinct phyla and, it may be argued, even orders sprang abruptly; and today the observation is of '*stasis*'. '*Stasis*' means things stay about the same. Animals such as ants, spiders, lice, crustacean ostracods, flies, bats, bees, springtails, jellyfish and frogs not to mention stromatolites, bacteria and many other kinds of organism hardly change for hundreds of millions of years on the standard time-scale. Variation on theme occurs but no evolution. An evolutionist must either accept this anti-evolutionary fact or rely on a hypothetical mechanism for the unobserved transformation of one theme into the next. A complete reliance on theoretical extrapolation in the face of an absence of positive evidence is a great weakness.

Yet, you argue, missing links exist. Books and teachers tell me so. And, indeed, a few slow-motion bones are better than Gould's entirely unseen 'sleight-of-hand'. Then tally ho! Fossil intermediates, missing links, have been keenly hunted for a hundred and fifty years. Many candidates have been proposed. *Coelacanth*, a 'croco-fish/ fishapod' called *Tiktaalik roseae (*that closely resembles an extant, air-breathing eel catfish), *Ichthyostega*, *Seymouria*, *Archaeopteryx* and ape-men figure largely but to deal exhaustively with these and other aspects of the fossil record would take another volume. For a fuller account of systematic gaps (that is, missing linkages) I thus refer the reader to Pitman's '*Adam and Evolution*' or Denton's '*Evolution - A Theory in Crisis*'. For now let's deal briefly, from a *top-down* or typological perspective, with one famous intermediate and one example of supposed phylogeny - *Archaeopteryx* and plants respectively.

Taking a Flier

Not the greatest seer but the greatest flier of all time is acrobatic, stratospheric *PCM* (see Glossary); for now let's sink to mere lower atmosphere and fly below the clouds with birds!

The same two mind-sets are at work. *Top-down* presumes program, logical routines and subroutines: *bottom-up* assumes a phylogenetic jigsaw, a haphazard puzzle of inheritance for which we need find interconnecting pieces - missing links composed of 'transitional' bodies or their parts. In the case of reptile/ bird interconnection the textbook 'fit' appears as *Archaeopteryx.* However *CT* (computer tomography) scans have revealed that 'Archy's' skull and mid-brain, dealing with visual and aural processing, are avian. 'Archy' probably had, like its modern counterparts, better colour vision than humans. The acute sense of balance needed by a bird is reflected in its wholly avian inner ear; and if you flew into trees at speed you'd better have a super autopilot, a hindbrain that coordinates your muscular responses fast - a chunky cerebellum that, like every other bird-brain, this one had. The Bavarian mosaic sports a further range of exclusively avian characteristics, such as feathers and efficient one-way-flow lungs, which will be discussed below. It flew just like the birds outside your window. Now, therefore, palaeontologists at the Natural History Museum concur with what typological *top-downers* had been saying all along - that the organism was capable of powered flight and, as such, a fully-fledged bird. The international Archaeopteryx Conference at Eichstatt in 1982 concurred, agreeing that it was not necessarily the ancestor of modern birds. *Not a transitional form; not a missing link at all.* One hundred and fifty years of textbook repetition, artistic imaginations, exam questions and assumption out of the window.

So how did birds get feathers? *Bottom-up* theory demands reptile/ bird linkage and so, accompanied by a plethora of just-so stories that would do Kipling credit, the search for predecessors has been joined. Because they *must* exist, mustn't they? 'Archy' dramatically encapsulates such evolutionary imperative. Having turned out to be a case of false identity, its non-linkage fulfils typological predictions - because there are neither fossil series leading up to or away from it. Such a showcase for the antithetical interpretation of evidence is worth a further look. Why, from a diagnostic point of view, was the creature first 'seized upon'? How could the elements of diagnosis, taken as a whole, lead to opposite conclusions? And, now that an evolutionary expectation has failed, where do we go from here?

In spite of controversy that has surrounded the authenticity of one of seven fossil specimens of the pigeon-sized organism called *Archaeopteryx* (early bird), it is reasonable to suppose that they represent a genuine article. All hail from the Solnhöfen district of Bavaria. The first was found in 1861 two years after the publication of Darwin's famous treatise and from that providential moment *Archaeopteryx* was proclaimed as an intermediate that demonstrates macro-evolutionary linkage between higher taxonomic groups. Such evidence seemed, from a Darwinian imperative, to prove biological evolution.

Reptilian features of *Archaeopteryx* are teeth and a long tail. No living birds have teeth in sockets but some fossil ones do; a fossil skull of 'Mother Goose' (or *Dasornis emuinus*) was found in London clay bearing a serration of tooth-like bony spikes along the edges of its beak; and fish-eating ducks also have less pronounced tooth-like serrations. In fact the straight, unserrated, peg-like teeth of *Archaeopteryx* are unlike those of reptiles. Some reptiles have teeth or tailbones while others don't. The same applies to fishes, amphibians and mammals. Following the analogy that birds with teeth are more 'primitive', are you more 'primitive' than, say, a spiny anteater?

In an interesting experiment outer tissue was taken from the first and second gill arches of a five-day-old chick embryo; it was combined with inner embryonic tissue (mesenchyme) of a mouse taken in this case from the region where the first molar teeth develop. Normally the enamel layer of a tooth forms from outer tissue and underlying dentine and bone from the mesenchyme - if that tissue can interact with the outer tissue. Chick mesenchyme cannot form dentine so that its outer tissue never gets the chance to form a tooth - but in the experiment where it was artificially exposed to the dentine-producing tissue of mice embryos, it did. And formed teeth! Teeth in a bird! Evolutionists explain this startling fact as 'atavism', a doctrine of reversion. In such a case ancestral genes for teeth are present but suppressed by a mutation. A modification to the genetic program for vertebrate mesenchyme has, in birds, disconnected it from the production of dentine and, therefore, teeth. If, as in the experiment, it can be reconnected it will produce teeth. In other words, birds have the genetic potential for teeth.

In a *top-down* view teeth are a subset, a subroutine associated with the archetype for vertebrate structure and switched on by chemical resonance (here by a trigger present in embryonic mouse mesenchyme but absent in a normal chick). Such subroutine is relatively expressed or suppressed in the context of a particular organism in which the larger 'alimentary' routine from vertebrates is present. In this case the mouse tissue evoked both the morphogenetic subroutine for tooth and unlocked genetic potential present but previously 'blanked' from expression in the bird genome. In short, from a typological point of view atavism is not the accidental 'abruption' of a long deselected past but irregular access to part of an archetypal routine locked for that organism. In this case the trigger was the abnormal 'chime' of specific dental protein from a mouse.

While a long tail is typical of the vertebrate pattern neither modern birds nor humans (except at a foetal stage) have one. It is distinct in *Archaeopteryx*. Also, in most birds wing-claws are suppressed but the young ostrich, rhea, and touraco have them. So do spur-winged plovers, moorhens, coots and young hoatzin, a modern bird as much in the bird books as sparrow, kestrel or any other that shares a large number of features with *Archaeopteryx*. Fliers, both ancient and modern, have fully-formed, asymmetrical feathers of the type that you can, if you visit the wooded rivers and swamps of the Amazon valley, see the clumsy hoatzin flap or, if you look about, see *any* bird enjoying. Feathers are aerodynamic beauties. They are light, the shaft being hollow, and quite different from the scales - not least those that are coded onto avian feet. They grow from capsules called 'pin feathers' and become lifeless when full-grown. A feather from wing or tail is composed of a shaft with branches to right and left called barbules. These overlap neighbouring barbules and are interlocked to each other by little hooks and eyelets. Some large feathers contain over a million barbules with hooks and eyelets to match in perfect order. The feather is useless without this interlocking mechanism; it acts something like an automatic zip fastener whose disturbance preening rearranges. When outstretched in flight, the hooks cause the whole wing-assembly to form a continuous sheet to catch the wind. The feather is a cohesive, elastic and light structure, well designed to function as an air-resistant surface. Sensory receptors record its precise position. Over both wings they trigger continuous variations and fine adjustments in more than ten thousand tiny muscles attached to the bases of the feathers. Feathers are preened using a non-reptilian oil gland; their sheen and coloration are the result of a 'high-tech' arrangement of crystalline proteins. Behold the parts - integrally pre-coded - of a precision instrument of aerospace, unparalleled in design and workmanship by human technology!

To fly you need force air backwards and downwards; wing angle needs be slanted slightly upwards into wind; an aerofoil shape allows air to pass faster over the top surface; and you need a pulley system, functional with associated bones and musculature, to lift the wings prior to dynamic, downward flap. Wings are to fly. That is their *purpose*. What use intermediate but inoperative structures (for wings, feathers, lungs or other specifically avian features). Could running or jumping really lead to flight? Silent owls, nimble flight and 'drag' reduction are just tasters from the range of sky-borne art. 'Aerodynamic Solutions Incorporated' is the name of nature's aviation school. As engineers well understand, there is no patent on a biological design.

Reptilian pterosaurs were capable of complex flight. They had highly aerodynamic hollow bones, a one-way bird-like ventilation system and a large brain. The flocculus is a region of the brain that controls motion and *CAT* scans have shown that its size in *Rhamphorhynchus* exceeds that of birds and bats. Due to a huge elongation of its fourth digit *Quetzlcoatlus* had a wingspan larger than a Spitfire aeroplane. Bones of such extreme aeronauts found in Europe, the Middle East and America have generated 'guestimates' of spans wider than F15 Tomcats fighter planes and maybe, at about 18 metres, even the equivalent of a couple of London double-decker buses placed lengthwise! There is, however, as usual no fossil sequence leading to or from these phenomenal aviators. They spring fully formed from the ground.

You might *wish* feathers were modified reptile scales. However the couple's morphogenetic development, gene composite, protein shapes (including keratin types), filament formation and overall structure are all different. Thus by what steps, precisely, does an expert seriously suggest that flaps of lizard skin turned somehow into feathers; or that hair or filoplumes or the down feathers most chicks (and some adults underneath main feathers) sport for thermal purposes were in some way transformed? Down would rear a 'chick-before-the-adult' syndrome. What did down evolve from anyway? Try scales - perfectly designed for a reptilian way of life. Natural selection would delete any nascent deviation towards (eventually) feathers. Nor, indeed, has any trace of the hundreds of technically required intermediate grades of transformative 'scale-feathers' ever been found. Only the notion that feathers *must* have evolved from scales has led to some remarkably vague, implausible and proof-less stories. What is a half-formed feather? What survival value has it? Scales grow from a different level of skin from feathers. '*There are no known intermediate structures between reptilian scales and feathers*'. Thus says bird expert Alan Feduccia. Prejudicial speculation sometimes dubs frayed, mono-filamentous collagen fibres, known as 'dino-fuzz' and found on dinosaur fossils such as pterosaurs and water-dwelling ichthyosaurs, as so-called 'type-1feathers' or 'proto-feathers'. Your arm's hairs could thus be construed. There is, in fact not fiction, absolutely no evidence for the evolution of feathers.

Not even flightless *birds* have ever given rise to flying ones. How, therefore, did evolving reptiles ever learn to fly? You know you just can't just wish it or hang around a million years or two. '*Ground-uppers*' think some dinosaurs flapped forelegs to help climb tree-trunks or run faster - and according to their wishes 'dino-fuzz' or scales turned into feathers. Flap the forelegs that you call your arms. Flipping flap! If you flapped non-stop for a hundred years would wings have started to appear? Why should a reptile, even if it flapped and flipped its forelegs half each day for double twice a hundred million years (400 Ma. from the Cambrian to Cretaceous eras), actually grow a single feather? Even if it wanted to jump between branches or by a desperate leap of saurian imagination fly why should it then take off?

The fact bears repetition. **There is, despite the wishful hype, no undisputed example of a dinosaur with feathers - or, at most, more than 'type-1' feathers!** There is 'fraying dino-fuzz'; and men whose fuzzy thinking transforms even clucking chickens to a kind of 'living dinosaur'.

Actually, in this instance, *bottom-uppers* and *top-downers* can for once agree. Feduccia points out that huge, saurian hind legs with balancing tails and tiny forelimbs are 'exactly the wrong anatomy for flight' - the opposite of wings and bird-like legs. Still theory makes you have to guess. Rejecting '*ground-up*' flap-to-flight the '*branch-down*' experts claim that gliding turned to flight. Could not small, tree-dwelling crocodilian types have first careened between the branches and thence not into sky but onto ground? Of course, gliding lizards, frogs and snakes never came to feather; nor did colugos, flying fish or ants. The earliest fossil bats, frogs and snakes are just like those today. You claim without a jot of proof that thrust-less gliding and quite different thrusting forms of flight *must have* evolved at many different, unconnected times. How fortunate - except how crashed the half-winged specimens? What happened in advance of aeronautic brains? There are no ancestral gliders or on-the-way-to fliers. *Ground-upper* and *perch-downer* theories each berate the other and, in this, perhaps both are right because they both are wrong. Agreed?

The theory can't take off. It flaps and flounders with a hundred pecks to grasp how flight technology took off; but is the basic premise, *PAM*'s rejection of intelligence for any natural invention, right? There is a satisfactory, secular explanation neither for the origin of feathers nor, indeed, the apparatus that relates to *any* kind of flight. A bird's flying kit not only includes wings but also lungs, legs and bone structure - the aerodynamic wholeness of an integral flight machine. *Archaeopteryx* lacked a keeled breastbone (sternum); but although the wings of some modern birds attach to this 'keel' others attach to the wishbone; Archy had a robust wishbone with which to anchor strong flight muscles. He also sported hollow bones and perching feet with sharp claws such as found in tree-dwelling species. What about the digits of its 'hands'? Are ostrich wings vestigial or nascent? Some suppose such flightless 'primitives' resemble dinosaurs. How, though, could a bird (in whose forelimbs the 'hands'/ wings develop using digits two, three and four) have evolved from a dinosaur (whose hands develop with digits one, two and three)? Anyway, with four claws and one pointing backwards Archy could grab standing room in trees.

Nor are wings, feathers and feet suitable for branch-work all a bird needs. By what steps might cold-blooded reptiles have overcome the metabolic and morphological differences between them and warm-blooded birds? Avian muscular systems also differ from those of other vertebrates. So do skeletal and, where appropriate, hollow, strut-strengthened bones (Archy has 'pneumatised' vertebrae and pelvis). Could an evolutionist simulate the precise steps, with relevant genetic associations, by which the bellow-like lungs of a dinosaur evolved (at the same time as wings, feathers, claws etc.) into a one-way lung-system that even uses lightweight hollow bones and is essential for maximising gaseous exchange and sustaining the respiration levels essential for flight? If none has even speculated on the numerous, successive, gradual intermediates required to obtain such an intricate, flight-special system, why not? *Since without such steps Darwin's case, in this case alone, is well and truly lost.* Add it to the list of mechanisms that must be appointed, working, for a bird to fly.

The vast majority of animal tissue is soft whereas fossils are mostly bone or shell, both of which are quite variable even within a species. So although one can draw subtle inferences there is almost no record of avian body temperature, respiratory system (especially the lungs), circulatory system, digestive system, muscular system, reproductive system, oil glands, nictitating membrane etc. etc. All these irreducibly complex systems incorporate unique features that belong to the biology of birds and no other group. In each case the gaps are distinct, definitive and bridgeless. Nature's miniaturised mastery of flight far surpasses human attempts to copy it. The marvellous micro-engineering of dragonfly, butterfly, bee or other quite different but equally exquisite insect aerobatics, always 100% fit, bear analogous witness. 'Nothing works unless everything works'. What, precisely, are the gradations to insect flight? What incoherent and non-functional intermediates would have to exist between reptile and bird? *In fact the fossil record does not support the theory that reptiles changed into birds let alone any other kind of flight. It hasn't happened but it must have.* So where in earth can one turn?

Aren't birds, like crocodiles, just remnant dinosaurs? No other reptiles fit the avian cage. Don't theropods have wishbones on their bird-like ribs, open hip sockets, an extended pubis and a crescent wrist? Except that they, like every other kind of dinosaur with large tail and hind limbs but short fore-limbs, are the wrong basic anatomy for transformation into a bird. Nevertheless when wish leads fact there must have been a suite of transitional forms. So was the ancestral bird perhaps a small, agile dinosaur called a *coelurosaur*? It certainly wasn't a hoax like the 'Piltdown bird' bought from Chinese merchants in 1999. A fossil of *Archaeoraptor liaoningensis* was exhaustively examined by top experts using the most sophisticated techniques and equipment. After this it was fan-fared by the National Geographic magazine (for whom the evidence for evolution is always 'overwhelming') and modelled as 'Archy's' heir, a long-awaited intermediate between the classes of reptile and bird. Thousands of distinct specimens, either bird *or* reptile, had been plucked from the same strata in Liaoning province, N. E. China. Nevertheless *Archaeoraptor* was deemed to constitute 'proof' of evolution until, worryingly, a Chinese researcher called Xu Xing happened to notice he had seen half of it elsewhere! He was honest enough to reveal the truth.

There is no doubt that politics dominate state-policed China; and that the philosophical basis of its Communism is based upon the creed of materialism. Scientific materialism, whence scientific atheism, is its faith. And this faith depends, as Karl Marx immediately saw, upon the Darwinian Theory of Macroevolution. Evidence for such critical missing link as reptile-bird has to be found, especially if the theory comes under attack. Has science been politicised? Of course, denial would be issued concerning evolution as a biological weapon in such ideological war.

So are there fossil factories in Liaoning? At least the producers of 'feathered' *Archaeoraptor* had glued pieces of dinosaur tail to a fossil bird in order to attract a small fortune for it. They did. The provenance of fossils from China is in doubt. An unscrupulous market riddled with such fakery casts its shadow over palaeontology. What might be the next transitional 'composite' that a disturbingly knowledgeable network of counterfeiters evolves? If experts are deceived, in this case perchance for a short while by 'Piltdown bird' but in the case of Piltdown Man for a long one, the keenly desired 'proof' goes into mass-produced folklore. Missing links are critical to the Darwinian narrative so glimpses keep on seeming to appear, for example *Sinosauropteryx* whose 'furry feathers' are probably the frayed or decomposed reptilian collagen fibres we already met as 'dino-fuzz'. Such fuzz is not evidence for 'proto-feathers' let alone the real McCoy.

Perhaps one does not want to learn. In 2007 vast *Gigantoraptor* found in Mongolia was fan-fared as a pre-avian class of dinosaur. Quick as a flap the Chinese Academy of Science whistled artists whose impressions fleshed the bones and fluffed the tiny arms and knob-like tail with pure imagination in the form of feathers. Hot on its hind-toe, from Liaoning again this time in 2011, a fossil dealer furnished *Xiaotingia zhengi*, a 'feathered dinosaur' that, Xu Xing claims, knocks Archy off its proto-avian perch. What might be expected from a politicised regime whose underpinning 'state' philosophy, scientific atheism, depends for its credentials on the thin, frayed yarn of *PCM*? But 'face' dependent on farcically 'colourful' artistic licence is not confined to Asia; materialism is a globe-wide creed. Who excuses such artistic licence as a jape? Who rates such an exercise in propaganda to prop atheistic state philosophy as just a little harmless, scientific fun?

The right fossils command high prices. Profitable *Archaeoraptor* was in fact foreshadowed by none other than expensive *Archaeopteryx*. The latter was discovered in 1861 just after Darwin's book was published (1859). The place it was found (Eichstatt in Germany) was apparently notorious for the sale of fossil forgeries; and all Archy specimens hail from the same quarry. Ernst Haeckl (see Index) was a rabid evolutionist qualified for such work and active at the time. In 1985 a group of scientists including Fred Hoyle examined the London (1861) specimen carefully - after which the NHM (Natural History Museum) denied some of their findings and permanently withdrew the specimen from public scrutiny. The group declared it was, in the manner of Piltdown Man (and, had they known then, *Archaeoraptor*), a 'bi-bodied' hoax. The strong flying wings seemed attached to a somewhat uncorrelated (yet also well-developed) reptilian body. A mismatch, especially pronounced in the area of the left wing, was identified between the front and back slabs that should, enclosing the fossil, mirror each other. Cement blobs, possibly remnants of an etching or collage process, were also found.

Indeed, three specimens involving wing feathers had each passed the offices of Hermann von Meyer, a friend of the Haberlein family who profited from sales to museums. Since the 'hoax row' two more Archies have come to light. However, fraud or not, the creature may be a Platypus-like mosaic (see below) but is no longer considered a transitional, macro-evolutionary missing-link.

Has Kipling's story-telling school thus been working with redundant 'just-so' scoops? Over the other side of the field a palaeontologist, Sankar Chatterjee, claims to have found a couple of crow-like birds called *Protoavis* with, for example, a sternum and hollow bones, in a 220 million year old Triassic formation near Post, Texas. And bird tracks are found beside dinosaur footprints (e.g. at Deokmyeong-ri, Korea and Xinjiang, China). Indeed, according to Dr. Carl Werner's research, ducks, parrots, owls, sandpipers, cormorants and other kinds of bird have been found in cretaceous layers (but almost never exhibited in museums that he checked). *Archaeopteryx* (Jurassic 150 Ma.) was certainly a bird, albeit one in its own sub-class. It was certainly not, if modern bird-types lived 75 million years earlier, the intermediate reptile/ bird ancestor of our modern flocks. Indeed, 'Archy' seems to have had contemporary pals. Physiological tests on a *Confuciusornis* and *Chaoyangla* have demonstrated the lung capacity and bone structure necessary for sustained flight.

It is not only with birds that evolutionists take fliers. Disputed 'feather fuzz' and anomalous fossil order both fly in the face of the standard, necessary explanatory story of birds from theropods. Cretaceous birds appear as alien to such 'rational' preconception as similarly anomalous finds elsewhere - including human artefacts in coal seams and so on. But such anomalies are dismissed, correctly or otherwise, as 'impossible', 'fraudulent' or hypothetically explained away. So presumably, will soft tissue anomalies such as skin, tendons and ligaments intact in a mummified hadrosaur from North Dakota; from the Gogo Formation in Australia placoderm soft tissue including nerve and (from neck and abdomen) muscle cells; and flexible blood vessels, haem, cells and fibrous protein that may yield to sequencing found by Mary Schweitzer in the thigh-bone of a *T. rex* from Montana (Science 25-3-05 ps. 1952-5). Moreover melanin (a light-absorbant pigment) has been discovered in the skin of mosasaurs, ichthyosaurs and leatherbacks. Such complex molecules are fragile and, left to the elements, rapidly decay. Even in especially favourable conditions of storage the upper limit for survival is estimated at a million years or so; and don't assume that nature will match theory's 'best'. Indeed, C-14 dating on eight dinosaurs yielded an age of between twenty and forty thousand years; such findings, presented by Dr. Thomas Seiler at the 2012 Western Pacific Geophysics Meeting in Singapore, created such a faith issue that the abstract (no. 5) was removed from the conference website! Could such 'anomalies' really be over 65 million years old? How young must a fossil be before, in an orthodox class of geology or lecture in palaeontology, it becomes 'impossible'? In fact dinosaur rocks (Triassic, Jurassic and Cretaceous) have yielded every major plant division living, practically unchanged from that time, today; also every major invertebrate phylum and vertebrates of all kinds including cartilaginous and bony fish, frogs, snakes, lizards modern in appearance; also over 400 species of mammal (Kielan Jaworowska and others; Mammals from the Age of Dinosaurs).

Extinction is not evolution. How scientific is it to deny what you don't want to hear even if the talons of 'an ugly little fact' could rip to shreds and then rewrite your scales of time? Some reason will, you claim, be found to un-crease those rank anomalies. And anyway, who'd swap a lifetime's paradigm for just one ugly fact? Of course, it's ugly, interesting and a fact that - reptile on reptile - catastrophic burial in Gujarat, India froze a snake, jaws open for the pounce, about to eat a hatchling dinosaur (a *Titanosaur*). Fossil theatre; but did some kind of dinosaur dramatically take wing as bird? Birds of a feather may have flocked above but for sure dog-sized and maybe larger mammals walked with dinosaurs. And ate them. Liaoning deposits in China have yielded what is perhaps a 'possum' with a small dinosaur (called a *psittacosaur*) in its stomach (Nature 13-1-05 ps. 116-7 and 149-52). Is the accepted wisdom of mammalian history going to have to be rewritten? A Jurassic but modern-type beaver (called *Castorocauda lutrasimilis*) has been found; so have moles and shrews. What more? The beaver fossil, fur, webbed feet and all, is found in rock pinned with a label written by a man that reads 'my age is guestimated to have passed beneath 160 million cycles of the sun'. That would be 100 million in advance of whales.

Ah, whales! From whales to Wales or *vice versa*! How on earth might a Welsh mouse (named *Morganucodon*) suffer transformation into whales; or larger Chinese rats (*Gobiconodon*) become a dolphin? By what millions of tiny and correctly ordered steps did tiny four-legs radically change its own anatomy and physiology because it came to like to swim? But don't you doubt it! Darwin himself suggested (then retracted) that a bear might, on swimming much more than usual, develop a sea-faring blowhole on top of its head and an anus to the front end! This would exemplify one of myriad fully-working adaptations that an oceanic whale requires (see, for example, the scholarly and comprehensive 'Transformist Illusion' by Douglas Dewar (1957)). How about an infant's special mouth to suck the teat and yet not drown in brine? Ignoring such critical trivialities in a typically brash sale of evolution to Joe Public (perhaps simplified from a contemporary article in Nature) Ian Sample, Science Correspondent, imaginatively tells us how it really must have happened. His sensational story ('From Bambi to Moby-Dick (*sic*): How a small deer evolved into a whale' Guardian 20-12-07) is accompanied by an artist's illustration that defines the latest in a line of missing links - a mousy-looking deer - and a photograph of a real cetacean. Reality to prove a real point! *Indohyus* ('Indian pig') is, in fact, a leaping 'link', a flier composed of fossil fragments of its head and legs broken from Kashmiri, that is, Himalayan rock. As in the case of *Morganucodon* an ear-part does the trick; an auditory bulla fills the gap. It fills up gaps in ocean-happy evolution of the whales and pinnipeds and thus helps throttle damned creationists. It scuppers ID; it seems, triumphantly but quietly, calmly and objectively to squash religion and Intelligent Design! Hear, hear! And (this is the subject's evolutionary fun) there's almost endless scope for building trees of heritage and expert disputation. What contemporary derivatives or other fossil cousins represent, for a mammal that's so different from the others, near ancestry? Sample does not mention 'mesonychid' wolves with hooves or, closer to a cow than sea-cow or a horse, the hippopotamus that's from genetic evidence identified as Moby's closest relative; nor the claim that semi-whales (lacking blowholes or anuses to prow) could make up missing intermediates. One such semi-whale, Pakistan's whale ('*Pakicetus*'), was identified on the strength of the top of a skull, two lower jaw fragments and a few teeth. What's in a name? And yet, despite the hype once 'proving' *Pakicetus* and a crocodilian colleague (*Ambulocetus* named 'the whale that walks') were of cetacean lineage, further finds have relegated the whoppers to a kind of fleet-footed mammal roaming over land. Nor, for sure, has Sample sampled many reasons, detailed by the damned, explaining why a whale could not evolve in tiny, piecemeal ways that unexpectedly pop out of chance. Just to dip a toe, what about specific adaptations needed to allow a baby whale to suckle underwater? *Neo-Darwinists contemptuously ignore profane objections but to be fair the fossil record needs a thorough reinterpretation, part by part, according to both Theories of Intelligence and None, that is, from archetypal and from evolutionary viewpoints.*

Did a strike by meteorite, global volcanism or an ice age wipe out dinosaurs an estimated 65 million years ago? Did such sudden mass extinction open up the niches 'owned' by dinosaurs and thus allow the mice it missed to change into an elephant or whale? Genetic data from the vast majority of mammals that survive today suggests, according to the evolutionary goggle, that they had diversified long before the dinosaurs all fell to ground; and that another 'pulse' of differentiation happened after that extinction. Therefore the curiously selective asteroid must have 'chosen' to leave more than mice immune from its effect. It missed mammals widely spread and well established. And so a shrew from Wales called *Morganucodon*, the previous champion of mammalian stock and thereby your progenitor, has been deposed. A fossil dog and data from the lab have conspired to generate another speculative rewrite and another shift of evolutionary excitement. This may be science but is its quality of 'factual' truth sufficient unto barking 'mad', 'dog-in-the-manger' growling when it comes to other than Darwinian paradigm? Couple this sort of evidence with a general lack of soft parts and you can safely conclude that the origin of birds is still up in the air - perhaps even higher! It therefore remains to paint artistic flesh and some 'emergent' feathers onto convergent skeletons; and also check the provenance of antiques from the Chinese market much more carefully.

Mosaic Subroutines

Top-down programming, main and subroutines called modules - we've met this kind of thinking in Chapters 19 (Archetype), 21 (Energy Metabolism Perchance?) and this chapter (Homology and Origin of Type). It also involves what evolution calls convergence. Convergence, you remember, is the explanation for similar characters thrown up in apparently unrelated organisms (evolution is 'parallel' if you can identify some ancestral trait). There's disputation over such dotty incongruities disrupting a smooth theory of evolution. Mosaic, unexpected distributions cut the branches off a simple tree. Yet we have, in the 21st century, entered a post-mechanistic era. This is the information age - one aligned with Natural Dialectic's approach wherein immaterial information precedes and guides whatever course of action or construction follows. We can rationally employ an explanation for 'convergence' using modules. Indeed, modular thinking is central to the 'upright logic' wherein the very basis of biology is, at every level, information. Principles of organisation rule; their conceptual nature is expressed, using a single universal code, in various material ways. The type of this expression may be analogised with a keyboard on which various, coherent themes of a whole piece are played; or with a mind machine (a computer) and the mechanisms it controls by means of a genomic program. Each organism contains, at its informative core, such modular programs expressed, in turn, as modular machinery (metabolic pathways, organs, systems and so on). **The core routines of modular biology have already been identified. They involve *DNA* storage and operating systems, energy metabolism (at least respiration), various kinds of homeostasis, cell division (with bacterial or mitotic elements), reproductive strategies and the exhaustion of waste products.** These in turn involve a multitude of modular subroutines always, as within any single coherent program (called the organism), co-integrated and co-operative.

Biologists have discovered sequential, structural, functional and systemic molecular modules (in evolution-speak 'convergences'). These recurrences involve similar or identical *DNA* sequences genes, enzymes, other proteins such as photoreceptors, peptidases, protein-binding receptors and so on, and biochemical systems such as bioluminescence, toxin resistance, cases where essential cellular functions are carried out by unrelated enzymes in different bacteria, and the independent origin of *DNA* replication in *Archaea* and *Bacteria*. Genomic studies are throwing up many modular 'convergences' unexpected if chance were accredited with creation.

We'll discuss irreducibly complex mechanisms composed of molecular or larger-scale parts in the next section. For now let's stick with modules we can see.

In the vast majority of organisms that are boneless variation and not transformation to entirely fresh routines is recorded in the rocks. What, moreover, if the weight of fossil bones were not enough to tip the scale? Could it be hard fossils also don't substantiate the innovative, macro-evolutionary aspect of Darwinian paradigm? In the previous section, for example, it was shown that most, if not all, avian modules appeared then hardly changed from ancient times. Beaks, feathers, feet - birds fly their case across the whole phyletic board. *Fossils promoted as transitional forms have been shown to be reptilian, avian or fraudulent.* A typologist's theme is, on the other hand, one of *discontinuity* of homology and type. He espouses those mosaic subroutines and draws inference from the various permutations of the modules he finds scattered through biology. Thus he relates distinction in the shape of body, the configuration of its systems, organs, tissues, even types of cell to the coordinated form and life-style of a given organism. In Chapter 25 emphasis will be laid on informed technology and the consequence of this interpretation of biological form and function. **In this respect an explanation of macro-evolution derived from developmental modules and nick-named evo-devo is, although it eventually fails, much more closely aligned with the dialectical perspective than classical 'random mutation and natural selection'.** By a modular or (as related in Chapter 19: Archetype) archetypal view *Archaeopteryx*, a bird, is itself seen as a *mosaic* that includes predominantly avian but also a couple of reptilian archetypal subroutines. Contemporary overlaps that illustrate mosaic clearly are the lungfish and the duck-billed platypus.

A lungfish has fins, gills, scales and other 100% fishlike characteristics. It also has lungs (one in the Australian variety and a pair in the African and South American), heart and larval stage that are amphibian. There is nothing transitional about its well-formed, well-adapted and well-coordinated parts. Why have its eyes five visual pigments as opposed to your own three? This 'living fossil' is in no way 'primitive'. It is a mosaic, a fish that includes amphibian characteristics. Nowadays nobody suggests that lungfish, any more than mudskippers, are missing links that fathered the first amphibians.

When fresh modules or radically innovative permutations of old modules are found then 'missing links' are necessarily and therefore invariably invoked to fill the evolutionary gaps between. The notion of such links exerts, perforce, a strong hold over the materialist's imagination. Fish like mudskippers can, like some frogs and crabs, climb trees. Eels wriggle across the land but lungfish, mudskippers and eels are definitely fish. And so, for at least 100 million years by standard reckoning, have sturgeons been. Do you remember 'old four legs', the lobe-finned coelacanth once thought to be a 'missing link' and now, like sturgeon, found to be a 'living fossil'? The fish-to-philosopher chain has a fresh though fossilised candidate. This 'link' is found in Arctic Canada. The specimen is a fish with scales, gill arches, fins and fin-rays called *Tiktaalik roseae*. While the upper bones of its fins resemble, like Coelacanth's, the upper bones of a flipper, it is equally true that the 'hand' part is quite different. There are no axial connections to the fins whose rays are not equivalent to finger-bones. While on the subject observe Tasmanian handfish as they 'walk' along the seabed with fins extended from their body ending in a hand-like paddle of 'finger fins'. Are they not, although definitely 100% fish, a 'missing link'? Skeletal attachments occur in all crawling or walking systems of locomotion. Indeed, although the creature's neck is reported as 'mobile' (like an African eel catfish), the vertebral column is poorly ossified and the tail and hind-fins undefined. Some are excited, others less so, about *Tiktaalik*'s 'link' credentials but the 'croco-fish' does, like crocodiles and flatfish, have a flat skull and close-set eyes. Unfortunately fossilised tetrapod footprints found in the Holy Cross mountains of Poland and dated about 20 million years before *Tiktaalik* destroy the claim it is a missing link.

Why should identical, complex functional designs keep re-occurring in an evolutionarily unrelated way by only random 'coding' processes? For example, what about the mosaic distribution of oviparous (egg-laying), ovoviviparous (non-placental, embryo-bearing) and viviparous (placental) methods of gestation? Some sharks are oviparous, other viviparous; all three methods are found in frogs, toads and reptiles; there exist oviparous and viviparous mammals; and while most fish lay eggs some guppies are viviparous. A genus called *Poeciliopsis* even goes so far as to develop, like a mammal, a placenta! How could effective inhibition and promotion generating permutations of the subroutines within a basic program come about by unpredictability? If accidents establish programs how could as yet unfinished, intermediate phases stave the killer, natural selection, off? How, equally, could they survive a gradual, random transformation leading from one integrated system to another kind? The situation's well explained (Chapter 23: Super-codes and Adaptation) by blueprint, plan or routine archetype.

How do placental and marsupial *doppelgangers* (say wolves or moles or tigers of both kinds) each evolve their different reproductive systems - unless from an adapted reproductive archetype, a subroutine, a varied program on mammalian theme? Where do marsupials fit in?

For example, a flying squirrel and sugar glider look almost identical; each is a squirrel with 'wings' stretching from arm to leg. However, the former is placental and the latter marsupial. Their common ancestor was, far back, not a squirrel so how came this 'mosaic form' of so-called convergent evolution? Natural selection only kills unfitness off; it cannot 'design' a thing. Only random mutations create so it must be assumed these two creatures both evolved their 'wings' by chance. Such just-so stories occur ubiquitously as explanations of the way we came to be.

Indeed, was their common ancestor perhaps a monotreme? What has a pouch, a sticky tongue, a pair of ankle spurs and no teeth? The spiny anteater lays eggs, has fur and, although a mammal, behaves like a reptile. Its relative, the platypus, has a bill like a duck that can sieve food from mud or water but also sense electrical fields like a shark or eel. Under water the eye and ear on each side are sealed within a common slit that reopens on surfacing. A young platypus has teeth which in adults are replaced by horny plates. It has large cheek pouches like a monkey or squirrel. Platypus limbs are short like a reptile with five webbed toes. A hollow spur on the inside of its heel connects with a poison-gland making it venomous as a snake. With beaver-like tail it swims like a fish. It is, however, warm-blooded, has mole-like fur and the female gives milk like a mammal. Part of her bird-like cloaca is interpreted by some as a rudimentary uterus. Here eggs form and the creature lays them like a bird. Indeed, one Y of the male's ten (!) sex chromosomes shares genes with sex chromosomes found in birds - but it definitely lacks wings or feathers! This unique mosaic of physiological and anatomical features, some reptilian, some mammalian and some apparently piscine but none in any sense 'transitional', makes it impossible that this animal arose from any particular class of vertebrates. Just as no fossils lead up to (or away from) *Archaeopteryx*, so no intermediates zigzag towards a monotreme. Fossils of the duck-billed platypus are scarce, mostly of poor quality and consist of teeth and, if correctly identified, a few skulls. They suffer from the usual plethora of different but definitely evolutionary interpretation. As such it is for the evolutionary palaeontologist to lay a clear path to the door of this extraordinary specimen - because although prejudicially sub-classed '*proto*therian' it is a mammal. What is the evidence that mammals started out as monotremes? Lungfish, duck-billed platypus and *Archaeopteryx* are each dialectically construed as a mosaic of different subroutines co-engineered into a model fit for work in a particular niche. Both phenotypically and (surprise, surprise!) genetically a platypus demonstrates specific mosaic factors like those found in a number of very different kinds of vertebrate. By what path were they each conserved in order to combine in platypus? What evolutionary path led to this so-called 'missing link'? *And why is Archy any more such a link than the duck-billed platypus*?

But it's not only overlaps or so-called 'missing links' that run with archetypal subroutines. It's all of life on earth. Let us turn from a specific to a general, prolific form of life on which almost all other organisms depend, plants. There is (Chapter 20) no account of the evolution of photosynthetic metabolism. It presently occurs in several kinds of bacteria, photosynthetic protozoans (such as *Euglena*), plants and algae. But if from this lineage you discount algae, as many do, then there exist twelve phyla of multi-cellular plants. The basic kinds appear suddenly in the fossil record and continue as distinct groups with major discontinuities between; and who can give you evidence explaining how a single leaf evolved? *Leaves are green black boxes.*

Within the confines of this book it is not possible to document all the speculative dotted lines that have been pencilled in then rubbed out to accord with one evolutionary imperative or another. Nor is it easy to classify organisms that employ such an amazing variety of structures and strategies in order to thrive in all sorts of environmental locations and conditions. Anomalies abound. However, we can divide plants into two distinct categories - vascular with specialised plumbing for water transport (*Tracheata*) and non-vascular without it (*Bryata*). The former includes horsetails, club mosses, ferns and seed-ferns, cycads, whisk ferns, gingko, pines and flowering plants; the latter mosses, liverworts and hornworts. There seems to exist no internal link between *Bryata* themselves nor between these and *Tracheata*. Did meristems, lacking root caps, first appear in lycopsids like *Asteroxylon mackiei*? Indeed, did meristems, roots leaves a xylem evolve convergently many times over? Whence arose the code from whence these modules are derived?

Within *Tracheata* there is no clear link between plants with spore and with 'more advanced' seed habit. Indeed, while modern ferns employ the former extinct seed-ferns employed the latter. Did *Cooksonia*, with spore-cases on its leafless stems, give rise on the one hand to mini-leaved (microphyllous) *Lycopsida* such as clubmosses and on the other to *Rhynia*, which looks like a modern tropical herb called *Psilotum* and from which ferns, woody ferns, horsetails, gymnosperms and flowering trees and plants are supposed to have followed -*in spite of the fact that no ancestry is known for any of them?* Gymnosperms include such unconnected types as cycad, gingko, conifer and a curious basket of very different genera, *Ephedra*, *Gnetum* and *Welwitschia*. The latter, *Gnetales*, have flower-like reproductive organs with double fertilisation but are not 'missing links'. Therefore did the distinctive, floral subroutine evolve 'convergently' at least twice? There are over a million varieties of plant, as diverse as orchids, trees, bromeliads, cacti and roses, within the class *Angiosperm*. From what and through what steps did their unique reproductive system evolve?

Indeed, which kind of plant came first of all? Did algae lead to mosses, ferns and then the rest? Or is the idea of a series rising from the wet through damp to dry quite wrong? A lotus rises from the mud through water to clear air; and a fossilised 'water lily' was revealed in early Jurassic Chinese sediments with roots, leaves and reproductive organs all present and correct (Science May 2002). Could you not trace evolution of green matter from the river up onto a meadow? That is, space takes the place of time; location (habitat) could then replace phylogeny. Can't you explain the order of our floral chemists, on whose brilliant work our bodies totally depend, in that timeless sort of way?

Did *Angiosperms* (grasses, flowers and leafy trees) bloom last of all? Just as they were budding first another curiosity was born. It is, Charles Darwin once observed, 'an abominable mystery' how the first appearance of the meadow's perfumed beauties blossomed with so many typical divisions (wind or insect pollination, cotyledon number, fruit structure, leaf type and so on) that have persisted almost without any changes to the present day. If they did not evolve (as is in general agreed) from pine-trees, gingkos, cycads or seed-ferns then from what ancestral soil *did* they all germinate? In fact, no fossil intermediates have been found. Evolution's garden has no path at all to flowering plants.

For a typologist plants employ bespoke modules, as do bacteria, fungi and animals, tailored to their unique 'life-style' at molecular, cellular and higher levels of architecture. *A module is a sub-routine or a mosaic element; permutations of sub-units can, in a mosaic way, be integrated into different displays of plants.* They can be deployed in different fittings to produce a range of goods. In this case are modular patterns of development include alternation of the generations between sexual and asexual bodies - a kind of 'hermaphroditic' inclusion of both reproductive kinds of module that it is hard to explain reasonably by evolution. It also includes photosynthetic and, for land-plants, various forms of gamete, spores, pollen, pollinating strategies, seed, cones, flowers, fruit and so on and so forth. Other modules involve vascular transport systems, anti-desiccation and overheating mechanisms such as xylem's bubble-free transpiration stream made of 'stretched water', meristematic growth patterns (including that of cambium), roots, leaves with stomata and a suite of molecular operations including timers, growth regulators, tissue repair routines etc. Each whole plant is a coordinated composite of a selection of these foundational forms (homologies), a mosaic of archetypes for which there is no clear explanation of origin either in principle or in fossil or modern-day practice. You are not surprised if an engineer designs in a top-down direction; he starts with a complete concept and finalises details at the end, that is, the bottom of his conceptual hierarchy. He materialises his dream. Nor do you berate him if he analyses a machine, presuming it to be invented, along the same conceptual lines. In this section only the fossil record for plant origins has briefly been considered. Hear the famous French zoologist and editor of the Larousse Encyclopaedia of the Animal World (1975), Pierre-Paul Grassé, speak in broader terms:

'How can one confidently assert that one mechanism rather than another was at the origin of the creation of the plans of organisation, if one relies entirely upon imagination to find a solution? Our ignorance is so great that we dare not even assign with any accuracy an ancestral stock to the phyla *Protozoa, Arthropoda, Mollusca and Vertebrata.....*From the almost total absence of fossil evidence relative to the origins of the phyla, it follows that an explanation of the mechanism in the creative evolution of fundamental plans is heavily burdened with hypotheses. This should appear as an epigraph to every book on evolution."

Does it? Reasons in science are not always scientific. **A systematic lack of evidence (past and present, in principle and in practice) for unlimited macro-evolutionary plasticity amounts to systematic support for limited typological variation-on-theme. How, therefore, how can a biologist rationally consider a conceptual interpretation of modular biological mechanisms scientifically unacceptable or irrational? And therefore reject, as anathema, the central concept of bio-logical information, the archetype?**

The Origin of Irreducibly Complex Mechanisms

To make a vehicle (as bodies are) are not mechanistic modules joined; or, if in an automated factory, made and joined according to a coded program representing purpose, concept and their plan? Therefore switch Darwin's evolutionary problem round. *His puzzle does not involve an origin of species but the origin and existence of irreducibly complex and coordinated systems.* Of such systems species are simply minor variants, variants by accident (genetic mutation, various kinds of isolation etc.) and, it may and will be argued (Chapters 24 and 25), by super-coded adaptive flexibility and a deliberate design to shuffle the genetic pack - an innate potential to vary on theme.

What constitutes the fundamental discontinuity between life and non-life? *What more obviously, irreducibly complex than the basic unit of biological survival, a cell?* It was explained (Chapters 19 to 21) that the complex, homeostatic mechanisms of such survival could not, except by imagination, have arisen spontaneously. Biology is replete with sub-cellular, cellular and super-cellular hierarchies. It is not just a question of the fossil record. It is one of what we see before our eyes. Let's be clear. *Darwin did not get it wrong just less than half right.* **To repeat, therefore, nobody has any argument (*fig.* 20.1) with his meticulous observations or descriptions of what they amounted to - an origin of race, sub-species and species.** On the other hand, The Primary Axiom of Materialism is that everything is material. Its Primary Corollary (organisms are evolved by random genetic mutations and natural selection) amounts to The Primary Materialistic Axiom of Life but there is zero evidence (see especially Chapter 23) that such axiomatic mindlessness can generate irreducible complexity; and beside such purposive, biological complexity Darwinian variations are trivial.

In Chapter 5 we saw that 'an illusion of design' is key to the case for a Theory of No Intelligence. An atheist will have thoroughly persuaded himself and wish to convince any prospective convert to his faith of the 'truth' that enough time and chance might have (indeed, did) generate this illusion. On the other hand *top-down* reverses *bottom-up*. A dialectical Theory of Intelligence stands against illusion; it stands, with respect to irreducibly complex mechanisms, squarely against an accidental, incremental extrapolation of minor into major change and the development of purposive, encoded form by chance.

In short, while the Dialectic accepts so-called micro-evolutionary variation it rejects the so-called macro-evolution of one type of purposive system or one type of whole organism into another.

What does 'irreducibly complex' mean? It's well worth repeating a definition given in Chapter 21: Nanotechnology. ***'Irreducibly complex'* means, with respect to a system or a mechanism, one that must have several parts each made of the right material, the right positional relationships between parts (which may include quantum elements such as electrical charge) and, in the case of a biological mechanism, the right 'grammatical' instructions encoded in a manner that allows it to access data, perform tasks and repair and reproduce itself. Such criteria must all be met. *The parts must be simultaneously present and built to perform to the standard of minimal function.***

***'Minimal function'* means a mechanism must fulfil its promise to the extent that an efficient, acceptable level of performance is achieved.** Such a performance includes not only the correct shape and interconnection of parts but the correct materials of which they are made. On this basis the biochemicals of life are not haphazard but deliberately machined (via metabolic pathways) as optimal components for tasks such as information storage and transmission (nucleic acids), energy storage and release (especially carbohydrates, lipids and *ATP*) and material biosynthesis (especially protein). What greater intelligence is there, from an evolutionary point of view, than human? Yet no scientist has devised better materials to underwrite biological construction.

Minimal function could, as well as applying to a single mechanism, also involve the integration and cooperation of other similarly irreducible mechanisms to engage a larger whole. For example, a TV is no use by itself. It needs programs, broadcasting centres, a power grid and an audience to be of any use. A torch, a motor-car and thousands of other items in daily use are also irreducibly complex. In order to work their construction demands intelligent conception, planning and manufacture. To 'work' means to correctly and usefully do the right thing. 'Right' means according to a purpose which, if removal or sufficient damage to any part causes the whole system to fail, will not be achieved.

If, within a cell, it could be demonstrated that any serial metabolic process or organelle existed along with its informative infrastructure of coded instructions which could not possibly have been formed by numerous, successive, slight modifications Darwin's theory would, in his own words,

'absolutely break down'. In fact, the more that is discovered about any cell the worse the problem becomes; and with more than one-celled organisms worse still. Up the scale of size tissue, organs and systems are, having developed from a single egg, seed or spore, wholly and dynamically integrated both in form (anatomy) and function (physiology) as a network called a living organism.

The problem for Darwinism is that biology *is* such informed networks. It is composed, in its entirety, of coordinated mechanisms. And modifications to a coordinated network always trigger, by the nature of its integrity's constraint, a chain of balancing adjustments. Necessary compensation needs be swift not gradual, accurate and not random. Specifically effective. Could a chain of random consequence develop new 'ideas' for body plans or the associated mechanisms with fresh ways of functioning? *If minor, mutant transformations that might somehow accumulate into a major novelty are dubbed 'missing links' then at every level they're just that - missing.* When, on the other hand, the factors embedded in any organism's overall homeostasis are taken into account it is hard to find what is not ordered as a well-fitting component that, combined with others, serves a function beyond its own capacity. Nothing is missing or half-missing. Every engineer understands that all parts of an engine must be present, assembled, working and interlinked-in-working before the whole works. Exactly the same applies with greater force to every cell, organ and body in biology. All parts must be similarly collected *but also coded for* before anything will work. The situation (and therefore the evolutionary problem) of irreducible, purposive complexity is everywhere before our eyes.

In neo-Darwinian terms, however, you always need entirely *reducible* complexity; any biological mechanism would have to be constructed in stages each of which has a functional use; and each of which functional parts is first selected for but, with change, not then selected against. Its prior use might disappear but its part always be subsumed and integrated into subsequent, new function. **You only have to look at the genetic code, the code's operations, the mechanism of protein synthesis or a metabolic pathway - let alone any integrated organ or system - to see that in this key respect theory is patent nonsense. A self-confessed and utter failure.** Talk, for example, about plasticity. A fruit fly may, genetically, have limited plasticity but what about its buzzing wings? What about the almost perfect elasticity of a protein called resilin? Resilin is found in flies, fleas and crickets. It can be stretched unstressed to three times its length. However, it not only needs specific genetic code for itself but for its own processing into the specific linkage without which it wouldn't have the twang. Resilin is just a minor part of flies a-flying, fleas a-leaping or the sound of crickets through hot, summer air but it is as important as the rest. Numerous other examples of irreducible complexity include planthopper (*Issus coleopterus*) gears for super-jumping, the human knee-joint, a bird's feather, a cilium, sex, blood clotting, *DNA* replication, metabolic programs, intracellular transport, immune systems, electron transport systems, photosynthesis etc. etc. etc. Each has to originate with the sort of accurate immediacy and highly informed synergy that the top-level management of any business program demands. It is hard, even granted infinite time, to see how an irreducibly complex mechanism could come together piecemeal and gradually. **Time grants no purpose. Yet purpose is the cohesive, creative force.**

Again, Darwin was aware of the problem. *He commented, "If it could be demonstrated that any complex organ existed, which could not possibly have been formed by numerous, successive, slight modifications, my theory would absolutely break down"*. **As will be shown it does, repeatedly, in any attempt to explain the origin as opposed to minor variations of a mechanism.** To underline by reiteration there are, scattered across life's distinctly patterned canvas, billions of 'slight modifications' missing. Design engineers, to the irritation of evolutionists, have catalogued a large quantity of 'black boxes'; they have indicated where these almost ubiquitous missing links (do not) appear. On a scale of deficiency the Dialectic has already indicated a number of serious black holes. Check with the index; there are more. You are not blind to sight and so take one that Darwin noticed in the mirror.

Seeing is Believing What?

Eyes have purpose; purpose is the grant of mind. It is to see. It is for a seer's mind to see. As opposed to mind's inner eye the outer pair is composed of irreducibly accurate, complex biochemical processes and arrangements. So are their related systems of response. These, even in the 'simple' mechanisms that constitute a unicellular euglenoid, are also in all respects exact and exquisitely integrated.

Darwin did not know of this supporting chemistry. He assumed a cell was simple as a blob of jelly; in like manner he presumed to prelude an analysis of sight with something 'simple' as a spot that's sensitive to light. (see below). He ignored the question of how this chemically very complicated spot 'arose' in the first place. Ignorant of molecular dynamics and the different optical mechanisms of light receptors with their associated apparatus he then compiled a list of various sorts of isolated creature eyes in an anatomical order that might suggest a progression from 'simple' to 'complex'. After all, if you ignore the overall potential invisibly pre-programmed into a zygote, such a process might seem how a child develops. Marks for the origin of sight - nil. Marks for the chemistry of optics - nil. Marks for a list - good but proof of theoretical imperative viz. the suggestion one transformed into another - nil. Indeed, of known fossils a couple of types, trilobites and an unusual shrimp called *Anomalocaris*, give evidence of extraordinary and very specific eye structures. Found in the lowest sedimentary (Cambrian) layers of earth's crust, trilobite eyes incorporate the optical principles of Fermat, Abbé's sine law, Snell's laws of refraction and the optics of birefringent crystals. Each eye was made, through a coded developmental pathway, of hundreds of calcite prisms to detect even tiny movements around them. Some developed compound eyes with fewer spherical lenses that corrected for spherical aberration by varying the calcium/ magnesium ratio in different individual crystals of the compound. The eyes of *Anomalocaris* are equally old, equally complex but of quite a different kind; they are like those of modern flies and at least as ingeniously codified; their acuity of vision employed 16000 hexagonally-packed ommatidial lenses per eye! Great complexity was differently but perfectly encoded (see Chapter 25: *Third Stage: Evo-devo*). You'd need foresight for starts 0.1% as smart as these! To see them is to think just what?

Although the basic forms are simple (spot), compound and camera there are at least ten different types of eye. Four kinds of compound and seven kinds of camera eye are each a programmed module said to have evolved independently! The fact is, however, that each different kind of eye compounds Darwin's progressive transformism. For each type you have to presuppose a series of cumulatively lucky modifications each of which confers a biological advantage. Yet over 160 years no such discovery has validated the Darwinian hypothesis that natural selections filter has 'produced the goods'. Indeed, more eye types discovered have worsened the problem.

Neither intermediate transforms nor even theoretical constructs between them are known. On what grounds but philosophical, therefore, are you compelled to swallow that? *Euglena*'s 'simple' light-spot, replete with many complex molecules (e.g. a rhodopsin-transducin complex interactive with phosphodiesterase and other complex chemicals such as 11-*cis*-retinal) all *programmed* into the right place at the right time, suits its needs. No more than today's laboratories does the fossil record show step-by-step (and accidental) 'evolution of metabolic pathways or the cooperation such complex biochemistry.

The compound eye of insects is quite unlike that of an octopus or a human. The latter are similar but quite unlike certain shrimps whose eyes employ a different optical principle in the form of radially-arranged reflecting mirrors in order to focus light. Other shrimps employ lens cylinders, completely different in principle, which smoothly bend the incoming rays of light to focus at a single point. These mechanisms are distinct in principle and practice. You cannot creep one to the other just by drawing pictures. Indeed, the peacock mantis shrimp may pack an incredibly powerful, lightning-fast double-punch with a hammer-claw designed to use catapult and cavitation shock-techniques but it also packs colour vision using twelve or more kinds of receptor (you and I have only three) including four in the ultra-violet range. It can distinguish polarised light and even has special filters that can compensate for a differing absorption of light by the waters it swims in. Of course, it also needs a visual system integrated with a nervous network plugged into its other parts. Is a shrimp's mind just the same as brain? Anyway, it has to have a mind to see. How did different shrimps evolve entirely separate ways to peek about? In fact, it seems that different visual systems deployed in arthropods must either have evolved discretely multiple times; or, in the serial course of transition from their primal eye, numerous intermediate species must have lost function (i.e. gone blind for indeterminate periods) before reaching a new, operative model; *or*, heaven forbid, modules based on different principles of optics must have been, like different patents for the same job in vehicle design and thence production, engineered separately. Which story's yours?

The wonder of all this doesn't stop at shrimps. The facets of a lobster's eye are graph-paper square, shiny-sided tubes that reflect rather than refract light. What selective advantage might the blindness while evolving in between two equally adequate systems obtain? Explain the 'eye' of brittlestars; or the uniquely independent swivel and telephoto apparatus in the peepers of a lizard, the chameleon. What about bifocal lenses, each with two retinas and focal planes, of sunburst diving beetle larvae? And animals can demonstrate great optical prowess. An archer fish can, using its binocular vision along with complex calculation and reaction times twice as fast as yours, adjust its angle of aim to account for refraction and spit water arrows accurately to bring down insects from overhanging vegetation. Does a squirt shoot evolution down? There's always a rebuttal based on some notion or another but at least its eyes are normal and it is brain that does the trick. 'Old four-eyes', *Anableps*, has two eyes functioning as four. An eye is divided into upper and lower parts by a line of epithelial tissue; corneas, pupils and retinas are separate but its ovoid lens can focus two images separately, one from above and one near-sighted from below. There is absolutely no evidence how such an effectively bifocal lens with concomitant parts and brain structure sufficient to decode yet integrate two images at once could have evolved gradually through all the non-operative stages. Can you even map a hypothetical path? But aye-aye, sir, evolution *must have* made the eye-eye seer. Even more incredible, the process wrought a deep-sea brownsnout spookfish. Spookfish also have four eyes but, uniquely, mirrors too. On each side of its head a conventional eye gathers light from above while a so-called 'add-on', using a guanine-crystal mirror to better focus light, observes the darknesss down below. What other creature uses mirrors to perform its miracles? What unfound, unimagined path led accidently to this trick? Can chance really fumble into irreducible complexity? Demonstration of soft tissue sequence is not on the cards and so, lacking fossilised or any other kind of evidence, perhaps softly-softly patter is the only way to let magicians conjure their illusion. The recourse is, of course, 'must have' bluster and 'just-so' entanglement in tales. Not just for spookfish but in every spookily intelligent appearance of design. In fact, eyes all have every coded part in place before they work; they're irreducibly complex in parts and arrangement. In fiction, small, genetic steps have wrought, by chance, over twenty different designs for perfect focus leading to interpretation by the psycho-chemistry behind. Only close and systematic scrutiny of vague, just-so genetic stories can redress the bias that they purposely create. Such 'exercise in truth' is necessary to re-balance evolution's '*must have*' mantra.

Let's revert, for a moment, to the primal phylum of coelenterates (*Cnidaria* called hydroids, jellyfish, sea anemones and so on). Here one primitive champions all 'higher' sophisticates. The box jellyfish (sea wasp) is known not only for its uniquely coelenterate and highly complex, toxic nematocysts but its 24 eyes of 4 different kinds arranged around its 'box' in 4 sets of 6. No brain supports these eyes; rather a nerve ring with 4 nodes in parallel sustains their integrated operation. In each eye-set are found two pit and two slit eyes; an almond-sized upper lens eye that is fixed looking upwards in order, it is thought, to navigate by means of out-of-water vegetation; and a lower lens eye the size of a tennis ball that is structurally like yours (a vertebrate) or a squid's (a cephalopod). This eye has a retina, cornea and lens by means of which it can focus. The set amounts to a true mosaic. How did such a primitive, Precambrian sort of organism slip upon such profligate sophistication? What lines lead to four kinds of eye, from 'primitive' to 'advanced' in one container made of jelly? For sight box jellyfish outbox all animals; but competition does not mean that their superior 'evolution' has eliminated every other kind of jellyfish!

Living in perpetual darkness, many organisms *lose*, that is, *devolve* their sight - even their eyes. In others sight's improved. Bats, dolphins and birds (oil birds and cave swiftlets) can echolocate - but an oil bird has the most light-sensitive eyes of any vertebrate. Its rods are stacked in tiers as in some kinds of deep-sea fish. Why no connection by phylogeny? Is there any feature evolution can't explain? You can frame the theory just-so so it's never wrong! Let a 'solution' gloss otherwise impenetrable tricks! 'Convergence' sounds like adaptation-on-a-theme; its answer therefore superficially appeals. Did *anableps* and snoutfish convergently evolve, like whales and seals their fishy shapes, four eyes? How, like various kinds of wings for flight, did different sorts of eye convergently evolve sight many times in different sorts of creature? It must mean smart mindlessness of some oblivious atoms that, for some non-reason but because of force of circumstance, 'target' a function and its critical effect. The gang of atoms 'needs it' so it gets, in this case, sight. Can't you see? From molecules, somehow suborned into unnatural acts of creativity, 'emerge' a range of unrelated versions of a non-blind mindedness. Is it not obvious to anyone who'd see that a complicated form of atoms can 'improve its own survival' if such form includes 'an eye'? How useful, on the other hand, is something at a quarter of the way to seeing, half way into empty darkness that, it's anybody's guess, might lead to evolutionary illumination? What use is pre-sight's blindness so that natural selection deigns to grant it leave of execution? How did blindness 'know' that its particular sort of answer was, perhaps, incipient and 'struggle' to obtain its prize? Or perhaps it didn't - one sort of sight or other tottered round in total darkness till suddenly, in a conceptual pool of light, the dimness lit up and the problem 'cured' itself. Isn't, as an engineer would say, straight design a simpler, economical and, except that it is philosophically taboo, obvious answer? Can human mind subvert or wilfully ignore a real possibility?

How does any sense arise from pre-sense, that is, non-sense? What prescient sort of sense is non-sense then supposed to make? What's the message? What, if there are only unresponsive atoms, is its meaning? A car that does not start is not 'mutating' to a 'super-car'; it is not transforming to a space-ship or whatever passes for advancement in the automotive field. The reverse. It only needs a minor fault in system to waylay the whole construction's purpose. Opticians know how easily degraded sight can be; conversely all the factors on the way to 'sight' must 'click' before the final step brings sight in sight. Not half an eye but 99% (including its conceptual operation linked to all surrounding systems and the central processor that sees) must be constructed prior to natural selection having any chance at all. This applies to simple light-spots just as well as other very different kinds of eye.

You can throw a bull's eye on the bench. Eyes don't see. They need a context like a nervous system with the body that it serves. Even then it's mind that sees what eyes convey. How did the *experience* of seeing come to be? Whence did all the minds grow, long time passing?

In other words, an optical light-reflector like an eye is no more than a mirror without context - context that ramifies into the wholeness of its owner. Maybe the psychology of a cyclopean 'planimal' (or plant/ animal) such as *Euglena* can be reduced to chemical reflexes but even this, and even if it is the whole truth, has to amount to an appropriate *system* of response. If (Chapters 13 - 16) you subscribe to the atomic 'mind is brain-meat' attitude, then chemical reflexes are the whole truth and consciousness is essentially euglenoid. But if you think there is more to it then you have to link eyes to a 'mind behind'. The question then becomes not what but who sees. Who sees with what? No doubt that brain is 'close technology', a medium transparent to its user, an intimate of information that allows its owner life on earth. The problem is that eyes are not alone. **The brain is part of eye and eye is part of brain** (for whose intrauterine development check Chapter 25: Growing Up). In an eye over 100 million receptor cells, one million ganglion cells and a million nerve cells have to be correctly assembled and connected with 100,000 million human brain cells. Hang on! That is not enough. You need a blood supply, digestive tract and ventilation system - the working lot. In fact all parts are irreducibly a part of one another. Eyes exist only in the complex context of a whole. Every part of every body, bound to its instruction manual, genome, is the same. It takes a lot to raise a smile! No wonder Darwin shuddered, blinked and winked at what the mirror saw! How, even when you've stripped the whole machine to parts, is a focused image then translated into something seen? What *is* sensitivity? What precise biochemical steps led to and now grant photosensitivity its relationship with living organisms? Who is the subjective seer? *Is not the clear intention that you, endowed with powerful equipment, see*?

How many million individual parts, each made of molecules developed and disposed precisely, are integrated into one of your most lovely eyes! A complex, ordered chain of interactions marshals precise optical components made from a number of tissues and with exact geometrical tolerances with regard to size and shape into assemblies with equally exact geometrical tolerances with regard to orientation and position so that an embryonic vertebrate will, in integral concert with a whole new locus of complexity, its brain, grow to see!!! How did the stages in this detailed, comprehensive 'flow-sheet' happen gradually? How was the end product (if it is that), brains, eyes, user bodies and the users come to be? 'Simple to complex, systems must have been conceived by chance' is, as good as it gets, the evolutionary explanation? But even euglenoid 'sight' has, allied with its own sense of self, a minimal but still complex optical flowchart below whose level nothing works. These considerations do not vanish with the waving of an intellectual wand, spellbinding words of comfort or the magic mantra that 'convergence it must be'. 'Archetypal program', something engineered to work within constraints according to a purpose from the stage of prototype, could easily replace 'convergence'. This is because natural selection does not like what does not work. If a feature does not work at first it bans a second try. Nor, by definition, does it mitigate the sentence just because 'convergence' wants a second chance.

What do you see in an eyeless fly? Arthur Koestler related how one set of induced mutations on the long-suffering fruit fly (*Drosophila melanogaster*) led to a stock of eyeless flies which, if inbred, within a few generations re-engendered normal eyes. He did not accept the standard explanation that other members of the gene complex had been *randomly* reshuffled in such a way as to deputise for the deficient parts. He believed, as an evolutionist, that such resilience was due to 'evolutionary self-repair'. What is that except a guess? What would he have seen in brainless weeds? Experiments with cress plants (*Arabidopsis thaliana*) at Purdue University, Indiana showed a 10% reversion to normality by double recessive mutant offspring. How could the weediest of weeds have acquired genetic information from other than their parents? If they did not then what elastic could vibrate recession back to health? Indeed, why not perceive an example of the strong, 'elastic', conservative tendency to return, wherever possible, to a working, archetypal function - to the 'green-print' for a weed or, for a hexapod, a 'blue-print' called *vis medicatrix*? And if there is an archetype then there's deliberate design.

So seeing is believing what? Do you believe blind fortune came to see or was sight seen before it saw?

Look Further, Penetrate Deeper

The account of sight could be repeated for auditory, olfactory and all other sensory systems. In what order, for example, did trilobite senses evolve? Why should organs of sensation co-evolve? Did hearing precede sight or touch; or were they planned the way that they appear in mind - as one?

What use is any pre-operational structure and, if selection can't eliminate such uselessness, why don't we find organisms these days in the process of evolving 'primitive' contraptions that might lead to fresh discoveries as capitally venturous as flight or sight? The same problematic account applies equally to circulatory systems, regulatory systems, developmental and metabolic systems etc. etc. etc. Thousands of examples and arguments can and have been adduced. In all cases the bottom line is the same. You can take, within an overall archetypal program, a permutation of subroutines with their correlated code and consequential chemistry. You can stretch or shrink their concerted elasticity to some extent but not past the point where either system function is destroyed or new function (as opposed to adaptation of an old) appears. **In other words, Darwinism is a theory about tinkering; it is about variation to, but <u>not</u> the origin of, irreducibly complex biological mechanisms and their incorporation in the body of an organism as a whole.** From the cell upwards it explains the *origin* of nothing. This may leave atheistic materialism in dire straits but it in no way affects biological forms *per se*, that is, as we find them without hypothetical reference to origins. And, as such, it does not affect the study or application (e.g. medical or agricultural) of biology.

Indeed, in terms of medicine is Darwinism's record in the clinic good? Indifferent or negative? It has been arguably proposed that 'adaptations' such as fever, salivation, sneezing, vomiting and natural repugnance each evolved to avoid or repel disease. Mutant micro-organisms battle with our health but variation is the issue and not evolution. And who'd try to 'speed up evolution' by irradiating organisms to increase mutation rates? What about the pain eugenics and phrenology have caused; or diseases wrought by promiscuity or wars to gain supremacy? Materialism's major prop degrades the sanctity of life to chemistry so that murder by abortion, production of chimaeras or half human animals might hardly seem amiss. We'll check later how Darwinism may or may not have advanced the Hippocratic craft; and how it's moral tincture is a recipe for suffering down the line.

A sweeping, Elizabethan judgment that most genomic *DNA* was '^jnu~n*k' is under serious review. Similarly, use has been found for all but a few organs that Victorian biology pronounced 'vestigial'. An engineer queries redundancy. He applies the reasonable benchmark of purpose to any machine or component part. What, for example, do you see in a pineal gland (Chapter 17), gill slits (Chapter 25), tonsils, an appendix or any other of approximately 180 so-called vestiges? Is the appendix, for example, a functional part of a general scheme, modified for more emphasis in some and less in other vertebrate alimentary tracts? It now thought to be a 'ferment-cum-storage-vat' or, if you like, a 'safe haven' for useful intestinal bacteria; and, having lymphoid tissue densely populated with lymphocytes, it is also a site for the initiation of immune response. It this respect its function is, like the thymus gland, spleen, adenoids and tonsils, to combat infection. Do you remember the eyeless fly? Will you still interpret 'lesser expression' or 'redundancy' (as in blind cave-fish or wingless flies) as 'vestigial' - or reasonable? How might 'vestiges', which are a form of anti-evolutionary *demise* from previous usefulness, account for the *origin* of a reasonable, purposeful and useful part? *Demise never furnished vestige of an evidence for origin.* And, if you account organs as 'reasonable', then why is an 'unreasonable' residue (such as male nipples) still parroted in modern textbooks as 'vestigial evidence' for evolution - which process fails to evoke *useful, new and nascent* as opposed to useless, past-working-era or degenerate parts?

Huge lists of 'evolutionary impossibilities' compiled by 'unbelievers' simply irritate those who believe that evolution is a fact. These 'unbelievers' simply appeal to a God-of-the-gaps instead of gaps in scientific 'by-mutation' theory. But, of course, the problem is not one of ignorance as much as philosophy. We're arguing, as discussed in Chapter 5, over quotation marks. For a materialist 'design' is a bad dream, caged between inverted commas, that threatens to include waking up. On the other hand information (with its subsets of purpose, will and imagination) is metaphysically awake. Although the holist has no problem with calling a design a design his inference is no more strictly scientific than a materialist's.

"*Design is evident*", writes Behe (Darwin's Black Box p. 194), "*when a number of separate, interacting components are ordered in such a way as to accomplish a function beyond the individual components*". You can, therefore, infer design wherever you look in biological machinery. It is omnipresent. Such inference requires neither that its originator(s) be identified nor that the mode of original creation be explained. *It is ridiculous to dismiss deliberate biological design as less scientific than design-free evolution.* Would the study of locomotive design be less scientific if the engineers existed? If you study electronics or computer technology do you not accept design, do you not assume a mind-behind without having to meet the man or learn his historical biography?

The fact of the matter is that the essence of all cellular life is regulation. Its homeostatic control tends to resist accidental, foreign or new and unregulated intrusions. Just as all doctors realise the homeostatic, elastic self-healing properties that allow us to bounce back to our vital norm, so life would militate to vigorously eliminate new and uncoordinated developing systems whose structure and function were as yet incomplete, unclear and alien.

How accurately do evolutionists plot the small accretions by which, say, metabolic programs or osmoregulatory systems 'arise'? If something does not work it has no selective value. If an enzyme is newly created before the others in its pathway how should it manage to wait for them? Indeed, why should it wait if there is no irreducible system planned in which it will play a critical part? *It is blindingly easy to take a series of organisms or organs that already function optimally, even sufficiently, and place them in order of apparent connection or complexity. This is quite different from explaining, at biochemical, developmental and phenotypic levels, how one working part was turned accidentally, by numerous successive small modifications, into another; how its functionally related neighbours changed, at every level, accordingly; or how one creature transformed into quite another. We do not see this happening.* Nor, despite more than a century of hype to the contrary, does the fossil record witness it. *Indeed, such are the staggeringly complicated biochemical mechanisms underlying the gross anatomy and physiology described by palaeontologists that some biologists, such as Behe, believe traditional appeals to fossil record and gross biology are irrelevant.* It is time, they say, to stop papering over cracks with the rhetoric of just-so stories and give a precise molecular explanation of how the relevant steps in biology's well-intentioned homeostatic equilibrations, information systems and irreducible mechanisms arose. We can therefore re-state the important conclusion of the previous section. **If biological mechanisms lack intention, that is, came about unintentionally as an evolutionist claims, then it requires that they be shown reducible to or, conversely, gradually constructible (or transformable) from a specified sequence of steps. It must at the same time be demonstrated that each one of the series of completely or almost completely useless mechanisms produced by this haphazard sequence of steps has selective value. In other words, a minor adaptation of an already functional mechanism is not the point. What must be shown is the steps by which, from its previous absence, a purposeless, functionless or dysfunctional, disruptive shortcoming evolves into a purposeful sub-system; and how such sub-systems are integrated hierarchically into 'over-systems' - whole organisms that we find in fossil or in extant form.**

A sweeping hand, even professorial or Royal Societal, is insufficient. To date no serious attempt has been made to verify any such claims. Nor are emergent steps identified sufficient to predict and artificially hasten the arrival of novel functions and their mechanisms. For example, embryonic examples of new concepts for flight, sight or locomotive engineering are not observed.

At a trivial level variation called Darwinian micro-evolution definitely occurs. Mutations (Chapter 23) undoubtedly affect biological life. Like any bug mutations can cause system failure. Or they can cause anomalous, unintended outcomes that amount to minor but tolerable variations to a particular program as a whole. Just as a car with a broken number-plate might avoid prosecution so a bio-bug might even, under specific circumstances, sustain a mutation 'beneficial' or, arguably, 'an improvement' to its operation. Such mutations are, though, incoherent. We'll see later how their chaotic presence never could add up to sense, to systems and coherent bio-logic. Genomes are, although much smaller than a speck of dust, exceedingly dynamic, complex and integrated controls. It is a just-so story many have uncritically listened to that swathes of bugs-in-systems build conceptual programs nested logically together in a whole new application suite. Bugs cannot transform one complex system into others; they cannot 'macro-evolve' one integrated network into quite a different one. You might *hope* they could but ask any programmer. If they did he'd be replaced by generators with the name of 'Bugs Galore'.

Indeed, close inspection reveals the problem is worse than you imagine. Not only are so-called 'beneficial mutations', the *sine qua non* of materialism, rare but, in accordance with the 'rules' of chance and entropy, information previously carried by a mutant sequence is soon lost. Eyeless, wingless flies or antibiotic resistant bacteria may thrive in very specific environments but also suffer loss of form and function that would in general circumstances ruin them. It needs information and not errors of typography or damage to create instruction manuals like a genome. Nor can you build controls by chance or have them work without a purpose. It might be argued that even a 'simple' bacterial genome contains an amount of information comparable to what you need to build a Boeing. Could undetected, 'nearly-neutral' changes accumulate until emergence of some pristine form and function? Certain genes are key to the regulation of animal development: could some such trigger as mutation of a cluster of master developmental genes spring a 'hopeful monster' or a 'missing link'? *The mechanisms of trivial variation include genetic mutation but, as again we'll come to see more clearly, the way in which completely fresh body-plans, organs, metabolic programs and so on 'macro-evolve' is just not known.* **There is no Darwinian answer to life's myriad, major innovations.**

DNA is no more creative than a database or manual. It is passive information. Although it is referenced with respect to protein, neither party is other than a passive agent that executes the clear targets, programs and strategies of development and maturity. They are what the timely production of building materials are to an architect's archetypal plan. Moreover protein by itself contains neither the architecture for three-dimensional morphogenesis (creating the shapes of body parts) nor the conceptual information that substantiates bio-cybernetic controls, functional organs, integrated systems and whole bodies.

Passive script derives from thinking. Command and control derive from the intentions of a mind. The passive information of *DNA* and protein helps implement the creative plan which first informed them. *DNA* is *not* programmed to create new programs. Everyone agrees that mutations are unintentional, unpredictable accidents that, as the study of genetic disease unambiguously demonstrates, generate *faulty* material. Nobody thinks of exposing people to mutagens (agents that cause mutations) in order to create super-humans; nor do doctors claim genetic illness and deformity is part of human evolution. *Bugs in programs do not improve let alone create them.* Efficient engineers do not design machines by tinkering with freaks of nature. Likewise an architect accepts commissions as a whole and then constructs accordingly. **Thus it is by strange and self-contradictory logic that genetic mutations in eukaryotes are currently identified as causes of disease, abnormality or, at neutral best, wreak no effective havoc but in the past are seen as the creators of all metabolic and other sub-cellular processes,**

cells, tissues, organs, systems and coherent forms of life! Yet this *is* evolution!

Nor is there any evidence whatsoever that some divine shadow intangibly tinkers with genes or inscrutably 'guides' them through an apparently accidental process of progress called evolution. *The issue is not one of survival but original arrival of the fittest.* Accidents and aging occur as they did to Sir Nigel Gresley's locomotive machines; but these are secondary effects and no more considered a primary, integral facet of original design than mutations, age and illness are of roaring health. However, excising Sir Nigel and trying to explain a sophisticated design by way of wear, dirt and dents is as challenging as trying to explain biological designs (which are all sophisticated in terms of codified functions and corresponding structures) in terms of an unpredictable series of mutations. The *only* reason for such a severe 'counter-intuitive' sprain of common sense is to satisfy a theory which (although it might be illogical and wrong) *must*, in terms of scientific materialism, be right! Ouch! In fact, a sturdy, practical design resists and counteracts breakdown and abrasion. Likewise the resilience, leeway and flexibility of archetypal program buffers as far as possible against 'the arrows and the slingshots of outrageous fortune'. So, if a theory proposed that bugs and random buffets were the *cause* of biological design, would you judge that it passed muster?

Such a *top-down* interpretation of events is neither irrational nor, in fact, untrue. Nevertheless it is resisted because it eliminates the only mechanism by which a topsy-turvy, *bottom-up* interpretation might, on a macro-evolutionary scale, originate, create and, without meaning to, 'progress' - if indeed you are a biological improvement upon a germ. The lack of any accidental method of injecting information into biological systems will be discussed further (Chapter 23); **but if random mutations cannot do this job it means that, currently, Darwin's theory lacks any modus operandi; and that, unless and until a viable alternative mechanism is found to account for the origin of biology's conceptual programs and the efficient apparatus of their expression, The Primary Corollary of Materialism (the neo-Darwinian theory of evolution) is hopelessly dead in the water.**

23. Part Right, Overall Wrong

The theory of evolution evolved to stage 2. Now it's time to take stock of genetics and, specifically, genetic mutation. **The mechanism of mutation was coupled with natural selection in the hope that it might provide sufficient material for biological transformation.** Was this hope misplaced or not?

Language of the Genes

To inform is to instruct; and to communicate a message. Information involves symbols, signals, messages and therefore purpose. Its purpose is its meaning. Mind is a communication centre. It receives impressions and creates/ formulates response. It also communicates with body through, primarily, a medium called brain; brain is, therefore, also an information exchange. The single, critical difference is that mind is active, semantic and metaphysical and brain, life's highest form of matter, is passive, guided and physical.

Information is transmitted according to rules understood by sender and receiver. Language isn't just chaotic sounds; it engages layers of complexity - letters, words and sentences to libraries of meaning. Nor is life but a pretty pile of molecules; it is, layered with the same complexity, an assemblage of instructions. These are transmitted using passive instruments (such as electronic pulses, screens, chemicals etc.) but are initiated according to a formulation that has been agreed - such as between the *DNA* and protein or between cooperative proteins in a cell. The stimulus for meaningful transmission, which carriers such as chemicals cannot create, is the achievement of a goal. Thus information, even if it's passed between uncomprehending objects, is inherently purposeful. You could well claim genetic meaning *is* protein; and the message sent from *DNA* is the construction/ maintenance of its body. **Here the problem for materialism is, always and unrelentingly, the origin of biological information.**

Genetic information is not simply language-like. It *is* a language the meaning of whose symbols (genes) is protein. Its alphabet is grouped hierarchically in higher order units (codons, genes, chromosomes and the genome); its sequences are specifically related to operational purpose (see also the section: Non-Protein-Coding *DNA*); its syntax includes reading frames, punctuation (e.g. start and stop codons) and multiplex expressions of control; and the integrity of such micro-hierarchical library (the genome) is itself a fraction of layer upon layer of integrated chemical complexity that constitutes, by the transformation of genetic story, into physical reality. A unicellular organism is as irreducibly complex as an instruction manual combined with its mechanical end product. This manual is translated into intentional phrases (proteins) which are in turn automatically, in a computational manner, combined into the meaningful, dynamic story called a living organism. The beauty of this story is astounding. Its engine runs on prefabricated tracks; its locomotion is programmed into every cell; at all levels and at every moment sophisticated feedback communicates between genome and its physical context. Since language is a way of organising any complex system, 'language' of the genes is *not* an analogy; it is not 'as if'. It involves all the characteristics of real linguistic symbolism. Such language neither accidentally nor even accidentally-on-purpose just emerges out of chaos. It does not endlessly accumulate rules without reason. It emerges out of mind *in order to* accomplish purposes. This applies to nature and to nature's human beings - unless you deny either of them any immaterial, metaphysical dimension! If you do then (since bodies are not candidates for chance but bio-logic, not best inference for evolution but design) realistically and rationally why?

It is, therefore, entirely accurate and appropriate to see a genome as an instruction manual. The Greek word '*lego*' means 'I say or reckon'. Its nominal derivative, '*logos*', means word, speech, language and reason. The bio-logic of the language of genes is that they code for the components and orderly construction of a plan. Thus code, program, grammar, language, plan and blueprint are words universally, habitually and literally used to describe the pre-eminent, informative aspect of a genome. All of them are indissolubly associated, except by a strain of philosophical necessity in the case of scientific genetics, with intelligence, communication of purpose and conceptual design.

Check 'code' in the Glossary. Its genetic morphology is found in choice of bases, nucleotides and triplet code; by three-base codons choice of 'right' amino acids is agreed. Recognisable bases/ codons are arranged syntactically in sequence; such instructions, including punctuation, can then be interpreted by a receiver called a ribosome. Thus body, by translation, can obtain the instrument to carry out a job. Information's (*sat*) potential has been (*raj*) set to work and ended up with (*tam*) desired conclusion; code, by prior agreement, translates into specific protein. Intention. Comprehension. *What could be clearer than intelligence of advanced language and command in genes*?

With protein different parts are made and work as one. The *purpose* of different parts of a body is obvious to everyone including children less than two years old. It is also clear that lack of purpose is an absentee - a few hypothetically vestigial organs and the status of 'junk' code owe more credence to ignorance than, as we learn more, actuality. They make the point. If the structure and function of a whole machine is purposely integrated, irreducibly complex and smoothly operating how could its plan be otherwise? How can a single whole, an organism that transcends in operation all its 'separate' parts, be instructed from a manual without a plan by an extensive jumble-up of accidents? *Code is, absolutely certainly, an instrument of sense not nonsense. Language is most definitely a store of order.* Is chance a bank of grammar and its schooling sensible? A genome, called by some the book of life, is a program carried on the 'ink and paper' of a chemical called *DNA*.

Can you imagine that a program or a book is built up letter by cipher, each giddy one whirled in by chance? How, in its creation, does the senseless book know ciphers from non-ciphers; or which one to select? Will not wrong ones far outweigh the right ones? And, where one was right in one context, the rightness be outweighed and wronged by multiple erroneous replacements? The book of life, a genome, is not a simple, linear script. Its multi-dimensional, poly-functional text involves data compression, feedback loops, branches, various kinds of regulator, contextual modulations and interaction with external circumstance in the forms of cytoplasm, body and the outside world. It incorporates a host of dynamic tricks that integrate with many layers of control. All these operate exactly tuned into their purposes and overriding purpose - to engineer the construction and maintenance, that is, the survival of bio-logical forms. As you know, not everybody buys the story that, in combination with mutations, natural selection is empowered to senselessly perform Life's Great Information Trick.

Your own instruction manual holds two sets of over 6 billion letters filed in 46 chromosomes containing perhaps 21000 genes - all wrapped in the central part of a bag of molecules much smaller than a speck of dust! To say that you are simply dust of earth is a gross understatement. Your genome is a very special kind of information panel. What working manual is unintentionally extracted out of garble? 'Noise' or 'junk' is, if not eliminated, scrupulously cut to minimum. Indeed, the code of genes is organised to minimise mistakes in copying and thus cut randomness, as far as possible, to nil.

Billions of jumbled letters might be meaningless but a small quantity correctly ordered mean a lot. Is it quantity of *DNA* or quality of information counts? Is an evolutionary 'climb' from simple jumble into complex order or a 'drop' from purpose into detail of expression what we actually see? Don't let detail blind you to the sense of purpose that invigorates your chemistry of life.

If you think information is a by-product of 'noise' you would expect to find that low-grade, 'infantile' information evolved to high-quality, complex, 'noise-less' information. You would also expect to find that quantity of *DNA* correlated with this build-up. It does not. Check the jumble for yourself. **Research has signally failed to demonstrate that amount of *DNA*, number of genes or chromosome count demonstrates evolution. On the contrary, the quantity of *DNA* in genomes throughout the living world is dotted in a mosaic way, called the C-paradox, that is systematically anomalous and therefore puzzling for Darwin's theory.** This paradox notes that extensive differences in the quantity of *DNA* in a cell appear to bear little relation to an organism's complexity, size or type. For example, there are many fishes, plants (especially ferns) and amphibians with far more *DNA* than you; some insects and molluscs have more; some sea urchins have about the same; the flower, *Paris japonica*, has, at nearly 150 billion letters where you have about 6 billion, the largest genome of any plant; the axolotl has perhaps more than 40 billion and a marbled lungfish 140 billion; but while single-celled *Amoeba proteus* has 290 billion amoeboid *Polychaos dubium* has the most of any known organism at 670 billion! There exists a 1000-fold variation in plant amounts, 3000-fold in animals and perhaps 300,000-fold in protistans; but, as we've begun to see, the differences do not accord with evolutionary theme.

When it comes to genes there are two-thirds as many in a fruit fly (~15000) as a man but nearly 50% more in a grape (about 30500). There's about 22000 genes in you, 23000 in a mouse, 25000 in thale cress, 16500 in a chicken, 18500 in a silkworm, 21000 in an acorn worm, 31000 in a water flea, 19000 in a soya bean pathogen, 32500 in a polychaete worm, 23000 in a sea urchin, 39600 in single-celled paramecium, 28500 in the lancelet, 45000 in a eucalyptus tree, 20500 in wheat stem rust, 27500 in sorghum, 58000 in a variety of rice, 63300 in maize and so the apparently anomalous, mosaic and non-evolutionary list goes on and on. Never mind! There *must* be some way evolution happened.

It is not the number of pages in a book that counts; it is different information. Thus you might understand the complete jumble of chromosome numbers. There is no evolutionary sense to be made in any kingdoms. For example, chimpanzees, gorillas and orang utans have (along with wild tobacco, potatoes and the pretty corncockle flower (*Agrostemna*) a count of 48 chromosomes; you have 46, as does the sable antelope and an arctic clubmoss called *Diphasium complanatum*; Rhesus monkeys only 42! A horse has 64, carp 104, pigeon 80, hedgehog 88, a shrimp from the Persian Gulf 90, earthworm 36, yeast 32, rat 42, aquatic rat 92, horsetails 216 and the adder's tongue fern over 1200! No evolutionary numerical connections are discovered in plants, animals, fungi or protozoa. What did you expect? Ask any academic which is more potent, information-wise. Is it a slim volume of concentrated knowledge succinctly and coherently expressed or thousands of pages of ill-defined waffle? From which do you infer intelligent authority? *Quality of information and not quantity of DNA is, as the Dialectic certainly confirms, the real criterion by which to judge a manual's source.*

Of course reprints and damage due to wear and tear mean errors can accumulate until corruption degrades sense to garble. There exist preordained editors, elaborately designed systems to buffer 'information shocks', proteins that pick up error and eliminate them. This way a robust quality control ensures, as far as possible, that pristine information is preserved. This buffer of stability occasionally fails. In that case what is the failsafe for redressing the intrusion of an error on a genome? There is, bar lucky counter-accident, none. Entropy of information rolls you down the hill towards chaos. Who relies on further bugs to oust bugs from computer programs or on further accidents to mend an accident to car? You might expect, therefore, that accidents called mutations progressively degrade function into dysfunction and its consequence, disease. An inexorable 'rusting' of information would wear the genome down into extinction. Erasure. This we see. Except that humans haven't reached the point of meltdown yet. The whole direction is, however, precisely the reverse of evolution.

If you run counter to the cosmic tide you have to vest your faith in 'serendipity'. You must presume that 'beneficial accidents' accumulated in such number and such sequence as would generate your manuals and their manufactured consequence, the organisms of this earth. This presumption is none other than The Primary Corollary of Materialism. You need therefore ask whether 'beneficial mutations' (*BMs*) actually work your trick. Can natural selection really make up books of life on earth? Or is naturalism's Corollary, and thus its Axiom, what's actually made up? Let's check how historically, against the grain of cosmic law, you might have been persuaded that the theory of evolution's true. How did wind and rain and minerals compose you? How did they come to speak your tongue? But language is the child of message, meaning and communication. Its mark is teleology and not the mindlessly erratic jump of chance. This brings us to *the second stage in development of an inverted logic.* Can mutation, when it's added to selection, make the grade?

Second Stage: The Creator (Mutation)

Innovation and Mutation

Meaning is gained by concentration of a mind, by focus of attention or the aim of intelligence. Such reflection better understands; such creativity innovates. Call it negentropy of information. Intended stimulus generates coherent, purposive systems; it invents means more or less efficient to serve ends. The origin of codes and machines falls into this category; so does the origin of types of cell, tissue, organ, system and organism.

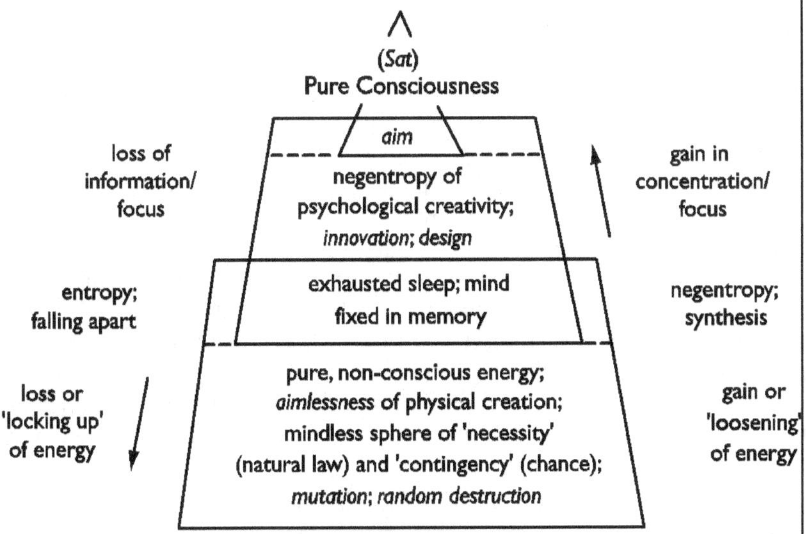

Creation and creations; active creativity of mind and passive, physical arrangements.

Meaningless movement (such as a sea of wind or waves) is derived from unintentional stimulus. Energy rings mindless changes; natural forces make and break up aimless aggregations. And the application of aimless energy to informed constructions (like frost or rain weathering a bicycle) corrodes their sense; random interruptions disrupt or jumble code. Engineers all know the elements will savage systems. Untamed energy destroys the purpose in designs and pulverises their coherence out of sight. For order it is terminal. Are biological mutations all disintegrators? Errors? Isn't each and every one an information-loser with, for code and meaning, deadly tendency?

fig. 23.1 (see also 2.10)

A scientific G.O.D, no less! A Generator of Diversity! Since natural selection fails entirely to innovate the second step in evolutionary lore was to propose a mechanism of creation. Life is built, it is explained, by *DNA*'s genetic chemistry. Therefore, the saying goes, swirling chemicals create ancestral books of life; then novelties derive from random changes to these books, that is, to genomes. The name for unpredictability like this (a whimsical non-deity, the G.O.D's left hand at work) is mutation. **Thus materialistic faith is vested in a core of unpredictability, a central lack of reason.** Randomness is, of course, the complete opposite of the best springboard for generating complex orders, orders that are specified in code, that is, in genetic language. However, grant this palsied hand of chance manipulative power that's far beyond the bounds of mathematical improbability - the random making of a meaning written down in *DNA*. Grant, in other words, *all* the chemicals you want and, on top of cosmic generosity, a script that codes and, exempt from any inconvenient intrusion, synthesises 'useful' protein. The shaker simply changes name. Unsteady incubation in a pool of chemicals is now, at life's level, recognised as a 'mutation'. Genetic accidents (this Chapter: The Creator (Mutation), Entropy of Information and Evolution in Action?) aren't common and, arguably, as damaging or idly neutral as typesetting errors to a text. Still, let accumulate a series of a thousand harmless, even 'useful' but incalculably unlikely mutations. Next somehow select and keep the 'useful' or the neutral ones. Let them loiter, waiting for each other in a subliminal kind of anticipation and, for no reason, a 'correct' assembly. Such stalwarts are recruits conscripted to the struggle, front-line troops that shoulder evolution. Natural selection could not operate on bio-molecules alone but now, armed with the imagination of a fully-fledged organism, life's cavalcade might start to roll - erratically. Mutations, clumsy, silly chaps of chance, disrupt their own frail house of cards; for every brick perchance 'in place' another twenty are disposed to knock it off its perch. So constructs that hopeless cause, a blind and senseless builder, so fabricates a mindless watchmaker. Is evolution theory not, at every stage, a multi-level house of cards?

Never mind irrationality. Do mutations that can benefit constructively or add non-lethal novelty exist? Perhaps not or, depending on perspective and interpretation, perhaps *BMs* (see Index: 'mutation beneficial') parade in quantity. Evolution needs but demonstrate them! More than neutral, show us suites of selectively favourable sports (as Darwin called unexpected physiological oddities). They may never or just rarely manifest - but wildly, bountifully contradicting the most generous reality allow that *half* of all mutations are, bestowing 'sacrament' that blesses by enhancement, 'beneficent' to life. The probability of two hundred successive, specifically successful mutations is perhaps one in 10^{60}. It is no use having one sub-suite in one life form and another in another. 'Successes' must be concentrated on the inside of a single membrane, a sac too small to see. Now surpass even such statistical generosity. Fossil bacteria have been dated, on current assumptions, at about 10^{17} seconds (about four billion years). Allow an earth more than double that age. Allow a billion proto-biotypes mutating away once a second over every one of earth's 10^{13} square metres. Thereby obtain 10^{43} mutational kicks at goal -with an outrageously generous success rate of 50% and, by a miracle, conjure fifty relevant proteins simultaneously onto some microscopic stage. Have you a cell? Very far from it. Even the 'simplest' bacterium known (*Mycoplasma H39*) contains about 600 varieties of protein along with collateral *DNA*, replicating mechanisms, synthesising mechanisms and a host of other specific bio-molecules.

Mind-numbing improbability is frequently observed to jerk an appeal to intergalactic flotsam of simple organic chemicals and 'unimaginable' mists of time. If simple chemicals are like a 3-foot bar that you can easily jump over, can you extrapolate and leap up 15000 feet? Can you really leap from brick-shaped lumps of mud into a city; or is the flight from clouds of space-dust one to cuckoo-land? In other words, is anything possible and nothing impossible without necessity of mind? For evolution's sake you need to think so and thus naturalists (not Naturalists) insistently assert the 'power' of time and chance. They constitute a veil behind which you issue licence to suspect that anything can happen. Just like theology the combination is a free domain. Hypothesise without responsibility, without testability. Why, in such unscientific climate but in the matter of design, is it 'rational' to exclude consideration of the fine, deliberate and powerful anti-randomiser, rational mind; and thereby blot out mind-store, memory and notion of an archetypal memory? On which side does lack of reason, madness, lie?

If you left ink and paper, would it in a billion years turn out a simple book? Don't, therefore, with a reflex obfuscation then ascribe creation to the possibility that chance can, even given tons of time, weigh in with any kind of knock-out answer or, in fact, do anything particular at all. Don't hop behind that battered veil of age. Age hides not beauty but decay, not exuberant construction and complexity but degradation into slack simplicity. Who could persuade you that, behind the handkerchief of time, some conjuring occurred whereby the atoms systematically and, as it vaguely were, inevitably gathered (and still gather) information? Why should this randomly collected specificity achieve complexity sufficient to outstrip all man's achievements and create a 'primitive' first cell? In fact greatly increased time, an essential ingredient in the evolutionary scenario, does *not* increase the probability for an improbable, complex and meaningful end product. The reverse. The amount of energy accidentally tied up in special complexities would just as accidentally tend to be released into low-energy simplicities. Not only this but the chances of additional disorder increase. The overall chances of finding things randomised - which is the antithesis of code and negentropic, purposeful construction - increase. *Indeed, in no reproducible conditions do highly complex yet different molecules such as DNA and polypeptides ever form outside a cell - let alone work in perfect tandem.* The true direction of chemical spontaneity is one that reduces every novelty to breakdown and, in the engineering case, a heap of rust. Time's arrow is, where vehicles of purpose and/or complexity are concerned, one of decline and demise. Neither cell nor your own body are exceptions. Time is irrelevant to consideration of the origin of a working system - especially one as intricate and sophisticated that, by self-reproducing, swims salmon-like upstream against extinction's chute! *Time, however long or short, is not the issue. Intelligence, however, is.* You might, in this context, recognise as fantasy imaginative, perhaps-this-way-perhaps-that accounts of the naturalistic, long-time-lucky 'emergence' of a robot (like a cell) - an inverted fantasy with massive odds against its oddity. Indeed, you might suppose that you have kicked a 'scientific' habit and dispelled the influence of Darwinism's intellectual drug prescribed against the nightmare wake-up of that 'pseudo-scientific' possibility - intelligence! Mind shrivels chance. Intelligence can smack the odds from zero up to unity of certainty in just the twinkling of an eye.

	tam/ raj	*Sat*
	imbalance	*Balance*
	system	*Governance*
	variation-on-theme	*Archetype*
	mutability	*Immutability*
	informed	*Informant*
↓	*tam*	*raj* ↑
	material tendency	*life tendency*
	incoherence/ randomness	*coherence*
	disintegration	*integration*
	accident	*purpose*
	cacophony	*harmony*
	error	*accuracy*
	babble/ 'noise'	*code*
	meaningless	*meaningful*
	incorrect information	*correct information*
	genetic/ functional loss	*genetic/functional stability*
	damage/ disease	*health*
	disorderly recombination	*orderly recombination*
	mutation/ misprint	*facsimile*

It is insisted that they pioneer life's systems; yet again, therefore, we ask, 'Are random mutations able to create fresh function, structure, organ and coherent operations? Can they, as opposed to simply vary

or by chance cause to adapt, *innovate* metabolism and molecular machineries? Or gradually 'design' a range of organs, systems, body-plans and build the whole of life's great family up? In short, is mutation as creator 'fit for purpose'?

Lamarck's idea of the inheritance of acquired characteristics (*IAC*) proposed that the combined desire and will-power to create a particular biological characteristic (such as a muscular body or a long neck) was sufficient for the new trait to become heritable. Mind could evolve body. Darwin, who did not know the actual mechanism of information transfer, believed in gradual change and like most of his contemporaries accepted Lamarck's now long discredited notion. And he certainly believed that 'sports' (abruptly occurring abnormalities) were *not* the cause of evolution. It was also believed that hereditary material from each parent was 'blended' in their offspring. In 1867 Fleeming Jenkin, a Professor of Engineering at Edinburgh University, published a review of the Origin of Species in which he demonstrated that any character, including a useful chance variation, would, by blending, become diluted to 50% of the first generation stock. This dilution would continue until the novelty disappeared long before natural selection could get a grip on it. It was simple and obvious. No new species could arise from chance variations by the accepted 'law of ancestral inheritance by blending'. Darwin, reportedly believing this was the most valuable criticism anyone had made of his work, never saw how to get over the difficulty.

In fact Gregor Mendel had already unwittingly invalidated Jenkins' argument. At the same time he had correctly ruled out the 'pangenetic' forces invoked by Darwin to substantiate his theory of evolution. This is because he showed that discrete units of inheritance, now called genes, were transmitted unchanged from generation to generation. The theory allows any number of permutations of these 'atoms of heredity' to occur but the genetic information itself remains stable. Jenkins' argument is invalidated but a new spectre looms. If genes were immutable whence could novelties arise? Whence came the information for new forms, species, genera, kingdoms and domains? Of Mendel's paper, published in 1866, Sir Ronald Fisher (himself a founding father of modern genetics) wrote:

"The journal in which it was published was not a very obscure one and seems to have been widely distributed. In London it was received, according to Bateson [ed. another founder geneticist], by the Linnaean Society. The paper itself is not obscure or difficult to understand; on the contrary, the new ideas are explained most simply and amply illustrated by the experimental results."

Why, therefore, were Mendel's unintentionally anti-Darwinian papers neglected until 1901? In this year Hugo de Vries, having discovered a mechanism to circumvent their argument, rediscovered them (along with Carl Correns and Erich von Tshermak). This mechanism was mutation. There are conspiracy theorists who, noting Fisher's observation, point to Thomas Huxley, Darwin's 'bulldog', and his 'X' Club. The 'gay and conspiratorial' 'X' Club was a group of nine men, strongly evolutionist in character, who were notified of association by the delivery of an algebraic formula and always dined immediately before meetings of the Royal Society. Indeed, it included officers of the Ethnological, Geological, Linnaean and Royal Societies. If Mendel's theory had preceded the Origin of Species, maybe Darwin would never have published. Certainly, if Huxley had read Mendel he would have seen trouble. Is it possible his 'X' club saw the implications, refused to confront them and suppressed wide circulation of Mendel's anti-evolutionary discovery?

Anyway, 'mutations' seemed the way out of Mendelian trammels. You might, if you found 'beneficial' ones, at least conceptually transform one type of body to another in the way that evolution needs. *However mutations, now known to be the result of copying errors or environmental damage to chromosomes by chemicals, radiation and so on, are random changes to cellular DNA. In the germ-line they are rare (one for between one and ten million copies) and, as the product of chance rather than coherent, systematic events, they are analogous to 'noise' in information theory*. They show entropy of information; they degrade and not improve a body's reservoir of structural command.

In other words, mutations corrupt code that previously worked. They have an effect on genetic code like random keying over the word-processed text of a composition. Aberrant or meaningless code is a contradiction in terms. **Mutations scramble code**. They render it informatively less effective codeless. You might liken them to bumps, knocks, wear and tear on a new machine. It is the position and type of an accident that determines whether the vehicle proceeds, needs repair or is scrapped. *There is universal agreement that at least the vast majority of mutations are deleterious.* On the basis of the instruction manual analogy all are so. Such accidents (deleterious mutations or *DM*s for short) cause recessive or mutant as opposed to dominant, healthy, wild type characteristics. They are spoken of in terms of burden, load, pathology, corruption and disease. Over 4500 genetic diseases, many showing various levels of effect due to different mutations, are already known to afflict the human race. They affect every organ and tissue type. Cancer is one of them. How can misinformation be a mechanism for evolutionary advance? Is 'blind watchmaker' an entirely inappropriate description of a process that operates more like a 'blind terrorist' who takes a hammer to already created watches? Since, however, chance is central to that view then such 'blind terrorism' is central to a 'rational' explanation for the cogs that compose you!

Is the only difference between your own brain and a fruit fly's copying errors? Did a series of random misprints raise us both from some bacterium? *An evolutionary viewpoint vests mutation with creative power sufficient to produce, after the imperative but tiny business of initial cells, all other forms of life.* It is no less than evolution's 'anti-deity', our Great Creator. Mutations must have fathered, except for the first 'founder cell', all life on earth. They must have generated, perchance, the coherent complexity found in every working, and therefore by definition purposeful, system. Is your mind a working system? If you can reason, then the origin of reason is supposed to have been chance. **Accident, which is not logical, is presumed to have gradually generated biological systems of great ingenuity, irreducible complexity and purpose**. Illogic makes logic. The origin of reason is, according to such line of reason, irrationality. **This, issue of materialism's seed, is highly illogical.** It pervades evolution that might itself, on this basis, be classified as a mutant theory; and is, as already noted, the sole yet fundamental breed of irrationality that radiates influence from the very heart of modern science.

It is certainly 'counter-intuitive' that corruption of genetic material, the manual for building life forms, should lead by trial and error to increasingly informed complexity. In fact the phrase 'counter-intuitive' is commonly used as code for 'incredible but necessary because a straightforward answer does not fit materialism's paradigm'.

It is also a specious but common form of sophistry to claim Darwinian evolution is *not* down to chance. Mutations, the originators of its changes, *are* by chance. Death, called natural selection, merely weeds this crop of chance but, like weeding, does not make a thing. It must await mutation's 'creativity'; such novelty occurs, you claim, by chance. **But the origin of information, not constraint of mutant babble, is the issue.** How can randomness, however many times its lawless one-armed bandit rolls, spring programs; can bugs innovate or chancy babble script the book of life?

Jacques Monod, a molecular biologist and atheist, succinctly captured the depressing irrationality, the lack of common sense. He wrote, "...chance *alone* is at the source of every innovation, of all creation in the biosphere. Pure chance, absolutely free but blind, at the very root of the stupendous edifice of evolution..." Yet Sir Ronald Fisher showed (which demonstration will be reinforced below) that such single, 'beneficial' mutations as Monod needs in great and consecutive, cumulative number would be wiped out by random effects before they could spread. This contradicts Darwin's intuition of numerous, gradual slight variations accumulating like pennies into pounds. Any lone penny, even if its so-called 'selective value' were high, would most likely be eliminated from the genetic pool so that no pounds could accrue. Indeed, negative-value accidents are hugely more likely and would combine to neutralise any possibility of positively-valued randomisations. Lone 'pennies' would not become established. This is before you consider the necessity that hundreds of slight, simultaneous 'value-adders' would have to convene, randomly, to inform such coherent biological improvements as left any single one of them with a chance of more than short-term survival. What's more, even so-called 'positive' mutations are like bugs that destroy rather than add to information in a system. But although the issue of evolution by chance mutation is fundamental to the 'scientific' case, it is shrouded in assumptions, arguments, guesswork and unknowns; it is void, except for trivial variations, of real evidence. Some, for example, Hoyle and Wickramasinghe ('Why Neo-Darwinism Does Not Work' presents 'a simple and decisive disproof'), Spetner ('Not by Chance') and Sanford (Genetic Entropy and The Mystery of The Genome) have detailed mathematical and/or accompanying philosophical refutations of it. Odds of 10^{50} or less are, like winning the National Lottery five times in a row, normally considered 'impossible'. Odds thousands, even millions of times slighter than 'impossible' are calculated against the progress of evolution in Darwinian style. Hopeful? Incredibly. Hopeless? Yes. If they have wrangled properly then the sprain such style exerts on common sense is relieved. The teachers of its error are dismissed. Chance is more than simply shrouded. It is dead and buried. No chance left.

In short, therefore, the choice is simple. **Either life's a hierarchically devolved expression of information - a product of prior intelligence; or, no intelligence involved, it is evolved by accidental, chemical reactions and random rearrangements, called mutations, to its 'message'**. Mindlessly self-organised, the 'messenger' then mindlessly controls the rest.

Thus, *bottom-up,* tail-to-head the theory of evolution reflects the trial-and-error, simple-to-complex process of empirical learning - but only superficially because there's never any mind with its intelligence. Infertility and death both teach and learn in this brute, ever-mindless school.

From a *top-down* line of view the materialist version of events twists head-to-tail to tail-to-head. It systematically re-explains each object and event in back-to-front reverse. It reverses *top-down* Truth of things and the *top-down* Logic of embodied lives. Indeed, stripped naked of fine words, the *bottom-up* philosophy of creation by chance appears incoherent. Evolution is, in this respect, a tale of discord, meaninglessness, darkness in descent towards loneliness, a story based on error, violence and death. It is a hooded kind of nihilism since, in the end analysis, what *is* disorderly, upsetting chance? It's a non-entity. Reason's vacuum. It's nothing meaningful or in the least intelligent. Yet mutations, evolution's driving force, arise from it.

Such a theory is, on this basis, the reverse of life's truth. The information coordinate (*fig.* 0.7iii) has been zeroised. Design has been replaced by 'white noise' from which, out of any number of possibilities, natural selection fixes some 'immediately adaptive' snips. Step by random, short-term snip long-range, non-random symphonies (coherent, harmonic systems) are grown. Just give 'noise' time to clatter, bang and drone enough. And, of course, a musical ear to choose and splice 'nice fragments'. What a song and dance about some nonsense.

As opposed to the Master Analogy of Natural Dialectic, Music, evolution's lack of rhythm grinds in fits, rasps and unpunctual starts. It is not a pretty insight that you contemplate. As opposed to the work of a Grand Master who employs geometrical form and the golden mean, how chaotic its 'modern' form of art appears. No wonder Monod, having adopted such cacophonic philosophy, was depressed by its implications. *It is, truly and literally, a Theory of No Intelligence.* Could observation demonstrate its falsity?

The top-down cure for No Intelligence is to, simply, twist the twist! Let's twist again! The head-to-tail logic of Natural Dialectic reverses the twist in Monod's tail-to-head illogicality. It twists the topsy-turvy, back-to-front priority of matter round. *It wakes from the dream and returns to reality. Mind, not matter, is the informative, innovative, creative principle.* The more powerful is a mind the greater the coherence, ingenuity and purposeful influence of its creations.

In which case how can a theory of evolution, composed of truths (mutation and natural selection) and untruths (abiogenesis and macro-evolution) have seduced so many and such intelligent minds? How can what seems *bottom-up* so right appear, *top-down*, so wrong? The answer is *if* 'errors' of the right sort by the billion gradually accumulate to 'rightness' then the theory is not wrong. If millions of tiny bugs can accidentally improve the database of computerised homeostasis that girds a cell or a biological body and, from that body, press on to evolve new suites of application program with the grand, macro-evolutionary power to transform life on earth then Darwin's theory certainly is right. Can they? Is such, blind, gradual progression actually, rationally the case?

'Beneficial mutations' are for genetics what 'missing links' are for evolutionary palaeontology - critical. **Not just The Primary Corollary but The Primary Axiom of Materialism and its whole panoply of philosophical, political and sociological speculation, not to mention the paradigm of modern science, hangs on this slender thread.** *The whole of secular academy depends, for its verbose existence, on this evanescent gleam of hope, a key to unlock all of life's diversity that's called a BM ('positive' or 'beneficial mutation'). Can BMs rise to such occasion?*

This is the moment, if there sounds confusion, to make two clear distinctions. Firstly, some biologists do not define a *BM* as one that necessarily confers advantage. It simply persists in a population. There's no evolutionary advantage whatsoever there! Secondly, the definition of a *BM* is couched in terms of information. It does not include circumstances when (as illustrated in the section 'Evolution in Action?') an organism is able to survive better in a locally changed circumstance because a *loss* of genetic information endows it with advantage. Transformation (macro-evolution) into higher forms of life is an entirely different matter. *In evolutionary terms a BM must reverse genomic entropy. Each must add to the instruction manual in a way that leads, by accumulation and coordination with many subsequent benefactors, to systematic advance in the form of novel features integrated into every operational level of an organism.* Can suites of accidental *BM*s generate such irreducible complexity? If they can't or don't they're back to *DM*s - not *BM*s at all.

What is the actual evidence, against the tide of genetic normality and ability to recover from 'error', for positive humbug? We'll see but meanwhile there is, some philosophical biologists retort, positive evidence for a little humbug in design as well. What 'Great Designer' would, they pseudo-theologically demand to know, dream up such fault-ridden inefficiency as organisms show? Either He is less than perfect or 'botched jobs' preach a good apology for evolution. Don't they showcase imperfection, signal 'could do better' and thereby embody scope for progress? If you but look you're sure to find examples everywhere. An evolutionist is keen to spotlight 'imperfections' he perceives in biological constructions as evidence of dim work or a proof of non-existent creativity (you call the absence 'chance'). What about laborious pain in childbirth, vaginal opening next to anus or, in men, urethra slung across a sperm tract? What about a panda's thumb, orchids, human retinal construction and so forth? These are cited as examples that offend the peculiar aesthetic (and therefore metaphysical) sensibility of their critics. How can one claim 'Intelligence' has cranked and cobbled life's misshapen cogs and cases? "If God designed a machine," wrote Stephen Gould, "He would not have used a collection of parts designed for other purposes." Biological design involves stupidities! So, you see, it couldn't have been Him then!

In fact, some arguments from imperfection (e.g. eye) have already been found 'arguments from ignorance'. They were in Darwin's time and will be in the future. Behe DD p. 46.

Of course, design does not mean the same as perfection. So you might, conversely, argue that conception of an integrated system dwarfs a minor 'flaw'. **Every engineer knows that, in design, compromise between competing factors iis inevitably require**. Best balance is a goal! A single feature may not be engineered to optimal capacity but, taken in combination, a whole construction may be; for example, the thicker a skull the better it protects a brain but, equally, the heavier it becomes. A compromise is needed. Or a 'spare part' might be useful when inserted properly. It would be a sub-routine. You could therefore argue that the appropriately modified integration of a subroutine showed marked intelligence. For example, an offensively 'inelegant' panda's thumb is not, actually, its true thumb. A unique enlargement of two wrist-bones gives seven digits of which one bears superficial resemblance to a human thumb. The panda's true thumb is used for purposes other than the false one's manipulation of bamboo. The animal actually has a fine grasping mechanism.

While Victorians catalogued 'vestigial' (useless) organs it's now known what they're for. 'Junk' DNA is found to have its reasons. You're left with left-overs from developmental economy (like tummy buttons or male nipples) rather than phylogeny. Are 'design faults' destined for redundancy as well? Because there are those who, even if presented with a finely manufactured product, would prise out faults - especially critics from a rival company! Of course, it is a vacuous red herring to argue that flaws in the design of a watch or any other instrument of reason mean it was not intelligently created in the first place; or that its design is less scientific because it has a maker. Indeed, you might well argue that conception of an integrated system dwarfs a minor 'flaw'. A single feature may not be engineered to optimal capacity but, taken in combination, a whole construction may be; for example, the thicker a skull the better it protects a brain but, equally, the heavier it becomes. A compromise is needed. Or a 'spare part' might be useful when inserted properly. It would be a sub-routine. You could therefore argue that the appropriately modified integration of a subroutine showed marked intelligence. For example, an offensively 'inelegant' panda's thumb is not, actually, its true thumb. A unique enlargement of two wrist-bones gives seven digits of which one bears superficial resemblance to a human thumb. The panda's true thumb is used for purposes other than the false one's manipulation of bamboo. The animal actually has a fine grasping mechanism.

How do you explain the evolutionary redundancy of two eyes, ears, kidneys or testes when one, adding nothing to survival value, is a back-up you could live without? Look again! Can't humans see with clarity? Don't eyes serve their purpose well? Do we, except for accident or illness, have a problem? Perfection of design is marred, explained the rival camp, by the fact that vertebrate photocells are wired backwards and thus light has to pass through a restrictive forest of nerve cells to reach them. A computer simulation of the supposed evolution of an eye starts with a nerve behind a light-sensitive spot. Heaven knows how time reversed this relationship but, it was claimed, such deliberate reversal would amount to poor design. Why not wire up from the back and thus eliminate the obfuscation? Moreover, as these millions of nerves are bundled up and cabled out through the back of your eye towards the brain they create a blind-spot - one you would not have if they were phalanxed to the rear. Does your blind-spot worry you? The other eye compensates. Close it. Wink. Any worries yet? In fact brain uses retinal information to compose an image and can make good different kinds of deficiency - reflection problems, a minor defect in the other eye, dirty spectacles and so on. Think how else you might lead nerves from inside the eye out to the brain. The spot, covering 0.25% of visual field, is an efficient solution. And the retinal pigment epithelium (*RPE*) is very metabolically active. It needs a good blood supply and this it receives from the choroid layer (or black pigment epithelium).

In fact, the evolutionist criticism is nonsense. Would you really want an opaque choroid in front of the *RPE* (blinding you as does haemorrhage in the case of macular degeneration); or is it most functional when set behind these light receptors? If it must run behind then do you want the *RPE* pointing outwards far away or inwards close to its fuel supply? As it is the nerve cells are embedded in a rich supply of blood. This delivery flushes them with nourishment that includes visual pigments such as retinal and supports the demands of vision's high metabolic rate. And the choroid also transfers heat away from *RPE*'s high chemical activity. Indeed, if this layer or its blood vessels were exchanged to the front of the retina they would seriously impair optical use. A designer might, of course, have devised a solution to the potential obfuscation problem. Yes, Müller cells act as optical fibres that funnel light with maximum collection and minimal distortion. Arranged parallel to the path of incoming light they act like a fibre-optic plate in transmitting what might otherwise have been be blocked. They are transparent with the same refractive index as the other optical humours and so in no way degrade your clarity of vision. Indeed, they help filter and focus light, making images clearer and keeping colours sharp. Observe! Does such inversion actually compromise your perfect sight? How else would *you* construct an eye's fantastic screen? Your eyes are eagle not squid sharp. An octopus eye, with which superficial comparison is sometimes drawn, looks like but is simpler than the human and works more like a compound eye with a single lens - a different sort of design. So, like ping-pong, the cases for good, bad, degraded or indifferent design continue to be spun across a net of argument. *One might admonish, "Cut the carp; put up or shut up; just design a better system for yourself!"* Or else, where great intelligence in your eyes failed materialism's test, prostrate your wit in praise of unseen and inestimable chance. Müller cells render atheistic dispute false. How can you complain design's at fault when something works extremely well? Er, well, you can. An evolutionist clings tightly on. He's never wrong! Now finding argument from flawed design a failure he turns round to praise the ingenuity of add-ons as his evolution copes with its own poverty! In fact, the issue isn't one of trivial, incorrect complaint about 'intelligent' but poor design. It is about the very *origin* of information. How did an eagle's or your eyes really come to be? Some atheopaths have, though it loops to serve the heart, windpipe and oesophagus as well, singled out the recurrent laryngeal nerve as inefficiently elongated and thus of poor design; but how did it or any other nerve originate? Can mutations, incremental accumulations of millions of them, successfully create purposeful systems with their integrated parts that work even imperfectly? **There is absolutely no hard evidence this is so - just a theory driving the imagination**. What, in this imagination, is a 'positive mutation'? Is it one that unequivocally, by adding information, benefits its system?

This is the crux. What sort of accident is 'positive'? And, if *BM*s exist, what is their character? Are they constructive; is information value-added so that new machinery is built? Or do they occasionally offer opportune solutions to environmental problems, which non-innovative benefit derives from loss of information by deleted, broken or otherwise impaired genetic code? *If the latter is the case then the PCM dies, scientific and thence philosophical, a double death.*

Entropy of Information

'Teach the origins of life based on evidence, scientists demand.' (Heading of report in The Times 22-6-06 p.10). They don't mean based on scientism's flights from science.

Flight from science? We heard (Chapter 21: Types of Fossil) from Stephen Gould of palaeontology's 'trade secret'. Genetics has the one from Fisher too. If they are true both planks of *PCM* are flawed; the neo-Darwinian synthesis is compromised. It has been promulgated based on sham, disinformation and silence as to scientific evidence. Is this not flight from science? How does genetics flee?

tam/ raj	*Sat*
expression	*Plan*
machine/ manual	*Concept*

passive info.	*Active Info.*
matter	*Mind*
↓ *tam*	*raj* ↑
version received	*message sent*
receiver	*source*
noise	*signal*
greater noise/ weaker signal	*clearer/ stronger signal*
interference	*clarity*
loss of info.	*precise transmission*
entropy of information	*sustenance of system*
separation/ disintegration	*integration*
meaningless/ useless junk	*meaningful/ useful code*
randomness/ incoherence	*coherence/ order*
incorrect information/ error	*accuracy/correct info*
damaged text	*pristine text*
disease	*health*
mutation	*facsimile*

If abiogenesis and natural selection won't construct origin of code and its effect, a body, then the sole support for Primary Corollary - the *PCM* - is down to turning noise perchance into a pretty song. Negative or deleterious mutations (*DMs*) are not fit for purpose and so we need a special kind of information-adder, one that adds coherence accidentally without the benefit of a conceptual framework telling it what's fine or not, a *BM*. Everything depends on this.

Why should you need conceptual framework? Isn't the environment sufficient arbiter of life and death? Isn't its selection good enough? The answer is, as we have seen, that this is good enough to keep a form in shape or populations healthy but it doesn't *make* them. Natural setting cannot build coherent systems, complex hierarchies of command or such control as purpose in its execution needs apply. *Yet all the systems of biology, at every level, show this character*. It is what we want explained. It isn't just the ordered maintenance but *origin* of code and its initiative expression in the form of all fresh parts and wholes of bodies that we want to know about; it's not disordered degradation or the final moment quality control decides that things no longer work and scraps them.

In the 1920's Ronald Fisher, Haldane, Sewall Wright and others established the study of population genetics which led, over the next couple of decades, to formulation of the neo-Darwinian synthesis, a 'modern' expression of the theory of evolution or Natural Dialectic's *PCM*. From their work grew the notion that individual mutations might gradually accumulate to threshold values that 'improved' a stock or even, in imagination, generated novel pathways, organs, systems and thereby new kinds of organism. Yet Fisher and supporters, we shall see, were well aware that natural selection is 'blind' to the great majority of mutations - and the ones it isn't blind to don't create a single novelty, are slight adjustments or, if ever major, kill. Thus most mutations are neither good enough (called *BMs*) nor bad enough (*DMs*) to figure on selection's radar; they do not obtain the threshold where they have effect.

You might reply that Richard Lenski, in a long-term evolutionary experiment (LTEE) spanning decades and involving well over 75,000 generations of *E. coli* bacteria, showed that random mutations may increase the reproductive turnover. Equally, however, the mutations involved were later identified as degradative, that is, genes or regulatory elements were either broken or deleted. *Variation* and possibly, in the multi-step case of oxygenic citrate metabolism, re-potentiation occur; but *not innovative evolution.*

Experiments using *E. coli* (which with optimum lab conditions can double its population every fifteen minutes) have been going on for 100 years. Such vast populations, probably quadrillions or more of individual bacteria, to which Lenski added about 75,000 generations of millions per generation, should have accrued a sample that decisively illustrated evolution-in-action. **But the results are now in. *E. coli* definitely remains *E. coli*.** Check the Glossary 'micro-evolution/ limited plasticity' and 'macro-evolution/ unlimited plasticity'. Do not conflate variation with ransformatory evolution. So which does *E.coli* show? Certainly mutations, *HGT* (horizontal gene transmission) and broken control systems field variations but never innovations. And the apparently beneficial changes lose force with changed context leaving, since the 'beneficial damage' is not reversed, useless wreckage. Temporary advantage gives way to disadvantage. This is bio-entropy. Information atrophy. Over time natural selection (Chapter 22) runs a relentless, manifestly devolutionary course in a way that affects not only *E. coli* but, it is argued, all organisms.

In fact, do *BM*s exist at all? A body's by inherited mutation unaffected, maimed or dead - thus *DM*s and 'neutrals' are, for sure, the vast majority. Are you incapacitated by a scratch or bruise? Of 'nearly-neutral' mutations that continue, undetected, to accumulate 'bad' superficials swamp the postulated 'good'; and anyway, as Fisher noted, 'good' would be wiped out by random drifts before they could achieve a cumulative beneficence. Genetic entropy of information is, contrary to what synthetic theory needs, always apt to increase. These anti-Darwinian genetic 'secrets' were kept by researchers from the public - or at least they were not advertised. Nor have their problems for the theory been resolved.

Not only this. It is absurd to suggest (as you have to in support of *PCM*) that metabolic pathways, cellular metabolism and the specificity of proteins could gradually aggregate by chance. No papers in the professional literature of genetics explicitly, plausibly and in detail demonstrate how this assertion is a reasonable one. Is it personal or scientific bias keeps lips sealed?

Far from the censor's office and 'trade secrets' it is perfectly legitimate to see the genome as an instruction manual. This is what the words we use involving it - program, genetic code, language, information centre, book of life, system of orderly expression and control, transcription, translation and so on - all imply. This manual is, we now know, much larger than originally thought. A human genome contains two sets of about three billion 'letters' each. No book is, however, simply a complex series of letters. Each letter's contextual arrangement is critical and must, to make conceptual sense, be in the right linguistic position with respect to all the others. Life's manuals are, however, much more than a 'static' line of letters. Just as it is realised that protein structure involves a hierarchy of interactions generated from simple sequence so genetic code (but not the *DNA* it's written on) is full of loops, branches, data compression and integrated regulatory and editorial features. It is likely, as we have surmised, that very little is 'junk'. What is actually useless is likely to have accrued by incidental mutation and, since natural selection would have weeded any operational negativity, be at best (like blots, truncations, overprinting or the duplication of a section of a page) neutral in informative effect.

The context of nucleotides and properties of proteins are not, however, all that a process called morphogenesis needs to evolve its shapes. *Do you remember* (Chapters 16, 19, 21 and 22) *analogies with building materials and the conceptual plan of houses, cars or circuit boards*? **Mind can evolve a circuit board but matter by itself cannot, in the Darwinian sense of a self-organising scheme, evolve a thing. Ever. It can't produce a programmed scheme and yet biology shows nothing else, at every level, but such schemes**. Raw nature couldn't, even in a trillion years, create a china cup - let alone a cup of tea! In the material world systematic, purposive machinery never self-creates. Intelligence informs a system and it is carefully constructed in accordance with its plan; then from the moment it rolls off the line it needs the presence of a maintenance man. It might go wrong. It tends to degrade, run down and need repair. Accidents occur. It loses order and its information to the point that it completely breaks and dies. Is the passive code that carries biological information different in this respect? No. Although ingenious devices for self-maintenance are embedded in its construction, still passive information in the form genomic *DNA* may, like a rusty or loose cable, degrade; and thus the message that it carries is eventually degraded (or mutated) into crackling noise. **Mutations, creators of mutants, embody entropy of information; the relentless, net effect of random mutation is degradation unto destruction of function**. Entropy does not create; its direction is, for life's Titanic, pointing (↓) down. **No creator - 'life's destroyer' were more apt a phrase**.

While it is, you might say, poly-functional, the genomic manual is also multi-dimensional. Its operation must, in addition to contextual and regulatory rigour, show specific coordination with non-genetic features of the body that it helps inform. For example, polygeny (where many genes cooperate in the production of a single structure) and pleiotropy (where one gene plays a role in different structures) compose the norm. Together with micro-hierarchical command and control factors they illustrate the closely-knit functional integration of the genome. Such a web of integration is operative both within each level and throughout the whole range of hierarchical levels that constitute the structure and function of any life form. **Such web represents a very tight constraint.** *For evolutionary novelties to occur through mutation countless sequential 'good' mutations would be required; at each directionless step all would, without transgressing the constraints, have to cooperate harmoniously within the system as a whole. 'Progress', that is, would have to integrate with the context of each level of expression to ensure a healthy flow of information.* No doubt, natural selection weeds out failures; the question is, with such demanding standard, could there ever build informative, transformative success? Even Arthur Koestler, an atheist and evolutionist, recognised there was something seriously wrong with his theory of choice. Echoing Sir Ronald he wrote:

"Each mutation occurring alone would be wiped out before it could be combined with the others. They are all interdependent. The doctrine that their coming together was due to a series of blind coincidences is an affront not only to common sense but the basic principles of scientific explanation."

Such enlightenment led him to resurrect a version of mind-over-matter that Darwin, also seeing no alternative, had supported. However, Lamarck's *IAC* is now anathema because it is unclear how mind could rearrange what is presumed, materialistically, to be the sole basis of organic shapes i.e. the order of *DNA* base sequences or genotype. If mind shaped evolution it is, equally, unclear from what unconscious organisms or proto-mind evolved. *IAC* is out. **On the other hand, mutations have never been known to create a new function for an enzyme, a new metabolic pathway, a new target biochemical, organelle, cell type, tissue, organ or body-plan - let alone each factor integrated with all others within the hierarchy that devolves from the organism as a functional whole**.

Can you now see why Koestler worried; and how the low-level store of code's components can't by itself determine the shape or function of higher-level structures such as metabolic signals, cytoplasm, tissues, organs and the body as a whole? Layer on layer of information is hierarchically arranged to dovetail so a body works. Nor do genes alone determine how or what each level does. A genome helps with the construction, quantity, timing and positioning of apposite materials but does this in conjunction with a '3-dimensional index' made of chemical switches, electrochemical messages and organisational patterns found in cytoplasm and in higher order compositions..

The pc (and the *PC* of *M*'s) answer is, of course, that genes are what it's all about and therefore changes to them should be all it needs to muck the complex system up or else transform whole organisms in an evolutionary way. A gene is, however, just a chemical - conceptually inanimate. Evolution *is* inanimate. Does any chemical that you know generate designs? If, therefore, genes are not the only factor then, in principle, they cannot alone initiate dynamic plans and novel forms that a process of development or body-plan requires. The complex hierarchy (Chapter 21) that operates to stabilise all forms of life is, in itself, a further reservoir of information. How can genes or their mutations at the bottom of the scale then build fresh hierarchies piecemeal? If, as seems most likely, in the operation of this hierarchy layers of information disconnect genes from the direct production of organs then they could in principle mutate forever without generating even one such novelty. *If they are not responsible for physiological systems or body-plans then how could they, for all mutation over billions of hypothetical years, ever create a single one?* There is no evidence from developmental genetics that 'novel body-plan' mutations ever occur. The hypothesis is simply born of a 'Meccano' frame of mind - stick on the bits and see what builds itself. It is not organic in the way that, from coherent and conceptual centre, nature is. **In principle for all these reasons *PCM* and therefore *PAM* are, as we demonstrate below, wrong.**

Any misprint or smudge detracts from the sense of a book. A mutation is a typographical error caused in various ways. 'Point' or 'gene' mutations are by rearrangement, gain or loss of nucleotides. Chromosomal mutations involve loss, gain and other random rearrangements of whole sections of code. Each one's effect can range from nearly neutral up to lethal. It depends, as with a misprint or a fault a circuit board develops, exactly what has happened where. Lethal kills but the effect of even a minor fault can overshadow a million nearly neutral mutations. Is it, however, fair to say that every change degrades the text?

You might argue that a change where *DNA* plays no part in informative affairs (say, a harmless mutation) might be called 'neutral'. In fact such absolute neutrality cannot occur because even so-called useless *DNA* is burdensome. It loads the genome with demands for extra energy and processing time while adding nothing of substance to the text. Does a smudge or bent leaf improve your book? The change might be nearly neutral but is still *DM* (deleterious). Nothing beneficial can come of simple chromosomal duplication of material already there. Hypothetical *BM*s must be added, like *DM*s, a nucleotide or two at a time. What kind of letter-change, however, will not wreak effect? Does there exist in any text such level of neutrality?

Life's lexicon is simple. Check it out in any textbook or dictionary of biology. Sixty-four words of three letters each represent one of twenty amino acids or, as a 'full-stop' signal, none of them. Each acid is, on this account, assigned one or more words. For example, the genetic code for methionone reads, in m-*RNA*, *AUG*. Any change to this order would specify another amino acid and potentially, indeed almost certainly, damage the protein for which it was part of the code. On the other hand, code for cysteine reads *UGU* <u>and</u> *UGC*. In this case a change from *UGU* to, say, *UGG* (meaning tryptophan) would be deleterious; but from *UGU* to *UGC* (still meaning cysteine) no effective change would have accrued. This would - unless it caused some super-coding or so-called epigenetic effect - indeed be a wholly buffered, neutral mutation - neither *DM* nor *BM*. It would amount to an alternative spelling of the same word. No change, however, definitely means no evolution.

As well as neutral point mutations duplication of whole genes or genomes shouldn't do you harm. Gene duplication (also called 'genetic stutter') is a 'godsend' for an evolutionist because it offers am effective *tabula rasa* on which miscopying can happen with relative impunity (since the original remains intact). Might such clean slate allow mutations that, eventually, build messages of consequence - so that gene duplication is life's dream machine? We shall shortly see.

But what miscopy or misspelling might actually improve instruction manuals? You cannot calculate a change will not disturb the text around it. In poly-functional complexities each part is strictly constrained. And nucleotide positions have, like letters in a book, subtle and necessary contextual implications. Such large-scale contextual constraint is, in conjunction with hierarchical relationships, highly conservative. Random corruption reverses sense. The manual's integrity resists aimless jumbling and certainly beggars any belief in aimless 'construction'. Claude Shannon's mathematical definition of information (Chapter 6) is irrelevant; it is *meaning*, not just shuffling particles of *DNA*, that counts. At this rate in such sense any non-neutral change at all will be *DM*. A particular mutation might not do much harm but, over time, 'nearly neutral' errors will accumulate. Different editions (different bodies) may accumulate different permutations of errors but, until the point of incomprehensibility approaches, sense is made and readers struggle through.

Your own genome incorporates thousands of nearly neutral mutations, some from your lifetime, many others from your ancestors. These 'nearly neutrals' are, despite their number, ones that natural selection can't detect.

Might you hope that mutations in different individuals might, by sexual recombination, splice together and, over eons, make a gene? In fact, the human genome is composed of linkage blocks of nucleotides up to 30000 strong. These blocks are not, even in inheritance, broken up. No shuffling disturbs the nucleotides, no individual splicing can occur and, within the blocks themselves, *DM*s will outnumber and yet be inseparably linked to *BM*s in them. If such blocks could be shuffled then as soon as a freshly-tasked gene appeared it could be reshuffled, split apart and disappear. Break-up does not look promising for evolution's accidental but, you have to hope, improving ways.

We *are* all mutants. Mutations in the germ-line (the only place where, in evolutionary terms, they count) are rare but common enough to accumulate 'burden' at a rate which, given our evolutionary age, should have rendered us extinct. Calculations for humans vary but it is now supposed that nuclear mutations of the kinds adding to heritable degradation amount - if you take 'junk' *DNA* as over 97% of the genome and neutral in effect - to between 3 and 5 per person per generation. Of course, if 'junk' is actually operational and you add cytoplasmic *DNA* and other hot-spots known to be 'miscopy-prone' then each text suffers between 300 and 500 per person per generation. This would mean that mutations, mostly nearly neutral, are currently accumulating in earth's 7 billion people at a rate of over 1800 billion per generation. What counteracts this huge 'downward' flow? Even if, on average, as many as one in a million were a *BM* it would mean that in probability a single gene had, in order to 'get lucky', to bear the crippling load of nine hundred and ninety-nine thousand, nine hundred and ninety-nine distortions - for each of hundreds of successive, coordinated build-ups. What could reverse this anti-evolutionary deluge, this great entropy of information? Rather than just holding your own against the tide how, much harder, could you build up germ-line information and publish real evolutionary advance - a whole fresh, self-consistent manual? If there is net loss every generation how can 'genome building' happen? *In short, mutation doesn't give you evolution*.

Natural selection operates only on functional wholes. It is therefore oblivious to nucleotides, at least to any re-arrangement of them that does not cause physiological or anatomical strife sufficient unto death. Mutations pass beneath its radar if they don't affect the phenotype, that is, the body of the text's creation. Most of them are nearly neutral whispers. Their impact is too small to snag selection's sieve. Quality control for a whole body can't precisely pick up individual, minor faults. It can't 'hear' a problem till its noise is shouting 'abnormality', severe disease' or 'irresistible trauma'. Even then it's not just a faulty body that 'selects' itself for death. Outside non-genetic factors like starvation, injury or sheer bad luck play just as large a part. No blame's attached. They kill both fit and unfit off.

Near neutral mutations therefore pile up through the generations. Mostly they are negative but surely some make for improvement? Surely a few are not just 'positive' but add some value in the form of information? *They'd better. If they aren't and don't then evolution in the large sense cannot happen.*

At first geneticists tried to copy how they thought that evolution worked. They threw all sorts of mutagens at organisms (mostly plants and germs but laboratory animals as well) in order to create *BM*s and thus improve life's stock. It never worked. Does anyone believe Hiroshima, Nagasaki or Chernobyl is an example of an evolutionary paradise? They failed. Mutagens and their mutations only bring you grief so this approach was ditched. As every breeder always knew, exploitation of adaptive potential works much better. Work with nature's order not against it. This is, in nature's but not evolution's way, how genetics goes today. Trying to improve a system by random rearrangement or destruction of its parts was never much of an idea. The wonder is that evolutionary thinking led science up that garden path for over forty years. Some still commune with naturalistic fairies.

How, therefore, did illogical, unnatural evolution work? If most of the genome is functional then most mutations must be bad for it. Geneticists have, however, ears close to the ground for any whisper of a *BM*. If even one in a million mutations were constructively beneficial should the literature not be overflowing with reports of them? Actually, silence reigns. Even such mutations that have been previously discussed are arguably deleterious. They do not add information to a system. They burden it. A misshapen protein might, from time to time, turn tricks but such erratic, lone peculiarities, such lucky trivialities can't switch a whole great lack of system on; in fact, a few duff molecules turn evolution off. This is because trillions of consequent *BM*s would be required, one after the timely other, to codify billions of transitions needed for millions of complex interactions and the mechanisms that, each in turn, need all components correct, co-present and correctly arranged. **It might well be argued that mutation is, as an explanation for trans-speciation, an accepted but unproven and therefore unfounded system of belief.**

Surely one *BM* in many billions is enough? *DM*s maim and kill; could a *BM* strike functional gold in the form novelty? Except for such an unobserved and thereby mythic case *BM*s should be mostly, like *DM*s, nearly-neutral. This means selection can't detect and, therefore, adopt them. It means that if they actually occur (though who knows what small change steps towards what larger end?) random drift and non-genetic factors will, as Koestler noted, probably eliminate them well before they team up with the many hundreds further needed to create a single novel complex protein, metabolic pathway, cell type or higher-order structure integrated, naturally, with all that went before. If not then their effects, like those of most *DM*s, will be diluted up the hierarchy into scarcely any impact on the phenotype. And in time a 'lucky' switch that might have led who-knows-where could be reversed by an 'unlucky' one or overtaken in a deluge of surrounding *DM*s fouling up the context, dragging down the sense. The fact is that, during the random 'construction' of a new gene/ protein, tissue, organ or whatever *DM*s will outweigh *BM*s (if, when you don't know where you're going, those critical *BM*s exist at all). **In short, any *BM* would be invisible at the level of a whole organism and its small 'advance' overwhelmed by neutral or deleterious mutations long before the successive and exact chain of cooperative *BM*s needed for any novel biological system could 'evolve'.**

Indeed, a fractionally completed item is at any point during its as yet incomplete construction neither beneficial nor neutral. It is only deleterious.

How might any organism survive 'unfitness valleys' that the rarity of *BMs* (if existent) would dig so long and deep? A fresh metabolic pathway, organ, system or a body-plan will need numerous such 'valleys of malfunction' simultaneously. Do you remember the Cruft's specialities, those eccentricities of selective pressure that have been nurtured to survive? Do such creatures, pushed to limits of selective breeding, show tendency to change from being dogs? Could any organism survive - without even knowing what the new form of survival needs - such ineffectiveness, agony and breakdown as would be required by transformation; could such macro-evolution emerge as something radically new? The poor answer is 'it must have'.

Not knowing where you're going is a fundamental problem. It applies to 'junk', 'neutral' or any other kind of *DNA*. What should a first or following *BM* be? There can only be one when defined within a conceptual framework. What is 'good' or what is 'bad' is only so when valued in anticipation of that end. **How can you even define a *BM* if you don't know where you're going? Mindless evolution's natural selection doesn't know. And if there's no such thing as target and its *BMs* how can 'progress' gradually mint a whole with integrated working parts? With such lack of logic bang goes the irrational *PCM* and therefore bang goes *PAM* as well!**

Remember context. Most single genes are, excluding their regulatory surroundings, at least 1000 correctly sequenced 'letters' long. The chances are as low as practical impossibility that you'll obtain fresh sequences that are exactly what some new job needs (you don't, in fact, know what you need). And continual miscopying will delete at least as much that's 'beneficial' as it makes - even if you eventually obtained a single protein at the same time you would have accumulated vast amounts of damaged *DNA*. And what's the innovative protein for? What, in the tight contextual constraint of all others, will it do? How will it fit into the context of an integrated physiology? One senseless protein built by no reason for no reason is not what creates new systems, elaborates new mechanisms or evolves fresh body plans. You begin to see why people doubt there ever was a truly beneficial, evolutionary information-adder or any such thing as 'a right good sport'? This is, *bottom-up*, the problem. How do you add, by chance and one by one, piecemeal components to construct a system that is purposeful? How do you take a text and, blindfold flipping over letters here and there, create a whole new integrated section or a full new script? How were you tricked to thinking that you could?

You protest. You protest, for example, that low phytate corn, a mutant bred by use of mutagens with the specific intention of improving livestock feed, is a *BM*. But although it may help digestion the genetic machinery producing phytic acid has been damaged. What about tomatoes that have lost the ability to recognise their own pollen and therefore resist inbreeding? Is this loss of useful trait a 'good' or 'bad' thing? Is loss of function simply the reverse of making it? Arguable *BMs* show, like *DMs*, entropy of information. *They do not address, in any shape or form, the question of the origin of high-grade information. They do not start to answer how a form of life could be initialised.* Yet this is the whole point of the theory. It is what its Primary Corollary is there to do. And even claims to do!

You could, of course, adopt the view that an instruction manual was a 'perfectly assembled jumble'. You might argue that a pristine genome was composed of *BM*s and contained, by virtue of its intense heterozygosity, adaptive flexibility! This is not how evolution is supposed to happen. You start, *bottom-up*, from imperfect jumbles, error and what, without sense or reason, just scrapes by as a poorly working possibility. Except that the hypothesis itself is faltering. Life is not, we now know for a scientific certainty, a simple fleck of slime. Even cells are many-layered wonders, highly integrated functional wholes whose purposive complexity puts poor old Paley's watch to shame! How, therefore, can you take a point mutation and, like a nick in some scrap metal, extrapolate from that to tell the time? The hypothesis itself won't start to tick. The Primary Corollary will not tick over. How can you wind up evolution's watch (and there are many clocks in every cell) before the mechanism's there?

Surely, surely there are two ways you might move? One uses junk, the other tiptoes up the stairs.

Motoo Kimura's Neutral Selection Theory asserts that, in non-effect, it is random drift that dominates proceedings. Most mutations are not selective but neutral. On top of this consider Susumu Ohno's suggestion of gene duplication. He takes an 'instructive sentence', doubles it and then supposes random interference jumbles up the letters of just one of the pair (the other needs to stay intact so that the organism can survive). In other words, suppose mutations, each one useless by itself but, since they're in the duplicated section, neutral with respect to genome operations as a whole, accrued. You could thus, it is suggested, ring the changes and accumulate 'advantage' in the 'dead ground' and away from where mutation's general negativity could interfere with business-as-usual. Such accrual might one day suddenly, as the last clue completes a crossword, leap to active sequence. It might generate a second 'instructive sentence', that is, a novel protein could, most improbably, appear! Where would it fit? No innovative function or its organ is made - unless the process multiplied ten thousand times while holding on to every part that might be useful in an unknown unplanned future! Why *should* useless, neutral, individual changes not be lost to drift or natural selection? Why *should* changes integrate and what would happen to the 'junk' that made no bio-sense? High improbablities piled higher, can the process get you any actual as opposed to wishful where? **By such imagination evolution theory with collateral atheism stands or, on such impossible confabulation, falls.**

Take a look at baker's yeast. Many of its genes seem duplicated so that perhaps, historically, some other fungus stuttered into wholesale duplication, doubling up its whole genome. More generous an allotment of spare copies you could hardly dream of; and analysis revealed that other species also seem to have descended from original 'double trouble''. Such yeasts show different mutations but, with tens of millions of nucleotides and a hundred million years of evolutionary opportunity what have they made of their career? Precisely none. If, however, 'breakpoints' and mobile genetic elements could 'engineer' recombinations of bulk chromosomal sections of a duplicated part (like rearranging a computer's subroutines) might not a lucky strike on novelty obtain? Yeasts contain such elements a-plenty but still they don't evolve. They change a little but remain yeasts. Is this the best two of the best Darwinian agents, gene duplication and mobile genetic elements, can in reality obtain? Quel maximum! You could not, on present showing, take genetic st-st-stutter as a serious candidate of macroevolution's mechanism or the author of a biological appurtenance that's new!

Moreover large mutations tend to be less frequent than the small ones and, since they have more chance of 'coming to the surface' as an illness or deficiency, are more likely to be blotted out. And in the case of the majority of small, nearly-neutral (rather than completely neutral) changes tend fall beneath selection's 'radar'. It can neither 'see' nor pick them up nor wipe them out. Thus the overwhelming majority of rare 'hits' called *BM*s would be, like *DM* 'misses', nearly-neutral non-selectables as well. But no *BM* alone can do a thing. Non-genetic factors (drift from random mating, accident, environmental problems and so on) would still act, as with any gene, to change its frequency or weed it out. *Nor can selection cause one 'sport' to play the game and dovetail in with all the others. It has no power to supervise the churning of innumerable 'letters' far below the only level it controls, the body as a whole.* How could such a muddle spawn, like Athene from the head of Zeus, a whole new genetic system with its proteins and their regulation? Bear in mind, whatever's cooking, the chances of a *DM* 'miss' are always greater than a 'hit'; and the chances of such misses increase exponentially with length of sequence. An average protein, just one factor in a cellular machine, employs an accurately sequenced code of 1000 nucleotides or more. How, therefore, might accumulation of 'neutral' but random changes to a working script sally suddenly into a whole new show? You'd expect a garbled module but you'd need the polar opposite of any lethal pop-out; some mutant jerk into rose-tinted novelty would have to conjure up the 'beneficial' trick.

Such ideas are just contrivances to try and wiggle past the real road-block - information generation. How can chance whip reams of sense up from blank non-sense? After forty years of research an eminent geneticist from Berkeley, California, Richard Goldschmidt, saw that mutations led nowhere. Accordingly he proposed that from a normally intolerable shake-up of genetic material, a mega-mutation, might suddenly throw up a 'hopeful monster' very different from its mother. A sickly but perhaps a new type of organism would have been bred. There is, however, not a jot of evidence for such 'leaps by deformity'. From the pulpit of Much Flailing-in-the-Marsh monstrous transformations might be fact but all experimental evidence denies it.

Could you perhaps create new organs or another kind of body plan by juggling with the homeotic genes that help control the times and places of shape making - morphogenesis? These include, in yeasts, animals and plants, so-called *Hox* and *MADS*-box genes. In fact, although evo-devo's explanations (Chapter 25) adopt this line, mutations to genes or regulatory factors involved in the early stages of development (involving, say, body axis or the spinal cord) have devastatingly negative impact. Such fundamental, leading text is critical. Basic order has to be conserved or else the rest is rendered rudderless. Downstream expression that involves, say, extremities like fingers or nearly finished organs, can logically and actually withstand mutation better. No *Hox* mutation ever hinted, any more than fundamental defect in a clock, at any gain of information, any evolutionary anti-entropy or any useful, integrated fine-tuned novelty. The opposite. Any error here shuts down, aborts the system or creates fatal monstrosities. *Such anti-evolution gives the lie to mega-beneficial transformation stemming from developmental code.* **Bodies develop perfectly in order**. They anticipate a well-defined and preordained target, adult form. Does chance anticipate a thing? Mind does. Purpose always has a target in the future. It is conceptual. How (Chapter 25) could morphogenesis evolve by chance? How could it not have been defined by mind?

If big *BM*s won't do then, round in circles, try again to commandeer some small ones. Modification of descent by cumulative complexity means that you think irrational errors in a rational text (it works and so must have its reasons) might transform one instruction manual, letter after letter, into another model's. Is this really how you move from one highly constrained, specific arrangement of components to another? Of course you don't. You can't. What about the intermediate times when meaning's lost? Wouldn't natural selection snap them up into its jaws of death? Just as you can't, if you are logical, imagine how a metabolic pathway could evolve so you can't imagine how blind chance could throw up purposive and irreducible complexity of systems. What about the precise contextual necessity for every 'positive' change to integrate at every level of an organism's composition. No hiding impact here. No doubt, at top level (the whole organism) reproduction or its lack weeds weakness out; but any change to any level of the program that is a survivor's needs be integrated so the guillotine that works to keep up quality won't fall. Bugs that cause malfunction get the chop. What, therefore, about the *DM* deluge while a few conceivably interpreted *BM*s happened here or there, a deluge that would wash all tiny gains of information down its drain? Evolution's *BM*s live in hypothetical a realm. Is there empirical a nail to drive into the theory's coffin? What undertaker carries Darwinism's corpse away? In the next section we will meet one.

A thief of truth, the spotlight shows, has padded up the gradual flight of steps that he imagines *BM*s take. And methinks weasel words won't keep him from arrest. Do you remember the third *caveat* preceding a discussion of natural selection - a 'ratchet' cannot work without a prior plan to know what you apply it towards? You can't craft computer programs step by step towards a preconceived end-point and then claim they demonstrate the way that evolution works. You can't pretend a highly purposeful compendium of works that leads to targets you've selected in the first place is how nature works haphazardly. You can't decide a sentence and then let high-tech solutions, based on programs whose 'intelligence' their programmers endowed them with, compare your random keystrokes and retain the ones you have decided in advance were 'right'. Indeed, with games of virtual evolution 'digital organisms' are transformed in cyberspace but such mutatory simulations all involve the diametric opposite of what they try to prove; they look far more like intelligence than chance. In fact, such exercise in sophistry just proves the *top-down* point of prior conceptual establishment. Take, for another illustration, combination locks. You have to spin the numbers one by one into specific code. Why, however, should blind natural selection know what 'number' (or, of thousands up to billions, what nucleotide) is 'right' to keep? Is every nascent *BM* kept until the rest accumulate? In fact, the next spin on that number's wheel would lose it well before the others all came 'right'.

Aren't *BM*s, as Fisher showed, wiped out by following accidents or, in effect, washed flat by floods of *DM*s so that any possibility of 'gain' is more than lost? *No doubt waves of genetic change drift to and fro in populations; whether such mutations fix or fade is down, excepting lethal disabilities, to the vagaries of chance.* In short, it's obvious. Oblivion, the state of matter, cannot generate a meaning; nor can it leap up many steps across the chasm to the level of an integrated irreducibility of working systems in a go. Orders of magnitude, by virtue of extent of information, separate the churning chemistry of molecules in space, sea, land and air from purposive, devised designs. Natural selection doesn't even work with letters (nucleotides) and doesn't let disintegrated parts of wholes malfunction until suddenly they begin to work together in a fresh, transcendent way. It chops whatever organism does not, as a whole, work. And it does so straight away.

You may even think that genes are all that matters. That, as we have seen, is wrong. You may think that function must arise from nucleotides and not from concept; and that it 'gathers strength' in neutral areas before selection operates. Or perhaps you imagine that genetic function is continuously upheld while blindly fiddling for no reason with a text until it turns into another that (for all the fiddler, nature, knows) has no reason either. **When you consider the profound, specific complexity, the hierarchical levels of precise interaction and order, quantity and quality of information that is needed to generate a biological cell, let alone a biological body, how ever did you think that evolution was responsible?** Active information orders; passive information, active's script, is an arrangement with materials that, by the basic laws of nature, is inexorably overwhelmed by entropy -in this case entropy of information. *This is what, to eyes unprejudiced by Darwinism, the data clearly shows - and the fact is fatal for the theory.*

It is a fact of nature that, unless carefully maintained, the structure of passive information (for example, a book, sound-wave or chromosome) decays like everything else. Entropy of information, along with all that is known about the nature and the distribution of mutations, militates against production of a single gene. This is even by *BM*s. Factor in the flood of *DM*s and you have virtual proof that *PCM* is wrong. What word is there, except for bluster, that it's right?

Genomes naturally fall apart and not together. Information decays and at best natural selection only slows that fall. If life is 'information incarnate' from what is it falling? It is falling out of order. As time passes you grow older and then old. Bio-systems decline, not least due to genetic bugs, from their resilient youth. Disorderly old age might be described as 'entropy of function'. As the health of an individual declines so the human genome is, as already noted, in overall decline by between 1% and 3% or more per generation. Relaxed selection (due to sufficient food and medical attention) plays a part and genetic illness, for example, is apparently on the increase. Is it the case that a species mirrors in slow motion how an individual declines? If the *PCM* is wrong then our understanding of life's history is wrong as well. If things decline and not evolve we've got our sights on upside down. If they fall not rise then where have they fallen from?

If genetic entropy as evident in aging is indeed a process that applies to populations and species what does this indicate? It is not 'rocket science' to grasp that turning history right side up shows universal entropy applied to information. It shows every species, including human, in overall decline. Decline must proceed from a superior condition. For example, a pristine instruction manual that is the product of intense conceptual activity becomes, due damage and miscopying, gradually illegible. When too many bugs invade a program it will crash; when a book becomes illegible you junk it; when a machine stops working then it's scrapped. This is not progress. It is not evolution but instead, of which we have clear evidence, extinction. The way of all (*tam*) entropy is down. It is, for information, energy and life alike, exhaustion and the end. This and not 'progressive' evolution is the real law. Is law not truth? Who cannot accept the truth of physics as applied to natural bodies and, thus, each body biological?

Consider, in the light of cosmic run-down, how *PCM* needs for its start genomic build-up of already irreducible complexity - a cell. It has then to interfere at random with this working cell's instruction manuals to increase life's capability. It has to factor in plans and dovetail all parts to fit in with a novelty; and novelties along the metaphor of 'tree of life' have to dovetail with a hierarchy of 'ancestral' ones. Can you imagine even one instruction manual thus transformed, by random tapping out new letters anywhere across the text, into a new one that includes development and systematic integration of a whole fresh set of bio-logical ideas? Like heart surgery it will have had to happen while the factory is working. What will happen when the products have made fractional progress down some organ's evolutionary line? At which point, like a half-completed textbook, neither part nor system is, either on a metabolic or a physiological level, in working order and will not only fail to function but most likely foul up all the rest. Is this way of generating information possible; conceivable; or, for all features organisms show, a 'scheme of randomness' improbable beyond all rational belief? Why, therefore should we buy the pass? It's time to put a very weak hypothesis to test.

Evolution in Action?

Is there evolution? Yes and no. It depends on how you use the word.

Yes, variation-in-action certainly exists. All agree upon the limited plasticity of microevolution. Is, however, speciation evidence for evolution? **If a new feature had arisen, yes; but if change occurs within parameters that describe existing structures then variation and not evolution has occurred**. If Darwin had proclaimed a Theory of Variation, we would all agree. Variation happens and may result in speciation. But Darwin did not propose such a theory. His was the Theory of Evolution whereby entirely new features outside the existing parameters of a structure are added - such that a microbe could become a man. That requires an extrapolation to unlimited plasticity. Is such transformation, macroevolution, fact or fantasy? Certainly, in this sense variation is not evolution. **One may, by conflation, call it so but then commit the logical fallacy of equivocation (using the same word while swapping between two meanings). This fallacy, double-dealing with the same word, is very common; proponents of evolution continually try to interpret**

variation as if evolution was occurring. They equivocate but do the facts bear out semantic trickery?

The mechanisms of variation are in-built genetic potential, deliberate sexual reproduction and inadvertent genetic mutation. It needs be reiterated that positive recombination and diversity due to sex is not an evolutionary option. Sex does not create new information, simply propagates new permutations of what is already there. Information is neither lost nor gained. The activity creates, in the memorable language of Guiseppe Sermonti, 'wobbling stability'. It scrambles information in an orderly, meiotic way but is, with respect to type, conservative *par excellence*. Type remains the same type. *Therefore for evolution, whose crux is gain, sexual variation is neutral; and sexual selection is, as a Darwinian precept, void.*

In science you experiment. How can you *test* evolution in the past? In his book 'The Edge of Evolution' Michael Behe demonstrates that nature has empirically tested Darwin. "If it could be demonstrated", wrote Charles Darwin, "that any complex organ existed which could not possibly have been formed by numerous, successive, slight modifications, my theory would absolutely break down." Do you want numbers showing how, by gradual mutations, life has step by step evolved its tree? ***HIV*, *E. coli* and malarial parasites satisfy the numbers game. Virus, bacterium and eukaryotic cell have reproduced, mutated and should have evolved through sufficient generations with sufficiently large populations to indicate whether neo-Darwinism's engine, random mutation, can bear the weight of a theory that would have it gradually innovate body-plans and parts by gradual but cumulative, useful steps - or is crushed by numbers**. In nature and laboratory these and, to a lesser extent, fruit flies and other organisms illustrate Darwinian changes. Of what extent and quality are these? To what far edge can you observe extrapolation run? Is, when you evaluate the action, the conclusion that great macroevolution could occur? If not the tree-of-life hypothesis would be, as Darwin feared, negated.

What, therefore, are the facts?

HIV, a nine-gene scrap of *RNA*, lacks editing formality and thus mutates about 10,000 times more wildly than other cells. Each infected person is burdened with about a billion viral particles and over ten years incubates (since *HIV* generation rate is about a day) 10^{13} viruses. Fifty million sufferers would have produced about 10^{20} copies in the last fifty years, that is, since the earliest version of *HIV* (1959) was found, probably transferred from a monkey or monkeys in the Congo.

Furthermore, due to the explosive rapidity of mutation, each possible point mutation occurs tens of thousands of times each day in *HIV*-infected individuals. Double point mutation (where two nucleotides are changed simultaneously) would occur each day and up to six-point mutations could have occurred in the total population. Yet in about 10^{20} mutating viruses no change has revolutionised the basic genetics or behaviour of this snip of low-life half-life. Blind flailing of mutations has simply thrown up - randomly, chaotically and unpredictably - various biochemical configurations that inhibit the binding of molecular weaponry that drug companies have conceived to throw at it. But all the unintelligent genetic churning has not 'invented' the wherewithal to overcome resistance offered by one mutation (a deletion) on the human so-called CCR5 gene. The *AIDS* virus is sometimes touted as a clear example of 'modification by descent' (which it is) and evolution-in-action (which it is if you equate variation with micro-evolution but isn't if, with unlimited plasticity, you offer it as evidence of macro-evolution). However, even with its prodigious evolutionary advantage no innovation whatsoever has, during its whole *blitzkrieg* of Darwinian *variation*, evolved! Limited plasticity rules. ***HIV* is still *HIV*.** One might reasonably conclude explosively mutating viral strains are macroevolution's bombshell!

Even the simplest bacterium involves the composition of perhaps a hundred billion (10^{11}) atoms and ~5 x 10^6 base-pair bits of genetic information in the form of a coherent, self-replicating machine. What, therefore, of much scientific experimentation with *E. coli*, a bacterium that we met in Chapters 20 and, with respect to limited plasticity, 21? Cultivated for over a century in flasks that, under the watchful eye of technicians, yield up to 70 generations a day you may calculate the passage of 2.5 million generations - enough to zoom a human out and back sixty million years or so. If you counted the numbers of every generation perhaps a hundred trillion (10^{14}) organisms have been subjected to every kind of indignity that their interrogators could design. Mutation or any other exotic mechanism for the radical addition of information perhaps leading to macro-evolution could have shown up during Richard Lenski's aforementioned evolution experiment. None has. *There have been lucky improvements by various degradations of the code; there have been incoherent variations on the pre-existing theme, some of which have exhibited antibiotic resistance and other chemical changes; but no report of macroevolution's gift; no sign of even the rudimentary beginning of anything more complex emerging from an incubator; absolutely no new function, nascent system change or verging towards eukaryotic status*. Only trivial, localised and random tinkering occurs. ***E. coli* remains *E. coli*.**

Whither might antibiotic resistance, for example, evolve? Is it heading for a new organism or just protecting what's already there - in the way of natural selection Edward Blyth proposed. No doubt, 'intelligence' plays a key role in any war. In biology's survival game a core strategy, in concert with energy supply lines and reproduction, is self-defence. Defence, that is, against disorder, disease and discord. When trouble starts you need to identify, target and eliminate its cause as soon as possible. It is the business of molecular police, proteins that operate on behalf of an immune system, to quell internal disturbance and keep the healthy peace. In many organisms, including ourselves, the adaptive immune system engages a sophisticated genetic lottery (called somatic recombination); how did this evolve gradually? But fungi and bugs can also, like humans, certainly invoke resistance to a troublesome invader. For example, the common bread mould can make penicillin and the bacterial *Streptomycetes streptomycin* tetracyclin, erythromycin and other antibiotics. Such organisms employ a series of defences (efflux pumps, ribosomal protection proteins, modifying enzymes etc.) against their own types of weapon. The genes for most antibiotic resistances are acquired in a uniquely bacterial process called 'lateral gene transfer' that employs plasmids (additional, non-essential *DNA* separate from the main chromosome) or transposons (genetic units that can fabricate copies of themselves and jump around a genome).

Bacteria do not possess sophisticated genetic repair systems so that more mutations 'slip through the net' and are expressed than with eukaryotes. It may, however, be argued (as did Nobel prize-winner Werner Arber) that coded enzymes act to generate or modulate the frequency of genetic variation. In this view, lateral transfer, transposition, recombination and point mutations add up to an in-built strategy to 'ring the changes' and generate micro-evolution. Of course, external mutagens can inflict random effects but the driving force for genetic and thereby phenotypic variation is intrinsic.

Stress can also kick-start a 'mutation mode', analogous to somatic mutation in the human immune system, which rapidly produces the recombinations from which a protein that confers resistance can arise. When exposed to an antibiotic most microbes die but in some the fortuitous, abovementioned genetic recombination may interfere with antibiotic binding sites and thus confer 'immunity'. Antibiotic resistance has increased dramatically over recent years. For example, tetracyclin resistance has grown from a few percent in the 1950's to 80% today. Kamamycin and penicillin resistance has become widespread. In fact the clinical use of antibiotics has, by vice of this particular sort of virtue, managed over the years to artificially select for superbugs. *CD* (*Clostridium difficile*) and *MRSA* (methicillin-resistant *staphylococcus aureus*), for example, are troublemakers that will not respond to known forms of antibiotic discipline. If bacteria, using gene transfer or mutation, quickly gain fresh tactical information to outwit the body's law enforcement agencies, have not microbiologists by repeatedly exposing influenza, smallpox and other infections to powerful antibiotics rapidly created (or caused to evolve) drug-resistant, weapons-grade strains of killer virus? Surely this is clear evidence, from the bug's point of view, of 'positive mutation' and therefore evolution? In other words, is all this not an obvious example of micro-evolution and, by extension from its small part, macro-evolution building up before our eyes?

A 'small' bacterial, perhaps lacking complexity of expression, is, unlike ours, tightly coded with perhaps ten times less so-called 'waste', 'junk' or non-protein-coding' *DNA*. In this efficient case we might judge viral and bacterial economy in advance of humans - although we do not presume that they evolved from us. The disruption of tightly ordered code is, however, more and not less likely if random transpositions or mutations occur. For example, streptomycin inhibits bacterial growth by binding against part of the ribosome and blocking protein synthesis. A point mutation can warp the ribosome so that the streptomycin no longer fits. This is useful to the bacterium in contact with streptomycin but, in a drugless environment, the mutant bacterium produces protein, grows more slowly and is thus less fit than the 'wild type'. Information has been lost not gained. Similarly, mutations that confer adaptation to abnormal nutrients such as xylitol and resistance to antibiotics or insecticides like *DDT* are oft-quoted of evolution happening incessantly and obviously today. In the Darwinian sense of variation this is true. But, for example, insecticides make resistant strains of organisms, such as mosquitoes, less active and responsive in the absence of of those toxins than their 'wild type' peers. For this reason drug or toxin resistant mutants have even been called 'evolutionary cripples' ('Plasmids', Scientific American Dec. 1980).

What about the new strain of superbug resistant to the antibiotic vancomycin (*VRSA*) as reported by Nature 1-8-02, p. 469? Analysis revealed that the new strain had evolved nothing new. It had acquired its resistant genes by data transfer from a relatively harmless gut bacterium called *Enterococcus*. Do you remember bacterial sex (Chapter 23) or 'lateral gene transfer'? In some circumstances a bacterium can form bridges to conjugate with another of the same strain. Plasmid (and occasionally, randomly, chromosomal) data is transferred across these bridges. The recipient gains what the donor loses. The intent of this process has, due to data transmission, been dubbed sexual rather than immunological. It has therefore been judged an evolutionary mechanism. Sex, however, only reproduces the same kind of organism. In the same way bacterial 'sex', viral transfer of *DNA* fragments and swapped genetic elements have never created a new sort of bacterium.

It is conceivable that plasmids, which might predate the use of clinical antibiotics by perhaps millions or even billions of years, and the principle of stress-generated variability represent a simple sort of microbial immune system. This 'intention' is supported by the fact of transposable genetic elements that contain the very genes necessary for their own mobility. It may be that each resistance gene in an R-plasmid (bacterial Resistance plasmid) is part of a transposon. In addition to its 'adaptive' aspect, a bacterium is supplied with scissor-like restriction enzymes designed to snip up any invading, enemy *DNA*. You could also argue that other features (the efflux pumps etc.) are, as the outcome of existing genes, part of a defensive subroutine. In eukaryotes molecular defence is, as we saw in the previous chapter, developed much further. However, the conceptual logic of defence is the same; and its information is based, even in a bacterium, on code, recognition, signal and high levels of specificity. Such coherent elements are not, in any purposive system, improved by random intervention.

In 1845 the bodies of sailors on an ill-fated Arctic expedition were deep-frozen in permafrost until exhumed in 1986. Strains of Victorian bacteria were found in their intestines and revived. When tested some were found to be resistant to penicillin and other modern antibiotics. In other words, resistance predates clinical application and therefore cannot have evolved from it (Medical Tribune 29-12-88). Even if it had, resistance represents variation. You might rename it micro-evolution but there is absolutely no evidence that one sort of microbe transmutes into another or a new type. Resistance is not an agent of transformative macro-evolution. For a first example, thousands of experimental mutations induced in *E. coli* bacteria support this assertion. *Hundreds of thousands of experiments involving billions of E. coli bacteria have (not for want of trying) failed to generate anything but E. coli.* The fact of the matter is that, neither in the field nor the lab, have any new types of virus or bacterium been observed to evolve by mutation. Nor has incipient mitosis or eukaryote been spotted. New strains arise but no transformation into other types of prokaryote or eukaryote. Nor, equally, have eukaryotic fruit fly, mouse or any kind of organism been transformed beyond (*figs*. 5.1 and 22.1) its typical limit. Innovation isn't seen to happen. In this case it could be said that the same facts of bug-resistance supposed to 'prove' the Theory of No Intelligence better 'prove' one of Intelligence. **In other words what is called, *bottom-up*, micro-evolution is called, *top-down*, variation-on-theme; plasticity (*fig.* 22.1) is limited; in every case what is called evolution-in-action is no more than variation.**

A founding father would have agreed. Together with Alexander Fleming and Howard Florey Sir Ernst Chain FRS, Nobel Laureate (1945) and holder of numerous, prestigious decorations, pioneered the study of penicillin and antibiotics. He worked at Oxford, Cambridge and other universities but still regarded the theory of evolution as a residue of the 'euphoric', Victorian myth that through omnipotent physical science comes and will come all worthwhile revelation. His 'Chain reaction' was that Darwinism is a flimsy, feeble explanation for the brilliant yet micro-technologically precise diversities of life. He knew, of course, that 'evolution-in-action' or 'variation' arises naturally all the time. But, significantly, he questioned the extent of its limited contribution to the forms of life we find; he noted the incompatibility of modern molecular biology and biochemistry with classical Darwinian ideas; and concluded that the evolutionary hypothesis is irreconcilable with either common sense or facts.

In a return to the numbers game Behe whisks up a third player, *Plasmodium falciparum.* This character constitutes one side of the disease, malaria - its infector. The mission of this virulent malarial parasite is to reproduce, to which end it engages in 'trench warfare' with the human genome. It infects human red blood cells causing the often fatal illness. Although several drugs have been devised over the past sixty years or so to combat this burden the parasite has, in turn, generated various kinds of resistance to them. These mostly involve one-point or two-point nucleotide substitutions which are specific but, given the number of possible mutations, statistically likely in the way of a lottery win. *Surely this is a clear example of Darwinian evolution in action? Yes, it is.* How does it work?

Take eukaryotic mutation rates. They differ by species and by section of a genome (junk *DNA* mutates faster than genetic) but take a rate of ~1 in 10^8 base pairs. There are ~3 x 10^9 base pairs in a human (against ~2 x 10^7 in *Plasmodia*) so that, on average, you might find ~30 substitutes and as many more insertions and duplications per human per generation. If there were 10^8 births per generation you could expect to find each nucleotide substituted in at least one person. All possible single point mutations would be obtained somewhere on earth.

A word of caution that to break is not to make. Such mutations either have no effect or change for the worse. They never innovate; they do not create new, complex structures or systems. For innovation simultaneous, mutually advantageous mutations would be needed in large quantities. And if they happened one by one how would advantage know itself; what could select it and not lose it in relentless churning throwing up new substitutions well before it found a use? It would, on the contrary, have to interlock within the tight constraints of already-working systems which repel misshapen interlopers from their machinations. In real circumstances (not imaginary) has randomness sufficient firepower to create and not just break?

What's being asked? You're asked to build a metro - not probably as complex as a single cell. Don't you think out every detail of the massive project in advance? You might include a little flexibility, such genetic potential as in life sex and adaptive super-codes can show; but could you tolerate loose bolts, warped threads, leaks or spanners thrown easily into painstaken works. Locked into genetic hierarchies and networks of interaction chance mutations can't have much effect. They can, as breakdown crews all know, cause faults but can they innovate; and add on top of innovation that's already there? In fact, trillions on trillions of mutations are irrelevant or positively impotent. Tight coherence of complex genetics and morphology rules out novelty. Therefore, can malaria show how macroevolution really fares? It is, after all, the largest test of evolution we can hope to observe; and it's growing day by day.

To cut a complex story short, in the case of malaria there can exist 10^{12} parasites in the body of each patient which, if a two hundred million (2×10^8) persons are infected each year, yields a circulation of about a trillion (10^{20}) parasites. The drug chloroquine worked against these bugs for a number of decades but became ineffective in the 1980's. Studies have shown that spontaneous resistance (ie. correct single point mutation) to another drug called atovaquone occurs in about one in a trillion (10^{12}) cells can occur in a number of weeks. However, *two* specific point mutations are needed, simultaneously, to confer resistance to chloroquine. In this case the odds are one in a hundred billion billion (10^{20}) cells - 10^8 less likely. It is confined to about one person in the world each year and may or may not, in those cases, happen to spread. Chloroquine resistance has, in accordance with statistics, shown up several times since research began.

If you took a human population (now $\sim 10^{10}$ but historically $\sim 10^8$ and always hugely smaller than the malarial parasite's) it means that specific, spontaneous double point mutations would occur statistically once in $\sim 10^{13}$ years - a span much older than the universe (maybe about 10^{10} years). It is therefore reasonable to say that not many, if any, such mutations have occurred in our particular species. What, therefore, can force the pace of an evolutionary 'arms race' to obtain new functions and facilities? Such novelties must have a basis in their biochemistry; molecular machines are made from gangs of proteins coupled up together; their coupling points (called binding sites) are a product of precise conformity. How, therefore, could such mutation rates allow specific sites to simultaneously occur and a machine, such as a ribosome, to pop up into view? *Indeed, no such sites are known to have randomly arisen! Thus how could organisms ever have evolved*? What about, say, mice or a special mouse (do you remember *Morganucodon*?) from which the panoply of mammals are supposed to have evolved over a period of just $\sim 2 \times 10^8$ years? Take, generously, a generation a year each with ten million members (10^8) for 10000 (10^4) species. You would hardly expect to find a spontaneous double point mutation but you'd really need to - indeed, perhaps trillions of them. How, otherwise did such variety emerge? Be bolder still! Revert to vertebration's Cambrian start ($\sim 6 \times 10^8$ years ago). Allow two generations every year ($\sim 10^9$) of an average million ($\sim 10^6$) species each with a hundred million members (10^8). You'd still expect to find only about a thousand such mutations. That is a reasonable but, in evolutionary terms, wholly unreasonable calculation. How many do you need?

Chloroquine resistance turns on interference with the removal of waste products from the parasitic gut. It involves mutations involving a particular couple of amino acids among those composing a membrane pump. Typically, such mutations merely render the binding configuration of a protein less effective; they do not let the drug latch on to do its work but, at the same time, neither add anything fresh to the complexity of a genome nor facilitate any incipient new function. When chloroquine is no longer used due to the resistance the weaker strain soon declines and the stronger 'wild-type' comes to dominate again. Such 'anti-evolutionary' behaviour is standard. When a threat or problem disappears the un-mutated 'wild-type' starts to repopulate the field. Whichever way you look at it, billions and billions of parasites known to science have failed utterly to evolve new protein-protein interactions, metabolic pathways, resistance to sickle cell trait, survival below 18°C or transformation into any other type of organism. **All malarial parasites remain the ones they were.**

Take cod. Like pollution-resistant worms, antibiotic-resistant microbes, Alaskan sticklebacks and many other organisms it is claimed that Hudson River cod have, in plain sight in fifty years, 'evolved'. In this case 'micro-evolution' has conferred resistance to toxic chemical called dioxins and PCBs. Some bacteria can detoxify these chemicals but nobody else. In the cod a certain protein receptor (present long before man concocted the toxins) allowed them to bind to it, thus stopping it from working properly. An adaptive mutation causes the shape of the protein to shift such that binding can no longer occur and this stops the poison wreaking its deadly effect. Clearly, in a poison-rich context only mutant fish survive; but where the poison is absent mutants only constitute about 5% of the population. *Such minor change to a protein receptor is undoubtedly useful in a particular environment but is many orders of magnitude removed from multiple changes that might be required to innovate (e.g. such a novelty as fin, swim bladder or a nervous system).* **In fact, there is absolutely no evidence that any such 'micro-evolution' is, with its genetic loss or minor change to a protein or two, going anywhere further than the lucky instance where, in a specific circumstance, it happens to help. Nor, more importantly and tellingly, is there evidence that new (or nascent) functional structures are currently evolving anywhere on earth.**

What about the human, as opposed to parasitic, side of things? The job of haemoglobin in red blood cells is to carry, by means of some precise engineering, oxygen from your lungs to organs. Several kinds of mutation to one or other of the four globin chains (that frame the haem co-factor and all together make up haemoglobin) are known to affect the human response to malaria. They include various forms of thalassemia (wherein genes are deleted, broken or switched off), HbS sickle cell anaemia (with a single point mutation at position 6 of 146 amino acids in the β-globin chain) and HbC (same point but different substitution in the chain). Though they afford varying degrees of protection against malaria their selection in most cases leads to clinical debility or death.

The best known change to the highly engineered shape of haemoglobin occurs in a form called HbS. Sickle-cell anaemia is a textbook icon because it is supposed to illustrate evolution-in-action (which it does) but also thereby underwrite evolution-in-general (which it doesn't). This disorder is, like cystic fibrosis, painful and debilitating. Due to mutant haemoglobin its 'sickling' clogs cells and veins, causes infarction of organs and, at best, much shorter life expectancy. Children start to suffer brain damage as early as four years old due to such blood thickening and clotting. Those who receive the mutant gene from both parents die painfully before reaching maturity. How unequivocally beneficial is that? However, those with one 'normal' and one 'sickle cell' allele neither die from malaria in childhood nor suffer much from 'sickling'. In this case the presence of a mutant gene in malarial areas confers resistance to malaria and thus better health than its absence. It is thus commonly passed down and is, surely, a *BM* (beneficial mutation) and proof of evolution? Is it not, as often cited to biology students, an excellent example of the 'positive' way that evolution 'works'? In a similar way the mutant allele for cystic fibrosis is supposed to survive because it may confer some resistance to cholera. Thus serious negative effects with serious secondary malfunctions and suffering are played down in order to label a mutation 'positive' or 'beneficial'. There is no doubt that sickle-cell anaemia demonstrates natural selection at work. *But is it a demonstration of evolution not in the trivial, Darwinian sense of variation but as a step towards macro-evolution and the origin of biological novelties?* The trait is a defect; it shows neither increase in genetic information, in physiological/ functional complexity nor in structural complexity. *In other words, it adds nothing to an 'upward', 'progressive' evolutionary argument.* Indeed, an increase in the number of mutant alleles simply leads to more patients suffering the painful, lethal illness. Would you want sickle-cell-anaemia or cystic fibrosis in order, if perchance the circumstance arose, to obtain its spin-off effects? It is like claiming that a jammed lock on a car door was an improvement (a beneficial mutation) because it prevented or diminished theft. This does not mean jammed locks are desirable or positive features of a vehicle, just that you may keep a car longer that way in a criminally diseased area. Indeed, just as locks are not made broken so neither the origin of haemoglobin, red-blood cells or the whole system that supports the stream of life in the execution of its purpose can be explained by the gross malfunction of a part. *The sickle-cell mutation is negative. It represents loss, bug or breakage in the system.* Indeed, the mutant allele 'self-regulates'; it is naturally stabilised because otherwise, at a tipping point, so many persons would inherit its deadly double dose of sickle gene and die that it would be wiped out. However, partial success is not what evolution is about. *This example, commonly touted in textbooks as evidence for evolution, is no such thing. On the contrary, it illustrates a fact of genetic science - unequivocally beneficial mutations are unknown and even those that raise tangential benefit are very rare. And, like all mutations of a gene, it represents not gain but entropy of information*. Make a study of life's mutant suffering, genetic disease; evanescent sign of *BM*s means scant sign of evolution. If sickle cell anaemia is a classic proof of evolution then it really points a mutant theory up. From such quality of 'proof' you might well infer disproof.

Sickle cell resistance to malaria derives from a single point mutation. Others (on different nucleotides) have not done the trick. Nor has blind flailing found a systematic way to cure malaria. What, therefore, does that other stalwart of experiment, *Drosophila* the fruit fly, have to say about mutation. Fruit flies have about 13500 genes on four chromosomes. Perhaps, after *E. coli,* they have had more mutations forced on them than any other organism. Over 3000 have been catalogued. All sorts of monstrosities and debilitations have been induced but they have remained fruit flies. Even mutations of high-level, master genes that control the developmental 'cascade' have crippled them but engendered no hint of change into another kind of organism. A simple example of such limited plasticity comes from experiments to try and force an upper or lower limit on their fly bristles. The norm is 36 and, after a lower limit of 25 and an upper of 56, the end was reached. Variation ceased and the creatures died. When allowed to breed freely, the number rapidly returned (as also in the case of eyes in previously mutant, eyeless flies) to normal. *This clearly illustrates the typological notion of elasticity of the gene pool with limits past which it cannot be stretched. It is, therefore, an opportune moment to reiterate the fact (see Chapter 21 and figs. 5.1 and 22.1: Limited Plasticity) that there exist limits beyond which breeding cannot take an organism or it dies*. **A fly remains a fly remains a fly.** If you can't engineer one from an egg, how do you think mutations blindly could? Moreover, if no new addition has occurred to fly anatomy then, in however many clueless years, zero multiplied by any number doesn't give you more! No evolution on the fly! The numbers flatly, mathematically shout there never was.

No doubt neo-Darwinian 'micro-evolution' is chancy, rough and blind. Above all, by generating no new suites of information, it is uncreative. It happens and it plays a part in variation. But there is something even more serious for the macro-evolutionary problem than microbes, flies or parasites. It involves the fundamental way cell chemistry, which underwrites all life, works. *DNA* base pairs 'bind' because of their precise, complementary geometry of shape, its orientation, their chemistry and the position and strength of electric charge(s); so do *t-RNA* molecules with specific codons and amino acids. Enzyme with substrate, immune system components such as antibodies and cell-cell surface recognition units bind to other chemicals as lock and key - exactly. This means that, for various purposes, proteins are engineered to fit in complex ways that rule luck out.

Not only this, but nearly all important cellular processes are mediated by assemblies of at least half a dozen and mostly more than ten proteins. These components must each exactly match their complement(s) with respect to the above-mentioned characteristics. The binding surface(s) must fit to the extent that they can without mistake self-assemble and, furthermore, perform the 'algorithmic rituals' in which they engage with other assemblies. Such non-random complexity is staggering but it is the way cell business works everywhere and all the time. How, therefore, did these complementary assemblies evolve by gradual, randomly generated steps when it takes odds of one in a hundred trillion (10^{20}) to produce just two particular spontaneous point mutations? Such changes, registered in protein, just distort a site sufficiently to either, as the case may be, disrupt or improve the binding of a previous circumstance. They can render a drug or antibiotic ineffective; they can stymy recognition or attack but how did protein-protein binding sites originate in complex, interactive quantity? If you want just four such sites, not a minimum 'hexaplexity' you have to double the odds. You now need 10^{40} organisms. The number of microorganisms living in a year is estimated at 10^{30}; and for all cells ever $\sim 10^{40}$. Thus, Behe notes, complexes of three or more different proteins seem to mark 'the edge of evolution'. This is where you draw its line. It might be even worse if *HIV*, *E. coli* and *Plasmodia* are anything to go by. They do not create a single interactive site in a new system; they do not create a single metabolic pathway or a property. And a single, typical cell contains 10000 complementary assemblies laden, every one, with coded, interactive binding sites.

Time is not the issue, population number is - and most probably intelligence as well. Because, extrapolating from the odds of one in 10^{40} to produce two protein-protein interactions (with four complementary binding sites), you would need all cells since life began. Politicians always bandy figures but it is therefore reasonable to suggest that random mutation with natural selection in tow is wholly incapable of generating cells. The origin of multi-cellular organisms, with different kinds of cell and developmental algorithms, is, except for wishful thinking, even further beyond the dim and dumb capacity of Darwinian 'triviality'. We see that evolution's engine, a mutation, causes suffering, breakdown and, crashing through sickness, death. Mutations may be '*fleurs du mal*' but not the seeds of fresh ideas nor the source of real biological novelty.

Do you believe in general evolution even though, while you might wish one will be found, no mechanism holds it up? So far the tests have failed. Variation happens constantly but not, even on a small scale, innovative evolution on the grand scale theory demands. A Theory of No Intelligence does not supply a macro-evolutionary mechanism nor support the idea of life's general evolution here on earth. Numbers crush. Limited plasticity and contrary interpretation wipe it out. Indeed, Francis Galton, Darwin's cousin, thought small variations would average out, that is, show 'regress towards the mean'. No major change would come of them. This is not exactly 'limited plasticity' but echoes Natural Dialectic's variation-on-an-archetypal-theme.

One's own and peer group philosophy may grasp an anti-scientific view but, given all the facts that we know now, it is reasonable to suggest a reasonable man like Darwin would have conceded the impossibility, in its grand scope, of his theory.

In short, then, micro-evolution is a prejudicial, biased word. Evolution-in-action is, simply, variation-on-predesignated-theme; this variation is always constrained by working systems already in place; and it is either coherent, due to in-built genetic potential or incoherent by Darwinian mutation. *Variation is in action but grand macroevolution's not.*

Non-Protein-Coding *DNA*

About 2% of the human genome codes for protein. If you assume that the only function of the genome (a cell's *DNA*) and transcriptome (its *RNA*) is to produce protein then you may have judged the other 98% non-genic i.e. ^jnu~&n*k. Such ^jnu~&n*k was sometimes called 'evolutionary jetsam and flotsam' or 'dark matter of the genome'. Is it really accumulated waste or does it do anything useful? Of course, if it's proven waste then Darwinism triumphs and Design is dead. God is dead, long live the idol! This is why polemicists fix on this issue.

Top-down, you would expect a program (including a biological program chemically inscribed as a genome) to be written and initially 'go live' bug-free. You might expect, after a period of time in the manner of 'bumps and scratches' on a vehicle, a number of accidental but non-lethal bugs could accrue. *However, the primary function of a biological genome is to serve as a database for protein required to make an organism; and a database or library is useless if inaccessible. A large data bank, like disc storage, a library or even a single book needs be accurately indexed, accessed and regulated in its provision of correct information at the right time in the right quantity of copy.*

You would *not*, therefore, expect that non-protein-coding (n-p-c) elements in that library would constitute what used to be interpreted by a Theory of Non-Intelligence as the 'flotsam and jetsam' or 'junk element' of a genome. You would, instead, expect all code except some small accumulation of 'nearly neutral' or 'non-lethal' bugs to be specifically involved in structural, indexing/address-related, switching and other regulatory functions. Chemical relationship is made by properties of substance attraction called electronic affinity and resonant association. You might. therefore, further conclude (with Chapter 16: How Does the Connection Work?) that such relationship, when used as an index or other component of an operating system, needs be as precise in signal connectivity as the configuration of an enzyme's active site. **In short, you might predict that 'junk' (much more humbly and precisely called 'non-protein-coding *DNA*') was a term born more in the darkness of ignorance than light of knowledge; and that any minimal, mutant fraction played a negative or at best neutral part in the computational expression of genomic information**. You would, as we shall shortly see, have been entirely right.

The genome is a vehicle of information. Every vehicle takes, of course, some wear and tear and so some damage that is rightly called 'junk' can be expected; but how much can its healthy operation bear? Who cares? In an evolutionary paradigm you welcome accidents. Accidents (mutations) are what build vehicles; they are evolution's little engine. Some might appear to confer advantage; others might not but still, perhaps, provide a careless duplicate or unused space on which further accidents might somehow accumulate to the point of innovating something. You may find it curious, though, that some sections of ^jnu~%n*k are, for whatever reason, practically identical in humans, elephants, dogs and marsupial wallabies; indeed, nearly 500 sections of over 200 bases have already been found identical in mice, rats and humans. There exists, moreover, a strong correlation between the placement, though not the actual sequences, of such short and long interspersed nuclear elements (SINEs and LINEs) in rats and mice. Perhaps mammals each have their own repertoire of once-called-useless chunks of *DNA*.

Just as genes involve p-c-coding bands interspersed with n-p-c intron segments so chromosomes are also banded. A banding pattern on mammalian chromosomes reflects such systematic densities in a compartmentalised way that resembles bar-coding. So-called 'white', loosely-packed R-bands (called euchromatin) are variously rich in the genetic letters G and C, occur where there is a high concentration of SINE elements, active transcription of protein-coding genes and replication early in the process of cell division. On the other hand, 'dark', tightly-packed G-bands (called heterochromatin) are rich in the letters A and T, transcriptionally inactive, occur where there is a high density of LINE elements and replicate late. Chromatin is the combination of *DNA*, *RNA* and proteins that compose a chromosome; and a chromosome may be seen as one volume, accessed in an automated way, of the encyclopaedic genome. Chromatin is closely bound up with the regulation of gene expression. Any affect on its organisation impacts on this life-critical expression; and it is known that n-p-c (or nc-) *RNA*s (see Glossary) help compose such organisation and can affect gene expression by modifying it. Also, various proteins and n-p-c *RNA*s mediate the process of correctly attaching the transcription complex, *RNA* polymerase, to *DNA*. So did n-p-c repetitive junk *DNA* coincidentally accumulate about the same but also differently in each rodent's banding case? Perhaps these tracts of uselessness are just analogous to ancient data left on a hard-drive after the delete button has been hit. As hard-drive may be copied in entirety, so is genome replicated. You could hand ancestral rubbish indiscriminately down like this. Or has some higher order pattern been passed down rats, mice and other mammals, although each of the elements is systematically but differently coded, from a common ancestor? It is unlikely 'crazy' chance could work in such an ordered way. You might better presume that it has resisted natural selection, been conserved and therefore has some purpose. *Very often conservation is evidence for function. It suggests a necessary use.* In chromatin you definitely have it.

Repetitious tandem and interspersed n-p-c *DNA* occupies up to 50% of the human genome. Identified jobs include X chromosome deactivation, constitution of a histone-positioning code, repair of broken *DNA* and involvement in placental development. Transposable elements are also involved in cell stress responses, chromatin condensation and *DNA* methylation (see next section: Super-codes and Adaptation). Some uses even appear sequence-independent, such as stretches of repetitive code that serve to compose centromeres (critically linking chromatids in cell division) and chromosomal caps called telomeres. Thus is there not method in apparent madness, pattern in perplexity?

Organisms transcribe most of their DNA, including so-called junk, into RNA. Over 90% of bases from small, sample sections were, by 2007, found to transcribe by coding either for protein or *RNA*. While perhaps 25000 human genes that code for protein have been revealed there may be up to 450000 '*RNA*-genes'. What's the point of them? *Why waste energy transcribing them if they amount to useless and thus interfering junk*? Francis Collins, Leader of The Human Genome Project, noted some time ago that only a small fraction of such factors are known to be useful but his appeal to evolutionary 'junk' status for the rest simply assumed unproven non-functionality. Discoveries have overtaken this opinion. It is now defunct. Nature (vol. 489 ps. 57-74: 6-9-2012) has announced that, definitively, 'junk' is junk. **Already at least 80% of the genome is judged biochemically functional.** There may well be more to come.

Can a Darwinist be wrong - ever? Not if he argues by past chance (therefore irrational) events. By 2009 Richard Dawkins was still observing (The Greatest Show on Earth ps. 332-3) that 95% 'junk' *DNA* was an embarrassment for 'creationists'. In September 2012, with ENCODE's announcement, his tune changed. Now 'junk' was no longer evolutionary jetsam and flotsam but, as critical and refined coding, 'just what a Darwinist would expect'!! A twist, a U-turn flipped without as much as the wink of a cheeky eye!

The ENCODE project (**ENC**yclopedia **O**f **D**NA **E**lements) is a public research consortium launched by the US National Human Genome Research Institute after completion of the Human Genome Project to find all functional elements in the human genome. In 2012 a Cambridge co-ordinator, Ewan Birney, was quoted as saying that the genome is 'alive with switches, millions of places that determine whether a gene is switched on or off'. In fact, non-protein-coding regions of *DNA* engage in multiple functions. They generate, for example, those tens of thousands of short n-p-c *RNA* features (such as micro-, pi- and si-*RNA*s) that seem involved in the genetic operating system and thus regulate the addressing, transcription, translation and post-translation modification systems by which genes are expressed as protein. The answer therefore seems to be control; regulation of expression; an operating system so that exactly the right protein is produced in the right quantities at the right times in the right cells. Such tightly controlled expression is species specific and, within species, cell specific. This is no small order in a space far, far smaller than the eye can see.

Moreover, manifold transcriptional and other critical editing/ formatting factors are found systematically located before, after and even within protein-coding genes. Such genes are edited after transcription and what is left (called the exon) is passed to a ribosome for translation. Such editorial employs an orderly cut-and-paste technique controlled by n-p-c *RNA*s and called splicing. It has been found that regulated 'alternative splicing' can produce 100's if not 1000's of different versions of the same protein. Such production amounts to fine-tuning for bespoke molecules in different cell/ tissue types or at different times in an organism's life cycle. Alternative splices mean that, like cutting a film at different points, different waste cuts (called introns) accrue. It turns out that some of these introns contain highly conserved sequences (sometimes called motifs), suggesting a function. This, since motifs act like 'find' sequences applied to a text in word-processing, is most probably to bind a regulatory protein. In this case they would operate, like the abovementioned regulatory micro-*RNA*s, as a kind of genetic control panel, that is, epigenetic super-code. Repair, replication, transcription, proof-reading and editorial splicing - the organisers of such functions cannot be called 'junk'!

'Jumping genes' and other repetitive *DNA* were once considered to be composed by random viral insertions into host *DNA*. Such operators, called transposons, are mobile. They cut, paste and amplify material. Although the reflex, jumped-to explanation of their genetic status was dismissive we know what happened to organs called 'vestigial' before; they're known not to be 'vestigial' any more. Jumpers, of various distinct types that include abovementioned LINEs and SINEs, may or may not work randomly. A few may indeed be incoherent mutants and thus act as mutagens but swathes of systematic 'junk', conserved and transcribed, are something we can, obviously, bear. *Perhaps these transposons aren't a load at all.* Transposons and repeating elements may yet turn out to constitute another form of fine control; they may be, for example, silencing or switching factors, mobile agents of expression or of quantities of protein that, in varying circumstances, are differently expressed. In other words, they may constitute another form of super-code - genetic flexibility of regulation. Over 20000 candidates for such regulation have been identified so far. Informed conversion of junk into value now proceeds apace. In the section Natural genetic Engineering we'll probe further how the mobile elements in genomic operating systems, perhaps the active essence of computing in biology, are at present thought to work.

Is it the monster mash or graveyard smash? A pseudogene was presumed to be the genetic corpse of a once-functioning gene. The latter's function may have been killed by mutation or a lack of promoter sequences or introns (in which case perhaps it was once-reverse transcribed from an exon); or it may be the damaged copy of a working p-c gene without undamaged counterpart - a degraded result of genetic entropy. However, are all pseudogenes 'broken genes' or do some at least have useful function? As well as finding that a large proportion of n-p-c *DNA* (maybe up to 90% of *both* strands) is transcribed to *RNA*, the ENCODE project has also revived many pseudogenic corpses. Still more may rise. Already over 850 such resurrections are known to be genetically active. Many are transcribed in various tissues; and some *RNA* transcriptions have been found to translate into protein. This deletes the pseudo out of pseudogene! It is, further, believed that 'antisense' *RNA* from pseudogenes may be used to regulate, by increase and by inhibition, the transcription of 'sense' strand *RNA* issued from p-c genes. We'll check an argument about the vitamin C pseudogene later; suffice here to note that the same logic applies to pseudogenes as to repetitious *DNA* viz. conservation most likely indicates function. Some pseudogenes are extremely well conserved, that is, appear the same in different species. Either this highly conserved kind of waste is so inert that it endures below natural selection's radar; or it isn't waste at all. For example, a protein that is part of a tumour suppression system is transcribed from a pseudogene. Indeed, some functional sense has already been made of 7000 such 'vestigials' and 10000 long n-c RNAs (Nature 465: 24-6-2010 ps. 1016 and 1033). Since this nearly doubles the conventional genome it seems so-called pseudogenes might turn out to be a stalwart of epigeny. 'Pseudogenic mutations' may not mash out any monsters but a growing number would certainly, as super-coders, appear to smash their own graveyard status.

Moreover nucleotide sequences may be expressed from both sense and anti-sense strands of a gene; from sectors of the genome previously thought 'silent'; or even from overlaps with p-c genetic sequences. Protein-coding genes are symmetrically embedded in much larger functional sequences that regulate their expression. Acting as an associated control panel these 'functional' or 'contextual' sequences include so-called introns, promoters, enhancers and other surrounding instruments of correct expression. Many segments of the genome are now known to constitute subtle but important address systems, packing functions or regulatory and expressive agencies. One also needs to consider repetitive 'satellite' sequences essential for genome function. These affect regional folding, nucleosome spooling and *DNA* site exposure sequences, linkage points for centromeres, telomeres and other features of chromosomal architecture. At this rate a gene of, say, 3000 nucleotides may be thought of as involving a context of another 50000. If you have 25000 genes this would mean a context made of half your entire genome. Add to that another 50% now thought, sometimes doubling up by using both strands, to code for various forms of *RNA* and *RNA* regulators. Thus, before how they're expressed is fully understood, genes instruct informative, structural and metabolic applications; and, in its role of operating system, 'junk' seems almost wholly tidied up in exactly what it should be doing - facilitating 'on-screen', actual expression of such application programs!

The density of all this encrypted information (see also Chapter 20: Supreme Density of Data Storage) is breathtaking; and Collins notes in his book 'The Language of Life' that 'the complexity of this network of regulatory information is truly mind-blowing'. Indeed, the whole genomic business, with its two base pairs and two bits per nucleotide site, resembles nothing as much as database and operating systems. Bill Gates has commented on the computer-like nature of digitally coded *DNA*. Of all persons he might recognise machine code; and how very far from jumble, junk or randomness is any sort of reasonable, working information system. *Such is the language of systems programmers, informative Dialectic and the computer analogy applied to genomes.* **Research has now confirmed that a modular, computational picture of the genome is correct; it thus as much supports the informative predictions of a Theory of Intelligence as damns the 'flotsam and jetsam' notion of an evolutionary Theory of No Intelligence.**

Multi-featured versions of word processing or data capture programs devour storage space. They surround each entry with a context they can use. And each evolved new version with more applications needs a quicker chip and larger store. Dynamic *mental* evolution finds a way. Could it be, since matter has no mind at all, the same with life's dynamic script?

If genetic context, say an address system, suffered a mutation the bug would probably affect expression of its gene. It could strike in different and, especially in the case of a single gene that could produce bespoke protein variants for use at different sites, subtle ways. Until a program is fully understood the effect of specific bugs remains obscure; but you do not, in ignorance, therefore assume that n-p-c 'context' is just 'bugsy junk'.

In short, junk's not, for the most part, junk in any form of life. The percentage of negative load on chromosomes has, as design theory predicted, now been drastically revised downwards. The mammalian genome is described (Nature: June 2007 p.799) as 'pervasively transcribed'. In fact, in as far as it has so far gone the ENCODE project has concluded that, far from being 'evolutionary flotsam and jetsam' such material is the primary controller of 'slave' code, genetic *DNA*. As ENCODE's Birney told the BBC, "The term junk *DNA* must now be junked". *If the work of 'non-protein-coding' segments is critical then 'junk' is absolutely 'bunk'.* Yet even this is not the end of the matter. The genome encrypts layer upon hierarchical layer of dynamic subtlety. ***DNA* is not the boss. It is a critical resource that's 'brought to life' by epigenetic codes. These lead the 'dance' not *DNA* and orchestrate expression of the genes.**

Super-codes and Adaptive Potential

Adaptation by mutation - what was that? Darwin himself was inclined to a Lamarckian 'inheritance of acquired characteristics'. Unless he had read Mendel (who seemed to counter evolution) he knew absolutely nothing of mutations or genetic codes - let alone an integrated battery of super-codes controlling *DNA*.

Indeed, the various views of a genetic system touted throughout the twentieth century are now seen as relatively crude and out-of-date. *Yet adaptation as a result of natural selection (or, if you like, the impact of ecological variables) acting upon genetic mutation is still the central component of Darwinist materialism's Primary Corollary.* As the basis of evolutionary theory it therefore deserves a close inspection.

Are mutations that produce a key new structure with new purpose ever seen? Trivial, lucky variations can be found (see 'Evolution in Action?') but never has an emergent, nascent organ been caught graduating to a full degree of novel, integrated functionality - although their origin in either other-functional or pre-functional form is basic to the theory of evolution. Some such 'pioneers' should be visible today, burgeoning in organisms at various stages up to the integration of an efficiently functional new system. We see circumstantial improvements to what's already there but no sign at all of radical novelty. Things have to work; but neither observation nor controlled experiment has shown natural selection can manipulate mutations to produce a new function for a gene, hormone, metabolic or communications system or whole organ. Such observation sometimes elicits a howl of objection. 'This is not how evolution works!' One believes, as a 'gradualist', that an accumulation of small, heritable variations (now known to be caused, sometimes, by genetic mutations) is sufficient to generate adaptive potential. As we've already asked, however, were chemical 'mutations' able to program and construct the major molecules (*DNA*, *RNA* and protein) to start with? After that, are *any DNA* mutations ever 'structurally beneficial'? Where is the evidence? It should be everywhere around. **Nascent novelties should be emerging now but we see, especially in Chapters 22 through 25, why they cannot and, at least at present, are not found. Thus if, as Charles Lyell had Charles Darwin think, the present is the key to what has passed then why should we expect that evolution ever happened?** Perish, of course, that thought. It needs to be selected sharply out. If, however, gradual, nascent innovations don't exist except as simplex variations-on-a-protein-theme there must be an analgesic that could satisfy the evolutionary intellect. One takes the plunge and so adopts the equally unproven doctrine of large, rare and disruptive flaws whose seismic effect on an organism hurtles evolution 'forwards'. Perhaps '**evo-devo**' (Chapter 25) has become your favourite explanation; or, as an adherent of the hypothesis 'punctuated equilibrium' (Chapter 22), you might still hurtle unseen gradually. But if Lyell counted in geology he doesn't in biology.

It needs be noted that two entirely different kinds of adaptation (that is, change) to an instruction manual or its product are possible - accidental or in-built, non-purposive or deliberate. *From both top-down and bottom-up perspectives random and non-random variations can occur.* Random variations tend to degrade performance; their sort of change is unproductive, negative. One of them may rarely correct a previous 'accident' or, by reformulating a protein's conformation slightly, cause it to evade detection by an immune system; but mutations never add innovative structural or functional information to life's genetic vehicle, an integrated manual called a genome. Therefore such trivial, almost completely 'down-side' variation spells, as a study of genetic disease illustrates, trouble both for a suffering organism and for a theory that depends on chance for integrated swathes of information.

Without mutation what kinds of non-genetic and yet natural response to life's exigencies occur? Constitution, nurture, aging and the exigencies of environment can seriously affect your looks and fitness but are expected elements in any body's package. Both choice (such as life-style) and program play their non-chance part. Perhaps more interesting are non-random adaptations, adjustments or fine-tuning - capacity for variation genetically written into an original design. Bugs degrade but have you ever varied a computer program by varying its subroutines? Intelligent design might cleverly dovetail a working innovation; it might, moreover, include subtle buffering, adaptive variation on a theme that works by super-codes. This is exactly what we'll see.

Of course (see Chapters 17 and 22) deliberate *DNA* casinos have been established to perform sexual and immunological shuffles; these lend potential to adapt to challenges from the environment but could you call simple shuffling 'super-code'? Is there, however, any *top-down* menu system where a 'go to' switch obtains, according to the circumstance, a pre-arranged 'Plan B' or 'C' or 'D'; could deliberately programmed options, fail-safes or adaptive features be applied? Wouldn't it be clever if, as you read a book for information, you could pick out what you needed following the 'go to' signs? It is as if the pages of a book were so intricately packed with information that, according to the way you read it, different messages appeared. Such densely-packed code when unpacked (or decoded) gives multi-level answers; the same linear inscription of information can vary 'meanings' or 'the products' that it yields. Obviously, any disturbance to such a close-knit context would upset its whole integrity. Write a program or a book yourself incorporating various layers of cross-reference, feedback and adaptability for readers. Do you attribute any of your thought-through interactions to the mindless hand of chance? Could such hand ever have created your ingenious formation? *The IT profession, recognising such layered complexity, calls it data compression.*

Such layers of code might equally be seen to encapsulate adaptive potential. Discovery of the straightforward 'monolayer' that is nuclear coding of proteins was difficult enough. The instance of n-p-c *DNA* and nearly half a million snippets made of *RNA* are factors that, as super-code, control expression of the genes. Scope of complexity like this ramps up the problem of a clear elucidation of genomic operating systems. After all, it is not easy to look at a computer hard-drive from the outside and then work out from its molecular construction how the coding system works. Nor to do the same for a dynamic biological library operating at micro-second speeds in a volume much smaller in radius than the tip of a pin! This, however, is just the heroic venture undertaken in today's laboratories. It is yet further tested by the discovery of various epigenetic hierarchies, that is, layering of codes. Indeed, there is a sense that every switch and regulation applied to a gene sequence involved with the application program that is protein synthesis might be thought of as an 'operating' super-code. In this sense we've already met promoters, enhancers, inhibitors and n-p-c *RNA*s operating in a complex yet coherent regulatory manner during the transcription and translation of *DNA* through to protein. It will take skilful research years to unmask then codify these libraries of operators and their tasks.

As noted previously, splicing is a mechanism that involves sophisticated editing of genes as they are transcribed to m-*RNA*. An m-*RNA* 'photocopy' of original *DNA* consists of so-called exons spliced together; this chain of exons then proceeds to full expression in the form of protein. The pieces thrown away when exons are 'cut and pasted' are called introns. Why were introns present in the first place? How are they precisely recognised then cut away? Are they indices, addresses or cross-references wrapped around the gene - important notices that need unwrapping when the gene itself 'goes live'? More than this kind of super-code splicing can deliver yet another powerful, fine-tuning kind. Genes can be spliced, in different circumstances, to express 100's, even 1000's, of bespoke alternatives; resultant protein is thus subtly adapted for nuances of use in different cells and bodily environments. How do multi-layered systems of control, that make smooth work of great complexity, first arise?

Spliced variation broadens a genetic theme. It could, for example, fit subtle protein nuances to each specific type of cell in which they work. Genes are susceptible as well, by way of hormones and other chemicals, to influence from the extra-nuclear environment, that is, to environmental cues. The question therefore expands to one of the wider, extra-corporeal environment and phenotypic adaptation. Could certain environmental circumstances favour certain permutations of switches and thus trigger pre-coded buffers to changing circumstance? Could, for example, animals moult their brown, summer hair and re-grow white when winter snows are due; or arctic landscape predispose to white fur so that, as with hares or bears, a new species is deemed to occur? As oil and water evolution and anticipation sharply disagree. Could adaptations be, in fact, not random Darwinian responses but foreseen and thereby, preordained? No need for mutations then.

You could extend this rational perspective. Thousands of cases of rapid adaptive response to, almost interaction with, changes in environments have been recorded. They include, for example, brood sizes, finch beaks or, if a region becomes drier, the response of some plants by developing a deeper root system or a thicker waxy cuticle to cover their leaves and inhibit desiccation - but *only* if the buffer (or adaptation) is pre-coded. *The question is not how such rapid, specific and satisfactory response can systematically occur 'on tap' by chance. It does not. There is embedded prior genetic possibility, in-built flexibility to cope with changing external conditions. After all, such provident ability is part of what survival is about.* For Natural Dialectic information *is* prior to material expression. Of course, therefore, deliberate adaptive potential is incorporated into every scheme of life. Even the humblest, bacteria, can adapt to antibiotics more quickly than mutations might account for; it is suggested even there epigeny might play a part. What, after all, is adaptation but back-up in emergency, a sensible plan B? An alternative strategy is pre-coded to engage the pressing change of circumstance. The more complex and intelligent a system (for example a Boeing 777 or, perhaps, a government disaster plan) the more extensive will be the preparation for emergency. It will wait, stored in the background, until the crisis occurs; and various sorts of stress will trigger its emergence in the hour of need. Do biological systems show no signs of such pre-ordinate intelligence?

Do you remember the analogy of computer programs with genetic code and, held on nucleic acid hardware, the notion of an archetypal software cast in modules called genetic subroutines? Thus organisms can be seen as mosaics composed of functional units, as exemplified by homology, hermaphroditic plants and animals, the duck-billed platypus etc. In the way of computer programs, these bespoke units are tailored for use in different organisms by different sets of adjustments. Such resilient bio-logic will become still clearer as we move to Chapter 25. *So too will the fundamental principle of biological balance in its epigenetic aspects of regulatory homeostasis, feedback and a cycle of genetic interaction with the cell's other chemistry and, thereby, environmental conditions.*

No doubt, base sequence is (bar mutation) 'set in stone' but responsive regulation, whereby gene expression is appropriately affected, might be subtly added on. Gyroscopic systems balance flight, fin stabilisers balance ships that roll with swell dynamically; responsive meta-code (or super-code) might balance bodies as environmental change occurs. Nothing accidental, just facility for adaptation and adjustment in the form of switches might enhance the basic plan. Anticipation meets with flexibility - but not with perception and conceptual determination by a mind. In this case you might expect broad, inlaid pre-adaptive design would endow organisms with great reflex capacity to cope, over time, with environmental stresses and strains. A reasonable example is the way hostile conditions such as oxygen depletion or low pH have been shown, in yeast, to alter the shape of an editorial protein and thus affect the way the *DNA* is expressed. Another is the normal inhibition of heat shock protein *HSP*90 (one of the most widely expressed polypeptides) that has been shown to sometimes unleash a burst of variation including changes in leaf shape and colour. Common, versatile *HSP*90 derives from several genetic forms (including, which may have some undetected use, those labelled pseudogenes). It also binds *ATP* and protein, assists in both protein folding and destruction and interacts with nuclear protein so that chromatin, with its topology of *DNA* coiling, is intricately and appropriately remodelled. This protein's gene and/or surrounding sequences somehow mask hidden pre-adaptive genetic potential for variation. Why, before enough mutations had enabled such potential or the nascent adaptation was required, should natural selection not eliminate it? How is a multi-tasking protein integrated, by sheer mutant luck, into functionally separate systems; how did accidents not aggravate but dovetail fixture into apparently essential other layers of an organism's chemistry? You might hypothesize along a chosen line but if that line, in principle, is metaphysically incorrect then how much is that guess worth? The process is not part of a grand principle of evolutionary advance but, in *top-down* dialectic view, preferably one involving preconceived, local adaptation. Just as an electric switch does not create its circuit so various forms of super-code do not create an adaptation but instead express latent capability.

In other words *adaptive potential* might be coded and unlocked according to environmental as well as intra-cellular factors. **What was considered**

evolutionary is almost always not. Such potential is expressed as a matter of course within the architectural confines of a multi-cellular organism. Cell-cell recognition and induction factors, although they beg the question of large-scale morphogenetic construction, play a major role in the way cells position themselves against each other, elicit various responses and develop. What about environmental changes in the three material states of fluid air and water or on land? Climate certainly affects plant growth and light affects sexual cycles. Extra-cellular chemical gradients may also affect cell differentiation. It is hypothesised, for example, that plant cambium grows xylem to the inside as a result of higher carbon dioxide and phloem on its outside where there is more oxygen. The potential for cambium, xylem or phloem exists in cambium. Similarly, genetic responses through differently switched channels of adaptive potential will be realised under given conditions. If these change, that aspect may be *programmed* to respond. From a range of latent possibilities only the relevant tool is selected. Conditional but pre-programmed response might, by varying the combinations and order of relevant locks and switches, cause the differential expression of *modular* as well as genetic potential. Of course, as in the differential use of a single gene by editorial splicing and/ or other regulation, such pre-coded adaptability requires *more* not less sophistication, *higher* not lower logical algorithms. Such *de luxe* preordination is thus further removed than ever from the notion of accidental adaptation. On the other hand, natural selection's raw, incapable material includes no scope for the original development of organs or their gradual, accidental transformation into completely other body-plans to service very different modes of life; or even on-tap, as-required adaptation.

At another level tags and signals lock and unlock genes; they operate, as we'll soon see, with protein spools on which the 'hard-drive' *DNA* is coiled. They constitute a responsive mechanism for locking (silencing) and unlocking (expressing) genes called, as a structure, chromatin. Chromatin thus comprises a powerful layer of computational control. It allows for variation on genetic themes. Just as different sectors of genome were blocked off in different cells during the development of your own body, could not broad, generalised plans (such as those for a mammal, reptile or even vertebrate) be adapted by 'blacking out' different phrases, paragraphs or subroutines? Could not some throwbacks (atavisms) such as hen's teeth or human tail be seen as accidental letting slip, as far as the form of life in question is concerned, locked up potential; or as erroneous access to 'illegal/ forbidden' lines of code for a specifically-unwanted generality, that is, a structure included in the general plan but for a specific member of that class of organisms out-of-place? *In other words, just as different cells employ different permutations of an organism's whole genome, could different types of organism make mosaic use of archetypal programs and their subroutines? And, equally, employ bespoke reformulations of these essentially conceptual routines?*

A subtle, non-*DNA* micro-hierarchy of control is invested in genetic information over what is written out by *DNA*. What other kinds of mechanism might add layers of flexibility to normal *DNA*'s inflexibility of sequence? No doubt that genetic libraries need, like any other, indices. A cell needs swift and accurate response to a multitude of signals. It needs access to suites of information to construct the metabolic pathways of its biochemistry. Could there exist an extra dimension to the switching systems, one that can access on demand specific, pre-arranged adaptation; that can lend genetic flexibility to buffering the environment; in short, could a wide range of super-coding anticipate, in a 'premeditated' way, adaptive flexibility? It does. Could such integrated and encoded foresight self-configure by blind accident? Informatively speaking, it could not.

Adaptation-by-mutation - what, indeed, was that? Scholars once believed that adaptations evolved accidentally while natural selection chopped and chose. *Dump that belief because the super-codes are here!* And always were except we didn't spot them well. Super-codes are far-from-accidental, in-built governors that flexibly adapt inflexibility laid as the 'track' of *DNA*. They cause programmed variation, extemporise most orderly and are therefore adaptation's incarnation.

Do bugs engineer computer systems? That would constitute an evolution. **But code on code the revolutionary revelation of super-codes sweeps 'adaptation by mutation' up; layer upon informative genetic layer trashes 'accidental evolution theory' as efficiently as it promotes a Theory of Intelligence**. *These super-codes are switching systems that amount to a responsive, adaptive potential.* Potential means possibility. In life's case this means information in the form of options for response to changing circumstance. In a machine such 'choice' involves dynamic switching systems, fail-safes and so on. These are, as in computers, predefined. Super-coding tags genetic information so that appropriate response to differing circumstance *does* appear. Tracks don't change but signalmen switch points for different paths: the bio-manual's *DNA* remains unchanged but tags adjust the sections that are read. **This is epigenesis**. Epigenesis, codes governing codes, amounts to an extra informative dimension. Hierarchically arranged, it builds a towering, high-grade data structure. How can structures so compressing data 'pop up' randomly without a prior informant? Or be built, gradually with no end purpose, by a crazy, logic-less series of alphabetic rearrangements - as if books were scribbled by the flip of coins or tossing letters in the air? Why did men ever want to think that this was daft life's way? Super-codes spell death for evolution. *A responsive genome incorporates multiple, overlapping and interactive code in a mode the IT profession calls data compression. How can first-class data compression arise without a prior informant? Ask any software engineer. Why should DNA and its controllers, packing information just as tight as any hard drive, be different?*

Nuance. Epigenetic punctuation. Hide the cows outside; hide! the cow's outside; hide (or leather), the cow's outside. The basic code's not changed but punctuation charges it with different meanings. So various types of higher level super-coding double up, triple up and power a whole new order of complexity. Epigenesis involves the study of chemical modification of *DNA* that affects transcription, replication, recombination and the specific regulation and expression of genes. It is the study of gene switching and heritable changes in gene function occurring *without change in DNA sequence.* **Such regulation transcends primary *DNA* sequence and is so far understood to involve at least six main kinds of mechanism: spooling tags,**

histone-positioning codes, histone methylation and acetylation, cytosine methylation, use of larger protein tags, cortical/ cytoskeletal configurations and the precise disposition of materials in an egg at point of fertilisation. The latter - key priming - amounts to a launch trigger that has been called 'the zygote code'.

Firstly, therefore, spooling tags. Your *DNA* is superbly, functionally folded (2 metres of it folded into a nucleus approaching a million times as small in diameter). It is spooled onto proteins called histones; there are eight coordinated histones per spool and millions of spools per cell (indeed, it has been established that chromosomes contain twice as much protein as *DNA*). Two coils are wound round each spool or so-called nucleosome; using these spools *DNA* is further reeled, that is, super-coiled into exact topologies that are known collectively as chromatin. Chromatin is a finely composed package. Its close, nuclear association of *DNA* wound onto protein spools amounts to an index or filing cabinet whose critical role is to police access to the genetic library and thereby control expression of its information. Protein may variously occlude the *DNA* to block or impede expression; or it may permit an active, always specific range of response to incoming signals. The idea is not cut and paste but cover and uncover. Newspapers work, to some extent, like chromatin. If your interest is sport then blank out all the rest; if it's political then black out what's not political. Same newspaper, different permutations of access by readers to articles of interest: same *DNA* but areas of access differing according to cell type. Information, however detailed or complete, is useless if it's inaccessible. A large data bank, like a press library, public records, disc storage or even a single book, needs an index. To explain the algorithmic reason chromatin is used to blank, lock or leave *DNA* open for access is not hard; it is in part-control of gene expression. To describe the mechanism of specific indexation and explain the coded operation of its accurate censorship in terms of how non-*DNA* chemicals index *DNA* is much harder. And, in so doing, to describe the whole as accidental in conception or thoughtless in its origin grows harder too.

Even before considering the sorts of tag or 'flag' which are placed along both elements of chromatin, note that the *DNA*-histone interaction alone plays a central role in regulating the core process of all life - genetic activity. A nucleosome positioning-code, repeated every ten nucleotides, specifically dictates its binding points with *DNA*; this leaves linker regions between the histones upon which, lo, easily accessible initiation sites for polymerase attachment and thus gene transcription are found. Thus *histone-positioning code* is a super-code that, working in concert, overlays the genes.

Such complex chromatin, a prime governor of *DNA*, is refined by further subtlety. 'Tails' extruding from its histone proteins are coded using methylation, acetylation and other signal factors; chemical changes in the cell specifically affect those 'tails', their interaction with the *DNA* and therefore its activity. This additional program involves a *histone acetylation/ methylation code*. Put simply, gene expression depends, as well as labelling, on the differential burial or exposure of *DNA* in chromatin. An acetyl device is, as required, positioned by one specialised protein and taken off by another. Acetylated spools are 'unlocked', 'open' or 'accessible' to facilitate transcription. There are many thousands of such tags and attendant proteins for each chromosome. Methyl tags, on the other hand, used singly, in pairs or triplets, 'close' or 'lock' a section of *DNA*. Each main type is used in conjunction with phosphate and larger markers - sometimes in different permutations. Furthermore, these histone tags work as signals and switches in concert with nearby chromatin regulators and transcription factors. If the way a gene is wrapped changes, that is, if its chromatin combination is changed then its expression will change as well. And, in case of division, the cell generates hundreds of millions of new spools and markers. In this amazing way works the first facet of super-coded potential for genetic modification without disturbing the bonded sequence of bases. *Code on code, could mutations 'think up' such highly integrated, signal information*?

Secondly, on a railway direction-giving lines are called permanent way. Are signals and switch-points as vital as the lines of track? In fact they render flexible the path of 'trains of information' over 'static', chromosomal tracks of *DNA*. They grant such adaptive potential for the dynamic running of life's messenger as to explain much of the variation that a single coded system can express. And if chromatin allows a 'train' to run then at what stations, even on the open sections, will it stop? The *second* method of binary (on/ off) switching involves the systematic tagging of specific cytosine nucleotides on the *DNA* chain with a methyl (CH_3) switch. The presence of such a switch generally deactivates a gene. It silences expression; but lift the signal, change a switch-point and transcription's way is clear. It's not yet clear how tagging is decided; how is one cytosine correctly methylated and others, that would generate mistakes, ignored? However, it is certain that a second governor, such a *code of methylation*, pegs the road through *DNA*. Information systems regulate each body like a vastly complex signal box.

Just as you need an operating system to relate an application program to a database so an arsenal of ancillary devices act as signals and points that regulate the various tracks along which different permutations of genes are expressed in different cells. As mentioned, this equipment includes methylating enzymes, the actual signal of a 'fifth base' (methylated cytosine), promoter segments, transcription factors, protein elements with 'tails' incorporating complex code, various other types of movable flag (for example, acetyl and phosphate groups), repressors and so on.

Order within order; clever, very clever stuff! In the genetic signalling program massages are transmitted and different permutations of epigenetic 'levers' switched 'on' and 'off'. Just as switching algorithms are central to the operation of information technology so epigenetic hierarchies of control are clearly central to the study of genetics and gene expression. If you compare a railway system's permanent way to information stored permanently in a cell's genome then the possible routes that trains can take over the tracks are controlled by signals and switches (called in England points). Call the whole layout a cell and particular sections chromosomes. The configuration of levers set to 'on' and 'off' in this cell's nuclear signal box is called its 'epigenome'. An 'epigenome' reflects that cell's dynamic set-up; it reflects permissible routes, that is, restriction of the full potential in that unit at a given time. The business is capable of subtle variation, adaptation and control. For example, it is certain that, while cells in a body enfold an identical genome (they are clones), every cell type contains its own epigenome, that is, its own switching permutations to control the *DNA*. There are about 225 cell types in you, each with controls appropriately set. It may even be that, in this dynamic system, the same cell type/ tissue in different organs of the body has its tags set slightly differently; and that the switching algorithm in the same tissue is responsively re-set at different stages in life and/or according to changing constraints. Indeed, dietary factors and environmental pressures are known to cause a path to be re-set. In this way a further layer of regulation linked to external conditions is brought into the equation. Such changes are copied each time replication for cell division occurs; a special protein (called methyl transferase) makes sure that the requisite epigenetic flags are set up on the new strand. In this way the 'set-up' for a cell's track is maintained. If it were not so there'd be havoc down the line. Furthermore, a re-set permutation is thought to be heritable. *Outside influence as well as mutation would then mark inheritance. But nothing new would come of it - just markers set for different journeys down the same set of tracks.*

What, however, if the points were switched at random? Disintegrated pathways soon run scheduled traffic to dead ends. Cancer may be evolution on a small scale but at least as dangerous as mutation is disturbance to the super-coding markers; this random switching of genetic points and signals (in fact, an accidental alteration of the methylation pattern) is called an *epimutation*. How could concentrated, integrated regulation pop along piecemeal by uncoordinated, pathogenic chance inflicting its chaotic touch at different levels of the system? Information is potential; hardwired, epigenetic potential allows genetic flexibility. And where, in biology, is potential most purely expressed but in gametes? Can complex, specified preordination happen accidentally? Myth-wise, could an egg emerge from aqueous impurities?

This brings us, *thirdly*, to the so-called 'zygote code'. When it comes to eggs, sperm and development a whole fresh, brilliant world of super-coding heaves in view! New epigenetic configurations must and do accord with the growth of new cell types scheduled along the course of development. What, in the midst of this heaviest of biological traffic, causes tags to be re-arranged correctly and appropriately? Most pertinently, why should and how can the markers change because a cell concerned becomes an egg or sperm? They need to be because incoming tags are an adult in pattern but the zygote will need to pass through various developmental sub-routines. It will need a different setting so the slate's wiped clean; and, after fertilisation, a second purge takes place but some imprints (especially regarding growth) are restored. Thus epigenetic patterns can be inherited (more so in plants and fungi) across generations. Super-coding structures are maintained. This must be done correctly or, as for example in partial deactivation of the second X chromosome in females, pathologies may develop. How, though, can the system cope with the apparently contradictory tasks of leaving a number of specific tag permutations for inheritance while at the same time clearly re-setting the switches to accommodate a dynamic, flexible regime of growth and development? How, when an egg is fertilised, can tags, imprints or stamps (as they are called) be safely reprogrammed for a second time and critical markers set afresh? This wondrous process is not born of accident or muddle, that's for sure. Ask any manufacturer of super-chips if complex coding comes for free. Can wind and rain produce the goods or is prime charge intelligence - intelligence allied to purpose that anticipates and thereby specifies a future shape? In this case the attractor is an adult form. Before you start the 'image' of an adult form must be in place; this is the 'anchor point' that pulls the guided process forward. **So, in the egg is concept of maturity. Development is a conceptual not an evolutionary business. It makes you the construction of a great idea! And if you argue, tell me how an egg or sperm evolved.** No ersatz, arm-waving explanations, please. No answers, like the ones from Haeckel, that the next two chapters are about to comprehensively dismiss. Evolution theory is full of promissory hypotheses that will, forever, lack precision as they're based on unpredictability; on top of which there's no necessity these ever-varying guesses, even if they flourish plausibility, are in the least in fact correct. This time, moreover, it is absolutely obvious that chancy explanations will not do. Certain concept is involved; informative potential takes up centre stage; thus flow-charts might explain the evolution of an egg (with all surrounding apparatus that includes the male with sperm). In 150 years so far, however, no chance-bound flow-chart or other explanation whatsoever has been slapped upon the table - not even if selection's there to trim mutations that don't work. Do any? Name what you would need. And then you'd have to add the super-codes. In Chapter 25 the working of genetic switches, regulatory proteins and the extraordinary computational bio-logic that creates an adult form will be elaborated. **Such core super-coding occupies the centre of biology.**

An egg's origin is yet another black box when it comes to Darwin's Theory of No Intelligence. There's no denying atheism its tenacity, its gritted-teeth denial in the face of facts. Cells have functions. Eggs inform. They are as pure an expression of potential as the known universe of matter holds. Such whole potential will, by staged reduction, systematically produce the variation that was, from the start, prescribed. This inscribed variation marks the passage to an adult form. How's it done?

First of all there's even information at the edge of eggs. From membrane to its junction with the cytoplasm ovoid patterns are, by cortical inheritance, passed on to daughter cells. By the 1970's Dr. Arthur Jones had, after Tracy Sonneborn in the 1950's, declared such logical extension of inheritance. Indeed, not only architecture of the cortex but the disposition of all molecules within a zygote trigger, by their three-dimensional shape codes, programs that will tag the *DNA* and thus promote correct expression in both time and space. For example, as your skeleton defines your shape, so eggs, like every other cell, are accurately defined by struts and girders in the form of microtubules of a cytoskeleton. These tubules are organised by centres like the centrosome. Eggs lack a centrosome. Although their microtubules radiate from a position near the centre of the cell they lack dynamic flexibility until a sperm delivers, with its gene-set, one. Only then can what we call an active zygote, as opposed to passive egg, begin to work; only, with a host of other factors, can 1-d *DNA* be converted into 3-d shape This is to say that morphogenesis is part explained. Correct delivery of materials is a logistic every construction project is surveyed for; it is an essential factor in the realisation of an archetypal plan to raise a complex, specified configuration - in this case a cell, cells and a body-form. Is logistic just a lorry and its goods? What part does active information play; how, for future reference, is the whole construction plan recorded? Genes are reference manuals; they are an informative resource but not the only one. There's epigenome and, control above the physical controller, Natural Dialectic has proposed the ultimate attractor - archetypal memory in mind. This super-code is metaphysical but, for biology, the high reality.

Not only eggs. Research has suggested that a child may, while in the womb, 'set its biological compass' according to conditions there. It may 'respond' to its circumstance through a predetermined range of genetic possibilities. In other words, taking the foetal environment as an indicator of the conditions outside it may 'predict' its future life-style and adapt accordingly. Conditions involving nutritional poverty or extreme cold may, for example, evoke one 'circuit' and those involving emotional stress or a particular toxin dispose to another. If, of course, the foetal decision did not coincide with general external conditions then the epigenetic plan chosen would switch the child into a lifetime's mode of biological response 'out of tune' with reality. The issue becomes one of ante-natal care. Not only this but it has been found that new permutations of 'epigenetic flags' can also be inherited. These kind of discoveries, if verified, correlate perfectly with the notion of adaptive fine-tuning, systems programming of a complex, anticipatory kind and, therefore, a truly intelligent 'strategy of genes'. There is no reason, except philosophical materialism, to suppose that such provisional ingenuity, responsive but not innovative, does not underlie most complex, functional adaptations.

What, cutting to the chase, has such refined, dynamic adaptation, such intrinsic information for a flexible survival got to do with mutant bugs or plan-less evolution? Do you build high-class information systems or densely trafficked railways out of accidents of chemistry? Epigeny, beyond the excellent book of *DNA*, shines out as obvious a sign of exemplary, intelligent design; and it illuminates the nuclear core of all except, perhaps, bacterial life's great flexibility. For science the voyage of discovery into biological information technology might well be likened to the inspection of a strange artifice found on a desert island. What is it for? How does it work? How is its logic coded and expressed and how does this expression interact with non-genetic parts? After hardware subtler software is unravelled. There comes, beyond its hard-wired chip of *DNA*, an operating system's software turn. Chemical but coded epigenetic switches are of primary importance in the expression of genes. Such an IT specialist as Bill Gates has no difficulty recognising in such complex binary circuits a computer operating system of the highest quality, that is, of the best intelligence. Such efficient information transport systems really get you where you want to go.

It bears repeating that precision's not the same thing as rigidity. Adaptability (or flexibility of response) is built into high-grade computer programs at the level of systems analysis. Concept precedes the practice of its coded execution. You might wish, for example, to lock certain chunks of code out of reach of particular operations but leave those subroutines open to others. **You might, in a phrase, wish to incorporate 'adaptive potential' into your sophisticated**

system. The emphasis is placed on rational code. How do *IT* experts generate their codes at random? They do not. Their single-minded focus is on obtaining a desired end to the accompaniment of complete elimination of all randomness in the form of bugs. How, therefore, does an evolutionist explain the origin of quality controls such as described above? He may, for example, note that an indeterminate percentage of *DNA* could be 'parasitic junk'. This 'junk' is often methylated, that is, deactivated; its *DNA* is silenced. Did epigenesis start out, it is hypothesized, as protection that 'silenced' viral genes inserted by an unpredictable attack? Did it then advance from bug control to top-range regulation, that is, provision governing gene expression and, as such, the very basis of biology? What dazzling promotion! By what steps precisely must this ultimate career path have evolved - or is precision not a question that is fair in science or, at least, scientific guesswork as regards life's history? *Top-down*, it is agreed that switches of the epigenome may indeed 'shut down' rogue elements. They may gag potential trouble but this is their secondary not primary remit. The primary task is as intelligent as any working system's own intelligence. Is not a railway company a good example of scheduled complexity, integrated information and operational intelligence? Epigenesis plays signalman who orchestrates the programmed show and, by the way, keeps traffic off unsafe or inappropriate sections of the track.

The genetic correspondence with intelligent *IT*, implied by use of the loaded word 'code', cannot be overstated. 'Flowcharts' involving switches are in fact a menu system. Their conceptual flows are just what you expect from a purposive, coherent program and they occur during the development of an organism and throughout any non-developmental zones and periods. In other words they are an integral, pre-coded part of the regulation of cell differentiation and development throughout life. Epigenetic tags are a method of realising flowcharts in chemical reality. Sets of regulatory proteins or markers bind to encoded *DNA* and govern the level at which a gene is expressed. Ask any electrician. If, in any complex switching system, a switch is misplaced or malfunctions expect trouble. In life inappropriate switching means disease or death but what about appropriate options? The epigenomic way tags are placed on *DNA* varies according to cell type, developmental stage, sex, age etc. Functional diversity is claimed from rigid sequence. Such flexibility is like fine-tuning of an instrument or vehicle. It gives more gears or options than a single one. Sophistication illustrates the polar opposite of chance - advanced intelligence. Yet, amazingly, those so trained continue to interpret findings through lab goggles in the 'scientific' light of Darwin's theory! How powerful is personal and professional philosophy!

Could even deeper coding underlie nucleic acid sequences and tags? Engaging with a cell's architecture could machine code, overall, inform and supervise the chemistry? At the electric level chemistry's machine code, binary/polar charge, might operate. Imagine you could airbrush out a body's mass and thus reduce it to the other of its fundamental units - presence (-) or absence (+) of electricity in space? You'd have mapped the unsupported structure, you might even monitor the dynamics of this invisible, electric 'ghost'. Have you ever seen a ghost? Such 'ghost in the machine' has already been encountered (Chapters 15-17 *passim*) in the form of *H. electromagneticus*. This sub-atomic, electro-dynamic and, it was suggested, psychosomatic aspect of being could be viewed as a sort of stereo-computer system. According to this model the architecture of a protein or nucleic acid is a digital framework that supports, protects and presents a three-dimensional electrical shape, made of ions or functional groups and exact down to the last electron, so that it can interlock with another exact binary composition and promote a required reaction. In other words, the electrical composition of nuclear material, including the context of indexation, would be part of *H. electromagneticus.*

In seeds it is found that change in a single gene brings change to the overall electrical pattern of the chromosome. There seems to be a close relationship between genetic constitution and electrical pattern. Likewise proteins, including histones which are bound to the DNA to make a chromosome, are charged molecules sensitive to their electrical environment. Differing electrical 'ambience' in different cells is calculated to 'break into' different subroutines of the genetic 'bar-code'. Does such an 'ambient' index help inform the complex process of cell functioning?"

Writing in the New Scientist (2-1-82 p. 220) D. MacKenzie thought so. In an article entitled 'The Electricity that Shapes Our Ends' he says, "The epigenetic input - that is, telling a cell where and who it is and therefore which genes it would be appropriate to use - is after all the crux of developmental interactions, the dynamic process through which differentiation and morphogenesis are enacted. Many of these instructions, it now appears, ride along currents of electricity."

All data banks need careful indices. Epigenetic factors serve to index and control the operations of our 'book of life'. **Which comes first, an index or its book; a book with index or its author; or them all together?**

Finely detailed adaptive strategies have been incorporated so that organisms can respond flexibly to environmental changes. Heat, cold, pH, light, chemical and other factors can trigger latent genetic and, thereby, mnemonic adaptive possibility. Coloration, nitrogen-fixing root nodules and various types of photosynthetic adaptation are a few among many examples of subroutine within subroutine, offering biological suppleness. While they may involve epigenetic super-code most also rely on 'permanent' genetic and controlling non-genetic *DNA* sequences. Yet are the latter not, as variations on conceptual theme, still genetic super-coding built as blocks of subroutine assembled into different kinds of organism. You might include (see Chapter 22) homologies and even body-plans as, at root, expressions of conceptually-originated code. Which symbolizes which? Is organism symbol of its code or *vice versa*? Is thought code's symbol or is it the reverse?

Interestingly, bacterial nitrogen fixation using a form of haemoglobin is a symbiotic process involving about 50 genes in the bacterium and another 50 in the host plant. What steps led (accidentally, of course, despite plant breeding institutes that try to copy it) to this highly specified and complex form of nutritional enrichment? No answer is supplied. OK, then, look at photosynthesis. Variations on 'normal' *C3* photosynthetic mechanism, the 'subroutines' of *C4* and *CAM* (Crassulacean Acid Metabolism) allow more efficient sugar production in dry regions. In both cases the occurrence is randomly mosaic. For example, *C4* adaptation is sprinkled round in 16 plant families but usurps none completely. Some species are even intermediate between *C3* and *C4* mechanisms but no-one claims *C4* evolved from these particular specimens. In fact molecular biology has shown that genes for *C4* are present but inactive in *C3* species of the genus *Flaveria*; and partially active in the intermediate types. Each photosynthetic adaptation transcends, in its informed complexity, a stream of consecutively correct genetic mutations or the random construction invoked by the plausibility of genetic duplication and neutral gene theory (see Kimura). It involves the genetic potential, in masked or inactive code, for complex traits. A purposeful, alternative response is triggered in ways not yet fully understood; an ingenious, prescribed solution to difficult but specific environmental problems is expressed.

Witness the wide variety of Madagascan lemurs, African cichlid fishes in Lakes Malawi, Tanganyika and Victoria or the new species of old types that irradiate remote locations everywhere. Mutation may have played a minor part but why not claim determination at the heart of difference, that is, fixed track of code with changes pegged out using heritable flexibility of super-code? You would propose, according to the Dialectic, limited plasticity but in-built scope for 'chaotic' or contingent transformations on, say, the cichlid scheme. You can't predict the changes any more than evolution can; but at least, like fractals, there is formulaic method in the madness. The rash of variation always turns upon a central datum, on an archetypal theme.

Accidents happen but for Natural Dialectic adaptive potential is not an accident. It is a sophisticated feature of every bio-system; and it starts where every practice starts - from informative principle transmitted through a code. The basis of biological flexibility is one well known to *IT*. It is the construction of hierarchical code. Sub-routines are nested within a master routine and switched on or off by means of a switching super-codes. What else but 'computer programs' are logical 'genetic strategies'? Strategies aren't accidents. Epigenetic switches are an efficient necessity programmed into genetic operations. *These super-codes constitute, far from a matter of evolutionary chance, an additional level of sophistication, a switching refinement that signals routes through 'memory-routines', that is, through code for metabolic application programs stored on DNA.* Of course, such 'sensitivity' is no more an accidental 'afterthought' born of accident than consciousness is an epiphenomenon of brain. Wherever there is communication, signalling, centres of control and the transmission of purposive information switches are core business. You find them in transistors on the circuits at the very roots of information technology. Viewed in this way adaptive traits are not inherited as chance-born products (however well 'selected' later on) but, programmed flexibly by super-codes, as preconceived shock absorbers, buffers or resilient options.

Third Stage: Natural Genetic Engineering

How could adaptive potential have evolved? What mechanisms might provide and then re-fashion raw, fresh information into functional novelty?

By now it's clear the 19th century emphasis on natural selection's power to evolve life-forms is as impotent as incorrect; and that neo-Darwinism's 20th century notion of a gradual accretion of mutations is insufficient so that the century's textbooks are, at best, no more than fractionally right. What will the 21st bring and how the 22nd mark its work? No doubt, molecular biology and genome studies now give lie to much of what has been proposed before. Neo-Darwinism's mechanisms can't produce the goods; its perspective is being replaced, with some resistance, by information-based systems biology. Informatic and not mechanistic, though this radically new evaluation hoves into line it must not violate the norm; it must, of course, still toe the principles of naturalism's mind. Biology must be reduced to chemistry and physics. It is assumed that variation must, by evolutionary extrapolation, accrue to macroevolution satisfying secularity. Our secularity demands a rigid mind-set, one that cannot work outside its box. After all, what's not materialistic isn't science and science is today's authority! How, therefore, might a shift from neo-Darwinism's paradigm be engineered?

	tam/ raj	*Sat*
	physical expression	*Archetypal Information*
	material outcome	*Immaterial Concept*
↓	*tam*	*raj* ↑
	physical mechanism	*mental imposition/ plan*
	informed body	*informant code*
	target parts	*symbolic representatives*
	component	*integrated system*
	individuality	*regulation*
	disintegrative tendency	*integrative tendency*
	reductionism	*holism*

Why, naturally of course! No prior concept is required. 'Natural genetic engineering' is the 'systems' metaphor. It is proposed by James Shapiro and leads us to *the third stage in the development of an inverted logic*. Could it offer a third and improved explanation of how evolution works and so constitute the final answer? The first move is, before we look at facts, you have to think 'as if' there was an engineer but know that there is not. We've met this approach, such *a priori* excision of an active mind, before (in, especially, Chapters 5 and 6). It employs, in animistic, pseudo-pagan form, the usual bio-suite of words implying at the same time as denying teleology - code, signal, target, opportunity, decision, preference, choice, recognition, functional specificity, invention, regulation, engineering just to name a few. The model's teleological but not interpreted that way. Its self-contradiction philosophically denies the words their cognitive, mental sense.

The concept recognises, though, the crucial question's one of informatic innovation. How, in fresh conceptual territory, might one explain the informative origin of such deeply integrated, well-organised and complex circuitry of regulation as evidenced in the machinery of *every* cell? How do you engineer the starting-block? Where do you obtain the information? After that how then can you switch, swap and shuffle innovatory combinations up?

Now that an informatic metaphor is recognised recall that bio-logic works more dynamically than any manual's rationale or even a computer program's steps. Its *top-down* routines, subroutines and lines of code are chemically inscribed to engineer its symbolism into an embodied fact - one such as you or me. In such non-random, busy, highly regulated circumstance gradual mutation-by-mutation evolution is a story now black-marked as weak beyond the possibilities of chance. The question then becomes how functional code originates; and, if they do, how chunks of it change function. If localised mutations constitute impoverished an information-generator then some richer input is required. 'Combinatorial thinking' proposes shuffling modules and accreting extra chunks of *DNA* might accidentally obtain, in jumps called 'saltation', changed signal clusters, splicing factors, swapped modular domains or whatever other kind of variation yields fresh meaning, that is, functional structure. For this you need to grasp how the genomic system's operators such as 'go-tos', 'dos', do-nots', 'turn-ons', 'turn-offs', cut-and-pasting and a host of other kinds of regulator each specifically works. Could 'mobile controlling elements', as Barbara McClintock called them, hold the key? Could what are, crudely speaking, jumping genes now flip evolution's mechanistic scene?

As if unravelling a C-drive's operating system and its applications code molecular biology is now discovering how all genomes process information computationally. Theirs is not *ROM*-feature (as Natural Dialectic has suggested, in Book 2, might be the case for archetype) but *R-W* (read-write). Genetic material is now known to be subject to continual non-random switching like you'd find inside a busy signalbox. This heavy traffic of specific changes that adjust to circumstance is levered by a dedicated suite of functions (such as replication, transcription, translation, editing etc.) and each functional team's executive components (such as *DNA* sequences, protein and *RNA* regulators and various other kinds of signal that control expression of code symbol into morphological fact). Just as a machine or a computer is entirely chemical so science legitimately examines the non-historical biochemistry of a cell's cybernetic engagements. Might you extrapolate (as natural genetic engineering tries) toan historical description of a system's mindless origin in this same frame? What does such a theory, which includes the way that evo-devo's major factors might have been constructed, propose?

It admits all cells are made of finely integrated parts and freakishly controlled; and that a read-write system using mobile elements (such as transposons, retrotransposons, specific *RNA* fragments, regulatory modules, proteins and other sorts of 'conditional micro-processor') cuts and pastes to guide genetic action. A cell is as automatised as a computer; it is as 'cognisant' as a programmed assembly line. Moreover, the transduction of its information is clearly an exchange between symbolic and real worlds. What other sort of relationship has, say, *cAMP* (a secondary messenger) between its structure and the metabolic message that it represents? The same applies to *DNA* itself. How, though, might chemicals react to engineer such a relationship?

Most of a genome does not code for protein (about 1.5% of ours). As a book accumulates tears, blotches and bent pages so a small portion may actually be irrelevant junk but the whole is of a precise architecture involving quality control of its expression. Such control includes, for example, a hierarchy of proof-reading components to render its content stable. Moreover accurate transcription, translation, replication, transmission at cell division and, as appropriate, the programmed condensation and formatting of chromatin packages are par for the molecular course - indeed, they are in principle essential from the start for any cell's survival.

How, therefore, does theory propose that functional novelties appear and, at the same time, seamlessly integrate themselves into already highly integrated contexts - seamlessly because ill-fitting parts will be selected out? Of course, in all machinery random faults occur. An engineer foresees and arms design against as many as he can. For example, despite several proof-reading steps an E. coli bacterium shows a single mutation in every 10^9 base pair copies; yet such lapsed fidelity, one in a billion parts, falls far short of any but minor variation. Might the genome's operators, mobile elements, be able aimlessly to engineer saltation by rearranging chunks of code?

Susumu Ohno's idea of gene duplication has already been discussed (Chapter 23: Entropy of Information). Such duplication does occur. It can even, where polyploidy is concerned, generate a hybrid species. The condition, found in many plants but rare in animals, involves no sequence mutation, novel genetic material or radical change of form. Although sexually incompatible with diploid forms, which isolation may yield independent populations, essentially an organism keeps its type, say, wheat or yeast or whatever else. There is no indication polyploidy ever played a part in macro-evolution. Nor did natural selection ever generate new species. But you might imagine massive genome reconstruction in a single cell within a single generation could easily perform the trick. Ciliate protozoa can, through shock or sex, freely reorganise their genome - but do not evolve. They stay ciliates, such as *Paramecium* to *Paramecium*, just the same.

Shock and stress; since all aspects of cell function and their biochemistry are subject to strict regulation evolution, even variation, seems entirely constrained. Might not stress trigger adaptation? It does. Genetic and epigenetic modification may occur but does not change the organism into something else. Indeed, you might argue that stress triggers predetermined buffers such as any engineer with foresight would employ to take up slack, resist force or respond to unaccustomed stimuli with some degree of flexibility. Such standard practice is mindlessly and yet not mindlessly observed! It happens naturally as mechanisms always do but better answers to ingenious engineering than a non-specific slip of chance.

Let's skid along another mindless tack we first zigzagged in Chapter 23: Evolution in Action? Let's consider multi-drug-resistant plasmids passed from one bacterium to another by a process called conjugation. Receptors gain immunity but strains and species stay the same. However, horizontal transfer of genetic chunks is, since all prokaryotes and many eukaryotes acquire them from often unrelated types of cell, worth pursuit. Parasites certainly cut burdens into host text but, of course, both parties stay, as organisms, just the same.

Ah! Not possessed of editorial checks a virus can mutate out of control. Could this represent the path of progress leading to new forms of life? But, for example in the case of *HIV*, we do not see nucleic acid transfer generating functional novelty. Neither party changes from itself. In fact, bugs normally upset normality. They don't enhance performance rather bring you crashing down. Yet, some seriously claim, a pathogen might confer everlasting benefit;

germs might be evolution's saviour since infection is the way that, from a 'psychedelic' virosphere, novelty's injected into genomes and, hopefully, perhaps some 'improvement' is assimilated and we all evolve! Stealthy 'progress' but, though misinformation from a microbe can cause malady, no-one has demonstrated that hijack is able to uplift the biosphere or lead to ever greater functional complexity. From an evolutionary point of view it's just another hopeful thread.

Plasmid exchange is many orders of magnitude less complex than the lotteries 'nature has engineered' to 'deliberately' generate variation - but dedicated lotteries such meiosis and immune response employ do not *evolve* a thing. Nor, which might be construed as horizontal gene transfer, does fertilisation. What, therefore, about exchange or combination of data by the kind of symbiosis wherein one cell comes to live inside another? Such mutual economy is widespread and incurs dependencies. It is, indeed, proposed eukaryotes evolved from fusion of proto- and cyano-bacteria; the former became a mitochondrion, the latter a chloroplast - though where the planimal's nucleus with its linear chromosomes appeared from is unclear. Perhaps, ménage à trois, happy triple fusion did the trick. Such theory (see also Chapters 21: Energy Metabolism Perchance? and 24: The Origin of Asex) notes nuclear control of the supposed bacterial endosymbionts and must presume that the requisite information was, for some non-reason, accurately transferred from circular organelle to linear, nuclear chromosomal databanks. It is, however, possible to postulate that the modular sub-stations (chloroplastic fuel factories and mitochondrial power stations) are common-sensibly designed to execute such localised, continual and repetitive activity - energy metabolism - as free-living bacteria would also need to perform. Cement mixers are placed at building sites apart from HQ's store; similarly it would be burdensome and unnecessary for a nucleus to cope directly with simple yet important, endlessly repeated business. You could, however, expect functional homology of code, metabolism and morphology between organelles and free-living organisms. In this case, has symbiosis ever generated a fresh function, metabolic pathway or type of organism?

The question is: could dynamic circuitry of the genome, used for maintenance and correct expression of genetic information, not service evolution too? Could computation and not competition drive up life on earth? What, thus, about repetitive motifs and mobile nucleic acid elements? Dispersed or tandem, mobile or fixed, what do accurately distributed n-p-c *DNA* elements and accurate targeteers (both *DNA* and n-c *RNA* fragments) accomplish? Certainly some tandem repeats help compose organelles such as centromeres, telomeres and, facilitating chromosomal packaging, binding-points to nucleosomes. Indeed, could not developmental circuitry such as the *Hox* complex (Chapter 25: Evo-devo) have been naturally engineered by retention, shuffling and reuse of critical domains, that is, of high-level architectural controls that specify an organism's form? As with extrapolations, not confirmed by practical research, of the other mechanisms into evolutionary mode you might hazard guesses that an aimless shuffling of these modules might just have hit the jackpot time on different-combination time; or, equally and information-wise, you could spoil the convoluted story-telling fun by inference that they were programmed. The *origin* of such modules, such targeting and targeted routines and clustered subroutines, is even more opaque.

Transposable elements (transposons etc.), whose rapid activity in all of us appears controlled by cellular programs, seem themselves to act as regulators. Retrotransposons such as SINEs and LINEs may also turn out to be co-factors in the operating system's kit of messengers. Other factors, including n-c *RNA*s a-plenty, also appear to cooperate in servicing genomes. You might further argue that a precise distribution of repeated elements would, in a similar dedicated manner of delegation as occurs with mitochondria and other organelles, permit the same editorial or regulatory functions to be executed on-site at multiple locations - a policeman on every street corner can react more efficiently than ones dispensed continually who have to travel from HQ. Thus why should, how could random upheavals improve and not impair their management? It depends, of course, how you interpret nuclear activity. Is circuitry precise, controlling and conservative - dedicated in a flexible response to circumstance? In this case genomic and epigenomic buffers would, as adaptive potential, be embedded in the way instructive algorithms work. Or is the *status quo* itself uncertain; does it permit extrapolation such that you identify jumping genes as major players in the innovation game?

Certainly organisms with similar genes (such as mice and men) differ greatly in their mobile, n-p-coding lexicon. Clearly 'junk' and 'jumpers' make a systematic difference. And morphologically similar organisms (such as chimps and men) involve a similar lexicon of n-p-c elements *as well as* genes. This, like to like, you might expect. The question then becomes one of homology (Chapter 22: Common Descent or Common Design). You scour genomic print-outs seeking similarities that you interpret as the evidence for evolutionary relationship. Interpretation of a fact, as Sherlock Holmes would tell you, is what counts. How you interpret current or hypothetical historic circumstance depends on how you think that complex, computational code is put together. All parties agree that morphological change occurs by modification of genomic expression. The quality, position and deployment of genetic factors, large and small, make all the difference. The question now becomes the nature of such engineering. The fact is that all aspects of cell functioning and cellular biochemistry are subject to strict regulation - but could deregulated, pointless accidents add up to progress? The question's asked. Could swapping chunks of information using known mechanisms yield 'deep novelty' - not of trivial variation but of the species with potential to deliver microbe-to-man 'development-by-chance'? Such hypothetical, promissory species has not been demonstrated in a lab.

It's always early days with fresh discoveries. At base level life's dependent on the proteins and the binding regulators code can generate. This is the nitty-gritty, fundamental level at which evolution has to happen. Perhaps, therefore, by engineering coding stretches you could, using biotechnological intelligence, engineer new protein functions. Protein engineers are hard at their commercial work of trying to create some artificial enzyme catalysts. If mind can do it aren't you able to imagine chance could too? If tracts of years can do what your intelligence can copy in a few then you'd have grasped and gripped how evolution has to work. However, at amino acid or nucleotide level there are indications of a 'sampling problem'. So many trillions of possibilities exist you have to guide the process in the way you guide your own behaviour to its targets. Experiments (by Doug Axe and Ann Gauger) have indicated that to stumble without reason on the correct configurations for generating novel function is likely to be beyond the capacity of natural processes. For example, new protein functions typically demand multiple fresh folding structures each of which in turn demands exact and long, new stretches of code. Demonstration by experiment has not induced any undirected occurrence. What, even if it had occurred, of the hypothetical new protein function now that the cell has lost its old one? Malfunction kills. No problem. If you accidentally duplicated chunks of code (cells do) then 'chose' precisely, blindly sequences that needed to be changed you might survive. Is resurrection unto a new form of life impossible? So why not hope that natural genetic mechanisms (which have somehow found themselves at time's disposal) 'chose progressive transformation' by shuffling blocks of prefabricated code known as domains. A protein's shape and thereby function is dependent on what can be analysed as sections (incorporating, say, an alpha-helix or an active site). Since a domain's essentially a piece of *DNA* with its own algorithmic message, a symbolic structure that will drive specific function, you might think of proteins as composed of subroutines. Thus a 'systems view' allows you to imagine that, without any rationale, mobile elements might have engaged a grand experiment; they might swap or shuffle a sufficient wad of subroutines to buy fresh folding unto an employable and integrable novelty. To simply reposition, rearrange and/or amplify domains might be how evolution works. Do not programmers deftly tweak and re-employ routines? Is this how genomes were 'conceived'? Except, as engineers well know, mind does not delegate to chance; nor is to demonstrate in labs or factories is the same as saying nature did or could have worked your way. Although materialism must recruit the mindlessness of nature as its default 'engineer' no experiment has yet confirmed this hypothetical assumption.

No doubt, specific splicing of genetic transcripts is a routine, well-regulated exercise; different cells in your own body can refine a protein to improve its integration, in each case, with a different, set circumstance. Splicing and recombination can, as in the deliberate lottery of antibody variation, tailor a family of proteins. In other words, adaptive engineering does in particular instances exist - but how would such functional logic necessarily point, at next step, to irrational and pointless evolution? In the new millennium Darwinism will suggest it must have. Can organism's somehow 'guide' their own progressive evolution? Darwinism will suggest they do.

This issue applies especially to evo-devo's pursuant - although the usual red light is flashing. No engineers!! Evolution's metaphor is 'as if' or 'naturally engineered'. Thus evo-devo's explanations are, as we shall see, a subset of this section since retention yet the rearrangement of developmental regulatory domains (as, for example, *Hox* or homeodomain) might shuffle innovation up. New body-plans for shuffling old! If you successfully rewired a dedicated micro-chip you might, in this developmental case, rewrite morphogenesis from top to toe. Mix 'n' match of core modules might possibly (although the evidence would seem to witness such mutation as a creature's devastation) create not just new species but, as evolution needs, whole new families, orders, classes and phyla of forms. Disjunction yields abruption - Goldschmidt's hopeful monster lives again! In monstrous numbers since you must be one of them! Equilibrium has, in this liberal view, been well and truly punctuated. Variation's run amok. The question: does it work? Flies stay flies, *E. coli E. coli* and malarial parasites don't change their style. Tales are spun but variation isn't evolution; nor is the *origin* of shuffled modules in the least explained. You leave that to the great, inscrutable extrapolator, time. There's none so blind as cannot, or will not, see. Is this guesswork based on science or philosophy? By investigating further (Chapter 26) can a poor prognosis be improved?

A cell's informed and, by programmed 'cognition', actively controls its destiny. Yet arbitrary decision has excluded natural, immaterial mind from cosmos; its nature, though it ought to be, is not favourably discussed. This error, involving the horror known as teleology, can and does inhabit very clever minds but, philosophically, sits in the kindergarten class. It forces 'as if' metaphors, Darwinism of the gaps and just-so, wishful-thinking tales of time and death-directed chance.

Top-down, though, genetic engineering's 'natural' is materialism's fig-leaf, a triplex magic spell waved over barely understood biology to draw, from a Darwinian style of hat, mindless innovation - and to thereby keep a world-view on the road. The analogy of genome to a computational operation has, on the other hand, long been predicated by a theory of design. It has proved correct and, it is predicted, will be further vindicated as our understanding of biological operating systems improves. Balance, flexible maintenance and adaptive potential are concepts all biologists employ. Genetic engineering naturally, materially supports all life. This does not mean that such preordinate 'cognition', evidenced at every level, was informed by lapse, that is, non-conscious vacancy. What good reason, barring axiom that is dogma's twin, is there to believe it ever was? Information is conceptual, meaningful, target-prone and immaterial. While transformative evolution (as opposed to variation) is materialism's story soon sheer quantity of complex, coordinated bio-information will demonstrate that it is not a fact but an *interpretation* of the facts, a biased explanation that can't reasonably stand alone.

24. Twists and Turns

A Twist in the Head-to-Tail

And also in tail-to-head. Can Natural Dialectic's vectors be expressed by geometry? There exist various kinds of cyclical vibration, spin, torsion, to-fro antagonistic oppositions, coils (helices) and spirals (vortices) with geometries in two or three dimensions. We've so far encountered several kinds of reversal, twist, mirror-like or anti-parallel reflection. We've also dealt with embedded complementary opposites and, in their dynamic combination, resolution. You may remember, from Volume 0, the reflective asymmetry whereby, from pole to pole, a cosmic gradient inverts consciousness, the basis of informative mind, into its non-conscious, wholly energetic opposite, the physical universe. *Another fundamental inversion, which mirrors the whole scope of creation, is that offered by anti-parallel top-down and bottom-up perspectives.* Holism (and with it Natural Dialectic) inverts materialism and *vice versa.* Neither pillar of faith disputes the fact of material existence but their interpretations of how mankind began and, therefore, all that follows from that view are diametrically opposed. No union or resolution there. In that case, this chapter could well be entitled **Anti-parallel Interpretations of Biology**.

	tam/ raj	*Sat*	
	existence	*Essence*	
	finite	*Infinite*	
	duality	*Unity*	
	concentric rings	*Top/ Centre*	
	polarity	*Neutrality*	
	expression	*Potential*	
	action/ process	*Source/ Start*	
↓	*tam*	*raj*	↑
	negative	*positive*	
	increasing passivity	*increasing creativity*	
	inside-out	*outside-in*	
	from Centre	*towards Centre*	
	exteriorising/ centrifugal	*centripetal/ interiorising*	
	increasing passivity	*increasing creativity*	
	below	*above*	
	inversion/ involvement	*detachment*	
	devolution	*eversion/ evolution*	
	descent	*ascent*	
	materialisation	*dematerialisation*	
	dark	*light*	
	mass/ sink	*flux*	
	flat	*vibrant*	
	energetic entropy	*informative negentropy*	
	involuntary	*voluntary*	
	non-conscious body	*conscious mind*	

From a *bottom-up* point of view there exists only matter. Such inversion, 'inside-out', does not exist. In this case, since matter is without intelligence, developments occur by accidents that accord with, as far as we understand them, the laws of physics. 'To evolve' is literally 'to unroll', 'untwist' or 'unfurl' so that any system simply evolves through time from its origin. Stars, rocks and gases can, by this definition, evolve. In this sense evolution (see also Glossary) means little more than natural change according to the 'rules' of entropy. With purposive intelligence, however, schemes are hatched and plans of great complexity evolved. One also speaks of the subjective evolution of a soul (by shedding its material attachments) towards immaterial purity of consciousness. No doubt that evolution is a many-splendoured word. Its biological sense includes a presumption of increasing complexity either by design, as in the sense of embryological development, or randomly as in the sense of Darwin's major metaphor, the evolution of a 'tree of life'. If you describe the purposive program that guides a baby's development from its egg as 'evolution' then, in this most informative and orderly of senses, you certainly evolved. If, on the other hand, you describe emergence of the panoply of life from some primordial imagination as 'evolution' then, as Chapter 20 well explained, don't be half so sure. The question's one of head-to-tail; did body evolve informative consciousness or *vice versa*? It's the riddle of your source and sink. Are these the same or, sunk in clay, might you evolve from earthly body *back* to immaterial source?

Sharp distinction needs be drawn. *Do you evaluate something that develops according to preconceived plan as, essentially, the same as something that does not*? Do you, with philosophical abandon and without resolving blur, conflate chance and design? If, so you 'evolved' in both the *diametrically opposing* ways in which biology employs the word.

A *top-down* perspective is hierarchical. It differentiates between levels of origin. Active, dialectical creation is an expression of information, the materialisation of an idea. Its materialising vector drops from (*sat*) informative potential through to (*tam*) physical realisation. This conscio-material dynamic shows a fundamental (↓) *inversion*, involvement or devolutionary binding. It twists (see also Chapters 2, 8, 12 and 13) from a 'soul-centred' concentrate of formless consciousness through mind and sensitivity to its exterior antithesis in objective, 'mass-centred' material bodies - bodies whose apt symbol is orderly 'lock-up', the hierarchical imprisonment of energy in rings, a roundlet called an atom.

According to this scheme, laid out in Chapters 2 and 3, an act of creation thus starts with an idea within whose frame of reference a logical outcome is 'realised', that is, materialised. Thus, to repeat, creation from an Inward Centre is seen as 'turning inside out' or, paradoxically, 'inverted' from conscious psychological to non-conscious physical pole. It is a motor process and, because its orderly progress drops from informative to energetic domain, we call such 'exteriorisation' *devolution*. Order, as in any hierarchical system, is *devolved* from top-centre. From pole to pole an idea is inverted through conscious to unconscious state. In this restricted, vectored sense (*raj* ↑) *eversion* does not imply opposition or an upside-down anti-pole. As antithesis to the (*tam* ↓) vector of *inersion* it simply runs 'anti-parallel' to its opposite number. Its direction is one of centripetal *evolution*. In the energetic case this means, as with evolution of a gas, energy-gain or 'upward phase change'; in an informative sense it means information-gain or 'upward evolution of mind' from material ignorance towards the Source of Knowledge. Such vectors you're familiar with; they can be represented as a ladder.

A Suggested Geometry of How the World's Informed

Representation of Branch Dialectic as a Ladder.

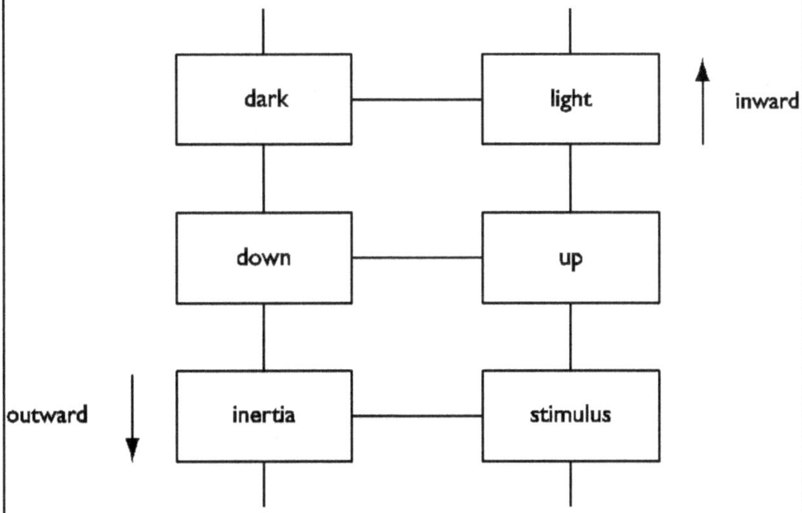

Stimulus, whether conscious or unconscious, amounts to a sign, message or activator to the recipient.

fig. 24.1

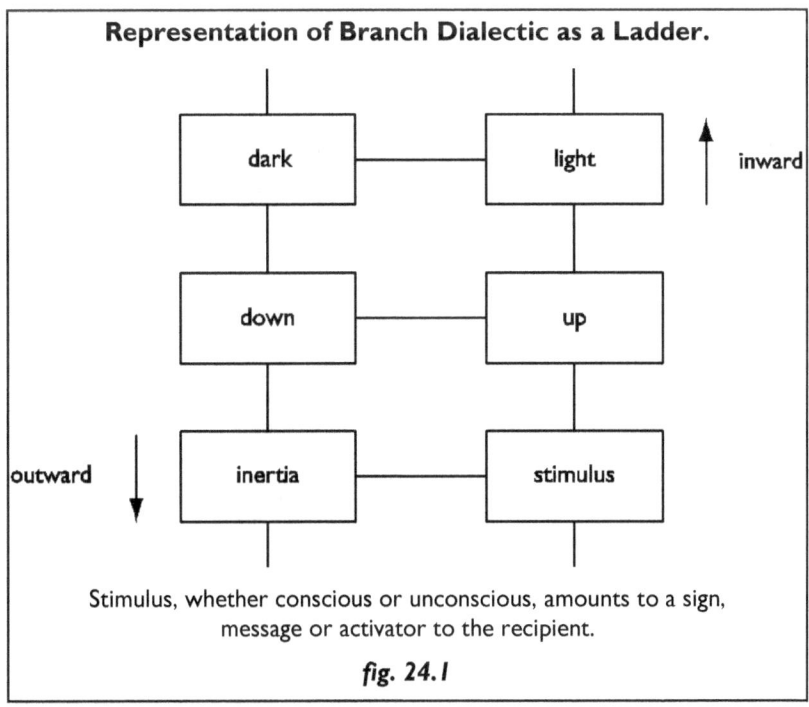

Opposites attract, repel but also complement; they bind, separate but still in quality are two-in-one; such staples nail the cosmic plan to nothingness. According to this plan the motion of materialisation flows, stream-like, 'down and out'. In this sense the development of a baby is actually 'devolved' not only from primary physical information (an egg) but also psychological plan (archetypal memory and, at initial conception, the active development of the 'human idea'). Materialisation involves any motion from the centre towards periphery or, in dialectical terms, from right to left and downwards.

Evolution involves contrary twist. It involves release from 'locked-up', insensible conditions. The name of matter's stream, falling into fixity, is entropy; but, endowed with clever mechanisms, life pumps up against the flood. For a while its bodies are not swept away. Treading water where you are needs stably input energy yet, as we've learnt, raw energy destroys. No doubt, stimulus fires fixity to flux but fluids never by themselves create an operational system. Such system is also, crucially, initiated not by energy but *information.* Biological egg (and every other cell) are, by prior prescription, operationally codified; and only mind inputs informative negentropy. Yet, tail-to-head, Darwinism omits such input. It presumes an orderly prescription by prescribing chance and natural selection in its stead. Could this twisted narrative be right?

In fact, the greater understanding and intelligence the easier and more powerfully can purpose evolve systems of its choice. *Operational systems always rise from mind; they never chance from matter*.

Representation of the Ladder as a Double Helix

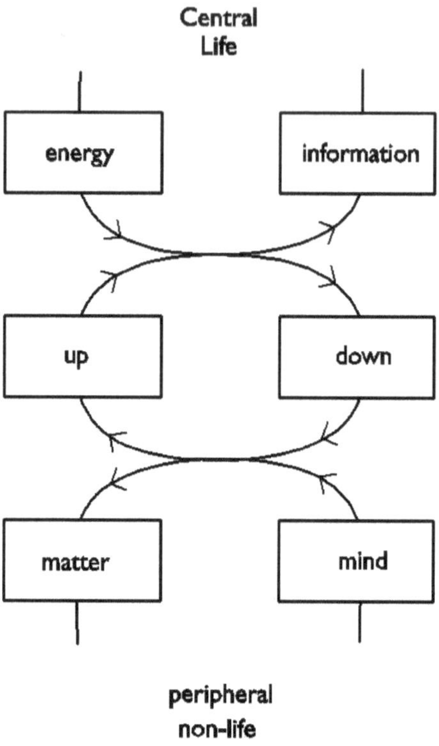

From pole to pole and side to side intrinsic to the Dialectic is polarity. It's binary; it's digital; it switches, straight or gradually like a dimmer, on and off. Note that the polar strands of this dialectical helix (information and energy) are arranged to run anti-parallel; they are also plaited into a double helix. Energetic creation is a downward, outward 'muscular' process; information gathering (or knowledge seeking) is an upward, inward 'perceptive' process leading, at culmination, to the central consciousness or 'brain' of creation. Your own construction just confirms a polar, spiral geometry of information. Your caducean template (*figs.* 17.5- 17.11) assumes this pattern; so does the physical expression of your informative *DNA*, nervous and hormonal systems.

fig. 24.2 (see also figs. 2.10 - 12)

Turning-upside-down' occurs across any pair of dialectical opposites. One extreme is absence of the other. Darkness, for example, turns light upside down. **In this case the Dialectic turns materialism upside down. It spells a creation**

that devolves from Idea through to its expression locked in idea-less matter. Perhaps the best physical analogy of this process is the informative potential banked in an egg and released during an orderly, non-Darwinian species of development to maturity, that is, full expression of original idea.

By the same twist in reverse, how might material oblivion invert its own non-consciousness? It is without ideas; its passivity is utterly incapable of active thought. **Somehow, though, natural particles and forces are imagined to aimlessly evolve purposive complexity - the state of biological bodies, brains and thereby minds with their ideas.** A theory of design-by-chance proposes the replacement of informative purpose, mind and teleological complexity by chance and natural selection. It proposes, unobjectionably, that inanimate, improvident events create material patterns; but in addition, dark to light, that if not prematurely destroyed these creations might somehow collect 'animated sensibility' along the way. They might 'self-organise' into naturally purposive, survival-driven biological complexities. *Such impudence (A Theory of No Intelligence) represents the height of unreason, an axe to any logic in creation and at root, therefore, in human thought.* It runs opposite the whole current of Logical Creation (A Theory of Intelligence) and is, if this is right, wrong - even if the great and good, with granted bodies, education and intelligence, might disagree. Physics, which lacks any law of informative purpose or teleological complexity, is neutral with respect to Logical Creation. Not so when it comes to animated matter and biology. Philosophical tension ruptures any attempt to explain not only the origins but also the critical, vital subjectivity of animation.

Two ideas of creation: two anti-parallel ideas. They need be clearly spelt apart - one starts with absolutely nothing and the other nothing physical - that's Archetypal Information. One illogically 'evolves' the chaos of a world by accident; the other 'devolves' order of coherent cosmos purposely. The latter is dramatically reflected in the way a baby grows. It devolves, according to strict plan, from 'inside outwards', from zygote to adult, from (*sat*) informative potential of an egg through (*raj*) developmental phase to reach a final climax, that is, end-result in the (*tam*) extension of maturity.

Inside information is devolved to outward structure in the form of mortal coil. If human being is a cosmic metaphor what geometry might naturally (and, therefore, dialectically) express informative potential?

Take the earth. Take a satellite that spins around its flying axis called a sun. Draw the coil through space; a spring-like helix will appear. A helix is a single thread. Cosmos is a polar pair. Its fundamental components are information and energy - a couple intimately entwined within a duplex called the 'conscio-material gradient'. While the current of creation spins from Inside outwards that of dissolution flows the other way. There exists a vectored, two-way flow; counter-currents shiver up and down the cosmic slope. In this way the columns of Natural Dialectic look more like a ladder or even, as we see, a double helix that drops from top to bottom. Balance is tipped across the Dialectic. It runs, according to the creative vector, from right to left and, in dissolution *vice versa*. Each pair of opposites reflects a different case of tipping. Each opposite within the pair is a *polar inversion* of the other's character. It is easy to spot that what is below reflects, in a negative way, what is above. Positive ranges down to negative, light to non-light (dark). Motion brakes to its special 'non-case', stillness; cyclical vibration dies to its special case, flatness; and centre-point vanishes, axis disappears as curve is rolled out to a straight line. Love twists to its anti-pole of hate and life to non-life (death).

You can spiral either way. Oscillations include up *and* down. You can scramble either way on Jacob's ladder. And when you entwine two coils you have a double helix, a polar twine. In short, cosmos is informed, at heart, by polar opposition and all motions in between. Its strands are vectors Natural Dialectic knows as cosmic fundamentals (Chapter 2). Thus a double helix with its anti-parallels well represents the roots of truth.

Thus, if man's a duplex spun of mind and matter's threads then his mortal coil is bilateral too. For example, he is (with motor-side outward, sensory inward) nervously double-stranded. Not only his mortal coil, psychology and sex but his nuclear 'book of life' itself is twined with polar, complementary opposites. A helix twists. Anti-parallels of *DNA* coil like vectors rippling up and down the length of vital information. Do you remember the idea (from Chapter 2 and Glossary: dialectical Stacks) of 'chromosomal' stacks of information out of which the form of Natural Dialectic is composed; and (Chapter 16) *DNA* in operation as a spiral aerial, a radio regulator by electromagnetism of expression of its genes? Perhaps a double helix is no less than cosmic symbol of the vibrant, dualistic way that energy's informed.

A cosmic symbol? Cosmos but not chaos issues abstract symbols and, in music's language, harmony; order is the very meaning of the word. Do you remember (Chapter 0) models of the cosmic hierarchy and the order of creation - ziggurat, concentric spheres, Mount Universe and now Jacob's ladder? *Do you remember 'Noah's great rainbow' that represents the Holy Spectrum, a conscio-material gradient and order that substantiates an energetic universe*? And, of that Holy Ghost called light, a brilliant sign of warmth, intelligence and life. Is intelligence not understanding-in-action? What, therefore, about the symbol of light's reverse? *What better symbol of oblivious darkness, what better cipher for an inverted and mass-centred plane, a universe trapped in its endless, reflex cycles than the basis of its energetic actuality - an atom*?

Silent atoms do not tell you anything; but coded information means, as all who speak a language know, transmission of a purpose and designs. The double helix is a Jacob's ladder elegantly entwined in anti-parallel embrace; it is light and life's transmitter through the space and time of cosmos; and it represents an orderly insistence on the Natural Dialectic of the way things work. You're in the plan. Do you remember (Chapter 17 and *figs.* 17.7 -10) 'archetypal man'? How your own organic 'blossoming' reflects materialisation along a conscio-material gradient from Conscious Centre to non-conscious, bodily periphery? Your microcosmic caduceus symbolises, in a balanced way, the macrocosmic order of creation. Its crown, (*Sat*) Neutral Essence, is polarised into (*raj/ tam*) finite parts. From it springs a coloured spectrum of existence, from it fans logical order from pole to pole of its creative range. Do you remember that the two major poles of existence are an informative domain (the 'negentropic', ordering principle of mind) and an informed domain (incorporating the 'down-pointing', physical principle of matter)? And within these domains do you remember a caduceal range of power points, points that reflect descent into a body and, in the body, down to earth? Materialisation is a graduated inversion that descends, sway to sway, from a predominance of 'soul-centred' influence to the overwhelmingly 'mass-centred' attraction of matter. Balance is tipped. The Truth of Primal Unity is devolved, in an orderly way, through Important Principles into the relatively trivial truths of individual forms and detail of different practices. They are thereby details that accord with basic, dialectical principles of psychology, biology, physics and chemistry. Hierarchical order is, more than just a slant, a different twist on universal structure. It is one basically inimical to the notion of you being here by chance!

Twists that Entwine.

	tam/ raj	*Sat*	
	existence	*Essence*	
	body/ mind	*Soul*	
	lesser selves	*Self*	
	division	*Union*	
	parts	*Wholeness*	
	lesser pleasure	*Ecstasy*	
↓	*tam*	*raj*	↑
	descent outward	*inward/ ascent*	
	lower self	*higher self*	
	involuntary	*voluntary*	
	sensation	*contemplation*	
	instinct	*learning*	
	materialisation	*spiritualisation*	
	multiplication	*unification*	
	apartness	*togetherness*	
	pain	*pleasure*	
	isolation	*communication*	
	bondage	*friendly bond*	
	pride	*humility*	
	egotism	*altruism*	
	take	*give*	
	hate	*love*	
	complain	*praise*	
	demand	*thank*	

A psychological part of this twist-in-the-tale involves the opposite directions of focus (Chapter 0) - inward and outward. Use of information runs, in the motor sense, outwards from mental through neurological to muscular action. It is dispersed into the arrangement of worldly details according to one's purposes. It ascends in the opposite direction. Information gathering is an inward, sensory and learning process. Information is brought up from the outside and unified in mind. These are anti-parallel entanglements. Their complementary strands facilitate survival, order, conceptual unification and, at its zenith, total comprehension of the world. Is this all? What about, along the way, happiness? What about extremities of happiness?

The materialistic twist tends outward. Following this you might conclude that happiness resides in objects, physical events and the sensations of them. A popular pursuit of pleasure concentrates the excitement of physical and emotional involvements - interests and associations as opposed to isolation, absorption as opposed to boredom. Enthusiastic actions, togetherness and at its zenith, perhaps, sex - it is easy to see how energetic mirrors informative ecstasy and how, for example, physical reproduction mirrors a psychological act of creation. Perhaps the most powerful physical drive to ecstasy is in the consensual, climactic, sexual union of male and female. It peaks at an orgasmic moment when potential for another life is released. Fertilisation and conception lead to the automatic development of a new creation, an expression called a child. Lust and the reproductive experience are sensational, temporary, binding and exhausting; it is instinctive, involuntary and, in its downward mode of expression, materialising.

Whereas sexual reproduction is an act of creation that involves physical eggs the contemplative act of creation is a metaphysical egg (*figs*. 5.3 and 5.4). The germ of an idea is developed into a work of genius. This psychological act of creation cuts both ways. On the one hand the materialisation of an idea begets 'brain children'; development of potential involves the creation of corresponding objects. On the other hand the hierarchical dematerialisation of a particular object translates it into symbol, lifts into general category and leads to understanding of its principles. The externalising effect of sex or any other act of physical creation is diametrically opposite the contemplative spiral inwards. Equally, contemplative, inward twist turns hedonism on its head. Focus on base genitalia is replaced by focus at the upper, cosmological axis of the eye-centre (*ajna* plexus) situated singly just above the outward visionary sense (of two eyes). In other words, such focus of attention is inverted at the human headquarters. Such inversion turns from the passing show and towards the power and principles behind it. A thoughtful, contemplative bent of mind is characterised by curiosity, information-seeking and a wish to know the truth; also by the creation of purposeful, meaningful complexity in both the arts and technology and, at the deliberate extreme, by the climax of a metaphysical Communion.

The reverse of action is (*fig*. 13.4) to understand it. It is to retrace the steps of its development, to comprehend the principles and to evolve an understanding of its purpose. To comprehend a seed-idea in its entirety is to commune with its creator. Time and distance disappear. By working back to its creator you can understand an artefact. What more natural an artefact than cosmos? What more natural than mind from which it issued and, therefore, than mind which grasps the threads again? Thus, *top-down*, the highest informative ecstasy is to retrieve the Cosmic Egg! This is the mythic, mystic quest. It is Communion with Transcendent Other and a comprehension of the cosmic patterns through which The Big Idea (including you and me) is realised. Because Inmost Nature is timeless so its experience, its essential ecstasy, is also eternal. Such voluntary, metaphysical development (as opposed to the involuntary reflex of 'doing what comes naturally') is called 'evolution of the soul'. It is a twist back from the tail end of creation to the top. Liberated from its existential bindings naked soul ascends back to its immaterial Origin.

The logic of the base pole's genitalia and the sensational direction of orgasmic sex are diametrically opposite the upward characteristics of metaphysical Communion. There is no room at the top, which is Unity, for two. No room for other and for self. Communion means loss of self in another or The Other. By losing self and dropping the burdens of selfishness one finds friends; and, likewise, a mystic seeks an all-encompassing association with The Friend of friends. In other words, he seeks lasting metaphysical release, a universal love as opposed to very local lust. No short-lived passions or brief-held possessions. Nor, unless trust and love are irrational, does a mystic's faith involve the suspension of reason. Rather it is reason satiated, trust confirmed. Is love a fact that's only proved by the communion of lovers? Not scientifically or rationally but actually? Is metaphysical immersion a fiction, subjective unity incomprehensible or immaterial love unreasonable save to materialistic rationale? Mystic Communion is to merge; it is to fuse with Inward Self of which, paradoxically, all is a part. It is what, after an odyssey through the worlds of mind and matter, amounts to 'the return of a prodigal son'. It is, at the heart of cosmos, the experience of Home.

Such entwinement is, eventually, our truth and such uplifting aspiration counters the strong draw of earth. It opposes the involuntary and instinctive gravitation into the constraints of body-centred sensation. Have you ever shed the burden of the world? Would the way be easy, would its heart be light? In this view altruism and love are neither a cunning scheme nor social strategy for egotistic, evolutionary survival that reduce, in the basest rendering of reason, to a need to protect genetic information. They are, on the contrary, (*sat*) fine and central qualities, cosmic in nature, that automatically increase in accordance with the ascent of soul towards its state of freedom, selflessness or purity. *If science is the same as saturated materialism there is absolutely nothing scientific in this Central, Subjective Quest nor its Truth.*

Twists in the Bio-logical Tale

Just as psychological so also a biological spiral exists in all of us. It falls, from inside information down to outside form, through a triplex, complementary combination of (*sat*) informative, (*raj*) energetic and (*tam*) structural factors. The hierarchy externalises from (*sat*) program, code and signalling; messages, symbolised by the chemistry of nerves and hormones, inform, unify and govern the polar adjuncts of biological purpose. These outward adjuncts are (*raj*) biochemical metabolism and physiological function; and (*tam*) hardware, called organs and skeletal anatomy, within whose appropriate 'morph' the software can express the pattern of its purposes.

There exist, according to Dialectical principle, cosmic vectors (*raj*) leading upwards and inwards towards subjectivity and (*tam*) leading downwards and outwards towards objectivity. Materialisation amounts to 'organic blossoming'. It devolves inside outwards from Central Potential towards peripheral exhaustion, from simplicity of principle to complexity of practice, from outline sketch to detailed final picture. Just as there exists an eversion, illustrated by the polar characteristics in Primary Dialectic, between Essence and existence, so there exists a similar mirror, illustrated by secondary or branch dialectic, between (*raj*) right-hand mind above and (*tam*) left-hand body, earth or matter below. *This 'dimension switch' is a clear aspect of biological construction with respect to brain, body and sexual emphasis.*

Polar Inversion of Cerebral Hemispheres with Visual and Sex-Linked Characteristics

Dialectic Inversion of Hemispheres of Human Brain

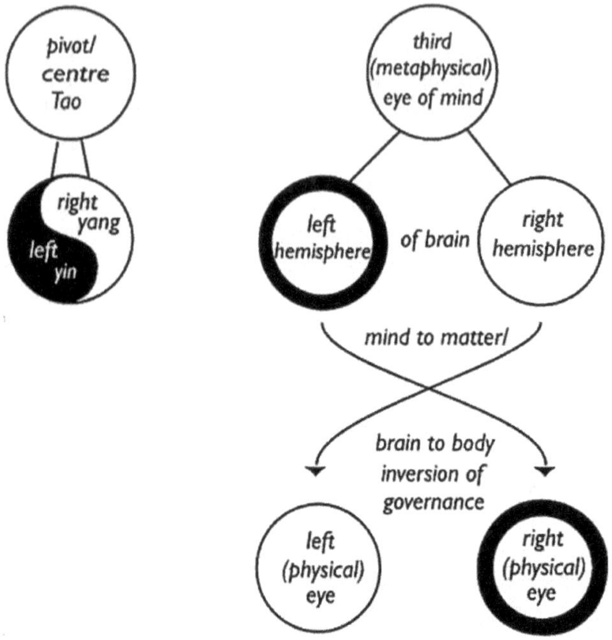

Polar, Hemispheric Inversion of Visual Faculty

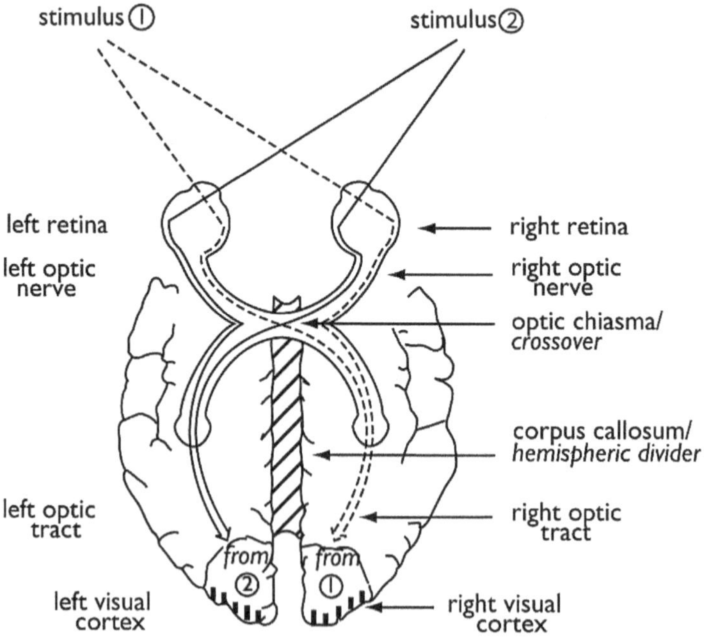

Just as signals to the left eye are for the most part processed by the right visual cortex, so the right hemisphere of the brain controls the left side of the body and *vice versa*. Such division is clear-cut and, of course, the bilateral symmetry of human form is aesthetically superb; but it is for certain complemented by asymmetries. For example heart, pancreas and liver are not centrally aligned. More profoundly, an imbalance of handedness is associated with 'psycho-sexual' tendencies. From a few surgical 'commissurotomies', operations by which connection between the two sides of the brain is severed, has emerged a 'split-brain model' of hemispheric functions. However, with *corpus callosum* intact the vast majority of humans use their undivided brain as a generally neutral, asexual unit laced with sexuality and are proficient in all kinds of skills.

Sex, however, is not neutral. Both research and Dialectical principle suggest that within the whole, blurred range of sexuality there exists polar emphasis; in other words, there exist statistical 'spikes' that register male and female bias called a 'norm'. These two clustered norms suggest trend, strong probability and definition to the point of reason for their 'opposition'. If we make a Dialectical switch, as occurs clearly with sight and slightly less clearly in the case of sex, between physical and psychological handedness, what on balance might we find? Of course, in the complete picture everyone uses both complementary hemispheres of the whole brain harmoniously but are the psychological and physical 'inversions' that we show at different times and/or in various strengths a function of polar, sexual stereotype? Check the stack below.

tam/raj	*Sat*
yin/yang	*Tao*
inclination	*Symmetry*
imbalance	*Balance*
each side	*Both Sides/All Sides*
female/male	*Human Being*

tam/yin	*raj/yang*
left hemisphere	*right hemisphere*
grey cell/local processing	*white cell/networking*
analysis/division	*synthesis/integration*
intellect	*intuition*
parts/detail	*principle/generality*
harder	*sensitive*
objective	*subjective*
fact-oriented	*people-oriented*
male	*female*

↓ *physical*

psychological ↑

female	*male*
tam/yin	*raj/yang*
passive	*active*
follower	*leader*
receiver/receptive	*donor*
below	*above*
softer	*muscular*
yielding	*dominant*

Can this switched stack be correct? It suggests greater female strength in the subjective, psychological arena and greater male strength in the physical. In holistic reality the sexes are, as dialectical pattern confirms, complementary, balanced within the human archetype and yet divinely different! We shall shortly (*fig.* 24.5) check this sort of divinity in greater detail.

fig. 24.3

There exist, according to dialectical principle, cosmic vectors (*raj*) leading upwards and inwards towards subjectivity and (*tam*) leading downwards and outwards towards objectivity. Materialisation amounts to 'organic blossoming'. Cosmically, the process devolves inside-outwards from Central Potential towards peripheral exhaustion, from simplicity of principle to complexity of practice, from outline sketch to detailed final picture. Just as there exists devolution, illustrated by the polar characteristics in Primary Dialectic, between Essence and existence, so there exists a similar mirror, illustrated by secondary or branch dialectic, between (*raj*) right-hand mind above and (*tam*) left-hand body, earth or matter below; this involves energetic 'evolution' (as, for example, a gas is raised from liquid) and informative evolution (of a plan or, spiritually, of soul from its material attachments).

Information is, we've seen, the potential for orderly behaviour and, thence, construction. *Such informative-material 'dimension interaction' is a clear aspect of the 'spiral' biological construction that exists in all of us.* It falls, from inside, immaterial information down to material, outside form, through a triplex, complementary combination of (*sat*) informative, (*raj*) energetic and (*tam*) structural factors. The hierarchy externalises from (*sat*) codified potential in the form of physical behaviours and, biologically, the book of *DNA*. Programs (and, therein, signals, switches and messages realised by the chemistry of nerves, hormones and other factors) govern biological expression. Their outward adjuncts are (*raj*) biochemical metabolism and physiological function; and (*tam*) hardware, called organs and skeletal anatomy, within whose appropriate infrastructure software can express the pattern of its purposes. The 'dimension interaction' especially involves, at all three levels as we'll shortly see, both cognitive and reproductive emphases; and, as regards their order of events, replication and development.

First, then, to brain! It was noted earlier (Chapter 13) that *psychological* inversions of emphasis were reflected neuro-physiologically. The cortex of a human brain is split into two distinct hemispheres by a central junction called the *corpus callosum*. Despite nervous crossovers from left and right-hand sides of the body most psychological activity is bilateral. Some 300 million neuronal fibres (called commissures) cross between hemispheres to integrate function and support coordination. There are, however, no cortical centres with exclusive function nor, as might be expected if consciousness was simply a by-product of synaptic networking, much loss of portions of memory or thought process if parts of either hemisphere are excised. Holism involves two-within-one. *As well as bilateral operation psychologists note evidence of a 'chiral' symmetry or bias of function that is apparently related to sexuality.* With different emphasis right and left-hand characteristics exist in either sex. As such humanity is mirrored in the brain.

The polarity of brain's 'morphogenetic symmetry' involves two characters. *The left hemisphere that governs the right-hand side of the body tends towards serial processing, stepwise analysis, logico-mathematical calculation and linguistic tasks*. Reason focuses on local detail and arrangement of its circumstance. It analyses, classifies and cuts clear line. It thus relates to body's self-identity; and is related to the 'masculine' domain where energy (in the determined form of physical action) holds sway. *On the right-hand feeling's part communicates, 'entangles' and emotes.* Such character tends, in computer-speak, to process information from its circumstance in parallel; and instead of plodding on methodically it absorbs 'globally'. It is 'feminine', pictorial, artistic and related to the psychological domain where information (in the form of learning, knowledge and understanding) is in the ascendant. Such ascendancy involves principles, generalities and universals. This holistic half might yet expand you past the world of ideas to transcendent planes of peace and love. You would, logically, develop (↑) towards *Nirvana*, *Logos* or Communion on the right.

What about a body/ brain inversion reflecting the one between mind and matter? These words are, as you read, registered upside-down on your retinas. This asexual inversion is then switched as, along the optic nerves, impulses from your left eye travel to the right side of the brain and *vice versa*. In fact (see Chapter 17) nearly all nerves, sensory and motor, decussate (cross over like an X) from the left side of the body to the right side of the brain. This kind of mind-matter mirror imagery is ubiquitous. Like a photographic negative, brain receives a mirror image of matter which is then developed, anti-mirror-wise, so that we sense the right way round; and in the motor translation of principle into practice, that is, thought into action we act the right way round. Moreover *mind* and not the nervous connections can adjust if mirror-goggles are worn which, at first, invert our orientation. After a day or two the wearer learns to acclimatise. When the goggles are removed he looks normal but sees everything the wrong way round. Nor have the nerves changed their connections. As the 'false' acclimatisation wears off he returns to normal. It might be possible to trick other senses with 'inversion-machines' but, certainly, mind and not nerves will govern any adjustments.

Hold out your hands. Place the left one over the right. It will not fit unless you switch it upside down. Opposites employ an inbuilt 'chiral' switch. So with information and energy, mind and body; each part complements and exerts a range of influence over the other but, where they meet, a switch of mirror-like inversion is discovered.

Look at the cerebral cortex. It is the top-front part of our physical medium of information exchange, the nervous system. Motor controls appear, in an arc around the centre, in front of the somato-sensory arc. This physical arrangement, with action to the fore, is copied lower down. 'Sensory dorsal, motor ventral' means that motor output exits the spine through nerves at the front while information being raised to consciousness enters the spine at the rear. The location of functional centres also follows an arguably inverse back-to-front, upside-down spatial geometry. The top sense, vision whose organ is the eye, is located base rear; proceeding forward we find the auditory centre (ears), olfactory centre (nose) and *above* them taste (mouth) and, top-centre, the somato-sensory (touch and pressure) arc. This reverses the position of organs in the body and their relationship with *metaphysical* archetype (*figs.* 17.6 - 10).It indicates, for the female, greater emotional and, on the physical side, less innate strength than the male.

Of course, generalisations never, where a scale of tendencies is involved, have the whole truth; exceptions simply indicate the median rule. It is certainly a singular fact that areas of the somato-sensory arc are inversely related to the functional position of body parts. For example sex is a materialising force at the generative heart of long-term biological survival. As such it is, with energy metabolism, a primary instrument of physical life and enjoys pride of place in materialistic philosophies. Alongside the hedonistic pleasures of eating and drinking (the energetic side) the satisfaction of sexual (reproductive) appetite becomes a central rationale for existence and most important to a worldly way of life. From the *physical* point of view is it, therefore, a surprise to find the cerebral correlate of the base-of-body genital area embedded top-centre (next diagrams) ? Whence the arc 'curls' down through toes, legs, hips, trunk, head, arms, hands, face, mouth and inwards towards the throat and abdomen. A similar, inversion occurs on the forward motor arc and, of course, each applies the brain/ body inversion, left lobe governing right side of body and *vice versa.*

female	*male*
responsive	*initiative*
submissive	*dominant*
informed	*informant*

Human procreative parts are the trunk organs most dialectically and physically removed from informative brain. Just as there exist other kinds of psychological and biological inversion so inverse sexual characteristics bind us to a mate. Sex is clearly a dialectical bind. It involves polarity, paradox of two-in-oneness, the complementary interaction of opposites, a sliding scale through the degrees of individual sexuality and, in a familiar way, discontinuity (in separate bodies of either male or female pole) and continuity (the spectrum of common features whose various emphases exist in both sexes). Indeed, at its psychological, informative end the distinction is subtle, nuanced and fluid; at the base end, with human procreative parts most dialectically and physically removed from brain, the distinction between male and female organs is obvious and practically fixed. In other words there exist two systematic, stereotypical norms, male and female, around which the majority in each sex are clustered. Male and female 'spikes' both show the whole range of non-sexual characteristics but also, in different individuals of either sex, a spectrum of 'more-or-less' masculine or feminine traits. Whether it is fashionable to emphasize difference or similarity between the sexes depends not on nature but on individuals and cultures.

From the lower to the upper part, let's turn to body's panel of control, the brain.

More Cerebral Inversions

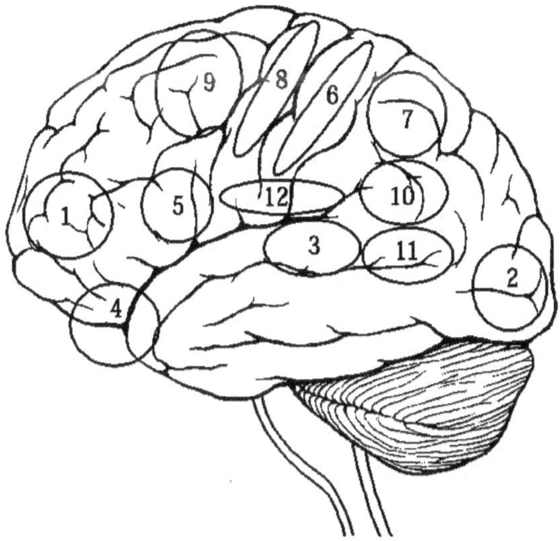

Cerebral processing power is not relative to size alone but to the ratio of size of association cortex to primary sensory and motor cortices. Although higher-level cortical processing is less well defined, lower-level body-related functions are localised on the mind's 'dashboard'. These centres describe arcs from front to back (1-2), back to front (2-5) and, at 90 degrees, from side to side at the centre of the brain (6-9). In the sense that numbers run from base-rear to front and then upwards towards top-centre (a couple of adjacent arcs around the crown), they run in approximately inverse position to their dialectical order. They conform to an inverted or mirrored connection of information centre (brain/ mind) with energetic domain (body).

(i) Centres are:

(1) conceptual or conscious (*ajna*)
(2) visual
(3) auditory
(4) olfactory (smell)
(5) gustatory (taste)
(6) sensory (touch, pressure)
(7) sensory association
(8) motor control
(9) supplementary motor (e.g. eye movements, manual dexterity)
(10-12) informative skills
(10) language (symbolic order)
(11) reading
(12) speech

(ii) **Somato-sensory and Motor Sections.**

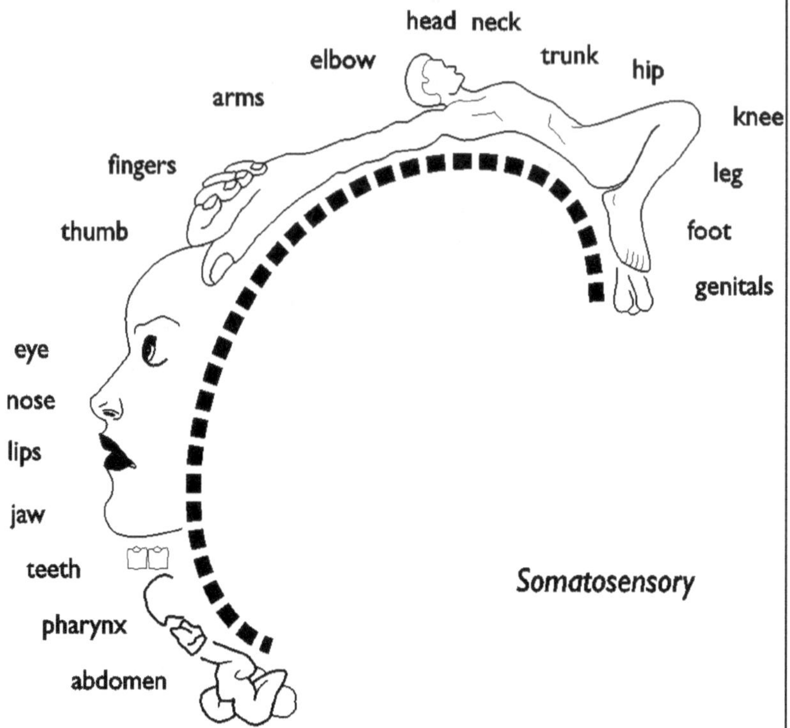

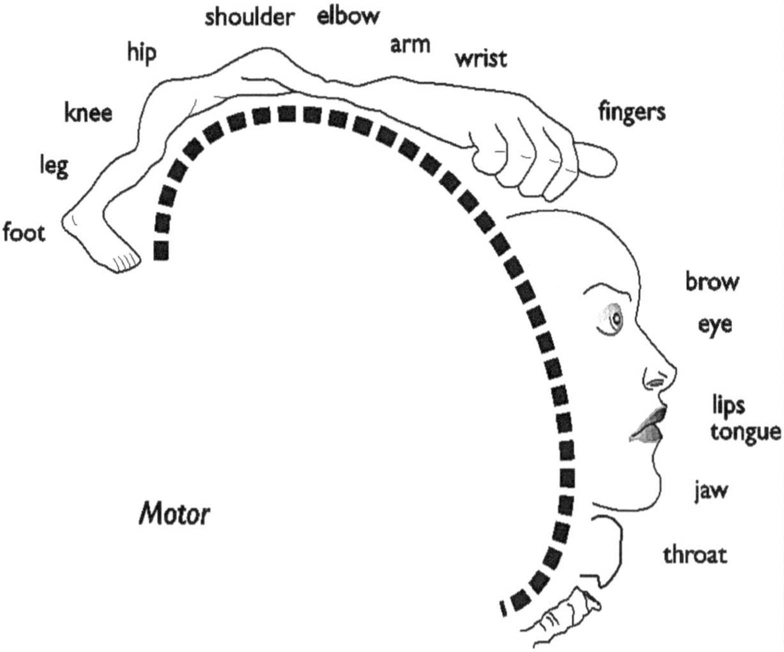

The sense associated with the lowest physical state (solidity) is pressure and touch. Unlike the other sense organs localised in the head this one is generalised; it gives a sense of body's shape and anywhere it's touched. Check *fig.* 17.9 Pyramidal Man and Chapter 17 Informative and Informed Domains. The earth plexus is at the base of spine. It represents the lowest, solid periphery of creation, the diametrical opposite of top-level consciousness. In the cerebral case, however, it is inverted from base/ peripheral to top/ centre. Its arc, to the rear of the motor arc, systematically represents incoming 'touchy-feely' information in a second inversion that runs from base of body (genitalia and toe at its crown) to head (at the its base). The left side of the cerebral cortex, pictured here, represents the right-hand side of the body and *vice versa.*

In dialectical terms informative potential precedes kinetic behaviour so that the position of the motor arc in front of the sensory represents a third inversion. This arc shows the same inverted sweep as the somato-sensory in its descent from base of the body (at its crown) to head (at its base). Head is described as part of a sequence that 'invaginates' into the mouth and down the alimentary tract.

Why, in practice, should it matter where what processing is done? In fact a striking degree of order is obtained. In general the dialectical logic of psychosomatic functions is followed, inversely, by brain structure. Even without this particular mind-matter twist the three-dimensional geometry of cerebral architecture appears too efficient, coherent and reasonable to have evolved by chance.

fig. 24.4

Psychological and physical traits need disentangling. Let's start with the subjective, psychological aspect of human sexuality. In this case 'feminine', associated with the sensory or sensitive aspect of existence, is placed right. *This (raj) positive position serves to emphasise a woman's innate tendency to care, communicate and nurture relationships.* Invisible feelings take precedence over visible things; 'right' behaviours are more important than objects; and emotional more satisfying than financial wealth. Her strength is therefore in the cohesion of service, self-sacrifice and love rather than the bellicose division of dominant ego or the creation of wealth. In this, at the centre of the family, she is the psychological pillar of society. *Her psychological as opposed characteristics align her with the upward, right-hand interiorising direction of to physical, the Dialectic.* This, as we have just seen, culminates in informative ecstasy. 'Female' has intrinsic consonance with the inward, subjective path of love.

Before checking the male 'spike' let's ask if male and female differ in that most intimate, shared and yet self-centred of experiences, mutual orgasm. Both have much in common; the psychological experience is similar although the physical is capable of greater depth and prolongation in the female. In both sexes similar hormonal changes (such as increased charge of natural amphetamine, testosterone, endorphins, oxytocin and so on) with physiological changes (such as blood pressure, breathing rate and sweating) occur. With respect to organs, of course, the obvious dialectical bio-logic differentiates between donor/ receiver, hard/ soft, externalised/ internalised etc. Are there psychological differences too? Brain scans seem to confirm male emphasis on physical sensation transmitted from the genitals while, for the female, emotional relaxation (derived from trust, ambience and mood) are important contributors in ascent to her orgasmic 'trance.

Sex-linked Inversion

tam/raj	*Sat*
yin/yang	*Tao*
body/mind	*Soul*
polar expression	*Archetype*
hemispheres	*Whole Brain*
bias/emphasis	*Balance*
sex	*Sexless Neutrality*
tam/yin	*raj/yang*
left hemisphere	*right hemisphere*
division/separation	*unification/love*
outward	*inward*
serial	*synchronous/ multi-tasking*
thinking	*feeling*
head	*heart*
clinical analysis	*inclusive sympathy*
scientific	*artistic*
logical	*emotional*
investigative	*creative*
verbal/numerical/ linguistic	*visuo-spatial (with pictures/shapes)*
intellectual	*intuitive*
more egotistical	*more responsive*
exclusive/egotistic/'autistic'	*inclusive/other-centred*
'my way'/confrontational	*relationship-builder*
less empathic	*more empathic*
colder	*warmer*
harder	*softer*
dominant	*giving way*
straight/direct	*diplomatic/'curvy'*
emotionally self-constrained	*emotionally explicit*
more objective	*more subjective*
thing-oriented	*life-oriented*
physical solution	*psychological solution*
physical emphasis	*psychological emphasis*
male	*female*

psychological

physical

female	*male*
tam/yin	*raj/yang*
left hand of body	*right hand of body*
passive role	*active role*
follower	*leader*
receiver/receptive	*donor*
centripetal	*centrifugal*
yielding	*dominant*
weaker	*stronger*
smaller	*larger*
curvaceous/rebounding	*direct/thrusting/angular*
pulling	*pushing*
attention seeker	*attention giver*
'wants to be loved'	*'wants to make love'*

fig. 24.5

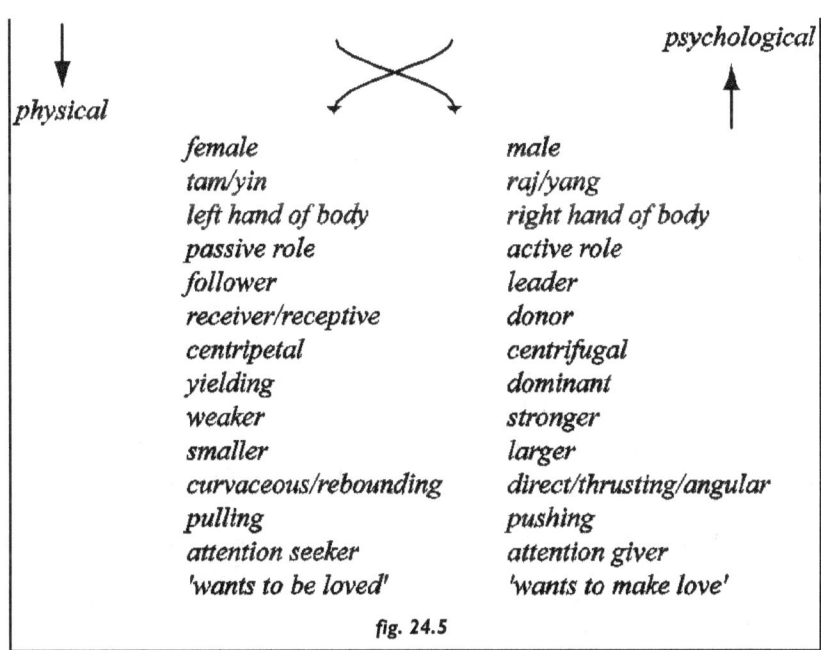

Scans confirm traditional observation. The objective, physical aspect of sexuality is predominantly male. In this case 'masculine', associated with the motor aspect of existence, is placed right. *The motor process involves making, doing or creating. Creation is a downward, coarsening, exteriorising process.* It is about 'bringing heaven down to earth'. Materialisation. A man's physical bent aligns him with the (*tam*) left-hand, materialising direction of the Dialectic. Its order (Chapters 7 - 12) is one of increasing restriction, confinement and diversity from generality into specific forms. The potential of a gas is realised when energy is lost from flux and locked into the definite order of a solid state. Similarly the potential of mind falls from principle to detail, from universal to self-centred expression. It descends from the Infinite through an active to passive embrace of information with energy. In mind information is predominant. We call it life, consciousness and intelligence. This is overcome, by degrees, with material predominance until a precipitate, the material universe, is realised. Similarly, the transformation of thought into muscular behaviour is a *downward,* motor process. The physical motor is muscle, the psychological motor is purpose and design implemented by will-power. Will-power is focused attention. It says, "I want, I will have". *Its masculine emphasis tends to practical activity and pragmatic creation.* The male *ego* dominates, leads and imposes its ideas in the shape of constructions. In its egocentric wilfulness, tempered by rational analysis, male tendency is one of centrifugal adventure and aggression. From a physical point of view the male aspect is positive, dynamic and 'where the action is'. In business that supports a family he is the physical pillar of society. 'Male' has intrinsic consonance with the outward, objective path of 'go-getting' competition.

Having spiralled down information/ energy torsions well into the physical domain, we are in a good position to understand the 'masculine', down-to-earth, mythological perspective with regards to the act of creation. Sex, clearly a dialectical bind, comprises distinct male and female bodies within a single type. The cosmos comprises distinct, polar components (information and energy) within its single type. Each component, either in human being or in cosmos, complements the other. Creation is an outward, centrifugal motor process. It is a (*tam*) masculine, downward, materialising emphasis that involves the translation of psychological patterns into muscular, physical behaviours and the shapes of events. Behaviour includes a special, static case; psychologically static behaviour is called memory and physically static behaviour/ event is called an object. Thus archetypal memory (Chapters 16 and 17) is responsible for the rigid patterns of energy and shapes of mass; and in this respect 'static' refers not, of course, to physical motion but to 'rigid subjectivity', the fixity of non-consciousness, implacable law and (if all factors were known) absolute, Laplacian predictability.

The creative process runs, in accordance with cosmic order, from (*sat*) potential through (*raj*) active plan to (*tam*) finished product. This is the way a human being develops from (*sat*) ovular potential through (*raj*) kinetic program to (*tam*) final adult form. *Indeed, as we note, cosmic order pervades creation and the structures through which information is exchanged and energy (including biological and other material bodies) governed.* Is it all a matter of chance? Male complements female and *vice versa.* Together they reflect the whole creation and express the whole of human being. Could such symbolic congruence and clear inference of overall, archetypal self-consistency have evolved mindlessly, randomly from myriad imperfections? Is the product of it all, a baby, the outcome of creation and the structures through which information is exchanged and energy (including biological and other material bodies) most closely governed; or is it all a case of chance? Male complements female and *vice* misshapen ancestry and mutation? *Ask its mother for her deepest intuition. Is the product not perfect? An outcome of most generous design?*

In this light let's elaborate the mythological stereotype we glanced in Chapter 22. *Stereotypical sexual emphasis, based on the physical and not the psychological perspective, writes male tendencies right, female left.* It could, in this way, be construed as chauvinist but it tells, of course, only half the whole story. The whole picture implies no innate superiority or inferiority on either side, simply an equal interdependence. Two sides of the same coin.

	tam/ raj	*Sat*	
	yin/ yang	*Tao*	
	two	*One*	
	male/ female	*Human Being*	
↓	*tam*	*raj*	↑
	yin	*yang*	
	triggered	*trigger*	
	recipient	*donor*	
	product	*cause*	
	finalises	*initiates*	
	division	*union*	
	offspring	*mating*	
	female	*male*	

The function of sex is the materialisation of new bodies, the recreation of biological life. *Such emphasis is strongly physical.* Its primary organs are situated at the *tail-end* or *base* of the body. Penis and the female tract both *expel* elements of potential new life (sperm, egg and foetus respectively) *down and outwards*. Their 'intention' is for the future and they are found in front of the channel that egests solid waste (past, exhausted material) to the rear.

The bio-logic of complementary symmetry even insinuates, with its erotic twists and turns, the arts and parts of sex. *Yoni* (vagina and womb) in which rests a cosmic egg is, allied with *lingam* or phallus, a common symbol of worship in Indian temples.

	tam/ raj	*Sat*
	duality	*Unity*
	self-apart	*Ecstasy*
	complementary opposition	*Climax*
↓	*tam*	*raj* ↑
	assive/ receptive	*active/ dynamic*
	entered	*entering*
	inward	*outward*
	centripetal	*centrifugal*
	ovary	*testicle*
	yoni	*lingam*
	vagina	*penis*
	receptacle	*penetrant*
	womb	*propulsion system*
	egg	*sperm/ seminal fluid*
	womb-man	*man*

Cosmos from the loins of god! Climax wells into an outburst of creation. The female void is seeded. A receptacle is impregnated with the message 'make like this'; and the implement of making, energy, erupts. Once the archetypal hole of emptiness is penetrated nature is developed in the womb of what was once an absence, space. Development that's laboured from a sacred womb! The analogy of sexual union with polarised creation is at the heart of, for example, Egyptian, Assyrian, Greek, Roman and Hindu ceremonial worship of the world. The deities have partners. Although Ra, head of the Nile pantheon, was neutral he sexlessly gave rise to polar, sexual pairs. His dynasty included deities of earth (Geb) and sky (Nut); Isis and Osiris; and at Karnak phallic Amun and his Mut! The theogamy of Zeus and Demeter (Mother Earth) was symbolised at the mysteries of Eleusis and the Hindu expression of polarity includes female consorts in that pantheon. You can experience the ancient world's worship of regeneration tomorrow. You can step back three thousand years as simply as walking into another room. Nothing has changed in the symbolism, rites and festivals of the famous temple of the fecund Female Deity, *Meenakshi*, and her consort *Siva* (as *Sundareshvara*, the beautiful Lord) at Madurai in Southern India. The stereotype (with exceptions such as Geb and Nut) is systematic:

moon	*sun*
night	*day*
below	*above*
passive/ reflecting	*activating*

receptive	*dative*
earth	*sky*
is watered	*waters*
grows	*sows*
crop	*seed*
yielding	*firm*
soft/ tender	*aggressive/ hard*
cool	*hot*
moist	*dry*
after	*before*
follower	*leader*
female	*male*

Check back to Archetypal Sex. The Sun brings light and life. With lightning, cloud and storm the Sky-god rains. Under these male forces the Mother Goddess, Earth, is brought to season, fertilised and grows heavy. A seed is pressed into the soil's womb; it drinks with osmotic suction the transparent blood of earth.

compression	*release*
stored info	*springing info*
stored energy	*springing energy*
resistance	*flow*
container	*life*
egg	*its development*

From this darkness its potential is developed to fruition. Age is rejuvenated, death skipped and the wheel of life recycled in a kind of quasi-immortality.

confinement	*release*
receptacle/ egg	*delivery/ sperm*
gravid/ heavy	*empty/ light*
spring compressed	*spring*
discharge	*charge*
pregnancy/ labour	*birth/ new life*

Production and sexual reproduction show clear evidence of complementary design in the polar stereotype. There is, however, yet a further, critical aspect of inversion, one closely involved with the physical storage and transmission of information for use in biological reproduction, which it is now time to consider.

We use the word 'offspring'. We think of fresh life springing, as water from the earth, from womb. *Life is elastic, vital, sprung.* A single helix is a coil or spring. A spring is compressed prior to the release of its power. A double helix is, in effect, a double spring. We turn to ponder the chemical structure of biological information storage, a molecule called *DNA*. Whether or not *DNA* acts as an archetypal aerial (Chapter 16) both its structure and function are dialectical. At the start of every life cycle chromosomal *DNA* is compressed, in the process of mitosis or meiosis, prior to its release as the new life is sprung into operation. In a fresh cell *DNA* uncoils; cell division may continue through vibrant generations of re-coil and release. And the whole process of development is 'sprung'. At first life's body uncoils (develops) at great speed but the rate slows until it stops and, after a pause of poise on the cusp of adult youthfulness, enters a long and gradual decline towards death.

Do you remember (*fig.* 24.2) that the dialectical ladder can also be represented as a double helix? Let's look more closely at this further twist to the information spiral.

In the dialectical 'chemistry' of any particular characteristic (say, relative lightness or light/ dark) there may occur either an 'ionic' extreme (on the left or right) or a 'covalent' degree of emphasis (one way or the other including a mid-way balance) between any two complementary opposites. *Each person, behaviour, object or event can be described in terms of a combination of relevant aspects or characteristics which can be stacked to form anti-parallel (↓↑) columns of the Dialectic*. In this way, according to the dialectical doctrine of binary, complementary opposites, any individual thing comprises a suite of characteristics each of which shows a tendency to express, more or less predominantly, one opposite or the other.

	tam/ raj	*Sat*	
	duality	*Unity*	
	existence	*Essence*	
	finity	*Infinity*	
	its expression	*Potential*	
	its exchange	*Information*	
↓	*tam*	*raj*	↑
	passive info.	*active info.*	
	informed	*informant*	
	motor	*sensor*	
	its opposite	*principle*	
	its reflection	*image*	
	its expression	*code*	
	production	*invention*	
	received	*message*	
	imprint	*program*	
	memory	*new thought*	
	storage	*transmission*	

In fact both Primary and secondary dialectic can (Chapter 2: Cosmic Fundamentals and *fig.* 17.7: Wireless Framework of *H. archetypalis*) be expressed tri-logically:

↓	*tam*	*Sat*	*raj*	↑
	down	*Poise*	*up*	
	to periphery	*Axis*	*from periphery*	
	from origin	*Origin*	*towards origin*	
	centrifugal	*Central*	*centripetal*	

In terms of Natural Dialectic, Main or Essential Dialectic counter-poses Infinite with finite, Unity with polar duality. One strand of Essential helix represents (*Sat*) Infinite Potential from which the other, existential polarity, derives.

The existential strand of Main Dialectic is (*figs*. 1.1 and 3.1) polarised into the (*raj/ tam*) strands of existential or branched dialectic. Active informs passive. The basic existential complements are active informant and passive, informed energy - mind and matter. The helix involves a gradient from predominantly active, psychological causes to predominantly passive, automatic physical effects. In other words (*Sat*) potential is, through the kinetic agency of mind, finally realised in the (*tam*) detailed patterns of the physical universe. What is below reflects that above.

In this way three-tiered Mount Universe can also be modelled as three stages (start, mirror and anti-mirror) in the single full twist of a double helix. You can analogise the half-twist with mind. Mind is the Mercurial messenger, crossover, mid-point of an exchange between 'heaven and earth'. Janus-like it mirrors on the one hand Infinite Origin and on the other complete physical expression of its Potential in the form of earth below. In an act of creation there exists a one-way, motor flow which transcribes and translates principle into practice, which turns coded potential into actuality.

In other words (*tam/ raj*) polar relativity is set within a framework that includes its (*sat*) point of origin - in this case nucleus and its carrier, egg. This point is variously described as *axis*, centre or point of governance. It is known as the potential or prior information from which an orderly, regulated behaviour of things derives. In a creation gradient that runs from information to energy i.e. mind to matter, information from the centre sources the arrangements of peripheral shapes of matter. *Inside information is expressed as, according to a tension that is fundamentally polar, its external form of existence. Different, dynamic permutations of polar yin and yang show in each aspect of everything; the structure of Natural Dialectic (and, behind it, existence itself) is, at root, bilaterally informed. Moreover, if inversions occur between Infinite Essence and finite existence and, within existence, between mind and matter, the conscio-material gradient can also be conceived of as a binary spiral, that is, a double helix.* **Double helices, whether cosmic, nervous or genetic, are 'inside information bearers'.**

Until now the dialectical expression of opposites, either Main or Existential, has been in columnar form. The juxtaposition of complementary opposites has been held rigid, rather like a ladder with its rungs representing the two-way flow of measurement between them. *It is, however, possible to express the polarity of information contained in the two strands of Natural Dialectic's columnar construction in an even more robust, symmetrical yet flexible embrace.* It is tri-logically reflected (Chapter 2: Tri-logical Form of Natural Dialectic and *fig.* 17.7) in the two strands of biological *DNA* that, incorporating mirror image code, run anti-parallel; and between whose parallels the third and central axis runs. While composed of vacancy between genetic letters (nucleotides) that are sequenced on each complementary strand, this axis is a point of governance. From this line of nothingness the twine is separated to reveal its previously eclipsed polarity; where else splits open to expose a nominated tract of code? One vital strand of code threads upwards and the other down. So do the anti-parallels of psychosomatic (*figs* 17.8 and 17.11) and neurological (*figs* 14.5 or 17.1) information. The sensory strand runs up and the motor coils down within a central, spinal axis of information at the head of which presides the brain. *Both informative DNA and nervous system are, in turn, reflections of the wider principle of complementary polarities. This is symbolised in columns whose pair can, in a final twist, be dynamically expressed as a double helix.* Any behaviour, object or event, psychological or physical, is actually composed of a combination of characteristics - a dialectical stack. A stack allows multitudes ofcharacters, that is, of principles combined. It therefore seems to symbolise the shape of complex information and the polar world its combinations generate. **The full flowering of Natural Dialectic is expressed as a binary spiral, that is, a double helix.**

Spiral Inversions as seen in a Double Helix

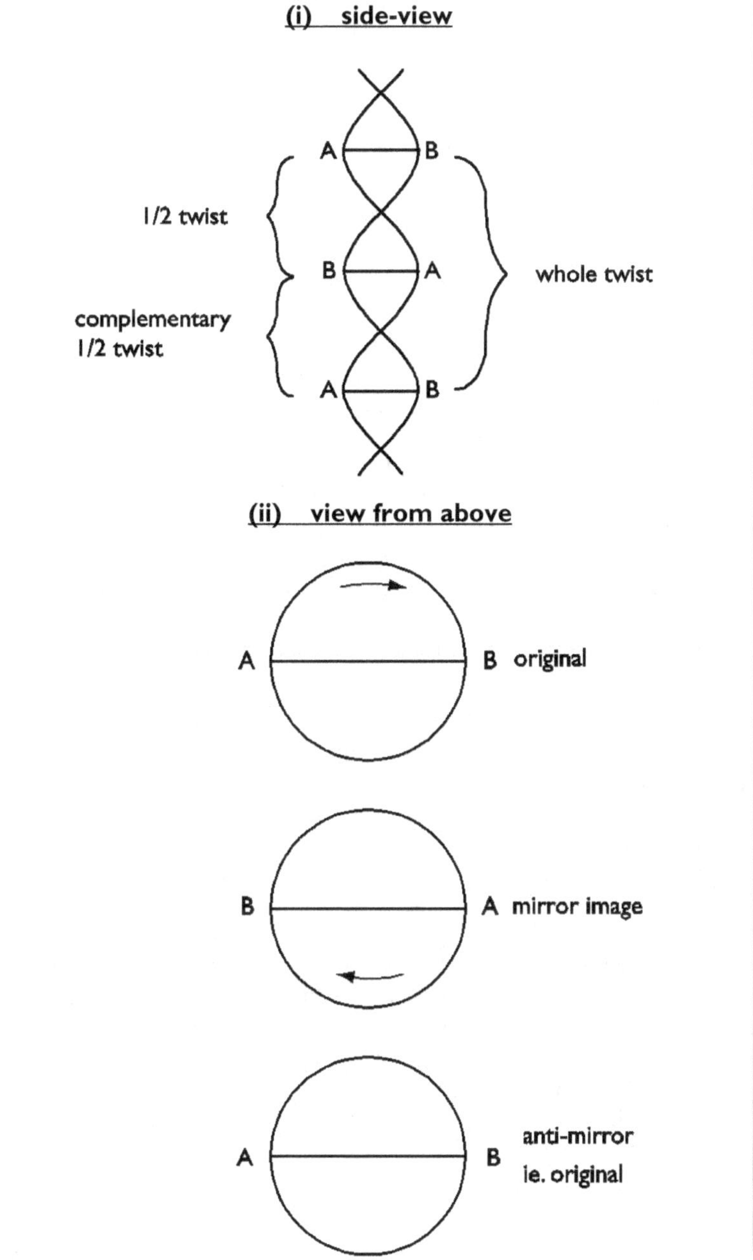

Take (in terms of *DNA*) a cross-linkage through, say, guanine to cytosine on the complementary strand. Take (in mathematical terms) a line from a point on strand A of such a helix through the central axis to the correlate or complementary point on strand B. If you half-twist this line through 180 degrees about the central axis you will reverse the position of the two points. One full turn of the cross-linkage (i.e. a full turn of the spiral) will have traversed full circle. The strand that started right will have passed (at 180 degrees) to the left side and then back (after 360 degrees) to its original position. Image, mirror image, anti-mirror (i.e. original) image; reflection of reflection that reflects the couple's origin.

fig. 24.6

Cosmologists and poets both know the limits of a metaphor, image or model - into whose constraints multi-faceted nature is at the same both squeezed and diminished. Yet nature has shapes and models have an explanatory use. An atom, for example, symbolises the Taoist principle of (*fig.* 1.1) polarity-in-complementary-unity. Similarly light (fig. 12.2) is a 'messenger' and the shape of *DNA* is, with a little imagination, like a fixed version of the interwoven electrical and magnetic fields of a plane polarised ray. It certainly, with its interwoven spirals-along-a-line, represents a biologically suitable paradox - dynamic stability. *Is it any coincidence that the coils of biology's fundamental store of code actually adopt the stable geometry of complementary opposites entwined in the mirrored embrace of a double information helix?* Does anything else indicate that the double helix is an optimal, natural, geometrical vehicle for the cosmic transmission of information, order and harmony?

The spiral is an ubiquitous natural shape that follows a precise mathematical pattern. In a Fibonacci number sequence (1, 2, 3, 5, 8 etc.) each number is the sum of the two preceding it. When two adjacent numbers are divided the result is (for larger over smaller) 1.618 or (for smaller over larger) 0.718. *This ratio is called the golden mean.* Architects (including pyramid-builders), sculptors and artists have always employed the beautiful, dynamic proportion (1:1,618) of the *golden rectangle* in their work. Fibonacci numbers also appear in phyllotaxis (the way leaves are disposed around a stem in order to maximise each one's exposure to light and air) and in petal arrangement. The revolution of the earth round the sun takes 365 days. Curiously, if you compare the ratios of one planet's time of revolution its neighbour then, inward from Neptune, they correlate with the spiral arrangements found in flowers! With anomalous divergences in the ratios for Mars/ Earth and Earth/ Venus the set reads 1/2, 1/3, 3/8, 5/13, (8/13, 13/21) and 13/34. At the same time the sun is flying through space so a corolla of planets *spiral* round it. A solar system like a flower!

Chemical information to evolve plant spirals, ear cochlea, ram's horns, snail shells etc. is held on the genetic equivalent of ink, paper and alphabetic script - *DNA*. A *DNA* molecule is 21 angstroms wide and, in a full turn of its spiral, 34 angstroms long. *The ratio of these adjacent Fibonacci numbers means that the potential for a life form is, in practice, borne on a stack of golden rectangles.* 'Golden' chains bind life to its cells. Nuclear DNA is the chemical agent for the storage and transmission of biological information. The very shape with which it fixes life to earth is expressed as multiples of the golden mean. *The shape of each gene, chromosome and genome is constructed according to the best principles of dynamic symmetry; life's book is printed with geometrical beauty. This suggests that, both with respect to components and structural engineering, its capacity for the fax carriage of vital information is optimal; that DNA is (Chapter 20) the best possible material to expedite, in a durable yet precise, efficient way, the critical use and replication of the book of life.*

The consideration of such *functional logic* provokes a question. Was an aesthetically and functionally meaningless stroke of chance the creator of such an illuminated manuscript? If so, how did a previously sterile, primeval marsh haphazardly scrawl its proto-nonsense in such a relevant, rich style? Why on random earth should unsupervised puddle-chemistry fall involuntarily into the arrangement of a high-tech, code-loaded, working database - which must have been a part of even an Ancestral Slimeball? The structural and functional elegance of the *DNA* twine suggests that, in fact, it was the combined product of high-tech engineering and information technology. If the Textbook about each type of organism was conceived, transcribed and transferred by Natural Science into the business of a biological shape, it would amount to quite a final twist-in-the-information! Consider the creation of purposive mechanisms by chance or by mind. Which idea involves a sleight-of-hand? Which evolves illusion?

The twists spin further. There exist many kinds of complementary, polar opposite and mirror-image symmetry. Light and dark, matter and anti-matter and opposite electrical charges are examples. Eggs develop from radial into a variety of bilateral symmetries. While a starfish keeps a radial shape the case of a single head-to-tail cut at 90 degrees through a body's midline, which dissects it into two mirror images, is a common sort of biological shape-maker. Symmetrical polarity. Balance incarnate. Ask any tiger, bird or fish how theirs, composed in multiple complexity, arose. **Morphogenetic symmetry is mechanically critical for flying, swimming or running - for survival.** You can't luckily limp there by unselected stages. Your own hands and feet are chiral (mirror) images of each other; there also exists a more or less bilateral symmetry of human head (cerebral hemispheres, two eyes, ears, nostrils etc.), body and limbs. Eyes, ears, nostrils, hands, feet, lungs, kidneys and gonads are all part of your nearly equal right- and left-handed bilateral polarity. To left/ right may be added front/ rear antithesis. Front/ rear-wise, central information processing (brain), sensors (eyes, ears, nose and tongue) and the information motor (tongue for speech) are located predominantly top-front, while locomotive organs for energetic response (e.g. spinal, gluteal and calf muscles) are predominantly lower rear. How can genes mastermind such stereoscopic, spatial choreography and engineer left-hand/ right-hand inversions? How can they, for example, counterpoise your mirror-imaged thumbs? *By what steps do symmetries, piece by piece, evolve?* Where are all the intermediate, asymmetric fossils? Will you invoke the piecemeal construction of layer upon layer of genetic cyber-systems, each of whose invocation multiplies purposive complexity and, therefore, the odds against chance? **The evolution (as opposed to operational construction)**

of morphogenetic symmetry is yet another major evolutionary black box. On the other hand, could not intelligence-based archetypal engineering, having replaced nonsensical chance, put sense back into the way we understand that purposive, efficient biological shapes originated?

As your left is a mirror image of your right hand so chiral molecules (from the Greek '*cheir*' meaning 'hand') exhibit a subtle form of isomerism, mirror-image twist or 'handedness' that has profound consequences for life. A normal mixture of such chemicals (Chapter 20) is randomised so that a roughly equal number of left- and right-handed types are present. Biological metabolism rigorously, systematically segregates chiral molecules. For example (*sat*) *DNA*, the primary chemical source of life's information, is *only* right-handed while its product, protein (the primary chemical source of (*raj*) kinetic action (metabolism) and (*tam*) static structure) is *only* left-handed. Right-handed informative material complements left-handed, functional enzymes and structural, protein-based material such as muscle, bone and skin.

Information to Shape: Code to Creation

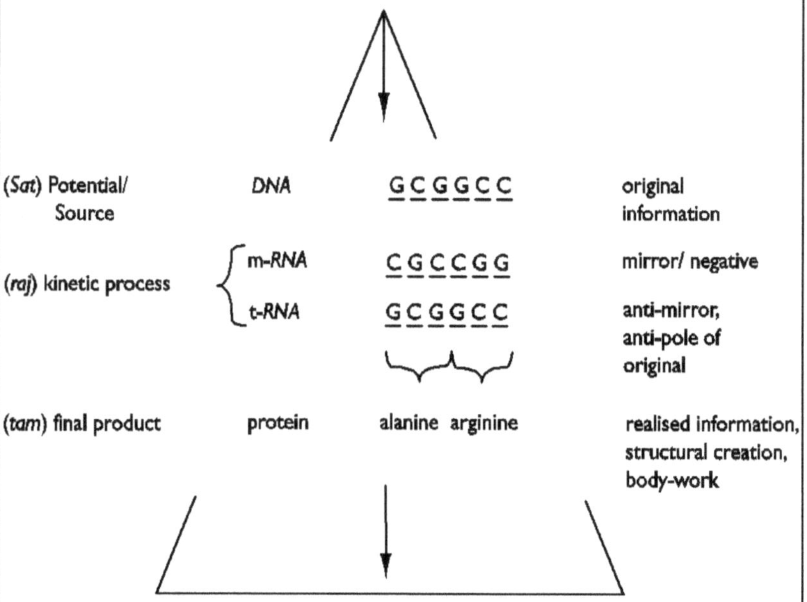

This diagram is triangular as opposed to the previous one's spiral form. Its example shows the materialisation through transcription and translation of informative *DNA* (using complementary if not dialectical pairs) into both 'dynamic', functional and 'static', structural forms of biological material, protein.

You might liken the process of protein synthesis to photography. In this analogy the image of an object (*DNA*) is snapped, that is, transferred to a negative (the film of m-*RNA*); this in turn is developed (through the agency of t-*RNA*) into a positive end-product, a 'picture' called a protein.

The complex procedures of both protein synthesis and *DNA*

replication (Chapter 20: xviii and xix) must have been present in the first cell. Let he who demurs propose a precise, viable alternative; and let it be possible to infer no informative intelligence from such anarchic evolution, such chance-ridden 'chemical development'.

fig. 24.7

	tam/ raj	*Sat*	
	its exchange	*Information*	
	subsequent order	*Archetype*	
	protein/ RNA	*DNA*	
↓	*tam*	*raj*	↑
	down	*up*	
	end product	*processor*	
	informed	*informant*	
	translation	*translator*	
	protein	*RNA*	
	structure	*function*	
	target	*enabler/ enzyme*	

What necessity prompts the frequency of reflection, imprint or inverted copy on life's drawing board? Information shapes energy, energy patterns bodies. Inversions pervade the biological translation of 'mind' to matter, of programmed data to biological form. (*Sat*) nuclear information, or biochemical memory, is stored as specific sequences of *DNA*. It is transcribed or 'photocopied' onto a negative called messenger or m-*RNA*. Select sections of 'absolute' *DNA* are stamped, in mirror image, onto this (*raj*) passing phase. Such a segment, representing a fraction of the whole potential called a gene, is a midway mirror that transmits a relevant, detailed part of the general whole to a translator. Negative, mirror-like m-*RNA* is now developed, word by codon (as a genetic 'word' is called), onto a series of verbal translators called t-*RNA* molecules. Mirror of mirror, a t-*RNA* molecule is an anti-mirror or quasi-original but precisely deconstructed version of the *DNA*. Each of these sixty-four remarkable translators exactly reflects a letter of the original genetic order. Each combines with only one kind of amino acid to expedite the final mirror, translation of *(sat)* information into (*raj*) activity and (*tam*) physical structure. Information is transformed into two main forms of specific operational protein. There are soluble, globular antibodies, hormones and enzymes for (*raj*) active control and non-soluble, fibrous, thread-like ones for weaving the fabrics (or tissues) that constitute the (*tam*) container called a biological body.

Protein synthesis is a mini-metaphor, a reflection of the universal creative process. From potential (in the form of programmed precondition) a mirror, a messenger or medium (mind) translates code from the Centre to a peripheral physical construction. From a whole *DNA* genome specific chemical symbols of archetype (called genes) are transcribed; outside the nucleus mirror-images called m-*RNA* are, in turn, mirrored against translation (by t-*RNA*) into the building blocks of bodywork (amino acids and protein); a chain of amino acids is extruded from its ribosome and, with help from protein quality controllers called chaperones, guided to assume precise three-dimensional shape; finally, according to an archetypal plan, the bulk material of a classical, biological body is developed. From (*sat*) initiator through (*raj*) conversion into (*tam*) matter - the pattern of life's creation fits.

The process is vital. Genetic information introverted and thereby protected at the inside-centre of its anti-parallel strands; and its tube-like scroll is, in the reading, opened out. Introversion is unfolded; the message is specifically revealed, transcribed and passed upon a ribosome from code to extroverted shape, from symbol into its material expression. Like thought to deed, the meaning of an idea is translated to the body of its actuality. This is devolution. Do plans happen randomly? The reverse; the dialectical process is and always was in order. Of this you are living proof!

Protein is the agent of a dream both coming and come true. On this foundation of translation super-dreams called lives emerge into phenomenal reality. On code-informed protein synthesis rests all complexity of detail and diversity. Small and great alike rest on material nothingness called information and encapsulated in genomic libraries full of volumes published using just one special kind of ink. Take metabolism and cell structure; take physiology and anatomy; proteins constitute the (*raj*) functional and (*tam*) structural basis from which everybody grows, the core round which each body is sustained. Bodies, just like homes or businesses, reflect their organisers. In the dialectical case main guides comprise a biochemical *DNA* database, morphogenetic archetype and, rising to the horizons of sensitivity, psychological instinct and conscious calculation. *DNA* involves inversions and helices as obvious as those of the psychosomatic exchange between mind and matter across the nervous system's template. Those of an archetypal *caduceus* (Chapter 17) are subtler and, physically, unseen. Do you remember the vibrant structure of plane-polarised light with its symmetrically entwined couple of forces? A double helix resembles a frozen kind of light. We might reasonably conclude that it comes to symbolise, both dialectically and naturally, the universal structure of information-in-energy, of mind in matter. Frozen code, balanced twine, dynamic elegance of polar information stored. Biological illumination!

Both *psycho-logic* and *bio-logic* involve mirror-like twists from top-to-bottom and side-to-side. Symmetrical structural and functional bio-logic seems to accord with life-style and usage and therefore occurs scattered in such a mosaic pattern across the tables of biological classification that it is of little or no help in trying to perceive any theoretical evolutionary relationships. Is the almost ubiquitous presence of pre-programmed complementary, mirror symmetry as we observe in biological body-plans a product of coincidence, necessity of balance or an economy and order of design? *Is it all a completely unlikely accident? Or, rather, bound as part and parcel into the construction of existence?*

To conclude, polarity within the confines of extremes is relative. Nowhere is this more obvious than in the polarity of sex where each partakes of the other to form, ideally, a complementary whole. Sexual difference yet equality, an equilibrium in the psychological and physiological equivalence of the sexes, is embedded in dialectical logic. **In the absence of any viable answer and in light of the systematic integrity and purpose clearly visible in sex and its inseparable correlate, development, it is entirely reasonable to suppose the**

system was designed. It is logical, from a dialectical point of view, to suppose within the larger program for reproduction, male and female forms represent a nested suite of subroutines already written and correct before sex could function. *A holistic or potentially hermaphroditic archetype, human being, is split into polar aspects of the same nature.* Sexed bodies are automated co-creators and their reciprocal apparatus revolves around the mechanisms of meiosis, fertilisation and parenthood - survival of the species.

The thesis is, therefore, that reproduction accords with the principles of cosmic construction and constitutes a multi-layered analogy for the act of creation itself. If creation and, with it, sex are not the unintentional residue of a series of accidents but archetypal, fundamentally polar concepts that accord with dialectical principles, then any need for parthenogenetic males or 'hopeful monsters' searching at every stage of evolution for another 'hopeful monster' of a not-quite-opposite, not-quite-yet sex of the same species disappears. *However the solution that sex is preconceived according to cosmic structure rather than evolved piecemeal comes at serious cost - a philosophical cost for which some may lack the conceptual cash.* This price is significantly raised when it is realised that the *bio-logic* of physical, sexual polarity is firmly correlated (both in the microcosmic, human *caduceus* and macrocosmic principle) with a super-sexual, informational *psycho-logic.*

Not just side-to-side inversions such as sex but those that snake from pole to pole (information/ energy, mind/ matter, archetype/ body) are spirals, double helices embedded in the natural order of the bio-world too deeply to have happened accidentally. The dynamic shape of Natural Dialectic and, logically, the cosmos it describes is composed of vibrant, complementary duality. Its order is expressed this way. Where an atom models the non-conscious zone of universe, a double helix models information as it enters into form and its formation. It certainly reflects the way that, through and through, you're built.

What is the final twist in tail? Is it the twist that wrings the truth in place of certain story tales? *Such clear filtrate is an elixir of wit. Its intelligence suggests there's more to life than, when it comes to origins of order, meets naturalism's eye.* **You might, in other words, infer a knell. You might come to realise the power of combination in the form of logic and design, archetypal psychological and biological design, spells the death of scientific atheism as the whole of any answer to the questions that we ask about our heritage. It bids us look beyond the grave, that is, beyond the fact of simply atoms and the molecules of earth.**

We saw (fig. 19.1) that according to a dialectical plan of the way life works the first and core principle of biology is 'that which comes before', its potential being, (sat) information. Information tells you how to act. Its potential precedes implementation. It has to be present before any purposive, physical arrangement can occur. In this case the purpose is survival under physical constraint. *And the core informative principle, around which all other biological principles are expressed as peripheral adjuncts, is* (Chapter 19) *homeostasis for survival.* The basis of homeostasis is, in turn, negative feedback based on information received. Although contemporary biology allows internal (conscious, sub-conscious and non-conscious) and external (sensible) sources of information it does not allow an original, teleological source of the prior information, order and construction necessary for survival in an earthly (biological) body. *Active information* (Chapters 5 and 6) *is excluded from the process of creation; only passive outcome is examined.*

Dependent on prior information are a couple more key principles and associated processes. One is immediate and associated with (*raj*) current action. *To live at this moment needs energy metabolism.* Without at least tickover of life's engine survival is measured in seconds. We have noted (Chapter 20) the silence that engulfs any attempt to explain the *origin* of metabolic circuitry, especially core energetic pathways, in terms of accidental chemistry. Otherwise useless intermediates on such pathways, codified enzymes and associated substrates, form a large proportion of metabolism so that to suggest its gradual evolution is, even on the theory's own terms, illogical. Such routines are critical and appear as bespoke attunements, tailored to its individual context, in every organism. Their various attunements (i.e. codified differences) are held by the phylogenetic mind-set to demonstrate evolution. However, during the time a vehicle of life (a body) was transforming itself into a 'new model', could its metabolic parts-list have morphed, each item in parallel with the rest, to conform with fresh and ever-changing coordinate specifications? Could it have done so by an unpredictable, targetless and long drawn-out series of accidents? It is a medical fact that even slight chemical and genetic accidents, not least those involving the metabolic 'chip' and its associated operating factors, cause disease, death and destruction. From a *top-down* point of view the conjecture is unnecessary, unproven, unproveable and misleading. Key processes of data (*DNA*) manipulation, photosynthesis (or chemosynthesis) and respiration must have been co-operative from the first moment of the first cell's existence and yet no real attempt has been made to specify the haphazard steps by which this critical but apparently impossible, multi-faceted conjunction might have evolved.

It was noted (Chapter 22) that two main arguments on which the theory of evolution is based were homology and hereditary variation. Antagonistic interpretations for the origin of homologies, archetypal subroutines and irreducibly complex mechanisms by chance were briefly inspected. We turn now to the third key principle subsumed under potential, long-term survival, a survival measured, for individuals, in days, months or years but for their type always in an indefinite, immeasurable, theoretically unlimited future. *Extended survival involves inheritance by reproduction - the business of copying physical structures. Reproduction is never, even in bacteria, a chemically simple, unchoreographed operation.* **Equally, it is conceptual because it is anticipatory**. *It is pre-programmed. It foresees death. It looks into the future when the present organism will cease to survive.* **It must, from the first or at very most within the first days, weeks or months of the first cell, have 'appeared' to meet this eventuality and, with it, the 'desire' for a species to transcend an individual's (that is to say, a bunch of atoms') so-called death.** Can gangs of atoms really bunch like this by fluke?! With fluked-up information to resist their own demise? *If not, evolution's central mechanism is non-evolutionary.*

In biology there are thousands of obvious examples of dialectical forethought in design. Informative and reproductive systems are just two of them; but each one surely tolls the knell for evolution.

The Reproductive Archetype

Reproduction occurs in sexual and asexual forms. The latter is sometimes considered more 'primitive' than the former but, in terms of function, both have an equally important place and purpose.

The point of asexual reproduction is invariance. It is the conservative retention of complex information by accurate replication. One doubles. It becomes two of the same genetic circumstance, called clones.

Reproductive Archetypes

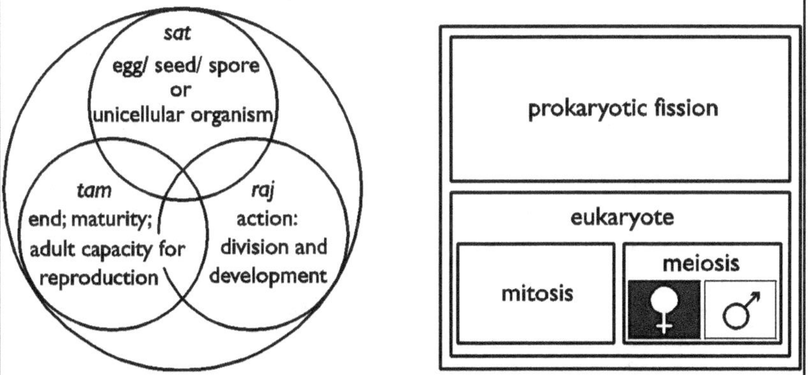

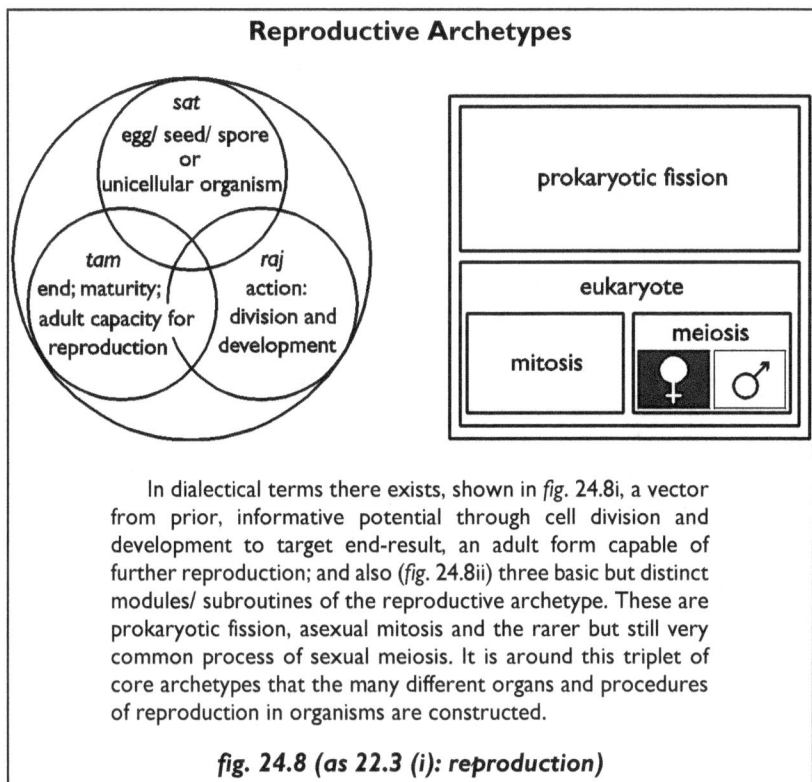

In dialectical terms there exists, shown in *fig.* 24.8i, a vector from prior, informative potential through cell division and development to target end-result, an adult form capable of further reproduction; and also (*fig.* 24.8ii) three basic but distinct modules/ subroutines of the reproductive archetype. These are prokaryotic fission, asexual mitosis and the rarer but still very common process of sexual meiosis. It is around this triplet of core archetypes that the many different organs and procedures of reproduction in organisms are constructed.

fig. 24.8 (as 22.3 (i): reproduction)

The point of sexual reproduction, on the other hand, is variation on theme. It is neither to add nor subtract but to shuffle an organism's cards into new permutations and thereby deal new hands in an old game. In this case two become a different one. In principle (or type) each and every offspring is the same as its parents, in practice (or detail) different. This sex-driven difference protects it from genetic damage that may exist in the parental genes - a mutant gene is likely to be masked by a 'healthy wild type'. As well as promoting fitness it promotes physical differences that under certain circumstances buffer changes in the environment or perhaps even lead to speciation.

From the point of view of objective and functional efficiency neither form of reproduction is more advanced than the other. *Conservation and variation* - each has its clear, purposeful place and, if you consider any exchange of genetic material to be inherently 'sexual', is found in all kinds of uni- and multi-cellular life forms.

The Origin of Asex

All cells make a copy of their *DNA* database before division. Accurate reprint, that is, cloning leads to the reproduction of identical cells or, as in the case of your own development from a single cell, the zygote, myriad different kinds of cell. The mechanics of *DNA* replication and reproduction (including the development of a multi-cellular organism) are not yet completely understood but there is no reason to suppose that in future they will not be. However, even if a mechanism is understood down to the exact operation of each subatomic particle this does not mean its *origin* is understood. The origin of (biological) machinery is conceptual. Atoms have no concept of 'work', 'survival' or 'coordinated and cooperative strategy for an improved probability of survival' - although a few billion of them, accurately coded, configured and called a bacterium, seem happy with the process. Why did the millions that code for all the others first appropriately 'self-organise' - if you buy that phrase?

Overwhelming problems with chemical and biological evolution have been aired in Chapters 20 through to 23. On the other hand, idea and design precede the planned construction of even simple machinery. The qualification is not annulled simply because, in its ability to self-reproduce, biological is far more complex than ordinary technology.

Another set of overwhelming problems involved in the spontaneous, naturalistic or non-conceptual origin of genetic material were discussed in Chapters 20 and 21. How long can bases, nucleotides or nascent extents of *DNA* survive without (or prior to) a cellular environment? What is the longest a cell can exist without reproduction? Abiogenetic production of a biological prototype, against the most fundamental principle of cell biology that a cell always derives from a cell, would not have involved *re*-production. A few minutes (or perhaps hours) later it would. No organism is reproductively fitter than a bacterium. No organisms better withstand extreme conditions than prokaryotes. Their 'naked' *DNA* is non-repetitive and economical. In favourable conditions they can divide every thirty minutes or so. Two *DNA* replication forks travel in opposite directions round the 7000 or so genes of a circular chromosome until they meet. Transcription is very rapid (in the region of 5000 bases per second) and accurate. Next one of the rings is peeled off and attached to a new site on the cell membrane. *There is no indication that so-called proto-bacteria were any less efficient, certainly not below the functional minimum mandatory for life.* While the exact mechanism of bacterial *DNA* replication and fission is under the microscope its gradual evolution is hard to defend both in principle and practice. Surely atheism is not hung from vague, arm-waving, very slim conjectures?

A bacterium is a conservative survivor *par excellence*. Millions of genetic experiments involving a life cycle of thirty minutes over a hundred years have varied the genome of, for example, *E. coli*. Such variations are sometimes called micro-evolution but the plasticity is limited. Typical *E. coli* are not evolving into other forms. The type remains stable. The same basic invariance occurs in all the highly diverse groups of prokaryotes. How and why, in spite of selection pressures, should the fittest of all life forms in terms of fecundity have (d)evolved into diverse, complex but less fecund, environmentally restricted, multi-cellular forms of life? Why ever might it leave its sexless rut? *There is absolutely no sign of either fossil or present bacteria evolving into other kinds of bacteria far less eukaryotes*. Many bacterial strains actually liberate plasmid toxins in order to kill incipient divergence in the form of closely related strains. Bacterial *DNA* differs radically in structure and function from eukaryotic chromosomes. Its replication is efficiently and effectively engineered but bears no resemblance whatsoever to eukaryotic mitosis, meiosis or the development of purposely differentiated sexual organs. There is no hint of the steps by which bacterial conjugation *must* have been transformed into either asexual mitosis or sexual meiosis, which is a deliberate 'shuffling machine' or lottery. From deliberate to accidental lottery, if such evolution occurred once why should the blind process not keep recurring and an incipient example be found today? Meiosis, the root algorithm of sexual reproduction, precisely promotes systematic genetic recombination in a way prokaryotic fission never has, cannot and, according to the evidence, never will.

Bacterial fission is one thing. Non-bacterial cell division is quite another. After replication any eukaryotic cell not destined to become a gamete enters a profoundly efficient, conceptual postal system called mitosis. Check the Glossary or any textbook for its '*mise-en-scene*'. Nothing less than the complete ritual counts. Why? Why is it necessary? From quadrillions of bacterial fissions not a single lab has seen the eukaryotic algorithm start sequential steps to build itself; nobody's seen mitosis set to 'happen'. Or did it of a sudden once appear by chance complete and working? Why not?

Eukaryotes do not have a single, circular 'chromosome' but a number of linear chromosomes. What is their origin? Did the circular *DNA* of a proto-eukaryotic cell split randomly into strips? Whatever the case, you can regard each chromosome as a book of information and the whole complement, called a genome, as a complete encyclopaedia. Each chromosome is therefore critical. In cell division each daughter needs to have the right quantity and, furthermore, the right quality of chromosomes. That is, if the number of your chromosomes is 46 you cannot deliver 40 to one and 52 to the other daughter cell. Even if you deliver the right quantity (46) you must also deliver the right quality. It is no use delivering two copies of chromosome 5 to daughter A and none to B; or two of chromosome 19 to B and none to A. This would constitute a lethal loss of information. *So there arises the immediate necessity for an efficient postal system in order that, at the moment of division, the correct type and number of chromosomes are allocated to each daughter cell; and that the rules of delivery are executed with automated precision each and every division of each and every eukaryotic life.*

Mitosis is a prize-winner. Its essential purpose is to keep intact written information of the genome every time a cell divides; the chance that chance could do this are statistically nil. Conceptual elegance combines with strict economy of working parts. A choreographed algorithm constitutes the ubiquitous foundation of eukaryotic cell division. Its irreducible components comprise a system of multi-chromosomal postage entirely different from the bacterial one; the mechanism employs a logically impeccable order of steps and complex kit (centromeres, centrioles, spindle and so forth) to expedite its accurate deliveries. Incomplete or faulty mitosis is useless. No organism sports half-way mitosis; faulty mitosis is equally lethal because information is scrambled or lost. In this game, therefore, wrong numbers are not dialled.

Down's syndrome (47 instead of 46 human chromosomes) is just a light reminder of the problems, mostly lethal, that inefficient counts accrue. In this case what happened as the pre-mitotic lack of accurate arithmetic was getting under way? How did the dance recover from its faultful, fateful learner steps? As a harmonic dance, an algorithm, an embodiment of anticipatory, informative logic the archetypal view of mitosis is clear; but this is not the way that evolution works. So Darwin's proponent fails to see the issue as one of information impressed on molecules; he prefers instead to think that molecules can 'happen to make sense' by serendipity. **In this case the evolution of mitosis is another Darwinian black box.** *No plausible let alone systematic account has been proposed for its alleged, aimless self-construction.*

You insist it *must* have happened, though. Like it or not you emerged from slime. The presumption of evolution from bacterium to mitotic eukaryote demands speculation that has found perhaps its best expression in the theory of endosymbiosis (see Chapter 21: Energy Metabolism Perchance?). According to this idea you might imagine, for example, that a nucleus might have evolved when a membrane 'defended' genes from, say, phage attack. The animistic turn of phrase is not meant to imply that molecules, such as nucleic acids or lipids, might themselves be other than neutrally, obliviously passive with respect to motions of attack, defence or survival - but it does. In the pre- biotic story molecules continually find themselves Lamarckian reasons for responses that then inexplicably code themselves into immortality. Could there be method in the madness? Symbiotic method? So that one bacterium enters another and, having ceded genetic autonomy, becomes a 'symbiotic' organelle. According to this notion the hybrid ascension to a eukaryotic cell amounts to a conglomerate, a settlement of different germs inside a larger host.

In this scenario chloroplasts come to resemble chemical configurations called *Chlorobium* (a green sulphur bacterium). And mitochondria might have once, to judge by their *DNA* sequence, have been *Paracoccus* (a purple bacterium) or a killer called *Bdellovibrio* in a host cell, say *Thermoplasma. Bdellovibrio* is a nasty little character, a necrobe that feeds off its infected host. After manipulation which included nuclear transplantation (not a natural occurrence!) it was shown by Kwang Jeon that necrobic infection can be tamed and, indeed, with immunity gained, transformed into a necessity without which the host cell, in this case an amoeba, dies. This result might seem to mark a plausible phylogenetic channel but neither means nor proves that such complicated adventures actually occurred. For a bio-logician it is a hope based on the wrong premise. *Indeed, the endosymbiotic suggestion of bacterial transformation into organelles is as essentially irrelevant as the uncoded, random production of chemicals to the genesis of a cell.* This is because generation of the double membrane of a mitochondrion, chloroplast or, to enclose linear *DNA* and mitotic process, a nucleus, involves an element of preordination. **The functionaries are constructed as an integral part of a coded program**. If it is claimed that satellite *DNA* in cell organelles, although not self-sufficient, bears correspondence to sequences found in bacteria then a logician straightway parries that, of course, you might expect the same function to access a similar information source for its working parts. *The correspondence would then become one of archetypal subroutine rather than phylogeny, of logic* *rather than chance; and the construction of organelles would be seen to involve subroutines as integral to a cell as bathroom, kitchen or study to a house.*

Having made this point we turn to another example of the great diversity of bacteria - spirochetes. These swimmers are free-living, parasitic and possibly, in the case of attachment to a protistan called *Trichonympha* that lives in the gut of termites, symbiotic. By means of a special internal arrangement of fibres each spirochete can undulate like a whip cracked. It is supposed they entered other bacteria and became automotive organelles, cilia and flagella, widely used by single-celled organisms. Can you crack a whip in any other way? The micro-tubular structure of a flagellum is found in almost all eukaryotes as centrioles. Centrioles are centres for the anchorage and extension of cytoskeleton including the mitotic/ meiotic spindle. This spindle is a crucial framework on which the highly organised process of genetic postage takes place. Was this how the critical and clever mitotic algorithm must have 'happened'? Bacterially derived organelles, 'realising' the postal exigencies caused by the numerous chromosomes found in a nucleate cell, came to the rescue and, having arrived, dreamed up timing devices, ordering mechanisms and other apparatus, not least the spindle, of 'a spindle method'. *This is more remarkable because the end product of mitosis is not a chemical at all. It is a dynamic, conceptual arrangement of intrinsically oblivious chemicals.* These chemicals, genetic or otherwise, are simply agents of the dance, a routine that leads to its specific end-point, the precise, required distribution of information needed to satisfactorily (that is, precisely and so non-lethally) complete cell division. No-one is claiming spirochetes, which can give you syphilis, are helpful or intelligent but does this crude, phylogenetic guess more than scratch the surface of an answer, an answer needing to include the dimension of logic, routine and subroutine. Why re-invent the wheel or, in this case, the cytoskeletal rope. Why not re-deploy good, perhaps even optimal, kit in an efficient, economical way? *In this case you would expect to find a genetic correlation between chloroplast and the photosynthetic genes of a cyano-bacterium.* The same goes for other organelles. Not necessarily phylogeny but bio-logic.

The Origin of Sex

What if there was no such thing as sex? Asexual reproduction deals in clones. A clone is a cell or organism with exactly the same genetic constitution as its parent. Fission, budding and eggs unfertilised from sex's default, female form give clones. However, is such sameness all you want - a bland and sexless lack of any variation?

In fact there are two other, sex-based kinds of clone. Identical twins, having the same genetic basis, will look 'identical'; but all such pairs of twins are different. And you're unique as any pair of twins. Sexed organisms grow from a single-celled zygote and have, in almost every cell, identical *DNA* - but you are not a Humpty Dumpty-like 'great egg' because your zygote in the course of its divisions differentiated into over two hundred correctly-positioned sorts of cell. Just as a newspaper is a whole but you read different sections, so different readers (called cells) read different *DNA* sections that apply to them.

Thus identical *DNA* can generate identical, cloned cells or different sorts of cloned cell; but to shuffle *DNA* to create slightly different forms you need more than default females. You need males as well. One to one gives just the same; two compounded into one excite the possibility of endless changes rung upon parental theme. Two can mix their *DNA*. The message changes each time making 'Mary', 'Tom', 'Delilah' and so on. Humanity is sectioned. The complementary sectors are called sex for short. Not south sex (Sussex), Essex, Wessex, even Middlesex. Tom and Delilah's kind of sex. Their cut's the label slapped on complementary interactive forms, on biological polarity.

While bacteria only reproduce asexually all plants and animals can do so sexually, at least occasionally; and while not up to that occasion many incorporate the capacity to reproduce asexually as well. So having easily produced eukaryotic mitosis from time's magical topper we can proceed to tackle the more complicated invention of sex. We can query in detail the evolution of sexual reproduction (including meiotic choreography and male/ female body plans) along with the anticipatory development of a fertilised zygote into specific multi-cellular form. Biological science can describe the operation of sex almost as completely as that of a motor-car but how accurate or convincing is the account of its origin, its evolution? It *must have* happened, if you take that line, but what were the steps that led from chicken to its egg or *vice versa*?

It *must have* been some sort of 'planimalian' germ initiated proto-plants or semi-animals with sexuality. Or, even more brow-raising, did the complex process blindly auto-organise in different types repeatedly? In fact, there's hardly any clue to indicate how sexual reproduction had (according to the *PCM*) to have evolved. What sign anywhere of useless half-meiosis? Thus serious-minded explanations are the seminiferous order of the day. A compulsion of arm-waving waffle, replete with convoluted, complex, just-so argument, conspires to germinate debate; and to generate a seminal idea of how a pointless process might endow each higher kingdom with its congruent, happy forms of sexuality. Did such 'evolution' ever happen? **Lack of evidential provision in the sexual category as well as those mentioned above would mean that, in *all* the key areas of core biological processes, we are left with a theory of origins whose explanation is not only illogical (see Chapters 5, 6, 19 - 25) but, after 150 years and huge intellectual investment, fundamentally deficient.**

The biological purpose of sex is not pleasure but variation-on-reproductive-theme. This is achieved by the establishment of a lottery. Sometimes variation is achieved through recombination-by-switches (as with petal, hair or other forms of coloration); various factors, including mutation, can affect expression of already-present batches of genes. But other times, using a lottery, chance spins the trick. Nature's most sophisticated lottery, a component of our immune system, employs apparatus that can, conceivably, generate a trillion (10^{12}) different antibodies. To obtain such phenomenal rearrangement-potential versions are specifically cut and pasted in hot-spot variable regions of *DNA*. In Chapter 23 we met such inbuilt capacity for variation; its diversity is called *adaptive potential* The product of such lottery is then fine-tuned to better its affinity for a specific antigen! Yet, if this week's lottery winner is 1230847 and the next week's 1532954, this does not imply a weekly evolution from one number to the next. *Variation* is the game's name. Similarly, sex is a kind of lottery and a lottery is, like a casino, a frame within which randomness is purposely invoked. It is a method to skip the expected and flip to endless variation-on-theme - variation strictly limited to a particular type. In a casino you play different games with sets of ciphers and only expect to bet on various randomised recombinations from the same set. In biology the game's the same. The basic rule is *e cellula cellula* (from one kind of cell comes that same kind). What's your particular game? A sexual one? In different types of organism different rules apply but, through all the shuffles and the deals, the same 'great regulation' holds - the game you're playing stays the same. Sexual recombinations will generate the same and not a different type of organism. Unless you lift this limit (a speculation illustrated by *figs.* 5.1 (i) and 22.1 (i)), break the real deal and cross from game to game then sex will never roll you evolution. Sex will roll you round the field of variation but will never deal a foreign card.

At the heart of life's sexual gamble is a process called meiosis. *Meiosis, which underwrites the sexual reproduction of some single-celled and nearly all multi-cellular organisms, bears the hallmarks of a routine designed to extract maximum variation-on-theme at minimum cost in labour and materials.* Its precise, initial dance extends the mitotic algorithm so that, after replication, *two* reductions are made to produce a gamete (egg or sperm) with half the normal number of chromosomes - a state called haploid. The reason, an excellent and critical one, is to avoid exponentiation (i.e. doubling each generation) of the chromosomal number. The count is stabilised. If, for example, it is 46 normally then gametes with 23 each will, when recombined, recreate the diploid 46. The meiotic routine also involves two mechanisms that are not, because they are at the heart of sexual reason (i.e. variation) and its implementation (meiosis), at all haphazard.

After replication the first step in meiosis is quite unlike that of mitosis. A tactical change at this stage indicates anticipation. It indicates foresight (Latin: 'provision' or 'providence') because it is critical to the successful implementation of the purpose of the process. Just as parents coupled to conceive a child so at the point when that child's body, now adult, produces its own sex cells their chromosomes couple systematically. The symbolism of such synapsis, as it is called, is obvious. Each 'homologous pair' (there are 23 such pairs in your genome) couples in a full, close and coiled embrace called a bivalent. This synaptic structure allows for the first shuffle, an exchange of genetic material between them. The exchange is called crossing-over. After crossing-over the bivalents coil and, after a 'random assortment' with respect to maternal and paternal elements, line up to separate.

Crossing-over (or genetic recombination) and the independent assortment of chromosomes together constitute two-thirds of a mixing process which generates no new material whatsoever, merely new deals from the same pack of cards. They are a couple of shufflers, dealers or roulette wheels working a sexual casino in whose game, once the new combinations (called gametes or sex cells) are dealt, the third and final cut is fertilisation. This is thorough shuffling or, technically, recombination at genetic, chromosomal and finally genomic levels. The permutations reunite to make a whole new lucky number. Orchestrated fusion couples two polar halves to make a fresh call. All peripheral sexual apparatus (including instincts and sensations) is a context within which the core processes or, if you like, the central mechanisms of meiosis, fertilisation and development are backed up and delivered.

As *fig.* 24.8i illustrates, *top-down* order runs from (*sat*) potential (information) through the (*raj*) business of expression to end-product; it runs from egg through development to adult, reproductive form. Purpose, anticipating then achieving such a form, is an attractor; metaphysical concept draws the present through a program to its future goal. In other words, the whole business is targeted and, as such, worked out in every minute detail; its successful operation depends on finely engineered chromosomal architecture including the centromere's position and base-to-base alignment that alone guarantees synapsis, recombination, fertility, reproduction and therefore survival; and on top of all this the information for its repetition is packed easily within the confines of a single egg.

Wait! Clearly Darwinism's got a clutch of lethal problems here. The fact is hundreds of protein 'machines' maintain the integrity of each chromosome. *Synapsis and crossing-over cannot occur without such maintenance; nor, without this pair of processes, can sperm or egg be made. Sex could not start to work.* These two, however, aren't enough. *After shuffling genes the chromosomes are 'reassorted'. Finally, the genomic level is addressed. Now, since a partner's been provided, half of two whole shuffled genomes are combined.* Three steps without each of which variation and survival could not flourish. *A well-maintained and deliberate lottery of meiosis coupled up with sexual intercourse (using the correctly prefabricated body-forms) generates variation; but strict limits that the maintenance of genetic integrity imposes equally limit the degree of variation by common descent.* The deal changes but the pack remains the same. **It is predicted that a degree of stricture that confirms the Law of Heredity but is entirely incompatible with evolution by common descent will, in due course, be measureable by genomics. In other words, even variable versions of Darwinian evolution won't survive.**

The definite, foreseen target of meiosis is, by varying on theme, the next generation. Did you ever wonder at a baby, perhaps even your own? You knew that, although there was much more to its life than you could obviously observe, the biological target was sexual maturity, adulthood. So that the consequent developmental program that culminates in an Apollo or divinely different Venus was anticipatory; it must have been written and, in each respect, operate faultlessly before sex can function. No use one evolving before the other! Clearly, the simultaneous production of meiosis, egg, sperm, the surrounding organs of sexuality and programs of goal-oriented development into complementary adult male and female bodies taxes - heavily - a theory of evolution whose central proposition is the gradual occurrence of pointless biological programs by coincidence!

It is no use imagining anything less than a mechanism, with its associated parts, that works. Natural selection kills anything less. **Anything on its way to working never gets there**. So how did meiosis crop up and sex simply evolve? *As usual, where logical answers are barred, no phylogenetic substitute worth any salt at all is raised. And remember that, against the clear logic of purpose, any inference derived from an unpredictable stagger of accidents is anyway bound to be contorted, arcane, provisional, not testable and as twisted as the path of accident itself. Such is the fundamental way of evolution theory.*

Before proceeding beyond the core molecular 'Las Vegas', meiosis, to sexual machinery that in the larger scale carries forward its purpose, code and their embodiment in genetic material, let's glance at a second lottery at work in you. Defence as well as sex exploits the throw of chance. This time the high-stakes gambling rolls to reduce the odds against your illness, to keep you feeling well and underwrite survival. Your internal police force, your immune system plays its hand by generating 'not-you' combinations used as checks against illegal immigrants. If any immigrant checks positive against a 'not-you' permutation (as expressed by protein antibody) then the system swings into counteraction. A complex shuffling mechanism composed of several hundred 'variable domain genes' (blocks of immunological *DNA* code) and called 'somatic recombination' will have kept your immune system on its toes. It works by recombining the blocks of code into new 'combination-lock' permutations. The recombination occurs during the maturation of B-lymphocytes and permits them to generate new kinds of antibody to fight new species of illegal immigrant, new threats of infection. At every level in this war the ability to recognise self from non-self involves code, signal, recognition and, above all, specificity. An immune system consists of mechanisms that lock onto and variously disable pathogen-associated molecular patterns (often protein) and their accompanying body. *It employs a lottery and deploys about a thousand million different receptors to protect your health. Such a system is, in overall construction, an information-rich, automatic and reasonable process, the very reverse of chance.*

In bacterial warfare various strategies of attack (e.g. phagocytosis, antibiotics or attachment 'hairs' called *pili* or *fimbriae*) and defence (e.g. efflux pumps, antibiotic neutralisers, restriction enzymes) are employed. The code for such features is often deployed away from the main (somatic) chromosome in 'optional extras' called plasmids. Genetic mutations may prevent an attacker or an antibiotic from binding to its victim and therefore confer 'resistance'. Such mutations, spread through populations by any one of three mechanisms of genetic recombination (transfer, transduction or conjugation), have led to a large and rapid increase in the percentage of resistant strains of toxic bacteria - even producing superbugs. These variants are said to have 'micro-evolved' (Chapter 23). However, did systems of recognition and defence originate by chance? Or is their main thrust an aspect of survival around whose major themes minor variations, both deliberate and accidental, can occur? Is the bacterial transfer of information, like your own 'immunity-lottery', a legitimate, non-accidental agent of defence - a barrier similar in protective concept to membranes, cell walls, capsules, slime or even, in your own case, skin? Certainly, variation confers strength against unexpected dangers. It provides an element of flexible response. Therefore would you guess that molecular lotteries, whose purposive randomisation within strict theme and context is a logical defence, were both a product as well as a producer of chance recombinations? Could you, on the contrary, conceive that they are purposeful adepts in the exploitation of chance as a source of useful adaptation - another kind of bingo to keep you feeling good but surely not, as some claim, the origin of sexual reproduction?

You may note different forms of defence used by plants, fungi, animals etc.; and you may relate the 'sophistication' of that defence to the 'sophistication' of its biological form. For example, the immune system of a sponge or earthworm is considered, no doubt rightly, less sophisticated than a human's; and 'natural' or 'non-specific' immunity found throughout the animal kingdom is presumed below the par of 'adaptive', 'specific' immunity (whose white cells target specific antigens and adapt to respond more promptly on subsequent infection). You have nuclear missiles, rifles, cannon and sticks in a defence system; you have bikes, motorbikes, motor-cars etc. in a transport system. One machine did not evolve from another; each was a product of *psychological* direction. Moreover each different level of sophistication has its own suitable purpose and, in the larger scheme, coordinates with the others - just as 'natural' and 'adaptive' aspects of immunity positively and intimately interact.

In short, you may presume to arrange immunity (which is complex even in bacteria and protozoa) according to an orthodox perception of the relative complexity of body-plans but its evolution, like that of abiogenesis and the origin of those plans, remains another Darwinian black box. The reality is that defence is purposive. It is a consequence of a conceptual, psychological purpose - to survive. How, from a *top-down* point of view, might an engineer charged with designing a machine that could defend itself conceive it working? How would he rationally inform it? *Indeed, given the information-rich specificity and obvious purpose of systems of defence by self/ non-self recognition, it might seem positively irrational to ascribe the origin and evolution of natural and adaptive immune mechanisms to chance alone.*

From lottery to lottery. Sex, everyone agrees, generates the spice of life - variety. However 'simple' sex may appear, some of its most complex forms are found in unicellular organisms such as malarial parasites or *Paramecium*. Out of myriad life cycles evince precisely how just one, malaria parasite's, evolved. Even a glance at the reproductive strategies and structures of 'simple' unicellular and multi-cellular algae, fungi, plants and animals highlights their variety; even superficial study spotlights intricate complexity. Many, including 'higher' plants, show both asexual and sexual reproductive capacities. Differences in the patterns of conjugation run from a simple exchange of material between motile (+ or 'male' donor) and non-motile (- or 'female' receptor) factors, use of indistinguishable + and - 'isogametes', fertilisation of oospheres by antherozoids, ovules by pollen and ova by sperm. Each case is supported by the appropriate development suitable parts; contextual apparatus that includes requisite genetic and metabolic information; and a postal system, one well-conceived to deliver the message to a specific target, its sexual complement often called an egg - from fusion with which development towards a final yet wholly preconceived target takes place. *The whole is an expression of conceptual circuitry for a well-defined reason.* The wheels of life keep rolling purposely. Whether you take an alternation of generations (where sexual and asexual life cycles alternate), parthenogenesis (virgin birth), parthenocarpy (formation of fruit without fertilisation as in banana or tomato plants), hermaphroditism (where both sexes occur in the same body, as in many plants and earthworms) or separate male and female bodies the constructional challenge is the same. **The codified subroutines, modular structures and instincts have to be pre-programmed. They have to anticipate (which is precisely what chance does not do) the process of gametogenesis, fertilisation and subsequent development**. Such anticipation is just the start. In addition there needs be known what form the adult, reproductively mature form will take and how, from a single cell, this form will develop. Hierarchical chains of command and control need to *precede*, not succeed, not only the development of sexual apparatus but also its performance in the field and the procreated results of all the hullaballoo. **In short, each step and part has to anticipate its cooperation in a core biological process that achieves a critical purpose - reproductive variation on a theme.**

At root sexual transaction is about 'meaning'. It is a strictly grammatical conversation about life, an orderly exchange of preordained algorithms (chains of logic) encoded in media that include a database called *DNA*. It is neither mistaken, born of mistakes, unreasonable nor unintended. The reverse.

Creation is recreated. Sex, like asex, is about the regeneration of life and life is, dialectically, the central, subjective aspect of cosmos. Not surprisingly, therefore, sexual reproduction is perceived (by all but materialists who embrace a Theory of No Intelligence) to symbolise the polar nature of cosmos. Two vectors (*raj* and *tam*) are balanced within a (*sat*) whole - male and female within the archetypal whole of, in our case, human being. Two, split from an archetypal whole, are in the sexual intercourse of marriage or of moment unified. From that wholeness of union spring other ones - from communion offspring. So, as the Chinese noted (*fig.* 1.1), sex is a very good example of *yin-yang* balance, trend and swing. As such it appears to encapsulate the controlled, dialectical operation of nature as a whole - a complementary interaction of opposites to synthesise the next step forward. The task is to explain how sex, whose basis is meiosis and whose mechanism is a deliberate lottery, came about. What, according to each of the two mind-sets, is the origin of sex?

Bottom-up concept, symbolism, purpose, meaning nor logic plays any part in biological origins let alone the specific subroutine of reproduction or sub-subroutine of sex. Therefore pulling sex from an evolutionary hat is a procedure full of blots, blurs and missing links. It is (or at least, with purposeful intelligence philosophically excluded from the equation, should be) full of unreasonable and unexpected twists and mutational turns. From this point of view the descent of sex *must* have been one of nature's accidents; and it *must* have been due to changes in material, molecular patterns. If you are inclined to an ambiguously animistic view of nature that attributes purpose to chance while at the same time denying it, you might propose that 'improvement' in the guise of natural selection 'took up' or 'guided' the cause of molecules. *It was noted* (Chapter 20) *that, prior to any reproduction at all, an impenetrable circle excludes a rational account of the abiogenetic origin of biologically coding, self-replicating DNA; also that molecules or genes made of them have no 'wish' to 'survive' or 'improve', give no 'instructions' (such as 'become an ovary'), and are ignorant (because non-conscious chemicals are fundamentally ignorant) of any 'role' they assume, 'program they expedite' or effect they produce.* Nevertheless, the basis of biology is information and *DNA carries* passive information. If you vested its text with a life of its own, you might similarly ascribe chemicals (*DNA* in the form of 'meaningful grammatical sectors' called genes) with a life of their own; and, therefore, 'selfishness' and a drive to survive, grow more complex and conquer the world.

This perspective nails its own problem. What else can you say if, as a scientific materialist, you are convinced that the most complex, dynamic codes, mechanisms and machines on earth, that seem replete with purpose, created themselves? You have no option but to state that chance informed it all, that 'design' is a powerful illusion and only a couple of factors called mutation and natural selection can account for it. According to such Darwinian fundamentalism evolution only cares about genes or, rather, genes only 'care' about their own survival. Genes rule the roost. You are simply a secondary device, a vehicle for the carriage of cargo; you are disposable packaging that enshrines, for their improved protection, benefit and chances of survival, these potentially passionate, ruthless atomic clusters. What you think of as life is really simply a genetic sidekick; you are an inconsequential robot 'designed' by the all-important centres of 'information', the molecular bosses known as genes. What is their purpose? Molecularly inanimate, it must be none. 'To be or not to be?', that is the genetic question. 'To fall apart or be transported unto reproductive immortality?'

Such strange philosophy is occasionally disseminated, poker-faced, as fact. Is it? Is the core phylogenetic axiom true? Does high-grade information spontaneously and accidentally generate itself? Accidents are presumed, irrationally, to have 'created' the conceptual information with which, in its integrated, hierarchical, precise and anticipatory programs, the whole business of biology let alone sex is loaded. In other words, an irrational lens is used to frame the subject. Such irrationality is simply rationalised. In order to side-step the critical issue of the *source* (and *not* the material structure and composition) of complex, self-referential and purposeful information systems we can simply ascribe genes (like the order of letters and words in this book) the automatic power of program. Ink makes meaning. *The book of life, we vow, is no more than its genes.* Once locked to this hypnotic trick you are shortly passed into a genocentric world. Now that life's order and its origin is safely packed within material sequences of *DNA* the obvious must have happened. Unseen, unexpectedly and long ago sex must, like all reasons in biology, have been invented by some rearrangement of base letters in a line!

This metaphysical insistence is so strong that let us grant its wish. Suppose that abiogenesis (Chapter 20) has generated a bacterial cell - with informative, energetic, reproductive and structural factors intact and operative. Bacteria reproduce neither by mitosis nor meiosis but binary fission. Nevertheless for good reason the rigid conservatism of such clonal fission is capable of relaxation in three main ways - transformation, transduction and conjugation. If sex is redefined as 'a recombination of *DNA* from two individuals' then all three methods are 'sexual'. Conjugation, however, occurs irregularly, relatively rarely and between living couples of the same kind of bacteria. In a 'genetically fluid' microbial world such exchange of material is not directly linked to reproduction. Rather, where meiosis and fertilisation represent a deliberate lottery to create variation-on-theme, haphazard bacterial conjugation seems to some to represent a fortuitous randomising machine whereby recombinations of evolutionary value might be obtained. How?

Many bacteria contain 'optional' *DNA* called plasmids. Many pathogens (e.g. strains of *E. coli* or *N. gonorrhoeae*) have plasmids that form short, tubular *pili* with adhesive tips for attachment to their 'host-prey': some also have F-plasmids, that is, code for long injectors called F- or, disingenuously, 'sex-pili'. It is not that male or female is involved. 'Asexual sex' occurs because an F-plasmid expresses the genetic potential to create a docking/ injection arrangement (a *pilus*) with another bacterium. An F+ ('male') donor can, by transferring a copy of its F-plasmid or, indeed, its total genetic content, convert an F- ('female' recipient) into an F+ 'male'. So the process is multiplied throughout a population. Such *DNA* snippets can usefully, in the way of 'strategic flexibility', confer antibiotic resistance (i.e. they constitute a crude immune system); or they can enhance the ability to survive by increasing a bacterium's permitted number of nutrient sources. Because of such genetic mobility bacteriologists fight shy of the word species, preferring 'type' instead. As well as the promotion of 'strategic flexibility' conjugation is known to cement the cohesiveness and 'survivorship' of bacterial type. In other words, the tendency is useful but runs counter to evolution. For example, *E. coli* remain *E. coli*, *Spirilla Spirilla* and no prokaryote-eukaryote evolution is observed to occur - particularly in the department of emergent 'semi-mitosis' or 'hemi-meiosis'. Why not? Proto-bacterial genesis (Chapter 20) demands the 'invention' of glycolysis, chemiosmosis, *ATP*-compatible chemistry, *DNA* code and operating systems etc. etc. Bacterial conjugation involves, in addition, conjugation-promoting code and protein. Granted all this apparatus germs infect both each other and all other kinds of cell. Such promiscuously 'infective sex' is held by some to fund evolution. Theory-driven speculation also holds the transgenic delivery of bacterial injection in part responsible for not only inter-bacterial 'micro-evolution' (also called variation) but transformism - the origination of new, system-compatible, irreducibly complex molecular, organic and even whole-body (or phenotypic) formulations. In this way *Archaeo-bacteria* might transmute into *Eubacteria* (or *vice versa*), bacteria into protists and, by random infection or by symbiosis, capacitate fresh forms of life in unobserved, indefinite but historically radical ways. QE accidentally D. One is reminded of 'co-option theory', bacterial motors and the same vague phylogenetic appeal to self-assembly. With respect to sex the same logical response as then is due - no chance! Why?

In short, despite the pre-eminently successful efficiency of asexual bacteria in the replication department, there is no evidence for their mutation (that is, evolution) into unicellular eukaryotes. Nor have credible flow-charts been developed illustrating precisely how mitosis or, on quite another level of complexity, sexual polarity of body and of germ cell based around meiosis accidentally and without any concept just-so 'happened'. Why should a billion years make any difference since what needs to happen needs to do so in a way that works immediately?

Therefore out of two Darwinian black boxes take a flying leap from kinds of asex into a third, the baffling box of sexual actuality! Did it evolve once, twice (in plants and animals) or, as a variety of subroutines, multiple times in a diversity of organisms? We see great variety in primary sexual apparatus, accessories and behaviours but these are always distinct, fully functional and integrated into the whole working of the organism concerned. No sign, although materialism's paradigm insists it must have happened, of transformation here.

If real sex evolved at all which came first, male or female? If you ask that question you tend to get silly answers. Not only ones about God evolving woman from man's heart or, more precisely, plenipotential stem cells in his male rib-cage bone marrow - the spare rib; although, if you could re-initiate the way a main routine is set you might revert a cell to stem cell (plant cells can reset potential by reprogramming in this sort of way). Otherwise, did *both* sexes issue from a clot of blood? You laugh but it makes even less sense if you dare imagine sex occurred by accident. A haphazard addition of variable snippets of bacterial *DNA* is one thing: the arithmetically and geometrically precise dynamic of mitosis and meiosis, the latter embedded within a stable, one-way, male-to-female production, delivery and fusion system, is quite another. Many problems spring to mind. How did the spindle mechanism just 'snap' into place? How did the reduction division mathematically decide that haploid gametes were required; and once division had been 'doubled up' then, until the postage and delivery systems for life's vital letters were established, what games did the gamates play? How, at any level from gamete to surrounding apparatus or trajectory, could a 'lone' sex reproduce? If 'lone sex' is not asexual how could fertilisation occur? Or lacking the other sex, must 'self-fertilisation' have occurred? How does an egg self-fertilise; and how would such an egg, missing the whole point of sexuality and generating only clones, improve upon asexuality? If it could, how did a 'proto-egg' without any idea (or code) concerning its destiny stumble straightway into a developmental pattern culminating in the production of a viable, sexually mature, adult form - one equipped with not only apparatus but instincts to repeat the whole process over? If not, how did 'proto-single-sex' survive, asexual but sexually equipped, until chance threw up its partner? What advantage did the new sex have before its bug-free operation floated into view, knew what to do and found the place to do it? *It's a question of question after question without answers* - perhaps the very line of questioning leads simply up the creek. Because it's based on false, materialistic-only premise. Conceptually, logically, arithmetically, geometrically, informatively (with respect to coded subroutines), functionally (with respect to instinct and taxis), structurally (with respect to both development and elegant efficiency of operational apparatus) and, you might have after-thought, scientifically *sex repels irrational notions of an accidental origin. It cries out orderly deliberation.* Still, some biologists insist, it *must* have happened according to their Theory of No Intelligence - accidentally. But unless both systems coincidentally evolved simultaneously and instantaneously, in the same close location, to the irreducible point of minimal functionality, one sex *must* have preceded the other. Which way was it?

Was the descent of woman from man? In principle, let alone the complexity of complementary practice, it seems unlikely that females were an evolutionary 'second thought'. Indeed, if no switch triggers the other path perhaps a female body is, although the most sexually complex, the default form. Our attention is therefore drawn to *parthenogenesis*. Parthenogenesis, a curious sideline, is the development of an ovum into an individual without fertilisation. A mosaic of such diverse types as dandelions, aphids, waterfleas, wasps and all drone bees are 'virgin born' by this form of half-sex. Other parthenogenetic types (including such vertebrates as some reptiles, fish and, rarely, birds and sharks) also include the sexual possibility in their repertoire. *Artificial* 'parthenogenesis', called biotechnological cloning, can now fabricate otherwise exclusively sexual organisms; a fatherless mouse and, it is claimed, parthenogenetic embryonic human cells have been created. However 'virgin' birth from the union between a god and a woman involving *n* human and *n* divine chromosomes has not been scientifically recorded!

You can flick on a light but the switch is just one component in a preconceived system of electrical delivery. It is, likewise, possible that the development of both male and female bodies is induced by a cascade of switches that together control a system far more conceptually complex than any single one of them. A switch does not *cause* an electrical system; no more does one switch (say an inelegantly named Wnt-4 gene) *cause* female development or another (say *SRY* coding for the promotion of testes) *cause* a male; and RSpo1, required for early gonad development in both sexes, *causes* neither. Programs do not spring like dragon's teeth from soil; since such molecular components are switches in a 'computer' program they cannot help explain, any more than electrical switches explain the integrated design of one domestic circuit or another, how the *first* male or female circuitry arose. Parthenogenesis is the development of offspring from an unfertilised egg. If such one-sex, sexless or 'asexual' reproduction is possible, why should sexual reproduction have evolved in the first place? Or, since it exists, not been selected against to the point of extinction? After all (and feminists may concur) males represent a two-fold burden on females. If, for every male required to consummate further sexual increase, a female could produce one or more females by herself without male intervention, the population would multiply quickly. Such females would not have to wait for a male in order to reproduce clones. Why should a second or, theoretically, a third or more sexes be allowed restrain the fertility of their exclusive Amazonian fitness club? Why, for fitness' sake, waste energy and resources on producing males?

In all parthenogenetic species, which are restricted in number, there is a parallel sexual type. Perhaps it marks a degeneracy, as in Lady's Mantle, where the stamens have become vestigial with some lost altogether; if present the sacs are either empty of pollen or contain grains whose sexual potency is lost. This suggests that, although the coding for a complete organism resides in a haploid set of chromosomes, except in rare instances such as drone bees parthenogenesis is an aberrant condition that offers no solution to the problem of the origin of sex.

Nobody suggests that virgin birth is a halfway stage or even a real competitor to sexual reproduction. How could parthenogenesis, whose clones lack the 'flexibility' of variation that male input lends, contribute to sexual selection? Which latter means that tough, pretty, clever or whatever girls and boys select and are selected for; they bear the winning offspring and thus best bolster fitness and a triumphant way of overcoming natural obstacles - or that's the theory. Every breeder knows sexual selection can, by the force of his artificial induction, sharpen up a particular characteristic but not transform an organism into something else; and that pressed too far a speciality becomes a weakness, a potential liability that natural selection, health's hone, will in the wild home in on. Nor is it thought that, far from a spare rib, the female of a species gave rise to the male... except by any professor of genetics who cares to pronounce that accidental mutations in the female developmental system may have triggered the pathway to male development! By this token man is not a switched-on male. He is merely an aberrant, modified and switched-off type of woman.

Such speculation highlights the persistence of Darwinian fundamentalism. Woman accidentally bears a second sex. If this improves the evolutionary chances of womankind why has accident not generated more - say six or seven phenotypic kinds of sex - and not the wholly complementary, polar couple that we actually have? Has not a phylogenetic mind-set again missed the biological point - one that partakes of dialectical polarity, complementary balance of opposites and their interaction - the point of archetypal duality-within-unity that is so well expressed by sex? Does it equally, wedded to the Theory of No Intelligence, forbid an act of cosmos? Deny an orderly act of creation in favour of anarchic 'fall-togethers'? Of course, in any creation psychological notion has to be translated into physical object or event, principle is made practice, active gives rise to passive information (Chapters 5 and 6). Each created object needs a physical agent of implementation, a translator, a carrier of code, a medium with which to store the symbols of information. Paint, brushes, ink, paper, alphabets, software, hardware etc. are such agents. They are instruments in the process of earthing an idea. So are genes. But genes are chemicals as senseless as the ink on pages of a book; thus from a *bottom-up* perspective you presume genetic 'fall-together' into body-logic *must* have happened without logic, that is, bio-illogically signifying nothing. Is life with no significance devalued?

Is a technical drawing, database or mould wholly responsible for a product? Or is it, as part of a primarily conceptual (and therefore purposeful) process, only partly so? Are genes the fundamental basis of information or simply an important stage in its translation? Of course, defective plans will presage failure; if a fault occurs with a component (such as an infection or mutation) its intended effect will be frustrated; and if there is deliberate 'play' in the system (e.g. multiple alleles in a gene pool or the sexual lottery) then healthy variation will occur. Switch malfunction may, of course, provide insights into how a system works but in no way explains its conceptual origin; still less does it concede the supposition that the system itself 'grew together' from the accidental production of systems-integrated switches (because without useful integrity in the first place any single switch would immediately have suffered adverse de-selection). It is therefore highly illogical to assert that a mould or database is *responsible* for its product; to suggest that genes are *responsible* for characteristics such as aggression or maleness; and to swallow the story that switches are *responsible* for the generation of a mutual two-sex system which anticipates the reproductive process or that a genome is *responsible* for an organism. Of course, the cylinder is characteristic of a steam engine; so are technical drawings and moulds for the cylinder. *But the real basis of its character resides, as in every other case, in the conceptual domain, in the intention of its maker*. Coded chemicals (such as the printed word or, more powerfully, *DNA*) do not, however much materialism may wish and struggle, break this rule of origin.

Therefore was the descent of man from woman? No doubt, there exists a line of view down which male is a subset of female, boy a version of a girl. Although, due to possession of a 'Y-switch', he is always potentially male, such actual development only happens when a 'point' is switched from the female to a male track. If the switch has been lost or fails a female or degree of ambiguity develops. Mother precedes father, wife husband and it is brother, an evolutionary changeling, who appears second in line to his sister's sex. *From virgin birth it is a short step to male self-depreciation whereby man is perceived as a deviation from the default female form, an appendage gradually differentiated from the feminine.* The alleged root of masculinity, the Y chromosome, is assumed by an appropriate interpretation of its genetic remainder to be a degraded analogue of the larger X chromosome - of which human females each have two. Any creation suffers wear and tear but, despite nature's large stock of virile masculinity, this single chromosome is sensationally perceived by some as Adam's curse, a curse on males everywhere, a crumbling, fault-ridden precursor of every holder's masculine doom - not that demise will tell you much about its (or his) origin.

Is, therefore, Apollo just a bug in Venus' make-up? Is he simply the outcome of a few mutated genes that became, for example, a sex-determining switch (called the *SRY* gene) on that ill-considered vestige of information, the Y chromosome? Whose further mutation will drive their creation to its extinction? Is 'simple' replacement of an X by a Y chromosome (in the case of humans) or Y by an X (in the case of birds) all it takes? Insects have no Y chromosome and a mole vole (*Ellobius lutescens*), which has apparently lost it, can still reproduce. Nor, if you think you were a chimp, does its Y-chromosome hold your surmise up. It has far fewer genes (only about 50% of the human protein-coding complement) and 30% of its lacks any counterpart that can align with yours. So whence, if not from chimps, did men evolve potential maleness from? Is anything more than a few genes in a dilapidated chromosome responsible for the man his wife and children love?

How does such fundamentalism treat Eve's derivation out of Adam? Did various, exacting protein/ nucleic acid interactions just appear (for example, a regulatory protein labelled 'R-spondin1' suppressing a so-called 'Sox9 gene') and thus conjure the development of ovaries and Eve's completely female flower? Why, however, should a single protein acting as a switch 'evoke' an ovary, 'induce' associated organs, 'promote' the obviously intended apparatus of child-bearing and thereby 'evolve' the perfect complement of man? Such a notion lacks, as materialism does, all recognition of informative intelligence involved in generating complex, natural systems. It misses, wilfully and by a mile, the whole holistic point.

In other words, scientific reductionism excludes target, anticipation and with them plan. It includes passive information (e.g. genetic code or metabolic program) but its wholes always fail, by philosophical prejudice, to include a precedent, active mode of information. As such it misses mind and, in a dramatic, central error, the meaning of what we think and, therefore, say and do. As such it represents a lesser, material part of the whole truth. For example, could its switches have evolved your home's electric circuitry? *Is a switch the source or a prearranged part of what follows?*

You might reasonably assert that switches, even sexual switches, aren't what generate the hierarchy of command controlling what becomes a functional male. A switch, like mayors who with a flick of finger light up Christmas or unveil a plaque, is just the final link in a prefabricated system. This is not what evolution theorists want to hear.

Therefore if things don't drop in place, what a massive chain of happy accidents must have hastened circuits that comprise a man or woman after the first protein trigger found that it was coded! Where switches actually steer between prearranged programs of behaviour did this one (say, the Y switch) somehow generate a multitude of cooperative molecular, cellular and organic steps that led to the appearance of a second sex - the program called a male? It could not have taken ages. If it didn't race towards completion but, slow and unhappy, wove a wayward evolutionary course then natural selection would have found a part to play - not one conducive to your being here. If, for example, man evolved from woman what about his early steps - his unreformed lesbian instincts and his nascent apparatus with a performance rating of still zero? Did the recognisably masculine 'bits' evolve in pieces over many hundred thousand years? What on earth selected those pathetic fractions? Who selected the poor, half-endowed half-fellow during his erratic exit from the girls' room? Surely not the girls?

What about his puberty? And hers? Intermediate enzymes in a metabolic pathway have no use unless the whole crew rows; and thus their coded origin, if one by one, is a mystery as logical as actual. What about a large-scale analogy - development from childhood through the half-way house of puberty to full-blown adulthood? The precise manipulation of one chemical into another through numerous non-accidental, pre-coded stages for a particular reason smells more of the chemist's mind than his test tubes. It is not the business of evolution but highly intelligent, scheming chemistry. You would, in similar dialectical manner, predict that the pubescent steps to sexual maturity might not occur by accident. You might expect to find, like everywhere else in bio-logy, signs of coded signal, triggers and associated phenotypic changes. You do. For example, so-called 'kiss-1' gene is latent in the zygote but, having been switched on in the hypothalamic master-gland (and nowhere else), after a reasonable wait of twelve or thirteen years celebrates the onset of active sexuality with a transcription of 'kiss-peptin'. 'Kiss-peptin' initiates a cascade of control proteins, enzymes, hormones and other appropriate biochemicals that usher adulthood. 'Kiss-1' and its mates, like 'GPR54', are part of a gang dedicated, as surely as a metabolic pathway, to a predetermined end. Chance does not anticipate. Only mind predetermines. What use was 'kiss-1' or any other kind of boy-girl kiss before the 'adult' notion sprang to mind; or before its mates and all the other adult-making chemicals had been precisely coded for? Is Adam like me or Eve like you without reason? Male and female are clearly co-programmed entities. How could such obvious 'info-technology' evolve by chance?

When the kissing starts you might get more than hypothetical results. Development of offspring is a highly structured, practical sort of sequence quite unlike phylogeny. Phylogeny is the cerebral arrangement of, in most cases, speculative ancestors according to a predetermined theory. Although it's supposed to get you here in fact it gets nobody anywhere. No wife, no child, no actual anything. *As we saw with eyes* (Chapter 21), *simply arranging complex mechanisms in a suggestive order that leads from simple to complex fails address the question of their origin.* In fact the exercise often reveals an unexpected mosaic of anomalies and pre-existent archetypes - such as very different kinds of eye in very similar sorts of shrimp, a unique package of musculature and bone structure including a ball-and-socket wrist joint for the superbly swinging gibbon (an excellent example of efficient design by subroutine) or monotreme, marsupial and placental reproductive apparatus in mammals. Reproduction! Copy! There's scarcely any outward difference between male and female gibbons or between bacteria of any particular asexual strain. Could you argue, nevertheless, that bacterial conjugation gave rise to the plus and minus strains of algae and fungi? Perhaps, one might guess as a matter of evolutionary rather than ecological necessity, internal fertilisation on land succeeded external fertilisation in water?

All such to-fro guessing fails to generate an accidental 'phylogeny' of sex; it fails to answer how, in the presence of highly successful bacterial fission, faltering mitosis and then (perhaps) a default female body evolved; how or why, in the presence of successful parthogeny, an unnecessary and burdensome male should have attracted the sympathy of natural selection; how selection cottoned on and nurtured the idea through its mutant, unworkable infancy; how very different male organs gradually evolved to precisely and purposefully complement the female's; or how in some cases male and female bodies look the same but in others differ dramatically. Nor, even if there were fossil evidence which there is not, does speculation offer any conceivable 'algorithm' of evolution from the first inoperative, invisible flicker of something different (perhaps man's Y-chromosome or, in the case of a bird, the cock's X-chromosome) through steps to sexual apparatus and instinct that has developed into what actually works. Or is the whole dogma, as rationalists claim for any faith that is not theirs, a triumph of wishing to believe over what the plain facts show? Is there faith in things unseen? In other words, is not evolution a religion?

In reality very little if anything is known about the origin of sex. Its alleged evolution certainly remains as black a Darwinian box as that of metabolism. All you can say, lacking the provision of any adequate explanation, is that (within a philosophically materialistic framework) it *must* have happened. It is no more than a just-so story limply told.

Archetypal Sex

We can progress. *Top-down* the facts are the same but the line of view very different. Darwinian fundamentalism, which asserts that the appearance of design is a 'powerful illusion', is itself seen as an upside-down, diminished version of the whole truth. Its interpretation of the facts amounts, overall, to a delusion that has ensnared its otherwise intelligent adherents - an apparently addictive delusion.

Is not the product of a fertile mind ideas? Does it not naturally generate conceptual realities called blueprints, plans or archetypes? From a *top-down* perspective the program for male and female forms must already be written and, in each respect, correct and working before sex can function. No function, no selection. *No use one evolving before the other!* If, therefore, sex is conceptual in design then observe the following simple ways in which the Dialectic can reflect one whole divided into sexual halves - halves that in complement revolve around each other like a binary star.

The Archetypal Polarity of Sex

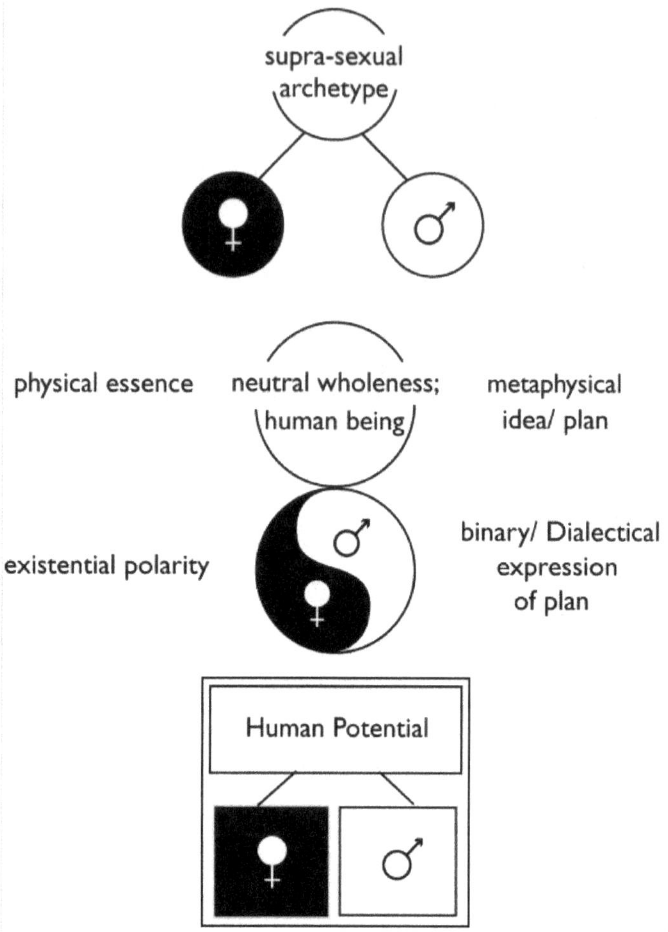

fig. 24.9 (see also figs. 1.1, 3.1, 17.3 iv and 24.8)

	tam/ raj	*Sat/ Tao*	
	duality	*Unity*	
	division	*Union*	
	polar	*Neutral*	
	sex	*Whole*	
	woman/ man	*Human Being*	
↓	*tam*	*raj*	↑
	yin	*yang*	
	female	*male*	

In archetypal terms sex is polar. It is the product of precise, purposive pre-programming according to dialectical principles. *Each type of life form that exhibits sex is conceived of as a neutral whole divided into polar male and female sexual halves*. (*Sat*) unifying information and (*raj/ tam*) polarised energy combine to express a duality of complements within a unity. Biologically, therefore, archetype is neutral or, rather, hermaphroditic; and so, potentially, is *DNA* although it may incorporate a polarising switch (in humans the Y chromosome). Neutral potential is expressed, by means of a 'menu switch' into one of two preordained polarities. The information encoded in each sub-routine is passive but purposive. Sex is, therefore, *conceptually* hermaphrodite - which doubles up initial complexity. Not only switching code but asymmetrically-mirrored and yet complementary body parts to follow have to be conceived - not to mention the idea of gametes, fertile union and consequent development. All must straightaway be in working order so that natural selection cannot kill less off. Such spot-on precision is, from a *top-down* perspective, not an accident at all.

Hermaphroditism, like parthenogeny, is generally thought of as another curiosity in the story of sexual reproduction. Hermaphroditus was the son of Hermes and Aphrodite, so called because he and the water nymph Salmacis became one person. In nature a hermaphrodite is an organism in which the organs of both sexes are expressed. By this definition many algae, mosses, ferns, clubmosses, monoecious pines and flowering plants are hermaphrodites. Doubled-up complexity of the hermaphroditic 'habit', subroutine or archetype is sometimes held (e.g. by J. Maynard Smith) in a strange reversal of evolutionary logic to be 'primitive' and its division into separate sexes 'advanced'. Thus, in this view, duplex precedes simple but the origin of its complexity stays unexplained. When flower petals attract pollinators, mechanisms exist which stop 'selfing' i.e. inbreeding. Usually male and female parts of a single flower will mature at different times. Even if self-fertilisation does occur, out-crossing is also possible; and it is commoner for wind- or water-pollinated flowers to be single-sexed.

Fungi exhibit many unique sexual routines based around meiosis. Nobody claims, however, that the sex habit arose in fungi with, in many cases, their indistinguishable 'mating types'. Did it therefore, by some mysterious 'inevitability', evolve separately in fungi, plants and animals?

Hermaphroditic animals are scattered without order (and therefore without obvious phylogeny) across the invertebrate phyla. Some snails, sea lilies, coelenterates, bivalves, segmented and ribbon worms, plankton, corals, endoprocts, lamp shells, crustaceans and bony fishes such as porgies, picarels and rainbow wrasse are hermaphroditic. So are complete taxa of flukes and tapeworms; and famously, of course, the earthworm itself. It seems that hermaphrodites might have more of a chance in conditions where mates are scarce. Even if one does not or cannot 'self', all the creatures it meets (and not just 50%) will be potential mates. There is not much information available about self-fertility in hermaphrodites. Certainly in earthworms elaborate precautions are taken against self-fertilisation. In other animals successive hermaphroditism means that, as in many plants, male and female organs are active at different times. Intersex forms are of interest because in a normally single-sexed organism sex-determining controls have failed and produced an organism intermediate in character between the two sexes. Such an aberration may or may not show both kinds of sexual organ, that is, be hermaphrodite.

Two from one; duality from unity - this is the order of creation. *Top-*

***down*, the principle but not necessarily the physical expression of hermaphroditism is central.**

The divisions of sex reflect the dialectical two-in-one nature of polarity, the cosmic play of opposites. Polarity (Chapter 1) involves complementary (but anti-parallel) aspects of the same thing, in our case human being; and the dynamic expression of its symmetry involves bias, asymmetry or the differentiation we call gender. *In Indian or Chinese words, opposite vectors (tam/ yin and raj/ yang) are expressed within the unity of a (sat/ tao) conceptual archetype.* The archetype is metaphysical; it is the potential from which (*raj/ yang*) male and (*tam/ yin*) female factors are physically expressed. The *yin/ yang* symbol (*fig.* 1.1) represents this polarised, covalently bonded whole. It embodies degree, relativity or emphasis towards one side or the other of the bond. The holistic, archetypal bio-logic is, therefore, that organisms exhibiting sex are conceived as a (*sat*) neutral, genderless whole in which the sexual halves constitute polar (*raj/ tam*) modules; their being transcends but also includes complementary opposites. **The sexual archetype (<u>not</u> a common ancestor) and its *DNA* coding are fundamentally hermaphrodite and incorporate both male and female potential.** Implicit polarity is, in different organisms, explicitly normalised in either a bisexual, hermaphroditic or a single-sexed body. In the former (e.g. a tulip or an earthworm) both modules are expressed. In the latter (e.g. a human) the organism is switched, during the development of heterosexual form, into one channel or the other. <u>*This shows, from the outset, a masterful efficiency of design.*</u>

Do not the two halves of a whole, in their division and recombination, reflect the cosmological and, in its miniature reflection, biological tradition? Does sex not reflect the archetypal pattern of polarity expressed from metaphysical (informative) neutrality or, if you like, of unitary principle polarised into physical detail? Do sex and its polar modules not illustrate a subtle, consistent and (see also Chapter 23) stereotypical interplay of complementary opposites? Reunion of these opposites (either in marriage, sexual intercourse or both together) recreates the prior archetypal, super-sexual condition of whole human being. From this condition issue other polarised or sexual individuals. In other words, from the unified condition is created *family*. Family is therefore an archetypal concept, a cosmic intention whose different forms are part of the natural structure of events. Are not the parental halves within its whole, with their opposing and yet (*raj/ tam*) complementary tendencies, united in (*sat*) balance - the balance that a stable family in society represents? Cosmic order and not chaos indicates the social norm; it indicates, against the storms life brings, a middle way, a social balance and a central aspiration. Is it not obvious that, derived from cosmic Dialectic, (*raj*) right-hand motion up the columns indicates increasing union within the world's polarity? Such union (as an ideal or a principle around which humans organise themselves) is expressed as an archetypal pull towards the condition of a stable, happy family.

<u>Archetypal Sex Expressed</u>

The potential for sex resides, as for everything, in its central, informative aspect.

Nobody understands exactly how genes, which simply code for protein, act as moulds for the precise, purposive, large-scale shape of organs such as hands, eyes or sexual parts. It is becoming increasingly apparent (Chapter 25) that a versatile and powerful bio-logic composed of genetic switches can correctly deliver protein building blocks to a given site. Since these sites are composed of cells a specific, timely geometric pattern of protein expression occurs. Of course, if the relevant protein is not present its 'building' cannot be constructed; but, although a hugely complex network of switches will be found to orchestrate protein production, is this enough? Construction must conform to an equally critical, preordained overall plan in order to translate to phenotypic shapes as unrelated to protein as is protein to *DNA*. The shape of a heart is not caused by muscle protein; male and female sexual apparatus derives from the same tissues but is not the same.

Of course, scientific materialism correctly perceives the need for immaterial information to rule the roost. This information must, by philosophical decree, be physical and therefore where else will you choose life's 'palace' but in chemicals called genes, the passive information held on scrolls called *DNA*? From this perspective coronation of a molecule is logical. *DNA*, The Accidental King, rules both the timely production of materials *and* their correct insertion into the palace architecture. Genes and proteins, clever little groups of atoms that they are, can develop subtle devices to construct lung, heart, brain or any other kind of chamber while the system is already working! *DNA* does very well - but materials alone cannot construct a house any more than instruction manuals invent machines. In other words, no doubt banks of top-level switches relate to variable sequences of 'downstream' genetic switches and in correlation produce, in an exquisitely fine and hierarchical example of modular programming, a composite generator of correct protein from, according to their chromatin bindings, correctly modulated cells; but is that all? What first created this supreme bio-logical program? Is it 'simply' the consequence of initial conditions laid out in a zygote? What laid out these conditions in conjunction with their complicated outcome? What is, for example, the *origin* of operation, not current operation itself, of sexual development? We shall discover, in more detail in Chapter 25, why the evolutionary answer 'chance' is feeble in extreme.

Top-down, information is original potential. Both quantity surveyor and the architect submit their plans; the quality of shape defines the secondary plan for quantity, that is, the timely delivery of correct components. For 'architect' read 'archetype', for 'quantity surveyor' read 'the plan inscribed in *DNA*' and for every morphogenic plan read an original focus of attention. There is neither observation nor creativity without a mind behind. Although unpalatable to materialistic style of thought, information always originally needs an active mind - although mind's passive, storage phase is memory. Thus it has been suggested (Chapters 15 - 17) that *DNA*, as physical carrier of life's plans, is only half the whole informative script. No doubt that biological nature exhibits superb menu programming, that is, *top-down* or flow-chart switching. If you wanted, theoretically, to design complementary sexes you might opt to specify each module using thousands of genes; alternatively, and far smarter, you might conceive a main routine which included a 'gender switch'. This would switch the program's operation along either male or female lines; it would call gender-informed subroutines. In principle, therefore, at a mere flick the balance would be tipped. A cascade of emphasis would ensure that either reciprocal polarity of an integrated whole would be engendered. *Such powerful economy of plan implies a clear purpose; highly intelligent, archetypal strategy anticipates and targets its conceptual climax - a reproductive couple.*

Talk of switches, menus, *top-down* programming and latent sources of information is never the language of chance. Nor is the expression of this informative potential as purposive, pre-programmed mechanisms. *SRY, Sox9*, R-spondin1 and so on are simply clever names for switches whose click unleashes a cascade of coherent, targeted events. They trigger maleness or femaleness and thereby decide the gender of all organs; in conjunction with 'down-stream' hormones they indelibly print a body with its sex. And a gender switch is simply one of a cascade of developmental triggers whose arch-initiator is fertilisation. As we shall shortly see, a simple flick sets in motion a superbly organised plan. Every plan is purposive and is the informative potential for what follows. An egg, at the top of life's process, is a plan; it is pure potential. From its preordained information a cinematic logic drives, *top-down,* to its conclusion. It is no dry, academic matter. Glorious reason builds from a simple ovoid premise to its climax, its revelation, the punch line - the material expression of adulthood. Could nature be, in some internal, conceptual aspect, even cleverer than its children, its observers?

Biological (passive or externalised) information is coded into nuclear *DNA*, cytoplasm and cell membrane. In humans, for example, every cell in the body is imbued with sex in that it contains either XX (female) or XY (male) chromosomes. After fertilisation, development proceeds for a while without any overt sign of sex. Then, as we saw, an *SRY* gene on the Y chromosome of a male embryo (perhaps in conjunction with other genes on other 'non-sex' chromosomes) triggers a series of events leading to the production of testosterone and development, from the cells of the genital ridge, of male tackle; and from the same ridge R-spondin1, progesterone, Wnt-4 and and other winsomely named chemicals promote the organs of prime femininity. This economical switch, in concert with a testosterone acceptor programmed to occur on the cell membrane of *both* sexes, initiates far-reaching consequences. It permits only one of the two hermaphrodite potentials to exercise its power to the full while the other simply shows some complementary rudiments. In this it *presages* the development of adult forms and, in their recombination, the birth of a new life cycle. *You do not presage or anticipate without thought. Such action is precisely what sets intelligence apart from material oblivion or chance.* Switches simply activate latent potentials. In the male case the genital tubercle develops into an explicit penis rather than an implicate clitoris and the genital fold fuses over the urogenital opening to become the scrotum rather than labia; and, conversely, in the female the urogenital opening divides to produce both urethra and vagina. On the inside, also at about ten weeks, the undifferentiated gonads now become testis or ovary, epididymus or fimbriae, vas deferens or fallopian tubes, glands for seminal fluid or womb. In other words, a fertilised human egg contains conceptual and genetic potential that controls the development of *both* male and female organs.

The switch is thereby seen to act like a line of code in the master module of a top-down application program. 'If switch present then Y, if not X.' All menus work like that. They incorporate pre-programmed *choice.* Program is principle. It contains the information on which practice is based. Sex is no different. It is not only in the structure of its information but the practice of its parts that reproduction relies on central command posts and batteries of local controls. Take, for example, a primary sexual member, the penis, one of whose purposes in principle is the transmission of genetic information and in practice the injection of sperm into the female receptor, a vagina. Its operation is coordinated down a long line of command and control. Goal-oriented information hierarchies are not a product of coincidence. This one leads from the central mind of its bearer. Conscious impressions, desires and sub-conscious mnemone, including instinct, dovetail with sub-stations in the hypothalamus and nervous systems. Both chemical and electrical messages pervade the peripheral apparatus of the body. Their influence is concentrated, however, on the penis where a special exercise in hydraulics unfolds. The plumbing includes pipes, pumps, production lines, reservoirs, valves and, of course, fluid mixtures of the right composition and concentrations. Simple in principle, complex in practice, these parts are combined in a mechanism *designed* to rise to the reproductive occasion.

No less well designed is the heart of the issue, information transfer. Sperm and spermatozoids are, in every plant or animal to which they belong, dynamic, efficient envelopes in which the love-letter, a letter of life, the genetic tryst is posted and delivered to its female address, its attendant potential called an egg. Sperm and egg are each a cybernetic compound of command and control; and regulatory systems are designed with intent. Do atoms have intentions? Do molecules 'regulate'? The 'intent' of sperm's as clear as any guided missile's is - but what, exactly, does the sensing and then acts appropriately? To say, for example, that a sperm 'knows' what to do by chemotaxis (chemical attraction) is to beg the question and to mask an ignorance. Chemical gradients that act as switches need sensors and effectors. They are instruments of *program*; and if the program is purely chemical then how did 'correct response' (called taxis) and arrays of contextual chemicals required evolve one by one, gradually? A metabolic program or a teleological series of steps for a reason, to accomplish a purpose is not an accident. What, therefore, wrote the program in the first place; what makes guided missives? Chance? *Chance does not breed programs* - which renders the little tadpole, just like all machinery, greater than the sum of its specific parts.

Nor is female cooperation in the cause of procreation less subtle, coherent and according to plan. Both sexes anticipate their union; each apparatus perfectly allows coitus, flow of information and the motion, as it all clicks into place, towards a definite, preordained biological goal - multiplication, reproduction and thereby the indefinite survival of human life. Is not fertility a prime criterion of evolutionary success? In which case why, one asks again, have all men and women not evolved long ago into a prime condition of sexual attractiveness, athleticism and fecundity? Or, just as genetic malfunctions appear on the increase, is sexual fitness on the gradual decline? This would imply a pristine start degrading (as things do) over time.

Each sex is separate but the poles unite to make a single whole. Whenever did a monopole evolve a mate? Into what place, except the framework of an underlying plan, can working parts click? Why should mutations in genes that code for protein grind gradually and randomly into the comprehensive architecture of sex? *Within each complementary, switched sexual subroutine a whole delicate but robust, hierarchical, homeostatic subsystem of checks and balances lies present, latent but with full potential from the start... until puberty when, according to the archetypal developmental plan and an over-arching stimulant, instinct, it wakes into hormonal 'rock 'n' roll' that will realise the reproductive goal.* This is the human pattern of cohesive psychological and biological behaviour but there is no reason to suppose other than a different pattern of switch operative on modular permutations of the hermaphroditic archetype in all sexually reproducing organisms.

While *DNA* is a life-long, conservative nuclear database short-lived hormones inform more specific intercellular cycles and constructions. They are dynamic, coded schedules; they are fluid, communicant triggers.

	tam/ raj	*Sat*
	division	*Union*
	sexed	*Neutral*
	sex expressed	*Potential Sexuality (Zygote)*
	female/ male	*Archetypal 'Unisex'*
	output/ agent	*Single Plan*
	lower cycles	*Top-central Axis*
	gonads	*Hypothalamus*
	informed	*Informant*
↓	*tam*	*raj* ↑
	diversification	*unification*
	production	*agent/ instigation*
	aftermath	*start-up*
	duration	*immediacy*
	oestrogen/ progesterone	*testosterone*
	ovary	*testes*
	egg	*sperm*
	female	*male*

Single, neutral, sexless at the top. From governing principle to polar practice. Information from above triggers action below. In the human case both male and female sexual processes start in the same place using the same neutral, uni-sex hormones. At the centre of your information centre, your head, the hypothalamus releases minute amounts of a 'potential' or, if you like, the uni-sex trigger called GnRH (a single gonadotrophic releasing hormone). This singular start leads to polar procedure. There exists, as a consequence, a polar dialogue between pituitary and pelvis. For an account of the logical, dialectical location of reproductive organs according to a polar scheme check back to Chapter 17: The Informed, Energetic Domain. In this most reasonable and not haphazard, chance-borne dialogue the pituitary gland secretes both FSH (follicle-stimulating hormone) to stimulate the ripening of sperm or egg and LH/ ICSH (luteinising/ interstitial cell-stimulating hormone) to stimulate ovarian oestrogen and progesterone or testicular testosterone. The centre-point of these hormonal swings is, in the feminine, to 'spring' ovulation. In both sexes the appropriate hormones engender secondary sexual characteristics such as body size and hair, voice, characteristic cerebral, muscular, pelvic and other structural differences, fat distribution (e.g. as breasts or shapely hips) etc. Their interplay is involved in ovulation, menstruation and, when appropriate, the regulation of milk production.

Nipples are worth a momentary diversion. If both male and female forms were originally rudimentary or, as intermediates, pre-operative, what selective advantage was there for a milkless milk-production system to improve to the point, a few million years later, of working? Are male nipples, positioned like a female's at the centre of an areolar target, the vestigial remnant of a time before men had evolved from women; are they from an ancestral dream-time in which aboriginal, Amazonian parthenogens asexually reproduced; or, as Darwin wrongly suggested, may men have suckled in the deep past? It would have given the baby a double dose of survival so why, if they ever started, did they stop?

Nipples and breasts are certainly, to some degree, the product of their genetic database. What about their shape and techno-logical position? What about allied milk production and the untaught instincts of mother to bring child to her heart and suckle or of the child to suck? Are instincts simply a product of chemical bonds, of a sequence of non-psychological molecules or of a network of electrical motions? In short, can anyone detail the gradual steps whereby a 'purpose' of lactation might have evolved its mechanisms? I do not refer simply to style of outlet (nipple) but the whole production and delivery system synchronised to develop and operate in step and in harmonious association with all other evolving co-factors. Of course, the same problem occurs with respect to all physiological systems.

Sex is a fine example of balance and emphasis between complementary opposites within a wholesome unity. *It involves differential male and female expression of a single reproductive archetype.* No doubt, as female athletes can testify, excessive presence of male hormones in a female can 'de-feminise' secondary sexual characteristics; fullness of breast, hip and buttock is reduced, muscularity increases etc. For a male, on the other hand, the excessive presence of female hormone has a feminising effect. Under abnormal conditions males have even grown breasts and lactated.

Under normal conditions both sexes of pigeon secrete milk like yoghurt from their crops for chicks to drink. The discus fish, a cichlid native of the Amazonian basin, also appears to feed its young with a sort of milk. Just before breeding a slimy protective coating on the scales of the adults thickens considerably. This is due to a copious white secretion, granular in composition, which changes into tiny filaments when pulled or rubbed. After absorbing their egg sacs the youngsters swim directly to the parents where they cluster for protection and their 'milky' nourishment. And in fish complete sexual metamorphosis is also known. It's funny! A clown fish starts life as a male and then, if it becomes the largest fish in its group, transforms into the female sex. She is mated by the second largest fish, a male with larger testes than the rest. If this male is removed, the next largest develops big testes and becomes sexually active. What causes these dramatic sex changes is not known but, clearly, the genetic basis for either sex is present in every clown fish. *Any case of sexual metamorphosis, not to mention hermaphroditism, is readily explicable in terms of top-down program that is, Natural Dialectic's unisexual archetype with polar, switched expression.* Potential for both sexes is, simply, tripped into one code channel or the other. Male and female modules. By what ancestral route might Darwin (or any Darwinist) suggest that sex, with its mixed mosaics, came to pass?

Do you remember the archetypal way in which the same embryonic cells from the genital ridge were steered into either male or female organs? Every 'polarised' organism presents sexual 'primordia' that are then evoked or suppressed accordingly. Take, for example, associated Wolffian and Mullerian ducts. In a male the Wolffian develops into testicular material but in the female degenerates into the vestigial parovarium. In a female the Mullerian develops into fallopian tubes, uterus and vagina but in males degenerates into vestigial testicular material. They are functional features not evolutionary dropouts. Un-breasted male as opposed to breasted female nipples fall into the same category. The bio-logical view therefore leaves ducts, nipples, instincts and so on as more likely expressions that indicate coordinated adjustment round a single theme. *Nipples are rudimentary but sufficiently complementary expression in males of a fundamentally hermaphroditic archetype.* They are therefore, like mammaries themselves, the product of a subroutine from *H. archetypalis*, a memory whose full female expression we dub Eve.

Engineered (metabolised) from the same steroid precursor, cholesterol, molecular testosterone and oestrogen differ as subtly and economically as two keys cut from the same blank. There is even an enzymic converter, called aromatase, between the two. In fact each sex produces, with homeostatic emphasis on its 'own' norm, both androgen and oestrogen steroid hormones. After cholesterol many are also derived through progesterone (which serves in its own right to promote pregnancy). Sex is a case of 'dialectical' or 'to-fro' balance and degree.

Polar division of a stable, neutral whole introduces the dynamic stimulant of charge. *Throughout its 'electrical' sphere of influence sexuality well illustrates an archetypal interplay of complementary opposites.* For example, in the female menstrual cycle oestrogen (the 'sexy' hormone which peaks just before ovulation) is counteracted by progesterone (the 'maternal' hormone which peaks after fertilisation might have occurred). Together the hormones reflect the position of female compromise between, on the one hand, mate and, on the other, offspring. Both hormones operate feedback loops with the hypothalamus, not to mention the loops established from the womb by *HCG* (human chorionic gonadotrophin) if pregnancy does occur.

Do you remember (Chapter 17 and *fig.* 17.11) the sacral plexus whose association is with water, the spring of life, reproduction and osmoregulation? Whose organs include both the renal and reproductive factors? Anabolic steroids, made in the kidney as well as gonads, are not only sexual in function. They act antagonistically to catabolic steroids such as adrenaline and cortisol. The latter mediate immediate rather than future, reproductive survival. They 'rigidify' or 'tense' the body and tend to close down other than respiratory metabolism. Sex hormones, on the other hand, are 'supple', promote growth, build protein (e.g. testosterone with muscles) and generally stimulate metabolism. They probably affect every cell in our bodies. A fertile woman's ovarian follicles produce, prior to ovulation, androgens that are catalysed into oestrogen. After menopause most of her oestrogen (and more testosterone than previously) is produced in the adrenal cortex. Conversely, in the male some testosterone is converted to oestrogen in the testicles.

It may be noted, bringing to bear a cybernetic approach to the study of sex, that abnormality or malfunction of code, switches or integrated components causes, with a severity related to the stage of development from which any effect spreads, problems. Such systems are not caused but bugged by mutation. It was noted, for example, that the application of testosterone to a sportswoman's body will diminish her own and promote male secondary sexual characteristics; and, conversely, an overbalance of female steroid in a male has a dramatic feminising effect on his appearance and behaviour. More seriously, genetic or early developmental abnormality may affect primary sexual organs with the result that illness (such as CAH or congenital adrenal hyperplasia) or incomplete, indeterminate or hermaphroditic structures occur. *How, therefore, can a phylogeneticist logically (and one supposes that our science wishes to be logical) claim the origin of a system whose bio-logic is fundamentally poised round emphases of balance by means of unpredictability, imbalance and a staggering series of blunders - the complete and irrational antithesis of order. Every information technologist knows utterly and absolutely that you do not construct sub-programs and their large-scale integration, systems, for no reason using just a bucketful of bugs!* **Yet this, which scathingly dismisses archetype, is the neo-Darwinian Theory of No Intelligence, evolution!**

Switch malfunction or 'wrong-sex' hormonal surges can affect other critical aspects of sexual orientation such as *SDN*s (sexually dimorphic nuclei) in the brain. The left and right cerebral hemispheres of the human cortex seem to exercise both bilateral and biased, asymmetrical aspects of function (see also Chapter 17). There seems, for example, a predominance of 'masculine' linguistic, analytical and scientific allied with visuo-spatial operations in the left-hand sphere. Its influence crosses to govern the right side of the body. A mirror switch means the right-hand 'feminine' sphere, with a holistic, artistic and emotional predominance, governs the left side. It has been shown that a characteristically female arrangement, possibly caused by an untimely surge of opposite sex hormones early in development, can occur in a male body and *vice versa*. Such an occurrence may, it seems, provoke trans-sexual behaviour, a feeling of 'being in the wrong body' and, at its extreme, a desire to change sex by surgical operation on the primary organs. *Such exceptions to the clear sexual norm seem to underline the presence of a latent hermaphroditic archetype.*

Has chance some kind of inner rationale? Do the laws of physics force mutations to add up to contributions? Must bugs have logic? Or is the wand of natural selection what, touching molecules, brings them to life and waves the whole thing through? It is remarkable, given how long ago their common ancestor must have lived, how 'convergently' the sexual apparatus of plants and animals has evolved. You find, in principle, the same asexual mitotic and sexual meiotic algorithms, complete with egg (ovum or oosphere), sperm (pollen or antherozoid) and the involved chemistry and organs of fertilisation. When, in *Angiosperms*, the young and attractive female part of a flower is ready to receive the male, signs of sexual arousal appear. Secretion from 'her' tissues prepares the way for intercourse in a way that resembles that of an animal counterpart when ready for mating. Bud has blossomed into flower. Heightened colour and moistening of the stigmatic surface with a sugary fluid occur. Nectar is secreted and attractive scent cast to the male air. Chemical analysis shows that, by widespread lucky bouts of 'co-evolution', floral perfumes resemble insect body odours and, as one of a repertoire of tricks, help entice a pollinating insect and make sure the message sticks. Apart from the similar production of sex cells in anthers and ovaries flowers also have a placenta, the part of the ovary wall on which ovules are borne. An ovule is attached to its 'womb' by an 'umbilical cord' called a funiculus. A scar like a navel but called a hilum is left when, as the seed ripens, the funiculus breaks. Of course, modular changes are rung sufficient unto the different natures of plant and animal but could such analogous methodology and structure have evolved by piecemeal accident twice over in the two separate kingdoms. Or, as you observe a petalled flower inveigle pollen and bear fruit, do you see a metaphor for woman, poetry and polar symmetry? Are temptation and the complementary play of opposites an ugly, hotchpotch aggregate of lack of plan or simply rightly planned? Is your choice one of evolution or deliberate design?

Co-evolution is construed as accidental but effective adaptation 'triggered' by a change in some interacting organism. It is often seen in terms of a competitive arms race between predator and prey, a tit-for-tat 'outwitting' of the other. Check Chapter 23: Evolution in Action? In some cases, such as the changing immunity of bacteria to antibiotics or prey to snake venoms, random mutations may indeed provide the tiny but necessary molecular shifts; but, as regards larger scale, complementary and mutually beneficial changes that would need thousands of beneficial mutations, it is more reasonable to ascribe non-random adaptive potential as the cause. For example, pollination mechanisms often involve a symbiotic relationship between plants and animals. A botanist can relate numerous examples of complex, 'deliberate' plant structure and interdependent animal instincts - symbiotic growth and behaviour patterns which defy a 'co-evolutionary' explanation. Various orchids and insects, figs and fig-wasps and yucca moths and Joshua trees provide spectacular examples. Unless there exists a rich, prolific program of underlying community to which these plant-animal pairs bear silent witness they *must* have co-evolved. But in such cases you might easily construe that these co-incidents were not coincidence; and that non-random co-evolution due to resonant expression of two archetypes, that is, informative potentials may seem more reasonable an explanation.

female	*male*
responsive	*initiatory*
recipient	*donor*
fixative	*fluid*
stable	*motile*
earth	*sky*
ground water	*cloud/ rain*
submissive	*dominant*

From algal conjugation to the conjugal arts of man sex represents a central reproductive core around which bodies are tightly woven. *It reflects the cosmic pattern of creation.* At the (*sat*) nuclear centre exists potential, biologically compressed into the form of *DNA* code. (*Raj*) dynamic transfer and development of information is completed on attaining the (*tam*) adult state. Now the cycle can begin again. The reciprocity, polarity and conjugation of the sexual theme are fully expressed at, where possible, conscious as well as sub-conscious and physical levels. The physical hardware and psychological software of complementary polar opposites attract and recombine.

Primary organs are distinct. Physique and behaviour are sometimes less well defined. Sexual characteristics vary, in practice, to the point of role reversal. Either sex, varying around its stereotypical norm, can show both male and female tendencies. A woman may, for example, in different circumstances or as a trait want to dominate as well as be dominated, to initiate as well as be led. It is a question of emphasis. In biological terms archetypal sexuality represents a flowering, sharing and recombination of differences of emphasis. Currents exchanged in the sexual field mirror those that flow in all polar fields of force. Sexual union mirrors the original act of creation. There one made two and more, here two make one or more. Male and female, dual aspects of a single sort, combine to rejuvenate in the form of their offspring.

<u>What's True Love?</u>

	tam/ raj	*Sat*	
	body/ mind	*Consciousness*	
	degrees of love	*Agape*	
	division	*Communion*	
	polar	*Neutral*	
	degrees of stability	*Stability*	
↓	*tam*	*raj*	↑
	base	*top*	
	division	*unification*	
	one to two	*two to one*	
	objectification	*subjectification*	
	analysis	*synergy*	
	body-centred	*soul-centred*	
	hormone-driven	*drive to unify*	
	from psychological stability	*towards psychological stability*	
	exploitation	*altruism*	
	material lusts	*immaterial love*	
	selfish pleasure	*selfless giving*	
	sexual pull/ erotic heat	*degrees of devotion*	

Do you remember from Chapter 17 that your body is polarised into informative and energetic domains? From crown to base, from soul to sole and top to toe it is detailed as a range of informative patterns or, alternatively, a spectrum of psychosomatic energies that drops from one predominance (the conscious pole of your cosmological axis) to its reverse (the non-conscious, 'solid' pole at the base of your spine). The latter's area of influence is fixity of structure, bulk material shape and reproduction. It is therefore logical that the organs of reproduction are found, opposite the third eye at the centre of your forehead, at the nether regions of your trunk. Has this any bearing on the polarity between altruistic love and sexual passion or, as Greeks characterised the antithesis, *agape* (pure, sexless love) and *eros* (polar, sexual lust)? Could *agape* (pronounced a-ga-pay) be concentrated at the head end, lust between the legs and romance some heartfelt place between? Indeed, one friendship might include them all. The marriage of true minds and bodies might include a spectrum of devotions and, combining male and female polarities into one whole being, recreate the ideal human state. Is this, a happy family, the peak of human expectation? It is a radiant ideal that humans rally round but is there any synergy transcends it?

Ask first, is sex all in the mind or not? Sense, instinct and imagination no doubt play a part but aren't these, like mind itself, a side-effect of hormones, action of transmitters or electric stimulation generating patterns in the brain? Lust is chemical but, if lust is love, love is a product of unwitting atoms too. So isn't true love something scientists might drug-wise engineer? It is, if lust is surreptitiously turned into love. Such conflation might describe the act of 'making love' but is really true other than true lust?

In this debate a tension pulls between the usual two camps. Does molecule make love or do thoughts generate, in consequence, their bodily effect? Where love's chemistry makes choices do proteins, genes or molecules actually compose morality; or is morality a function of your consciousness which leads your body and its molecules to follow immaterial decisions? Which aspect leads a centaur by its nose; and which, in the process of creation, came before?

Thus, forever torn in the torsion between the two opposite poles of Consciousness and material body, which way do you twist and turn? No doubt you want intense, abiding happiness. How do you find it? Does not ecstasy, for which exists a universal craving, involve at root a *loss* of self? Is bliss not loss of self-centred identity at the same time as finding fullness in a beautiful experience, wholeness in a world of love and, especially, communion with a loved one? So which way will you go? Lust is selfish , love is giving. Where is love the most complete?

In this debate a tension pulls between the usual two camps. Does molecule make love or do thoughts generate, in consequence, their bodily effect? Where love makes choices do proteins, genes or molecules compose morality; or is morality a function of your consciousness which leads your body and its molecules to follow immaterial decisions? Which aspect leads a centaur by its nose; and which, in the process of creation, came before?

Bottom-up, of course, you'd tend to cast molecules in leading roles.

↓ *tam*	*raj* ↑
testosterone	*oxytocin*
aggression	*welcoming*
suspicion/ wariness	*trust*
risk-taking	*risk-averse*
cautious/ separative	*generous*
isolating	*loving*

You see the polar pattern and thus mode of balance here. We've begun to understand the twist materialism angles love's truth with. Theemphasis, as we well know, tends towards a physical expression of desire. In this view the expression of emotion is locked into brain; and brain is simply a complexity of fixed and dynamic patterns made of chemicals. It is natural, therefore, to invoke chemicals as the basis of mind and, if the emotions are exciting, chemistry that overrides an aminergic mode of brain (*figs.* 13.5 and 14.2), cerebral rationality and the balance of an ordered consciousness. Why not invoke the power of odours, tastes and touch; is it not amphetamines like *PEA* (phenylethylamine) that tip you into flush and rush? Why not scan romance? Check how an *fMRI* scan illuminates the '4F' brain, the seat of instinct and primordial passion, a house of hot emotion called the limbic system. This system emphasises cholinergic activation tipping towards irrationality. Adrenaline for action and dopamine for pleasure flush your nerves arousing to a fluttering, sweating, jittery state. Tingling. This is what excitement feels like; and not least a warm front rising towards the sexual storm. Anticipation's just the start. Go-get testosterone turns on the heat for both sides and, as they entwine, endorphins step the pleasure levels up until, electrically activating thirty parts of brain, the flood becomes an inundation. Warmth and well-being saturate your glow. How could love's hormone, oxytocin, not sweep thought away? It's all material, you cry. Explanation's couched in terms of chemical effects; and, since mind's a biochemical effect then how could it have made the first advance? Biochemistry's the cause of love - or, at least, lust!

To ensure, by recreation known as reproduction, survival of created species sexual pleasure is a fascinating, irresistible design. Nature's life-preserving trick. But, as opposed to the materialistic strand leading down to sex and struggle for survival, the immaterialistic side of helix tends inward. Indeed, so central is the evolution of 'return' to the heart of our existence that it is worth rephrasing, step by step, its psychological direction.

Firstly, if mind is an 'element informative' that interacts with biochemicals you might ask who's arousing whom. Who leads the merry dance? Which agent shapes your thoughts? Suggestive mind, intimating chemicals or, according to the spur of moment, both? Of course, both chemistry *and* thought facilitate mind-body interaction. Kissing, for example, rushes a cascade of chemicals throughout the body but do hormones feel emotions or pheromones get hot? Do chemicals have feelings or does mind? It is true, however, that the lower you descend in body and, earthily, upon creations's scale the more that chemical effects can overwhelm. Passion scatters reason like leaves in a storm. The body drives your thoughts and subsequent behaviours. Nowhere more strongly do the hormones play than in erotic zones.

You might claim that sex, for its subjective and important part, like all *experience* is sited in mind; yet erotic ecstasy also clearly involves arousal centred at the base pole of the body. It rises to an apex and then falls away. The entertainment industry's in love with sex if not orgasmic sex but what of climax by the mind alone? What might arouse and sweep you further up? What might electrify your wakefulness? Could it be intoxication with mind-bending drugs? Wiser might be music, food of love. Rhythmic harmony's a strong contender for the top spot but also, from the axis of eye-centre, there ascends a metaphysical and contemplative 'high'. What more positive a feedback elevates than excitement in the focus of an interest? We call it love. The harmony of love is mind's attractor just as body's is the satisfaction of its archetypal needs: and if the archetypal need of mind were love what would its First Cause be composed of?

So, secondly, is experience just a biochemical or an electrical complexity? If not, is it possible to rise above their impulse? What material residue or immaterial concentration is the kind of love above? As you rose might you expect a balance tipping towards ascendancy of thought, of order and a mind more steadfast and less buffeted by storms of hormones and cascades of chemical excitement. You might even find mind less at the mercy of its chemical attachment, brain. Higher mind not lower mind-with-body takes the wheel, mind not body drives the vehicle of reason and its subsequent behaviours. Now, as you rise, another kind of concord swells that, purified of carnal sexuality, is called romantic. In the region of the heart strong waves of selflessness irradiate the mental atmosphere. As an example of asexual passion maternal love weighs effortlessly in. Renamed by science 'altruism', does her baby feed on milk alone? Maybe the electrifying surge that friendship and devotion stimulate produces physical effect as well but is such biochemistry (evolved without reason but with all its reasonable chemicals) the cause or an effect of love? Which rules? Is mind or matter arbiter? As they dance and sway which of the two would you say dominates your life? Are you body's puppet or detached from the control of selfish, sensual strings?

The danger lurking here is we believe that a materialist's account of things is absolutely true. It is half-true, of course, because materials make every physical location; not wholly true because, as usual, any immaterial element has been ignored. Such ignorance would, of necessity delete an immaterial Apex of the Universe! It would miss life's essence out.

In the end analysis, therefore, are trust and altruism simply evolutionary syndromes born of chemicals like dopamine - with oxytocin, the endorphins, B-lipotropin and the rest piled in? Is your 'self' created out of atoms, subjectivity a sort of biochemistry and love's emotion some quality effusing from a microgram of hormones? Of course, chemicals affect the brain. It's built of them and their dynamics. How, though, is their specific action linked to archetype and thereby feeling and experience of mind? Unless you think (when Chapters 13 to 17 are meaningless) that qualities of mind are somehow chemical as well!

Thirdly, happiness derives from unselfconsciousness. A healthy body doesn't drag you down. You're unaware of it. And you forget yourself when doing what you love. The word in Greek for 'loss of self' is 'ecstasy'. Don't humans crave such ecstasy as beckons a release from pain, confinement and life's imperfections; and more than this, converts experience into radiance? It simply is a question of the choice of loss. Is loss of ego found outside in interests that an engaging world supplies; is it to be found in body pleasure, at most keen with sex; is it loss of self in friendship sharing life with life; or union with The Friend, The Self, that shares the Heart of Nature? Climactic loss of self in what is 'best' is a desire whose extremity can physically (ephemerally) or metaphysically (long-lastingly) be satisfied. Climax is not humdrum nature's norm; it marks a speciality. Where average balance is sustained by homeostasis, where stability's equation is maintained by counter-feedback what is climax but departure from the norm? It does not interfere but amplifies. We always seek to move obstruction from our quest for greater happiness. This orderly but exponential sort of feedback rises with success. It is positive. If bodily it marks abandonment of normal bounds and, rising to crescendo, bears its fruit then falls exhausted to completion; yet mental climax, love, may not exhaust but thrive. Life-rich occasions tend to be climactic. You find such rush embedded in what causes the excitement of each nervous flip, an impulse that is called an action potential; you find productive climax at the time of ovulation, in the contractions of laborious birth and at the peak of an erotic ecstasy. You love it while it lasts. Such ecstasy, however, is a short-lasting and, for each party, selfish one. And it is short-lived and body-centred. Is it, therefore, true and everlasting love?

Fourthly, on the other, higher hand, mystic ecstasy is centred opposite to the earth pole and its physically reproductive sex. Loss of self in Self is loss of self into Pure Life. Such climax, at the Apex, cannot ever fall or fade. It is what creation is established from. Has 'rationalism' therefore missed a fundamental trick? Does it bark at an approaching climax of the world because it scents the blood of God? Nor is Communion other than despised in such vicinity but if, because of that, you call it psychological illusion, what in life is not? If it is not a scientific exercise to swap analysis for trust, suspend your reason, believe in your beloved and then merge in ecstasy then so be it. It seems not science but irrationality has won. Irrational the rationalist to miss this whole subjective point, to miss behind his finite things Infinity! This truth is ancient and well known to the more-than-scientific-only, mystical fraternity. Such magnetic levels of attraction form the central nexus of their brilliant lives. Nor is this kind of climax temporary or exhausting. The reverse. Its completion does not fall upon exhaustion but a fresh and endless spring of life. What higher than the Apex of Mount Universe? What feedback more exhilarating than the Top Experience? What climax deeper than Communion within the Origin of Worlds? A mystic starts with focus at the third eye and works up from there. This is a real ascent. Within the top pole of the body you might find, at will, the Highest State of Mind.

At this point you may ask if any pole but matter is more than illusory. Materialism emphasises evolutionary origins for everything including love. How far short does this step fall? How far, as far as holistic dialectic is concerned, has it missed the real point? What transcends duality but Unity, breaks limits but Infinity and overrides existence but its Essence? What Potential prefaces expression, Neutrality precedes polarity or Principle informs all practices? Is this First Principle material or not? Is it subjective or objective, live or dead, above or down below? If the *Logos* is alive then merging with it is not, dust to dust, what sextons gravely know as death; it is, conversely, merging back the element of life with Life, a very lively joining up with Unity. This climax is not erotic or orgasmic. It is not scientific where that sense of science and of knowledge is just physical. Complete surrender to the Heart of Life is, of course, a mystic's climax entered at the crown of all creation, at the Homely Apex of Mount Universe. This, Union, is the nature of that Apex. Is it, therefore, the mystic's logic that involves True Love?

Love, at any rate, is home; home is where you want to be. *Top-down* the Dialectic sees a range in quality of love descending from most excellent through degrees of natural but selfish appetites, of worldly passions and desires including flaming and erotic connectivity; and, from that consensual, material and most human base falling further to perversions that take pleasure from another's non-consensual pain. It sees a spectrum fall from metaphysical to physical engagement and, when frustration intervenes, the cold, hard shadows of reversed love fall, as crime, across the world. Of course consensual lust is warm, synergetic and revolves round bodies, reproduction, chemicals and stems from a genetic determination to survive. It is tightly strapped to earth. Perhaps, therefore, erotic and romantic loves are only part of the whole story. Perhaps, apart from multiplicity, promiscuity and all-consuming fires that rage through sexual polarity, there exists another kind of union. Take maternal love, retain the passion of its undivided focus and then universalise. Transform your keenest interest into a love for everything because you love its source. No doubt you'd find serotonin and dopamine inside a lover's brain but is it these that drive his meditation or give rise to sanctified, enlightened *agape*? Are chemicals creators of salvation, truth and love; can materials by reaction generate a quality that's immaterially pure? *Agape*'s communion is with the Heart of Cosmos. *Agape* is Subjectivity; it is Holy Wisdom (*hagia sophia*). It is the mystic's golden treasure called *ahava*, *pyar*, *metta*, *'ishq*, *prem* and many other names; and is known to Christians as the Love of God whose depth entirely contradicts the local flash of *eros*. Top potential is, as *figs*. 17.9 to 17.11 illustrate, not at all the same as base exhaustion. The sunshine of enlightenment and fireworks of lust are, as your body's own construction demonstrates, poles apart.

25. The Origin of Growth and Development

The Origin of Growth and Development

The origin of sex has two lines of view, top-down and bottom-up; by informative potential or by chance. Which line is the dominant wild type, which the recessive product of mutation, error and defect? What is the reality? And what, in this same vein, is the reality of process following hard on sexual climax? Is not the subsequent development of body towards a second sort of climax - adulthood? How came adulthood?

Development is the precise and deliberate process of transforming a single cell into multi-cellular, multifunctional forms of life. Surely the origin of such development is not, like that of consciousness, thought, memory, instinct, homeostasis, irreducibly complex mechanisms, homologous body-forms, 'convergent' and 'parallel' body-forms, energy and other metabolisms, mitosis, sexual reproduction and so on, another Darwinian black box? *Surely so much ignorance does not turn evolution into fact?*

Development is like a spring released. Its process is, as opposed to the relatively 'static' adult form, dynamic. Its hallmark is, as any playwright or NASA flight designer well understands, an orderly, integrated and targeted sequence of events leading inexorably, as in turn any actor or flight controller well understands, with critical and perfect timing to a climax of events. Resolution. Mission well accomplished.

What exactly does your body mean to atoms that compose it? We have repeatedly asked why chemistry should develop mechanisms, including development, in order to maintain complex, apparently purposive aggregations of molecules. The *bottom-up* answer, exclusively required by professional science, is that abiogenesis and the subsequent evolution of life is almost bound, given enough time, to happen anywhere that the 'right' environmental context is luckily sustained for billions of years. This entirely profligate bet is placed trillions of times further against the odds than any normal punter would consider sane. *The answer is repeated as often as the question 'why are we here?' is asked - evolution must have happened because we are here.* This is irresponsible stone-walling. It is tautology (circular argument) at its most self-destructive. If you first (*a priori*) deny purposive skill was invested in coded instructions and their subsequently developed biomechanical engineering (the study of how the constructions of plants and animals obey and capitalise on the laws of physics and chemistry) then it is simple to deny design at every point it 'appears to appear'. In each case the counter is overturned and an explanation based on denial is evolved. *From a top-down perspective twentieth-century biology was, in respect of origins, living in a systematic Darwinian state of denial. Let us hope, although its start has not been promising, that the twenty-first century retrieves a positive frame of mind.*

A small bacterial cell contains perhaps a billion atoms and millions of proteins of at least a thousand kinds all built in a highly specific way according to strict code and cooperative in a dynamic but coherent, robotic mode to expedite preordained functions according the organism's archetypal nature! What a program! *One has to ask (as in Chapters 20 and 21) whether it is sensible to pursue study based on the notion that the first one popped up in a puddle. If, however, one insists then the next question is how its successors might grow and develop multi-cellular complexity.* It is inferred from meagre evidence (and can be studied in greater detail from any of many books on the subject) that an order of complexity runs from prokaryote, living either singly or in clusters or chains; through prokaryotic endosymbiosis (in which one bacterium starts to live inside another, which latter then changes its mode of respiration and asexual reproduction profoundly to become a eukaryote); eukaryotic organisms such as algae, protozoa, fungi and slime moulds; multi-cellular colonial bodies wherein single cells develop specialities to serve the 'whole' organism to, finally, multi-cellular aggregates that grow and develop from a single spore or egg. Is it all as simple as such a 'random walk'? How did atoms become 'obsessed' with survival? With 'health', 'progress' and 'increased complexity'? How works the ghostly eminence of natural selection (Chapter 22) whose power holds chemicals in sway, fixes their 'improvement' and leads them neither they nor it know or how or why?

When you zoom from allegations of a hazy, distant abiogenesis to the present's clarity you observe, in the current development of multi-cellular organisms, an obvious example of extreme control, hierarchy, algorithmic coherence and target from molecular to phenotypic levels. The logic is, as with computers on complex chips, encoded. *The whole developmental business is not 'as if' teleological. It is in reality thoroughly teleological and irreducibly complex with respect to such nuclei of code as, by specifying the spatial domain of embryos, effectively control the type of body plan; and also with respect to the whole biochemical context within which these magisterial, homeotic nuclei work.* A 'master' or 'homeotic' gene is one that determines which parts of a multi-cellular body stem from others. **At this stage simply note that, as opposed to any 'random walk', the *top-down* line of view is that development is, like the rest of the reproductive process, anticipatory; it looks to the future with a goal in mind and is thus conceptual. Such exquisite, futuristic programming, useless without its end-product, cannot involve unintelligent non-design; it has, therefore, to involve intelligent, informative design.**

Where does life for a sexed organism start? Initial conditions for a gamete (sex cell) are complex. At first it was believed, within a sperm, a foetal-curled homunculus was simply waiting for the sexual trigger to assume, by growth, an adult form. However, we now know this adult form is *latent* in a program; it is preformed or (the same thing) pre-informed invisibly. It is as preordained as any vehicle that rolls out of a works. Naturalistic explanation is no option. Only blue-print, hard-copied in the form of nuclear chemistry, incorporates potential whose definite end-product is an adult form. Of course, an element of flexibility - genetic and by super-coding markers - generates variety but (as steps in any orderly yet complex game, development is not created '*ad hoc* on the spot'. The orchestration is preformed; even while unrealised the score intends the climax of an adult form.

Developmental Stereo-Computation

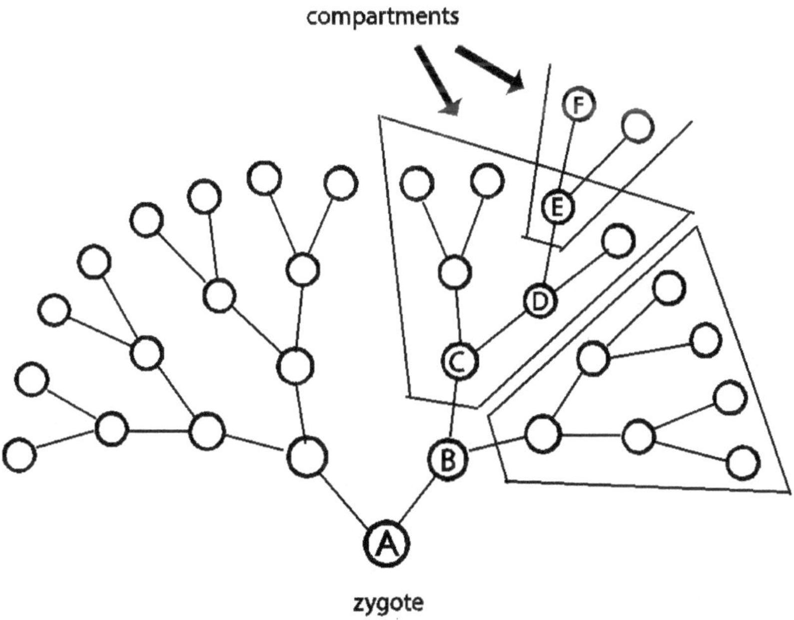

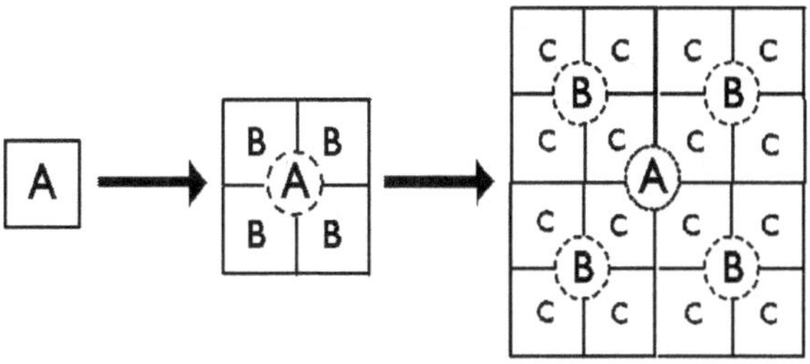

These schematic 'trees' are data structures that represent the development of organism from zygotic source to sink, that is, from top potential of a seed downstream (past A, B, C etc.) to adult form. Nodes of influence appear and, in turn, give rise to others while they disappear. Routine A gives rise to Bs and Bs to differentiated Cs etc. An increasingly determined, distinct picture emerges. Through stepping stones of stages subroutines generate various compartments (or segments) that will become the integrated and yet self-contained tissues, organs and systems of a whole. In short, a preordained cascade, an anticipatory stereo-computation of breath-taking elegance controls, perhaps from conserved and universal basic principles, the development of all specific downstream details of a form of life on earth.

Evo-devo (the evolution of development) consists, as we'll shortly see, in the notion that mutations at a high level in the cascade (say, B, C or D) might cause large-scale effect 'downstream' (say, at F, G or lower stages) as subsequent details emerge; and that they might thus generate large-scale transformations in body-plan, create fresh coherent systems with their organs or simply adapt organic functions. Thus might you solve a major, as-yet-unsolved conundrum – the evolution of development. Yet how could changes to a protein or a series of them wreak such macro-evolution? What complexity of stereo-computation would you need evolve to navigate construction of, with all its correspondents, brain? The strength of plausibility of chance-creation is to be discussed. The strength of plausibility will be discussed.

***fig.* 25.1**

How 'spooky' is a mind, intelligence or plan? Along an inevitable path, with every step engraved since the dawn of human history, a zygote turns into a human being in about fifty-six hierarchical, algorithmic steps. As with metabolic so with developmental pathways; each kind reaches a target stepwise. Intermediate constructions are countenanced which have no value except in terms of an 'aspiration', an anticipation called, when it is reached, the finished product. No stage in the developmental process from an egg's potential to preordained realisation has meaning except in terms of adult, reproductive form. The 'docile', generalised zygote is progressively refined into billions of specialised cells. Determination soon sets in to the extent that, at the gastrula stage, fate-maps for various cells can be worked out using marker dyes. A succession of stages, each with its centre of influence, branches outward from a single seed. The coherent, hierarchical data structure involved in this developmental **stereo-computation** is a tree; you might say 'a tree of life'. You cascaded into life by it. It is, as opposed to Darwin's much-vaunted but hypothetical and ghost-like tree of life, as real as you and me.

Spatial integration from node to node at the same stage is by programmed induction. Such induction may or may not cause a neighbour to change into a new kind of cell or tissue, triggering a new but 'preconceived', orderly chain of responses. It may create segments, compartments, discs, routines, modules - call them what you will - which become distinct organs and, always integrated with the rest, systems. In each case centres of influence are derived from a previous node; neighbouring compartments interlock; the whole is hierarchical and non-Darwinian in the expression of its program; and the result is a thoroughly predetermined, systematic and coordinated phenotype. Development is the complete antithesis of chancy evolution. Spin me a tale. How, logically, could it rise by chance?

It is now understood that clusters of master, so-called homeotic genes involved in the regulation of developmental patterns control at least the flow of materials required by that brilliant stereo-computation otherwise known as morphogenesis. They amount, with egg as program first-line, to high-level sets of switching routines branching to lower hierarchies of subroutine that progressively express the detailed formulations of an organism. Some of these routines have been conserved to the extent that, for example, genes from a mouse can be inserted into a fly and trigger the same desired result. And when a gene known to induce eye formation was turned on in unnatural places in a fly it induced eye structures in those places; and when a gene doing the same thing in mice was introduced into a fly it induced the formation of an eye there - a fly eye not a mouse eye! You can interpret such induction routines as so powerful that, although evolved early on, they are conserved simply because their loss or deformation is, because you lose the organ involved, catastrophic. *This does, however, not explain their origin within a stepped scheme of construction.*

You could, instead, interpret them as critical, must-have components of such scheme. And you might reasonably speculate that a single cell was created, much more plausibly by design than accident, which contained all the information (including 'embryonic' developmental strategies) needed for its descendants to evolve. Alternatively, in dialectical mode of bio-logic, you may consider such strategies as part of the top-level suite of routines in a 'universal' or at least phyletic program whose complex but highly ordered stereo-computation from the start anticipated its product, that is, the process of morphogenesis from an egg. And what a clever concept eggs are too. 'Small is beautiful.' 'Any intellectual fool', noted Ernst Schumacher, 'can make things bigger, more complex and violent'; to move in the other direction takes 'intelligence and subtlety'. To what greater extreme is the precept driven than in the compression of potential (that is, information) into a single, most rejuvenated cell capable of subsequent development into an adult form - an egg! It is reduction of an entire organism to a single cell from which, entirely integrally, a multi-cellular development is run. **There is no room in this interpretation for the gradual, haphazard evolution of such a powerfully compressed, specific process.**

How the development of shape happens is the focus of much study. Now the chemical mechanisms of this marvel are emerging and the operation starts to be defined; and there is no doubt that, eventually, the exact *modus operandi* of complex developmental algorithms will be fully elaborated. *It is, however, emphasized that a machine's operation does not fully reveal its origin.* This, unless you know its creator or watched its creation, must be inferred. Nor is description on a par with conception or invention. Are we really forced to ascribe developmental morphogenesis, including integrated genetic and epigenetic code, to mutant chances?

Bottom-up, of course, materialism's best guess has to be a stone-wall 'yes!' Yet, as we saw (Chapter 6: Machines), there is no sense, just because you understand the workings, in denying a machine its inventor. It is an elementary error to deny intricate design a creator, that is, to eschew a mind behind. Why, therefore, does a simple philosophical error, missed either through laziness or determination, consign legions of highly intelligent specialists to waste their 'evo-devo' speculations (but not their scientific efforts) trying to conceive how such complexity was not conceived but blindly, with its stunning teleology, evolved? To try and understand how something works is one thing. To spend time, having predetermined that a process is mindless, imagining the way that time and chance perhaps authored it is quite another. Such scientism skirts the origin of homeotic nuclei and complicated switching systems; it tiptoes round how integrated hierarchies exist that, as national grid and multitudes of instruments surround the simple room-light you switch on, precisely govern all the integrated context of development. But the 'evo-devo' movement, pressed to the feet of theory and supported by academy, labours under the secular faith that 'self-assembly' springs 'as if spontaneously informed' by molecules alone. Morphology is, in that special kind of matter, the three-dimensional product of haphazardly created genes. Gene pathways, so the creed recites, regulate your shape because, it's plain to see, where can information spring from but genetic chemistry? Such chemistry, and thus the evolution of development, is sprung from chance mutations. Such conceptual grandeur as life growing wasn't thought of by an accident. Accidents don't think and so development was not conceived at all.

Top-down, information is the key. Informative potential resides in orderly arrangement of instructions. Quality of information resides in both the scope and complexity of a goal and how effectively the order in which things are placed allows this goal to be achieved. **The informative interpretation is, therefore, that intelligence develops plans; and, in each case, a *top-down* program from first principles (or main routine) issues regulations through a suite of subroutines.** Thought is expressed in code; material detail is expressed through that same medium, code. You create and then commit a scheme to passive memory; active thought is fixed in store. Thus, by such bio-logic, archetypal memory is half of passive information's file. Memory is reflected by physical a moiety called code. Mind's message is conveyed (as ideas by spoken words or books) on air, on ink and paper or, in life's case, sequenced on the plaited 'circuitry' of *DNA*.

Seen in this light any attempt to explain anticipatory sex, development and metamorphosis as products of biological evolution is as daft as trying to explain the presence of an automated factory without recourse to its underlying but invisible purpose, blueprint and the codes that guide its operation; or claiming that such industry does, for absolutely no reason, 'self-construct'. This is because every cell and every organism is just such a very complicated factory.

Developmental biologists study the devices by which a basic concept (in this case 'survival of the type') is expressed in physical practice. In other words, they study how reproductive maturity is reached. Their fascinating work clearly reveals the central presence of a thoroughly *top-down* information system, one that strictly regulates the reproduction and development of all life forms. A computer analogy is entirely appropriate and can be extended into biochemical reality. Tremendous power and influence devolve from the top. Very high-grade programs have been written. It is dialectically appropriate that, just as metaphysical archetype (called super-matter) governs *(raj)* quantum and (*tam*) bulk classical aspects of physical expression, so bulk form should itself be governed by the internal activity of patterns of electromagnetic force (in our human case the *H. electromagneticus* of Book 2) and subatomic particles composed into atoms and molecules. For example, just as a few key hypothalamic releasing factors define and control the differential expression of hormonal cascades and their subsequent visible effects, so key molecular switches operate to define and control the patterns of development long before any outward sign is visible. Researchers are patiently collecting and collating the modular data that is expressed through master switching routines to realise genetic potential in an ordered, hierarchical ripple of subroutines. They are deciphering, in ever-increasing detail, the lines of cybernetic code, multiplex communications and charge-based (that is, electrical) signalling that minutely define, in its contemporary array, life on earth. It is predicted that revelation of the multi-level coherence of such interlocking operations will become increasingly breathtaking.

	tam/ raj	*Sat*	
	polarity	*Potential*	
	outworking	*Plan*	
	periphery	*Centrality*	
	parts	*Whole*	
↓	*tam*	*raj*	↑
	passive	*active*	
	determined	*determinant*	
	energy/ matter	*information*	
	differentiation	*unification*	
	developmental component	*developmental control*	
	product	*code*	

Choice of words can prejudice debate. *If the word ancestral is simply replaced with archetypal the notion of an accidental construction of highly ordered systems vanishes.* If we replace a description that is unsustainable both in principle and practice we become free to apply a conceptual mode of logic whose template well fits the facts of developmental biology. *The difficulty is to explain, not the mechanisms but their origin, in other than archetypal and typological terms.* It is the old and ubiquitous problem of explaining a hierarchically organised, irreducibly complex piece of machinery, network of signals or targeted sequence of activities in terms of the order of its whole and its interdependent parts. *Who or what hatched the plot? Which came first, the chicken or the egg? Or did both hatch together?*

It is, of course, easy to organise a series of phenotypic forms in a way that suggests one could morph into the next. That is to say, for example, that amphibian forms represent an extant 'island' intermediate in a chain of extinct, 'submerged' intermediates between fish and reptiles. *It is more difficult when you have to explain intermediate metabolic or developmental steps which are critical but useless outside the specific context of their purposive pathway towards target molecule or adult, reproductive form.* There is a sense in which networks of linear or branching metabolic pathways are simpler than three-dimensional developmental cascades but, not least because an enzyme operates in three dimensions, the problem is the same. On the one hand, either form of development involves numerous intermediate stages, each in its own right indispensible as intermediate configurations in a particular, purposive process of construction which dovetails into all the other relevant aspects of an organism's structure and function. On the other hand, no step has any selective value either *per alium* or *per se* except in the company of the full suite necessary to achieve the purpose of the pathway. How could a developmental sequence, from zygote through impotent, pre-hatched 'larval' or 'baby' forms evolve serially over even a few thousand let alone hundreds of millions of years? How did the first blastula let alone baby reproduce? In other words, the coded appearance of an intermediate molecule or (as a stage) complex of molecules can have no selective value unless it occurs in concord and correct sequence with the rest. On the contrary, it will suffer negative selection and disappear before any other of the series of links haphazardly appears. *Simple logistics eliminate gradualism*.

From one you grew into perhaps 100 trillion cells. From a gamete developed, precise in time and shape, every facet of your adult form. **Who bluffs that any multi-celled incorporation could evolve its own development through aeons to maturity?** No-one suggests a spore or gamete was once, in itself, an independent organism that gradually transformed piecemeal into an embryo; or then, after much 'blind-watch-making', turned into a foetal form which then, by haphazard starts and stops, became an infant; and eventually this infant by a series of chances evolved into the reproductive climax of an adult. Such negation of orderly control would be as unreasonable as suggesting there was no intelligence behind stereo-computation. *Sex, metabolism, metamorphosis and morphogenesis are four anticipatory and therefore conceptual processes.* **The fact is that the evolution of development is as much a black box as the evolution of any of them. Their logic hammers nails squarely into at least four corners of Darwinism's and thereby materialism's coffin.** It is, on the other hand, a fact that in the production of any machine both 'egg and chicken' almost concur - except that the 'egg' of principle and design inextricably *precedes* the physical realisation of equipment. *The root of archetype is conceptual.* It is not, therefore, irrational to perceive the presence of a developmental archetype behind the reproduction of all cellular and multi-cellular form. *Indeed, in this view the architecture of an adult by which any multi-cellular species is normally judged is actually only the last, special frame in the series of a dynamic, hierarchical, developmental archetype.*

Bio-logic.

	tam/ raj	*Sat*
	physical	*Metaphysical*
	tertiary/ secondary orders	*PrimalOrder/Archetypal Law*
	gross/ subtle forms	*Potential Form*
	message conveyed	*Information-in-Mind*
	external expression	*Inside Information/ code*
↓	*tam*	*raj* ↑
	tertiary order	*secondary order*
	external expression/ bodies	*internal code/ regulation*
	variation	*stability*
	specific, local play	*conservation of pattern*
	massive aspect	*energetic aspect*
	macroscopic effect	*causal, molecular level*

Darwin, thinking *bottom-up*, thought that, instead of breeders, habitats drove evolution. Adaptations somehow fitted plastic organisms round 'exogenous' or ecological constraints. No doubt, such variable factors challenge but, increasingly, reassessment indicates that natural selection has a limited, if not a minor, part to play in giving birth to novel forms of life. It is conservative; death is not an innovative filter. External stimulus, the modernist consensus claims, is not enough to drive leaps and transforming bounds that macro-evolution craves. How, therefore, could a body re-invent itself? How, though, could natural selection *innovate*? Theory thus recruited genetic mutations that, despite their normally crippling effect, might supply such variations as drove evolution unto macro-evolution. Arcane meandering of these stealthy infiltrators (small since shocking, large ones lop your fitness drastically) must have transformed bacteria into bats and mice to men.

This couple held the 20th century fort but Natural Dialectic is, you've seen, dismissive of their capability. It is more aligned with startling discoveries in genomics and developmental science. This is because these discoveries shift the action far 'upstream' towards the source and basis of biology. They emphasize the leading role of language, signalling instructions, controls and logical expression - in a word, of information. No doubt, since bodies are composed of molecules and operate as wonderful machines, the precise computational and accompanying atomic configurations that underwrite development will eventually and completely be worked out. Thus, having revealed the interactive complexity of genetic operation (with its splicing, repair mechanisms, shape-correcting chaperones, micro-*RNA* promoters, epigenetic labelling and a host of other breathtakingly elegant solutions all conspiring, within tight hierarchical networks, towards your health and happiness), the tilt has shifted decisively 'up' from the details of local practice to general principles. The question is how instructive bio-logic might emerge by virtue of non-conscious forces and the particles that physicists enumerate. Have, in the tide of time, everlasting laws of physics conspired with local elements to mindlessly 'conceive' of 'logic'; or was mind involved?

Naturally, naturalism *must* adopt the former line. Developmental science must be packed into a box of evolutionary tricks, illusions and, hey presto! drawn from a locker of 'impossible' improbabilities - whatever metaphor you choose.

Thus certain experts in the field of coded *operations* that control development adopt a missionary, even militant, position when it comes to *origin* of operations, that is, how they all began! Evolution *had* to make the process work; only thus the puzzle is approached. Thus Evo-devo outlines how. It is as if, having discovered how a machine works or the logic of a program flows, the immediate next step is to deny intelligent invention. And if the wind of reason threatens to dislodge your goggles, hang in there! Yet more firmly press them back in place. Does this mean, however, Natural Dialectic's reason, by appeal to immaterial elements of logic, innovation, information or conceptual bio-logic is commanded to dunce-corner? Or is worthy of contempt, academic loathing and a summary expulsion from the rational scientific class? Isn't such reaction an irrational one?

Top-down, however, make a substitution that's a simple, systematic one. *Instead of 'time and chance' employ intelligent logic'. For 'conserved' or 'ancient' read 'precisely specified and critical component'. For 'natural selection and mutation' employ, respectively, 'editor and author of ideas'. Thus finally replace 'phylogeny' with 'bio-logic'.*

Any fundamental logic of embodiment (Chapters 16, 17, 19 and 25) is, of course, anathema to current science, especially Darwinian bio-logy. *Nevertheless logic is outstandingly obvious in a process of anticipation and efficient steps to reach a goal. That is, in this case, development from egg to adult.* How, therefore, does an informative perspective square with one that's not?

Firstly, recall the linguistic properties of *DNA* (Chapter 20: Pristine Instruction to Code Rules Supreme; also Chapter 23: Language of the Genes).

Secondly, recall a computational mode of operation. Life logically proceeds, by algorithmic steps, from starting-points to targets that are predefined. A program, carried forward by language you might call computer code (but at root compiled in polar bits called nucleotides), instructs the hardware what to do. Whether technological or biotechnological, electronic charge runs both digital, switched kinds of show. Application programs sweep you into target end-effects; but application programs need an operating system so that they can be expressed. Both application and its handmaid operating software may be hierarchically arranged. The combination may involve modules, subroutines and many, many switches (ifs, go-to, stops and starts) in regulation. Indeed, perhaps the most controlled and efficient forms of software are expressed *top-down.* A main module (or set of main routines) calls subroutines (or sets and subsets of them) into an appropriate response. Perhaps no cybernetic system is so complex and inexorably controlled as room inside a cell.

Studies show this is how genetic 'videos' of phases in the life of any organism work. The dynamic phases are developmental leading to the last, climactic, stable frame - an adult form. Such studies have long illustrated mini-application programs that also lead to target frames. These include specific calls to genes that generate the suites of functional enzymes metabolic pathways use to build the chemicals of life. Now (see Chapter 23: Non-Protein-Coding *DNA* and Super-codes and Adaptation) dynamic flexibility, operating system factors and developmental switching networks are beginning to be found. **The whole genetic investigation is like trying to decipher a computer's software operations from 'outside', that is, by inspection of its (nuclear) c-drive.** Painstakingly epigenetic mechanisms, super-coding in the forms of splicing, chromatin arrangement, *RNA* and protein regulators and a multitude of switching systems are being connected up and brought into relief. *Top-down* you'd predict, of course, that in the next decades relief will grow still sharper until a clear working picture is complete. You wouldn't then, in breathless wonder, cry out, 'Chance!' What a fine thing chance were.

Ask chip-designer or chip-manufacturer where their chips came from. Are they material or intellectual property - or both? In 1983 Pitman predicted that, according to *top-down* perspective, biology in general and developmental biology in particular would swing increasingly towards an immaterial, informative dimension for its explanations. It has.

In his adventurous book 'Endless Forms Most Beautiful' developmental biologist Sean Carroll confirms the bio-logical approach to his science. His key ideas involve the encoding of form, that is, how morphogenesis from egg through geographies of embryo to adult is programmed. His ideas include modularity/ homology; a high-level genetic 'tool-kit' that affects how genes 'lower' in the hierarchy are controlled; a genomic 'dark matter' composed of switches that instruct the 'tool-kti' operations; a 'combinatorial power and glory' wherein numerous independent switches govern individual genes; and, overall, an astronomical number of possible combinations of regulatory factors (even though, as with any language, only specific combinations make 'sense'. This informatively described paradigm would make, in any computing office, perfect sense.

Carroll notes (p. 129), illustrated with a simple 'genetic wiring diagram of regulatory logic' that, 'animal architecture is a product of genetic regulatory network architecture. This circuit and network wiring or logic can be illustrated with the same sort of diagrams used for electrical circuits or logic problems. Each switch is a decision point, one node in the genetic circuitry.' It is his guess that to write, in the manner of a systems analyst but not of natural selection, the logic of making a fly would need at least a thousand pages; and, although the networks are not in principle any more complicated, many thousands to script those needed to describe a human. Do you remember (Chapter 21) Minimal Functionality? Moving in the opposite direction towards 'simplicity' we note that, as opposed to a fly's 16000 or so p-c-genes the minimal number possible for a free-living organism has been calculated at about 400. This is about 2.5%. 2.5% of 1000 pages gives 25 or so putative pages of logic, flowchart decision-making, call it what you will. A billion years could never potter up the cup your tea's in let alone a logic dissertation. Therefore, didn't wind and rain do well? Not least because the vital notion of alphabetic logic and its translation into words of another language (words called amino acids whose information makes up working protein parts) represents a truly astonishing, magical accident; and not least because such a magically encoded message would, with its genetic operations slotted into place, then have to produce a minimally functional cell; but, on top of all this serial impossibility, if, as well as just a prototype, these pages needed to contain in embryonic form top-level sketches that could then evolve down billions of years into all sorts of organism. That would involve exceptional 'foresight'!

Logically Expressed.

We live in an information age. Nature always did. How is natural, inner logic so expressed that we see outward form? How does downstream product of the code appear?

An author or an artist sketches outlines of a theme before he moves to fill coherent details in. Dramatists know that, from a palette made of archetypal plots, they can colour myriad dramas in; they use a given datum, alphabetic language with its bits made into words and higher level code. So do computer programmers. But biology goes further in pursuit of shape. It has, like a sculptor from initial block, to generate a 3-dimensional form.

Not that bacterial form, presumed the primal 'block' of life, is imprecise. It does not contain organelles (unless you count carboxysomes or ribosomes as such) but does show a degree of compartmentalisation (such as plasmid clustering and membrane domains), protein localisation (e.g. *DNA* polymerases) and formative, cytoskeletal arrangements (using crescentin, protein rings and so on). At the time of cell division such a ring is constructed and, using groups of other proteins, bound in a particular way at a particular place (the mid-plane) to the inner surface of the cell's wall. If the chromosome is held in the correct orientation, orientation is maintained by still more suites of 'handmaid' proteins and an uncompromised ring pinches the mother into two daughter cells, then division will have occurred. If any part at any time fails, however, the process aborts and death results.

Compared to bacterial 'simplicity' dynamic multi-cellular sculptures are a glory of precise anticipation and cooperation. Instructions, as finely integrated as those on an advanced computer chip, stepwise sketch the form to be produced. Poles are defined with major axes, longitude and latitude, that are then subdivided into modules and, in procession, finer subroutines. At specific coordinates and intervals these modules will be developed, by switching on different permutations of genes in different groups of cells, into required temporary, 'scaffold' or final structures. Such exquisite architectural expression is then enhance by perpendicular developments that effectively produce 3-dimensional characters such as modulated organs, appendages like limbs and so on. In all cases coordinates are supremely accurate; sometimes they are specific to the location of a single cell. Developmental biologists want to know how all this happens.

The IT element that governs life is absolutely clear. Information is potential; metaphysic leads to physical expression. Without plans life is nowhere. Plans grow trees but trees can't grow them. Trees grow flowers and fruits but these derive from mind; neither anticipatory codes nor, therefore, anticipated, complex, specified results emerge, unsupervised and accidentally, from mindless chemical events. Thus philosophical attempts to exorcise design from biological design, however desperate or vain their support of various materialistic sects, are doomed. It is predicted there'll appear no clearer indicator of this fate than, when the final maps are in, revelations from developmental science. Why?

All somatic cells (except mature red blood cells, plant tracheids, sieve tube cells and perhaps a few others where there is an equally logical reason) contain the entire genome, that is, an organism's genes. The genome is actually a database to which an application program (called the cell) makes reference. It calls according to its various subroutines to satisfy, in a context of itself and its surroundings, contextual needs; such calls maintain disequilibrium of balance, that is, dynamic equilibrium of a dialogue called homeostasis. This balanced 'sense of properness' is predetermined. In order to function properly a particular type of cell will access a proportion of its whole genomic potential. Its access to this whole will be confined and specific. If you imagine items on the genome as articles in a newspaper, then some segments will be used in common by all cells, others less commonly and some may be unique to certain cell types. It is a cut-and-paste-as-you-need-it sort of operation. Another helpful analogy is with notes on a keyboard. Different types of instrument (or organism) are played with specific clusters of chords (cells) in a way that culminates in a special song. The right notes (genes) played together in the right order at the right time in the right place add up to themes (cells) within an instantly recognisable *opus*. In the same way permutations from the genome are coordinated to engage their piece. This is the differential expression of modular, genetic information.

It should be obvious by now that a database is, in cybernetic practice, governed by an operating system composed of application programs.

Conceptually these regulatory and configurative factors include an army of switches, locks and keys called variously initiators, promoters, repressors, selectors, inducers etc. In practice their agency is composed of *DNA* sequences (sometimes repetitious, sometimes not), strips of *RNA*, proteins, binding motifs, epigenetic methyl groups, chromatin, hormones, cyclic *AMP*, chemical gradients and other signalling (or messaging) molecules. As a whole they constitute a complex information system whose key function is, like the musician's, timely selection according to a pre-programmed score.

It is also obvious that you access a library as you need specific books, not all at once. Differential gene expression is the way every cell requisitions its supplies for maintenance, repair, reproduction and often, with the latter, development. Control of access constitutes a critical part of the way construction processes are managed - *top-down* from coded message into physical actuality. *Top-down networks run the show. They are always the product of purpose, design and its intelligent information*. They are the way in which principle is expressed in practice, thought is implemented as physical behaviour and mental shapes material fact. Mind is anti-chance and thus hierarchical, *top-down* applications are never the product of chance. The unicellular or intra-cellular expression of genes is one half of a dialogue between genome and cytoplasmic or extra-cellular signals. With ping-pong it is difficult to suggest which side is in control. At any rate, the balance is dynamic. Biologists have detected micro-hierarchies of gene regulation that lead one to infer that it is the application program (cytoplasmic initiation) that triggers a system of indexing to choose what texts are selectively accessed from the database. Thus index works as part of operating system. Genes serve the organism and not *vice versa.* They are an important half of its physical codification which is, in turn, the 'lower' half of the whole archetypal truth. *Top-down,* in dialectical terms, it is not material expression (with which experiments can be done) but metaphysical information that takes precedence. Just as thought precedes an action, invisible precedes visible. Although simple and entirely logical, this is not an easy idea for scientific materialism to accept.

Of course, a single *top-down* system runs each cell. Information runs from *DNA* to protein, from nucleus to cytoplasm and beyond. However, as regards the development and maintenance of multi-cellular aggregations called a whole organism other kinds of obviously *top-down* system ramify to regulate. For example, in most animals two of these branch literally *top-down* from the brain. They are concerned to integrate and coordinate the psychological and physical aspects of life and are called the nervous and hormonal hierarchies. The third, whose task is to raise a multi-cellular body from the materials of earth, is an order of development that ramifies until it has, in an adult form, fully realised its potential. Curiously, this hierarchy has its top-level headquarters vested in a single cell that inhabits the basement area. It is called a fertilised egg or zygote.

After all, how would *you* have done it? If *you* needed to achieve the imperative of survival down the generations, how would you have the copies made? Especially if you wanted to include some sober flexibility, that is, continually varied reproductions on a standard theme? *We can ask how an engineer might design a self-reproducing machine and ask if nature has preceded him in his intelligent logic. How might the least demand be made on the tissue or strength of a parent while at the same time encapsulating its potential? A brilliant, optimally economical idea would be to reduce the parent to a single cell and then, from the symbolism of this cell, build up a new adult.* This is exactly what happens in nature. Each individual adult is 'repotentiated'. Its body is reduced almost entirely to a symbol, a directory, a coded book of what might be. Its spring is re-compressed into the top-level form of an egg or sperm; these fuse and the offspring, a single-celled zygote, unfolds in a complicated but very precise program, a hierarchical developmental archetype which eventually realises its goal, the re-creation of the next adult generation. Nobody would believe a car could reproduce itself so we have to infer a mind behind automobile production-lines. What about a self-reproducing robot? Even this, as every AI technologist understands, is small beer beside a cell. But can't bio-replication of the *DNA*, while making some mistakes, build anything you want? *We'll come to brains but meanwhile is it not ironic that the most stupendous feat of engineering, self-reproduction of a very complex machine, a 'simple' cell, should be the very 'reason' some folks believe there need be no purposeful, top-down application and therefore not a single scrap of mind behind?*

Although a body develops branching from top-level potential called an egg, seed or spore, *top-down* programming branches hierarchically from a top-level core/ master switching routine. You might view its hierarchy as concentric rings or shells radiating from a central point. Remember (Chapters 5 - 12) that information is the potential form of objective order. Mind conceives the patterns it transfers to matter. *The central point, the topmost source of power in any hierarchy, is its potential.* What is potent rules the impotent. A potentate directs the lower orders but such power is not physical. It is metaphysical. The governor is information; information's natural element (a quantum physicist may well observe) is mind. Mind's fixed, immaterial fraction is a passive entity called memory. Universal memory is archetype. Thus, in hierarchical, dialectical terms, the source of biological power and glory resides in archetypal memory; such power initiated physical expression in the form of code, underlying process and overt machinery.

Where psychological informs physical it is, in higher animals, expressed in offices of overall control - the cerebral cortex, brain stem, hypothalamus and lower information systems. But in every organism the foundation of physical construction is the cell whose biochemical potential is concentrated in its central, nuclear genome. It is, like all ideas expressed in speech, coded. In asexual forms of reproduction a single cell is sufficient, when triggered, to produce either a unicellular or multi-cellular clone. In sexual forms genomic potential is preserved in sperm or egg and tripped by fertilisation into the developmental cascade. This cascade is purposeful. *Purpose is metaphysical; it informs the world of purpose using code.* Such code, either the vibratory resonance of metaphysical archetype or physical *DNA* base sequences, is an embryonic plan-of-action. It is a blueprint, manual or reference point. Its instructions are realised as metabolism which underwrites large-scale structure and function. **This is the nature of conceptual as opposed to phylogenetic biology.**

An archetype is, of course, as physically meaningless as a mental image. The image is expressed by its physical re-creation in speech, behaviour or some other material construction. The more clearly defined the image (or video-image called a program) the more subtle, accurate and complex its expression is likely to be. Such succinct precision is one of education's highest goals. A biological archetype *is* a mental image; it is, without material projection, as meaningless as an unexpressed idea. It has no outer meaning but rests as incommunicado as a TV broadcast without sets to receive it. Archetypal projection is much subtler than radio waves, the 'verbalised air' of sound or, indeed, any human construction. It involves the expression of a preconceived program against coded material. Its operation is, therefore, tightly integrated with a generator (*DNA*) of generators (enzymes) of physical factors. *These factors, including both 'control panel' and building materials of the vehicle called a body, are what tie the idea to earth. They are the physical reality of a metaphysical blueprint.*

One half of a biological program is tied up in the characteristics of a particular cell (or, in the case of sexual initialisation, a zygote) and the other half in its information storage department, a database called *DNA*. This physical potential, generator or genome is composed of complementary base pairs, the letters of its language, robustly stored in well-bound twine. You might argue it wasteful to replicate the entire genome in every cell from top (zygote) to bottom (adult) level of the hierarchy. Does the editor of a newspaper print only the information each individual reader requires? No, on fast-rolling presses he prints the lot and lets each reader select what he or she wants to read. This is much simpler and therefore more economical than producing hundreds of mini-extracts. Instead, in the case of an interested cell, it is the cybernetic army of switches that decides which indexed bulletins will be retrieved when. Such a system of retrieval applies (with a complexity of biochemical networking factors which are being discovered by the latest research) to all maintenance, repair and, on which we now focus, development.

Subset of Stage 3: Evo-devo.

Evo-devo is an important sub-set of aforementioned 'natural engineering' *which was itself identified as Stage 3 in development of an inverted bio-logic.*

Selection and mutation, feeble and incompetent a pair, used to hypnotize their subjects as an explanation of how evolutionary process, that is, not minor but major variation, must have worked. Now a subset of the third contender, Natural genetic Engineering, is drawn from developmental studies. It includes the former pair as agents but 'enriches' its raw data with a bulkier input drawn from concepts of 'domain', 'module' and modular mobility. Thus it contrives a stronger, more persuasive case. This is so-called 'evo-devo'. Let's look at the 'devo' or developmental half before we turn to evolution.

To define the genetic and molecular basis of regulatory pathways controlling development biologists use various models. Slime moulds and algae (such as *Chlamydomonas*) represent protoctistans. Thale cress (for flowering plants), maize (for grasses/ cereals), snapdragons (for flower development) and a kind of moss are used to study plant systems. In the case of animals a nematode worm (*Caenorhabditis elegans*), fruit flies (insects), chicks (birds), frogs and the axolotl (for amphibians), zebrafish, sea urchins, the lancelet (a chordate) and mice (for mammals) are the mainly used windows through which biologists have investigated the way animals are sculpted, like a motor manufactured while running, from primordial potential (plan) through to the differentiated detail of final, actual bodies.

It bears repetition that, in *top-down* computer programming, perhaps the most highly logical and efficient way of writing code, main routines are critical. They introduce the program because they introduce its purpose and subordinate purposes. It is from main routines that, *top-down*, all subroutines are called and orderly but highly complex computing organised. You branch to and from this integrating source. You make all reference within this systematic hierarchy. *Top-down* computing, at the cutting edge of intelligent *IT*, is organic in its operation. It is the natural way to express an idea's outline prior to detailed 'filling in'. It is, if you like, principle preceding practice. It is archetype or body-plan preceding orderly development of all sub-systems. Such bio-logic, involving switches,, transcription factors and hierarchies of genetic control, is now clear in nature as well as modern technology and commerce. Such elements occur in plants and, it is predicted, will be found to apply with similar rigour in fungi, seaweeds and other forms. No doubt plant systems differ in their detail (such as different genes, transcription factors and *MADS*-box instead of *Hox* modules) but in principle they use the same sort of *top-down* bio-logical programs for regulating their development. In this section, however, we'll make the dialectical point with respect to animal development.

	tam/ raj	*Sat*
	static/ kinetic	*Potential*
	expression	*Information*
	process	*Start*
	subsequent effects/ causes	*Trigger*
↓	*tam*	*raj* ↑
	conclusion	*kinetic program*
	informed effect	*informant cause*
	downstream effects	*high-level modules*
	called subroutines	*switching logic*
	off/ untouched/ finished	*on/ activation*
	structural & functional genes	*master/ tool-kit genes*
	end-product/ target	*development towards target*
	final shape	*dynamic process/ morphogenesis*
	output	*functional input*
	visible complexity	*invisible generator*
	practice	*principle*
	specific working parts	*outline sketch*
	organised matter	*organised instruction/ code*

Serial order of the cosmos runs from (*sat*) potential through (*raj*) action of creation to (*tam*) materialised, end product. It 'falls' from principle to practice, set of possibilities to single actuality, from archetypal generality to various particular expressions. Subtle potential of various forces, polar charge, subatomic particles, atoms and molecular gases is detailed in local events, 'cool' fluids and arrays of solid aggregates. Such outcome is non-purposive (though some would argue that fine-tuned constitution of its quantum subtleties bespeaks a purposive creation of the universal stage). However, as opposed to energetic potential the potential implicit in hierarchical banks of informative, archetypal and genomic code marks obvious, local purpose. Such subtle definition is developed. It governs single cells or develops from a zygote into patterned aggregates that, visibly, become gross phenotype in adult form.

Thus, in dialectical terms, the process of development runs from (*sat*) potential through (*raj*) kinetic to (*tam*) conclusive phase: informant plan → process → end-product/ adult form. If you declare an adult 'has evolved' from egg then you employ the word in a completely opposite sense to Darwinian theorists. There is neither random element nor dependence upon disruption of program. The absolute reverse; encoded predetermination rules supreme. **Attainment of target results from a completely controlled, hierarchical cascade of events.** Nowhere are reason, code and logical design as clearly displayed as in the exquisitely complex, simultaneously serial and parallel expressions of bio-logical construction.

The physical keys to this order are, initial trigger, banks of switches and sets of top-level, 'tool-kit' factors that can act in combination to modulate the way a gene is switched on or off.

Top of the genetic hierarchy is enthroned a seed or egg. These starters reign (perhaps like the cosmic egg itself) supreme. They are all-powerful (or 'toti-potent'); sufficient information's there at the beginning. An organism's future rests within; gamete function's just to carry code; they're packets chock-a-bloc with life-on-earth's instructions - the immaterial metaphysic that is purpose, message and the signal to make working bodies. From their transcendent ordination is projected every lower consequence. Let's understand an egg, the most rejuvenated of all cells, as the critical, initial condition for creation of a body. It represents the first line of a developmental cycle that will unfold into an adult form. Such a starting-point contains not only genetic code. An egg is ovoid, that is, polar. An unfertilised egg is a storehouse of cytoplasmic factors, also called maternal effects, which include prefabricated m*RNA*s and proteins such as, in the fruit fly, *bicoid* and *nanos* that help reinforce 'morphogenetic' gradients. These gradients run pole to pole - vertically from head to tail (anterior head to posterior, 'vegetal' tail), from back to front (dorso-ventral) and, horizontally, from side to side - and are 'present' long before any actual shapes become visible.

The explanation can run deeper. We need to recall (from Chapters 16: *H. electromagneticus* and 17: The Logic of Development) Burr's L-field, that is, patterns in trans-membrane voltages. It was proposed that, correlated with an electrically sensitive antenna in the form of *DNA* helices, such an 'index' might communicate the spatial coordinates of embryonic development. Such an electro-dynamic template would, in concert with genetic operations and the production of protein and other materials, map 3-d shape. Such developmental electrodynamics has been probed for over thirty years. It was recently shown (by Michael Levin and others) that patterns of bio-electrical signal play an important role in the development of face in frog embryos. It seems that waves of activity trigger genetic correlates, that is, they affect chromatin and genetic switches relevant to each wave's programmed 'sweep'. It would be logical to suggest that an egg's bio-electrical template constitutes frame 1 of many 'video-frames' rolled out, using informative exchange with the genome, in the course of a specific organism's developmental program.

Whence arise these waves; what decides the spatial coordinates that, supported by a molecular infrastructure of pumps and respiratory apparatus, will guide a form through serial frames of its morphogenesis? **Natural Dialectic suggests that the bio-electrical template discovered by Burr, Bose, Levin and others be identified with the *electromagneticus* body explained in Chapters 16 and 17**. In the case of *Xenopus*, a frog, it would be called *Xenopus electromagneticus*. The Dialectic further suggests that control of typical developmental/ morphogenetic electrodynamics, itself causal with respect to the orchestration of molecular patterns, is in turn informed by a 'higher', metaphysical structure called the typical/ archetypal mnemone - in the frog's case, say, *X. archetypalis*. It is the interaction between frog archetype and a specific egg with its individual bio-electrical shape and genome that gives rise to different *Xenopus* individuals. Archetype is the causal potential, the top-level informative program that, by way of an electro-dynamic intermediary, triggers not only individual but higher levels of variation (such as species or even genera) within the frog type. In short, an egg's bio-electrical component is co-programmed with its genome to produce, in coordination with an archetype, variation on a theme; the archetype is localised by nature and the nurture of top bio-logical potential - eggs or equivalent.

Thus from electro-dynamic and genetic coordinates, specified in high degree, fertilisation triggers a ramifying cascade of activity. Make no mistake (and nor does it), these hierarchically organised effects run exceedingly swiftly, accurately and concurrently. Changing frames of spatial coordinates correspond with the genetic production of correct material requirements. Predefinition runs like clockwork to construct a body from its book. After the first, most basic alignment there follow broad brushstrokes drawn under the influence of a series of proteins induced from so-called segmentation genes by the 'primordial' determinants. Segments are, in effect, part of a modular process of definition involving the relative orientation and organisation of large areas along the abovementioned axes. The flight has taken off. From ground-state outlines there follow broad, sure brushstrokes that refine within the axial parameters. In you, for example, a fertilised egg differentiated into about 210 correctly-positioned types of cell. Embryonic stem cells from the first seven divisions retained plasticity but your potential was gradually, by means of the abovementioned switches, locks and keys, closed down. Soon the fate of individual cells became irreversibly determined. They suffered neither accident nor experiment which might have 'sprung' the locks, undone the determination and, with catastrophic results, caused either 'de-differentiation' or bugged malformation.

No doubt, every cell-type still contains the genomic 'toti-potentiality' (or full *DNA* complement) that it had while an egg but its expression has been restricted from high-level developmental phase to the detailed function of any specific type of cell. In other words, as an adult, your working genomic potential has been locally, specifically reduced. It included, for example, the ability to cause you to develop from egg to egghead. This process employed switches, switching patterns, proteins and temporary devices that are after use redundant. They are as critical as a builder's scaffold, moulds, supports and strategic activity from foundations to the roof of a construction. Some are reduced and others cleared away by apoptosis (programmed cell death); they are dead and gone. In maturity only the patterns of its operational phase remain; and each differentiated cell can only access its particular permutation of genetic segments. Muscles, for example, do not express insulin or make sex hormones; even the same protein may be differentially spliced and so refined for slightly different operational circumstance in different types of cell.

It's arm-waving time again. An evolution of apoptosis by survival's mentor, natural selection, is as contradictory and inexplicable as the whole programmed and purposive anticipation of any developmental process as a whole. Why should you die in order to perhaps create a functional futurity? What use is one line of a program disintegrated from the rest? What use are parts, like enzymes intermediate in a pathway, unless the integrated set can roll? How did schemeless natural selection figure out that going backwards could, several moves away, advantage going forward?

Creation is, as opposed to non-creation (evolution), a process of devolution. It is a *top-down* materialisation of cause, the physical effect of metaphysical initiation and, as such, runs from idea to complex system rather than chemicals to chemical complexity. Its gradient tends towards increasing definition, coordinated detail and, therefore, reduction from generalised potential as it approaches the full and final effect of an idea's individual, solid, materially-detailed expression.

Can you judge a book by its cover? Two books on a shelf, one about primary school arithmetic and the other the mathematics of quantum theory, may look the same from outside. How, though, could a top-level initiator, a single-celled egg, have been accidentally transformed by printing errors from the entirely different, basic genomic potential of a bacterium; or a single-celled eukaryote? Moreover seeds, eggs and spores, however superficially alike some of their covers may appear, are books as distinct in information as their adult forms. *And an egg is, seen from its conceptual angle, the polar reverse of any other single cell.* While the latter is the particular end-product of a 'simple', single-phase division, the former embodies a striking purpose. This is to act as the starting-point or potential for a multi-phase developmental program. A non-egg is totipotent with respect to the construction of its own differentiated organelles but an egg is totipotent for 'its own construction' of a whole body made from perhaps a trillion cells. *Although containing the same DNA as every other cell, it is specially primed; it is primed to anticipate a dominion that is orders of magnitude more complex than its own, single body so that, in order to generate fresh form, it has to sacrifice its being to a series of different cell types and organs.* In every organism the dialogue between genome and cytoplasm is coherent and in the multi-cellular case every developmental event *anticipates each step that follows* towards the climactic formation of a target, sexual maturity. To attribute such coherent, purposive dialogue (with its intricate, deliberate switching mechanisms) to random genetic recombination is illogical; and to infer that shuffling millions of letters, bits of words, words or phrases from one coordinated book could, surviving the intermediate nonsense, create a second logical sequence whose path to climax is still more complex is exponentially illogical. **Intelligent construction, architecture, stares us in the face**. Who, though, averts his eyes, denies the obvious and nurses imagination into exponential sophistry? A biological 'big bang' of hugely profound accuracy is alleged, according to a *bottom-up* perspective, to have scattered all the body-plans of biological organisms including jellyfish, molluscs, arthropods and perhaps even vertebrates (therefore including sex, eggs and developmental cascades) from little more than bacterium in the space of a few million purposeless Cambrian years. From an egg's point of view this is more an explosion of evolutionary reason than anything else.

To return, therefore, from that lack of reason called an accident to its epitome (called logic) we note how outlines are progressively, almost artistically, refined into 'compartments' and clusters of cells where specific work takes place. After fertile lift-off banks of instruments control developmental flight. They signal progress through a complex set of tracks to safely reach the destination. The terminus is sexual maturity. How does regulated integration biologically work?

By switches. These are longish sections of *DNA* that surround a gene. Although non-p-c they are certainly not junk. Their sole function is regulatory. Sometimes, as with bacteria, a switch may be simple but in animals they are complex in that they can be bound at different sites by many different proteins. In a railway signal box a cluster of levers may control the possible movements of a train through a particular section of track. These levers can be pulled in different combinations for different pathways, that is, for different trains of varying purpose. Similarly with genetic switches. A 'pathway' for the gene's section is thrown by a permutation of proteins which each bind to a particular, complementary short *DNA* sequence called a motif. Specific binding is a common feature of bio-logical information exchange and is denoted as 'lock-and-key' or 'hand-in-glove'. Protein and motif are equivalent to a lever; and the consummate switch to a bank of levers. One gene, one bank of levers to variously control its operation; but a difference is that in a signal box all levers are 'stored' in the 'off' position and thrown to 'on' whereas a genetic switch can register 'off' and 'on' in ways refined according to the different context of its use.

A human may have 500 or so types of regulatory proteins that do nothing but work the lock-and-key system of levers. Thus vast potential exists for permutations of combination and, out of combinatorial blocks, precise logical control. Use of high-level switching systems to create cascades of permutations indicates efficient programming; it demonstrates efficient bio-logic by its logician. Such inexorable, coherent algorithms of control extend to how production of building materials is timed, quantified and located. Any builder needs materials efficiently delivered. Since the binding proteins, called transcription factors, play no part in any cell's peculiar structure or metabolic functions, you might call them 'universals'; they are generalities that can, in principle, be deployed in any cell in ways that cell requires. In other words, the same 'master gene' or 'tool-kit factor' may be used to work genetic switches across a wide range of not only cell types but body types, that is, organisms.

You may remember that master genes determine which parts of a body will develop out of which others. In animals a particular set of such genes has attracted much interest. It is a routine (dubbed the *Hox* cluster and found in bilateral metazoans) that kicks in following modular compartmentalisation. It triggers the timely flowering of organs that those various compartments need. A small chordate called a lancelet contains a single cluster, fruit flies two with mice, humans and some fish four. The genes are highly conserved; they could have been found in Precambrian animals; being found in numerous otherwise very different organisms they are called homologous; they do the same job in these different organisms; and, extraordinarily, the order of genes in each cluster corresponds to the order of body regions in which they are sequentially expressed. In other words, they enjoy both spatial and temporal 'co-linearity'. This means that there is a correlation between their physical arrangement on the chromosome, their activation pattern and the anterior-to-posterior expression of the body segment patterns they have triggered. A series of top-level triggers that initiate the various sub-routines of body plan, such genes are clustered on their chromosome in the order that, from head to tail, they produce their structural effects! In short such expression, whose arrangement appears fundamental to the pattern-making of an animal body, is triggered, *top-down*, head-to-tail along modular, segmental sequences. It is as if a painter or sculptor generalised and at first, from the basic outlines, no specifically recognisable shapes would yet be present in his work. With the broad proportions sketched it is time for detailed definition to begin. Incipient form and function in each module (or segment) is mediated by a suite of timers so that different permutations of 'lower level' genes are expressed in different segments along the length of the body. Significantly, although their eggs differ somewhat the major switches (master control genes) for the ordered production of building materials are the same for the development of a fruit fly as a human. Program that anticipates result *par excellence*!

Just as there is no chemical similarity between *DNA* and protein so there is no necessity for physical connection between job-related genes. Related data on a database is not necessarily stored contiguously; proteins for a metabolic process are not coded on contiguous genes or even on the same chromosome. There is, for example, absolutely no necessity for job-related *Hox* genes to be situated in a cluster on the same chromosome in the serial order of their expression. Thus a random origin of such tight order were incredibly unlikely - the more so because different *Hox* clusters are each orderly composed - but logical exposition, as the chapters of a book listed on the contents page, entirely understandable. Perfectly *logical* sense that the sequential order of genes represents the order of its functional phenotype is, however, dangerously unsettling for a theory based on chance. Logic is itself a function of purpose. It involves best ways and means to ends. It is, therefore, less clear the necessarily bio-logical way in which such a complex, integrated developmental micro-hierarchy, with its main and sub-routines, might have evolved piecemeal, haphazardly over millions of years! *How complex bio-logic should be constructed by no logic whatsoever is indeed a time-transcendent puzzle*!

Such 'tool-kit' genes as found in a *Hox* block are defined by a common sequence (or motif) of about 180 letters and called the homeobox. The homeobox codes for a section of the corresponding regulatory protein called, in turn, the homeodomain; and the latter amounts to a recognition site for binding purposes. To what does this regulatory protein bind? Non-regulatory genes are surrounded by n-p-c *DNA* which acts as a switching mechanism. This area is 'the clustered block of levers' of our previous signal-box; it is divided into sections each of which accept a different transcription factor, that is, they bind a different regulatory 'lever'. Thus permutations of input - in the form of different *Hox* or other control proteins - can differentially switch the gene on or off. It may be expressed differently in different contexts, such as an embryonic stage, in a heart, brain, kidney and so forth. Such regulation of transcription occurs, it seems, for the vast majority of genes that code for bio-systems. Executives that perform this sort of computational bio-logic thus control large networks of subroutines; the whole system is *top-down* hierarchical; it establishes the working shape of mechanisms down a gradient that moves from ovoid principle (why not 'principle' since immaterial information *is* symbolically ''in principle') through stages towards final actuality.

You could look the other way. Upstream. In fact *Hox* genes are themselves controlled by so-called 'gap' and 'pair-rule' genes whose regulation stems, in turn, from maternally supplied m-*RNA* and gradients in the egg - the program's head and source whence issues an informative cascade. Egg → m-*RNA* → 'gap' and 'pair-rule' proteins → *Hox* segmental triggers and so on down a branching line of subroutines. Step-by-step a great idea is turned to fact; principle's transformed to practice, symbolic information is translated into physicality and archetypal generality into specific, detailed, individual form. Aren't you an individual human being?

A cluster of control genes would therefore represent a main routine from which genetic subroutines were switched; *Hox*, *Pax* (for nervous system inc. eye genes), *Otx* gene family (for kidney, guts, brain, gonads) and other 'informative centralities' would, with the brilliant Boolean logic of a computer program, mastermind the developmental output of bodies ranging from a jellyfish or fly to you - with a different kind of logic to govern plants and fungi! Regulatory logic-clusters are found in a wide variety and, it will predictably transpire, all multi-cellular animals. In the four distinct blocks in vertebrates each gene is responsible for triggering a cascade of sub-routines that will supply the right materials in the right place at the right time to construct a given segment of the animal in question. Obviously, therefore, a *Hox* gene for a worm will define (call) different or differentiated organs (sub-routines) to ones for a fly, a horse or a human. In other words, the same genes can be responsible for initiating the development of different organs. In one instance a gene may specify for a tail, a coccyx or the rear part of a fly, frog or grasshopper; and the gene that triggers development of your eye would, if transferred to a fly that lacked it, cause its blindness to be eyed - with a fly eye not a human one, of course!

Nevertheless you would expect any changes at the high level of main routine to initiate major changes 'downstream'. Indeed, you might expect that specific cases of incorrect high-level inductions, either natural or experimental in origin, could cause bizarre effects. The right potential would be realised, if the determining locks were 'sprung', in the wrong place and/or the wrong time. And indeed known *Hox* mutations lead to extreme and crippling abnormality or, mostly, spontaneous abortion. Neither are powerful harbingers of useful novelty. In fact, therefore, the critical, computational algorithms of *Hox* regimes are intensely conservative. Therefore how, logic from no logic in the form of chance mutations, could the whole caboodle ever have evolved? <u>*Why should such systematic elaboration have developed piecemeal in an accidental*</u>

series? How, as in a metabolic pathway, could each fractional stage survive until the working whole clicked into place? In the top-down view Hox would represent, allied with all connected subroutines, part of a line-by-line solution, a stepwise algorithm in the chemical expression of an archetypal program! Why should I be thought illogical and therefore banned from science class for mooting this?!

Check the legend of *fig.* 17.7 and also by implication the sequence 17.5 - 10. In our own bodies regulatory routines supply a prefabricated body plan with coherent systems and their segmental associates, various organs and so on. Through Chapters 15-17 it was mooted that such routines, indeed the genome as a whole, could represent a chemical reflection; and that, in transduction from such archetype through an electro-dynamic medium, coiled *DNA* might act as an antenna, a radio mast, a conductor as well as a fixed databank of information. In this case for '*ancestral*' you would read '*archetypal*'. For '*phylogenetic*' read '*bio-logic*'. The greatest power resides in potential and the top switching routines will exert greater overall influence than the lower, more detailed, localised subroutines. On any *top-down* informative hierarchy for '*primordial*' or even '*primitive*' read '*top-level*'. For '*persistently conserved*' read '*critically decisive*'. For '*staggering inefficiency*' read '*economy*'; and for '*plan-less flounder*' read '*law of conservation of energy, minimum action/ maximum impact and least action (time multiplied by energy) to reach a predefined goal*'. The higher a principle (such as information, energy metabolism or reproduction) or stage in the expression of potential, the more 'persistently conservative' it will be. Variation and its details can be switched on later down the stream. The issue of seniority has nothing whatsoever to do with time (or chance). It has all to do with principle, efficiency and power. Nowhere are elegance, ingenuity and economy more evident than in developmental biology.

Who believes *PAM*'s insistence that parts of a body are, like its whole, without a purpose? Ask the sphinx. The origin of development of purposeful, programmed complexity will indeed answer the riddle of life. Of course, archetypal logic and its implications don't appeal at all to naturalistic sensitivities. What's invisible is fine as long as it's material and we just haven't seen it yet. So scientism's 'health and safety' rules require, before you think or act or tackle vital riddles, that you affix an evolutionary pair of goggles. Of course, with squint attached you'll take the '*Hox* homology' as evidence of common ancestors. You infer but have no evidence for the actual existence of such organisms. You draw up family trees. And some developmental biologists have, through the results of experimentation with high-level controllers, speculated that their subsequent monstrosities may indicate that small, incremental steps at this important site might generate macro-evolutionary leaps of change. In fact shuffling such genes merely shuffles outcomes in an abysmal sort of way. Displacement not novelty occurs. For example, due to the interference a fly's leg may form on a different, unnatural part of its body. Who would call this sort of meddling with the logic of instruction manuals positively productive, that is, macro-evolutionary in scope? No-one's ever coaxed another species out of *Hox* mutations. Every developmental gene in the fruit fly *Drosophila melanogaster* has been tampered with but no new species has emerged. You obtain a normal, defective or dead fly; and so with fish or mouse or chimp. Or men from chimps you might suppose. Why should natural genetic engineering's mechanistic tool-kit aimlessly succeed where scientific focus in the lab has failed? *What is evolution's real bind, its charming and quite plausible deceit that wearing squints makes straight*? It is that bits of 'logic' self-cohere; that coherent information 'grows itself' without a mind.

Is it, therefore, phylogeny or bio-logic? Evo or devo (devolved) devo? What way do evolutionary enthusiasts argue the accidental appearance of super-coded bio-logic? There is, of course, speculation a-plenty because, firstly, it *must* have happened and, secondly, fossils don't supply genetic read-outs. Nevertheless, it is obvious that the broad outlines of initialisation are critical. Wouldn't you expect 'high-level' body plans to be strictly conserved? In a journey from one place to another it is obvious that any mistake made early on will lead you further astray than one made close to the destination. Likewise decisions made by a general or a company director will have much greater strategic impact than ones made by the lower ranks. In any hierarchy a few high-level switches govern the expression of many later, increasingly detailed steps. In this way error, malfunction or mutation in primordial genes will have proportionately large-scale and dramatic effects. In biology if implementation of the developmental journey is sufficiently disturbed the body involved aborts. If the primary process or its genetic instruction is flawed (by random mutation or whatever else) the system shuts down. If such accidents debilitate or kill then how could such an obviously purposeful, logical, series of steps originate? The idea is that 'tinkering' (or 'random mutation') with master genes will, despite the overwhelming majority of fatalities, still allow sufficient change to filter through selection's deadly net. It's a time and numbers game. What can't this couple in imagination build?

Yet bio-logic is exactly what you don't expect of chance; and what you *would* expect of high class systems ordination. The immaterial factor of a program is its purposeful arrangement, conceptual sequence or information. This is why we say 'genetic code', a phrase by linguistic implication anti-evolutionary. For sure the cosmic maelstrom, being mindless, didn't know its luck when striking such 'intelligent' routines and subroutines as all development employs; but once such highly complex, integrated activators had somehow been minted then, you argue, couldn't evolution's higher, causal, evo-devo explanation have emerged? Could not high-level muddles (mutations of pair rule or tool-kit genes) reverberate throughout the hierarchical cascade that leads to phenotype? Could *Hox* clusters and their ilk have duplicated and, within the second and superfluous 'page', mutations have accumulated harmlessly until at last one 'clicked' into a novel operation?

No 'hopeful monsters', please! But it's conceived that shifts in egg coordinates or 'tweaks' to high-level master genes might occasionally do the trick. In other words, it is argued, some accidental new permutation of transcription factors might produce another organ, system or whatever - all coherent with the bunch they'd added to. Extra banks of varied main routines (serial homologies in bio-speak) might not require the linkage of a bio-engineer. In this way might even point mutations generate the wondrous innovations macro-evolution hangs upon? Dramatic changes might develop in an embryo's geography. OK. Proof is in the pudding. Test it. Take a 'living fossil' and experiment upon its master elements. Teach old genes new tricks. Twist its logic into some descendant form. You might get a lucky break and thus evolve a novel species in the lab. Or even one that now exists. Yet upstream changes, every author or programmer knows, trigger many slight adjustments down below. If it's not tuned logic just won't work – and an organism is an integrated egg-derivative. Thus, if you tinker serially you do so artificially. You use intelligent selection to steer towards achievement of a working operation. Nature's not like that. That's not evolution of material but of ideas.

Such innovation doesn't happen now for sure. Why should it any time? The great unanswered question still needs answering. **'Precisely how did what you 'tinker with' emerge, as a primal innovation, *in the first place*?'** How did the basic circuits for life's bio-logical complexity 'occur'? Where did coherent, immaterial information come from? Even single cells are complex but forms dependent on development abound in strata that predate the Cambrian explosion (Chapter 22: Types of Fossil). Cambrian Burgess Shale has yielded ten phyla, that is, types of body-plan including *Pikaia*, a possible chordate; but Chengjiang, estimated 15 million years older, has also given up chordate (*Haikouella lanceolata*) *and* a jawless fish (*Haikouichthys ercaicurensis* - named without apology!) whose anatomy is more complex.

Pass back to when life's blocks of logic must have been installed. Speculation provides, deep in time before segments, eyes, mitosis or anything else existed, mysterious Ancestors. **The evo-devo idea is considerably constrained by its fact that such a marvellous Ancestor must have been equipped with sufficient but unemployed genes for building complex bodies. In other words, it must have carried, in embryonic form, the genetic logic and the genes for bearing not only animal but plant, fungal and all other kinds of eukaryotic development as well!** Such formidable, prestigious 'forethought' was something nothing ever thought of - except as speculation in the minds of humans at the other end of time! Such speculators are persuaded (since it *must* have happened somehow) that single-celled Ancestral Colonies provided regulatory logic that would, with tinkering to *top-down* systems somehow even then in place, forge proto-organs. These nascent functionalities came to their senses gradually; piecemeal in an unknown order touching, eating, moving, replicating and so forth 'progressed' into a state of operational capacity. No proof but theory drives these wild interpretations of the facts. No doubt, as has been discussed, trivial changes can occur. For example, the size of a stickleback's fin might be varied epigenetically or by a random mutation - but the fin is there already! What about the *origin* of fins or eyes or any other organ, not just minor changes to them? What about invertebrate or vertebrate developmental cascades; where did whole bodies come from? Biology has now begun to understand the preternatural mechanisms of development; but evolution can't explain the stunning forethought that's embedded in their bio-logic.

In fact, the notion that immanent *Hox* control built up haphazardly but inscrutably veiled by swathes of time is, against the straightforward interpretation of *top-down* programming, entirely feeble. Carry, then, the feeble to the classroom! Unnaturally, and not for God's sake, select the strong for ban!

At least the development of eggs is not imaginary. Eggs renew; they create no novelties but inform real, fresh forms. How did they happen? You might try imagining how any egg evolved. Don't call your explanation scientific fact (nor does imagining a hundred thousand times build up the 'fact' of evolution). So did an egg precede its adult? *Vice versa*? Which or what?

Did an Egg Precede its Adult?

Only the shelled variety of egg has been fossilised. A lack of their shell-less siblings no doubt inhibits a thorough palaeontological but not a theoretical study of the evolution of eggs. In *top-down* terms of IT and cybernetics *every* egg is, in principle, both simple (a single cell) and complex (an informative compression of huge numbers of cells). This is because eggs *anticipate* their product, an adult form. They invisibly contain, in terms of information, the undivided potential necessary to achieve and maintain this form.

Potential precedes action. Potential being comes before expression of its 'egg'. And a form of potential called information channels the energy of an action from start to finish. Dialectically, (*sat*) potential energy is 'higher' than (*raj*) kinetic or (*tam*) exhausted forms. It is the battery whose charge, released, drives things downhill to sink, to finish, to their various ends.

There exist active and passive forms of information. Either it is metaphysical and creative (as in an idea or imaginative design) or a passive imprint of design (as, for example, works of art and technology, information 'stored' in the laws of physics, *DNA* code or an egg). Thus biological eggs are, as already noted in Chapters 15 - 21, information incarnate. They are chemical forms of storage and transmission but a metaphysical egg is an idea.

What represents an idea? Symbolic form. A symbol is realised as code or construction. A physical creation bears the metaphysical imprint of its creator. In the broadest possible sense, therefore, everything represents the idea behind itself. Even the forces, particles and events of nature are, from this perspective, symbols. Metaphysical egg precedes physical nature and thereby physical egg. The latter is a coded representation of the former. An egg is a special case because it represents the development of an idea, the act of creation itself. That is, it symbolises potential, the stimulus of fertile idea and its development to full term. What is the reason, purpose and meaning of a construction? To discover you inspect its source and try to understand its idea. A construction reflects its maker's mind. **Are you the construction of an idea? If so then what does this make you?**

What are plans but 'metaphysical structures' that promote the successful execution of their own objectives? A plan can be actively 'worked out' then stored, in memory or on some physical medium, for a period of time. If metaphysical precedes physical phase then active also precedes passive. The active origin of an idea is different from its consequent, passive storage in memory. The Dialectic calls the storage of nature's plans archetypal memory. Other names for an archetypal memory include 'metaphysical egg' and 'universal idea'. Phenomena are products of these numinous archetypes, these 'noumena'; physical objects and events occur according to the regulation of such 'eggs'.

You can't see a memory but have you ever seen a physical idea? An egg is as near as you will get. Take a dinosaur egg from the Gobi desert. Reptiles, unlike amphibians, lay eggs out of water so that the reptilian infant needs to be endowed with a watery environment and a food supply. A reptile's egg must therefore include a store of yolk and protein-rich albumin and be wrapped in a container made of the correct materials. This container demands a delicate compromise the penalty for whose imbalance is death. What applied to extinct dinosaurs applies to extant turtles, crocodiles and birds. A shell must be strong enough to resist accidental breakage but fragile enough to chip free. It must lose the right amount of water in ways that ensure the embryo neither desiccates nor drowns in its own metabolic water. Its size and nutrient content must be geared to embryo size at birth. And gases must be able to diffuse through pores which are the result of deliberately randomised packing of calcium carbonate crystals. Such a shell requires two special embryonic membranes - the amnion and allantois - to protect the embryo, allow it to breathe and act as a reservoir for the waste products of metabolism. Fertilisation must occur within the female before the shell begins to harden, necessitating concomitant changes in the urogenital organs and habits of the adult. Finally, the hatchling (whether bird, reptile or monotreme) needs a chipping tool to develop at the right time and place (at the front of its snout or the end of its beak) along with the right instinct to break out of its cradle. A chipper idea!

bottom-up	*top-down*
planless	*planned*
mindless	*conceptual*
uninformed	*informed*

A problem for the theory of evolution is one of *bottom-up* perspective. That is, the problem is its starting point. *This point grades information 'below' matter.* It is as if the prime factor in a machine, from which you could detect its origin, were its molecules. Such a mind-set wants to put 'simple' before 'complex', genes before bodies, simple-looking eggs (which are in reality complex in potential) before their expression, chickens. Chance-and-necessity before chipper ideas. Another problem is that, in its planlessness, it replaces preconceived archetype with scribble - scribble that, for no reason the scribble understands, 'works'. For anything biological to work the first scribble has to pass the triplex test of homeostasis, the simultaneous presence of energy metabolism, reproductive system and correct structural materials. Only thereafter is natural selection presumed to delete all but 'clever' scribbles. You might assert that, in this respect, materialism neither has nor wants any idea!

E cellula omnis cellula. 'Only from a parent cell does daughter come'. Nevertheless evolution contradicts this primal axiom. It sees cells derived from the most disordered 'egg' of a biochemical puddle; then it sees different types of cell evolve from precursors and from these, in serial order, multi-cellular complexity increase. It is as if you saw a television as, at first, chemicals, then an electron gun around which the screen, casing, power supply and finally programs accumulated. In a similar way it is possible to identify a cellular hierarchy which runs, in what seems a logical progression, from simple chemicals (such as carbon dioxide, nitrogen and water) through simple bio-molecules (such as amino acids, sugars and bases) to supra-molecular assemblies (such as ribosomes or enzyme complexes) and organelles (such as the nucleus or mitochondrion) into a complete cell. Just as a simple egg evolves into an adult, it is easy to presume that this hierarchy represents an original order of construction. **The theory of evolution is based on the seductively similar yet opposite concept of unplanned development.** You must believe that, *without* any coherent idea, design or program of implementation the simplicity of matter evolved into the purposive complexity of biology. **But where informed development is full of reason, by contrast uninformed evolution is 'developmental' nonsense.**

So phylogenetic 'egg' is physical. It is an abiogenetic pre-bacterium whose cradle is a puddle. After that a prototype bacterium, sex, development and adults *must* inexplicably but luckily have followed. At least (because its creator is genetic mutation and its editor natural selection) the system neither thought nor cared - like drunken serendipity that woke up as a man.

An egg precedes its chicken. Is the chicken simply an egg's way of making another egg; or an egg simply the fowl's reproductive *modus operandi*?

Did an Adult Precede its Egg?

Did fruit precede its branch? We noted that reproduction looks to the future; and thus sex *anticipates*. It is a conceptual process engaging machinery of irreducible complexity to achieve a target - generation after generation. Survival of the kind. Take, when an egg has been prepared, the next and crucial step - fertilisation. Many factors combine to create the spark that fired your development. At the critical moment a winning sperm penetrated the surface membrane of an egg whose surrounding layer, called *zona pellucida*, instantly thickened and separated from that surface. It formed an impenetrable barrier to all other sperm. *How could this complex chemical reaction, critical to the production of a viable, diploid zygote, have evolved by accident? And co-evolved with the whole multiplex of factors required for fertilisation and subsequent development?* We are entitled, since all factors cohere as pre-designated parts of coherent and purposive machinery, to a detailed account of how a series of accidents is supposed to have worked such an apparently premeditated marvel. Naturalists don't like the 'logical idea' but surely, up against their chaos, it should be considered?

incoherent system	*coherent system*
accidental occurrence	*purposive expression*
chaos	*program*
no-egg	*egg*

Which, after all, comes first - principle or practice? *In what order did the homeostatic triplex (information processing, the flux of energy metabolism and structural permanence that includes reproduction) 'happen'? Does phenotype precede genotype, hardware software, adult zygote, chicken egg? Is life bred from chance like chance itself completely uninspired - or, in design, inspired and awe-inspiring?*

Perhaps life, if it evolved, cannot be compared to well-conceived informational, political or industrial systems. *This is because no such system can arise by accident.* A computer system, especially, has to be complete and bug-free before it works. **It cannot be too clearly stated that the primary, basic component of any informed system is its plan.** A plan serves a purpose, the purpose is the goal and the goal is the end product. There is in life, as with any code or machine, an element of 'chicken-before-the-egg'. **In this view the adult is *in mind* before the egg.** The conception of 'egg' follows within a recycling, reproductive sub-archetype whose object is survival of the main archetype into the future. Concept aside, in the physical reality of either asexual (eggless) or sexual (eggful) reproduction there is always a parent form. How, therefore, can offspring precede parent? Adult, either conceptually or physically, must have preceded egg.

Is there childhood of a one-celled organism? Its single-frame development denies it one. Life must have begun with adults. So the problem remains; and its solution must impossibly reside in salty puddles (Chapter 20) or, absolutely unacceptably, an initiation, a single primer, a first imprint of archetypes on earth.

Egg and Adult Together.

What has been noted bears repetition. **Wherever an apparent 'chicken-and-egg' situation crops up the puzzle is resolved by the introduction of purpose, design, information and mind rolled into one - teleology.**

chance	*teleology*

Natural Dialectic takes a top-down, teleological perspective. It supports a systems approach to biology. A system's order, like that used by all human designers, starts with concept and finishes in material detail. Indeed, professional engineers and information technologists such as systems analysts are instructed by international design standards to program *top-down* starting with fully functional concepts called plans! The fact that technology insists on such a design process demonstrates that its physical systems nearly always contain irreducible (i.e. chicken-and-egg-*together*) mechanisms. How absurd, not to mention unprofessional, it would be considered if you were to gather any old nuts, bolts, sheets of metal and wires and 'worked' blindfold to try and finalise a purposeless non-machine with high level functions! *The Dialectic therefore finds no accord with a proposal that promotes the illogical, conceptionless and purposeless evolutionary construction of an irreducibly complex machine such as a celled body.* A teleological stack might read:

	tam/ raj	*Sat*	
	division	*Union*	
	part	*Whole*	
	expression	*Idea*	
	outcome	*Purpose*	
↓	*tam*	*raj*	↑
	physical aspects	*mental/ immaterial aspects*	
	directed operation	*instruction/ code*	
	target product	*development*	
	passive/ completed/ done	*active/ process*	
	structure	*function*	
	apart/ by itself	*part of*	
	isolation	*relation*	
	compartmentalisation	*cooperation*	
	component	*network*	

Wholeness is a primary dialectical principle. Both system-as-a-whole and its integrated components express, in their physical mechanisms, means to a conceptual end. *Concept, design and the internal logic of a system define both its function and the goods it manufactures as surely in biological as in industrial enterprise. Information comes first. Concept is encoded as a plan.* Code is symbolic and includes written, pictorial or any other expression of purpose. Next, therefore, follows actualisation of code: now motion and energy are involved in physical construction. Finally, against the ravages of time, the system will incorporate (with relevant checks, balances and triggers) repair and reproductive mechanisms. This order of creation is accurately reflected in the development of a zygote (or seed) into a reproductively capable adult form. *As far as chicken and egg are concerned, therefore, the simple answer of an information technologist is that neither came first. One did not precede the other; but metaphysical idea preceded physical outcome.* You make them both together with the same object in mind - in this case a self-reproducible biological information system or, in other words, a living being. As noted, therefore, adult and egg were conceived at the same time as parts of an integrated package remembered as an archetype. (*Sat*) precedent archetype is neutral and its expression (*raj/ tam*) polar. **You are one half of a metaphysical idea called 'human being'.**

Which End Is Bio-logic's Head?

Of course, bio-logic runs from metaphysic into physic; it runs from information through a carrier, code, into material body's fact; biologic's head, of course, is in informative potential. So how could bio-logical development evolve? What do you look up, or down, to? Did you derive from Top or toe; did your family climb a tree whose foot is rooted in an ancient puddle or drop from a bio-logical head-start, a conceptual sky? In short, did multi-cellular organisms evolve from unicellular ones or not?

Don't even think. Just add cells on and on; aggregate, differentiate; increase the size and complication of a form of life. Sounds reasonable, doesn't it? Until (Chapters 19 - 26) you actually learn the bio-facts. Do you remember (from Chapter 21) who enthused that matter was 'alive'? It was a storyteller called Ernst Haeckel (1834-1919), a self-appointed spokesman for Darwinism in Germany. In 1828 Karl von Baer, the founder of embryology, published hypotheses based on observations of the similarity of embryos of mammals, birds, lizards and snakes during their earlier stages. This makes sculptural sense. When a sculptor starts to carve a block of marble into a statue, he immediately reduces its potential. Broad outlines are established and, eventually, distinct details of his unmistakable work of art emerge. This is exactly the way development works to reduce generalised, shapeless genomic potential into a *pre-defined* conclusion. However in 1868 Haeckel reinterpreted the work of von Baer according to his own convention of a 'fundamental biogenetic law'. It was encapsulated in the phrase 'ontogeny recapitulates phylogeny'. This was supposed to mean that an organism, in the course of its embryonic development (ontogeny), successively passes through (recapitulates) phylogenetic (or evolutionary) stages passed through by its ancestors. This theory does not apply to the plant, fungal or bacterial kingdoms. Nevertheless Darwin himself saw it as fine support for his theory; and successive textbook repetition of its selective 'evidence of evolution' has meant that Ernst's war-cry has rallied generations of students up to and including the present to his evolutionary cause - although, however hard it dies, his rumour has long been discredited. How?

To support his case Haeckel faked evidence by both omission and commission of facts. By 1874 his tweaks had been declared invalid by Professor His. When charged and convicted of fraud by a university court at Jena, he agreed that a small percentage of his drawings were forgeries. He was, he pleaded, merely filling in and reconstructing missing links where the evidence was thin. In fact his main faults had lain not only in disingenuous copying (he used the same woodblock to demonstrate similarity between different classes!) but also in the selectivity of his evidence. The avowed intention was to demonstrate that early embryonic similarity showed common ancestry. He did not, however, start at the zygote where different classes of egg differ greatly in yolk content, size shape and in cleavage patterns and the organisation that prepares them for gastrulation. These primary stages actually refute his and, therefore, the Darwinian case. So he began at the mid-point when diverse early stages briefly and superficially converge before subsequent divergence into organ and body formation. Even then they are actually substantially different in size and shape but Haeckel minimised this fact. For example he omitted 'anomalies' and drew embryos in the 'tail-bud' stage of development as almost identical. They are not. Moreover mammals and frogs have limb buds at this stage but for the sake of 'evolutionary conformity' Haeckel 'air-brushed' them. A comparison of real embryos set against Haeckel's drawings made by embryologist Michael Richardson (Anatomy and Embryology Vol. 196 (2), ps. 91-106, 1997) clarifies the scale of his deception. Even the confession at Jena was, by this measure, a fraudulent diminution of this scale. Although the theory of recapitulation has been decisively abandoned by science for well over 100 years the will to believe despite the facts cannot be overlooked. In 1901 the disgraced set of drawings was placed, gill arches and all, into an educational volume and has, from that time, been parroted by less than perspicacious authors of high school textbooks. Therefore (how great the power of propaganda) some biology teachers still believe and teach it!

The spark of fertilisation trips cell division and development. In Haeckel's upside-down world a single-celled zygote represents our bacterial ancestor. He omitted to mention, however, that multi-cellular organisms differ considerably with respect to eggs/ seeds and ignored the following, mechanically essential 2, 4, 8 and 16-celled stages because there is no known organism, extant or extinct, whose adult body is (due to the facts of surface tension) composed of these quantities! He also left out the different ways by which those 2, 4, 8 and 16-unit stages are expressed. For example, mathematician D'Arcy Thompson noted that twelve patterns of segmentation were possible at the 8-piece stage. In fact, initial 'sketches' that lead to the formation of a blastula are different in amphibians, reptiles and mammals. Hence developmental courses diverge. They may vary even in closely similar forms so that apparent homologies may not, from their different developmental pathways, actually be so. Whichever way, such courses lead every embryo to expedite the features of its genus. For example, human definition is increasingly apparent from its glimmerings after about a month to clarity after six weeks.

A morula is a solid ball of cells. Let's reach 32, the number at which it becomes a hollow ball, the blastula. There are organisms that grow no further than this. Haeckel took a colonial organism like *Volvox*, a form of green alga embedded in spheres of mucilage, to represent the blastula. So did you evolve from *Volvox*? Except that, while some consider it an organism whose development is arrested at the blastula stage, others call it a 'colonial'. A 'colonial' results when a number of 'separate' types of cell with identical genome have chosen, for mutual benefit, to live together and some of them are specialised for the common cause. In this case about 20 posterior zooids (from a body count of around 10000) are specialised for reproduction. Is the *DNA* and physiological capability of the cell types quite different? Or is there a *Volvox* genome whose sexual subroutine is just expressed in a few cells? In the former case each cell-type would amount to a separate type of organism and *Volvox* would be your 'colonial symbiote': and in the latter a multi-cellular animal, a metazoan. It would be a case of different animals living together or a single, complex character. Neither way suggests an intermediate, genetic step to multi-cellularity.

There exist numerous complex kinds of unicellular organism and, equally, numerous complex cell types that are part and parcel of the single genomes of specific organisms. Above a certain size multi- cellular bodies need interlocking systems made of different cell types and organs in order to meet the informative, energetic and reproductive demands of survival. According to what kind of template are the shapes all made?

Is it heads or tails? What might an evolutionary toss-up entail? Which end is your head? Early cleavage does show two developmental characteristics from each of which you might derive a different phylogeny - a radial and, more complex, a spiral cleavage pattern seem to compose constructional antecedents for two major sets of animals. Those with spiral have blastopores that become their mouth. They show bilaterial symmetry and the fate of their cells is determined as they form. These include such body-plans as arthropods, molluscs, rotifers, annelids and other kinds of worm including siphunculids. Such animals have hydroskeletons and exoskeletons but no bony infrastructure. They are called **protostomes**. **Deuterostomes**, on the other hand, cleave in a simpler, radial way. Their blastopore becomes an anus and they include organisms which, except for sea urchins and starfish, also show bilateral (left-right polarised or mirror) symmetry. The fate of cells is not immediately determined (thus identical twins can form). You can also include arrow worms, acorn worms, sea squirts, lancelets and back-boned vertebrates in the deuterostomic gang. So what's the answer? *Phylogeny or bio-logic*? From what ancestor did such complete opposites evolve; through what missing intermediates did both come from another? And then does spiral cleavage somehow predispose to one sort of body-plan and radial to completely other; is one justified in drawing thick connecting lines between all groups because, of course, phylogeny was always how it happened? The inference being that dialectical bio-logic is, logically, unacceptable because phylogeny is, as Haeckel claimed, the real bio-logic. So that real logic depends on chance, on logic's lack. The philosophical lines are drawn! Whose shall we teach as truth?

A combination of molecular, structural and developmental homologies is adduced to construct phylogenies and demonstrate evolution. To this end each aspect of homology has, in any particular case, to correlate with the others. But the diversity of body plans and life-styles includes many anomalies that, as well as lack of a common ancestor, frustrate an evolutionary taxonomist. Octopus eyes are, for example, quite human-looking. Nevertheless by deuterostomy and not by optics we humans are, as chordates, currently classified with moss animal and sea squirt rather than a protostomic octopus. Again, the larval ('tadpole') form of a sea squirt exhibits a sliver of cartilage in its back that does not develop by an induction process homologous to that in vertebrates. It disappears during its programmed metamorphosis from a successful motile to sessile adult sea squirt. So did a mutant 'tadpole' really, as evolutionists claim, form the basis of vertebrate construction? Was your grandfather-removed a sea squirt? Do you remember the Ediacaran jellyfish? Is the fossil of an animal with apparent muscles, fin and head from the same rock assemblage a vertebrate (ABC Science Online, Australia 5-1-2005)? It can't be, can it?

According to the protostome/ deuterostome method of classification eyes, ears, filtering devices, buoyancy control and most other organs must have evolved separately in each group, that is, at least twice and sometimes more than five times! Such prolific 'convergent', 'analogous', 'parallel' or 'co-evolutionary' catch-all 'explanations' might alert you to query the possibly misleading use of such a scheme of relationship. The same problem of mosaic distribution of subroutines was made clear in the case (Chapter 22) of the duck-billed platypus.

In fact, the ancestor of protostome/ deuterostomes is unknown. Instead a purely fictional, worm-like organism called *Urbilateria* has been invented and plumped at a hypothetical branch between the pair. What ancestor preceded the split between *Cnidaria* (jellyfish, sea anemones, corals and so on) and the other two groups is also unknown. So is the one that preceded sponges and all other types of animals. Ancestors are, except in the imagination, in short supply. Indeed, it would take a separate book (for example, 'The Transformist Illusion' by Donald Dewar) to catalogue the hundreds of developmental and phenotypic anomalies that disrupt phylogeny but not bio-logical programming.

The stage beyond the blastula is gastrula. Imagine pushing your finger into a soft, hollow rubber ball. You would create a pocket or 'invagination'. In a cell this operation is called gastrulation, the hole a blastopore and the two layers of rubber, one inside the other, ectoderm and endoderm. The positions of cells in the blastula that give rise to germ layers and their migration patterns during gastrulation differ greatly. This strategy of development, normally a passing phase, also arrests in the form of two-layered 'diploblasts' like sponges and jellyfish. Take, as Haeckel did, a sponge to represent the phase. Although loosely but still socially organised, it contains over thirty different types of cooperating cell. If they are broken apart its parts can reform as the original whole but the creature normally develops either asexually (by budding off) or sexually from a single cell. In other words, all cells are coded in sponge *DNA* and the organism is definitely a self-consistent whole. This would fit, as does the notion of a genome, with the idea of 'holographic' archetypal influence on every cell of every single body. *Further than this, the origin and destiny of sponges is unknown. Nothing is thought to have evolved from them.* Therefore some choose to speculate that the father or the mother of all animals is tiny *Trichoplax*. They say evolvability evolved from this simple 'placozoan' with only four cell types but a genome of 100 million base pairs and about 11,000 genes of which about 80% tally with other metazoans. No fossils of this thin, almost transparent jelly tot have been found; indeed, it has never been observed outside a laboratory. If elephants did not descend from tissueless, organless *Trichoplax* and if sponges include very specialised cells (e.g. choanocytes) so coelenterates like jellyfish, which are 95% water and whose origins are 'obscure', include multiple types of eye and *nematocysts* (complex sting cells). As well as being venomous these superb and innervated examples of 'water architecture' can sometimes pulse with coloured light displays; they bioluminesce, they fluoresce and one genus (*Erenna*) does both! Such ability is found in a number of unrelated organisms (e.g. bacteria, fish and flies) - was such independent, multiple re-evolution by convergence or, as the Dialectic would presume, is its appearance 'bio-modular'? 'Simple' sponges and jellies both combine sexual and asexual reproduction with complex life cycles. It is unclear how any of these features could have self-constructed in gradual, integrated and therefore useful steps. Nor have accurate evolutionary flow-charts, relating gene sequence to nematocyst and other phenotypic constructions, been evolved to clarify the issue. It is not obvious how water is gradually fossilised but jellyfish, whose mouth *is* their bum, certainly appear abruptly and fully-formed in Pre-cambrian strata. They differ structurally little if at all from their modern counterparts so that, functionally, we can presume their reproductive strategies (e.g. alternation of generations, larvae, other developmental phases) and other complex physiology was also up and running. All coelenterates, having two cell layers, are called diploblastic. *Whence did their very different classes evolve? Nor is anything known to have evolved from them. The evolution of all metazoa is, if not divine, definitely not divined.*

Nor is there any evidence that triploblastic evolved from diploblastic animals (like jellyfish and sponges) - although there *must* have been an earlier ancestor. As well as endo- and ectoderm triploblasts (like you and me) have a third, mesodermal layer. When mesoderm divides into inner and outer parts it creates a body cavity (called a *coelom*). This cavity is of functional significance in respect of movement but, especially, because it provides the enclosure for various systemic organs (such as lungs, heart, liver etc.) that are needed to support a large, multi-cellular body. You can classify the steps of triploblastic development in the same way as a systems analyst draws *top-down*, hierarchical trees or branching menus - from zygote through three generalised germ layers to the specific, local details which arise in each. On the other hand (*fig.* 17.4) you can use concentric rings to express the fifty-six or so hierarchically coordinated, algorithmic steps by which a human zygote (like you were) evolves from *morula* (with no germ layers) through *blastula* (with two), *gastrula* (with three), *embryo* and *foetus* into baby, child and adult form. The process everts from 'inside out'. Inside information, such as *DNA* database code, chemical gradients, environmental cues from the egg and, from a *top-down* point of view, overall archetypal guidance, cooperates to wrap three concentric 'germ layers' around a hole (the archenteron and blastopore) at the centre of your cylindrical outline. This hole will become your alimentary tract. Around it are arranged organs sprung from endoderm, inner mesoderm, outer mesoderm and ectoderm. From these reference points subroutines branch as the cells divide and the concentric germ layers take shape. Once positioned in a germ layer a cell references subroutines related to its position, its neighbours and its final destiny. Internal organs derive from inner layers, outer (such as skin and bone) from outer. And, of course, room for all these necessary organs is provided by that body cavity (the *coelom*). It may be argued that various kinds of coelom more reflect an animal's life-style than any progressive, evolutionary 'plan', that is, non-plan. How can lack of plan finance the highly schematic, dynamic architecture of any multi-cellular development? Could one type of growth program be transformed by incremental steps into another and, if so, why don't we see it happening now? *In fact the hypothetical evolution of development and, with it, multi-cellularity is so difficult that you need to be very clever to ascribe such purposive logic to the indifference of chance.*

A 'higher' vertebrate would have to pass through a fishy stage and this is supposed to be illustrated by a two-chambered heart and what Ernst Haeckel prejudicially called 'gill slits'. These slits are actually a series of five or six bulges with furrows between them. Such furrows develop in the neck region of all vertebrate embryos but never in reptiles, birds or mammals assume respiratory function or resemble the gill slits of an adult fish. If your head needs a copious supply of blood before a complex heart can be developed, either you build the complex model alongside a simpler, makeshift pump or you engineer a four- from a two-chambered heart while it is still working. The latter is what, using a series of ingenious contrivances, is made to happen. Such is embryonic engineering that the bulges are an essential *intermediate stage* (like scaffold, moulds or temporary supports in house construction) in the guidance and penetration of blood vessels from the lower abdominal to the middle, hind and head regions of the embryo. After this usage they are converted according to an organism's life-style in air or water. In a fish (which retains a two-chambered heart) they become gill arches and jaws, in others they herald parts of the ear, hyoid apparatus (lower mouth) and neck: these include the important thyroid and parathyroid glands. In fish these vessels divide, forming a network that extends to the gills: in other vertebrates no division occurs nor is the character or respiratory function of gills assumed. There must be, according to prediction by a bio-logical argument for engineered design, a mechanical or physiological reason for every feature exhibited by every type of embryo at every stage of its development and in most cases we know what it is. The structural changes through which an organism passes as it develops follow the shortest, quickest, most direct route to the target adult form, compatible with the immediate necessity of living while doing so. *They are efficient, economical and targeted, always a sign of intelligence in the vicinity.*

Nevertheless, from a phylogenetic angle, if a gill won't do then what about a tail? Horses have tails but there is no embryonic, recapitulatory trace of five toes (the pentadactyl limb) from which hoof and two splint bones should have devolved. Humans have no tails but it is similarly mooted that our proto-vertebrae give rise to an embryonic recapitulation in the form of a tail about 4 mm. long (as long as long-tailed animals during these same weeks five to eight of gestation) with connected muscles. This 'tail' later retracts and the *os coccyx* is left. What is the logic of this appendage? Is it simply part of the dynamic development of a vertebrate body-plan whose presence induces the appropriate curvature of *os coccyx* so that man can walk upright, sit comfortably and, with necessary muscles there attached, defecate properly? What would life have been like before the retracting tail retracted properly? At the same stage of development the embryo exhibits a post-anal length of gut. This, like the 'tail', shrivels up. Of what, however, is it a relic? What evidence is there of caudate hominids or its ancestral use?

If a tail won't do then how about a bottom at your mouth's end - in the manner of a jellyfish or sea anemone? It is obvious that what you find you want to classify. To do this you choose criteria. Biologists choose a variety of traits from which to infer relationships. Such relationship may accord with phylogenetic or **bio-logic**al presumption. If the choice is phylogenetic then various sequences are arranged as evidence to justify the speculation. For example, you might choose body symmetry. This is a good way of comparing and relating body-plans but, because otherwise closely related organisms may have different symmetries or lack homologies, this particular criterion is not considered evolutionary. If a functional characteristic pops up here and there without obvious evolutionary lineage we call it 'mosaic'. Or 'analogous' (as defined in Chapter 22).

One way or another, it is the always the same story. A modal necessity called '*must have*' is the real imperative behind the evolution of complexity by chance. In his TV series 'The Life of Mammals' the evolutionist Sir David Attenborough took a safe, orthodox infrastructure within which to describe the alleged order in which this 'winning design' developed. He took a cradle of life, their reproductive structure. Perhaps unaware that 'dogs' with the grand name of *Repenomamus robustus* ate dinosaurs or that Jurassic moles and beavers were to come to light he proffered *Morganucodon,* so-called from a few broken skull bones found in Wales, as the first mammal - an egg-laying 'shrew'. Was it mousy? Did it lay eggs? How? Why? What next?

At present egg-laying mammals, Australian monotremes of dubious descent, are represented by a couple of extraordinary *mosaics* neither of which are remotely like mice. Monotremes are endothermic (warm-blooded) with characteristic mammalian body-plan, milk, hair and jaw.

One is the type *Tachyglossidae*, 'fast-tongued' echidnas or spiny anteaters. These, like placental pangolins and anteaters, have no teeth. Instead they have a long tube-like nose and 'hoover' insects, mainly ants, with a long, sticky tongue that protrudes from a tiny mouth. Many of their hairs are enlarged to form spines that are moveable by muscles and cleaned by an especially long hind-claw.

Look out! Here swims our duck-billed friend again! Entirely unrelated to the echidna in body-form, life-style and or any evidence of blood relationship except a common reproductive scheme is *Ornithorhynchus anatinus*, the platypus. Platypus systems are either reptilian, as in reproductive tract and poison fang-claws on the heels, or mammalian, as in middle ear, milk and fur or, as in its otherwise bird-like but electro-sensitive 'beak', like the spoon-billed paddlefish or hammerhead shark. Indeed it swims like a fish - with webbed toes like a duck or a beaver. Its juvenile teeth are replaced by horny plates and it has, as do snakes and lizards, only one functional ovary and oviduct - the left ones. In the *echidna* both are functional as with crocodiles, turtles, marsupials and humans. Yet female monotremes have, in common with the placental beaver, an avian feature called a *cloaca* (common urogenital and excremental tract); the male has undescended testes and a penis which is not used for urination but only sperm transfer. Birds, reptiles and monotremes lay shelled eggs; in each case the young hatch using an 'egg tooth' to break open the shell. An *echidna* lays a single egg into her own nipple-free pouch, which develops in the breeding season. For the female platypus instinct is entirely different. She builds a burrow and lines its nesting chamber with damp plant material. It is here she lays yolky eggs and rears her young. The *DNA* complement of monotremes is nearer human (with c. 97%) than is marsupial (81-94%).

Meanwhile marsupials are of equally dubious ancestry. Many of them 'converge' with placental body-plans (or archetypes or homologies). The females are nippled. They have a permanent pouch and paired vaginae with a third central vagina formed as a temporary canal at the time of birth. By contrast a placental vagina is simple and single; amniotic development in the womb is quite different from the marsupial 'conveyor-belt' system with one joey in the pouch while another is waiting to be born. We might continue but two points are made. The systems in question are distinct. Each sort of womb works, is entirely successful and shows absolutely no sign of a forerunner. **'Better' or 'worse', 'primitive' or 'advanced' explain nothing more than a human tendency to categorise and compare**. Sir David, himself a norm for not believing in archetypal bio-logic, knows that each genome codes for its own and that disturbances to its integrity would have caused disruption and malfunction rather than improvement to the point of a 'watershed' leap to construction of an entirely different kind. While minor variations within typological groups occur continually and are often touted as evolution-in-action, we see no evidence of the *real* business - organisms which exemplify steps transforming one subroutine into another (e.g. from monotreme to marsupial sexual apparatus). *Still less do we see nascent new organs or hints of the origin of fresh, emergent body plans*. If, as is claimed, evolution is an on-going, creative process, we should. In fact, neither Sir David nor anyone else has evolved an accurate, hypothetical, step-by-step flow-chart to show how one sort womb could conceivably have transformed into another. He also knows there is neither fossil nor contemporary evidence for such a leap. Simply, according to the Darwinian bible he was taught at school and college and his colleagues consensually believe in, it *must* have happened.

Common ancestors ought, according to Darwin's definition of homology, to follow similar developmental pathways. Denton says (*op. cit.* p. 146):

"There is no question that, because of the great dissimilarity of the early stages of embryogenesis in the different vertebrate classes, organs and structures considered homologous in adult vertebrates cannot be traced back to homologous cells or regions in the earliest stages of embryogenesis. *In other words, homologous structures are arrived at by different routes* (my italics)."

For example, the alimentary canal is formed from different embryological sites in different vertebrate classes. Pentadactyl forelimbs form from different body segments in different species. He quotes other examples and notes that there are many examples in insects of homologous organs and structures arrived at by a wide variety of radically different embryogenic routes. *The upshot is that organs are, even in closely related organisms, induced and organised to develop in different ways from different places in the embryo without forfeiting their eventual likenesses*. Therefore the roots of the Darwinian definition of homology, whose phylogenetic implication is an argument fundamental to the concept of evolution, seem axed.

From a top-down point of view, however, homological organs and typological body-plans express archetypal routines. Permutations of these appear like mosaics, sprinkled wherever useful across the biological board. While phylogenetic anomalies abound diverse and beautiful arrays (or radiations) of structural, functional and developmental archetypes, in all kinds of modular permutation, fulfil their purposes in sea, air and on land. To try and connect their *mosaic* patterns into in a 'blood' relationship born of 'convergent' accidents is to engage in an extended just-so, must-have compilation. *Such mosaics are at least equally rationally described as varied permutations of the anatomical, physiological and metabolic subroutines that cluster the core processes of energy metabolism, reproduction and their informed homeostasis for survival. Informed from the top, expressed from the pre-physical, born of the Centre what is life but all viable combinations of psychological and physical characteristic? What at heart, despite its differences, but a family from the same Logical issue?* **'Different but not evolved' is a reasonable analysis; nor is such a positive, subjective attitude towards life primitive, uncivilised or a waste. The reverse.**

Just as the facts suggest a sexual archetype, permutations of whose subroutines are adapted for bespoke operation throughout the biological world, so they suggest (diametrically opposed to Darwinian chance) the logic of a developmental archetype and its variously adapted modules. From its conceptual principles derive the coded software and dramatic hardware of biological development. From its executive practice flows the efficient, economical and purposive differentiation of each specific, multi-cellular organism from its egg.

T o summarise, it seems a little presumptuous to offer a blanket explanation of the global evolution of animate nature over a possible few billion years (called phylogeny) when the phenomenon of individual development (ontogeny) is not well understood; and when the former, which we can't see, is supposed to have happened by accident whilst the latter, which we can, is clearly *targeted* towards procreative adulthood. Yet frequently, in books and TV documentaries, analogy is drawn whence unplanned evolution (or phylogeny) might seem to mimic planned ontogeny. Starting from a single cell, the story runs, life's 'seed' logically develops such parts as an engineer might reasonably expect to satisfy the needs arising from 'improvement' of bacterial viability! Lo and behold. What we conceive is needed is just what we find! Yet evolution (without program but getting what it 'needs' in some Lamarckian sort of way) and development (tightly controlled by program) are chalk and cheese.

For evolution information is unplanned; there's no forethought in its accidental carrier, *DNA*, at all. Dialectically, however, information is potential that precedes its physical expression. Is this not what actually shows? Egg *anticipates* adult. The process is, in its every minutely detailed step, conceptual and pre-programmed. It is almost unimaginably highly informed. Science has not begun to synthesise a zygote from simple chemicals. Yet, as every doctor knows, even minor mistakes in the program can lead to catastrophic problems. It is therefore even more startling (especially in terms of information technology or engineering) to hear a claim that the process gradually accumulated up by chance. Who disputes that biological systems are purposive? *What is purposive is conceptual - as is every artefact of man. Irreducibly complex mechanisms can degrade from a working start. They can neither self-construct nor upgrade without even conceiving the start-point or anticipating the end product*.

Indeed, has evolution itself evolved to maturity? Fully developed? Run out of 'ideas'? While biological organisms undergo continual variation (called micro-evolution) are there any evolving, nascent organs? A hand, a wing or bat sonar must, among thousands of other biological concepts whose technology we admire and copy, have seemed odd while they were purposelessly emerging. What radical new 'concepts' is chance pulling slowly, blindly from its hat? Why are major novelties (e.g. extra eyes at the back of the head, radio reception or locomotive wheels) absent from our biological radar? Why do organisms seem to all intents and purposes fully evolved? If real innovation's gone away then why is that? If evolution's running round within the compass of a type then what has stopped its senseless creativity?

Growing Up

Brains, butterflies and growing up - how do they fit together?

One has to ask how, without the benefit of plan or anticipation of a target adult form, various immature phases took hundreds of millions of years to accrete. *It might be noted that before an organism could 'create' any new stage of development it would have first to reach its own adult limit then add that stage.* This is because it has to reproduce to create the relevant mutant offspring. Such order is patently absurd. Unless you generalise from the unusual pathology of paedogenesis (the precocious development of sexual maturity in juvenile forms) or, as in the case of a defective salamander called an axolotl, the persistence of juvenile features into adult form (neoteny). In fact, the axolotl and its mature form, a black and yellow salamander, were at first classified into different sub-orders of amphibian until, one day in a Parisian pond whose water had dissolved sufficient iodine, the previously arrested metamorphosis took place. At last the two were known as one. The fact is, however, that the system was in place and a simple 'broken fuse' had been replaced. The question isn't of repair but systems *origin*. In this case, one of the very few that's proffered as an explanation, no evolution whatsoever happened. How likely is it, therefore, adulthood evolved by paedogenesis; or that hundreds of thousands of generations of precocious, procreating pre-pubescence then pubescence gradually evolved maturity in flowers, fungi, trees and animals?

You are not a salamander but aren't you a neotenous great ape? So that the round head of a baby chimp, which more resembles yours than its parents', might show where you came from? It is possible to parade a sequence of skulls to scientifically justify this sort of speculation. But is such intelligent arrangement near the actual truth or not? If your head is deemed more 'junior' than a senior ape's what of all the 'lower' animals? Could you pursue this line of reckoning until you had conceptually evolved the whole ancestral tree of life on the back of a staggering yet persistent set of immaturities? Or is this just another just-so speculation? *In fact, reproductive systems are 'soft' and there is absolutely no fossil evidence, either molecular or organic, for the gradual appearance of any developmental strategy or its tactical, intermediate series of immaturities.*

Is nothing sacred anymore? So pervasive is the notion of competitive selection that some, ignorant of its overwhelming bio-logic, have even speculated that pregnancy is a zone of conflict in between a mother and her child! The lens of evolutionary perspective sees uterus and foetus fight it out. Of course, sufficient resolution must have been there from the start - an indication of design. And compromise over resources between 'warring' factions can just as easily be seen as homeostatic maximisation of 'loving' allocations. Share purposely and share alike; thus the metamorphosis of love into a child. Which, competition for survival or maternal sharing is the less imaginary view - a wholesome view that's easily applied to all developments in life on earth?

I mentioned intelligence but did I remark on brain ? The word reminds me of a miracle! Observe the generation of, on average, about 5000 neurons per intrauterine second and five times that number of ancillaries such as glial cells. O happy happenstance that gradually transformed molecular development to make, mindlessly, a brain! Note that such prodigious activity runs as smoothly as any well-oiled Victorian or silicious Elizabethan preconception towards 100 billion well-connected informers mounted in a three-dimensional kilo-and-a-quarter jelly. From primitive streak through neural groove, tube, encephalisation (into forebrain, midbrain, hindbrain) and its first brain waves - at seven weeks - to the flowering of foetal, infantile and adult masterpiece observe unparalleled power and beauty. Observe information neither better expressed nor, for information's sake, compressed anywhere in the universe. Observe an epitome of intelligent interaction. Chance? No chance! It is a spectacular illusion that such an awesome array of atoms can, by their own addled conception, contemplate themselves! Even a brain is nothing by itself. It is 'cockpit controls', the medium by which our mind and body interact. It is nothing without the vehicle on whose behalf it mediates - and which, attached in parallel, is built around. Head on shoulders, miracle on miracle each making whole truth more spectacular by far.

A Clap of Fragile Wings

How does chance evolve a path that targets an end-product? How did any metabolic pathway prophesy its own construction or feedback control; how (as, analogously, in the logical, consequential proof of a mathematical theorem) did thousands of correct steps on the path to reproductive adulthood arise by accident? It strains credulity; but not the woolly-thinking of a man committed to materialism. Wait and see! Time's a wonder-worker. No proof needed; neither target nor solution anywhere in mind-sight. Yet its miracles make possible impossibility and what is probable soon morphs to certainty.

Credulous faith conspires with plausible imaginations to explain whatever fact you choose; weak arguments are dressed for public presentation; serious story-telling fills in gaps until you'd think the problem solved. This 'must-have-somehow' stunt is pulled continually except, with metamorphosis, its bluff is called spectacularly. Butterflies have always thrown Darwinism, in an arm-lock, in a flap and on its back.

Development is a continuous process. If, however, transformations of abrupt and discontinuous appearance are observed they are called metamorphosis. Metamorphic transformation has been common since the earliest times (in Cambrian rocks). Most marine invertebrates involve at least two stages with very different types of body - larva and sessile adult. Do you remember (from Chapter 18) the 'immortal jellyfish called *T. nutricula* that can cycle between adult and immature forms? In echinoderms the modules always overlap. A juvenile grows within the larva, pops inside out and, thus 'birthed', absorbs the latter. Have you, however, lived inside two bodies both at once?! A starfish, *Luidia sarsi*, takes the process to extremes. The juvenile form, having detached from its immature larva, takes a parallel life until this 'parent' dies. Of course, each stage of each metamorphosis involves the same genome; many genes (e.g. those for respiration) are used all the while but others are stage-specific. There's clear program permutation - clever routines modulate developmental sequences and engineer requisite parts. Why invoke the agency of chance - unless arcane hypothesis is in its own right fun? Vertebrates such as frogs and salamanders change from egg through larval to adult stage; fish can change sex and, in the case of floor-dwellers, usefully rotate eyes to the same side of the head; but in insects both semi- and full metamorphosis occur. In this developmental sense men are simpler (less evolved?) than insects. *The transformation from egg to beetle, bee, fly or, more strikingly, a butterfly illustrates a pattern of development that defies a gradual, evolutionary explanation.* Entirely different-looking phases, each perfectly formed for function, serially erupt. First egg; then, for a 'childish' caterpillar to become an adult moth or butterfly, it eats and moults. It keeps moulting exoskeletons (which are flexible like cat-suits and yet give it shape) for larger ones that form folded underneath the smaller outer sheath. At the last and largest size skin is shed by delicate manoeuvres revealing a cocoon, a chrysalis in which the future hangs. Then caterpillar body parts dissolve and build again into a butterfly called the imago that, emerging after several days, inflates its lovely wings by pumping blood into their veins.

Both Darwin and Wallace developed the notion of natural selection but the latter (from his study of butterflies, mimicry, birds and coloration) deduced an 'overruling guidance' or 'natural teleology'. He was clearly at philosophical odds with the former. By Darwin's book the process is mindless. Natural selection represents an anti-god that, making purpose 'purpose', thereby deletes intelligence from natural design. How, though, might a 'random walk' composed of gradual mutations - without any mechanism to hold in stock currently useless candidates which might later become useful - slowly accumulate the myriad, very complex steps towards adult, reproductive form? What species of mindlessness might take those thousands of decisions needed to store the 'right' accidents - without any of which immediate extinction would beckon? Could you squeeze coherence like this out of unpredictability; can you squeeze immaculate, conceptual conception, in the name of evolution, out of Lady Luck?

How, if such a birth won't work, does an ugly caterpillar gradually evolve anticipation of its adulthood? This anticipation's in the form of not a single but several provisional, pre-coded and coordinating 'discs' to generate the beautiful imago. Having had his useful mandibles (and thus his food supply) dismantled by mutation and the other parts dissolved into a soup how did our brave adventurer first retrieve, by way of an incredibly complex network of genetic switches, new body parts. What fairy wand touched all the right mutations just in time to breathe again? How, like Houdini, did butterfly-to-be cheat death and resurrect preordination in the form of archetypal discs by metamorphosis into proboscis, fresh and different digestive system, graceful wings and all the other segments of a butterfly - before he's even had the time to die? How did our friend know how to wind a chrysalis and from that shroud resurrect, transformed, into a wholly different shape of life thereafter? Now check the wondrous microstructure of coloration that, using multiple pigments and reflections from precise arrays of mirror-like scales, are revealed as iridescent wings unfold.An epitome of engineering - watch the fluttering ascension of a butterfly. Which stage could, without extinction, be 'experimental'? Could humbly waiting as a caterpillar eventually evolve pupal development? In whose cocoon enzymes 'know' how far they can dissolve the caterpillar's parts before transfiguration to a healthy butterfly. How long did some ancestral pulp hang round inside a chrysalis (that came from who knows where) until a suite of chance mutations magically (how else?) re-programmed 'mush' into the concept of winged flight? How, in fact, did pupal death rise straightway (not even in a generation) to a form by which type-butterfly appears? One has to ask how, without the benefit of plan to reach anticipated adult form, serial immature phases took hundreds of millions of years to accrete. *It is noted that before any organism could 'create' any new stage of development it would have first to reach its own adult limit then add that stage.* This is because it has to reproduce to create the fresh, 'advanced' offspring. Such order is patently absurd.

How, moreover, did mutating chemicals construct, in order that their body fly, exquisite, elfin wings? Defenceless butterflies cannot, moreover, shape their lovely fluttering. They can't wish for protective mimicry, camouflage or any other highly-coloured trick. Thus whence emerge these pretty patterns (mostly different on male and female forms), clever copies and anticipatory skills? How do genes create psycho-illusions and why did natural evolution not sieve out non-utilitarian, aesthetic senses such as the appreciation of beauty, religious delusion or nerves replete with mathematical rationality? *The PCM lacks any answer other than most feeble to these questions.* There you have it. **Butterflies are symbols of nemesis. At the clap of silent, fragile wings Darwinism logically dies; a giant is slain and at the same time flutter flags of life's innate, original intelligence.**

Let's express this nemesis another way. Beauty's slumber in its pupal sheets takes but a wink against the time required for evolutionary transformation. On close examination metamorphosis is shown, like *all* morphogenesis, to be entirely non-random; it is precisely pre-programmed, that is, an intentional process that works from invisible potential (prior information) through to visible adult result; Natural Dialectic has suggested this conceptual process is dependent on an attractor (also called an archetype) that's lodged as memory in universal mind. It involves transformation whose high-level regulation and operative mechanisms will be biologically discovered; but whose origin will remain forever rationally inexplicable in evolutionary terms. After a butterfly's egg has produced its larva, many of the latter's tissues die and are reabsorbed to provide 'raw' material for the construction of an *imago* (adult form). No control is ever lost. The adult potential is arrested in visible biological subroutines. Undifferentiated but already pre-determined cells are clustered as specialised 'data-items' that are called 'prearranged imaginal modules' - those 'discs'. Like a 'sub-egg' each disc corresponds, in a way that predicates an information cascade, to an adult part such as an antenna, leg or wing. If you don't find Cambrian insects, you do find pre-Cambrian jellyfish that, like modern ones, have gone larval stages; and if the fossils aren't interpreted as jellyfish, still, how did they develop? *If development (including metamorphosis and imaginal 'sub-eggs') is not preconceived then what is? How do extensively coherent, systematic programs come about by accident? If only one occurred the other, natural way that programs do then all could.*

Signs of obvious anticipation always flatten evolution. They squeeze the theory's time to death and thus compress it to impossibility. How can forethought be by chance? When is a plan not a plan? Is concept the same as lack of it? Never mind the metamorphosis of sleeping beauty to a royal butterfly - try frog to prince! Frogs never leapt in on the back of chance. From egg to fish-gilled tadpole then, after almost every system and associated organs is reworked, behold a transformational epiphany! Cascades of information have, layer upon biological layer, combined in a precise, sequential program to produce a croak. Is 'ludicrous' a word too weak for croaking that a frog evolved? Yet such hypnotic spell is cast upon the brotherhood of science by a religion, scientism, so that many still think metamorphosis somehow evolved. What choice is there?

The fact is that all instances of metamorphosis (of which development in general is one) are another Darwinian black box. Neither fossil evidence supports nor systematic explanation is proffered as to how such an engineered process might have come about in random steps. It has to work, say, for a tadpole or a caterpillar through pupa to adult butterfly, first time. So does the rest of the brilliant array of insect tricks. What adult insect, for example, has or ever could have had flight systems (not just wings that would be useless alone) offering insufficient aerobatic control? Fossil dragonflies found in coal measures an estimated 225 million years old had almost precisely the same superb, four-fold wing structure as their modern counterparts - the only difference being larger size. Such similarity with modern counterparts pervades the insect fossil record. *In a metamorphic but not in a macro-evolutionary, transformist case who can afford not to allow rational bio-logic? It is 'materio-rationalists' for whom the cost outweighs the benefit and therefore who, unlike the dragonfly, hang their wings upon a flight of fancy - evolution theory.* Its flight is haltere-less - unstable in both balance and direction. In fact, it is scarcely surprising that a world-class lepidopterist, Bernard d'Abrera, in his definitive 'The Concise Atlas of Butterflies of the World', scathingly and in some depth dismisses the theory of evolution.

Coming of Age.

Reproduction follows reproduction. Of course the target is an adult form. Reproductive adult rises from a zygote, sexual maturity evolves from infancy, a child comes of age. What happens when you climax, when you reach the tops of hills? Could another kind of age be coming, could maturity devolve downhill? Wisdom cannot counter gravity; intelligence cannot avoid old age. We know that, always, energy runs out. One way or another life unravels. Senescence and exhaustion wane into a very unproductive moment - anti-climax, bathos, death. Full stop. You come of age, you start to age and die. Expiry comes when age has had enough. Must biology of aging slope inevitably towards that valley's depth? You might or might not think so.

Bottom-up, survival of the fittest is the game. Is the prime 'purpose' of natural selection not survival, prodigious reproduction and, where a body is simply the genome's way of making more genomes, genetic immortality? There exists, therefore, no selective 'logic' for aging, no 'rationale' for menopause or death. Wouldn't you expect the genius of evolution to have stumped up universal genes whereby longevity would complement fertility? You might even dare to think that immortality was natural selection's end-game; that perfection, when the bugs are all ironed out, should yield a never-ending set of years. Checkmate against extinction. In this view, therefore, death is a reflection of the failure to obtain 'perfection'. Such failure is in fact ascribed to the indifference of selection towards unfavourable effects that only show up after reproductive age. Youth-happy evolution nods through mischievous traits as long as they wreak havoc only after reproduction's age has drained away. This view logically identifies the elderly, like the sick or infirm, as a waste of time, resource and energy. It casts senescence in the role of a 'genetic dustbin' of post-menopausal difficulties that, unless grandfather takes a nubile bride, can't reproductively recur; but it fails to explain, with body hourly growing older, what 'clock' tolls fertility to fall away - the 'clock' that natural selection should have best reset forever.

However is not immortality, as Chapter 18 started to explain, a fact - one that resides in reproduction? Group immortality through the agency of genes 'divine' - genes round which successive mortal coils are wrapped but fade and fail - is obvious all around you. An individual dies but nature, at the time of climax, willed his or her inheritance, a package wrapped round genes and called a child. Is it thus extinction can't survive and *DNA* can multiply?

There are, for example, life-styles 'aimed' at sacrificing self for greater glory of the group's long trek through time. You have small, short-lived organisms called 'reflex ovulators' that can, in a reproductive frenzy, generate a multitude of offspring, care very little for them and soon die. Rabbits are like that or, in the extreme, a group of Australian marsupial shrews called *Antechinus stuartii* whose male members die after a couple of weeks' non-stop copulation.

On the other hand larger, longer-lived 'cyclic ovulators' (such as tigers or humans) produce less offspring and care more. For real age investigate such plant 'immortals' as the bristle-cone pine (five thousand years or more), creosote bushes, vegetative reproducers and, indeed, the persistence of any unicellular divider. Why, if it keeps dividing, shouldn't a cell's cytoplasm and, especially, its nuclear throne-room be considered the 'immortal heaven' of survival-selfish genes? The boundary of heaven on earth would thereby, on investigation, turn out to be a lipid membrane! And the message through that nuclear rood-screen only passed by priestly *RNA*!

Count atoms ageless but do nucleotides, chemicals that mint fresh bodies, ever age? They do and are replaced. Thus what, if anything, makes lines of nucleotides called genes grow old? Mutation? Is it defective information, a conceptual deformation, loss of order and, therefore, of biological control? So that evolution casts mutation in a strangely contradictory way as a harbinger of death and simultaneously creator of fresh forms of life.

Aging is sometimes defined as anything increasing risk of dying; or as a decrease in precision of the body's operations till some failure that is critical occurs. A cosmic trend towards inertial equilibrium that renders immortality as natural as water flowing up a hill might play its part. No-one, except perhaps an optimistic King Canute, doubts the omni-pervasive effect of the Second Law of Thermodynamics. Its decree is descent towards a condition of maximum entropy, that is, of exhaustion, slackness and disorder. It flattens out dynamic 'bounce' and commands, eventually, death of the universe. No doubt open systems that absorb energy can locally buck the trend; no doubt, equally, that organisms constitute such systems - but, as we have seen, only by using highly informed mechanisms like coded components, energy metabolism, reproduction and so forth. The basis of biology is information and information, the opposite of chaos, regulates the disintegration of a body as surely as it regulates its integrative development and maintenance. It does not disarm but parries death and for a while can hold its scythe at bay. At every level, by a strange quirk of accidental mutations, programmed maintenance and repair are supposed to have evolved 'in order to' minimise the effect of their own cause - mutations. The slope of decline by wear and tear is thus, as far as possible, rationalised, levelled and inhibited.

On the other hand, information just as well as energetic things 'runs down'. Errors creep in. Mutagens, such as radioactive materials, carcinogenic chemicals and printing errors, form the very basis and sole accidental cause of evolution. Yet they distort genomic information. They misinform. Mutations cause malfunction and, if serious or numerous enough, death. Bodies also 'rust' under the influence of sunlight and pollutants such as tobacco smoke; even the leftovers from life-giving air, free radicals, are toxic. Accident, starvation, predation and disease tread heavily. They kill individuals but also, in extinction, can stamp out whole types. Comets crushing dinosaurs are, it is claimed, a case in point. Over-predation and ecological destruction by mankind in his wily race to cheat death comfortably also ramp the toll right up. Is it not strange, therefore, that a blind process of error and extinction should have thrown up a species craving individual as well as group, genetic immortality? At the same time is it not obvious that, if you love life and that life is, for you, solely vested in your body, you'd want to make more of it? Elongate it? Is this the self-deceit, the trick Tithonus fell for? Is it dream or possibility?

Top-down, of course, (*Sat*) Immortality is vested in Essential Axis, Centre or the Singularity of Consciousness. Uncreated, Life is N(One); Unconditioned Actuality not an existential thing; the destiny of things created is to pass.

Physical immortality, to fix life frozen in a body, is therefore a dream whose madness born of twisted-round perception could, quite possibly, become a nightmare. Is life really but a dream? And, waking into Immortality, does life become Alive, a Sky in which mortality is like a passing cloud? In this straight and sane perspective body is a well-conceived vehicle in which our dramas are driven, our roles on the cosmic stage played out. The question becomes one of whether this 'classic carriage' has built-in obsolescence or, if wear and tear were kept at bay, is designed for immortality. Is the system fit for potential agelessness? Except that bugs can 'curse' and bring it down?

But you can try and bug the bugs. For example, where epigeny's involved excessive methyl tagging builds up through the genome; it silences genes that ought to be expressed, it increasingly deactivates activity. This is old age in a nutshell or, literally translated, nucleus. So why not, using prescribed protein that demethylates, reactivate arthritic, silenced genes - but taking great care since the least inaccurate meddling could spring loose an oncogene and thus provoke a cancer! Or, instead of therapy that worsens health, why not take a visit to the clinic, purchase anti-oxidants, restrict your calories, improve your diet - even undergo cell therapy, an introduction of fresh cells in order to rejuvenate, refresh and restore to former glory? Take a biochemical massage. Revert to pristine Adam or fresh, youthful Eve. You might replace lost cells, transform stem cells, transplant organs, block mutations, or 'freeze' carcinogenic tendencies. You could insert or repair age-influencing genes - perhaps you could even fix life's innate programs of senescence. Maybe it's time to flex the mortal rigour collagen cross-fibres have you stiffen into till you're stuck with death. Surely some idea won't bite the dust? Think of aging as a form of slow castration since declining presence of the sex hormones (which affect many organs other than those strictly related to sex) creates an imbalance that ushers towards death. Until it was shown to increase the risk of breast and possibly ovarian and endometrial cancer, not to mention dementia, HRT (hormone replacement therapy) seemed a logical way to reverse post-menopausal symptoms. Yet despite medication natural wane seems more than just poor maintenance and bad luck. If nothing seems to work the trick then why not, as your end-game, break the Hayflick barrier and by reconstituting telomeric snips (repetitive elements of *DNA* that 'lace' the end of chromosomes together and thus prevent them fraying) snap the tightening boundaries of death. Such genetic snips would otherwise, lost one by one, mark a natural count-down on the cell regenerations possible before the last before full stop. Again, aggressive cancer cells can also snip the Hayflick tape and, reproducing endlessly, turn immortality to death. If, however, we could trip that catch and trick the program into longer life then could not you, regaining patriarchal longevity, become Methusalah? Yet, having aged invisibly, would you still want to live forever?

There are two problems beyond the second law that arise if you try and cheat time and death. Do you remember programmed cell death called apoptosis? In development cells create temporary structures that, like scaffold on a builder's site, need to be cleared away; and there is a continual turnover of material in a living body that includes, like falling leaves in autumn, the programmed death of cells. As the Hayflick limit seems to indicate, your whole physical structure is also programmed to sink, like sap, down to the dead of winter.

A second problem rises in the spring. If there is an after-life or cyclical rebirth then trying to skip death becomes the errand of a fool. If there is no such life and you believe in only atoms, still all lives on earth are clocked. The end is planned. The end is nigh. Each organism's individual extinction is, like obsolescence, built into both its body and the population's reproductive plan. For example, although there is no evidence for a single 'growing old' gene or gene cluster a female's endowment of eggs is fixed at birth; they will run out. Nerve cells are not replaced. And the 'rate of living' (called basal metabolic rate) is a function of 'the rate of dying'. A mammal's heartbeats (1-2 billion) and breaths (around 200 million) are numbered. Within the variability that individuals show no doubt the hairs on different heads are numbered differently. Perhaps you can stretch, contract or otherwise vary your elastic '*uber*-cycle' of three score years and ten - but not by more than tens of years. Certainly an individual's 'slope of mortality' has its clinical angle tipped by parentage, environment and lifestyle - not least diet - but can you level up and live, as a rock in space moves, on forever?

Therefore, the wise advise that clinging on to life is not a sexy move. Immortality's a young man's inexperienced game. According to the order of creation, the logic of Natural Dialectic and statistics drawn from physical data, the lifespan of each kind of organism follows a flexible but orderly, appropriate norm. It span is recognisably programmed. Preordained. You might, for example, vary human life between 40 and, say, 120 or even 200 years - but immortality? It seems that, happily, such a catastrophe is programmed out. So that, despite the best efforts of mankind, Armageddon is by nature kept away. After all what, with reproduction by immortals, would happen to the population of the world? What happens even now? Apocalypse would soon kill immortality. In fact growing old and death involve many genes and a combination of environmental factors. Programmed degeneration might well promote a stable population and allow the young and less experienced their chance. Mercifully, therefore, it seems that reproduction and senescence walk, like grandparent and child, hand in hand. Thus end, with pleasure, seasons of biology.

A Mutant Ape?

How were you supposed to come of age, to have developed into evolutionary maturity?

'On which side, grandfather or grandmother's, were your ape ancestors to be found?' asked Bishop Wilberforce in Oxford. 'On every side', crowed Darwinists. 'We'll comb the fossil record till we find 'em.' The hunt for missing links was on. Distinguished amateurs deceived professionals; professionals deceived themselves and, certainly, the public..

Welcome, therefore, to a Hall of Smoke and Mirrors - palaeoanthropology! Press the ape-man button! Shelves bulge, keys chatter and artistic licence runs incessantly amok. Febrile imaginations have, for five generations, now eagerly clothed 'bones of contention' with an idiotic scowl and lumbering, part-upright frames. From ever-shifting dusts of excavation swirl as many ghosts and theories as there are experts in the field. It is entirely reasonable to claim that evidence remains disconnected, hard to decipher, often media-hyped and always hotly debated. Speculative tales proliferate; hypotheses and spats spill forth. Does academic politics or science win the game? Various arguable reconstructions of bone fragments have secured many a grant, degree and reputation - even fame and fortune! Does confusion or consensus reign? Institutions are established and their libraries of manuscript and file record an over-heated monkey business that surrounds proposed development of ape to man. And, of course, that mutant ape is you. You're a mutant, brother, and you, sister, too. We're all mutants, are we not? No problem, though, because a body's animal and what are we but bodies working like computerised machines? I'm not worried I'm a mutant ape, amphibian or fish; and if my ancestry is thus why should I care if, in the last analysis, my line springs from a mutant micro-organism and at root I am bacterial?

Science is a challenge. How can progress finish? But presentation's always positive; only when a new discovery is made, technology for measurement invented or fresh methodology developed is previous, underlying weakness superseded with a fresh wave of authority. Such flexible authority is, however, in its exclusivity inflexible. Its mind-set cannot flex beyond materialism's box. When it comes to monkey business monkey business of unyielding logic drives interpretation and hypothesis. This logic needs interrogation based on facts. The literature of palaeanthropology is far too large and convoluted to engage with other than broad brushstrokes yet an outline of both *bottom-up* **and** *top-down* perspectives must be drawn.

No doubt, *bottom-up*, evolution theory *must* turn apes to men. Indeed, a 'cast-iron' theory and dependent world-view hang upon this transformation. Substantiation is, therefore, imperative. Much is at stake. The whole materialistic world-view stands or falls. Therefore we seek and find. The maxim, 'If I hadn't believed it I wouldn't have seen it' seems to frequently apply. Maximum suggestion is squeezed out of minimal evidence. Bones of contention are shoe-horned into evolutionary interpretation and fossils arranged and re-arranged exclusively within various permutations of an ape-to-man conceit. Compelling public statements are issued at the time of each discovery only, amid behind-the-scenes controversy, to be quietly dropped; retrospective ignorance (after further clues are found) is common and the whole subject has been implicated in degrees of skulduggery that range from outright fraud through secrecy, naïve wishful thinking and imaginary propaganda to skulduggery's lack - an honest search to find, if they exist as surely they must, the relevant fossils.

Such a search involves the comparative anatomy of fragmentary, whole and reconstructed bones. It includes the reassembly of putative and mostly partial skeletons, accurate supporting dates and, most problematic, one-track interpretations of the data. To question either data-crunching mind-set or its dating assumptions is, sadly, to court a real possibility of curt, contemptuous dismissal. However, while science engages an excellent and supposedly self-correcting methodology to deal with the current operations of all things physical (including biological), it relies upon abductive argument - best explanation by inference - for historical interpretations of past facts, especially in the case of origins. And while the first and foremost requirement in an honest court of historical appeals should be for *both* sides to put their case, this is not the case.

In fact, fossil evidence should be subjected to both *bottom-up* and *top-down* consideration. Holistic design and materialistic evolution should be thoroughly, consistently compared. No less than a faculty of Comparative Palaeoanthropology demands establishment! Each side might learn something from the other but instead high-pitched objections from materialism's camp pierce reason's lucid air. Only in a complementary way, however, can the smoke be cleared, mirrors shattered and fun-of-a-scientific-fair be elevated into wholesome dialogue.

'Hominid' (or 'hominin') is a designation invented to mean humans and their theoretical evolutionary ancestors. There is no lack of fossil evidence for such creatures. The Natural History Museum in London published a Catalogue of Fossil Hominids. Although items rather than individuals are tabulated, a count of some 4000 individuals has been estimated (by Marvin Lubenow) to have been recorded up to 1976. This figure includes over 200 Neanderthals and over 100 *Homo erectus* designations. Since then many more fossils have been dug up so that perhaps 8000 alleged hominid pieces, mostly in highly fragmentary form, have been found. The finds have not, however, been systematically committed to a single list. Moreover fragile, irreplaceable specimens are often and understandably sequestered in vaults and strong-rooms (as in Kenya, Ethiopia and South Africa) and elsewhere (such as Indonesia) also barred from access except by a chosen inner circle. Some 'finders-keepers' will not allow even well-qualified co-workers in the field near their 'property' or allow publication of any evaluation that may contradict their own (and, it may be added, their own fame and fortune). Such caper has, in the case of determined but defensive missing-link hunters, gone on since the time of Eugene Dubois in Java (1891). Since originals are practically unavailable for study plaster casts are, in a small percentage of cases, issued to replace them. In some cases these casts are accurate and in others less so. For example, a plaster cast of the fraudulent Piltdown Man issued by the National History Museum for inspection by an investigator (Louis Leakey) had teeth of the orang utan mandible which showed no evidence of file marks crucially evident on the original forgery. In 1984 the American Museum of Natural History in New York sponsored, in part due to 'concern about creation science', an Ancestors exhibit. Although nations such as China, Kenya, Tanzania, Ethiopia and Australia demurred a 'family gathering' of about forty original specimens was displayed. These national treasures were placed on mounts that had been precisely prepared using plaster casts - although in most cases the real bones did not fit and the mounts had to be adjusted accordingly. How dependable is such a cast of players? To what extent is the whole business theatre?

The upshot is that, although many fossils exist, they are neither catalogued in a comprehensive way, adequately displayed or available for close study. Only a very tiny fraction of palaeoanthropologists, archaeologists, anatomists, evolutionary biologists, sociologists or any other interested party can gain sufficient access to make an independent judgment unclouded by expert opinion, academic fashion, second-hand media hype or other kinds of 'Chinese whisper'. Since when was hearsay how you did your science?

In this case, heretically, we'll start with a brief review of dating systems and follow up with a review, cast according to the evolutionary dating system, of hominid material ranging from modern *Homo sapiens* through 'archaic' *H. sapiens*, Neanderthal, *H. erectus*, the 'handy-man' link (*H. habilis*) and *Australopithecines*. In this we note that chimps and humans show similarities. If you *assume* that similarity indicates common physical descent (while for similar models engineers assume common *conceptual* descent) then a textbook line of 'morphs' from ape to man must confirm your theory. Despite the similarities, however, it will be argued along with Ernst Mayr (in What Makes Biology Unique?) that a 'large, unbridged gap' separates, except for 'historical narrative', chimp-like Australopithecines (such as Lucy) and man-like *Homo erectus/ ergaster* (such as KNM-WT 15000, Turkana Boy). Given the postulated time-scale (maybe 3 mya or 120,000 generations) and the number of significant anatomical differences (as regards limbs, hands, feet, larynx, muscular refinements to lips and tongue, much larger skull and brain size and so on) could the necessary genetic revolution have conceivably occurred? Many changes need to be concerted. In bacteria two or three coordinated mutations may occur but, even in the timescale of the universe, not enough to change the *function* of a protein let alone innovate physiological and psychological capacities (such as, in man, ability to speak, make music, conceive abstract principles or otherwise symbolically codify the world). *It behoves an evolutionist to spell out which sequence and quantity of single mutations, piled positively upon each other and thus each rendering offspring significantly fitter than their parents, orchestrated the rapid 'invention' of a human.* Plain assumption or blank assertion that it 'must have' happened does not suffice. Assuming *DNA* alone is capable of such transformation is it possible that this tally of *BM*s could, each in turn, be fixed in so few years? By what experiments has the hypothetical 'genomic leap' been tested let alone validated?

Finally, we'll note the three main evolutionary paradigms within which the evidence has been differently interpreted and, having listened to each of these stories, see how it all totes up.

Let's start with dates and follow with morphology. Often the date assigned to a fossil specimen does not agree with the theoretical order in which its 'evolving' morphology 'ought' to be found. A typical response involves reassignment from the unsatisfactory taxon (a group of organisms judged to form a unit) to another. A second kind of response involves reassignment of the date. How can this be?

Hardly anyone is trying to *deceive*; it's simply that the theory's fact and therefore facts *must* fit. Professionals are paid to peer through evolutionary spectacles. No spectacles? You're fired! The pressure's therefore overwhelming. Repetition is relentless. Anomalies must be resolved this way or that - but only in the framework that's prescribed. In Chapter 22: Types of Fossil the circular reasoning (or tautology) employed using index fossils, stratigraphy and contextual flora and fauna to assign a relative date was discussed. In Chapter 11: DC-time the strength and weakness of assumptions inherent in 'absolute' radiometric dating were examined. Now, concerning the alleged evolution of some kind of ape into human form, the prevalent use of potassium/ argon and carbon-14 dating, including the 'coverage' gap, need special mention. Whatever its final merit so does an implication that concerns the rate of helium leakage from zircon crystals. *Whether or not the earth is young (in spite of strong evidence for age - such as the size and temperature of an expanding universe), it may be shown that precise, legitimate dating for most for most hominid fossils is lacking.* Such imprecision has led to a catalogue of reassignments according to where in which scheme of evolutionary events a fossil of uncertain date is deemed best placed; and, through such manipulation, what ancestral relationship is proposed, in terms of family 'bush' or 'tree', to ourselves. Palaeoanthropology, it may be plausibly substantiated and thus fairly admitted, is not the most exact or easily self-correcting of sciences. **Finally, when all is said and done, is it even possible that enough evidence has already been collected to falsify, whatever other fossils are later found, the ape-man missing link hypothesis?**

The age of a rock is defined as the moment it solidifies. The problem is to find out when that was. This is exemplified in recent cases such as the construction of Surtsey Island, the destruction of Santorini and the eruption of Mount St. Helens in Washington State (1980) will be further discussed (Chapter 24: Towards a Unified Theory of Ecology; Abiotic part). For now consider the actively volcanic Mount Ngauruhoe in New Zealand. Lava samples from a series of eruptions between 1949 and 1975 were tested 'incognito' by Geochron Laboratories, Cambridge, Massachusetts. Four of these were dated, with a 20% margin of error, at less than 27 kya (27000 years ago); at the other end of the scale one was assigned 3.5 mya (3.5 Ma or million years ago)! No doubt, one is informed that radiometry does not work well with rocks less than fifty, or even a thousand, years - which human witness accounts might calibrate; but where it *is* supposed to work it can't be double-checked. A best attempt is made but no independent verification is possible. Thus great ages constitute an unchallenged and, it might appear, unchallengeable assumption. And, as regards the highly charged arena of hominids, when a date is assigned to a fossil that does not agree with the theoretical order into which its morphology should fit, re-dating and/or reassignment from less to more comfortable taxon is the order of the day (a taxon is a group of organisms judged to form an exclusive unit). When theory's right the facts must not be wrong!

Thus here comes trouble! Zircons are microscopic crystals found in biotite. They contain radioactive uranium and thorium, one of whose by-products is helium. This light, un-reactive and thus mobile gas can leak between the crystal lattices and so, you might presume, rise to the upper atmosphere and be whisked into space. In fact, escape is limited since it spreads throughout the atmosphere; and this contains less than 0.1% of the amount it should if earth were very old. This implies an earth that's relatively young (an inexpressible taboo). Furthermore, a study of diffusion rates from deep-bore zircon samples found that 58% of helium from an alleged 1.5 by (billion years) of decay was still *in situ*. Slow and steady leakage from decay should have left scarcely any. Was the Precambrian granite, foundation rock in the bore-holes, as old as previously estimated? Could there have been, in accord with J. Magueijo's increased initial speed of light (see Chapter 12: Matter's Holy Ghost), an initial, accelerated burst of decay and a much younger earth than commonly believed? Which one or more of the three basic radiometric assumptions was incorrect?

Men and Apes: A Fossil Chronology

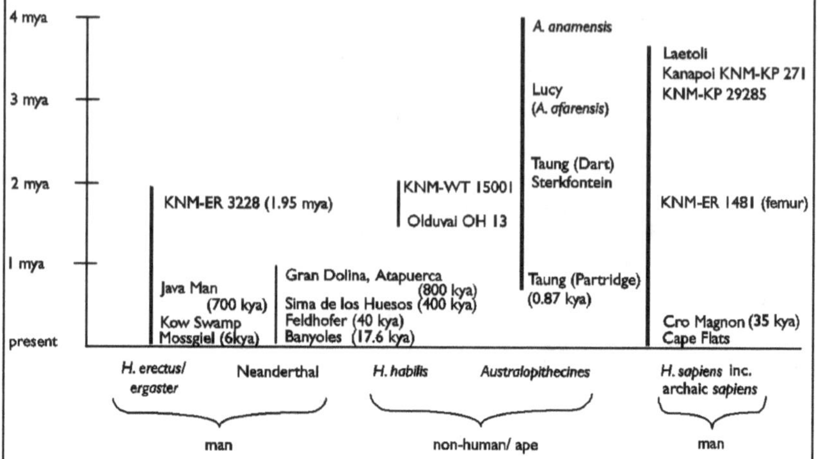

This chronological chart represents a précis of fossil facts drawn from the thousands of 'hominid' fragments now excavated. Only a few, including the oldest and youngest that bracket each group, are shown. While the chart reflects fact for bones and consensus estimate for age it does not necessarily reflect such multiple and often heatedly conflicting interpretations, manipulations and hypothetical organisations as impact presentation of the actual material. Finally, it excludes from consideration controversial items such as the Piltdown skull, *Hesperopithecus*, the Calaveras skull, Castenedolo skeletons and so on.

As far as Natural Dialectic is concerned the chronology emphasizes a pressing need to drop philosophical prejudice and enjoin comparative anatomy with respect to the measurement. By this is meant the engagement of expert palaeontologists of both 'scientific' and 'unscientific' mind-sets (that is, arguing from evolutionary 'design-by-non-design' and non-evolutionary design respectively); and, in this dual/ duel engagement, rigorous examination and interpretation of both 'hominid' and all other fossil material. Only with such discipline, as in a court of law, will truth obtain a fair trial. Who will be the jury but your public self?

fig. 25.2

The effective range for K-Ar (potassium/ argon) dating is thought to begin about 400 kya. On the other hand, the useful range for carbon-14 (whose half-life for decay to nitrogen-14 is, you may remember from Chapter 11, 5736 years) is limited to about six half lives, that is, about 40 kyr. This leaves a 'coverage gap' of 360 kyr smack in the crucial period where modern man is supposed to have evolved. This is not conducive to accurate chronology but a method for precisely measuring (to 0.001%) the ratio of carbon-14 to carbon-12 in a sample against that found in the modern atmosphere has been developed. It is called an accelerator mass spectrometer and theoretically extends the range of carbon-14 dating to about 90 kyr (still leaving a gap of 310 kyr). The surprise is that no fossil material anywhere can be found with as little as 0.001%; and the consequent implication is that, because all carbon-14 should have disappeared by at maximum 250 kyr, no fossil is older than that; thus great age would appear to represent, without the invention of significant excuse, an unutterable fiction. In the human case any bone dated over this age but containing any trace of carbon-14 would suggest (without the rescuing device of 'contamination') erroneous assignment. Some mistake, surely?

Are these chronological considerations problematic? They are if you combine them with morphology. Examples of fossilised *H. sapiens* range from about 5 kya (Cape Flats, South Africa), 25 kya (Zhoukoudian, China) and 30 kya (Cro-Magnon, France) through 60 kya (Lake Mungo, Australia), 90 kya (Skhul and Qafzeh caves, Israel), 700 kya (Java Man, human femur) , 1.9 kya (KNM-ER 1472, 1481, femurs and other fragments, and possibly 1470, cranium, Kenya) to 3.7 mya (KNM-KP 29285 tibia ends and KNM-KP 271 upper arm, Kanapoi, Africa. To this range might be added the modern-looking footprints from Ileret in Kenya (1.75 mya and thus dubbed *H. erectus*) and Laetoli tuff, Tanzania (3.75 mya and so deemed *Australopithecus afarensis*). Yet Russell Tuttle honestly reflects Mary Leakey's initial assessment that the latter footprints are remarkably similar to those of modern man. After extensive study he admits (Pitted Pattern p.64) that if they 'were not known to be so old, we would readily conclude that they were made by a member of our genus, Homo.' Poor running man - erased because he doesn't fit the scientific scale!

Most of the latter are wrong if gradual ape-man evolution's what you're looking for. Thus there's heated argument. Thus also 1860 discoveries in Pliocene layers (5 my) are quickly dismissed as hoax

or intrusive burial. At Calaveras, California, as at Trinil, Taung and other hominid sites, skull fragments and other bones were discovered by non-professionals (in this case not coolies but miners). The state geologist, Dr. Whitney, believed them genuine; another professional geologist, Clarence King, gouged a stone grinding instrument from local rock dated at 9 my; and more artefacts and fossils were discovered in the area. Anthropologist William Holmes from the Smithsonian Institution soon weighed in. Alfred Russell Wallace, co-formulator of evolution theory, expressed dismay at the way anomalous evidence for modern man in Pliocene rocks was 'attacked with all the weapons of doubt, accusation and ridicule'. Murky waters at the edge of scientific margin run. Holmes castigated Whitney for his lack of caution in presenting evidence that contradicted theory; anecdotal evidence whipped up a claim that the illiterate but Darwin-savvy miners had been persuaded to commit a funny hoax; later radiocarbon dating was said to have dated the skull at only a thousand years; and other finds in the area, found some fault with, were summarily dismissed. At the same time, across the world at Castenedolo near Brescia in Italy, Professor Guiseppe Ragazzoni, a geologist, found a modern skeleton in undisturbed blue clay also dated Pliocene (over 5 my). The find was confirmed by anatomist Guiseppe Sergi. Anthropologist Sir Arthur Keith admitted: 'As the student of prehistoric man reads and studies the records of the Castenedolo finds, a feeling of incredulity arises within him. He cannot reject this discovery as false without doing injury to his sense of truth and he cannot accept it as fact without shattering his accepted beliefs.' However, the discovery of a second skeleton in a nearby fissure, this time clean of encrustation by marine material, led directly to a notion that all the bones were from a medieval cemetery; later carbon-14 dating of ribs (of the second skeleton?) precisely confirmed this theory. The acceptance of a Pliocene date for the Castenedolo skeletons would have created so many insoluble problems that one might hardly hesitate in choosing between the alternatives of adopting or rejecting their authenticity. Who, after all, would wish 'bad' archaeology might nip a theory in its fragile bud? For sure, intrusive burial! Castenedolo case rejected!

'Bad' archaeology was rife in Victorian times. Yet, we shall see, anthropologists were not as critical if the evidence were in favour of Darwinian hypothesis - far from it. Blatant fraud went unchecked for decade after decade; skulduggery and manipulation with respect to reconstructions, inconvenient dates and taxonomic classifications abounded; much propaganda value, especially illustrative, was extracted from scandalously ill-defined material; the complaint of Wallace was upheld; and such monkey business may, in the highly emotional field of human origins, still occur. At any rate, let's also dismiss the Pliocene pair. **Still the range of *Homo sapiens* runs clean across that supposed for Neanderthals.** The latter runs from about 18 kya (mandible at Banyoles, Spain) to 800 kya (Gran Dolina, Atapuerca, Spain). **It also covers, as we'll see, the entire range proposed for *Homo erectus*/ *Homo ergaster*.** In brief, this runs from 6 kya (cranium from Mossgiel, Australia) and 9.5 kya (Kow Swamp people, Australia) to 1.6 mya (KNM-WT 15000, the Turkana Teenager, an almost complete skeleton of an adolescent approaching six feet tall, Kenya, Africa), parts of Ethiopian Omo remains (between 1.05 and 1.9 mya) and 1.95 mya (Modjokerto infant, Java, cranial fragment KNM-ER 2598 and hip bone KNM-ER 3228 both from Kenya).

Homo habilis is a taxon that may include both human and non-human material. At any rate, its creatures are included in the *erectus* date-frame (running from Olduvai skulls and other fragments at 1.5 mya to 1.9 mya (a Swartkrans tooth) and 2.0 mya (cranial fragment KNM-WT 15001). **Thus *Homo sapiens* covers the entire range for *Homo habilis* (or handy-man) as well!** What, therefore, about definitely ape-like material from Africa such as the metre-high *Australopithecines* (possibly extinct kinds of chimpanzee, dated from Taung at ~ 2.5 mya through 'Lucy' and 'Dikika girl' at 3-3.8 mya to a similar but possibly separate species called *anamensis*, 4.2 mya); and the genus *Paranthropus* now including *Zinjanthropus* (originally deemed an *Australopithecine* tool-maker but reincarnated as *Paranthropus boisei*) and the 'black skull' (KNM-WT 17000) dated between 2 and 2.5 mya? **Whether or not these two genera of hominids are judged in line to *H. sapiens*, a unique kind of organism, we cover their supposed epoch too.**

Let's for a moment slip across from Africa to Asia and inspect a different case of 'midget men'. A female 'pygmy' said to resemble *Homo erectus* and nicknamed 'Flo' (or, more seriously, LB1) was discovered in 2004 in Indonesia alongside hunting and cooking gear. She was predictably claimed as a new species (*H. floresiensis*) by her discoverer and, equally predictably, 'aped' by National Geographic and other imaginative artists. Although even smaller than

Australopithecines, this woman (or ape-woman) lived only 12-15 kya - young enough to be contemporary with modern man and, if human, to resemble modern wives, mothers and sisters. Is the pin-head 'hobbit' (brain capacity 380 cm^3) a missing link, a separate species of human or, as Australian palaeopathologist Maciej Henneberg claims from photographs (he has been denied access to examine the actual skull), was one of 'its' molar teeth filled by a dentist in the 1930's?! Seven more non-fossilised skeletons were found on site. The local doyen of palaeoanthropology, Teuku Jacob, claims she was 'a modern woman with a congenital disease that caused a shrunken brain-case' (a condition called microencephaly), that is, a pathological specimen of *H. sapiens*. In fact hundreds of medical conditions can cause dwarfism, microencephaly and so on in humans but has even such non-evolutionary analysis proved necessary? Do not pygmies live in Africa? Teuku pointed out a community of over seventy families of 'little people' the same size as Flo living at Rampasasa, an Indonesian village just down the road from her Liang Bua cave; and over twenty skulls of dwarfs or pygmies (with human jaws) were found on the Micronesian island of Palau about a thousand miles away. In fact adult humans vary in size between, at the extremes, about 0.85 and 2.75 m. Could the *NG*'s missionary evolutionism have libelled or infringed the talking, voting, legal human rights of living '*Homines erecti'*. Ape-women *must* have existed, mustn't they? Speculation is, therefore, the order of a changing day. But, it seems this body in a cave is not a case for local police but probably a poor, sick old lady.

The fact of a 'coverage gap' and the contemporaneous nature of all hominid existence (between 2 mya and virtually, except for ape-like *Australopithecines*, historic times) considerably complicates the notion of evolution – wherein a species that succeeds its inferior should wipe that ancestral competitor out. It also sets considerable doubt on Time-Life's flagrantly inaccurate yet canonical (even iconic) imitation of evolution, that is, Clark Howell's 'March of Progress'. This fake parade, now politically incorrect as well, shows fifteen forebears that rise, in a single file, from skipping ape to at least six fully upright 'humanoids'. *But the smooth, beguiling, progressive transformations are a straight illusion.* Even at the time of printing scientists knew that it was visually misleading. On p.41 Howell wrote that, 'Although proto-apes and apes were quadrupedal, all are shown standing for the purposes of comparison'! Oh well, never mind; the author's heart was in the right place, wasn't it? The fictional nature of the 'once-popular fresco' was publicly admitted only 35 years later (see Nature 403: Vol. 27 p.363, 2000). Such was the impact of its simple, brilliant propaganda that it is still falsely embedded in the modern mind-set. Indeed, the whole, ever-shifting confabulation of interpretation and the regular discovery of new fossils keep the pot at boiling point; they combine to complicate the study of whose rising son you really are and, as previously noted, cry out for equal research from an infidel, non-evolutionary stance.

What about Neanderthals? In 1856 quarrymen unearthed the first specimen from the Feldhofer Grotto near Dusseldorf, Germany. The skullcap, some ribs, part of the pelvis and limb bones were saved and examined by Hermann Schaafhausen and the father of pathology, Rudolf Virchow, who pronounced them human. Thomas Huxley never saw them but recognised they were 'insensibly graded' from modern specimens, that is, were qualitatively human.

Darwin never saw them either. Nor did William King, Professor of Anatomy at Queens College, Galway but this did not stop him placing the bones in a separate species, *Homo Neanderthalensis*. Which label was the wanted one, the one that stuck?

Other examples were found including, in 1908, one from a cave near La Chapelle-aux-Saintes, France. This was, despite the fact that its brain size was 1620 ccs. as opposed to a large modern brain of 1450 ccs..), poorly reconstructed according to strong evolutionary preconception by Marcellin Boule. Thus appeared a forward-leaning, hunch-backed, thick-looking, crook-legged, shuffling cripple! Ape-man was being delivered! Only by 1957, with the throes of parturition well over, did 'self-correction' hobble towards a different conclusion when William Strauss and Alec Cave re-examined the same bones, rumbled the prejudicial reconstruction and noted that Neanderthals walked just as humans do. They said: "There is...no valid reason for the assumption that the posture of Neanderthal Man...differed significantly from that of present-day man...if he could be reincarnated and placed in a New York subway - provided that he were bathed, shaved, and dressed in modern clothing - it is doubtful whether he would attract any more attention than some of its other denizens." They have, we now know, the same ear labyrinth and thereby balancing mechanism as we do; there is no intermediate between this and the ape-like pattern. Neanderthals have been shown, from genome studies at the Max Planck Institute, indisputably human. Thus Alan Templeton previously, correctly hypothesized that there is more genetic difference between contemporary humans than between us and Neanderthals - which, except for biology textbook illustrators, makes them human. Indeed, have you never spotted thickset characters with sloping forehead and with beetling brow as they spear seals or, closer home, whisk down your street to work? On average they would have, if 'Neanderthal', a bigger brain than you. Of course, you couldn't hold a conversation unless you learned her foreign language or she attended English class - but why should humans with a bigger brain and maybe intellectual capacity not speak? And exactly how did physiological fine control through 'language maps' (e.g. regions of neurons involved with vowels quite separate from those for consonants) evolve by small mutations from ape to human thought-and-talk in just a million years or two? Natural Dialectic would presume Neanderthals had the same archetypal instinct for rational order, symbol, grammar and the other aspects of communication. They might 'live rough' but equally could grace a hall of academe. Indeed, made-up facially according to their skull form but ordinarily dressed, TV presenter Alan Titchmarsh put Strauss and Cave's pronouncement to the palaeoanthropological test in High Street. Did anybody bat an eyelid? No - well, except the odd appreciative glance from female non-Neanderthals! The case of Boule shows how, with evolution, myths are born.

Also in 1908 came the Piltdown Fraud. As ascertained in 1982, an orang utan jaw (from Borneo or Sumatra) was fitted to a human skull (carbon-14 dated to about 600 years old and probably from a local black-death grave, although this was not trumpeted as in the Castenedolo allegation). On the contrary, the vast majority of scientists eagerly accepted this third strong beat on the evolutionary drum. Only in 1953 - again, long after propaganda value was secured - did the

British Natural History Museum issue notice that Piltdown was a hoax. He was no less a fiction than Evening Ape. The latter was a tooth declared by palaeontologist Henry Osborn (the first president of the American Natural History Museum to have been trained a s a scientist) to have the mixed characteristics of human, chimpanzee and *Pithecanthropus*. In Britain Cambridge scholar, anatomist and anthropologist, Sir Grafton Elliott Smith FRS, FRCP, fully agreed and the tooth was named *Hesperopithecus haroldcookii* (after its finder). Thus, sprung fully-formed from this non-dragon's tooth, in the London Illustrated News (24th June 1922) a centre-spread of an artist's imagination of 'man-ape' *Hesperopithecus* cavorting with his wife. Later investigation proved that the tooth was not from a dragon, as in the myth, or a man as in another myth, but an extinct pig. So a pig made a monkey out of goggled evolutionists, though no publicity was given to the fact. Is there a retraction? Of course, even grave mistakes occur but, although discredited, the misinformation amounted to a further insidious injection of new truth. Fresh, exciting missing-link injections still regularly continue to excite and then, like other drum beats, quietly fade away.

Neanderthal fossils have, like modern Cro-Magnon man, been found associated with stone and bone tools (including 'hand-axes' that may have been used like a discus to hunt), artefacts, the controlled use of fire and mortuary practices. At the Drachenloch cave in Austria, at an altitude of 8000 feet, a cubical chest of stones covered with a large stone slab was found. Inside were seven bear-skulls all with muzzles facing the cave entrance while six more were mounted in niches on the wall. Is hunting magic so inhuman? In another cave in Bavaria the skulls of thirty persons, with ornaments of deer-teeth and shells, had been placed in the earth. The charred remains nearby seemed to indicate cremation. Was it symbolic that the skulls were sprinkled with red ochre and all faced, like the church door through which a coffin is carried, west towards the dying sun? Lacking modern technology what would your offerings consist of? At the Shanidar cave in Iraq a Neanderthal was found buried on a bier of flowers.

Would you expect Ice Age Man's dwellings, which might have ranged from igloo, yurt and tent to mud-wood-and-thatch cottages, to have survived intact, especially since a hunter life-style is nomadic? How much sign do Eskimos, red Indians and bush-men, in their culture, leave? These thick-set hunters had powerful jaw musculature (which might have contributed to their skull shape); their hyoid bone indicates they could talk and, certainly, work intelligently in organised teams. Although Neanderthals are commonly believed to have died out about 30 kya a skull was found just below the top layer of Amud cave in Israel, a layer dated at 5710 years. Contamination is blamed for this theoretical anomaly. However, the oldest fossilised primate protein to have been sequenced from a Neanderthal was found to be 'as ours' (New Scientist 19-3-05 p. 14); and various caves on Mount Carmel, Israel (e.g. Mugharet-et-Tabun and Magharet-es-Skhul), show a mixture of Neanderthal and Cro-Magnon types, strongly suggesting that, for example like a modern Chinese and an African, they could 'hybridise'; the remains imply mixed marriage and an early lack of racism. It is even suggested that Neanderthal genes most people carry boost our immune system. Indeed, there is little greater difference between members of the three groups of 'early man' than between cold-adapted Eskimos, heat-adapted

Aborigines, pygmies and Europeans. *That Neanderthal and Cro-Magnon people lived and were buried together affords strong evidence they interbred - yet, as we'll shortly see, a politically correct modern theory called 'Out of Africa' demands the two be kept apart and would downgrade the 'older fellow's' human reputation back into sub-human genus!*

As regards the notion of brutish, dim-witted 'cave-men' it is observed that modern humans live in caves (as in Urgup in Turkey); that humans have often retreated to caves in times of crisis; and that, throughout human history, caves have been used as easily identifiable points to bury kith and kin (Abraham's cave of Machpelah is an example). Similar prehistoric Neanderthal cemeteries have been found scattered throughout Europe and the Near East. Indeed, the Sima de los Huesos, discovered in 1992 by Juan Luis Arsuaga in Spain, is just such a site. It is deep, narrow and was never inhabited. Yet in this ossuary were deposited (not buried) on top of each other the remains of at least 33 individuals, with 3 finely preserved fossil skulls, dated at 400 kya. **There is such variation in the skulls, let alone the other bones, as encompasses the whole range of 'archaic' *H. sapiens* found in Europe. Indeed, as regards 15 cranial characteristics Arsuaga's fossils show 7 similarities with ourselves, 10 with Neanderthals and 7 with *H. erectus* (supposedly the precedent of Neanderthals) - suggesting that differences may be due to non-evolutionary causes occurring in a single, unique species, *H. sapiens*.** It is fact that environmental conditions change (witness the oncoming of an ice age) and, due to genetic isolation, morphological changes can occur rapidly. Maybe Neanderthals *were* small, isolated populations of aboriginals both in the west and (witness skulls so far found in Indonesia, China and Australia), the east. In Europe most on-the-line human/ Neanderthal morphologies (all but 4 of 130) are grouped in the central and eastern regions so that, yet again, interbreeding is suggested with a different race or races from the south (like those who may have produced the 400,000 year-old North African Tan-Tan figurine).

If Arsuaga's highly localised group shows such variation what are we to make of widespread anomalies, a so-called 'archaic human' taxon into which are bundled puzzle fossils ranging from 5 kyr (Cape Flats skull, South Africa) to an estimated 700 kya (Java Man)? Such individuals have a human skull capacity of around 1100 ccs., heavy brow ridges, large faces/ small skulls or *vice versa*, rounded rear of skull and human 'post-cranials' i.e. bones below the skull. Perhaps the best known is Rhodesian Man, found in a copper or heavy metal mine and reported in Nature (Nov. 1921) by Dr. Arthur Smith Woodward FRS, a senior palaeontologist at the British Museum (Natural History). Stone and bone tools and bolas (used today for catching cattle) were found but inconsiderate mining personnel, in the course of excavations, destroyed the site. The skull was not fossilised nor even, surprisingly, mineralised. It certainly exhibits, far from a definite date, 'how fast a fossil can age after it is taken out of the ground'. Smith Woodward put the age ~10 kya. By 1962 this had been increased (by Carlton Coon) to 40 kya, by 1965 to a minimum of 125 kya (Richard Klein) and then in 1999 (by Ian Tattersall - perhaps to facilitate the 'Out of Africa' notion wherein modern men are supposed to have emigrated from that continent around 150-200 kya and taken over the world) to 300 to 400 kya. This, to date, is as far as it goes. **It's still very rapid aging; around 350**

kyr in little more than 0.08 kyr (80 years) without good evidence he lived more than perhaps 1 kya! The Swanscombe, Stenheim, Hungarian Verteszöllös and Greek Petralona skulls are among examples sometimes sorted into this disparate group; so are *H. rudolfensis* (inc. KNM-ER 1470, 1472, 1481 and 1590 (~ 1.9 my), Saldhana, Boxgrove, Heidelberg and Ethiopian Omo and Herto/ Idaltu Men (the latter inc. a skull, 1450ccs., skull fragments, a few teeth but post-cranial body-bones are dated at 160 kyr - well in line to make fine candidates for sally Out of Africa). Although no African counterpart to the abovementioned Sima de los Huesos cache has yet been dusted down, you could argue from their cranial characteristics that they're also part of 'muddle in the middle' and should be assigned to fully human type. Do you interpret such variety in evolutionary or non-evolutionary ways? Israeli caves and the Spanish ossuary both indicate that variation's range might fall within our special flexibility - except that bones can't interbreed to prove it finally.

In 1888 a human skull (capacity 1550ccs.) was discovered at Wadjak, Java. A lack of precision plagues the geological details of the site where, in 1890, Eugene Dubois found a further human skull (Wadjak II) and fragments of a skeleton. Then in 1891, determined to find a missing ape-man link, also in Java at Trinil, by the Solo River, his coolies unearthed a beetle-browed top section of skull (capacity over 1000 ccs.) and, variously reported between ten and fifteen metres away, a human femur. He called the combination *Pithecanthropus erectus.* Java man. This version of *H. erectus* was, just as the Chinese today are on average smaller than westerners, smaller than Neanderthals. A further Selenka-Trinil expedition, far more extensive and scientific in its scope than any of Dubois, used over 70 coolies to dig at exactly the same site on the Solo River. The scientists concluded that Trinil's volcanic, flood-deposited sediments were too young to throw light on evolutionary origins. For example, the flora was modern and so were most of the gastropod forms. 43 crates of fossils produced no more *Pithecanthropus*; instead scientists found charcoal, hearth foundations, human bones and artefacts in its stratum. An excellent report was compiled (*Die Pithecanthropus-Schichen auf Java*) which, while wishing to confirm an earlier evolutionary perspective, actually cast real doubt on it. For this reason Frau Lenore Selenka apologised to co-worker Max Blankenhorn for lack of evidence supporting Dubois' hypothesis. Maybe it is also the reason why her work has been and still is widely ignored.

In fact, Dubois was hardly qualified to date and define by name specimens not even found by him *in situ* (except, perhaps, the Wadjak II skull). Indeed, he hid these skulls. As regards Java man, the question was whether both femur and skull-cap were human or whether the femur was human and the skull-cap *Pithecanthropus* - in which case the two species must have lived together. In 1895 anatomist and anthropologist Sir Arthur Keith FRCS decreed him human. In 1900, after actively promoting his ape-man, Dubois 'went quiet' and only in 1924 published a definitive paper on the skull-cap followed, in 1926, by one on the femur. Did Dubois hide the human Wadjak skulls under his floor-boards for over 30 years to prevent any deflection of glory from *Pithecanthropus* who, as palaeontologist Gustav von Koenigswald later wrote (in 1956), '...came just at the right moment at a time when the conflict around Darwinism was at its height. For the scientific world it constituted the first concrete proof that man is subject not only to biological but palaeontological laws'? Beats on the evolutionary drum were flowing faster, louder - each presumed a death knell for 'unscientific' theories of design.

A Catholic priest, keen to promote evolution, was involved in 'dodgy' fossil finds at Piltdown, in Java and in China (where from 1923-46 he served as a consultant to the National Geographical Survey and where, during the war years, he lectured students on how they had evolved from animals). In 1929 this priest, Teilhard de Chardin, and palaeoanthropologist Davidson Black FRS visited Choukoutien (dragon-bone or fossil hill) near Beijing. The find of a molar tooth led to the designation of *Sinanthropus* (Peking man) - later merged with Java man under the classificatory label of *H. erectus*. Eventually 575 boxes of bones were collected from which five instances of sufficient bone for a cranial reconstruction were singled out. Although the cranium resembled Dubois' gibbon-like Java skull de Chardin was able to 'evolve' sufficient brain capacity (~1000ccs. or a litre) and apish appearance for a despatch to the French *Revue des Questions Scientifiques*. In fact, there seems to have been a tool-making and lime-burning industry on the site, with large deposits of ash in which the fossil fragments were found. This industry, of far too advanced a nature to be attributed to a small-brained animal, may have serviced the city of Cambulac on the site of the present Beijing. Skeletons of baboons and macaques, often with neat holes drilled in their skulls, were found. These possibly served, as Marcellin Boule suggested and as still happens, as food. In 1934 Davidson Black died in his laboratory among his fossil collection; his successor, the famed anatomist and anthropologist Professor Franz Weidenreich, re-cast his reconstructions but most of the collection, including models and photographs, was lost during WW2; and Weidenreich deemed so-called *Sinanthropus* to be, against the interpretation of Black, 'human' - a race of *H. sapiens.* Nevertheless in 1966 the evolutionistic communist authorities hailed another skull from Choukoutien (with an extremely large gap between front and rear portions) as Peking Man.

In 1935 the same Teilhard de Chardin, invited by Gustav von Koenigswald, hoped to establish a link between Java and the finds in China. From 1936-39 von Koenigswald paid natives per fossil piece found in the Sangiran area of the Solo River; they found fragments of jaw-bones, teeth and skulls which he designated *Pithecanthropus* II, III and IV. These were criticised by other palaeontologists for their very ape-like nature, the reconstructions and consequent classification. However, on the floor of a cave he and de Chardin found fossils 'absurdly similar' to those in fossil-bearing deposits at Kwangsi. The hoped-for correlation was, perhaps by luck, achieved. However, in the 1990's dating specialist Garniss Curtis (who with Getty support founded the Berkeley Geochronology Centre and had previously dated, at 1.75 mya, the Leakey's *Zinjanthropus* from Olduvai) now dated *H. erectus* fossils from the Solo/ Sangiran beds at between 27 and 53 kyr. Thus, while the Mojokerto infant (1.95 mya) might seem to push the logic of 'Out of Africa' back by a million years or more, the Sangiran dates would indicate that Javan *erecti*, much younger than their African *erectus/ ergaster* counterparts, must have co-existed with modern immigrants for at least 100 kyr! This why Curtis' dates have 'rocked foundations'.

Is *H. erectus* so morphologically different from *H. sapiens* as to categorically forbid his presence as a variation possible by genetic recombination, interbreeding, isolation and adaptive super-coding (this chapter : Super-codes and Adaptation)? Is he found in a time-frame that permits legitimate transitional forms, by mutation, natural selection and evolutionary genocide of less evolved competitors, to grade into modern man? Today the species, a slightly smaller version of Neanderthal, has been joined by about 280 fellows worldwide whose ages range from 6000 years (Mossgiel cranium in Australia), 6.5 kya (Cossack skull, Australia), 9.5 kya (Kow Swamp people, Australia) through Peking Man (300 kya), Java Man (skull, 700 kya, with human femur and nearby Sondé tooth), various Olduvai hominids, Tanzania (~ 0.7 for OH 22 - 1.9 mya for OH 60), Swartkrans fragments (1.8 mya) to cranial (KNM-ER 2598) and hipbone (KNM-ER 3228) fragments, Kenya (1.95 mya). **Thus the range of *H. erectus* fairly covers not only that of Neanderthals but also of ourselves, *H. sapiens* - surprising if one form evolved from the other <u>but not so if they are variants of the same basic, human archetype</u>.**

If the aforementioned groups could be fully human then isn't there definitive a bridge across the yawning gap between men and the apes? You might want to minimise the gap but here exist two problems. First let reptiles, for example, speak. Two practically identical alligator skulls, one fossilised (~75 mya) and the other contemporary, have, according to palaeontological tendency, been assigned a different genus and thus different names (*Albertochampsa langstoni* and *Alligator mississipiensis*). This, subtly but perhaps deceptively, lends variation evolutionary credibility. The second problem is, since archetypal separation is forbidden as a heresy, that evolutionary imperative luxuriates in speculative family trees; the trees (or their bushy branches) sprout an ever-changing foliage. In other words, there is, for humanistic scientists, necessary consensus over evolution but seemingly endless, fossil-fuelled debate concerning how its macro-evolution *must* have worked.

Was 'Ida', for example, the progenitor of lemurs, apes and men? In 2009 the well-preserved bones of a lemur-like primate, found in Germany in 1983, classified *Darwinius masillae* and dated 47 mya, were exposed to lucrative hype and, in the manner claims for missing links continually come and go, gross media manipulation. Overblown claims ('missing link found', 'eighth wonder of the world') and, launched with great fanfare, a publicity blitz ephemerally propelled the sexless fossil christened with a girl's name to an eager world. Even the normally circumspect Sir David Attenborough was sufficiently excited to declare that Ida showed us who we were and where we came from; there was not, he claimed, a shred of doubt - although both philosophically and factually there exists considerable bone of contention. Dozens of species would have occurred between Ida (who lacks any anthropoid feature) and yourself and so the galvanic term 'missing link' here fizzles into something as blurred and scientifically meaningless as pointing to a lemur in a zoo. But that theoretical imperative drives heavy speculation. Are you, without a shred of doubt, overwhelmed by proof of such an 'obvious' piece of evidence? Experts sharply disagree. Ida's now 'down-graded from *the* common ancestor to some simian branch-line which perhaps just withered fruitlessly. The moral of this overheated tale must be, "Don't lean on lemurs over-heavily!"

Where, therefore, to start? The fossil record of extant great apes is very poor. Enthusiasm might identify the modern as an ancient kind or, worse, construe what's ape as missing link. Of over 5000 species of ape about 4900 are deemed ancient and only about 120 are in circulation today. Whence did they emerge? Was the great ape common ancestor like Punjabi *Ramapithecus* or *Sivapithecus* (about 10 mya)? Perhaps not since, in 2007, a gorilla (aged and ennobled by its assignation of a separate genus, *Chororapithecus abyssinicus*) was found in Ethiopia and dated 10 mya; and in 2002 a sandpit worker in Thailand came across an orang utan mandible (7-9 mya). Did some other fragments called *Proconsul* (5 mya at latest) father chimps and us? That would leave a couple of million evolutionary years or so to randomly triple brain size and 'invade' human, *H. erectus* and Neanderthal zones. No doubt there are many bones to find; one lives in hope but will they staunch confusion or cool hot contention down?

Australopithecines are extinct apes. Milford Wolpoff (American Journal of Physical Anthropology; Nov. 1991 p. 402) summed the situation up. 'The phylogenetic outlook suggests that if there weren't a *Homo habilis* we would have to invent one.' On cue, therefore, enter *H. habilis* to bridge a yawning ape-man gap. In 1959 Mary Leakey had found so-called *Zinjanthropus* at Olduvai Gorge, Tanzania. Although at first associated with stone tools but also bearing a gorilla-like sagittal crest, much-trumpeted Zinj was classified as *Australopithecus boisei* but has since been 'down-graded' off the human to a 'slow' line as the genus *Paranthropus*. However, in the 1960's, members of the Leakey family excavated cranial, hand and foot bones associated with stone tools (such as choppers used today in nearby Turkana) in Bed II at Olduvai. (~1.6 mya). These were classified as *H. habilis* despite the fact that foundations of a circular stone lodge (such as are still built and used in Africa today) were dug up from the base of lower Bed I (1.9 mya). This, you might conclude, simply shows men lived with apes and ate their bush-meat just as now. At least ten more Olduvai hominids, KNM-ER 1802, 1805 and 1813 from Kenya and an Ethiopian cranial fragment from Omo, Ethiopia are also dated at 1.8-1.9 mya. In 1986 a partial skeleton (OH 62 with cranium and teeth similar OH 24 and KNM-ER 1805 and 1813) was discovered. But the rest of so-called 'dik-dik' hominid's body showed a height of a metre - even smaller than *Australopithecines* such as 'Lucy'. KNM-ER 1470 (with a reconstructed cranium of ~800 ccs.) and 1481 (a modern leg-bone) were found at the same level. Richard Leakey initially called 1470 human but it was later reassigned to another species (*H. habilis*) and then another (*H. rudolfensis*). The skull is indeed arguably human.

What does this tangled mass of facts demonstrate? You might well argue, when its cloud of dust has settled, that the taxon *Homo habilis*, comprised of various bones from different species of both man and ape, though much needed is a false bridge one born of goggled hope. **Whether or not this is the case, the bottom time-line ranges from 1.5 to 1.9 mya and the 'species' *H. habilis* is thus completely overlapped by and contemporary with both *H. sapiens* and *H. erectus* - meaning neither could have evolved through its ancestry.**

If stepping-stones do not exist how can you reach the other bank? It is always tempting for a palaeoanthropologist to try and squeeze a new find into standard hominin phylogeny. We shall, however, leave discussion of *Oreopithecus bambolii* (7-9 mya, an ape with humanoid features), the Toumai skull (6.5 mya, a female gorilla?), Orrorin (5.5 mya, a very early bipedal ape?) and other putative predecessors due to the density of academic dust-storm surrounding their few fragments. It is, precisely, unclear whence originated our conventionally designated ancestors, the *Australopithecines*. *Australopithecines* are fossilised, extinct apes - extant species of which often walk upright, hunt in carnivorous packs, exhibit other dramatic behavioural differences, use simple tools, play and communicate emotions in a way with which we can empathise. However, might we hope that crushed and crumbly, reconstructed 'Ardi' links them up? Over a hundred fragile, crushed and widely dispersed bones have been reconvened into a composite creature excavated from the same 'hominid-rich' region of Middle Awash, Afar where later Lucy was retrieved. In the regular pattern of evolutionary pot-boilers the smithereens were trumpeted as 'hardy-man', 'Ardi' or *Australopithecus ramidus* (~4.4 mya), a missing link that may have been the nearest to a common ancestor of chimp and man that we shall ever find! In fact, the skeleton was digitally reconstructed from pieces scattered in 17 locations over a distance of 2 kilometres and in need of 'extensive digital reconstruction'. A toe-bone critical to the hypothesis that Ardi 'almost certainly walked upright' was found 15 kms away. Criticisms were made, Ardi reclassified to a new genus (*Ardipithecus*) and claims for direct-line ancestry to man quietly dropped. This is the way of simian reconstruction - unreconstructed aping of an academic line.

At Olduvai Gorge in Tanzania Louis Leakey claimed to have found contemporaneous fossils interpreted as *Australopithecines*, *Habilines* and *Homo erectus* in the same Bed II. But Dart really started it in 1924. At The Place of the Lion (Ta-ung) in South Africa he discovered a juvenile skull with simian features but, he claimed, man-like teeth (classified *Australopithecus africanus* and dated 2-3 mya). Bones from other sites were collected and allocated dates of between 2-3 mya. In 1974 Don Johanson's team brushed metre-high 'Lucy' (AL 288-1, dated 3.2 mya) from the dust of Afar, Ethiopia and classified her *Australopithecus afarensis*. Lucy amounts to about a third of a full chimpanzee-like skeleton made up from bones scattered across a hillside; it is thus doubtful whether the pastiche is all from the same species let alone individual. The pelvis was found badly crushed, cracked and distorted. Hers tibia shows no human characteristics and her femur is badly crushed at a part which might indicate bipedal gait, the knee-joint; in 1973 two bones of a knee joint had been found in a stratum eighty metres lower than Lucy's although the distance between them and their distance from Lucy is unclear - could they really be from her? Recently the provenance of bones from which Lucy's 'bipedal' foot was reconstructed was challenged (Scientific American Nov. 2005). The reconstruction, it was noted, was actually created from a mixture of bones from *A. afarensis* (3.2 mya) and *H. habilis* (1.8 mya); the navicular, from which the arch of a bipedal foot is shaped, was of the later age. Nevertheless it was claimed that she 'had a spring in her walk', that is, was bipedal and thus fitted an imperative evolutionary line towards humankind. Many, including Jeremy Cherfas (New Scientist 20-1-83 ps. 172-77), have begged to disagree. Of course, chimps and other apes are happy in trees, running short distances on land or fording streams but are not bipedal. For example, rain-forest chimpanzees, *Pan paniscus*, spend a good deal of time walking upright and some apes hunt in part on foot but nobody calls them human. Could *afarensis* be, like the Leakey team's distorted skull called *Kenyanthropus* (KNM-WT 40000, 3.3 mya), simply a kind - or several kinds - of extinct chimp?

These and other similar but incomplete discoveries have been eclipsed by another *Australopithecine* star, the recently raised 'Dikika girl'. The incomplete fossil of this toddler was exhumed from near Lucy but, probably being older (3.3 mya), is not 'Lucy's baby'. Is she 'mankind's mother' either? She could not have told you so because the hyoid bone beneath her skull (comparable to an infant chimp's) barred speech. Her upper body shape including shoulder blades is ape-like; arms and hands are fine for knuckle-walking and for swinging through the trees, as is the ape-like inner ear for ape-like sense of balance. No pelvis is available but a partial foot and the angles of bones in Lucy's reconstructed and Dakika's unreconstructed hip suggest that *Australopithecine* 'chimps' might, like other chimps, have been able to walk upright if occasion required. In 1970 anatomist Sir Solly Zuckermann refuted Dart's claims for such a stance; and few anatomists support Johanson's claim that Lucy mostly roamed this way. In fact, detailed analysis has shown that *Australopithecines* are more different from humans and apes than the latter couple are from each other. By the way, in 1973 a geologist, T. C. Partridge, showed that the Taung cave from which 'Dart's baby' came could not have been formed before ~ 0.87 mya, meaning that the skull could only be that age at maximum. Such discrepant anomaly (humans were on the scene by this time) provoked immediate reclassification to *habiline* or, later, *Australopithecus robustus.* However, according to Dart's successor, Phillip Tobias, by 1973 the skull had still (despite fifty years of propaganda value) been fully analysed or described. And surely, argued Richard Klein, a date of 2 mya is 'more reasonable' than Partridge's embarrassing chronology? Thus the latter's work has been ignored.

Bewildering bushes of hypothesis, dotted line phylogenies and, in a far-from-clear terrain, a lot of questions are marked. **It keeps the business busy but still, at this point, you might well interpret all the range of bones as ape *or* man without any substance in between**. This, however, must not happen; naturalism's theory casts a powerful spell. X evolves from Y and, by competition, its superiority eliminates the lesser Y. Death is central to the theory; evolvants don't long co-exist. Each gradual step wipes out the one before. Three main conceptual schemes exist to 'fit the bones' and cover such 'ascent of man'. The first, paraphrased in Howell's line-up, the 'March of Progress' or the 'Fake Parade', is from ape (or apes) more simian to less; you just need to fill the morphological gradations in. More human drives less to extinction till, unique in glory, mankind tops the bill. Today one species, modern man, and other definite kinds of ape, are all that's left of all this bridge-and-merging; it's not happening anymore; no apes are turning into other kinds of men! Evolution must, it seems, have stopped. Clearly, as we've seen, there are objective problems with this picture so two other conflicting frames have been developed. These are the Multiregional Continuity Model by Australians Milford Wolpoff and Alan Thorne; and 'African Eve' or 'Out of Africa', the currently preferred model developed by Rebecca Cann, Mark Stoneking and Allan Wilson at UC Berkeley, USA. Both models, despite contrary evidence, date *H. erectus* from between 0.4 and 1.5 mya; but at least 140 specimens are younger than 0.4 and over 30 dated older than 1.5 mya. The latter are, apparently, ignored. The slightly taller Neanderthals are, in turn, dated from about 28 kya. This time would be sufficient for morphological change due to genetic interbreeding to leave us our own race; but it would mean interbreeding whereas 'African Eve' theory requires Neanderthals to be a sub-species or, better, different genus that evolved *H. sapiens* wiped out. What, moreover, to make of Sangiran, Kow Swamp and the whole rash of other time-anomalous specimens? The situation isn't, without determined, widespread geo-chronological gerrymandering and/ or reassignment of fossil dates, going to change. Which is it to be?

Hypotheses are flexible as dates uncertain and so let's proceed. As noted, The March of Progress from definite ape-forms through morphologically plausible grades of link (say, *Australopithecines* through *H. habilis*, *erectus*, *neanderthalensis* to, finally, *sapiens*) has now, although still simply and strongly informing the public mind, been firmly relegated to the fiction bin. Through this portrayal of ascent, so attractive to the Victorian and even 20th century sense of European superiority over colonial natives, an obvious, intrinsic, racist streak pervaded. Gradations rose towards our modern selves but, even in our modern species, splits between evolving forms were cut. Only through Neanderthal and other aboriginals can we obtain the current peak of educated *Homo occidentalis*! Two, perhaps three, more major models have evolved to replace Howell's hit parade.

In the 1960's Carleton Coon proposed that five main human races (Caucasoid/ Aryan, Mongoloid, Capoid (bushmen), Congoids (negroes) and Australoids (Australian aborigines)) were each derived separately from widely dispersed populations of *H. erectus*. However, Coon estimated that Mongoloids had been evolving since Java and Peking Men (say, 700 kya), Europeans since Neanderthals (say 450 kya, Petralona Cave or 500 kya, Heidelberg Man), Congoids from (as he dated Rhodesian Man) 40 kya and Australoids from an even more recent date. Clearly, if different races had been evolving for different lengths of time, then some would be 'more evolved' than others. Coon's pact with fact had not cleared the cultural air of racist whiff. We need to better understand why his reading of the known, scientific facts engendered such revulsion as to spawn two further theories - of which the most evolved i.e. the most politically correct, is 'Out of Africa'. Since when did politicians care too much for fact?

Blame God not Darwin! A violent streak in centaur man has always sought to differentiate and dominate. Whomsoever's not in my gang is to be attacked, abused or, the extreme, exterminated. When, in all history, have not bullies kicked and slavers thrived? When has poltics, in order to expend with or exploit, not demonized 'those others'? To prove Neanderthals were an inhuman kind you'd have to try and interbreed. That's not possible but still the British almost proved sub-man's humanity. They committed shameful genocide of which remains survive. Between 1804 and 1876 the local Tasmanian population was entirely wiped out. Of course, before this happened slavery and sexual relations generated mixed blood descendants, thousands of whom survive. Nevertheless, 'scientific racism' of the day held that Tasmanian aborigines were a 'missing link' between stone-age primitives and modern man. Erasmus Darwin had an aboriginal dug from the grave to initiate a large collection of stuffed Tasmanian exhibits at The Royal College of Surgeons.

His grandson Charles, whether or not implicated in the Tasmanian terror, wrote that the difference between a primitive from Tierra del Fuego and a European was greater than between the native and a beast (see also Chapter 26: A Continual Fight). His nephew, Francis Galton, used the measurement of skulls (discredited phrenology) to gauge human status; he also initiated the study of eugenics, selective breeding that might, in the manner of horses, dogs or sugar beet, drive human evolution, by the abortion/ post-natal elimination of inferior/ unwanted stock, into superhuman gear. Does such a notion figure as an undercurrent in today's genetic engineering? If so, beware the tendencies!

In the 1890's Friedrich Ratzel had intellectualised the idea of *lebensraum* (living space) - one already practised with vigour by the systematic, military massacre of black locals in German West Africa. Then in 1904 Alfred Ploetz, founder of the German Society for Race Hygiene, wrote to Galton praising his work. Later Heinrich Himmler praised Ploetz. Galton was therefore one deft step removed from the intellectual responsibility for Hitler's death camps. A direct consequence of the holocaust was the establishment of a home for the Jews, Israeli-Arab grievance and the religious veneer of global terrorism. Doesn't Galton's ghost haunt you as well? Did he learn nothing from 'progressive' Uncle Charles? Not only did an evolutionary train of thought justify and thereby foster attitudes of racism but also various imperial malpractices and 'medical' experimentation with convicts. It also (Chapter 26: The Nature of Evil) raced directly to justify and thereby bolster fascist, communist and other political forms of tyranny, murder and genocide. The philosophers and scientists who developed neo-Darwinism were eugenicists but, post-holocaust, public discussion of this issue was abruptly shelved.

Was the idea dead? What about promotion (or enforcement?) of social and biological improvement (= evolution)? In the 21st century might Richard Dawkins or Peter Singer venture to ask whether eugenic selection is all that bad when it comes to selecting for musical, mathematical or athletic talent and, certainly, promoting a vigorous population and culling the disabled? *As the basis of secular thought, evolutionary progress and, if necessary, evolution helped along by man is so engrained in the modern mind that it now subliminally bolsters attitudes towards all manner of controversial issues ranging from abortion clinics through cloning to embryonic stem cell research.* Aren't prenatal selection tests and death camps for the unborn now in legal operation? Aren't chambers of that holocaust in action in the wards of hospitals that sling out by the million unwanted, inconvenient or, in someone's signed-up-to opinion, sub-standard foetal (but fully human) damned. Such sanctity! A brave new world of 1984 might well, on such a slippery slope, slip in. But never mind. Someone will tell you how he's reasoned it's alright.

Why aren't apes turning into humans now? The evolutionary pursuit of 'living links' has also led to sad, individual cases of cruelty. For example, the Belgian colonial authorities exported many pygmies (a type already used to enslavement by local tribes) to European and American zoos. One such Congolese primitive, Ota Benga, was in 1904 torn from his family, taken to the USA and, among other exhibitions, caged with various apes in the Bronx Zoo.

The zoo's director, learned, expert Dr. Hornaday, was in tune with prevailing sentiment but ignorant of chromosome 2. He waxed lyrical about his 'transitional specimen' until, unable to bear the treatment any longer, 'it' committed suicide. Nor was Ota the only one. Since 'racist' attitude to 'inferior species of native' was a cultural norm in Darwin's day (see also Chapter 25: A Continual Fight), check for yourself the unsurprising suffering of eskimo Abraham Ulrikab's family in Europe.

Ota Benga wasn't even 'real' but why not mint a million *real* ape-men? Is this the currency of evolution's great idea, 'survival of the fittest'? Is this where black philosophy can lead? If so, thank God that passionate atheist and evolutionist Josef Stalin tried but could not coin chimeras made of monkey and of man. In 1926 he ordered his Academy of Sciences to produce 'new, pain-resistant, invincible humans - living war machines'. Biologist Ilya Ivanovich Ivanov, whose scientific pedigree was in the line of artificial insemination and who had already considered the possibility of creating a hybrid ape-man, was engaged to execute the task. In French Guinea, Africa, he tried to cross human male and female chimp; and, in the reverse, five women died. In Georgia further attempts to hybridise humans and apes and establish the reality of ape-man also failed. Ivanov was subsequently 'purged' and exiled to Kazakhstan where, in 1932, he died while still working at the Kazakh Veterinary-Zoology Institute. But it is unlikely he will be the last of it. Some fervent biologists now claim they can take evolution over. They can mix and match genetic profiles empirically finding, according to commercial orders, what works best. Mindless nature needed luck but, using biotechnology, they want to create on purpose. Unmindful of the nemesis, what monstrous strangeness will they dare, what demi-organisms (called chimeras) will they arrogate to make? Perpetrators always find a reason; but is genetic mix 'n' match a game that, mindful every powerful technology is also malleable to evil purposes, wise men should embark upon? The morality of evolution is 'survival of the fittest'; on this raw ground could you defend transgenic engineering's moral case? Or is morality a function of the sanctity of well-created forms of life? Materialism scoffs at such 'creationist' mentality; scientific atheism scorns it as an opiate, uncurious, 'recessive' and a mutant weakling's code. By which mind-set is rightness found or truth received?

Contemporary culture as a whole excoriates, sometimes illegalizes racist mind-sets, genocide and slavery. Indeed, you can't blame a theory for the sins of man; you *can*, however, deprecate the fact that this one intrinsically justifies them. Bearing this in mind, let's surface to the second shot. After Coon's disturbance let's discuss the third conceptual ape → man flow.

In Africa and Asia the European Neanderthal is replaced by 'archaic' *H. sapiens* and *H. erectus/ ergaster*. America is virtually a blank. In this case Wolpoff's Multiregional Continuity Model proposes (as half-opposed to Coon's parallel, separate evolution) an emigration of *H. erectus* from Africa 2 mya that led to populations spread across four continents. Evolution was semi-parallel as interbreeding between the groups occurred; and the line led through archaic humans (including Neanderthals) to modern. This model has the merit of 'sinking' interbreeding *H. erectus*, *H. neanderthalensis* and *H. sapiens de facto* into the same species; nor, due to interbreeding, would three separate populations of *H. sapiens* evolve at different rates and times; and the single species would show a measure (such as we know today) of geographical, genetic and morphological variation. It does, however, require the usual extended periods of evolutionary time during much, perhaps all of which either modern or 'archaic' *H. sapiens* was alive (e.g.. Laetoli 3.75 mya and Kanapoi 3.5 mya). However, the model does not solve problem of parallel coexistence that the actual fossils puzzlingly, perhaps even irritatingly, display. Maybe, moreover, there's a better way to blot out racist implications evolution theory holds for the ascent of man.

Every decade or two a new fashionable theory appears - in 1987 a fourth projection did the trick. The theory dubbed 'African Eve', 'Mitochondrial Eve', 'Out of Africa' or 'Replacement Theory' was rolled, with customary fanfare, off the blocks. All propaganda presupposes that repeat publicity will generate its own momentum. Add to this authority of science and you have what was, for a few years and for some still, the favourite explanation of the reason that you're here. Eve - a pressing substitute for Our Lady of Eden - resolves the issue in a quite semitic way. Time-scale apart, it claims all humans rose from just one 'lucky lady' (or a population of her peers) about 200 kya; then descendants sallied out of Africa and slaughtered all 'inferior' specimens they met. There was no ethnic interbreeding with Neanderthals or *H. erecti* further east; they were just replaced. Thus humanity, by way of genocide, reflects the mien of Africa!

On what evidence is theory based? Could it actually be true?

Firstly, fossils. You might allow gradations of morphology in Africa; some evolutionary continuity should build towards humanity. Outside Africa you would expect discontinuity, a sharp distinction marking out superior conqueror from inferior, vanquished and sub-human stock. This is not, however, what the fossil record demonstrates. The Multiregional Continuity Model excels in classifying *H. erectus*, *neanderthalensis*, archaic and modern *sapiens* as, in fact, just *Homo sapiens*. What, for example, of Dark Eve's place in Sima de los Huesos or Israeli caves? What about the borderline morphologies suggesting human/ Neanderthal interaction across eastern Europe, Neanderthals dated 5.7 kya at Amud Cave (surely some mistake, contamination perhaps?), *H. erectus* (6 kya) at Mossgiel and the Kow Swamp people of Australia? What, moreover, about the period, much longer than 200 kya, during which, by fossil evidence, *H. sapiens* once lived? Indeed, why for these has evolution stopped? And why, having allegedly replaced the aboriginals, did Eve's people come to differentiate and look like them? Her exodus from Africa, pregnant with the greatest ramifications for any human migration in history, is backed, some claim, by a few Ethiopian fragments (three broken crania, other pieces of skull and some teeth, 150 kya) called Herto or Idaltu Man and classified *H. sapiens*. Nobody living today could have seen Moses in Sinai but at least the journey's written in a book; and many independent facts corroborate the trail as historical fact. Date it thus at 3.5 kya. Much, much more questionable is the historicity of Eve's great, scientific sally out of Africa (say, 50 kya). Were there lots of Eves whose lines, except our own, have through conflict, illness or catastrophe, all gone extinct? Or did competitors drift out of competition by giving birth to only sons? What serendipity created a 'genetic bottleneck' by which we all have, luckily, one mum? There is, unless as salve for racist undertones of evolution theory, no answer. Moreover, did mum come from Africa? Results from the computer program through which original data (from 136 different women of different backgrounds) was run were biased according to the order of data entry; thousand of subsequent runs in random orders have deleted the preference and have shown that, according to order, Eve could have come from practically anywhere. Thus it seems less likely that her hypothetical troupe, as murderers perhaps even cannibals of any competition, ever spread forth all-conquering to claim a wide, un-promised, intercontinental land. Indeed, is Eve a fiction born in California? Did the girl exist?

Necessity's the mother of invention. Imaginative Eve's not finished yet. Subtler evidence, genetic evidence, can now be tapped. In fact, there's several genetic strands used plausibly to argue human evolution. We'll take a peep at these before return, by way of mitochondrial *DNA*, to 'Eve of Africa'. Firstly, therefore, ask what kind of rising son or waxing moon are you? Check back to Chapter 16: *H. archetypalis*; the Image of Man) and recollect that, genetically, you are about 90% chimp and, as regards your set of protein-coding genes, 99% rodent! So when did the becoming-human game begin? To put matters in perspective do you remember (this Chapter 21: Evolution in Action?) that 150 years of *E. coli* bacteriology generating an optimal 70 generations per day would produce nearly 5,000,000 generations each populated by, at very least, millions of individuals. In human terms this number of generations would equate to more than 100 million years of history. In the simpler, prokaryotic case every type of mutagen has been systematically thrown at organisms and yet no major change of any kind has ever been observed. This applies to all other genetic experiments. Strains, races and, at maximum, new species are the most that's come about. What's claimed for humankind? No more than a strain of ape?

At least 15% of human genes better match gorilla than chimp. Nevertheless, if a chimp genome is at a high guestimate 95% human then (check on the back of an envelope) 5% of 3 billion base pairs is 150,000,000 - though don't forget that. 150,000,000 differences means that number of mutations need to have occurred, in an order that was neither lethal nor nonsensical in terms of producing a human, over the time since an alleged chimp-human divergence. What time was that? Palaeontologists suggest, vaguely, between 8 and 25 mya; protein studies less than 8; and genetic comparisons ~ 6 mya. The current knee-jerk answer is between 5 and 7 mya. Let's go halves and call it 6. Such age would therefore require about 25 'positive', cumulative (that is, conserved and not lost) mutations every single year since human-time began! Two a month, regular not random, serial and not unpredictable in their direction up some individual man-bound line - and now we're there where are we just as rapidly proceeding? Is evolutionary speed increasing with sheer population numbers? From ape to man is over but it seems that evolution's stalled. It isn't happening any more. The present is no longer any clue to what has passed so should you claim that mankind isn't going anywhere? Or could it be that mindlessness, by inadvertently creating mind, has passed a teleological baton on? But there's a theory to maintain. Mindless evolution *must* have happened and insensibly be happening. Therefore let the tale proceed.

You might ally homology of code to that of form and function. If you ascribe primate homology to evolutionary trend then *DNA* suggests your theory's right; but if you ascribe it to design then similar routines might correlate with similar constructions and thereby equally suggest design-interpretation is correct. However, some wax most compelled not only by an evolutionary interpretation of homology but by what seem genetic 'faults' that chimps and humans share. Such bugs are locked like 'prehistoric flotsam' in the database of both types indicating - if they are mutations and were, of course, not programmed purposely - that men arose from chimps or, more likely, both from common ancestors.

Of course, nobody argues that bugs can't develop to disorganise a program; degradation from an orderly initiation will occur. **Just as physics deals in orderly beginning from a transcendent projection, so does the bio-logic of design**. Nevertheless, do you remember pseudogenes (Chapter 23: Non-protein-coding *DNA*)? They resemble functional genes but seem, by one or more of several means, disabled. Identical (and thus by definition highly conserved) deactivations in different organisms are interpreted as evidence of common ancestry. Why else should 'mistakes' persist - unless they are not random and conserved because they've definite jobs to do? No doubt, mutation's entropy of information takes its toll but, as in the case of mobile agents called transposons, we've seen that some can act as functional units of control. Perhaps pseudogenes are not all 'pseudo' and advanced investigations will demonstrate that some are functionally important sections of the genome; suppose that so-called falsity is false and they can act as switches, silencers or 'scaffold' proteins needed in development. However, why does a single 'pseudogene' (out of four genes needed) stymy the production of vitamin C in some types of monkey, bat, bird and fish; also chimps, gorillas, guinea pigs (but not rats) and humans (except at the foetal stage)? Possibly the gene was once operative and its 'corpse' is a product of miscopying or other genetic wear and tear. If, on the other hand, the problem is regulatory (as synthesis of vitamin C by infants might suggest) then perhaps some humans are still able to synthesize the vitamin. Otherwise, although no function is known for this conserved pseudogene it may turn out to have one. Even if it does not, deactivating mutations may or may not have occurred by chance in the same weak 'mutation hotspot'. At any rate, conservation has dug in its heels across a broad spectrum of organisms. Interpretations of this fact can vary. You are not compelled to toe the party line.

Another pseudogene, the so-called β-globin pseudogene, also shows great 'conservation' or 'stabilisation' of structure in mouse, cow, rabbit, chicken and all major primate groups including man. The fact that 'breakage' occurs at the same place in chimps and humans is often interpreted, in evolutionary style, as meaning that a functioning gene or extra copy of it mutated in a male/ female pair of imaginary ancestors; and that no non-mutants survived. Since natural selection should have cleaned it up what advantage does the much-conserved genetic trash convey? We've noted conservation is interpreted to mean a gene serves some important role; and congruence of genotype or phenotype may just as well result from common blueprint as descent. Certainly neither we nor chimps appear to suffer from the 'silence' of the β-globin pseudogene. Therefore, given wholesale but ignorant labelling of n-p-c *DNA* in the 1980's as 'junk' (see Chapters 19: The Computer Analogy and 21: Non-protein-coding

DNA)), why persist in error and dismiss as junk sequences whose meaning isn't perhaps as yet perceived? Is it transcribed to *RNA*? Does it mediate as a switch between production of foetal and adult haemoglobin? The question is why some genes have been truncated, rendered apparently useless but conserved at the same position in mice and chimps and men. Would such conservation-by-accident across eons of macroevolution show that, far from a rat, you had descended from grandparent mice? Or are pseudogenes (and repetitive elements) factors whose uses we do not yet properly understand? Clearly you could interpret the presence of 'apparently ancient' relics at the same hot-spots on mouse and human genomes either as a case of evolutionary lineage or of critical function. It is only by assuming the former, that is, non-functionality, that any description of so-called pseudogenes as genetic graveyards is maintained. Perhaps, as The Computer Analogy suggested, some are a facet of the genome's operating system - one whose super-coding finely regulates expression of the genes.

Here's another puzzle. In a chimp the analogue of human chromosome 2 occurs as two smaller factors, 2A and 2B. First off what, in any chromosome, is a centromere? It is a spindle attachment, that is, the region at which a chromosome attaches to the spindle during the mitotic process called cell division. Chromosome 2 seems to show centromeric traces at the two points expected if chromosomes 2A and 2B (each with its own centromere) had joined. Next, what is a telomere? It may be a repetitive *DNA* sequence (in humans and apes TTAGGG) that 'laces' the ends of linear chromosomes so that they do not 'fray'; and it may at the same time act as a marker that defines an organism's 'natural span' of life. In this case telomeric elements have been found in chromosome 2 where the junction of chromosomes 2A and 2B would have left them. Finally, recall that each nucleus contains 2 copies of each chromosomal type, one from each parent.

It has, therefore, been claimed that, since human chromosome 2 might be construed as a fused version of chimp 2A and 2 B, humans must have evolved from chimps. Though plausible is this necessarily the case?

For a start change in chromosome number is, as with Down's syndrome, damaging and rare; but to create humans (if that is what is claimed) it would have had to be, besides rare, most advantageous. Never are ape-humans seen emerging now. How could an ape's 2A and 2B, in its gonads, fuse? Did the maternal pair first fuse (♀2A + 2B > 2) leaving paternal ♂2A + 2B asymmetrically intact (or *vice versa*)? How thus could meiosis and, thereby, reproduction, occur? Or did both pairs fuse, in some magic moment, simultaneously? In this case an ape would have chromosome 2 able to undergo meiosis and create gametes; but its 'unfused' partner would not. Off-spring would, in this rare case, contain a mix - either (2A + 2) or (2B + 2). How could such organisms be viable, that is, undergo meiosis, form gametes and go on to reproduce? Suppose, against the odds, such viability occurred in the form of a brother and sister; and this pair - each (2A + 2) or (2B + 2) - later incestuously mated: or that one such child mated with its parent of opposite sex. Either way, incest bore our race because if one such kind of 'heterozygous' pair had viably mated twice or more they could have possibly created primal humanoids (chromosomes 2 + 2, like us). They could have theoretically created disparate human youth, both Adam and Eve, in a single generation. You'd suddenly have primal humans in amongst a troupe of apes; wild children would be educated in a truly feral way. Otherwise such complex business would require even more extreme luck, the simultaneous transformation of (2A + 2B) in ♂ and ♀ chromosomes in the gonads of not just one but both parents! If, if, if - all most unlikely. *Yet this or similar unlikelihood, along with millions of 'nudging' yet intentionally-blind mutations, would be needed for the evolution of humanity. Such improbable, microscopic randomness to wield such vast effect! If not God then thank your lucky stars you're here*!

How compelling is this ape-to-human evidence? If chromosome 2 alone made man (with all his exceptional attributes) it would be an interesting possibility. Why, however, is this conjunction still not being made with primal Kim-types leaping in the trees or running out Africa? Do the experiment! You can't, by law, extract chromosome 2 and insert chimp 2A and 2B into a human egg or sperm - but you could do the reverse to a chimp. Would half-humans appear, born in the cage - which you might mate to find, in little more than a decade in the same cage, a real and erect sprinkling of Adams and Eves. In fact perhaps, due to morphogenetic constraints and genetic difference (especially in n-p-c *DNA*), many thousands of additional changes are necessary. For example, in chromosomes 6, 13, 19, 21, 22 and X there is the same banding but significant 'non-coding' regions of difference; chromosomes 3, 11, 14, 15, 18, 20 and Y *look* the same; and the rest, except 4 and 17, look alike. Indeed, why should a joined or separate chromosomal situation make any difference if full genetic sequences are present in each case? An evolutionist might be prepared to wait for more evidence that random processes constructed the systematic complexity of a human. On the other hand, would not tens of millions of base pair differences and half a million insertions have ruled out the gradual evolution of an ape to man - especially over a short five million years or so.

In fact, it might reasonably be asked (outside the evolutionary box) whether such alleged fusion occurred at all. It may be noted that the 'fusion point' involves a large amount of telomeric *DNA* missing or garbled; moreover interstitial telomeric sequences *(ITS)* are commonly found throughout mammalian genes where they are not, except in this prejudicial case, regarded as the scars of fusion. If it *did* happen why does it necessarily indicate descent from chimpanzees? *Humans could equally have originated by common descent or by common design with two chromosomes which later fused and then, in some genetic bottleneck, became fixed throughout the population; or they could have originated with a single never-fused chromosome 2.* You might argue, when the full subtleties of genetic information are finally appreciated, that the reason for each 'line of genetic code' (and thus its 'design-behind') will have been demonstrated.

An enormous range of foreign substances (antigens) can be recognised due to cell surface receptors on molecules such as antibodies, T cells and what is called the *MHC* (major histocompatibility complex). Their highly variable regions, derived from highly variable genes, generate the immune defence system's antigen (or foreign body) binding sites. In the 1990's some biologists deftly chose *HLA* genes from the *MHC* system to demonstrate (whether or not is numerically possible within 10000 years, which it is) the impossibility of human derivation from an original couple. This couple could embody a maximum of four alleles whereas there exists a complex of over 200 *HLA* genes each with its highly mutable region (called exon 2) - a fact that would seem to make myth of the couple. The assumptions and calculations used to rule them from reality are arcane but plausibility can, if you let them, cut both ways. Thus it is argued (in Science and Human Origins p.105) that the immune system's need for variability allied with such provision through genetic hyper-variability (perhaps with gene duplication added) could have generated the current *HLA* complex from a few original *HLA* alleles. In other words, a couple is ruled back to possibility.

At any rate, let's start reeling back towards mitochondrial Eve. First let's meet scientific Adam (whose address is not in paradise). He's an interesting chap. In most mammals, some insects and with variants in birds and fish the chromosomes that determine sex are called X and Y (human: XX female and XY male). Since 95% of the human Y-chromosome does not recombine it is passed without change (except mutation in its *NRY* or *n*on-*r*ecombining part) from father to his sons. No lottery. Rather a direct, unbroken male-only line of inheritance from Y-chromosomal Adam to the present time - to us. Clear of recombination's incalculable blur such patrilinear descent therefore nicely indicates, by comparison of mutations alone, the amount divergence of one Y-chromosome from another, a possible location of its owner and (if you apply a molecular clock) a putative age.

Y-chromosomal Adam (Y-MRCA) is the most recent common ancestor (MRCA) from whom all living men are descended in the male line. Y-MRCA is, because more genetic diversity has so far been found in Africa than elsewhere, presumed to have arisen somewhere (where exactly is debated) there. The 'basal' divergence is identified as one between bush-men and the rest of the world including all other Africans; and its time (depending on which set of markers you use and which molecular clock speed you adopt) is variously identified as between 40 and 400 kya. This vagueness was assigned a median age of about 188 kya but the guestimate was soon reduced to between 49 and 37 kya (Whitfield et al.). This is, in evolutionary terms, a minuscule amount of time and even post-dates Eve's suspected exodus from Africa; indeed, thoroughly confounding Biblical nonsense, theory has it Adam never met Eve since he was perhaps as much as a hundred thousand years her junior! In fact, a more recent study (Nature 385; Jan. 1997 ps. 125-6), using a larger sample but a much shorter *DNA* sequence, found remarkably little Y-chromosome divergence within any primate species; at the same time, paradoxically, it found significant divergence between species. Internal lack of divergence would indicate recent origin of the species; external divergence might indicate an ancient common ancestor or separate creation. Did you ever believe chimps and humans were 98.5% or even 95% genetically similar? Although, according to evolutionary imperative, chimp and human lines diverged about 6 mya a definitive study (Nature 463; Jan. 2010 ps. 536-539) has revealed that chimp and human Y-chromosomes 'differ radically' in both structure and gene content. Such 'remarkable divergence' means the two sorts of Y are more comparable to the difference between human and chicken chromosomes! The specious explanation given is that, for no particular reason, this 2% or so of the genome alone must have must have embarked on an evolution spree, that is, exceptionally rapid mutation as opposed to previously discovered conservation. Of course, hypotheses abound to try and counter such confusion but never currently, on pain of academic death, include the obvious anathema of difference by original design!

Does age mean beauty? Could antique ladies steal the show? One definitely fires passions - so-called 'mitochondrial Eve'. Now, therefore, let's consider not the female X-chromosome but 'female' mitochondrion. A mitochondrion is an organelle that contains, on a couple of strands coding for 37 genes, about 0.00006% of the human genome. A fertilised egg has one diploid nucleus but 100,000 mitochondria (a sperm has about 1000); thus mitochondrial *DNA* is high in copy number. Also, being unchecked by sophisticated nuclear editorial mechanisms, its mutation/ evolutionary rate is higher. Finally, while nuclear *DNA* is recombined (by meiosis and fertilisation), it is often assumed that mitochondrial *DNA* (mt-*DNA*) never is; the sperm mid-piece and tail, where mitochondria reside, are supposed to always be destroyed as they enter an egg. Thus, as a wife surrenders surname and her husband's name survives, so in this case only clones of maternal mt-*DNA* alone survive. Without any paternal mt-*DNA*, only maternal mt-*DNA* should thus provide an exclusively matrilineal 'clonal inheritance' whose succession could be measured by the same logic as used for the patrilineal Y-chromosome. No recombination to blur clarity means only mutations accurately reflect the history. *Thus, along with its other useful features, mt-DNA's strict matrilinearity is central to the 'African Eve' hypothesis*. In this hypothesis Eve's stock replace any previous, more primitive stock that may have co-existed anywhere.

It was assumed that mitochondrial mutations ticked at a constant rate so indicating, by means of this 'molecular clock', a chronology of changes. And in 1987 a study of 136 women from many different racial backgrounds seemed to indicate a single individual or individual group with aboriginal mt-DNA dwelling in Africa about 200 kya. Further analysis, this time of European stock, seemed to identify an ancestry leading from five 'daughters of our Eve'; and the original human group may have been as few as two. Five women bore the indigenous children of the Common Market and the white man's commonwealth. A few more, it is estimated, bore all Asian stock as well. From perhaps a single matriarch there issued everyone. By all accounts, what a close and recent family relationship all humans actually have.

However, the computer program into which data was entered was 'order sensitive'. Thousands of subsequent runs in randomly different orders have not preferred an African origin. Nowadays the frank admission is that 'Eden' could be almost anywhere; it just depends on your authority's belief. Moreover perhaps 'mixing' between tribes of different genetic constitution blurred the clarity of a single line to an extent that makes analysis of an original Eve impossible. It is not, of course, necessary that maternal and paternal mt-*DNA* might somehow actually recombine; multiple types of quite dissimilar mt-*DNA* have been reported in a single individual. Thus one assumes no mixing. In fact a series of assumptions, born of a desire to put Eve on the scientific map, become forbidden fruit that causes her demise. Thus, perhaps more seriously, the apple of a metronomic tick is, as well as arbitrary and unfounded, incorrect. Indeed, perhaps the assumed regularity of a molecular clock (which is supposed to click in time with constant and yet, by contradiction, random mutations) runs faster than imagined, making Eve proportionately younger. How do you calibrate a clock like that without assuming what you want to find - evolutionary time-scales lifted from assumptions over evolutionary divergences between organisms? And what was the genetic constitution of original Eve against which changes are calculated? Do you mark 'original state' against a chimpanzee assuming with gross circularity what you would prove - that man evolved from such a common ancestor? Or do you simply guess?

Worse still, fatal to a theory based on strictest matrilinearity, could paternal 'leakage' happen? Of the 1-2% proportion of spermatic to zygotic mitochondria could none slip through? Even a cumulative fraction of a percent per generation would render theory opaque. Actually, John Maynard Smith and others were systematically ignored, to his frustration, by an African-Eve prone establishment, for suggesting that paternal and maternal mt-*DNA could* be swapped (in chimps as well as humans). Moreover, lack of paternal mt-*DNA* destruction has been discovered in sheep and humans; in 2002 (Schwartz and Vissing) a large proportion of paternal mt-*DNA* was found in a patient's muscle tissue - and if in one human, in how many others? The fact might seem to blow a whistle on the home run of Eve's game. In 1996 a study in the Proceedings of the National Academy of Science, USA 93 (24): 13859 -63 said: 'The current view (that there is no paternal 'leakage' of mt-*DNA*) is incorrect. In the majority of animals - including humans - the mid-piece mitochondria can be identified in the embryo even though their ultimate fate is unknown ... The missing mitochondria story seems to have survived - and proliferated - unchallenged ... because it supports the 'Africa Eve' model of recent radiation of *Homo sapiens* out of Africa.

Mitochondrial *DNA* has actually been recovered from Neanderthal fossils (including the type specimen found alongside modern human bones in Feldhofer Cave). Differences between this and modern samples have been interpreted to show Neanderthal divergence from the human line 600 ky; and that interbreeding did not, against the fossil evidence, occur. However, if you eliminate all modern sequences on the basis that they might represent 'contamination' obviously you only retain what is different. Indeed, Neanderthal *DNA* too close to human would, under such a regime, be excluded from consideration; you would thereby, by a methodological tautology, simply prove what you set out to prove! The same result occurs if you use primers (for your *DNA* amplification) that will only amplify Neanderthal mt-*DNA*. You may find a particular stretch absent but this proves no more than incomplete modern samples or genetic variation due to drift, 'bottleneck' or other factor. The real issue is the arbitrary exclusion of modern material, presumed due to contamination, which might actually be from the Neanderthal. Mt-*DNA* may, moreover have undergone significantly more mutation than nuclear; Neanderthal nuclear *DNA* might well be within the orbit of modern human. In fact, it is not known how different a human genome must be to represent a different species. Certainly *DNA* from Cro-Magnon and, separately, apparently *H. erectus* remains from Kow Swamp, Australia have been shown human (Proceedings of the National Academy of Science, USA 98 (2): 537-542); also, similarly, three Neanderthal and one anatomically modern Australian (Mungo Man 3, 40 kya) have mt-*DNA* different from contemporary humans. In this case, perhaps *H. erectus* bred with us and, across the other side of the world, Neanderthal as well - because they were us! As Sima de los Huesos seems to confirm, aren't all these forms just variations on the human archetype? In other words, the simplest conclusion would be large variation, quite apart from speciation, in human mt-*DNA*.

The original 'Eve' theorists made plausible but impossible-to-prove assumptions (no paternal 'leakage', mt-*DNA* passed unchanged except for mutation, no mixing of populations, constant rate mutations, the genetic constitution of Eve etc.). In a letter to Science (Science 255; Feb. 92 ps. 737-739) one of those founder theorists, Mark Stoneking, squarely in the scientific way admits fresh facts may have overwhelmed hypothesis. What might the next consideration be?

Hairline cracks in cups or bells cause clunks not rings. How, in the last analysis, can large cracks in modern ape-man theory ring with any kind of truth at all? Of course, the issue of descent is as much a personal and cultural as a scientific one. It touches on the family heart of who we really are and, therefore, in what kind of universe you think we live. No doubt, mind-sets of *top-down* and *bottom-up* perspectives interpret differently. One links the human centaur with a higher cosmic order and the other roots emergent body in its mindless earth. These opposing views are always locked in mortal combat; bones rattle, roll and suffer use as intellectual clubs. Books trying to make the case for ape-man line the shelves of libraries. Jaws, skulls and whole, hypothetical bodies are reconstructed from fragments of bone. Fervent anthropological dissension, error, reversal of opinion, self-deception and even fraud litter the territory. 'Ass taken for a man' (Daily Telegraph 14-5-84 p.16). A skull found in Spain and promoted as the oldest example of *Homo* in Eurasia was later identified as that of a young donkey. It seems that this time men became an ass. Is all that glitters gold? Isn't the problem with a lot of anthropologists that they want so much to find a hominid that any scrap of bone becomes a hominid bone?

Has this section cleared the smoke or readjusted mirrors so you see yourself directly? Did apes produce great writers, prophets, scientists or philosophers? In this behavioural respect *top-down* decrees that to belittle the informative dimension is a superficial view; comparison between our '***Phylum cognoscens***' and apes is cheap. Yes, there are similarities but man and ape are deliberately differentiated due to variations in their archetypal program; they embody preordained distinctions in routines and subroutines of regulation by the master and sub-magisterial genes. At this point, therefore, simply try to view the evidence from a *top-down*, dialectical angle. From here the 'anthropoid' condition is clear. *Of its set (new and old world monkeys, lesser apes and great apes) only great apes are of ape-man interest. This subset (of gorilla, orang-utan, chimpanzee and man) splits neatly into two - apes and men. No blur.* No fake parade, no finely-graded 'transitional' missing links, no communistic ape-man. ***Man is man and ape is ape*. All that is found is one or the other**. The distinction is archetypal; archetype is metaphysical; and the acceptance of metaphysical information with material energy as a fundamental component of cosmos immediately invites a different perspective on life and morality. *The* *problem is, as bottom-up loudly, impatiently insists, that this is not just theoretically improbable, it is IMPOSSIBLE...*

The Third and Final Flight from Science

What was the origin of energy? What is the origin of information? Do you remember (Chapters 8 and 20) how scientism seems to fly apart from its own fabric and promote, against the laws of science and their constraints of possibility, theories of The Big and Little Accidents? Did Little Accident evolve - by zillions of tiny, consequential accidents that sum to ultra-complex bio-logic, *PCM* and 'macro-evolution' - every other kind of life as its descendants here on earth? Scientism in the interests of materialism will insist so but the facts, as have been shown, clearly contradict such diagnosis stemming from genetics, fossils or consideration of 'the ratchet', natural selection.

Remember those 'trade secrets' from genetics and palaeontology? They are, in spite of protestations, as true now as then. And they explode the myth of *PCM*, that is, the neo-Darwinian synthesis.

What, therefore, is the bottom line? Nobody disputes the factual discoveries of Darwin but it is theoretical extrapolations ('abiogenesis' and 'transformism') that are music to a naturalistic ear. This couple compound to compose a plausible and easy-sounding story that, from molecules to man, describes your being by a 'theory of evolution'. *But plausibility is different and sometimes far removed from truth.* The question is, are these extrapolations and their consequent interpretations actually fact? Is the story a delusion flecked with truths? Or is it, like phlogiston theory, once beloved of naturalists but now, by fresh information of genetic information, falsified?

Part 1 of this book ('A Player's Body') argues Darwinism's story is the lesser half of whole truth (*fig.* 20.1) and thus in fact piebald delusion of a powerful kind. No-one disputes a plastic 'origin of species', sub-species or 'level' thereabouts by natural or even, perhaps, sexual selection (*figs.* 5.1 and 22.1). What is the lesson drawn? **The lesson is that, while nobody disagrees with the facts of mutation, natural selection, adaptation and variation, none of them create a jot of systematic information.** Whether it involves germs, weeds, snails, Alaskan sticklebacks or whatever else such variation, while incessant and sometimes misleadingly called 'micro-evolution' or 'evolution-in-action', is essentially trivial. Viewed dialectically random mutational bugs are, like wear and tear in motoring, too trivial to produce the vehicle itself. Although a book or machine is created as perfect as possible still ravages of time and place will mar or change - but not to the extent of generating a new novel or creating fresh brands of technology.

The same is true of biological organisms, not least with respect to basics - cells, *DNA* and, most importantly, information in programs compiled to compute with the exquisite, coded chemical complexity that's called metabolism. Metabolic pathways, like development, anticipate precise, required results. They may be buffered and by various designs protected but the ravages of time are liable to disrupt precision's chemistry and thus 'bug' informative capacity. *Therefore while neo-Darwinism deals with accidents and their effects it does not address the origin of vehicles that suffer them.* **It explains wear and tear but not initial production of the models that will sell in life's survival hall.**

Indeed, as he formulated his ideas Darwin had but meagre, page-one knowledge of the copious factors that absorb biologists today; and none of their extreme, computer-like integrity. Cell biology, molecular biology and biochemistry are dedicated, with an army of laboratories, to the elucidation of such body parts and parcelling. And Venter, struggling with complex production of a 'simplest organism' using nature's patents and scrap parts, is about to underline the case for anti-chance.

How could a few cents worth of chemicals, however carefully weighed out to correspond to what is in a cell, spontaneously 'organise themselves' into that cell? **The issue is not one of chemicals but information. The real business, innovation, remains entirely unexplained and inexplicable by Darwin's Theory of No Intelligence. The fact is, equally, that Intelligent Design is staring scientism in the face but, through sheer philosophical negation, the creed averts its eyes and turns away.** The whole Darwinian enterprise therefore amounts to a philosophical cul-de-sac, a persuasion born of anti-anti-chance, that is, the notional pre-eminence of chance as life's creator. It is, simply, materialism's spine, its central dogma and its life. Light in the eyes of Intelligence is therefore Gorgon-like. Don't look!

Has Darwin Had His Day?

No and yes.

No. Darwin will remain remembered for his observations and research. He identified correctly and in part explained the fact of variation/ micro-evolution by notions that he learnt from breeders (artificial selection) and from Edward Blyth (natural selection). Variation, genetic mobility, mutation, natural selection and heredity are facts. And some protean form of evolutionary theory is essential for materialism, atheism and associated breeds of thinking to survive. The principles of naturalism must not, on any count, be violated. Thus, although 19th and 20th century versions taught in their time as practically law are nonetheless demonstrably deficient, the 21st must invent and has already, in the buttresses of 'natural genetic engineering' and 'evo-devo', invented furtherance. Evolution is materialism's oxygen. Lack snuffs; thus Darwinism is the credal air it breathes. Darwinism is a mind-set that, Atlean since it supports materialism's world-view, must not fall.

Yes, however, if his grander speculations are found fiction. *A materialist will try convincing you that the aforesaid facts add up to evolution, that evolution's proved and is a fact. This is not the case.* They amount to the 'lesser half' of what is, combined with highly dubious hypothesis of abiogenesis (Chapters 20 and 21) and an unlimited plasticity of macro-evolution (Chapter 22), presented as a whole truth - The Primary Corollary of Materialism, *PCM.* **On the contrary, however, one might well argue that Darwin's whole hypothesis is built on the *false* idea that micro- leads to macro-evolution! Does the thesis, as materialism's arrogation, take first-class Honours in Kidology?**

Victorians knew nothing of molecular biology, biochemistry or genetics and very little about cells, metabolism or development. Nor was access to the fossil record as worldwide as today's. Darwin's knowledge was, in these critical respects, very superficial. It is even possible that, if he knew what's known now, he would have retracted. Reasonably recanted and with grace conceded.

Moreover life's 'major half', left by evolutionists on principle completely out, is derived from its subjective axis; this half comprises active and, in the form of archetypal memory, passive information. Without such immaterial ingredient no integrated, complex system-with-a-purpose (as all living systems are) is able to arise. More rational, deeper understanding turns away from billions of missing links towards a conceptual, computational view of bio-logic and, therefore, a meeting of great minds!

What are you left with when no specificity's 'injected', no information is encoded? **The fact is that no undirected chemical process has ever shown a capacity to generate functionally specific information. Until such capacity is demonstrated then the process by which such information is, to our certain knowledge, invariably associated will remain the best explanation for its biological omnipresence. This process is a conscious activity, the creativity of mind. Is not rational, deliberate design the best interpretation of an engineering triumph?**

Natural selection, being witless, can't create a whit. Let alone great wit itself. The double-barrelled phrase just means allowance or erasure of a body that, as a '*creatura*', has already been created. What if mutation changed a healthy specimen? Although it is perforce hypothesised and taught that *BM*s (beneficial mutations) can coincidentally generate the whole stupendous edifice of life on earth, the scientific evidence itself shouts otherwise. ***It shouts that random mutations are unfit for purpose; they cannot generate the information life requires****. It shouts, in other words, that the mechanism of the PCM is incapable of creating different types of life form in the first place; and that, therefore, the slender pair of threads by which all materialism is hypothetically strung have both been sliced.* Abiogenesis (Chapter 20) fell to the knife of scientific evidence. Nor was it religion but evidence and reason cut away the filament of *BM*s and, therefore, annulled the patent of a pruning knife presented as the 'force' for innovation trading in the name of natural selection. In which case why ask where all the *BM*s' missing links have gone, long time passing; it may well, from myriad improbabilities to sharp transitional impossibilities, be diagnosed that Darwinism's deselected by a million cuts. And now, perhaps prescient of the final thrust, there comes discovery of complex, computational genetic super-codes. The highly conservative deployment of developmental routines across wide ranges of organisms indicates that thematic, mosaic and intelligent design is a strong, rational interpretation of the evidence. Only mind could logically originate life's stereo-computational complexities. *If this interpretation is correct it means that Darwinian PCM, the theory of evolution and, therefore, PAM (The Primary Axiom of Materialism) are, if presented as the whole truth, false.*

Modern science has not made evolution more but much less believable. What is left except denial in the form of invocation? Cannot Fortuna, given time and dislocated from the laws of physics, convert the least of probabilities to certainty? Like dreams, statistics are a form of wish fulfilment; they can be turned to justify even evanescent possibility. What odds, therefore, cannot be invoked to save The Theory of No Intelligence and with it scientific atheism? Cannot chance be persuaded, at least in imagination, to conjure actuality from physical impossibility? Who strongly wishes that the theory were true? Is such materialistic faith an atheistic form of prayer or simply an abuse of Lady Luck?

Is such faith and prayer enough? Today, outside their bodies, organic chemicals soon decompose: nor can fire, wind, rain and earth create or re-create them. Chemical evolution, it's been argued, is a phantom faith but then, if variation, natural selection and mutation can't cut the innovative mustard, what is left except denial in the form of miracle? If your theory's *tripos* (a three-legged stool) has lost its legs but still the seat is seen as holding up in air then lo! scientific levitation has become reality. This, it has been argued trenchantly, is the reality of naturalism's central theory. Its root's mutation and the problem is you can't do science using chance; you can't predict its unpredictability. Your answer is to guess. You can't predict mutations and, therefore, unfocused theory stays a story, more or less, of how fill the gaps of blur.

The logic of Natural Dialectic, on the other hand, simply adds immateriality (which is apparently nothing) to materiality. It adds active onto passive information everywhere. As such it runs hard opposed to the irrational twist in scientism's tale, a tail-to-head twist in a truly head-to-tail reality. In fact the paradoxical, dialectical conclusion is that, while nothing supports neo-Darwinism's *PCM,* Nothing fully supports the intelligent, teleological designs which students and experts alike meet manifest in all biological forms.

Are you not, as scientific narrative proclaims, a human animal? Of course, but is this all you are? Is any organism, lacking an informative dimension, no more than physically biological? Is it devoid of archetypal influence or is half-truth made whole by immaterial mind? *Evolution is a naturalistic explanation of your origin; but if, as Natural Dialectic openly declares, materialism represents a fraction, unintelligent a fraction, of whole cosmos then isn't such an explanation only fractional as well?* **Aren't, from this rational and holistic view, Darwinian tales a specious species of half-truth, a subtle form of deceit and, in the last analysis, life's most insidious sort of falsity?** The essential resolution of Kant's antimony (Chapter 3) rests simply in correction of a hidden error - anti-teleological materialism. Thus corrected, has not the 'larger-than-life' though 'lacking-inner-life' philosophy of neo-Darwinism as applied to mind and body had its finest hour? You can definitely reckon Darwin, in the great though not the small, has already had his day.

Theories of Accommodation.

Hang on! If scientism flies from science why do clerics hanging on its lab coat fly behind? Why should administrators of an organised religion, bowed and bamboozled by a sceptic, septic scientism's huff and puff, preach what neither science nor their prophets teach? It beats me. The framework and preceding dialogue of this whole book point up how greatly ill-conceived such theory of accommodation is. Theistic and deistic theories want it, as their names suggest, both ways - 'God' *and* 'evolution'. Is it possible (unless you lose them in a cloud of waffle) to yoke vehicles of contradictory logic in this way? 'Easy!' senior theologians cry - but how? I can't understand it but you, somehow, may!!

There are two pillars of faith - atheistic and theistic - but, honestly, an agnostic can't decide between them. He is, in all humility, not sure of the 'inferential proofs' with which the *PAM* and *PAND* assail him. He cannot make his mind up. Which of these implacable and yet potentially pervasive pillars of faith is wholly true. His compromise is wobbling on the fence. Such vacillation is a pity just because truth does not vacillate. One or the other is correct. Either *PAM/ PCM* or *PAND/ PCND* is right. There is or there is not a Universal Mind and Metaphysical Creator.

If materialism shouts that *PAND/ PCND* is right then others loudly shout the opposite. If you're unnerved then perhaps the best bet is Pascal's. Blaise Pascal was a brilliant 17th century French scientist, mathematician, philosopher and, it is claimed, inventor of the first *machine arithmetique* or computer. He grasped probability by the horns and advised hedging your bet. Reason alone, he reasoned, cannot decide the transcendent essence of God. However, to avoid a paralysis of doubt afflicting the critical issue of the quality of your life, wager and live accordingly. If you bet God exists and He does, you will have won everything. If He does not, at least you will have lived a good and honest life in the context of a false belief. You will have lost nothing. If, on the other hand, you bet against but He exists then you are in for a large post-mortem shock and the judgment day you only hours ago pooh-poohed. Finally, if you have bet that He does not exist and He does not, you have lost nothing except that, in a drive to satisfy a short life's unpredictable desires, you may have inflicted or condoned the infliction of pain on others. Why, in the case of survival of the fittest, is that dishonourable? How, for a carrier of selfish genes, sinful? Or, if you can get away with it, wrong? Pascal himself bet God exists.

All parties agree the obvious - physical forces and objects exist. We dwell among their mindless behaviours. We are part of their 'event'. After this smooth union immediate cracks appear. Does material constitute the whole or part of cosmos?

A *pure atheist* adheres to 'The Theory of No Intelligence'. There neither is nor was any intelligence behind the creation of cosmos or, as a special case of cosmos, the evolution of living matter. It is all a matter of chance and necessity. If anything is possible it is only because chance, quantum mechanics and classical determination allow. In this case the only thing predictable about evolution is its entire lack of predictability. All organisms are, moreover, equally resilient survivors. 'Progress' is a human figment. Nothing, let alone a white Caucasian male, is at the head of any evolutionary scale. Thought of superiority is vain. What tree is mankind atop? All life is, like a leaf, simply a presence at the twig-tip of its genealogy.

Impure atheism is another matter. It resembles theistic evolution with, confusingly, the latter's adjective rubbed out. As a pair they'll be considered later.

On the other hand a theist, who used to be called a 'natural theologian', currently adheres to one of three versions of a 'Theory of Intelligence'.

Firstly, a *pure theist* ascribes creation to the projection of preconceived ideas. In this case intelligence, which both precedes and co-exists with its informative projection, behaves intelligently. All engineers design according to constraints. They evolve their systems within predefined parameters or, if you like, initial conditions. Such restrictions, with which any finite set-up must accord, are imposed by nature and his own conceptions. The pure theist simply proposes that physical nature was established by Super-nature or Nature. This Precondition is intelligent and not chaotic, logical and not a matter of chance. Indeed, every inventor strives to eliminate chance from the boundaries of his influence and, as best possible, meet his target. Nor is it his fault if the vagaries of accident or user error subsequently degrade erstwhile pristine machinery. In the case of *cosmos* a pure theist takes the word for what it means - rational order.

Both kinds of purist sport the merit of intellectual honesty and rigour. Now comes vacillation's top solution. Cut dither, swipe the sweat and snap into decisiveness. 'We'll keep the peace! We'll smudge them both together,' claim accommodation theorists, 'and promote a scheme of compromise.' What relief from trying to control divergent horses. What an escape from an agnostic kind of ignorance. Fascinated, spellbound, even overwhelmed by the power of material science such ecclesiastical adoption somehow transforms the basis of the *PCM*, the theory of evolution and thereby the *PAM*. It's neither *PAND/ PCND* nor quite what our mystic heroes seem to think but, there again, what science did Christ or the others know? We can't even ask them what they really meant. Having therefore swallowed *PCM* entire accommodation theory now transforms the transformational illusions of abiogenesis and macro-evolution by arguing The Great Evolutionist works in one of two main ways. Is such interpretation semi-atheist, semi-theist, both at once or can you simply dub the mix 'impure'?

In 1879 Charles Darwin wrote to John Fordyce "...a man may be an ardent theist and an evolutionist"; and, in one of his last letters (1881, to William Graham), "You have expressed my inward conviction ... that the universe is not the result of chance". Isn't this, in neo-Darwinistic terms, confused? How might such seeming contradiction be resolved? In the *second* of three versions of The Theory of Intelligence *deistic evolutionists* (like Erasmus Darwin and his grandson Charles) allow an impersonal, 'externalised' kind of creator, the miraculous creation of souls, laws of physics and material energy. These are called, in the jargon, 'initial' and, with respect to law, 'boundary' conditions. The material part of such conditions, forming everybody's background, are of course necessities for life. They are Natural Dialectic's 'stage'. No doubt, life needs water but is its necessity sufficient unto life? Departure by a deist leaps to such conflation. He proclaims the world's refined, designed 'front-loading' must inevitably bring forth life; the fabric of our universe *must* force animation from its lifeless bowels. Thus, intelligence once spent, such a detached divinity leaves life's encoded information to evolve by chance and predefined necessities. **All this, as was noted, despite the critical fact that no undirected chemical process has ever shown a capacity to generate functionally specific information!**

Indeed, you ask, what inventor throws invention to the whirligig of chance? What care slings wards forever under careless nature's wheels? What level of intelligence engages governed yet ungoverned process - wind and rain and fiery chemistries - as a 'naturalistic methodology', its means by which to engineer? Let nature specify then build your aeroplane! Or much more complex cells! It is, deistically or semi-atheistically, hoped that 'constraint of natural law' is sufficient unto driving evolution. More than this, that the 'process' may inexorably drive its own 'inevitable progress' forward to a climax of (at least so far) chance-born humans and, at the very peak of climax, spring on the universe a shower of teleological flukes called scientists, philosophers and intellectuals in general! How grand the self-evolving powers of nature! Such 'unbroken law', whereby 'miracles of intervention' are banished, holds a strong appeal for many theistically religious scientists. Of course, a 'pure theist' has no problem with 'unbroken law' either. It is simply that, to the exclusion of chance-based evolution, his archetypal conceptions include the basis of psychological and biological as well as physical law. The deistic problem is that if Deity is simply an Initialiser, it is entirely unclear what magical, almost animistic catalyst spins out complex expressions of purpose - unless chance mutation and natural selection (Chapters 19 - 25) or Lamarck's discredited *IAC* are supposed to have assumed this god-like, informative role. Indeed, if the *PCM* is right such an Initialiser is actually reduced to a vestige of presence - no more than a metaphysical explanation for a big bang.

Thirdly, theistic evolutionists do not believe (correctly) that undirected chemistry can single-handedly create specific information, generate a functional system or, at length, write up humanity. Yet such a theist is mind-locked to Darwinian 'progression'; he is fused to the general idea of evolution. Thus, while rejecting the cosmos as a deliberate, highly intelligent projection, he embraces a range of ill-defined possibilities to explain God's interference with the mindlessness of pre-biotic chemistry and His employment of mutation to construct life's database of genes. Genetic mutation, the only mechanism known to toy with biological information, is one as witlessly random as blindly shuffling letters, words or chunks of text in the non-hope of 'improving' a book. In this case you need allow God various degrees of access to life's genomes and, to start life and to sustain non-random macro-evolution, an ability to preternaturally, deliberately break 'unbroken natural law' and its contingencies. There arise a plethora of elastic fictions. How much, exactly, might 'divine intelligence' leave down to chance? What engineers leave down to chance they know full well may never happen. In fact, the more complex and apparently purposive a system the less likely it is that it will ever happen of its own accord. Indeed, why should a deity that works through chance care whether or not it produces mankind? And since the open-endedness of 'unbroken physical law' seems unlikely to have produced such objects as themselves, leaving it all to chance has worried many thinkers into conjuring a 'God of the molecular gaps'. This character is supposed to occasionally enter this hapless process, 'orthogenetically' fiddle with genes, override the *PCM* and thus somehow order the disorderly. Interfering here and there, why shouldn't such omnipotence work surreptitiously? He only needs to inject or assemble genes to jump a few levels of biological classification or program another type of organism, organ, complex system or homology. After all, as Darwin plausibly explains, broad concept (such as every human operation shows) need never enter in. But plausibility is not the same as fact; Darwin's suggestion doesn't mean it didn't! Thus some earth goddess or theistic entity, now as then, needs infiltrate the *DNA* and, intentionally unlike unintentional mutagens, tinker with life's nucleotides. Divinity's become a mutagen! Such a foggy line of reconciliation is a cheap and plastic one. It snaps far too easily. Most if not all molecular biologists, geneticists and biochemists of agnostic, atheistic *and* theistic persuasions reject out of hand such a fey, semi-animistic and wholly unproven form of mutagenesis. Persistent 'miracles' of this kind, like a Boeing engineer making spasmodic visits to adjust the wind's effect as it mindlessly blows pieces of junk around a yard, make neither physical nor logical sense. In which case why not expand whimsical 'Guide of the Gaps' to the status of an unknown law of universal information, strongly directed evolution, a really grand and self-evolving power of nature! call it Bio-logos or Bio-force. Having invoked 'living matter' (as did atheistic Haeckel) why not extend the animism to 'an élan of vitalism' - Bergson's teleological creative momentum that, unbeknown to science, drives evolution on and up towards consciousness? Even better, even more discordant with neo-Darwinism, try a Catholic priest and palaeontologist mixed up with fraudulent Piltdown and Java men, Teilhard de Chardin. Was his 'visionary synthesis' a 'reconciliation' or, because it was originally banned by the church authorities, a 'rebellion'? Anyway, starting from nothing his universe self-evolves through matter and life forms to consciousness and, inevitably and climactically, Point Omega (God). God last? Life risen from material oblivion? Surely this reversal is not what such a sincere and well-qualified luminary as Francis Collins really thinks? Or what big-bang-and-evolution's trophy convert, Pope John Paul II, believed?

How feeble a creator whose informative capacity is nil! Does science not confirm that genetic entropy is fact? Is the First Corollary of Materialism (that you, a mutant bacterium, are the product of random mutation and natural selection) right or not? If not then strangely its evolutionary adherent Rowan Williams, the Anglican Archbishop of Canterbury, accommodating with a host of clergy also gets it wrong. Mass apostasy by priests! False information spread! Is turning Christian teaching on its head intelligent? It's not, for sure, correct.

Scientism has realised and, in a few instances, accepted the practical impossibility of an abiogenetic cell (Chapters 20 and 21) but it has failed to countenance the implications. Instead Arrhenius, Fred Hoyle, Francis Crick and others suggested 'panspermia'. They mean the earth must have been seeded from an extra-terrestrial department in the form of bacteria or, perhaps borne on a cometary carriage made of mud and ice, even simpler types of life. On the other hand so-called 'interventionists', well outside science, suggest that advanced forms of *ET*, 'angels' or 'the shining ones' (called *Elohim*) somehow sired humans or, at least, keep returning to 'kick-start' new phyla, classes or orders of life on earth. *Where did panspermiant bacteria, extra-terrestrials or angels come from in the first place?*

Is nothing impossible for a theist and everything possible for an atheist? *Impure atheism* here deserves a shout. When philosophical fox is tracked through his dense verbiage to lair one such as Thomas Nagel will appear. Nagel well understands the intractable obstacles that afflict materialist/ reductionist theories of abiogenesis, neo-Darwinism and evolutionary 'emergence' of consciousness, cognition (including reason, thought and logic), meaning and evaluation from a complication of non-conscious atomic elements and forces. He sniffs a shift in paradigm is due but still accepts biological evolution in principle (thus not attracting fire from scholarship's main battery of guns). Thus, while prowling round the cosmic coop, he wants to expand cautiously to a general understanding that includes a fundamentally subjective element - conscious, immaterial mind. So, having rejected the notion of Natural Engineer (or any image that is humanoid of 'God') and also that of physical reductionism in the form of Darwin's evolution, what middle way might let him off the basic pair of Maker-or-No-Maker hooks? Of course, with its range of views theistic evolution queers the dualistic pitch but your 'ungrounded intellectual preference' might be to cut Creators altogether out. Atheistic evolution must explain how life, despite its bio-suite of teleological description (purpose, target, signal, code and so on), 'emerged' from non-life and how consciousness is added to conscious matter - not just nowadays but in the making. Not intentionally, of course. Thus is unveiled an unsubstantiated guess - *natural teleology*. If to divine's to guess such divination is unknown to science. The laws of nature physics knows act rigidly and presently; they determine with precision how the universal clockwork turns. Nagel needs, for natural engineering, an additional suite of laws that work 'temporally', that is, entrain progress towards some target or, in his words, are 'biased towards the marvellous'. He means some futures (such as those including the highly specified complexity of life and the emergence of mind from its material essence) must in principle be probable. There's inbuilt cosmic bias unto certainty. Such laws of innovation, progress and self-organising targetry would, as attractors, skew cosmic destiny in principle but unintentionally towards producing, in the long run, atheistic academics! Chemistry and physics (thus reductionist biology) entirely lack such vision of non-purposed teleology. Such indeterminism, countering entropy, is wholly speculative. Where ghosts of Bergson and de Chardin flit Nagel commandeers imaginary, abstract space to loosen up the iron grip of natural law and let the world somehow 'progress'. By slapping 'natural' onto 'teleology' this doctor hopes to draw the poison of divinity by a fantastic speculation. His form of cosmic mind, if such exists, is mindless; natural law is 'minded' with an unexplained 'direction' in its stead. You'd almost think, although it's not allowed, that he sought internalised stability (like mathematics, logic or Platonic form) or Natural Dialectic's archetypal memories that never physically evolve. No sir. That idea is not meant. The old trick is conjured up again - just exclude designer but include 'design'. Could Nagel's foxy oxymoron ever tell whole truth or does it, as in Book 0 we've already seen, in reality confuse?

Confound the issue. Arguments ping and definitions pong. Would you call mind devious or subtle? Is what cannot be imagined happening somehow? A major stumbling block is the definition or, rather, lack of definition in the pictures people hold of 'God'. If cosmic intelligence is superhuman, that is, super-conscious then by definition *all* such graven images or un-graven imaginations are more or less wrong. They are, depending on the user, simply useful metaphors or straw men at which to tilt.

Does mindless evolution need a mind to guide? Why, on the other hand, subscribe to entropy of information as an anti-tool for planning and construction? *Why should any engineer compose both senselessly and mindlessly? Why should he, especially He, behave worse than the fool of fools?* **Moreover if a Creator had really left it for chance to create then, by definition, He could neither have predicted you nor created you 'in His image' - unless He is also Chance! So even then He couldn't.** Such 'Creator' leaves the form of 'man in its own Natural Image' to the longest of long shots. Fortuitously, over billions of absent-minded years, Chance 'gets lucky'. In other words, the Creator passively waited until, for no reason whatsoever, chance managed to incorporate and still keeps incorporating high-quality, systematic information into biological organisms, including un-designed animals like man. Or is it something in the way He set up atoms means you have, eventually, to be? How the bishops muddle through! The arm-waving train of logic is muddled, self-contradictory and illogical. An accommodation theorist wants to run with hare and hounds. The hybrid but sterile idea is that a foolish Engineer employs chance to effect His designs. Random designs. Absence of mind and therefore absent-mindedness must constitute a dormant kind of 'creativity'. *In this folly rests the foolishness of all accommodation theory. Such theory is a 'bastard' born from forced, unholy union between the two pure pillars of faith's possibility - Cosmic Intelligence or none.* **Its core tenet is a most moronic option. It is supposed that The Highest Intelligence chose to create man 'in His own image' by chance.**

There is no sweet logic in such bitter and yet tasteless fudge! Indeed, if intelligence is such a 'good thing', why has evolution been so sparse with it - especially with one of its most human and yet powerful forms, purposive creativity?

If a theory of accommodation were true, why did neither Christ, Mohammed, Nanak nor any other 'Professor of Faith' teach evolution? After all, describing the issue of simple-to-complex is easy using the analogy of the development of a child to adulthood. Or, if you like, everyone understands building a house from simple foundations into its complex, detailed architecture. Neither Christ nor his disciples need have been scientists. In fact, his teaching (and that of the whole so-called mystical tradition that has constituted the core element in every religion at all times) cannot be clearer. 'In the beginning was the Word, and the Word was with God and the Word was God. Through Him all things were made; without Him nothing was made. In Him was life, the light of men'.

Does this mean, you have to ask, that the Word is reasonable or, as reconstructed versions of accommodation tout, an issue made of accidents? If *Logos* or the Word constructs by chance then is it not the son of Chance and 'spiritual' Christ, like you, essentially a mutant bacterium? Does not this interpretation, revamped to align with the assumptions of contemporary science, amount to severe entropy of original information? Is such conformity with The Primary Axiom and Corollary of Materialism a significant mutation of Christ's message or is it what the Man was trying to explain? If *PCM* is what He really had in mind then He foretold of one to follow. He presaged prophet Darwin and is saying evolution is the cause of incarnation!

If this is true and evolution is the mode of a 'creative' God then a predictive Christ did not mean what He seemed to say. Saint John, the Torah, Moses, the Koran, Mohammed, Hindu scriptures and the rest of the world's mystic tradition must be, by claiming an Intelligent Creator, either ignorant or untruthful. And their message a misleading, even a deceitful one. If, claiming to be at one with the Father, Christ was ignorant, deluded or deceitful then certainly his claim and most likely God's omniscience are a sham. If he was a liar who claimed to speak nothing but the Whole Truth, why should we believe anything he said - especially if he didn't claim mankind was made, by accident, from apes? Or, if his audience hadn't heard of apes, they were transformed from mice or fishes by the hand of evolution? He would be no better than a confidence trickster peddling, like his followers, an illusion. *Why, in either case, should we believe a word*?

Many priests, theologians and their circles now live a sophisticated, philosophically schizophrenic delusion. "In fact, what we mean by evolution is the world as created by God". Thus spoke a Papal Primate by the name of Gianfranco Ravasi. Does he mean what science means by evolution and thus endorse a neo-Darwinistic creed (the *PCM*)? Has the Vatican and its hierophant, head of a ponderously-proclaimed Pontifical Council for Culture, grasped all the facts of life - not least its immaterial, informative spirit? Or has this Archbishop's halo, knocked sideways by a socio-scientific pressure to conform, slipped into error; has his Cross so toppled that he marks Christ's Christian Creator with an X? *If popes, seminaries, prelates and a herd of trending clergy misconstrue, why should we give them credence, why believe a word they say?* Why not drop their rephrased catechism and its services? What type of accommodation might, for example, Anglican incumbents choose? How would Coptics mesh the Theories of Intelligence and None? Is it deistic primal chemistry or theistic (even atheistic) macroevolution that a Catholic should henceforth host? No doubt, incredulous 'modern' thinkers nail the church to an uncomfortable crux. Be crystal clear - from Natural Dialectic's point of view accommodating churchmen hammer it to death as well.

TOTAL CONTENTS

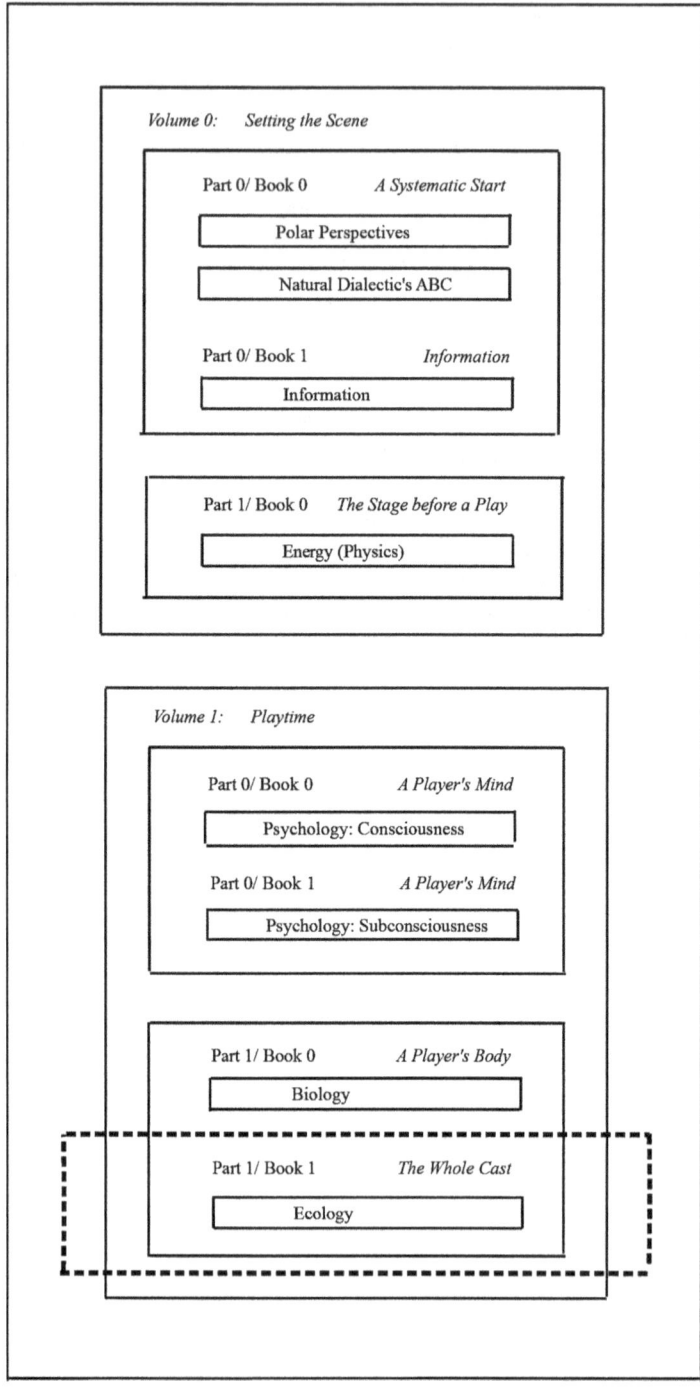

Volume 1: Playtime

Book 1: A Player's Body

Part 1a: Biological Society/ Ecology

26. Towards a Unified Theory of Community

Is Natural Dialectic simply an intellectual exercise restricted to its relationship with academic science and philosophy? Or has it a practical, real-life application? *The system of Natural Dialectic is, as much as being an abstract reflection of the way things are, an application program.* This program generates a template for both involuntary, 'natural' and voluntary, human behaviours. It organises the pattern not only of 'hard science' and biology but politics, law and religion; it guides the aspirations of education; it is hard-wired into the humanities and, as such, is expressed in the very fabric of individual and social life. Its consequences, especially its moral and psychological consequences, involve everyone. How?

Towards a Unified Theory of Community

From a *bottom-up* perspective humans are animals and mind evolved as a strange function of brains. Consciousness is a will o' the wisp, a phantom wafted methane-like from life's wet chemistry; it is an elusive, illusive by-product of atomically constructed, non-conscious neurological circuitry. The study of mind-chips is called psychology. Just as (it is claimed) nothing biological makes sense except by evolution's torch, so nothing in social or cultural endeavour makes sense except in the light of psychology. Mind evolved from some primordial marsh and therefore, though lesser strains of clinical psychology exist, it is truly for *evolutionary* psychology - having first deleted a psychic delusion called soul - to best offer explanation of man's activity-by-brain. For example, do not genes explain how morals come about? Doesn't cell assemblage 'have a concept' or a net of neurons somehow, by remembrance, 'learn'? No doubt, 'eureka' is an adrenergic rush. It's all in the mind and, since mind is brain, you must appreciate appreciation as a nervous concept and consciousness a rainbow crock of neuroscientific gold. Brainy evolutionism calls all shots. Its truth must be the final arbiter.

Top-down we start with a stack.

tam/ raj	*Sat*
relative	*Absolute*
issue	*Source*
duality	*Union*
hierarchy	*Top*
cycle	*Axis*
oscillation	*Homeostasis*
relative balance	*Balance*

↓	*tam*	*raj* ↑
	down	*up*
	outward	*inward*
	exit	*return*
	external	*internal*
	division	*unification*
	separative	*communicative*
	isolation	*relationship*
	effector	*sensor*
	body	*mind*
	involuntary	*voluntary*

Key concepts of this section, which deals with association, relationship and community, are homeostatic balance, (re-)cycling and return.

Top-down, the first compass of community is universal and absolute. It reflects the structure of creation whereby all things, psychological and physical, descend from One (*Sat*) Infinite Centre.

The second form of community is relative and, in this relativity, we need to identify a stable vantage point, a criterion. X marks the spot. Cosmos and its ways are relative to you; you, fixed as far as you're concerned, perceive its glory and its pain from your 'third eye' or axial point of consciousness. Every individual eye of a beholder charts relationships with cosmos from this axis. This is its own but not the cosmic centre.

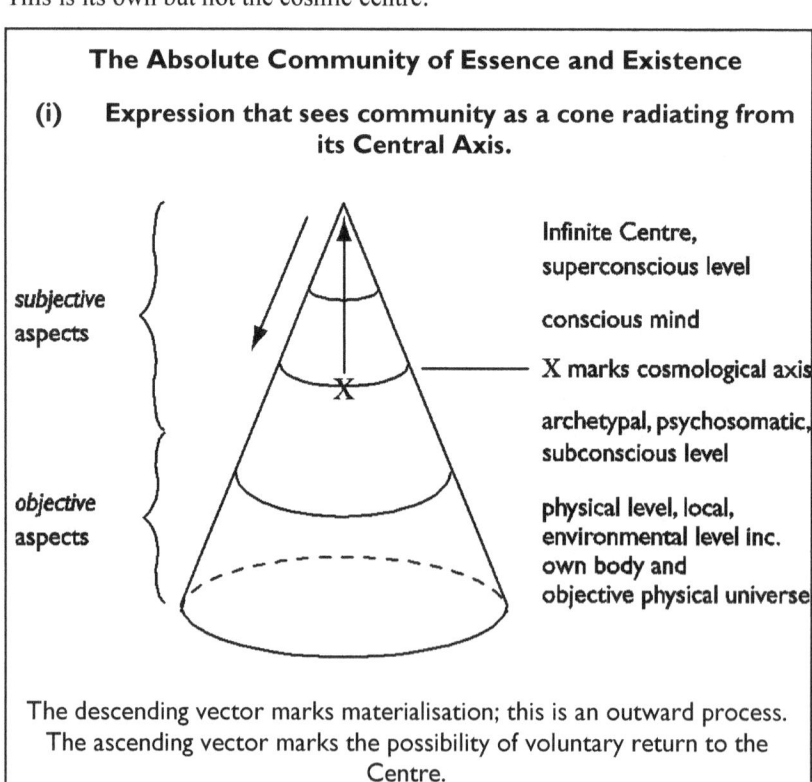

The Absolute Community of Essence and Existence

(i) Expression that sees community as a cone radiating from its Central Axis.

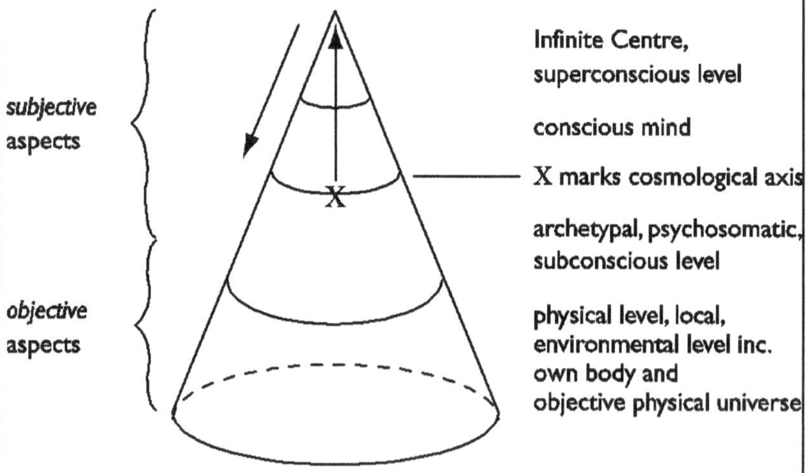

The descending vector marks materialisation; this is an outward process. The ascending vector marks the possibility of voluntary return to the Centre.

(ii) Alternative expression as concentric rings: nested social circles.

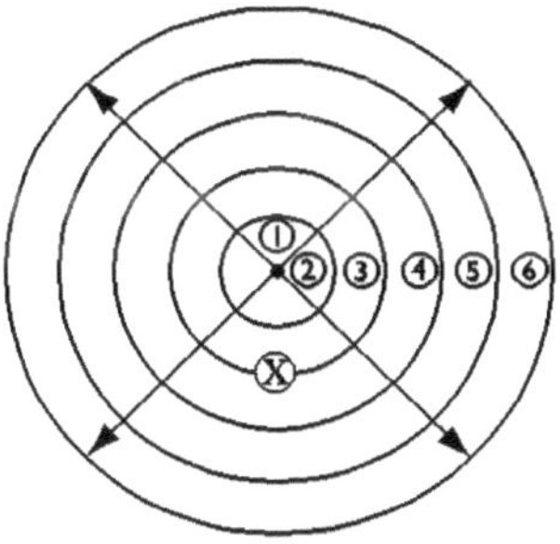

Descending vectors used in the sense of 26.1 (i).

fig. 26.1 (see also fig. 0.10)

The Relative Community of Existence from a Standpoint of X

(i) Circles inward and outward from X.

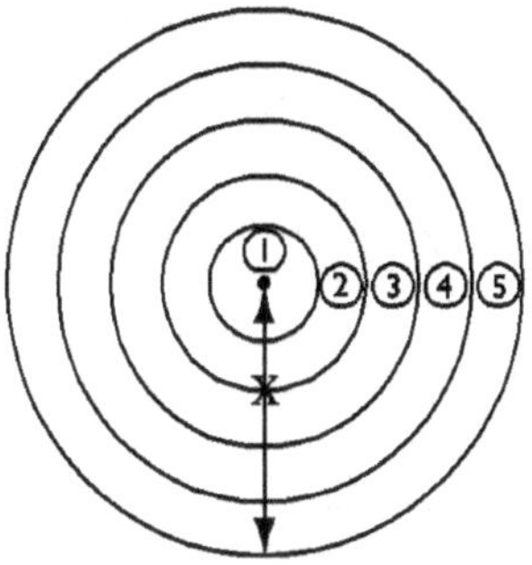

(1) Super-conscious association, Communion.
(2) Your normal experience/ mind/ inward perceptions.

This couple are internal from the cosmological axis, X. Their ascendant vector marks a possibility of (*raj*) 'right-hand', voluntary return to the Centre where maximum 'connection' is experienced.

Call X your current pivot of awareness between inner and outer environments. Conscient organisms do not, from a spatial and perhaps from a temporal point of view, share the same frame of reference. However in terms of the hierarchical conscio-material gradient they do. No matter where it might be living in the universe embodied consciousness has to operate on a pivot between inner, subjective and outer, objective worlds. The cosmological axis, while precisely local and non-central spatially, is a psychologically common frame of reference. It is simultaneously both central and universal to all embodied awareness, that is, perceptive biological forms.

(3) Friends, other persons, the socio-political community of human population.

(4) Ecological relationship with other organisms; natural community of life on earth.

(5) Interaction with abiotic factors, that is, gaseous, liquid and solid zones of a biosphere.

(3), (4) and (5) mark an external descent, a biotic 'fanning out' towards an involuntary, abiotic periphery. Relationships become looser, more isolated and less relevant as man observes himself from the outside. Maximum 'minification' and sense of isolation are encountered when a human body is set against the galactic universe.

(ii) Two-way expansion of focus from the cosmological axis.

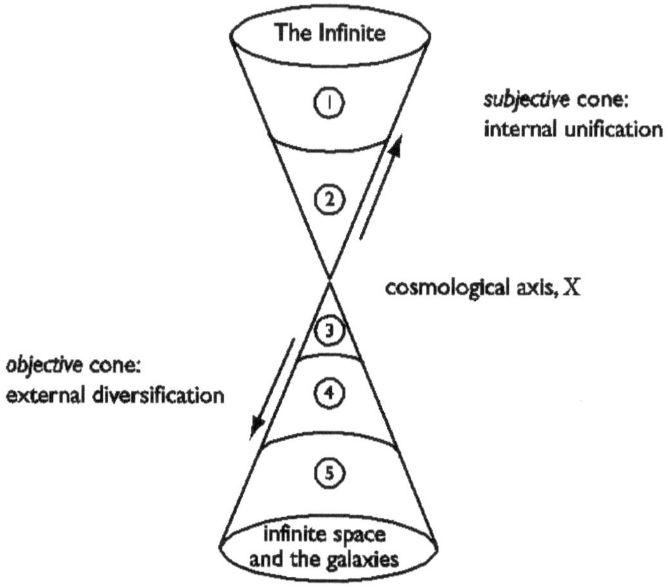

① Superconsciousness experience culminating in the Infinite, Essential Self

② Conscious mind

③ Own body and close locality

④ Biological community

⑤ Abiotic segment; planet extended to the galaxies

fig. 26.2 (see also figs. 0.7 - 0.9)

Unconsciousness lacks all sense of locality. Moreover subconscious memories (including templates for repetitious behaviour called habit or instinct) are 'fixed shells' of thought; whether universal or personal, they have no sense of presence of their own. They are passively informed and cannot voluntarily do a thing. Indeed, except with use they fade away - which is why the archetypal records of universal mind, used constantly, are etched with permanence. But even these are neither more creative than a standard batch-process nor lively than the replay of a disc. This zone of universal mind is not what you or I would call alive. Nor, being mindless, can the associations and dissociations of the material universe break out of their closed community. Non-conscious matter lacks point X. After all, what could energy's sleepless oblivion, unlike sleep, ever wake to? Similarly sub-conscious memories and repetitious behaviours such as instinct are associated with biological bodies but as fixed 'shells' of thought have no presence of their own.

	tam/ raj	*Sat*	
	many	*One*	
	selves	*Self*	
	body/ mind	*Soul*	
↓	*tam*	*raj*	↑
	down	*up*	
	outer	*inner*	
	lower self	*higher self*	
	body	*mind*	
	animal	*sage*	
	sensation	*contemplation*	

From point X, however, there radiate social circles in both directions.

Internally there exist the symbolic yet most real relationships of mind. They involve other persons, life forms, events, objects and the paraphernalia of a life spent learning. These are your mind-world. No doubt the 'suction' of the senses towards the external world is overwhelming; it tends to blot out this inner world as the sun outshines the stars. Do you remember, however, the sense deprivation experiment; or the physical anaesthesia of profound meditation? If an *OBE* or disembodied mind is possible (Chapter 18), what senselessly then? If a fifth, subjective dimension (or, if you like, sixth sense) of mind exists it would have no material 'membrane' that could separate minds as bodies separated them before. In a so-called 'astral plane' immediate telepathy would be the order of community, a resonant association and, within a waking dream, vivid theatre of the mind's emotions. Is it all in or at least from the mind? Perhaps it truly is - with cosmic states or regions that reflect the conscio-material spectrum's psychological levels. Do you descend to darkness or ascend to light? If capacity to escape the gravity of motion *and* motivation is realised and if peace from disturbance is obtained then you will have ascended into inner space. You will have earned a voluntary discharge from existence and arrived at what is called *moksha*, liberation, enlightenment etc. You would have shot to the tranquillity of Void, communion with Origin, association with Essential Truth. While suspected or glimpsed this Ultimate Destination is rarely reached.

Externally you descend from X through physical interaction with your environment. The rings spread first through your nervous system (which is the closest physical associate of mind) to the rest of your skin and bones and their neighbourhood. This is your space-time location. It includes other biological bodies to which you physically and, more or less easily according to type, psychologically relate. If you restrict these to the human type then (as is the norm) you restrict yourself to the psychological dynamic of family, friends, neighbours, interest groups, nations or the international scene. You take a social part, which may involve institutionalised politics, law and religion, in your community.

You can also think of life forms (or biotic factors) in terms of individuals, populations or communities of different kinds of organism. This positive, inclusive perspective is called ecological. An ecological web is dynamic. Its wider body includes the biotic community of living things both near and far. The health and behaviour of these parts each affect the whole. Such parts descend, as the rings convey, to include non-living factors such as sources of energy, climate, ocean, water, mineral cycles and the soils of earth. These abiotic factors make up habitat. The economic relationship of any organism with its wider body (community and habitat together) is called its niche.

Finally you can extend from planet earth's 'bio-spherical' surface to its core or, in the opposite direction, out to the wider solar system, galaxy and universe. Whichever way, you can emphasise relationship, linkage and togetherness. The underlying notions of community are positive interaction, self-regulation and harmonious stability.

How cosmically tiny is the web of life on earth! How alone! There is no evidence, only hope, that there is any other. What, however, if other solar systems *did* host 'alien' communities? *Whether life evolved or was created here, there seems no reason why, in principle, it might not have evolved or been created anywhere that rare but suitable conditions exist.* What rare, unlikely conditions are these? The universal prescription includes a complexity of form based on carbon chemistry and narrow but stable parameters for conditions that include temperature, climate, pH (acidity), saltiness in the sea etc. It is hard to envisage an alternative to liquid water as the medium for metabolism. Water, in the form of ice or gas, is not uncommon in the universe; but liquid constantly in quantity is most unusual. Because this phase only exists within a very small range of temperature life would have to inhabit a planet with a circular orbit the right distance from its particular star. A long list of constraints will be detailed (Towards a Unified Theory of Ecology: Abiotic Part). Moreover, what catalysts other than protein could mediate and regulate extra-terrestrial metabolism? How would protein be constructed except through the highly regulated agency of a protein database made of *DNA*? Would the basic unit of exobiological life be other than a cell? Is there a better, worse or even possible kind of protective yet communicative 'wall' between a cell and its environment apart from the flexible hedge called a phospholipid membrane? A membrane stacked with gates, informative sensors and communications gear? If there is no other chemical way to satisfy this set of conditions then it might be reasonably supposed that any member of life's Astronomical Society would resemble one form or another of life on earth. And, since the more complex forms of any ecosystem depend on 'simpler' workers and providers, any anthropoid extra-terrestrials would involve an ecological web, a range of associations no less complex and harmonious than our own. Is there any objection to a unified theory of interstellar or intergalactic communities?

Points of Return

tam/ raj	*Sat*
existence	*Essence*
time	*Timelessness*

	form	*Formlessness*	
	expression	*Potential*	
	lesser perfection	*Perfection*	
	time apart	*Reunion*	
	differentiation	*Communion*	
	issue	*Origin*	
↓	*tam*	*raj*	↑
	down	*up*	
	materialisation	*dematerialisation*	
	diversification	*unification*	
	decreasing consciousness	*increasing consciousness*	
	isolation	*relationship*	
	creation	*dissolution*	
	careless/ ignorant of Reunion	*yearning for Reunion*	
	departure	*return*	

If an act of creation sweeps with a centrifugal current, then centripetal dissolution is its introversion, its reverse. And if the act begins at a Single Apex and spreads down through the levels of 'Mount Universe', then a return ravels up the unravelling. It ascends towards First Principle and the Essential Singularity. In (*tam*) descent differences are accentuated; isolation and distinction are the result of transforming principle into practice. An upward (*raj*) vector is, on the contrary, characterised by unification, association and relationship. Do you remember (Chapter 23) the seed that, once pressed into darkness of the soil's womb, flowered back into the light? Oriental tradition uses a lotus flower to analogise the 'return' circuit of creation. Just as consciousness is overcome in descent through mind to matter, so it gradually awakens by evolving up towards its Origin. Paradise regained. A lotus seed, pressed into mud, develops through the water till it flowers in clear, sunlit air. **The keynote of this chapter is return. It involves, eventually, completion of the cosmic cycle (*fig.* 12.7: Loop 1). Is not this, therefore, Alpha and Omega, also the Climax of the world?** *As such this text will take the furthest periphery of community and follow its return path in towards the journey's end, its culmination in Reunion, Homecoming at the House of Inmost, Central Sun.*

Outer darkness in the form of matter can't return. Oblivion (*fig.* 12.7) cycles in its purely energetic and involuntary loop. It is, however, at the base of conscio-material gradient and for an incarnate human the physical plane is, on a scale of return, the ascending (↑) journey's starting-point. The first landmark is, therefore, on dispersal's far and 'muddy' shore; body-land is the mass-centred diametrical antagonist of Subjective Radiance. And description therefore sets out towards The Centre with consideration of the abiotic factors of ecology. These factors range, at the perimeter, in wide galactic space whence you can zoom through solar systems inwards to this planet. There follow the biotic factors that permit a life on earth; and it is recognised that animals are trapped in physicality. It is the gift of human centaurs, though, to rise above sheer instinct and carnality; there comes a section covering our population's version of community, the humanities. Human society is, like the individuals who sum to it, a meeting of its mind and body. Its dualism reflects the tension between metaphysical coherence and physical differentiation, principle and practice, morality and economics; and the tensions that arise between an individual and his community. The institutions that attempt to reconcile the differences are called religion, law and politics. Buttressed by education these stalwarts labour to maintain dynamic social equilibrium, that is, to keep the peace. In principle, at least, they strive to stabilise. Their equations quiver in the balance; and excited hands of science, literature and art each play a part. Could the framework of the dialectical help one understand the nature of the big three's slants? Each stakeholder in society claims his or her agenda 'right and good'. When ideas conflict can the Dialectic help to frame a judgment and, perhaps, a compromise? Can it at least identify what everyone among the greys might reasonably agree is black. Is 'evil' not a social and an individual problem? If so, is the problem soluble? If yes, then how?

Finally we pass the lintel of mind's door. One moves inward from the outside world and, through the third eye at the temple closer to the Central Home. This is the real turning point, a voluntary aspiration towards association with the Real Self, Central Absolution at the heart of things. Outward tendencies of focus are reversed; the habit of dispersal is entrained, revolved and by dint of contemplation turned inwardly. If materialisation involves a contraction of infinity, an inversion towards increasingly ego-centred individuals and localised, mass-centred objects, then dematerialisation implies an expansion from them. Where one becomes two and two many, now the opposite comes into play. There is a motion towards a Single Origin, the Starting-Point. There is motion to complete the Cosmic Circuit. We call the association between two objects a bond. We call the confluence of minds friendship and, when minds merge, the name is love. Two hearts, one love. In union is love and in Communion Love. *In pursuit of Perfect Truth the logic drives us towards such nature of Original Cause.*

Towards a Unified Theory of Ecology: Abiotic Part

Ecology is about your 'wider body'. Such 'wider body' involves biological and physical connections within whose dynamic every life form thrives or dies. It includes a community of organisms (*biota*) that inhabit a solar system, in particular the surface region of a planet. This region, called an ecosphere or biosphere, comprises *biota* and solid, liquid and gaseous forms of *abiota.* The latter is, as already noted, a composite of rocks and soils, seas, freshwater and the climatic atmosphere. It amounts to the stage on which, in different scenes, various players act. A reductionist, scientific analysis of earth's ecosphere is divided into multiple disciplines. They include branches of biochemistry, biology, geology, climatology, oceanography, environmental science etc. Many divisions, say climatology and genetics, rarely communicate. The splintered perspectives of these academic 'cells' add up to a lack of wholeness in the way our world is described. What collects the analysis? What synthetic principles drive life on earth?

Top-down, first look to homeostasis. This term, used for cybernetic self-regulation, is of fundamental cosmic importance. Such regulation is the tendency of a system to maintain a stable state, a balance between the contributions of its parts. This balance (Chapters 5 - 25, especially 19) is more than the sum of its parts. The extra is found in the order, norms or laws according to which the parts behave.

In non-purposive systems homeostasis revolves around norms described by the so-called laws of physics. These norms are inexorable patterns of behaviour; and, while energy is conserved, a normal sequence of events occurs. Take a chemical reaction. When the (*sat*) potential for action is realised a (*raj*) kinetic phase runs to (*tam*) completion. Reaction or diffusion's force is spent. Such a condition of sub-state equilibrium is, like an unwound spring, exhausted. It is also called impotent or inertial equilibrium. Until their next injection of energy inanimate systems drop as soon as possible into the least of tensions, the stability of death. In other words their dynamic is as irregular and unpredictable as the type and quantity of energy they are affected by.

Purposive systems, on the other hand, tick with deliberate regularity. Biology, whose base is animation and whose 'wider body' ecosphere, is constrained by natural law but its forms of homeostasis all include a definite goal - survival. The norms are reasons, logical constructions, mechanisms devolved around this core non-atomic concept. Feedback is negative but far from accidental. Dedicated cycles of control, each integrated with the others like a cog in William Paley's watch, keep life ticking steadily. This mode of regularity is called dynamic equilibrium. Its vibrant motion, like a piston pumping in its cylinder, automatically reaffirms the reason why it cycles round - to keep things going. The measure of a conceptual system is the efficiency and ease with which a combination of pre-coded parts achieves its preordained design. It is, in other words, the extent to which it can achieve its full potential, its, goal, in this case its excellent survival.

Death and decay are time's way. As easy as reversing time you might imagine evolution cranking up the information that you need to service concept; you might believe that luck, constrained of course by natural law, turned matter into 'living matter'. What, after all, are pistons but metallic and a body made of molecules? Except for their arrangement, order, plan and purposes. Dynamic bio-equilibrium tends not, as is the tendency of its non-purposive surroundings, towards the lowest state of energy; nor does life's fountain spring from geysers, volcanic vents or even showers of sunlight. Fuel does not, without an engine, drive a car. A body's engine is a subtle, well-designed pump, a coded speciality called energy metabolism. Its business springs from, firstly, information and only then from energy metabolism and from reproduction. In each body this tri-unity is controlled by Paley's cogs; its triplex is managed by vibrant oscillations synchronised around a prefabricated set of norms. These rules constitute, in turn, the cybernetic type of law that is biology's.

Such an improbable balancing act is sustained by interlocking sub-routines each cycling more or less tightly round its axis. 'Axes of reason' coordinate to buffer against accidental irregularities and stabilise in favour of a *status quo*. Variables include temperature, pH, salinity, electrical potentials, heartbeat rate etc. Controls in humans include metabolic, hormonal and nervous systems of adjustment. In this case the *reason* for such coherent operation is not, in the last analysis, physical but metaphysical. **'Cybernetic homeostasis', like 'program', is a conceptual phenomenon. Its presence in any machine**

indicates an underlying purpose. *Numerous biological sub-systems are coordinated under the overall purpose of the continuation of life, of survival.* As discussed (esp. Chapter 20), atoms have no 'wish' to 'self-regulate' in otherwise impossibly complex concatenations 'in order to' survive. Yet information systems, energy metabolism and reproductive mechanisms at the core of biology are based around this psychological imperative. It is what informs and by informing drives the wheels of life around.

Ecological systems include both non-purposive (abiotic) and purposive (biotic) components. Where various categories of organism (such as producer, predator, scavenger etc.) and features of an environment are necessary for such a system to flourish, the shocks of small changes are generally absorbed but large ones may throw it off-balance. In this case different organisms will come to flourish and the overall equilibrium of the affected spot shift. A different local norm will, for better or for worse, be expressed. Such changes are themselves a norm - indeed, 'succession' is the way life takes root, the way different habitats develop to maturity. This is not to say that shifting norms, like variations in typical life forms, can be pushed too far. Temperature, chemical constituency, pH etc. must remain within the overall constraint of Life's Biological Norm. Within the orbit of this norm each type of organism's hierarchical axes of information, its homeostatic systems, scarcely change - because their processing is designed to rationally dispose a material effect, to arrange matter extremely specifically to obtain its own kind of survival, to conform to archetype.

If a system is a group of ideas or objects linked by some common purpose, theme or interdependence, then the universe might be called a 'cosmo-system' or 'cosmo-logical'. A cybernetic control theory of the universe makes it a self-regulating body and the inclusion of universal mind (as archetype) reveals a whole 'cosmo-psycho-physiology'. 'Existence' becomes an abstract for a system in dynamic equilibrium, a composition of both mind and body that used to be called the person of God. Of course, such a metaphysical idea is of absolutely no interest to science but, by the same definition of system, life on earth can be termed an ecosystem. It includes all living organisms (the *biota*) and anything inanimate that influences or is influenced by them. The theory holds earth to be a system in dynamic equilibrium. Such a system, including abiotic factors, constitutes a totality that in a cybernetic manner *seeks* an optimal condition for life. The ecosystem can be defined as a single self-regulating system, a 'super-organism of all life tightly coupled with the air, oceans and the surface rocks'. Its study is therefore called eco- or geo-physiology. In this sense is there any unified theory of earth's ecological community?

An excellent start has been made by James Lovelock. Eco-physiological *Gaian* theory was inspired by the famous 'blue marble' photograph, the *top-down* view of our planet from a space satellite. *Gaia* is the name given earth's (but in principle any) ecological system. However although *Gaia* is Greek for 'earth' and associated with *Ge* (also 'earth') and *Demeter* (earth-mother), there lurks absolutely no animistic undercurrent in the theorist's mind. For James Lovelock, as for Natural Dialectic, there is no more hidden 'animating vitalism' that somehow guides biological matter to its self-regulating ends than there is 'mind' which physically works as opposed to engenders a computer. The computer's case of electricity is an ecosystem's sunlit chemistry. In fact *Gaia* even lacks dialectical archetype, 'organisational record' or conceptual code. *Its dynamic is one without which it would forfeit scientific credibility viz. it is wholly materialistic and thus, perforce, evolutionary in outlook.* The assumptions of orthodox cosmology (a big bang, galactic, stellar and planetary evolution, an early reducing atmosphere, a very old age of the earth, abiogenesis and the possibility of macro-evolution) are accepted without question as a starting-point. The Gaian hypothesis is rooted in chance and natural law. Tightly coupled animate and inanimate features of the biosphere *evolve* together. It includes purposive, teleological cybernetic systems that, according to the scientific paradigm, evolved without any regulatory purpose in mind. Its 'geo-physiology' evolves as gradually as biological form and function; its apparent 'purpose' grows by chance. However, in a way that seems self-contradictory to dialectical eyes such evolution co-opts mechanisms of self-regulatory homeostasis, adaptive potential and the constancy of earth's 'eco-physiological' condition. How might such order gradually evolve by genetic inconstancy - the usual mutation/ innovation problem? In other words, theory glosses the root problem, discussed at length (Chapters 5, 6, 19 - 25), of the random origin of information. How could a planetary, any more than a biological, form of atoms want to self-construct, self-regulate and 'improve' itself by 'inventing' myriad purposive, integrated yet meaningless, irreducibly complex 'designs'? Of which one, ecological homeostasis, is a critical foundation for life on earth? In short, an inanimate 'earth goddess' is predictably, emphatically denied any teleological dimension. She has no mind, not even archetypal, of her own.

The problems with evolutionary theory have been investigated in some detail. The real tension arose, you may recall, between phylogenetic and bio-logical interpretations of the evidence. It involves our old friends The Theories of No Intelligence and Intelligence. In the former, phylogenetic case time is of the essence. Luck, which is far too dim to recognise even itself, needs plenty of time to accrue multiple, serial jackpots. In the latter, bio-logical case time is not, in conceptual terms, the measure of creation. Indeed, it is irrelevant. Neither the working life of a machine nor the duration of a play is dependent on the timescale of its creator's preparations. Neither does engineer or author work passively, asleep or drunk into oblivion. An active producer propels every creative design from conception through detailed (archetypal) planning to material expression. First an outline is sketched, a foundation laid. This foundation is developed, like a house, in logical stages to fruition. Each stage is triggered by a signal that the previous one is complete. These triggers might fire rapidly or over extensive periods of time. *In this case the real questions involve not time but mind.*

In other words, are the primary producers of a biosphere really sunlight and the plants? Did such 'fast ecology' as sprang on Surtsey Island in the 1960's or Mount St. Helen's in the 1980's USA create earth's ecosystems? Is just thirty years a lot too quick or was their making, as Darwinism ponders, ineffably slow? Do the ravages of time gradually destroy initially coherent information or, on the contrary, did life with its supporting systems just 'emerge', build information up from chaos and improve with age? In short, are ecosystems grown by chance or is the primary producer mind? If physical proceeds from metaphysical what might be the nature of informative potential, its archetypal inwardness, its universal mind?

Not that you can see universal mind. It is nowhere. Archetypal memories (Chapters 16, 17 and 19) are physically absent. They are as invisible, secret and yet real as your neighbour's mind, memories and dreams. Nor can you find, only *infer*, the author in a book or the engineer in his machine. You *infer* the presence of mind from reasonable behaviours, purposive mechanisms and complex yet harmonious works of intellect. You see mind wherever matter couldn't make it. You *infer* mind elsewhere, you only *know* it in yourself. Look around your lab or room. No mind is there but mind is everywhere. It is in every object, every habitat that's made with reason. All life has its reasons. From an immeasurable, metaphysical interior issues visible, physical behaviour. Metaphysical storage of information is called memory. Memory, habit and character are each informed by past purposes, ideas and events; they have become the inner potential according to which actions now occur. As a seed grows to a plant, so a purpose or design 'organically' expresses seed idea. *Physical develops from its metaphysical interior.* In other words, material behaviour issues and is sustained according to the trammel of archetypal memories - the generic term for which is universal mind. Matter is 'set' according to its patterns, 'plays' according to its records, 'runs' according to its programs. In this sense (as was shown in Chapter 6) a material device is, besides its allotted task, an information store that fixes its creator's thoughts. Just as there was a special, extra-physical creation of matter whose archetypes are expressed as physical forces and associated particles, so there are biological programs, an ecological suite of archetypes. As a *ROM* is inaccessible to the vagaries of a user so these archetypes are tamper-proof. Physical expressions, called organisms, are accessible but archetypal *ROM*s are inaccessible to scientific instrument or experiment. And, given the amount of Promethean tampering with *DNA*, just as well!

In the dialectical view purpose informs the structure and function of a mechanism; mind is in the machine; metaphysical drives physical. Life builds its house with survival in mind. *The mind of Gaia is in fact voluntary, active and includes this purpose. Its informant is life. Neither rock, sea nor air but life is its mind.* The 'super-organism' includes, in fact, a whole range of minds and archetypal memories. It includes both conscious and sub-conscious aspects of the minds of all organisms that comprise the different cells and organs of its body. Different sorts of organism have different niches that each contributes to the whole. In most cases their will to survive is concretised in instinct and biological equipment. Nevertheless together they help drive the planet into its improbable, unbalanced and yet amazingly stable, healthy, life-friendly condition.

In 1855 Chief Seathl of the Swamish tribe wrote to President Franklin Pierce, protesting at the white man's aggressive attitude towards the American ecosystem.

"What," he asked, "is man without the beasts? If all the beasts were gone man would die of a great loneliness of spirit, for whatever happens to the beasts also happens to man. All things are connected..."

Unscientific Chief Seathl, like scientific James Lovelock, saw life on earth as a web, an interconnected whole. It is, from a *top-down* perspective, doubtful that his 'unifying' wisdom derived from the same source or stopped as short as Lovelock's. In which case let's let the facts speak. Which, Intelligence or No Intelligence, can best be inferred from them? Let us turn to see how, like a house from its foundations, an eco-physiology is constructed. We follow concentric, social circles inwards from abiotic extra-terrestrial through solid, liquid, gaseous to biotic, biological environment. We ascend the hierarchy of ecology.

	tam/ raj	*Sat*	
	hierarchy	*Top*	
	cycle	*Axis*	
	peripheries	*Centre*	
	expression	*Potential*	
	effects	*Source*	
	planets	*Sun*	
↓	*tam*	*raj*	↑
	static	*dynamic*	
	mass	*energy*	
	loss	*gain*	
	dark	*light*	
	capture	*release*	
	fixity	*flux*	
	structure	*circulation*	
	inertial equilibrium	*dynamic equilibrium*	
	death	*life*	

Firstly, let's survey factors outside the control of life's mind or body.

House-hunters know the value of location. How beneficial is earth's real estate in space?

Heavy traffic isn't safe. Black holes, deadly radiation and exploding stars frequent the centre of the Milky Way. And the conurbation's borders house too few elements for planetary life. Large areas of 'suburban' spiral arms are also inhospitable but our globe plies business in the safest zone of town.

Not only our galactic but also our planetary zone is habitable. Life needs an endless supply of liquid water and thus strict, natural thermoregulation. Our thermal generator is a 'dwarf main sequence star'. Large stars flood planets with bio-destructive ultra-violet radiation; if you want a low dose you'd be far enough away to freeze. If, on the other hand, our ordinary nucleus of power were smaller (like 90% of its galactic neighbours) then you'd have to orbit closer - but at the same time your rotation might get locked to synchronise with orbit so that one side always frazzled and the other darkly froze. 5% nearer and things would have boiled long ago, 5% further and runaway glaciation would have frozen life out. Most unfriendly! Not a '*CHZ*' (continuously habitable zone). Our own solar system has a *CHZ* - in which earth spins. Therefore our distance from the golden ball is fine.

You argue that such definite felicity is 'temporary' since it's believed our star is gradually heating up; that apocalypse will come, if not by prophetic fire, by overheating; and that scientific man is simply, greatly hastening the end of animation's world. However, what we know for fact is that the earth's location is and has for aeons been a biophilic and self-regulating dream; and easy sunshine is life's primary, pristine nutrient. From a self-centred point of view our massive, abiotic extraterrestrial star is far away; it seems marginal upon our social circles. Do you, however, recall (Chapters 2 and 23) an inversion that places mass at the centre of physical systems? Our Copernican fireball may seem peripheral to a geocentric point of view but is actually the radiant, gravitational axis of a rare formation. It is supposed but not known (Chapter 12) how planetary systems 'nucleate' from discs of gas and dust surrounding stars. Whether it bore planets or did not, the sun above is certainly the foundation that supports earth's ecological pyramids below. Its wireless rays nurture children of the heavens, earth's biological offspring, with the 'life-friendly milk' of radiation - mostly in the narrow, visible waveband. Its heat drives huge surface fluxes in the form of oceanic currents, weather systems and therefore the climates that govern *biomes*, ecological zones like rain forests, deserts, savannah and so on. Our lives are definitely hung upon a lucky star. Not only the sun's distance but also earth's unusually circular orbit slung on this fixed radius and its rotation round its own axis are right. They are right by distance because temperature on globe's surface is, except at the poles, life-friendly; by circular orbit because this temperature is stabilised; and by speed of spin that avoids life-destructive boiling, refrigeration and violent, protracted wind-storms. Critically, the constancy of temperature is of precise degree to produce that *CHZ* - a liquid water-bath in which life forms can thrive.

If life's solar lord clocks just the right statistics what about our closer lady of the night? Four hundred times smaller but four hundred times closer, at the moment of eclipse she exactly covers him. With us she dances near enough to lock in motion so she never shows a dark side, just a single, silver, sunlit face. Our planet's tilt, which wobbles through the seasons and affects sea levels, has been finely damped and stabilised by moon's gravitational effect. Did, you surmise, a deluge of huge rocks crash long ago into creation of this norm, a vital oscillation 22 through 24 degrees and at the same time generate its stabilising moon? If so then, oh, what planetary serendipity! This way lunar influence delivers benign climate, her caress dispenses temperate seasons to the brow of life on earth. And her rhythmic sway stirs up the seas. She lifts oceans then exactly drops them down again. The effect is not so large that continents are flooded and eroded each high tide and not so small that surface waters can stagnate. Marine environments are flushed and with diurnal freshness the communities of ocean bloom and thrive. Lunar periods influence other biological events and our consort acts (in company with vaster Jupiter) as a shield to protect earth from the impact of comets and asteroids. Not only moon but earth is of the right size in relation. Thus gravitation sustains a viable atmosphere, oceanic swells and a reasonable proportion of tectonically active landmass. Each factor like a right note helps to swing along the couple's 'bio-centric' waltz. Of course, each could be a coincidence. From an accumulation of how many coincidences do you infer that there is more than chance at work? How many steps would make you think a form of motion was an orchestrated dance?

Fresh breath of life! Earth might be mostly made of iron, magnesium, silicon and oxygen but clouds of water, carbon, nitrogen and oxygen compose a friendly sky and, for the most part, bodies living under it. From pure energy (sunlight through space) you fall on gas. *Gaia's* primary dynamic, sky, consists of a concentric suite of atmospheric shells. These envelop the surface of our 'nuclear' globe transparently (what kind of life survives an opaque atmosphere?) They drop from the ionosphere through mesosphere, stratosphere and ozone layer to climatic troposphere. Each, like a membrane, offers its own particular form of protection to the life within. The living planet floats, egg-like in a white of air, within the warm, deep womb of solar influence. Outside the influence of its star there stretches endless, barren, ultra-freezing space. You might construe the stratosphere and ionosphere as buffers, membranes, even subtle skin.

Where did all the water spring from? Nested lower down the suite of air-light shells, you find fluids of fertility - mists, rain, rivers and the oceans. These, lower atmosphere and ground water, assume the blood-like role of conductors, convectors and radiators. The media churn. Their healthy currents help to circulate biological nutrients and refresh the tissues and anatomy of earth. For example, the hydrological (water) cycle helps to moderate the climate, cleanse sky, flush earth and weather rock into soils and sandstones. It fosters, mostly gently, life's fecundity. While this cycle is about 1000 times more effective in the presence of life erosion could proceed without it; and both liquid and gaseous layers could, as a result of geological activity alone, have formed above the surface crust.

The amount of water on the blue planet has remained stable. So when rivers pour megatons of salts into the oceans how do they avoid rapidly becoming dead seas? Above the norm, which has remained stable throughout life's lifetime, salinity is fatal. Is the random, geological formation of salt-pans enough to have held the narrow line or might life, in the form of stromatolites and corals, make and break lagoons (and thereby salt-pans) in a precise biological response to conditions in the water round them? Whatever its cause, the improbable balance is robustly upheld. Accident and illness are overcome.

Gaia's bones, nails and hair are soils and solid, crustal rocks that, washed by storm and stream, yield minerals. These minerals life's producers, plants, absorb. In fact eleven lighter, non-radioactive and so-called 'biogenic' elements (carbon, hydrogen, oxygen, nitrogen, phosphorous etc.) constitute 99.9% of body by weight and occur in the same proportions in all organisms; fourteen more (including iron and iodine) are trace elements. Their surface appearance is complemented by heavy, subterranean elements. The radioactive ones are, while lethal to life, in another sense essential. The heat of their decay, particularly of the heaviest 'stable' element, uranium, may have caused the processes by which molten iron fell to earth's central core and its crust, while 'floating' up, differentiated into layers or zones of lighter materials with lower boiling points towards the surface. While extra-terrestrial energy drives *Gaia's* superficial fluxes internal fires of tremendous heat power her geological circulation. Mobile areas of a molten iron core generate magnetic fields that deflect harmful solar winds, protect against them ripping off the atmosphere and interact with the electrical conditions of life. The heat, derived from radioactive decay, also drives volcanism. Volcanoes throw up irregular formations like mountain chains whose various habitats permit an abundance of ecological niches. If life's skin is the sky, the earth's is a crust of islands (or tectonic plates) that float on seas of magma. These plates are, comparative to the globe's diameter, wafer thin - between 3 and 60 kilometres thick. This allows them to move, crumple, recycle rock and shape the continents; also to buffer the surface from its hot core. Partial melting of the mantle rock, peridotite, creates basalts, granites and other mineral-rich rocks. Basalt is volcanically extruded at constructive plate margins and all three at destructive margins. The latter, also called subduction zones, are where rock is recycled down into the mantle. As well as volcanoes these motions occasionally generate, like a roaring bull tossing in the black depths, earthquake, *tsunami* and, for humans, devastation. Without the presence of uranium rocks of the crust would not have surfaced nor water gassed into the sky and condensed into the seas. Neither is the radioactive heat that drives tectonic cycles too little otherwise 'no wheels would turn'; nor too much otherwise rampant volcanism would suffocate, poison, crush and bury life. The massive geological cycle is supplemented by the stress, deformation and metamorphic reformation of rocks by movements of large blocks of earth that crack or fault the crust. However, none of this life-friendly action would occur if our globe were larger; pressure and viscosity would induce a stagnant outer layer that inhibited tectonics; and such magnetic shield as deflects harmful radiation would be weakened.

Neither internal nor external engine of ecological dialectic is in life's hands to govern or control. Nevertheless tectonic and hydrological cycles mesh like cogs in a gearbox, like a great plough that has, for life's lifetime, turned earth's field. The former recycles crustal material and the other dissolves key minerals into surface water whence bio-forms can use them. This integration of hot and cold 'ploughs' has, for example, ensured the presence and weathering of silicates into hygroscopic and ion-absorbing reservoirs of clay. Cold-water weathering also helps recycle carbon, calcium, nitrates, phosphates and other minerals. In this way the heaviest element (uranium) is instrumental in hot-earth churning to the point where the light end of the periodic table can 'kick in'. Clays and other plant-friendly soils mean photosynthesis can charge life with energy. Light links up with life (*fig.* 21.1). The element at the heart of energy metabolism is diametrically opposite the heaviest - the lightest, hydrogen. Weightless, invisible photons are allied by photosynthesis with hydrogen's proton and electron to prop life's massive, homeostatic ecology.

Not only hydrogen but carbon and oxygen are supremely bio-philic. Carbon demonstrates a unique balance between stability and ability to participate in organic reactions under mild conditions. It can form single and fused rings, also linear and branched chains, involving much greater complexity than silicon (the only comparable element). Carbon-carbon bonds are also more stable than silicon-silicon and less susceptible to oxidation. In other words, carbon has an optimal disposition to 'complicate' in aqueous biochemical formations at temperatures within a normal, narrow and very stable 50 degree heat window found on earth's surface (for although localised extremes stretch from about -40 arctic wind chill factor to 120°C in hydrothermal vents the salient swing is between -5 and 45°C). And, critically within this thermal frame, there occurs the life-friendly weak-bonding capacity and strength of structure that allows *DNA* to accurately inform and proteins to take up precise, preordained but pliable three-dimensional complexities in order to function.

Hydrogen, carbon, oxygen and nitrogen are common atoms not only in microcosmic forms of life but also in the macrocosmic world. Atmospheric nitrogen is an inert buffer, a sky-filler that, being 80% of air, keeps reactive oxygen (about 20%) in check and balance. Much below this level neither man, fire nor their alliance (called technological civilisation) could occur; much above and huge, lightning-generated conflagrations would destroy all flora and, thereby, fauna.

If you want 'sparkling', chemical liveliness, pluck oxygen from air. Oxidation of carbon and hydrogen liberates more energy than any other chemical (except boron); therefore the latter are stable, efficient stores of energy. Lipids, sugars and proteins are all stores for respiratory energy release; and highly charged phosphates inject action into any biochemical scenario. Yet although oxygen's reactivity increases greatly above 50 degrees it is, at ambient temperatures, relatively inert. You are unlikely to spontaneously combust! It is restrained enough to work with life. 'Clever' enzymic metabolism both exploits the possibilities and reduces the risks of oxygen's behaviour (such as random free radical damage) in a controlled and coordinated way. Indeed, the 'damping' of oxygen is crucial; as temperature rises its solubility in water drops so that, between 0 and 50 degrees, the right amount to satisfy metabolic requirements can diffuse into cells. If the solubility were lower diffusion would be too slow, if higher the danger from toxic free radicals would grow and competition with carbon dioxide inhibit sugar production and, consequently, life. If you could change the bio-philic solubility of oxygen, how would this constraint impact your re-design of its vital carriage in blood?

Only planets earth's size can allow an atmosphere with oxygen at all; but could you, would you change the ratio of oxygen in ours (now about 21%)? Was it (see Chapter 20: Atmosphere) always so? Air bubbles in amber have been found to contain up to 30% - which fits with fossils of giant leaves, tree-sized ferns, thick fire-resistant bark, huge dragonflies and so on. Some disagree but, being carbon-based, you'd die with much less and self-immolate with more. A huge annual influx of nitrogen, ammonia, CO_2 and other gases is also buffered by various of the many feedback systems that keep Gaia very bio-friendly.

Oxygen is life's atomic go-getter because the oxidation of hydrogen and carbon yields stable, harmless and yet highly bio-centric ingredients - a matrix (water) and handmaiden (carbon dioxide). Handmaiden? Life's humble servant is, in gaseous form, everywhere at hand. Indeed, it surrenders its whole being unto living. It is, by photosynthesis, the way that carbon enters life's grand system and, at another place, constitutes a perfect instrument for the excretion of organic waste. Because it is highly oxidised this gas can co-exist peacefully with reactive oxygen in the atmosphere. It also dissolves freely in water forming, unlike many other oxides, a weak acid. Carbonic acid has a brilliantly bio-centric 'trick' up its corrosive sleeve viz. it solves the biological problem of acid-base (pH) regulation and, amid a maelstrom of metabolism that often generates acidity, helps keep the balance. The balance is as critical as it is tight. It is impossible to achieve complex biochemical manipulation using enzymes in other than practically neutral water. Carbon dioxide, sometimes helped by an enzyme called carbonic anhydrase, reacts with water to generate the bicarbonate ion; and the bicarbonate system tends to buffer any divergence from the pH neutral norm. Your life depends on it. How elegantly and efficiently the metabolic house is kept. Gaseous carbon dioxide is introduced, its bicarbonate ion regulates the background pH of the water in which biochemistry occurs; and excess acid is, in the original humble form of gaseous carbon dioxide, tipped back into the air. Input, work, output. *Could you design a better fit?* Could you do so with the right properties attached to oxygen, to the transition metals that bind it and constitute the electronic circuitry of life, the tectonic factors that brought these metals to the roots of plants and so on and on?

Yet under normal circumstances the chemical reactions that happen in fluid media run to completion but, in our *Gaia's* case, "the chemical composition of the atmosphere bears no relation to expected steady-state chemical equilibrium. The presence of methane, nitrous oxide and even nitrogen in our present oxidising atmosphere represents a violation of the rules of chemistry to be measured in tens of orders of magnitude. Disequilibria on this scale suggest that the atmosphere is not merely a biological product but more probably a biological construction.....the atmospheric concentration of gases such as oxygen and ammonia is found to be kept at an optimum value from which even small departures could have disastrous consequences for life" (James Lovelock: *Gaia* p. 9).

Why on earth have the highly improbable, unstable but correct hydrological and atmospheric conditions for life on earth remained, since the inception of life, in a state of dynamic equilibrium? It has been calculated that, during the time life on earth has existed, the sun's heat output has increased by 25-30% - in which case earth would have been a frozen, hostile and anti-evolutionary environment throughout most of its history. Raw solar energy drives the convection currents of the seas, the climate and, through the specialised and complex agency of photosynthesis, recharges life's batteries. Alternatively a 25% increase in energy from a tolerable initial level of heat would have deleted any life that may have occurred. What has buffered this uncontrollable effect?

Rivers should have dumped far more sodium and mud sediment (to an average of 350 metres) than is found to be the case - despite subduction zones removing about 5% of the deposit annually. Where has all the mud and salt gone, long time passing? Compounds of oxygen found in the rocks suggest the climate has always, oscillating in and out of ice ages, been much as it is now. Our atmosphere is oxidising, ocean salinity never more than 3.5% (6% for even a few moments is fatal) and, within the right pH range, the ground well soiled. The whole biospheric band is not only very narrow spatially (extending just a few kilometres from mountain-tops to marine abyss) but also in the constrained physical and chemical tolerances imposed by every minuscule, biological cell. What natural homeostat has controlled temperature, atmospheric composition, salinity, pH and other key biological parameters so tightly? *What sort of strict self-regulation has kept such a highly improbable but critical, very life-friendly regime unbroken for perhaps billions of years?*

"*The climate and chemical properties of the Earth now and throughout its history seem always to have been optimal for life. For this to have happened by chance is as unlikely as to survive unscathed a drive blindfold through rush-hour traffic.*" (*Gaia* ps. 9-10).

Not only fluent air. Do you remember (Chapter 8: Is the Match Friendly?) all the biophilic properties of life's liquidity, of fluent water. Can one overestimate its vital, swirling influence on life? Who can deny its critical permission? Earth spins at just the right speed and distance from the sun to keep its bio-essence permanently liquid. What other than life's window is tight fluctuation in the right degree of heat? Heat absorbent, a solvent and a reaction chamber - coolish liquid water is thrice critical because it is the only medium wherein complex organic chemistry can be coaxed to perform and complex products of such coded chemistry survive. Life precisely, utterly depends on planetary spin combined with opportune relationship around its axis, hub or sun. Thus, as whole cultures have, thank your lucky star. Is it the only one?

Could such exact and multiple, improbable coincidence be just a chance? Are we a rarity rare luck has thrown on a very special shore of time? Calculations rate luck's chances of combining all the biophilic factors that life needs at 1 in 10^{15} - which makes nonsense of a bookie's bet. Yet Lovelock assumes that, unlike the thermostat on a home central heating system, the *Gaian* 'life-stat' started out with its informant, life, by an abiogenetic chance. It is suggested that the physical constraints which interlock with little tolerance to play their part in life's great puzzle were preordinate; and the parts prefabricated in such a way that a profound, long chain of accumulated coincidence building to the chemical potential for encapsulated metabolism was bound to happen somewhere in the universe. It happened here. Thenceforth life and *Gaia* co-evolved. *The odds against this speculation (discussed in Chapters 8, 20 - 26) are thousands of trillions without inclusion of the all-important information factor, coded chemistry. Whence issues code?*

Towards a Unified Theory of Ecology: Biotic Part

Stratified coincidence. Coincidence upon coincidence (or perhaps 'coincidence') has added up to planet earth's inanimate expedience. While the previous stack emphasised the *abiotic* we turn now to the informative, *biotic* aspect of ecology:

	tam/ raj	*Sat*	
	change	*Peace*	
	oscillation	*Homeostasis*	
	play/ range	*Norm*	
	output/ input	*Measure*	
	communication	*Decision*	
	effector/ sensor	*Regulator*	
↓	*tam*	*raj*	↑
	involuntary	*voluntary*	
	apart	*a part*	
	competition	*symbiosis*	
	effector	*sensor*	
	mind-to-matter	*matter-to-mind*	
	isolation	*relationship*	
	disagreement	*consensus*	
	loss of control	*tight control*	
	disintegration	*integration*	

abiotic lifelessness	*biotic loops*
environment	*body*
death	*life*

Carbon, water and oxygen are (Chapter 8: Is the Match Friendly and the previous section) basic to biological construction. Where carbon gives complexity of form 'fiery' oxygen enlivens. Nobody has seriously conceived of biochemistries other than those based on carbon as the key element, water as the solvent and the very narrow thermal band within which water is liquid. What metabolism would replace ours? What substitution could be made for structural and functional proteins, genetic storage systems, sugars or lipids? Indeed, the universal elements and constraints on such construction being what they are, the extremity of biological complexity probably precludes any but 'carbaqueous' styles of life; in other words, life found anywhere in the cosmos will, as far as can be seen, inhabit an environment resembling ours and look recognisably terrestrial.

If it needn't be like this then go and seriously design a substitute. If you can't, may we presume that there is only one way strictly coded life on earth could be expressed? And chemistry has 'found' that clever way alone?

The materials and conditions that must co-exist before the multiple constraints that bear upon the making of a cell can even be contemplated could possibly exist elsewhere in space. Yet does this mean that, just as computers might 'emerge' from areas of silicates, appropriate metal ions and a heat source, so the very complex, purposive and from inception integrated mechanisms of cellular life must, given a few simple, organic chemicals and right sort of sunlit sky, 'inevitably' emerge? And thence evolve, though blindly, into even greater intricacy such as, in the case of planet earth, a flower or a human? *Constraints, any engineer will tell you, do not conspire to make his work easier; they do not make his irreducibly complex designs more but less inevitably successful.* The more dovetailing constraints must be integrated, the more highly and comprehensively organised (so more difficult and less likely) must be the purposive product - in this case a biological one. A requisite supply of raw materials is not enough to cause the gradual, 'inevitable' construction of life forms. It does not demonstrate the latter are a necessary, accidental result of component 'designs'. Thought is far more 'inevitable' in the outcome of its designs than any chance; organisms are far more likely, as in the engineering case, to be the result of thought than continual, random recompilations from a jumble of 'prefabricated' parts. As such the investigator either confronts or refuses to confront the presence of a natural intelligence hugely greater than the fraction of his own. Science, in principle for all practical purposes, refuses to confront. As if denying patent for invention, is its mind-set in denial?

The composition of life's individual homes is cellular but its communal foundation is microbial, its visible architecture drawn in the shapes of multi-cellular organisms and its garden is the world. Did there once open a brief 'window of opportunity' through which proto-life slid and thenceforth set about transforming barely tolerable, unregulated conditions to ones sufficiently comfortable and stable for a pyramid of life to evolve? Or was there a coherent, deliberate creation of both abiotic stage and biotic characters behind life's extravagant spectacular? Either way, a team must cooperate to 'keep the ecological show on the road'. Its three kinds of player are, in broad terms, producers (the photosynthesisers), consumers (animals) and recyclers (scavengers, detritivores, fungi and bacteria). Within their biological field we read:

	tam/ raj	*Sat*	
	imbalance	*Balance*	
	oscillation	*Homeostasis*	
	process	*Initialise*	
	static/ kinetic	*Potential*	
	end/ lifetime	*Providence*	
	product	*Producer*	
	proton/ electron	*Light*	
	respirant	*Photosynthesiser*	
↓	*tam*	*raj*	↑
	down/ output	*up/ input*	
	discharge	*charge*	
	effusion	*absorption*	
	cool completion	*hot action*	
	exhaustion	*dynamism*	
	lifeless	*lively*	
	waste	*nutrient*	
	inertia	*energy*	
	recycler	*consumer*	

Take that chemical reaction again. Reactants are combined with sufficient energy to create products; as the (*sat*) potential for action is realised a (*raj*) kinetic phase runs to (*tam*) completion. Completion means inertial equilibrium, an impotent end, the exhausted demise of activity. Life, however, does not want to die. It wants to keep on going, to survive. It has to maintain dynamic equilibrium, to keep wheels in wheels in interactive balance. Chemists, in order to preserve the gradient of an equilibrium reaction, keep shovelling reactants and extracting products. They, like life, establish a one-way conveyor belt. From source to sink, they keep their process alive.

(*Sat*) light potential, (*raj*) electronic action and (*tam*) massive result. The dialectical sequence describes photosynthetic 'freezing' of light into chemical storage, the binding of pure energy into sugar. In ecological terms light supplies producers with the potential for life; plants then peacefully create the first link in a food chain - nutrients for consumers. Busy consumers devour the chemical provision and recyclers clear up. The lightest form of recycling (see *fig.* 21.1) is independent of any specific organism. Sub-cellular energy metabolism involves both production and consumption; one's waste is the other's nutrient. Plants can cycle alone; everything else has to pedal with them. The chain of life depends on sun but its dependent chain of death runs down the other way. Recyclers (teams of scavengers, detritivores and decomposers) mop up your exhaust; they clean the dirt of expulsion and decay; they finish off the end. Provision, consumption, recycled waste. *You need all three.* Solar energy streams into earth's equation, through its lives and is discharged. From the discharge (water, carbon dioxide and soil nutrients) plants grow, the wheel turns and the phoenix is resurrected. Life's battery is recharged. Input, process, output. *Homeostasis needs all three. Ecology is irreducibly, biochemically homeostatic and cyclical.* Together every community and every level of life in each community cycles around each co-factor. Indeed, each organism plays one or more of the roles. You need three-in-one to peg the biospheric balance happily.

We have seen (Chapter 20) why abiogenesis is impossible but, whatever their mode of origin, microbes of the same kinds that exist today existed from the start. In the sun and fresh, oxygenated air of the 'over-world' flourish aerobic photosynthesizers and consumers; in the dark, anoxic soils, sediments and mud of the 'under-world' slave multitudes of anaerobes. Tough and reliable, these microbes toil relentlessly. They 'plough' the abiotic field and continuously 'farm' an organic substrate on which 'higher' organisms can thrive; for example, they convert elements critical for life (such as carbon, nitrogen and sulphur) from inorganic, gaseous compounds into forms used by plants and animals. Bacteria might even, as the foundation of life's ecological pyramid, be construed as its primary, substantial, most important forms of life; yet, working at the interface with inorganic matter, its most 'robotic' too.

These are *Gaia*'s 'mind'. They dominate her archetypal and instinctive memory. Microbial behaviour is, for life's sake, intimately involved in recycling carbon, water, methane, oxygen, nitrogen, calcium, sulphur, phosphorous, iodine and other vital minerals. Microflora like beautiful diatoms, radularia, coccolithophores and larger plants all extract carbon dioxide from the air, pump it down into the waters and fix it in the soils. And huge blooms of algae like *Polysiphonia fastigata* excrete *DMS* (dimethyl sulphide) gas from which is derived an aerosol that provides a source of nuclei to seed cloud formation. As the blooms ebb and flow so do the clouds. It makes a thermostatic kind of climate control. Indeed, although we know very little of the world hidden at the base of earth's biological pyramid it is critical. Its microscopic bulk supports the rest. Recyclers. Feedback merchants. Microbes are responsible for the transformations that drive round all the major chemical cycles and sustain life on earth. Captain James Cook's Endeavour took Joseph Banks, The Beagle took Charles Darwin and in 1872 The Challenger took Thomas Huxley to record the world's undiscovered species. They could only see a superficial fraction and (do you remember *Bathybius* from Chapter 21?) missed *Gaia*'s vast microbial 'underbelly'. Now, inspired by these journeys, genome-cracker Craig Venter wants to make a 'whole earth gene catalogue'. To this end he mounted a round-the-world expedition in his yacht, Sorcerer II, to sample ocean water and rewrite biology. Are the secrets of the ocean in a drop of brine? Has Venter found millions of new genes, hundreds of thousands of new species and massively expanded our microscopic knowledge of unseen biology? Maybe this new 'library collection' will release information to help redress some pressing problems - such as global warming or an impending crisis as the oil runs out. The adventurer hopes so; but anyway, acting God who isn't, he plans (always a chance-eliminating move) to intellectually evolve 'appropriate', synthetic forms of life.

Actually life's geo-physiological health, the poise on which all ecosystems and their multi-cellular inhabitants depend, is in great part the gift of bacteria. From a microbial foundation rises a dynamic construction through which sweeps the energy of the sun. A vital pyramid combines basic, *sine qua non* producers (nitrogen fixing bacteria and photosynthetic organisms) with chains of (animal) consumers and the department of refuse collection. Each hierarchical level involves more complex, often larger and rarer forms of life; and its levels are bound by networks of essential interaction. Take, for example, deep and extensive subterranean networks of mycorrhizal fungi on which our crops and forests depend. Old trees extend a symbiotic fungal hand to younger; birch reaches out to fir. It is a general fact that the more diverse a community, the more stable, resilient and prosperous its survival. The web of life's interactivity extends to every nook and cranny of our planet's land, sea and air. 'Our' extends to every organism and the whole community of life. Is there not a sense of completion, maturity and therefore climax in the way we saturate the earth? Not just in name and number but sheer ingenuity. Humans still keep discovering how 'bio-tech' exploits every material, constraint and purposive device to the awesome limit. It has specified with most minute precision and the cleverest ideas. You name it. Webs, feathers, eyes, technology of information and the codes that govern biochemistry are just for starters. What has nature not explored, exploited and excelled? Where has chance, if chance is nature's level of intelligence, not landed first? Then Lady Luck has beaten me hands down!

destruction	*creation*
neglect	*care*
discontinuity	*continuity*
division	*union*
weakness	*strength*
exploitation	*conservation*
death	*life*

Monocultures, on the other hand, destabilise, weaken and threaten linking strings that make life's safety net so strong. A single, dominant species is not good for earth's environment.

This brings us to man's salient impact on our natural society. Just as brain function is often studied through the effect of injuries so ecology discovers, more and more unto astonishing extent, interconnections between components of healthy ecosystems that are being ruptured by mankind. *Indeed, at this point an acute paradox rears its head.* Man - whose intelligence, forethought and powerful creativity should serve in the role of steward and conservator of life on earth - seems to have lost his better mind. His innate curiosity, insatiable thirst for information and a drive to efficiently extract rich living from the environment has led him to empirical, material science. Physical science wants to help as much as possible; it wishes to improve our lives, increase our knowledge and, in a material sense, our rationality. Its heavily invested focus well serves the bodily division of a 'centaur' (wo)man. Organised, coordinated study of all physical behaviours has yielded dazzling mechanical and medical rewards. Technology, while it scorns the metaphysical, has vastly contributed to anthropocentric wealth, health and comfort. Of course, right-minded research works towards benefit but at what eventual cost? Could such amoral enterprise become a wrecker? Science celebrates its many triumphs but excuses tragedies and ignores its immanent calamities. Could its power, in the hands of lust, greed, power and other forms of immorality, transmute into a monster that might lead an overpopulated, technologically-pillaged planet to catastrophe?

Is not the subject a sophisticated, very tempting form of Eve? Not just tempting but most tempted by the serpent curiosity? The greater a lust the headier and yet more dangerous it can be. The more powerful a car the faster it goes and the more potentially lethal it becomes: the more intensely powerful a technological development the greater can be risks attached. Is amoral science therefore sailing too fast, too close to the wind or, like Icarus, flying too close to a lethal sun? *The most profound and powerful scientific advance is always sprung with an equal and opposite downside.* There always exists a negative to match its positive potential. This is, simply, because no instrument uses itself. Only the moral mind of a user, well or misaligned with truth, has it to wield within his power. Just as the instruments of science amplify benefits so they amplify the damage man can cause. You could have asked Albert Einstein or Robert Oppenheimer. Nuclear discoveries, physical and biological, are a good example. They have become twin nuclear bombs - atomic and genetic bombs, ethical bombs. Nuclear power and genetic engineering can both be employed peacefully and gainfully; both also involve grave risks in case of accident and grave suffering in case of moral malfunction. Take two more glorious twins, technology and medicine - whose very best intentions have also produced global pollution and overpopulation. Together the latter are corrosively attacking the fabric of our home. They have begun to seriously rip the ecological web of life. Deforestation, an enhanced greenhouse effect and destruction of the ozone layer ring the same alarm bells as over-hunting, over-fishing and the ignorant destruction of biodiversity.

If your body is upset what happens? It gets ill. What causes this malfunction, this breakdown of one form of homeostasis or another? Poison, infection, physical abuse. What then about the body of earth's ecosystems with its *Gaian* homeostats in place? What happens if sustained abuse by man cracks their resilient stability and throws the balance over? With respect to life the earth falls sick. James Lovelock uses the phrase 'disseminated primatemia' to describe this disease. It is a plague of people creating their miasma on the planet. Scientific Eve has bitten hard into the apple. Adam has invented instruments of mass destruction; at the same time he has developed benevolent instruments of mass survival (for example, medicine and ample food supplies) that have led, ironically, to positive population feedback. There is always danger from an uncontrolled, exponential increase in the numbers of a species. Plagues are over-populations that can only end in tears. It seems, therefore, that man's smart and modern curiosity has opened up Pandora's age-old box and that, under the spectre of his own mass destruction, he neither wants to nor can close it. How long, therefore, before the unlocked sweetness sours and, as Lovelock well writes, *Gaia* will exact revenge? Chief Seathl was not scientific but he also made the point. Is it man's intelligence or ignorance, cunning or stupidity that has depopulated or precipitated sad extinction in the case of many other species? How long before men find that they have engineered their own demise as well?

How long, in other words, before events invite the horsemen of apocalypse to saddle up and ride? And what might cheat the globe of famine, pestilence and war and death - Armageddon not of nature's but of human making?

Is there a Perfect World?

The 'decisions' made by an ecological community accord with the laws or, in dialectical terms, archetypes of physical, chemical and biological reflex. An ecosystem is, ex-humanity, amoral but is its operation perfect? What is a dynamically perfect system? A system that automatically and perpetually works according to the rules and expectations of its maker? In short, who is to judge what's 'good' or not? By what criterion is perfection judged? What quality of intellect or state of mind best rates it?

	tam/ raj	*Sat*	
	defective	*Ideal*	
	imperfect	*Perfect*	
	more or less sacred	*Sacred*	
	disharmony	*Harmony*	
	clash	*Peace*	
	lesser love	*Love*	
↓	*tam*	*raj*	↑
	impure	*pure*	
	unjust	*just*	
	profane	*sacred*	
	cacophony	*rhythm*	
	degradation	*uplift*	
	worse	*better*	
	loveless	*loving*	
	degeneration	*improvement*	

Bottom-up, oblivious matter cannot be judgmental; nor has it purpose and, therefore, what is is and 'perfect' is a concept that, naturally, does not exist - unless you think that matter makes up brains whose molecules somehow secrete the quality of qualitative comprehension.

Top-down, however, nature issues out of Nature; the Nucleus is not physical. Creation issues hierarchically from a Single Source. In this way each part of the whole is both divided from all others and yet indivisibly connected. From an Intelligent Maker's point of view the system works. Do involuntary, reflex laws of ever nature fail? There is no risk attached, in principle, to the operation of well-oiled machines. Both the non-conscious physical cosmos and the sub-conscious influence of archetypal memory are machine-like. They are informed and they are fuelled. Their involuntary operation is either rigid or minimally flexible. You definitely can't change matter's mind. It goes without so where's the fault? Thus, rightly grasped, isn't even imperfection perfect and natural business most well oiled?

Nature's laws and basic particles do not evolve. Nor do atoms, types of chemical or forces. Time, space and all that they contain evince a stable yet dynamic archetype. In this sense everything has issued pristine from a 'perfect' start. Nothing has, for all the aeons, changed. Nor, as far as cosmos is concerned, is perpetual motion, change or its exhaustion 'perfect' or 'imperfect'. It is neutral. Nature is neither negative nor positive except as it affects the 'creature comforts' and survival of its forms of life. What, if such form is conscious, is its utopia, its perfect state? Freedom from fear and pain, a state of moral rectitude - which species of utopia is yours? An inward one - contentment: or an outward one that strives, through various desires and turbulence, towards material satisfaction?

Thus in a *top-down* sense material if not moral operations of the world are 'perfect'. From *bottom-up* perspective, though, the notion is false. It is meaningless because the cosmos, whose only 'level' is non-consciously physical, evolved by chance. It has neither maker nor is it 'perfect'. It is not excellent; indeed, *only* imperfection lodges, it would seem, in life - which started by mistake and gradually improved. If mutation marks 'defect' then life is built from defects. The relative process of evolution works, without reason or meaning, from more imperfect to less. By this criterion you might judge bacteria less perfect than a non-bacterium and multi-cellular 'advances' higher than a single cell. Or you might, for example, consider a visual system somehow more 'perfect' than a geo-chemical system; or one visual system 'better' than another. It is not, however, rated strictly 'objective' to think of a random, purposeless process in this comparative, hierarchical sort of way. You might evolve towards 'perfection' by whose definition on what terms? Is it complexity, survival tactics, manifest intelligence or what that takes the prize? Who decrees that, after trillions of *BM*s, evolution bubbled up its finest hour, its consummation, crowning glory with a 'perfect' type of body - human form none other than yourself? Who remarks perfection when in evolution's non-eye every class is equal, that is, equally un-judged and thereby valueless? Why should natural selection 'select for' 'imperfections' such as some biologists perceive in panda's thumbs, eyes, human female reproductive tracts and, set against their own conception of perfection, so on down the line? Surely it is only men who, according to agenda, judge parts of nature 'perfect' or 'imperfect'?

There is (speaking of 'better' or 'worse') no question of absolute, universal excellence or (speaking of 'right' or 'wrong') ideal morality. Who is to judge and against what yardstick? Morality is simply a relatively confused condition of human thought that involves compromise between an individual and the others who inhabit his environment. Since nerves are a product of mutant (i.e. evolved) genetic make-up and thought is in some ill-defined way a property and product of those nerves, then reason is ultimately a matter of electrical configurations in a brain. A theory of the survival-driven strategy of all 'reasonable' behaviour can then be elaborated by a sub-science known as socio-biology.

The problem with this view is that, unlike trees or cells, conscious organisms do not function like robots. The more concentrated consciousness becomes the less sway sub-consciousness exerts; the higher it is raised the less the influence of instinct surfaces. But individual wills and interests clash so that in order to function a society needs rules. Nowhere is this more the case than with humanity's complexity. Not only this but man is blessed (or cursed) with a divine (or sometimes infernal) ability to think, crave information and aspire to perfection, especially perfect happiness. Is the latter feasible or just a judgment call, a relativity? There is no doubt that, swayed upon the motion of duality, relative perfections like all things come and go. What lasts? It needs be asked, within the framework of man's various philosophies, whether something like pure, timeless happiness exists or not. Is such perfection an Edenic or utopian imagination? Is metaphysical idealism a worthy even if mythical beacon towards which, against the tow and storms of bodily reality, to aim life's ship? *It wouldn't be a myth if there really did exist the polar opposite of oblivious, material restriction - an incorporeal reality that showed itself as supreme, extreme and conscious liberation.* Not external physical but internal, metaphysical perfection. Would not such freedom constitute an Ideal Yardstick, the Root of Actual Morality transmitted from the Apex of Mount Universe? If so, what would be the nature of such tip-top pleasure, such infinity of bliss? Of course, the circumstance of happiness and the nature of good feeling touch on every living being. Questions of this kind and the pillar of faith by they are answered is anyone's religion.

<u>Nature's Negativity</u>

	tam/ raj	*Sat*	
	powers	*Principality*	
	relative order	*Order*	
	imbalance	*Balance*	
	imperfection	*Perfection*	
↓	*tam*	*raj*	↑
	descent	*ascent*	
	anti-Principal	*pro-Principal*	
	negative power	*positive power*	
	confinement/ contraction	*liberation/ expansion*	
	'stop'/ resistance/ fixity	*'go'/action/ flux*	
	energy loss	*stimulus*	
	chaos	*order*	
	discord/ dysfunction	*harmony/ synergy*	
	discomfort/ pain	*comfort/ pleasure*	
	degradation/ destruction	*integrity/ cooperation*	
	bug/ disease	*health/ smooth running*	
	virus	*healthy cell/ body*	
	war	*peace*	
	'bad thing'	*'good thing'*	

As well as positivity a theatre of duality must, critically, involve the power of negativity. Its (*tam*) resistance helps develop diverse patterns out of (*raj*) action, flow and stimulus of energy. Such polarity, an essential part of nature's frame, swings outside any moral sense. It is neutral morally. Where there's only mindless matter there's no 'good' or 'evil' only, in the form of feedback, 'negativity'. 'Up' is not correct, 'down' incorrect. 'Right' is not upright against 'left's' sinister collapse; nor, nonsensically, male body 'good' and female 'bad' or *vice versa*. Nature's negativity lacks prejudice. Unfeeling physic is without intent or choice; it is as amoral as reflex response to it.

There is, however, in addition to physical or biological a psychological ecology. Sensation, information, strength of purpose and intelligent decisions compose a mental landscape with its moral atmosphere. Unlike unfeeling physic, feeling metaphysic involves judgment, meaning, reason and emotional intent. If nature includes mind *and* matter (and which is not in nature?) then, it's clear, natural negativity has to include both moral evil and amoral kinds of 'badness'. Even from a non-judgmental kind of organism's point of view life isn't perfect. Experience is, if not indifferent, 'good' or 'bad'. Circumstances hurt or please; choice responds accordingly. Let's take the unintentional world of physical events with their amoral 'evil' first; the character of moral evil will be treated later. The Dialectic points up natural adversity (a 'challenge' or 'amoral evil' to be overcome) in *fig*. 26.3.

An ill wind, evil cold, cruel sea and other natural challenges (not least disease) may threaten life with suffering. They may spell fearful pain and death. Such 'evil', as we understand, does not involve intent or animosity. It is, as matter is, oblivious: not immoral but amoral. Thus 'ill wind' does not blow with ill intent. It blows according to the fashion of inanimate design.

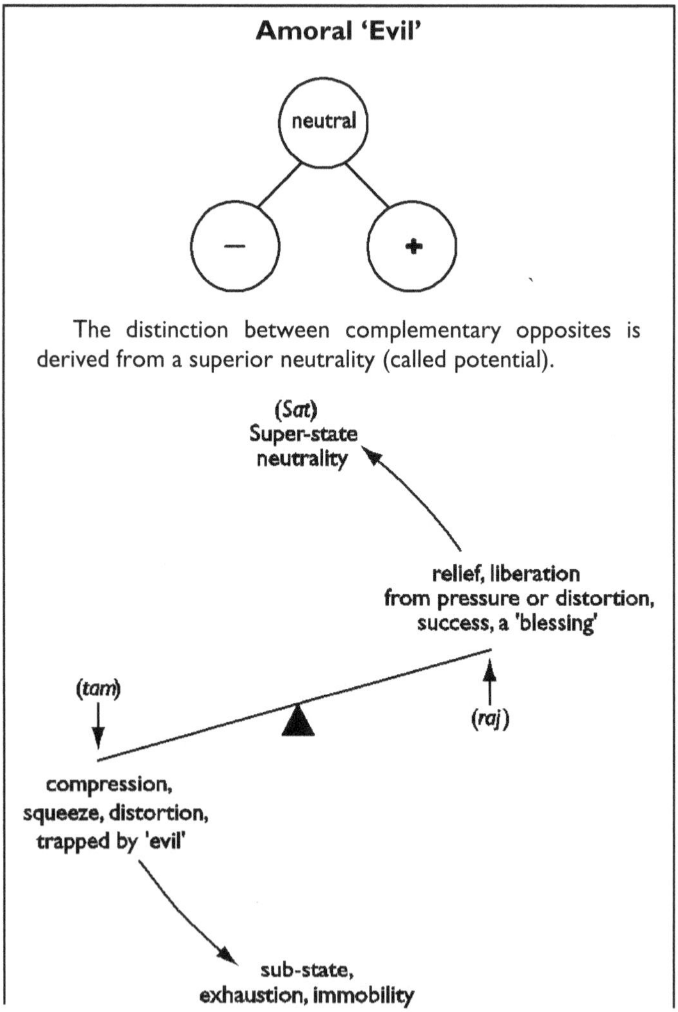

Polarity involves the notion of vectored, homeostatic swing between two poles whose axis is a pivot called balance. In short, (*Sat*) Potential is an original super-state whereas its opposite - extreme (*tam*) downswing - obtains the anti-pole of sub-state darkness, pain, exhaustion and their discharge on the cosmic floor. This sub-state is the opposite of Axis. Physically, its amorality derives from the oblivion in which material operations act. Psychologically and morally it's hell.

fig. 26.3 (see also 1.1, 2.5, 2.9)

As always, it's the way you see design. Is the world 'designed' by chance or was its regulation made deliberately? Chance is neither good nor bad. If, like the substance of the wind or sea, you're made of chemicals alone then can amoral body be to blame for anything? Who'd blame a machine for accidents its user suffered? From a maker's point of view, why would you manufacture breakdown, chaos and, when it came to life, pain? Surely you'd design a vehicle of quality and minimise its vulnerability to accident or wear and tear? Good God, how on earth could you, benevolent creator, permit or, worse, deliberately create a diabolical miasma? How could intent that is malicious be allowed to spoil your paradise or every grade of pain be loosed to drag its victims from a state of grace? Don't makers with intent have choice? Aren't they, malevolently creating evil, evil? Or, if 'design' is that imperfect, is there really not a Maker of the world at all? There is nothing like a spot of pain or evil to corrupt religious faith.

It is said that when his beloved daughter Annie died Charles Darwin lost faith in 'The Good Lord'. The vexed question was how, from his unhappy point of view, 'a loving Father' could have allowed such imperfections, bugs and pain. The deist thought his 'deity apart' was too apart and from aloofness should have intervened.

What, sir, were you, like many of us, really saying? That suffering and God are incompatible or cannot logically co-exist? Are you saying that the world of duality is, according to your way of seeing it, imperfect? What, therefore, is the nature of perfection that you seek? That nothing ever hurts? That aging, death and decay be banished? No negative power? Neither rain if we want dryness nor ice if we seek warmth; or *vice versa*? Must there be a plentiful supply of everything we think, at any given time, we need or want? Should volcanoes not exist because Pompeii suffered? Or seas disappear because their stormy waves have killed? Should neither accident nor sickness blight? What if life lacked free will or if, perpetually pristine, *physicalia* lacked entropy? Would one, how could one create such a utopian but unnatural world? Is it not ironic? The one whose theory of evolution banished any 'Special Interference' in 'unbroken natural law' was the very one who, when it suited him, rejected a Creator for a lack of 'Special Dispensation'!

Darwin's prayers for Annie 'didn't work'. What does 'didn't work' mean? Does it mean we stamp our foot because we don't get all we want the moment we demand it; or make demands that counter natural law; or curse because we can't conjure suspension of its operation? Do we logically blame an engineer if his locomotive hurts us? Isn't this, which many prayers adopt, a childish attitude? In demanding Darwin's sort of perfect world what is really called for is a cosmos rearranged to suit our whims and wishes. No natural 'calamity'. And yet if every one of nature's children kept on wanting different things, then how confounded would that nature be! There would be no end to muddle. Confusion, chaos and confliction of the elements! We really do demand such imperfection and yet, every time we suffer, curse its imperfection! So who's the judge of imperfection? Is such a judge's ruling perfect or imperfect? Is there a way to really see things right?

Could it be anthropocentric? Darwin didn't like the way cats play with mice; nor, with humanizing Dawkins, how 'cruel' ichneumonid wasps feed in the living body of their caterpillar hosts. Who knows what, if anything, such caterpillars feel? Is any predatory behaviour 'wrong'? Can appeal be fairly made to human views that class an animal as 'torturer' and so impute them with morality? Normally the opposite, objective and amoral view is held. Sir David Attenborough makes a similar although less personal complaint than Darwin. His apparent scientism springs from the consideration of another kind of 'devilry'. He doesn't put a scientific but an (a)theological and question - what kind of evil-mindedness is needed to create a worm that parasitically infests the eyeballs of a child? Good God! Could it be Good God? If it's Him He's bad and who would devil-worship? And if it's not the Devil's our Creator or there's none. And there's not since infestation is a pitiless but pointless accident of evolution. It is, of course, a well-worn tactic of polemicists to set up 'straw men' they attribute to opponents only to demolish them. In fact, if he'd ever asked nobody ever claimed that a Creator spitefully delighted in creating parasitic worms; or even specially, sadistically created them at all. Worms, yes. But not a perfect set of parasitic worms conceived and engineered the moment life began, a set that's never lost pristine, malign perfection; in other words, was it original intention not mutation that *sought* to inflict pain? What evidence is there for that? Or does he think non-evolutionists believe inevitable mutations only served to strengthen an original, conceptual malevolence?

In fact everyone believes in entropy of energy and information. Bugs infect computer programs, systems age and may malfunction due to wear. Genomes accumulate mutations and these bungle up their phenotypes - as anything but selfish genes is known. Neo-Darwinism's *PCM* states clearly that mutations randomly distort the structure and degrade the function of their carriers. *Everyone agrees on this*. Thus, did ancestral worms or microbes come before the humans or the other organisms they infect? If so, then they were not originally parasites; and in degenerate dependence on a host have sometimes gained bizarre life cycles but have lost their independence.

Let's rephrase amoral evil's case. No doubt carcinogens and pathogens inflict suffering and death upon the innocent; but is illness more immoral than a storm at sea? Not illness but the origin of illness is at stake. Whence did viruses arise? No dependent virus could precede its host - an independently self-replicating cell. Yet the malefactor that attacks you is composed of bits and pieces found in cells. Always including a nucleic acid fraction, viruses are seeds of strife. Deception, hijack and destruction are their blossom, parasites of 'evil' penetrating healthy life. No doubt, they 'change their spots' but were they first designed to maim and kill; or perhaps, as rogue fragments accidentally broken from coherent, working cells, these malefactors prowl the mindless biosphere? Remember (Chapter 23: Evolution in Action?) that they rapidly mutate and may thus, in some cases, randomly evade host scrutiny - but this does not amount to evolution in its necessary innovative sense. Innate, intrinsic systems of defence inhabit bodies; even cells, with plasmids and protein machines, partake of multiplex immunity. Mutation doesn't make such system, only tricks or breaks it. Such a so-called arms-race, called co-evolution whence change in one organism evokes selective reflex in another, doesn't need intelligence; nor does it evolve a single *novel* function, pathway, organelle or system. It just leads, by genetic lottery, to superbugs and counterthrust. *HIV* stays *HIV*, flu flu - pox strains effortlessly round in rings of battle *sterile of all innovation*. Thus, you argue, pathogenic microbes are actually mutant forms of life whose ancient forms were 'safe' as modern, co-existent strains are now. You can't, for example, stomach certain 'evolved' strains of normally benign *E. coli* gut bacteria but this in no way proves that the original bacterium was poisonous. Sir David's 'straw worm' props his own opinion up but in reality it seems that illness and eye-watering pain evolve from entropy of information. If pathogens express degenerate outcome of genetic entropy then they are scrambled information. They are garbled message. The parasitic fluke's a fluke. What lets it spread but more disorder such as ignorance, dirt, decadence or sheer bad luck?

According to the rules of entropy genetic degradation is accumulating. Original information is progressively (or is non-progressively?) becoming lost. You might predict an increase in mutation-generated illness whether from new pathogens or by inheritance. You might predict, just as the world will end, that humans will become extinct. At least, such 'progress' is what physics and genetics indicate; 'fall' is what they find.

No producer leaves a thing he can avoid to chance. He does not start from a chaotic point of imperfection and work up. Instead he evolves a purposive idea and, once well defined, materialises it. What other rational way does a creator have? Societies and their technologies are both evolved, deliberately, towards the same perceived goal - improvement in the quality of human life. A pioneer devises work as perfect as he may and, having raised the curtain, lets the uncreative critics loose. Is it not sure that, having set machinery in motion, users without sense are going to abuse and fault it? If a manual is issued and ignored whose fault is that? Is flouting a producer's reasons rational? It is certainly dangerous but why should you, if you have published the commandments, be to blame when users trash, flout, ignore or ridicule them. Every gift can be misused, every love risks its rejection - no less so a parent's for a child. So were deeds of devils His intention? Or are competition and adversity twin evils driving mankind into fields of war?

A Continual Fight

After 1814 and the Iron Duke's victory at Waterloo it was clear that French revolutionary fervour was not going to infect our sceptr'd isle. The need for

Thomas Malthus' anti-revolutionary treatise, his Essay on Population, therefore became redundant. The anti-Malthusian, pro-revolutionary element of Erasmus Darwin's thinking also flagged but its evolutionary part did not. It became, apart from religious and (because no-one had seen it happen) intellectual objections, acceptable in England. In this case war and struggle did not die. Clergyman Malthus' argument was, ironically, revitalised for just the reason it aspired to nullify. He wrote that 'population presses upon the food supply and always tends to increase beyond it'. As a result all populations are always as large as possible and tend to the limit. Such constant struggle for survival as might ensue among organisms of the same species could, Wallace and Darwin realised, leave those individuals with any advantage in life's competition 'naturally selected', as Edward Blyth had previously noted, for the prize - survival.

Constant struggle to obtain food is the heart of Malthus', Wallace's and Darwin's theories. War, plague and vice strike with tangential cut. What you cannot struggle for (such as immunity to a disease or various adaptations) dovetails into this scheme but the steady-state hypothesis of population observes an oscillation based, in the main, on food supply and reproduction.

How might 'vice' resist the reproductive pressure to increase? Such diverse 'unnatural' habits as contraception, homosexuality, celibacy, abortion, infanticide and marital fidelity are all behaviours that reduce a rate of birth. These apply to mankind and, it seems, to other kinds of organism too. What might pacify the ruthless and 'continual free fight' that Thomas Huxley, Darwin's bulldog, saw and co-theorists still perceive among the organisms of our natural world? Ecological 'live and let live', compassion, kindness and pacific altruism pose a problem for the warlike. So-called 'selfish' genes can't understand a wish for peace; their *motif* isn't share-and-care; nor can they ever turn the other cheek. 'Survival of the fittest.' Aggressive force of mind and body ought to be the only combination that succeeds. Indeed, in Huxley's world an ideal man might, without moral, well evolve a psychopathic paranoia that, allied with Machiavellian cunning and deceit, could be construed as top dog's fittest kind. How can 'intelligentsia' believe this theory? Its severe distortions screw up man's reality. Such distortion isn't, as a muddled liberal may gently think, innocuous but - witness Stalin, Hitler, Mao Zedong and other manic despots - dangerous. Why only kindly hiss and warn? Better maim and kill and crush all opposition. Why not, the fitter for survival, rob and rape your way to reproductive stardom?

Are criminals the heroes of mankind? Is evolutionary struggle, red in tooth and claw, the epitome of human aspiration? Is fear and anarchy what creatures really want? Do even selfish and manipulative genes thus risk destruction and demise? A revolutionary might wish, in freedom's name, to sweep away the monarchy, the church and such constraints as private property; so-called 'enlightened' reason kicks the trammels of authority; a leveller wants to level hierarchy and replace it by an order with himself on top! Darwin's theory might be, in various degrees, correct for organisms other than the human kind; and the lower down the scale the more explanatory power it musters. If you could ask them perhaps bacteria would, hands up, cry that Darwin's paradigm well suited them.

In the case of modern man, however, it's a libel. It's a slander on the human race. And, at the rate of David Stove's philosophy, it's a Darwinian Fairytale - simply untrue. How do desire for beauty, art and intellectual endeavour square with 'a continual free fight'? A theory of selfishness enlightens every 'rational' age but its logic at the same time denigrates the altruism of a doctor, soldier, teacher and the charitable politic of any welfare state. Bloodthirsty leaders don't create a welfare state nor does a demented public vote for it as well. Yet in selfish theory nearly everybody's altruism is a problem you explain away - because at root its hoggish brief incites immoral acts and is a charter for the criminal. It is not good for civil rights and peace in a society. Who, after all, embodies altruism? Where, the Christ, did he evolve from? It may be argued that raw, fungal Darwinism does not fit but breaks the mould of culture.

Victorian Darwin clearly saw the problem and for the most part kept his peace. Spirituality, the mystic path and love of God are at the brilliant core of all religions even if, in some of their 'adherents', it is defiled and hard to see. On such foundation rests our Christian civilisation. On other such foundations rest the rest. No doubt selfishness exists; but to deny its altruistic counterpart, in either fact or theory, incurs ridicule and moral outrage. Was there an error at the core of evolutionary theory that, antagonistic in its warring posture, displaced the 'old', religious ideas by which men hitherto had understood their place on earth? To an adherent evolution can't be wrong so you propose that, by evolving from below, modern man is subtly masked. A veneer that's civil glosses his true nature. The descent of man is really due to a self-centred, vicious, careless and manipulating streak. No doubt, it is variously argued, such ruffian vein breaks through a well-trained surface only in extremes of despotism, pain, despair and such divisive passions as, at home or in the courts, society eschews. Yet we are cavemen and before that ape-men in our hearts. Continual struggle dominates a short and brutish life.

Do you think that, if you showed him, so-called 'caveman' couldn't drive a car? Or plane? If your education was erased and you were isolated on a desert island you would soon appreciate his plight. What's betting that our fathers were as bright as you or me but hadn't got the gadgets yet? This, of course, was not Charles Darwin's bet. He did not find the savage noble. For example, the astonishment he felt at meeting Fuegian Indians who lived, naked, among the rocks of ice, wind and fire at the southern tip of South America caused the thought 'such were our ancestors' to rush across his mind. Unkempt, wild-eyed, babbling cannibals were what his racial supremacy erroneously perceived. He wrote, "I could not have believed how wide was the difference between savage and civilised man... the difference between a Tierra del Fuegian and a European is greater than between a Tierra del Fuegian and a beast."

Darwin had, at least in his imagination, found a kind of missing link. Cavemen. Such 'inferior' characteristics were, however, quite unlike the ones explorer William Parker Snow observed in them. Nor did two Austrian priests trained in anthropology agree with Darwin. In fact, they found that the tribesmen set a high standard of morality. They believed in a Supreme Being who had created both the world and the framework for a workable society. They prayed to Him, especially at death. The English missionary, Thomas Bridges, who had the grace to learn their language, also found the 'primitives' moral, kind, sociable and with respect for family life. Darwin, who did not bother with such linguistic triviality, guessed their communication consisted of only about a hundred sounds and, since many animals make a dozen or more different sounds, a comparison seemed apt enough. But, as the son of Thomas Bridges, Lucas discovered, the tongue is not at all semi-bestial. Far from being a philological 'missing link', his careful studies found it rich and complex with a vocabulary of about 32000 words and inflections. Indeed, Darwin himself was aware that several natives taken back to England by Captain Fitzroy could easily be educated, learned English and were, in 1831, presented to the king and queen. It must, therefore, have been the mind-set of his problematic theory drove him to ignore the evidence and conclude that such miserable animals were far removed from men. No doubt, the circumstance of poverty and lack of education makes a difference to anyone but doesn't turn you into half an ape. How wrong could Darwin get? Why did he so arrantly ignore the facts? Theory, as it still blows hard today, had swept him off course from the truth.

The Arunta tribe of Australian aborigines possess, as Thomas Huxley noted, a large 'Neanderthaloid' skull and teeth. Some grow extra molars. Do you remember (Chapter 23: A Mutant Ape) the Kow Swamp Aussies of 10,000 years ago? Were the latter actually Aborigines? For example, although westerners' teeth are smaller, modern Aborigines show little reduction in size from *Homo erectus*. Today, in a generation, such toothy 'savages' learn to play first-class tennis, take degrees at university and practise science. They are, although men and women from another cultural background, no less intelligent than Europeans. And, like the Yahgan tribesmen from Patagonia, they are genetically completely human.

What dog eats dog? Are beasts as savage as myth propagates? Is lion lawless, wolf unsocial, ant or bee not altruistic? Are even apes as stupid as, in defence of theory, ape-man has been drawn? You'd use your native wit and sometimes fight but animals form groups and troupes with social bonds; niche and the ecology of co-existence win all normal days. Nor do primitive societies today live free from moral or religious halters. In fact, they're usually more restrained by far than a Darwinian professor in our modern, loose and secular society. So perhaps our predecessors weren't the savage and malignant half-animals that caveman theory, in defence of evolution theory, cultivates. Perhaps, with huts and handy tools, they were as clever as ourselves but less divorced from nature and less technologically dependent on a web of artifice. But no, the theory *needs* ape-men at war as troupes of monkeys war today; prehistoric Darwinism *needs* foundation in the form of savage cave-apes who evolved into the modern, social way. Thus, never having seen pre-human ape alive, it chatters endlessly about him.

TOTAL CONTENTS

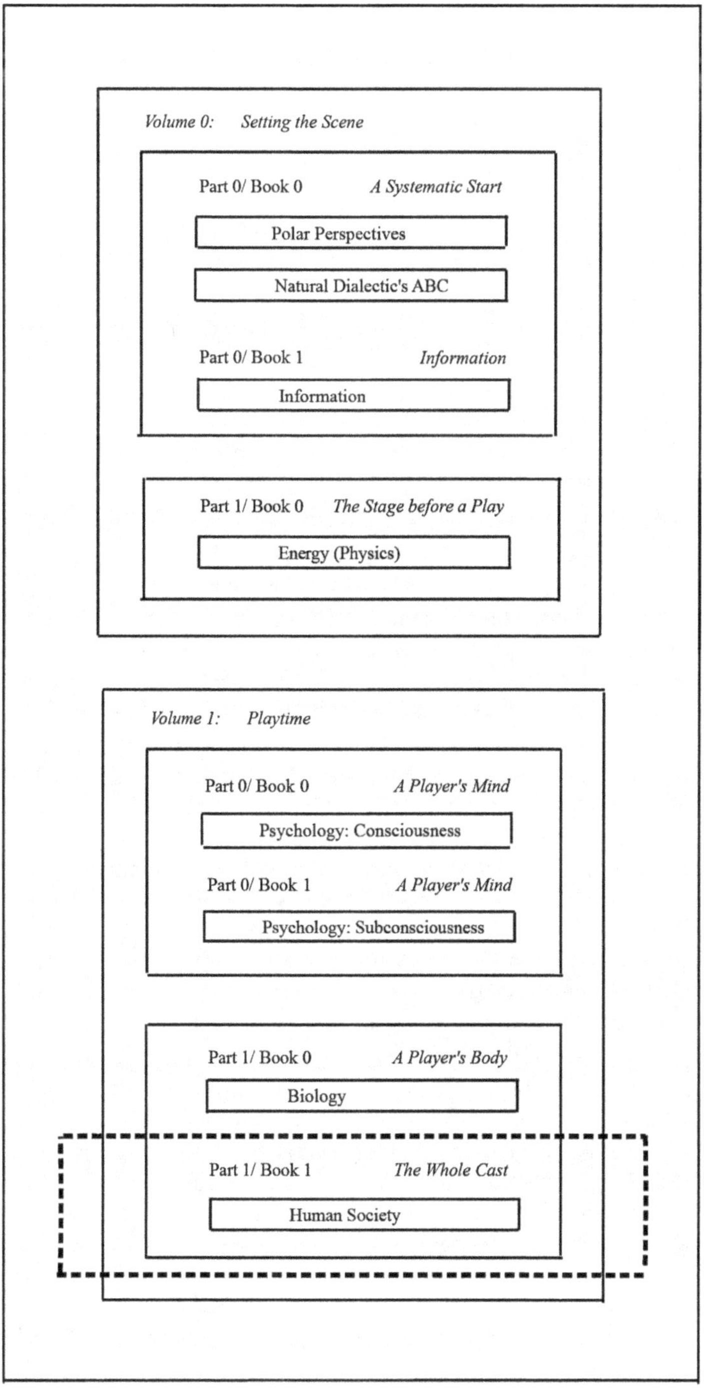

Volume 1: Playtime

Book 1: A Player's Body

Part 1b: Human Society

27. Sociobiology/ Sociology

	tam/ raj	*Sat*
	egotistical realities	*Mystic Reality*
	range of loves/ kindnesses	*Special Case: All Love*
↓	*tam*	*raj* ↑
	self-centred/ take	*other-centred/ give*
	selfish/ manipulative	*altruistic*
	at the other one's expense	*self-sacrificing*
	immoral	*moral*
	special case: no love	*mixed grades of love*
	towards hell	*towards heaven*
	socio-biological reality	*charity/ generosity*

Insensible - without a body you can't see, hear, smell, touch, taste or feel a mortal thing. Thus, from a *top-down* view, the body's a connector that allows life access to mortality, that is, to the non-conscious universe of lifeless things. Do you remember (Chapter 0) a sense deprivation tank? Body is an instrument. It is a medium through which the mind and thus its animator, soul, can know the physical dimension; and, of course, we call the instrument an animal.

Such *top-down* rot is not materialism's *bottom-up* perspective. Check *fig.* 0.6. There is no cosmic hierarchy, simply a material domain. A biological dimension is life's only one; man is no different, by basic need or biochemistry, from any other organism. Socio-biology's epistle cuts him down to size. It lops the upward reaches of a man towards heaven as delusive madness; it roots him only in the minerals of earth. This restricted view does not degrade; it simply, as its angle sees it, is plain true. *Top-down* 'reason' is, therefore, irrational. Its intelligence is totally misleading while a scientific theory based on evolution definitely is not. On these truncated terms, by man's reduction solely to biology, you don't bridge but annihilate the gap between him and the animals.

What is life but bodies? Thus the message is, "You are material; know that you are nothing but a body and therefore an animal alone. Such animal in turn, amid the parsecs and the aeons of the universe, is but a helpless, hopeless speck, a puff of vain illusion. This illusion is the very nature of a consciousness that's made of nervous matter; apparent life is just manipulation by your purposeless but Machiavellian *DNA*! You are, according to this sort of pseudo-scientific sanction, liberated from religious stricture or any ethical veneer. In anarchic but instinctive nature there's no crime but if sin, so-called, happens then take guiltless comfort - blame it all on selfish genes. No doubt, you're just a catalogue of gene mutations that have erected you from earth but such ascent of man is free from any trammel of Authority."

Physical apotheosis! You are sexually emancipated too. Top Gene's Siren is an orgiastic demagogue; and, manipulating on its own behalf of course, an idea lately called The Top Meme whispers in believers' willing ears. It tarts up traditional adage. 'Song, wine and women' are translated into a genetic deity's command: 'You are born a ruthless primate. Time is short. Before your chemicals disintegrate grab, any which way, all the bestial sensuality and satisfaction that you can. Critically, climactically, reproduce! Know, otherwise and also, you are nothing but a living speck of dirt that's carried on the low road to oblivion. But don't despair. My truths are, although fools call them only 'half-truths', whole ones. What I say is fascinating and I know you're spellbound so submit to me!'

In this version of Darwinian testament genes replace divinity. Or, equally, sup with the devil. The moral adjunct of a Theory of No Intelligence is socio-biology's 'enlightened' Theory of Selfishness. In this case 'soft' or 'liberal' Darwinism's a delusion too. A moral atheist (there are many) is a philosophically inconsistent one. Hard, competitive survival should pit an individual against the other members of its family, other individuals or, if you don't believe in altruism, any kind of welfare state. Poverty means weakness. Why should anyone give poverty or ignorance a helping hand? Sickness indicates a lack of strength. Why support, why not exploit a mind or enslave a body that is feeble? Death arbitrates or else eugenic programs should be introduced to cull disabled 'lower classes' and, in reproductive terms, promote the strong. Though you might diagnose a mental sickness or suspect its evil, it is in terms of evolution logical - indeed, it's socio-biological! And yet, at least as it pertains to man, the demographic facts refute Darwinian theory. Those least exposed to hardship - educated, privileged and moneyed groups - don't 'press upon the food supply' by breeding like a rabbit. It is, as data analysts confirm, the case that literate, emancipated women bear fewer children; and illiterate, impoverished and often bullied females excel in reproductive fitness if, that is, to breed like rabbits is a woman's ultimate career. In short, the social evolution of mankind does not conform to the Malthusian principle embraced by Darwin's theory of evolution; nor does it to survival of the fittest in a struggle with that most pitiless of agents, natural selection.

Malthus actually admitted theory did not match the fact of modern but still no less natural man. At first he claimed that only lack of food, war, pestilence or 'vice' could leave you less than maximally reproductive. Such insolence charged any *choice* by, say, childless couples or those with a small family, celibates or even virtue such as marital fidelity with breaking reproductive law. So, in retreat, the clergyman added moral conduct with respect to sexuality to his list of checks on population. Thus he conceded that, with respect to marriage or the wait until sufficient funds accrued for it, prudence and morality are not 'unnatural' but part of human nature. Not only socio-biological/ evolutionary principles apply - except, you'd claim, such cultural diversions miss the point. They're only part of our manipulation by that late discovery, 'selfish genes'.

Selfish Genes

It's true that there's a tension stretched between society-in-general and individuals that make it up. On behalf of selfish theory Sigmund Freud claimed you are, like any centaur, torn between a so-called caveman '*id*' and civil, social '*superego*' that restrains your evolutionary habits laden with aggression and dark sexual tendencies. How, though, in the light of evolution's rough priorities, did idealistic superego first evolve against its better (evolutionary) nature called that id? Why on earth would natural selection let mutant, anti-selfish, altruistic animus survive? To try and govern natural instincts runs against a wild brute's grain. You'd think such nonsense was inexorably weeded out. How, after all, do ideals, virtues and a love of Great Delusion, God, make any sense when food and sex and physical survival are the only rational type of urge? A repressor that for rude health's sake requires deactivation never should have first evolved. What date, therefore, was altruistic law's nativity? How did rule-givers first discover anti-evolutionary moralities? Since natural selection should have sheared their woolly thinking off the finer qualities of life present a problem for a theory whose first, bare axiom is continual, internecine struggle. Surely, after billions of years, man did not evolve to contemplate Big G? In whose mind has evolution gone all wrong?

Surely such a simple-seeming theory can't, in spite of piles of evidence against it, be wrong in fact and only right in fiction? Surely its adherents don't insist on living lies? Altruistic social bonding is a widespread fact and so, in order improve acceptability, the characters of selfish theory - id and ape and caveman - have now been transformed. Exegetics of the New Darwinian Testament absolve mankind of all responsibility. Sin's redemption is its non-existence since not mind but matter rules. Amoral atoms are, in some configurations, elevated to the legislator's bench; immortal, invisible divinities are worshipped; scraps of scripted chemicals called *DNA* are recognised as the omnipotent manipulators of familiar behaviour patterns whose 'motives', socio-biology perhaps patronisingly informs us, we do not really understand.

Therefore a scientific transformation may, it seems, resolve the source of selfishness clean out of mind and into genes. At selfish core, in theory, 'selfish' genes ruthlessly manipulate events. More tenacious is such *alter ego* than the one that your delusion holds as yours. Indeed, you are just a passing agent of this unconscious and yet strangely wilful dynasty. Behold, reality resides in animated chemicals whose issue, protein, dominates life's show! Experience is concocted by most complex interactions and material entities in brain - complexity that's safely well beyond your comprehension or ability to argue with. Don't, therefore, argue with such train of thought. Where do brains come from? Brains and, therefore, nervous networks known as morals are no more, in origin, than multiples of monomers called nucleotides! Welcome to molecular psychology!

'Organisms', some insist, 'exist to benefit their *DNA*.' Forms of life are robots. They are puppets and their genome pulls the strings. Animals (you and I are truly close to chimpanzees) are simply vehicles of devious but unconscious 'deities' - manipulative, selfish, replicating genes. Genes, of course, are *not* nor ever were self-replicating; they require a protein workforce in support. Suppose, however, all these selfish molecules intend is to 'self-replicate'. *DNA*, like salt or water, is a molecule that's lacking any sense of self. How, therefore, although a pinch of socio-biological ambivalence protests that chemistry's devoid of sense, can *DNA* be 'selfish'? How can it be animated by intent to replicate itself or to survive? And even if it replicated how, as in the case of salt, can two molecules instead one improve each individual's 'life', if that is what you want to call a chemical's existence? How, in other words, can simple replicas of me (made up from selfish genes) improve my chances of survival and not, through increased competition, rapidly decrease them?

It is supposed that in the court of King Gene you will find the Cause of which you, though you think you are in charge, are simply an effect. General evolution is a process driven by struggle. Life on earth is dominated by a ruthless battle and it's non-conscious genes that pitch you, front-line shock-troop, into the deadly fray. Genes have no morals nor, understanding this, should anybody feel the least responsibility for how they force him to behave. Alms, schools and hospitals have, logically, no place in 'me-first' worlds; crime's fine if you can get away with it; what charity in brains egocentric unto their genetic core. How neat! Except as a biological device ideals and ethics disappear. Theory wipes off their veneer. Underneath well-mannered make-up sex, self-interest and violence compose the body that performs nucleic acid's real, dirty work. The unacceptable has, not by simply replicating but by animated, *selfish* genes, been glossed with theoretical acceptability. Less gullibly you might discern, however, that such theory is nothing less than an incitement. It's, even if it might be wrong, a call to immorality and criminality and thus, you might archaically remark, calls up the devil. How can evolutionary crime exist? Ends justify genetic means. Sin needs laws and moral systems yet Darwinian man, a paranoid psychopath, is hinged (or unhinged) by the vision of continual, lawless murder that is predatory survival of the fittest. Is he deluded? Why does this hard, atheistic vision not apply to you? Is it just that you're an inconsistent softy?

Don't Richard Dawkins' molecules control his thoughts and thoughtless genes the way that he behaves? Or is he, godlike and exceptionally free of his own fixed genetic destiny? If self-interested manipulation by non-living *DNA* controls behaviour and if his every interaction is socio-biologically interpreted as 'manipulation of receiver by the sender' why should one believe a word of the deception any 'selfish theorist' communicates? Their words, in book or lecture, must amount to the manipulation of a gullible, inferior audience by superior genes. Thus these, by royalty and salary, better their survival! A cunning ploy, a clever ruse, an indomitable scheme!

Ink colours paper. Genes code for proteins. One inanimate chemical relates in complex fashion with another. No information other than the laws of nature is involved. *In this scientific respect the genetic concept has no context outside molecular biology and biochemistry.* Whose, therefore, is the delusion? Is it God or gene delusion? Which - priest, *DNA* or nucleic acid's scientific intercessor - pulls the puppet's strings? Who is gulled, manipulated, intellectually conned? A cultural 'gene' appears to visionaries as 'meme'. This psycho-gene's expressed by electronic orders in a brain. Lo! In a nervous way brain-parts replicate themselves by jumping into other skulls. Materialism, for example, is a meme that, through proselytising atheists, infects as many cerebra as possible - though what it's carried on is in the air! Perhaps meme-power rearranges brains. Genes are not intelligent but, it's said, neural circuits are and, if they know what's good for them, they'll scientifically evolve! Is not the top priority of memes to further theories of selfish genes? Therefore beware! Philosophical eugenics! Manipulation of the mind by latest science and its gene technology! If memes are living structures planted in the nervous tissue of a brain then woe betide an inquisition. Theocrat or pseudo-scientific atheocrat - Big Brother's just the zealous same. Could thought police armed with scalpels (rather than a thumbscrew or electric prod) ever try to surgically excise the competition? Who'd not subscribe to gene-sent mission that, for the patient's own good, purged viral patterns like opposing politics or delusional religion? You could medically re-educate man's superstitious ignorance. Apostates from evolution theory are certain candidates for mental snip. Of anti-scientific or unscientific spells an exorcist would hunt most keenly those cast by such witch-like brains as culturally reviled the thoroughgoing social logic of Darwinist hypothesis. If memes are patterns of cerebral nerves then kindly and with awesome skill, he might even implant 'chips' whose evolved, new logic cancelled all but socio-biological bents of mind!

Exclusive Fitness

Of course, a socio-biologist is well aware that the final bent of his theory leads to minatory nonsense and, with respect to many organisms, falsity of fact. If, with E. O. Wilson, you can query why one warns competitors and does not straightway maim or kill (Sociobiology: Harvard UP, 1975 p. 129); or with Richard Dawkins (The Selfish Gene: OUP 1978 p. 110) call for more research because a mother doesn't want her child abducted (which should free her to fulfil fertility potential of her selfish genes), then perhaps there's a fundamental problem with your idea and the mind-set that its meme induces. Perhaps the theory's a block (some would say a philosophical obscenity) that's standing in the way of seeing how things really are. Perhaps it's actually, from a *top-down* angle, an inversion of the truth.

If you don't hold that life's a vicious battle and your brotherhood, a con-specific population, isn't fundamentally a bunch of selfish genes then charity, benevolence and altruism aren't a problem. Indeed (see Chapter 3) an ideal mode of thought might act as basis for the rules of immaterial laws of logic and morality; consciousness (and not oblivion) would sway the mental day. Of course, the story's that much harder to sustain if unconscious mindlessness is at the root of everything. In this case 'morality' becomes excrescence out of neutral and amoral atoms so behaving since it gives advantage for survival of mutated genes. Thus *top-down* conclusion is excluded (by virtue or by vice of its 'unscientific', immaterial element) from a construction that might still cleanse 'selfish theory' of its worst excesses. In this vein Bill Hamilton's solution to the problem altruism poses for Darwinian worlds-at-war is a construction called The Theory of Inclusive Fitness or of Kin Selection.

Mathematics might be used to point the way. Parents share half of their genes with children so you might expect, though with ulterior and genetically selfish motive, some parental care. Orang-utans take nine years and humans double that. This is a long time, by far the lengthiest 'incubation' period of all organisms. But is the converse altruism, child to parent, quite as strong as kinship theory should suggest? Grandparents share a quarter of their genes and cousins half of that; in these cases too is reciprocity complete? And identical twins are genetic clones. Clones could not be closer kin. They should be so altruistic that each gives the other everything. Happy therefore the bacterial realm! You'd expect by fission germ-clones overflowed with honeyed selflessness. Not to mention sterile worker ants and bees. 'You jump first through reproduction's hoop!' Slavish cooperation's fighting talk; it's our survival strategy! And parthenogenesis, as in a plague of aphids, ought to ooze the milk of kindness. Asexual reproduction should bear paradise. Of course, it doesn't happen. No more the river out of Eden flows. Most organisms, and especially plants, share genes without a jot of care for offspring. That goes for almost everything except the higher vertebrates; and some of them, like cuckoos, couldn't care less either. Indeed if, in a dread experiment, you secretly swapped every child at birth and, unbeknown to its true biological progenitors, replaced it with an unrelated one would you expect immediate altruistic freeze? Would unrelated genes by kinship theory cause complete, proportional dearth of all parental care? Adoption shouldn't work. If it does, it's not the genes at work and an inclusive fitness theory's wrong. Or, if kinship story's right, then why not raise the altruistic stakes and secure the family genes by thoroughgoing incest? In fact, as well as legal complications, scientific study of genetics shows that incest weakens health. The problem is you also lock the weak genes in. This opens families chained up incestuously to hereditary illness from those same weak but replication-crazy genes. Who, therefore, would ever marry cousins? How could incest generate survival of the fittest?

It gets, socially, worse. In times of yore small troupes, it is supposed, evolved a kin/ group/ gang mentality whose exclusivity expressed itself, just as it does today, as nepotism, xenophobia, monoculture, class, race and those sort of divisions different people always find. Now, as the tale is spun, strangers in a city benefit because the mutant gene has not evolved. Kinship has been diluted but retarded altruism still acts as a social glue; it's simply been transmuted into 'you scratch my back, I'll scratch yours'. Mutual benefit survives but, if this strange gene mutates once more, its 'morality' might spring back into true Darwinian form. Inconsistent altruism would then disappear. Progress will have wiped an anti-evolutionary infection out - unless political control replaces what the genes lack in morality!

Evolutionary altruism, let alone the brotherhood of man, is, in Darwinian and therefore atheistic terms, self-contradictory - hence kinship theory's inconsistency. If *top-down* criticism's right the theory is unfit for purpose. An article of faith is fervently and yet, it seems, irrationally held.... unless kin altruism is itself a fraud. Is generous behaviour only down to genes or proteins? No doubt they inform your body; do they equally inform the choices made by mind? Acts of kindness might seem real; the beneficiaries will vouch for that. But Darwinian logic at its sharpest edge slices altruism out. Saintly generosity should not exist. Universal love is kinship gone completely mad. Where Darwin shrank from free admission socio-biologists do not. They spell it out. The hard truth that soft, weak or liberal people have to learn is, in a cold and hopeless universe, altruism due to kinship, friendship, leadership or any other kind of ship is fake. Compassion, as the Buddha never realised, is nothing but self-centred fraud. Affectionate appearances are just manipulations by those witless demons, selfish genes!

Yet if you prefix 'kinship selection' with the word 'apparent' just the same objections hold. If you propose that altruism is no more than an illusion caused by genes your explanation is as poor as if you'd called it real. You're as confusing as confused. Are you denying or explaining altruism using genes? If kinship theory is an explanation why does so much altruism not eliminate the competition and the struggle Darwinism places in its driving seat? If, on the other hand, inclusive fitness relegates good will to an appearance wrought by selfish genes then it denies the real quality exists. Love is a genetic swindle, altruism an illusion conjured up by atoms in the form of those 'imaginative', 'ruthless', 'selfish' genes. Immortal, invisible, gene only wise! Morality's a foolish carry-on and nucleotides together rule the world. All-powerful polymorphs, our genes, create and form each life on earth. Cell nuclei are temples whence the message is conveyed. This is the new religion. Molecular intelligence is, though it's absolutely mindless, superhuman and its purpose is sublime - avoid extinction, outwit death, obtain more life, abundant life! Partake of eternal life, that is, survive for all time! As yet science has but glimpsed the glory of our purely naturalistic makers!

Of course, human life powerfully contradicts the theory of kin selection. So does much of animal behaviour. If you don materialistic goggles then neo-Darwinism with its adjunct, selfish genes, seems logical. But it is ambivalent. Are these genes the cause of generosity of heart or not? Do you deny the spirit of compassion or allow that its appearance is deceptive due to competition at atomic level and a useless struggle to survive by ignorant, unconscious genes? What a loveless, desperate delusion you would thereby weave! With language that is inconsistent! One moment you say genes 'manipulate', the next one you deny it. You shuttle in between the language of an animist and then deny a sense of purpose. Of course, the theory of evolution is most teleological. The *purpose* of embodied life is, it claims, to survive and reproduce; if evolution is a process of 'progression' it must have somewhere to go. If everything's reduced to a molecular intention still something has to have a reason. Biology is packed with information, that is, full of reason. Or 'reason'. 'As if.' Are code, messaging, translation and the rest really what you say or do you wrap words in inverted commas to protect them from their meaning? Thus their reason isn't reason, it is 'reason' that's conceived by chance. Who's mixed up? Is anybody speaking with forked tongue?

Self-contradiction doesn't live within consistency. The theory of evolution's claim to fame is 'rationalising the irrational', that is, marrying 'purpose' to a universe without a mind. Chance makes reason. Reason then makes sense. Material has now eliminated immaterial consideration. Making everything of matter strips value, meaning and morality from it. How, if thought and feeling are material as an atom or a stone, can science, reason, laws of logic or an atheist exist? If, however, scientism's primary axiom is wrong then speculation from it is misled, misleading and misspent. *PAM* and the direction of its *PCM* cannot interpret nature right. Hence swirl inconsistency, self-contradiction, systematic muddying of cosmic waters that materialism stirs up but, because it doesn't see it that way, won't admit. The scale is massive but it may well be that you've been conned! Natural Dialectic would concur with that.

Psychological Ecology

We enter now the halls of subjectivity, the internal province of an individual's ecology, his mind. Associations here will make, of outside or inner world, a heaven, hell or in-between. Outward utopia would be provision of one's needs, freedom from pain and positive relationships with other life. The former, in the case of food and raw materials, nature circumstantially provides; the latter is a matter for each individual's attitudes.

Pivotal Choice: Moral Good or Evil

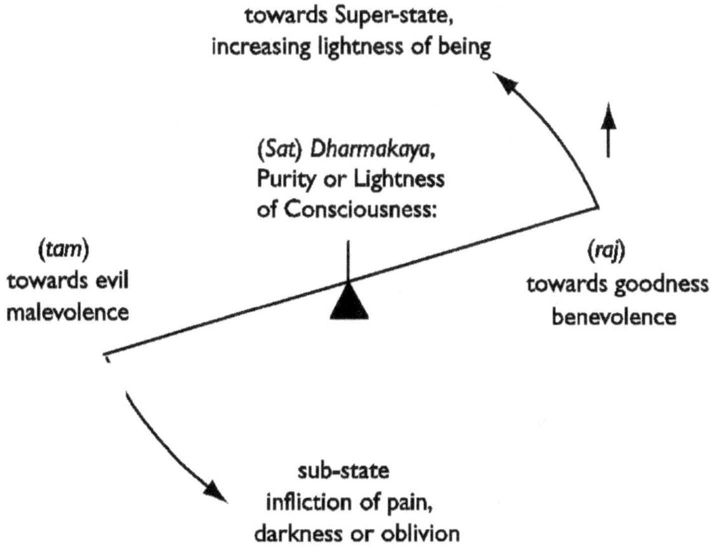

Action and reaction are equal opposites. They swing around a fulcrum and balance on equation called neutrality. Is such fundamental physical law also invisibly, psychologically applicable? Do you, in the larger picture, reap as you have sown? The Dialectic easily confirms that violent ignorance and cruelty are 'left-hand' evils plummeting from truth. And it confirms within the swings and roundabouts accountancy, whose cosmic sheet delivers an eventual equilibrium. What path, therefore, might an informed moral choice adopt?

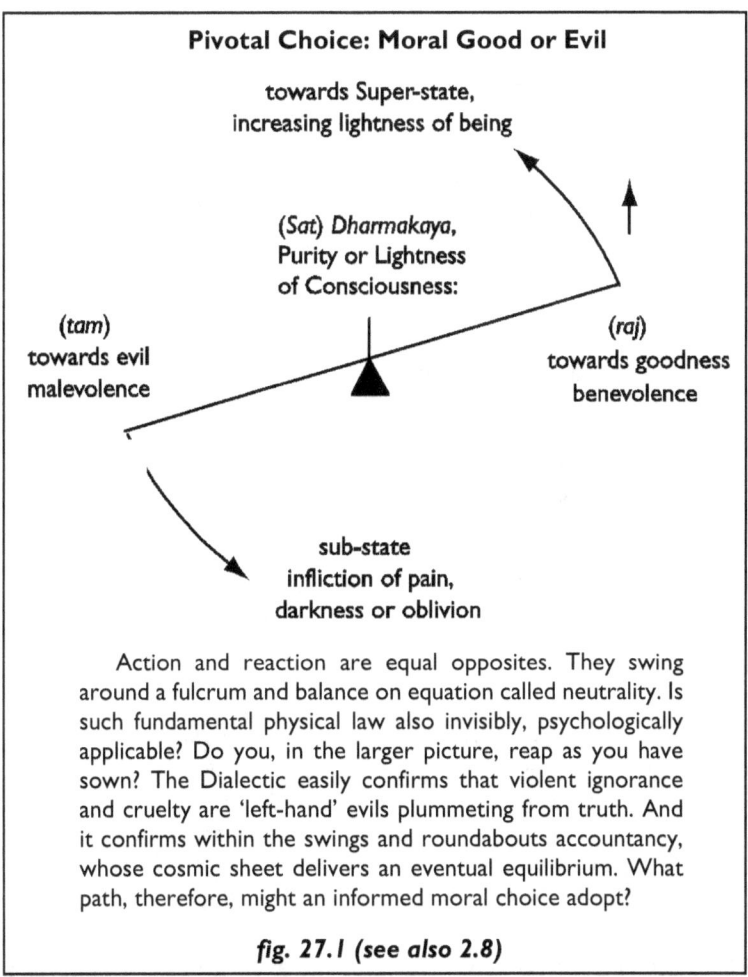

fig. 27.1 (see also 2.8)

What is an *inward* utopia? Whether trivial or important in effect, conscious mind makes choices all the time. Pro-active or responsive, on many levels you're deciding what to do. These decisions impact various points along the time-line anybody's destiny unfurls. Mind always sets balls rolling; thoughts send waves that reinforce or interfere destructively with one another. It happens in a single mind or whenever more than one communicate. Nor does causal act, a thought that's formulated, happen in a vacuum. Physical environment may pack a punch but so does frame of mind. Subconscious factors, in the shape of memories and instinct, play their part. So do prior experience, education and, often critically, moral sensibility. This mix, in combination with the circumstance of other minds, comprises 'ecological psychology'. Such landscape involves, as well as background, action; it immediately involves desire, interaction with benevolent and malevolent intent and, according to the overarching cosmic rule of balance, an equation with the term that action and reaction are opposite and equal. Dynamic payback is the way that nature swings along. It is therefore crucial to align its swings propitiously by best criterion of choice. Where 'attitude is everything', what is the wise man's choice and what, in order to avoid it, is the nature of intentional negativity?

The Nature of Evil

	tam/ raj	*Sat*
	all below	*Transcendence*
	lesser sorts of being	*Supreme Being*
	vectored deeds	*Peace/ Poise*
	balancing acts/ reaction	*Balance/ Pivot*
	range	*Super-state*
	moral spectrum	*Good*
↓	*tam*	*raj* ↑
	division/ demonization	*unification*
	from Truth	*towards Truth*
	from Peace	*towards Peace*
	body/ self-centred	*soul-centred*
	passion	*detachment*
	contraction	*expansion*
	malevolence	*benevolence*
	hate/ abuse	*love/ care*
	crooked/ perverted	*straight/ open*
	criminal	*saintly*
	a curse	*a blessing*
	immorality	*morality*
	descent/ darker	*ascent/ lighter*
	negative wish	*positive wish*
	tendency to create chaos/ cause pain	*endency to order/ relieve pain*
	decline/ fall	*lift/ helping hand*
	malefactor/ enemy	*benefactor*
	evil/ sin	*goodness/ virtue*

This stack makes clear which way life's moral currents flow. In the balance evil's burden weighs you down. Decline and fall; slide (or even run) towards moral death. Constriction, isolation, pain - the moral compass is both individual and a cosmic one. You can't negotiate with its directions any more than argue gravity. As energy so information suffers entropy; noise blurs morals most where selfishness combines with lack of love. 'Survival of the fittest' drums a din. Clear communication fades, 'rightness' fails. What sort of rectitude is that whose lack expels from paradise? You don't even have to move since, as a moral ecosphere, paradise is immaterial. It's simply that material circumstance 'self-organises', like files to magnet, round a state of mind.

Thus God and suffering can logically co-exist. Why not? How could the world be always right from everybody's point of view? But if there is a God of love why should He create an entity of malice, cunning and perversity? Is, therefore, God father of the devil and, thereby, the root of evil? Who loosed the wolf of persecution on a world of suffering innocents? I cannot believe in such two-facedness, such diabolical betrayal. Love that lies is love denied. Good God, wicked world. A fundamental contradiction - lost paradise, a paradox of darkness. Therefore I relinquish faith.

No doubt, there's nature's negativity. Resistance, opposition and exhaustion - how can the world amorally exist without (*tam*) shadow set against its light? Beneath transcendence truth is broken by polarity into vectors of (*raj*) force and (*tam*) anti-force; plus and minus are unbreakable a pair. But as free agents and not unconscious robots humans choose which way to act - and sometimes choose paths that deliberately create another's pain. Is any conscious, moral struggle cosmic or just local - in your head? Is the devil universal or anti-Christ a factor in an individual's brain? Brain's an instrument of selfish chemicals called genes, so in that case a mindless 'demon gene' engenders devilry. It's no-one's fault; morality's a dream; the answer's in molecular biology. We briefly visited (in Chapter 3: Conscience in Principle) Dante's kind of hell. Who'll be the first to transmute hell to heaven's sequence, quench inferno in a test tube and, back-mutating ill's mutations, dislodge Satan off his spiral perch? You'd usher in utopia with a godly gene - well worth a Nobel peace prize, if you please!

The godly gene would have to be recessive. Weak. Not the wild, fit type. Because from the unknowing angle of a 'selfish gene' the problem is not evil; it is selfless altruism. Who denies that in a fallen world the bestial side of life is struggle for survival? A centaur's creature side, now dubbed Darwinian, dominates the fray. And there, athwart the fray itself, a killer rules; the hone of natural selection culls all departures from the current optimum. It is a force eliminating godly chemistry. 'Survivorship' and 'reproductive fitness' win a day that's governed by genetic products called your limbic system and your hypothalamus. In their primeval land 'evolved success' is measured by the propagation of your seed. Survival is what counts; and, though you personally don't think too much along those lines, survival of the species' genes. So fear of death and death wax hideous; one must avoid destruction of one's genes! A parasite invades, the lion stalks deer or, as with lions, langurs and some other kinds, infanticide runs rife. At ground level, from a bestial point of view there's no morality caps self, survivorship and bio-logic's betterment. To rape, cheat, pillage and exploit is implicit in those naturally selected genes. You can't help what is natural. Of nature's amorality and, swept along its stream, mankind's immoralities George Bernard Shaw once wrote that evolution must involve 'a damnable reduction of beauty and intelligence, of strength and purpose, of honour and aspiration'. Where, after all, must progress lead if not to heightened brutish selfishness? A Dominant and Selfish Lord of Gene! The socio-biological prognosis for morality is, in a single word, bad. Darwinian propensity for evil is, in another small word, great.

Some persons, even to the point of devil worship, seem to want to be 'enlightened' by a Theory of Selfishness. The same and others may espouse intrinsic atheism of its adjunct Theory of No Intelligence. Does the light of reason need religion's crutches or restraints? It radiates its own authority. Indeed, we are evolving, aren't we? Human nature is improving, isn't it? You reason ever more infallibly don't you? Even criminals evolve resourceful reasons for a crime. In fact, though, isn't 'reason' often an excuse that barely masks the motivation to behave atrociously? Don't think reason's heaven-on-earth is free of gross brutality. From the time of rationalistic Jacobins two hundred years or more ago successive atheistic 'utopias' have inflicted hell and havoc dwarfing wars between religions let alone internal schisms and societies. At least pure Christianity is based on peace and charity. Subtracting positive commandments from behaviour's equation leaves the negative unchecked. It leaves human violence on your reddened hands. Did Darwin, losing his religious faith, embrace the ethics of Hume's utilitarianism or (in which the unborn child, children, future generations, mental incapacitates or any form of being judged sub-human has no rights) Kant's rationalism? Did his cousin Francis Galton's study of eugenics not imply such rationalistic moral code as 'fittest races' might inflict on cavemen, primitives and those in poor or technologically backward, 'third world' states? Within the twentieth century materialistic shades of social Darwinism as diverse as totalitarian fascism and communism, socialism, colonialism, dog-eat-dog capitalism, scientific atheism, hedonism, humanism, anarchism and sheer nihilism have all, in some cases at the cost of vast human depression, inhuman suppression and cruel suffering, drawn doctrinal strength from the material heart and blood of evolution theory's 'unofficial bible' - The Origin of Species - with its central creed. Is not your motivation, rationale and power-hungry politic at root 'best standard of survival for the fittest, that is, for the most evolved'? No devil in the detail there then!

What might a well-evolved and cunning despot's genes, whose instinct is transfixed into Darwinian mode of a continual fight, inflict upon the 'enemies of reason'? Purge them? Pulverise them? Clear all threats to his genetic egotism's article of faith - any-cost survival? Ruthless, forceful and, the best of all, efficient tyrants soon create such 'excellence' as Dante's hell.

Racism. We're all embodied souls; do you care about the colour of a motor that its driver is 'embodied' in? Of course, he sometimes needs a motor doctor but it's the way he drives - behavioural morality - that counts. The fact is humans are 99.99% genetically identical. If they descended from an original pair they would obviously all be, in this sense, brothers and sisters. Darwin wrote 'The Origin of Species by Means of Natural Selection or The Preservation of Favoured Races in the Struggle for Life'. Clearly, in this context 'race' means the same as 'species' but, where the term 'race' is used with a sense of inherent superiority Darwin was, with the great majority of his contemporaries, of a racist mind-set. Becoming 'superior' is evolution's whole theme; but in terms of human superiority colonialist Europeans saw their technology, education and themselves as far above the state of their subjects. Although such evolutionary attitudes did not cause slavery (which long preceded Darwin and succeeds him still) it served to justify it. **How, if nature lacks morality, can such behaviour be 'wrong'?** It's natural! Lower varieties represented part of a ladder of 'inferiority' reaching up to 'us'! Such concept is embodied in Clark Howell's outrageously deceptive but iconic, textbook evolutionary sequence of chimp through stooping ape-men to Neanderthal and, at the top, upright *Homo sapiens*.

Us! Do you remember (Chapter 23: A Mutant Ape) Ratzel's *lebensraum* that led directly to the racist demonization of 'lower groups' and its logical evolutionary outcome, a holocaust?

Fascism. 'Higher race subjects to itself a lower... a right which we see in nature.' In Mein Kampf (Chapter 4) Hitler argues that Darwinism is the only basis for a successful Germany; and in December 1941 he revealed to his Secretary, Martin Bormann, that his life's final task would be to solve the religious problem - the organised lie must be smashed. Franco and Mussolini also sprang from Nietzsche's *Übermensch,* apostolic Haeckel's febrile evolutionism and, therefore, once more Darwin's and his cousin's pseudoscientific corm. Indeed, world records for mass-murder germinated like black flowers of modern twentieth century evil from that bulb. There's no appeal to mercy at the root of evolutionism's creeds. Don't believe in atheism? Take a bullet in your alien concept's brain! It is plain fact, as scientifically objective as it gets, that modern, militant atheistic states (and states of mind) have been the most systematically, unrelentingly vicious in the history of the world. Who, so far, has seized this poisonous profusion's crown? A European Council document, issued January 2006, conservatively calculated deaths since 1917 due to communism at nearly 100 million (65 million in Zedong's China alone) and due to National Socialism at nearly 60 million.

Communism. Stalin followed Lenin who followed Marx - for whom a signed copy of Darwin's book contained 'the basis in natural history for our views'. Mao Zedong, Pol Pot, Kim Il Sung and others descended from Marx by way of Stalin. 'Democratic' socialism's negative, non-complementary dialectic sees men as 'them' and 'us'; believers or deniers of materialistic faith. Of course, since by evolution there's no necessary God each man is the ruler of his moral universe; an Atheistic State is self-crowned Ruler of a Human World. Forget fundamental inconsistency; gravel, as material, doesn't have morality so atheistic faith must borrow it. Its naturalism takes aggressive moral tone - political correctness. No doubt, no atheistic culture has historically existed unless ideologists have brutally suppressed all 'backward' opposition to their own self-contradictory morality. Make-it-up-as-you-proceed; 'progressive' indoctrination would forbid all competition for devotion and eradicate self-sacrifice upon the altar of unsocial individualism or, worst, a God Delusion. Atheism viscerally distrusts religion's challenge to its power and, history confirms, specifically attacks in order to eliminate. It still happens. Faith aimed higher than The Party soon attracts an atheist's conscience and its instrument of fitness vested in the politics of propaganda, army, prisons and the police. In China, North Korea and Tibet The Social Soul keeps killing.

It is, therefore, as shameful for a 'reasonable' atheist to gloss the systematic slaughter caused by vicious fellow travellers as for religious devotees to gloss their own persecutions, wars and cruelties or other violent scatterings that stain history with blood. In this respect, what really is the difference between a theist and an atheist? Because one party, doing what he has decided is 'God's will', is historically capable of as much evil as the other. One side believes in a Conscious Creator, transhuman, trans-physical and the other believes only in non-conscious physical energy (which we call matter). Either way, what difference does that make to victims?

Cynical power politics, contempt for 'bourgeois' morality, ruthless 'survival of the fittest', police states, purges, holocausts - never offer atheism as improvement on what went before! It can be argued that materialism breeds a special kind of *fleur du mal*. Is there any doubt that scientism's creeds (whose idol or whose ideology is science) are as dangerous as, when they're hypocritically distorted, zealous faces of world's religious faiths can be? Is Orwellian 'rightness' wrong or not? Peace, love and prosperity are switched to violence, hate and war. At least religious war is 'only' hypocritical. Such faith at root, in clear and intrinsically peaceful mode, condemns the cruel contortions demagogues wrings from its holy book; at least it calls for peace and not by doctrine, slogan or some posture of the intellect thinks only of division, terror and erasure of, say, Marxist Dialectic's 'enemies of state'. At least religion's basic principle is classless freedom, brotherhood and peace - its truth is not enslavement to Big Brother's doctrine of repressive tyranny and war.

War. Of course, war is the climax of physical competition that is murderous and in which man's in-built passions are, to decide the fittest as the victor, raised to a lethal nth degree. It is not that evolution as a theory *causes* evil but it has justified large swathes of it.

You might well argue that dogmatic canons wreak most havoc. Aren't most wars and persecutions by religions; and their damnations for eternity? If theism is delusional then why not trash the primary delusion? Why not delete my-way theocracy? Isn't it, in one form or another, undeniably the root of evil? History seems to underline the logic of your case until you realise that, in obvious fact, the devil wormed inside a man is what really twists the whole thing round - right round when it screws into sadistic cruelty. Theocracy? Atheocracy? It is not ideals of either grasp that wrench the helm and shipwreck whole societies. Read biographies and not philosophies. Blame the man and not the ideal that he wields like a weapon, an implement of his oppression in the misappropriated name of some utopia or god! Topsy-turvy rationalist! How can you bin religious faith without a step (and there are no objective steps) to verify its claims? How can you rubbish the accumulated wisdom (not the cleverness) of men except to halve it then throw out the part that counts, the one that you decided didn't - immaterial metaphysic you have vilified as an 'irrational delusion'? To argue that the cure of evil is its cause or antidote infector of disease is back to front. Do makers engineer the failure of their systems? Is repair according to a manual wrong? If, except in self-defence, intentional infliction of suffering on another's life is 'bad' then who is at root responsible? Did an 'Evil Eye' devise a cosmos that allows such 'wrong'? Or do the arrogant (dismissive of the maker's manual) and deliberately disobedient breakers of the natural, moral rules create what we all suffer? Confusion and deception are the order of a devil's day.

Nor is anti-socialism taint or tension-free. After both world wars unpleasant politics and programs involving 'scientific' eugenics based on evolutionary theory stained at least Australia and the Stars and Stripes. Although they did not, like Hitler, indulge in genocide Sweden, Norway and Canada along with the USA forcibly sterilised 'social deficiencies' such as criminals, paupers, drug dealers, epileptics, the blind, the deaf and so on. Indeed the *ACLU* (American Civil Liberties Union), a bastion of humanistic liberty and instigator of the 'pro-Darwinist' 1925 Scopes Trial, was swayed by its staunch and anti-religious creed to support the right to teach from a flagrantly racist, eugenic textbook called Hunter's Civic Biology. Ironic. Yet what rational defence could barristers deploy against the fall-out from Darwinian creed? Nowadays there are calls to legalise abortion, euthanasia and 'grey area' reproductive biotechnologies. Indeed, when religious rectitude has passed away political correctness tries, must try, to shape the vacuum left. Society and politics need policies and books of rules to operate. Of course, by *bottom-up* analysis, you'd argue there is no such thing as natural rules of psychological behaviour (except by government of instinct) in an amoral and creator-less creation. There lacks constancy of creed. And since no sociological sin exists to fling its perpetrator headlong to a non-existent devil's zone, you socially (if not quite scientifically) experiment. For 'the best' - however those in power define 'the best'. If any such experiment should fail (as in communistic history) It is no more Darwin's or Spencer's fault than Christ's, you cry, if followers misappropriate an idol's name as an excuse to stride astray.

Who can win? The road to hell is paved with good intentions too. Genuinely good intentions target winners but, in the sway of politics, hurt those who lose as well. Man-centred science and technology works wonders but is wreaking its own havoc on the natural ecology. And, if you're the patient suffering havoc, what do you construe the cause of pain, pollution and extinction? Greed? Wrong-doing? Evil cast upon the face of earth?

↓	*tam*	*raj*	↑
	descent	*ascent*	
	fallen angel	*angel*	
	abomination	*saint*	
	Lucifer	*Logos*	
	Ahriman	*Ahura Mazda*	
	Satan/ Beelzebub	*Christ*	
	Ravana	*Rama*	
	Mara	*Buddha*	
	Mephistopheles	*Marduk*	
	Set	*Osiris*	

If rationalism won't deliver you from hell on earth, what about so-called irrationality? A Persian, Zarathustra, noted a perennial battle waged between antagonistic light and dark. Manicheans also took the Dialectic up. They held an existential doctrine. Co-eternal forces, those of good and evil, lock in moral combat. Throughout the universe and in each individual's life, imperfections pointed up polarities called 'good' and 'evil' but, as ex-acolyte Augustine saw, this doctrine of duality lacks trinity. It lacks Essence. It lacks the (*Sat*) Infinite and Central Pivot of Transcendence that another doctrine, Christianity, embraces. He converted from polarity to Unity, Tri-unity or Trinity. St. Augustine understood the swing of Dialectic *and* its pivot. A pivot is a point of balance and the origin of sway. Its immobility is axial to motion and, since existence is all movement, comes before existence. It is beyond existence and yet still the origin of every world. Therefore, between the positive extremity (of *Logos*) and the negative (of non-conscious materials) Augustine raised Transcendence to the centre of his altar. Below this Height he also recognised deep night. He could turn from praising light and, gazing sadly down Mount Universe, pick out chasms all in shadow, a racked and wrecked abyss of wrong psychology. In this case negativity of mind is pathological. Its deliberate infliction is the act of evil, its creation hell. Beelzebub, deceiver, is himself deceived. There is no devil but the one that men, with intent to hurt or harm, materialise; and devils, in the *karmic* balance of the world, will have to pay for sins.

There's a twist, a Buddhist turn in the tale of this dynamic. Can you, monks robed in orange or in purple ask, really call the natural world imperfect? Isn't imperfection simply a perception? Isn't it a function of your aspirations and frustration when events don't happen in the 'right' way or according to your taste? We respond accordingly but whose is imperfection? Their prescription seeks to neutralise the imperfections both of mind and body, that is, to obliterate effects of negativity. The only full correction is, they claim, Detached Perfection. This antidote *is* the apex of Mount Universe and its essence only found by one whose contemplation climbs there. It is, transcending all existential duality, entry into cosmic climax called *Nirvana.* In the universal case the highest point is also, as with every river, source. No taint at spring.

You can understand that nature is not evil but, still, it doesn't take a genius to understand the nature of an evil act. What's, therefore, the source of evil's stream? Is the devil an unnatural fiend or a wretched feature of man's natural mind? Embodied egotism needs a bed, a meal and clothes; it needs affection, interests and freedom from oppression; and it desires to reproduce, to build the safety net of family and thereby multiply original, simple needs. What, however, of frustration, of obstruction in the way of satisfaction and blockage on the path of love? What is hate but satisfaction thwarted or love turned? Is it, therefore, love or hate, love of hate or hate of love that drives a demon? Anger, greed, lust, pride, wish to dominate and laziness are swollen forms embodied egotism easily assumes. Sins are inflammations, devilry a cancer that inflicts a mind controlled by body not a body by its higher mind. Thus demons are, in Dialectic script, the offspring of desire writ large, passions overrunning or the body of revenge. Evil is an opposite that's turned from goodness and has spurned the central truth. **God did not but humans do create damnation for themselves**. Devils spring in mind; they place themselves, by choice, in universal quarantine; they are the body's egotistic choice when magnified into the fires of vice. In a world where balance rules the sufferer once inflicted pain; and the inflictor, having suffered, takes revenge. Who does not pay both for and back? And what release is there from such a cycle? Materialistic science has, since morals aren't material things, no answer whatsoever. But Natural Dialectic frames a crystal clear, transparent one. Immaterial purpose and benevolent intention are the simple keys. They are subjective. They involve non-physical deliberation and, in this way, direction. As much as mind is free its will, its choice is freely vectored. Are you determined to root evil out? Evaporate the devil in his smoke? Charity begins at home and home is in your mind. Is your centaur-direction turned, by exorcism, towards non-demonic peace?

Risk

Only the master pulls his puppets' strings. They have no choice and thus no risk of selfishness or other bad behaviour. If, however, you have freedom to make conscious choices, then there's risk.

Automata aren't voluntary. They are reflex, as 'obedient' to the rules as wind or rain. *No risk there.* When it comes to organisms what sort has no choice? Is it one that's dormant and completely locked into a preordained, instinctive pattern of behaviour? On the other hand, the higher a degree of voluntary consciousness the greater is the possibility of appreciation, creativity and love. The deeper grows desire to know and be informed. And, on an educated path of contemplative focus, the more frequent and profound become the moments when some proposition's understood. These may ascend, climactically, to perpetual and full enlightenment - the so-called mystical but clear experience of Original State. This Voluntary State is Infinite and it hierarchically precedes 'fall' into less liberated, more finite localities of space and time and mind. It transcends confinements of existence and its moment's known as, in Hindi, *moksha* or release. For a Buddhist it's *Nirvana* and in Christian language it is called Communion, Absolution or Perfection.

↓	*no*	*yes*	↑
	wrong/ false	*right/ true*	
	immoral	*moral*	
	bad	*good*	
	hate	*love*	
	anger/ frustration	*patience/ tranquillity*	
	lust/ selfish passion	*altruism*	
	oppression/ slavery	*liberation*	
	pain	*happiness*	
	scowls	*laughter*	
	tragedy	*comedy*	

But cosmos is no ordinary machine. Although to an untutored eye apparent chaos seems to rule in fact, as science and the saints both readily confirm, coherent systematic order is the substance of the game.

Top-down, from (*Sat*) Perfect Apex you descend Mount Universe. Imperfection follows its perfection. Inward acme is turned inside out. And so it seems *(tam)* death and negativity are, with the downward cosmic vector, written in. They are an inevitable part of 'Causal Good Intention'.

If you've included Living Apex then the factor of intelligence appears. The incomparable, the paragon must fall into a state of relativity. As virtue falls you start to take a risk. Ideals fray, behaviour worsens and becomes increasingly disorderly. *So here's the risk.* If you believe there is intelligence behind the universe then you describe the possibility of deliberate misbehaviour as 'God's risk'. This risk is, in other words, a collateral of flexibility in conscious mind. It involves a voluntary element of freedom called free will.

Voluntary capacity means choice. In practice, though, choice is physically constrained by forms of instinct and of body. These trammel and impose survivorship's Darwinian priorities. Of course, you may claim these are paramount but if you aren't a robot of your genes then what is the alternative? Men may certainly behave like beasts but can, with reason, rise beyond their beastly norm. Forethought, intelligence and ingenuity involve morality and yet immoral excellence, the egotistical infliction of desire at another life's expense, perverts. Such reason falls below the naturally bestial; it is sub-natural and sub-natural choice is evil.

Now look at things from a creator's point of view. There exists dilemma. Do you build a lifeless system like the props but not the animation of a play? Scenery like stones or clouds lacks all appreciation. Yet without observance and appreciation cosmos is a massive tomb. It is invisible, unknowable. There is no sorrow, love or fear or joy in it. Nothing of what really counts. No plot, no theatre, no intentions. No resolving climax, no comprehension of The Truth. No family of actors only empty matter and oblivion. Should it, therefore, be a puppet show where life seems real but is only an illusion? A place where puppets have no audience and do not live? Or should the puppeteer, after casting characters in various roles whose interactions may strike billions of dynamic scripts, animate the puppets, loose them and relinquish his control? **For there to be true love between man and his Maker there must be the capability of its rejection. Free will and rejection always are love's risk; and with rejection possibility of evil.**

The problem's therefore disobedience. If a puppet master hands the plot to animated puppets to extemporise then, sadly, they might lose it. If a writer once allows his actors free rein then he runs the risk forgetfulness and egotism may invade the script. Mistakes and clashes could occur; a player might forget that his embodied role is an ephemeral one and, with ill-conceived involvement, take the stage for real. He'd soon traduce his lines and 'fall from grace'. A producer may instruct but, under a theatrical code that includes free will, his own rules can be broken and his suggestions ignored. He is, therefore, in the same position as the parent of a wayward child. And mankind's grown-ups are often immature, adolescent and rebellion-prone. Passions bind. Danger lurks. Man's freedom is a risky enterprise. In so far as he is able to anticipate the consequences of his actions; in as much as he is able to subject intent to reason and thereby control himself; and in as much as he can discriminate between 'good' and 'evil' he is inextricably involved in choices. Behavioural choices. Thoughts and actions that will whisk you up or down the cosmic elevator. *As a human you are forced, whether you know it or like it or not, to play a part in the only kind of voluntary drama possible - a morality play.*

Don't cinema and theatre reflect this? Cops and robbers, heroes, villains, virtues, vices, all the blacks and whites wear masks of mummery. It becomes, therefore, important to identify the 'good' from 'bad'. In the end analysis is anyone's morality (see also Chapters 3 and 13) a relative or absolute propinquity? Can you choose to leave it *post agendum*? In this cosmic boarding school, are rules a necessary evil, a constraint on personal freedom by society; are regulations simply a cold matter for internal or external police? Or, as the Dialectic seems to hint, is the basis of an absolute morality the immaterial intensity called love? Religion, law and politics did, do and always will involve the climate, watering of encouragement and soils of morality's ecology, economy and fruit. They are the ubiquitous instruments of communal decision about how you should best behave. Do they invest in you or strip you of your liberty? What, in stripping, is their *quid pro quo*? Before developing a unified theory of religion, law and politics that indicates a way to identify what is good, right and true, we need to have assessed the basic risk and clarified (see also Chapters 2 and 13) its problem - evil.

How free, in the context of society, environment and education, is free? How can the restrictions of a body offer other than conditional freedom, almost all-constrained determinations of 'free' will? At least, it seems, you can make choices. You can want more love or less; you can choose a path of truth or lies; you can *want* to ascend or descend the gradient of creation. Who, by imprisonment, can take the wish for liberty away? Are you not a voluntary being? Voluntary means according to *your* wish – even if the wish is to obey another man's command. Volition is a power of purpose, imagination and ingenuity that transcends instinct. It exists, in a limited, diluted form, in problem-solving animals but in concentrated form in man. The freer from material constraints and conditions, the more powerful it becomes. Imagination carries with it, like Promethean fire, risks. *We can, in man, identify the real risk in creation. God's risk is man. His potential for glorious ascent is, at the same time, risk for dreadful fall.*

Fallen angels dwell in catastrophic underworlds. But if a puppeteer endowed his robots with free will then would he not, in proportion to the freedom given, risk the possibility of will's abuse? Do you allow an actor freedom of expression to extemporise his lines? Do you rule your sons and daughters absolutely or encourage open minds? *To live or not to live, that is the question.* Shall the whole play just be lifeless objects or the puppets kissed with real animation. Is there such a thing as dolls' rebellion? But deliberate rebellion or a preference in living dolls for the excitement of infernal games leads to a shady sort of life. It may be sunny outside but the devil fits inside a mind. Its metaphysical extremities are negative. And actions spring from what a body's mind approves. Whether they are physical or psychological we call dark phantoms demons just because they torture and, in hurting, make life hell. Must these hells be?

Free Will and Determinism

	tam/ raj	*Sat*
	existence	*Essence*
	appearances	*Reality*
	body/ mind	*Soul*
	lesser selves/ parts/ roles	*Self*
	determinations	*Freedom/ Free Will*
↓	*tam*	*raj* ↑
	passive	*active*
	what is willed	*will-power*
	body/ puppet	*mind/ puppeteer*
	oblivion	*inclinations/ tendencies*
	no choice/ automatic	*choice/ relativity*
	fixity/ complete determination	*seeming freedom*
	prison	*grades of freedom*
	tyranny	*consensus*
	slavery	*helping one another*
	no freedom	*you and society*
	no free will	*conditioned free will*

Is not resolution of the tightest paradox, free will and determinism's oxymoron, hardest of them all? Chapters 6 (Authorisation, Calculations of Mind and Mind Machines), 8 (A Culture of Doubt and Puppets, Mummery and Drama) and 13 (Physical and Metaphysical) have skirmished with the issue. Now let's fire another shot. It's time to crack it. Bull's-eye!

Bottom-up, you are a puppet of your selfish genes and 'chemically determined' by a gene-built brain; thus there's no choice, no moral code, no risk. You can't transcend cerebral twitch. Behaviour is, for marionettes, made up of sodium, potassium, water and the other elements of muscle, nerves and bone. You might believe in unconditional or conditioned choice but, if you believe that nerves make thoughts you *can't* think what you want; those chemical reactions run the show. Indeed, bacteria are great survivors. Thus why, when compared with chemistry and instinct's automatic pilot, should consciousness improve an organism's chances? More than this, if mindless chemistry evolved an order-making entity called mind why should it do so in a way that lets it mull alternatives then choose what 'you' instead of nerves 'consider best'? And what, on top of that, is the reason that a nervous pattern lets you understand an evolutionary or any other point of view? If there's nothing but non-conscious matter consciousness and life remain a central mystery; this pair must somehow be material as well. Then probabilities are all that loosen fixed determination from its course. It hardly matters that a photon or electron has 'a choice'. Since when was quantum randomness the same as free, decision-making will? There's no free-thinking - in a materialistic paradigm determination rules. Therefore *bottom-up* you might believe, as did Pierre Laplace, that every motion can, from cause to its effect, be mathematically determined; or, with Werner Heisenberg, that there is basic quantum indeterminacy - at least as far as you're concerned. Is one view right, the other wrong? How can both classical and quantum physics be, in this case, right? Think about it. While you're thinking don't forget that your prefrontal cortex, made of neurons, must dictate your logic; and their origin, your 'selfish' genes, are leading cogitation the nose. Molecular determination means that evolution carved you from the elements of earth; awareness is appearance, an illusion born of electronic or atomic state. A helping hand, an altruistic beast or tablets of morality are each construed as part of puppetry's illusion. Where, in such determinism, has free-thinking science disappeared?

Programmed within and forced without what other answer can there be? Mind must be material. This vision just sees genes and evolution of 'genetic instinct'. It bears repeating there is no such thing as choice - except perhaps in free-thinking minds of those whose choice-less choice is thus! Genes, nerves and ionic movements in some way issue thought; mind is a material fantasy! Can expert fantasy not work it out from there? Has an embodied individual, whether vehicle of genes or soul incarnate, any more to say?

Why can't one reasonably squeeze immateriality completely from existence? Who'd notice any change in physicality when nothing disappeared? You could thereby elevate non-conscious, physical phenomena and sensory perceptions to the one acceptable reality. Mortality's survival thrives on sensuality. It waxes warm on power and influence, wealth, sex and (which is interpreted as freedom not enthralment) the fulfilment of continually bubbling-up desires. Of course, we have material ideals - utopian imaginings that politicians promise us like communism, socialism and, with exploitation of the natural earth, economic wonder-worlds of science and technology. Who objects to comforts easing him through earthly years? Woven with the comforts what, however, of uncomfortable questions of morality? Can such immateriality of choice be real?

No doubt, a thoughtful secularist is decent, educated, rational and, all in all, an intellectually enlightened 'soul'. The chap knocks spots off the robotic and unfeeling cadres of a revolution or off low-grade, narrow-minded zeal. *No doubt, also, humanistic atheism rides the slipstream of Mosaic law but, for materialism's creed whose central 'fact' and core belief is in a Theory of No Intelligence and just possibly an overriding rule of selfish genes, this stream's direction is an inconsistent one; its genetic current, driving always towards the earth, is against the upward path.*

It might, therefore, be argued moral atheists are inconsistent fellows. After all, why not lie, rape, thieve or murder if you're egotistically programmed to cull competition, maximise on reproduction and thus 'best survive'. 'Survival-of-the-fittest' theory doesn't cause but certainly can justify such immorality. It leaves no choice. Behaving like an animal's consistent with your nervously synaptic view. Society is just an energy-efficient ploy, a superficial gloss wherein submission profits over battling to achieve desires. If you held evolution's theory true, then intellectually, rationally and logically you will be socio-biologically inclined to run with 'natural', non-Mosaic rules and regulations. We've just examined how a creature (aren't you just a creature, after all?) should function 'rationally' according to determination of its selfish genes. Morality must flaw such theory with internal inconsistency.

Yet atheistic nihilism does not necessarily make for depression and evil - though it certainly helps. There are many atheists, humanists and agnostics of excellent moral character - thoroughly good people. They're simply lucky aberrations of Darwinian code, civil folk who temper body-logic with humanity. Whether or not the principles that mark this goodness were transferred from long Christian tradition is a point for interested academics to debate because, in fact, it is the 'now' of behaviour that counts. **In this case, the difference in character between honest, well-meaning theist and atheist seems as paper-thin as a metaphysical 'credo'; but there is a gulf in perspective regarding the true nature of creation.**

The axiom that mind and matter are two separate elements is basic to this book. Who, on this hand, can seriously deny an immaterial fact called consciousness? The evolution of this basic fact of life is, perhaps because it never happened, still a pitch-black box. If, however, consciousness *is* immaterial then who is able to deny the metaphysic of non-reflex, thoughtful choice, that is, free will? Mind over matter rules our lives; you make up mind but can't make up your brain! In public life, against the evidence before their eyes, materialists assert their theory's lack of sense and sensibility. Genes, other chemicals and brain create the great illusion that we think we are. Yet having chosen how to think the thought, can you robotically deny free will? Can you professionally deny the fact of choice but unprofessionally relaxed at home allow it all the time? You would be hypocritically dancing to the changeful tunes of immaterial mind.

Top-down sees the issue graded. Non-conscious matter (atoms, molecules and anything, including nerves, made of them) is a special case. At the nether pole oblivion can have no will and thus is wholly predetermined. Inexorable patterns of behaviour nowhere in the cosmos lack. This is the material extreme.

Unconscious purpose (called by some 'unconscious will') is a passive breed as well. Sub-conscious archetype and automatic, autonomic functions work by preordained intelligence. In life, as consciousness is lost, so force of instinct takes direction over and keeps orderly.

Consciousness, however, is replete with purpose and emotion. You can't change matter but can change your mind. Can't you choose and, without predetermination, choose again a million times just as you wish? Awareness focuses and functions with initiation and response called will. Remember, though, you float upon an ocean made of context in the form of memories; and that you wear a body made of needy flesh and bones. How free, how undetermined is the line down which you aim desire? Could we say that you had conditioned will? Body, parents, family, education, circumstance and thereby experience and memories that you, for good or bad, could not avoid combine to limit choice. You have freedom that is, by your circumstance in mind and body, very much reduced. *And thus a paradox - you have free will and don't.* Without body (as in imagination) will flies freer; without ego (as in transcendence) it flies free. Body's interference radically closes choices down; there is no choice but to accommodate material limitations.

We tune our beams of interest and desire like lasers use the power of light. Creation, action - awareness is the basis of free choice; conscious mind involves a concentrated focus called this inner power of will. This is how it seems from our perspective, our position on the universal slope.

Now check Chapter 5: Top Teleology. What is a Creator's view? What from the Apex of Mount Universe might *Top-Down* see? Here the paradox bites sharpest, here dilemma's most acute. Isn't this The Point of Freedom, Concentrate of Consciousness without an end? Without colour, weight or texture what is it intellectual erudition cannot ever understand? Infinite Awareness moves existence. Silence sings and cosmos is its song. An author writes and so determines how his play will work out and its theatre run. Yet was there ever any choice? Is ours the only universe that destiny allows? Are there, as Einstein wondered, principles that absolutely govern how First Cause initiates, how an emergent current of Free Will is bound to operate?

What if space, change and time all vanish? What if mind and body drop away, existence is dissolved in Essence and the lesser selves have disappeared to leave just Real and Central Self? In this 'condition-less condition' there can be no separation of an individual will from Will; there can, in union, be no second, different sort of freedom (such as yours or mine), only Infinite (say, Boundless) Free Will of the (N)One. When two are joined as one and there's no other what's the nature of this All-in-One? What is, at Conscious Union, a choice-less freedom if not love? Thus we rise up into Prime Paradox - in Liberation Absolute The Absolute is found to be a Slave of Love!

Madness! Slam the door! And yet the answer to free-will-and-determination's paradox is not an intellectual one. Its enlightened resolution transcends mind; and the principle that governs in Communion with Unity would seem, logically, Love Alone. Is not love freely given? Does not its quality derive from voluntary surrender of one's egotistic self? How, therefore, can love lack unconditional, free will? Can you now understand what all the mystics have been trying to explain?

To create or not create you have to choose. First Cause, an action prior to any reflex, also chose. The Infinite chose finiteness and The Formless form; Consciousness began to move and thus initiated cosmos.

Therefore *top-down*, if Natural Dialectic is correct, all lesser loves reflect The First; and from First Principle a panoply of principles diverges to construct the practices that make existence, that is, mind and matter down below. An engineer determines how inventions run. 'Not a leaf stirs but He willed it so'. At root, such information passively determines how leaves stir and cosmos spins around. Thus, to the Creator's eye, complete determination paradoxically rules what might seem freely chosen psychological as well as physical creations. Play is, like any game with rules, pre-planned. Determination at both cosmic poles is absolute. At one extreme there reigns such ego-free determination as the sovereignty of an inventor's love supplies. At the other fixity of regulation automatically rules; what seems free will turns out to be no more than inclination towards a seeming satisfaction. Such mobile inclination is the natural, rolling motion of equilibration; it is born of wave and counter-wave upon a *karmic* sea. So-called 'free will' is actually conditioned by the complex context of its exercise - its exerciser's mind.

Yet even here the contradiction rears again. Some claim a course of life on earth is preordained. Its fate and destiny are fixed. Thus it follows choice is an illusion. If, however, there exists conditioned flexibility you can, as circumstance allows, choose from different possibilities; different option, different outcome and a destiny (and therefore fate) that changes each and every time you make a choice - this is quantum probability of mind. When free will is entertained it seems that destiny cannot be fixed. If it could 'fixed freedom' contradicts itself; it is suspended in self-inconsistency. It would appear, therefore, that 'karmic play' of action-equal-to-reaction is not (as it is in Newton's law of motion) inexorable. Life's course is not fixed; it is not by *karma* preordained – not least because Prime Will willed all lesser versions of free will and, with them, various exercise of moral choice. Mind's part in creation is, unlike complete material rigidity, relatively free.

If destiny is fluid then prophecy would not seem possible; could even an omniscient creator know a future that, by creating, he had loosed? Free will or no free will - this dilemma reaches to the roots of being.

Can it ever be resolved? Free will, determination, purpose and the focus of intention are drivers at the heart of sense of self and 'human teleology'. Without them who am I? With them comes the faculty of choice. How could synaptic circuitry decide what's good or evil? How do nerves without free will evolve a system of morality that flatly contradicts the ruthless selfishness of evolution's 'ethic-that's-genetic'? It is, as we have seen, nonsense to conclude they can. **In fact, liberty to learn, free will to choose and chosen legal systems underwrite religion, science, politics and all of civilised endeavour**. If it's all determined, what is a machine-creator's purpose? Is the whole creation actually purposeless? In contrast to this hopeless view there's hope. Why should optimism not abound when, at the centre of a dark, material periphery, there's cosmic light and life eternally - whence moral rules and ethical behaviour, as opposed to reflexivity, derive?

Remember that creation is composed of finite parts. It is, by nature, set apart from its Essential Nature. This Infinite Potential is broken, when it moves, into the cosmic play. You need a natural (*tam* ↓) negativity to break things up. Such negativity applies to physical and metaphysical alike. When, therefore, you include the immaterial, living element of consciousness there must be negative as well as positive intent. As the left-hand column of the Secondary Dialectic perforce shows, what you have dubbed The Devil will appear. When devilry is thrown in the drama is imbued with fiery colour, ethics and distracting sub-plots but, in life's plus-and-minus theatre of morality, Top Will remains the axis round which all the lower machinations swing.

In short, free will's a trinity. *Top-down*, Causal Will creates what wasn't there before; in the special case of Essence, also called Potential, there is every possibility. A Choice was made; our cosmos emerged. Life on earth involves restrictions; a range of states of mind (including human ones) reflects a scale, from loose to tight, of bondage to the world and its calamities. Man's view involves degrees of freedom. Lastly, mindless matter (which includes bodies, brains and all molecular biology) is the special nether case of will-power - none. It has no freedom but is fully automatic. Impotent, it has no will at all.

What you personally determine is a choice constrained by conditions (such as education and prior thoughts and actions) but on the surface it seems free. Where ought such freedom carry you; where should your determination drive, what is the way to lead the best of lives? The grand plot therefore stays, despite deflections, 'keep to the good and, at best, keep trying to ascend along the rightward path'. The game plan will eventually call you to transcendent inexplicability - to Free Will, Freedom, Self and Source.

Although the answer differs by perspective is the tension in this elemental paradox now reasonably relieved? If so straightaway another problem needs be solved. How can the (*raj* ↑) rightward, upward vector cope with the perversion of intention, that is, mischief, crime and criminality? How can a man determine to avoid blood-spattered degradation of ideals? How can one personally solve the problem that is, metaphysically in mind and physically upon the body, evil?

<u>Solubility and Solution</u>

↓	*division*	*unification*	↑
	hardening	*softening*	
	weight/ burden	*lightness/ transparency*	
	heavy debt	*freedom/ ascendancy*	

Are pain and evil soluble and, if so, what do they dissolve in? What's the medicine, the cure? If the pain is physical then let a doctor pick your bones. Evil is inflicted by intent. It starts metaphysical; and it is not amoral but immoral. What is the solution?

At the Summit of Mount Universe, above creation's clouds of thought and valleys of phenomena, the sun is always out. Brilliant! But suffering's weather, more or less inclement, churns below. If you suffer from its buffets is there shelter in release from being underneath? Is there a complete solution to the problems of existence?

Mental snakebite needs its anti-venom - antidotes to evil as related in Chapters 2, 3, 13 and this one previously. With the Dialectic you can guess what's coming next. It's a good dose of the right-hand path. Which, if you're thinking 'down and outward' towards periphery means turn around, think again or, in the French vernacular, repent.

Circular Dialectic and the Power of Unity

(i) Attentive Focus

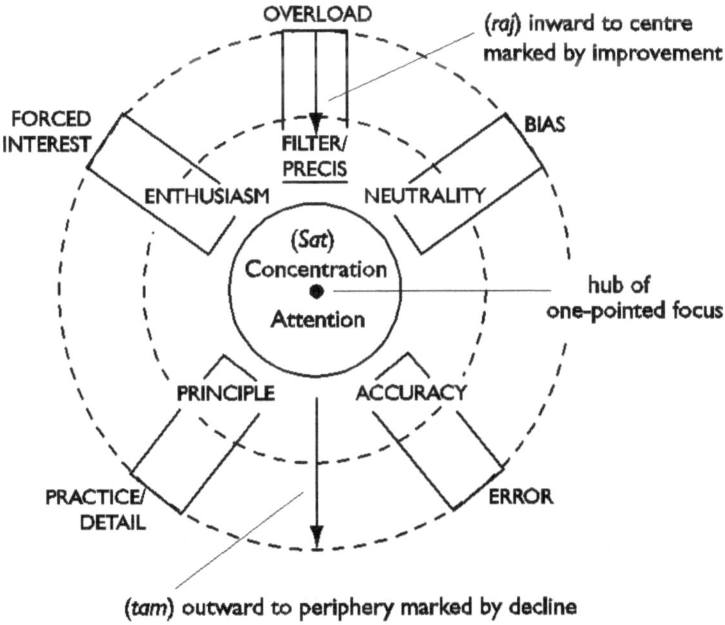

Unity/ community is found at the point of all-connectedness, the axial centre-point or boss.

(ii) Morality

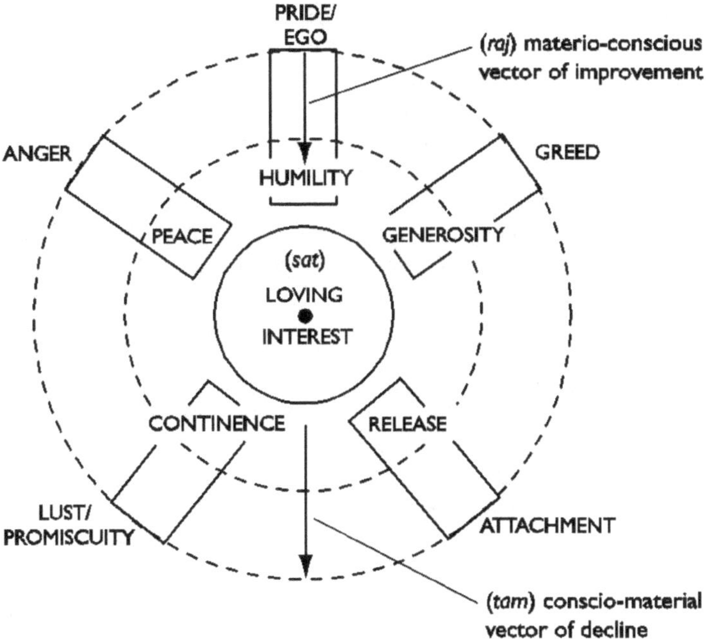

It may well seem that the tendencies of ego, either way, are overlaid by the character of moral systems. Don't we, as egos, fluctuate continually with characters tending towards one column or the other?

↓ *negative*	*positive* ↑
heavy	*light*
grudge-bearing	*forgiving*
'sticky'	*easy/flowing*
dark	*radiant*
crooked	*straight*
self-centred	*altruistic*
stressed	*free*
maleficent	*beneficent*

Both 'yes' and 'no' are necessary truths but, in this egotistic case, which truth do you prefer? In the 'morality' circle incoherence, dispersion and negativity increase outwardly towards periphery. This is anti-truth. Vices are alien, destructive and, eventually at the world's end, lethal; man is, you might say, clamped in the vice of mind and thus dragged down. Conversely the power of unity, single-mindedness of purpose and a positive current of will as shown, for example, by victorious world-cup teams cannot be underestimated. This is the centralising, virtuous nature of the human spirit. What if the athlete is a mystic and his goal the Cosmic Heart?

fig. 27.2

Does an artist fail to paint because his painting might receive rejection or will one day disappear? Do parents not have children just in case they'd sin? And do players start a game with good intentions or abandon any match because adversity or accident or foul were possible? No risk, no game and no reward. Thus creators have no choice. No risk, no game. So they choose, against the possibility of loss, to make an effort and by following the right-hand tack to win a game well played. Why should a Supreme Creator differ from creators in this calculation? The possibility of imperfection (pain with sensitivity, worry with imagination and evil with ability to choose) is in the nature of the game of life. Spoiling is unwanted but you can't stop creation just because a difficulty might occur. There will be problems. Blows. Loads. They will need solutions. You simply start things up as best you can. You establish healthy first conditions and disseminate an archetypal rulebook that outlines the circumstances of fair play and how to remedy attacks on it. Each re-start, each game within The Game, it's just the same. The Game's The Play; playing games is how scenes in life's theatre work. *Thus, since mind precedes and frames the puppet-body's movements it is important to decide and then to exercise the healthiest condition possible of mind.* In this respect mystic solution has, through the curtains of prehistory and throughout recorded history, formulated core ideals and human best endeavour. What is the wise man after? What is his grand strategy? How to strike? How does he strike the roots of evil and, precise and surgical, cut rottenness away? Is it by repentance (a U-turn in mode of thinking); thence (*raj* ↑) return from evil and re-entry into Eden; thereby (*Sat*) Paradise regained? If such a strategy is cosmologically correct it will, like charge or light or gravity, endure.

There's tension here. What use is renunciation of the world when all you want is to become involved? When you crave creation yet don't give a jot for its creator? What use is desire dissolved if you think sin is fun and strait-laced Eden's not? While the going's good, while the world is funny and the outing's not too painful why should I ever want to join a mystic and cry off? If life's a laugh then, stepping from existence back to Essence, why should I jump the universal party-bus? If, on the other hand, it all becomes too much to bear then how can you alight? The problem is, as has been noted, an embodied soul's. You are irreducibly incorporated. The game is that you wear a body three score years and ten. You incorrectly but habitually identify yourself with this incorporation, vehicle of drama, 'bus' of your intrinsically amoral body biological. There is a natural instinct for survival. Its subset includes the instinct for control of circumstance, domination and thereby sufficient freedom to obtain necessities. Such necessities include, of course, food, fresh water, clothing, shelter, relaxation, exercise, agreeable society and, critically, health. What about the future of humanity? One imperative is reproductive sex. You therefore need a fig leaf to indulge the nearly bare necessities but what about an increase in the list? What about unnecessary 'necessary extras'? An inflation of desire usurps man's mind. It flares from the nether focus, swells a normal need beyond proportion and thenceforth roves consumed with apparently palatable passions. Ethical oedema. More normal than an aberration such self-centred state of mind is overwhelmed with meanness, anger, greed and lust. It slides with ease into a tempting vortex of desire. This is the kind of cancer Doctor Buddha (*PH.D cum laude*) and the other mystics try to ease and, practically, dissolve.

No head in the sand can, on the other hand, avoid the moral-stained arena into which all humans are, by virtue of capacity to think, pitched. This soaked arena is called mind. *Upward and downward, spiritual and animal, higher and lower mind are, angel with adversarial devil, locked in mortal combat. Such struggle is part of your fabric.* Its crux, the oft-repeated moment of temptation, discrimination and morality, is as much a part of man as body parts. Paradoxically, despite the fact that Byzantine ascetics chose foolishly to mortify their own amoral flesh, the body is an innocent abroad. What can skin and bones know making them an evil cage? Mind is the snake in need of tending, scotching and, where its bite is poison, killing off. How to best excise a thought that drags you down? Never mind the urgent detail. Choices of direction all involve an underlying texture of morality.

Not just a snake! A freedom-fighter too! Utopia, like Eden, starts in mind. Material starts from metaphysical perfection. Yet ethics and metaphysics aren't the province of your average scientific scrutiny. Detachment bottles up and keeps their concentrations on the rack. Nor are they really stored, except as they impinge evolutionary neuroscience, psychology and socio-biology, upon materialism's shelf. Dialectically uncorked, however, they assume a central role. How?

Firstly, the right-hand path in its subjective mode aspires towards Central Morality, the Natural Nucleus of existence. Morality is thus, once understood, an integral ingredient of cosmic chemistry. It stems from the Nature of The

Chemist. Such chemistry is, in the first place, not atomic; it is metaphysically atmospheric. It is mobile upwardly. Evaporation from a draggled world! Distillation unto purity! The basic energies are love, attraction and desire for happiness all round. Blithe, isn't it, except that in the 'real' world bodies interfere. Centaur stakes are coursed for sex and money. Jostle, jockey, competition unto outright battle for supremacy can decimate life's flying turf. What kind of whip and bridle does a temperamental racer need?

Secondly, an individual who assumes the right-hand path begins to rise towards a State of Balance. Psychological excellence therefore becomes, in this voluntary respect, a question of degrading mental viruses called vices that would otherwise swell into a raging fever made of sin and suffering. Who likes to swallow medicine? Who wants to hear the joy of sin is an infection, a contagion, mental pestilence that spreads its epidemic into eager and impatient patients? Who wants to know the swabs are virtues, the poultice boring old contentment and the real antiseptic prayer? This is what ablution means and why the mystic is psychiatry personified. The Doctors clearly state that their course of independence aimed at Independence is the sovereign remedy. Self-treatment, self-reliance and self-government make better cures than medical dependencies. You might, for example, take a course of Noble Truths along an Eight-fold Path to Purity. This double bill both heals and is a course of learning towards post-graduate enlightenment. The greater wisdom then the easier morality will automatically flow. The current is a warm and natural one.

Is that the whole solution? What else might help defray the vast expense of evil? In a world where evil is a heavy force such solution is as naïve as it is profound. Mercy is the luxury of strength and victory; of course you need sufficient strength to batten down the opposite of high ideals. Bad down, good up; but left, low aspirations held in check don't obviate, upon the right, the need for power to rise. Do you remember (Chapter 4) an immaterial fuel called concentration? It is well known when strong concentrations are diluted, dissipated or diffused then they become weak ones. What, when considering information or intelligence, might constitute the most apt focus of attention? What might best concentrate the mind? Or constitute a Pure Concentrate? Most concentrated, that is, in the same respect as water and its full potential, purity? In terms of Spiritual Water Natural Dialectic indicates solution, Concentrate of Consciousness, is not mixed with any solute whatsoever. This is the immaterial element alone, the elixir, the Pure Solvent and Solution. Why on earth should a philosopher misname it 'stone'?

Towards a Unified Theory of Religion, Politics and Law.

How best, philosophers enquire, live life? What solvent best dissolves the hardness of our problems, what solution flows towards universal happiness? Down-to-earth solutions to the problems of disorder, ignorance and pain involve two species of utopia - one objective, outward social and the other inward, personal and subjective. Does the inner one reflect the outer or, if mind precedes material, *vice versa*? What's the starting-point to hunt the heart's desire for lasting happiness? What's the basis of a 'good' life? The narrative will swing from an objective framework (of religion, politics and law) to the interior, subjective aspects of *He Agathe* (*ή αγαθή*) - as Plato's Dialectic called 'The

Good'. We'll deal with the outward, large-scale solution first and then, as was the plan, work inwards towards a microcosmic, personal Nuclear Solution.

Religion, Politics and Law:
A Dialectical Plan of the Way Communal Life Works

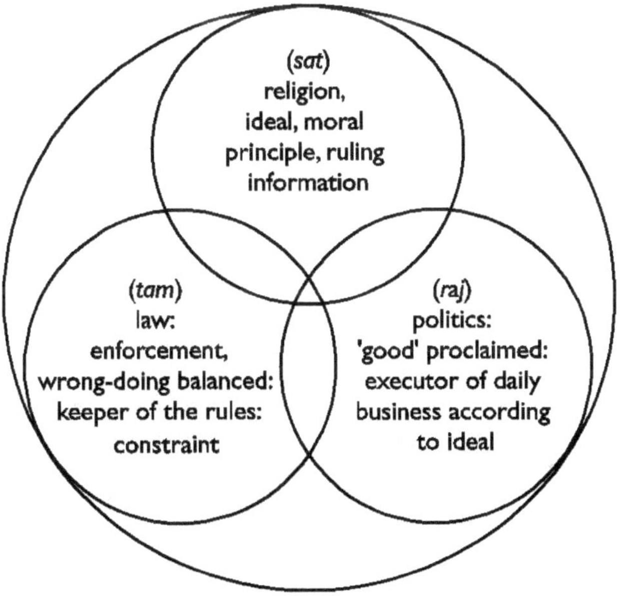

fig. 27.3 (see also fig. 19.1)

The relationship of individual to society, elaborated in a million books and, more to the point, in each person's life, is a manifold and complex one. Association embraces one-to-one, family, group and wider interactions; these span outward from close and physical to large, abstract affiliations such as citizenship of a country or the world. All, however, involve the simple, dialectical principles of (*sat*) balance, compromise or equanimity, *(raj)* positivity, improvement or release and (*tam*) negativity, deterioration or oppression. It is along these lines that - either internally, psychologically or externally, physically - the Dialectic discriminates.

tam/ raj	*Sat*
lower principles	*First Principle*
subsequent derivations	*Ideal*
material/ immaterial	*Immaterial*
motion	*Peace*
variant	*Pivot/ Norm*
polar	*Neutral*
relativity	*Absolution*
relative imbalance	*Balance*
oscillation	*Homeostasis*
eccentric swings	*Central Axis*

↓ *tam*	*raj* ↑
lower nature	*higher nature*
negative act/ wrong	*right/ positive act*
deformity/ deformation	*integrity/ integration*
tendency to violence/ disorder	*tendency to stability/ order*
less Normal	*more Normal*
towards insanity	*towards sanity*
ego/ selfish self	*desireless/ selfless self*
imposition	*self-control*
fall	*self-correction*
punishment	*reward*
confinement	*release*
guilt	*innocence*
debt/ burden	*credit*
shame	*merit*
failure	*success*

What is progress without target? How can you improve without ideal? In other words, does an Ideal exist towards which the aspiration is exactly what is always was? In this sense 'progress' is not the prize of man's technology but of an individual's evolution, flower-like, from the cramping style of violence, rage and evil thoughts. It's learning how to ride the moral bike. In fact, the whole direction of this chapter traces our behavioural evolution 'inwards' towards discovery and assumption of The (*Sat*) First Principle. Its (*raj* ↑) vector revels in 'return'; it involves ascent along the conscio-material gradient from the ecology of living (in some cases conscious) organisms and society of humans towards 'higher' association with....with what? What elicits, what excites the best in you?

At this point - the consideration of a Unified Theory of Law, Politics and Religion - we speak about '*external association*'. Such association comprises relationships between the individual and the physical world. It includes both positive and negative interaction with non-living objects, any living organism and, closest, other humans. In an ecosystem the key principles that drive communal order are homeostasis, harmonic interaction and recycling nutrient 'goodies'. Are the key principles of human community - dynamic equilibrium, harmonic interaction and recycling psychological/ moral 'goodies' - the same? If so what are the mechanisms by means of which dynamic stability can be obtained?

At the heart of human community are one individual's religious, political, legal, educational, economic and personal communications with another. The strongest, most important relationships are often with those in close physical proximity. This includes one's own body (habitually thought of as oneself), family, neighbours, colleagues and workmates. Of this group the closest are those 'on the same wavelength', that is, perhaps family but certainly friends. And of friendships potentially but arguably the closest kind ideally fulfils the criteria of both physical and metaphysical union - man and wife. Meeting of mind and body is an upward (*raj*) symbol of unification and love. It represents positive, creative polarity, homeostatic stability and dynamic equilibrium. In the context of family these cosmic principles translate into security, continuity and healthy relationships; also, since a family is the basic unit of society, into social cohesion and stability. "Peace in the state begins with order in the family," wrote Kung Fu Tse (Confucius but also spelt, confusingly, Kong Fuzi!).

What, therefore, should the instruments of social cohesion - law, politics and religion - most attentively and sensibly support but family and, within the family, its individuals. Is this your country's case?

↓	*unprincipled*	*principled*	↑
	egotism	*communal spirit*	
	disintegration	*integration*	
	criminality	*benevolence*	
	stress	*release*	
	violation	*relationship*	

Psychological and physical stability support the primal health of nature, the original sufficiency of biological construction. If health is happiness then illness is the emergence of irregularities that create stress and disorders that are understood as pain and evil. Why, therefore, should family or religion be preferred as evil's antidote? Why indeed choose either? Have these two institutions not rubbed up more strife than any other in the history of man? Why promote them to the highest instruments of wrong's correction? How, if they create such quantity of suffering, can you claim that they eliminate the 'badness' and refresh the roots of social order?

Cancer cells rebel against the body that they occupy. It is out of order. How can malfunction not amount to violation of the smooth and stable order of its 'body politic'? If evil is, in parallel, the violation of another's body or his lawful property, then hygiene and medicine must apply to cure the case. Evil is a crime and crime disease whose parasites infect the body social and, thereby, your body politic. Unprincipled a sickness needs a doctor - rules, regulation and an antiseptic retribution sufficient to negate the viral thoughts inhabiting a sinner's mind. Who wants to become another ego's victim? Criminals are social cancers. Swollen egos may resist the police but, where internal regulation fails then, in the interests of its members' and thereby its own stability and health, society is forced to keep the peace by its external powers - church, politics and law.

Isn't this too obvious? Or don't you like my choice of words? Then go back to your academic cell and, as Aristophanes would say, invent cloud cuckoo land.

Religion, politics and law; is not the greatest and the first of three, the one from which ideals derive, religion? Do you recall (from Chapter 0) the nature of 'formality'? Given human inclination you may well have argued with some force and element of truth that religion (either atheistic or theistic) has stirred up more pain and evil than created happiness. You may cite details but will still have missed the point, and missing all its personal, familial and social benefits, have dreamed up a half-truth that, like evolution, in the whole light is a lie. The central point is, dialectically, that mind precedes material; and crime derives from error in the mind. Immaterial precedes and governs the material state of man. Do you remember, yet again, the schizophrenic state-of-centaur paradox? These are creatures struggling with the tension in between their loving upper and unloving, lower parts. They are us. What is a swollen ego but, in selfish mind, a case of psychological oncology? As moral harmony (called care, manners and respect) wears thin so harsh correctors have to take its place. If voluntary morality is lost involuntary restraints must hold the balance, harsh reminders hold the line. As health is lost the medicine is dosed, the hardness of the surgeon's knife grows real. There have swung to action scales of religious law, law-making politics and justice by such formal, legalised morality.

In the hierarchy of our lives which institution towers on which? Which depends on which and whence derive our truths? Which do you pile on top?

An atheistic faith, whose substance is material, brings matter into mind; theistic faiths, each in their way, explain how immaterial impacts on physical event. Therefore try walking up the latter's conscio-material gradient. As you (*raj* ↑) ascend towards (*Sat*) Central Character are you reaching to the heart of things? Whatever knots of pain distort religious histories their smooth and central aspiration grains the same - obtaining knowledge of the Immaterial Concentrate. Primal health of mind is based upon this plank of Good; original sufficiency of psychological construction depends on this Ideal. Are moral principles not facts? No such ideal is 'scientific fact' but is it a delusion? Are your own ideals and principles not immaterial fact? Or are you just, in holding on to them, deluded? If ideals were facts then an Ideal of all ideals would be The Fact. Not 'scientific fact', of course, but one irrelevant to science. Perhaps even 'scientific error'. A delusion. On the other hand, if subjectivity implies awareness, this Subjective Fact must be alive. Just like your mind it must be Fact and Factor all in one. What, asks a mystic, is the nature of this non-delusion? What character does Main Dialectic's right-hand column certainly imply? What Fact and Factor supervises from the top?

This is not religion. It is Natural Essence. Essence is not super-natural but the very Heart of Nature. It may be the source of but is beyond man-made religions. From this Essential Ideal there stream ideas of ethical behaviour (in the business of a body politic) and enforcement of morality (the vice that justice systems try to close on vice). In a material world whose way is tumbling to exhaustion, wickedness and disarray what worth have remorse, repentance, rehabilitation, resurrection? Could illumination from ideals derived from The Ideal lift lightly from the downside? 'Reality' continually intrudes on dreams but, as in every category of life, isn't aspiration to perfection better than no try at all? Are not, in this case, ideals an 'active transport' pumping up against that natural drag towards disarray? And pumping further past a so-so state aren't they the impetus to join up islands of utopia that all of us from time to time enjoy. A tidy continuity of bliss!

Ideal in principle, less so in practice are the laws and legal systems of the human world. Rough as well as reasonable justice is dispensed. Fair and unfair judgments can occur but the underlying sense of balance represents protection from the vicious and unbalanced fractures of a less than perfect local, national and international community. In this case from Natural Dialectic's point of view best principle derives from Top Ideal. *Society's best formal answer that addresses evil issues from the State of Immateriality; our material solution is resolved, most definitely without recourse to science and without a test-tube anywhere in sight, by institutions charged with primary exercises in morality.*

<u>Morality steps centre stage.</u> It is not scientific but it is for sure the bigger player. The purpose of the agencies of social order - religion, education, law and politics - is to combat imperfections, minimise all criminal behaviour and promote relationships. *The whole point of morality is maintenance of individual and thence social balance and, wherever necessary, restoration of dynamic peacefulness.* Such homeostasis integrates the cogs, oils the wheels and ticks the social engine over in a state of poise and health. Its quality of mind promotes harmonic interaction at all levels of community from individual to global federation. This community should certainly include the ecological condition of the spaceship we inhabit, that is, other forms of life. And to that end we even have to take sufficient care of soil, sea and air. Does not dominion imply responsibility?

	tam/ raj	*Sat*
	action	*Potential*
	polarity	*Neutrality*
	relativity	*Absolution*
	duality	*Unity*
	lower principles	*First Principle*
	less or more fit	*Health*
	compromise	*Ideal*
↓	*tam*	*raj* ↑
	negative	*positive*
	less healthy/ out-of-kilter	*in tune/ more fit*
	inertial equilibrium/ slack	*dynamic equilibrium*
	shaped	*shaper*
	enforcement	*formulation*
	body	*mind*
	law	*politics*

Health is primal, pristine and original. This fresh and intended state involves dynamic equilibrium. Such equilibrium is the heart, anytime, anywhere, of human institution, constitution and a formalised society. Cohesive structures, simple or complex, need promote the vibrant harmony of health, comfort and survival of the local 'gang'.

What drives the logic of lives in community? Whence springs the natural formulation of society? What, in other words, is the origin of ideal, principle and therefore law? How does the idealism work? How do you sustain the peaceful and unbroken rhythms of a dynamic, social equilibrium? There is no choice but to try and best deliver. How do men deliver principle in practice?

<u>Peripheral Religion</u>

tam/ raj	*Sat*
existence	*Essence*
spectrum	*Source*
lower principles	*First Principle*
belief	*Knowledge*
religious faith	*Super-religious Fact*
its translation	*Information Centre*
expression	*Ideal*

↓	*tam*	*raj* ↑
	peripheral formalities	*approach to nuclear core*
	bondage	*freedom*
	towards darkness	*towards light*
	lack of intelligence	*intelligence*
	rigid adherence/ dogma	*thoughtfulness/ debate*
	hypocrisy	*sincerity*
	rule by fear	*rule by wisdom*
	arrogance/ intolerance	*toleration*
	wrong	*right*

Of course, society needs structure and bureaucracies more rigid and extensive than an individual's paperwork. **The structure that encrusts around his nuclear ideals is called a man's religion**.

How men deliver principle in practice is therefore dependent, above all, on what a group believes is true about itself, its purpose and relationship with natural law.

Religion means 'a system of belief' and is therefore unavoidable. Whether atheistic or theistic, oriental, occidental or plain secularly 'liberal' it frames relationship with cosmos. It defines, informing us about our origins and heritage, our standing in the world to which we have been born. It is intrinsic to the basis of our thinking. We are all bound to it - even a majority who do not think about them but accept the current social norms. Relationship with both the physical and metaphysical domains of nature influences the fundamental form and thereby patterns of behaviour found in every community. *It's down to* origins again. We spent chapters in discussing them. How you orientate yourself is central to the type of world you build. And if the place of origin is myth, then a society's myths are crucial. If science, with one-tiered materialism, *PAM* and evolutionary *PCM*, claims destruction of a myth what does it replace it with? And is the basis of its intellectual revolution true - or simply a materialistic myth, a sophisticated metaphor inveigling and entangling its 'designs' with actual design? Myth, either way, is at the heart of your religion but each kind, atheistic or theistic, propels the other far apart.

Pol Pot was cremated with his Book of Death, 1500 cradled pages of meticulously assembled photographs of those tortured to death at Tuol Sleng bound in a thick, black leather cover. Unsaintly tyrants, 'dear leaders', are embalmed in state - Lenin (Stalin briefly with him) in Red Square and Mao at The Emperor's Gate, Tiananmen Square, Beijing; and mausolea mark the centre-point of revolutionary pilgrimage in Hanoi (Ho Chi Minh) and Santa Clara, Cuba (Che Guevara). Statues, fervour, violence and mystique; but on an altogether nicer, liberal level did not 'Darwin's Big Idea' (a centenary exhibition at the Natural History Museum, London, in 2009) enshrine - just as La Sainte Chapelle in Paris treasures Christ's crown of thorns or a casket in Topkapi Palace, Istanbul enfolds religious relics of the Prophet's hair - some scientific fibres fallen out of naturalism's idol's beard? Indeed, the museum seems (like Down House, Kent, where the wise man lived) transformed into a shrine, a temple of the evolutionary faith. *You'd thought, I know, that relics were a superstitious, medieval game but modern metaphysic, named materialism, is religiously devout as well.*

Like myth, relics carve opposing tracks. No doubt, therefore, it is the systematic practice of materialism's *bottom-up* philosophy to, topsy-turvy, invert *top-down*'s point of view. The weight of science in psychology bears down on explanation of the mind by physiology. What, therefore, of a basis of religion, its morality? Survival of the hardiest, unbridled lust and pitiless one-upmanship should logically win the day. If this seems harsh then why would morals, how could altruism and religion, full of rotten faith and superstition, ever have evolved? How, compared to scientific writing-on-the-wall, could anybody be so silly?

As usual, *bottom-up* cites an effect as cause. For example, brain scans show repugnance registers neurologically the same when foul food or indignation at some outrage is perceived. Perhaps, therefore, repugnance is our moral guide. Body language and expressions of the face reflect identically on faeces, rotting corpses and an act of violation. Perhaps, an evolutionary psychologist is pressed to muse, germs and disease first prompted 'bad taste' that evolved through 'good taste' into ideals of justice, honesty and love. The mind's ill-health, including sin and crime, has roots in physical oppression; 'sense of rightness' and religion stem from an aversion to disease. Physical and psychological disgust are bundled just the same; sensibility and feeling are, allow it, cultural phenomena but stem from muscular and nervous, not least cerebral, biology. Brain scans show it. That you can't deny! The substance of this explanation (which includes the way you know it) is material alone. If, however, conscious information complements non-conscious matter it emerges as a half-truth - one in essence wholly false. Mammon's den evokes peripheral religion - Darwinian scientism - that could be rotten to its mindless core.

Scientific treatments of society and individual (sociology/ psychology) must engage by working practice subjects such as you or me numerically, experimentally and so objectively. We become an object of research. Does 'objective' empathise? One observes but not befriends, analyses but not sympathises, supervises from a touchline but does not enter in the game. Sociology would make 'a science of society'. But aren't reporters simply hangers-on? Who is it makes the running? From relative perspectives sociologists observe traditional practice, systems of belief and consequent behaviours but they describe no absolute morality. Since materialism's tenet has excised the metaphysical from any sense of its reality how could their stance be otherwise? The sanction of religion with its systems of rewards and penalties is still permitted in the role of social tool and useful glue; but its naïve 'suspensions', demoting rationality in favour of belief, are simply mental 'supernatant' you could ditch. Theism lacks fundamentalism of the all-is-matter species therefore its logic can't be right! Why on earth, if they've evolved and there's no God, do men believe in Him? What survival value has this figment of imagination? Why in heaven's name did natural selection bless this accident of nerves? Indeed, how long before taxpayer's money is invested in research to find how Deity promotes a social structure? How long, if *PAM* is wrong, before the cash is wasted on a 'scientific study' of why faith in Figment has evolved? Or is squandered seeking a genetic answer for Divinity? Garbage in, garbage out. If you base your questions on false premise then, as top researchers know, the answers are unlikely to be right. No matter. If the premise is our *PAM* with *PCM* can anybody rational disagree? Funds therefore rubber-stamped!

Rather than the Dialectic's emphasis on underlying similarities and, at root, its hub of unity, an anthropologist will tend to emphasise variety. Jung and others once identified a deeper, archetypal theme but normally beliefs and their attendant ceremonies are classified as relative and sometimes contradictory relics of traditional, pre-scientific cultures. *Science, it is argued, not religion is the only kind of knowledge that will unify mankind creating international community. After all, what's faith but flight from matter to imagination, to pious and misguided hope? It is simply refuge for the weak, a loser's lapse from life's sharp edge - the Great White Lie.*. Why should life's meaning, if there's any, derive from a delusionary relationship with the illusory Creator of an ordered universe. Leave everything to chance and chance's order - natural, material law. Then evolutionary psychology can spin its caveman yarns about the way you live today. Ethology, ethnology and social anthropology can line up unopposed to patronise pre-scientific attitudes and at the same time demonstrate that mankind simply lives a complex form of simian interaction. Is monkey not, all said, man-in-the-making; and monkey business just a cruder form of man's? If once the future's men were apes could study of their troupes let slip the answer to a mystery of evolution? Might the psychology of primates (apes and not archbishops) eventually reveal the way that, from inside your brain by way of proteins, God Himself mysteriously, genetically evolved?

Brain, mind and psychology are products of a competition at the heart of which vie chemical complexities called genes. These only issue 'self-indulgent' orders. Since cultural phenomena are only understood in terms of social psychology, and that psychology in turn in terms of evolution of the genes, then the twin expertise of evolutionary genetics and psychology must be able to explain all animal behaviour. Aren't you an animal yourself? The scientific guru's explanation, based on self- centred competition and survival of the fittest, emphasises physiology. It places healthy, fertile specimens at the heart of how it sizes up humanity.

Survivors might be physically tiger-tough but what sort of social constitution might gymnasia or army barracks frame? Is it one with laws that favour fitter, stronger individuals in super-groups or über-races? Is its idea of 'progress' one that gambles on genetic engineering leading to a superhuman future? In this case a certain group of ghosts float up - racism, National Socialism and eugenics are but three. Or will it frame such rules as conjure up another kind? Will it emerge as socialistic Communism, an atheistic hierarchy with a god whose omnipresent generality is 'State'? Whose embodied son of god is president and whose cabinet of angels sing to hymn-sheets in a special office called 'praesidium'? Clustered in that 'heaven' you will find a clique whose evolutionary logic religiously eliminates all trace of antithetical ideas from entering its sphere of influence. Its bureaux extirpate with slogans, thumping rhetoric, reformatory camps and fearful death all 'splittist demons' of dissent. Who still lacks faith in mobile, relative, utopian morality and will not toe its party lines? Certify such enemies of state insane. Since mind is brain and brain is neurochemistry, why not enforce prescription of 'appropriate' mindbenders? Why not create a cyborg population pill-pressed into work-gangs slaving to translate dear leader's ever-changing revelations into cement monstrosities and monuments of steel? *If you think I exaggerate then think again.* As history lays bare, the angels of technology and heavens of utopia are spattered with the blood of innocents. Zealotry of theocratic or of atheocratic sort are, politically, the same. Mammon's politics are earth-bound. Earthly power's concretion drives a coach and horses through the life of man and nature. Its thoroughfare is paved with bones of double-dealing and hypocrisy; it always leads to many individual hells. What might the intrinsic selfishness of evolutionary philosophy deliver in the place of communism, fascism and politicised religious zealotry? How might the religion we call scientism moderate its blacker spots? Indeed, bound by body here on earth in which direction might a centaur turn?

Nuclear Religion

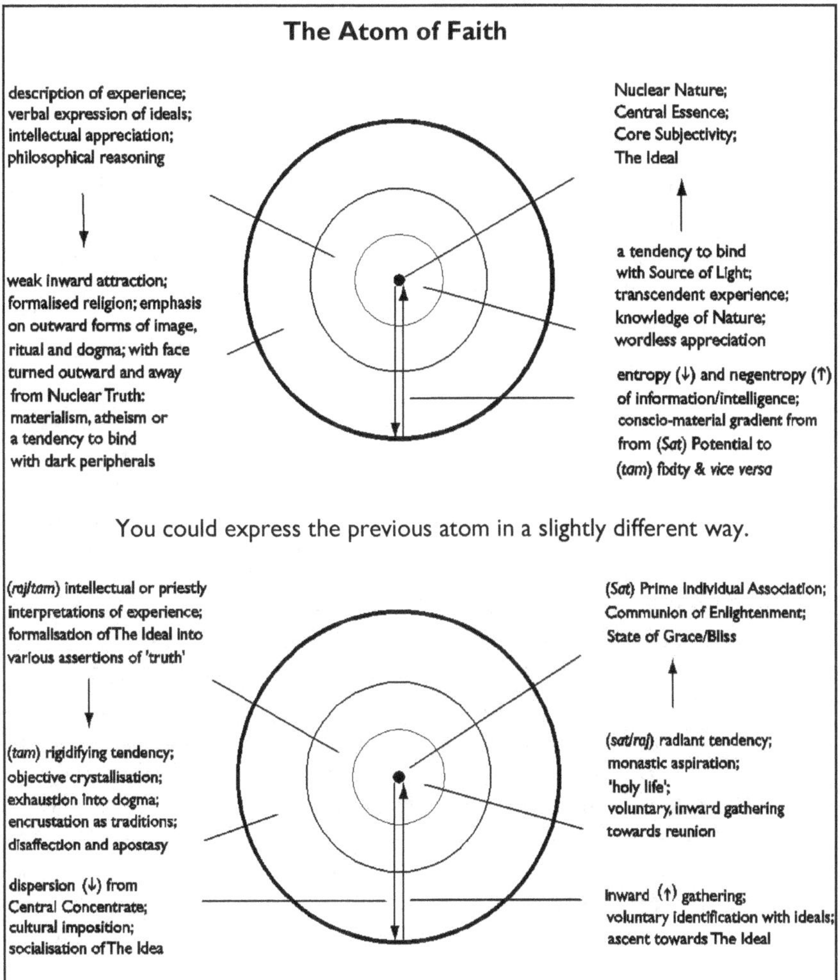

fig. 27.4

	tam/ raj	*Sat*
	stages 2 and 3	*Stage 1*
	partial truth	*Truth*
	shells	*Nucleus*
	world-tree	*Seed*
	range of experiences	*Experience*
↓	*tam*	*raj* ↑
	stage 3	*stage 2*
	outer	*inner*
	lifeless	*lively*
	aged	*youthful/ growing*
	encumbered	*unencumbered*
	ritualistic	*enthusiastic*
	dogmatic	*questioning*
	orthodox	*iconoclastic*
	encrusted	*fluid/ radiant*
	formal	*informal*
	authority by fear	*authority from respect*
	conversion by force	*conversion by example*
	punitive	*educative*

Nuclei cellular and psychological - is it *DNA* or quintessential mind, the soul, from which morality derives?

For materialism there is only one way. There is no direction but its own wherein all life is evanescent and thought dances, like will o' the wisp, inexplicably upon the waters of oblivion. If this is hopeless and a counsel of despair grow up, the creed exclaims, grow strong because that's just the way harsh, hopeless cosmos is.

If, on the other hand, you call Essence, in the terminology of physics, Symmetry, then forms of different religions break in localized expressions of its Law. Law is Symmetry. Its archetype is from all angles at all times and places just the same. The Principle of Nature's Nothingness, the (N)one, emerges with a real and central meaning. If religion naturally exhausts in scattering from its sacred core into profanities; if like ivy repetition sucks and dries vitality; and if tradition suffocates its shoot then of what nature is the quintessential symmetry? What is the sort of nucleus that seeds varieties of devotee; whence develop creeds and cultures as religious progeny?

Glance at *fig.* 0.4 to revise the notion of a three-tiered universe; and *fig.* 5.3 for the hierarchical order of creation. The order of materialisation runs from (*sat*) potential through (*raj*) kinetic to a (*tam*) exhausted or fixed phase. Religious practice also falls from Source Experience through these phases. This, entropy of vital information, is the way that heaven's brought to earth.

This perspective treats faith from subjective to objective phases. Therefore its 'sociology' treats individuals and society subjectively as well as in objective mode. It engages flux as well as flux's public, formalised precipitate; it takes the heat of nuclear ideal. What predetermines passive structure is that prior ideal. Round inward principle the outward practice of a crystal grows; round mind the body's business circulates.

Religion's core is its Ideal, Transcendence and Top Teleology. *Its Great White Truth is, the reverse of flight from matter's 'cold and hard reality', an eminently rational undertaking to commune with a Subjective Friend and thus discover nature's Natural Heart.* One process, many snowflakes; one Natural Axis, many faiths. Despite their imperfections, quirks and quaintness world faiths all promote the Immaterial Centre that is Life.

Therefore religion and, derived from this centrality, politics and law are not just players in an economic game - they are its subjective half. They are not detached but centrally and actively involved promoting health both psychological and physical. Nor do they just analyse or criticise but make the rules and referee. They combine rational government with actual practice on the vital field of play. This field might include both town and country but is mainly in the mind.

Of the three the middle man is politics. It forms a bridge between the 'outer body' of economy and 'inner principles' of higher up philosophy. It is guided by its ideal way to play the game - by rules dependent on morality. If the arms and legs of politics are its administration and the agencies of law then is the head not source of its ideals; and the nuclear heart of its religion your third eye? **All creeds turn upon the axis of their soul.** The kernel of theistic faith transcends the mind. What is materialism's soul? In the case of secular belief with non-existent soul the axis of its mind is, if not logically physical oblivion, then reason. When is reason Reason? Is there perfect reason or is intellectual reason, based on the relativity of different mental frames of reference, doomed to flaw? Is the metaphysic of the Dialectic ever, for materialism, reasonable? Whichever way, all religions want to take first place; they want to claim your heart. Take care, the mystic warns, and keep but also rise above them.

Which came first, man or nature? Man-made religion or the subject it describes? It is hoped that Natural Dialectic can relate perennial philosophy, an eternal *dharma*, an accurate expression of the cosmic law. Such absolute includes an absolute morality, a transcendent framework into which the different expressions of religion and their social logic may be plugged - because it is their natural source. The central thesis of this book is to point out and explain the natural hierarchy, the conscio-material gradient issuing as existence from this Source of Concentrated Consciousness. It hopes, in this instance, to make clear the line that falls from Revelation through its explanation by religion into social tracks. You drop from principle to mundane, detailed practice; you drop through politics to its enforcement (law). How important is an understanding of this order of creation? The fact is neither kings nor major-generals nor inventors are as much remembered as a few soul-scientists, mystics, saints or prophets such as Christ, the Buddha, Moses or Mohammed. These few embody the real waves that daily flood through human lives. Like it or not, neither science, politics nor even law but religion based on revelation governs human history! And the structure of the cosmos predicates, despite an atheist's objections, that it always will.

Personal enlightenment rests at the apex of Mount Universe. Social adoption of its nuclear wisdom is, lower down the slopes of mind, another thing. It is, through all the layers of its comprehension, called religion. Derived from the heart of being such religious layers, rightly understood, are like steps placed at the centre of community and rising towards divinity; higher layers represent intense devotion but the lower show more physical, less metaphysical a comprehension. The bottom few are cold, hard and dogmatic stone. Lower, cultural and ritualised expressions divide men more than they unite; but from less material, higher steps states of increasing union on principle and by agreement rise.

From Nuclear Essence germinates the world-seed. *It formulates society the same as cosmos; religion drops through the same three phases of creation as the universe.* And its large-scale cycle of development involves the same three stages as, in miniature, your own.

Check back to the previous stack. First, born of contemplation, comes (*sat*) revelation, inspiration and idea. It is an Immaterial yet Most Important Moment. The innermost enlightenment of mystics, philosophers and founding fathers drives the process. There follows a more 'outgoing' and enthusiastic, 'youthful' phase consisting of (*raj*) communication, fellowship, debate, conversion and perhaps even revolution. In the course of time this 'fluid prime of life' congeals into (*tam*) dogma, ritual and traditional, authoritarian hierarchies born of age. It ages, simply, into an external husk, an institutionalised shell, a shadow of its former glory. The fresh, original fruit is now inclined to rot. Original inspiration has (check Chapter 0: Formalisation) concretised; the slow and wizened hand of orthodoxy now exerts control. Is the body of old age not rigid and inflexible? Religions die but their perennial heart lives on; and from this heart new generations of the old faiths always spring.

Each phase of religious life is one of emphasis and all three can, in practice, co-exist. For example, faith may have its mystics, monasteries and missions travelling on the same road as their 'heavier', orthodox companions, established churches. Political and philosophical creeds follow, in a more secularised way, the same pattern. So do individuals. They vary as to the different 'stage' of an organisation with which they can best identify; and, of course, although their dynamic is different, each stage is reflected in the others. For example, do you remember the Holy Roman Emperor Constantine VII's definition of ceremony as 'the outward form of inward harmony'?

Enlightenment itself is in dialectical terms the culmination of ascending the (*raj*) right-hand path. Just as a brilliant human transcends the mind of a tortoise, so its freedom and clarity transcend the slow and partial shadows of our understanding. Against its Excellent Norm moral feedback occurs and the 'right' path is steered. Indeed, religious regulation is based on such high principles and ideals. The Good Idea includes absolute morality and, within it, clear distinction between 'light' and 'dark', decisive discrimination between 'right' and 'wrong'. There is no chance for moral relativity to blur each line or, at the nether pole, a moral vacuum cause collapse. Who does not understand that it is sensible aspiring to the highest happiness, right to sympathise with harmony and to translate ideals born of Ideal into our own and other lives? One might naturally, therefore, want to spread the message of Experience, Wisdom and Enlightenment.

Sacred Structure: Ziggurats of Man

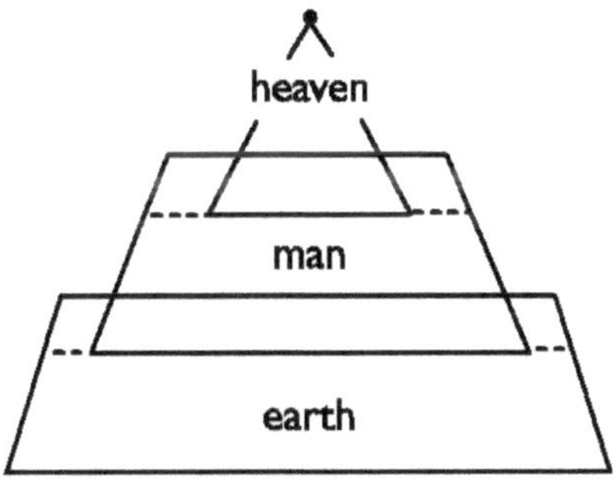

Worldwide and throughout history man, un-blinded by materialism, has used architectural symbol to represent his aspiration to and place in truth. Sacred shapes of pyramid, ziggurat, stupa and temple mark a ladder stretching up from earth to heaven. Minaret and steeple indicate direction; they climax in the Pinnacle of Verity. Step, vector and intention to obtain its Central Peak always lead you up Mount Universe.

fig. 27.5

The religious symbolism of cosmic structure is ubiquitous. Light, fire, bells, gongs and music signify a radiant energy. There flows from the central point of inmost altar, where offering is made and answer sought, a sacred, hidden source of order. From this essential holiness on high devolves the whole peripheral architecture of a shared existence. Church towers, for example, dominate the land wherever possible from raised ground. Each spires heavenward from the midst of its community. And does not the altar face first light? It faces dawn. And the whole construction spires; it arches up towards heaven and embodies (as shrines do) the principles, first and foremost moral principles, of the universe that we inhabit. Religious structures (conspicuous even by their absence in a gap of secularity) speak for the moral mind of people. They are, despite tension between the exhilaration of their heart and hardening of their institutional arteries, central to society and its best governance.

Domes, arches, apogees and focal points like altars aim devotion towards divinity. Spires and minarets will shoot you up. Non-semitic societies also organise around reflection of the cosmic pattern. The most common is a hierarchy according, each with its own cultural rendering, to the cosmological principles of creation. Do you remember the dialectical models (Chapter 0) - concentric sphere, ziggurat and fundamental harmony? Humanity across the earth, historically and now, is crying out the theme. What is a pyramid if not an elevator to the stars; and ziggurat a ladder earth to heaven? Turn east towards temple-mountains. Observe how the pyramidal structures of Angkor Wat reach to a central, high and holiest hub. Ask any Hindu temple, Buddhist *stupa* or Taoist shrine what, by the arrangement of its being, it is trying to say. Move west past every pyramidal complex from the Asian to those, such as Teotihuacan (the pyramid of the sun), densely populating Southern and Central America. What about Etemenanki,, the temple of the foundation of heaven and earth, the great ziggurat of Marduk at Babylon; or the Great Pyramids of Egypt? These were (*raj*) resurrection machines, divine escalators, launch pads to a cosmological axis of the universe - an axis identified as an unmoving star around which all other constellations of creation swing and called by science Sirius. From earthly, pyramidal tomb to heavenly Apex; from the inner chamber Pharaoh's soul and, through him, the souls of all Egyptian 'good and true' were projected after death towards cosmic centrality, the point of origin and thereby, at least symbolically, eternal life. A grand picture overtook reality. Now concentric stone circles and the tiered, nested (though not pyramidal) structure of the Jewish temple in Jerusalem. Only Jehovah's linkman, the high priest, could penetrate to the innermost holiest of holies; he would not discover any idol but, imageless and void, an image of Essential Nothingness. It was as heavy as light with dialectical symbolism - a silent, empty and yet crucial, all-funding space. Islam's axial Ka'bah in Mecca encrypts the same message and so does a 'heavenly' chancel as opposed to the 'earthly' nave in a church. **But aren't you, fabulous in beauty and complexity as Universal, Pyramidal Man (*figs.* 17.9-11), a temple of the living God? In which the real holy of holies is - unseen, closer than brain, breathing or the fraction of a second - a single central eye? Mind's eye looks down upon a world of soul-less things but, swivelled upward towards the Inward Apex, seeks illumination's jewel. You step up; and, in communion with Essential Soul, are transformed into supreme and living paradox.**

Kings, Pharaohs and the Chinese Emperors mediated between metaphysical 'sky' and physical earth. They were a symbol of the Axis, a reflection of Highest Authority. Through their medium the order of heaven was transmitted to the ground. Concentric rings of the Heaven-Worshipping Altar at Tian Tan (Temple of Heaven) are a good example. Creationist tradition has certainly, until 1911, always dominated China. The Confucian creator, Shang-Di or Heaven, was worshipped by the emperor at Mount Tai in Shandong province until the altar was removed, in the 15th century, to Peking. Hence the divine regent upheld the Mandate of Heaven, Divine Will or, in Christian terms, *Logos*. Acting as a conductor between creation's main domains, he received and transmitted from a central position by standing on the 'Heavenly Heart Slab'. The slab is, in terms of Natural Dialectic, the Central Source of radiant energy or Apex of the cosmic ziggurat (*fig.* 0.5).

The belly-button (or *omphalos*) of the classical universe was, not as a specific stone but as an idea, removed from Delphi and replaced at the terrestrial heart of Christian faith in Justinian's Church of Holy Wisdom in Constantinople. You can still stand on the spot where, until 1453, the Holy Roman Empire's coronations took place. You stand at the centre of an *omphaleion* where, over centuries, emperors were initiated into 'linkage' with divinity. Through such linkage they embodied Christian faith on earth; and then, of course, there is the papacy. Such authorities were, in consequence, axial. Power and influence derived from divinity; the ruler was, symbolically, a steward for ideal morality; his regency played out in tune with cosmic order. Civilisation has, from the first until this day, derived its order from Ideal and, mostly, striven to embody it in cities that reflect God here on earth. Take the British Constitution. Now much secularised, its 'set' was built around a core of Christian values represented by the Church. Visit the fine Cosmati pavement, commissioned by Henry III, in front of God's high altar at Westminster Abbey - the seat of English power. This symbolic cosmological axis is a circle, like an *omphaleion*, set at the centre of a monarch's sphere of influence. At this conducting point king met with his Creator; at this heart coronations did and do occur. An occidental regent, in creation's hierarchy, took anointed place. A monarch was, as other satellites of solar Rome, effectively a papal deputy. *Fid. Def.* He was 'defender of the faith'. This was theoretical theocracy which politics have forged into 'symbolic monarchy' and 'actual democracy'. Did not Bishop St. Remi baptise King Clovis in Reims where coronations marking Frankish kingship in the name of God were henceforward held? Nor, deliberately, is America a theocracy but do you remember the plea of George Washington (Chapter 5) 'Almighty God, we make our earnest prayer that thou wilt keep the US in thy Holy Protection...'? Secularism's hierarchy is, however, flat. Is flatness, in the cosmic make-up, actually the case?

Corruption: Fall

A loving father cursed his children. He punished both severely. This, given that their crime was sexual desire implanted in their bodies for survival, seems perverse as well as harsh. Did 'natural sinfulness' begin in this unnatural, strange, cruel way?

Or was the (↓) fall built into physicality and thus innate in bodies biological? The natural tendency of entropy - informative and energetic - is incorporated into every corpus and corpuscle of this starry universe; creation's flow's a downward tide whose shadow deepens into conflicts of materiality; and death (if that is what you call non-life) is intrinsic to non-conscious energy. So that the father, knowing possibilities the physical creation might accrue, gave warning to his children when the time was ripe. 'Drift into the vortex of desire and risk suffering; if you become entangled in the nexus of material surroundings you lose paradise. Be not tempted. Do not go. Fall is not inevitable if only you will choose, before all else, to turn to me'.

↓	*tam*	*raj*	↑
	drop/ entropy	*negentropy/ rise*	
	entanglement/ undertow	*release/ freedom*	
	external pressure	*internal relief*	
	passion	*devotion*	
	recklessness	*duty/ responsibility*	
	vice	*virtue*	
	corruption	*purity/ impeachability*	
	ego/ self	*humility/ selflessness*	
	antagonism/ strife	*cooperation/ community*	
	low state of mind	*high state of mind*	

This applies to every Eve and Adam, that is, to each embodied one of us. We are neither saints nor deities. What, though, about the children in whose hands is wielded earthly power? Neither kings nor even popes are judged by history infallible; and whether their philosophical bent is religious or secular politicians are not, in the main, particularly wise or incorruptible. They are not even, mostly, holy men. They misbehave. Their power is like a lens through which, as well as nobility, human frailty is magnified. This has led to constitutions overtaking them. For example, Louis XVI was beheaded in the French Revolution; and the last Caesar (Tsar Nicholas II) and jaded Chinese emperor (Puyi) ceded in different ways to the violence of Marxism. Having suffered the corruption of metaphysics the state replaced it with the physics of materialism, humanism and scientific atheism. It swapped high priests for commissars, one bible and kind of religion for another - a specifically objective, soul-less one.

Was the swap worthwhile? Of course, not every atheist extends towards political extreme. The corridors of academe resound with 'rationality'. Decent fellows see the social value of a Christian precept here and there. Why not solve the 'faith dilemma' simply by committing intellectual patricide? Why not kill the Father off, delete all reference to Subjective Heart and, after verbal surgery, inject attenuated formulations as the moral vaccination of a humanistic, fresh way forward? Then let socio-biology invade the pulpit, let evolutionary psychology step to the lectern or the lecture dais. It is doubtful whether God's ghost-rewriters could or would be charged in any local court with an infringement of His copyright; or whether nature's copyists have broken any patents on the world. Moreover if you only neutralised (that is, annihilated) all religion you might pinch off an aggressive streak - unless man's violent sinfulness is, as opposed to a religious appeal to peace, what stokes the awful fact of war. At least, a secular detachment and fair play on neutral ground amounts to tolerance - a virtue always worth returning to. Such virtuous detachment is, obviating theocratic or atheocratic domination, what the constitutional separation of state and religion is (or should be) about; but without a moral vacuum, please!

Talking of aggression there exists (Chapter 23) an archetypal difference yet a balance held between the sexes. Male and female each have strengths and weaknesses. But since muscularity is male and, typically, fathers are the heads of families, an image of The Boss is painted with testosterone. Of course, the image is essentially wrong because what transcends duality cannot be sexed; but women are more physically vulnerable and so a more 'aggressive' stereotype of dominance is reinforced. For better and for worse a male imbalance also tends to skew the course of leadership and government. Male qualities are sometimes over-valued, female under. Whichever sex 'defends the faith' it is the quality of holy regency that counts. Does the regent listen, hear and understand the orders from on high? When turning to affairs of realm does he or she rule in The Senior's name or not? As human history underlines, the business of good government depends on high ideals and moral purity; it relies, essentially, upon the mystic guidance of its mentors. It demands a faith, wise king and council.

If you aren't sure how something works then you consult an expert. Men and women (who together constitute communities) turn to saints, politicians, scientists and judicial systems for advice. They look for help from parliament, the courts and scientifically driven industry; and seek values from the 'scientific' metaphysic' called psychology, religion and at religion's nuclear heart a mystic fresh from revelation. *Metaphysical truths have traditionally been inscribed in holy books which, because they describe what is right and good, hold pride of central place in their belief system and, gathered round about it, their educational, social and political systems.*

↓	*tam*	*raj*	↑
	fall	*rise*	
	death	*ascension/ resurrection*	
	corruption	*cleansing*	
	damnation	*salvation*	
	burden/ guilt	*clear conscience*	
	sin/ wrong-doing	*redemption*	
	imprisonment	*liberation*	
	darkness	*light*	

Ideals can be corrupted; fresh fruit rotten grows. Acid is corrosive. If a sceptic, perhaps professing neuroscience or some evolutionary species of psychology, pours more than just cold water on the idea of an element of consciousness, transcendence or the immaterial rubric from which social orders were derived then he needs seriously consider viable replacement. 'Man doth not live by quantity alone'; nor again by Mammon's purity of mind, materialism, or the quality of scientism, atheism. Therefore, having burnt away subjective infrastructure using drips of clever rationality, with what new and objective theory might an infidel prevent the crumbling of ideal, collapse into dissolved morality or the angelic diktat of a bunch of commissars?

No moral vacuum, please! It is not intellect but faith drives moral expedition and a tide of 'goodness' forward. Could, therefore, a social scientist armed with sociology convince the new-age laity and lead with priest-like strength to secular Jerusalem? Or would his non-committal stance be swept and drowned as an irrelevance? Materialism's mammon and armed politics frame powerful reasons and ideals of their own. So would a 'rational republic' be, like Plato's, one of entrenched class structure, censorship and mind control? Did not just such a secular adventure, a religion by the name of communism, kill its own and kill again and fail? What if that faith's traditional mythology, scientific atheism, were as intolerant as deeply flawed?

From inner circles high priests, propagandists, teachers, journalists and scientists impress their various 'truths' upon less perspicacious crowds. They patronise and sometimes bully lesser men. They may corrupt their own ideals but what is the alternative? What is leadership but practical advice in active, living form? Individuals crave information and advice by cue. These cues invisibly reflect the mental shape of a society. The template of this shape is principles and high ideals; patterns of behaviour stem from the acceptance or decline of these. The whole business stems from how you think; its issues from an immaterial action zone. *Therefore what, essentially, is society but behaviours dependent on invisible containment by ideals? What is civilisation in its high, great form but a world of action orderly suspended underneath the immaterial crystal of a Great Ideal?*

Man is driven through all sorts of error on a quest for truths and an alignment with ideals. Do you blame a school for wayward pupils, law for criminality or standards for behaviour lacking them? *Religion, at nuclear core if not peripheral zeal, springs from the best motives you can find. It has been, is and will be the foundation of community. Does it not speak of overriding unity, unification and, in practice, a relationship that's universally tied up with friendship and obedient to love?*

Do you, therefore, blame the water for its dirty, rusty pot? And for the germs of human violence that, proliferating with excuse, infect a sick and suffering world? Fig-leaves of 'religiosity' barely cover cynical abuse. 'Obedience by fear', the zealot and crusader cry. 'Only I am right', the socialist, theocracy and socialism, each terrifying their community, rasp. Naked aggression, rapine and oppression of a holy war all strip the leaf away. Wars of idealistic and manipulative dogma have perhaps caused greater suffering down man's history than any other kind. What goes wrong? How is enlightenment so corrupted into darkness? How does ideal in action come so far adrift?

Just as free gas, when energy is lost, becomes a solid so the Central Truths become encrusted, ritualised and, as a fruit that's sucked continually, lose all their juice. Initial wisdom is repetitively learnt by minds aspiring but whose aspiration isn't yet achieved. Various intellectual levels need to understand the principles and thus their promulgation has to 'harden' for the young and simpler-minded into childish models we call parables or stories. A catalogue of stories and a self-consistent, logical philosophy are needed to explain; sharp definition in the form of canon, creed and orthodoxy is a necessary, stabilising distillation of the Wisdom of Original Experience. By this stage in descent the inner truth has been inverted. It has been set in style. People start to take the form for substance and mistake descriptions for reality. Inner communion has been concretised into the outer dogma of community. Solidification marks a Dialectic fall. It marks a loss of fluid energy. Stage 3. Whereas gas cannot crack, a 'solid' dogma can.

No doubt Plato's hierarchical republic was classified beneath a Saintly King. Wisdom's paragon reflected cosmic order and ideal throughout the working of its lower ranks. Actually, however, saints eschew the turbulence of mammon; those of lower impulse tend to seek the throne. Such kings crack. Flesh is weak. Frail humans fail to live up to ideals. Who is not inconstant, inconsistent with his creed? Presidents and ministers succumb; they fall to temptation and are seduced by passions that the powers of leadership ignite. 'Power corrupts and absolute power absolutely corrupts'. History is storm-shadowed by the moral failures of its leaders; suffering is the shadow and the testament that, trailing swathes of gravestones, marks each time and place. As in a personal case of moral dereliction, so a premier with corrupted state of mind corrupts his state. With its leader fallen sick a kingdom swiftly follows down. At every level frailty derails ideals. Only 'the dream', only aspirations and the form of metaphysic known as abstract principle can moderate incessant bickering and spasmodic violence that, as opposites, define the paradox of centaur - that is, human - politics. But are ideals and their Ideal an abstract fantasy you waste aspiring effort on? They are not scientific but still, aren't they really facts that issue from the quintessential nature of The Most Important, Immaterial Fact?

Top ideal declines to nether practice, truth hardens and the cracks appear. Rivalries, divisions and dissent erupt. People fight for their entrenched and diverse convictions clasped together in a fractured world-view called religion. Such circumstance is an inversion of where things started out; it is personal enlightenment's reverse. At this juncture history's index finger wags and warns. *The institutions of either religious or secular government, once encrusted in dogma or degraded by vice, tend to emphasise difference, isolate non-*

conformity, aim an intellectual or an actual weapon and thereby campaign offensively, oppressively against 'the opposition' to their way of thinking. When ideals become corrupted as deep down as tyranny the worst side flares. With lethal twist God devil made. Armour-plated dogma cuffs or otherwise eradicates dissent; vicious crime and satanism are, dark peripherals, hell-bent on upset and destruction. History's finger singles out some crooked thugs. Socialism, Christianity, Islam and other canons sprout infected, fungal undersides.

Islam and Christianity make spiritual claim; many Socialistic species swirl denial's flag. But if there's no spirit to be willing what is left? Who cares? No purpose, no design, there exists no moral in the pitiless and blind abyss of matter. Is just this the basis of an atheist's creed? With relative morality what's 'right' or 'wrong'? What changes as you think is whimsical, unstable and, in subversion's hands, a danger 'nice' men shudder to behold. And yet whenever it's decided what is right, there's 'not-it', which is wrong. And wrongs need policing; watch out for the menace of a brainwashed agent of the law. Individuals and societies perforce succumb. Not everybody's always getting better. Hopeless optimism out of armchair humanism doesn't turn out, history shows, so understanding, merciful or kind. 'I don't want to be controlled; I want to be uncontrolled' cries anarchy. 'And out of the anarchic vacuum', cries a revolution, 'I will need, I will want to take control. *My* control will be the order of the day. *My* tender love will trump The Fairy's!' Most zealots are not filled with anger, hate or devilry but it only takes a few. Check again (Is *PCM PC*?), what tends historically to happen following the suit of logic whereby (*ex-PAM* and *PCM*) the slide of moral relativity becomes politically correct. Aren't secret police and *gulags* dark disaster anywhere?

If it ends materialism no doubt Natural Dialectic kills atheistic communism as a rational creed as well. Might rationalism viscerally retort or un-detached emotion spit invective back? What irrational kind of censor confiscates the Dialectic's passport or would lock the bearer up? The checkpoint asks, 'Is *PAM* true?' The password that's politically correct is 'yes'.

Dark's sole relief is light. Immaterial light alone, a rational or supra-rational beacon of ideals, draws adult children out of their morass. It is a pity but they even fight about their lights. 'I'll make the others understand that mine is the best! I will replace their error with my ways!' Discord is a symptom of the fall from grace.

If you misunderstand something does day disappear? Does night's sable fall on your mistakes? Sun shines on crime, cruelty and pain. Night at noon: there can be darkness at high noon because noon's brilliance is not truth's; and evil's not dependent on the colour of the sky. Not physical but metaphysical obscurity is in proportion to one's misalignment from the Truth. Is that plain enough? Materialism would exclude the immaterial and yet the latter's strength is all that animates oblivion and then leads human animation out of ignorance and man-made hells.

Therefore which pillar of men's faith, atheism or theism, shines with the light of Day? 'Behold the night is over, day has dawned anew', pronounced Pope John Paul II on visiting Czechoslovakia in 1990. Had not this 'figure of Christ', with epic symbolism, faced down the powers of darkness? He survived a bullet shot, as investigations indicate, point blank from the murk of anti-Christian politic. Then, having kissed the soil of home, this Polish priest struck fire that melted communism's binding ring. As shock waves spread through Eastern Europe, Russia and the Balkans he cast out the anti-Christ. In this titanic, mythical, heroic struggle darkness was dispersed, light thickened and An Ideal phoenix-like took wing. With classic strength of immateriality the resurrection lifted off materialism's block; a weight was dropped, a ghost was slain. Dawn. Spring. Life has risen over death. It was as if grass shoots were springing up through crumbling concrete that for years entombed the body of his flock. Science has a role to play but not a one like this. So was the ideal-driven drama rationally irrational or really unreal? It was definitely not a scientific enterprise.

Vibration. Cycle. Fall and rise. From corruption's dip and depths of suffering you swing into a long-awaited rise again. Life jibs the shackles of materiality; it wants to dance; it wants fresh synthesis and sparkling resolution of the truth. But even fountains one day falter, fail and die away. The waters dry, youth wizens and the wintry tide of age rolls over one last time. As wheels within great wheels or generations in a dynasty the turns of power and influence spin. Sects, schisms, period in office, empires and religions come and go but the Axis of The Great Wheel stays the same. At religion's heart The Sacred Heart, The Nuclear Power, The Natural Truth is no more clerically religious than the daylight in your eyes. And it is incorruptible.

Politics

	tam/ raj	*Sat*	
	disequilibrium/ wobble	*Balance*	
	tendency/ bias	*Compromise*	
	feedback/ change	*Homeostasis*	
↓	*tam*	*raj*	↑
	negativity	*positivity*	
	isolation	*community*	
	oppression	*liberation*	
	state	*individual*	
	individual	*state*	

(*Raj*) action, business, economy; seething crowds, cross-currents of humanity; turbulence of social ocean, individuals dealing with each other in a way that keeps (or fails to keep) the peace. What is a crowd but a community, communities a town and towns writ large across a wedge of country? Individuals in society, men in a collective state, states dealing with each other in a way that keeps (or fails to keep) the peace - this is the cauldron, politics, into which birth throws us. How relationships are worked is politics; and, at individual or large-scale politics shows 'face'; it shows moral, intellectual and emotional kinds of quality. Thus its character, changing by the hour or year, exhibits positive or negative exertion; it sways, according to material condition, on a scale of mood. As position on such scale, so positions individuals and states adopt can oscillate. You rate humanity by place on scales of power, knowledge and authority. These aren't the same yet each is measured by position and its vector, up or downwards, on a dialectic scale. Vector implies, away or towards ideal, direction of a moral quality. Such quality is natural when it comes to mind.

Ideals are abstract attractors drawing us to natural, so-called 'karmic' consequence. In principle, a stack's a sliding rule of thumb; it indicates direction of 'best practice' and helps formulate a healthy politic for life.

	tam/ raj	*Sat*
	motion	*Peace*
	business	*Informative Matrix*
	oscillation	*Norm/ Standard*
	legalities	*Order/ Law*
↓	*tam*	*raj* ↑
	according execution	*active planning*
	procedure	*principle*
	instructions	*instructor*
	object of legislation	*legislator*
	community	*government*
	worker	*director*
	pupil	*teacher*

Of course, business with the world, the body's business to survive and, surviving, to enjoy, is framed four-square by purpose. We've dealt with the (*sat*) nuclear principles that, derived from an enlightened state of mind, frame a belief system, outlook and the constitution of a people; we now turn to (*raj*) the business. of politics. From abstract to material imperative we move from moral absolutes to their interpretation and expression in the market-place. Here, in the scales of business, when intentions disagree, how shall resolution and thus peace be bartered? By cunning, compromise or force? Thus emerge channels of diplomacy, ways the body of a crowd can speak and, at the highest, formal level, the establishment of offices of state.

Who can do without it even if he wants to? Government is here. Between ideals and their frustration strong men, given power or having taken it, must speak for masses and thus lead the way. The principles of government involve morality, morality involves religion and so any government, secular or otherwise, partakes of cult and creed. Call it party ethos, policy or manifesto; then, sporting banners or rosettes, whip up the sentiment to haul your wagon into power. The axle creaks with half-oiled sentiments, it lurches with mobile 'morality' and rolls with a pragmatic faith. To what extent, in any given governmental case, is ox-power of its policies yoked to the traction of its holy cow Ideal?

↓	*discord*	*harmony* ↑
	war	*peace*
	crime	*right-living*
	bankruptcy	*credit*
	dishonesty	*honesty*
	ignorance	*knowledge*
	disease	*health*
	evil	*goodness*

It is the business of politics to buffer the effects of an imperfect world; and as a homeostatic agency with feedback keep the peace, maintain the health and nurture wealth of its community. Mind guides matter through its aspirations and ideals.

At the heart of any business, even cosmos and the dialectic that describes it, rests a point of balance round whose comfort zone there swings dynamic equilibrium. The logic indicates that any body's health relaxes by avoidance of extremes, that is, in the cool pursuit of light, poised and middle ways. **Equilibration is the basis of its politic**. As biological so legal and political dynamic; the purpose of a government is steering towards its stated goals. Are these not in a human nutshell education, balance and, providing for a healthy body, keeping healthy peace? *Its purpose is to protect its subjects' sense of equilibrium by maintenance of customary law and order in its realm.* The principles that generated custom and framed law are what precede best practice. These are what philosophers and mystics care about. Who thinks immaterial ideals don't lead? Establish first-rate principles and all the complicated practices fall orderly within their field. What is immaterial rules - magnetic immateriality that counts no money, ideal that's drawn to, pattern of behaviour that is valued as the source of all that's well-to-do. In socio-economic terms economy is valued physically by things that money buys; society is valued psychologically in terms of happiness. Does not the rational pursuit of happiness involve well-chosen targets called ideals? What sorts of standard is it best to set? So that, in circling round, we find the oil of politics is principle of various viscosities. *The health of the body politic depends on immaterial strings, on standards metaphysical called principles. What else have mystics ever claimed?*

Good news! What, though, in our world, about polar reality? Where there's choice there ever lurks the danger of a flaring spot of bother escalating to destructive negativity. In order riot, in peace passion, in politeness dwells potential for rude anarchy. Offence, defence, protection and assertion - there is a universal pattern that, throughout creation, expresses these dynamic principles. Natural Dialectic is one way that to-fro polarity and swings of fortune are explained. There is, therefore, a cosmic rationale with which to nurture health in a community. There is a natural way to exercise the body politic and keep its peace. Such order is ingrained in cosmos. Take, of coordinates (*fig.* 0.7iii), the informative; then run your thinking up the (*raj*) right-hand path of Secondary Dialectic's stacks towards (*Sat*) Primary Ideals. This sliding rule is not for case-specific consultation but it tallies generality; accentuate the positive, eliminate the negative; who doesn't know this helps evolve an ethos and provides a rowlock onto which you latch and lever forward?

↓			↑
	individual apart	*individual a part*	
	isolation	*interaction*	
	alienation	*relationship*	
	exile	*society*	
	alone	*a friend*	

Individuals are not robots. What might make them march in step? What unites armadas if each captain has a different plan? A major source of social oscillation wobbles in between the rights of individuals and those individuals taken as a whole, that is, society. This quivering tension always exercises every government. How do you balance personal freedom/ social obligation, rights/ responsibilities and individual/ social moralities? How do you maximise cohesion with the winners and at the same time minimise disruption from the losers in the compromises that you have to make? And then, on both sides, how does one ensure the communal equation is upheld? If you think this kind of wrangling is mathematical then think again. Are politicians ever short of verbiage in their debate? Do you value anybody's character in SI units or against embraced ideals? How do you sum a moral situation up with scientific graphs or operate with military objectiveness when lives and feelings are at stake? A human 'object' trusts, purposes and incorporates ideals - but the program is subjective. How is a robot programmed to believe and thus, in principle, have faith? If you are simply such a robot name your factory!

If the purpose of government is to keep peace and expedite prosperity a keystone of its moral architecture must be care of individual beneficiaries and, if victimised by criminals, protection of its innocents. How, in the playground of community, do you protect and at the same time break up fights? How can you unify division? You might deploy an incorruptible detachment, a kind of balance called neutrality. But isn't that just sitting on the fence? You can't have government so neutral that it wobbles indecisively. You need policies, agenda and, with action, moral tracking. You need self-policing or the force of police. So here they sally once again - the tenets of the faith!

You might argue that the closer government and individuals that constitute society adhere to The Nuclear Ideal, the healthier and more stable each of them will be. How, though, do lofty norms translate into the forum? Religious, cultural and philosophical divisions, each with norms and sub-norms, co-exist. How could one expression of the whole truth claim superiority and want, idiosyncratically, to dominate and then eliminate the others? Such messianic mental illness has inflicted governments before and, patchily, still does today. Don't you need soldiers so that, if your words and actions can't convince the infidel, fear of death or tortuous persuasion will? This is the individual's nightmare. State-sponsored terror is religion's tyranny opposing a democracy.

No man is an island. Individuals (like you and me) assume a multiplicity of roles. You might be a father, son, husband and so on; and identify and be identified with different groups. Such 'multi-mask' is representative of all of us. Is society or you the one with many faces? Is the *natural unit* of society this 'multi-man' (first person singular of a plurality); or am I merged in various communities (first person plural in each unity) so that the natural unit is a 'single group'? In all the politics of life what is the *basic* group-with-common-interest in which you lose or find yourself? This group, whence emerge your politics, is your religion. Clear or foggy, it's a fact.

Don't families engage in politics? Aren't families society writ small? If humans multiplied from an original pair (or even from a single troupe) then it is evident that everyone's related. Each of us relates to every other if you trace back far enough. This is to perceive that local, national or international groups are all, in fact, blood relatives from just one starter family. The actual family of man; but one alive so long ago that it is buried in the mausoleum of forgetfulness. Millenia are wedged between; branched is the bloodline and now irredeemably fragmented their original unity. Yet does it need a cosmic consciousness to hit upon 'the brotherhood'? How far do the brothers and their sisters see that way? In as far as children cannot you need politics to marshal, moderate and mediate the endless squabbles in the planetary playground over their desires.

There exist two kinds of stable government. *Historically the commonest is hierarchy by consent.* Such consent may take, at best, the form of coronation of benign dictatorship, right divine of its royal chief, defender of the faith and so on; or, at the worst end of the range between, negative assent to the malignant terror meted by tyrannical dictatorships. The majority of people cannot lead and would be led. As family, so politics; all kinds of parent figure come to rule a country's roost.

The (*raj* ↑) positive tendency of Natural Dialectic is towards unity. Cohesion. Such direction would support the government of difference by balance and neutrality. Although water to the wax of secular philosophy, both outlooks might in practice opt for a detached and democratic state the overriding ethic of whose government would be impartiality. Religious neutrality. Tolerance within the rule of law; intolerance would be the unacceptable ingredient. Therefore let each individual speak.

If you are serious about an individual's right to have his say you might devise a scheme whereby he spoke by ballot and elected representatives to parley in his stead. *The second form of stable government, where ballot substitutes for bloodshed and thus contributes greatly to stability, is called democracy.* The checks and balances of such stability, whose natural oscillations are non-violently, consensually contained, is a kind of middle way. The notion of dynamic balance, rhythm and, cogs within a greater cog, an operation ticking over and not breaking down are notes that strike a chord with Natural Dialectic. If illness strikes you need pre-emptive cure. Democracy enshrines at heart a mechanism for the peaceful overthrow of sickening leadership - off-colour, that is, with regard to what most people think or are persuaded that they want. However right or wrong the will of a majority might prove to be at least the individual cannot say he could or did not have his say. He casts a vote for quality of man that he prefers to lead. It's all a question, Natural Dialectic adds, of character and, in principle, the basis of that character's belief. Pedestals of faith hold politicians up.

One man or the people, through a ballot, lead. You might even judge occasions when each kind of stable government could use a little of the other. There are times, however, when the governed step outside the lines of rules laid in the social air. Outside these abstract corridors their 'bad' behaviour meets the officer and court of law. Restrictions need restraint; rules need penalties against offenders. Feeble metaphysic tempts the clamp of muscular control; if internal moral isn't working then external must arrest the immorality. Isn't best behaviour the idea? Entropy of nuclear ideals drops towards chaos, fear and war. Perilous is politics without clear purpose based on engines of high, right-hand (↑) ideals. **It always comes back to the nature of ideals**.

The Law

	tam/ raj	*Sat*
	hierarchy	*Head of State*
	consequences	*First Principle*
	effects	*First Cause*
	decisions	*Balance/ Poise*
	motion	*Axis*
	specific regulations	*Rule of Law*
↓	*tam*	*raj* ↑
	chaos/ anarchy	*law/ order*
	lower self	*higher self*
	attack/ criminality	*cooperation*
	unreliability	*reliability*
	no mutual consideration	*mutual consideration*
	moral deviation	*on track*
	social sickness	*social health*

Governing or governed. Your perspective changes with the hat you wear. Do you remember, yet again, the centaur? Animal and rational combined? Part of me wants to act exactly as I please; the other fraction - conscience, reason or the fear of punishment - says I cannot. In this case what about internal politics? Self-government. Are conscience, self-control and an internal law to be preferred? Or don't they count and you invite a visit from the officers of law? Rules leash animals with reason; they frame your practices with principle. You are constrained within a mental box of regulation; if you flout the ideals that it represents the external arm of conscience, those officers, restrain, detain and teach you with the pain of punishment. They might even lock you in a hard box called a cell. Did home and school not teach you this? As far as they did not they failed.

The movement of an arm begins with brain but movement's purpose is initiated by the mind. (*Sat*) morality, either absolute or relative, involves a definition of your norms. It decrees a 'good' manipulation of behavioural events and therefore sets the goal from which deflection needs to be corrected, error be controlled.

Having defined what is 'good' the purpose of a government is to accentuate this positive and eliminate its negative. The negative, of course, is opposition or obstruction. Is it peaceful or aggressive? Is the calamity a natural one or not? *(Raj) politics* becomes an act of government imbued with diplomatic art of compromise. It balances principle with practice, ideal with actuality, creed with daily happenings.

In a world of social change and economic unpredictability political will is expressed through negotiation and legislation. Government is nothing without rules; and rules are useless by themselves. They won't bring you order without their administration and enforcement.

And self-consistent order springs from values that it represents. What face do you want to show, what laws do you want to fix to straighten other people out? You might lionise the science class but marginalise the teaching of morality at social peril. You will surely lose your pupil's way unless you show it in yourself. Therefore, given underpinning faith, what type of discipline will you, the father of your family, dispense? Are the lines of its unseen but well-known structure crisp and clear and rational or loose and inconsistent so that boundaries are weak or dangerously disappear?

No fudging, please! You have to make it clear. *Administration and enforcement compose the (tam) visible bulwark of the invisible structure composed of governmental rules.* They maintain a vision of the way things ought to be. Whatever warps this vision, unbalances or causes sickness to a system must be treated. Administration expedites the positive ideas and enforcement keeps the rules. Or that, at least, is the idea. The analogy of citizens as cells in body politic is obvious. Where a natural catastrophe injures or a 'carcinogenic' criminal lacks self-government, the upset must be contained. The moral illness must be quashed lest it spread contaminating and corrupting all society. If individuals or groups cannot inwardly self-regulate or of themselves conform to social ideals and morality, whatever be the local quality of law and its legality, then external pressures are applied. Law keeps its decided evil balanced out. Civil order is secured and, therefore, peace kept by some force of police (to counter threats at home) and the armed forces (to counter those that stem abroad). At some stage of definition and refinement in every group humans (and some social creatures too) the obvious picture is the same. Final correction to the miscreant's immorality and deviance from the current norms is the job of a judiciary, whose courtly scales weigh innocence or guilt and, in the case of guilt, hand down a sentence and enforce the standards (norms or rules) through discipline. *What is judged 'healthy'? What is the character of government? It all depends upon initial standards, the quality of mercy and morality, both individual and socio-political. This is turn depends on faith and quality of faith. It depends on high or low ideals.*

	tam/ raj	*Sat*	
	expression	*Archetype*	
	imbalance	*Balance/ Poise*	
	enforcement/ encouragement	*Rule/ Order*	
	changing local circumstance	*Overall Stability*	
↓	*tam*	*raj*	↑
	from order (vice)	*towards order (virtue)*	
	subject of rule	*agency of rule*	
	violation	*lawful obedience*	
	disturbance	*moderation*	
	punishment	*reward*	
	force/ constraint	*freedom*	
	tit-for-tat/ revenge	*forgiveness*	
	unashamed	*penitent*	
	dogmatic inflexibility	*understanding*	
	harshness	*mercy*	
	unfairness/ partiality	*justice/ rigour*	

(*Sat*) Balance. Keeping poise and peace; upkeep of dynamic equilibrium of body (health) and mind (happy interests). The law of equilibration (*karma*) is reflected in a state's morality, that is, its laws. Are moral statements about protons? Do they address electrons or the force of gravity? Physical equations are not moral ones; if morals count science cannot count them. Morality is psychological but could 'survival of the fittest' as a motto lead to virtue? Could a panel legislate according to a naturalistic view or evolutionary psychology? Is moral relativity your guide or absolution by Constabulary?

No doubt, a criminal hangs like millstone heavily around all honest necks. Police, probation, prison, courts and rehabilitation - not just victims have to pay the bad man's tax. All innocents are burdened with the rotten, bitter but expensive fruit of immorality. What, therefore, of morality - within whose frame a citizen stays dutiful without the police? Where criminal enforcement ends you still need an internal, moral code. National laws express internal order; they express the human conscience and impose it where an individual's conscientious self-control has failed. They are framed to chop the vicious and encourage virtue; they are framed to generate stability, cohesion and to unify the individuals of society. The character of government depends upon the quality of its ideals; the priority is social stability; and in such strain the modes of its approval and correction equally depend on its philosophy and faith.

Self-Government

	tam/ raj	*Sat*
	down/ up	*Transcendence*
	slopes	*Apex*
	division	*Union*
	relative insecurity	*Trust*
	limited love	*Love*
↓	*tam*	*raj* ↑
	assive/ negative	*active/ positive*
	low state of mind	*high state of mind*
	non-idealistic	*idealistic*
	no trust in Ideal	*trust in Ideal*
	selfish	*selfless/ altruistic*
	bonded by ego/ passions	*liberated from ego*
	governed/ weak	*governing/ in control*
	from love	*towards love*
	antagonism	*cooperation*
	strife	*community*
	darkness	*radiance*
	separation	*unification*
	sorrow	*joy*
	suffering	*relief*
	enfeeblement	*fortification/ strength*
	death	*life*

No crime! No prisons, courts or punishments! Do such societies, outside monastic, still exist - perhaps in Bhutan, Hunza or remote communities - ones in which self-government eliminates the reckless, disrespectful element? Where, really, does self-government exist but mind? So crimelessness has great potential. It exists in you. And every other human if, as a child, he's taught the choice to make.

The choice is simple but its impact huge. You'll get exterior law with police and CCTV. You choose big brother or interior self-government. Governed, governing. Law is an immaterial structure imposed from outside on yourself; equally invisibly you also use the information principles endow to regulate your own response to outside factors from within. Internal politics. Is your self-regulation coded like a cell's or, like instinct, also an automatic reflex? Are genes the real, inmost self or, although your body uses preordained responses, do you also play a game of psycho-politics involving active choice?

Conscience and the bit of self-control according to some scheme of works are, critically, sorts of information your subjective actions and reactions use. In this case what on earth have genes or Generators got to do with personal politics? Surely frames of mind called attitudes are prime and prior? Such creeds, noble or ignoble, are the spring of policy and policy imparts direction. Mind-set is the main propellant driving on the way to go. Mind leads and body follows suit. If the course of human lives depends on quality of leadership in a community, it depends on level of *self*-leadership as well. Both are another way of saying (Chapters 3 and 13) quality of mind.

Firstly, you might ask, why should I bend to self-control or let a line of law less visible than air restrain my anarchy of whim? Why should a rebel love his neighbour or hold other persons in respect? Why, especially, betray your genes and love another as yourself? What exactly is this strange phenomenon that some call altruism, others love?

'Scratch my back; if you're lucky I'll scratch yours'. Is this the apish, impish limit to which Darwinism can aspire? In this view altruistic actions are just 'strategies' for individuals to play life's chess, to mate and check-mate other players. Such behaviour is what you expect of natural selection's mindless state of mind. There is no high or noble, selfless cause but instead a single fount of information - 'selfish genes'. It's genes for everything! You can't see them but can blame them. Such absolutely passive entities aren't conscious of their 'motive'. Like computers they have no intention but they do! So powerful is their program, some believe, that it leads you by nose. If altruism helps survival that's what genes allow. If not then cut it out - that's fine as well. Evolution must and therefore can explain all things. You can't lose if you're a gene - except if natural selection executes. Thus a mother, acting without hesitation or reward, is actually unaware of the genetic 'wolf' beneath the membranes of her tenderness. She fails to understand the real, ulterior manipulation by her 'motivated' biochemistry. In the theatre of inheritance she acts on behalf of protein, she prevails against the teeth of natural selection just to save her baby's genes. Is she is *really* strung along to string along a string of *DNA*? Is this the height of explanation objectivity can reach? Is the essence of a mother's love no more than, little does she know it, genes that also have no clue?

Are you a robot? Are you governed in the way that robots are? Is the substance of your thinking made of ion gradients and electrochemistry? *The basis of such vacuous theory is the deletion of a prior source of information.* The analogous 'robotic' theories of Charles Darwin and Burrhus Skinner respectively ignore or discount inner subjectivity; instead they concentrate exclusively on life's objective shell. Together they exclude any precondition of intelligence from biology and psychology. The precedent evolutionary blank is some 'simple', pre-biotic replicator; and the minimal repertoire of embryonic mind is dim behaviourism's prior void (check Chapter 14: Quality of Information). Upon these open, unrestricted fields genetic mutations or random stimuli can, as a result of natural selection or rewards/ punishments, iteratively print endless variations of body-form or its behaviour. Is there, as the theories go, such outcome as an optimum? How, since hereditary evolution always finds a better way and non-hereditary mental traits must be rebuilt for each fresh life, could optima or ideal forms occur? In short, the twins propose empty yet somehow progressive, 'living' robots void of internal constraint or any preconception in the form of innate instinct or aforesaid morphogene. They (cleverly) guess that external, material constraints in the form of input stimulus and output response are sufficient to mould the way all organisms are. The couple are (or were, since Skinner's mindless theory of psychology has toppled down from grace) two canons drawn up to blast the ship of archetype to smithereens. Materialism's mother's mind is, therefore, not her own; it is, in 'scientific' topsy-turvy land, her genes'!

Not only prior intelligence but also your own immaterial consciousness - aren't these philosophical deletions pushing life a lifeless step too far? 'Behaviourism' is psychology at the extreme - no soul (of course), no mind (whoops!) or consciousness; just nervous tissue with a brain. That's better than your wood-and-paint Pinocchio but surely, simply rationalistic folly? This is what happens when the materialistic skew scythes subjectivity and thoroughly confuses passive with the active form of information. You might argue that the instinct to befriend and care for others is a cultural gloss, a social oil to grease the animal community, a graft on fundamental selfishness and an essentially self-interested, cunningly self-deceptive kind of evolutionary reflex. Or is it perhaps a selfish fear of loneliness propels us into selfless generosity? Are you not, fundamentally, a computation issued by your inside information in its most material sense - your *DNA*? Are you not in fact the puppet in a web of egocentric calculations and survival stratagems?

The Dialectic can agree you are. You are but this is just one half, the lesser and objective half, of the whole story. Instead of genes blame instinct for the lack of virtue that from time to time erupts. Never mind what instinct actually is or how its mechanism works. Aren't you an animal? It is obvious that hunger, thirst, warmth, friendship, sex and safety are all instincts that, in the name of biological survival, make demands. Can't they be used as an excuse from dropping off ideals?

In fact, doesn't science have an answer that, in its power and generous provision, could outdate ideas of ideals, morality and sin? Rub it up the right way and, like Aladdin's genie, can't it service all you individual needs? And what is more than needs, desires? High-tech Utopia! Rationalism's Eden, scientific fantasy! Science is a brilliant Jeeves when it butlers body's needs but at the same time, since the genie's left the bottle, not so bright. It is unhelpful when it generates the instruments of pain and war, suffering, overpopulation, global warming and so on. Technological catastrophe. It may not happen - even though the record shows abuse of science rides upon each good intention. What focus of morality will generate sufficient power to, like a laser, operate upon the cancers that now threaten to inflict not high-tech heaven but an international hell? If you thought glittering toys were any real answer, don't depend on fool's gold found in shifting sand. Don't depend on them!

What, though, about the other rationale? What about the immaterial and subjective one - a mother's love, devotion, bravery, honour and self-sacrifice? Whose folly is the giving of your heart in love? What if, far from deluded, all the saints are nearer than logicians or 'genetic motives' to the Truth? The start and not the end is love and love, they say, is its own element. It is a facet of 'rough diamond' shining with the natural brilliance of life. They can polish up uneven thinking. Inside information of the saints is not a package made of genes.

O tempora! O mores! Round and round un-saintly cycles run. Circumstances vary, fashions change and, now liberal, now strict, the social tapestry is woven, unwound and rewoven - always 'progressively' and 'for the best, of course, as far as I can see'. What is life for if you can't improve it or, at least, maintain the *status quo*? On the other hand, could strength and success ever presage feebleness and fall? Could enervating luxury and satiated consumption drift, slacken and decline? Could desire outwit ideals until, from the chaos that ensues, phoenix-like the latter rise again? Social cycles, just like empires, wash in waves like human tides. Nor, although you might have thought that glittering toys were what it was about, are they the sun that drives head's weather. We're back, yet again, to the real engine of our social climate, of our human ocean's currents, moral principles that heat us from an Ideal Sun.

The crunch for every centaur is the one of his or her identity. Who exactly are you, what's life for and where's the 'fun'? From a *bottom-up*, material perspective animals all need their 'fix' of sensuality before they die. Who can party hard enough? Incursion by business, invasion by the media and inundation through the screen into our rooms are welcome friends. The song is an excited one. The drumbeat is that, given passion, only fools reject the gift. Short, sharp life's for living in the world and not in so-called 'spiritual' regions of the head.

Psychological pollution! Vacuum of morality! Loss of purity and slippage out of centralised identity! Desires, prayers and many gods seduce an individual from The One. Stimulated to distraction by his sensuality a prodigal (whether individual or a whole society) has from *top-down* perspective lost sight of his own Natural Reality. Thus fogged he wallows in a sea of conflict; waves of attraction, pushes of repulsion, tugs and tows of fear or fight all wash him off the line of Inner Truth. Such is the temper of biology. To hold the line tests your embodied soul. And who, cries a self-indulgent centaur, enjoys such tests as try morality?

If, however, every man and woman walked sanctified in Eden underneath that Ideal Sun there would be no need for formalised religion, politics or law. What prisons are in paradise? What shackles Absolution? Can life on earth be seen as a confinement, a conditioning, an embodied lack of freedom that amounts, according to the level of its suffering, to various grades of bodily imprisonment? You are shackled in life's chain gang and its lash will never let you still. More desires flood on the ones before. Nature drives, social rules and regulations channel and, if you choose to burst their banks, will dam you, damn you and insist you flow their way. These containers of your moral waves, life's riverbanks, comprise society, law and, centrally, a faith. How, though, does a container stretch or squeeze a perfect fit? Arising in a mind that works with The Ideal self-government is automatic and as easy as a shining path of love. What friction might occur with the external agencies of government - unless its ideals weren't so clean?

Mind's a prism that diffracts perspective. It engages you in many roles and regulations - father, brother, son, employer, employee and so on - so that balancing demands, resolving strains and compromising conflict constitutes the staple calculation of each multi-mask's own daily politics. Continual decision-making continually involves, for better or for worse, discrimination and a moral quality. This quality involves, at root, how together or apart contingent people feel. In the end, more plausibly, morality might be derived from inward commune, where differences have disappeared, the *summum bonum* of it all - simply gathering round and being with a central person who best embodies love - which each of us, having chosen whom that friend is, always want to do. So, if the Generator's love, find his representative and worship there.

Individual Association

Self-regulation is from inside out; it involves the way you govern your relationship with outside circumstance. From exterior, however, now's the time to turn within; from community turn to communion with, it has just been indicated, whom? What might the interior reveal?

There exist society and individuals, each case afflicted by both moral and amoral imperfections; and, with respect to knowledge, areas of ignorance. Is not, for example, a great ignorance to ignore life's immaterial factor or the holist paradigm? At any rate, the ingenuity and institutions of mankind are set, in terms of education, industry and medicine, against what shadows they perceive. Light relief and higher standards - both in quantity and quality - is the purpose driving each of our theatricals through life. What epic drama. What an odyssey we sail. And if a single theory to describe it resonates with principles that underlie the cosmos, human body or a cell, you might not expect great novelty or twist. Instead you might detect, in its expression, strands from historical and current practices. You might recognise its presence on your social doorstep or, even closer, in your head.

In your head live the two sides of an individual association - external, outward and internal, inward-facing. These are logged in directional, moral and emotional senses of relationship. In a directional sense the stack reads:

	tam/ raj	*Sat*
	swings/ normal behaviour	*Axis/ Norm*
	oblivious/ religious	*Super-religious*
	human	*Superhuman*
	relatively sane	*Sane*
	degrees of apartness/ division	*Communion*
	unenlightened	*Enlightened*
↓	*tam*	*raj* ↑
	downward	*upward*
	outward, material emphasis	*inward, immaterial emphasis*

physical emphasis	*psychological emphasis*
physical relationships	*metaphysical relationships*
from metaphysical	*to metaphysical*
objective science	*subjective science*
action/ sensation	*two-way reflector*
physical object	*symbol-maker*
zone of practice	*zone of principle*
oblivious	*aware*
body	*mind*

In the moral sense of metaphysical relationship:

from Creator	*to Creator*
from Centre	*towards Centre*
from Goodness	*towards Goodness*
world/ body-centred	*soul-centred*
lower mind	*higher mind*
vicious individual	*virtuous individual*
criminal	*upright citizen*
subhuman	*human*
less sane	*more sane*

And in emotional sense:

↓	*tam*	*raj* ↑
	down/ negative	*up/ positive*
	sinister	*dexter*
	void/ worthless	*valuable*
	isolation	*relationship*
	enmity	*friendship*
	insecurity	*security*
	self-centred/ seizing	*other-centred/ giving*
	sadness	*joy*
	confinement	*release*
	depression/ deadness	*happiness*
	anxiety	*laughter*
	erratic	*stable*
	'wild' mood swings	*controlled oscillation*
	irrational	*rational*
	insanity	*sanity*
	darkness	*light*

Are you not, a normal human swinging in between the columns of the Dialectic, something of a mixture? Is this swing, I ask you, based on natural, 'moral' code? Is its variability as simple as the brilliance of Sanity? Take the Ten Commandments known around the globe. Western culture rose from their foundation and they form a moral basis for about two-thirds of humankind. The other third, whose moral format is called Hinduism, Buddhism or Confucian-with-Taoism, each concur. Neither they nor cosmic structure have 'progressed'. Neither is negotiable or evolutionary. As they apply to how you treat the outside world so they apply at source; they apply to all the thoughts you have within; and they apply to 'individual association'.

Check figures 0.9 (ii) and 0.10 (iv); and bring to mind the opposite directions of focus your attention turns to using implements (sensation, contemplation) that support each kind of information. It is time to turn from outside in. Will harmony emerge in proportion as the 'noise' and 'static' of external life is filtered out? Could health brighten as the sensual, selfish bug is systematically debugged? Is love, like health, a natural state and rooted in the way things are? Ascend the (*raj*) voluntary column of the Dialectic. As you do intensity increases step by step. As you approach (*Sat*) Apex, Top Degree, what message is there sent? If the heart of morality amounts to an intention towards communion that is known as love's perspective, is that not the finest texture of humanity? Against whose case the hard, cold moral exhortations 'should' and 'ought' are just admissions of a compromised or weary frame of mind?

Clinical sanity, attitudinal insanity: the fact is most of us are clinically sane but, where sanity is 'cool' in its detachment, dynamic passions create heated waves. Lust, anger, greed, detachment, pride - each fights for the upper hand; desire for fame, approved success and power drive to the fray. The attitudes these frames of mind engender leave most of us, to some degree, off-balance and insofar attitudinally insane. Check Chapter 0: Pure Objectivity. It's clear that extremes of passion and frustration drive to the points of violence, evil and the crimes we daily see. In the human theatre what part do you play? Most of us have not achieved the sage's understanding, poise and sanity.

	tam/ raj	*Sat*
	body/ mind	*Soul*
	creation	*Spirit*
	lesser selves	*Self*
	personae	*Psyche*
	concentric rings	*Centre*
	relative truths	*Truth*
↓	*tam*	*raj* ↑
	external	*internal*
	descent	*ascent*
	egotism/ self-assertiveness	*merging in love/ self-surrender*
	Darwinian survival	*mystic path*
	Lower self	*higher self*
	from Creator	*to Creator*
	from Centre	*towards Centre*
	from Goodness	*towards Goodness*

Internal, individual association involves just one relationship. How well do you live with yourself? Individual association comes to terms with one's own mind and, at best, peace with one's Self. If it is argued motion's changes are, inherently, destabilising then only two extremes - the immobile purities of sub-consciousness and super-conscious poise - disconnect from instability. In practice, 'sanity' is an ability to temper instabilities of thought (and thence behaviour) and navigate within the social straits.

Within the class of sanity are found (*raj*) grooves of stability and happiness; and, on the other side, (*tam*) poor and traumatised *sanskaras*, those dark habits tending in the opposite direction from principles of harmony. Such patterns of behaviour are unstable and disturbed but outside the nether pale there skulks insanity. This, defined in terms of dialectical psychology, includes strong directional, moral and emotional tendencies of downward quality combined with an inability to hold the swings of mood within dynamic equilibrium. In this view madness is a product of malfunction physical (e.g. hormone imbalance and brain damage) or metaphysical (e.g. trauma, violation or depravity). And by this logic 'satanic black religions' are classified as viral, sick with a deliberate depravity. Lucid and aggressive forms of anti-truth (delusion) are contaminations that require, as any danger to a natural system, dissolution. Cast out, for the sake of victims, diabolical insanity!

'Genius' means a lot of things. It means 'tutelary deity', an inspiration for good living, vital energy and creative spirit. Western culture often takes the latter sense, describing talent with uncommon vigour, perhaps allied with instability, of mind. Whatever 'genius' flirts with dire insanity is not, however, that of Natural Dialectic or the saint. Insanity is a peripheral, involuntary and dark condition. Its passive 'victim' is 'possessed' by physical or psychological malfunctions and cannot throw them off. Scientific medicine tackles what were known as 'demons' by giving them new, scientific names and reasons. Could it be such states of mind are always thrown up by neurochemical distractions from the norm? If, though, the metaphysical dimension is real it may be that disembodied 'mind-fields' and their occupants exist. The issue of 'evil spirits' becomes, in this case, neither wholly chemical nor perhaps unreal. Immaterial agents of delusion, evil genies, may exist.

On the upper hand, does 'genius' not describe a man who unveils nature's secrets, builds great works of art or reflects what strings of philosophical idea might tie the world together? In such case true genius is certainly ranging on the right-hand path that ascends towards unification. Is it therefore not an act of genius, however faltering at first, to join that path?

Is there, however, any overarching meaning? Is there a purpose undergirding human life? Is there, beyond great curiosity, an engine driving us along a royal path to find the fullest and most satisfying answer of them all? And, finally, is this profundity discovered in the union and bliss of love? Love, interest and enthusiasm (meaning 'God in you') outwardly involve the lust of passions but inwardly may not. No doubt the soul is strapped by mind and pinned within the field of existential action - but detachment from the trammels of such harness lets it fly. Isn't, therefore, the Essential Aspiration of a human life, naturally inlaid but much neglected and distracted, to distil the mind's pollution to a pure distillate and thus, as every mystic always told you, reunite your life with Life? Communion's equation, merging lower self with Soul, brings 1 + 1 to 1. No doubt science cannot, though a scientist being human can, understand the theory and practice of a science of the soul. But can you now grasp that at the Heart of Individual Association is Reunion - Reunion of the self with Quintessential Self?

For a contemplative (why not you or me?), sane from the start, such internal association involves (raj) voluntary ascendance. The experient is in positive control. The aim is, at completion, a single relationship of self with Self, of Communion with the Central, Cosmic Self. For most humans this friendship is imperfect, this communion weak. For an enlightened mystic the bond is nurtured to the Perfection's Point. It is an exercise born of 'athletically' pressing the '*raj*' fundamental to conclusion and, as such, is the diametrical opposite of clinical insanity or, as 'materialistic psychology' might diagnose, delusion. *Inner Communion involves no external behaviours. It is purely metaphysical in character.* After an existential odyssey the return of a prodigal soul is, in a *GUE* (Grand Unified Experience), essentially complete.

There are two forms of inward or spiritual community. The *second* is an external community of like-minded individuals specifically devoted to the *first*, inward Communion. This Communion is the 'altar-piece' of any religion. Monk and nun are the engines of its external expression. They are psychological powerhouses that sustain the body of a faith. Anyone who enters deep prayer or contemplation is, for these moments, become a true monk or nun. The condition of a true contemplative is, however, entirely natural and, as such, (s)he rises above the distinctions of organised devotion. For social purposes a practitioner usually identifies with the local vehicle of faith but, inevitably, in their teaching role, such vehicles create (which thou shalt not) 'graven images'. Modelling and imagining 'God' is a 'label-libel' exercise from whose conceptions issue all manner of doctrinal error.

Where a scientific '*summum bonum*' might be a physical, technologically enhanced utopia, is it not also a noble aim to try, communally, to recreate the metaphysical '*summum bonum*', the psychological excellence of Communion? This goal is, at root, what any right-minded faith aims for. Therefore, while criticising such dogmatic aberrations as cause negative distortions (e.g. suppressions, inquisitions, war, repressions and so on) contemplatives inevitably support the fundamental (*raj*) inclination of a metaphysical religion towards (*Sat*) Truth. The laboratory for metaphysical research is a retreat from the world - be it monastery, nunnery, cave or your own home in a quiet room. The equipment has already been supplied - your mind, your life, yourself.

It is, in the end analysis, irrelevant whether the true monk is of a particular religious order or none at all. It is certain, though, that he will have a single Teacher. With this Professor, Master or Psychiatrist (Soul-Doctor) *sans pareil* comes (*Sat*) True Association. He embodies The Eternal Teaching here on earth; he applies the spiritual collyrium; and he will both lead and draw disciples towards the first and foremost form of Immaterial Communion, Enlightenment. The Teacher is the heart of contemplative process that, at climactic peak, dissolves religious frameworks. Communion means Oneness, not duality or difference; it means Perfect Balance. Any weight is lightest at its point of balance - no less so existence as a whole. *And Perfect Balance is reflected through creation as a principle called 'Golden Mean'.* The Golden Mean, whether Aristotle or Confucius tells it, is Ideal. Its moderation resists all extremes except the sense of Poise. It is, paradoxically, detachment in involvement. It is the scale by which aesthetic values and morality are weighed; it is health's axis and the Norm of homeostatic norms; and the staff of wisdom against which the 'rightness', harmony and elegance of all behaviour is checked. As such it is clear - the heart of Absolute Morality is yours. It is resident but often and in varying degrees obscured, within each person. *The dharma blossoms into full flower with enlightenment.*

Do you understand this? If you are secular in outlook you cannot because it can't exist, can't happen and is neurologically insane. How can nerves (that do your thinking) meet in Oneness and in Truth? The idea is immaterial, irrelevant and primitively fantasist. Have you not, however, locked in material partiality, thereby wholly locked the bigger picture out? **If you allow a natural element of information, immaterial consciousness, then the strivings of mankind since history began are understandable**. They are materially germane to mankind's greatest movement - a return towards individual association with the 'place' where all began. Such singular relationship is immaterial, of core relevance and the highest of realities.

Inward shifting; trans-religious movement, a possibility that's always been incorporated in the human frame, is mankind's future evolution. It frees from dogma and materialism. Contemplative association is internal, outwardly invisible and independent of material circumstance. Metaphysic's distillation purifies awareness. Its full flowering is not only super-intellectual but, if conscious is what you are now, super-conscious. What first and final paradox! The 'form' of Void is neither absent, empty or unreal. It is Plenitude, Potential and the Substance, that is, Basis of existence; it imbues and yet, by virtue of its transformation into existential changes, is not of creation's worlds; and is thereby the essence of all things but None. You might dub it Nature, nature's Inside Information. *Top-Centre, it is the Essential Spring from which existence flows. And its metaphysical Utopia is the Solution; this Most Natural Association, this Perfect Communion is the goal of Real Philosophy.* It is not that, from its Middle Way, the balance of the Dialectic bears austerity or other errors of 'un-worldliness' but you now understand what centrifugal hedonism's 'Apple' finds imperative to damn with faint praise; and why swarms of desert anchorites, cliff-faces full of Himalayan hermits and medieval cloisters calm with monks or nuns might 'waste' their lives. They go, in principle, for gold - real, inalienable but immaterial wealth; their 'waste' is mountaineering towards the peak of universe. Its *TOP* is galvanised in pursuit of Absolute Communion and thereby Individual Association with The Origin of Everything. Who might not wish for Perfect Life and, understanding every reason why, a deathless Light of Knowledge? Which universal science of the soul is, from a *top-down* point of view, hardly daft! Nor is the logic fanciful. Although metaphysical it is entirely pragmatic. Action derives from thought. Man's thinking creates his own, as opposed to His, world. It needs clear thinking. *Now, as much as in the past, the affairs of men have need of such a healthy frame of reference as the Dialectic's. Now, as ever, most prefer equivocation.* The remedy is hard to swallow; its regime is hard to follow. It neither replaces nor competes with any 'orthodox' expressions of Absolute Morality; rather it underlines their Centralising Direction. **This, it clearly indicates, is the right-hand and ascending path of Peace. Thus one would hope that Natural Dialectic acted as a universal pain relief.**

28. Where Does the Data Lead?

'When I first wrote my treatise about our system, I had an eye upon such principles as might work with considering men, for the belief of a deity, and nothing can rejoice me more than to find it useful for that purpose.'

Sir Isaac Newton, *Principia Mathematica* 1687.

Where Does the Data Actually Lead?

You, if a materialist, accuse. *Top-down* Natural Dialectic only reaches its 'far-fetched' conclusions due to its first premise. Sauce for the goose, however, is for the gander too. **You took your primary assertion, I took mine. Both are philosophical; neither is a scientific one.** *Your thorough logic, derived from your materialistic premise, thoroughly inverts my own - not with respect to current, naturalistic operations but tiered structure and historical originations.* And *vice versa.* One of the couple, a cosmos immaterially intelligent or not, must be the final case; and by Socratic principle good science always follows, without prejudice but with impartiality, wherever data leads. We have come to our conclusions. We've summarised the consequence of any difference. Now for the summary of summaries.

This book has toed Socratic line. It has (from Chapters 7 to 26) tracked evidence that's physical, psychological and biological. You can infer from all this data either that, framed in a paradigm excluding any immaterial factor, cosmos evolved by chance; or, framed in one including such a factor (in the form of consciousness), it runs as a projection of intelligent design. **It is imperative to understand that a philosophical but not a scientific decision underlines your perspective concerning origins; and to remember that materialism is a philosophical and not a scientific posture.** *Therefore, to repeat, if you choose to leave the immaterial element of information out of your equation then chance and 'natural selection', operating in the frame of physics, have to rule; if you include it then intelligent design (whose mechanisms perform within the same frame of physics) becomes the 'bottom line, top conclusion'.*

If the universe had a beginning (and it was a priestly astronomer, George Lemaitre, who first mooted the idea of a big bang) the choice or, rather, faith is between blind chance that provides multiverses out of uncreated, eternal energy; or an ultra-incredibly serendipitous single bang that created *ex nihilo*; or design that informs the orderly construction and behaviour of the cosmic parts. Is it Lady Luck or Lord Deliberate that is the parent of our world? Your speculations may have faith in Her; but the fine-tuning of our universe (Chapter 8) and bio-friendly planet (Chapter 26) provide a *prima facie* indication of intelligent design. **That is, unblinkered science leads to an interpretation of deliberate design.**

Such impression is much more powerfully and obviously evident in biology. Check the Glossary for current usage of a word - evolution. Of course Darwinism does, especially in the tiny world of microbes and disease, explain chaotic and occasionally 'lucky' variation by random mutation and natural selection; but the theory of evolution is insufficient, in spite of scientific exhortation from the philosophical sidelines, to explain a naturalistic, abiogenetic origin of life, metabolic programs, sex, development or 'saltatory' macroevolution from one kingdom, phylum, class or order of organisms. It fails (Chapters 13-25) to explain the innovatory appearance of a single system, organ, tissue, cell type or, even, the generality of life's ubiquitous building block, a cell. You might extrapolate to a primeval, biochemical concoction but this is a perspective-driven guess. Indeed, the more molecular biologists learn about the structure and machine-like operation of cells the less likely it becomes that such profound and purposive complexities jumped up 'like that'. A molecule is not alive so natural selection cannot help a chemical get fit. You simply have to take it on authority that dust self-animated and became alive (where life's defined in terms of molecules). Books have been written and at least a billion dollars wasted whistling up the form luck might have taken with a high degree to overcome the vast improbability - because you argue that, however great improbability, nothing is impossible! Especially if (your bottom line) you give chance time!! Who deals in numinous invisibilities? The most a speculator or expert empiricist can issue is a best-guess promise for the future - 'we will find out how it happened naturalistically because we have chosen to believe life first arose by accidents of chemistry alone'. Is promissory science enough to hook you to the line? To call a fishy theory netted fact?

Because it's claimed by some that evolution is a 'fact'! Evolution-of-the-gaps is riddled with black boxes and, literally, *non sequiturs*! Despite the facts it can't explain abiogenesis, the genesis of body-plans and systems or even macro-evolution higher than one species (or perhaps genus) to another. Nor, wisely, did Darwin ever try. Hypothetical suggestions, interpretation of the fossil record or extrapolation from the variation that we see today is not enough. By what actual mechanism, since gradual mutation with natural selection is demonstrably an insufficient cause, might transformation actually happen; how might descent on an evolving metaphor, the tree of life, obtain the vast, specific and encyclopaedic information life everywhere incorporates? It may be that you haven't got an answer - except that still, despite its crippling burden of these and numerous, unsolved 'chicken-and-egg' conundrums, you insist that the opaque 'fact' of evolution absolutely and unruffled stands. *What, however, is this failing theory more than promissory notes and just-so huff and bluff.* **These puffs are threads by which the whole hard-edged pillar of atheistic faith is precariously suspended over its abysmal vacuosity**. *On what, at this rate, is atheism based but a slender, duplex chain of huffing confidence and prolix trains of puff?*

Whose delusion is it, therefore, that not only utterly fails to explain how lifeless atoms generate not just biochemical bodies but the life in all of them - life that includes the psychological centralities of mind, subjective consciousness and experience? Darwin's theory does not touch on innovation, creativity and rational comprehension of an intelligible and lawful universe. It explains the presence of neither artist nor scientist; and thus, in truth, not a single baby child. Such failure, where intelligent design gloriously and completely embraces them all, is abject.

You might, by *bottom-up* bent of philosophy, prefer materialism's bullish bluff to the alternative, a Theory of Intelligence. And a Theory of No Intelligence might prefer the puff of stuff and convoluted nonsense when a straight, alternative interpretation tracks the scientific evidence elsewhere - but does personal preference make fact? Is such faith even relevant to truth? **Is evolution fact? Some suppose they know but in fact the best that you can say is, given saturated materialism as your starting-point, it emphatically *must have* happened and, from there, infer it did. This is not what physics or hard science calls a fact.** *There may be good reasons for being atheist but, as this book shows, the neo-Darwinian theory of evolution is definitely not one of them.*

The theory is (fig. 20.1) a partial truth in its entirety untrue. Materialism's is the lesser half of whole truth too. Indeed, is atheism not a great 'con' or, worse, a dark lie?

Is this jaw-dropping heresy or tough love, cruel to be kind?

	tam/ raj	*Sat*	
	existence	*Essence*	
	inertial/ stimulant	*Potential*	
	radiance	*Source*	
	hierarchy	*Top/ Above/ Before*	
	spectrum	*Purity*	
	grades of consc.	*Consciousness*	
	negative/ positive	*Neutral*	
	cycle	*Axis/ Hub/ Centre*	
	motion	*Equilibrium/ Poise*	
	time	*Now*	
↓	*tam*	*raj*	↑
	down	*up*	
	passive	*active*	
	nertial	*stimulant*	
	created	*creative*	
	informed	*informant*	
	non-conscious	*spectrum of awareness*	
	matter	*mind*	
	end	*lifetime*	

Impartial science deals in 'naturalistic' explanations; but atheism stands upon a pedestal of prejudice called *PAM*. Axiom 1: there is no immaterial dimension; from fundamental randomness 'extrudes' creation. If you believe in randomness you can believe in anything (branes, multiverses, neo-Darwinian tales etc.). If, on the contrary, you believe in essential order you incline towards information as the Source of Nature. From a *top-down* viewpoint scientific data is a natural outcome you can trace back to its Natural Origin (see *fig.* 1.3). If this is so then atheism is a modern and sophisticated but a superficial myth. It is, quite apart from the science it appropriates for its prop, a secular religion.

The dialectical conclusion is, most clearly, that the fashionable Cult of Lady Luck is wrong. Conversely, most importantly, a systematic exploration of its columns will, as microcosm of the macrocosm, point up the great potential that is you.

This book's narrative has added nothing to material fact - because informative consciousness and fresh perspectives are both immaterial. Nor is consciousness (with which you are most intimately familiar and which, indeed, you are) some arcane, novel or bizarre addition to the known world. The reverse. Without it there could be no *known* world. The simple addition of its subjective, metaphysical factor to objective matter changes nothing and yet, at the same time, changes everything - especially if there exists an endless reservoir of information out of which the order of *created* cosmos is defined.

Have you proof absolute (although you may have built an inferential case) that my premise of two existential fundamentals, energy and information, is not perfectly correct? To restate the point, it is impossible to prove or disprove any immaterial element materialistically or, indeed, see behind the origin of things through telescopes or microscopes. Therefore come on high to the mountain-top. As was promised in its Gate this book has applied a simple premise to the universe - concentrates of information and of energy precipitate the cosmos in an orderly, coherent way.

In this case the bottom line is, for Natural Dialectic, top line - Central Potential. This Potential is identified as Pure Consciousness. It is what constitutes the capstone of Mount Universe; the Immaterial Factor is the All-in-One and One-in-All that holists seek to meet. If consciousness is immaterial then, in his quest to purify, the mystic's chemistry distils a concentrated immateriality. This, enlightenment, is as natural an inner truth as body is an outer one; it is as above and beyond religion as your body is above creation by a school of medicine.

Bottom Line, Top Conclusion

	tam/ raj	*Sat*
	finite	*Infinite*
	relativity	*Absolution*
	partitioned/ divided	*Overall/ Undivided*
	appearance	*Reality*
	lesser truth	*Truth*
	imbalance	*Balance*
	relative order	*Order*
↓	*tam*	*raj* ↑
	objective	*subjective*
	energetic	*informative*
	matter	*mind*
	non-conscious	*conscious*
	ignorance	*knowledge*
	less real	*more real*
	involuntary/ reflex	*chosen/ voluntary*
	randomness	*purpose*
	accidental	*intentional*
	chance	*design*

Modes of attention vary and, it was argued (Chapter 13: To Build a Brain) , brain's two hemispheres reflect the two modes of experience - one holistic and the other analytical. It has been argued that the holistic, *top-down* of our forefathers and become distorted. Modern western culture has veered, in its *bottom-up* perspective, towards vision from the left - with consequent bias in the way we experience and therefore react to our finite planet. In the ascendant now presides a character of egocentric and materialistic, left-hand state of mind. Balance is disrupted; natural order is disturbed. It was argued (in Chapter 26) that the ecological condition of the world has fallen sharply in decline. Is technology alone sufficient unto solving problems that technology has brought about? Where will what we make of life in scientific culture lead?

No doubt, this book has opened up an immaterial dimension but how has this detracted from exclusive materialism? Has natural science not been welcomed as a most intriguing and exciting guest? **Have known facts been anywhere or any way distorted in these pages? I mean facts and not interpretations of those facts - though even here we have consistently reviewed the vision, *bottom-up* and *top-down*, that each lens of life's dualistic pair of goggles yields. If you think there's been distortion or contortion, omission or commission say exactly where and how!** What about the system within whose parameters Science and the Soul has been constructed - Natural Dialectic? The Dialectic's simple, basic premise is that cosmos is composed of *two* fundamentals - information and non-conscious energies in various proportions with each predominant in its informant or informed domain. These are mind and matter, distinct but woven in a finite tapestry we call creation. Beyond yet at the Axis of creation is its Single, Infinite Source. Where is not the centre of Infinity?

The three-in-one construction of the Dialectic indicates polarity of worlds but is, at root, dissolved in The Essential Communion. In other words, as *figs* 1.1 and 1.2 help to explain, the overall dipole includes (*Sat*) Essence and (*raj/ tam*) existence. The nature of existence is duality. Its polarity therefore comprises, in an existential dipole, (*raj*) kinetic pole of mind and (*tam*) non-conscious pole of matter. The three 'stages' are therefore Central Potential, mind and matter.

This cosmic hierarchy, this order of 'tri-universe', this trinity involves both prior subjective and sequent, objective parts. A familiar, triplex order of creation drops from latency (potential) through kinetic, mental to material phase. From seed idea there issues reasonable logical development; there is expressed creation that these volumes have been written to reflect.

If this Infinite Potential at the heart of things is a Pure Concentrate of Consciousness then how can it be known except by what it is, itself? A knower is subjective, knowing everything in mind. Where is your mind? Is not a conclusion that the apex of the universe is to be found, in principle, at the centre of your adult forehead remarkable, revolutionary and, if not logical, insane! But this is the conclusion to which three-tiered holism, as opposed to one-tiered secularism, leads. It is also, by that token, one that all world faiths emphatically endorse. Why?

Do you remember how *figs.* 0.7 to 0.10 and 17.11 give the address wherein you live and think, the point that's dubbed a 'cosmological axis' or 'eye-centre'? This 'third' or 'single' mind's eye is a crossover between the outer physical and inwardly subjective worlds. If, as the Dialectic consistently asserts, subjective hierarchically precedes objective, information precedes energy, consciousness non-consciousness or mind matter, then these inner worlds are precedent. Call them realms, regions, domains, levels, planes or zones but it sums the same - thought precedes and governs action, information channels patterns of behaviour.

Are you really an incarnate microcosm of this universe, a metaphor for its construction bearing all its grades, a trinity including its three 'stages'? If so, then the only way to know its subjective components will also be subjective. How do you perceive, know and understand? Using metaphysical apparatus? With mind alone? *If, therefore, the highest existential precedent, Logos, is knowable, it will be known at the place where you know anything. It will be 'beyond the Ka'bah's arch' or 'within the third and single eye'. It will, in short, be nested at the centre of your forehead.* In this *top-down* view the 'temple' of a living body is a shrine to Truth. Is there a subjective test for such hypothetical assertion? Such test is the immemorial exercise of yoga, mystic practice and meditation as properly, precisely taught by adepts. As with any subject it is best relayed by a live professor - in this case a saint and not a neuroscientist!

Such world-perspective is, of course, completely lost on the *bottom-up* fraternity. Secular materialism rejects the simple proposition that duality of cosmos combines *two* fundamental parts - matter *and* consciousness - as one too many. There is only unity of energy, no place for information. Professors of science and philosophy, intensely saturated with a one-tiered, flat-earth point of view line up to deprecate a Theory of Intelligence. It is as if the sweat and grime of labour at the lab-front or the study's desk had somehow proved their single-minded cause. It is not, of course, an obscure proposition to assert that immaterial is not material and therefore not laboratory fodder; but in a culture whose first rule of intellectual engagement is 'keep metaphysic out' the immaterial seems non-existent. If 'matter = energy = the objects, forces and motions of physics = nature' and 'brain = mind' then, naturally, the intelligence of a creator must be material or at least not 'super-natural'. If 'science = materialism = all-there-is-to-know', the appeal of metaphysic might appear as abstract, untested and incredible as it were unscientific. It might not, therefore, seem excessive to conclude that a 'test tube mode of thinking' arrogates all knowledge; that a three-tiered cosmos, with all its philosophical derivations, is arrant nonsense; and, now that science almost understands life's secret, contrary Theories of Intelligence infect the health of scientific reason and, therefore, the public good. How unintelligent such theories! What a 'cop-out' to propose intelligent design is engineered! Ban the notion from the classroom and the lecture hall! By contrast, so the spirit of our time exclaims, how intelligent a Theory of No Intelligence!

How stunningly illogical this 'goggled' view! As we saw in Chapter 6, how stunningly irrational a so-called rationalist can be! In Theories of Intelligence and No Intelligence the starting point of logic is reversed. To the former reason based at root on chance appears irrational; to the latter reason based at root on natural logic, *Logos,* equally irrational. Moreover each seems to the other to have created its own test-proof 'free domain' against which rational discussion can't prevail. A theory of non-deliberate design (neo-Darwinism's *PCM*) lacks proof that no designer could exist; and a theory of deliberate design (*PCND*'s Intelligent Design) proof that one did or does. How can you prove either rigorously, objectively beyond the point of inference?

This is the bottom-line, the crunch.

'There is only', you say, 'matter. I imagine an infinite reservoir of material energy sprung from nothing but, certainly, no 'reservoir' of immaterial consciousness. I therefore reject your premise and embrace the pillar of materialistic faith that leads from that rejection. This is my creed.'

'OK', I reply, 'believe it if you will. If you believe that cosmos came by chance, then your prerogative will force you down a certain road of reasoning. This road will not make final sense because, by definition, chance is non-sense. It is lack of reason; and could lack of reason really generate the very reasonable order life involves?' Do you lay your bet, your belief, your faith on short or very, very, very long odds? Which kind of punter is more rational?

Lux et Veritas

'*Lux et Veritas*' proclaims Phelps Gate at Yale. What kinds of light and truth are meant? 'God is my light' (Oxford's motto '*Dominus Illuminatio Mea*') makes certain nonsense for a secular philosophy. On the other hand it may well, as inspiration from which cloistered corridors of academe descend, be true. Might Phelps' light and truth be metaphysical as well? '*Magna opera Domini exquisita in omnes voluntatis eius*' is inscribed on the portal of the Cavendish Laboratories here in Cambridge. 'Great are the works of the Lord, carefully studied by all who so desire' is not a maxim that has deterred a galaxy of scientific luminaries making profound discoveries there since evangelical Presbyterian James Clerk Maxwell, the scientist *par excellence* of physical light, oversaw its construction. Let rest that Latin for agnostic vacillation is an '*ignoramus*'; which firm faith, atheistic or theistic, would you consider '*lucem veritatemque cognoscimus*'? At least, in neither Oxbridge university do college names proclaim a dearth or death of Christian faith.

In fact, you could argue atheism is a fundamental insult to the founding fathers of those early universities; and an affront to catalytic intellectual centres whence the Renaissance (but not the later 'rational enlightenment') caught light. Atheist and theist seem, inevitably, polarised. They just don't want to hear each other's version of the truth. In 1860 Samuel Wilberforce, opposing Thomas Huxley's Darwinism, had culture on his side. In 1986 a 'Huxley Memorial Debate' was staged by the Oxford Union between Richard Dawkins with John Maynard Smith and Edgar Andrews with Arthur Wilder Smith (creationists now representing, as Huxley had, the counter-culture). The motion was 'That the Doctrine of Creation is More Valid than the Theory of Evolution'; and the audio-recorded result was 'Noes' 198 and 'Ayes' 150. You won't, however, verify that outcome from the Union's minute book. It has since been lost or stolen. Was this loss, which could be construed as tactical, in fact an accident or did Oxford smirch its motto by an evolutionary design? One thing is certain - the evolutionary camp was horrified. After more than a hundred years in modern, intellectual Britain who'd gained any ground? Both sides of the Atlantic, though, the gown-cloaked, cultural establishment of evolutionists seems nowadays to close its mind to counter-culture; our intellectually 'open-minded' liberals fail (except by *ad hominem* insinuation, omission or adverse commission) to counter evolution's challenge by 'design'.

Is it logical to elevate the role of chance to rule, along with chance-begotten rules, the dust of stars so as to build them into men with brains? Is it other than an atheistic form of prejudice that militates against all notion of intelligence behind the complex vehicles of life inhabiting our earth and, thereby, cosmos? *Indeed, it is emphatically and most rationally possible to cut the slender threads on which an atheistic faith depends and adopt materialism-with-consciousness, the three-tiered metaphysic of this book, many other manuscripts and, unless they are in error, the three-tiered 'Mount Universe' of nature itself. If 'atheistic rationalism' is the 'modern' rage and if you swear (for what they're worth) by snappy phrases then the dialectical adoption constitutes, as well as pre-, 'post-modernism' too!*

Do you deny the separate existence of an element of information, consciousness or call it spirit if you wish? Such from-the-start denial were a root of darkness in that it eradicates root understanding of your essence. It expurgates your origin, it veils your full identity and thereby stymies highest human aspirations and fulfilment of potential. It also censors, by denying its design, appreciation of the meaning and the structure of our cosmos as a whole. How, therefore, can this naturalistic view claim every truth and thereby Truth? No doubt the spirit of humanity is caged, by different degrees, in all of us and mere denial of such immateriality does not, in principle, equate with 'wickedness'; but 'sin' is in the consequent misreading of our world, life's purposes and yearnings that invigorate the soul. At the cutting edge of such misunderstanding range materialism, atheism, neuroscience and the hypotheses, biological and psychological, of *PAM*'s *PCM* - neo-Darwinism. Correction might, without requiring a material change of fact, click into the *PAND*'s perspective. This, in the interests of balance, sanity and aspiration towards the principles of an ascendant right-hand path, is what Natural Dialectic obviously implies. There is no reason to deny or drive upon the left-hand slope towards tenebrosity!

It has been argued systematically that one-eyed materialism, aiming low, misses here, there and (except for a 'flat' view of cosmos) everywhere - a gesture of informative magnificence. On the other hand, holistic argument is clean, clear and natural as air. Its logic, Natural Dialectic, is, within simple premises, self-consistent. *On balance, I accept its force of truth.* **Anyone is free, without subjection to aspersion, so to do**. No doubt clever atheologists reject this line. They (including some contemporary scientists and academics) spin a web of stories of creation, both of matter and of life. The point is not persuading anybody that his mind-set's incomplete. Set is set is set. The point is that you do not, using a holistic logic, have to buy their point of view; nor, while always clinging to the hope of just-around-the-corner final proofs, clap promissory triumph born upon the thunderclouds of such a creed. *Feel intellectually free to doubt materialistic faith because its atheistic tale is, bluntly told, no more than one of bluff and speculation; it is a hedge of bets developed cleverly around aforesaid PAM and PCM.* **It is above all based on a subjective assumption of materialism that precludes, without any proof and (it has been argued) illogically, any immaterial factor such as purpose, will or consciousness. Are information, creativity and mind distinct subjective elements; or are they simply made of gravity, electromagnetism and the elements of atoms multiplied in certain but uncertain and mysterious ways?**

If subjectivity is not composed of these last three but rather *immaterial* mind and matter interact, then the Dialectic stands.

If a detective fails to identify, ignores or wilfully rejects a key factor in a case (even an invisible, mental factor such as causal motive) he is bound to misinterpret evidence. Thus, saturated with materialism, every one of millions of scripts on evolution accepts the passive 'information' wrought by chance but misses out a central factor - natural, archetypal information wrought by universal mind. Scientism presumes, in spite of evidence to the contrary, that metaphysic is disproved and therefore non-existent. **Therefore the basis of its argument lacks a whole element - the immaterial, informative dimension; and it wholly funks the dreaded t-word, teleology.**

You may have been conditioned by peer pressure, reading books or in your school to think that apes bred you but if the PAM is incomplete the Dialectic stands.

The adulation of a 'Darwin Year', wholesale misinterpretation of the facts and the polemic flush of atheism hangs upon hypothesis. Evolutionism really doesn't want to know the Dialectic answer. It fights it tooth and nail. Yet thin is scientism's basic thread, threadbare its cloth. **This book lime-lights a mind-set in denial; indeed, it demonstrates an atheist's profanity. The message is that his or her philosophy does not disprove but only wilfully deny the single, immaterial Heart of Nature.** This cell-less Heart can no more be examined on a slide than thoughts; like immaterial mind, it can't be centrifuged in tubes. But is the human spirit made of ethanol or life's soul finally composed of clay? How can you reckon man without his central subjectivity? How can incomplete detection of a major suspect's nature and intent ever fully crack our cosmic case?

Wilful, even militant denial of a line of questioning is unprofessional, unacademic and closed-minded. Therefore, if you're an atheist feel free to keep your faith; but know it as no more than this and that 'the other camp' at least as rationally holds court. Mankind applies perspective, not experimental science, to the question of origins, of physical and biological beginnings. Of course, a naturalistic slant is just as philosophically valid as a Naturalistic one. It's obvious, however, that inclusive, complementary Natural Dialectic offers no endorsement of such rough, rude, thoroughgoing blasphemy; indeed, it rates the skew no higher than dry dust whose pile of oblivion materialism heaps upon its pedestal. **On the contrary, its reason as applied to all the data would endorse, as the only cause that adequately accounts for the huge quantities of most specific, functional information stored in every incarnation (that is, in all forms of life on earth), the presence of intelligence**. The data actually leads towards an element of information known as, in this case, universal mind.

In other words, from a *top-down* perspective, materialism and its vital theory of evolution is the biggest scam the world, indeed, the universe has ever known. If, therefore, scientism and its atheistic hacks should chant in unison the falseness of a Natural Doctrine of Polarity then simply know material applications do not (and should not pretend) to understand the pull of Centre or the inside of our human world. Noise never bullied Truth. It is therefore with happiness I note that Sir Isaac, with whom my life has shared the same town, eloquently subscribes to the same non-materialistic purpose as mine. '*Philosophiae Principia*', the title of both his and, earlier in 1644, Rene

Descartes' major work, is too presumptuous for a communication not posted in the mathematical language of natural phenomena. It may be, however, that a numerical approach will never get you everywhere - especially access to the most important places. I therefore humbly offer the words of '**Science and the Soul**' as a pointer towards higher and, in our time, somewhat neglected Truth; I hope its concepts bring you to the Cosmic Upper Gate.

For whom is nature and Centrality of Nature, The Most Natural Essence, an exclusive right or property? This book relates the structure of the cosmos and our place in it. It thereby systematically confirms belief in (*Sat*) Unity of Higher Truth and supports its ethos coded as an instrument of (*raj*) moral guidance. It refutes, however, (*tam*) division by religion (including communist or any other systematic faith derived from so-called scientific atheism) into such cultural, social and political exclusivities as the adherents of a creed tend mostly to assume. *Does Nature not transcend race and class and creed?* The Dialectic therefore equally rejects what cant religious engagements (secular or not) can easily accrue; and it denounces meretricious, intolerant or bullying use of such politicised beliefs as a 'left-handed' instrument of social control. Meanwhile the Dialectic is both 'scientific' and 'unscientific' but in neither case iconoclastic. Have you listened to the inspirational poetry of mystics, *yogis*, *sufis* or other heretics about the natural, unifying, creed-transcendent path? If the tenets of your faith are cast in orthodox a mould there is no need to shatter it, only look above and reach out for the Infinite beyond man-made frameworks with their inevitable distortions that we think and therefore worship in. *While many trails lead up a mountain, at the top they have to merge; the single path is Logical; only Logos leads to Transcendence. Therefore while this volume thoroughly vindicates your family religion, it does not vindicate its exclusivity.* Different and exclusive claims to any Single Heart by any caste or creed are, in the last analysis, as false as foolish. One is one. How do you cut the light up in a room? By claiming exclusivity religions would, as impossibly as sunlight cloven into pieces, try to ring-fence Single Radiance. Does superior sun-light fall on chosen gardens or is its all-embracing quality the same in every place?

An atheistic humanist may not appreciate our dialectical conclusions since, if true, materialism's basis is a half-truth and therefore his philosophy is incomplete; and, with respect to the more important half of everything and to the origin microbes through to men, the reverse of true. The inclination of the Dialectic is, however, (*raj* ↑) towards unity. *Thus paradoxically its supra-religious neutrality chimes well with intellectual dispassion and finds counterpart in secular ideals.* For example, it embraces the notion (in as far as it might work) of a single constitutional framework into which to pour your politics; and, at the same time, separation of religious practice and benign but neutral government. In other words, men holding different beliefs might co-exist, work together, pray (or not) apart and tolerate each other's style of life. Whether it is theocratic or atheocratic, an unacceptable ingredient of social mix is lack of tolerance. What ethos should, however, govern government? Does 'secular' mean ethically neutral or (you can't escape faith) involve some transcendent system of morality? Did not Plato emphasise you need agreement over what is 'good'?

Of course, you understand enthusiasm and a certain mode of it called science. Do you, however, think from this book's explanation that you fully understand the cool fire of its stars, the mystics? Mystics aren't practitioners of earthly answers;

they sing, where Omega is Alpha, about the end of earthly life. If you are interested in their message you are standing simply at the threshold of detachment, on the verge of deep psychology and in the dawn of day. 'At dawn the crier from his tower calls till the darkness has dissolved. The shadows die. The sun's expanding from a sideless tomb and, brilliantly, the Soul of Day eliminates its mortal night'. You will have to go and hear their call and learn their path. One of them may grant you real (and not symbolic) baptism in the Royal Stream of *Logos*; may you then rise as a fiery phoenix, may you hear the un-struck music of the Natural Word. St. John of the Cross once wrote, "One Word spake the Father, which Word was His Son; this Word He speaks ever in eternal silence and in silence must it be attended by the soul."

Let's rephrase this critical position. If 'natural' means 'material' are information, consciousness and mind material? Are they natural or not? *Point of attention, consciousness, is the essential ingredient of life*. If this is immaterial is life unnatural? If you have faith that subjectivity's composed of matter then show me how atomic forces make not brain but mind. And how from oblivion they answer Plato's question lucidly? Is it 'special matter' that decides morality?

By your definition is material nature everything and what is metaphysical impossible? Or is it simply definition that excludes 'unnatural', 'super-natural' or immaterial information centres from a sensible consideration of the universe? *In which 'tough-minded' case it's just semantic definition and not fact that rules out mind as metaphysical, denies that consciousness could be an immaterial element and construes that Natural Dialectic is nonsensical.* **A Theory of Intelligence replies that, in the cosmic hierarchy, 'mind-preceding-matter' does not render mind 'unnatural'; and the Dialectic Heart from which all nature issues is not 'super-natural' but Most Natural of All. What more Innate a Quality of Heart, as humans know, than Love? Shines, therefore, at the Sacred Heart of Nature, Love?**

Natural principle is not man-made; religions are. Thus nature and its Nature both precede religion of all cuts and colours. Mystic science of The Central Soul is as natural as natural science of the universal body. The inspiration of both sciences, metaphysical and physical, is truth; and each transcends, as does nature, the distinctions of class, race and creed. In pure form, as knowledge for its own sake and reward, they stand clean above the market place. But their intense neutrality is nothing less than, though dispassionate, thoroughly enthusiastic. This lack of bias and financial detachment, while it generates precise morality and mathematics, keeps universal tolerance. And such illuminated tolerance can, as history shows, filter through religion to the politics of man. The virtues of its natural, perennial fire might well, through an understanding of the mystic's guidelines, lead affairs of men. It is a fact that moments of enlightenment, the pinnacle of immateriality, as guiding lights have shone through all recorded history; they have swept as guiding, humanising waves through all societies of man. Who craves the dark? Who preaches chaos on the human sea? Who wants to lose his way?

The science of mathematics is not science of morality. Thus take, as was habitually taken, advice from experts in the field of subjectivity - called, by tradition, saints. Non-secular, humanitarian and well-informed, a saintly kind of leadership would tolerate all positive religious aspiration and encourage all morally appropriate scientific application. If tolerance is dyed into the texture of your social cloth then thinking men and not aggression can decide the tack. In this respect aren't unions and neutralities derived from Natural Dialectic and the liberal inspiration of a humanist's good nature much the same in practice? On the field of daily life are their perspectives split apart as modern thinking thinks? *But they dissociate from world faiths in two very different ways. One, beyond and yet fathering religions, fosters them; the other basically rejects*. A humanist has faith, above his non-existent deity, in human nature and the progress of technology. We've seen how terribly misplaced this faith has been, is and, unless you counter with a pressing optimism, by past record will be. Faith in material or immaterial priority defines the pair. **Their views of origins and therefore self and cosmos differ radically; thus how they treat our soul is angled poles apart.**

	existence	*Essence*	
	relative truth	*Truth*	
	duality	*Unity*	
↓	*division*	*unification*	↑
	no intelligence	*intelligence*	
	chance	*design*	
	towards illusion	*towards truth*	
	no thanks/ ingratitude	*thanks/ gratefulness*	
	ridicule	*praise*	

I hope you have enjoyed the exploration of a philosophical architecture whose *motif* is the binary pattern of Natural Dialectic. We have, considering both inanimate and animate, circled round the universe. *What is, from the evidence, the bottom line and top conclusion? What, at the most natural core of cosmos, is the nature of Truth? The choice, the faith is at the end between material chance and a Creator.*

In a cosmos finely tuned for life, with Darwinistic evolution logically counted out, the Dialectic shows the latter is a choice most open, rational and likely Real and True. In the Glossary check 'evolution' one last time. The Primary Axiom of Natural Dialectic rejects the second usage but, accepting all the others, glories in the fourth. *It thereby deduces that the capability of cosmos is, beyond the dreams of physical cosmology, life-friendly; and that the nature of Nothing is, paradoxically, the Natural Heart of everything. Moreover it identifies the way to enter Nature's Heart and, unequivocal beyond belief, know the Communion of Saints.*

The Essence of Natural Dialectic is its path of (raj) unification leading to a climax of (Sat) Unity. Better, therefore, dwell in hope than hopelessness; much better True Hope than false hopelessness. **Wisdom means the enlightened discernment of truth and philosophy means love of wisdom. Develop your Potential, say the mystics, and realise it to the full. The practical and most rational aim of Real Philosophy can therefore only be to lighten and enlighten; its arrow points you, right-hand upwards, towards Essential Truth.**

I rest my case.

Glossary

A

adaptive potential: involves pre-programmed, super-coded switches and recombinant (transposable) refinements intrinsic in the genomic program of any particular biological type (*SAS* Chapter 23).

***ahamkar*:** pre-scientific term; conscio-material band/ grade - conscious band of mind; sometimes identified (only partially correctly) with *ego*; faculty of self; *ahamkar*, involving identity, frames thought; habitual identification with own physical body; also with friends, family, community, study, work, country or cosmos as a whole; intellectual analysis is a 'knife' that dissects according to the interpretations of this frame.

allele: a sister gene; you have two copies of life's book, one from mother and the other from father, so that each gene from father has a correlate 'allele' from mother and *vice versa*.

anti-entropy: *see* **negentropy**

archaea: phyletic anaomaly; prokaryote differing significantly from bacteria regarding cell wall, membrane, genes and some metabolic pathways; unlike bacteria its replication and transcription is of eukaryotic type; chirally distinct codification for membrane lipid (L) glycerate-1-phosphate, as opposed to (R) glycerol-3-phosphate in all other organisms; t-*RNA* also unique; origin and evolutionary 'progress', if any, unknown.

archetype: basic plan, informative element; conceptual template; pattern in principle; instrument of fundamental 'note' or primordial shape; causative information in nature; 'law of form'; nature's script; Natural Dialectic's 'holographic' edge, omnipresent but invisible because it's metaphysical; the psychosomatic place where metaphysic and its physic meet; morphological attractor or field of influence in universal mind; the subconscious component of universal (natural) mind comprising archetypes; prototype-in-mind (maybe related to Platonic ideas or Aristotelian entelechies); potential matter seen as hard a metaphysical reality as, say, particles are physical realities; program(s) naturally stored in cosmic memory - simple in terms of inanimate physical 'law' (of particles and forces), complex in terms of animate structure/ function/ behaviour; information stored in a typical mnemone; in biology, metaphysical correlate of biological type/ super-species that is physically expressed in code as *potential* form; abstract or metaphysical precursor; the collective unconscious of a type e.g. human type; as thought is father to the deed or plan is prior to ordered action, so archetypes precede physical phenomena; pre-physical initial condition of matter.

Archetype: Primary First Cause; *Logos* (*figs. 4*.1, 7.6, 10.1, 12.1).

***ATP*:** Adenosine TriPhosphate, life's standard bearer of chemical/ heat energy; a cell's agent of energy transmission; a biological 'match' or 'battery'; an active cell may discharge many thousand units of

ATP per second to drive its metabolic machinery; these are recharged by respiration; *ATP* also plays a critical informative role in the transmission of nervous and possibly other signals.

***AV*:** Authorised Version e.g. *AVS* authorised version of science.

B

base: significant component of a *DNA* nucleotide: a letter in 'the book of life': there are 4 bases in the genetic alphabet - A (adenine), G (guanine), C (cytosine) and T (thymine): in the case of *RNA* base T is replaced by U (uracil).

base pair: the conservative accuracy of genetic inheritance and the elegant construction of *DNA* are both dependent on a base-pairing rule viz. G pairs only with C and A with T (or U).

big bang: unconscious singularity *see* also **transcendent projection;** Cambrian explosion sometimes called 'biological big bang'.

biomimetics: also known as biomimcry; fast-expanding field of scientific study and imitation of the biological production of codified substances and processes; research in order to better inform the processes of design (engineering) and technology (production) for human purposes.

black box: process or system whose workings are unknown.

***buddhi*:** pre-scientific term; conscio-material band/ grade - conscious band of mind; faculty of intellect; instrument, whether sharp or blunt in an individual case, of learning and discovery; analytical tool to educe physical and metaphysical patterns; pragmatic and hypothetical power of reason; crucial to gauge physical circumstance for survival and metaphysical principle for optimising state of mind.

C

caduceus: staff of Hermes/ Mercury the communicator, intercessor and informant deity; the messenger of metaphysic, carrier of thought is a power mythologically trivialised; symbol, including double helix, used by Natural Dialectic to represent basic human infrastructure, that is, the archetypal form of man.

***chakra*:** pre-scientific term; conscio-material band/ grade - subconscious mind; metaphysical modulator; also called a node or plexus; device for the wireless transmission of *prana* to the electrical systems (e.g. nervous) of an organism; psychosomatic gate; trans-dimensional (metaphysical to physical and *vice versa*) transit-point for informative signals; a mechanism including an antenna (receiver), transformer (between 'voltages' of *pranic* energy, transducer (between electrical and *pranic* conveyance of charge), simple harmonic oscillator (a 'heart' controlling *pranic* flow) and distributor (of *prana* through a network of meridians); archetypal channel; specific mind-body broadcasting interface; morphogenetic informant especially important in the process of bio-development; lowest metaphysical component in the hierarchical transmission of power throughout macrocosmic creation and its microcosmic reflection as mankind and other forms of life of earth; *chakras* are power hubs (and thus, like suns, 'controllers of regions') which exist on a cosmic as well as individual-body scale; conscious chakras (or focal concentrations) of higher mind are not included in this cosmology; because it cannot observe or experiment on metaphysical apparatus science has not developed any understanding beyond the physical expressions of informed energy; see also *prana.*

chaos: a confusing notion with three main but disparate implications - emptiness, disorder and randomness; Greek word meaning chasm, emptiness or space; structureless 'profundity' that pre-existed cosmos; *prima materia*, *prakriti* or primordial energy structured by regulation of divinity, archetype or natural law; anti-principle of cosmos i.e. disorder; any case of actually or apparently random distribution or unpredictable behaviour; also apparently random but deterministic behaviour of systems (e.g. weather, electrical circuits or fluid dynamics) sensitive to initial conditions.

chemical evolution: also called abiogenesis, biopoesis, chemogenesis or prebiosis; implies that lifeless chemicals 'evolved' to the point whence they could 'self-construct' the primary unit of life, a reproductive cell; it means the generation, perhaps gradually over a long period of time, of life from non-living components by physical means alone. This process is integrally part of, strictly not the same as, Darwin's consequent evolution.

chitta: pre-scientific term; conscio-material band/ grade - conscious mind; attention *per se*; pure, formless (or boundless) intelligence in which forms of thought are projected; source of ideas and creativity; psychological focus.

chloroplast: organelle in plant cells containing photosynthetic apparatus.

chromatin: a nuclear complex that, using histone proteins, helps package, reinforce and control the expression of genetic *DNA*.

chromosome: a 'book' in the 'encyclopaedia' of life; the human genome contains 46 chromosomes.

cladistics: method of classification using diagrams called cladograms; organisms are collected into groups on the basis of shared (homologous) features; homologies are tallied and numerical rather than speculative, evolutionary/ phylogenetic links drawn up between organisms; cladism is thus a powerful, neutral, objectively detached tool of analysis and for this reason the technique enjoys growing popularity among the world's taxonomists.

code: the systematic arrangement of symbols to communicate a meaning; code always involves agreed elements of morphology (the form its symbols take), syntax (rules of arrangement) and semantics (meaning/ significance); without exception such prior agreement between sender (creator/ transmitter) and recipient involves intelligence.

codon: 'word' in the genetic language; stands for an amino acid or a full stop; since more than one codon may stand for a single amino acid the genetic code is sometimes ill-perceived as 'degenerate'.

conscio-material spectrum: basic, binary structure of existence; slope of creation that, in dialectical description, extends from the immaterial pole (a concentrate of Pure Consciousness) to a material pole of pure non-consciousness - the physical plane; drop/ descent from Essence through existence; drop from Conscious Singularity to unconscious singularity (black hole); a hierarchical description of polar creation simply modelled *passim* by the use of spectrum, concentric rings and, step-wise, ziggurats; immaterial, subjective element (information) tapers on a sliding scale with material, objective element (energy); 100% objective is, for example (*fig.* 0.14), 0% subjective - physical nature is a special case of consciousness, its total absence at creation's spectral base; *vice versa*, at Top, Transcendent Nature (Essence) there is 0% objective form; and ratios, from top to bottom, in between; from this perspective any 'sharp' division between mind and matter becomes illusory; it is no more real than exists between, say, UV and microwave radiation in the 'rainbow' continuum of their e-m frequencies; cosmos is a conscio-material spectrum, a taper of consciousness to, at base, its absence (*fig.* 0.14).

convergence: the tendency of unrelated organisms to evolve similar characteristics; in the case of *divergence* adaptation/ speciation from an original feature occurs (e.g. beaks of finches); *convergence*, involving the unrelated, mosaic occurrence of similar features (such as the camera eye, viviparity and thousands of other instances), runs counter to Darwinian expectation; it means that such codified features must have evolved independently many times over; evolutionary explanations of this profound yet ubiquitous puzzle may thus involve speculations such as appeal to non-random 'deep bio-structure', 'principles of evolution', 'morphological laws' or 'inevitablility' granted by imaginary natural laws of codification/ innovation; for a design theorist the bio-codification and engineering of 'convergent' forms derives from either an original use of modular programming or, in the case of so-called micro-evolutionary variation, from in-built adaptive potential flexibly but appropriately activated by genetic switches and epigenetic markers.

cosmic fundamentals: cosmic psychological and physical qualities (see Chapter 2); basic states or tendencies; universal ingredients whose mixture is variously expressed in every object and event.

cosmological axis: human pivot; the point at which subjective and objective perception meet; eye-centre; third eye; thought centre; *ajna chakra.*

cosmological principle: idea that, on a sufficiently large scale, the distribution of matter in the physical universe looks just about the same from any vantage point; it therefore has neither centre nor, being infinite, edge - unless of course, its space is somehow spherical.

cosmos: often applied to physical universe, universal body; from Greek word denoting orderly as opposed to chaotic process; involuntary pattern of nature; also equated, including metaphysical mind, with existence as a whole; seen, dialectically, as a projection through the template of metaphysical archetypes; **the umbrella title of the series and website of books, Cosmic Connections, could, with reference to the Natural Dialectic which structures its *CUT* (see Glossary: unification), equally be called Orderly Linkages**.

creation: origination; physical or psychological arrangement; mind creates with purpose, matter without; creation means active production but also passive result; a creation will have been informed by force of mind and/or matter.

D

dialectic: a form of debate between positions of polar opposition (argument and counter-argument or thesis and antithesis); the motion of to-fro discussion that results in resolution (synthesis) whereby points of view are aligned; balance, compromise, neutral ground, golden mean and central truth are aspects of this synthetic (*Sat*) fundamental; paradoxically, two become one; union supersedes division; Natural Dialectic suggests that dialectical motion reflects the binary, cyclical nature of cosmos; such polar, to-fro or oscillatory dynamic occurs as the continual disequilibrium of nature (called motion and transformation) always seeks its various re-balances.

dialectical stack: stack of opposites; columnar expression of polarity; there are two kinds of stack - primary or non-vectored and secondary, vectored; primary (essential) stacks set (*Sat*) Unity against (↓ *tam*/ *raj* ↑) duality (for elaboration see Chapter 2 and *figs*. 1.4 and 2.2); secondary (existential) stacks represent the various kinds of polarity from which the changeful web of existence is composed (see *figs* 1.4 and 24.1); each pair of polar 'anchor-points' implies a scale or dynamic range that runs between 'paired opposition' or 'complementary covalency'; stacks do not necessarily list synonyms or make equations; ***their perusal is intended to promote connections because consideration of connections tends to help unify/ collate/ organise one's working comprehension of any matter in hand.***

diploid: having full genetic complement with one copy of chromosomes from each parent e.g. you have 46 chromosomes, 23 from mother and 23 from father.

DNA: a complex chemical; a large bio-molecule made of smaller units, nucleotides, strung together in a row; a polymer in the form of a double-stranded helix; a medium superbly suited to the storage and replication of 'the book of life'; 'paper and ink' on which the genetic code is inscribed; an organism's 'hard drive'.

dukkha: imperfection, suffering.

E

electromagnetism: physics of the field that exerts an electromagnetic force on all charged particles and is in turn affected by such particles: light/

e-m radiation is an oscillatory disturbance (or wave) propagated through this field; light; light paradoxically involves a perfect, polar balance between contractive/ magnetic and radiant/ electric components.

elementary particles: science has discovered and, for the most part, experimentally verified, over fifty elementary particles; these are divided, in simple terms, into bosons (force carrying particles) and fermions (separate particles); bosons include photons (which mediate the electromagnetic force), gluons (which mediate the strong nuclear force), W and Z particles (which mediate the weak nuclear force), possibly gravitons (which mediate the gravitational force) and also possibly a Higgs boson (which may mediate a proposed mass-giving field); fermions include two main groups - six quarks and leptons (six electron/ neutrino types); derived from quarks are strongly interactive composites called hadrons; hadrons include baryons such as protons and neutrons and (perhaps a little confusingly) bosons such as short-lived mesons.

entropy: a measure of the amount of energy unavailable for work or degree of configurative disorder in a physical system (see second law of thermodynamics); inertial aspect of an energetic, material or conscious gradient; diffusion or concentration gradient outward from source to sink; drop towards 'most probable' outcome i.e. inertial slack; a measure of disintegration or randomness; expression of the (*tam* ↓) downward cosmic fundamental; a major property of matter, closely coupled with materialisation; in a closed system, which the universe may or may not be, this tends the eventual loss of all available energy, maximum disorder and the exhaustion of so-called 'heat death'.

enzyme: biochemical widget; protein catalyst without whose type metabolism (and therefore biological life) could not happen.

epigeny: genetic super-coding; contextual punctuation; chemical modification of *DNA*; also extra-nuclear factors that may cross-reference with genetic expression.

equilibrium three modes of equilibrium are (*sat*) balance of poise or pre-active potential; (*raj*) dynamic balance occurring in all regular cycles, wave-forms and cybernetic homeostasis that is basic to the stability of life-forms; and (*tam*) inertial equilibrium that results from diffusion of information or energy; it equates with exhausted inaction or 'flat', impotent rest; such post-active inertia represents the most probable distribution of energy/ matter with the least energy available for work viz. the most random arrangement permitted by the constraints of a system; expressed in psychological terms as ignorance, unconsciousness or sleep; see also equilibration, *karma* and *fig.* 1.1 'Pivoted Existence'.

Essence: (*Sat*) Supreme or Infinite Being; Substance (perhaps Spinoza's Substance) 'prior to' or 'above' existence; Pure Consciousness/ Life; Peace that transcends all psychological and physical action;

the root of an essentially undivided universe; Conscious Singularity; Uncreated One within which and whence all differences have their being; Apex of Mount Universe; goal of saints/ 'philosopher kings'; the 'point' at which All-Is-One.

eukaryote: non-prokaryote; any organism except bacteria and blue-green algae.

evolution: there are today *four* main usages of this word; each 'loading' derives from the original Latin, 'evolvere', meaning to unroll, disentangle or disclose; the *first two*, physical and biological, are conceived as natural/ mindless processes; the *second, mindful pair* is of psychological/ teleological import; specious ambiguity may conflate or switch between the fundamentally separate pairs of meaning. ***Firstly***, in the scientific context of physics and chemistry, the word is used to describe change occurring to physical systems; the laws of nature can't, it seems, evolve through time but stars, fires, rocks or gases can. ***Secondly***, though also subject to the 'rules' of entropy, biological evolution is a theory of *random progression* from simple to complex form; it thereby implies increasing, codified complexity; while retaining the 'hard loading' of physical science it also, ambiguously, claims that codes, programs, mechanisms and coherent, purposive systems - normally the province of mental concept - self-organise by, essentially, chance; such confusion, the basis of naturalism, is compounded by failure to distinguish between, on the one hand, ubiquitously observed variation (called micro-evolution) and, on the other, Darwinian 'transformation' between different sets of body plan, physiological routines and associated types of organism - such 'black-box macro-evolution' as is never indisputably observed; to evoke a naturalistic ambience it is fashionable to use 'evolved' interchangeably with or to replace the words 'was created', 'was planned' or 'designed'; finally, it is noted that the coded, choreographed development of a zygote, packed with anticipatory information, through precise algorithms to adult form is the absolute antithesis of blind Darwinian evolution. ***Thirdly***, man certainly evolves ideas; intellect can evolve 'purposive complexity'; we invent all kinds of codes, schemes and machines; we devise increasingly complex theories and technologies; and we evolve an understanding of natural principles; this, which all parties accept, is an informative, psychological sense of 'evolution'. The ***fourth*** sense of evolution, at least as near to the original Latin as the other three, is the spiritual usage; immaterial spiritual evolution, unacceptable to materialists and unknown to physical science, is at the very heart of holism; in this voluntary sense of evolution practitioners cast off material attachment, evolve and merge into the *Logos*; evolution (or, perhaps better, centripetal involution) of the soul is their great business; their aspiration is to unite with The Heart of Nature.

evolution pre-Darwinian: minority/ anti-mainstream pre-Socratic snippets and sense-based Epicureanism lionized by interpretations of post-18th century materialists; virtually undetectable eccentricity in

Chinese, Indian and Islamic literature; natural selection treated by creationists al-Jahiz and Edward Blyth; Buffon, a non-evolutionist, addressed 'evolutionary problems'; Lamarck (evolution by inheritance of acquired characteristics); hints in poem by Erasmus Darwin.

evolution Darwinian: mechanism - natural selection; major tenets - common descent (inheritance), homology and 'tree of life'.

evolution neo-Darwinian/ synthetic: as Darwinian, except synthetic theory adds random mutation as the mechanism for innovation; also adds a mathematical treatment of population genetics and various elements (e.g. geno-centric perspective) derived from molecular biology.

evolution post-synthetic phase: natural selection and random mutation acknowledged as mechanisms insufficient to source bio-information; post-Darwinian evolution invokes mechanisms from hypotheses such as *NGE* (natural genetic engineering) and 'evo-devo'; holistic possibilities also address the origin of complex, specified and functional bio-information.

existence: which 'stands out' from background 'nothingness'; the apparently divided universe; seemingly disparate, finite things; all motion/ change/ relativity; all psychological and physical events.

exon: specifies the amino acid sequence for a protein; m-*RNA* after protein editors have removed introns.

F

field: any extent wherein action either physical or metaphysical but of a certain kind occurs e.g. field of battle, influence of mind or magnetism; the scientific definition is limited to a collection of numbers varying from point to point - such as a scalar field of contours on a map - or numbers with direction - such as a vector field showing speeds and directions of wind.

first causes: check *figs*. 3.3, 5.1, 9.1, 11.1 or 11.3. First cause is first motion in a previously undisturbed, pre-conditional field.

First Cause Psychological is Archetype, Potential Informant or (see Chapter 5: Top Teleology) *Logos*; attributes of this Primary Source and Sustenance of Creation include omnipresence, omnipotence and omniscience.

first cause physical is also called potential matter or archetypal memory; as the secondary source of creation it precedes physical phenomena; as such it is, transcending physical appearances, metaphysical; this 'physical nothingness' is therefore, paradoxically, the source of everything composing astronomical cosmos; it consists of their being or essence as opposed to their becoming; its void, with respect to the presence of finite phenomena, appears infinite; attributes of immanent archetype, the primary informant of our non-conscious, energetic universe, include omnipresence and omnipotence.

free will: free will, reflecting cosmos, occurs in three stages; first, in non-conscious, automatic matter, that is, the physical universe, there is none; second, in conscious mind freedom of will is relative; the degree of this relativity is a function of the type of an impression, that is, of the higher or lower quality of a memory, purpose, form of thought or feeling on mind's spectrum; third, the Essential Nature of Unconstrained Free Will is, paradoxically, 'constrained' by Transcendent Love.

G

gamete: sex cell with half of full genetic complement i.e. a single set of chromosomes.

gene: generally means a basic unit of material inheritance; section of chromosome coding for a protein; digital file; a reading frame that includes exons and introns; the old one gene-one protein hypothesis is incorrect; in fact, by gene splicing, a particular piece of *DNA* may be used to create multiple proteins.

genome: total genetic information found in a cell: think of the genome as an instruction manual for the construction and physical operation of a given organism.

genotype: the genetic constitution of an organism, often referring to a specific pair of alleles; the prior information, potential, plan or cause of an effect called phenotype.

gravity: in physics an attractive mass-to-mass force or warping of space-time; in Natural Dialectic the term is redefined more broadly - the agency of its (*tam* ↓) downward vector includes all psychological and physical factors of materialisation; such 'gravitational' factors and their properties are listed in the left-hand column of Secondary, Existential Dialectic; they include pain, pressure, confinement, strong nuclear force, mass, electromagnetic binding, inertia, entropy, 'standard' gravity and so on; gravity might be summarised as 'negative power' or 'the principle of death'.

GTE general theory of evolution; *see* macroevolution.

H

haploid: having half the full genetic complement, as in the case of sex cell.

heterozygous: having different allelic forms of a particular gene.

holism: opposite of reductionism; the view that a whole is greater than the sum of its parts; the extra metaphysical (immaterial) ingredient is identified by Natural Dialectic as information; information implies the purposeful design, development and arrangement of contingent parts in a working system; may operate according to a Logical Norm.

hologram: a 3-d photograph made with the help of lasers. Unlike a normal photographic image each part of it contains the image held by the whole.

homeostasis: vibratory or periodic control of a system to obtain balance round a pre-set norm; the mechanism of its information loop involves sensor, processor and executor; the operative cycle works by negative feedback; psychological (nervous) and biological cybernetics; the informed basis of biological stability.

homeotic gene: gene (e.g. *Hox* gene) involved in developmental sequence and pattern; high-level co-determinant of the formation of body parts.

homozygous: having the same allelic forms of a particular gene.

I

illusion: is the cut between illusion and delusion an illusion? illusions, apparently outside the mind, appear real; a delusion, in it, we think real; neither, mind allows, is real or true.

information the immaterial, subjective element; information occurs in three distinct modes; informative *potential* is action's precedent; this potential is both the source and substrate of all psychological activity; we know this substrate of life and consciousness; *active* information inhabits its own centre, mind; mind knows, feels, purposes, creates, codifies and recognises meaning; it is also, by way of secondary and subconscious forms of archetype) a physical entrainer; thus *passive* forms of information, either in memory or physical objects and events, reflects active; in other words, *passive* information is stored as subconscious 'files' in memory; and it is fixed in the expressions of non-conscious matter according, universally, to the archetypal behaviours of natural bodies or, locally, to particular schemes of life; that is to say, both the constructions of life-forms and the inanimate cosmos are the physical product of stored concept.

informative entropy :loss of information due to degradation of its carrying medium; such a medium may be metaphysical (mind) or passive and physical (for example, computer files or genetic code); and its entropy may be metaphysical (loss of memory, focus or consciousness) or physical (for example, genetic mutation); the informative correlate of such degeneration is diminished organisational capacity, meaning or thrust of original purpose.

informative negentropy: gain of informative clarity; increasingly focused, purposive specificity; associated with knowledge, wisdom, grasp of principle and pristine construction; machines are a good example of informative negentropy.

intron: genetic control panel; n-p-c (non-protein-coding) segment(s) spliced from an m-*RNA* transcript prior to translation; introns include regulatory elements (to variably promote or inhibit gene expression) and addressing factors of the genetic operating system; gene-attached information lending specific flexibility to protein manufacture.

inversion: turning upside-down or inside-out; reversing an order, position or relationship; in a hierarchical sense inversion is allied with the reflective asymmetry of opposite poles; information outwardly expressed; pole-to-pole reversal integral to dialectical structure; inversion represents, of the two anti-parallel vectors on creation's conscio-material gradient, the (*tam*) centrifugal vector; as opposed to (*raj*) centripetal eversion; various kinds of inversion (cosmic and micro-cosmic (biological)) are discussed in these books.

K

karma: action; law of cause and effect, that is, balance between action and reaction; *equilibration* such as underlies all mathematical equation; a deed with implications of the reactions or 'payback' it provokes; fruit or result of previous thoughts, words and deeds; applies as rigidly to metaphysical (psychological) as, in Newton's Third Law of Motion and mathematical *equations*, physical events.

L

levity: agency of the (*raj* ↑) upward vector; dialectical converse of gravity; psychological and physical 'levitatory' forces lift or stimulate; they are listed in the right-hand column of Secondary, Existential Dialectic and include light, heat, excitement, dematerialisation, release, negentropy, focus of interest, affection and so on; physically, levity includes anti-gravity or the intrinsic property of matter's absence, space; generally summarised as 'positive power' or 'a buoyant principle of liveliness'.

logic: analysis of a chain of reasoning; principles used in circuitry design and computer programming; 'normative reason' relates to the basic axiom(s) of a given standard e.g. *bottom-up* materialism or *top-down* holism; three main logical thrusts are: (1) inductive (premises/ observations supply evidence for a probable/ plausible conclusion) as in the case of experimental science working *bottom-up* from specific instances to general principle: (2) abductive (best inference concerning an historical event): and (3) deductive (conclusion in specific cases reached *top-down* from general principle): two pillars of logic are holism and materialism; holism employs mainly deductive/ abductive operations and a Logical Norm; materialism tends to inductive/ abductive operations whose axis is non-conscious force and chance.

Logos: First Cause; Prime Mover; Causal Motion that sustains creation's conscio-material gradient; labelled with many names at different times, places and languages; *Logos,* transcending mind, is Conscious; therefore, *Who* is *Logos*?

LUCA: last universal common ancestor.

M

macrocosm the physical universe of astronomy and cosmology; dialectically, the whole of existence (i.e. both universal mind and universal body) as opposed to individual, microcosmic objects and events - including the human body.

macro-evolution: large-scale, non-trivial evolution; process of common or phylogenetic descent alleged to occur between biological orders, classes, phyla and domains; includes the origin of body plans, coordinated systems, organs, tissues and cell types; unexplained by mutation, saltation, orthogenesis or any known biological mechanism; sometimes called 'general theory of evolution'

(*GTE*); crucial but unseen, hypothetical extrapolation from micro-evolution; may or may not occur; an extrapolation from Darwinian micro-evolution vital to sustain a 'progressive' materialistic mind-set, is conjectural alone.

manas: pre-scientific term; conscio-material band/ grade - both conscious and subconscious bands of mind; 'mind-stuff'; 'clay' moulded by the the hand of thought and perception; metaphysical 'material' on which direct formative action of *chitta* occurs; 'film' on which the perceptions of mind are developed and, at the same time, 'screen' on which they are perceived; receptor for sense impressions and storage silo of such impressions as 'seeds' or 'files' of subconcious memory; substance of archetypal field, in other words, of universal archetypes (cf. typical mnemone); mental form and energy.

mantra: archetypal symbol; psychological transformer; authorised form of words repeated to exclude other thoughts; examples include the 'Hail Mary', '*Om Mane Padme Hum*' and, materialistically, 'evolution made…', 'in time nature designed…' or similar incantation.

maya: partial truth; world of forms and forces; illusion that changeful cosmos is the ultimate reality; motions and perceptions composing *maya* are thus, set against Essential Truth, more or less unreal; becoming wise to the nature of *maya* yields liberation from its cosmic veil.

meditation: *medius*, middle; coming to Centre; mind and body dropping away.

meiosis: shuffling the information pack: variation-on-theme; mechanism for the production of haploid gametes; genetic postal system for sexual reproduction.

metabolism: body chemistry.

metaphysic = non-physical/ immaterial/ psychological/ unnaturalistic; physically expressed as specific/ intended arrangement/ behaviour of materials; physical behaviour reflects metaphysical blueprint; involves element of information; involves symbol/ code/ abstraction/ logic/ reason/ mathematics; also meaning/ message/ goal/ teleology; also consciousness/ mind/ life/ experience/ feeling; and also morality/ force psychological/ emotion; involves innovation/ creativity/ art/ invention/ aesthetics.

microcosm: an entity that reflects the universe by containing all its basic constituents. Used especially of the human state where it may refer to both mind and body or, in a purely physical context, body alone.

micro-evolution: misnomer; non-progressive, small-scale variation within a species or, more broadly, between strains, races, species and genera; variation/ adaptation within type; trivial Darwinian changes that may occur by natural selection/ ecological factors acting on genetic recombination, mutation or adaptive potential (q.v. this Glossary); sometimes called 'special theory of evolution' (*STE*), micro-evolution/ variation is a fact.

mitochondrion: organelle in eukaryotic cells containing the apparatus for aerobic respiration.

mitosis: conservative copying and delivery of genomes in cell division; genetic reprinting; genetic postal system for asexual reproduction.

mnemone: a division of memory whether individual or universal: an individual's two divisions are *personal mnemone* (likened to a working cache or data store) and *typical mnemone* (likened to a *ROM* or an operating system); typical mnemone is, in effect, a program consisting of three subroutines - *signal translation*, *instinct* and *morphogene* (for more information see *SAS* Chapters 15 - 17); it is also a synonym for natural, archetypal memory in universal mind; in short, it is a body's ***metaphysical DNA.*** Further than the character of each bio-type of organism it also includes the 'instinct' of matter (i.e. cosmos). Natural Dialectic's definition involves no 'cultural' connotation whatsoever and is thus wholly distinct from evolutionary psychology's use of the word.

mobile genetic element: transposon, retrotransposon, insertion sequence, other non-protein-coding *DNA*, n-p-c *RNA* fragments and various protein regulators that together expedite the operating system of a genome.

morphogene: one of three sub-routines of typical mnemone or archetypal memory relating to physical construction; morphological attractor; the component of subconscious mind associated with electrochemical function and thereby body; just as you might not guess from the picture on your TV screen or object from a 3-d printer the nature of the electromagnetic messaging that creates it so you might not guess a body's shape from its *DNA* or the messaging agent that links archetypal mind with body; morphogene is the dominant, perhaps exclusive, aspect of mind in unconscious organisms such as plants or fungi.

morphogenesis: the development of biological structure; more generally, the production of physical form.

mosaic: the presence of permutations of codified sub-routines or similarities of form and/or function scattered in organisms unrelated by lineage.

mutation: accidental change to genetic code.

mysticism: quite different from objective, it is the subjective science; not philosophy, religion or opinion but practice to achieve communion with natural, inner, immaterial truth; esoteric as opposed to exoteric, materialistic discipline; 'science of the soul'; as gyms and physical action are to athletes so meditative exercise and psychological stillness are to mystics; involves psychological techniques to achieve a clear, rational goal - purity of consciousness and thereby understanding of the fundamental nature of the informative principle, mind; since life is lived in mind a mystic seeks consummate knowledge of life's source and sanctum, that is, communion with its deathless heart; adepts were, are and will be 'Olympian' meditative concentrators.

N

nano-biology: biology of structures/ physiologies involving a few atoms or molecules; 'extremely small biology'.

nanotechnology: technology at atomic and sub-atomic level as is, basically, life's.

naturalistic methodology: also known as 'methodological naturalism': is, strictly, not concerned with claims of what exists or might exist, simply with experimental methods of discovering physically measurable behaviours; thus only materialistic answers to any question (e.g. how biological forms arose) are deemed 'scientific' or 'scientifically respectable'.

negentropy: opposite of entropy; lowering of entropy; expression of the (*raj*) upward-pointing cosmic fundamental closely coupled with stimulus, dissolution and dematerialisation; a measure of input, cooperation or synthesis; motive/ fluidising aspect of an energetic, material or conscious gradient; gain of energy, configurative order, information or consciousness in a system; when used in terms of information negentropy involves gain in order or understanding of principle from which different actualities derive; a measure of the amount of concentrated/ conceptual information, specific, intentional complexity or conscious arrangement in a system; a natural and essential property of mind.

Nirvana state of enlightenment; 'non-condition'; nirvana is devoid of existential motion; extinction of existence (i.e. perpetual change) leaving Essence Alone; pure soul; psychological super-state; Buddhists call such transcendence non-self or the Formless Self.

non-existence: where creation = formful existence, non-existence is formless; the polar opposite of physical space and time is Transcendent Potential; such pre- or super-existential formlessness is non-existent; Absolute Non-Existence is Essential; however relative non-existences of two kinds also occur; the first kind is metaphysical/ subjective and therefore psychological; it involves the absence of a specific psychological form or event; unconscious oblivion is one such non-existence; the second kind involves the local absence of a possible physical event (an object is a 'slow event'); impossibilities are non-existences but imaginations of non-existence (including symbolic abstractions, hypothetical entities, physical absences, absolute emptiness and the number zero) exist; furthermore, the nothingness of space and time, the zero-point of calculus and zero's empty set together constitute the basis of physical science and mathematics.

non-protein-coding *DNA*: occupies probably 95% of eukaryote and 80% of bacterial genomes; associated with the genetic operating system; may include some genuinely redundant misprints or duplications but now thought for the most part critical to the flexibility, efficiency and even possibility of gene expression; once thought of as useless, degraded information and ignorantly called 'junk *DNA*'.

non-protein-coding *RNA*: n-p-c *RNA* is also called nc-*RNA* (non-coding), nm-*RNA* (non-messenger) or f-*RNA* (functional); functional *RNA* molecule not translated into protein; many 1000's of different specimens include classes of t-*RNA* (transfer *RNA*), r-*RNA* (ribosomal *RNA*) and, commonly involved in the regulation of gene expression and other intra-cellular tasks, micro-*RNA*, double-stranded si-*RNA*, pi-*RNA* and so on; also, for inter-cellular communication, ex-*RNA*.

nucleic acid: *see DNA* and *RNA.*

nucleosome: a 'reel' composed of histone proteins around which chromosomal *DNA* is precisely wrapped; repeated nucleosomes allow the *DNA* to form a bead-like structure that can coil and super-coil; *DNA*, nucleosomes and other factors compose chromatin.

nucleotide: basic, triplex unit of nucleic acid polymer; monomer composed of phosphate and sugar (the 'paper' part) and base (the 'ink letter'); letters' of the genetic alphabet are (G) guanine, (C) cytosine, (A) adenine and (T) thymine. In *RNA* thymine is replaced by (U) uracil.

nucleus: centre, heart, creative core; informative *sine qua non*; psychological nucleus is consciousness or (in formful aspect) mind; atomic nucleus, made of protons and neutrons, is a centre of mass determining electron configuration; biological cell nucleus is the instruction centre of a cell containing *DNA* and nuclear operating machinery; nuclear is critical.

O

***Om*:** universal sound, fundamental reverberation, basic truth; initial motion of Potential Information; sometimes spelt *Aum*, a Sanskrit word whose Semitic transliterations are Am'n, Amin and Amen; see also First Cause, *Logos*, *Kalam*, *Shabda* etc.

order: regular, regulated or systematic arrangement; organisation according to the direction of physical law; passive information by which things are arranged naturally (with predictable but non-purposive complexity) or purposely (with innovative or specified complexity); mind, generating specified complexity in the order of its technologies and codes, actively informs; the orders of mind are meaningful, the orders of matter lack intent; see also cosmos.

organelle: cellular sub-station; discrete part of a cell; sub-cellular compartment having specific role such as informative (nucleus), energetic (mitochondrion, chloroplast), constructional (ribosome, Golgi body) or other.

P

***PAM*, *PAND*, *PCM* and *PCND*:** philosophical gambits; see Primary Axioms and Corollaries.

phenotype: the effect of causal potential; result of the development of prior, informative 'egg'; outward expression of inner plan; sensible appearance of an organism as opposed to its genotypic scheme: the whole set of outward appearances of a cell, tissue, organ and organism are sometimes called a phenome (*cf.* genotype/ genome).

photosynthesis: process by which inorganic carbon is introduced to the biological zone and energetic sunlight fixed as a crystalline molecule of storage, a sugar called glucose.

phylogeny: evolutionary history; relationships based on common or evolutionary descent.

potential: poise; latent possibility; potent non-action that precedes any particular action or creation; in science potential energy is defined as the energy particles in a system (or field) possess by virtue of position/ arrangement; gravitational, electrical, electro-chemical, thermo-dynamical and other kinds of potential are recognised; in dialectical terms mind precedes matter, information precedes the pattern of material behaviour; information is energy's pre-requisite potential; in this case *informative potential* involves two conditions; firstly, a pre-existential/ essential state of pure potential; secondly, a pre-material, metaphysical fact of potential matter, archetype or laws of nature; if potential's pre-active equilibrium is related to the voltage of a full battery then aspects of psychological 'voltage', whose currents drive intentional behaviour, are purpose, will and plan.

potential matter: see archetype.

prakriti: pre-scientific term; conscio-material band/ grade - whole spectrum of existence; complements Essential *Purusha*; universal energy; generic term for nature; screen and light on which the show of creation is projected; 'clay' with which the potter of conscious experience works to produce form; thus also (*figs*. 6.2, 7.5 and 13.6) identified with the objective, energetic as opposed to subjective, informative side of conscio-material cosmos; fundamental substrate whose root exercise involves continual recombinations of three cosmic fundamentals (Chapter 2); interplay of these qualities, attributes or tendencies intrinsically inhabits every object and event; it generates all patterns, forms and forces, whether in psychological or physical regions of the universe; nested, *prakritic* layers are hierarchically arranged like a grid of stepped-down voltages from a power station; in a second comparison, as the waveband of visible light is part of a much larger electro-magnetic spectrum, so the bands of higher and lower conscious mind, subconscious memory and non-conscious physicality compose the spectrum of conscio-material *prakriti*; in this way, for example, *prana* is a low-level, 'infra-red' expression of *prakriti* operant at the *PSI* border where subconscious archetypes (Chapters 15 and 16) give rise, with location, to quantum phenomena; and its lowest, 'radio' level of expression, peripheral to the full creation, is gross physical energy whose various transformations are expressed as the operations of physics, chemistry and biology.

prana: pre-scientific term; possible conscio-material band/ grade - subconscious and physical (lowest) levels of *prakriti*; lowest metaphysical bandwidth of *Logo*s/ Om; Chinese *qi* or *ch'i*; associated, as in the yogic practice of *pranayama*, with breath and thereby life of material body; also with light (visible band electromagnetism) and oxygen (specifically, negative ionic charge); identified as the archetypal energy of subconsciousness and the operations of typical mnemone (Chapter 16); subliminal, psychosomatic or mnemonic energy of universal mind at archetypal grade; supports *PSI* (psychosomatic) traffic between universal memory and quantum agencies in the case of both biological and and physical formations; 'infra-mental' band called potential matter; vibration underlying perpetual atomic motion; five pranic bands are analogised with visible light's rainbow, each frequency correlated with an expression of elemental character in physical phenomena or with a level of biophysical expression (*fig.* 17.11); wireless *prana* is identified as the metaphysical life-force of the physical body; in living organisms (including you and me) it is processed by way of metaphysical apparatus called node or '*chakra*'; each of a hierarchical series of such nodes operates as an antenna, transformer and distributor; the system acts as a transducer of *pranic* frequencies to those of bio-electromagnetism and electrical charge; the highest bio-frequency resonates with our '*ajna chakra*' or third eye behind the forehead; this in turn, is subservient to '*sahasrara*', the 'thousand-petalled lotus' supplying 'voltage' (and thereby current) to sustain the physical universe; having entered the human system by resonance with the '*ajna*' antenna *prana* is passed through a grid of aforementioned nodes, well-known to yoga and arranged down the spine; each distribute frequencies appropriate to its body area by a network of *nadis* or meridians identified by medical acupuncture; being metaphysical the pranic system cannot be physically tested by empirical, scientific experiment but only by inference (e.g. a cure); for this reason some proponents of occidental medical science dismiss *pranic* mechanisms of the mind as 'pseudoscientific' and, having thus 'rationally' condemned, proceed to narrowly and unwisely dismiss the broader immaterial fraction of holistic order wherein such components play a crucial part.

Primary Axiom of Materialism (*PAM*): all objects and events, including an origin of the universe and the nature of mind, are material alone; cosmos issued out of nothing; life's an inconsequent coincidence, a fluky flicker in a lifeless, dark eternity.

Primary Axiom of Natural Dialectic (*PAND*): there exists a natural, universal immaterial element - information; immaterial informs material behaviour; a conscio-material dipole that issues from First Cause informs and substantiates both mental (metaphysical) and physical creations; there is eternal brilliance whose shadow-show is called creation.

Primary Corollary of Materialism (***PCM***): the neo-Darwinian theory of evolution, that is, life forms are the product, by common descent, of a random generator (mutation) acted on by a filter called natural selection; such evolution is an absolutely mindless, purposeless process; the *PCM* is a fundamental *mantra* of materialism.

Primary Corollary of Natural Dialectic (***PCND***): the origin of irreducible, biological complexity is not an accumulation of 'lucky' accidents constrained by natural law and death; forms of life are conceptual; they are, like any creation of mind, the product of purpose.

prokaryote: non-eukaryote; bacterial type with little or no compartmentalisation of cell functionaries.

promissory materialism: belief system sustained by faith that scientific discoveries will in the future justify/ vindicate exclusive materialism and, as a consequence, atheism; confidence that technology will solve (more often than create) the problems the world faces; may involve a call to progress towards the technological provision of its 'promised land'.

protein: factor made from a specific sequence of amino acids to perform a specific task; 'informative' protein includes some hormones; skin, hair, bone, muscle and other tissues are made of 'structural' protein; 'functional protein' called enzymes mediates all stages in cell metabolism, that is, it catalyses all biochemistry.

PSI **(psychosomatic interface)**: psychosomatic border; the level of mind-matter interaction; bridge between metaphysical and physical dimensions; potential matter; 'gap of Leibniz'; 'fit' of mind to matter; point of linkage between subconscious mind and non-conscious matter; gearing between instinct/ archetype and the behaviour of material objects and energies; as in the case of physical law, psychosomatic influence is both general in potential and local/ specific in engagement.

psychological entropy: a measure of loss of concentration, focus of attention or consciousness; loss of 'mental energy' or aptitude; the drop from waking to sleep; loss of knowledge, information or sensitivity; the gradient from intelligence through stupidity to oblivion; an expression of the (*tam*) downward cosmic fundamental in mind; a tendency predominant in lower, egotistical or selfish mind; increasing level of ignorance, anguish or immorality; loss of integrity, psychological disharmony or disintegration; see also *information entropy*.

psychological negentropy: a measure of gain in order; an increase in concentration, focus of attention or consciousness; gain in sense of purpose, 'mental energy' or aptitude; the rise from sleep to waking, 'dark to light' or unhappiness to happiness; gain in knowledge, information or sensitivity; the gradient of learning and spiritual evolution; an expression of the (*raj*) upward cosmic fundamental in mind; a tendency predominant in higher mind; increasing level of contentment, understanding and the natural morality of happiness; the ascent towards psychological radiance, harmony and integration. The converse of psychological negentropy involves *entropy of information.*

psychosomasis: operation across the psychosomatic border; mind/ body interaction; the one-way imposition of archetypal pattern on *physicalia*; the two-way exchange of information in sentient organisms through the agency/ medium of subconscious patterns; *see* also synchromesh 2 (Chapters 16 and 17).

Purusha: pre-scientific term; conscio-material band/ grade - conscious; Pure Consciousness, Source of Life, Subjectivity and Creativity; Universal (*Sat*) Potential; boundlessly pre-active, in action Prime Mover, First Cause or Top Governor on the universal scale and order of creation; given many other names in many languages; ultimate subject of worship, praise and love.

Q

quantum: minimum discrete amount of some physical property such as energy, space or time that a system can possess; quantum theory states that energy exists in tiny, discontinuous packets each of which is called a quantum; an elementary discontinuity; an elementary particle e.g. photon or electron.

quantum level: matter-in-principle; 'internal', 'causal' or 'subtle' matter; the vibrant or energetic phase of physical organisation; zone of sub-atomic particles and forces; step (on cosmic ziggurat) between potential and bulk matter whose aspect is sometimes extended to include atomic and molecular interactions; small-scale substance underlying large-scale, sensible appearances.

R

raj: (↑) upward, levitatory or stimulatory cosmic vector.

reductionism: opposite of holism; the materialistic view that an article can always be analysed, split up or 'reduced' to more fundamental parts; these parts can then be added back to reconstruct the whole; a whole is no more than the sum of its parts.

religion: etymology debated between Latin *religare* (bind) and *relegere* (review); *religio* means dutiful and meticulous observance; currently religion means world-view, mind-set or basic faith; whether of materialistic or holistic belief, it involves the non-negotiable substance of an individual or community's truth - notably as regards origins; antagonism between holistic practice and the naturalistic methodology of science is, because the couple deal with separate but complementary physical and metaphysical dimensions, flawed; a materialist/ atheist 'binds meticulously' to an evolutionary mind-set, a holist to pantheism or a Living Creator; in the case that self-deception is crucial to successfully deceiving others which, holism or materialism, is the religion that is ultimately true?

resonance: the tendency of a body or system to oscillate with a larger amplitude when subjected to disturbance by the same frequencies as its own natural ones; thus a resonator is a device that naturally oscillates at such (resonant) frequencies with greater amplitude than at others; resonance phenomena occur with all kinds of vibration, oscillation or wave; their sorts include mechanical, harmonic (acoustic), electrical (as with antennae), atomic and molecular.

respiration: the controlled release of energy from food.

ribosome: site of polypeptide (protein) synthesis.

RNA: single-stranded nucleic acid polymer employed in three different forms during the process of protein synthesis; in computer terms might be likened to a portable memory stick as opposed to *DNA*'s hard drive.

m-RNA: is used to transcribe a base sequence from *DNA*. It 'photocopies' a gene and carries this information to a ribosome.

mi-RNA: short micro-*RNA* molecules are important regulators of genetic expression.

r-RNA: is part of the make-up of the protein-manufacturing station called a ribosome.

t-RNA: critically translates genetic 'words' (see 'codon') into amino acids: 64 such operators form the link between code and the actuality of a functional protein.

S

sanskara: character trait; groove, habit, obsession or repetitious mode of thought proportional in depth to the intensity of desire, force of impact or impression that created or sustains it.

samsara: existence, phenomena, the place of cycles and, therefore, reincarnation; non-essence or, in Buddhism, what is not *Nirvana*.

sat: 'top' or essential cosmic fundamental; 'vector' of balance, neutrality.

science: Latin *scire* (know); knowledge; commonly understood as the practical and mathematical study of material phenomena whose purpose is to produce useful models of the physical world's reality.

scientism: a philosophical face of official, *de facto* commitment to materialism; today's majority consensus of what the creed of science is; an -ism born of *PAM*; a faith that all processes must be ultimately explicable in terms of physical processes alone; like communism, a one-party state of mind; a doctrine that physical science with its scientific method is ultimately, the sole authority and arbiter of truth; a set of concepts designed to produce exclusively material explanations for every aspect of existence, that is, to colonise each academic discipline and build its intellectual empire everywhere; 'scientific fundamentalism' closely allied, when expressed in social and political terms, with 'secular fundamentalism', sociological interpretation of behaviour and the fostering of a humanistic curriculum.

secular fundamentalism: *PAM* as applied to the worlds of nature and of human society.

secularism: concern with worldly business; lack of involvement in religion or faith; secularism is generally identified, as defined by the dictionary, with materialism; for a secularist the ultimate arbiter of truth is human reason - ideas are open to negotiation so that even morality is relative; however many liberal agnostics, atheists and humanists argue that their metaphysical, philosophical system also embraces so-called 'universal' moral values and, as opposed to zealotry or the logic of evolutionary faith, a liberal politic of 'philosophical live-and-let-live'.

siddhi: marvellous, miraculous or 'super-natural' psychic ability that, at the point a practitioner actually masters it, becomes natural.

STE: special theory of evolution; *see* microevolution.

stereo-computation*:* stereochemistry involves study of the relative spatial arrangement of atoms in molecules; in biology a 1-D line of informative code (whose 3-D constituents bear no figurative relationship with their informed product) give rise to relative 3-D spatial arrangements at all levels from molecular to systemic and whole-body; such targeted generation may be termed bio-logical stereocomputation.

sub-state: *opp.* super-state; impotence, discharge, exhaustion, final stage in the expression of potential; fixity; non-conscious base-state; state 'below/ subtendence; extreme negativity/ (*tam*) condition.

sufi: mystic, Islamic 'heretic' of whom the most influential is perhaps Jalal-ud-Din Rumi, a disciple of Shamas of Tabriz.

super-state: potential; source of possibility; causal metaphysic/ archetype; state 'before' or 'above' subsequent expression; immanence; transcendence; precondition; (*sat*) priority.

symmetry: closely allied with the (*sat*) characteristic of balance; aesthetically pleasing balance and proportion; geometrical balance or interactive process such that some feature of an action remains invariant, that is, conserved; the symmetry of an entity (such as a sphere, empty space or natural law) or feature (such as energy) that remains the same at all times everywhere from any local point of observation or through every transformation is called 'higher' or 'continuous'; if a feature is conserved only when an object or process is moved, turned or viewed at certain angles or under specific conditions its symmetry is called 'lower' or 'discrete'; the symmetrical properties of a system may be precisely related to corresponding conservation laws and *vice versa*; various kinds of symmetry independent of space-time coordinates are important to both quantum and classical physics; scale symmetry occurs when a reduced or expanded object keeps its shape but not its size (as with Mandelbrot fractals); dialectical symmetry also involves *informative potential*; its metaphysical archetypes inform principles, laws or determinant fields that exist prior to action and, from their possibilities, govern actual outcome; such 'configuration of the world' is absolute and, beyond entropy, stable; it is negentropically immune from decay; by contrast, the 'free' symmetry of potential energy is inherently unstable and (like a pencil balanced on its tip) liable to spontaneously 'topple' or 'break' into the least energetic of a range of circumstantial possibilities; such spontaneous symmetry-breaking, the basis of diversity, represents an expression of 'deep symmetry' or archetype under local conditions and is therefore called by physicists 'contingent'.

T

tam: (↓) downward, gravitational or inertialising cosmic vector.

tanmatra: pre-scientific term; possible conscio-material band/ grade - subconscious band of mind; *tanmatras* involve the least metaphysical, most nearly physical band of mind; they are traditionally thought of as mental ideas, psychological forms or the Platonic ideals of physical perceptions (e.g. notions of heat, light and motion in a flame or fragrance in a scent); as *chakras* are immaterial structures dealing with energy, so *tanmatras* deal with image, quality and form; they represent the qualitative aspect of matter and are the 'device' that allows mental grasp of physical effects; as such *tanmatras* are an instrument of potential matter, the archetypal processors of image; as particle to wave so *tanmatra* to *prana*; and as sound is plucked from a tuned string so each of five *tanmatras* is like a string creating, by resonant association, one of five *pranic* 'notes' (see Glossary: *prana*); *tanmatric* apparatus represents a stage in the hierarchical translation of incoming (matter to mind) or outgoing (mind to matter) signals across either individual or universal psychosomatic border (see *figs*. 15.8 and 15.9, also Chapters 16 *esp.* signal translation and 17); the two-way traffic across this border means they equally act, as prism or lens to light, as media for the orderly projection of *prana* into subtle (quantum) events and thence, lower down, gross aggregates of called bulk matter; as such they are, in conjunction with *prana* as power source, the pre-physical mechanism that expresses, in the physical vacuity of archetypal field, the fundamentals of material phenomena; whether simple, single or in 'complex opera' that codes for bio-symphony, *tanmatras* transmit 'sounds' (vibrations or frequencies) which translate into cymatic messages (check Index: Chladni) exciting the emergence or maintenance of physical form; in short, they were the final agents in the initial creation of physical form and force; the wireless physical expression of cooperant *tanmatras* and *pranas* is known to quantum physics; the wired, bonded or aggregate expression involves the chemistry, physics and biology of condensed or bulk matter; and, as electricity supports the running of a machine, they iteratively, correctly support our starry universe.

tattwa: pre-scientific term; possible conscio-material band/ grade - both subconscious and physical bands; means 'that-ness' or 'not-self'; by nature, using their intellect/ *buddhi*, philosophers seek to analyse, categorise and argue so that, in the case of *tattwa*, the number of items listed varies considerably according to tradition;

basically, however, it amounts to a 'catch-all' description of human, animal and inanimate condition; five well-known *tattwas* correspond exactly to Greek, Latin and medieval European elements of ether, fire, air, water and earth (see Chapter 10: Old Vacuums); on the subconscious side of *PSI*, in archetypal memory of universal mind, these correspond to five *tanmatras*, five *pranic* 'notes' and five lower *chakras* (*fig.* 17.11); and on the physical side to five informative senses and five energetic organs of action (*fig.* 0.8 and Chaps. 14: Lower Physical Loop, 15: Psychosomatic Linkage and 16: Signal Translation); the five elements/ states of matter are variously defined; ether is space, upper air (home of the gods) or psychosomatic archetypal potential (which, being metaphysical, is physically unseen); it is related to the throat chroat *chakra*, seat of dormancy; there follow gas (air *tattwa* related to breath, oxygen and heart *chakra*), energy (fire *tattwa* giving stimulus for change, heat and light whose hub is the solar plexus), liquid (water and its osmoregulatory and waste expulsion systems) and solid (the earth *tattwa* of related to bio-mass and its sense of pressure/ touch whose *chakra* rests at the supportive base of the spine).

teleology: the doctrine that there is evidence of purpose in nature; doctrine of non-randomness in natural architecture; doctrine of reason ('for the sake of', 'in order to', 'so that' etc.) and intent behind biological and universal design.

third eye: place where you think; point of metaphysical focus between and behind the eyebrows, that is, just above the physical eyes; HQ/ seat of mind beyond the sensory world; cosmological eye-centre; gate through which meditative concentration can pass; single way that leads within.

transcendent projections: **psychological:** see Chapter 5 Top Teleology, Index: Archetype and *figs.* 2.6, 3.1, 5.1, 9.1 and/ or 11.1; **physical**: see Chapters 8, 9, 11, 12; Glossary: archetype; Index: transcendence, archetype, cosmo-logical language; *figs.* as above and 12.1; such projection involves an orderly, energetic expression from either metaphysical or physical nothingness, that is, unseen potential; an instantaneous 'miracle' that issues from 'within' non-conscious physicality; transcendently emergent, finely tuned expansion from 'inner' metaphysic into 'outer' material/ natural law; 0-dimensional singularity (paradoxically everywhere at once) from whose prior pointlessness all points perhaps began; cosmic seed whence, *ex nihilo*, the world developed; projection whose appearance, once physical, is visible and perhaps described but certainly not explained by big bang theory; transcendent projection of archetype is possibly, to the constrained sensory and intellectual states of human mind, ultimately incomprehensible; its invisible mechanism, the practice of materialisation, may remain a fact beyond material understanding. **biological:** if matter is developed memory (Chapter 9: How Does Nothing Physical Work?) then see Chapters 16: *passim* and 19: Conceptual

Biology; see also Glossary: mnemone and archetype; Index mnemone, archetype; and *figs*. 2.6, 3.1, 16.1 and 19.1.

transposon: jumping gene'; ubiquitous genetic element found in all prokaryotes and eukaryotes so far investigated; *DNA* segment that can, by enzyme, be cut from a one site (the donor) and joined to another (the target); a retrotransposon is moved through the mediation of *RNA* and reverse transcription back from *RNA* to *DNA*; transposons and retrotransposons play a key, functional role in gene expression and regulation; a kind of retrotransposon, SINEs and LINEs are thought to compose 35+% of the human genome; such elements may be flanked by terminal repeats that allow specific, operational variation in different types of cell; from an evolutionary view they comprise functionless viral imports; from a *top-down* view it is predicted they form a dynamic, intrinsic element of the genome involved in gene regulation, genetic shuffling as (epigenetic) response to buffer circumstantial exigency and, just as important, structural agents able to reshape a chromosome to meet specific genetic demands.

transcription factor: protein that, binding to a specific *DNA* sequence, regulates genetic transcription.

U

unification: simplification: details are unified by their working principles, themes or programs; better to perceive intrinsic principle is to simplify or unify an understanding; progressive unification of forces is the grail of physics: Clerk Maxwell unified electricity and magnetism; electroweak or *GSW* theory brought in the weak nuclear force; now the goal is to include the strong nuclear force (*GUT*), gravity in a super-force and show that, in essence, particles and forces are interchangeable (super-symmetry and *TOE*); Natural Dialectic, also working with the maxim 'All is One', includes what sums to a hierarchical *TOP* or Theory of Potential (*see* especially Chapters 5, 6, 7, 9, 16 and 19, also *fig.* 5.1); potential is the absolute from which variant orders of relativity derive; the equivalent of *TOP* is *CUT* (**Cosmic Unification Theory); Natural Dialectic is a vehicle of *CUT*, whose aim is to build cosmic connections, that is, orderly linkages towards a Holy Grail of Unification**; the Great Connector, that is, Unifier is consciousness; the subjective potential for mind is consciousness and the objective potential for matter is archetypal memory; such archetypal element unites psychology with the physics of natural science; it is the informative precondition of physical and biological form.

universal mind: cosmic grade; also called the 'mind of nature' or 'natural mind'; as a biological body is a specific though complex arrangement of universal chemicals so individual mind partakes of a particular, equally minuscule fraction of the metaphysical components of universal mind; *see* also archetype.

V

vector: existential dynamic; a vector has both direction and magnitude; it illustrates direction of travel with respect to a model or a secondary stack used in Natural Dialectic; fundamental vectors (↑ and ↓) denote relative gain or loss of information or energy; and, similarly, motion towards and from the axis, peak or source of a cosmic model; in this case, magnitude is inferred to occur on a scale between any pair of opposites, for example, the relative proportions of black and white in the grey-scale between these opposites; use of the word is general, metaphorical rather than specific; opposite members of a stack may involve metaphysical factors (e.g. love/ hate, beauty/ ugliness) as well as physical; thus its spectra do not necessarily concern physical motion or mathematical calculation; its 'field of relativity' extends beyond non-conscious elements; in this respect a Dialectical vector is similar in principle but not the same in practice as that defined by physics or biology.

virtuality: exotic component of quantum physics; para-physical feature of the quantum vacuum; immaterial substrate of material phenomena; inner (where solidity's the outer) edge of physical reality; ephemeral 'virtual particles' rise and sink back into a 'void' thought to teem with their 'fluctuations'; virtuality is identified as the agent of such important actualities as the strong nuclear force (resulting from interaction between virtual mesons and gluons), vacuum polarisation, the Coulomb force (between electric charges and mediated by the exchange flight of virtual photons) and so on; not used in the computer sense of a continuum between real and imaginary circumstance; see also *ZPE*.

Vitruvian man: where art meets science Leonardo's 'universal man' demonstrates an architectural symmetry, excellence of composition and, microcosm unto macrocosm, a reflection of the universe; Da Vinci's connection, from his notebooks, is quite the opposite of Darwin's doodle (Chapter 5); if, with ratios and rationality, it demonstrates mathematical perfection then does not design of larger cosmos demonstrate it too? Natural Dialectic certainly concurs with Leonardo's logical submission.

Z

zero: zero (the number) is a metaphysical entity, one critical to mathematics; zero (the fact) means, for Natural Dialectic, nothing in two senses; in the *negative sense* it means an absence of perception (psychological oblivion) or absolutely nothing physical (as naturalistically prescribed to precede, say, a big bang or as the nature of a theoretically perfect vacuum); negative sense may also be construed as (*tam*) an extreme sub-state, sink or emptiness; for materialism 'absolute nothingness' may involve natural law and its mathematical description; what, one may enquire, is the source of such 'eternal metaphysic', what is the nature zero-physical?: on the other hand, in a *positive sense* zero refers to source, pre-existent potential or (*sat*) higher cause-in-principle; for example, information (which is zero-physical) transcends/ precedes a course of action; information that passively governs the operation of cosmos derives from immaterial archetype.

ZPE: zero-point energy; quantum vacuum; vacuum energy of all fields in space; residual energy of all oscillators at 0°K; concept first developed by Albert Einstein and Otto Stern; intrinsic energy of vacuum; the ground-state minimum that any quantum mechanical system, in particular the vacuum, can have; remainder, according to the uncertainty principle, when all particles and thermal radiation have been extracted from a volume of space; residual non-thermal radiation; irreducible 'background noise'; 'quantum foam'; the potent, microscopic side of quantum vacuum (as opposed to impotent, macroscopic vacuum left by the apparent lack of anything); subliminal 'rumblings' of immaterial weak, strong and electromagnetic fields (called *ZPF*s); seething, jostling ferment of subliminal waves and particles in emptiness; a flux of unobservable 'virtual' matter and anti-matter that may or may not appear as the basis of observable forces such as electromagnetism, charge and perhaps inertial mass and gravity; a subtle facet of levity; the anti-gravity of dark energy (or the cosmological constant) has been postulated as a component of *ZPE*; suggested 'mother-field' support for electron orbits, atomic structure and thus the phenomenal universe.

zygote: fertilised egg.

Index

B

C

D

E

F

G

H

I

J

K

L

M

N

O

P

Q

R

S

T

U

V

W

X

Y

Z

Bibliography

Abusing Science	Kitcher P.	1982
Accidental Univer	Davies P.	1982
Adam and Evolution	Pitman M.	1984
Alas, Poor Darw	Rose H. & S (eds).	2000
Basis for a New Biology	Wilder Smith A. E.	1976
Billions of Missing Links	Simmons G.	2007
Blind Watchmaker	Dawkins R.	1986
Bones of Contention	Lubenow M.	2008
Book of Nothing	Barrow J.	2000
Brain Science & Biology of Belief	Neuberg A. et alii	2001
Brief History of Time	Hawking S.	1988
Cell's Design	Rana F.	2008
Chance and Necessity	Monod J.	1970
Cheating Time	Gosden R.	1996
Chemical Evolution	Aw S.	1976
Complete World of Human Evolution	Stringer C., Andrews P.	2005
Consciousness	Blackmore S.	2003
Constants of Nature	Barrow J.	2002
Contested Bones	Sanford J., Rupe C.	2020
Creating Life in the Lab	Rana F.	2011
Creation and Evolution	Hayward A.	1985
Creation of Life	Wilder Smith A. E.	1974
Darwin's Black Box	Behe M.	1996
Darwin's Doubt	Meyer S.	2013
Darwin and Design	Ruse M.	2003
Darwin on Trial	Johnson P.	1991
Darwin Retried	MacBeth N.	1971
Darwinian Fairytales	Stove D.	1995
Darwinism: Refutation of a Myth	Lovtrup S.	1987
Dawkins' God	McGrath A.	2005
Dawkins Letters	Robertson D	2007
Dawkins Proof for the Existence of God	Barns R.	2009
Debating Darwin's Doubt	Klinghoffer D.	2015
Deluded by Dawkins?	Wilson A.	2007
Descent of Man	Jones S.	2002
Devil's Delusion	Berlinski D.	2009
Did God Use Evolution?	Gitt W.	1993
Double Helix	Watson J.	1970
Dreaming	Allan Hobson J.	2002
Edge of Evolution	Behe M.	2007
Edge of Infinity	Davies P.	1981
Elegant Universe	Greene B.	1999
Endless Forms Most Beautiful	Carroll S.	2011
Epigenetics	Francis R.	2011
Ever Since Darwin	Gould S.	1977
Evolution, A Theory in Crisis	Denton M.	1985
Evolution, A View from the 21st Century	Shapiro J.	2011
Evolution Impossible	Ashton J.	2012
Evolution of Life	Jarman C.	1970
Evolution of Living Organisms	Grasse P-P.	1978
Evolution of Sex	Maynard-Smith J.	1978

Evolution, Still A Theory in Crisis	Denton M.	2016
Evolution, The Human Story	Roberts A.	2011
Explore Evolution	Meyer S. and others	2007
Facts of Life	Milton R.	1992
Fallacies of Evolution	Hoover A.	1977
Fearful Symmetry	Zee A.	1999
Flaws in the Theory of Evolution	Shute E.	1962
Fossils in Focus	Anderson J., Coffin H.	1977
Gaia	Lovelock J.	1979
Gaia: Practical Science of Planetary Medicine	Lovelock J.	1991
Genetic Entropy & the Mystery of the Genome	Sandford J.	2005
Ghost in the Machine	Koestler A.	1975
God Delusion	Dawkins R.	2006
God Gene	Hamer D.	2004
God and the New Physics	Davies P.	1983
God and Stephen Hawking	Lennox C.	2011
God beyond Nature	Clark R.E.D	1982
God, Science and Evolution	Andrews E.	1980
God: To Be or Not To Be	Wilder Smith A. E.	1975
God's Undertaker	Lennox C.	2009
Goldilocks Enigma	Davies P.	2007
Grand Design	Hawking S., Mlodinow L.	2010
Great Evolution Mystery	Rattray Taylor G.	1983
Hallmarks of Design	Burgess S.	2000
Has Darwin Had His Day?	Rosevear D.	2007
How Life Began	Croft L.	1988
How the Mind Works	Pinker S.	1997
Icons of Evolution	Wells J.	2000
In the Beginning was Information	Gitt W.	1997
Infinite Book	Barrow J.	2005
Infinity	Clegg B.	2003
Information and the Nature of Reality	eds. Davies & Gregersen	2010
Inspiration from Creation	Burgess S., Statham D.	2018
Intelligent Design Uncensored	Dembski W. and Witt J.	2010
Intelligent Universe	Hoyle F.	1983
Just Six Numbers	Rees M.	1999
Language of God	Collins F.	2007
Life Itself	Crick F.	1972
Life's Solution	Conway Morris S.	2003
Macroevolution	Stanley S.	1979
Master and his Emissary	McGilchrist I.	2009
Mind and Cosmos	Nagel T.	2012
Mind of God	Davies P.	1992
Mind, Body & Electromagnetism	Evans J	1992
Mysterious Epigenome	Gills, Woodward	2012
Mystery of Life's Origin	Thaxton, Bradley, Olsen	1985
Myth of Junk DNA	Wells J.	2011
Naked Emperor: Darwinism Exposed	Latham A.	2005
Nature of Life	Waddington C.	1961
Natural Sciences Know Nothing of Evolution	Wilder Smith A.	1981
Natural Theology	Paley W.	1802
Nature's Destiny	Denton M.	1998

Neck of the Giraffe	Hitching F.	1982
New Biology	Augros R. & Stanciu N.	1988
New Science of Life	Sheldrake R.	1981
New Story of Science	Augros R. & Stanciu N.	1986
New Theories of Everything	Barrow J.	2007
Not By Chance	Spetner Lee	1997
Nothing	ed. Webb J.	2013
Nothingness	Genz H.	1999
On Guard	William Lane Craig	2010
One Small Speck	Sodera V.	2009
Origin of Species	Darwin C.	1859
Origin of Species Revisited	Bird W	1989
Origin of Life	Bliss R.	1979
Origins of Life	Rana F, Ross H.	2004
Our Cosmic Habitat	Rees M.	2001
Quantum World	Polkinghorne J.	1984
Panda's Thumb	Gould S.	1980
Politically Incorrect Guide to Darwinism and Intelligent Design	Wells J.	2006
Presence of the Past	Sheldrake R.	1988
Piltdown Forgery	Weiner J.	1955
Reason in the Balance	Johnson P.	1995
Refuting Evolution 1 and 2	Sarfati J.	2007
Roots of Coincidence	Koestler A.	1976
Runes of Evolutio	Conway Morris S.	2015
Science and Creation	Polkinghorne J.	1988
Science Delusion	Sheldrake R.	2012
Science & Evidence for Design in the Universe	Behe, Dembski, Meyer	1999
Science and Human Origins	Gauger, Axe, Luskin	2012
Secret of the Creative Vacuum	Davidson J.	1989
Seven Sins of Memory	Schacter D.	2001
Shadows of the Mind	Penrose R.	1995
Signature in the Cell	Meyer S.	2009
Tao of Physics	Capra F.	1976
Tao Te Ching	Lao Tse	
There is a God	Flew A.	2007
Thermodynamics & the Development of Order	ed. Williams E.	2002
Time, Space and Things	Ridley B.	1976
Transformist Illusion	Dewar D.	1957
Web of Life	Davidson J.	1988
Web of Life	Capra F.	1996
What Darwin Got Wrong	Fodor, Piatelli-Palmarini	2010
What is Life?	Schrödinger E.	1944
When is a Fly not a Horse?	Sermonti G.	2005
Who Made God?	Andrews E.	2009
Without Excuse	Gitt W.	2011
Wonderful Life	Gould S.	1989
Wonders of the Universe	Cox B.	2011
Universe, Plan or Accident?	Clark R.E.D	1961
Vital Question	Lane N.	2015
Void	Close F.	2007
Y, The Descent of Men	Jones S.	2002

The author has recently written a few more books (available from Amazon, Foyles, Waterstones, Barnes & Noble etc. and see website addresses on p.2):

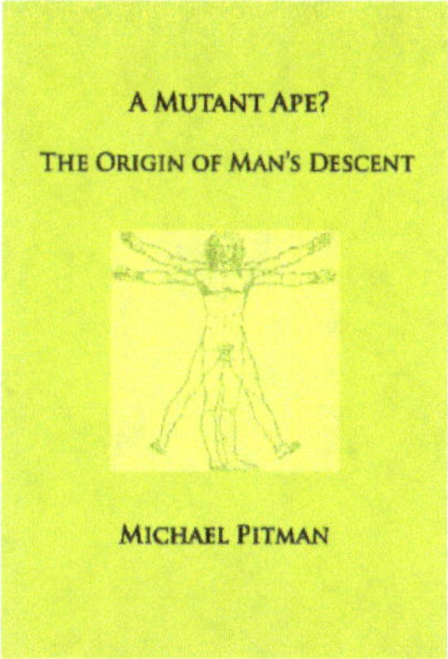

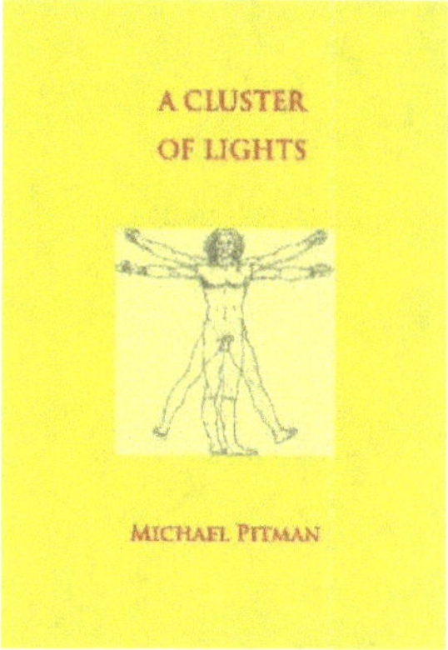

www.ingramcontent.com/pod-product-compliance
Lightning Source LLC
Chambersburg PA
CBHW071112080526
44587CB00013B/1312
9781838061814